INTERFACIAL DYNAMICS

SURFACTANT SCIENCE SERIES

FOUNDING EDITOR

MARTIN J. SCHICK
1918–1998

SERIES EDITOR

ARTHUR T. HUBBARD
Santa Barbara Science Project
Santa Barbara, California

ADVISORY BOARD

DANIEL BLANKSCHTEIN
Department of Chemical Engineering
Massachusetts Institute of Technology
Cambridge, Massachusetts

S. KARABORNI
Shell International Petroleum
Company Limited
London, England

LISA B. QUENCER
The Dow Chemical Company
Midland, Michigan

JOHN F. SCAMEHORN
Institute for Applied Surfactant Research
University of Oklahoma
Norman, Oklahoma

P. SOMASUNDARAN
Henry Krumb School of Mines
Columbia University
New York, New York

ERIC W. KALER
Department of Chemical Engineering
University of Delaware
Newark, Delaware

CLARENCE MILLER
Department of Chemical Engineering
Rice University
Houston, Texas

DON RUBINGH
The Proctor & Gamble Company
Cincinnati, Ohio

BEREND SMIT
Shell International Oil Products B.V.
Amsterdam, The Netherlands

JOHN TEXTER
Strider Research Corporation
Rochester, New York

1. Nonionic Surfactants, *edited by Martin J. Schick* (see also Volumes 19, 23, and 60)
2. Solvent Properties of Surfactant Solutions, *edited by Kozo Shinoda* (see Volume 55)
3. Surfactant Biodegradation, *R. D. Swisher* (see Volume 18)
4. Cationic Surfactants, *edited by Eric Jungermann* (see also Volumes 34, 37, and 53)
5. Detergency: Theory and Test Methods (in three parts), *edited by W. G. Cutler and R. C. Davis* (see also Volume 20)
6. Emulsions and Emulsion Technology (in three parts), *edited by Kenneth J. Lissant*
7. Anionic Surfactants (in two parts), *edited by Warner M. Linfield* (see Volume 56)
8. Anionic Surfactants: Chemical Analysis, *edited by John Cross*
9. Stabilization of Colloidal Dispersions by Polymer Adsorption, *Tatsuo Sato and Richard Ruch*
10. Anionic Surfactants: Biochemistry, Toxicology, Dermatology, *edited by Christian Gloxhuber* (see Volume 43)
11. Anionic Surfactants: Physical Chemistry of Surfactant Action, *edited by E. H. Lucassen-Reynders*
12. Amphoteric Surfactants, *edited by B. R. Bluestein and Clifford L. Hilton* (see Volume 59)
13. Demulsification: Industrial Applications, *Kenneth J. Lissant*
14. Surfactants in Textile Processing, *Arved Datyner*
15. Electrical Phenomena at Interfaces: Fundamentals, Measurements, and Applications, *edited by Ayao Kitahara and Akira Watanabe*
16. Surfactants in Cosmetics, *edited by Martin M. Rieger* (see Volume 68)
17. Interfacial Phenomena: Equilibrium and Dynamic Effects, *Clarence A. Miller and P. Neogi*
18. Surfactant Biodegradation: Second Edition, Revised and Expanded, *R. D. Swisher*
19. Nonionic Surfactants: Chemical Analysis, *edited by John Cross*
20. Detergency: Theory and Technology, *edited by W. Gale Cutler and Erik Kissa*
21. Interfacial Phenomena in Apolar Media, *edited by Hans-Friedrich Eicke and Geoffrey D. Parfitt*
22. Surfactant Solutions: New Methods of Investigation, *edited by Raoul Zana*
23. Nonionic Surfactants: Physical Chemistry, *edited by Martin J. Schick*
24. Microemulsion Systems, *edited by Henri L. Rosano and Marc Clausse*
25. Biosurfactants and Biotechnology, *edited by Naim Kosaric, W. L. Cairns, and Neil C. C. Gray*
26. Surfactants in Emerging Technologies, *edited by Milton J. Rosen*
27. Reagents in Mineral Technology, *edited by P. Somasundaran and Brij M. Moudgil*
28. Surfactants in Chemical/Process Engineering, *edited by Darsh T. Wasan, Martin E. Ginn, and Dinesh O. Shah*
29. Thin Liquid Films, *edited by I. B. Ivanov*
30. Microemulsions and Related Systems: Formulation, Solvency, and Physical Properties, *edited by Maurice Bourrel and Robert S. Schechter*
31. Crystallization and Polymorphism of Fats and Fatty Acids, *edited by Nissim Garti and Kiyotaka Sato*
32. Interfacial Phenomena in Coal Technology, *edited by Gregory D. Botsaris and Yuli M. Glazman*
33. Surfactant-Based Separation Processes, *edited by John F. Scamehorn and Jeffrey H. Harwell*
34. Cationic Surfactants: Organic Chemistry, *edited by James M. Richmond*
35. Alkylene Oxides and Their Polymers, *F. E. Bailey, Jr., and Joseph V. Koleske*
36. Interfacial Phenomena in Petroleum Recovery, *edited by Norman R. Morrow*
37. Cationic Surfactants: Physical Chemistry, *edited by Donn N. Rubingh and Paul M. Holland*
38. Kinetics and Catalysis in Microheterogeneous Systems, *edited by M. Grätzel and K. Kalyanasundaram*
39. Interfacial Phenomena in Biological Systems, *edited by Max Bender*
40. Analysis of Surfactants, *Thomas M. Schmitt*

41. Light Scattering by Liquid Surfaces and Complementary Techniques, *edited by Dominique Langevin*
42. Polymeric Surfactants, *Irja Piirma*
43. Anionic Surfactants: Biochemistry, Toxicology, Dermatology. Second Edition, Revised and Expanded, *edited by Christian Gloxhuber and Klaus Künstler*
44. Organized Solutions: Surfactants in Science and Technology, *edited by Stig E. Friberg and Björn Lindman*
45. Defoaming: Theory and Industrial Applications, *edited by P. R. Garrett*
46. Mixed Surfactant Systems, *edited by Keizo Ogino and Masahiko Abe*
47. Coagulation and Flocculation: Theory and Applications, *edited by Bohuslav Dobiáš*
48. Biosurfactants: Production • Properties • Applications, *edited by Naim Kosaric*
49. Wettability, *edited by John C. Berg*
50. Fluorinated Surfactants: Synthesis • Properties • Applications, *Erik Kissa*
51. Surface and Colloid Chemistry in Advanced Ceramics Processing, *edited by Robert J. Pugh and Lennart Bergström*
52. Technological Applications of Dispersions, *edited by Robert B. McKay*
53. Cationic Surfactants: Analytical and Biological Evaluation, *edited by John Cross and Edward J. Singer*
54. Surfactants in Agrochemicals, *Tharwat F. Tadros*
55. Solubilization in Surfactant Aggregates, *edited by Sherril D. Christian and John F. Scamehorn*
56. Anionic Surfactants: Organic Chemistry, *edited by Helmut W. Stache*
57. Foams: Theory, Measurements, and Applications, *edited by Robert K. Prud'homme and Saad A. Khan*
58. The Preparation of Dispersions in Liquids, *H. N. Stein*
59. Amphoteric Surfactants: Second Edition, *edited by Eric G. Lomax*
60. Nonionic Surfactants: Polyoxyalkylene Block Copolymers, *edited by Vaughn M. Nace*
61. Emulsions and Emulsion Stability, *edited by Johan Sjöblom*
62. Vesicles, *edited by Morton Rosoff*
63. Applied Surface Thermodynamics, *edited by A. W. Neumann and Jan K. Spelt*
64. Surfactants in Solution, *edited by Arun K. Chattopadhyay and K. L. Mittal*
65. Detergents in the Environment, *edited by Milan Johann Schwuger*
66. Industrial Applications of Microemulsions, *edited by Conxita Solans and Hironobu Kunieda*
67. Liquid Detergents, *edited by Kuo-Yann Lai*
68. Surfactants in Cosmetics: Second Edition, Revised and Expanded, *edited by Martin M. Rieger and Linda D. Rhein*
69. Enzymes in Detergency, *edited by Jan H. van Ee, Onno Misset, and Erik J. Baas*
70. Structure–Performance Relationships in Surfactants, *edited by Kunio Esumi and Minoru Ueno*
71. Powdered Detergents, *edited by Michael S. Showell*
72. Nonionic Surfactants: Organic Chemistry, *edited by Nico M. van Os*
73. Anionic Surfactants: Analytical Chemistry, Second Edition, Revised and Expanded, *edited by John Cross*
74. Novel Surfactants: Preparation, Applications, and Biodegradability, *edited by Krister Holmberg*
75. Biopolymers at Interfaces, *edited by Martin Malmsten*
76. Electrical Phenomena at Interfaces: Fundamentals, Measurements, and Applications, Second Edition, Revised and Expanded, *edited by Hiroyuki Ohshima and Kunio Furusawa*
77. Polymer-Surfactant Systems, *edited by Jan C. T. Kwak*
78. Surfaces of Nanoparticles and Porous Materials, *edited by James A. Schwarz and Cristian I. Contescu*
79. Surface Chemistry and Electrochemistry of Membranes, *edited by Torben Smith Sørensen*
80. Interfacial Phenomena in Chromatography, *edited by Emile Pefferkorn*

81. Solid–Liquid Dispersions, *Bohuslav Dobiáš, Xueping Qiu, and Wolfgang von Rybinski*
82. Handbook of Detergents, *editor in chief: Uri Zoller*
    Part A: Properties, *edited by Guy Broze*
83. Modern Characterization Methods of Surfactant Systems, *edited by Bernard P. Binks*
84. Dispersions: Characterization, Testing, and Measurement, *Erik Kissa*
85. Interfacial Forces and Fields: Theory and Applications, *edited by Jyh-Ping Hsu*
86. Silicone Surfactants, *edited by Randal M. Hill*
87. Surface Characterization Methods: Principles, Techniques, and Applications, *edited by Andrew J. Milling*
88. Interfacial Dynamics, *edited by Nikola Kallay*

ADDITIONAL VOLUMES IN PREPARATION

Computational Methods in Surface and Colloid Science, *edited by Malgorzata Borówko*

Adsorption on Silica Surfaces, *edited by Eugène Papirer*

Fine Particles: Synthesis, Characterization, and Mechanisms of Growth, *edited by Tadao Sugimoto*

INTERFACIAL DYNAMICS

edited by

Nikola Kallay

**University of Zagreb
Zagreb, Croatia**

MARCEL DEKKER, INC. NEW YORK · BASEL

Library of Congress Cataloging-in-Publication Data

Interfacial dynamics / edited by Nikola Kallay.
    p. cm. — (Surfactant science series ; v. 88)
    Includes bibliographical references and index.
    ISBN 0-8247-0006-6 (alk. paper)
    1. Solid-liquid interfaces. 2. Electrolyte solutions. 3. Surface chemistry. I. Kallay, Nikola, 1942– II. Series.

QD509.S65.I58 1999
541.3′3—dc21   99-051473

This book is printed on acid-free paper.

Headquarters
Marcel Dekker, Inc.
270 Madison Avenue, New York, NY 10016
tel: 212-696-9000; fax: 212-685-4540

Eastern Hemisphere Distribution
Marcel Dekker AG
Hutgasse 4, Postfach 812, CH-4001 Basel, Switzerland
tel: 41-61-261-8482; fax: 41-61-261-8896

World Wide Web
http://www.dekker.com

The publisher offers discounts on this book when ordered in bulk quantities. For more information, write to Special Sales/Professional Marketing at the headquarters address above.

Copyright © 2000 by Marcel Dekker, Inc. All Rights Reserved.

Neither this book nor any part may be reproduced or transmitted in any form or by any means, electronic or mechanical, including photocopying, microfilming, and recording, or by any information storage and retrieval system, without permission in writing from the publisher.

Current printing (last digit):
10 9 8 7 6 5 4 3 2 1

PRINTED IN THE UNITED STATES OF AMERICA

Preface

The aim of this book is to present up-to-date knowledge in the field of interfacial phenomena. Many of the chapters concentrate on the solid–liquid interfaces but other systems are also presented. The book covers fundamentals, special topics, and applications. It consists of 20 chapters written by numerous internationally recognized experts.

The book starts with an introductory chapter on the equilibria in electrolyte solutions. The next chapter describes the dynamics of interfacial water. Since the properties of interfaces are influenced and determined by both phases in contact it is essential to understand their individual properties. The next two chapters describe the theoretical models for treating the equilibrium at solid–liquid interfaces. The following chapters are devoted to the determination of the enthalpy of interfacial reactions: application of electrokinetic data in the interpretation of the adsorption equilibria, interfacial layer in nonaqueous systems, and immersion phenomena. Kinetics of adsorption/desorption processes of ions and proteins is covered in two chapters, followed by the microcalorimetric studies of biomaterials at the interface.

The next problem covered in this book is related to crystal growth and dissolution. The chapter describing the general knowledge on the kinetics of these processes is accompanied by two chapters describing special topics: the dissolution of hydroxyapatite and the dissolution of metal oxides in solutions of organic acids. Special topics include vapor adsorption isotherms, thermosensitive gels, electrical behavior of thin films, and conductivity of microemulsions. The two remaining chapters describe the application of real systems and include examples of how the fundamental knowledge may be applied in solving practical problems. The topics covered include environmental problems in soil, as well as the production of printing ink.

In preparing this book I enjoyed great cooperation with specialists in interfacial chemistry from all over the world. This is the beauty of science. Also, the team from Marcel Dekker, Inc., was very cooperative during the preparation of the volume. To everyone I would like to express my sincere gratitude.

Nikola Kallay

Contents

Preface *iii*
Contributors *vii*

1. Equilibria in Electrolyte Solutions 1
   Vladimir Simeon

2. The Significance of Interfacial Water Structure in Colloidal Systems—Dynamic Aspects 35
   Miroslav Čolić and Jan D. Miller

3. Thermodynamics of Adsorption at Heterogeneous Solid–Liquid Interfaces 83
   Władysław Rudziński, Jolanta Narkiewicz-Michałek, Robert Charmas, Mateusz Drach, Wojciech Piasecki, and Jerzy Zając

4. Surface Complexation Model for Solid–Liquid Interfaces 163
   Marek Kosmulski, Ryszard Sprycha, and Jerzy Szczypa

5. Enthalpy of Surface Charging 225
   Nikola Kallay, Tajana Preočanin, and Suzana Žalac

6. Interpretation of Interfacial Equilibria on the Basis of Adsorption and Electrokinetic Data 249
   Nikola Kallay, Davor Kovačević, and Ana Čop

7. Electrical Interfacial Layer in Nonaqueous Solvents 273
   Marek Kosmulski

8. Immersion Phenomena 313
   Jean Marc Douillard

9. Ionic Adsorption/Desorption Kinetics 351
   Kazuaki Hachiya, Yoshiko Moriyama, and Kunio Takeda

10. Quantitative Analysis of Protein Adsorption Kinetics 405
    Vladimir Hlady, C.-H. Ho, and D. W. Britt

11. Microcalorimetric Studies of Biomaterials at Interface   419
    Wen-Yih Chen, Fu-Yung Lin, and Ching-Fa Wu

12. Kinetics and Mechanisms of Crystal Growth in Aqueous Systems   435
    Ljerka Brečević and Damir Kralj

13. Dissolution of Calcium Hydroxyapatite   475
    Philippe Gramain and Philippe Schaad

14. Interfacial Chemistry of Dissolving Metal Oxide Particles: Dissolution by Organic Acids   513
    J. Adrián Salfity, Alberto E. Regazzoni, and Miguel A. Blesa

15. Ion Transport and Electrical Conductivity in Heterogeneous Systems: The Case of Microemulsions   541
    Federico Bordi and Cesare Cametti

16. Electrical Behavior of Inorganic Thin Films in Electrolyte Media   565
    Juana Benavente, José Ramos-Barrado, and Sebastián Bruque

17. Thermosensitive Gels   591
    Kimiko Makino

18. Vapor Adsorption Isotherms of Medium-Sized Alcohols on Various Solids and the Behavior of Adsorbed Molecules   621
    Akira Nonaka

19. Interactions at Colloidal Soil and Sediment Interfaces   651
    Bernd Dieter Struck and Milan Johann Schwuger

20. Application of Surfactants in Liquid Printing Inks   699
    Ryszard Sprycha and Ramasamy Krishnan

Index   737

Contributors

Juana Benavente Department of Applied Physics, Universidad de Málaga, Málaga, Spain

Miguel A. Blesa Department of Chemistry, Atomic Energy Commission, Buenos Aires, Argentina

Federico Bordi Sezione di Fisica Medica, Dipartimento di Medicina Interna, Università di Roma "Tor Vergata," and Istituto Nazionale Fisica della Materia, Unità di Roma 1., Rome, Italy

Ljerka Brečević Department of Materials Chemistry, Ruđer Bošković Institute, Zagreb, Croatia

D. W. Britt Department of Bioengineering, University of Utah, Salt Lake City, Utah

Sebastián Bruque Department of Inorganic Chemistry, Universidad de Málaga, Málaga, Spain

Cesare Cametti Dipartimento di Fisica, Università di Roma "La Sapienza," and Istituto Nazionale Fisica della Materia, Unità di Roma 1., Rome, Italy

Robert Charmas Department of Theoretical Chemistry, Maria Curie-Skłodowska University, Lublin, Poland

Wen-Yih Chen Department of Chemical Engineering, National Central University, Chung-Li, Taiwan

Miroslav Čolić Research and Development Division, ZPM Inc., Goleta, California

Ana Čop Faculty of Natural Sciences and Mathematics, University of Zagreb, Zagreb, Croatia

Jean Marc Douillard LAMMI, Montpellier University of Sciences, Montpellier, France

Mateusz Drach Department of Theoretical Chemistry, Maria Curie-Skłodowska University, Lublin, Poland

Philippe Gramain Hétérochimie Moléculaire et Macromoléculaire, CNRS—Ecole Nationale Supérieure de Chimie, Montpellier, France

Kazuaki Hachiya Department of Mechanical Engineering, Okayama University of Science, Okayama, Japan

Vladimir Hlady Department of Bioengineering, University of Utah, Salt Lake City, Utah

C.-H. Ho Department of Bioengineering, University of Utah, Salt Lake City, Utah

Nikola Kallay Faculty of Natural Sciences and Mathematics, University of Zagreb, Zagreb, Croatia

Marek Kosmulski Institute of Catalysis, Polish Academy of Sciences, Krakow, and Department of Electrochemistry, Technical University of Lublin, Lublin, Poland

Davor Kovačević Faculty of Natural Sciences and Mathematics, University of Zagreb, Zagreb, Croatia

Damir Kralj Department of Materials Chemistry, Ruđer Bošković Institute, Zagreb, Croatia

Ramasamy Krishnan Technical Center, Sun Chemical Corporation, Carlstadt, New Jersey

Fu-Yung Lin Department of Chemical Engineering, National Central University, Chung-Li, Taiwan

Kimiko Makino Faculty of Pharmaceutical Sciences, Science University of Tokyo, Tokyo, Japan

Jan D. Miller Department of Metallurgical Engineering, University of Utah, Salt Lake City, Utah

Yoshiko Moriyama Department of Applied Chemistry, Okayama University of Science, Okayama, Japan

Jolanta Narkiewicz-Michałek Department of Theoretical Chemistry, Maria Curie-Skłodowska University, Lublin, Poland

Akira Nonaka Institute of Applied Physics, University of Tukuba, Tukuba, Japan

Contributors

Wojciech Piasecki Department of Theoretical Chemistry, Maria Curie-Skłodowska University, Lublin, Poland

Tajana Preočanin Faculty of Natural Sciences and Mathematics, University of Zagreb, Zagreb, Croatia

José Ramos-Barrado Department of Applied Physics, Universidad de Málaga, Málaga, Spain

Alberto E. Regazzoni Department of Chemistry, Atomic Energy Commission, Buenos Aires, Argentina

Władysław Rudziński Department of Theoretical Chemistry, Maria Curie-Skłodowska University, Lublin, Poland

J. Adrián Salfity Department of Chemistry, Atomic Energy Commission, Buenos Aires, Argentina

Philippe Schaad Department of Chemistry, Institut Universitaire de Technologie, Illkirch-Graffenstaden, France

Milan Johann Schwuger Institute of Applied Physical Chemistry, Research Centre Jülich, Jülich, Germany

Vladimir Simeon Faculty of Natural Sciences and Mathematics, University of Zagreb, Zagreb, Croatia

Ryszard Sprycha Technical Center, Sun Chemical Corporation, Carlstadt, New Jersey

Bernd Dieter Struck Institute of Applied Physical Chemistry, Research Centre Jülich, Jülich, Germany

Jerzy Szczypa[†] Department of Radiochemistry and Colloid Chemistry, Maria Curie-Skłodowska University, Lublin, Poland

Kunio Takeda Department of Applied Chemistry, Okayama University of Science, Okayama, Japan

Ching-Fa Wu Project Management Office, Sintong Chemical Industrial Company Ltd., Taoyuan, Taiwan

Jerzy Zając LAMMI, Centre National de la Recherche Scientifique, Montpellier, France

Suzana Žalac Faculty of Natural Sciences and Mathematics, University of Zagreb, Zagreb, Croatia

---

[†] Deceased.

1
Equilibria in Electrolyte Solutions

VLADIMIR SIMEON Faculty of Natural Sciences and Mathematics, University of Zagreb, Zagreb, Croatia

I.	Solutions and Mixtures	2
	A. Mixtures	2
	B. Solutions	3
	C. Ionic solutions	4
II.	Ionic Interactions	5
	A. Ionic cloud models	6
	B. Advanced theories	11
	C. Ion association	15
III.	Hydron Transfer Reactions	21
	A. Brønsted's acids and bases	21
	B. Equilibrium constants	22
	C. Definition and interpretation of pH	26
	D. Complex formation	32
	References	33

Equilibria in solutions of electrolytes is a topic covered in all undergraduate chemistry curricula, so every chemist feels more or less familiar with it. Nevertheless, some parts are rather tricky, requiring a comparatively high level of thermodynamic knowledge. Fortunately, owing to the efforts of many individuals and international bodies, during past two decades the concepts of chemical thermodynamics have been made more transparent. The first part of this chapter contains exact and, we hope, understandable definitions of basic concepts and physical quantities relevant to the subject. The style conforms to the current international recommendations [1], except in some minor details.

The central part of this chapter is a review of ionic interactions, which are still the subject of active research (though this research is not nearly as intensive as it used to be in the first six or seven decades of this century). Besides two classical theories (Debye–Hückel, Bjerrum–Fuoss), which are still the most frequently used by interfacial chemists, some newer theories are also reviewed. Having in view the profile of this book and its presumed readers, almost all mathematical details are omitted. Starting assumptions of each theory are thoroughly expounded while, in most cases, mathematical techniques are only mentioned by name; the main results of the theory are quoted, commented on, and correlated to experimental findings.

As the hydronation/dehydronation* equilibria and the concept of pH are of particular importance in interfacial and colloid chemistry, these topics constitute a significant part of this chapter, although most of the material is not new. Nevertheless, in view of not infrequent misunderstandings, it was deemed important to offer a systematic and fairly rigorous (but not rigid) account.

It is hoped that colleagues working in the fields of interfacial and colloid chemistry will find some of the ideas useful, especially mathematical and numerical techniques (e.g., Monte Carlo and molecular dynamics) used in newer theories of electrolyte solutions.

I. SOLUTIONS AND MIXTURES

As the main subject of this chapter is electrolyte *solutions*, it will be useful to expound some basic concepts and definitions, however trivial they may seem. For example, it is amazing how many chemists are still unaware that the distinction between solutions and liquid mixtures, although conventional [1a], is deeply rooted in the conceptual apparatus of chemical thermodynamics.

A. Mixtures

All components of *mixtures* are treated in the same way. The relative amounts of the components, n_A, n_B, n_C, ..., n_K, ... are usually expressed as mole fractions; for example, the mole fraction of the component B is defined as

$$x_B = \frac{n_B}{n_A + n_B + n_C + \cdots} = \frac{n_B}{\sum_K n_K} \tag{1}$$

The *standard state* of each component of a liquid mixture is defined as the state of pure component, under the standard pressure $p^\circ = 100$ kPa. Note that here, unlike gases, for example, the standard state is experimentally attainable, at least in principle. Values of many thermodynamic quantities, for instance the activity and activity coefficient, depend on the choice of standard states. The *activity* of a component, B say, is defined in terms of its chemical potential, μ_B, *relative* to standard chemical potential:

$$\ln a_B = \frac{\mu_B - \mu_B^\circ}{RT} \tag{2}$$

(therefore the full name of a is 'relative activity' [1a]). The activity coefficient γ_B of a component of mixture B is defined as

$$a_B = x_B \gamma_B \tag{3}$$

When all components of a mixture have unit activity coefficients, the mixture is called *ideal*.

---

* According to IUPAC recommendations [2], the name "proton" should be used specifically for $^1H^+$.
When the isotopic composition is unknown or unimportant the term "hydron" should be used.

Equilibria in Electrolyte Solutions

The activity equilibrium constant for the reaction

$$a\mathrm{A} + b\mathrm{B} + c\mathrm{C} + \cdots \rightleftharpoons p\mathrm{P} + q\mathrm{Q} + \cdots \qquad (4)$$

taking place in a mixture is defined as

$$K_a = \prod_{\mathrm{K}}(a_{\mathrm{K}})^{\nu(\mathrm{K})} \qquad (5)$$

ν_{K} being the *stoichiometric numbers* of the components K = A, B, C, ... , P, Q, ... :

$$\begin{aligned}\nu_{\mathrm{A}} &= -a \\ \nu_{\mathrm{B}} &= -b \\ &\vdots \\ \nu_{\mathrm{P}} &= +p \\ \nu_{\mathrm{Q}} &= +q\end{aligned} \qquad (6)$$

When $p = p^{\circ}$, the activity equilibrium constant becomes the *standard equilibrium constant* K°.

The corresponding stoichiometric equilibrium constant is usually called the *rational equilibrium constant*:

$$K_x = \prod_{\mathrm{K}}(x_{\mathrm{K}})^{\nu(\mathrm{K})} \qquad (7)$$

B. Solutions

The essential feature of a solution is the distinction between its major component, the *solvent*, and the minor one(s), the *solute(s)*. Although most frequently the solvent is just one component, a liquid mixture of constant composition can also be used as a solvent. The relative amount of solvent is expressed as its mole fraction, *not* the concentration. The standard state of a solvent is defined in precisely the same way as for components of liquid mixture.

The most frequently used stoichiometric quantity for solutes is the *concentration*,

$$c_{\mathrm{B}} = \frac{n_{\mathrm{B}}}{V} \qquad (8)$$

However, many physical chemists prefer to use *molality*,* for its important advantage of being temperature independent:

$$b_{\mathrm{B}} = \frac{n_{\mathrm{B}}}{m_{\mathrm{A}}} \qquad (9)$$

(B denotes the solute, A the solvent and m_{A} the mass of solvent).

---

* Instead of the recommended [1] symbol m for molality, the ISO approved symbol b is used, in order to prevent confusion with the symbol for mass.

The choice of *standard states* for solution components is asymmetrical. While the standard state of solvent is defined in precisely the same way as for the components of a mixture, the standard state for a solute is defined as its state in a *hypothetical* ideal solution of standard concentration, $c^\circ = 1$ mol dm^{-3} (or, alternatively, standard molality $b^\circ = 1$ mol kg^{-1}) at the standard pressure of 100 kPa. The solute's activity is defined also by Eq. (2), but the defining formula for the activity coefficient is different:

$$a_B = \frac{c_B \gamma_B}{c^\circ} \tag{10}$$

The solution of solute B is called *ideal* if γ_B equals 1 and does not change upon dilution.

As the solute's standard state ($a_B^\circ = c_B/c^\circ = \gamma_B^\circ = 1$, $p = p^\circ$) is not accessible to experiment, standard thermodynamic quantities are usually determined by using some extrapolation procedure.

The *activity equilibrium constant* for a reaction taking place in solution is defined by a formula derived by substituting Eqs. (3,10) into Eq. (5). The *standard equilibrium constant* is defined as

$$K^\circ = \left\{ (x_A)^{\nu(K)} \prod_K (c_K)^{\nu(K)} \right\} \cdot \left\{ (\gamma_A)^{\nu(A)} \prod_K (\gamma_K)^{\nu(K)} \right\} \cdot (c^\circ)^{-\Sigma} \tag{11}$$

where Σ stands for the sum of stoichiometric numbers, K = B, C, ... , P, Q, ... are labels of the solutes, and A denotes the solvent, which is here, for the sake of generality, assumed to take part in the reaction. (Note that the above equality contains the mole fraction of the solvent, *not* the concentration.) If the solvent does not participate in the reaction, the factors containing A are equal to 1.

The *stoichiometric equilibrium constant** is equal to the left-hand factor enclosed in braces. For sufficiently dilute solutions, x_A is close to 1 and can be omitted.

C. Ionic Solutions

It is well known (but still too often overlooked) that some thermodynamic properties of individual ionic species (notably chemical potential, relative activity, and activity coefficient) are not accessible to physical measurement.

One basic reason for this is the strength of the electroneutrality principle. The chemical potential of an electrically charged species I at constant T and p is formally defined as

$$\mu_I = \left(\frac{\partial G}{\partial n_I} \right)_{T, p, n(J \neq I)} \tag{12}$$

It is easy to show that an immeasurably slight disturbance of the electroneutrality of the solution generates a tremendous electrical potential. For instance, the hypothetical removal of a minute amount of potassium ions, $\delta n(K^+) = 10^{-10}$ mol, from a flask containing 0.5 dm^3 of a KCl(aq) solution ($c = 1$ mol dm^{-3}) would result in a potential of as much as $+87$ kV (at the distance of 1 m). In such circumstances, the attempt to

---

* This name is more appropriate than the frequently used term "concentration constant," which does not cover the cases when the solvent participates in the reaction.

measure the accompanying δG would be devoid of any sense. The same applies to the relative activity and activity coefficient, which are closely related to the chemical potential [*cf.* Eqs. (2,10)].

An alternative way of measuring the ionic activity would be to measure the potential difference between an electrode and a solution, for instance Ag(s) and Ag$^+$(aq). As generally known, such a measurement is also impossible [3].

In contrast to the chemical potential (or activity, or activity coefficient) of an ionic species, the properties of an electroneutral combination of ions are measurable, at least in principle. Of these, the most frequently used in the *mean activity coefficient*, which is defined as the geometric average of the (hypothetical) ionic activity coefficients. For instance, the mean activity coefficient of the ions in a K$_2$SO$_4$ solution is

$$\bar{\gamma}_\pm = \sqrt[3]{\gamma^2(\text{K}^+) \cdot \gamma(\text{SO}_4{}^{2-})} \tag{13}$$

The formula for mean activity is completely analogous:

$$\bar{a}_\pm = \sqrt[3]{a^2(\text{K}^+) \cdot a(\text{SO}_4{}^{2-})} \tag{14}$$

and the (seldom used) mean ionic chemical potential is the arithmetic average of (hypothetical) ionic chemical potentials.

Nevertheless, there is no harm in using the chemical potential (or activity, or activity coefficients) of an individual ionic species in theoretical work, provided their mean remains unchanged throughout.

If the chemical equation of an ionic reaction is written in a form like

$$\text{Cu}^{2+}(\text{aq}) + \text{Cl}^-(\text{aq}) \rightleftharpoons [\text{CuCl}]^+(\text{aq}) \tag{15}$$

i.e., so that the charge numbers do not sum to zero (this is often done when the counterion is not deemed important), the standard equilibrium constant contains the activity of at least one individual ionic species and *sensu stricto* is only a hypothetical quantity. Nevertheless an acceptable estimate of it can be obtained if the mean coefficient is known:

$$K_1^a = \frac{[\text{CuCl}^+]c^\circ}{[\text{Cu}^{2+}][\text{Cl}^-]} \cdot \frac{1}{\bar{\gamma}_\pm} \tag{16}$$

II. IONIC INTERACTIONS

One of the most important aims of early theories of electrolyte solutions was to predict their thermodynamic properties, notably activity coefficients of solutes and osmotic coefficients of solvent (in fact, the latter have been the most frequent actual observables, due to the wide applicability of isopiestic and osmometric measurements).

With the development of diffractometric methods of structure determination (x-ray and, especially, neutron diffraction), a new observable emerged, *viz.*, the pair-correlation function (called also the radial distribution function [4a]). This function $g(r)$, defined as the density of probability of finding two particles at a distance r,

$$dP(r) = 4\pi r^2 g(r) dr \qquad (17)$$

contains important structural information. Also, Friedman [5] has shown how thermodynamic quantities can be calculated from radial distribution functions. Unfortunately, the measurement of radial distribution functions is still comparatively difficult and expensive.

Classical theories of ionic interactions have been based on two apparently different approaches. One of them is the *ionic cloud* model, upon which the celebrated Debye–Hückel theory has been founded, and the other is Bjerrum's concept of *ion association*. The newer theories, mostly based on Mayer's cluster expansion of osmotic pressure, can encompass both approaches, at least in principle.

A third option, the quasi-lattice model, was proposed long ago by Ghosh [6]. In his model, the solution of a 1 : 1 electrolyte is thought of as an expanded, solvent-filled *quasi*-crystal of the NaCl type, with a lattice parameter (mean interionic distance) of $(2Lc)^{-1/3}$ [*cf.* Eq. (26) below]). The idea was resuscitated in the late 1960s, when it was noticed that, in the concentration range from ≈ 0.001 to ≈ 0.1 mol dm^{-3}, log activity coefficients of some alkali halogenides were linearly related to $c^{-1/3}$ rather than to $c^{1/2)}$ (as predicted from Debye–Hückel theory). Since quasi-lattice models were hardly reconcilable with some experimental findings (e.g., ion association, radial distribution functions), they will not be discussed any further here, and the reader is referred to Ref. 7.

A. Ionic Cloud Models

1. Primitive Models

Most of the theories of electrolyte solutions, though differing in many respects, are based upon a common so-called *primitive model* picture: ions are thought of as hard (incompressible) electrically charged spheres dispersed in the solvent—a hypothetical isotropic, structureless dielectric fluid whose macroscopic properties are equal to those of the actual solvent. This model picture has been used in static electrochemistry since the beginning of this century. It has been the basis of the Gouy–Chapman electrical double layer model and of the first theory of ionic interactions by Milner [8a]. Born made use of it in his classical theory of ion–solvent interactions, Debye and Hückel in their celebrated theory of ion–ion interactions and, finally, Bjerrum in his theory of ion association. In spite of the simplicity of the underlying "primitive" picture and some minor inconsistencies, some of these early theories have proved to be qualitatively correct, some of them even approaching the semiempirical level (under favorable conditions). The primitive model is still widely used, both in current theoretical work and in numerical (Monte Carlo and molecular dynamics) calculations.

2. Debye–Hückel Model

In the Debye–Hückel [8a] theory, the nonideality of electrolyte solutions is ascribed solely to electrostatic ion–ion interactions. As each of the ions in the solution is a source of electric field, acting on other ions and being acted upon by them, it is natural to expect that the electrical potential is not uniform throughout the solution but varies from point to point; the same is true for local number densities of ions and charge densities ρ. In other words, ρ is a function of configurational coordinates of ions, as is the local electrostatic potential φ. Under the "primitive" model (spherical ions, isotropic solvent),

Equilibria in Electrolyte Solutions

the only relevant internal coordinates are the interionic distances. As the ions are thought of as hard spheres, there should be a lower limit to the interionic distance; the distance of closest approach will be denoted by σ.

Let α and β be two ions, chosen at will from an arbitrary configuration of ions dispersed in a solvent whose electrical permittivity is ϵ. The fundamental approximation of Debye and Hückel can be expressed as

$$W_{\alpha\beta} = z_\beta e \varphi(r) \tag{18}$$

Potential energy $W_{\alpha\beta}(r)$ can be interpreted as the work required to bring ion β (charge $= z_\beta e$) from infinity to a point at a distance r from the ion α, while $\varphi(r)$ denotes the electrostatic potential due to the presence of *all* ions other than β.

According to the classical Boltzmann–Maxwell statistics, the charge density at a distance r from α is

$$\rho(r) = \frac{1}{V} \sum_\beta z_\beta e \cdot \exp\left(-\frac{W_{\alpha\beta}}{k_B T}\right) \tag{19}$$

V being the volume of solution and k_B the Boltzmann constant. It is easily seen that relations completely analogous to Eqs. (18) and (19) are obtained by averaging over all possible configurations of ions complying to the constraint N, T, V const.* The symbols $\varphi(r)$ and $\rho(r)$ in the subsequent text should therefore be understood as *averages*, $\langle \varphi(r) \rangle$ and $\langle \rho(r) \rangle$.

Since Eqs. (18,19) define a relation between $\varphi(r)$ and $\rho(r)$, they can be combined into the so-called Poisson–Boltzmann equation:

$$\nabla^2 \varphi(r) = -\frac{e}{\epsilon_0 \epsilon_r V} \sum_I N_I z_I \cdot \exp\left(\frac{-z_I e \varphi(r)}{k_B T}\right) \tag{20}$$

The summation here goes over ionic species I instead of individual ions as in the preceding equation; ϵ_0 and ϵ_r denote electrical permittivity of vacuum and relative permittivity (dielectric constant) of solvent, respectively. However, as it stands, Eq. (20) is not quite consistent: it is a hybrid of the Poisson equation (based upon the linear law of superposition of potentials) and the nonlinear (exponential) Boltzmann term on the right side. This contradiction can be avoided by limiting the scope of the theory to the special case when the potential energy of ionic interaction is much lower than the energy of thermal motion of the ions:

$$W_{\alpha\beta} = z_\beta e \varphi \ll k_B T \tag{21}$$

The exponential can then be expanded into a series whose first term vanishes, due to the electroneutrality principle, and (nonlinear) terms higher than the second are neglected. Thus Poisson–Boltzmann equation assumes the form

$$\nabla^2 \varphi(r) = -\frac{e\varphi}{\epsilon_0 \epsilon_r V k_B T} \sum_I N_I z_I^2 \tag{22}$$

---

* All possible configurations constitute a Gibbsian canonical ensemble. A very thorough account of the Debye–Hückel theory can be found in the book by Fowler and Guggenheim [8a].

which does have physically consistent solutions. Using the Poisson equation—a classical electrostatic relation holding true for infinitely divisible ("smeared out") electrical charges rather than for real, discrete charged particles—introduces the other important limitation to the theory. Onsager and Kirkwood [8b] have shown that the errors generated by this approximation will be less serious for low concentrations, large ions, and small ionic charges.

Omitting further and rather lengthy derivation, only the most important formulae will be quoted and briefly commented on. The solution of the Poisson–Boltzmann equation is

$$\varphi(r) = \frac{z_I e}{4\pi\epsilon_0 \epsilon_I r} \cdot \frac{\exp\{-\kappa(r-\sigma)\}}{1+\kappa\sigma} \tag{23}$$

where κ is a characteristic parameter of the theory, having the physical dimension of reciprocal length:

$$\frac{1}{\kappa} = \sqrt{\frac{\epsilon_0 \epsilon_r \kappa_B T}{e^2 \Sigma_I N_I z_I^2}} = \sqrt{\frac{\epsilon_0 \epsilon_r RT}{2F^2 I_e}} \tag{24}$$

F denotes the Faraday constant and I_c the ionic strength (based upon the concentration):

$$I_e = \frac{1}{2} \sum_I c_I z_I^2 \tag{25}$$

Hence the Debye length $1/\kappa$ is directly proportional to $\epsilon_r^{1/2}$ and $T^{1/2}$ and inversely proportional to $I_c^{1/2}$. It was shown that the charge density $\rho(r)$ should have a maximum at $1/\kappa$ [8a]. Therefore the Debye length is usually interpreted as the "mean radius" or "thickness" of the (counter)ion "atmosphere" around the ion, or as the "most probable [ion–counterion] distance." However, at sufficiently low concentrations (for aqueous solutions at 25°C, $c < 1.15 \times 10^{-3}$ mol dm^{-3}), $1/\kappa$, which is proportional to $c^{-1/2}$, becomes *greater* than the average interionic distance, which is proportional to $c^{-1/3}$:

$$\langle l \rangle = (2Lc)^{-1/3} \tag{26}$$

Because of this contradiction, the interpretation of $1/\kappa$ as the "most probable" or "mean" ion–counterion distance should be used with great caution, if at all.

The most important result of the Debye–Hückel theory is the formula for ionic activity coefficient:

$$-\lg \gamma_I = \frac{A_c z_I^2 \sqrt{I_c/c^0}}{1+\kappa\sigma} \tag{27}$$

where A_c(water, 298 K) = 0.509 is the first Debye–Hückel constant and σ is the distance of closest approach, i.e., the sum of ionic radii (see below). By means of Eq. (13), activity coefficients of cation(s) and anion(s) can be combined into the mean activity coefficient

$$-\lg \bar{\gamma}_\pm = \frac{A_c |z_+ z_-| \sqrt{I_c/c^0}}{1+\kappa\sigma} \tag{28}$$

which, unlike γ_I, is a measurable quantity. Due to solvation, as well as to the approximations introduced in the derivation, σ is greater than the sum of crystallographic ionic radii and has to be treated as an adjustable parameter.

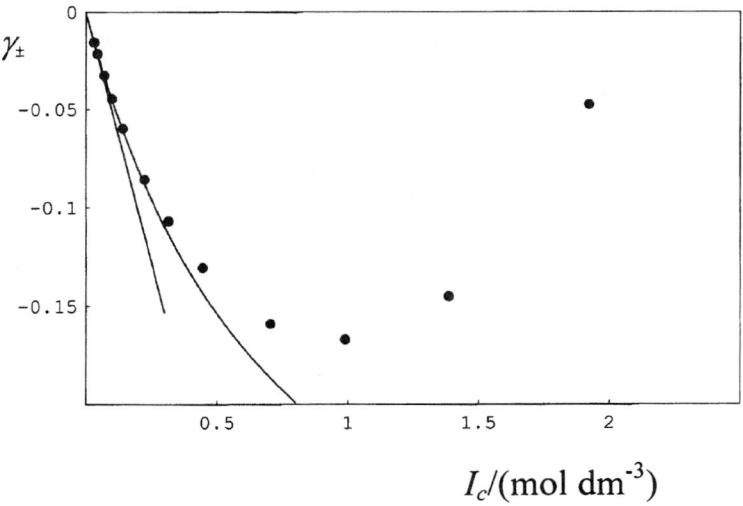

FIG. 1 Mean activity coefficients of NaCl(aq): Stokes and Robinson's [8e] experimental values (points) and predictions from Eqs. (29) (line) and (28) (curve).

In very dilute solutions ($I_c \leq 0.001$ mol dm^{-3}), $\kappa\sigma$ becomes small compared to unity, and Eq. (28) reduces to the well-known Debye–Hückel limiting law containing no adjustable ion-size parameter:

$$-lg\,\bar{\gamma}_\pm = A_c \mid z_+z_- \mid \sqrt{I_c/c^o} \tag{29}$$

The limiting formula for γ_I can be obtained from Eq. (27).

As is generally known, Eqs. (28,29) are very important formulae in experimental chemical thermodynamics and are used for extrapolation of various experimentally determined quantities to $I_c \to 0$. The agreement of these two formulae with the experimental activity data for NaCl(aq) at 25°C is depicted in Fig. 1. It is seen that the limiting law Eq. (29) fits the data correctly for ionic strengths below, approximately, 0.001 mol dm^{-3} (unfortunately, there are rather few reliable data for $I_c < 0.001$ mol dm^{-3}), while the validity of Eq. (28) extends up to ca. 0.1 mol dm^{-3}, provided a suitable value of σ is chosen. It is not surprising that Debye–Hückel formulae cannot predict the minimum in $\gamma_\pm$ appearing, for most strong electrolytes, at $I_c \geq 0.3$ mol dm^{-3} where the basic assumption of Eq. (21) is no longer true.

3. Extensions of the Debye–Hückel Model

(a) "Exact" Solution of the Poisson–Boltzmann Equation. Several amendments to the theory of Debye and Hückel have been proposed, one of the earliest being the attempt of LaMer et al. [8b]. If, instead of linearizing the exponential in the Poisson–Boltzmann equation (20), the sum on its right-hand side is split into two terms, cationic and anionic, one obtains an expression containing $\sinh(ze\varphi/k_BT)$. However, as shown by Fowler and Guggenheim [8b], the solution of the thus transformed Poisson–Boltzmann equation necessarily leads to inconsistent results, a consequence of the inherent contradiction of the Poisson–Boltzmann equation discussed above.

(b) Virial Expansion of Debye–Hückel Formula. The other, more successful, approach is based upon Brønsted's principle of specific interactions of ions: "Ions are uniformly influenced by ions of their own sign and specifically influenced by ions of opposite sign" (quoted from Ref. 8c). In other words, due to the much greater probability of encounters of oppositely charged ions, the interactions of ions bearing like charges are much weaker and, as an approximation, can be neglected. Taking as reference a (real or virtual) 1 : 1 electrolyte obeying Eq. (28), Guggenheim [8d] proposed a virial expansion of Eq. (28). For a mixed electrolyte the virial formula has the form

$$\ln \frac{\bar{\gamma}_{MX}}{\bar{\gamma}_{ref.}} = \sum_{X'} \beta_{MX'} c_{X'} + \sum_{M'} \beta_{M'X} c_{M'} \tag{30}$$

β being the empirical "coefficients of specific ion interaction." For a single-electrolyte solution, Guggenheim's formula reduces to

$$\ln \frac{\bar{\gamma}_{MX}}{\bar{\gamma}_{ref.}} = 2\beta_{MX} c \tag{31}$$

in accordance with the well-known Hückel equation.

However, one should be aware that the range of validity of Eq. (30) and similar linear formulae is limited, even under the primitive model. Therefore Pitzer and associates [9] in the 1960s and 1970s introduced additional cross-terms into the formulae intended for use at higher ionic strengths and/or with mixed electrolytes, in order to allow for nonlinear chemical effects. As the extension of Eq. (30) and similar equations requires introducing additional empirical parameters, the practical user will frequently be confronted with the lack of necessary empirical data. In spite of this, Guggenheim's and Pitzer's method of virial expansion is a logically consistent and practically useful phenomenological approach. Besides, Pitzer's virial coefficients can be, at least approximately, derived by methods of statistical thermodynamics, i.e., interpreted in terms of ions and molecules (see Section II.B.1).

(c) Solvation Model. The importance of ion hydration has been stressed long ago, in pre-Debye–Hückel time, by Bjerrum. Several experimental techniques (e.g., neutron diffraction, NMR, compressibility measurements) can yield rough estimates of the hydration numbers, i.e., the numbers of solvent molecules in primary solvation sheaths of ions. For univalent ions in water, hydration numbers generally range from 2 to 8; the values for Na^+ and Cl^- were estimated to be 5 ± 0.6 and 2 ± 0.8, respectively [10a].

Consider the following example: the mole fraction of "free" water in 1 molal aqueous NaCl, taking "naked" Na^+ and Cl^- ions as stoichiometric entities, is $x_w = 55.5/(55.5+1) = 0.98$. However, if the above estimates of hydration numbers are true, the effective mole fraction of "unbound" water is substantially lower, $x'_w = (55.5 - 5 - 2)/56.5 = 0.86$, which means that the activity of water should be lower than would be expected from simple stoichiometry.

Stokes and Robinson [11] devised a simple yet apparently powerful formula for the mean activity coefficient of an electrolyte containing v ions that 'bind' N_{hyd} molecules of water with an interaction energy significantly larger than $k_B T$. Besides the Debye–Hückel term of Eq. (28), their formula contains two terms allowing for the effect of hydration:

$$\lg \bar{\gamma}_\pm = -\frac{A_c |z_+ z_-| \sqrt{I_c/c^o}}{1 + \kappa \sigma} - \frac{N_{hyd}}{v} \cdot \lg a_w - \lg[1 - (N_{hyd} - v) b M_w] \tag{32}$$

here b denotes molality of the solute, $c^\circ = 1$ mol dm^{-3}, and M_w is molar mass of water; $a_w(b)$ is the experimentally determined activity of water. Although the Stokes–Robinson formula contains only two adjustable parameters, *viz.*, σ and N_{hyd}, it was found to reproduce experimental values of solute activity coefficients of 35 single electrolytes (charge types 1 : 1 and 2 : 1) in wide molality ranges (up to 1 mol kg^{-1} or more), all absolute mean deviations being below 0.005. Estimated hydration numbers ranged from 0.6 (RbI) to 20 (Zn(ClO$_4$)$_2$), and distances of closest approach from 3.6 Å (RbI) to 6.2 Å (Zn(ClO$_4$)$_2$). Most of the Stokes–Robinson estimates of hydration numbers are lower than the values compiled by Bockris [10a].

The use of the Debye–Hückel term in Eq. (32) has been criticized from the theoretical viewpoint because at high ionic strengths (≥ 1 mol dm^{-3}, say) the concept of smeared-out "ionic atmosphere" becomes utterly unrealistic [10b]. An obvious explanation for the predicting power of the Stokes–Robinson formula seems to have been overlooked in the literature: the activity of water, a_w, used as the predictor of $\gamma_\pm$ in Eq. (32), contains exactly the same information as $\gamma_\pm$ (because of the Gibbs–Duhem equation). This means that the formula is a kind of tautology.

B. Advanced Theories

1. Mayer's Virial Expansion of Osmotic Pressure

Mayer, the author of successful theories of real gases and liquids, proposed in 1950 [12] a more systematic and general approach to the problem of ionic solutions, also based upon the primitive model. His theory, encompassing—at least in principle—the two earlier, more specialized approaches, has exerted a profound influence on most of the subsequent work in the field. Because of mathematical complexity, only the fundamental ideas and assumptions of Mayer's theory will be sketched.

Analogously to his theory of real gases, Mayer assumed that clusters of ions are formed in solution, their sizes being $v = v_M + v_L$. Applying, *mutatis mutandis*, his and McMillan's treatment of liquid mixtures [13], he developed the concentration dependence of the osmotic pressure π of the solution into the virial series:

$$\frac{\pi}{k_B T} = C_M + C_L - \sum_{2}^{v}(v_M + v_L - 1)B_v C^v \tag{33}$$

where C denotes the number concentration. The vth virial coefficient, $-(v-1)B_v$, refers to all possible clusters containing v ions. Each B_v may be expressed as an integral of a function depending on the coordinates of the v ions. Assuming that the potential energy of a v-cluster W_v can be approximated by the sum of the pair potential energies $w(R)_{ij}$,

$$W_v = \sum_{i=1}^{v}\sum_{j=1}^{i-1} w(R_{ij}) \tag{34}$$

and neglecting the interactions of three or more ions, the integrand of B_v will be a sum of products of the functions

$$f_{ij}^* = \exp\left[\frac{-w(R_{ij})}{k_B T}\right] - 1 \tag{35}$$

The pair potential can be approximated by the function

$$w(R_{ij}) = \frac{z_M z_L e^2}{4\pi\epsilon R_{ij}} \exp(-\alpha R_{ij}) + w^*(R_{ij}) \tag{36}$$

where the second term represents the contribution of short-range forces, which, under the primitive model, is

$$\begin{aligned} R_{ij} \leq \sigma_{ij} & \quad w^*(R_{ij}) = \infty \\ R_{ij} > \sigma_{ij} & \quad w^*(R_{ij}) = 0 \end{aligned} \tag{36a}$$

The electrostatic term in Eq. (36) contains the function α which is similar to the Debye–Hückel κ but takes into account the concentration dependence of the solution permittivity. Introducing the exponential factor into the expression for electrostatic potential energy ensures the convergence of the integrals constituting the virial coefficients. For $i \neq j$ such an integral has the form

$$B_{ij} = \frac{1}{2} \int_\sigma^\infty 4\pi R_{ij}^2 f_{ij}^* \, dR_{ij} \tag{37}$$

This smart stratagem has its physical justification because the exponential factor in Eq. (23) accounts for the screening effect exerted by ions other than i and j.

The rest of Mayer's exceedingly massive and complex derivations will be omitted. Suffice it to say that, in order to render the problem solvable, various additional assumptions and approximations (e.g., truncations of infinite series) had to be introduced, lowering the upper limit of the theory's applicability. On the other end, the formulae of Mayer's theory for $\gamma_\pm$ at low ionic strengths reduce to Debye–Hückel Eqs. (28) and (29).

2. Mean Spherical Approximation (MSA)

Several theories of liquid-state physics have been applied to electrolyte solutions. Some of these theories also use expansion into series of powers of density or concentration, but instead of osmotic pressure (used by McMillan and Mayer), the pair correlation function, $\ln g_{ij}$ is expanded. The results of various expansions are integral equations that can be solved by introducing some approximations (known under the names of Percus–Yevick, Ornstein–Zernike and hypernetted-chain); for basic information see Refs. 4b, 10c, and 14. Their solution requires the introduction of further simplifying approximations, such as the *mean spherical approximation*, introduced by Waisman and Lebowitz [15], who found the analytical solution of the Ornstein–Zernike equation under the restricted primitive model (symmetrical electrolyte, $r_+ = r_- = \sigma/2$). Their formula for the pair correlation function can be written as

$$g(r_{ij}) = -\frac{z_i z_j e^2}{4\pi\epsilon_0 \epsilon_r k_B T} \cdot \exp(-2\Gamma r_{ij}) \tag{38}$$

Γ being a parameter related to the Debye–Hückel κ. For ions of the same size, Γ is given by

$$\Gamma = \sqrt{\frac{\sqrt{1+2\sigma\kappa} - 1}{2\sigma}} \tag{39}$$

Fawcett and Tikanen [16] estimated separately the electrostatic and hard-spheres contri-

Equilibria in Electrolyte Solutions

butions to $\gamma_\pm$ in solutions of alkali halogenides by

$$\ln \bar{\gamma}_\pm = \ln(\bar{\gamma}_\pm)_{es} + \ln \bar{\gamma}_{hs} \qquad (40)$$

The *electrostatic contribution* was calculated by means of the formula

$$\ln(\bar{\gamma}_\pm)_{es} = -\frac{|z_+ z_-|e^2}{4\pi \epsilon_0 \epsilon_r(c) k_B T} \cdot \frac{\Gamma}{1 + \Gamma \sigma} \qquad (41)$$

taking into account the concentration dependence of relative permittivity $\epsilon_r(c)$, which is an almost linear decreasing function of concentration.

The *hard-spheres contribution* to the activity coefficient was obtained from the Percus–Yevick model for noninteracting hard spheres:

$$\ln \bar{\gamma}_{hs} = 6\frac{\eta}{1-\eta} + 3\frac{\eta^2}{(1-\eta)^2} + 2\frac{\eta}{(1-\eta)^3} \qquad (42)$$

where the packing fraction η is defined as

$$\eta = \frac{\pi \sigma^3}{6} \sum_i c_i \qquad (43)$$

It should be noted that the Fawcett–Tikanen γ is defined on the concentration, rather than molality, scale.

Fawcett–Tikanen predictions of γs for a number of alkali (Li^+, Na^+, K^+, Rb^+, Cs^+) halogenides (F^-, Cl^-, Br^-, I^-) agree well (within ± 0.005) with the experimentally determined values in broad concentration ranges: maximum 3 mol dm^{-3} for KBr, minimum 0.5 mol dm^{-3} for LiI and CsF (MSA cannot be expected to work well when the ion sizes are very different). Thus their formula, yielding about as good predictions as Stokes and Robinson's Eq. (32), has three important advantages: it is theoretically consistent, it has only one adjustable parameter σ, and it needs no empirical data. An important feature of MSA is, probably, that the finite sizes of *all* ions are accounted for, not just that of the central ion.

3. Higher Level Theories

Although the primitive models are still being actively investigated and gradually improved, the work at higher levels of sophistication is also progressing. The first higher level of models was called by Friedman "Born–Oppenheimer (BO) level" [17]. In this class of models the solvent is no longer treated as a dielectric continuum but is usually thought of as a collection of (either hard or soft) polar spheres, and the ions are modelled as (either hard or soft) charged spheres. There has been some initial work on the next higher levels, called "Schrödinger (S) levels" where the solution is thought of as consisting of more realistic entities, such as either (a) electrons and atomic kernels (*i.e.*, nuclei and inner electrons) or (b) electrons and nuclei. This work falls outside the scope of this short review; the reader can seek more information in the texts by Pitzer [9], Bockris [10], and Friedman [17].

4. Numerical Simulations

The horrifying complexity of mathematical apparatus used in advanced theories and the steadily growing power of electronic computers motivated efforts to get predictions from physical models by numerical simulation instead of trying to solve hardly soluble or insoluble mathematical problems. Obviously, the simulation approach is well suited to the problems of statistical mechanics. Most numerical simulations have been done under some kind of primitive model.

(a) Monte Carlo Method. The total interaction energy of a system consisting of a sufficiently large number N of particles (ions or molecules) can be computed provided the energies of interactions among them are known or assumed in advance [for example, under the primitive model, pairwise interaction energies are given by Eqs. (36, 36a), and higher interactions are usually ignored]. If a sufficient number, M of *random* configurations of the N particles is generated, the total interaction energy U_j, of each configuration, can also be evaluated. The same is true for any physical property Y that is, a known or assumed function of particles' coordinates. Therefore the statistical expectance $\mathcal{E}(Y)$ can be estimated by means of the well-known Botzmann formula:

$$\mathcal{E}(Y) = \sum_{j=1}^{M} P_j Y_j = \frac{\sum_{j=1}^{M} Y_j \exp(U_j/k_B T)}{\sum_{j=1}^{M} \exp(U_j/k_B T)} \tag{44}$$

However, such an elementary approach (sometimes called the crude Monte Carlo method) is computationally very inefficient because too many of the randomly produced configurations will have some particles lying on top of one another and consequently will have an infinite energy. The true Monte Carlo algorithms (like the one first proposed by Metropolis) contain efficient sets of rules for accepting (or discarding) randomly generated configurations (see Ref. 4c for additional information).

Among the early Monte Carlo (MC) studies, the paper by Card and Valleau [18] has received much more attention than the preceding works by Poirier and Shaw [19] and Vorontsov-Vel'yaminov *et al.* [20] Card and Valleau studied a "typical" 1 : 1 electrolyte ($\sigma = 4.25$ Å) using, besides the MC method, an extended (nonlinear) Debye–Hückel equation and two approximate solutions of integral equations (see Section II.B.1). Their MC prediction of $\log \gamma_\pm(I_c^{1/2})$ function had a qualitatively correct shape (i.e., it possessed a minimum). Contrary to some interpretations in the secondary literature, the range of quantitative agreement of the MC predicted activity coefficients with empirical values (up to ≈ 0.1 mol dm^{-3} for LiBr, < 0.1 mol dm^{-3} for LiI and NaI) was not much broader than the range of validity of the Debye–Hückel formula (28).

(b) Molecular Dynamics. The initial configuration in this method is usually a regular array of particles that are assigned the momenta in a random manner. Assuming the form of interparticle forces, the calculations are carried out by numerically integrating coupled differential equations of motion. The influence of the initial configuration dies out after some time and, from then on, the results (configurational coordinates and momenta of the particles) are accumulated and used to compute the radial distribution function or any other observable.

C. Ion Association

1. Bjerrum's Theory

A different approach to the problem of charged hard spheres in an isotropic dielectric continuum has been proposed by Bjerrum [8e], who, instead of using the Poisson–Boltzmann equation, took into consideration only the interactions between the neighboring ions, attempting to evaluate the probability of finding an ion, α say, at a distance r from another one, β. This probability should directly depend on the number concentration of *all* ions, $C = cL$, and inversely (in accordance with Boltzmann's law) on the potential energy of electrostatic interaction, $W_{\alpha\beta}/k_B T$ (ignoring the effects exerted by other, farther ions). The core of Bjerrum's formulae is his tentative expression for the density of probability of finding an ion at a given distance from another ion:

$$C \cdot \left\{ \exp\left(\frac{-W_{\alpha\beta}}{k_B T}\right) \right\} \cdot 4\pi r_{\alpha\beta}^2 = C \cdot \left\{ \exp\left(-\frac{z_\alpha z_\beta e^2}{4\pi \epsilon_0 \epsilon_r r_{\alpha\beta} k_B T}\right) \right\} \cdot 4\pi r_{\alpha\beta}^2 \qquad (45)$$

(the reader will note that Bjerrum in fact modelled the pair distribution function introduced in Eq. (17) at the beginning of this section). Bjerrum showed that this density of probability has a minimum at a characteristic "critical" distance usually denoted as q:

$$q = -\frac{z_\alpha z_\beta e^2}{8\pi \epsilon_0 \epsilon_r k_B T} \qquad (46)$$

for example, in water at 25°C, $q = 3.6$ Å. Although the function (45) does not meet the criteria for a probability density (its integral diverges when $r \to \infty$ so the normalization constant C does not exist), Bjerrum, focussing his attention onto small interionic distances, $r \leq q$, arrived at reasonable results.

The pair of oppositely charged ions whose distance lies in the range $\sigma \leq r \leq q$ (i.e., between the distance of closest approach σ and the "critical" distance q) was defined by Bjerrum as an *associated ion pair*. Bjerrum's expression for the fractional concentration of the associated ion pairs in a symmetrical binary electrolyte can be written as

$$\alpha = 4\pi C \int_\sigma^q r^2 \left\{ \exp\left(\frac{z^2 e^2}{4\pi \epsilon_0 \epsilon_r k_B T r}\right) \right\} dr \qquad (47)$$

The integral is anything but elementary but can be solved numerically, after a suitable variable change, $2q/r = y$. It can be seen that (a) α is proportional to the total ionic concentration, (b) α is inversely related to the dielectric constant (relative permittivity) of the solvent and to (c) ionic radii that determine the lower integration limit σ. Once α is known, the concentration stability constant $K_c(\text{assn.})$ of associated ion pair is easily computed:

$$K_c(\text{assn.}) = \frac{\alpha c^o}{(1-\alpha)^2 c} \qquad (48a)$$

At sufficiently low concentrations, where Bjerrum's neglect of the effects of "remote" ions (i.e., those whose distance from the "central" ion is greater than q) is an acceptable approximation, the concentration stability constant should be close to the standard thermodynamic constant; more precisely, $c^o K_c(\text{assn.}) \approx K^o(\text{assn.})$.

Attempting to allow for the effects exerted by ions lying at $r>q$ and thus extend the range of the validity of Eq. (48a), several authors [8e,10c] introduced activity coefficients. For a symmetrical electrolyte, the activity coefficient of the associated ion pair (an electric dipole, in fact) can be approximated by unity, ignoring the effect of its electrostatic field. The mean ionic activity coefficient can be estimated theoretically, e.g., by means of the Debye–Hückel (Eq. 28) or a similar formula, taking into account only nonassociated ions when computing I_c and κ. Because of the limited range of validity of the available formulae for estimating $\gamma_\pm$, the modified formula

$$K^\circ(\text{assn.}) = \frac{\alpha c^\circ}{(1-\alpha)^2 (\bar{\gamma}_\pm)^2 c} \tag{48b}$$

also has a limited range of applicability.

Although some experimental data supported Bjerrum's theory (see Section II.C.4.a), his choice of q as critical distance for distinguishing between associated ion pairs and free ions has repeatedly been criticized for its alleged arbitrariness. However, having in mind the statistical nature of ion association, it is readily seen that the choice of a dividing radius between associated and free ions must be arbitrary to some extent. Moreover, Fuoss (see below) convincingly demonstrated that Bjerrum's choice of critical distance was reasonable.

Because of the already mentioned divergence of Bjerrum's probability integral, making its normalization impossible, Eq. (47) may yield physically impossible results, even for moderately concentrated solutions.

2. Fuoss's Revision

Fuoss [8e, 21] devised a smart definition of an ion pair (associated or not): a cation and an anion, whose center lies at a distance between r and $r + dr$ from the cation, are considered as an ion pair, provided no other unpaired anion lies within the sphere of radius r centred at the cation. Fuoss's revised version of Bjerrum's theory, built upon this definition, will be summarized in the following paragraphs.

Let $G(r)\,dr$ be defined as the probability dP that an anion j lies at a distance between r and $r + dr$ from the other member of the pair, cation i. The probability density function $G(r)$ can be expected to be proportional to C_j, to $4\pi r^2\,dr$, to the Boltzmann factor from Eq. (45), and finally to the probability that an unpaired anion is not already present in the sphere of radius r centred at i, which is, obviously,

$$1 - \int_\sigma^\infty G(x)\,dx \tag{49}$$

It should be pointed out that here also the screening by other ions is neglected altogether. Omitting further derivation, the expression for Fuoss's probability density $G(r)$ can be written as

$$G(r) = C_j \cdot 4\pi r^2 \cdot \exp\left(\frac{2q}{r - 4\pi C_j \cdot \int_\sigma^r x^2 e^{2q/x}\,dx}\right) \tag{50}$$

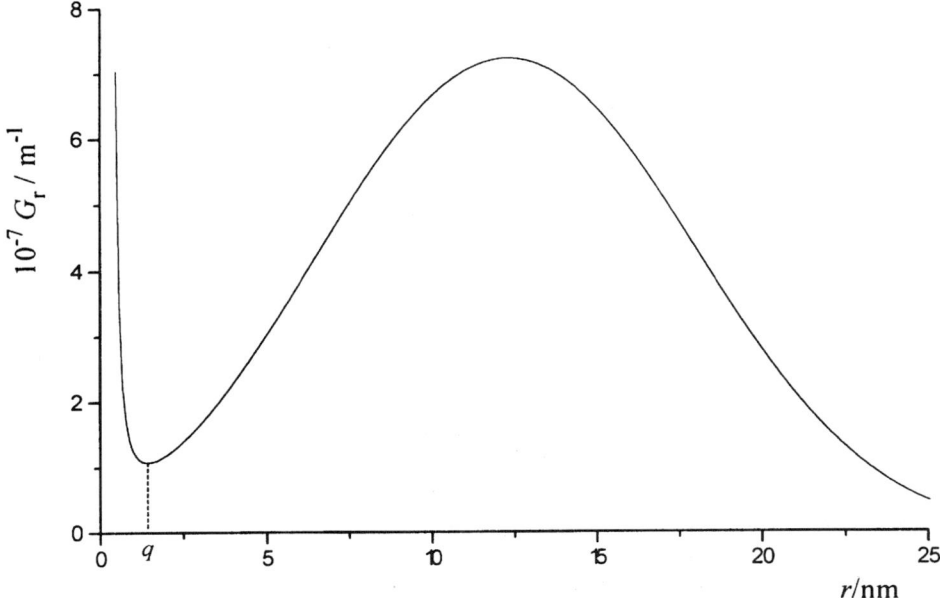

FIG. 2 Fuoss's probability density of distance between two ions, Eq. (50).

(again, σ is an adjustable parameter). It may be shown that (unlike Bjerrum's) Fuoss's probability density function is normalized:

$$\int_\sigma^\infty G(r)\,dr = 1 \tag{51}$$

so that the fraction of associated ion pairs can be computed by integrating $G(r)$ over the range $[\sigma \leq r \leq q]$ (however tedious the integration may be). A graph of the $G(r)$ function is shown in Fig. 2.

$G(r)$ has a maximum at approx. $(2\pi C_j)^{-1/3}$ and a comparatively deep minimum near q. It can be seen from the graph that in the neighborhood of $r \approx q$ there are very few ion pairs, compared with either the number of associated pairs (near σ) or that of nonassociated ones (near the maximum). Obviously, the choice of dividing radius is not critical, and the allegations for arbitrary choice of q (however frequently heard) are groundless. Unfortunately, Eq. (50), being difficult to handle numerically, found little practical use.

3. Second Fuoss Model

A couple of decades after the above contribution, Fuoss [22] proposed another model of ion association, without using *any* critical distance like q. He redefined the associated ion pair as an *immediate contact* of two oppositely charged ions, irrespective of its duration. One of the ions in contact is thought of as a virtual charged sphere whose radius is equal to the sum of ionic radii, $\sigma = r_+ + r_-$, while the other is represented by an opposite *point charge* located at the distance σ from the center of the first ion. According to Eq. (23) (borrowed from Debye–Hückel theory), an anion in contact with a cation (taken as the "central" ion) experiences the electrical potential

$$\phi(r = \sigma) = \frac{z_+ e}{4\pi\epsilon_0\epsilon_r\sigma(1 + \kappa\sigma)} \tag{52}$$

and its potential energy is

$$W = -\frac{z_+ z_- e^2}{4\pi\epsilon_0\epsilon_r\sigma(1 + \kappa\sigma)} \tag{53}$$

The calculation of the fraction of associated ion pairs is based on the following thought experiment. Suppose that in a solution of a 1 : 1 electrolyte containing Z cations and Z anions in volume V, there are $Z_+ = Z_-$ free ions and $Z_\pm$ associated pairs. Upon adding a minute (but finite) quantity δZ of cations and an equal number of anions into the solution, part of them will associate so that

$$\delta Z = \delta Z_+ + \delta Z_\pm \tag{54}$$

The number of nonassociated cations can be expected to be proportional to the number of ions added δZ and to the available volume $V - Z_+ v_+$ ($v_+ = 4\pi\sigma^3/3$ denotes the volume of a cation under the Fuoss model). The probability of ion association is proportional to the amount added δZ, to the volume occupied by the cations $Z_+ v_+$, and to the Boltzmann probability $\exp(- W/k_\text{B} T)$. The rest of the derivation may be skipped, with the remark that the available volume is taken to be approximately equal to V [this simplification, though not necessary, is justified once Eq. (52), valid for dilute solutions, has been adopted]. The final formula for the association equilibrium constant,

$$K_\text{assn.} = \frac{4\pi\sigma^3 L}{3} \cdot \exp\left[\frac{2q}{\sigma 1 + \kappa\sigma)}\right] \tag{55}$$

is frequently written in the approximate form

$$K_\text{assn.} = \frac{4\pi\sigma^3 L}{3} \cdot \exp\left(\frac{2q}{\sigma}\right) \tag{56}$$

valid for dilute solutions where $1 + \kappa\sigma \approx 1$. The results computed by means of Eq. (55) or (56) are similar to those obtained by using the expression

$$K_\text{assn.} = \frac{4\pi\sigma^4 L}{q} \cdot \exp\left(\frac{2q}{\sigma}\right) \tag{57}$$

derived by Fuoss and Kraus [23] from Bjerrum's formulae (47,48) for solvents with low permittivities. However, the simple Fuoss formulae (55) or (56), having a sounder theoretical basis, are to be preferred.

4. Experimental Evidence for Ion Association

Both theories of ion association, Bjerrum's and Fuoss's, are focussed on the interaction of *only two* oppositely charged ions,* largely neglecting the effects exerted by the remaining, more distant ions. Although this model picture is certainly exaggerated, there is ample experimental evidence of its undeniable virtues:

> The phenomenon of ion association has not been only inferred from conductivity data but was unequivocally detected by direct methods: vibrational and electronic spectroscopy, x-ray and neutron diffractometry (see below).
> The extent of association and stability constants of associated ion pairs, measured in a large number of instances, were found to depend on charge type (2:2 > 2:1 > 1:1) and solvent permittivity as predicted by these theories, at least qualitatively.
> Ultrasonic absorption, besides giving the evidence of ion association, made it possible to distinguish between solvent-separated and contact ion pairs [24, 25].
> Kinetics of many chemical reactions could hardly be explained without assuming ion association mechanisms [26].

(a) Conductometric Data. Bjerrum's concept of ion association was shown to be in reasonable agreement with experimental data. James *et al.* [27] found that the dependence of $K_{ass.}$ of La[Fe(CN)$_6$] on ϵ_r (57 < ϵ_r < 79) in a series of dilute solutions in partly aqueous solvents was in accordance with predictions from Bjerrum's theory. The same is true for the results of Fuoss and Kraus [23] with tetraisoamylammonium nitrate in water/dioxan. Patterson *et al.* [27] explained the high-field Wien effect exhibited by the solutions of MgSO$_4$, CuSO$_4$, ZnSO$_4$, and La[Fe(CN)$_6$] by assuming the formation of Bjerrum's associated ion pairs.

(b) Electronic Spectra. Shifts of UV and/or visible absorption bands with changing concentration have long been observed for a number of solutions of inorganic salts with polyatomic anions, either in water or in organic solvents [28, 29] and tentatively interpreted in terms of electronic structure of cations and anions [30, 31].

Applying a factor-analytical method to nitrate bands in UV spectra of LiNO$_3$(aq) and Co(NO$_3$)$_2$(aq), Tomišić and Simeon [32, 33], besides detecting the formation of associated ion pairs of 1:1 stoichiometry, were able to determine the equilibrium concentrations, as well as to isolate the spectra of ion pairs, in each solution where association took place [>0.8 mol dm^{-3} for LiNO$_3$, >0.2 mol dm^{-3} for Co(NO$_3$)$_2$]. Apparent stability constants were found to steadily increase with increasing salt concentration, reaching approx. 0.5 mol^{-1} dm^3 at highest concentrations. As the absorption bands of [LiNO$_3$]^0 and [Co(NO$_3$)]$^+$ were shifted in opposite directions (relative to unassociated NO$_3^-$), they concluded that the former was more like a classical ion pair and the latter to a complex.

---

*Although there is some experimental evidence pointing to the possible formation of ion clusters (triples or even quadruples), the phenomenon was reported only for solvents with low relative permittivities (<15).

(c) Raman Spectra. Because of its applicability in aqueous systems, Raman spectroscopy has been widely used in studies of ion association [34]. Various vibration bands of polyatomic anions have been found to change their positions with the changing total salt concentration [35]. Some workers attempted to determine the equilibrium concentrations of associated ion pairs [36], most of them by *a priori* assuming the band shapes of spectrally active species. Tomišić and Simeon [32], applying a similar factor-analytical method as in the case of UV spectra (see above), determined the concentration dependence of apparent stability constants of (also 1 : 1) associated ion pairs and isolated the spectra of free and associated anions in aqueous solutions of $LiNO_3$, $Co(NO_3)_2$, $Mg(NO_3)_2$, Li_2SO_4, and $MgSO_4$. By comparing UV and Raman results it is seen that, surprisingly, the ranges of existence of associated species differ substantially, although the apparent stability constants are of a similar order of magnitude ($\sim 10^{-1}$ mol^{-1} dm^3). For instance, the association in $LiNO_3$ solutions was detectable by UV spectrometry at 0.8 mol dm^{-3}, while Raman spectra indicated the emergence of a new species only near 2 mol dm^{-3}. The explanation of these findings may require a thorough reconsideration of the concept of "spectrally active species."

(d) Ultrasonic Absorption. In their classical study of the absorption of ultrasound by aqueous solutions of 2 : 2 electrolytes, Eigen and Tamm [24] concluded that the ion association was a multistep process. In the associate formed in the first step, both ions retain their solvation sheaths; in the next step they associate more closely and share one or more solvent molecules; and in the last step a contact ion pair is formed. They were able to compute not only the equilibrium constants but also the rate constants for all these reactions. Gilligan and Atkinson [25] studied the ultrasonic absorption in the solution of sulfates of alkali metals.

5. Association vs. Coordination

Many, if not most, chemists use the terms "complex" and "coordination compound" interchangeably, as synonyms, although the first term is also used with quite different meanings (for instance, "activated complex" in chemical kinetics). Therefore many inorganic chemists prefer the latter, more specific term. In spite of its importance and the huge size of this class of compounds, the term "coordination compound" still seems to lack a concise definition [37]. Although the classical Lewis–Sidgwick definition of coordinative bond ("donation" of electron pair(s) by a Lewis base to a Lewis acid) covers a majority of coordination compounds met with in usual practice, some important cases are left out, for instance delocalized bonds, "back-donation" by π-bonding ligands, and complexes of cations with noble-gas electron configuration ($[Ca(edta)]^{2-}$, cryptates). A practical definition of a coordination—or complex—compound, just for the needs of the present discussion, would be the following: it is an entity containing one or more metal atoms or ions, each of them in a close ("bonding") contact with at least one nonmetal or semimetal atom or ion.

Experimental (mainly spectroscopic) results commented on in the preceding section suggest that ion association should not be regarded as a pure electrostatic phenomenon: besides electrostatic (coulombic, van der Waals, London) forces, chemical (i.e., quantum-mechanical exchange) forces are also contributing to the mutual affinity of oppositely charged ions. Therefore associated ion pairs can be considered as a limiting special case (of negligible chemical bonding) within the broader class of coordination compounds. In this context, it should be pointed out that different experimental methods detect different interactions in solution. For instance, the conductometric stability constant of $[MgSO4]^0$ ion pair in aqueous solution was found to be $\geq 10^2$ [38], while Raman spectrometry yielded values of the order of 10^{-1} [32,34,39]. As another example, UV and Raman results obtained by Tomišić and Simeon with $LiNO_3$ and $Co(NO_3)_2$ [32,33] suggest that the existence of a characteristic spectral pattern need not necessarily mean that a new chemical species has been formed.

Quite often Debye–Hückel and Bjerrum–Fuoss approaches are interpreted as two opposed theories, although in fact they are complementary. It can easily be seen that Mayer's theory (Section II.B.1) relies on the concept of ion clusters whose number and sizes are not limited *a priori*. If, in a particular case, there are only one or two dominant clusters it will probably be useful to adopt the concept of ion association. In the opposite case of no prevailing cluster size, activity coefficients should be appropriate.

Although the primitive models can still be improved to some extent, future research in electrolyte solutions will probably have to be promoted to higher, Born–Oppenheimer or Schrödinger, levels, which is a tremendous task.

III. HYDRON TRANSFER REACTIONS

Acid–base equilibria often play an important role in the processes taking place at solid|electrolyte solution interfaces. The solution itself may contain acidic and/or basic components whose interactions with the surface of the solid phase may be dramatically different; in such situations a detailed knowledge of actual concentrations of dissolved species is necessary. In other, more complex cases, the solid may have acidic or basic functional groups interacting with the components of the solution.

As most of the work on solid|electrolyte solution interfaces is done in water or other solvents that can donate and/or accept hydrons, the Brønsted–Lowry definition of acids and bases was deemed to be a sufficiently broad basis for the present discussion. It has the additional advantage of being much more familiar to every potential reader than the broader Lewis definition.

A. Brønsted's Acids and Bases

According to the well-known Brønsted–Lowry definition of acids and bases, the ionization equilibrium of a weak acid in aqueous solution,

$$HL(aq) + H_2O(l) \rightleftharpoons H_3O^+(aq) + L^-(aq) \tag{58}$$

is the result of the competition of two Brønsted bases, L^- and H_2O, for the hydron. Analogous equilibria take place in all solvents with basic properties. Reaction (58) is usually written in a shorter form:

$$HL(aq) \rightleftharpoons H^+(aq) + L^-(aq) \tag{59}$$

Although economical, the short notation may be misleading in some cases because it suggests a dissociation, rather than the actually occurring process of hydron transfer.

Acidic solvents (like acetic acid) will, in agreement with IUPAC recommendations [2], be termed *hydronogenic* (="protogenic"). For basic solvents (liquid ammonia and diethyl ether, for instance) the terms *hydronophilic* (="protophilic") will be used. Those solvents that exhibit neither acidic nor basic properties (hydrocarbons, for example) will be called *anhydronic* (="aprotic").

As water is not only a Brønsted base but also a Brønsted acid, the hydron can be transferred to its conjugate base, the hydroxide ion:

$$HL(aq) + OH^-(aq) \rightleftharpoons H_2O(l) + L^-(aq) \tag{60}$$

Solvents that, like water and alcohols, exhibit both acidic and basic properties are called *amphihydronic* (= "amphiprotic" = "amphoteric"). Their characteristic feature is the equilibrium between the hydronated and dehydronated forms, e.g., for water,

$$2H_2O(l) \rightleftharpoons H_3O^+(aq) + OH^-(aq) \tag{61}$$

or, in shorthand notation,

$$H_2O(l) \rightleftharpoons H^+(aq) + OH^-(aq) \tag{62}$$

The traditional name for this process, "autoprotolysis," can be translated into the new terminology as *autohydronolysis*.

A situation frequently met in interfacial chemistry is the heterogenous equilibrium between an acidic or basic solid phase and an electrolyte solution, for instance,

$$\equiv EOH(s) + H_2O(l) \rightleftharpoons H_3O^+(aq) + EO^-(s) \tag{63}$$

or

$$\equiv EOH(s) + OH^-(aq) \rightleftharpoons H_2O(l) + EO^-(s) \tag{64}$$

E standing for the atom of any element capable of such reaction(s). Although the processes described by Eqs. (63) and (64) are essentially analogous to those represented by Eqs. (58) and (60), respectively, the analogy is not complete: $\equiv$EOH in Eqs. (63) and (64) denotes only one of many acidic sites fixed on the surface of a solid macroscopic body, while in Eqs. (58) and (60) it refers to mobile microscopic particles.

B. Equilibrium Constants

1. Thermodynamic Equilibrium Constants

The thermodynamic equilibrium constant for the dehydronation of HL (Eq. 60), defined in terms of equilibrium activities, depends only on temperature and pressure. If $p = p^o = 100$ kPa, the activity constant is called the *standard equilibrium constant*,

$$K^o(a) = \frac{a(H_3O^+) \cdot a(L^-)}{a(HL) \cdot a(H_2O)} \tag{65}$$

(a) standing for acid. As hydronation/dehydronation equilibria are mostly studied in condensed phases, the pressure dependence of the equilibrium constant is usually negligible.

By definition, the standard equilibrium constant is closely related to the standard reaction Gibbs energy

$$\Delta_r G^o = -RT \ln K^o(a) \tag{66}$$

(subscript r is a standard symbol for "reaction" in general).

By inserting defining relations (2), (3), and (10) into Eq. (65), one obtains the exact formula for the standard dehydronation constant,

$$K^\circ(a) = \frac{[H_3O^+] \cdot [L^-] \cdot (\bar{\gamma}_\pm)^2}{x(H_2O) \cdot \gamma(H_2O) \cdot [HL] \cdot \gamma_{HL} \cdot c^\circ} \tag{67}$$

where the square brackets denote equilibrium concentrations. As the activity coefficients of charged species (H^+ and L^- in this instance) are not measurable quantities (see Section I.C), their geometric average $\bar{\gamma}_\pm$ is used instead.

In sufficiently dilute solution $x(H_2O) \approx 1$ and $\gamma(H_2O) \approx 1$, so Eq. (67) reduces to

$$K^\circ(a) = \frac{[H_3O^+] \cdot [L^-]}{[HL]} \cdot \frac{1}{c^\circ} \cdot \frac{(\bar{\gamma}_\pm)^2}{\gamma_{HL}} \tag{68}$$

If HL is electrically neutral, its activity coefficient is closer to unity than is $\bar{\gamma}_\pm$ and can, as an approximation, be set to 1, provided [HL] is not higher than, say, 0.1 mol dm^{-3}. Therefore in sufficiently dilute solutions, the standard dehydronation constant can be approximated by

$$K^\circ(a) = \frac{[H_3O^+] \cdot [L^-]}{[HL]} \cdot \frac{(\bar{\gamma}_\pm)^2}{c^\circ} \tag{69}$$

2. Empirical Equilibrium Constants

The first factor in formula (69) is the commonly known *stoichiometric constant* of dehydronation (= "deprotonation"), which is most frequently denoted by

$$K_c(a) = \frac{[H_3O^+] \cdot [L^-]}{[HL]} \tag{70}$$

By using the above formula, Eq. (69) can be simplified to

$$K^\circ(a) = \frac{K_c(a)}{c^\circ} \cdot (\bar{\gamma}_\pm)^2 \tag{71}$$

For simplicity, the cologarithm of $K_c(a)$ is very often quoted instead of $K_c(a)$:

$$\begin{aligned} pK_a^\circ &= -\lg K^\circ(a) \\ pK_a &= -\lg(K_c(a)/c^\circ) \end{aligned} \tag{72}$$

Unlike the standard equilibrium constant, the stoichiometric constant depends not only on temperature but also on total concentrations of the components. Despite this limitation, stoichiometric constants are useful for at least two reasons: (a) in many cases they can be determined experimentally, and (b) their physical meaning is obvious, so they are easy to use in practical calculations. Therefore, stoichiometric constants are the most important kind of empirical equilibrium constants, at least in solution chemistry and related areas. Most often, though not necessarily, they are determined by the constant-ionic-strength technique. The constants determined in a medium containing a constant excess of an "inert" electrolyte can be, at least approximately, treated as standard thermodynamic constants valid for a specific "solvent." For example, if the determination of the K_a value for a weak acid is performed in an aqueous solution containing, besides the acid ($c \approx 0.001$ mol dm^{-3}, say), a constant excess of KNO$_3$

($c = 0.1$ mol dm^{-3}), then 0.1 molar KNO$_3$ can be taken as a specific solvent and the stoichiometric constant can be interpreted as a standard thermodynamic quantity in that solvent. In other words, the solution of weak acid in 0.1 molar KNO$_3$ is treated as ideal. However, the constant-ionic-strength technique has some serious limitations that are discussed in Section III.C.7.

Another kind of empirical constants, called "mixed" or "Brønsted" constants, defined in terms of *conventional* pH (see Sections III.C.4, 5), is very frequently seen in chemical literature, especially in routine characterization of acidic or basic compounds:

$$K^B(a) = \frac{10^{-pH} \cdot [L^-]}{[HL]} \tag{73}$$

Very often the reliability of mixed constants is below reasonable standards. Therefore their use should be limited to qualitative or, at most, semiquantitative work.

3. Polyhydronic Acids: A Rational Formalism

The equilibrium stoichiometric calculations can be considerably simplified if, instead of dehydronation constants, hydronation ("protonation") constants are used, especially in the case of polyhydronic acids. Throughout this section (as well as in Section III.D) it will be assumed that the use of stoichiometric equilibrium constants is legitimate, and the notation will be accommodated to that customarily used by chemists working in the field of solution coordination chemistry.

For a monohydronic acid, the hydronation constant is simply the reciprocal of $K_c(a)$:*

$$K_1^H = \frac{1}{K_c(a)} = \frac{[HL]}{[H][L]} \tag{74}$$

It is easily shown that the fractions of hydronated (α_1) and dehydronated (α_0) forms of a monohydronic acid in equilibrated solution are uniquely determined by [H$^+$]:

$$\begin{aligned} \alpha_0 &= \frac{[L]}{c_A} = \frac{1}{1 + K_1^H[H]} \\ \alpha_1 &= \frac{[HL]}{c_L} = \frac{K_1^H[H]}{1 + K_1^H[H]} \end{aligned} \tag{75}$$

Consecutive hydronation of a base that can bind two hydrons, for example L^{2-} anion, can be written as

$$H^+ + L^{2-} \rightleftharpoons HL^- \tag{76a}$$
$$H^+ + HL^- \rightleftharpoons H_2L \tag{76b}$$

---

* Charge numbers are omitted from mass-action-law and similar algebraic formulae, for the sake of simplicity.

the respective hydronation constants being

$$K_1^H = \frac{[HL]}{[H][L]}$$
$$K_2^H = \frac{[H_2L]}{[H][HL]} \tag{77}$$

Consecutive hydronation constants are related to the dehydronation ones by

$$K_c(a, 1) = \frac{1}{K_2^H}$$
$$K_c(a, 2) = \frac{1}{K_1^H} \tag{78}$$

However, reactions (76) can alternatively be written in the "cumulative" form

$$H^+ + L^{2-} \rightleftharpoons HL^- \tag{79a}$$
$$2H^+ + L^{2-} \rightleftharpoons H_2L \tag{79b}$$

The above equilibria are described by *cumulative* hydronation constants (in solution coordination chemistry the cumulative constants are customarilly symbolised by β):

$$\beta_1^H = \frac{[HL]}{[H][L]}$$
$$\beta_2^H = \frac{[H_2L]}{[H]^2[HL]} \tag{80}$$

The relations among the three kinds of constants, $K_c(a)$, K^H, and β^H, can easily be derived:

$$\beta_1^H = K_1^H = \frac{1}{K_c(a, 2)}$$
$$\beta_2^H = K_1^H \cdot K_2^H = \frac{1}{K_c(a, 2) \cdot K_c(a, 1)} \tag{81}$$

The formulae for equilibrium fractions of L^{2-}, HL^- i H_2L species ($j = 0, 1, 2$) assume the simplest possible form when written in terms of βs:

$$\alpha_j = \frac{[H_jL]}{c_A} = \frac{\beta_j^H[H]^j}{1 + \sum_{j=1}^{2} \beta_j^H[H]^j} \tag{82}$$

It should be noted that here again the αs are uniquely determined by the value of $[H^+]$.

The generalization to acids containing an arbitrary number of hydrons ($j = 1, 2, \ldots, J$) is straightforward. The jth cumulative constant is given by

$$\beta_j^H = \frac{[H_jL]}{[H]^j[L]} \tag{83}$$

and the equilibrium fractions of H_jL species are

$$\alpha_j = \frac{\beta_j^H[H]^j}{1 + \sum_{j=1}^{j} \beta_j^H[H]^j} \tag{84}$$

4. Equilibrium Constant of Autohydronolysis

The standard equilibrium constant for the reaction of autohydronolysis of an amphihydronic solvent like water (see Section III.A),

$$H_2O(l) \rightleftharpoons H^+(aq) + OH^-(aq) \tag{62}$$

can be written as

$$K_w^o = \frac{[H][OH] \cdot (\bar{\gamma}_\pm)^2}{x(H_2O) \cdot \gamma(H_2O)} \approx K_w \cdot (\bar{\gamma}_\pm)^2 \tag{85}$$

where

$$K_w = [H][OH] \tag{86}$$

is the stoichiometric constant of water autohydronolysis. As the standard equilibrium constant of water autohydronolysis is very small ($K_w^o = 1.008 \times 10^{-14}$ at 298 K, $pK_w^o = 13.996$), the equilibrium concentrations [H] = [OH] are fairly low, and the approximation on the rightmost side of Eq. (85) is excellent.

5. Weak Bases

The hydronation of a weak base in an acid or amphihydronic solvent can be described in an exactly analogous way as had been done for weak acids. For example,

$$NH_3(aq) + H_2O(l) \rightleftharpoons NH_4^+(aq) + OH^- \tag{87}$$

The stoichiometric constant of NH_3 base, $K_c(b)$, is

$$K_c(b) = \frac{[NH_4][OH]}{[NH_3]} \tag{88}$$

By introducing the acidity constant of $NH_4^+(aq)$,

$$K_c(a) = \frac{[NH_4]}{[NH_3][H]} \tag{89}$$

the well-known relation among $K_c(a)$, $K_c(b)$, and K_w, viz.

$$K_c(a) \cdot K_c(b) = K_w \tag{90}$$

can easily be derived from Eqs. (86,88,89). As $K_c(a)$ and $K_c(b)$ are equivalent, there is no need whatsoever to use $K_c(b)$ for reactions taking place in amphihydronic solvents.

C. Definition and Interpretation of pH

Although the potentiometric determination of pH is a very popular analytical procedure, many if not most chemists find themselves in trouble when asked by a layman, What is pH? Some would answer by quoting Sørensen's [40] 1909 definition of what he denoted by the symbol p_H:

$$p[H] = -\lg \frac{[H^+]}{c^o} \tag{91}$$

However attractive because of its simplicity, this definition is impractical because the

Equilibria in Electrolyte Solutions

relation of p[H] with the e.m.f. of the glass electrode (or hydrogen electrode) cell is too involved. Others will try to define pH as the cologarithm of the activity of hydrogen ions, which, taken literally, means

$$pa_H = -\lg a(H^+) = -\lg \frac{[H^+]\gamma_H}{c^o} \tag{92}$$

However, this formula defines a quantity that cannot be measured (see Section I.C). Indeed, there seems to be no definition of pH that would be both simple and true. One of the very simplest definitions, which is correct but too vague, would be: pH is a *conventional* measure of acidity that is closely related—but not equal—to the decadic cologarithm of equilibrium hydron concentration, $[H^+]$.

1. Basic Concepts

The fundamental device by means of which the reference values of pH are assigned is the hydrogen electrode [40a].* Consider the following galvanic cell consisting of a hydrogen indicator electrode and a silver chloride reference electrode:

$$\text{Pt(s)} \mid \text{H}_2(\text{g}) \mid \underbrace{\text{HL(aq},c_1) + \text{ML(aq},c_2)}_{\text{Soln. I}} \vdots \underbrace{\text{MCl(aq},c_3)}_{\text{Soln. II}} \mid \text{AgCl(s)} \mid \text{Ag(s)} \tag{93}$$

(M^+ denotes a simple cation, *e.g.*, Na^+ or K^+). The overall cell reaction is

$$1/2\,H_2(g) + AgCl(s) \rightleftharpoons H^+(aq, I) + Cl^-(aq, II) + Ag(s) \tag{94}$$

The concentration dependence of the cell e.m.f. is given by the Nernst equation:

$$E = E^o - \varpi_N \lg \frac{a(H^+)_I \cdot a(Cl^-)_{II} \cdot a(Ag)}{a(AgCl) \cdot \sqrt{a(H_2)}} + E_j \tag{95}$$

where

$$\varpi_N = \frac{RT \ln 10}{\nu F} \tag{96}$$

denotes the Nernstian slope (for ν exchanged electrons; here $\nu = 1$) and E_j is the liquid-junction (or diffusion) potential. The pure solid phases (Ag, AgCl) are in their standard states (or very nearly so), hence $a(Ag) = a(AgCl) = 1$. For the sake of simplicity, let it be assumed that $p(H_2) \approx p^o$ and that H_2 is ideal at p^o; thus $a(H_2) \approx p(H_2)/p^o \approx 1$. By using hypothetical ionic activities one obtains

$$E = E^o - \varpi_N \lg \frac{[Cl^-]\gamma_{Cl}}{c^o} - \varpi_N \lg \frac{[H^+]\gamma_H}{c^o} + E_j \tag{97}$$

---

* Spectrometric and "colorimetric" pH measurements by means of indicators, which still find use in work with non-aqueous media, will be shortly commented on in Section III.C.6.

If [Cl⁻] is constant, the second term is constant, and this relation can be further simplified to

$$E = E_o - \varpi_N \lg \frac{[H^+]\gamma_H}{c^o} + E_j \tag{98}$$

Unfortunately, this relation cannot be directly used as a basis for potentiometric determination of either p[H] or pa_H because it contains two quantities, γ_H and E_j, that cannot be measured and that depend on the kinds and concentrations of all species present in the two solutions. Even when the chemical composition is precisely known (a situation seldom met with in practical applications), γ_H and E_j can be theoretically estimated with acceptable accuracy only for sufficiently dilute solutions containing precisely known amounts of simple salts.

2. Bates–Guggenheim Convention

The problem with γ_H was solved in extensive and extremely careful studies of the cells without liquid junction but otherwise similar to (93), for example

$$\text{Pt(s)} \mid \text{H}_2\text{(g)} \mid \text{HL(aq)} + \text{ML(aq)} + \text{MCl(aq)} \mid \text{AgCl(s)} \mid \text{Ag(s)} \tag{99}$$

The e.m.f. of such a cell is given by a relation similar to (97), i.e.,

$$E = E^o - \varpi_N \lg \frac{[Cl^-]}{c^o} - \varpi_N \lg \frac{[H^+]}{c^o} - \varpi_N \lg(\gamma_H \gamma_{Cl}) \tag{100}$$

containing the product $(\gamma_H \gamma_{Cl})$ which, unlike its factors, *is* measurable:

$$\gamma_H \gamma_{Cl} = (\bar{\gamma}_\pm)^2 \tag{101}$$

If one of the two ionic activity coefficients, e.g., γ_{Cl}, were assigned a value *by convention*, the value of the other, γ_H, would be uniquely determined. Bates and Guggenheim [40b] proposed that γ_{Cl} should be computed by means of a slightly modified Debye–Hückel formula (27),

$$-\lg \gamma_{Cl}^{BG} = \frac{A_c \cdot \sqrt{I_c/c^o}}{1 + B\sigma \cdot \sqrt{I_c/c^o}} \tag{102}$$

in which the $B\sigma$ parameter is arbitrarily set to 1.5, which corresponds to the effective ionic radius $\sigma_{Cl} = 4.6$ Å (in water: $B = \kappa I_c^{-1/2} \approx 0.33$ Å^{-1} at 298 K).

By introducing activity coefficients as defined by the Bates–Guggenheim convention (denoted by BG superscript), Eq. (100) becomes

$$\begin{aligned} E &= E^o - \varpi_N \lg \frac{[Cl^-]\gamma_{Cl}^{BG}}{c^o} + \varpi_N \lg \frac{[H^+]\gamma_H^{BG}}{c^o} \\ &= E^o - \varpi_N \lg \frac{[Cl^-]\gamma_{Cl}^{BG}}{c^o} + \varpi_N \text{pH} \end{aligned} \tag{103}$$

This formula defines the conventional pH value for hydrogen electrode cells without liquid junction.

3. Liquid-Junction Potential

Liquid-junction potential E_j is mainly determined by the differences in mobilities and concentrations of ions present in the two solutions in contact. As H^+ and OH^- ions have exceptionally high (though different) mobilities, E_j will be quite high on the boundaries of solutions greatly differing in acidities. When the solvents of analyte and bridge solutions are dissimilar, the differences in ionic mobilities are enhanced, and E_j values may amount to as much as several hundreds of millivolts.

Fortunately, E_j can often be largely reduced by using a suitable salt bridge—a concentrated solution of salt(s) whose cation(s) and anion(s) have about equal mobilities. If the bridge solution is saturated KCl(aq), the potential at its boundary with aqueous analyte solutions will seldom exceed several millivolts (provided the analyte solution is not too acid or too alkaline). For example, E_j on the HCl(aq, c) KCl(aq, satd.) boundary is estimated to be 14.1 mV for $c_{HCl} = 1$ mol dm^{-3} (corresponding to the error $\Delta \text{pH} = 0.24$), 4.6 mV for $c_{HCl} = 0.1$ mol dm^{-3} ($\Delta \text{pH} = 0.078$) and only 3.0 mV for $c_{HCl} = 0.01$ mol dm^{-3} ($\Delta \text{pH} = 0.051$) [40c], which can be tolerated in many applications.

4. Operational Definition of pH

Today's conventional pH scale is defined [34] by five aqueous buffer solutions whose compositions are:

1. Potassium hydrogen tartrate (saturated at 25°C)
2. Potassium hydrogen phthalate ($b = 0.05$ mol kg^{-1})
3. Potassium dihydrogen phosphate ($b = 0.025$ mol kg^{-1}) + disodium hydrogen phosphate ($b = 0.05$ mol kg^{-1})
4. Potassium dihydrogen phosphate ($b = 0.008695$ mol kg^{-1}) + disodium hydrogen phosphate ($b = 0.03043$ mol kg^{-1})
5. Borax ($b = 0.01$ mol kg^{-1})

The values of pH(S) assigned to these primary standards (in the temperature range from 0 to 95°C), derived from measurements on cells without transference, on the basis of the Bates–Guggenheim convention, are listed in Table 1 [40d,41].

Hydrogen electrode cell similar to (93) can be calibrated by means of one of the standard buffer solutions, applying the formula

$$\text{pH} = \text{pH(S)} + \frac{E - E_s}{\varpi_N} \tag{104}$$

which summarizes the so-called operational definition of the practical pH value for this cell [40a].

However, the hydrogen electrode is almost never used in practice. Therefore the calibration formula has to be accommodated to the properties of the glass electrode that is today generally used for pH measurement. Although the mechanism of response of a glass electrode is rather complex, a simple linear calibration formula of the form

$$E = E_o + \varpi \cdot \text{pH} \tag{105}$$

can still be used, keeping in mind that the value of the "constant" E_0 slowly changes with time and, besides, depends on the electrode's history. The slope parameter ϖ is usually sub-Nernstian (usually $0.95 \leq \varpi/\varpi_N \leq 1$) and also subject to slight fluctuations. For these reasons glass electrode cells need relatively frequent recalibration (at least two or three times during a working day). At least two buffer solutions are required for calibration,

TABLE 1 Values of pH(S) for Primary Standard Buffers

T (K − 273.15)	pH(A)	pH(B)	pH(C)	pH(D)	pH(E)
0	—	4.003	6.984	7.534	9.464
5	—	3.999	6.951	7.500	9.395
10	—	3.998	6.923	7.472	9.332
15	—	3.999	6.900	7.448	9.276
20	—	4.002	6.881	7.429	9.225
25	3.557	4.008	6.865	7.413	9.180
30	3.552	4.015	6.853	7.400	9.139
35	3.549	4.024	6.844	7.389	9.102
38	3.548	4.030	6.840	7.384	9.081
40	3.547	4.035	6.838	7.380	9.068
45	3.548	4.047	6.834	7.373	9.038
50	3.549	4.060	6.833	7.367	9.011
55	3.554	4.075	6.834	—	8.985
60	3.560	4.091	6.836	—	8.962
70	3.580	4.126	6.845	—	8.921
80	3.609	4.164	6.859	—	8.885
90	3.650	4.205	6.877	—	8.850
95	3.674	4.227	6.886	—	8.833

Source: Refs. 40 and 41.

and the parameters are computed by means of the formulae

$$\varpi = \frac{E_2 - E_1}{\text{pH}(S_2) - \text{pH}(S_1)} \tag{106}$$
$$E_0 = E_1 - \varpi \cdot \text{pH}(S_1) = E_2 - \varpi \cdot \text{pH}(S_2)$$

E_1 and E_2 denote e.m.f.s measured with buffer solutions S_1 and S_2, respectively).

5. Interpretation of pH

It is now obvious that the conventional pH value, measured according to the operational definition [Eqs. (105) and (106)] is *not* an exact measure of either [H$^+$] or a(H$^+$). A quantitative interpretation of pH is possible *only* for dilute aqueous solutions of known composition, containing simple solutes, i.e., for solutions of salts and/or buffers similar to those chosen to serve as pH(S) standards. In this, unfortunately rather narrow, class of solutions, pH may be taken as an acceptable approximation to $-\lg([\text{H}^+]\gamma_\pm)$ [40e]. Nevertheless, one should not try to interpret pH values measured in solutions of low buffer capacity, since these may be grossly inaccurate.

6. Nonaqueous and Partly Aqueous Solvents

(a) Potentiometric Measurements. According to what was said in the preceding paragraph, pH values measured in concentrated aqueous salt, acid, or alkali solutions ($I_c > 0.1$ mol dm^{-3}), suspensions, and colloids *cannot* be interpreted simply as $-\lg([\text{H}^+]\gamma_\pm/c^\circ)$. A specific interpretation is required in each case. The same applies *a fortiori* to solutions in nonaqueous or partly aqueous solvents.

Equilibria in Electrolyte Solutions 31

Many chemists working with nonaqueous or partly aqueous solutions used to calibrate their pH-measuring cells with aqueous buffers, tacitly assuming that the relation of e.m.f. to pH remains linear, like Eq. (105), with (almost) unchanged slope, ϖ, so that only E_0 is affected by the solvent change (probably, the most important factor affecting E_0, besides the liquid-junction potential, is the work of transfer of H^+ from water to the actual solvent). Even if this assumption is true (which is seldom verified), the "pH" scale in the new solvent is a relative one, its relation to $[H^+]$ being anything but transparent.

(b) Indicators. Because of the importance of determinations of p[H] or related acidity functions in nonaqueous media, a number of nonpotentiometric procedures have been proposed. Several early methods have been based upon the use of minute amounts of a suitable indicator, more specifically on the (usually colorimetric or UV-Vis spectrometric) measurement of the equilibrium extent of indicator (de)hydronation in the solution under study, which is uniquely determined by pH:

(unchanged base) $\qquad H^+ + \text{Ind} \rightleftharpoons \text{IndH}^+$ (107a)

(unchanged base) $\qquad H^+ + \text{Ind} \rightleftharpoons \text{IndH}^+$ (107b)

One very popular acidity function was introduced by Hammet:

$$H_0 = pK_a^o(\text{IndH}^+) + \lg \frac{[\text{Ind}]}{[\text{IndH}^+]} \qquad (108)$$

(subscript zero indicates that the indicator base Ind is uncharged; if the base bore a negative charge, Hammett's function would be denoted by H_-). The ratio of equilibrium concentrations, [Ind]/[IndH], can be measured spectrometrically and H_0 can be obtained provided the pK^o value is *independently* known. Requiring the use of a number of indicators, the method is rather laborious. Theoretically, it relies upon the assumption that the relative strengths of indicators of the same charge type are independent of the medium. Unfortunately, there are many cases where this assumption is very far from the truth.

Although Hammett's function was applied with success in a number of instances (for example in mixtures of sulfuric, perchloric, hydrochloric, or nitric acids with water and in anhydrous formic acid), it has been the object of serious criticisms [40f]. Classical work on pH and related acidity functions has been reviewed by Bates [40g]. Recent techniques of pH assessment in partly aqueous and nonaqueous media are reviewed in Chapter 7 of this book.

7. Calibration in Terms of $[H^+]$

If a large (at least about 50-fold) excess of "inert" strong electrolyte is added to the solution(s) under study, the ionic strength will remain essentially constant, no matter what processes take place. Therefore it is reasonable to believe that γ_H and E_j in Eq. (97) will also remain essentially constant and add to the value of E_0. As a result, the e.m.f. of a cell with constant ionic strength is expected to be an approximately linear function of p[H], within acceptable error range (typically ca. ± 0.01 [42–45]).

This stratagem, which can be dated as far back as 1905, is very popular among solution chemists [45–47]. Most of the equilibrium data (hydronation constants, complex stability constants, ligation enthalpies and entropies, etc.) have been determined at constant ionic strength. However, the validity of this procedure critically depends on how

D. Complex Formation

The formalism introduced in Section III.B.3 for polyhydronic weak acids can easily be generalized to cover complex formation equilibria (the term "complex" is used here in the meaning defined in Section II.C.5). Let a polynuclear (de)hydronated complex consist of q metal cations $M^{z(M)}$, of n molecules or ions of ligand $L^{z(L)}$, and of p hydrons, so that its formula is $[M_qL_nH_p]^{q \cdot z(M)+n \cdot z(L)+p}$ (which will be written in abridged form, $[M_qL_nH_p]$). It is convenient to introduce, in analogy to Eq. (81), the (stoichiometric) cumulative stability constant

$$\beta_{qpn} = \frac{[M_qL_nH_p]}{[M]^q[L]^n[H]^p} \tag{109}$$

referring to the complex formation equilibrium

$$qM^{z(M)} + nL^{z(L)} + pH^+ \rightleftharpoons [M_qL_nH_p]^{q \cdot z(M)+n \cdot z(L)+p} \tag{110}$$

Obviously, the equations for acids $[H_pL]$, simple mononuclear $[ML_n]$, and polynuclear complexes $[M_qL_n]$ are special cases of Eqs. (109,110). An advantage of using cumulative stability constants is that the equilibrium concentrations of a complex species can be expressed in terms of the concentrations of simple constituents (M, L, and/or H^+), which are usually easier to measure than concentrations of various possible complex species:

$$[M_qL_nH_p] = \beta_{qpn}[M]^q[L]^n[H]^p \tag{111}$$

Some complexes may undergo dehydronation (for example, complexes of glycylglycine with transitional metal ions). The loss of a hydron from the $[M_qL_n]$ complex is conveniently denoted by a negative stoichiometric index:

$$qM^{z(M)} + nL^{z(L)} \rightleftharpoons [M_qL_nH_{-1}]^{q \cdot z(M)+n \cdot z(L)-1} \tag{112}$$

Note that the ratio of cumulative stability constants, β_{qn0}/β_{qn-1}, is equal to the hydronation constant of $[M_qL_nH_{-1}]$ species (or, equivalently, to the reciprocal dehydronation constant of $[M_qL_n]$):

$$\frac{\beta_{qn0}}{\beta_{qn-1}} = \frac{M_qL_n]}{M_qL_nH_{-1}][H]} \tag{113}$$

More information on the stability constants of metal–ion complexes and their determination, use, and interpretation can be found in the monograph by Martell and Motekaitis [45], which contains also several useful computer programs.

REFERENCES

1. I. Mills, T. Cvitaš, K. Homann, N. Kallay, and K. Kuchitsu, *Quantities, Units and Symbols in Physical Chemistry*, 2d ed., IUPAC and Blackwell Scientific Publications, Oxford, 1993, (a) pp. 41–54, (b) p. 62.
2. G. J. Leigh. ed., *Nomenclature of Inorganic Chemistry. Recommendations 1990*, IUPAC and Blackwell Scientific Publications, Oxford, 1992, pp. 103–105.
3. E. A. Guggenheim, *Thermodynamics*, 6th ed., North-Holland, Amsterdam, 1967, pp. 270–275.
4. J. N. Murell and E. A. Boucher, *Properties of Liquids and Solutions*, John Wiley, New York, 1982, (a) p. 11, (b) pp. 48–51, (c) p. 75.
5. H. L. Friedman. J. Phys. Chem. *66*:1595 (1962).
6. J. C. Ghosh. J. Chem.Soc. 449, 707 (1918).
7. J. O'M. Bockris and A. K. N. Reddy, *Modern Electrochemistry*, 1st ed., Vol. 1, Plenum Press, New York, 1970, pp. 267–273.
8. R. H. Fowler and E. A. Guggenheim, *Statistical Thermodynamics*, 2d ed., Cambridge University Press, Cambridge, 1956, (a) pp. 383–394, (b) pp. 405–410, (c) p. 417, (d) pp. 415–420, (e) pp. 410–415.
9. K. S. Pitzer, in *Activity Coefficients in Electrolyte Solutions* (K. S. Pitzer, ed.), CRC Press, Boca Raton, FL, 1991, pp. 75–91.
10. J. O'M Bockris and A. K. N. Reddy, *Modern Electrochemistry*, 2d ed., Vol. 1, Plenum Press, New York, 1998; (a) p. 119, (b) p. 297, (c) pp. 314–329, (d) pp. 333–337, (e) p. 264.
11. R. H. Stokes and R. A. Robinson. J. Am. Chem. Soc. *70*:1870 (1948).
12. J. E. Mayer. J. Chem. Phys. *18*:1426 (1950).
13. W. G. McMillan and J. E. Mayer. J. Chem. Phys. *13*:276 (1945).
14. H. L. Friedman. Ann. Rev. Phys. Chem. *32*:179–204 (1981).
15. E. Waisman and J. L. Lebowitz. J. Chem. Phys. *56*:3086 (1972).
16. W. R. Fawcett and A. C. Tikanen. J. Phys. Chem. *100*:4251 (1996).
17. H. L. Friedman. Annual Review of Physical Chemistry, *32*:179 (1981).
18. D. N. Card, J. P. Valleau. J. Chem. Phys. *52*:6232 (1970).
19. J. C. Poirier, in: B. E. Conway, R. G. Barradas (eds.), *Chemical Physics of Ionic Solutions*, Wiley, New York 1963.
20. P. N. Vorontsov-Velyaminov, A. M. Elyashevich, A. K. Kron. Elektrokhimiya *2*:708 (1966).
21. R. M. Fuoss. Trans. Faraday Soc. 967 (1934).
22. R. M. Fuoss. J. Am. Chem. Soc. *80*:5059 (1958).
23. R. M. Fuoss and C. A. Kraus. J. Am. Chem. Soc. *55*:1019 (1933).
24. M. Eigen and K. Tamm. Z. Elektrochem. *66*:107 (1962).
25. T. J. Gilligan and G. Atkinson. J. Phys. Chem. *84*:208 (1980).
26. R. G. Wilkins, *Kinetics and Mechanism of Reactions of Transition Metal Complexes*, VCH, New York, 1997.
27. G. Kortüm, *Lehrbuch der Elektrochemie*, 2d ed., Verlag Chemie, Weinheim, 1957, p. 198.
28. L. I. Katzin. J. Chem. Phys. *18*:789 (1950).
29. L. I. Katzin. J. Inorg. Nuclear Chem. *4*:187 (1957).
30. B. J. Hathaway and A. E. Underhill. J. Chem. Soc. 2257 (1962).
31. D. E. Irish and A. R. Davis. Can. J. Chem. *46*:943 (1968).
32. V. Tomišić, Ph.D. thesis, University of Zagreb, Zagreb, Croatia, 1997.
33. Vl. Simeon and V. Tomišić. Phys. Chem. Chem. Phys. *1*:299 (1999).
34. D. E. Irish and M. H. Brooker, in *Raman and Infrared Spectral Studies of Electrolytes* (R. J. H. Clark and R. E. Hester, eds.), Vol. 2, Heyden, London, 1976.
35. D. E. Irish, A. R. Davis, and R. A. Plane. J. Chem. Phys. *50*:2262 (1969).
36. A. R. Davis and B. G. Oliver. J. Phys. Chem. *77*:1315 (1973).
37. F. A. Cotton and G. Wilkinson, *Advanced Inorganic Chemistry*, 5th ed., John Wiley, New York, 1988.
38. E. M. Hanna, A. D. Pethybridge, and J. E. Prue. Electrochim. Acta *16*:677 (1971).

39. R. M. Chatterjee, W. A. Adams, and A. R. Davis. J. Phys. Chem. *78*:246 (1974).
40. R. G. Bates, *Determination of pH. Theory and Practice*, John Wiley, New York, 1964, (a) p. 63, (b) p. 43, (c) p. 41, (d) p. 76, (e) pp. 91–92, (f) pp. 167–171, 206, (g) Ch. 8.
41. A. K. Covington, R. G. Bates, and R. A. Durst. Pure Appl. Chem. *57*:531 (1985).
42. H. M. Irving, M. G. Miles, and L. D. Pettit. Anal. Chim. Acta *38*:475 (1967).
43. Z. Kralj and Vl. Simeon. Kem. Ind. Zagreb *28*:299 (1979).
44. P. May and D. R. Williams. Talanta *29*:249 (1982); *30*:899 (1983).
45. A. E. Martell and R. J. Motekaitis, *The Determination and Use of Stability Constants*, VCH, New York, 1988, pp. 7–31.
46. F. J. C. Rossoti and H. Rossoti, *The Determination of Stability Constants*, McGraw-Hill, New York, 1961, pp. 27–30, 38–46, 58–62, 86–97.
47. H. Rossotti, *The Study of Ionic Equilibria*, Longmans, New York, 1978.

2
The Significance of Interfacial Water Structure in Colloidal Systems—Dynamic Aspects

MIROSLAV ČOLIĆ Research and Development Division, ZPM Inc., Goleta, California

JAN D. MILLER Department of Metallurgical Engineering, University of Utah, Salt Lake City, Utah

I.	Introduction	36
II.	Physical Chemistry of Bulk Water and Water Clusters	37
	A. Introduction	37
	B. Experimental techniques for the study of bulk water	38
	C. Spectroscopic studies of water clusters	41
III.	Role of Interfacial Water in Interparticle and Intermolecular Forces	43
	A. Introduction	43
	B. Rheological and colloid stability studies	44
	C. Direct surface force measurements	54
IV.	Spectroscopic Studies of Hydration	57
	A. Introduction	57
	B. Spectroscopic studies of colloid hydration	58
	C. Spectroscopic studies of hydration in biological systems	60
V.	Hydrophobic Hydration and the Role of Dissolved Gases in Colloidal Phenomena	62
	A. Introduction	62
	B. Water structure in the interfacial region	63
	C. Role of dissolved gas in colloidal phenomena	64
	D. Short- and long-range hydrophobic forces	66
VI.	Influence of Electric, Magnetic, Electromagnetic, and Sonic Fields on Interfacial Water	69
	A. Introduction	69
	B. Experimental studies on the effects of fields on interfacial water	70
VII.	Theoretical Studies of Bulk and Interfacial Water	73
	References	75

I. INTRODUCTION

The structural and dynamical properties of water have long been studied but as yet are not fully understood [1–3]. Water is a natural solvent for colloidal and biological systems. Therefore understanding its structure and dynamics is of the foremost importance. In many common and relevant situations water is not in its bulk form but instead is attached to some substrates or filling small cavities [4]. The importance of such interfacial or confined water is significant to numerous technological and biomedical problems such as corrosion and scale prevention, mining and mineral processing, oil recovery, waste water treatment, drinking water purification, pharmacology, and health care. Bound water has attracted considerable interest in recent years. Despite fast progress, old controversies that plagued this field for decades still exist [5]. Israelachvili and Wennerstrom stated in a recent review in *Nature* that extended layers of bound interfacial water do not exist and that water is just another small molecule [5]. Significant evidence collected by other authors points in another direction [6]. Yet it is still uncertain how important a role water indeed has in control of intermolecular and interparticle forces.

In particular, the assessment of the perturbation of liquid water structure and dynamics near model hydrophilic and hydrophobic interfaces attracted the attention of multidisciplinary scientific teams [4]. Hydrophilic silica surfaces and hydrophobic octadecyltrichlorosilane (OTS) coated surfaces are only two such examples. Hydration of proteins and other biological systems exemplifies another popular research topic [3]. As Finney pointed at the recent Faraday Society Discussions overview lecture [3], the overall approach to trying to understand hydration processes over the past decade might be summarized by asking the following questions.

> First, in a given system, where is the water? What is the structure of the hydration water and of that further away from the interface? Can we characterize the detailed hydration geometry at the interface? Secondly, how does the water move? What is its dynamics? Over what range of timescales can we characterize the dynamics, and what timescales (picoseconds to hours?) are relevant to understanding a particular process or function? How do the water dynamics couple to that of the macromolecule? Does one control the other, and if so, which? Thirdly, once we have some understanding of these structural and dynamic questions, how can this understanding explain the behavior of a particular system, not only at the molecular and macroscopic but also mesoscopic level? And increasingly important, how can we exploit this understanding to modulate interactions and modify operation in a controlled manner?

Recent progress in resolving such issues will be reviewed in this chapter. A variety of spectroscopic methods that directly probe the structure of bulk and interfacial water, such as Raman spectroscopy, hole burning, OHD RIKES, sum-frequency generation (SFG), attenuated total reflectance spectroscopy (ATR), second-harmonic generation spectroscopy (SHG), dielectric spectroscopy, NMR relaxation, 2D and 3D NMR, and vibration-rotation tunneling spectroscopy of water clusters, will be reviewed. Mathematical modeling [Monte Carlo simulations, the hypernetted chain (HNC) method, etc.] methods have provided significant new insights into the dynamic behavior of interfacial water clusters.

Other experimental techniques to measure surface forces and their dependency on hydration also have contributed to an improved understanding of water/solid interfaces on the microscopic and mesoscopic level. Specifically, atomic force microscopy (AFM) and surface force apparatus (SFA) measurements have been used for this purpose.

Macroscopic colloidal techniques such as rheology measurements or colloid stability measurements [photon correlation spectroscopy (PCS) or light scattering] also have provided useful information regarding water/solid interfaces and their interactions with ions, surfactants, and polymers. Hydration of silica and silica interaction with ions in aqueous solutions is still a lively and controversial issue that will be discussed. Significant new progress has been made in this area by application of the above-mentioned novel techniques.

The role of gas in describing the characteristics and behavior of the interfacial water region appears to be quite important. Hydrophobic hydration and its relation to surface forces between hydrophobic interfaces in water is also an emerging issue with many new problems to be resolved. Combinations of all the above-mentioned experimental techniques are being used in this important area of research. The origin and nature of the hydrophobic effect, though, are still unknown and are subjects of some controversy [3].

Finally, the interactions of electric, magnetic, sonic, and electromagnetic fields with interfacial water will be discussed. This is also a relatively new area with many current and possible biomedical and industrial applications, but fundamental mechanisms associated with these processes are not well understood. Sonoluminescence, the generation of visible light by sonic treatment of a gas/liquid interface, is one of the most exciting areas of modern chemical physics, but despite great interest in the scientific community, its origins are still unknown.

II. PHYSICAL CHEMISTRY OF BULK WATER AND WATER CLUSTERS

A. Introduction

Understanding of bulk water structure and dynamics along with hydration of ions and molecules is a necessary prerequisite for improved understanding of the interactions of water with colloidal systems. While major progress is being made in this field, the 25-year-old remark of Felix Franks [7] that water is the most studied but least understood liquid still probably holds. That intermolecular hydrogen bonds play a major role in the behavior of water is widely accepted. Yet the microstructure of water clusters in liquid form and the supramolecular architecture (if any) of such clusters is still unknown. Numerous proposed models can be divided into three groups. (1) There is the continuous model, according to which water can be pictured as a continuous network of tetrahedrally bonded molecules, in which the bonds are more or less distorted, giving rise to a continuous distribution [8]. (2) There is the discrete model, according to which a discrete number of species that differ from one another in microstructural architecture and number/geometry of hydrogen bonds exist [9]. (3) A third model proposed by Teixeira and Stanley more recently [11] is, in a sense, of intermediate character [12]. In their model, each water molecule is assigned to one of five species, according to the number of hydrogen bonds (one to four) it forms with neighboring water molecules. The problem is then treated in terms of percolation networks. Four bonded water molecules are given special attention. Water molecules that form four hydrogen bonds with their neighbors, form structures similar to that of ice, although they are still in the liquid state. Clusters of such water molecules tend to form finite regions (patches) whose structural properties are different from that of the remainder. Hydrogen bonds are not considered intact or broken in such a model [11,12]. A threshold exists, above which the hydrogen bond is so distorted that it is considered broken. It is also possible that such a threshold is different for different physical transformations of water. This model, though, failed to explain the behavior of supercooled water (liquid water below 0°C). Such water is of particular interest for modelling the hydration of colloids since, like water of hydration, it is also of restricted mobility and possesses more structural order. Attempts have been made since then to describe supercooled water as a mixture of two components, one an "open" icelike water network that would correspond to tetrabound water molecules, and another a "closed" water molecule network that would correspond to the remainder. In addition, at low temperature, a small quantity of true ice would exist as a heterophase fluctuation. Of course, the lower the temperature, the more ice heterophase would exist in water [13].

Recent evidence by Stanley and coworkers [14] showed that in supercooled water systems two different forms of amorphous solid water coexist: high-density and low-density forms. While final proof is still needed with different techniques, this is the first time that phase transformation and coexistence of two phases of noncrystalline pure substance have been observed. Even more exciting was a finding of Li and Rose that two kinds of hydrogen bonds exist in crystalline ice [15]. Biochemists proposed that hydrogen bonds of different strength can be involved in the exceptional success of enzyme catalysts [16]. This hypothesis, though, has been strongly criticized by the mainstream biochemical community [17].

The model of the two different hydrogen bond types in liquid water is actually not so recent [18]. Giguere proposed in 1984, based on the results of the Raman spectroscopy of water, that two different types of hydrogen bonds exist in water: stronger linear hydrogen bonds, predominant in cold water, and bifurcated hydrogen bonds, prevalent at higher temperatures. Some hydrogen bonds are so bifurcated that they are considered nearly free! Reorientation between these three states occurs through the rotation of water molecules about one of their OH bonds [19]. The bifurcated hydrogen bond model can account much better than earlier models for the thermodynamic anomalies of water. It is also this model that is used in the interpretation of FTIR, Raman, sum-frequency generation, and other vibrational spectra of bulk and interfacial water.

The role of defects in a totally connected random network of hydrogen bonds in liquid water was recently scrutinized by Stanley and coworkers [20]. They used molecular dynamic simulations to show that even a pure water random tetrahedral network is perturbed geometrically by the presence of the fifth molecule in the coordination shell or topologically by the presence of bifurcated hydrogen bonds. They showed that the mobility of water molecules is significantly influenced by such defects. Tanaka used simulation studies to show that hydrophobic cavities (defects) are particularly important in the perturbation of water structure [21]. Ninham pointed out that the role of dissolved hydrophobic gas might be much more important in the colloid chemistry of water solutions than ever expected [22]. Bunkin and coworkers discovered the presence of nanobubbles in nonoutgassed water [22a]. Recent spectroscopic studies by the same group indicated a difference in the structuring of water in the presence of dissolved gas [23]. All those findings slowly enhanced the conclusions that different types of hydrogen bonds are present in water. The continuum model of water, though, still cannot be discounted. Water chemistry seems to be one of the more exciting areas of science awaiting many answers in the next millenium. Some of the simulation and experimental techniques recently used in studies of the dynamic of bulk water will be reviewed in this chapter.

B. Experimental Techniques for the Study of Bulk Water

Raman spectroscopy along with the less powerful FTIR was historically used in the early studies on bulk water structure. High-frequency Raman spectroscopy has been recently supplemented with powerful low-frequency depolarized Raman spectroscopy (DRS).

DRS proved to be a valuable tool in dynamic studies of water structure. Dielectric relaxation in microwave and far-infrared optically heterodyne detected Raman-induced Kerr effect spectroscopy (OHD/RIKES), infrared hole-burning spectroscopy, and the time-dependent fluorescence Stokes shift technique (TDFSS) helped to improve significantly our understanding of ultrashort relaxation phenomena in water. Earlier techniques such as inelastic neutron scattering and nuclear magnetic resonance (NMR) relaxation time measurements were empowered with the advent of high-resolution techniques and the implementation of supercomputer molecular dynamic simulation for the interpretation of the results.

High-frequency Raman and FTIR spectra are generally interpreted with the earlier-described bifurcated hydrogen bonds model [24]. It is not the purpose of this review to discuss the applicability of different models. Rather, we shall describe spectral characteristics as accepted by the majority of the water chemistry community, with the clear notice that all models have some drawbacks.

Three different bands centered around 3600, 3400, and 3200 cm^{-1} have been found in Raman and FTIR mid-range spectra of liquid water. The OH stretching bands have been observed to shift from above 3600 cm^{-1} at high temperature to below 3200 cm^{-1} in the spectra of ice at low temperature. Spectral features in that range are generally attributed to the hydrogen-bonded OH stretch with different degrees of bond ordering [24–26]. For example, the peak at about 3600 cm^{-1} is attributed to the non-hydrogen-bonded or very strongly distorted hydrogen bonds vibration. The water peak at 3400 cm^{-1} has been assigned to the somewhat distorted hydrogen bonds (bifurcated hydrogen bonds). The peak at 3200 cm^{-1} is generally ascribed to the symmetric OH stretching mode associated with tetrahedrally coordinated water molecules, as in ice. In any event, these spectral features in the OH stretching region can be used as indicators of the extent of hydrogen bonding, in other words, the degree of structure present in the water being examined.

Zhelyaskov and coworkers recently developed an improved direct method of spectral deconvolution of these three bands in water spectra [27–29]. They performed a detailed analysis of the temperature influence on intermolecular coupling and Fermi resonance constants in the Raman spectra of liquid water, heavy water, liquid water–heavy water mixtures, and liquid water–hydrogen peroxide mixtures. The different spectral lines discussed above were still clearly present in the deconvoluted spectra.

Low-frequency Raman spectra (0–340 cm^{-1}) are more informative with regard to the dynamic properties of water. While the relaxation time effects are studied through the comparison with other techniques and yield rather indirect results, the observed spectra are still informative. The existence of two spectral bands near 190 and 60 cm^{-1} have been attributed to the O–H–O stretching and bending vibrations of hydrogen bonds [30]. Later, Walrafen reassigned these modes to longitudinal (dilation) and transverse (shear) type acoustic modes in disordered systems, such as glass. Recently, a broad relaxation mode was observed in a region below 20 cm^{-1}. This relaxation was considered to correspond to the duration time of the tetrahedral structure of bulk water. Tominaga and coworkers performed a detailed depolarized low-frequency Raman spectra study of aqueous solutions of lithium, sodium, potassium, and rubidium as well as calcium and magnesium chlorides. Lithium and sodium increased the relaxation times of water, while rubidium and potassium did not appreciably change it until very high concentrations were reached, and even then increased it. The result of their study correlated directly to the viscosity measurements but did not correlate to the NMR or depolarized Raman scattering results. NMR measurements of proton relaxation time are believed to be influenced by the first coordination shell of tightly bonded water. On the other hand, viscosity is more sensitive to the behavior of bulk water and to the less tightly bound water in outer shells around ions. This study strongly suggested that the structure making ability of ions also influences bulk water. Small, strongly polarized ions such as lithium or sodium seem to have the ability to organize water strongly. Such so-called structure makers promote the ordering of water in their vicinity. The opposite is true for large ions such as rubidium or cesium. They promote the disordering of water molecules in the immediate vicinity, as observed with NMR and depolarized Rayleigh scattering studies. Yet, all ions studied by Tominaga and coworkers increased the relaxation time of water. It seems that bulk water is perturbed by the addition of ions in a different way. The relaxation time for water around lithium ions was calculated to be around 1 ps. This is close to that of bulk water, which additionally suggested that low-frequency Raman spectra probed bulk water molecules farther away from the ions used. It is also important that magnesium and calcium influenced the structure of bulk water and the relaxation time much more dramatically. The calculation also showed that the uniformity of water clusters was destroyed by these ions.

Other recently developed dynamic measurement techniques were used more directly to probe the ultrashort relaxation phenomena in water. Several groups used the OHD/RIKES technique to study the dynamics of liquid water. In this technique, birefringence is used to measure the nonlinear polarization. This kind of study gives two relaxation times: 0.5 ps and 1.7 ps at 298 K as well as 0.40 ps and 1.16 ps at 295 K [31]. As mentioned before, depolarized low-frequency Raman spectra yielded one uniform relaxation time, unless high concentrations of magnesium or calcium were added.

The TDFSS technique [32], which is assumed to give a linear response from the solvent, has recently been improved to give femtosecond-time-resolved water measurements. In TDFSS a suitable solvent molecule is excited by a short laser pulse, thereby changing the charge distribution. The subsequent reorientation of the solvent is monitored by measuring the time-dependent fluorescence. Water response below 55 fs was measured with this technique. Spectral hole burning with a powerful infrared laser fired with the pulse duration of 11 ps also identified a population lifetime of 8 ps for a bifurcated hydrogen bond [33]. Yet the same groups used faster lasers and recently also confirmed ultrashort relaxation phenomena in water with structural relaxation times of 1.5 ps at 273 K and 0.8 ps at 343 K. Picosecond and subpicosecond relaxation phenomena in liquid water seem to be a reality. Another ultrafast technique giving a linear response is terahertz time domain spectroscopy (THz/TDS). This technique is based on ultrashort far-infrared (terahertz) pulses that are generated and detected by small photoconductive antennas driven by fs laser pulses [34]. The bandwidth of the THz pulses extends from 50 GHz to several THz. The broad bandwidth, the high signal-to-noise ratio, and the frequency covered make THz spectroscopy an excellent ultrafast technique. While the initial experiments have been performed in transmission mode, the reflection mode was needed for strongly absorbing water molecules. As in the case of Raman studies, the influence of temperature on the dielectric relaxation of water in the THz range was followed. Subpicosecond relaxation times were again measured (0.2 ps for single-molecule reorientation at room temperature). However, molecular dynamic studies and fitting of experimental data indicated that a model with two different hydrogen bond types could not describe the observed data. Also it was observed that no cooperativity was needed to be assumed during hydrogen bond formation. This was also recently proposed by Luzar based on her molecular dynamics calculations [35].

Interfacial Water Structure

Despite the significant progress in advanced dynamic spectroscopic techniques that has occurred in the last ten years, it is still not possible to get a realistic picture of liquid water. As indicated in this last paragraph, even similar techniques still show results that can be better interpreted by either the existence of different types of hydrogen bonds in liquid water or more continuous bond distribution. Better techniques with even higher resolution and reproducibility are needed to teach us more about relaxation phenomena in liquid water. More sophisticated simulation techniques will also help in clarifying the currently still foggy picture of the structure and dynamics of liquid water.

C. Spectroscopic Studies of Water Clusters

Scientists often resort to simple model systems in order to understand more complicated phenomena that are not fully accessible to a direct molecular analysis approach. In the case of bulk water studies, the natural choice is to study small gas phase clusters with 2 to 20 water monomers. Theoretical and spectroscopic studies of such clusters seek to explain structure, dynamics, and thermodynamics in terms of the underlying potential energy surface and the form of the intermolecular interactions [36]. Global analysis of such clusters is often performed theoretically by rigorous analysis of potential energy surfaces. Changes in the most favorable cluster morphology can be qualitatively understood as a function of the range of the interparticle forces. Thermodynamic properties can be calculated from a representative sample of local minima on the potential energy surface. The prediction of dynamics, on the other hand, requires also the knowledge of transition states and reaction pathways [36]. Minima in the potential energy surface are like the basins in a mountain range. Any change in the structure of a cluster with potential energy being in such a minimum is an uphill energy increase process. A transition state, however, is also a stationary point but corresponds to mechanical instability and is the highest point between two basins representing different minima [36]. For any multiatomic molecule, the number of different minima and transition states can be enormous. Global optimization theory readily yields the answer for situations where energy surfaces "funnel" towards the minimum. When this does not happen, the number of different minima exponentially increases with the number of involved atoms. Needless to say, such an analysis is possible only for small clusters. The lowest energy morphology for clusters bound by pairwise additive isotropic forces is determined by a balance between the total number of nearest-neighbor interactions (stabilizing) and the strain energy (destabilizing) [36]. The strain energy reflects the deviation of nearest-neighbor contacts from their most favorable value.

Water clusters with their strong directional intermolecular bonding and the capacity of the water molecule to act as both a double donor and a double acceptor of hydrogen bonds result in great deal of complexity in explaining the behavior of even the smallest water dimers. Nevertheless, recent dramatic progress in the spectroscopy of water clusters and a theoretical analysis of potential surfaces has resulted in insights into not only the structure but also the details of the dynamics of small water clusters. A recent development of far-infrared vibration rotation tunneling spectroscopy to study jet-cooled small water clusters at a temperature of only 5 K led to an understanding of the structure and the dynamics of water dimers, trimers, tetramers, pentamers, and hexamers in such detail that comparison with theoretical calculations yielded excellent agreement. Moreover, to improve the interpretation of splitting in obtained spectra, a global view of the potential energy surface was needed. Splittings in the recorded spectra were correlated to the splittings in the vibration rotation energy levels that result from rearrangements of the hydrogen bonding network. If the minima in the potential energy surface were close enough and the barrier separating them were not too high, interference would occur between the localized state of each minimum, resulting in the splittings of the localized states [36]. These are the splittings that are observed spectroscopically. The effect is a result of quantum-mechanical tunneling through the energy barriers. The challenge for theoreticians was now not only to calculate the minima in the potential energy space but also to predict the mechanism of cluster rearrangements and transition states energies. They responded with calculations of the different minima and predictions of the transition states involved, but direct calculation of all splittings is still not possible. However, it is possible to assign features in a spectrum to a particular mechanism without these calculations.

It was now possible to analyze the structure and dynamics of water clusters as large as the hexamer. As expected, dimers, trimers, tetramers, and pentamers had two-dimensional structures. The hexamer, however, turned out to be a 3D cage. Moreover, to compare spectra and theory, Gregory and Clary [38] had to incorporate zero-point fluctuations in their model. The fact that the hexamer has a 3D structure is important since this might explain the exceptional solvent ability of water to some extent. Preliminary observations and calculations for larger clusters suggested 3D structures for all of them.

Theoretical studies of the dynamics of water clusters suggested two transition states with acceptable energy. The first is a single flip [39]. Second, a more complicated rearrangement involved bifurcated hydrogen bonds and was consequently named bifurcation tunneling. Both splitting types resulting from such transition states have been observed experimentally. Furthermore, it was possible to estimate the O–O distances for each of the clusters well enough to yield a quantitative experimental measure of hydrogen bonding cooperativity. If what we learn from water clusters can be extrapolated to bulk water, cooperativity of hydrogen bond rearrangements and bifurcated hydrogen bonds will have to be accepted as a preferred model. The very detailed analysis of water dimers [40] will also yield the better pair potentials needed to improve our molecular dynamic simulations of liquid water. Preliminary observations on many-body interactions, which are so characteristic for cooperativity in any system, were also achieved. Currently, simple pair potentials used for liquid water molecular dynamic simulations were not helpful in describing gas water cluster dynamics. Many-body interactions in liquid water seem to be more complicated, as expected.

Another weakness of far-infrared vibration rotation tunneling spectroscopy is that it cannot yield the relative concentrations of different water clusters. Saykally and coworkers [41] developed another powerful technique to help resolve this problem. As mentioned earlier, mid-infrared absorption spectroscopy was used to study structuring in bulk water. Such was also the case in early cluster studies. Unfortunately, perturbations of the water clusters by the environment during measurement were severe and prevented quantitative analysis. Mid-infrared cavity ringdown laser absorption spectroscopy was developed with the intent to study isolated jet-cooled water clusters with the infrared absorption technique [42]. The technique yielded the expected result, namely that at lower pressures in the vacuum chamber only very small clusters existed. As the pressure was increased, even hexamers and heptamers were observed. Surprisingly, at any pressure studied, more trimers than dimers were observed. This strongly indicated the cooperativity between different hydrogen bonds. The authors hope that it will be possible to identify even larger water clusters. Castleman and coworkers [43] used mass spectrometry to analyze supersonically cooled water clusters and were able to identify clusters with average monomer number as high as 20. Once again, how applicable such data are to liquid water is still not fully clear.

III. ROLE OF INTERFACIAL WATER IN INTERPARTICLE AND INTERMOLECULAR FORCES

A. Introduction

The structure and properties of interfacial water are of great importance in many chemical, biological, and technological phenomena and processes such as stability of biomolecules, solubility of ions and macromolecules, dispersion of powders in water-based solvents, flocculation, adsorption, flotation, and emulsion formation and breakup. Recent research suggests that the structural features of interfacial water determine the nature of interparticle and intermolecular forces that control the aforementioned phenomena [44,45]. Wetting phenomena seem to be controlled to a great extent by the interaction of materials with water. Hydrophilic particles are wetted completely and can be fully suspended in aqueous-based solvents. Nonpolar (hydrophobic) particles, on the other hand, show a tendency to phase-separate from water. Many surfaces of intermediate nature are also observed. In this section we will concentrate on interactions of water with the hydrophilic particles and molecules. This section will deal with the role of interfacial water in interparticle forces. In the next section, we will discuss spectroscopic evidence for the existence of interfacial structured water.

Classical theories of aqueous colloidal stability and interparticle forces have for the most part ignored the role of interfacial water structure. Total interparticle force was traditionally described by the Derjaguin-Landau-Verwey-Overbeek (DLVO) theory [46,47]. According to DLVO theory, the total interaction force between two particles is obtained by summing the electrostatic double layer forces with the van der Waals forces. Yet an analysis of the results of colloid stability, rheology, and direct interparticle or surface force measurements indicated the existence of at least one additional force. This force seemed to correlate to the structure of interfacial water and was named the hydration force. Electrostatic double layer and van der Waals forces are long-range forces that act at distances of tens of nanometers. The hydration force, on the other hand, turned out to be a short-range force and was felt only a few molecular layers away from the interface.

Colloid stability of aqueous suspensions and consequently sedimentation and rheological behavior at low ionic strength are controlled and well described by the DLVO forces. For particles with the same sign of charge (positive–positive or negative–negative), the electrostatic double-layer force is repulsive. On the other hand, the van der Waals forces are, under most practical circumstances, attractive. Total interparticle forces are obtained by summation of those two components. If the total interparticle interaction force is repulsive, particles repel one another, and stable colloidal suspensions are prepared. Systems with attractive forces, on the other hand, are unstable. This results in higher viscosity and lower sedimentation stability. Charged colloidal surfaces can be neutralized with the addition of the oppositely charged counterions. DLVO theory predicts that when salt exceeds a certain concentration, the so-called critical coagulation concentration, the double-layer repulsions are reduced to such an extent that the van der Waals potential causes particles to form touching networks. At even higher salt concentrations, the electrical double layer is completely compressed, and further salt addition is not expected to influence colloid stability or rheology of suspensions. Recent colloid stability and rheology measurements, which will be described in this chapter, indicated that this is not true. The addition of salt concentrations up to 5 M decreased colloid stability and increased the viscosity of studied slurries. Moreover, the effect of nonadsorbed indifferent electrolyte ions on colloid stability was even observed at the isoelectric point of metal oxide, where particles are not charged, and no effects of added salts on colloid stability were expected. Furthermore, the size of monovalent indifferent electrolyte ions influenced colloid stability and rheology of studied systems. Finally, rheological measurements indicated that slurries whose charge was neutralized by the addition of salt had lower viscosity than slurries whose charge was neutralized by adjusting the pH to that of the isoelectric point. All those facts have been interpreted with the structure of interfacial water as one of the controlling parameters.

Direct surface force measurements between mica plates, silica, alumina, and rutile surfaces also indicated the possible presence of an additional force. That force was also of a short-range nature. As surfaces approached one another, the force was changing in the oscillatory manner with a period comparable to that of twice the size of water molecules.

Mathematical models have been developed to interpret this observed behavior and correlate it to the structure of interfacial water and its relation to the adsorbed counterions. Significant controversies still exist regarding the interpretation of the rheological, colloid stability, and direct force measurements with regard to hydration behavior. The current situation will be described in this section.

B. Rheological and Colloid Stability Studies

1. Interpretation of Colloid Stability Behavior

(a) DLVO Model. As was mentioned earlier, the total interaction between any two surfaces must include at least van der Waals attractive forces and electrostatic interactions [46,47]. Unlike the electrostatic forces, the van der Waals forces/potentials are largely insensitive to variations in pH, ionic strength, or concentrations of the adsorbed indifferent electrolyte ions. For any solute/solvent system, the van der Waals attractions are usually considered fixed. If the radii of the interacting particles and the distances between them are known, van der Waals forces can be calculated provided that the so-called Hamaker constant is known [46]. The theory of van der Waals forces is well presented elsewhere and will not be discussed here. The Hamaker constant can be calculated from theory, measured directly, or obtained as an adjustable parameter, e.g., from coagulation measurements. The values of the Hamaker constant for most condensed phases are known and lie in the range of 0.3–4×10^{-19} J [47]. Note that different methods yield different values of the Hamaker constant for the same system. Systems with different geometries also yield different values of the Hamaker constant (sphere–sphere, sphere–surface, two cylinders, two crossed cylinders, etc.). The deviations of the values obtained by different methods are due to the use of the approximate theoretical models. For example, the primitive models used in calculation of the electrostatic contribution to the interaction energy would be reflected in the adjusted value of the Hamaker constant.

Electrostatic forces between surfaces in liquid can be manipulated by changing the composition of the liquid medium, i.e., by changing pH, ionic strength or amount of phys-adsorbed molecules/ions. Electrostatic forces exist only between charged interfaces. There are actually numerous ways in which a surface can become charged. Ionization of the surface groups, such as carboxy groups, yields a charged interface. Adsorption of ions from solutions onto a previously uncharged interface will also yield a charged interface. Adsorption of water onto anhydrous oxides with subsequent protonation and/or deprotonation of the hydroxyl groups is a common method of charging metal oxides. Surface charge is balanced by an equal, but oppositely charged, atmosphere of counterions. Counterions are in rapid thermal motion in the region close to surface, termed the diffuse layer. Counterions are also bound directly to the oppositely charged surface groups at the interface [47a]. Different models of the electrical double layers are described elsewhere in this book.

The Poisson–Boltzmann (PB) equation is used to calculate an ionic concentration profile away from a surface [47]. Also, when calculating the forces between particles, one should consider the overlap of two identical or different electrical interfacial layers. In doing so one cannot avoid the application of the approximative procedure. In the so-called primitive approach, like classical DLVO theory, one meets two kinds of approximations. Mathematical approximations are related to the solution of differential equations, e.g., by linearization of the PB equation. The computer era makes such kinds of approximations unnecessary. However, one cannot avoid physical approximations related to the simplifications of the real systems [47]. The primitive model used for calculations of the electrostatic interaction energies assumes a smooth, smeared out surface charge. This is not true in real systems where surface charges are discrete. The finite size of counterions is also neglected in the primitive model. Image forces and ion–ion correlations are also neglected in such a model. Counterion charge is also smoothed out, and the discrete nature of counterions is not taken into account. Finally, the most serious problem of the primitive model is that it neglects the structure of the interfacial water/solvent as well as solvent–surface and solvent–solvent interactions. The interactions of ions with bulk and interfacial water are also not taken into account. In addition to the above problems, the reversibility of two approaching electrical interfacial layers is questionable [47b]. In practice, two limiting considerations are used: the constant potential model is assumed to apply for the reversible case, while the constant charge model is related to the irreversible situation. These two extremes are related to the relaxation of the electrical interfacial layer in the course of particle collisions.

Despite unrealistic idealization, and the significant differences between theoretical models and real systems, the primitive models may be successfully applied for numerous systems providing that conditions are appropriate. For example, the constant potential model explained the effects of electrolytes on particle detachment from surfaces, while the constant charge model disagreed with experimental findings. This result indicates that the electrical interfacial layers were in equilibrium during the detachment process. Also, the lyotropic sequence effect in the coagulation of colloidal particles was explained in the case of rutile and hematite [47c,47d]. In these studies, the electrostatic potential at the onset of the diffuse layer was obtained on the basis of the surface complexation model using electrokinetic, surface charge, and adsorption data. In using equations describing the electrostatic interactions, one meets the problem of so-called surface potential. It is clear that in this context the surface potential is the potential at the onset of the diffuse layer (ϕ_d), which is not measurable. One solution is to use the electrokinetic zeta-potential and calculate ϕ_d by the Gouy–Chapman theory, assuming certain slipping plane separation (15–20 Å) [47e]. Another approach is to use the surface complexation model and to calculate ϕ_d. These procedures result also in a decrease of ϕ_d by electrolyte addition. In conclusion, one may say that primitive models can in most of the cases successfully explain the coagulation abilities of counterions, but often the exact values of, for example, critical coagulation concentration cannot be predicted.

(b) Advanced Models of Interparticle Interactions. As it was described above, our understanding of interparticle forces/potentials can be improved only through development of more realistic models of the electrical interfacial layer. Historically, the Stern–Grahame model (SGM) [47f] was the first attempt to correct some of the assumptions of the primitive model. In the SGM the finite size of counterions/surface groups is taken into account, resulting in the introduction of the plane, a short distance away from the surface. Counterions cannot approach the interface beyond that layer. This also results in the introduction of an additional capacitance and potential. Another layer was introduced in the so-called triple layer model (TLM) [47g]. According to the TLM, some ions can bind directly to the surface; others are farther apart, and the rest of the surface charge is neutralized in the diffuse layer. Such a model requires two capacitances and three different charges and potentials, which are commonly obtained as fitting parameters. None of the above models takes into consideration the structure of the interfacial solvent or the hydration of ions. Recent models that take into account the structure of the interfacial layer will be discussed in the last section of this chapter.

2. Experimental Rheological Studies and Their Interpretation

The relation between interparticle potential (force) and rheological and colloid stability behavior of slurries has received considerable attention [48]. The interparticle pair potential is a convenient way of viewing the interaction of particles in suspensions. Many of the properties of particle networks are controlled by the depth of the potential well in which particles reside. The total interparticle potential is commonly plotted versus particle distance. The barrier in such a plot is considered to be the repulsion potential that particles have to overcome in order to form touching networks. When repulsions are overcome by the attractive potential due to van der Waals contributions, particles form touching networks with much higher viscosity and lower sedimentation stability. The depth of the potential well is controlled by the van der Waals potential and the distance to which the particles can approach. The distance to which particles can approach one another depends on the size of the adsorbed species. In the absence of any specific adsorption, this distance is defined with the thickness of the interfacial water layer and adsorbed indifferent electrolyte counterions. At the isoelectric point this distance is determined by the thickness of the interfacial water only. Particles that approach one another at the isoelectric point fall into a deep potential well. The thickness of a hydration layer is also expected to be shorter at the isoelectric point. Water molecules are dipoles and are structured more efficiently near the charged interfaces away from the isoelectric point.

Most colloid stability and rheological studies were performed with metal oxides. The surface of oxide particles in water is covered with a chemisorbed water layer that dissociates yielding surface OH groups. Such groups can either accept or donate a proton to the solution depending on pH. At the pH of the isoelectric point (iep), the surface concentration of positive and negative groups is equal and the surface is neutral. At pH's below the iep, the surface is positively charged, and counterions are negatively charged anions. At pH's above the iep, the surface is positively charged, and counterions are cations. The surface can be neutralized by the addition of counterions. Usually, at concentrations higher than 1 M of monovalent nonadsorbing electrolytes, surfaces of metal oxides are fully neutralized. The effects of further salt addition are then studied and correlated to the ability of salt to perturb interfacial and bulk water structure.

Velamakanni and coworkers [49] studied the effect of surface hydration on surface forces by measuring the rheology and particle packing density of alumina powders in aqueous suspensions. It was discovered that alumina slurries coagulated by a high amount

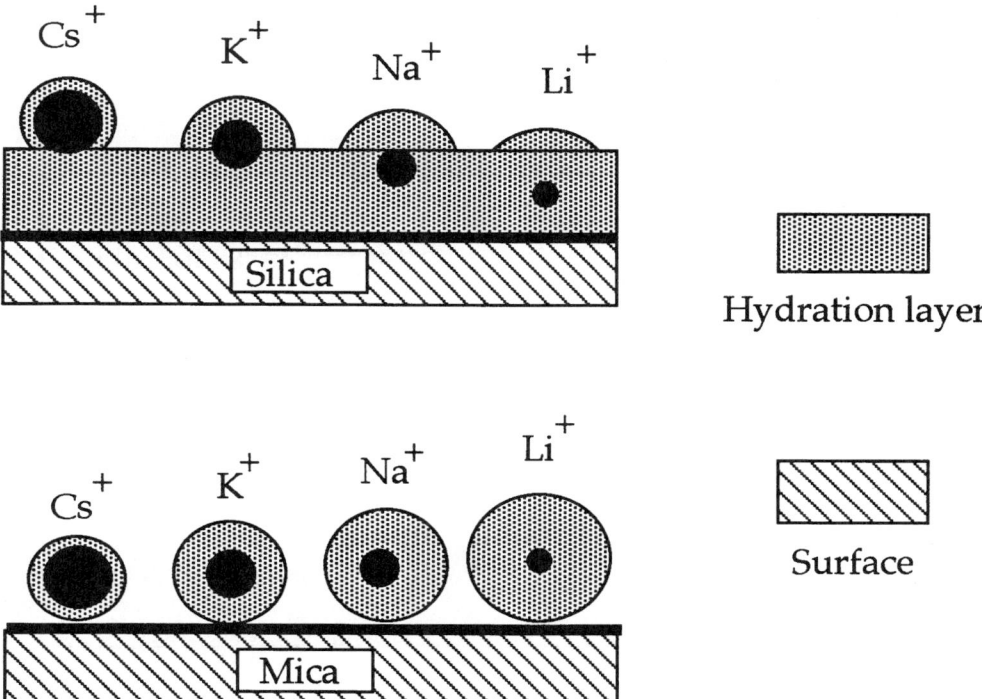

FIG. 1 Schematic representation of the structure of the oxide/water interface (pH>pH (iep)) and the mica/water interface with various counterions adsorbed to the surface. Small counterions with a greater affinity for water penetrate deeper into the surface hydration layer of oxides than do large ions. At the mica surface, ion exchange of fully hydrated counterions for the lattice potassium ions occurs. (From Ref. 55.)

of an indifferent electrolyte, ammonium ion, exhibited an order of magnitude lower viscosity and significantly higher packing densities when compared to the slurries prepared at the iep. On the basis of conclusions reached by Pashley and coworkers [50], based on their surface force apparatus measurements on mica, it was speculated that the additional short-range repulsive force (potential) might originate from the force needed to dehydrate counterions that crowded the surface at pH away from the iep. Using the surface force apparatus (SFA), Ducker et al. [51] confirmed the existence of a short-range repulsive force between basal saphire plates when excess salt was added.

This research was significantly extended by Čolić and coworkers [52–56]. They speculated that if structuring of interfacial water and its interactions with the adsorbed counterions is controlling the observed behavior, results of the ion size sequence experiments would yield a different sequence than if dehydration of the adsorbed counterions were controlling the colloid stability of metal oxides at high ionic strength. Pashley and coworkers [50] observed that short-range hydration forces with lithium counterions were of the longest extent, and those with cesium, of the shortest extent. This was observed with mica plates which are poorly hydrated. Čolić and coworkers speculated that if primary water hydration is perturbed with the adsorbed counterions, an opposite sequence would be observed, as shown in Fig. 1. Lithium would penetrate into a hydration layer of colloidal particles deeper and closer to the surface than cesium. This would yield

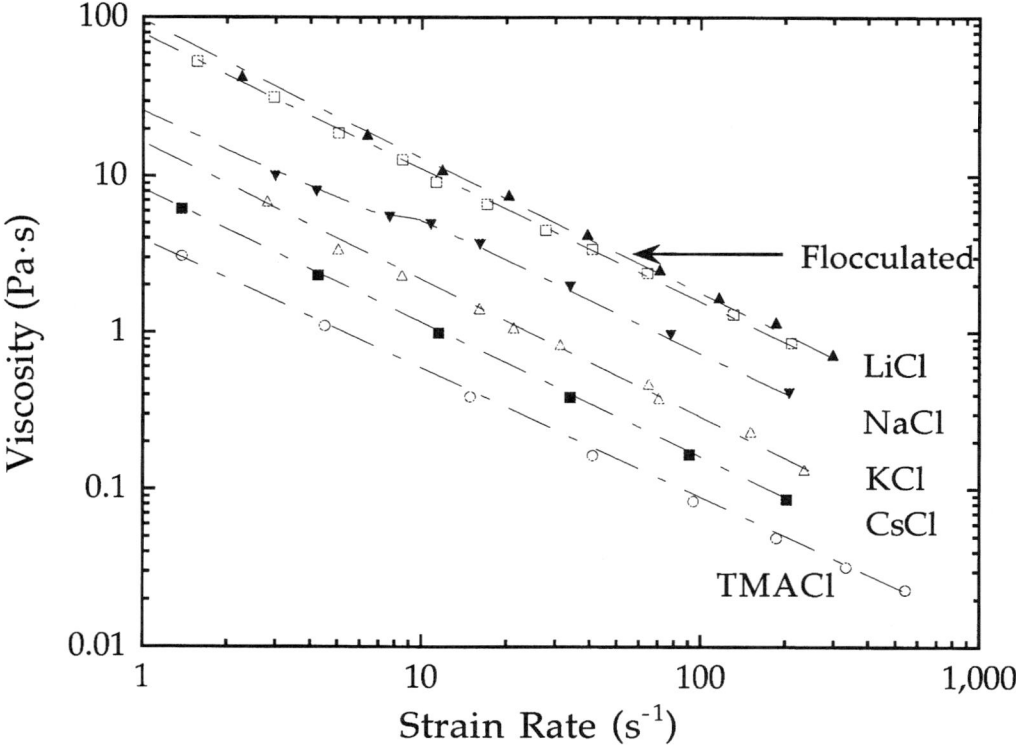

FIG. 2 Viscosity versus strain rate for alumina slurries (0.20 volume fraction solids) formulated at pH 12 with 0.5 M chlorides (as labeled) compared to the flocculated slurry formulated at the iep (pH 9) with no added salt. (From Ref. 52.)

short-range repulsive forces, the shortest extent with the lithium slurries and the longest extent with the cesium slurries. This was indeed observed with the positively charged alumina slurries at pH 12 and cation concentrations of 0.5 M or higher, as shown in Fig. 2 [52]. Similar results were observed with the consolidation of the slurries and the flow stress behavior of the consolidated bodies.

At this point we will explain the correlation between interparticle potentials, rheology, and packing behavior/flow stresses of the consolidated slurries. The relation between interparticle potential and the rheological behavior of slurries has received considerable attention. The viscosity, the shear modulus, and the apparent yield stress have been measured and related [57] to the potential between particle pairs. These rheological results can be correlated to the strength of the particle network, which is controlled by the interparticle pair potential as well as the volume fraction of particles within the liquid. Suspensions with repulsive interparticle potentials exhibit nearly shear rate independent (Newtonian) viscosities, while attractive particle networks exhibit shear thinning behavior (viscosity decreases with increasing shear rate) [58]. The viscosity of an attractive interparticle network, as well as the shear rate dependence, increase with the depth of the potential well. The force between particles at any separation is the negative derivative of the potential with respect to distance. The force between particles is zero at the equilibrium separation distance, whereas a maximum force required to pull particles apart occurs at the inflection point between the equilibrium separation. The yield

Interfacial Water Structure

stress of the network is related to the force needed to pull particles apart. The second derivative of the potential versus distance relationship at the equilibrium separation is proportional to the spring constant between the particles, which is related to the modulus of the network. As expected, interparticle potentials with deeper minima produce the higher shear moduli, yield stresses, and viscosities.

The consolidation behavior of slurries as well as the flow stress of the consolidated bodies are also a function of the interparticle pair potential [52]. The densities of bodies consolidated from dispersed slurries (long-range repulsions) are at a maximum for the particular powder and are relatively insensitive to applied pressure. The packing densities of bodies consolidated from flocculated slurries (at the isoelectric point where only attractive van der Waals forces exist) are low and strongly dependent on the applied pressure [52]. For bodies consolidated from slurries possessing weakly attractive interparticle potentials (short-range repulsive potentials), the packing density is intermediate to well dispersed and flocculated with the depth of the potential minimum controlling the packing density and flow stress of the consolidated bodies. Deeper minima, viz., stronger attraction, results in bodies with lower packing density and higher flow stress during compressive mechanical testing. It is well known that particles that form networks due to attractive interparticle potentials must rearrange to increase their packing density under an applied pressure. This process is believed to be correlated to the repulsive, lubricating interparticle force and interparticle friction. The deeper the interparticle potential well (created by short-range repulsive forces of lesser extent), the greater the friction will be and thus the more difficult it is for particles to rearrange and pack well [53]. This results in lower packing densities and stronger pressure dependence (more force needed for rearrangement) for bodies consolidated from slurries with deeper potential wells. Such behavior is presented in Fig. 3.

Čolić and coworkers also performed a detailed study on the influence of ion size and its affinity for water on the stress-strain behavior of consolidated water-saturated ceramic bodies [53]. The results of their investigations are best described in the abstract of their manuscript [53].

> The effect of ion size on the force necessary to remove counterions from the surface when particles are forced into contact (primary minimum) was investigated. Alumina slurries coagulated at pH 12 with 0.5 M of Li^+, Cs^+ and TEA^+ (tetraethylammonium$^+$) chlorides were consolidated by pressure filtration. Uniaxial compression tests were performed to determine the plastic-to-brittle transition pressure for each counterion. When particles are pushed together to form a strong touching network brittle bodies are produced. The force required to push particles together is related to the slope of the repulsive potential barrier, the steeper slope requiring greater force. The consolidation pressure necessary for brittle behavior was greater for slurries with smaller counterions such as Li^+ as compared to Cs^+ and TEA^+. The results presented here show that the resistance to pushing particles into contact is related to the size of the bare counterion in weakly attractive (salt coagulated) slurries. That is, the slope of the potential barrier produced when the slurry is formulated with smaller ions is steeper. This result is consistent with the idea that smaller counterions are more strongly bound to the surface as well as the fact that the repulsion due to the finite size of the ion occurs at smaller interparticle separation distances for the smaller counterions.

Such behavior is schematically presented in Fig. 4.

In short, this study further reinforced the idea that smaller counterions with stronger affinity for water penetrate deeper inside the hydration layer and closer to the surface. This yields slurries with higher viscosity and consolidated bodies with lower packing density and less plastic, more brittle behavior. Moreover, when particles are pushed together with

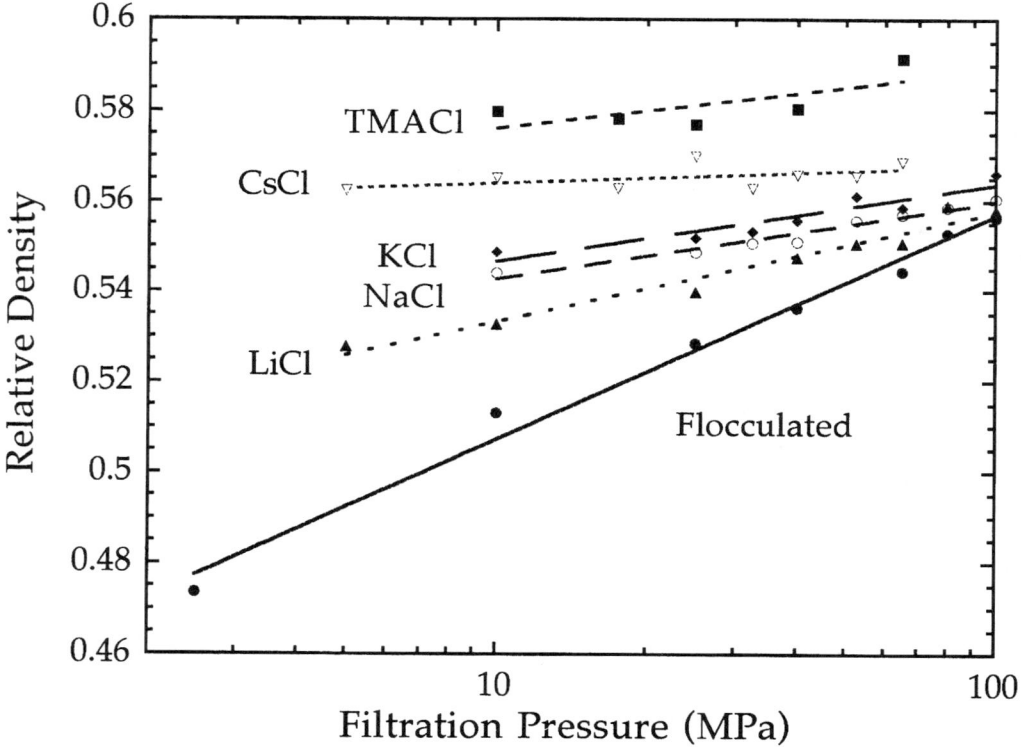

FIG. 3 Relative density versus applied filtration pressure for alumina compacts consolidated from coagulated slurries (0.20 volume fraction, pH 12, and different chlorides) compared to the flocculated slurry at pH 9. (TMACl = tetramethylammonium chloride.) (From Ref. 52.)

the brute force used in the mechanical testing machine, it was more difficult to push particles together with the bodies produced by consolidation of slurries prepared with 0.5 M of lithium ions. This strongly suggests that lithium ions are adsorbed closer to the surface than, say, cesium ions. Some "barrier" seems to be present between indifferent adsorbed electrolyte ions and surface groups on metal oxide. While it is quite probable that that barrier is water, direct spectroscopic studies should be performed to test this hypothesis.

In an attempt to explain further the effects of structure-making and structure-breaking ions and water hydration on interparticle forces and the rheological behavior of slurries, Čolić and Fisher [54] studied the influence of excluded (nonadsorbed) ion size and affinity for water on viscosity of alumina and silica slurries at the isoelectric point (iep) as well as below it, where cations are coions and are not expected to adsorb in the absence of the specific adsorption. Leberman and Soper [59] recently compared the effects of pressure and ion addition on water structure and discussed their results in lieu of a Hofmeister series for ion size effects on protein stability. Parsegian even proposed that the effects of ion size will soon be found for not only adsorbed but also excluded ions [60]. Parsegian also proposed that effects of nonexcluded ions on protein or colloid stability will be correlated to an ion's ability to structure water. Yotsumoto and Yoon [61,62] studied the effect of NaCl additions on the stability of silica and rutile sols at

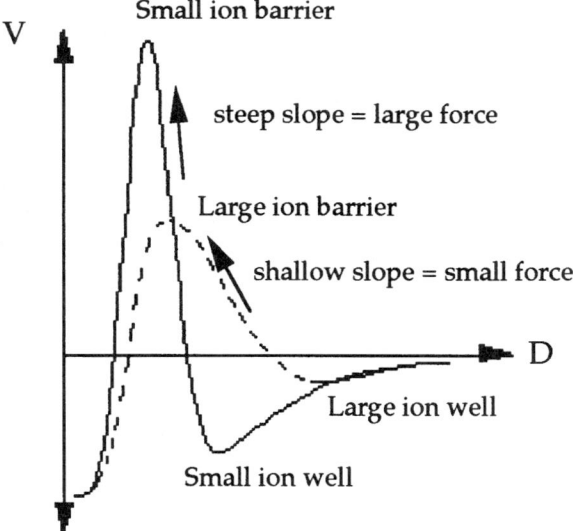

FIG. 4 Schematic of weakly attractive pair potentials when the counterion concentration is large and the Debye length has collapsed to its minimum size. The solid line is the pair potential for small counterions and the dashed line is for large counterions. The smaller counterion is more strongly bound to the surface (higher potential barrier) and produces a deeper potential well located closer to the surface, which results in a steeper slope of the potential barrier and thus a greater force to push the surfaces together. (From Ref. 53.)

the iep. Although the ions were not expected to adsorb, a decrease in colloid stability was observed above 1 M of added salt for presumably well-hydrated silica sols. The opposite effect was observed for presumably less well hydrated rutile sol.

The results of Čolić and Fisher [54] gave an even stronger hint that the ion's ability to structure water directly correlates with its influence on colloid stability at high ionic strength, where the electrostatic double layer forces are insignificant in the absence of specific ion adsorption. The authors realized that at high ionic strength (2 M or higher) slurries prepared with smaller ions with higher affinity for water always had higher viscosity than those prepared with larger ions with a weaker affinity for water. Sedimentation rate studies showed that slurries prepared with lithium always had the highest sedimentation rate, those prepared with cesium the lowest. If solution viscosity were controlling the total viscosity and sedimentation behavior, this would not be the case (particles would sediment at a slower rate in a solvent with higher viscosity).

Silica is a complicated system, and its behavior is still not fully understood. Yet the results with alumina at pH 4, five pH units below its iep, clearly showed that the ion size sequence for cations, which are excluded coions at this pH, exists. While alumina behavior might also be complicated, with gelation at high pH, such behavior has never been observed at pH 4. Specific adsorption is another possible explanation for the observed behavior. Healy and coworkers [63] observed that larger specifically adsorbed ions yielded slurries with lower viscosities. To resolve this problem, Čolić and Fisher studied another ion series. Tetraalkylammonium ions (alkyl = methyl-butyl) also differ in size and the ability to structure water. While they structure water in a different way, conveniently, the largest ion in this series, namely tetrabutylammonium ion, has the highest affinity for water [64]. If specific adsorption were controlling the behavior of slurries, the systems

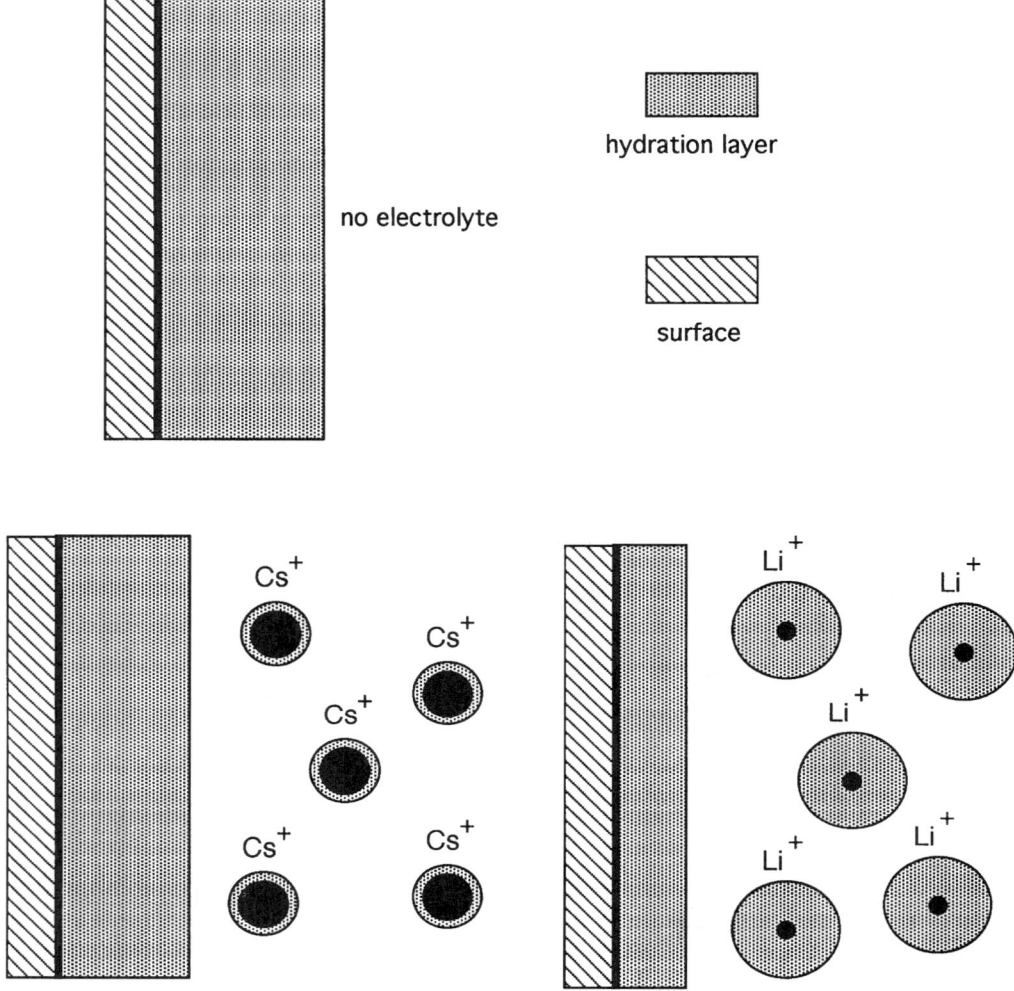

FIG. 5 Schematic presentation of the interaction of nonadsorbing structure promoting (lithium) and structure breaking (cesium) ions with the hydrated metal oxide surface at its iep. (From Ref. 54.)

prepared with tetrabutylammonium ion would have the lowest viscosity. The opposite was observed. The schematic model for interaction of structure-making and structure-breaking ions with well-hydrated interfaces is presented in Fig. 5.

While the experiments of Čolić and coworkers and Yoon's group strongly suggest the presence of thicker hydration layers on oxide interfaces and their interactions with nonadsorbed (excluded) ions, more direct spectroscopic proof is needed. The spectroscopic techniques used to study interfacial water will be discussed later in this section.

The behavior of the silica water system, which has been of particular interest for colloid chemists for many decades, still presents many challenges. Čolić and coworkers [55] also studied the influence of ion size on the rheological behavior of silica slurries. Once again, at 4 M of salt, lithium slurries always had the highest viscosity and cesium slurries the lowest. As expected, silica slurries at the iep (pH = 1.8) had, in the absence of salt, almost as low a viscosity as at pH 10 where silica is highly charged. This is the so-called anomalous behavior of silica [65]. More peculiar, even the addition of 4 M of salt at the isoelectric point did not produce slurries with non-Newtonian rheology, characteristic for coagulated systems. Another anomaly was observed. For other oxides, the farther the pH is away from the pH (iep), the larger the difference observed in the effect of ion size on viscosity. For silica, on the other hand, the opposite behavior was observed. In any case, Čolić and coworkers observed that at high ionic strength (4 M) and high volume fraction of the suspended silica, lithium slurries once again showed the highest viscosity and cesium slurries the lowest. This cannot be explained with the ion adsorption strength, since cesium adsorbs stronger than lithium at the silica/water interface. As in the case of alumina, Čolić and coworkers [55] proposed that the reason for such a lyotropic sequence observed at high ionic strengths and volume fractions of oxides lies in the strong competition for water needed to hydrate ions and colloids. Such findings strongly indicate the presence of thicker hydration layers at the silica interface. Other groups, which mainly used AFM or surface force apparatus direct force measurements, suggest that there is only one layer of organized water near the interface [66].

So what can be the reason for the difference in the observed behavior in surface force apparatus, AFM, or colloidal stability studies with the pyrogenic silicas and precipitated silica obtained from Nissan Chemicals and used in work of Čolić et al.? Ninham and coworkers [67] recently found that all silica surfaces produced by flame treatment or cleaned by plasma eventually gelled, given enough time. Such surfaces always carry some intrinsic silicic acid, which indeed controls their behavior. It was also realized by Čolić et al. that by adding silicic acid to silica samples the viscosity can be lowered. Yet all such samples gelled in 1 to 4 days. The silica sols obtained from Nissan Chemicals and used in the work of Čolić and coworkers have not gelled in more than three years. NMR analysis of these silica samples did not identify any silicic acid on the surface.

Čolić and Morse [68] recently realized that silica samples treated with plasma or electromagnetic radiation always contained some silicic acid on the surface and ultimately gelled with time. It was also observed that plasma, flame or electromagnetic radiation produce free radicals inside silica. Atomic hydrogen can be entrapped inside silicas for years [69]. When such silica samples are submerged under water, atomic hydrogen produces protons and free electrons, which are ultimate structure breakers [70]. The structure-breaking properties of electrons probably produce more active, smaller water clusters that dissolve some silica even at pH's as low as 3. Consequently, water penetrates inside the silica structure and gels the surface. When two gelled surfaces come together, they will ultimately adhere strongly, if given enough time.

The above-described results, we believe, will help explain a long dispute about what is causing the unusual stability of silica samples in water solutions. Both the adsorbed layers of silicic acid and the extensive hydration of the surface seem to contribute to that stability, depending on the sample, experimental conditions, and time given for the equilibration. Finally, other possible explanations for the observed phenomenon such as interparticle bridging [71] or inadequacy of the linear summation of electrostatic double layer forces and van der Waals forces should be considered. Such possibilities were recently scrutinized by Ninham and Yaminsky [72].

The topics described in this section clearly show the need for a comparison of bulk colloidal measurements with the more direct spectroscopic measurements. Only a combination of bulk, spectroscopic, and calorimetric data can actually yield new insight into the hydration of colloidal systems. Such spectroscopic measurements will be described later.

C. Direct Surface Force Measurements

As mentioned in the previous section, classical theories of aqueous colloidal stability and interparticle forces have for the most part ignored the role of interfacial water structure. Total interparticle force was traditionally described by the Derjaguin-Landau-Verwey-Overbeek (DLVO) theory [46,47]. The total interparticle interaction forces were generally considered in terms of the attractive van der Waals and repulsive electrostatic double layer forces. In recent years, direct measurements of surface forces identified other non-DLVO forces, including the so-called hydration force. This force was attributed to the solvent structure near the interface [73]. Unlike DLVO forces, which are of long-range nature and are felt up to 30 nm or more, the hydration forces are short range and do not contribute to the total interparticle forces at distances longer than 3 nm.

Several techniques have been developed to measure directly the forces between smooth solid surfaces, lipid bilayers, or model biomembranes containing proteins or other biological molecules [74]. In the surface force apparatus, the forces between two surfaces immersed in a liquid are measured from the deflection of the spring supporting one of the surfaces, while the distance between the surfaces is measured using an optical interference technique [74–76]. Until recently, these techniques could be used only with molecularly smooth mica sheets. Recent developments made it possible to use other materials such as silica, alumina, or rutile. In atomic force microscopy (AFM) surface force measurements, a particle is attached to the cantilever of the AFM. The force is measured as the cantilever approaches a smooth surface of the same or a different material [76–78]. The particle can be in the micron size range. Great care has to be exercised when attaching a particle to the cantilever in order to avoid contamination by the glue. Also the opposing surface should be as smooth as possible to avoid artifacts in the measurements. A variety of materials can be used with the AFM technique. On the other hand, the obtained force vs. distance curves do not yield as much information as SFA measurements. The osmotic pressure technique is particularly suitable for biological systems. Measurements of surface forces between lipid bilayers, proteins, or DNA molecules were performed [79–82]. Water is osmotically withdrawn (usually with polyethyleneglycol) from a dispersed stack of molecules studied (say lipid bilayers) and the change in the lamellar repeat spacing, for instance, is measured with x-rays. The interfacial gauge technique [83–85] is particularly suitable for experiments with freshly molten glass (silica) balls. This instrument is also capable of measuring interparticle forces vs. distance curves, but with the mechanical spring constant up to 500 N/m. This is stiffer by many orders of magnitude than a torsion balance or even an AFM cantilever. It is hundreds of times stiffer than the spring of an SFA apparatus [83].

So, what did we actually learn about the hydration of colloids from the direct surface force measurements? This will be the topic of this section.

Israelachvili and Adams studies the interparticle forces between mica sheets immersed in aqueous KNO_3 solutions and discovered an additional non-DLVO short range force. This force was acting when surfaces were less than 3 nm apart and prevented mica sheets from coming into direct adhesive contact [86]. This behavior has been ascribed to the hydration effects at the mica/water interface. Earlier, it was proposed that hydration effects on mica should be ascribed to the hydration of potassium ions present in the natural crystal lattice of mica [87]. This motivated Pashley [88] to study the effect of counterion size on DLVO and hydration forces between mica sheets in lithium, sodium, potassium, and cesium solutions. Deviations from the DLVO behavior was observed only after a certain cation concentration. Lithium suspensions always yielded short-range repulsions with the highest extent, cesium with the lowest. "Hydration" forces according to Pashley arise when hydrated cations adsorbed on mica are prevented from desorbing as two interfaces approach. Dehydration of the adsorbed cations then leads to a repulsive hydration force [88]. Lithium has the highest affinity for water and therefore produces systems with the short-range repulsions of the largest extent. Such hydration is actually better described as "secondary hydration." "Primary hydration" is a different phenomenon and describes the structuring of water at the interfaces [89]. As was discussed in the previous section, hydration of interfaces and adsorbed ions are often interdependent, and a combination of both effects controls short-range repulsive forces.

As discussed in the last section, one of the most interesting and unusual systems studied by colloid chemists is colloidal silica. The surface and interfacial properties of silica are important in the production of optical and electronic materials, adsorbents, polymer filters, composite materials, chromatographic columns, fillings, etc. Surfaces of numerous important high-tech materials such as silicon, silicon nitride and silicon carbide are covered by native silica film when exposed to air. Despite great interest in silica interfaces, we still do not fully understand the suspension behavior. The unusual stability of silica sols attracted the attention of the surface force measurements community [90–94]. Several earlier explanations for the colloidal stability behavior of silica have been advocated by different authors, and no clear solution to the problem is in sight. However, some recent contributions seem to have introduced clarification to at least some of the underlying problems.

The determination of some short-range repulsive forces is common to all surface force studies involving silica systems. It is the *nature* of that force, which is proposed to be responsible for the high stability of silica sols, that is still unclear. Different researchers have proposed at least three models to account for such stability behavior. Perturbed water structure near a silica surface with long-range ordering of water molecules is one of the more popular models [95]. "Hairs" of polysilicic acid growing from the surface and giving a kind of steric stabilization have also been proposed to account for the unusual colloid stability of silica sols [96]. Surface gelation and the thickness of the surface gel layer have also been proposed as the controlling factors for the behavior of aqueous silica systems [97].

We will first review reports on surface force measurements on silica in which the presence of a thick hydration layer is suggested to stabilize silica systems. Horn and coworkers [90,91] pioneered surface force apparatus measurements with silica. They prepared thin sheets of silica with smooth surfaces, suitable for use in SFA. The method involves melting the end of a tube of high-purity silica, using a hydrogen/oxygen torch and attaching a small remelted area of silica to freshly cleaved mica sheets. Different techniques are then used to clean the surface of impurities and hydroxylate silica. The exposure of silica to a water steam, ammonia, water RF plasma, or UV plasma have all been used. Horn and coworkers showed that regardless of surface treatment, short-range repulsions were always present and did not depend significantly on the type of treatment [91]. The authors claim that the results presented clearly indicated the absence of any significant silicic acid or gel layers. Their arguments [91] are summarized as follows: (1) the range of short range repulsions did not depend on the time of immersion of silica in water (up to 24 hours); (2) the additional repulsion was not affected by water vapor steaming or even ammonia treatment, which are definitely shown to produce surface gel; (3) covalently bonded propylamine or aminosilane removed most, if not all of the repulsion. If excess non-DLVO short-range repulsions were due to the presence of a surface silicic acid layer or gel, an additional repulsion would be observed after the binding of such short chain molecules. On the other hand, if surface silanol groups are masked by a covalent binding of those molecules used, no strong hydrogen bonds with water can be formed, which should disturb the ordering of water; (4) the repulsions did not depend on salt concentration up to 0.1 M of salt, which indicates that secondary hydration effects observed with mica are not responsible for additional short-range repulsions with silica.

The work of other authors on SFA measurements with silica seems to corroborate the model of Horn and colleagues [92–94]. Grabbe performed detailed experiments with sylylated silica surfaces and showed that covalent binding indeed removed all short-range repulsions [92]. Chapel [93] studied the effect of ion size on hydration forces between silica surfaces. Chapel's results indicated the ion size effect opposite to that observed by Pashley on mica. This also strongly suggested that the origin of short-range repulsions on mica and silica are different. While ions enhanced the short-range repulsions on mica, they destabilized silica. As discussed earlier, recent rheological data also seem to support Horn's model. Detailed spectroscopic studies on silica will be discussed in the next section. Spectroscopic studies also appear to give support to a model in which the unusual behavior of silica in controlled by a thick surface hydration layer.

Israelachvili and coworkers performed detailed adhesion, friction, and interparticle force measurements in air and aqueous salt solutions [98]. They deposited thin silica films by electron beam evaporation, plasma enhanced chemical vapor deposition, or flame glass blowing from the melt. The authors interpreted their short-range force measurements on silica in a different way from Horn and coworkers. Israelachvili and colleagues advocate the model in which the unusual stability of silica is caused by the presence of surface "hairs" of silicic acid, and possibly, a layer of surface gel. Their arguments are as follows:

> (1) The adhesion of silica surfaces in air increases slowly with contact time, especially in humid air where the "contacting" surfaces become separated by a 20 Angstrom thick layer of hydrated silica or silica gel; (2) the friction of two silica surfaces exhibits large sticking or "stiction spikes," whose magnitude increases in the presence of water and when the surfaces are kept in contact longer before sliding; (3) the non-DLVO repulsion commonly seen at short range (<40 Å) between silica surfaces immersed in aqueous solutions is monotonically repulsive, with no oscillatory component, and is quite unlike theoretical expectations and previous measurements of forces due to solvent structure; (4) dynamic contact angle measurements reveal time-dependent effects which cannot be due to a fixed surface chemical heterogeneity or roughness.

The authors further claim that this surface layer of silicic acid in the presence of water is responsible for chemical sintering, adhesion, and friction which are time dependent. They suggest that the surface gel layer shifts the outer Helmholtz plane outward and adds a monotonic short-range polymer-like steric repulsion to the DLVO interaction. Similar models that involve silica particle sintering and the presence of a silica gel layer on the interface have recently been proposed by Depasse [99] in order to interpret silica behavior. It is important to note that the plasma techniques used to deposit silica in the work of Israelachvili and coworkers is different from the techniques used by Horn, Grabbe, and Chapel.

Yaminsky and coworkers reported [100] that plasma treatment modifies the surface of silica so that any surface will gel, given enough time. The possible role of plasma in modifying silica solubility, gelation, and surface properties was discussed in the previous section. All said, it is clear that the real cause of unusual silica behavior is dependent on sample treatment and history. The real picture might be somewhere in between (both silicic acid and the gel layer, as well as hydration, might influence silica stability). More work remains to be done in the future to understand this interesting colloidal system.

Direct force measurements were also used to study the hydration and short-range repulsions of biological systems. Lipid bilayers and model membranes are the most popular systems studied along with some recent work performed on DNA and proteins. Surface force apparatus, atomic force microscopy, and osmotic technique were used with lipid bilayers [101–104]. The obtained results, as in the case of silica, are interpreted in different ways. Israelachvili and Wennerstrom [104] once again claim that there are no extensive hydration layers and that the observed short-range repulsions are due to the protrusion of the surface groups and the dynamics of the flexible interface as they approach one another. Parsegian and colleagues, on the other hand, advocate the presence of thicker hydration layers and their role in short-range repulsions. Spectroscopic experiments that will be discussed in the next section seem to show that extended layers of ordered water are a reality. Much more work with many novel techniques will be done in the near future to understand biological systems and the role of hydration in their behavior.

IV. SPECTROSCOPIC STUDIES OF HYDRATION

A. Introduction

As discussed in the introduction section of this chapter, there are some simple molecular questions that we have to answer in order to understand the hydration of colloids. First, in a given system, where is water? Second, what is the orientation of the water molecules with regard to the surface? Third, how fast can water molecules reorient themselves and how fast do hydrogen bonds break up and reform? And finally, how do molecular changes of the hydration process influence the bulk macroscopic and mesoscopic behavior of the system? While these questions seem simple, sophisticated molecular techniques are needed to yield answers. Spectroscopic techniques seem to be among the most powerful tools needed to answer these questions.

The adsorption of molecules at interfaces presented a long-lasting challenge to spectroscopists. The surface concentrations are commonly very small and yield only a very weak signal that is difficult to detect. The signal-to-noise ratio in any measurement is very low. Moreover, unabsorbed molecules yield an overwhelming signal from the bulk suspension. This is especially obvious in water adsorption studies. The water concentration in the bulk is around 55.5 M. Only a very small fraction of that amount (millimoles or usually much less) is bound at the interface. Many techniques such as x-ray diffraction do not even see protons directly.

To circumvent these problems, many ingenious techniques have been developed and tested during the last ten years. Some of them are particularly suitable for inorganic interfaces; others are more specialized for biological systems with their high order. Attenuated total reflectance (ATR) FTIR is one technique that is particularly useful in interfacial studies. Nonlinear optical techniques such as second-harmonic generation (SHG) and sum-frequency generation (SFG) revolutionized our ability to study interfaces. Pulsed NMR as well as 2D and 3D NMR techniques also yielded significant new insights into hydration and the dynamics of adsorbed water molecules. Scattering techniques (both neutron and x-ray) also significantly improved our understanding of hydration. Finally, some indirect probe techniques such as fluorescence spectroscopy also helped explain the dynamics of hydration. Dielectric spectroscopy and time-dependent reflectometry also enhanced our ability to study hydration. Other techniques such as Raman spectroscopy sporadically contributed to our understanding of the hydration process. Those techniques will be discussed in this section.

B. Spectroscopic Studies of Colloid Hydration

As mentioned earlier, spectroscopic studies of the interfacial water were troubled by many problems including a low signal-to-noise ratio and the small amount of absorbed water as compared to bulk. Some of the early studies relied on vacuum techniques in which metal samples were dried, evacuated, and then tested for any strongly bonded water that could survive such treatment. Consequently, the in-situ analysis of such systems was not possible and will not be reviewed in this chapter. One of the first attempts to study colloidal interfacial water with ATR/FTIR spectroscopy was made by Anderson and coworkers [105,106]. In this work, an "inert" zinc selenide internal reflectance element was used to enhance multiple internal reflections and enhance the signal-to-noise ratio. A suspension of goethite in heavy water (D_2O) was used as a sample. The authors analyzed different spectra in the mid-infrared region. The difference between the goethite + D_2O spectrum and the pure solvent (D_2O) was attributed to interfacial water structure. These authors observed shifts in the D_2O frequencies and concluded that the shifts were caused by stronger hydrogen bonding to the interface.

Miller and coworkers used internal reflectance elements as a model surface [107]. Silicon wafers with a nanometers-thick surface layer of silica were used for this purpose. This approach yielded band position, intensity, and shape. Additionally, depth profiling was possible by using different angles of probing beam incidence. As in the case of bulk water, characteristic water bands were observed in the mid-infrared region. As expected, the increase in temperature resulted in a decrease of bands assigned to strongly hydrogen-bonded water, while the intensity of bands for partially hydrogen-bonded water increased. Also, as the depth of penetration decreased, the band for icelike hydrogen-bonded water became more dominant. The closest distance to which data could be collected was ca. 153 nm. While most of the signal intensity comes from the surface, this also raises the question of how much of the contribution to the band intensities comes from the bulk water. Also, as in the case of bulk FTIR and Raman spectroscopies, deconvolution has to be performed to interpret the signal.

Another novel spectroscopic technique was developed to probe directly those first few layers near the interface. Second-harmonic generation in the UV and visible ranges was used to probe water near the silica interface. While the presence of interfacial water was identified, little data could be extracted from this work owing to the low amount of information about water molecules available in that spectral range [108].

To circumvent this problem, a similar nonlinear spectroscopic technique, namely sum-frequency generation (SFG), was developed and applied to interfacial water studies. Optical second-harmonic generation is the nonlinear conversion of two photons of frequency ω to a single photon of frequency 2ω, which, in the electric dipole approximation, requires a noncentrosymmetric medium [109]. Interfacial water is just such a noncentrosymmetric system. Bulk water is not, and no second-harmonic signals are obtained from the bulk. In the sum-frequency generation technique, two laser beams of visible and infrared frequency are combined, and the reflected SFG signal from the interface is detected and mathematically analyzed. The beauty of this approach is that while water can be probed in the mid-infrared region, detection is done with the very sensitive visible light detectors. The underlying physics is complicated, but the bottom line is that only interfacial water is identified.

The results of SFG studies of the silica/water interface gave further support to the old models, which predicted that a charged interface promotes the structuring of water. As the pH of the solution was increased from 1.8 (isoelectric point of silica) to 3.8, 5.6, 8, and 12.3, water became more ordered, and the number of layers of interfacial water

Interfacial Water Structure

increased. At high pH (12.3), water at the silica interface was almost as ordered as interfacial ice. The SFG results also showed that at least one layer of interfacial water exists at the isoelectric point of silica and that as the pH is raised to 12.3, up to seven layers can be observed. The change of orientation of the adsorbed water molecules when changing pH from 1.5 to higher pH values (which also resulted in the positive surface becoming a negative one) was also observed. The change of molecular orientation was identified by measuring not only the value but also a phase of the SFG signal and its components. Such a change was also proposed earlier by Bockris and coworkers for metal surfaces. Water molecules adsorbed at positively charged interfaces seem to present electronegative oxygen to the interfacial surface groups. On the other hand, negatively charged interfaces attract electropositive hydrogen atoms closer to the interface. At the intermediate pH's, where the surface is not strongly charged, intermediate water conformation is possible. Such water molecules are not as strongly attracted to the interface and are also less oriented. In such cases, ordering of water molecules in second or additional layers is not possible.

As mentioned in the last section, silica is a very complex system and not yet fully understood. To avoid any artifacts coming from the complicated chemistry of aqueous silica solutions, Miranda et al. [110] studied the effect of pH on the ordering of water under a Langmuir film of carboxylic acid, hexacosanoic acid. As in the case of silica, the increase in surface ionization resulted in an ordering of water and the presence of multiple layers of organized water underneath. At high pH's, water was once again organized almost as well as in ice. Shen and coworkers also studied the ordering of water underneath alcohols with nonionizable OH groups. Surprisingly, water was highly ordered [111]. The authors explained their findings with the proposal that some epitaxial ordering occurred.

Scattering techniques are not a spectroscopy in the technical sense but are often described along with other molecular/spectroscopic methods of measurements. Recent results of incoherent quasi-elastic and inelastic neutron scattering spectra from confined water has been analyzed by numerous authors [112–115]. In particular, Bellisent-Funel and coworkers [112] analyzed results obtained from water contained in the micropores of Vycor glass. The short time self-diffusion constant, the elastic incoherent structure factor, the residence time for jump diffusion, the rotational correlation time, and the proton density states were obtained as functions of temperature and degree of hydration. As expected, water was more confined at lower temperatures and hydration extent. The values for the hydrogen bond lifetime were very close to that of bulk water, namely they varied between 1.1 and 3 ps. On the other hand, the residence times of water molecules at local sites was longer. The intermolecular motion of water molecules close to the surface was also hindered. The hindrance increased with decrease in temperature and extent of hydration. The authors expect that at 25% hydration, bending modes of water vibration at the surface would completely disappear. In the future, the authors contemplate performing new incoherent inelastic neutron scattering experiments at low temperature and degree of hydration to study intermolecular density states of confined water. Such experiments would then be correlated with new high-resolution NMR experiments and computer molecular dynamics studies.

All said, what did we learn from the recent progress in surface spectroscopic and scattering techniques? It seems that confined water is a reality, and that up to at least four layers of structured water do exist near charged interfaces. The orientation of water molecules depends on the polarity of the surface. As expected, the confinement of water is dependent on pH, ionic strength, temperature, and level of sample hydration. On the other hand, water near interfaces still has a significant freedom with hydrogen bond lifetime being close to that of bulk water. For the time being, it seems that multiple layers of organized structured water near the interface are reality. In the future, more direct experimental techniques will be used to test our current knowledge. The proposed models will have to survive the scrutiny of many different techniques used to test them, for instance, SFG, NMR, and neutron scattering along with computer molecular dynamics.

C. Spectroscopic Studies of Hydration in Biological Systems

Water–protein, water–DNA, and water–lipid interactions have long been recognized as major determinants of chain folding, stability, internal dynamics, and binding specificity [116]. Historically, it was thought that water molecules form multiple layers on macromolecule surfaces and are excluded from the interior, mostly nonpolar parts of, for instance, proteins or membranes. New 2D and 3D NMR measurements along with oxygen-17 and deuteron spin relaxation rate measurements, as a function of frequency, indicated that most of the immobilized water signal actually comes from the protein interior. Such water can be deep inside, the so-called buried water, located in the clefts, or be present as the hydration water of metals which are part of the protein structure. Surprisingly, surface water was fairly free with short residence times on the local surface groups, though some slowing in comparison to the bulk water was definitely observed. Neutron scattering [117] and computer molecular dynamics studies [118] complemented and confirmed the results of the NMR measurements. Dielectric spectroscopy measurements gave further support to the above-mentioned techniques [119].

Theoretical and experimental characterization of biomolecular hydration and hydration dynamics has proven to be a difficult problem for many decades. This field was plagued with interpretation controversies and the lack of adequate techniques to perform high-resolution studies [116]. Modern NMR techniques such as multidimensional nuclear Overhauser enhancement spectroscopy (NOESY) have helped observe individual water molecules indirectly via their dipole couplings with protein protons [120]. These techniques demonstrated that most water molecules at the surface have a mildly reduced residence time that is still in the picosecond range. However, buried water molecules with very long residence times ranging into milliseconds or longer were identified [116].

The dynamics of protein hydration have been traditionally studied with the nuclear magnetic resonance relaxation dispersion technique (NMRD) [116]. This technique lacks the special resolution of NOESY and was also plagued with hydrogen atom exchange and molecular interpretations. Recent studies in which oxygen-17 and deuteron nuclei were used to study protein hydration dynamics helped resolve some of these ambiguities. As mentioned above, the results of such studies clearly indicated that NMRD is not due to the relaxation of the long-lived surface water layer but rather to the small number of well-defined molecules trapped inside cavities or deep clefts or coordinated to the protein-bound metal ions [116]. Such studies, combined with protein engineering to remove some of the buried water molecules, showed that some of the slowest water molecules had residence times in milliseconds and most others somewhere between 10 picoseconds and 10 microseconds at room temperature. The proteins investigated ranged from the small bovine pancreatic trypsin inhibitor to the mid-sized lysozyme or myoglobin [116]. This study also confirmed that surface water molecules have relatively short residence times (10–100 ps) and are not significantly slowed even with the presence of charged carboxylate groups. This indicates that the simple analysis of hydrogen bond strength and the counting of hydrogen bonds cannot explain the observed behavior. To understand the dynamics of water molecules bound to proteins, it is clearly necessary to understand the cooperativity of the hydrogen bonding, which is involved in the fast redistribution among local molecules as local defects in the hydrogen bonding propagate through space [116]. Long-lived molecules that we classify as immobilized water, structured water, etc., are actually shielded from the local solvent and buried inside cavities or clefts. This precludes the cooperative exchange mechanism. It seems that in addition to the strong hydrogen bonding of water molecules to the protein, their isolation from the rest of the solvent is also needed to make them long-lived. This means that even a water molecule buried inside a hydrophobic cavity can have a very long exchange time, if it is isolated from the outside solvent.

Neutron scattering techniques have also been used to complement NMR measurements in understanding the hydration dynamics of proteins and other biomolecules. As in the case of Vycor-silica studies, the obtained results suggested only small but statistically significant changes in the residence time of water molecules as compared to the bulk [121]. The effects of temperature and the extent of protein hydration were followed. As expected, water lateral diffusion was slower at lower temperature and extent of hydration. The dynamics of hydration were described in terms of the relaxation model used to describe kinetic glass transitions in dense supercooled water. The authors proposed that the somewhat slower dynamics of hydration water inside and at protein surfaces might be the reason for the slower dynamics of protein conformation rearrangement itself. They also proposed that protein hydration dynamics might be the source of the anharmonicity, which is required in order for proteins to function.

Dielectric spectroscopy has been successfully used to study the interactions of macromolecules with water and hydrated counterions [122]. Results obtained at partially hydrated DNA samples were of particularly good resolution. DNA molecules carry negatively charged phosphate groups on the outside and behave similarly to polyelectrolytes. Positively charged counterions are strongly attracted to such charged surfaces. Both ions and phosphate groups are hydrated. The total interaction of ions with hydrated surfaces will therefore display the complex interplay between long-range electrostatic forces and short-range ion–water, surface–water, and water–water interactions [122]. In the absence of water, the ions were localized to charged surface groups and restricted to the immediate vicinity of the phosphate groups. Hydration above the critical limit produced a population of delocalized ions, which is consistent with the polyelectrolyte behavior and the existence of a condensed counterion phase. While the addition of water to sodium–DNA caused more or less all counterions to become delocalized, a different behavior was observed with magnesium. Magnesium is divalent and can bind to two surface groups. It also has a very strong affinity for water. In the magnesium solutions, some water has always been more strongly bound between ions and the surface, and ions were also more localized. The authors proposed that the formation of multiple hydrogen bonds to other functional groups from DNA, such as bases or sugars, can be a reason for such behavior.

The solvation dynamics inside small biological molecules such as gamma-cyclodextrin have also been studied. In one particular study, femtosecond to nanosecond solvation dynamics in pure water and inside the cyclodextrin cavity were compared. Time-dependent Stokes shifts of Coumarin probes were followed using femtosecond fluorescence upconversion and time-correlated single-photon counting techniques [123]. The observed relaxation times ranged from less than 100 femtoseconds for the relaxation processes in pure water to nanoseconds inside the cyclodextrin cavity. Once again, water molecules inside cavities where they were separated from the bulk showed a significantly slower relaxation behavior.

Protein engineering was also utilized to probe the role of hydration in, for instance, DNA-homeodomain protein binding studies. Such proteins are very important transcription factors involved in the control of the development of many organisms, including humans. Mistakes in DNA binding of homeodomain proteins often result in developmental errors, death, or serious diseases such as cancer [124,125]. Too much or too little binding, or binding to a wrong site, are common observed errors. In one particular study, the fact that water is more organized near strongly charged groups was utilized. NMR studies showed multiple water layers between DNA and bound homeodomain protein [124]. The authors speculated that if a phosphate group in DNA is modified to nonpolar methylphosphonate, water will disappear from the binding site, and a significant change in protein binding will occur. Such behavior was indeed observed. NMR studies along with molecular dynamics simulation indicated the very complex role that the hydration water of an interfacial intermolecular cavity plays in DNA protein binding. These results support the hypothesis that the hydration of an interfacial cavity functions as a noncovalent extension of the DNA interface [124].

To conclude, it seems that the hydration of charged and polar sites plays an important role in the behavior of biological molecules. While other contributions such as electrostatic, steric, and van der Waals forces may have a dominant role in the stability control of biomolecules, the role of water cannot be neglected. Organization of water molecules around nonpolar molecules, the so-called hydrophobic hydration, might play an even more important, if not dominant, role in the stability of proteins and other biological macromolecules. Hydrophobic hydration and hydrophobic forces will be discussed in the next section of this chapter.

V. HYDROPHOBIC HYDRATION AND THE ROLE OF DISSOLVED GASES IN COLLOIDAL PHENOMENA

A. Introduction

Hydrophobic interactions are still not very well defined or understood phenomena. The name "hydrophobic interactions" was given to the entropic driving force that tends to bring together two nonpolar interfaces or groups in aqueous solutions. These interactions are significantly temperature sensitive [126]. The traditional definition of this driving force is that it originates from some kind of modified 'ordering' of the water molecules in close contact with the nonpolar surface or group [127]. According to this model, when two such interfaces/groups are brought together, part of this "ordered" hydration shell is expelled to the bulk. Subsequent changes in entropy favor such a process, in which hydrophobic entities in water tend to bind together and minimize a contact area with water.

Hydrophobic interactions are found to play a major role in many fundamental processes such as protein stability, molecular self-assembly, adhesion, wetting, film stability, dispersion/aggregation of colloidal suspensions, and particle bubble interactions/flotation [128]. Surface forces between the hydrophobic entities in water are significant and even larger than the attractive van der Waals forces. This has been confirmed with direct force measurements [129]. An early evidence that hydrophobic interactions between macroscopic bodies extended over ranges considerably larger than chemical bond distances was provided by measurements of the film thickness between a gas bubble and a methylated silica surface. Destabilization of the film occurred at separations greater than 60 nm. Since charges on hydrophilic and hydrophobic silicas used were similar, and no such effects were observed with the hydrophilic silicas, the effect was assigned to the interaction of the bubble with the hydrophobic surface [130]. Yoon and Yordan recently showed a direct correlation between the critical rupture thickness of thin films and the hydrophobicity of the surface [131].

The direct measurements of forces between macroscopically hydrophobic surfaces in aqueous solutions revealed strongly attractive forces, 10–100 times stronger than the expected van der Waals attractions [132]. The range of interactions reported varied from 10 to 300 nm. The decay length also varied from 1 to 50 nm [132–138]. The results of experiments performed on similar systems by different groups often varied. The reasons for such behavior will be discussed in this section.

The origins of these hydrophobic interactions/forces remained controversial, in spite of considerable experimental and theoretical research in the last two decades. Theoretical models proposed to explain why hydrophobic forces are fundamentally different. Proposed explanations included van der Waals attractions [139], hydrogen bonding [140], hydrodynamic correlation between fluctuating liquid–solid interfaces [141], electrostatic interactions arising from surface-induced perturbations in the fluid next to a hydrophobic surface [142], or a density fluctuation model based on the elasticity of water [143]. All proposed models have some difficulties.

One of the more acceptable hypotheses is that water film between hydrophobic surfaces may be metastable. Such a perturbation may give rise to the formation of a cavity when the surfaces come into contact, thus causing attraction [144,145]. The cavitation mechanism has been inferred both theoretically [146] and experimentally [147]. Related to that issue, it was proposed that dissolved gas might enhance long-range hydrophobic forces. And indeed, it was experimentally observed that the length and range of hydrophobic forces did depend on the amount of dissolved gas [148–150] and on factors that influence gas solubility [151]. Laser cavitation experiments directly showed that the probability of cavitation near hydrophobic surfaces was much higher in the presence of dissolved gas [152–155]. Recently, it was even directly spectroscopically observed that dissolved gas accumulated near the hydrophobic interface [156]. It was also shown that dissolved gas influenced the emulsion stability and emulsion polymerization process [157]. The effect of gas on filtration of hydrophobic particles and hydrophobic coagulation was also experimentally observed [158,159]. Spectroscopic evidence even indicated the effect of dissolved gas on the structuring of bulk water [160,161]. All these exciting phenomena will be discussed later in this section.

B. Water Structure in the Interfacial Region

Knowledge about interfacial water structure near hydrophobic surfaces appears to be crucial for an understanding of hydrophobic interactions. As mentioned earlier, a traditional way of explaining hydrophobic hydration is to assume that one hydrogen bond must be broken near the nonpolar interface. To compensate for the enthalpy of the hydrogen bond breakage, water then rearranges itself into a clatharate-like structure with some defects but an overall increase in order [162]. Frank and Evans proposed long ago that water close to hydrophobic solutes must rearrange itself to compensate for the lost hydrogen bond in such way as to make a bond with another water molecule, a bond that is even stronger than the initial one [163].

Consequently, spectroscopic evidence was needed for the formation of broken hydrogen bonds as well as the formation of more ordered hydrogen bonds in the vicinity of the nonpolar group/surface in aqueous solutions. Sum-frequency generation (SFG) [164–166], Raman [167], and FTIR [168] spectroscopies of water structure near nonpolar interfaces have been performed. SFG results clearly indicated the presence of a broken hydrogen bond near a hydrophobic interface. Air/water [164], hexane/water [165], and octadecyltrichlorosilane OTS/water [166] interfaces were studied. SFG experiments at the water/CCl_4 interface also clearly indicated the presence of more ordered icelike water [165]. SFG studies of the air/water interface at different temperatures indicated a very small temperature dependence. Also, the addition of an alcohol monolayer at the air/water interface promoted icelike ordering of interfacial water. Neutron scattering experiments indicated the presence of broken hydrogen bonds near a hydrophobic interface but did not show any increased ordering of nearby water. This discrepancy is still unresolved.

An alternative mechanism of hydrophobic hydration and hydrophobic interactions is proposed by Wiggins [169]. This author proposed a particular rearrangement of water near the nonpolar interface that might have the effect of decreasing both entropy and enthalpy. Such a process relies on the cooperativity of water–water hydrogen bonding. It is proposed that the loss of a whole hydrogen bond by a water molecule near the hydrophobic surface is propagated as weak bonding through several molecular layers that, relative to normal water, have high enthalpy and somewhat increased entropy. The author claims that the experiments with cellulose acetate films and polyamide gels showed that appreciable volumes of water at hydrophobic surfaces have decreased density, changed solvent properties, and increased viscosity. It was suggested that this resulted from the collective response of hydration waters to a state of increased enthalpy, an expansion to decrease the local chemical potential. Such a change in the local strength of hydrogen bonding should also change other fundamental properties of water near hydrophobic interfaces. The author claims that this results in the existence of low-density water and normal-density water. It was also proposed as an explanation of the lyotropic series of ion adsorption and exclusion near different interfaces.

C. Role of Dissolved Gas in Colloidal Phenomena

It is clear from numerous studies that trace amounts of impurities influence the properties of solid-state systems such as semiconductors. The amount of dissolved gas in water is of the same order of magnitude. The concentration of air in air-saturated water is ca. 0.005 M. Yet until recently, few studies of the role of dissolved gas on the behavior of liquids and suspended colloids were attempted. Lord Rayleigh [170a] demonstrated long ago that cavitation of liquid mercury between glass plates occurred at distances of microns. As mentioned earlier, Kitchener and coworkers [170b] showed that a film of water between a bubble and hydrophobized silica ruptures at distances up to 60 nm. If water had the tensile strength predicted by theory, such ruptures would occur at distances of a few nanometers. It has actually long been known that the tensile strength of water and other liquids is orders of magnitude lower than predicted by any accepted theory. For solids, such disagreements between theory and experimental results were often attributed to impurities present in the material. For liquids, such impurities might originate from dissolved gases.

Recently, Bunkin and Bunkin [170c] proposed a theory in which the existence of stable or metastable submicron bubbles in liquids is responsible for lower tensile strength. The authors also proposed that such bubbles are stabilized due to adsorbed ions. Such bubbles form fractal aggregates and can coagulate and yield macroscopic clusters. Another group also proposed the existence of submicron entities in water and their role in gas nucleation. Bowers et al. studied gas evolution reactions in water and concluded that no current theory could explain the observed kinetics [171]. The observed kinetics could be explained when it was considered that gas molecules that preexisted in water were aggregated in small clusters. The authors proposed that such clusters are stable because they do not have a well-defined interface. Consequently, an extremely high surface tension, which would cause the collapse of such clusters, would not exist in such system. Bowers et al. called such clusters "blobs."

Some preliminary experimental results seem to confirm the above-mentioned theories. Bunkin and Lobeyev used small-angle laser scattering in doubly distilled water and identified the presence of some scattering objects [172]. Similar results were obtained with neutron scattering [173]. More recently, small-angle light scattering and optical breakdown experiments were performed with doubly distilled and well outgassed doubly distilled water. The results clearly showed the influence of dissolved gas on the small-angle light scattering. Moreover, an optical breakdown could not be induced after degassing [174]. The interior structure of degassed water was studied by Rayleigh using four-photon polarization spectroscopy and compared to water that was air saturated [175]. Clear differences in the obtained spectra were observed. Moreover, the spectra of outgassed water were closer to that of ice than the spectra of air-saturated water. It seems that outgassing increased the ordering of liquid water. The authors also proposed that the destruction of quasi-rotational lines in the spectra of air-saturated water might show that submicron bubbles are actually charged, the local electrostatic field being responsible for the destruction of these spectral lines. Kikuchi and coworkers studied water collected from the cathode area after electrolysis of 100 ppm of NaCl [176]. Such water was supersaturated with molecular hydrogen. The authors used photon correlation spectroscopy to follow the size of hydrogen bubbles. Surprisingly, they were able to identify submicron bubbles as small as 5 nm. Submicron bubbles were stable for at least 48 h. They were coagulating and redispersing so that the average bubble size varied between 5 and 400 nm. Therefore it would be better to describe such submicron bubbles as metastable.

The role of dissolved gas in the behavior of hydrophobic systems is of particular interest. The long-range nature of hydrophobic forces is one phenomenon that might involve dissolved gas. As mentioned earlier, one theory of hydrophobic interactions of surfaces immersed in water involves the metastability of the water film due its confinement between incompatible hydrophobic walls [177]. Cavitation was experimentally observed when two hydrophobic surfaces were separated from contact in water [178]. Several research groups also showed that dissolved gas was involved in long-range hydrophobic forces.

Optical laser pulse cavitation experiments in thin films bounded by hydrophilic and hydrophobic surfaces have been performed [179,180]. Thin films of water were placed between silica plates and a laser beam of low energy was used to produce optical cavitation at energies that were not high enough to produce the plasma breakdown of water. The probability of optical cavitation was significantly enhanced at hydrophobized surfaces. Outgassing significantly decreased the probability of cavitation near hydrophobic surfaces but had only a mild influence on cavitation near hydrophilic surfaces.

Recently, Miller and coworkers [181] used ATR/FTIR measurements to show the role of dissolved gas in the structuring of interfacial water. The authors studied the adsorption of hydrophobic gas butane near hydrophobic silicon and hydrophilic silica surfaces. As in the case of cavitation experiments, no accumulation of gas near hydrophilic surfaces was observed. On the other hand, a significant amount of butane was adsorbed at the hydrophobic silicon interface. Moreover, analysis of their spectra indicated that butane that was accumulated at the interface was in the form of aggregates rather than single molecules.

As already mentioned in the introduction to this section, the role of dissolved gas in emulsion stability, emulsion polymerization, filtration of hydrophobic solids, and coagulation of hydrophobic colloids was also observed.

The relationship between dissolved electrolytes, bubble coalescence, surface tension, gas solubility, and electrolyte ion hydration was also extensively studied [182–185]. Ninham and coworkers observed a peculiar phenomenon that while some electrolytes stabilized macroscopic bubbles, other did not influence bubble coalescence at all. Moreover, a combination of some cations and anions would always produce stabilization, while a combination of others would never stabilize or destabilize bubbles. After careful analysis of the problem, Weissenborn and Pugh [185] offered the explanation that gradients in surface tension vs. electrolyte concentrations correlate with gas solubility and bubble coalescence. The authors proposed that macroscopic bubbles are surrounded with the gradients of submicron bubbles and that such gradients play an important role in the coalescence of macroscopic bubbles. Electrolytes, which are successful in decreasing the solubility of gas, decrease the concentration and gradient of submicron bubbles and therefore stabilize macroscopic bubbles. On the other hand, the electrolytes that did not influence solubility of a particular gas did not influence the stability of macroscopic bubbles to an appreciable extent.

D. Short- and Long-Range Hydrophobic Forces

Wetting behavior of aqueous solutions of hydrophobic solids gave a first indication of the existence of an additional non-DLVO attractive force with relatively long range when compared to that of DLVO forces [186]. While both electrostatic double layer and van der Waals forces theories and their combination, such as DLVO, predict that a thin film of water on a hydrophobic surface should be stable, this is obviously not what is experimentally observed. An additional attractive force that can disrupt the film and form droplets had to be invoked. As mentioned earlier, studies of the disruption of the water films between bubble and hydrophobized silica indicated that the destabilization of the film occurred at separations greater than 60 nanometers [187].

The improvements in the surface force apparatus technique (SFA) and recent use of an atomic force microscope (AFM) for the direct force measurements enabled the application of direct force measurements between macroscopic hydrophobic bodies in aqueous solutions [188–192]. These direct measurements strongly indicated attractive forces 10–100 times stronger than the expected van der Waals attractions. Attractive forces extending between 10 nm and 300 nm were later measured by numerous researchers. How could the range of the observed forces vary so much, when studied by different groups? It seems that a large variety of techniques were used to hydrophobize surfaces. Different parameters such as the nature of the surface, ionic strength, and temperature also varied.

Craig and coworkers [193] recently classified different methods of surface hydrophobization as follows: (1) Phys. or chem. adsorption of (usually) single chain surfactants from solutions to produce a hydrophobic monolayer [188,189] that is often incomplete [194,195]. Cationic surfactants and negatively charged surfaces at concentrations below critical micelle concentration were used. (2) Langmuir–Blodgett monolayers of insoluble double-chain surfactants adsorbed to mica surfaces [196,197] or single-chain alkanethiols on gold and silica surfaces were also successfully used in direct force measurements [198]. Monolayers of double-chain surfactants have also been deposited from organic solvents [199]. Polymerized surfactant monolayers have been deposited from aqueous solutions [200,201]. (3) Chemical modification of surfaces by reaction with silane coupling agents in organic solvents [202,203]. Such modifications produce very robust hydrophobic layers. Unfortunately, unless great care is taken, multiple layers form, particularly when surfaces with high hydrophobicity are desired [202]. (4) Solid polymer surfaces such as polypropylene or thin polymer films have also successfully been used in direct force measurements. [204,205].

The range of hydrophobic attractions measured does appear to be dependent on the method used to prepare the surface, with a few recent exceptions [206,207]. Early experiments performed with adsorption of cationic surfactants on mica and subsequent SFA measurements [188,189] gave rise to forces with ranges of less than 30 nm. Forces extending up to 100 nm were observed with the Langmuir–Blodgett deposited films. Silanated surfaces gave rise to forces with ranges between 50 and 100 nm. Clean polymeric surfaces exhibit low to moderate ranged forces (12–40 nm). Finally, the longest ranged attractive hydrophobic forces were measured with Langmuir–Blodgett deposited polymerized surfactants (up to 300 nm).

The early attempts to explain such long ranges of hydrophobic forces relied on interfacial water structure, hydrogen bonding, or even van der Waals forces. However, it was very clear that none of these effects have a range, that is even close to that needed to explain experimental observations. Electrostatic interactions are capable of giving rise to forces of the range observed in direct force measurements. The following electrostatic interactions have been proposed to be involved in the rise of hydrophobic forces: the interactions caused by the correlated fluctuations of polarization due to the state of the water molecules adjacent to the hydrophobic surface [208], the interactions caused by the fluctuating electric field induced by the lateral mobility of adsorbed ions [209], and the electrostatic interactions caused by the electric field generated by the ordered structures of the adsorbed surfactant films [210]. As in the case of electrical double layer theories, the dependence of the observed forces on the addition of electrolytes is a good test of the theory. The change in force decay length with increase in ionic strength is also an important test of any theory utilizing an electrostatic mechanism. Most measurements performed to test this theory indicated only a very weak dependence of the hydrophobic forces and their decay length on ionic strength up to 0.01 M of electrolyte. Reduced force, but unchanged decay length, was found in some experiments with Langmuir–Blodgett deposited films. The experiments with inert fluorinated or polymeric surfaces did not show any dependence of hydrophobic forces on ionic strength whatsoever. Some authors found that if the attraction is divided into short-range and long-range components (>25 nm), only the long range is dependent on ionic strength [211,212]. Others found that a small increase in adhesion energy follows the increase in ionic strength [213,214]. Kokkoli and Zukoski recently proposed that this is probably due to a change of the chemical potential of water [214]. In any case, the electrostatic theory of hydrophobic forces does not fit the observed experimental data.

It is apparent that the phenomena occurring at the water/hydrophobic surface are responsible for the observed long-range attractions. Yet how it is possible to explain the long-ranged nature of the observed phenomena? Ruckenstein and Churaev [215] suggested that due to the incompatibility between water and the hydrophobic surface, the planar interface becomes unstable to thermal and mechanical perturbations, and fluctuating gaps are generated between the two. The fact that experimental observations on the flow of water through hydrophobic capillaries cannot be explained by theory gives indirect support to such a proposal. The flow rate of water through a capillary is greater than expected and can be increased by deareating solutions [216]. This suggests the presence of some kind of gap between water and the hydrophobic interface and the increase in size of such a gap upon removal of air/gas. According to Ruckenstein and Churaev [215], the fluctuations of the water interface generated by the instability of the hydrophobic–hydrophilic interface propagate via a pressure field from one interface to another. A long-range force is then generated due to hydrodynamically correlated fluctuations of the two water interfaces. Bridging cavities between surfaces can then form if the fluctuations are large enough. Yaminski and Ninham proposed a similar model [216]. A modified version of Ruckenstein's model was recently used to explain the unexpected significant dependence of the hydrophobic force on temperature [217,218]. The obtained equations predicted the temperature dependence of the hydrophobic force due to the temperature dependence of viscosity. This observation brings indirect support to the proposal that the hydrophobic forces might have a hydrodynamic origin. The author concludes with the statement that short- and long-range hydrophobic forces may indeed have different origins, namely hydrodynamic fluctuations and submicron bubble bridging.

This brings us back to the role of dissolved gas in the behavior of colloidal systems. Rabinovich and Yoon [219] reported in 1994 that the outgassing of a solution significantly reduced the extent and decay length of hydrophobic forces between hydrophobized silica. Similar observations have been made since then by several other researchers [220–222]. Ninham and colleagues also observed that outgassing produced more stable emulsions and modified emulsion polymerization processes [223]. Zhou and coworkers showed that dissolved gas influenced the filtration rate of hydrophobic solids and the coagulation of hydrophobic coal [224,225]. It seems that at least long-range hydrophobic forces are indeed influenced by the presence of dissolved gas. In the bridging bubbles model, this is explained with the bridging of nanosized bubbles from the approaching surfaces. Bridges can be formed far away from the surface and produce attractions of very long range, depending on the size of the submicron bubbles. Such a model is schematically shown in Fig. 6.

In the real world, attraction often happens between dissimilar surfaces. It was only recently that Yoon and coworkers performed a detailed study of the direct forces between a glass sphere and a silica plate hydrophobized to a different extent [226]. It was found that the power law force constant for asymmetric interactions is close to the geometric means of those for symmetric interactions. Thus hydrophobic force constants can be combined in the same manner as the Hamaker constants. A plot of the power law force constant versus water contact angles indicated a direct relation between hydrophobic force and the contact angle. This was the first clear experimental correlation between hydrophobic effects and contact angles of the macroscopic surfaces. It now appears that hydrophobic surfaces with larger contact angles result in hydrophobic forces of longer extent. This might explain a lack of long-range hydrophobic forces in experiments involving surfaces with low contact angles. Earlier, Rabinovich and coworkers showed that surfaces with higher contact angles contain hydrocarbon chains that are more ordered and eventually end with a methyl rather than a methylene group [227]. The role of the end-group of the hydrocarbon chain in the mechanism of hydrophobic effect remains to be clarified in the future.

The mechanism or mechanisms of hydrophobic interactions are still not fully understood. Progress is expected to occur in the near future. Collaboration between experts in different fields and simultaneous applications of numerous different techniques to study a well-defined model system will be needed. The future of this exciting area of scientific research will, most certainly, be full of surprises.

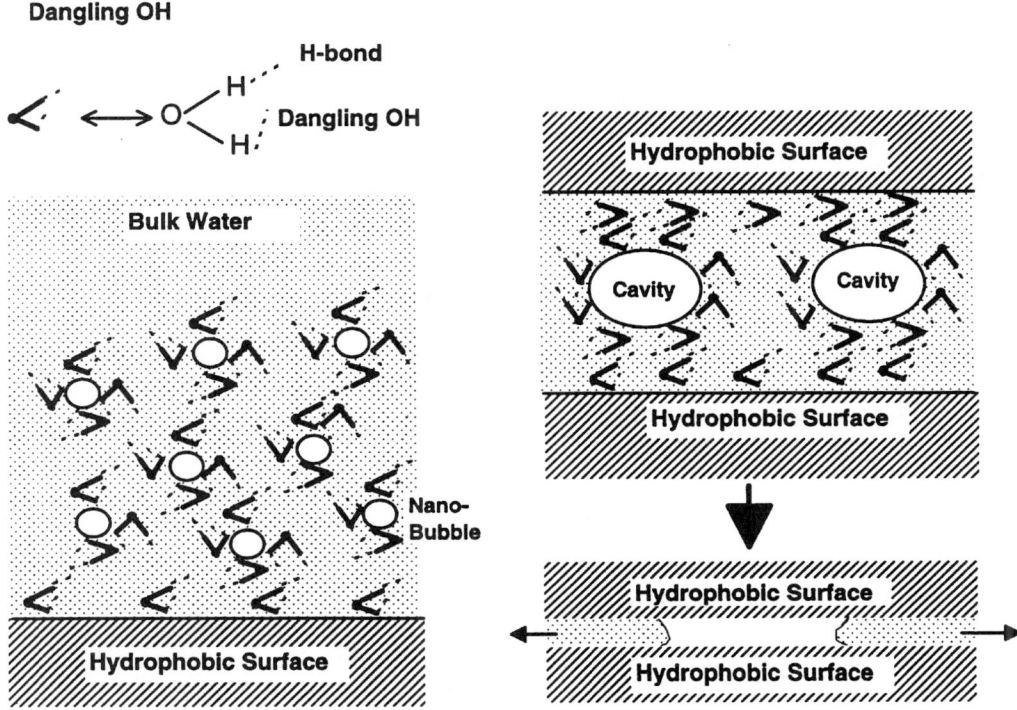

FIG. 6 *Left*: Nanobubbles (ca. 10 nm) are formed near a hydrophobic surface by dissolved gases, and a liquid/gas interface is established in the interfacial region. Water molecules near a hydrophobic surface (bubble and/or solid surface) form a H-bonding network in which each water molecule is engaged in three H-bonds with one dangling OH bond. *Right*: The slippage of water from nanobubble surface due to H-bonding interaction from the dangling OH bond with adjoining water results in coalescence and the formation of cavities. Water molecules still show the H-bonding network with one dangling OH near the cavity surface. When two hydrophobic surfaces approach each other the net attractive H-bonding interaction from the dangling OH bond gives rise to the displacement of water molecules and the coalescence of cavities—the long-range hydrophobic force. (From Ref. 156.)

VI. INFLUENCE OF ELECTRIC, MAGNETIC, ELECTROMAGNETIC, AND SONIC FIELDS ON INTERFACIAL WATER

A. Introduction

The effects of different fields on water solutions and suspensions have been noticed early in this century. It was claimed that, for instance, magnetic treatment of water helps increase the growth rate of irrigated plants, reduce the deposition rate of scale, or modify the hydration rate of cement [228]. Later, similar effects were described for water treatment with electromagnetic radiation in the radio-frequency, microwave, millimeter wave and far-infrared regions of the electromagnetic spectrum [229]. In the recent years some of these claims were confirmed by serious scientific studies [230]. The observed phenomenon was appropriately called "the magnetic memory of water." The observed effects on water solutions and suspensions persist hours after water treatment. Most relaxation phenomena in water occur on the time scale of milliseconds to seconds. The question to be answered is, How then can water treatment effects last for hours? We are only beginning to answer this question, and the proposed models are still speculative.

Another old phenomenon that received much attention in recent years is so-called sonoluminescence. When water is bubbled with air or some other gases or, alternatively, is in the presence of a single levitating bubble, application of resonant ultrasound in the kilohertz range produces visible light. This is at least eleven orders of magnitude amplification [231]. Moreover, while the ultrasound used has rise time in microseconds, the observed light pulses occur in the picosecond time scale [232]. Numerous models are proposed to explain the observed phenomena but none can explain all observed experimental results [233]. The observed light pulses are accompanied by the splitting of water into hydroxyl radicals and atomic hydrogen. Hydroxyl radicals then recombine to yield hydrogen peroxide, which can be identified after the sonochemical treatment. Recent studies of the electromagnetic and sonic water treatment will be described in this section. The proposed models will then be presented.

B. Experimental Studies on the Effects of Fields on Interfacial Water

The effects of electric, magnetic, electromagnetic, and sonic fields on aqueous solutions and suspensions [234–237], colloidal particles [238–240], biological systems [241–244], and industrial processes [245–247] have been studied in various fields of science and engineering. The effects observed with magnetic colloids and solutions are understandable with the current knowledge of colloidal forces and magnetic and/or other fields [248–251]. However, the effects of various fields on nonmagnetic materials could not be explained with the current state-of-the-art theories. As mentioned earlier, the long-term effects that persist after field treatment of aqueous solutions and suspensions present a real challenge to water researchers. The existence of the so-called magnetic memory of water was a rather anecdotal phenomenon until recently. Industrial and biotechnology applications of such treated water were documented. Water can be treated with electric, magnetic, electromagnetic, or sonic fields and used minutes or hours later for the desired applications. A reduced rate of scale deposition and an increased rate of cement hydration or plant or animal growth are only some of the described uses of such technologies.

Recent contributions from several laboratories assisted in quantifying the magnetic memory of water as a real and fairly reproducible phenomenon. Chibowski and coworkers [238–240] studied the effects of radio-frequency (rf) electromagnetic fields (44 MHz) on pH, conductivity, and zeta potential of colloidal suspensions. Oscillations in zeta potential hours after the rf treatment were observed for diluted suspensions of numerous oxides such as rutile, zinc oxide, hematite, and silica. Higashitani and coworkers [241] also measured the effects of magnetic water treatment on the precipitation of calcium carbonate from calcium chloride and sodium carbonate solutions. For minutes or hours after treatment, the solutions were left separated and then mixed to precipitate calcium carbonate. Nontreated solutions yielded mostly calcite; treated solutions produced a mixture of calcite and aragonite. Particles precipitated from the treated solutions were also larger. Higashitani and colleagues [242,243] also observed the magnetic memory effect with measurements of colloid stability, electrophoretic mobility, and interparticle forces of the treated solutions. The same laboratory also observed the magnetic memory of water effect at the molecular level. The magnetic treatment of water solutions of hydrophobic fluorescence probes resulted in changes in the measured fluorescence, while no such change was observed with the hydrophilic fluorescence probes [244]. The observed changes lasted for at least 8 h after the treatment.

Fesenko and coworkers [252–254] realized that when water is treated with microwave electromagnetic fields and later added to cells, the same effects on the behavior of membrane proteins was observed as when the cells were indirectly irradiated with the same signal. The microwave treated water was analyzed with capacitive discharge measurements [254], and it was concluded that some kind of unspecified changes in water structure occurred and lasted for hours. Fesenko pointed out that water was always in contact with air and that long-term effects on water solutions probably resulted from the modification of the gas/water system.

Motivated by the above-described results, Čolić and Morse [255–259] recently studied a multitude of phenomena involving the effects of electromagnetic and other fields on aqueous solutions and suspensions with the aim of understanding the mechanism of the magnetic memory of water. Colloid chemical and bulk physicochemical techniques were used along with molecular techniques to quantify the water treatment effects. It was also concluded that water structure was somehow changed and that this change lasted for hours. Following Fesenko's suggestion that electromagnetic water treatment perturbs dissolved gas in water, Čolić and Morse realized that outgassing of water prevented the effects of electromagnetic fields. Detailed study of the behavior of the gas/liquid interface revealed that it is the primary target of the electromagnetic field action and the memory effect. Gas/liquid interfaces seem to relax much more slowly than pure water when perturbed.

Other authors performed spectroscopic studies on the effects of electromagnetic fields (EMF) on water structure. Raj and coworkers [260] observed the change in UV and the fluorescence spectra of water (in other words, the appearance of some bands) after magnetic water treatment. They attributed it to the enormous changes to the bond angle of water. However, after careful examination of their spectra, one cannot avoid noticing similarities with the spectra of ozone and hydrogen peroxide. It is quite tempting to identify spectra of water treated with one magnetic pole as that of oxidizing species (ozone) while that of water treated with the opposite pole as the spectra of hydrogen peroxide (reduction product). After all, Lorentz forces can produce small currents inside solutions, producing the same effect as electrolysis of water. Perhaps atomic hydrogen is also present as a product of magnetic water treatment, and not only of the electric field treatment.

Ozeki and coworkers [261] studied the effects of the magnetic field on the adsorption/desorption of water vapor on hydrophilic and hydrophobic materials. Electron spin resonance (ESR) spectroscopy measurements identified the presence of some long-lived paramagnetic species at the silica/water interface. SQUID (superconductivity quantum interference device) measurements of the magnetic susceptibility corroborated the presence of long-lived paramagnetic species at the silica/water interface. The possibility exists that they have identified encaged atomic hydrogen. Other groups have shown [262,263] that hydrogen encaged inside silicates is extremely stable and long-lived. Ozeki and coworkers also observed that water adsorption at the hydrophobic surfaces was more strongly affected by the magnetic field treatment.

Hayashi [264] measured the size of the water clusters with ^{17}O NMR before and after water electrolysis. This author analyzed line width and noted that the average cluster size in the bulk water collected from the cathode area decreased from 10–13 to 5–6 water molecules per cluster. The author claims that water with smaller cluster size is more reactive.

Two questions still had to be answered. (a) How do the changes in the water/gas interface and the properties of water influence the observed changes with bulk colloidal systems? (b) Why are changes in the dissolved gas concentration or state influencing the properties of water? As mentioned earlier, numerous authors described changes in shape, morphology, and phase of the precipitated calcium carbonate when solutions of calcium chloride and sodium carbonate were pretreated with magnetic or electromagnetic fields. The effects of electromagnetic fields on the gas/water interface can influence calcium carbonate precipitation in two different ways.

First, since concentration and diffusivity of carbon dioxide are changed after the treatment, precipitation of calcium carbonate can be influenced through the formation of hydrogencarbonate. Calcium hydrogencarbonate is much more soluble than calcium carbonate. Changes in the amount of dissolved carbon dioxide also modify the pH and conductivity of solutions. Carbon dioxide reacts with water yielding carbonic acid. Dissociation of carbonic acid decreases the pH of water. Removal of the dissolved carbon dioxide results in the increase in pH of the solution. Changes in pH are probably responsible for the change in zeta potential and colloid stability after rf or magnetic water treatment. Indeed, it was found by Čolić and Morse that outgassing of water solutions before electromagnetic water treatment resulted in the absence of any effects of the field on water solutions or suspensions [255].

Second, calcium carbonate precipitation, zeta potential, and colloid stability of suspensions might also be influenced by the water structure. If water hydrates ions to a higher extent, it is then more difficult to dehydrate the precipitation nuclei. Consequently, larger nuclei that form result also in larger precipitated particles with different morphology. If ions are hydrated to a lesser extent, dehydration occurs more easily, and smaller nuclei start to form smaller precipitate particles. Both phenomena were experimentally observed [256]. Increased thickness of the hydration layer around colloidal particles would also result in the higher colloid stability and modified zeta potential due to a change in the location of the shear plane.

Another puzzling issue was that electromagnetic fields influenced solubility and diffusivity of dissolved gases. If gases were in the form of neutral single atoms or molecules, no apparent reason exists why such systems would be modified by oscillatory electromagnetic fields. This question might have been answered with the work on gas nanobubbles in water that was described in the previous section of this chapter. Suppose that the nanobubbles model of Bunkin [265,266] and Ninham and coworkers is correct. According to this model, dissolved gases aggregate into small nanobubbles that are charged and form fractal aggregates. Such bubbles are like impurities or defects in water, and their removal or aggregation might result in improving the structuring of water. On the other hand, dissolution or formation of gas-in-water solutions would increase the concentration of defects in water. Gases that promote the formation of clathrate-like structures in water, such as noble gases, hydrogen and carbon dioxide, would particularly influence the structuring of water. Such gases would then be influenced to a higher extent than others such as oxygen. This was indeed observed in the electromagnetic treatment experiments of Čolić and Morse [256]. Dissolved gases also tend to accumulate near the hydrophobic interfaces. Indeed it was observed by several authors that the behavior of hydrophobic systems is more affected by electromagnetic and magnetic fields than that of hydrophilic systems.

Researchers who study the effects of electric, magnetic, and electromagnetic fields on water solutions and suspensions have made another interesting observation. Various fields produce free radical species and products of their recombination during water treatment. What could be the mechanism of this process? Electric fields simply electrolyze water and produce numerous reactive species such as ozone, hydrogen peroxide, and atomic hydrogen. While some such species are short lived, others like hydrogen peroxide or ozone can persist for hours or even days in the absence of decomposition catalysts. A magnetic field can produce similar species through Lorentz forces and induced electric fields. On the other hand, it is more difficult to explain how high-frequency oscillatory fields such as radio-frequency fields can produce the splitting of water and the production of free radicals. This is where some parallels between oscillatory electromagnetic fields and resonant sonic fields start. Both phenomena require a gas/liquid interface and produce some free radicals, with the resulting hydrogen peroxide as the long-term product. In sonoluminescence, a collapsing pulsating bubble results in very high temperatures, and the resulting splitting of water might be a high-energy process. On the other hand, rf water treatment is a low-energy nonthermal process. Eberlein recently proposed a quantum vacuum radiation model that predicts that an oscillating low-energy electromagnetic field could produce energy that in turn would split water [267]. The origin of such a phenomenon is in the dynamic Casimir effect [268]. In this model, the two interfaces with different polarizabilities (water and bubbles) are treated as a possible two-photon state source during excitation by electromagnetic fields. The virtual two-photon state becomes real when the dielectric moves due to an electromagnetic field. An oscillating electromagnetic field can influence the bubble/water interface simply by producing coherent oscillations of hydrogen-bonded dielectric–water molecules.

Researchers involved in multiple bubble sonoluminescence studies also realized that the nature of the dissolved gas influences the observed emission spectra and the amount of free radicals and hydrogen peroxide produced [269]. While sonication of water saturated with helium yielded barely measurable emissions and trace amounts of hydrogen peroxide, water saturated with neon, argon, and xenon, respectively, produced sequentially higher emissions and amounts of hydrogen peroxide. Moreover, sonoluminescence was observed only after water was saturated with the respective gas. In single-bubble sonoluminescence experiments, it was also observed that water saturated with oxygen or nitrogen did not yield appreciable emissions. The addition of even trace amounts of argon significantly increased the observed emissions. Noble gases do promote clathrate-like structuring of water to a higher extent than oxygen [270].

It is important to note that our understanding of the role of dissolved gas in the behavior of aqueous solutions and suspensions is still at the rudimentary level. Our understanding of the mechanisms of the effects of magnetic, electromagnetic, and sonic fields on aqueous solutions and suspensions is also not satisfactory. Much more work remains to be done to understand these exciting phenomena. As we improve our understanding, current models described in this chapter will certainly be modified, and in some cases rejected.

VII. THEORETICAL STUDIES OF BULK AND INTERFACIAL WATER

This chapter reviews mostly experimental studies on the dynamics of water and the hydration of colloidal systems. Theoretical studies will be mentioned only when relevant to the interpretation of the experimental research. Ohmine and Tanaka recently reviewed theoretical studies of fluctuation, relaxation, and hydration in liquid water, along with the dynamics of hydrogen bond rearrangements [271]. As mentioned at the beginning of this chapter, different models of the structure and dynamics of liquid water and hydration were proposed. Those models succeeded to some extent in interpreting the thermodynamic properties of water but failed to improve our understanding of dynamical properties of liquid water and the hydration process.

Historically, the first molecular dynamics (MD) calculations on liquid water by Rahman and Stillinger [272] were a great step forward in computer modelling of liquid water structure and dynamics. The authors realized that some kind of cooperative molecular motions are involved in liquid water behavior. However, due to the limited capabilities of the computers used in the early 1970s, it could not be clarified what spatial, time, and energy scales cooperate in the dynamics. Stillinger and Weber [273] dealt with this problem by developing new computational methods in which the separation of the fundamental structure component from the vibrational component (thermal noise) was possible. The authors constructed a total potential energy surface and found multiple minima, which they termed "inherent structures." This approach was later used to analyze the cooperative nature of the hydrogen-bond network in water dynamics and its time evolution [274].

Molecular dynamics and related Monte Carlo simulations were later also used to describe the hydration of small solutes, macromolecules, and colloids. In principle, such calculations first describe the basic interactions between all atoms involved and then let the system evolve according to Newtonian physics. Such studies require two sets of data: intermolecular and intramolecular pair potentials for all molecules involved [water–water, solute–solute, and water–solute (or water–colloid)], and a prescribed procedure for following movements of all atoms involved through time (molecular dynamics) [275]. Such calculations produce a sequence of configurations like frames in a movie. Each atom moves through time in a series of discrete steps. During each step, forces exerted by other atoms force a particular atom to accelerate, which in turn changes its velocity. If forces are constant during that time step, Newton's laws apply and we can calculate the new velocity. This updated velocity is then used to calculate the new position of the atom. For strongly interacting systems such as water molecules in a hydrogen bond network, the times steps have to be very short, somewhere in the low femtosecond range. The movie produced as a result of such calculations is very explicit, showing all rotations, vibrations, and movements of water molecules such as lateral shifts. Current computers do not allow such simulations for more than a few nanoseconds. In short, the results of such computations showed that water at the surface of biological or colloidal systems is not substantially slowed and that most rearrangements occur in the 10–100 ps timescale. Results of such simulations with proteins [276], hydrophobic surfaces [277], and hydrophilic silica and surfaces with electrical double layers with mobile counterions [278] indeed indicated that surface water molecules did not become slowed down for more than a factor 2–5.

Recent experimental results also seem to present a serious problem to any theoretician interested in explaining hydrophobic hydration and hydrophobic forces. Indeed, as discussed in previous sections, spectroscopic and neutron scattering studies did indicate broken hydrogen bonds near the hydrophobic entities in water, but no structural enhancement was observed. As already mentioned previously, Wiggins proposed a model in which no clathrate-like structuring of water was needed to explain hydrophobic hydration, hydrophobic forces, or protein folding [279]. In this model, the true hydrophobic interactions occur due to incompatibility of low-density water and oil. Low-density water is perturbed water inside the double layer. In this model, hydrophobic forces are felt in the systems where amphipathic molecules or surfaces are present in water. Hydrophilic groups promote low-density water formation (particularly when ionized), and hydrophobic entities try to avoid such more ordered water. Consequently, hydrophobic entities sequester out of contact with low-density water. In this model there is no need for any strong structuring of water. Instead, even smaller changes of the dynamics of water at charged or hydrophilic polar interfaces are felt far away due to hydrogen-bond cooperativity. On the other hand, Luzar and Chandler [280] recently proposed a model in which hydrogen bond cooperativity is completely absent. Head-Gordon [281] also presented some arguments for and against clathrate-like ordering of water around hydrophobic groups. Finally, it was recently shown that even the word hydrophobic is often misused. Hydrophobic effects exist in water–nonpolar molecule or interface systems, which are two-phase systems. A single molecule or a group is better termed nonpolar. Haymett et al. [282] actually recently showed that there is an attraction between water and nonpolar dissolved molecules. The hydrophobic force/effect has different origins. There is no repulsion between water and nonpolar groups. Overall, the molecular origins of hydrophobic effects are still unknown.

Other models of a static nature have been developed. Long ago, Kjellander and Marcelja have modelled the orientation of water at colloidal interfaces and its relation to short-range forces [283,284]. Hydration of lipid bilayers was also a popular subject of more static models [285]. Such models were a starting point for the development of more recent mathematical approximations used to describe the interactions of ions with hydrated surfaces. Early diffuse double-layer models did not take into consideration the ion size or the presence of the solvent (water). Integral equation theories based on the hypernetted-chain approximation (HNC) [286,287] can be viewed as a weak coupling treatment of surface–ion interactions. Such models were more suitable for solvent treatment as distinct molecular species than the modified Poisson–Boltzmann theory. Linearized HNC was not particularly successful. Modified versions of this model, the so-called reference hypernetted chain (RHNC) and the asymmetric hypernetted chain (AHNC) were developed to circumvent these problems [288,289]. According to RHNC, water near any polar interface is more structured than in the bulk. Smaller ions with an affinity for somewhat structured water, such as sodium, will then preferentially adsorb closer to the interface. Larger ions such as potassium will prefer to stay outside, closer to the bulk, where water is less organized. According to such a model, even indifferent electrolytes can reverse a charge sign of a colloidal particle, if it has a strong affinity for interfacial structured water. This was recently observed with strongly hydrated aluminum hydroxide samples at high concentrations of lithium [290]. The results of the rheological experiments of Čolić et al. described earlier in this chapter also seem to fit such a model [291].

Our models of the hydration and structuring of bulk and interfacial water are still primitive, and much work remains to be done to develop realistic models of the simplest systems. Better and more realistic pair potentials for interatomic interactions are needed. Faster computers would allow longer simulation time of the dynamic behavior, which is currently limited to a few nanoseconds. More experimental data with powerful time-resolved spectroscopies or other dynamic techniques are also necessary to be able to compare simulation results with experimental data. Realistic molecular models of hydrophobic hydration are particularly needed. We hope that developments in the next millenium will improve our understanding of water and of the hydration of colloidal and biological systems.

REFERENCES

1. F. Franks, ed., *Water—A Comprehensive Treatise*, Plenum Press, New York, 1970, p.1.
2. D. Eisenberg and W. Kauzmann, *The Structure and Properties of Water*, Oxford University Press, Oxford, 1969.
3. J. L. Finney. Faraday Discuss. *103*:1 (1996).
4. W. Drost-Hansen, J. Colloid Interface Sci. *58*:251 (1977).
5. J. Israelachvili and H. Wennerstrom. Nature *379*:219 (1996).
6. M. Čolić, M. Fisher, and D. Morse. Submitted to J. Colloid Interface Sci.

7. F. Franks, *The Physics and Physical Chemistry of Water*, Plenum Press, New York, 1972, p. 1.
8. J. A. Pople, Proc. R. Soc. London A *202*:323 (1950).
9. H. S. Frank and W. Wen. Discuss. Faraday Soc. *24*:133 (1957).
10. H. E. Stanley and J. Teixeira. J. Chem. Phys. *73*:3404 (1980).
11. A. Geiger and H. E. Stanley. Phys. Rev. Lett. *49*:1749 (1982).
12. R. L. Blumberg, H. E. Stanley, A. Geiger, and P. J. Mausbach. Chem. Phys. *80*:5230 (1984).
13. G. D. Arrigo. Nuovo Cimento B *61*:123 (1981).
14. O. Mishima and H. E. Stanley. Nature *392*:164 (1998).
15. J. Li and D. K. Ross. Nature *365*:327 (1993).
16. J. Peter Guthrie. Chem. Biol. *3163* (1996).
17. S. Scheiner and T. Kar. J. Am. Chem. Soc. *117*:6970 (1995).
18. P. A. Giguere. J. Raman Spectrosc. *15*:354 (1984).
19. I. Ohmine and H. Tanaka. Chem. Rev. *93*:2545 (1993).
20. P. H. Poole, F. Sciortino, U. Essman, and H. E. Stanley. Nature *360*:324 (1992).
21. H. Tanaka. Chem. Phys. Lett. *282*:133 (1998).
22. B. W. Ninham, K. Kurihara, and O. Vinogradova. Colloids Surf. *123*:7 (1997).
22a. N. F. Bunkin and V. B. Karpov. JETP Lett., *52*:669 (1990).
23. A. F. Bunkin, N. F. Bunkin, A. V. Lobeyev, and A. A. Nurmatov. Phys. Lett. A. *225*:349 (1996).
24. P. A. Giguere. J. Raman Spectrosc. *15*:354 (1984).
25. J. R. Scherer, in *Advances in Infrared and Raman Spectroscopy* (R. J. Clark and R. E. Hester, eds.), Heyden, London, 1978, p. 149.
26. J. R. Scherer, M. K. Go, and S. Kint. J. Phys. Chem. *77*:2108 (1973).
27. V. Zhelyaskov, G. Georgiev, Z. Nickolov, and N. Miteva. J. Raman Spectrosc. *21*:203 (1990).
28. V. Zhelyaskov, G. Georgiev, Z. Nickolov and N. Miteva. J. Raman Spectrosc. *20*:67 (1989).
29. V. Zhelyaskov, G. Georgiev, and Z. Nickolov. J. Raman Spectrosc. *19*:405 (1988).
30. K. Mizoguchi, Y. Hori, and Y. Tominaga. J. Chem. Phys. *97*:1961 (1992).
31. E. W. Kastner, Y. J. Chang, Y. C. Chu, and G. E. Walrafen. J. Chem. Phys. *102*:653 (1995).
32. W. Jarzeba et al. J. Chem. Phys. *92*:7039 (1988).
33. H. Graener, G. Seifert, and A. Laubereau. Phys. Rev. Lett. *66*:2092 (1991).
34. C. Ronne, C. Throne, P. O. Astrand, A. Wallquist, K. V. Mickhelson, and S. R. Keideing. J. Chem. Phys. *107*:5319 (1997).
35. A. Luzar and D. Chandler, Nature *379*:55 (1996).
36. D. J. Wales. Science *271*:925 (1996).
37. K. Liu et al. Nature *381*:501 (1996).
38. J. K. Gregory and D. C. Clary. J. Phys. Chem. *100*:18014 (1996).
39. D. J. Wales. J. Am. Chem. Soc. *115*:11180 (1993).
40. S. C. Althorpe and D. C. Clary. J. Chem. Phys. *101*:3603 (1994).
41. J. B. Paul, C. P. Collier, R. J. Saykally, J. J. Scherer, and A. O'Keefe. J. Phys. Chem. A *101*:5211 (1997).
42. J. J. Scherer, J. B. Paul, A. O'Keefe, and R. J. Saykally. Chem. Rev. *97*:25 (1997).
43. X. Yang, X. Zhang, and A. W. Castleman. Int. J. Mass Spectrom. Ion. Proc. *109*:339 (1991).
44. J. M. Delgado-Cabro Flores et al. J. Colloid Interface Sci. *189*:58 (1997).
45. H. Yotsumoto and R. H. Yoon. J. Colloid Interface Sci. *157*:434 (1993).
46. B. Derjaguin and L. Landau. Acta Physiochim. URSS *14*:633 (1941).
47. E. J. W. Verwey and J. Th. Overbeek. *Theory of Stability of Lyophobic Colloids*, Elsevier, Amsterdam, 1948.
47a. M. A. Blesa and N. Kallay. Adv. Colloid Interface Sci. *28*:111 (1988).
47b. J. N. Israelachvili, *Intermolecular and Surface Forces*, 2d ed., Academic Press, New York, 1991.
47c. N. Kallay, M. Čolić, H. M. Jang, D. W. Fuerstenau, and E. Matijević. Colloid Polym. Sci. *272*:554 (1994).

47d. M. Čolić, D. W. Fuerstenau, N. Kallay, and E. Matijević. Colloids Surf. 59:169 (1991).
47e. N. Kallay et al. Croat. Chem. Acta 63:467 (1990).
47f. O. Stern. Z. Elektrochem. 30:508 (1924).
47g. J. A. Davis, R. O. James, and J. O. Leckie. J. Colloid Interface Sci. 63:480 (1978).
48. L. Bergstrom, Surface and colloid chemistry, in *Advanced Ceramic Processing*, Surfactant Science Series, Vol. 51 (R. J. Pugh and L. Bergstrom eds.), Marcel Dekker, New York, 1994.
49. B. V. Velamakanni, J. C. Chang, F. F. Lange, and D. S. Pearson. Langmuir 6:1323 (1990).
50. R. M. Pashley. Adv. Colloid Interface Sci. 16:57 (1982).
51. W. A. Ducker, Z. Xu, J. Israelachvili, and D. R. Clarke. J. Am. Ceram. Soc. 77:437 (1994).
52. M. Čolić, G. V. Franks, M. L. Fisher, and F. F. Lange. Langmuir 13:3129 (1997).
53. G. V. Franks, M. Čolić, M. L. Fisher, and F. F. Lange. J. Colloid Interface Sci. 193:96 (1997).
54. M. Čolić and M. L. Fisher. Chem. Phys. Lett. 291:24 (1998).
55. M. Čolić, M. L. Fisher, and G. V. Franks. Langmuir 14:6107 (1998).
56. M. Čolić, M. L. Fisher, and D. W. Morse. Manuscript in preparation.
57. J. A. Yanez, T. Shikata, D. S. Pearson, and F. F. Lange. J. Am. Ceram. Soc. 79:2917 (1996).
58. J. W. Goodwin, R. W. Hughes, S. J. Partridge, and C. F. Zukoski. J. Chem. Phys. 85:559 (1986).
59. R. Leberman and A. K. Soper. Nature 378:364 (1995).
60. V. A. Parsegian. Nature 378:338 (1995).
61. H. Yotsumoto and R. H. Yoon. J. Colloid Interface Sci. 157:426 (1993).
62. H. Yotsumoto and R. H. Yoon. J. Colloid Interface Sci. 157:434 (1993).
63. Y. K. Leong, P. J. Scales, T. W. Healy, D. Boger, and R. J. Buscall. J. Chem. Soc. Faraday Trans. 89:2473 (1993).
64. M. Talukdar and K. K. Kunda. J. Phys. Chem. 96:9702 (1992).
65. L. H. Allen and E. Matijević. J. Colloid Interface Sci. 31:287 (1969).
66. J. Israelachvili and H. Wennerstrom. Nature 379:219 (1996).
67. V. V. Yaminsky, B. W. Ninham, and R. M. Pashley. Langmuir 14:265 (1998).
68. M. Čolić and D. Morse. J. Colloid Interface Sci. 200:265 (1998).
69. R. Sasamori, Y. Okaue, T. Isobe, and Y. Matsuda. Science 265:1691 (1994).
70. P. Han and D. M. Bartels. J. Phys. Chem. 95:5367 (1991).
71. J. Depasse. J. Colloid Interface Sci. 194:260 (1997).
72. B. W. Ninham and V. Yaminsky. Langmuir 13:2097 (1997).
73. R. M. Pashley. J. Colloid Interface Sci. 83:531 (1981).
74. J. Israelachvili and G. E. Adams. J. Chem. Soc. Faraday Trans. 1 74:975 (1978).
75. R. G. Horn and D. T. Smith. J. Non Cryst. Solids 120:72 (1990).
76. G. Vigil, Z. Xu, S. Steinberg, and J. Israelachvili. J. Colloid Interface Sci. 165:367 (1994).
77. W. A. Ducker, T. J. Senden, and R. M. Pashley. Nature 353:239 (1991).
78. L. Meagher. J. Colloid Interface Sci. 152:293 (1992).
79. S. Leikin, D. C. Rau, and V. A. Parsegian. PNAS 91:276 (1994).
80. C. Reid and R. P. Rand. Biphys. J. 72:1022 (1997).
81. S. Leikin, D. C. Rau, and V. A. Parsegian. Phys. Rev. A 44:5272 (1991).
82. F. Volke, S. Eisenblatter, and G. Klose. Biophys. J. 67:1882 (1994).
83. V. V. Yaminsky, B. W. Ninham, and R. M. Pashley. Langmuir 14:3223 (1998).
84. V. V. Yaminsky, B. W. Ninham, and A. M. Stewart. Langmuir 12:836 (1996).
85. V. Yaminsky, C. Jones, F. Yaminsky, and B. W. Ninham. Langmuir 12:3531 (1996).
86. J. Israelachvili. Faraday Discuss. 65:20 (1978).
87. R. M. Pashley. J. Colloid Interface Sci. 80:153 (1981).
88. R. M. Pashley. J. Colloid Interface Sci. 83:531 (1981).
89. W. Drost-Hansen, in *Biophysics of Water* (F. Franks and S. Mathias, eds.), John Wiley, Chichester, 1982, pp. 163–169.
90. R. G. Horn, D. T. Smith, and W. Haller. Chem. Phys. Lett. 162:404 (1989).
91. A. Grabbe and R. G. Horn. J. Colloid Interface Sci. 157:375 (1993).
92. A. Grabbe. Langmuir 9:797 (1993).

93. J. P. Chapel. J. Colloid Interface Sci. *162*:517 (1994).
94. J. P. Chapel. Langmuir *10*:4237 (1994).
95. G. Peschel et al. Colloid Polym Sci. *260*:444 (1982).
96. R. K. Iler, in *The Chemistry of Silica*, John Wiley, New York, 1979, p. 640.
97. J. Depasse and A. Watillon. J. Colloid Interface Sci. *33*:430 (1970).
98. G. Vigil, Z. Xu, S. Steinberg, and J. Israelachvili. J. Colloid Interface Sci. *165*:367 (1994).
99. J. Depasse. J. Colloid Interface Sci. *194*:260 (1997).
100. V. V. Yaminsky, B. W. Ninham, and R. M. Pashley. Langmuir *14*:3223 (1998).
101. J. Israelachvili and H. Wennerstrom. J. Phys. Chem. *96*:520 (1992).
102. J. Marra and J. Israelachvili. Biochemistry *24*:4608 (1985).
103. T. J. McIntosh and S. A. Simon. Biochemistry *32*:8374 (1993).
104. J. Israelachvili and H. Wennerstrom. Nature *379*:219 (1996).
105. I. M. Tejedor-Tejedor and M. Anderson. Langmuir *2*:203 (1986).
106. E. C. Yost, I. M. Tejedor-Tejedor, and M. Anderson. Environ. Sci. Technol. *24*:822 (1990).
107. M. R. Yalamanchili, A. A. Atia, and J. D. Miller. Langmuir *12*:4176 (1996).
108. S. Ong, X. Zhao, and K. B. Eisenthal. Chem. Phys. Lett. *191*:327 (1992).
109. Q. Du, E. Freysz, and Y. R. Shen. Phys. Rev. Lett. *72*:238 (1994).
110. P. B. Miranda, Q. Du, and Y. R. Shen. Chem. Phys. Lett. *286*:1 (1998).
111. Q. Du, R. Superfine, E. Freysz, and Y. R. Shen. Phys. Rev. Lett. *70*:2313 (1993).
112. M. C. Bellisent-Funel, S. H. Chen, and J. M. Zonatti. Phys. Rev. E *51*:4558 (1995).
113. P. M. Wiggins. Prog. Polym Sci. *13*:1 (1988).
114. D. C. Steytler and J. C. Dore. Mol. Phys. *5*:1001 (1985).
115. M. J. Benham et al. Phys. Rev. B *39*:633 (1989).
116. V. P. Denisov and B. Halle. Faraday Discuss. *103*:227 (1996).
117. M. C. Bellisent-Funel, J. M. Zannoti, and S. H. Chen. Faraday Discuss. *103*:281 (1996).
118. M. Gerstein and M. Levitt. Scientific American, November 1998, pp. 101–105.
119. H. R. Garner, T. Ohkawa, O. Tuason, and R. L. Lee. Phys. Rev. A *42*:7264 (1990).
120. K. Wutrich et al. Faraday Discuss. *103*:245 (1996).
121. M. C. Bellisent-Funel, S. H. Chen, and M. Zannoti. Phys. Rev. E *51*:4558 (1995).
122. R. S. Lee and S. Bore. Faraday Discuss. *103*:59 (1996).
123. S. Vajda et al. J. Chem. Soc. Faraday Trans. *91*:867 (1995).
124. L. Labeots and M. A. Weiss. J. Mol. Biol. *269*113 (1997).
125. C. O. Pabo and R. T. Sauer. Annu. Rev. Biochem. *61*:1053 (1992).
126. J. L. Parker, P. M. Claesson, and J. L. Parker. J. Phys. Chem. *96*:6725 (1992).
127. C. Tanford, *The Hydrophobic Effect*, John Wiley, New York, 1980.
128. J. Israelachvili, *Intermolecular and Surface Forces*, 2d ed., Academic Press, London, 1992, Chapter 13.
129. M. Hato. J. Phys. Chem. *100*:18530 (1996).
130. T. D. Blake and J. A. Kitchener. J. Chem. Soc. Faraday Trans. 1 *68*:1435 (1972).
131. R. H. Yoon and J. L. Yordan. J. Colloid Interface Sci. *146*:565 (1991).
132. R. M. Pashley and J. Israelachvili. Colloids Surf. *2*:169 (1981).
133. J. Israelachvili and R. M. Pashley. J. Colloid Interface Sci. *98*:500 (1984).
134. J. Israelachvili and R. M. Pashley. Nature *300*:341 (1982).
135. R. M. Pashley, P. McGuiggan, B. W. Ninham, and D. F. Evans. Science *229*:1088 (1985).
136. P. M. Claesson, C. E. Blom, P. C. Herder, and B. W. Ninham. J. Colloid Interface Sci. *114*:234 (1986).
137. H. K. Christenson and P. M. Claesson. Science, *239*:390 (1988).
138. Y. A. I. Rabinovich and B. V. Derjaguin. Colloids Surf. *30*:243 (1988).
139. C. J. Van Oss, D. R. Absolom, and A. W. Neuman. Colloid Polym. Sci. *258*:424 (1980).
140. C. J. Van Oss, M. K. Chaudhury, and R. J. Good. Adv. Coll. Interface Sci. *28*:35 (1987).
141. E. Ruckenstein and N. V. Churaev. J. Colloid Interface Sci. *147*:535 (1991).
142. P. J. Attard. J. Phys. Chem. *93*:6441 (1989).
143. V. V. Yaminsky and B. W. Ninham. Langmuir *9*:3618 (1993).
144. E. Ruckenstein. J. Colloid Interface Sci. *188*:218 (1997).

145. D. R. Berard, P. Attard, and G. N. Patey. J. Chem. Phys. 98:7236 (1993).
146. V. V. Yaminsky, V. S. Yuschenko, E. A. Amelina, and E. D. Schukin. J. Colloid Interface Sci. 96:301 (1983).
147. P. M. Claesson and H. K. Christenson. J. Phys. Chem. 92:1650 (1988).
148. L. Meagher and V. S. Craig. Langmuir 10:2736 (1994).
149. O. I. Vinogradova et al. J. Colloid Interface Sci. 173:443 (1995).
150. V. V. Yaminsky and B. W. Ninham. Langmuir 9:3618 (1993).
151. V. S. Craig, B. W. Ninham, and R. M. Pashley. Nature 364:317 (1993).
152. N. F. Bunkin et al. Colloids Surf. 100:207 (1996).
153. N. F. Bunkin and F. V. Bunkin. Sov. Phys. JETP Engl. Transl. 101:512 (1992).
154. N. F. Bunkin et al. Langmuir 13:3024 (1997).
155. N. F. Bunkin and F. V. Bunkin. Laser Phys. 3:63 (1993).
156. J. D. Miller, Y. Hu, S. Veeramasuneni, and J. Lu. In press in Colloids and Surfaces.
157. M. Karaman, B. W. Ninham, and R. M. Pashley. J. Phys. Chem. 100:15503 (1996).
158. Z. A. Zhou, Z. Xu, and J. A. Finch. J. Colloid Interface Sci. 179:311 (1996).
159. Z. A. Zhou, Z. Xu, J. A. Finch, and Q. Liu. Colloids Surf. 113:67 (1996).
160. N. F. Bunkin and A. V. Lobeyev. Phys. Lett. A 229:327 (1997).
161. A. F. Bunkin, N. F. Bunkin, A. V. Lobeyev, and A. A. Nurmatov. Phys. Lett. A. 225:349 (1996).
162. T. Head-Gordon. PNAS 92:8308 (1995).
163. H. S. Frank and M. W. Evans. J. Chem. Phys. 13:507 (1945).
164. Q. Du, R. Superine, E. Freysz, and Y. R. Shen. Phys. Rev. Lett. 70:2313 (1993).
165. D. E. Gragson and G. L. Richmond. Langmuir 13:4804 (1997).
166. Q. Du, E. Freysz, and Y. R. Shen. Science 264:826 (1994).
167. Z. S. Nickolov, J. C. Earnshaw, and J. J. M. Garrey. Colloids Surf. 76:41 (1993).
168. J. D. Miller, Y. Hu, S. Veeramasuneni, and Y. Lu. In press in Colloids Surf.
169. P. M. Wiggins. Physica A 238:113 (1997).
170a. Scientific Papers of Lord Rayleigh, Dover, New York, 1964, Vol. 4, p. 430.
170b. T. D. Blake and J. A. Kitchener. J. Chem. Soc. Faraday Trans. 1 68:1435 (1972).
170c. N. F. Bunkin and F. V. Bunkin. Sov. Phys. JETP 74:271 (1992).
171. P. G. Bowers, K. B. Eli, and R. M. Noyes. J. Chem. Soc. Faraday Trans. 92:2843 (1996).
172. N. F. Bunkin and A. V. Lobeyev. JETP Lett. 58:91 (1993).
173. N. F. Bunkin et al. JETP Lett. 62:685 (1995).
174. N. F. Bunkin and A. V. Lobeyev. Phys. Lett. A 229:327 (1997).
175. A. F. Bunkin, N. F. Bunkin, A. V. Lobeyev, and A. A. Nurmatov. Phys. Lett. A 225:349 (1997).
176. K. Kikuchi, H. Takeda, T. Okaya, and Z. Ogumi, paper presented by K. Kikuchi at the 2nd Functional Water Symposium, Tokyo, Japan, 1997.
177. E. Ruckenstein. J. Colloid Interface Sci. 188:218 (1997).
178. P. M. Claesson and H. K. Christenson. J. Phys. Chem. 92:1050 (1988).
179. N. F. Bunkin et al. Langmuir 13:3024 (1997).
180. N. F. Bunkin et al. Colloids Surf. 100:207 (1996).
181. J. D. Miller, Y. Hu, S. Veeramasuneni, and Y. Lu. In press in Colloids Surf.
182. V. S. Craig, B. W. Ninham, and R. M. Pashley. Nature 364:317 (1993).
183. V. S. Craig, B. W. Ninham, and R. M, Pashley. J. Phys. Chem. 97:10192 (1993).
184. P. K. Weissenborn and R. J. Pugh. Langmuir 11:1422 (1995).
185. P. Weissenborn and R. J. Pugh. J. Colloid Interface Sci. 184:550 (1996).
186. J. Laskowski and J. A. Kitchener. J. Chem. Soc. Faraday Trans. 1 29:670 (1969).
187. T. D. Blake and J. A. Kitchener. J. Chem. Soc. Faraday Trans. 1 68:1435 (1972).
188. R. M. Pashley and J. Israelachvili. Colloids Surf. 2:169 (1981).
189. J. Israelachvili and R. M. Pashley. J. Colloid Interface Sci. 98:500 (1984).
190. J. Israelachvili and R. M. Pashley. Nature 300:341 (1982).
191. R. M. Pashley, P. M. McGuiggan, B. W. Ninham, and D. F. Evans. Science 229:1088 (1985).

192. P. M. Claesson, C. E. Blom, P. C. Herder, and B. W. Ninham. J. Colloid Interface Sci. *114*:234 (1986).
193. V. S. J. Craig, B. W. Ninham, and R. M. Pashley. Langmuir *14*:3326 (1998).
194. V. V. Yaminsky, B. W. Ninham, H. K. Christenson, and R. M. Pashley. Langmuir *12*:1936 (1996).
195. V. V. Yaminsky, C. Jones, F. Yaminsky, and B. W. Ninham. Langmuir *12*:3531 (1996).
196. H. Christenson and P. M. Claesson. Science *239*:390 (1988).
197. H. K. Christenson, J. Fang, B. W. Ninham, and J. C. Parker. J. Phys. Chem. *94*:8004 (1990).
198. E. Kokkoli and C. F. Zukoski. Langmuir *14*:1189 (1998).
199. Y. H. Tsao, S. X. Yang, and D. F. Evans. Langmuir *7*:3154 (1991).
200. K. Kurihara, S. Kato, and T. Kunitake. Chem. Lett. *50*:1555 (1990).
201. K. Kurihara and T. Kunitake. J. Am. Chem. Soc. *114*:10927 (1992).
202. J. L. Parker, P. M. Claesson, and P. Attard. J. Phys. Chem. *98*:8468 (1994).
203. J. L. Parker, D. L. Cho, and P. M. Claesson. J. Phys. Chem. *93*:6121 (1989).
204. M. E. Karaman, L. Meagher, and R. M. Pashley. Langmuir *9*:1220 (1993).
205. J. Wood and R. Sharma. Langmuir *11*:4797 (1995).
206. P. Kekicheff and O. Spalla. Phys. Rev. Lett. *75*:1851 (1995).
207. R. H. Yoon and S. A. Ravishankar. J. Colloid Interface Sci. *179*:391 (1996).
208. J. C. Eriksson, S. Ljunggren, and P. M. Claesson. J. Chem. Soc. Faraday Trans. 2 *85*:163 (1989).
209. R. J. Podgornik. J. Chem. Phys. *91*:5840 (1989).
210. P. J. Attard. J. Phys. Chem. *93*:6441 (1989).
211. H. K. Christenson, P. M. Claesson, J. Berg, and P. C. Herder. J. Phys. Chem. *93*:1472 (1989).
212. H. K. Christenson, P. M. Claesson, and J. C. Parker. J. Phys. Chem. *96*:6725 (1992).
213. J. L. Parker and P. M. Claesson. Langmuir *10*:635 (1994).
214. E. Kokkoli and C. F. Zukoski. Langmuir *14*:1189 (1998).
215. E. Ruckenstein and N. J. Churaev. J. Colloid Interface Sci. *147*:535 (1991).
216. V. V. Yaminsky and B. W. Ninham. Langmuir *9*:3618 (1993).
217. E. Ruckenstein. J. Colloid Interface Sci. *188*:218 (1997).
218. J. L. Parker, P. M. Claesson, and P. Attard. J. Phys. Chem. *98*:8468 (1994).
219. Ya. I. Rabinovich and R. H. Yoon. Colloids Surf. *93*:263 (1994).
220. L. Meagher and V. S. Craig. Langmuir *10*:2736 (1994).
221. J. Wood and R. Sharma. Langmuir *11*:4797 (1995).
222. A. Carambassis, L. C. Jonker, P. Attard, and M. W. Rutland. Phys. Rev. Lett. *80*:5357 (1998).
223. M. Karaman, B. W. Ninham, and R. M. Pashley. J. Phys. Chem. *100*:15503 (1996).
224. Z. A. Zhou, Z. Xu, and J. A. Finch. J. Colloid Interface Sci. *179*:311 (1996).
225. Z. A. Zhou, Z. Xu, and J. A. Finch. Colloids Surf. A *113*:67 (1996).
226. R. H. Yoon, D. H. Flinn, and Ya. Rabinovich. J. Colloid Interface Sci. *185*:363 (1997).
227. Ya. I. Rabinovich, D. A. Guzonas, and R. A. Yoon. Langmuir *9*:1168 (1993).
228. J. S. Baker and S. J. Judd. Wat. Res. *30*:247 (1996).
229. M. Čolić and D. Morse. In press in Colloids Surf.
230. M. Čolić and D. Morse. Manuscript in preparation.
231. S. Putterman. Scientific American, February 1995, pp. 47–51.
232. P. Bradley, P. Barber, and S. Putterman. Nature *352*:318 (1991).
233. R. Hiller, K. Weninger, S. J. Putterman, and B. P. Barber. Science *266*:248 (1994).
234. E. Chibowski and L. Holysz. Colloids Surf. *101*:99 (1995).
235. K. Higashitani, A. Kage, S. Kotamura, K. Imai, and S. Hatade. J. Colloid Interface Sci. *156*:90 (1993).
236. K. Higashitani, K. Okuhara, and S. Hatade. J. Colloid Interface Sci. *152*:125 (1992).
237. K. Higashitani, H. Iseri, K. Okuhara, A. Kage, and S. Hatade. J. Colloid Interface Sci. *172*:383 (1995).
238. E. Chibowski, S. Ghopalkrishnan, M. A. Busch, and K. W. Busch. J. Colloid Interface Sci. *139*:43 (1990).

239. E. Chibowski, L. Holysz, and W. Wojcik. Colloids Surf. *92*:79 (1994).
240. L. Holysz and E. Chibowski. J. Colloid Interface Sci. *165*:243 (1994).
241. D. J. Muesham and A. A. Pilla. Bioelectrochem. Bioenergetics *35*:71 (1994).
242. C. H. Grissom. Chem. Rev. *95*:3 (1995).
243. R. A. Luben, in *Electromagnetic Fields: Biological Interactions and Mechanisms* (Martin Blank ed.), American Chemical Society, Washington, D.C., 1995.
244. I. J. Lin and J. Yotvat. J. Magn. Magn. Matl. *83*:525 (1990).
245. D. Morse et al., U.S. Patent 5,606,273 to ZPM Inc. (1997).
246. B. G. Welder and E. P. Partridge. Ind. Eng. Chem. *46*:954 (1954).
247. O. Sohnel and J. Mullin. Chem. Indus. *11*:356 (1988).
248. M. Ozaki, K. Nokata, and E. Matijević. J. Colloid Interface Sci. *131*:233 (1989).
249. M. F. Haque, N. Kallay, V. Privman, and E. Matijević. J. Colloid Interface Sci. *137*:36 (1990).
250. N. Kallay and E. Matijević. Colloids Surf. *39*:161 (1989).
251. M. F. Haque, N. Kallay, V. Privman, and E. Matijević. J. Adhesion Sci. Technol. 4205 (1990).
252. V. I. Geltyuk, V. N. Kazachenko, N. K. Chemeris, and E. E. Fesenko. FEBS Lett. *359*:85 (1995).
253. E. E. Fesenko, V. Geletyuk, V. N. Kazachenko, and N. K. Chemeris. FEBS Lett. *366*:49 (1995).
254. E. E. Fesenko and A. Y. Gluvstein. FEBS Lett. *367*:53 (1995).
255. M. Čolić and D. Morse. Langmuir *14*:783 (1998).
256. M. Čolić and D. Morse. Phys. Rev. Lett. *80*:2465 (1998).
257. M. Čolić and D. Morse. J. Colloid Interface Sci. *200*:265 (1998).
258. M. Čolić, A Chien, and D. Morse. In press in Croat. Chem. Acta.
259. M. Čolić and D. Morse. In press in Colloids Surf.
260. S. Raj, N. N. Singh, and R. N. Mishra. Med. Biol. Eng. Computing *33*:614 (1995).
261. S. Ozeki, J. Miyamoto, S. Ono, C. Wakai, and T. Watanabe. J. Phys. Chem. *100*:4205 (1996).
262. R. Sasamori, Y. Okaue, T. Isobe, and Y. Matsuda. Science *265*:1691 (1994).
263. M. Pach and R. Stosser. J. Phys. Chem. *101*:8360 (1997).
264. H. Hayashi. *Microwater, the Natural Solution*, Water Institute, Tokyo, 1996.
265. N. F. Bunkin et al. Langmuir 13:3024 (1997).
266. N. F. Bunkin et al. Colloids Surf. *100*:207 (1996).
267. C. Eberlein. Phys. Rev. Lett. *76*:3842 (1996).
268. R. Golestanian and M. Kardar. Phys. Rev. A *58*:1713 (1998).
269. Y. T. Didenko and S. P. Pugach. J. Phys. Chem. *98*:9742 (1994).
270. H. Tanaka and K. Nakanishi. J. Chem. Phys. *95*:3719 (1991).
271. I. Ohmine and H. Tanaka. Chem. Rev. *93*:2545 (1993).
272. A. Rahman and F. H. Stillinger. J. Chem. Phys. *55*:3336 (1971).
273. F. H. Stillinger and T. A. Weber. Phys. Rev. A *25*:97 (1982).
274. F. H. Stillinger and T. A. Weber. Science *225*:983 (1984).
275. M. Gerstein and M. Levitt. Scientific American, November 1998, p. 101.
276. C. Y. Lee, J. A. McCammon, and P. J. Rossky. J. Chem. Phys. *80*:4448 (1984).
277. S. H. Lee and P. J. Rossky. J. Chem. Phys. *100*:3334 (1994).
278. P. Linse. J. Chem. Phys. *90*:4992 (1989).
279. P. M. Wiggins. Physica A *238*:113 (1997).
280. A. Luzar and D. Chandler. Nature *379*:55 (1996).
281. T. Head-Gordon. PNAS *92*:8308 (1995).
282. A. D. J. Haymet, K. A. Silverstein, and K. A. Dill. Faraday Discuss. *103*:117 (1996).
283. R. Kjellander and S. Marcelja. Chem. Phys. Lett. *120*:393 (1985).
284. R. Kjellander and S. Marcelja. Chemica Scripta *25*:73 (1985).
285. T. J. McIntosh and S. A. Simon. Rev. Biophys. Biomolec. Struct. *23*:27 (1994).
286. J. M. Eggbrecht, D. J. Isbister, and J. C. Rasaiah. J. Chem. Phys. *73*:3980 (1980).
287. J. C. Rasaiah, D. J. Isbister, and G. S. Stell. Chem. Phys. *75*:4707 (1981).

288. S. Marcelja. Nature *385*:689 (1997).
289. G. M. Torrie, P. G. Kusalik, and G. N. Patey. J. Chem. Phys. *91*:6367 (1989).
290. W. N. Rowlands, R. W. O'Brien, R. J. Hunter, and V. Patrick. J. Colloid Interface Sci. *188*:325 (1997).
291. M. Čolić, G. V. Franks, M. L. Fisher, and F. F. Lange. Langmuir *13*:3129 (1997).

3
Thermodynamics of Adsorption at Heterogeneous Solid–Liquid Interfaces

WŁADYSŁAW RUDZIŃSKI, JOLANTA NARKIEWICZ-MICHAŁEK, ROBERT CHARMAS, MATEUSZ DRACH, and WOJCIECH PIASECKI Department of Theoretical Chemistry, Maria Curie-Skłodowska University, Lublin, Poland

JERZY ZAJĄC LAMMI, Centre National de la Recherche Scientifique, Montpellier, France

I.	Thermodynamics of Adsorption from Binary Mixtures of Miscible Liquids at the Solid–Liquid Interface	83
	A. Introduction	83
	B. Measurements of adsorption from binary mixtures onto solids	85
	C. Theoretical interpretation of surface excess and heat of immersion	91
II.	Simple Ion Adsorption at Oxide/Electrolyte Interface	106
	A. Isotherm of adsorption of ions forming different surface complexes	109
	B. Accompanying heats of ion adsorption	116
	C. Results of our investigations	124
III.	Adsorption of Cationic Surfactants on Hydrophilic Silica: Effects of Surface Energetic Heterogeneity	131
	A. Introduction	131
	B. Conceptualization of ionic surfactant adsorption on charged solid surfaces	133
	C. Theoretical model for the adsorption of cationic surfactants on a hydrophilic oxide surface	136
	D. Comparison of the model predictions with experimental data	144
	List of Symbols	153
	References	157

I. THERMODYNAMICS OF ADSORPTION FROM BINARY MIXTURES OF MISCIBLE LIQUIDS AT THE SOLID–LIQUID INTERFACE

A. Introduction

In this section we shall consider the thermodynamics of adsorption of low-molecular-weight uncharged adsorbates from binary solutions at the solid–liquid interface. The discussion will be limited to the simple case of mixtures of liquids that are miscible in the bulk phase over the entire range of compositions. The mixtures of practical interest show low or moderate deviations from the ideal solution behavior.

The molecules of the two components may have unequal sizes. In general, interactions of such adsorbates with the solid surface should be in some ways similar to each other. In consequence, the change in composition of the interface is a result of the adsorption of both components, and the overall process has a pronounced competitive character. The phenomenon appears to be very sensitive to any deviations from the ideal behavior. In particular, heterogeneity of the solid boundary and nonideality of mixing in the bulk phase and in the interfacial region play an important role in determining the changes in thermodynamic functions upon adsorption.

The practical use of adsorption from liquid mixtures onto solid substrates has long been appreciated, but the fundamental understanding of this phenomenon has developed relatively slowly. Early attempts focused on dilute solutions, and consequently the associated theories were constructed by simple analogy with gas adsorption. The development of more versatile and more precise methods of measuring adsorption from liquid mixtures over the whole concentration range and the pioneering works of Kipling, Everett, Schay, and Nagy [1–4] drew attention to the inadequacy of this approach. The competitivity of solution adsorption was fully recognized, and the rigorous thermodynamic treatment could be formulated [5,6].

The breakthrough in the understanding of broad principles governing adsorption from solution at the solid–liquid and the liquid–vapor interfaces has stimulated widespread theoretical studies on the phenomenon. One of the simplest and most widely used approaches is to make the assumption that the adsorbed layer is monomolecular [7–11]. Multilayer adsorption from solution is less common than in gas adsorption. This is due to a marked screening of interactions in condensed media. The formation of a multilayer may be encountered rather in the case of adsorption from solutions of solids close to saturation, and from incompletely miscible liquid pairs close to the miscibility limit. However, according to Lane [12], multilayer effects are negligible only for adsorption from nearly ideal bulk mixtures. A multilayer model of adsorption was considered by Guggenheim [13,14], Lane [15], and others [16–19]. All mono and multilayer models are generally based on either a lattice model of the adsorbed and bulk phases [e.g., 13–15] or a more phenomenological mass action approach [e.g., 1,2,18,19]. Most of them include intermolecular interactions in both the bulk and the adsorbed phases, while the effects of surface heterogeneity on the thermodynamic properties of the adsorption system are usually omitted.

Surface heterogeneity has often been expected to have a substantial effect on adsorption from solution. The importance of this factor was shown in some experimental papers [20–23] and was also quantitatively discussed by Everett [2]. The first systematic development of a heterogeneous surface model for adsorption from binary mixtures is generally attributed to Rudziński et al. [24]. A more advanced theoretical treatment of the combined effect of heterogeneity and solution nonideality on the surface excess isotherm can be found in works of Dąbrowski, Jaroniec, and Koopal [25–31]. Sircar [32–34] and Rudziński and co-workers [35–44] have addressed the combined effect on both excess isotherm and heat of immersion.

There are still insufficient experimental data available with which to carry out detailed tests of the various theories. In particular, the theoretical interpretation of the existing experimental results rarely goes beyond the fitting of adsorption isotherms. An adsorption isotherm is a necessary but not sufficient way of describing the thermodynamics of adsorption. The lack of other thermodynamic data reported on the actual adsorption systems has left most of the developed models for adsorption from binary mixtures essentially untested. Full description of the phenomenon requires knowledge of mutual interactions between all the components of the system. Such opportunity is offered by flow and batch liquid adsorption microcalorimetry. The importance of calorimetric measurements was emphasized by Zettlemoyer, Everett, Groszek, Sircar and Myers, and Woodbury and Noll [45–50].

B. Measurements of Adsorption from Binary Mixtures onto Solids

Adsorption at the solid–solution interface is usually determined by the measurement of the change in the composition of a binary liquid mixture that is allowed to attain equilibrium with a given sample of solid. The amount adsorbed and heat effects accompanying the phenomenon can be precisely defined in operational terms by considering two types of experiment: the immersion experiment and the flow displacement experiment. The former is the basis for static experimental procedures [46,50–52], and the latter underlies a variety of dynamic methods [50,53–60].

1. Immersion in a Pure Liquid

Before immersion the system separately contains an outgassed (i.e., free of adsorbed liquid and vapor impurities) solid sample of a given surface area A_s in a vacuum and pure liquid at temperature T and pressure P. Then the solid is immersed in the solution under conditions of fixed T and P [e.g., 61,62].

In most practical cases, the adsorbent remains thermodynamically neutral, which means that the thermodynamic properties of the bulk of solid are hardly affected by the presence of foreign molecules in the surroundings. The occurrence of side phenomena, like absorption, dissolution of the solid in a liquid, or swelling of the support, is excluded from further analysis. The process is carried out in an excess of liquid, high enough for the enthalpy change in the bulk liquid to be neglected during immersion. Moreover, adsorption at the solid–liquid and the liquid-vapor interfaces can be ignored for a one-component liquid phase. In consequence, the total enthalpy change is ascribed only to the solid–liquid interface. If the Gibbs convention for the interface is used, the enthalpy of immersion can be written as

$$\Delta_{\mathrm{imm}} H = A_s [H_{\mathrm{SL}}^{\sigma(n)} - H_{\mathrm{So}}^{\sigma(n)}] \tag{1}$$

where $H_{\mathrm{SL}}^{\sigma(n)}$ and $H_{\mathrm{So}}^{\sigma(n)}$ are the reduced excess enthalpies (per unit area) for the solid–liquid and the solid–vacuum interfaces, respectively.

The experiment of immersion can be performed under completely different conditions [61,62]. Consider a solid in equilibrium with the vapor of the immersional liquid at the saturation pressure p_0. The adsorbate forms an adsorbed film on the solid surface. When solid is subsequently immersed in the liquid, the enthalpy change, called the *enthalpy of immersional wetting*, $\Delta_w H$, will be different from $\Delta_{\mathrm{imm}} H$. If the immersional liquid spreads out spontaneously over the solid surface (i.e., wets out the solid), the final states in both experiments are completely equivalent and we can write

$$\Delta_w H = A_s [H_{\mathrm{SL}}^{\sigma(n)} - H_{\mathrm{SV}}^{\sigma(n)}(p_o)] = A_s [H_{\mathrm{SL}}^{\sigma(n)} - H_{\mathrm{So}}^{\sigma(n)}] - A_s [H_{\mathrm{SV}}^{\sigma(n)}(p_o) - H_{\mathrm{So}}^{\sigma(n)}] \tag{2}$$

where $H_{\mathrm{SV}}^{\sigma(n)}$ is the reduced excess enthalpy (per unit area) for the solid–vapor interface. The first term on the right-hand side of Eq. (2) is the enthalpy of immersion, while the second one refers to the formation of an adsorbed film on the adsorbent surface, $\Delta_{\mathrm{ads}} H(p_0)$. So far as the enthalpy is concerned, the interface between the saturated adsorbed film and the equilibrium vapor phase at p_0 may be identified with the interface between

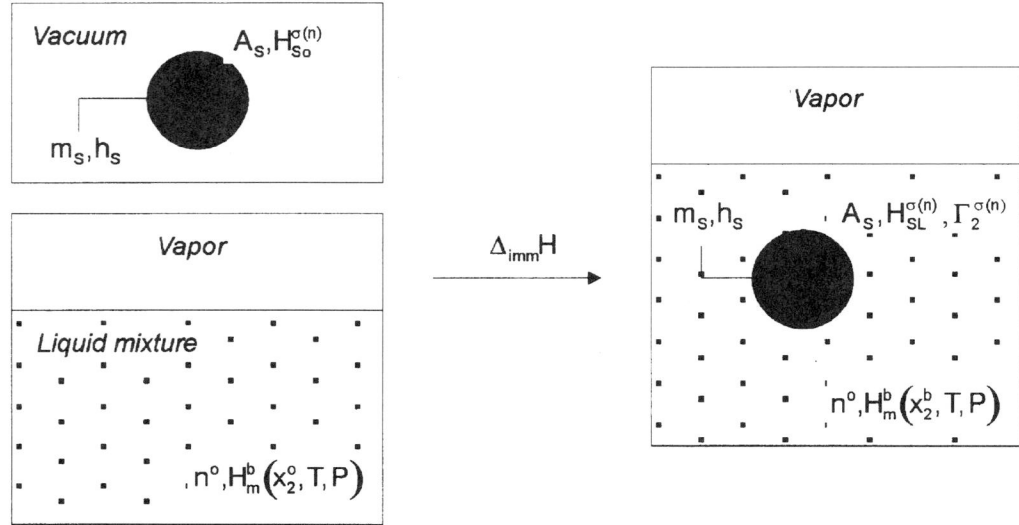

FIG. 1 Schematic representation of the immersion experiment.

the liquid and its own vapor. Thus

$$\Delta_{\text{imm}} H = \Delta_{\text{ads}} H(p_o) - A_s H_{\text{LV}}^{\sigma(n)} \qquad (3)$$

where $H_{\text{LV}}^{\sigma(n)}$ is the reduced excess enthalpy (per unit area) for the liquid–vapor interface.

2. Immersion in a Binary Mixture

In this case, the solid sample with a clean surface is immersed in n^0 moles of binary mixture of composition x_2^0 under temperature T and pressure P. After the attainment of adsorption equilibrium, the mole fraction of component 2 in the bulk solution has a new value of x_2^b because one of the solution components is usually preferentially adsorbed at the solid surface. The immersion process is schematically represented in Fig. 1.

The experimentally measured *surface excess* of component 2 (assumed to be preferentially adsorbed) is calculated as

$$\Gamma_2^e = \Gamma_2^{\sigma(n)} = \frac{n^0(x_2^0 - x_2^b)}{M_S} = \frac{n^0 \Delta x_2}{M_S} \qquad (4)$$

where $\Gamma_2^{\sigma(n)}$ is the reduced adsorption of component 2 and $M_S = A_S$, the surface area of the adsorbent. When the latter parameter is unknown, the quantity of adsorption is reported as the surface excess per unit mass of the solid sample (i.e., $M_S = m_S$)

The enthalpy of immersion in a binary liquid can be expressed in the form

$$\Delta_{\text{imm}} H = A_s[H_{\text{SL}}^{\sigma(n)} - H_{\text{So}}^{\sigma(n)}] + n^0 [H_m^b(x_2^b) - H_m^b(x_2^0)] \qquad (5)$$

where $H_m^b(x_2)$ is the molar enthalpy of the liquid mixture corresponding to the mole fraction x_2 at temperature T and pressure P.

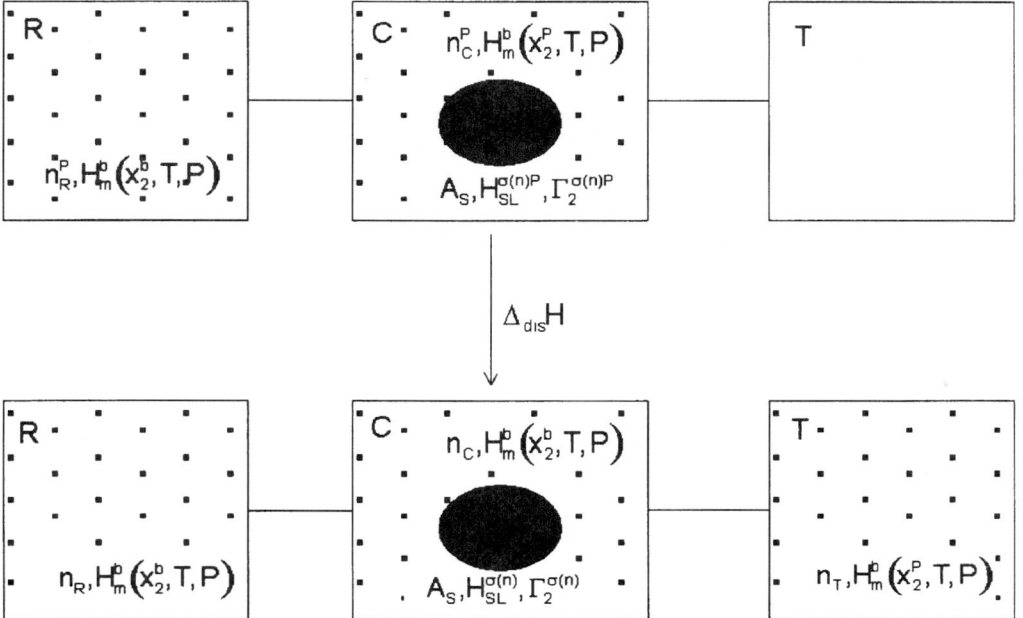

FIG. 2 Schematic representation of the flow displacement experiment.

3. Flow Displacement Experiment

In the flow displacement experiment, the model experimental system is composed of three parts: the reservoir (R), the cell with the adsorbent (C), and the trap (T). The adsorbent is initially immersed in n_C^P moles of binary mixture, the composition of which (sufficiently far from the solid surface in the cell) is x_2^P. The initial and final states of the flow system are illustrated in Fig. 2.

As the flow occurs, the interface between the solutions of different compositions remains sharp, and the effects of adsorption or desorption change the position of the interface according to the ideal chromatographic behavior [50]. In the final state, the number of moles in the trap is n_T and its composition is x_2^P, whereas the composition of liquid mixture in the reservoir and cell (far from the solid surface) is equal to x_2^b. The temperature T and the pressure P are assumed to be uniform throughout the system.

From the law of conservation of mass in a closed system, the surface excess of component 2 after displacement is given by

$$\Gamma_2^e = \Gamma_2^{\sigma(n)} = \Gamma_2^{\sigma(n)P} + \frac{(n_T - n_C^P)(x_2^b - x_2^P)}{M_s} \tag{6}$$

where $\Gamma_2^{\sigma(n)}$ and $\Gamma_2^{\sigma(n)P}$ are the reduced adsorptions of component 2 in the final and initial states, respectively.

The enthalpy change, $\Delta_{\text{dis}}H$, for the flow displacement process can be written as

$$\Delta_{\text{dis}}H = A_s(H_{\text{SL}}^{\sigma(n)} - H_{\text{SL}}^{\sigma(n)\text{P}}) - (n_\text{T} - n_\text{C}^\text{P})[H_\text{m}^\text{b}(x_2^\text{b}) - H_\text{m}^\text{b}(x_2^\text{P})] \tag{7}$$

where $H_{\text{SL}}^{\sigma(n)}$ and $H_{\text{SL}}^{\sigma(n)\text{P}}$ are the reduced excess enthalpies (per unit area) for the solid–binary mixture interface in the final and initial states, respectively.

The flow displacement experiment can be carried out in steps, starting with pure component 1 in the cell, replacing that by a dilute solution of 2 in 1 [50,53,54]. If component 2 is preferentially adsorbed on the solid surface, the molecules of 1 will be displaced from the interface and replaced by molecules of the second type. The system equilibrates so that the surface concentrations of 1 and 2 will be appropriate to their concentrations in the bulk liquid. Then the enthalpy of displacement becomes

$$\Delta_{\text{dis}}H_1 = \Delta_{\text{dis}}H(0 \to x_2^\text{b}) \approx \Delta_{\text{imm}}H(x_2^\text{b}) - \Delta_{\text{imm}}H_1^\text{o} \tag{8}$$

where $\Delta_{\text{imm}}H(x_2^\text{b})$ is the enthalpy of immersion in a binary liquid mixture corresponding to the equilibrium bulk phase of composition $x_2{}^b$, and $\Delta_{\text{imm}}H_1^0$ is the enthalpy of immersion in pure liquid 1.

This procedure can be repeated with a more concentrated mixture of 2 in 1 until the solid surface is in contact with pure component 2 only. The enthalpy change associated with each step, $\Delta_{\text{dis}}H_d = \Delta_{\text{dis}}H_d(x_2^\text{b})$, is defined by Eq. (7). The total enthalpy change, $\Delta_{\text{dis}}H_\text{T}$, occurring when the solid adsorbent in contact with pure component 1 is exchanged for the solid in contact with pure component 2, is the sum of such contributions. From Eqs. (7), (8), and (1), we obtain

$$\Delta_{\text{dis}}H_\text{T} = \sum_{x_2^\text{b}=0}^{x_2^\text{b}=1} \Delta_{\text{dis}}H_d(x_2^\text{b}) = A_s[H_{\text{S}2}^{\sigma(n)} - H_{\text{s}1}^{\sigma(n)}] = \Delta_{\text{imm}}H_2^\text{o} - \Delta_{\text{imm}}H_1^\text{o} \tag{9}$$

where $H_{\text{Sj}}^{\sigma(n)}$ is the reduced excess enthalpy (per unit area) for the solid–liquid j interface ($j = 1, 2$). The enthalpy change on complete displacement of components is equal to the difference in the static (batch) enthalpies of immersion.

4. Classification of Experimental Curves

Experimental procedures lead to the determination of the surface excess of component 2. For a given amount of solid sample M_S, the amount adsorbed at equilibrium will depend on T, P, and the composition of the bulk solution. The curve given by the equation

$$\Gamma_2^\text{e} = \Gamma_2^\text{e}(x_2^\text{b}) \qquad T, P = \text{const} \tag{10}$$

represents the so-called *composite* or *surface excess isotherm*. The transfer of component 2 to the interface must be accompanied by the transfer of an equivalent amount of component 1 to the interior of the solution. For a two-component system $\Delta x_1 = -\Delta x_2$, and so the surface excess of component 1 is just the negative of the excess of component 2. In consequence, only the composite adsorption is measurable.

Schay and Nagy [4] have classified the observed experimental isotherms into five types. Nevertheless, only three types, as shown in Fig. 3, are adequate to describe the general behavior of the adsorption system. The first group of systems is characterized by the U-shape isotherm (type I): the surface excess Γ_2^e has positive values over the whole adsorption range. This isotherm is presumably consistent with preferential adsorption of component 2 in systems exhibiting small departures from the ideal behavior (a slightly

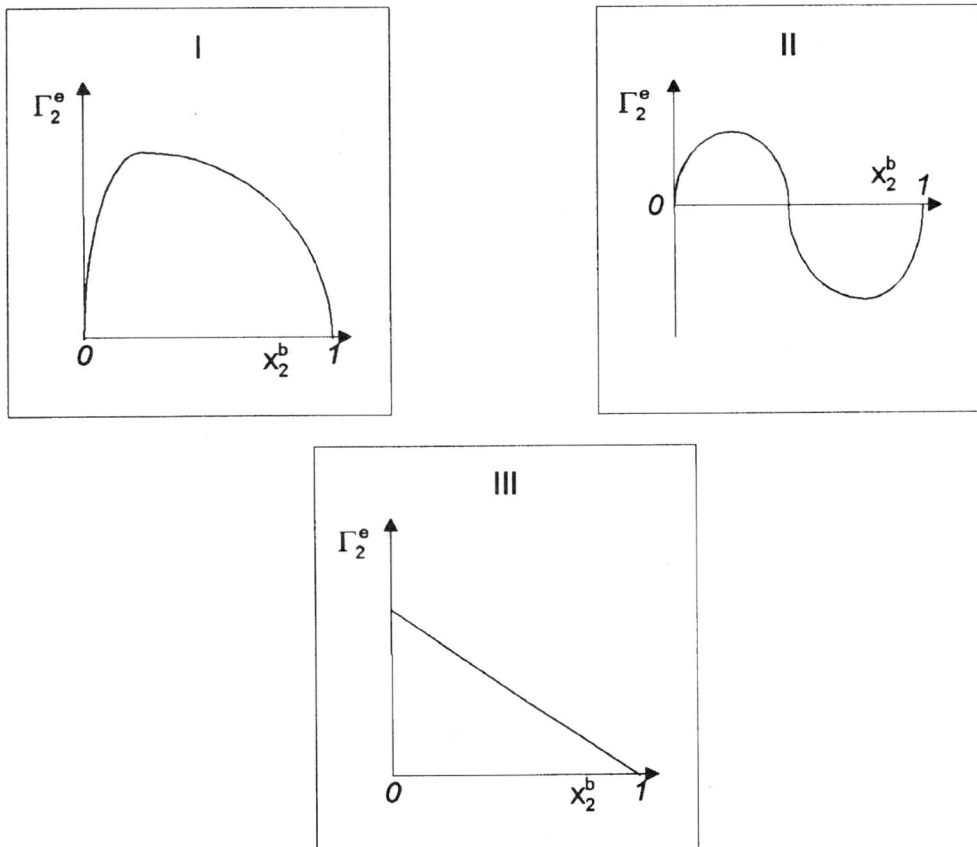

FIG. 3 Classification of composite adsorption isotherms.

heterogeneous solid surface and a perfect liquid mixture). The detailed shape of the isotherm depends on the degree of preferential adsorption of component 2 and on the experimental conditions. The second type of composite isotherm is S-shaped (type II), with a change in the preferentiality of adsorption. This adsorption behavior has been proposed as indicative of a competitive adsorption on strongly heterogeneous surfaces from nonideal solutions [1,2]. The linear isotherm (type III) corresponds to the molecular sieve effect during adsorption of components having different molecular sizes: only one component is adsorbed within the pores of the adsorbent. For example, the linear isotherm has been observed for the mixture benzene(1) + n-hexane(2) in contact with Linde molecular sieve 5 A: the benzene molecules are excluded from the zeolite porous structure [63].

The enthalpy of immersion is readily accessible by immersion experiment. It can be also determined from the appropriately designed flow measurement, as indicated by Eqs. (8) and (9). The enthalpy change on immersion is negative. By convention, the heat of immersion is quoted as a positive value. Experimental heats of immersion are generally reported in the form of heat isotherms showing the enthalpy of immersion, $\Delta_{imm}H$, taken with the opposite sign, per unit mass of solid sample (i.e., $M_S = m_S$) or unit area of adsorbent surface (i.e., $M_S = A_S$) as a function of the mole fraction of component 2 in the equi-

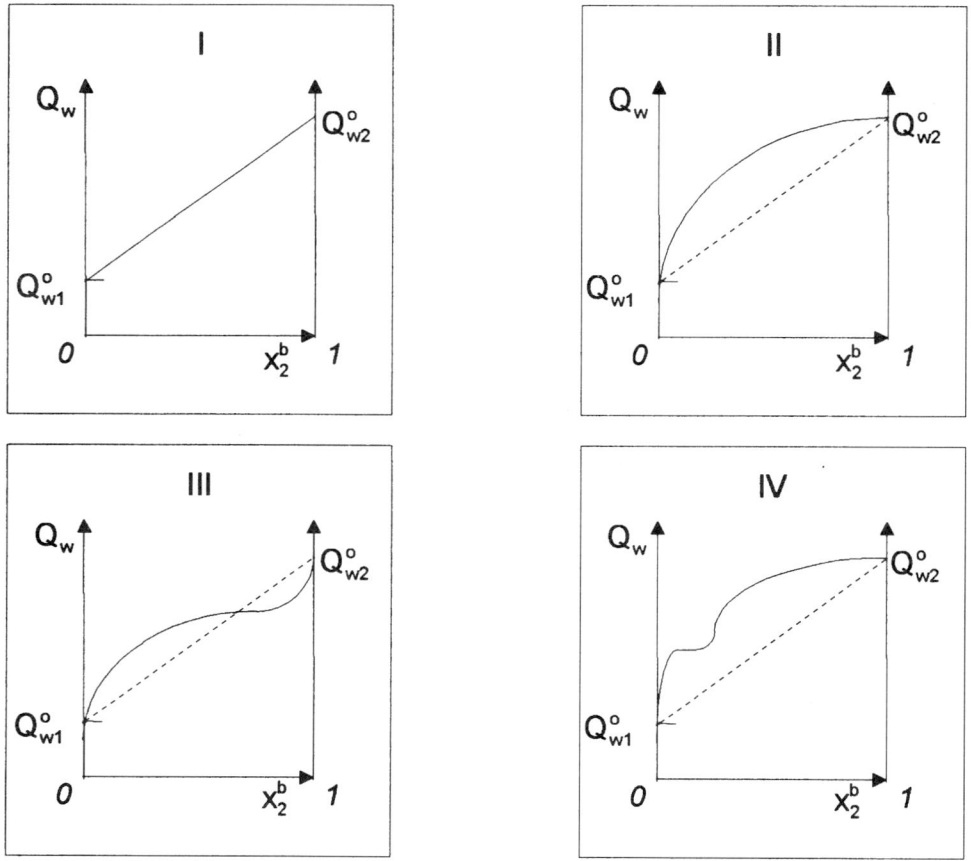

FIG. 4 Various types of isotherms of heat of immersion plotted against the mole fraction of component 2 in the equilibrium bulk mixture. Q^0_{wj} denotes the heat of immersion in pure liquid $j (j = 1, 2)$.

librium bulk phase, namely,

$$Q_w = Q_w(x_2^b) \equiv -\frac{\Delta_{\text{imm}} H(x_2^b)}{M_s} \qquad T, P = \text{const} \tag{11}$$

The possible classes of such curves are shown in Fig. 4. Examples of curves of types I and II have been observed for nearly ideal systems [59], the last two types, for nonideal systems [58,60,64]. Line I is the trivial case, which would be expected for a homogeneous surface immersed in an ideal liquid mixture of similar size molecules when there is no preferential adsorption. Curve II has been obtained for preferential adsorption of component 2 over the entire concentration range. Curve III is comparable with an S-shape adsorption isotherm and can be observed for competitive adsorption on heterogeneous surfaces from a nonideal binary mixture. Curve IV is said to indicate a change in the orientation of adsorbed molecules or the formation of a second adsorbed layer [58].

C. Theoretical Interpretation of Surface Excess and Heat of Immersion

In the previous paragraphs, adsorption from binary liquid mixtures on solid substrates has been described in terms of phenomenological thermodynamics. One of the advantages of this approach is that all ad hoc assumptions about the structure of the adsorbed layer are avoided. The two characteristics of the solid–solution interface appearing in Eqs. (4–9), i.e., the reduced excess enthalpy, $H_{SL}^{\sigma(n)}$, and the reduced adsorption of component 2, $\Gamma_2^{\sigma(n)}$, can be determined from the appropriate measurements of the surface excess and heat of immersion. However, the phenomenological treatment is too abstract, and a molecular model of the interface has to be constructed to make further progress in the understanding of the phenomenon.

It is possible to establish a link between the experimental procedures and the theoretical treatment of the adsorbed phase. Firstly, the role of solid substance is reduced to the creation of the external potential field for the adsorbing molecules and to the supply of the specific surface area. Secondly, the concentrations of both liquid components equal zero within the solid phase. With these comments in mind, the Gibbs dividing surface is located between the liquid mixture and the solid, i.e., it coincides with the surface of the adsorbent. The reduced adsorption of component 2 and the reduced excess enthalpy can now be expressed in terms of the total quantities of the adsorbed phase [42]. Therefore, from Eqs. (4) and (5), respectively,

$$\Gamma_2^e = \frac{n^s(x_2^s - x_2^b)}{M_S} \tag{12}$$

and

$$Q_w = -\frac{x_2^s}{M_S}\Delta_{imm}H_2^o - \frac{x_1^s}{M_S}\Delta_{imm}H_1^o - \frac{n^s}{M_S}[\Delta_{mix}H_m^s(x_2^s) - \Delta_{mix}H_m^b(x_2^b)] \\ -\frac{n^o}{M_S}[\Delta_{mix}H_m^b(x_2^b) - \Delta_{mix}H_m^b(x_2^o)] \tag{13}$$

where n^S is the total number of moles in the adsorbed phase, x_j^S is the mole fraction of component j in this phase ($j = 1, 2$), $\Delta_{mix}H_m^s(x_2^S)$ is the molar enthalpy of mixing of the two components to form a surface mixture of composition x_2^S, and $\Delta_{mix}H_m^b(x_2)$ is the molar enthalpy of mixing for the bulk mixture of composition x_2.

In ideal bulk solutions, the value of $\Delta_{mix}H_m^b$ is zero. The enthalpy of mixing is thus considered as a convenient measure of the solution nonideality. This observation may be extended to the case of adsorbed phase: the surface binary mixture is ideal when $\Delta_{mix}H_m^s = 0$. The last dilution term on the right-hand side of Eq. (13) is relatively small, and consequently it can be neglected.

Eqs. (12) and (13) can be used for the molecular interpretation of adsorption. The final form of expressions describing surface excess of component 2 and heat of immersion will depend on the molecular details of the model, and especially on how the nonideality effects are treated in the surface mixture.

1. Quasi-Crystalline Lattice Model of the Adsorbed Phase

The general thermodynamic treatment proposed here follows the lines of the thermodynamics of solutions, in that the properties of a real system are compared with those of the *ideal adsorbed phase* formed on a planar heterogeneous surface. The effects of nonideality in the adsorbed phase will be considered by introducing the concept of surface activity coefficients and the molar enthalpy of mixing for the surface binary mixture. The description of the model is very brief. More details can be found in Refs. 36–44.

Since diffusion in solutions is a cooperative process and many adsorbate molecules adhere strongly to solid surfaces, both components of the mixture are supposed to be localized on some adsorption sites supplied by the solid adsorbent. In consequence, many available models of the solid–solution interface are based on a lattice model of the surface solution [e.g., 5,13–16,35–44]. Although it is well understood that such a model may to some extent overestimate the local order in liquids, it has received some support from the Monte Carlo simulation for the solid–solution interface [65–67]. The lattice model of an interface requires the simultaneous use of the same model for the bulk mixture.

A two-component mixture being in contact with an insoluble, inert solid adsorbent is pictured as a collection of lattice planes stacked parallel to the planar solid surface. Each plane comprises the same number of lattice sites, and each lattice site has z nearest neighbors. The fraction l of the total nearest neighbors lies in the same lattice plane, and the fraction m, in each of the adjacent planes, i.e., $2m + l = 1$. The lattice structure is treated as rigid and not affected by the presence of the solid phase. The first lattice plane adjacent to the solid is identified with the adsorbed layer one molecule thick. Here, the adsorption centers constitute the lattice sites. Any adsorption site has only $l + m$ nearest neighbors in the first and second layers, respectively.

During adsorption, the components 1 and 2 compete with each other for the adsorption centers. The adsorption process is thus represented by the quasi-chemical equation of the phase exchange between the surface mixture and the equilibrium bulk phase,

$$\frac{1}{r}(1)^s + (2)^b \Leftrightarrow \frac{1}{r}(1)^b + (2)^s \tag{14a}$$

where the parameter r has been introduced to account for unequal adsorbate sizes. According to the parallel layer model [5], each molecule of the greater component (say component 1) is divided into r segments or submolecules of about the size of the other component. Each molecule of type 2 occupies one adsorption site, whereas each molecule of type 1 occupies r connected adsorption sites in the adsorbed layer. The conditions of adsorption equilibrium for both mixture components at constant temperature T and pressure P can be summarized by

$$\frac{1}{r}(\mu_1^s - \mu_1^b) = (\mu_2^s - \mu_2^b) \tag{14b}$$

where μ_j^s and μ_j^b are the chemical potentials of component j in the surface and the bulk mixtures, respectively.

For unequal adsorbate sizes, the total number of adsorption sites (expressed in moles), n^a, is different from the total number of moles in the adsorbed phase, n^s. It is convenient to introduce the *surface fraction* of component 2, ϕ_2^s, defined as the fraction of adsorption sites occupied by this component; therefore

$$\phi_2^s = \frac{n^s x_2^s}{n^a} = \frac{x_2^s}{x_2^s + r(1 - x_2^s)} \quad \text{and} \quad n^a = n^s[x_2^s + r(1 - x_2^s)] \tag{15a}$$

The total number of adsorption sites is constant and all of them are always occupied.

Correspondingly, the volume fraction of component 2 is defined as

$$\phi_2^b = \frac{x_2^b}{x_2^b + r(1 - x_2^b)} \tag{15b}$$

The changes of the chemical potential on mixing for the adsorbed phase are expressed in terms of the surface fraction of component 2 (see Refs. 5, 13, and 14):

$$\mu_1^s - \mu_1^{o,s} = RT[\ln(1 - \phi_2^s) - (r-1)\phi_2^s] + RT \ln f_1^s \tag{16a}$$

$$\mu_2^s - \mu_1^{o,s} = RT[\ln \phi_2^s + \frac{r-1}{r}(1 - \phi_2^s)] + RT \ln f_2^s \tag{16b}$$

where $\mu_j^{o,s}$ is the chemical potential of component j in the one-component adsorbed phase and f_j^s its activity coefficient in a binary surface mixture. When the surface mixture is ideal, $f_j^s = 1$ ($j = 1, 2$). For the equilibrium bulk mixture, the surface quantities $\mu_j^{o,s}$, ϕ_2^s, and f_j^s are simply replaced in Eqs. (16) by their bulk counterparts, i.e., the bulk potential $\mu_j^{o,b}$, the volume fraction ϕ_2^b, and the bulk activity coefficients f_1^b and f_2^b. Equations (16) and (14b) can be combined and on rearrangement give

$$\exp\left[\frac{\frac{1}{r}(\mu_1^{o,s} - \mu_1^{o,b}) - (\mu_2^{o,s} - \mu_2^{o,b})}{RT}\right] = \exp(K) = \frac{\phi_2^s f_2^s}{\phi_2^b f_2^b}\left[\frac{(1 - \phi_2^b)f_1^b}{(1 - \phi_2^s)f_1^s}\right]^{1/r} \tag{17}$$

Equation (17) implicitly gives the actual concentration of component 2 in the surface monolayer as a function of the composition of the equilibrium bulk mixture at constant T, and P. For that reason, it is called the *individual isotherm* equation for adsorption from binary liquid mixtures on solid surfaces.

There arises the problem of how to describe the activity coefficients and the molar enthalpy of mixing in the adsorbed and bulk phases. The total potential energy of lateral interactions between molecules of both components is often calculated as the sum of contributions arising from all contacts between neighboring elements (i.e., molecules of component 2 and segments of component 1). The potential energy of interaction between unlike elements is characterized by the *interchange energy*, which is the energy gained on mixing to create one mole of pairs composed each of a monomer 2 and a neighboring segment of type 1. For the adsorbed monolayer, the appropriate interchange energies may differ from their bulk counterparts on account of perturbations in the adsorbate–adsorbate interaction due to the presence of the solid. In the bulk lattice planes, the dimensionless interaction energy parameter is w^b. It is denoted as w^p when referring to the interaction within the first plane and as w^v when considering the interaction between the first and second planes.

According to the Flory–Huggins approximation of completely random mixing within layers [5,13,14], the components 1 and 2 are miscible in all proportions in both bulk and adsorbed phases when

$$l \cdot z \cdot w^p < \frac{1}{2}\left(1 + \frac{1}{\sqrt{r}}\right)^2 \quad z \cdot w^b < \frac{1}{2}\left(1 + \frac{1}{\sqrt{r}}\right)^2$$

$$z(w^b - m \cdot w^v) < \frac{1}{2}\left(1 + \frac{1}{\sqrt{r}}\right)^2 \tag{18}$$

The bulk activity coefficients corresponding to the Flory–Huggins convention can be expressed as

$$f_2^b = \exp\left[z \cdot w^b(1 - \phi_2^b)^2\right] \qquad f_1^b = \exp\left[r \cdot z \cdot w^b(\phi_2^b)^2\right] \tag{19a}$$

Correspondingly, the bulk enthalpy of mixing per one mole of lattice sites is

$$\Delta_{\text{mix}} H_m^b = \Delta_{\text{mix}} H_m^b(\phi_2^b) = RT \cdot z \cdot w^b \phi_2^b (1 - \phi_2^b) \tag{19b}$$

On account of the competitive character of adsorption from binary liquid mixture, each elementary act of adsorption is described by the *energy of adsorption* referred to one mole of adsorption sites,

$$E = E_2 - \frac{1}{r} E_1 \tag{20}$$

where E_j ($j = 1, 2$) is the energy of adsorption for a single molecule of the jth component (the potential energy difference between the lowest energy state of the molecule in the gas phase and its lowest energy state in the adsorbed phase).

When the molar partition functions of both components in the adsorbed phase, q_j^s, and in the bulk phase, q_j^b, are used directly to determine the related chemical potentials $\mu_j^{o,s}$ and $\mu_j^{o,b}$, the constant K in Eq. (17) becomes

$$K = \frac{E}{RT} + \ln\left[\frac{q_2^s}{q_2^b}\left(\frac{q_1^b}{q_1^s}\right)^{\frac{1}{r}}\right] = \frac{E}{RT} + K_o \tag{21a}$$

From the relationship between $\Delta_{\text{imm}} H_j^o$ and $\Delta_{\text{ads}} H_j^o$ given by Eq. (3), the first two terms on the right-hand side of Eq. (13) may to a good approximation be expressed in the form

$$\begin{aligned}
x_2^s \cdot \Delta_{\text{imm}} H_2^o + x_1^s \cdot \Delta_{\text{imm}} H_1^o &\approx \Delta_{\text{imm}} H_1^o - n^a \phi_2^s E + n^a \phi_2^s RT^2 \left(\frac{\partial K_o}{\partial T}\right)_v \\
&= \Delta_{\text{imm}} H_1^o - n^a \phi_2^s (E - \lambda)
\end{aligned} \tag{21b}$$

In general, solid surfaces are to some extent heterogeneous [30,68]. To include heterogeneity in the formal calculation of the properties of the adsorbed layer, the adsorption energy distribution must be known; namely,

$$\chi = \chi(E) \qquad E \varepsilon \Omega \tag{22}$$

where Ω is the set of the allowed values of E. In this case, all macroscopic quantities of the adsorbed layer should be averaged over all accessible E values. The following analytical expressions for χ have been used in theoretical treatments of adsorption [24–44].
(1) Dirac Δ distribution for an ideal solid surface (i.e., homogeneous surface)

$$\chi_{\text{DI}}(E) = \delta(E - E_o) \qquad \Omega = \{E_o\} \tag{22a}$$

(2) Gaussian (GS) distribution

$$\chi_{\text{GS}}(E) = \frac{1}{\sqrt{2\pi(D_{\text{GS}})^2}} \exp\left[-\frac{(E - E_o)^2}{2(D_{\text{GS}})^2}\right] \qquad \Omega = (-\infty, +\infty) \tag{22b}$$

(3) Gaussian-like (QG) distribution

$$\chi_{QG}(E) = \frac{1}{D_{QG}} \frac{\exp\left[\dfrac{E-E_o}{D_{QG}}\right]}{\left\{1+\exp\left[\dfrac{(E-E_o)}{D_{QG}}\right]\right\}^2} \qquad \Omega = (-\infty, +\infty) \qquad (22c)$$

(4) Gamma (GA) distribution

$$\chi_{GA}(E) = \frac{(E-E_1)^n}{n!(D_{GA})^{n+1}} \exp\left[-\frac{(E-E_1)}{D_{GA}}\right] \qquad \Omega = [E_1, +\infty) \qquad (22d)$$

For the special case of $n=0$, this reduces to the exponential distribution function. (5) Weibull (WB) distribution ($n \geq 2$)

$$\chi_{WB}(E) = \frac{n(E-E_1)^{n-1}}{D_{WB}} \exp\left[-\frac{(E-E_1)^n}{D_{WB}}\right] \qquad \Omega = [E_1, +\infty) \qquad (22e)$$

For the special case of $n=2$, this reduces to the Dubinin–Radushkevich distribution function.

In the above equations, E_0 is the most probable energy of adsorption and E_1 is the smallest physical value of E for a given adsorption system. Parameters D and n (n is integer) determine the shape of the distribution function. The smaller the parameter D, the narrower is the distribution function. At the limit of $D \to 0$, all the distribution functions (with the sole exception of the gamma distribution for $n > 0$) reduce to the Dirac δ distribution.

An additional important aspect of the adsorption system to consider along with surface heterogeneity is the spatial arrangement of adsorption sites characterized by the same value of energy E. Two extreme cases are usually compared in theoretical studies [24–44]. These are patchwise and random topographical models of heterogeneous surface.

The *patchwise* model allows a natural extension of the theory developed for an ideal solid surface to adsorbents composed of a number of unisorptic areas (patches), which, for simplicity, are considered as being independent of each other. At adsorption equilibrium, the adsorbed molecules of a given species on each patch will have the same chemical potential. The surface concentrations to achieve this equality will be different on different patches. Therefore, the potential of average force acting on a molecule confined to a certain patch depends only on the local surface concentration for this patch.

The combination of Eq. (17) with Eq. (21a) and the Flory–Huggins approximation provides the relationship $\phi_2^s = \phi_2^s(E, \phi_2^b, T, P)$, which describes the fraction of adsorption sites with a given value E occupied by molecules of component 2. The quantity ϕ_2^s is thus called the *local surface fraction* of this component. Since

$$\frac{\phi_2^s \exp[z \cdot l \cdot w^p(1-2\phi_2^s)]}{(1-\phi_2^s)^{\frac{1}{r}}} = \exp\left\{K_o + \frac{E}{RT} + \ln\left[\frac{\phi_2^b}{(1-\phi_2^b)^{\frac{1}{r}}}\right]\right.$$
$$\left. + z(w^b - m \cdot w^v)(1-2\phi_2^b)\right\} \qquad (23a)$$

the local surface fraction is an implicit function of the energy of adsorption for the patchwise topographical model.

All macroscopic quantities relating to the overall adsorbed phase are obtained by averaging the related local functions over all accessible values of adsorption energy E. The *global surface fraction* of component 2 is

$$\langle \phi_2^s \rangle_P = \int_\Omega \phi_2^s(E) \cdot \chi(E) \, dE \tag{23b}$$

where the local surface fraction $\phi_2^s = \phi_2^s(E)$ is given by Eq. (23a). The subscript P refers to the patchwise model of surface topography. The global surface activity coefficients and the global enthalpy of mixing per one mole of adsorption sites can now be written as

$$\langle f_1^s \rangle_P = \exp\left\{ r \cdot z \left[l \cdot w^p (\langle \phi_2^s \rangle_P)^2 + m(w^v - w^b)\phi_2^b + m \cdot w^b (\phi_2^b)^2 \right] \right\} \tag{23c}$$

$$\langle f_2^s \rangle_P = \exp\left\{ z \left[l \cdot w^p (1 - \langle \phi_2^s \rangle_P)^2 + m(w^v - w^b)(1 - \phi_2^b) + m \cdot w^b (1 - \phi_2^b)^2 \right] \right\} \tag{23d}$$

and

$$\langle \Delta_{\text{mix}} H_m^s \rangle_P = RT \cdot z \left\{ l \cdot w^p \int_\Omega [\phi_2^s (1 - \phi_2^s)](E) \cdot \chi(E) dE + m \cdot w^v \right.$$
$$\left. [\phi_2^b (1 - \langle \phi_2^s \rangle_P) + (1 - \phi_2^b)\langle \phi_2^s \rangle_P] - m \cdot w^b \phi_2^b (1 - \phi_2^b) \right\} \tag{23e}$$

For the random arrangement of equal-energy sites (i.e., *random* topographical model), the picture of the adsorbed phase is quite different. The local concentration of the surface mixture is the same throughout the whole adsorbed monolayer and therefore is equal to the average surface concentration. The potential of average force acting on any element localized on one adsorption site will be a function of the global surface fraction of component 2, which is defined as

$$\langle \phi_2^s \rangle_R = \int_\Omega \phi_2^s(E) \cdot \chi(E) \, dE \tag{24a}$$

where the subscript R refers to the random topographical model. The local surface fraction ϕ_2^s is not only a function of the adsorption energy E but also a function of the global surface fraction $\langle \phi_2^s \rangle_R$. This dependence is given by

$$\frac{\phi_2^s}{(1 - \phi_2^s)^{\frac{1}{r}}} = \exp\left\{ K_o + \frac{E}{RT} + \ln\left[\frac{\phi_2^b}{(1 - \phi_2^b)^{\frac{1}{r}}}\right] + z(w^b - m \cdot w^v) \right.$$
$$\left. (1 - 2\phi_2^b) - z \cdot l \cdot w^p (1 - 2\langle \phi_2^s \rangle_R) \right\} \tag{24b}$$

Adsorption at Heterogeneous Interfaces

The global surface activity coefficients and the global enthalpy of mixing per one mole of adsorption sites are defined by the expressions

$$\langle f_1^s \rangle_R = \exp\left\{r \cdot z\left[l \cdot w^p(\langle \phi_2^s \rangle_R)^2 + m(w^v - w^b)\phi_2^b + m \cdot w^b(\phi_2^b)^2\right]\right\} \tag{24c}$$

$$\langle f_2^s \rangle_R = \exp\left\{z\left[l \cdot w^p(1 - \langle \phi_2^s \rangle_R)^2 + m(w^v - w^b)(1 - \phi_2^b) + m \cdot w^b(1 - \phi_2^b)^2\right]\right\} \tag{24d}$$

and

$$\langle \Delta_{mix} H_m^s \rangle_R = RT \cdot z \Big\{ l \cdot w^p \langle \phi_2^s \rangle_R (1 - \langle \phi_2^s \rangle_R) + m \cdot w^v [\phi_2^b(1 - \langle \phi_2^s \rangle_R) \\ + (1 - \phi_2^b)\langle \phi_2^s \rangle_R] - m \cdot w^b \phi_2^b(1 - \phi_2^b) \Big\} \tag{24e}$$

Taking account of all the above relationships, the two fundamental observables for the solid–binary mixture interface, i.e., the surface excess of component 2 and the heat of immersion given by Eqs. (12) and (13), can be now written quite generally as

$$\Gamma_2^e = \frac{n^a}{M_s}\left\{ \langle \phi_2^s \rangle_T - \left[\langle \phi_2^s \rangle_T + \frac{1}{r}(1 - \langle \phi_2^s \rangle_T)\right] x_2^b \right\} \tag{25}$$

and

$$Q_w = Q_{w1}^o + \frac{n^a}{M_S} \int_\Omega E \cdot \phi_2^s(E) \cdot \chi(E) \, dE - \frac{n^a}{M_S} \lambda \langle \phi_2^s \rangle_T \\ + \frac{n^a}{M_S} RT \cdot z \cdot w^b \phi_2^b(1 - \phi_2^b) - \frac{n^a}{M_S} \langle \Delta_{mix} H_2^s \rangle_T \tag{26}$$

where the global surface fraction of component 2, $\langle \phi_2^s \rangle_T$, and the global enthalpy of mixing $\langle \Delta_{mix} H_m^s \rangle_T$, are given by the set of Eqs. (23) for the patchwise topographical model or by the set of Eqs. (24) for the random topographical model.

2. Effects of Surface Heterogeneity and Solution Nonideality on Adsorption

An illustrative model investigation can be carried out to show how various model parameters influence adsorption from binary mixtures on heterogeneous solid surfaces. For this purpose, the model surface excess and heat of immersion isotherms

$$n^E = n^E(x_2^b) = \frac{\Gamma_2^e M_s}{n^a} \qquad q_w = q_w(x_2^b) = \frac{(Q_w - Q_{w1}^o)}{n^a RT}\sqrt{M_S} \qquad T, P = \text{const} \tag{27}$$

are generated by using Eqs. (25) and (26). There are several parameters that must be taken into consideration for the meaningful interpretation of the data.

The bulk and adsorbed phase nonideality is characterized by three dimensionless interaction parameters: $W^p = z \cdot l \cdot w^p$, $W^v = z \cdot m \cdot w^v$, and $W^b = z \cdot w^b$. The unequal adsorbate sizes are reflected in parameter r. Surface heterogeneity is described by the type of distribution function Eq. (22) and by the model of surface topography (i.e., patchwise or random). In order to quantify the effect of heterogeneity on the model curves, the appropriate characteristics of distribution $\chi = \chi(E)$ should be considered. For comparative purposes, it is reasonable to use the expectation, ESP, and the standard deviation, DEV, of the random variable E with density function χ. Both variables are related to individual parameters n, D, E_0, and E_1 appearing in Eqs. (22). Therefore

$$\text{ESP}_{\text{DI}} = \frac{E_o}{RT} \qquad \text{DEV}_{\text{DI}} = 0 \qquad (28a)$$

$$\text{ESP}_{\text{GS}} = \frac{E_o}{RT} \qquad \text{DEV}_{\text{GS}} = \frac{D_{\text{GS}}}{RT} \qquad (28b)$$

$$\text{ESP}_{\text{QG}} = \frac{E_o}{RT} \qquad \text{DEV}_{\text{QG}} = \frac{\pi D_{\text{QG}}}{\sqrt{3}RT} \qquad (28c)$$

$$\text{ESP}_{\text{GA}} = \frac{E_1 + (n+1)D_{\text{GA}}}{RT} \qquad \text{DEV}_{\text{GA}} = \sqrt{n+1}\,\frac{D_{\text{GA}}}{RT} \qquad (28d)$$

$$\text{ESP}_{\text{WB}} = \frac{E_1}{RT} + \frac{\sqrt[n]{D_{\text{WB}}}}{RT}\frac{\Gamma(1/n)}{n} \qquad \text{DEV}_{\text{WB}} = \frac{\sqrt[n]{D_{\text{WB}}}}{RT}\cdot\sqrt{\frac{2}{n}\Gamma\!\left(\frac{2}{n}\right) - \frac{1}{n^2}\Gamma^2\!\left(\frac{1}{n}\right)} \qquad (28e)$$

where Γ is the gamma function. The deviation, DEV, is now a quantitative measure of the surface heterogeneity. Adsorbent heterogeneity increases as $\text{DEV} \to +\infty$. For an ideal solid surface, $\text{DEV} = 0$. The expectation, ESP, also has an important physical meaning. The value of ESP determines the position of the adsorption energy distribution along the energy axis and consequently the difference in affinity for the surface between the two components. This parameter reflects the degree to which component 2 is preferentially adsorbed at the solid–solution interface.

Many features of the monolayer adsorption model, described in the previous paragraph, are illustrated in Figs. 5–10. These examples have been selected to show that the theoretical model can mimic all the shapes of the experimental curves included in the classification systems presented in Figs. 3 and 4. More extensive parametric studies have been reported in Refs. 36, 37, 39, and 42.

Figures 5 and 6 show the effect of the bulk nonideality on the model curves relating to a homogeneous solid surface [Dirac δ distribution, Eq. (22a)] and an ideal surface mixture ($W^p = W^v = 0$) of adsorbates of similar sizes ($r = 1$). In Fig. 5 the affinity of component 2 for the solid surface is much greater than that of component 1 ($\text{ESP} = 2$). It is clear that the adsorption of component 2 is preferential over the whole concentration range and even strong bulk phase nonideality cannot reverse this trend. However, the degree of preferentiality is to some extent decreased by negative deviations from Raoult's law. This picture is different for a small value of ESP, as shown in Fig. 6. The model curves show a change in the preferentiality of adsorption depending on the sign and the value of the bulk interaction parameter W^b. It is worth noting here that the quasi-chemical approximation has also been analyzed [42]. It appears that the differences between both approximations are small and practically disappear for great values of the coordination number z.

The type of the adsorption energy distribution and the degree of heterogeneity are two important aspects of the adsorbent heterogeneity that should be analyzed, at first, in separation from the bulk and surface nonideality. When component 2 is more strongly adsorbed than component 1 (i.e., for great values of ESP) on surfaces with low or moderate heterogeneity, the detailed shape of the distribution function is not critical in determining the shape of the excess isotherm and the heat of immersion curve. The effect of the distribution type may be pronounced, if both mixture components are bound to the surface with comparable mean adsorption energies (i.e., for small values of ESP), as shown in Fig. 7. The shape of the model curves undergoes great changes as the adsorbent heterogeneity increases. Figure 8 illustrates such behavior of the adsorption system for the gamma distribution function, Eq. (22d).

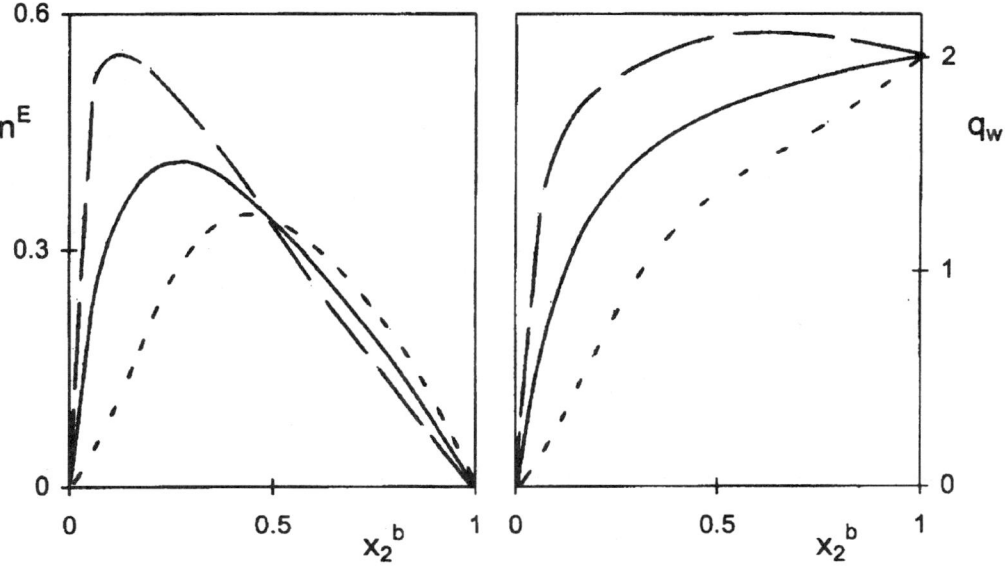

FIG. 5 Effect of the bulk interchange energy, W^b, on the model surface excess, n^E, and heat of immersion, q_w, isotherms generated by assuming an ideal surface mixture, similar adsorbate sizes, and a homogeneous adsorbent surface with ESP = 2: $W^b = 0$ (———); $W^b = 1.44$ (— —); $W^b = -1.44$ (- - - -).

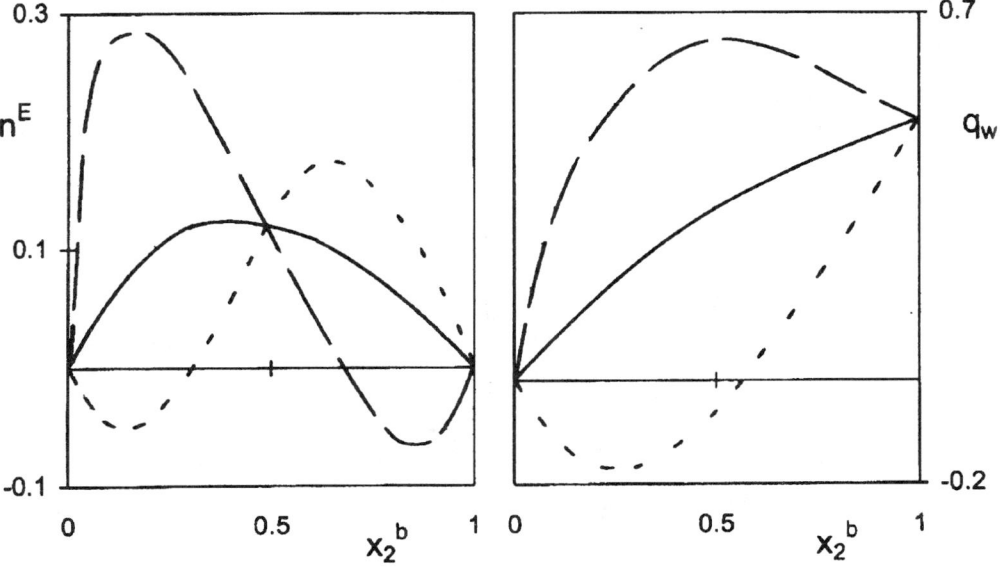

FIG. 6 Effect of the bulk interchange energy, W^b, on the model surface excess, n^E, and heat of immersion, q_w, isotherms generated by assuming an ideal surface mixture, similar adsorbate sizes, and a homogeneous adsorbent surface with ESP = 0.5: $W^b = 0$ (———); $W^b = 1.44$ (— —); $W^b = -1.44$ (- - - -).

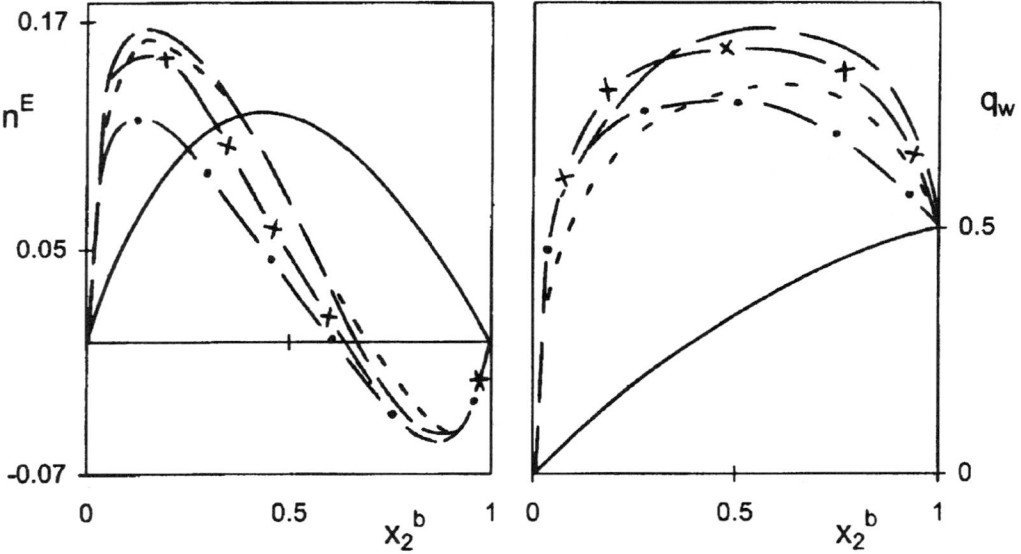

FIG. 7 Model surface excess, n^E, and heat of immersion, q_w, isotherms corresponding to the various distribution functions, similar adsorbate sizes, and ideal bulk and surface mixtures with ESP = 0.5 and DEV = 2: homogeneous surface (———); Gaussian distribution (— —); Gaussian-like (- - - - -); Dubinin-Radushkevich (—×—); exponential distribution (— • —).

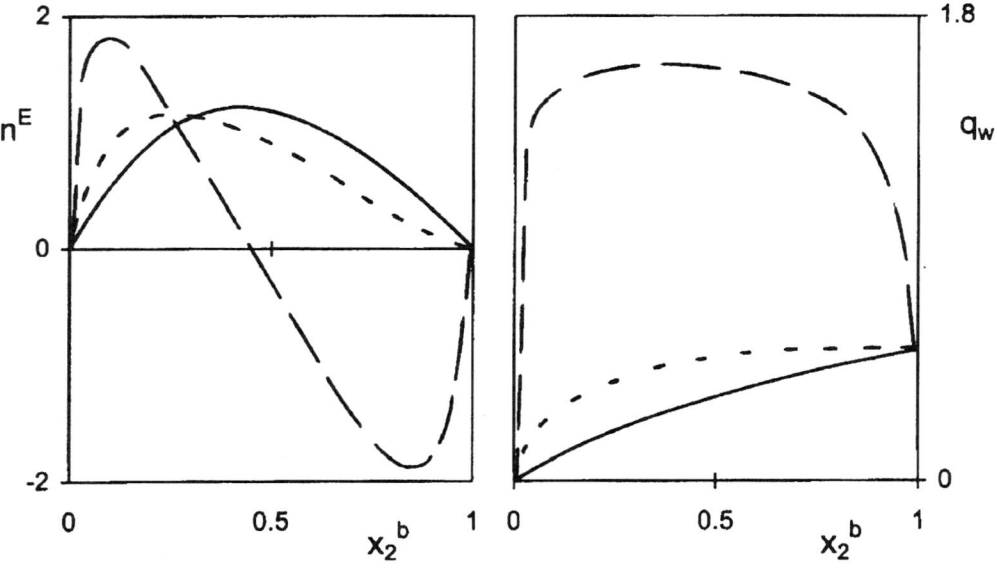

FIG. 8 Effect of surface heterogeneity on the model surface excess, n^E, and heat of immersion, q_w, isotherms generated by assuming ideal bulk and surface mixtures, similar adsorbates sizes, and gamma distribution with $n = 1$ and ESP = 0.5: homogeneous surface (———); DEV = 1 (- - - - -); DEV = 4 (— —).

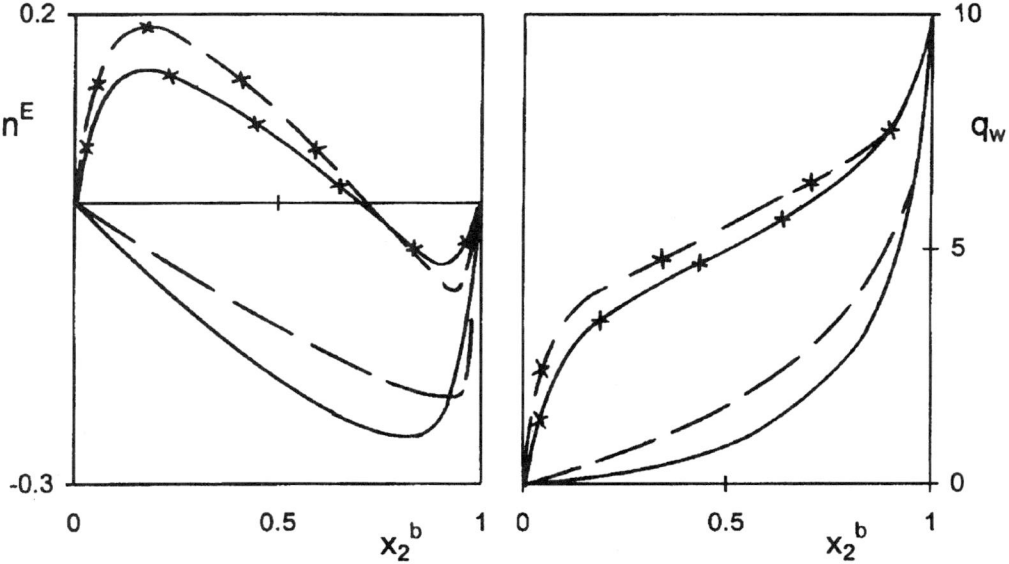

FIG. 9 Effect of the degree of surface heterogeneity and surface topography on the model surface excess, n^E, and heat of immersion, q_w, isotherms generated by assuming negative deviations from Raoult's law in the bulk and surface mixtures ($W^p = -1$, $W^v = -0.25$, and $W^b = -1$), unequal adsorbate sizes ($r = 2$), and Dubinin-Radushkevich distribution function with $E_1 = -5RT$: patchwise topography (———); random topography (— —); ESP = -2.5, DEV = 1.9 (simple lines); ESP = 0.06, DEV = 3.8 (lines with crosses).

The combination of surface heterogeneity and surface mixture nonideality leads to complexity in details, but some simple generalizations can be made. Both physical factors have a marked influence on the shape of the model curves. The effect of surface heterogeneity is to some extent compensated by negative deviations from Raoult's law in the surface mixture and intensified by positive deviations. There are only quantitative differences between the model curves based on the patchwise and random topographical models: the shape of the model curves remains essentially unaltered. Figures 9 and 10 present two sets of surface excess and heat of immersion isotherms evaluated for nonideal bulk and surface mixtures containing adsorbates of different sizes and two distribution functions Eqs. (22c) and (22e). Parameters ESP and DEV are clearly dominant factors. It should be noted that the heterogeneity effects on adsorption may be masked by a strong preferential adsorption of one of the mixture components. For the analysis of some experimental data, the choice between the homogeneous surface and the heterogeneous surface models will be indeed very difficult.

3. Experimental Adsorption Systems

The theoretical equations Eqs. (25) and (26) describing both observables depend on numerous parameters. One has to be extremely careful in designing and carrying through any data-fitting procedures that involve so many disposable variables. It seems probable that the fluctuations in the various variables are correlated when fitting too many of them at once. It is necessary to assess the sensitivity of fitting to variations in each variable. Finally, when values of some parameters can be evaluated from independent information,

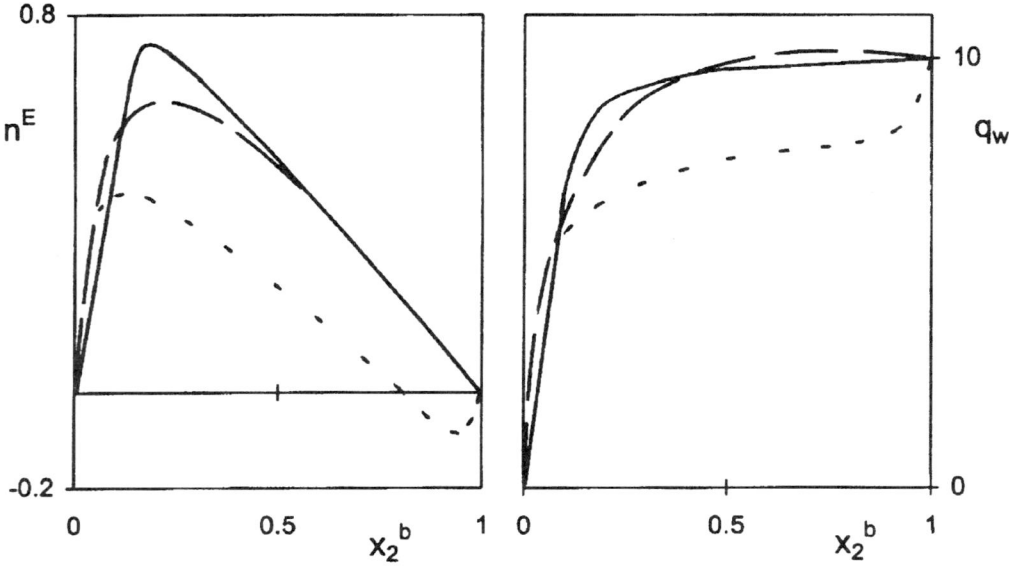

FIG. 10 Effect of the degree of surface heterogeneity on the model surface excess, n^E, and heat of immersion, q_w, isotherms generated by assuming positive deviations from Raoult's law in the bulk and surface mixtures ($W^p = 1$, $W^v = 0.25$, and $W^b = 1$), unequal adsorbate sizes ($r = 2$), patchwise topography, and gaussian-like distribution function with ESP = 2: homogeneous surface (———); DEV = 0.9 (— —); DEV = 3.6 (- - - - -).

they should be best inserted rather than left floating. Another difficulty is caused by the way in which the solution of the integral equations Eqs. (25) and (26) is obtained. Only exact methods would be free from any assumptions, which may additionally limit the model validity. However, this can be done only in a few favorable cases. The use of numerical computer-aided procedures is time-consuming and may be a source of round-off errors in computation. The best research strategy is to approximate the surface excess and heat of immersion by relatively simple analytical formulas that are accurate enough to make a quantitative analysis of the experimental data feasible. The solution is then obtained with considerably less effort. The detailed treatment of the problem can be found elsewhere [35–42].

The precise measurements of the amount adsorbed and the related heat of immersion over the entire composition range are rare in the literature. Therefore it is difficult to find experimental data suitable for making a quantitative assessment of the model. The experimental results of adsorption and calorimetric measurements reported for the mixture cyclohexane (1) + benzene (2) adsorbed on silica gel at 273.15 K and for the mixture n-heptane (1) + benzene (2) adsorbed on silica gel at 298.15 K fulfil all the formal requirements of the theoretical model [39]. The explicit values of the bulk mixture parameters can be estimated from independent literature data [68]. Cyclohexane and benzene have similar molar volumes so that $r = 1$. For the n-heptane + benzene mixture, benzene is the smaller molecular species and thus r is an adjustable parameter. In both cases, benzene is preferentially adsorbed on silica gel, and the shape of the experimental adsorption and heat isotherms may be explained on the basis of the monolayer adsorption model. Silica gel is a common heterogeneous substrate, and the concept of surface heterogeneity should be taken into consideration when fitting theoretical expressions to experimental data.

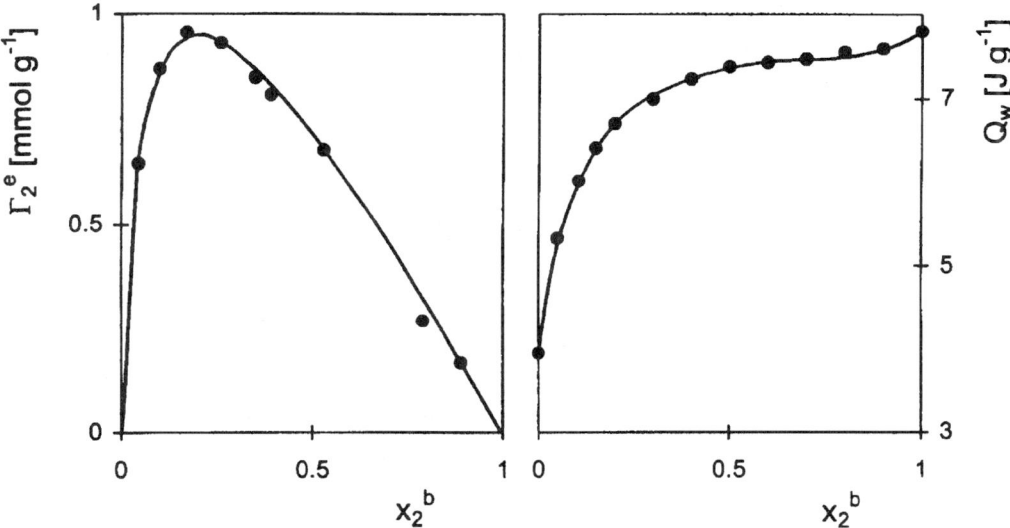

FIG. 11 Comparison of the models surface excess, Γ_2^e, and heat of immersion, Q_w, isotherms with the experimental data for the system [cyclohexane (1) + benzene (2) on silica gel] at 273.15 K. The experimental data are shown by filled circles. The solid lines represent the model isotherms for the best-fit parameters: $n^a(M_S)^{-1} = 3.1$ mmol g^{-1}; $r = 1$; ESP$_{QG} = 2.6$; DEV$_{QG} = 3.5$; $W^p = -1.0$; $W^v = 1.9$.

The best goodness-of-fit was obtained for gaussian-like distribution Eq. (22c) of the adsorption energy and the random configuration of equal-energy sites. The agreement between theory and experiment is shown in Figs. 11 and 12. It can be seen that the model curves fit the experimental data very well.

The mean values of the adsorption energy are very similar for both adsorption systems. They confirm the strong preferential adsorption of benzene on the surface of silica gel. Surface heterogeneity is higher for the cyclohexane + benzene mixture, which may be ascribed to differences in the adsorbate size and in the molecular geometry.

There is an anisotropy of mixing in the adsorbed phase. The best-fit values of interchange energy parameters suggest the negative deviations from the ideality within the adsorbed monolayer and a worse vertical mixing. The bulk and surface activity coefficients of both components are compared in Fig. 13. The picture of the nonideality is different in the surface mixture. This result seems to be surprising in view of the fairly ideal behavior of the appropriate bulk liquid mixtures. The assumption that the adsorbed phase is more ideal than the bulk solution has been exploited by many authors [70–73]. In the theoretical model described by Eqs. (25) and (26), the surface nonideality is a combined effect of molecular interactions and surface heterogeneity. For numerous adsorption systems, the latter can be the main factor governing the behavior of the solid–binary mixture interface. In consequence, the models of adsorption from binary liquid mixtures based on the assumption of an ideal surface mixture formed on a strongly heterogeneous surface may sometimes lead to more realistic results than those developed by neglecting the concept of surface heterogeneity.

Figure 14 shows experimental data for the adsorption of methanol (1) + benzene (2) by active carbon Carbo Medicinalis No. 5 at 298.15 K. Note that component 2 has been chosen to have the larger molecules in the mixture so that the current value of the adsorb-

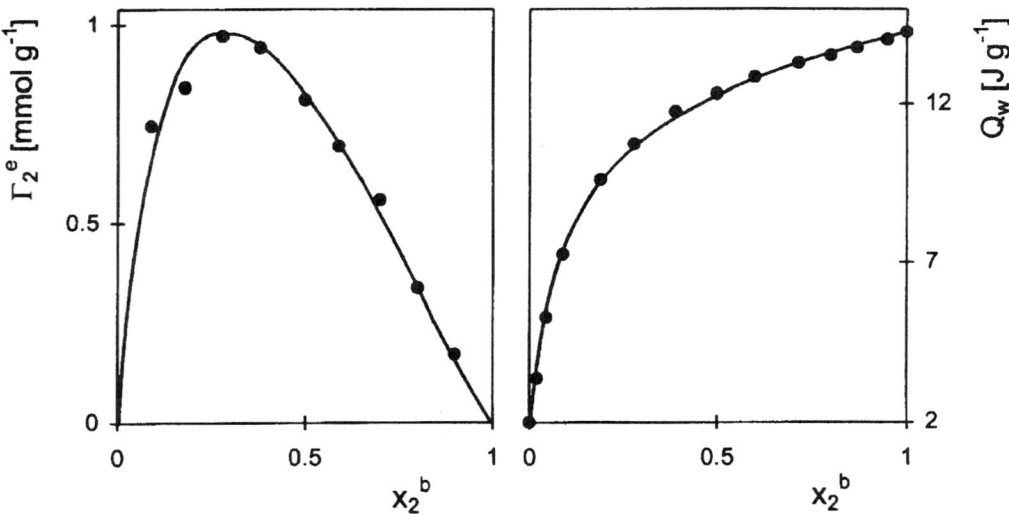

FIG. 12 Comparison of the model surface excess, Γ_2^e, and heat of immersion, Q_w, isotherms with the experimental data for the system [n-heptane (1) + benzene (2) on silica gel] at 298.15 K. The experimental data are shown by filled circles. The solid lines represent the model isotherms for the best-fit parameters: $n^a(M_S)^{-1} = 2.6$ mmol g^{-1}; $r = 1.4$; ESP$_{QG} = 2.1$; DEV$_{QG} = 1.8$; $W^p = -1.8$; 8; $W^v = -0.1$.

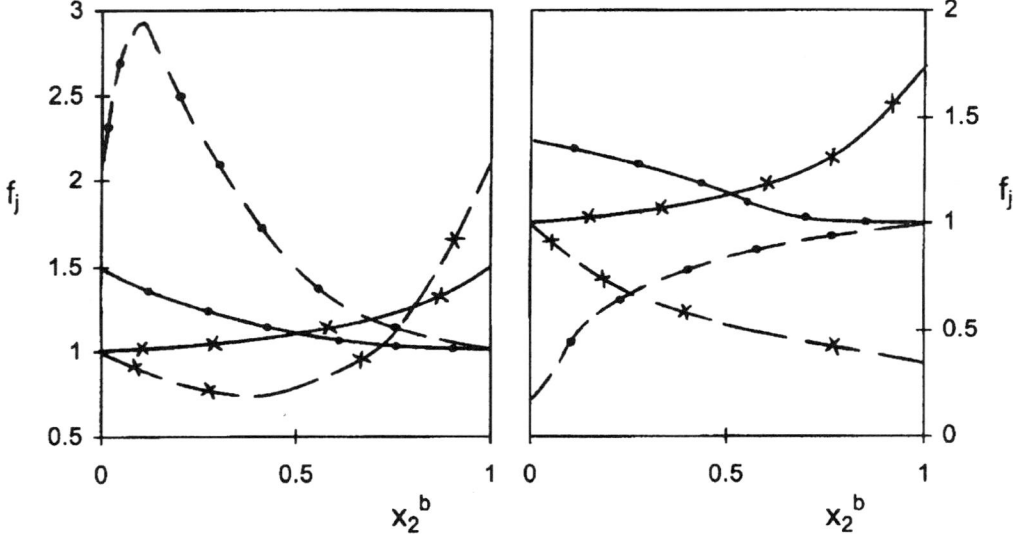

FIG. 13 Comparison between bulk (———) and surface (— —) activity coefficients for the system [cyclohexane (1) + benzene (2) on silica gel] at 273.15 K (left panel) and for the system [n-heptane (1) + benzene (2) on silica gel] at 298.15 K (right panel). The crosses correspond to component 1 and the circles refer to component 2.

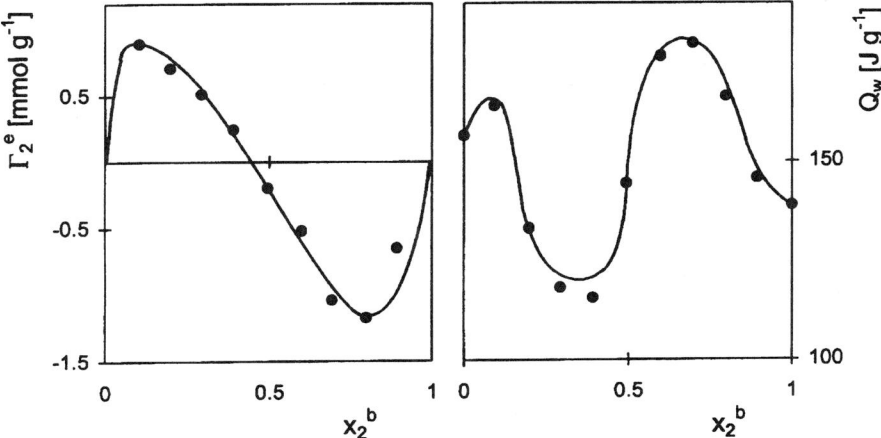

FIG. 14 Comparison of the model surface excess, Γ_2^e, and heat of immersion, Q_w, isotherms with the experimental data for the system [methanol (1) + benzene (2) on Carbo Medicinalis No. 5] at 298.15 K. The experimental data are shown by filled circles. The solid lines represent the model isotherms for the best-fit parameters: $n^a(M_S)^{-1} = 15$ mmol g^{-1}; $r = 0.5$; ESP$_{WB} = 0.5$; DEV$_{WB} = 0.8$; $n = 2$.

ate size ratio r is the reciprocal of the parameter defined in Eqs. 14. Benzene is preferentially adsorbed only over a part of the mixture composition range. At $x_2^b \approx 0.44$, the surface layer has the average concentration equal to x_2^b, and $\Gamma_2^e = 0$. By analogy with the bulk mixtures, the term *surface azeotrope* can be applied here. Above this azeotropic composition, methanol becomes the preferentially adsorbed component.

On comparing the enthalpies of immersion of the active carbon in pure liquid components ($Q_{w1}^0 = 156$ J g^{-1} and $Q_{w2}^0 = 139$ J g^{-1}), we expect the mean energy of adsorption from a binary mixture to be fairly low. As a consequence, heterogeneity effects may manifest themselves clearly in the surface excess and heat of immersion. The surfaces of active carbons consist primarily of graphite-like basal planes, which are relatively uniform in nature [74]. Various substances are adsorbed on this surface mainly under the action of van der Waals forces, though interactions with π-electrons are expected to occur as well. The presence of nonideal aromatic rings in the graphite crystal structure and oxygen functionalities located at edge defects endow heterogeneity of the active carbon surface. The evolution of the heat of immersion with the bulk mole fraction of benzene, illustrated in Fig. 14, indicates that the behavior of the adsorption system is indeed dominated by surface heterogeneity.

Theoretical expressions for Γ_2^e and Q_w, obtained from Eqs. (25) and (26) by assuming an ideal solid surface, completely failed to fit the results of calorimetric measurements. In contrast, the heterogeneous surface model with Dubinin–Radushkevich distribution of the adsorption energy and $W^p = W^v = 0$ gave a reasonable fit to both types of experimental data. In Fig. 14 the related theoretical curves are shown by solid lines.

Examination of the present example provides a convincing argument for the high sensitivity of the thermal effect of adsorption to the nature of the solid–solution interface. To illustrate better this observation, the surface heterogeneity component of the heat of

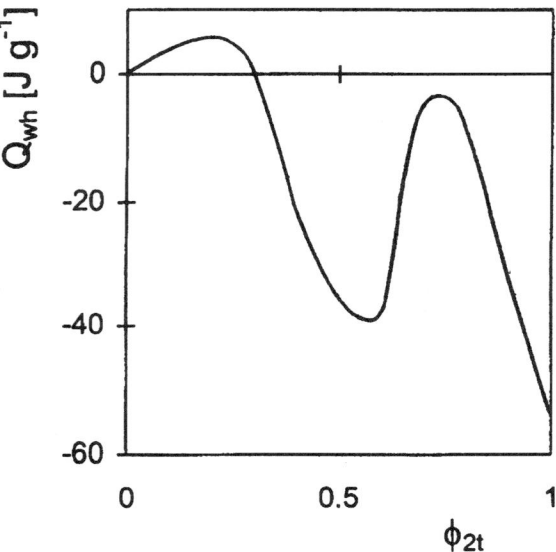

FIG. 15 Contribution to the heat of immersion arising from the surface heterogeneity plotted as a function of the global surface fraction of component 2, $\phi_{2t} = \langle \phi_2^s \rangle$, for the system [methanol (1) + benzene (2) on Carbo Medicinalis No. 5] at 298.15 K.

immersion, Q_{wh}, has been calculated and is presented in Fig. 15 as a function of the global surface fraction of benzene, $\langle \phi_2^s \rangle$. It is evident that the surface heterogeneity makes a significant contribution to the heat of immersion.

One important conclusion that emerges from the analysis of the adsorption systems studied is that much more reliable estimates for the model parameters are obtained while fitting simultaneously Eqs. (25) and (26) to two types of experimental data.

II. SIMPLE ION ADSORPTION AT OXIDE/ELECTROLYTE INTERFACE

Studies of the surface heterogeneity of oxide surfaces have a long history in the area of adsorption and catalysis. In catalysis, the surface defects in oxide surfaces are believed to be the catalytic centers for various catalytic reactions. For many other reactions, the variation of the acid–base strength on the oxide adsorption sites is of crucial importance. So it is no surprise that the surface heterogeneity of oxides was the subject of dozens of papers published by scientists working on catalysis.

Oxides are also very important adsorbents used in many technological processes employing adsorption phenomena occurring at the gas/solid interface. Here, the fundamental studies of surface heterogeneity of oxides are probably most advanced and they are included in the book of Rudziński and Everett [68]. Papers discussing surface heterogeneity effects in adsorption at the oxide/nonelectrolyte interface were also published by Jaroniec and Madey [30]. However, this is the oxide/electrolyte interface, which is most important for life, science, and technology. This is also the largest natural interface existing on our planet. So it is no surprise that adsorption at the oxide/electrolyte interface was the subject of hundreds of papers, several reviews, and a number of monographs.

Many of them were devoted to a theoretical description of adsorption phenomena occurring at this interface. Of course, the most pronounced of them is the formation of the electrical double layer at the oxide/electrolyte interface. Except for a few, all the theoretical descriptions were based on a physical model of an energetically homogeneous oxide surface. In terms of the most popular surface complexation approach, it means that all surface oxygens capable of forming surface complexes have identical adsorption properties. This is striking in view of the well-established image of the heterogeneity of really existing oxide surfaces. This is obvious for the amorphous oxide sample. However, even the interior of an oxide exhibits a regular crystallographic structure; its surface shows usually much less organization. Bakaev's computer simulations [75–77] of oxide surfaces suggest that even in the case of oxides having a well-defined bulk crystal structure, the degree of surface disorder may be larger than is generally believed.

The next surprise is that various theoretical descriptions neglecting the surface heterogeneity of oxides led to a fairly good agreement with experiment. This fact bothered Sposito [78] who wrote, "the surface complexation models are, in a sense, too successful, i.e., several different models can represent the same set of adsorption data equally well with corresponding chemical parameters in the model taking quite different values." Our publications [79,80] seem to provide an explanation for this striking situation.

Namely, the various theoretical approaches were tested by fitting the experimental titration curves—the most commonly measured adsorption characteristics. This surface charge isotherm is a composite one, so there are many factors affecting its behavior. As is to be expected in such a case, there will be also several compensating effects. Some of the physical parameters may be highly correlated. In other words, the surface charge isotherms, i.e., the titration curves, will not be sensitive to a physical model underlying the corresponding computer fit. We have demonstrated this in one of our publications [80]. Thus the theoretical and corresponding numerical analyses of titration curves do not create much chance for distinguishing between a proper adsorption model and a less adequate one.

In the instances where single (individual) isotherms of surface complexes could be measured, the failure to fit them by isotherm equations, corresponding to a model of a homogeneous oxide surface, was reported first. These were the studies of trace adsorption of such ions as Cu^{2+}, Zn^{2+}, Cd^{2+}, and Pb^{2+} carried out as early as the beginning of the 1970s. In those and all later studies, only the isotherm equations, corresponding to a model of a heterogeneous surface, were applied to correlate the experimental isotherms. As the end of the 1970s, Garcia-Miragaya and Page [81,82] and Street et al. [83] reported a successful correlation by the Freundlich equation of the data of trace Cd^{2+} adsorption by both clay minerals and soils. Benjamin and Leckie [84] found the same for the trace adsorption of Cu^{2+}, Zn^{2+}, Cd^{2+}, and Pb^{2+} onto amorphous iron oxyhydride. The use of the Freundlich equation was also suggested in works by Sposito [85,86.]

In the theories of gas adsorption the applicability of the Freundlich equation was long ago associated with the energetic heterogeneity of the adsorption sites on the actual solid surfaces. It was also known that the Freundlich equation is a simplified form of a more general isotherm equation that is now commonly called the Langmuir–Freundlich isotherm. Thus in 1980 Sposito [86] suggested that it was high time it were used also in the case of ion adsorption at the water/oxide interface.

Benjamin and Leckie [84] were among the first who initiated the studies of the surface heterogeneity effects in 1981. They reported that the adsorption of Me^{2+} metal ions onto oxyferrihydride could be described only by assuming a large dispersion of adsorption site affinities. Two years later Kinniburgh et al. [87] tried to correlate such adsorption isotherms by using other empirical equations employed earlier to correlate experimental adsorption isotherms for single gas adsorption onto heterogeneous solid surfaces. Theoretical studies of surface heterogeneity effects on ion adsorption within the electrical double layer were much advanced by Koopal, Van Riemsdijk and coworkers [88–92]. Rudziński et al. [93] developed the theory reproducing the behavior of the log–log plots of the experimental adsorption isotherms of bivalent ions on oxides at low surface concentrations. These experimental adsorption isotherms show a transition from a linear log–log (Henry's) plot with a tangent equal to unity to a Freundlich log–log plot with a tangent smaller than unity.

Electrokinetic curves are also related merely to one kind of adsorption isotherms. On the acidic side, this is the adsorption isotherm of the doubly protonated (positively charged) surface oxygens. On the basic side, it is the adsorption isotherm of cations of the basic electrolyte that determines the amount of the free (negatively charged) surface oxygens. The electrokinetic properties are related to the balance between these positively and negatively charged surface sites. So it is no surprise that striking differences were usually observed between the theoretically predicted and the experimentally monitored electrokinetic curves. At the same time, a usually good agreement could be seen between theoretical and experimental titration curves (for the same physical model and set of parameters).

Radiometric methods making measurements of single adsorption isotherms possible have provided a large body of experimental evidence for the necessity of improving the existing theoretical models and descriptions. In one of our recent publications we have shown that introducing the concept of a heterogeneous oxide surface substantially improves the agreement between theory and experiment [80].

Nevertheless, the obtained agreement was still far from being satisfactory, considering the level of accuracy achieved in the radiometric experiment. At the same time, as usual, it was no problem to fit the corresponding titration curve, obtained in the classical titration experiment. At that time we had no particular explanation for this.

Of all the models published so far, the most frequently used is the so-called triple layer model (TLM) [94–98] along with the 2-pK charging mechanism. All the assumptions accompanying the construction of the TLM were discussed in detail in the paper by Robertson and Leckie [99]. Just recently, TLM was subjected to rigorous analysis by Sahai and Sverjensky taking into account many experimental systems [100–102].

The modifications of TLM published in the literature assumed different directions. One group of scientists has challenged the charging mechanism of the surface oxygens. They invoke Pauling's principle [103] of local neutralization of charge to show that the protonated surface oxygens will, in general, have fractional values of charge. This hypothesis is commonly called the 1-pK charging mechanism [104]. Moreover, they argue that various surface oxygens may have different 1-pK charging mechanisms, depending on their local coordination to cations. That model is known in the literature as the multisite complexation model (MUSIC) of the oxide electrolyte interface [105–108].

Another group of scientists has accepted the classical 2-pK charging mechanism but emphasized the importance of a small degree of crystallographic organization of the surface oxygens, compared to the situation in the interior of an oxide crystal. That smaller degree of surface organization should lead to a different status of the surface oxygens, resulting in what is called "surface energetic heterogeneity" [80,87–89,109–111].

The third refinement of the TLM is the four-layer model (FLM). Its idea was introduced to the literature by Barrow et al. [112]. A new layer (the fourth one, as the name indicates, but situated as the second, next to the surface layer where protons are adsorbed) was reserved first for the bivalent metal ions or anions of multiproton oxyacids. Cations and anions of basic electrolyte were still placed in the same layer as in the TLM. Another kind of FLM was launched by Bousse et al. [113], who presented in their paper a diagram of FLM, in which there are only ions of basic electrolyte and the potential-determining ions H^+. They argued that anions and cations of the basic electrolyte are not located in the same layer (as in TLM) but in two separate ones. A rigorous thermodynamic description based on this physical model, and providing theoretical expressions for all the experimentally measured physicochemical quantities, has been recently published by Charmas et al. [114–117].

In this subchapter we want to show our results obtained using various modified TLMs with the 2-pK charging mechanism.

A. Isotherm of Adsorption of Ions Forming Different Surface Complexes

1. Triple-Layer Complexation Model

The starting point of our consideration will be the triple-layer model proposed by Yates et al. [94] and Davis et al. [95,96]. Below, we are going to repeat briefly the principles and definitions that will be necessary for our further consideration. When a metal oxide is brought into contact with an electrolyte, the outermost surface oxygens adsorb one or two protons, a cation, or an aggregate composed of two protons and an anion. In this way various surface complexes are formed. Thus we consider the dissociation reactions of the surface proton and the coadsorption of anions A^- and cations C^+ as the surface reactions that cause the formation of the following surface complexes:

$$SOH_2^+ \xrightleftharpoons{K_{a1}^{int}} SOH^0 + H^+ \tag{29a}$$

$$SOH^0 \xrightleftharpoons{K_{a2}^{int}} SO^- + H^+ \tag{29b}$$

$$SOH_2^+ A^- \xrightleftharpoons{{}^*K_A^{int}} SOH^0 + H^+ + A^- \tag{29c}$$

$$SOH^0 + C^+ \xrightleftharpoons{{}^*K_C^{int}} SO^- C^+ + H^+ \tag{29d}$$

The equilibrium constants K_{a1}^{int}, K_{a2}^{int}, $^*K_A^{int}$, and $^*K_C^{int}$ of the above reactions were defined in the literature [95,96].

For our purpose, it will be useful to consider also the following equivalent reactions:

$$SOH^0 \longleftrightarrow SO^- + H^+ \tag{30a}$$
$$SOH_2^+ \longleftrightarrow SO^- + 2H^+ \tag{30b}$$
$$SO^- C^+ \longleftrightarrow SO^- + C^+ \tag{30c}$$
$$SOH_2^+ A^- \longleftrightarrow SO^- + 2H^+ + A^- \tag{30d}$$

for which the equilibrium equations are the following [79]:

$$K_{a2}^{int} \exp\left\{\frac{e\psi_0}{kT}\right\} = \frac{(a_H)\theta_-}{\theta_0} \tag{31a}$$

$$K_{a1}^{int} \cdot K_{a2}^{int} \exp\left\{\frac{2e\psi_0}{kT}\right\} = \frac{(a_H)^2\theta_-}{\theta_+} \tag{31b}$$

$$\frac{K_{a2}^{int}}{*K_C^{int}} \exp\left\{\frac{e\psi_\beta}{kT}\right\} = \frac{(a_C)\theta_-}{\theta_C} \tag{31c}$$

$$K_{a2}^{int} \cdot *K_A^{int} \exp\left\{\frac{e(\psi_0 - \psi_\beta)}{kT}\right\} = \frac{(a_H)^2(a_A)\theta_-}{\theta_A} \tag{31d}$$

when we introduce the notation: $\theta_0 = [SOH^0]/N_s$, $\theta_+ = [SOH_2^+]/N_s$, $\theta_A = [SOH_2^+A^-]/N_s$, $\theta_C = [SO^-C^+]/N_s$, $\theta_- = [SO^-]/N_s = 1 - \sum_i \theta_i$ ($i = 0, +, A, C$), and

$$N_s = [SO^-] + [SOH^0] + [SOH_2^+] + [SO^-C^+] + [SOH_2^+A^-] \tag{32}$$

In Eqs. (31a–d), a_H is the proton activity in the equilibrium bulk phase, and a_A and a_C are the bulk activities of anion and cation respectively. Further, ψ_0 is the surface potential and ψ_β is the mean potential at the plane of specifically adsorbed counterions.

The surface charge δ_0 is defined as [79]

$$\delta_0 = B_s[\theta_+ + \theta_A - \theta_- - \theta_C] \quad \text{and} \quad B_s = e \cdot N_s \tag{33}$$

where N_s is the surface sites density (sites/m^2).

The relationships between the capacitances, the potentials, and the charges within the individual electric layers were presented in our papers Refs. 79, 80, and 93.

The nonlinear equation system Eq. (31) can be rewritten in the Langmuir-like form [79]

$$\theta_i = \frac{K_i f_i}{1 + \Sigma_i K_i f_i} \quad i = 0, +, A, C \tag{34}$$

where

$$K_0 = \frac{1}{K_{a2}^{int}} \quad K_+ = \frac{1}{K_{a1}^{int} K_{a2}^{int}} \quad K_C = \frac{*K_C^{int}}{K_{a2}^{int}} \quad K_A = \frac{1}{K_{a2}^{int} \cdot *K_A^{int}} \tag{35}$$

and where f_i ($i = 0, +, A, C$) are the following functions of proton and salt concentrations:

$$f_0 = \exp\left\{-\frac{e\psi_0}{kT} - 2.3\text{pH}\right\} \quad f_+ = f_0^2 \tag{36a, b}$$

$$f_C = a_C \exp\left\{-\frac{e\psi_0}{kT} + \frac{e\delta_0}{kT c_1}\right\} \tag{36c}$$

$$f_A = a_A \exp\left\{-\frac{e\psi_0}{kT} - \frac{e\delta_0}{kT c_1} - 4.6\text{pH}\right\} \tag{36d}$$

where c_1 is the first integral capacitance.

To express ψ_0 (pH) dependence, which occurs in the equations for the individual adsorption isotherms, we accept here the relation used by Yates et al. [94], by Bousse et al. [118] and by Van den Vlekkert et al. [119],

$$2.303(\text{PZC} - \text{pH}) = \frac{e\psi_0}{kT} + \sinh^{-1}\left(\frac{e\psi_0}{\beta kT}\right) \tag{37}$$

where β is given by

$$\beta = \frac{2e^2 N_s}{c_t kT}\left(\frac{K_{a2}^{\text{int}}}{K_{a1}^{\text{int}}}\right)^{\frac{1}{2}} \tag{38}$$

and where c_t is the linear capacitance of the double electrical layer that can be calculated in the way proposed by Bousse et al. [118,119].

The way of solving the set of equations Eqs. (34–36) for TLM, to obtain individual adsorption isotherms θ_i's and the surface charge δ_0 as a function of pH, was shown in our previous papers [79,80,93].

The experimental titration curves corresponding to different concentrations of the basic (inert) electrolyte very often have a common intersection point (CIP) in the point of zero charge (PZC), where $\text{PZC} = \text{pH}_{\delta_0=0,\psi_0=0} = -\log H$, and where H is the activity of protons in the bulk solution at PZC. It means that the PZC value for a given system of oxide and electrolyte does not practically depend on salt concentration in the bulk solution. We were the first to draw readers' attention to the fact that such independence of PZC of salt concentration can be formally expressed as the set of two equations when we assume that $a = a_C = a_A$ [79]:

$$\delta_0(\text{pH} = \text{PZC}) = 0 \quad \text{and} \quad \frac{\partial \delta_0(\text{pH} = \text{PZC})}{\partial a} = 0 \tag{39}$$

The above formal criterion applied for TLM or for our modifications of TLM presented in this subchapter decreases by two the number of equilibrium constants K_{a1}^{int}, K_{a2}^{int}, $^*K_A^{\text{int}}$, and $^*K_C^{\text{int}}$. Such interrelations are important because they reduce the number of the independent equilibrium constants (best-fit parameters) determined from fitting the suitable experimental data.

The interrelations between the equilibrium constants inhered in TLM have been known in the literature for a long time [95,96]. They are easy to obtain without the criterion of Eq. (39):

$$H^2 = K_{a1}^{\text{int}} \cdot K_{a2}^{\text{int}} \quad \text{and} \quad H^2 = {}^*K_C^{\text{int}} \cdot {}^*K_A^{\text{int}} \tag{40a, b}$$

Their logarithmic forms are

$$\text{PZC} = \frac{1}{2}(\text{p}K_{a1}^{\text{int}} + \text{p}K_{a2}^{\text{int}}) \tag{41a}$$

$$\text{PZC} = \frac{1}{2}(\text{p}^*K_C^{\text{int}} + \text{p}^*K_A^{\text{int}}) \tag{41b}$$

where $\text{p}K_{ai}^{\text{int}} = -\log K_{ai}^{\text{int}}$, $i=1, 2$, and $\text{p}^*K_i^{\text{int}} = -\log {}^*K_i^{\text{int}}$, $i = \text{C, A}$. The same interrelations of Eq. (40) are obtained by applying the criterion of Eq. (39). However, for more complicated models discussed below, the interrelations can be obtained using only the criterion of Eq. (39) because of the complex form of $\delta_0(\text{pH} = \text{PZC}) = 0$ equation.

2. Modifications of the Triple-Layer Complexation Model Taking into Account the Factor of Energetic Heterogeneity of the Surface

In a previous paper, Ref. 109, we showed the way of rewriting the "intrinsic" constant K_i's,

$$K_i = K_i' \exp\left\{\frac{\varepsilon_i}{kT}\right\} \qquad i = 0, +, A, C \tag{42}$$

where ε_i is the adsorption (binding) energy of the ith surface complex; K_i' is related to its molecular partition function.

The surface heterogeneity will cause the variation of ε_i across the surface, from one site to another. Then, for the reasons explained in our previous publication, Ref. 79, we accept the random model of surface topography. According to this model, adsorption sites characterized by different adsorption energies are distributed on a solid surface at random. Also variations in the local coulombic force fields ψ_0 and ψ_β [120] will occur, but we will neglect them because coulombic interactions are long-ranged. In the case of random topography, this will cause "smoothing" of these variations over the local structure of the outermost surface oxygens. Thus, as in the model of a homogeneous oxide surface, we will consider ψ_0 and ψ_β to be functions of the average composition of the adsorbed phase. Chemical binding forces are short-ranged, so the local variations in ε_i must be taken into account.

The experimentally measured adsorption isotherms have to be related to the following averages, θ_{it},

$$\theta_{it}(\{a\}, T) = \int \cdots \int_\Omega \theta_i(\{\varepsilon\}, \{a\}, T) \chi(\{\varepsilon\}) \, d\varepsilon_0 \, d\varepsilon_+ \, d\varepsilon_A \, d\varepsilon_C \tag{43}$$

where $\{a\}$ is the set of the bulk concentrations a_H, a_C, a_A; $\{\varepsilon\}$ is the set of the adsorption energies ε_0, ε_+, ε_C, and ε_A; Ω is the physical domain of $\{\varepsilon\}$; and $\chi(\{\varepsilon\})$ is the multidimensional differential distribution of the number of adsorption sites among various sets $\{\varepsilon\}$, normalized to unity:

$$\int \cdots \int_\Omega \chi(\{\varepsilon\}) \, d\varepsilon_0 \, d\varepsilon_+ \, d\varepsilon_A \, d\varepsilon_C = 1 \tag{44}$$

Now we have to consider the fundamental physical question of whether the variables ε_i are totally independent, in other words, whether the value of ε_i is correlated in a way with $\varepsilon_{j \neq i}$ ($i, j = 0, +, A, C$) when they change from one site to another. The degree of the correlation between ε_i and $\varepsilon_{j \neq i}$ will affect the result of the integration in Eq. (44).

In our investigations, we considered three models taking into account correlations. The first model assumes that for all the adsorption sites the difference between the adsorption energies ε_i and ε_j is constant and equal to Δ_{ij}—we called it the model with high correlations between the adsorption energies of different surface complexes. The other model assumes that the energies ε_i and $\varepsilon_{i \neq j}$ are not correlated. However, this assumption was subjected to a rigorous analysis by Rush et al. [121], who showed that it corresponds to the situation when certain small correlations exist between the adsorption energies of various components (complexes). The third model was based on the assumption that no correlations exist between the complexation energies of the surface complexes SOH^0, SOH_2^+, and SO^-C^+. As the anion in the surface complexes $SOH_2^+A^-$ is kept merely by simple coulombic interactions, we assumed that the complexation energies ε_+ and

Adsorption at Heterogeneous Interfaces

ε_A are correlated in the same way as in the first model of high correlations. We will call such a model as one with partial correlations between adsorption energies of different surface complexes.

(a) The Model Assuming High Correlations Between Adsorption Energies of Various Surface Complexes. We begin by considering the first model-assuming high correlations between the adsorption energies ε_i and $\varepsilon_{j\neq i}$, i.e.,

$$\varepsilon_j = \varepsilon_i + \Delta_{ij} \tag{45}$$

This physical situation was considered by Koopal and coworkers [122]. In our previous publications [80,93] we used the following function to represent the differential distribution of the number of adsorption sites, among various values of ε_i, normalized to unity, $\chi_i(\varepsilon_i)$,

$$\chi_i(\varepsilon_i) = \frac{\frac{1}{c_i}\exp\left\{\frac{\varepsilon_i - \varepsilon_i^0}{c_i}\right\}}{\left[1 + \exp\left\{\frac{\varepsilon_i - \varepsilon_i^0}{c_i}\right\}\right]^2} \tag{46}$$

This is a gaussian-like, fully symmetrical function, centered around $\varepsilon_i = \varepsilon_i^0$. The heterogeneity parameter c_i is proportional to the variance of $\chi_i(\varepsilon_i)$, i.e. $\pi c_i/\sqrt{3}$.

Then, after rigorous thermodynamic treatment [80,93] it follows that θ_{jt} in this model takes the explicit form

$$\theta_{jt} = \frac{K_j^0 f_j}{\sum_j K_j^0 f_j} \cdot \frac{\left[\sum_j K_j^0 f_j\right]^{kT/c}}{1 + \left[\sum_j K_j^0 f_j\right]^{kT/c}} \qquad j = 0, +, A, C \tag{47}$$

where $c_0 = c_+ = c_C = c_A = c$, and

$$K_j^0 = K_j' \exp\left\{\frac{\varepsilon_i^0 + \Delta_{ij}}{kT}\right\} \tag{48}$$

The problem of establishing the relations between the intrinsic constants K_{a1}^{int}, K_{a2}^{int}, $*K_A^{int}$, $*K_C^{int}$ using the criterion of Eq. (39) leads in this model to the equations [80]

$$K_{a2}^{int} = \frac{H^2}{K_{a1}^{int}}\left(\frac{H}{K_{a2}^{int}} + \frac{H^2}{K_{a1}^{int} \cdot K_{a2}^{int}}\right)^{\frac{kT}{c}-1} \quad \text{and} \quad *K_A^{int} = \frac{H^2}{*K_C^{int}} \tag{49a, b}$$

The two above equations are independent. The first one is a nonlinear equation with regard to K_{a2}^{int} and can be solved by means of an iteration method.

(b) The Model Assuming Small Correlations Between Adsorption Energies of Various Surface Complexes. Now let us consider the second model of adsorption, assuming that small correlations exist between the adsorption energies ε_i and $\varepsilon_{j\neq i}$. This assumption led us to an equation of the type that has been known for a long time in theories of mixed-gas adsorption on solid surfaces and was called "the multicomponent Langmuir–Freundlich

isotherm." Its form, derived [80,93] for ion adsorption on the oxide surfaces, is

$$\theta_{it} = \frac{[K_i^0 f_i]^{kT/c_i}}{1 + \sum_i [K_i^0 f_i]^{kT/c_i}} \qquad i = 0, +, A, C \tag{50}$$

where

$$K_i^0 = K_i' \exp\left\{\frac{\epsilon_i^0}{kT}\right\} \tag{51}$$

and where the c_i's are the heterogeneity parameters, whose values can be different for different surface complexes, $i = 0, +, A, C$.

Applying our criterion of Eq. (39) for this model, we arrived [80] at the set of equations

$$\left(\frac{H^2}{K_{a1}^{int} - K_{a2}^{int}}\right)^{kT/c_+} + \left(\frac{kT/c_C}{kT/c_A} - 1\right) \cdot \left(\frac{{}^*K_C^{int} a}{K_{a2}^{int}}\right)^{kT/c_C} - 1 = 0 \tag{52a}$$

$$\frac{kT}{c_A}\left(\frac{H^2 a}{K_{a1}^{int} \cdot {}^*K_{a1}^{int}}\right)^{kT/c_A} + \frac{kT}{c_C}\left(\frac{{}^*K_C^{int} a}{K_{a2}^{int}}\right)^{kT/c_C} = 0 \tag{52b}$$

For given values of the heterogeneity parameters kT/c_i ($i = 0, +, A, C$) and some given values of K_{a2}^{int} and $*K_C^{int}$, one can easily calculate first the value of K_{a1}^{int} from Eq. (52a) and then $*K_A^{int}$ from Eq. (52b).

(c) The Model Assuming Partial Correlations Between Adsorption Energies of Various Surface Complexes. We have show in our previous publications [80,93] that the model assuming the existence of high correlations between binding-to-surface energies of all the surface complexes is not able to explain two fundamental features of ion adsorption: the difference between PZC and IEP, and the transition from the Freundlich to the Henry log–log plot in the experimental isotherms of Me^{2+} ions, at some low ion concentrations. On the contrary, the model assuming small correlations was able to predict the existence of the two above-mentioned adsorption phenomena. The small correlations between SO^-C^+ complexation energies, and those related to proton binding, can be understood by taking into account the expected differences between the cation-to-surface oxygen (chemical) bonds, and the proton-to-surface oxygen bonds. Even more questionable seems to be the assumption that small correlations exist between the complexation energies of the surface complexes SOH_2^+ and $SOH_2^+A^-$. As the anion in the surface complexes $SOH_2^+A^-$ is kept merely by simple coulombic interactions, the assumption that the complexation energies ε_+ and ε_A are not correlated (or small) needs probably a serious reconsideration. By the "simple" coulombic interactions we understand the interactions that do not change significantly the electronic structure of interacting chemical individuals.

We have decided, therefore, to study at first to what extent the questionable assumptions discussed above may be responsible for the moderate success of the equations developed by assuming the fully symmetrical gaussian-like adsorption energy distribution of Eq. (46). A major effect is to be expected by assuming that from all ε_i only ε_+ and ε_A are fully correlated.

When $\varepsilon_A = \varepsilon_+ + \Delta$ and there is no relation between ε_i and $\varepsilon_{j\neq i}$ for $i = 0, +, C$, we obtained [123] the expressions for θ_{it}'s ($i = +, A$) for the particular case of the adsorption energy distribution, Eq. (46),

$$\theta_{+t} = \frac{K_+^0 f_+}{K_+^0 f_+ + K_A^0 f_A} \cdot \frac{X_{+A}}{1 + X_{+A}} \tag{53a}$$

$$\theta_{At} = \frac{K_A^0 f_A}{K_+^0 f_+ + K_A^0 f_A} \cdot \frac{X_{+A}}{1 + X_{+A}} \tag{53b}$$

The related expressions for θ_{it}'s ($i \neq +, A$) are

$$\theta_{it} = \frac{[K_i^0 f_i]^{kT/c_i}}{1 + \sum_{j=0,C}[K_j^0 f_j]^{kT/c_i}} \cdot \frac{1}{1 + X_{+A}} \qquad i = 0, C \tag{53c}$$

where

$$X_{+A} = \left[\frac{K_+^0 f_+ + K_A^0 f_A}{1 + \sum_{j\neq +,A}[K_j^0 f_j]^{kT/c_j}}\right]^{kT/c_+} \tag{53d}$$

In the case of the partial correlations, the condition CIP expressed by the criterion of Eq. (39) leads to the following relations between the intrinsic equilibrium constants:

$$\left(\frac{H^2}{K_{a1}^{int} K_{a2}^{int}} + \frac{H^2 a}{K_{a2}^{int} * K_A^{int}}\right)^{kT/c_+} - \left[1 + \left(\frac{*K_C^{int} a}{K_{a2}^{int}}\right)^{kT/c_C}\right] \cdot \left[1 + \left(\frac{H}{K_{a2}^{int}}\right)^{kT/c_0} + \left(\frac{*K_C^{int} a}{K_{a2}^{int}}\right)^{kT/c_C}\right]^{kT/c_+ - 1} = 0 \tag{54a}$$

$$\frac{kT}{c_+} \frac{H^2}{K_{a2}^{int} * K_A^{int}} \left(\frac{H^2}{K_{a1}^{int} K_{a2}^{int}} + \frac{H^2 a}{K_{a2}^{int} * K_A^{int}}\right)^{kT/c_+ - 1} - \frac{kT}{c_C} \frac{*K_C^{int}}{K_{a2}^{int}} \left(\frac{*K_C^{int} a}{K_{a2}^{int}}\right)^{kT/c_C - 1}$$
$$\cdot \left[\left(\frac{H}{K_{a2}^{int}}\right)^{kT/c_0} + \frac{kT}{c_+}\left(1 + \left(\frac{*K_C^{int} a}{K_{a2}^{int}}\right)^{kT/c_C}\right)\right] \cdot$$
$$\left[1 + \left(\frac{H}{K_{a2}^{int}}\right)^{kT/c_0} + \left(\frac{*K_C^{int} a}{K_{a2}^{int}}\right)^{kT/c_C}\right]^{kT/c_+ - 2} = 0 \tag{54b}$$

3. Modifications of the Triple-Layer Complexation Model into the Four-Layer One

The refinement of TLM into FLM is that the cations C^+ and the anions A^- of the basic electrolyte are located separately in two different layers: the anions are situated within the layer of potential ψ_A (with the charge δ_A), whereas the cations are within the layer of potential ψ_C (with the charge δ_C). According to the TLM, both anions and cations are situated within the same layer of potential ψ_β. The definition of the surface charge δ_0 and the diffuse layer charge δ_d remain unchanged as in the TLM. Such a schematic picture of FLM adsorption within the oxide/electrolyte interface was first published in the paper by Bousse et al. [113], and then the thermodynamic relations were derived by Charmas et al. [114,115]. As we showed in our papers [114,115], from all the equations forming the equation set (34–36) for the TLM only the functions f_C and f_A are different. Now, they have the following form [114]:

$$f_C = a_C \exp\left\{-\frac{e\psi_0}{kT} + \frac{e\delta_0}{kTc_+}\right\} \tag{55a}$$

$$f_A = a_A \exp\left\{-\frac{e\psi_0}{kT} - \frac{e\delta_0}{kTc_+} - \frac{e\delta^*}{kTc_+} + \frac{e\delta^*}{kTc_-} - 4.6\text{pH}\right\} \tag{55b}$$

where

$$\delta^* = \delta_A + \delta_d = B_s(\theta_- - \theta_+ - \theta_A) \tag{55c}$$

and where c_+ and c_- are the electrical capacitances constant in the regions between the planes: the surface plane and the cation location plane, and the surface plane and the anion location plane, respectively [114].

The way of solving the set of equations (34, 35, 36ab, 55ab) for FLM, to calculate the individual adsorption isotherms θ_i's and the surface charge δ_0 as a function of pH, was shown in our previous papers [114,115].

Using the criterion of Eq. (39) we arrived for FLM at the relations [114]

$$H^2 = K_{a1}^{int} \cdot K_{a2}^{int} \cdot Y^* \quad \text{and} \quad H^2 = {}^*K_C^{int} \, {}^*K_A^{int} \cdot \frac{1}{X^*} \tag{56a, b}$$

or in logarithmic form

$$\text{PZC} = \frac{1}{2}(\text{p}K_{a1}^{int} + \text{p}K_{a2}^{int} - \log Y^*) \tag{57a}$$

$$\text{PZC} = \frac{1}{2}(\text{p}^*K_A^{int} + \text{p}^*K_C^{int} + \log X^*) \tag{57b}$$

where

$$X^* = P^* + a\frac{\partial P^*}{\partial a} \quad \text{and} \quad Y^* = a^2 \frac{\partial P^*}{\partial a} \cdot \frac{H^2}{K_{a2}^{int} \, {}^*K_A^{int}} + 1 \tag{58a, b}$$

$$P^* = \exp\left\{\frac{e\delta^*}{kT}\left(\frac{1}{c_-} - \frac{1}{c_+}\right)\right\} \quad \text{and} \quad \delta^* = -B_s \frac{{}^*K_C^{int} a}{2K_{a2}^{int} + H + 2{}^*K_C^{int} a} \tag{58c, d}$$

$$\frac{\partial P^*}{\partial a} = P^* \ln P^* \left(\frac{1}{a} \cdot \frac{2K_{a2}^{int} + H}{2K_{a2}^{int} + H + 2{}^*K_C^{int} a} + \frac{d\ln(1/c_- - 1/c_+)}{da}\right) \tag{58e}$$

Determination of two equilibrium constants from Eqs. (57ab) for given values of two others is not as simple as in the case of the TLM. We presented it in detail in our paper [114].

B. Accompanying Heats of Ion Adsorption

It has been also known for a long time in adsorption science that calorimetric effects of adsorption are much more sensitive to the surface energetic heterogeneity than the adsorption isotherms. The reported experimental heat effects accompanying single gas adsorption could not, as a rule, be reproduced by theoretical expressions corresponding to various models of adsorption on a homogeneous solid surface [68]. Thus it is expected that the calorimetric studies of ion adsorption at the oxide/electrolyte interface may put more light on the role played in these systems by the energetic heterogeneity of oxide surfaces.

The first studies of enthalpic effects of ion adsorption, based on temperature dependence of adsorption isotherms, started 30 years ago. It was Berube and de Bruyn [124] who studied the effect of temperature on PZC. Studies of this kind were next conducted by Tewari et al. [125,126], Fokkink et al. [127,128], Schwarz and coworkers [129], Kuo and Yen [130], Akratopulu et al. [131,132], Blesa et al. [133], and Kosmulski et al. [134]. The influence of temperature on the surface charge vs. pH profiles was studied by Blesa and coworkers [135,136], Brady [137], and Kosmulski [138]. Valuable information concerning this problem has been collected in the review by Machesky [139].

First direct calorimetric experiments were reported by Griffiths and Fuerstenau [140] and Foissy [141], who measured the heat of immersion of an outgassed solid sample into the solutions of changing pH.

The titration calorimetry is a better experiment for theoretical interpretation. So far only a few experiments have been reported where, in the course of titration, also the accompanying heat effects were measured upon the addition of an acid or base. The first experiments of that kind were performed by Machesky and Anderson [142] in 1986, Mehr et al. [143] in 1989, De Keizer et al. [144] in 1990, Machesky and Jacobs [145,146] in 1991, Kallay et al. [147] in 1993, and Casey [148] in 1994. After introducing an outgassed solid sample into a solution, the pH of that solution is measured, and sometimes also the concentration of other ions in the equilibrium bulk electrolyte. Then a titration step is carried out (from the base or acid side), and the evolved heat is recorded. The accompanying measurements of ion (proton, cation, anion) adsorption are useful for further theoretical interpretation. This is the way in which Machesky and Anderson [142] and Machesky and Jacobs [145,146] carried out their experiments. De Kaizer et al. [144] monitored the heat effects accompanying ion adsorption as a function of the surface charge.

Kallay et al. [147,149,150] have recently proposed a titration experiment aimed at determining the experimental conditions under which the obtained experimental data could be free of the coulombic contribution to the measured heat of adsorption.

It seems, however, that progress on the experimental side has not been accompanied by progress in theoretical description of the enthalpic effects accompanying ion adsorption. Interpretation of the experimental data was carried out mostly on a qualitative level.

The first quantitative interpretation was proposed by De Keizer et al. [144] in 1990. One year later, we published our first theoretical papers concerning this problem [109]. Based on the triple layer model, we developed corresponding equations for the heats of adsorption on both a homogeneous and a heterogeneous oxide surface. Using these equations we analyzed enthalpy changes measured directly by using immersional adsorption calorimetry. Recently an exhaustive theoretical study of these enthalpic effects has been published by us [151–155]. This study also takes into consideration the data obtained by using titration calorimetry. A theoretical study based on phenomenological thermodynamics has also been published recently by Hall [156].

The already published experimental and theoretical studies of the enthalpic effects accompanying ion adsorption at the oxide/electrolyte interface have shown that these studies may contribute to our understanding of the adsorption mechanism in these systems. These studies are also very important for scientists active in the area of soil science and environmental science. The reason is the following. As we have shown in our previous publications [151–155], theoretical description of the temperature dependence of ion adsorption involves introducing the same functions and parameters that appear in the theoretical description of the enthalpic effects accompanying ion adsorption.

In large areas of our planet, which are important for growing plants for food, the changing seasons are a source of temperature changes in soil by as much as 40 K. Of course, such temperature changes must affect strongly the adsorption of ions, so calorimetric studies of ion adsorption should be an important source of information about the above-discussed temperature effects. As surface energetic heterogeneity is known to affect strongly the calorimetric effects of adsorption, it is essential that the relevant theoretical studies should be based on a realistic model taking into account the surface energetic heterogeneity.

Theoretical interpretation of the 'batch adsorption calorimetry' is rather difficult, as we showed in our previous publication [151]. Much easier for interpretation are the heat effects measured in the titration calorimetry experiment [142–148]. Here, after introducing an outgassed solid sample into a solution, the pH of that solution is measured, and sometimes also the concentration of other ions in the equilibrium bulk electrolyte. Then a titration step is carried out (from the base or acid side), and the evolved heat is recorded. That heat effect is given by the expression

$$M \int_{\mathrm{pH}}^{\mathrm{pH}+\Delta\mathrm{pH}} \left[Q_0 \left(\frac{\partial \theta_0}{\partial \mathrm{pH}} \right)_T + Q_+ \left(\frac{\partial \theta_+}{\partial \mathrm{pH}} \right)_T + Q_C \left(\frac{\partial \theta_C}{\partial \mathrm{pH}} \right)_T + Q_A \left(\frac{\partial \theta_A}{\partial \mathrm{pH}} \right)_T \right] d\mathrm{pH} \qquad (59)$$

where Q_0, Q_+, Q_C, and Q_A are the molar differential heats of the formation of SOH^0, SOH_2^+, SO^-C^-, and $SOH_2^+A^+$ surface complexes, and M is expressed in moles.

Depending on the reported experiments, the consumption (adsorption) of protons and the ions of the inert electrolyte is monitored, accompanying the change of pH by $\Delta\mathrm{pH}$. And this is the way in which Machesky and Jacobs [145] carried out their experiment. They ascribed the heat evolved on changing pH solely to proton adsorption (consumption) also monitored in their experiment. The heat effect, Q_{pr}, considered by these authors is described by the equation [153]

$$Q_{\mathrm{pr}} = \frac{\int_{\mathrm{pH}}^{\mathrm{pH}+\Delta\mathrm{pH}} \sum_i Q_i \left(\frac{\partial \theta_i}{\partial \mathrm{pH}} \right)_T d\mathrm{pH}}{\int_{\mathrm{pH}}^{\mathrm{pH}+\Delta\mathrm{pH}} \left[2\left(\frac{\partial \theta_+}{\partial \mathrm{pH}} \right)_T + 2\left(\frac{\partial \theta_A}{\partial \mathrm{pH}} \right)_T + \left(\frac{\partial \theta_0}{\partial \mathrm{pH}} \right)_T \right] d\mathrm{pH}} \qquad i = 0, +, C, A \qquad (60)$$

where Q_i is the molar heat of formation of the ith complex.

A few years ago, Kallay and coworkers [147,149,150] started a theoretical and experimental study aimed at determining the experimental conditions under which the obtained experimental data could be free of the coulombic contributions to the measured heats of adsorption and refer only to the nonconfigurational heat values. This would eliminate the necessity of carrying out a complicated theoretical–numerical analysis. In one of their papers the authors postulated that the "average" heat of proton adsorption should correspond to the heat of titration from PZC-$\Delta\mathrm{pH}$ to PZC+$\Delta\mathrm{pH}$. The consumption of protons from the bulk solution was monitored in the Kallay experiment, so the "average" molar heat of proton adsorption, Q_{Av}, is given by [154]

$$Q_{\mathrm{Av}} = \frac{\int_{\mathrm{PZC}-\Delta\mathrm{pH}}^{\mathrm{PZC}+\Delta\mathrm{pH}} \sum_i Q_i \left(\frac{\partial \theta_i}{\partial \mathrm{pH}} \right)_T d\mathrm{pH}}{\int_{\mathrm{PZC}-\Delta\mathrm{pH}}^{\mathrm{PZC}+\Delta\mathrm{pH}} \left[2\left(\frac{\partial \theta_+}{\partial \mathrm{pH}} \right)_T + 2\left(\frac{\partial \theta_A}{\partial \mathrm{pH}} \right)_T + \left(\frac{\partial \theta_0}{\partial \mathrm{pH}} \right)_T \right] d\mathrm{pH}} \qquad i = 0, +, C, A \qquad (61)$$

The conditions under which Kallay's method may lead to reliable estimation have been discussed in our recent publication [154].

Adsorption at Heterogeneous Interfaces

The molar differential heats of adsorption Q_i are "configurational," as they must depend on the concentrations of the surface complexes θ_i. Their detailed explicit form can be developed by applying the appropriate thermodynamic relations [151],

$$Q_0 = -k\frac{\partial}{\partial(1/T)}\left[\frac{\mu_{\text{SOH}^0} - \mu_{\text{H}}^{\text{b}}}{kT}\right]_{\{\theta_i\}} = k\left(\frac{\partial F_0}{\partial(1/T)}\right)_{\{\theta_i\}} \tag{62a}$$

$$Q_+ = -k\frac{\partial}{\partial(1/T)}\left[\frac{\mu_{\text{SOH}_2^+} - 2\mu_{\text{H}}^{\text{b}}}{kT}\right]_{\{\theta_i\}} = k\left(\frac{\partial F_+}{\partial(1/T)}\right)_{\{\theta_i\}} \tag{62b}$$

$$Q_C = -k\frac{\partial}{\partial(1/T)}\left[\frac{\mu_{\text{SO}-\text{C}^+} - \mu_{\text{C}}^{\text{b}}}{kT}\right]_{\{\theta_i\}} = k\left(\frac{\partial F_C}{\partial(1/T)}\right)_{\{\theta_i\}} \tag{62c}$$

$$Q_A = -k\frac{\partial}{\partial(1/T)}\left[\frac{\mu_{\text{SOH}_2^+ \text{A}^-} - \mu_{\text{A}}^{\text{b}} - 2\mu_{\text{H}}^{\text{b}}}{kT}\right]_{\{\theta_i\}} = k\left(\frac{\partial F_A}{\partial(1/T)}\right)_{\{\theta_i\}} \tag{62d}$$

where $\mu_{\text{SO}-\text{C}^+}$, $\mu_{\text{SOH}_2^+\text{A}^-}$, μ_{SOH^0}, and $\mu_{\text{SOH}_2^+}$ are the chemical potentials of these surface complexes, whereas μ_j^{b} ($j = $ H, C, A) are those in the bulk phase of proton, cation, and anion of the basic electrolyte, respectively. The analytical form of the functions of F_i depends on the investigated model. So above we wanted to show the derivation results for some of the models presented in the previous subsection A.

1. Triple-Layer Complexation Model

In particular cases for a homogeneous oxide surface model considered here, the analytical forms of the functions F_i ($i = 0, +, $ A, C) were developed elsewhere [80]. Then, from Eqs. (62), we arrive at the following expressions for the configurational heats of formation of different surface complexes, Q_i [80,151]:

$$Q_0 = Q_{a2} - e\psi_0 - \frac{e}{T}\left(\frac{\partial \psi_0}{\partial(1/T)}\right)_{\{\theta_i\},\text{pH}} \tag{63a}$$

$$Q_+ = Q_{a1} + Q_{a2} - 2e\psi_0 - \frac{2e}{T}\left(\frac{\partial \psi_0}{\partial(1/T)}\right)_{\{\theta_i\},\text{pH}} \tag{63b}$$

$$Q_C = Q_C - e\psi_0 - \frac{e}{T}\left(\frac{\partial \psi_0}{\partial(1/T)}\right)_{\{\theta_i\},\text{pH}} + e\frac{\delta_0}{c_1} + \frac{e\delta_0 T}{(c_1)^2}\left(\frac{\partial c_1}{\partial T}\right)_{\{\theta_i\},\text{pH}} + k\left(\frac{\partial \ln a_C}{\partial(1/T)}\right)_{\text{pH}} \tag{63c}$$

$$Q_A = Q_A - e\psi_0 - \frac{e}{T}\left(\frac{\partial \psi_0}{\partial(1/T)}\right)_{\{\theta_i\},\text{pH}} - e\frac{\delta_0}{c_1} + \frac{e\delta_0 T}{(c_1)^2}\left(\frac{\partial c_1}{\partial T}\right)_{\{\theta_i\},\text{pH}} + k\left(\frac{\partial \ln a_A}{\partial(1/T)}\right)_{\text{pH}} \tag{63d}$$

where Q_{a1}, Q_{a2}, Q_{aC}, and Q_{aA} are the nonconfigurational heats of appropriate reactions presented in Table 1,

$$Q_{ai} = -k\frac{d \ln K_{ai}^{\text{int}}}{d(1/T)} \quad i = 1, 2 \tag{64a}$$

$$Q_{aC} = -k\frac{d \ln (K_{a2}^{\text{int}}/{}^*K_C^{\text{int}})}{d(1/T)} \tag{64b}$$

$$Q_{aA} = -k\frac{d \ln (K_{a2}^{\text{int}} \cdot K_A^{\text{int}})}{d(1/T)} \tag{64c}$$

While calculating the derivative $(\partial c_1/\partial T)_{\theta_i,\text{pH}}$ we assumed [153,154] that there are two different values of the c_1 parameter, depending on the pH value: one for the acidic region

TABLE 1 Surface Reactions, Equilibrium Constants, and Heats of Adsorption

Reaction type	Equilibrium constants	Heats of reaction
$SOH^0 + H^+ \leftrightarrow SOH_2^+$	$-pK_{a1}^{int}$	Q_{a1}
$SO^- + H^+ \leftrightarrow SOH^0$	$-pK_{a2}^{int}$	Q_{a2}
$SOH^0 + H^+ + A^- \leftrightarrow SOH_2^+ A^-$	$-p^*K_A^{int}$	$Q_{aA} - Q_{a2}$
$SO^-C^+ + H^+ \leftrightarrow SOH^0 + C^+$	$-p^*K_C^{int}$	$Q_{a2} - Q_{aC}$
$SO^- + 2H^+ \leftrightarrow SOH_2^+$	$-pK_{a1}^{int} - pK_{a2}^{int}$	$Q_{a1} + Q_{a2}$
$SO^- + 2H^+ + A^- \leftrightarrow SOH_2^+ A^-$	$-pK_{a2}^{int} - p^*K_A^{int}$	Q_{aA}
$SO^- + C^+ \leftrightarrow SO^-C^+$	$p^*K_C^{int} - pK_{a2}^{int}$	Q_{aC}
$SOH_2^+ + A^- \leftrightarrow SOH_2^+ A^-$	$pK_{a1}^{int} - p^*K_A^{int}$	$Q_{aA} - Q_{a1} - Q_{a2}$

(pH < PZC) and another one for the basic region (pH > PZC). Then, following Blesa's and Kallay's recommendation [135,157], we treated c_1 as the linear functions of temperature,

$$c_1 = c_1^L = c_1^{L,0} + \alpha_1^L \cdot \Delta T \qquad \text{pH} < \text{PZC} \qquad (65a)$$

$$c_1 = c_1^R = c_1^{R,0} + \alpha_1^R \cdot \Delta T \qquad \text{pH} > \text{PZC} \qquad (65b)$$

which may be considered as the formal Taylor expansions around $T = T_0$ so that $c_1^L(T_0) = c_1^{L,0}$ and $c_1^R(T_0) = c_1^{R,0}$.

We have shown in our previous publication [82] that the derivatives $(\partial a_i / \partial(1/T))_{\theta_i, \text{pH}}$, $i = C, A$, occuring in Eqs. (63) can be expressed as

$$\frac{\partial a_i}{\partial(1/T)} = -2.0172 \cdot 10^{-8} \cdot T^4 a_i \ln \gamma_i \qquad i = A, C \qquad (66)$$

Let us consider finally the derivative $(\partial \psi_0 / \partial(1/T))_{\text{pH}}$ occuring in Eqs. (63). After evaluating it from Eq. (37), we arrive at the explicit expression

$$\left(\frac{\partial \psi_0}{\partial(1/T)}\right)_{\text{pH}} = \psi_0 T \left[\frac{\beta}{\beta + t} + \frac{t}{\beta + t} \cdot \frac{Q_{a2} - Q_{a1}}{2kT}\right] + 2.3 \frac{kT}{e} \frac{\beta}{\beta + t} \frac{\partial \text{PZC}}{\partial(1/T)} \qquad (67)$$

where

$$t = \left[\left(\frac{e\psi_0}{\beta k T}\right)^2 + 1\right]^{-\frac{1}{2}} \qquad (67a)$$

In Eq. 67 the derivative $(\partial \text{PZC}/\partial(1/T))$ is replaced by the term $(Q_{a1} + Q_{a2})/4.6k$, according to the result of the formal differentiations of PZC in Eq. (41a), and the definition of Q_{a1} and Q_{a2} in Eq. (64a),

$$2.3 \frac{\partial \text{PZC}}{\partial(1/T)} = \frac{Q_{a1} + Q_{a2}}{2k} \qquad (68a)$$

Similarly to Eq. (41a), the same differentiation with respect to $1/T$ in Eq. (41b) leads to the result

$$2.3 \frac{\partial \text{PZC}}{\partial(1/T)} = \frac{Q_{aA} + Q_{aC}}{2k} \qquad (68b)$$

Adsorption at Heterogeneous Interfaces

in view of the definitions of Q_{aC} and Q_{aA} in Eqs. (64b,c). From Eqs. (68a) and (68b) we have

$$Q_{aA} - (Q_{a1} + Q_{a2}) = Q_{aC} \tag{69}$$

Except for the functions Q_i, the analytical forms of the derivatives $(\partial \theta_i / \partial \mathrm{pH})_T$ from Eqs. (60) and (61) for the homogeneous model of TLM were presented in our paper Ref. 79.

2. Modifications of the Triple-Layer Complexation Model Taking into Account the Factor of Energetic Heterogeneity of the Surface

Just recently, we have published [155] our results dealing with the calorimetric effects accompanying ion adsorption, taking into account two of those heterogeneous models presented in the previous subsection, i.e., models assuming high correlations and small correlations between adsorption energies of different surface complexes.

Thus, for the considered model assuming high correlations to exist between the adsorption energies of surface complexes, taking into account the explicit expressions for the F_i functions, the molar differential heats of adsorption $Q_i^{(c)}$ are given by (Ref. 155)

$$Q_i^{(c)} = Q_i - c \ln \frac{1 - \theta_{-t}}{\theta_{-t}} \qquad i = 0, +, C, A \tag{70}$$

where the superscript (c) stands for high correlations, and Q_i is the heat for the model of a homogeneous oxide surface, Eqs. (63), and where $\theta_{-t} = 1 - \Sigma_i \theta_{it}$ is the function of the free surface oxygens. From Eq. (49a) we obtain

$$2.3 \frac{\partial \mathrm{PZC}}{\partial (1/T)} = \frac{\dfrac{Q_{a1} + Q_{a2}}{2k} - \left(1 - \dfrac{kT}{c}\right)\left[\dfrac{Q_{a2}}{2k} + \check{r}\dfrac{Q_{a1}}{2k}\right] - \dfrac{1}{2}T\dfrac{kT}{c}\ln\left(\dfrac{H^2}{K_{a1}^{\mathrm{int}}K_{a2}^{\mathrm{int}}\check{r}}\right)}{1 - \dfrac{1}{2}\left(1 - \dfrac{kT}{c}\right)[1 + \check{r}]} \tag{71}$$

where

$$\check{r} = \frac{H}{K_{a1}^{\mathrm{int}} + H} \tag{71a}$$

Next from Eq. (49b) we have the same relation as (68b) for the homogeneous model. From Eqs. (71) and (68b) one can eliminate one of the four parameters, Q_{a1}, Q_{a2}, Q_{aC}, and Q_{aA}, as we did previously in the case of the homogeneous surface model.

To calculate Q_{pr} defined in Eq. (60), we still need to know the form of the derivatives $(\partial \theta_i / \partial \mathrm{pH})_T$ for this model. Their explicit forms were presented in Ref. 155.

We performed the same transformations for the model assuming small correlations between adsorption energies of different surface complexes. The related expressions for the molar heats of adsorption $Q_i^{(s)}$ take now the form [155]

$$Q_i^{(s)} = Q_i - c_i \ln \frac{\theta_{it}}{\theta_{-t}} \qquad i = 0, +, C, A \tag{72}$$

where the superscript (s) stands for the small-correlated adsorption energies, and Q_i are still the expressions for a homogeneous solid surface, Eqs (63).

From Eqs. (52ab) we can eliminate H to obtain the following interrelation of the adsorption equilibrium constants [155]:

$$G = \ln\left[K_{a1}^{int}K_{a2}^{int}\left[1 - \left(\frac{kT/c_C}{kT/c_A} - 1\right)\left(\frac{*K_C^{int}a}{K_{a2}^{int}}\right)^{kT/c_C}\right]^{c_+/kT}\right]$$
$$- \ln\left[\frac{K_{a2}^{int*}K_A^{int}}{a}\left[\frac{kT/c_C}{kT/c_A}\left(\frac{*K_C^{int}a}{K_{a2}^{int}}\right)^{kT/c_C}\right]^{c_A/kT}\right] = 0 \qquad (73)$$

Now, we calculate the derivative $\partial G/\partial(1/T)$ to arrive at the interrelation of the nonconfigurational heats of adsorption:

$$\frac{Q_{a1} + Q_{a2}}{k} - \frac{T}{kT/c_+}\ln\left[1 - \left(\frac{kT/c_C}{kT/c_A} - 1\right)\left(\frac{*K_C^{int}a}{K_{a2}^{int}}\right)^{kT/c_C}\right]$$
$$- \frac{kT/c_C}{kT/c_+}\left(\frac{kT/c_C}{kT/c_A} - 1\right)\left(\frac{*K_C^{int}a}{K_{a2}^{int}}\right)^{kT/c_C}$$
$$\cdot \frac{T\cdot\ln\left(\frac{*K_C^{int}a}{K_{a2}^{int}}\right) - \frac{1}{a}\frac{\partial a}{\partial(1/T)} - \frac{Q_{aC}}{k}}{1 - \left(\frac{kT/c_C}{kT/c_A} - 1\right)\left(\frac{*K_C^{int}a}{K_{a2}^{int}}\right)^{kT/c_C}} - \frac{Q_{aA} - \frac{kT/c_C}{kT/c_A}Q_{aC}}{k} + \frac{T}{kT/c_A} \qquad (74)$$
$$\ln\left(\frac{kT/c_C}{kT/c_A}\right) - \left(1 - \frac{kT/c_C}{kT/c_A}\right)\frac{1}{a}\frac{\partial a}{\partial(1/T)} = 0$$

And, again, from Eq. (74) we can eliminate one of the four parameters: Q_{a1}, Q_{a2}, Q_{aC}, and Q_{aA}.

The derivative $\partial PZC/\partial(1/T)$ is to be evaluated from the equation system Eq. (52ab). This can also be done by solving this equation system first to eliminate one of the equilibrium constants, $*K_C^{int}$ for instance:

$$\left(\frac{H^2}{K_{a1}^{int}K_{a2}^{int}}\right)^{kT/c_+} + \left(\frac{kT/c_C}{kT/c_A} - 1\right)\frac{kT/c_A}{kT/c_C}\left(\frac{H^2a}{K_{a2}^{int}*K_A^{int}}\right)^{kT/c_A} - 1 = 0 \qquad (75)$$

From Eq. (75) we evaluate now $\partial PZC/\partial(1/T)$:

$$4.6\frac{\partial PZC}{\partial(1/T)} = \frac{\frac{kT}{c_+}\left(\frac{H^2}{K_{a1}^{int}K_{a2}^{int}}\right)^{kT/c_+}\left[\frac{Q_{a1}+Q_{a2}}{k} - T\cdot\ln\left(\frac{H^2}{K_{a1}^{int}K_{a2}^{int}}\right)\right]}{\left(\frac{H^2}{K_{a1}^{int}K_{a2}^{int}}\right)^{kT/c_+} + \left(1 - \frac{kT/c_A}{kT/c_C}\right)\left(\frac{H^2a}{K_{a2}^{int}*K_A^{int}}\right)^{kT/c_A}} +$$
$$+ \frac{\frac{kT}{c_A}\left(1 - \frac{kT/c_A}{kT/c_C}\right)\left(\frac{H^2a}{K_{a2}^{int}*K_A^{int}}\right)^{kT/c_A}\left[\frac{Q_{a2}}{k} + \frac{1}{a}\frac{\partial a}{\partial(1/T)} - T\cdot\ln\left(\frac{H^2a}{K_{a2}^{int}*K_A^{int}}\right)\right]}{\left(\frac{H^2}{K_{a1}^{int}K_{a2}^{int}}\right)^{kT/c_+} + \left(1 - \frac{kT/c_A}{kT/c_C}\right)\left(\frac{H^2a}{K_{a2}^{int}*K_A^{int}}\right)^{kT/c_A}} \qquad (76)$$

The explicit forms of the derivatives $(\partial\theta_i/\partial pH)_T$ for this model were published in the paper [155].

3. Modifications of the Triple-Layer Complexation Model into the Four-Layer One

We also published a comparative study of the heat accompanying ion adsorption predicted by both homogeneous TLM and FLM [117].

Transformations similar to those outlined in the case of homogeneous TLM lead to the following expressions for the configurational heats of formation of complexes SO^-C^+ and $SOH_2^+A^-$:

$$Q_C = Q_{aC} - e\psi_o - \frac{e}{T}\left(\frac{\partial \psi_o}{\partial(1/T)}\right)_{\{\theta_i\},\text{pH}} + e\frac{\delta_0}{c_+} + \frac{e\delta_0 T}{(c_+)^2}\left(\frac{\partial c_+}{\partial T}\right)_{\{\theta_i\},\text{pH}} + k\left(\frac{\partial \ln a_C}{\partial(1/T)}\right)_{\text{pH}} \quad (77a)$$

$$Q_A = Q_{aA} - e\psi_o - \frac{e}{T}\left(\frac{\partial \psi_o}{\partial(1/T)}\right)_{\{\theta_i\},\text{pH}} - e\frac{(\delta_0 + \delta^*)}{c_+} - \frac{e(\delta_0 + \delta^*)T}{(c_+)^2}\left(\frac{\partial c_+}{\partial T}\right)_{\{\theta_i\},\text{pH}}$$
$$+ e\frac{\delta^*}{c_-} + \frac{e\delta^* T}{(c_-)^2}\left(\frac{\partial c_-}{\partial T}\right)_{\{\theta_i\},\text{pH}} + k\left(\frac{\partial \ln a_A}{\partial(1/T)}\right)_{\text{pH}} \quad (77b)$$

where the nonconfigurational heats Q_{a1}, Q_{a2}, Q_{aC}, and Q_{aA} are still given by Eqs. (64). The form of molar heats of adsorption Q_0, Q_+ and the derivatives ψ_0 and a_i with respect to $(1/T)$ are the same as in the homogeneous TLM, but not $\partial\text{PZC}/\partial(1/T)$, because PZC is expressed now through Eqs. (57).

Assuming that c_+ and c_- are the linear functions of temperature, we have two new parameters α_+ and α_-,

$$c_+ = c_+^0 + \alpha_+ \cdot \Delta T \quad (78a)$$
$$c_- = c_-^0 + \alpha_- \cdot \Delta T \quad (78b)$$

These linear functions may be considered as the formal Taylor expansions around $T = T_0$, such that $c_+(T_0) = c_+^0$ and $c_-(T_0) = c_-^0$.

Now, we take the advantage of having two expressions for PZC(T) to develop interrelations of some temperature derivatives of interest. Thus we calculate $\partial\text{PZC}/\partial(1/T)$ from two equation systems (57a,58) and (57b,58) and as a result we obtain [117]

$$\frac{\partial\text{PZC}}{\partial(1/T)} = \frac{\dfrac{Q_{a1} + Q_{a2}}{4.6k} - \dfrac{1}{2}\dfrac{\partial Y^*}{\partial(1/T)}}{1 + \dfrac{1}{2}\dfrac{\partial Y^*}{\partial\text{PZC}}} \quad (79a)$$

$$\frac{\partial\text{PZC}}{\partial(1/T)} = \frac{\dfrac{Q_{aA} - Q_{aC}}{4.6k} + \dfrac{1}{2}\dfrac{\partial X^*}{\partial(1/T)}}{1 - \dfrac{1}{2}\dfrac{\partial X^*}{\partial\text{PZC}}} \quad (79b)$$

where the derivatives

$$\left(\frac{\partial \log Y^*}{\partial\text{PZC}}\right)_{\frac{1}{T}}, \quad \left(\frac{\partial \log Y^*}{\partial(1/T)}\right)_{\text{PZC}}, \quad \left(\frac{\partial \log X^*}{\partial\text{PZC}}\right)_{\frac{1}{T}}, \quad \left(\frac{\partial \log X^*}{\partial(1/T)}\right)_{\text{PZC}}$$

have pretty complicated explicit forms, given in Ref. 117.

Finally, from Eqs. (79ab) one can arrive at the following interrelation of the nonconfigurational heats of adsorption [similar to Eq. (69) for the homogeneous TLM]:

$$Q_{aA} = (Q_{a1} + Q_{a2}) \cdot \frac{1 - \frac{1}{2}\frac{\partial \log X^*}{\partial \text{PZC}}}{1 + \frac{1}{2}\frac{\partial \log Y^*}{\partial \text{PZC}}} + Q_{aC} - 2.3k \frac{\partial \log X^*}{\partial (1/T)}$$

$$- 2.3k \frac{1 - \frac{1}{2}\frac{\partial \log X^*}{\partial \text{PXC}}}{1 + \frac{1}{2}\frac{\partial \log Y^*}{\partial \text{PZC}}} \cdot \frac{\partial \log Y^*}{\partial (1/T)}$$

(80)

One should remember that $(\partial \log Y^*/\partial(1/T))_{\text{PZC}}$ is a function of Q_{aA}, and the above equation can be solved as an implicit one using the iteration method.

C. Results of Our Investigations

The theoretical expressions developed in the previous sections make it possible to analyze both the temperature dependence of measured surface charge and the calorimetric effects accompanying simple ion adsorption at the oxide/electrolyte interfaces. The theoretical description of the temperature dependence of ion adsorption, and of the calorimetric effects accompanying ion adsorption, involves introducing the same parameters and functions. We start our consideration with a description based on a model of an energetically homogeneous surface. The developed equations presented in the previous sections were applied to reproduce the temperature dependence of some experimentally measured surface charge isotherms [151].

For the purpose of illustration, we have chosen the data reported by Blesa et al. [135]. These are their titration curves $\delta_0(\text{pH})$ for magnetite, measured at three concentrations of the inert electrolyte KNO_3: 10^{-3} mol/dm^3, 10^{-2} mol/dm^3, and 10^{-1} mol/dm^3. Every titration curve was measured at three temperatures: 30°C (303 K), 50°C (323 K) and 80°C (353 K). As a matter of fact the experimental data of that kind have rarely been reported, and Blesa's data are probably those containing the largest number of experimental points. Blesa and coworkers [135] proposed also an interesting theoretical interpretation of their data, which in part inspired us to publish the paper [151].

While fitting the experimental data of Blesa et al. [135] to our theoretical equations, we arrived at a reasonably good agreement between the theory and experiment, which is shown in Fig. 16. The parameters found in these calculations are collected in Table 2. Some of them could be determined independently, some of them were calculated from others through certain interrelations, and some of them were found by computer as the best-fit parameters (for details see Ref. 151).

We would like to draw attention to the calculated individual adsorption isotherms of ions, presented in Fig. 17(a). They show that the adsorption isotherm of anions, θ_A, is pretty much temperature independent, whereas the individual adsorption isotherm of cations, θ_C, is affected strongly by temperature. This property seems to be a common feature of these systems. Figure 17(b) shows the same effect found in adsorption studies, where the isotherms of cations and anions were directly monitored experimentally by using the radiometric method [138]. One can see that the adsorption of cations Na^+ on alumina is strongly affected by temperature, whereas the adsorption of Cl^- ions is pretty much temperature independent.

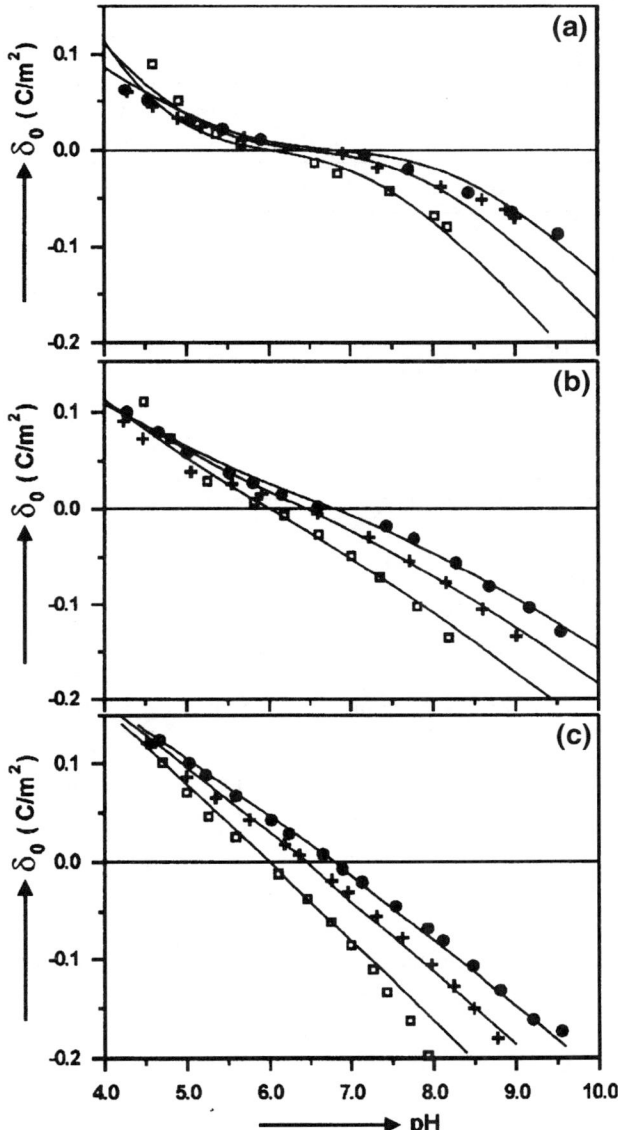

FIG. 16 The comparison of the theoretical titration curves calculated by us [151] (solid lines) using the best-fit and calculated parameters collected from Table 2 with the experimental data found by Blesa et al. [135] from the KNO_3/magnetite system, at three temperatures: (●) 30°C, (+) 50°C, (□) 80°C, when the concentration of the inert electrolyte KNO_3 is (a) 10^{-3} mol/dm^3; (b) 10^{-2} mol/dm^3; (c) 10^{-1} mol/dm^3.

TABLE 2 The Collection of the Best-Fit and Calculated Parameters Used by Us to Fit Our Equations to the Blesa Titration Curves Measured at Three Temperatures and for Three Concentrations of the Inert Electrolyte

N_s sites/nm^2	Q_{a1} kJ/mol	Q_{a2} kJ/mol	Q_{aC} kJ/mol	Q_{aA} kJ/mol
6	20.5	45.0	0.0	65.5

T	pK_{a1}^{int}	pK_{a2}^{int}	$p^*K_C^{int}$	$p^*K_A^{int}$	PZC
303 K	4.40	9.20	6.60	7.00	6.80
323 K	4.18	8.72	6.12	6.78	6.45
353 K	3.90	8.10	5.50	6.50	6.00

c mol/dm^3	α_1^L F/m^{2}deg	α_1^R F/m^{2}deg
0.1	0.003	0.0
0.01	0.005	0.0
0.001	0.060	0.0

	0.001 mol/dm^3		0.01 mol/dm^3		0.1 mol/dm^3	
T	C_1^L F/m^2	C_1^R F/m^2	C_1^L F/m^2	C_1^R F/m^2	C_1^L F/m^2	C_1^R F/m^2
303 K	1.40	1.80	0.95	1.15	1.10	1.30
323 K	2.60	1.80	1.05	1.15	1.16	1.30
355 K	4.40	1.80	1.20	1.15	1.25	1.30

Source: Ref. 151.

Our theoretical calculations and the above-discussed experimental findings show the risk of using the simple relation [139]:

$$Q_{\text{ion}} = -k\left(\frac{d(\ln a_i)}{d(1/T)}\right)_{pH} \qquad i = A, C \qquad (81a)$$

or [138]

$$Q_{\text{ion}} = -k\left(\frac{d(\ln a_i)}{d(1/T)}\right)_{\delta_0} \qquad i = A, C \qquad (81b)$$

to estimate heats of adsorption, where there is a tendency to treat Q_{ion} as a nonconfigurational quantity. The temperature dependence of cation adsorption might suggest the existence of considerable heat effects, accompanying cation adsorption. Meanwhile, that behavior of the cation adsorption isotherm was found here by us by assuming that the nonconfigurational heat effect accompanying the cation adsorption, Q_{aC}, is nonexistent. Our calculations [151] show that the temperature dependence of cation adsorption is due to the temperature dependence of the (first) proton adsorption isotherm, θ_0, which is the adsorption process strongly competitive to cation adsorption.

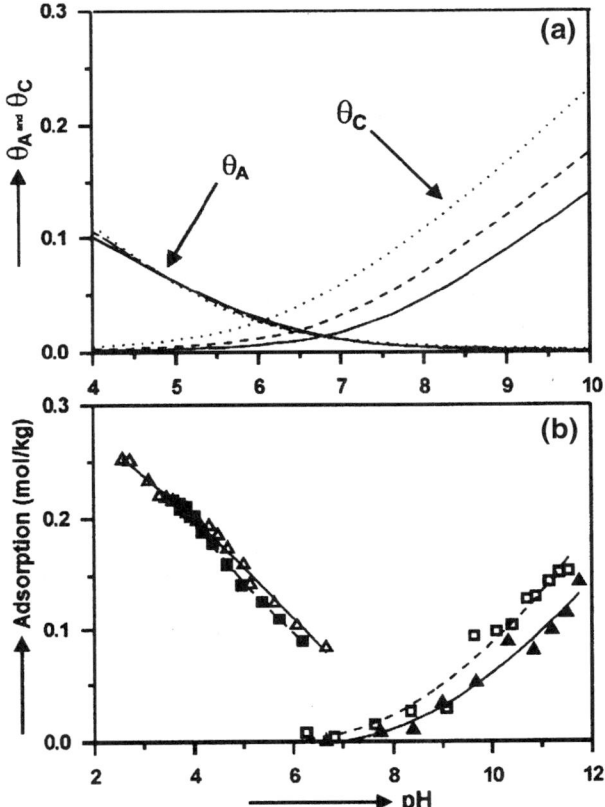

FIG. 17 (a) The individual adsorption isotherms of anions and cations of the inert electrolyte, θ_i ($i=$ A, C), calculated by us [151] for three temperatures: (———) 30°C, (- - - - -) 50°C, (......) 80°C, when the concentration of the inert electrolyte KNO_3 is 10^{-2} mol/dm^3. (b) The individual adsorption isotherms of Na^+ and Cl^- ions on alumina measured by Kosmulski [138], using radiometric methods. Here (△) and (■) are the individual (experimental) isotherms of Cl^- ions measured at 15°C and 35°C respectively. Then (▲) and (□) are the individual (experimental) adsorption isotherms of Na^+ ions measured respectively at 15°C and 35°C. The solid and broken lines are the second-order polynomial approximating best experimental data, and drawn here to help the eye.

We also used [154] the parameters collected in Table 2 to our model calculations to see under which conditions the data obtained in the Kallay titration calorimetry experiment [147,149,150] could be used to determine easily the heats of proton adsorption, Q_{a1} and Q_{a2}. The behavior of the calculated function Q_{Av} defined in Eq. (61) for the parameters collected in Table 2 is shown in Fig. 18. One can see that when the concentration of the inert electrolyte increases, the values of the function Q_{Av} tend to the value $(Q_{a1} + Q_{a2})/2$, as was assumed by Kallay and coworkers. However, our present model calculations suggest that in order to arrive closely at the value $(Q_{a1} + Q_{a2})/2$, the Kallay titration experiment has to be carried out at possibly high concentrations of the inert electrolyte. Meanwhile, Kallay proved first his hypothesis by comparing his experimental heats Q_{Av} for TiO_2 [147] with the values of $(Q_{a1} + Q_{a2})/2$ determined from Eq. (68a). The agreement was very good. (For recent results see Chapter 5 of this book.)

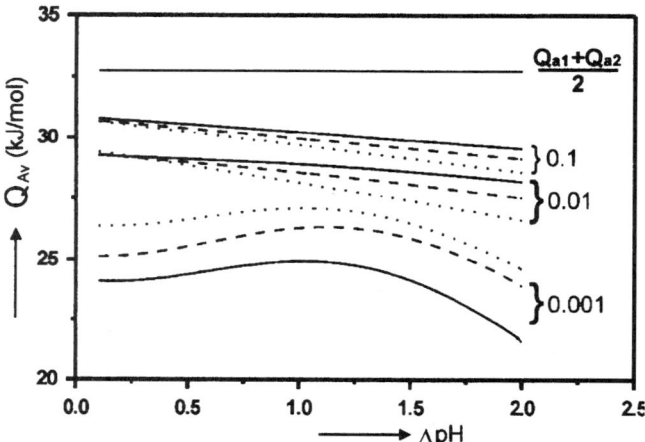

FIG. 18 The behavior of the calculated functions Q_{Av}, depending on ΔpH, at three temperatures and for three concentrations of the inert electrolyte KNO_3 [154]. The solid line (———) is for the lowest temperature, 30°C, the dashed line (- - - - -) is for the temperature 50°C, and the dotted line (......) is for 80°C. The numbers 0.1, 0.01, and 0.001 at the three families of the curves (—; - - -; ...) denote the concentrations of the inert electrolyte (in mol/dm³) for which the three functions Q_{Av} were calculated. All the calculations were done by using the same set of parameters as in Fig. 16.

In the paper [154] we were trying to find out under which conditions the Kallay calorimetric (titration) experiment will yield the value $(Q_{a1} + Q_{a2})/2$. It seems that the symmetry of the titration curves may be a practical guide to a successful application of the Kallay procedure of estimating $(Q_{a1} + Q_{a2})/2$. Of course, the validity of that practical guide deserves further theoretical studies. Such model calculations will require first a full analysis of the temperature dependence of adsorption (titration) isotherms or a full analysis of accurately determined heats of titration in a wide range of pH's. Unfortunately, the small body of experimental data reported so far in the literature, and certain problems related to their accuracy, create, for the moment, serious limitations to an exhaustive study of such a kind.

In our recent paper [117], we used the equations from the previous sections and presented a comparative theoretical study of the heat accompanying ion adsorption, predicted by both TLM and FLM for homogeneous surfaces. The expressions developed are successfully applied to describe quantitatively the heat effects accompanying ion adsorption at the TiO_2/NaCl solution interface recorded by Mehr et al. [143]. Figure 19 shows a reasonably good agreement between the experimental and theoretical heats of proton adsorption, Q_{pr}, calculated by using Eq. (60), for both TLM and FLM.

However, the application of the FLM allows for more accurate estimation of the nonconfigurational heat of the second proton adsorption, Q_{a1} (over 15% difference in the obtained values for TLM and FLM) having the experimental values IEP and PZC for the adsorption system under consideration [117]. Figure 19, presenting the calculated heats Q_{pr} vs. pH the for TLM, distinctly shows discontinuity in the point pH = PZC caused by the stepwise change of the values c_1^L into c_1^R and α_1^L into α_1^R going from low to high pH values through PZC.

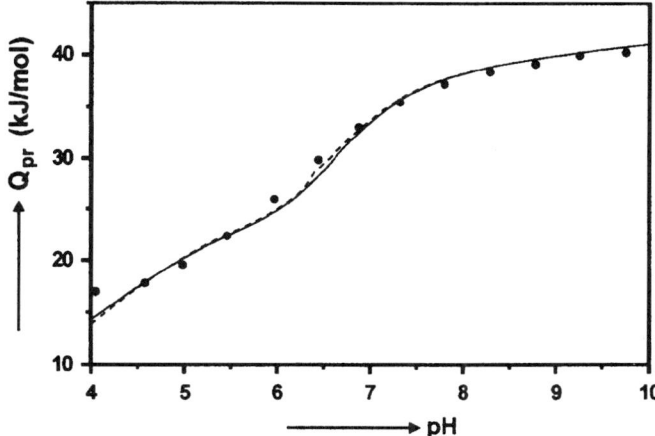

FIG. 19 Comparison of experimental and theoretical heats of proton adsorption [117]. The black circles (●) are the arithmetic averages of the experimental values of Q_{pr} recorded by Mehr et al. [143] in the course of acid and base titration. The broken line is the best fit obtained by accepting TLM, and the solid line is the best fit obtained with FLM. The parameter values are collected in the tables in Ref. 117.

On the contrary, the FLM model is characterized by continuity of all thermodynamic functions. Moreover, the FLM predicts that even small differences between PZC and IEP values may affect strongly the heat effects below PZC. Our computer exercises showed that there is not much room for maneuvering while choosing the parameters. A slightly different choice of any of them will destroy substantially the agreement between theory and experiment.

The difference between both surface complexation models accepted in this paper [117] becomes more evident in the calculation of isotherm derivatives in the interpretation of calorimetric titration experiment. As for real values of nonconfigurational heats of adsorption of ions as well as temperature derivatives of capacitances it seems that due to sensitivity of theoretical predictions (a small change of parameters causes a great change of theoretical heat look), the accuracy of the experimental data of measured heat must be taken into account. Obtaining such data is not easy and requires perfect knowledge of equipment and experimental technique as well as taking into account various heat corrections from the direct experiment as presented very well in the paper by Machesky and Jacobs [145].

Our theoretical studies of the enthalpic effects accompanying ion adsorption were extended to take into account the energetic heterogeneity of the actual oxide surfaces [155], where two different models of that surface energetic heterogeneity have been considered by us. One of them assumes that the binding-to-surface energies of different surface complexes vary from one to another surface oxygen but in much same way, i.e., these energies are highly correlated. The other model of surface heterogeneity assumes that only small correlations exist between the energies of adsorption of various surface complexes, when going from one to another surface oxygen. For both models of the surface energetic heterogeneity the corresponding equations are presented in the previous sections for the isotherms of ion adsorption, and for the enthalpic effects accompanying ion adsorption.

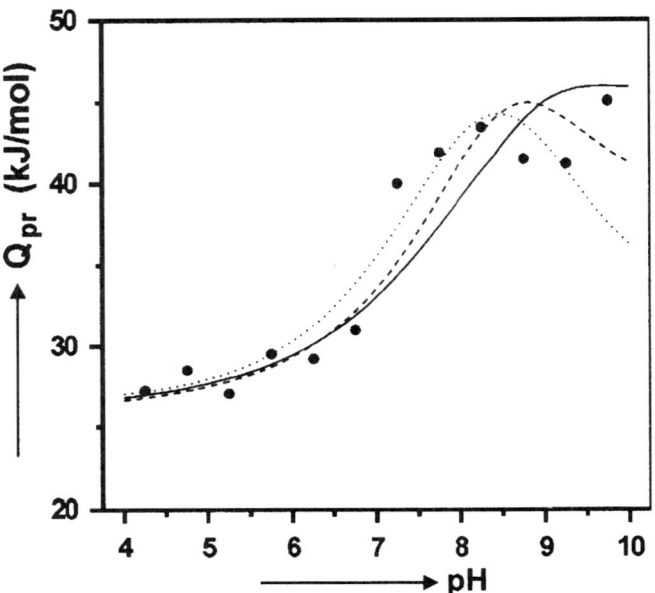

FIG. 20 Example of the comparison between the experimental (●) and theoretical heats of proton adsorption [155], calculated by assuming that $kT/c_+ = kT/c_A = 0.9$ (———), $kT/c_+ = kT/c_A = 0.8$ (- - - - -), and $kT/c_+ = kT/c_A = 0.7$ (......) and assuming that $kT/c_0 = kT/c_C = 0.9$. Other related parameters are collected in the table in Ref. 155.

We accepted for our theoretical investigations the experimental data of the alumina/NaCl interface reported by Machesky and Jacobs [145,146]. The choice of alumina as a good example to study the effects of surface heterogeneity was dictated by the fact that five kinds of surface oxygens may exist on the alumina surfaces [146], so the effects of surface heterogeneity should be clearly detected in this system.

As auxiliary experimental information, the radiometrically measured isotherms of Na^+ and Cl^- adsorption on alumina, reported by Sprycha [158], were also taken into consideration. Thus the theoretical expressions developed by us were used to fit simultaneously, applying the same set of parameters, the three sets of experimental data:

The titration curves δ_0 [145]
The enthalpy changes accompanying the titration steps [145]
The individual isotherms of Na^+ and Cl^- adsorption on alumina, reported by Sprycha [158]

Our numerical exercises showed that the best simultaneous fit of all those experimental data is achieved by using the expressions developed for the model of surface heterogeneity assuming small correlations between the adsorption energies of various surface complexes. The examples of theoretical curves of heat of proton adsorption are presented in Fig. 20. The fits of experimental data for the case of the model assuming high correlations between the adsorption energies of various surface complexes and for the case of energetically homogeneous solid surface are much poorer. In paper [155] we performed the modeling exercises that show as a result that a good simultaneous fit of all three kinds of experimental data can be obtained, with a small uncertainty for the values of the estimated adsorption parameters. A simultaneous fit of the measured enthalpic effects appears to be an especially strong criterion for a proper choice of adsorption parameters.

Now we would like to refer again to the papers by Machesky and Jacobs [145,146], where they attempted to draw some qualitative or semiqualitative conclusions concerning the enthalpic effects of ion adsorption. Their considerations were based on a somewhat different model of charging the surface oxygens, being a generalization of the 1-pK charging model for the case when a number of different surface oxygens exists on an oxide surface (MUSIC). Quantitative conclusions were drawn concerning the free energies of adsorption, but qualitative conclusions were drawn only about enthalpies of adsorption predicted by the model of charging the surface oxygens. This is because the related equations for enthalpic effects, corresponding to this charging mechanism, have not been developed yet.

Our theoretical considerations have been based on the 2-pK charging model, which is still the most commonly accepted in the literature. The question of which model represents the charging mechanism better, 2-pK or 1-pK, requires further extensive fundamental studies. As the definition of enthalpic effects depends on the accepted charging mechanism, the semiquantitative conclusions by Machesky and Jacobs cannot be directly compared with the quantitative conclusions drawn by us, accepting the 2-pK charging mechanism.

It seems, however, that many scientists still using the triple-layer model will appreciate having at their disposal a theoretical description of the enthalpic effects based on that triple-layer model. Even if more refined models gain great popularity in the future, it is worth knowing the full possibilities offered by the classical triple-layer model.

III. ADSORPTION OF CATIONIC SURFACTANTS ON HYDROPHILIC SILICA: EFFECTS OF SURFACE ENERGETIC HETEROGENEITY

A. Introduction

The term *surface-active agents*, usually contracted to *surfactants*, is used to designate amphiphilic molecules composed of two moieties, one highly soluble in water (*hydrophilic head-group*), and one oil-soluble (*hydrophobic tail*). The general behavior of such molecules in the aqueous solution and at the water-containing interfaces reflects the opposing tendencies of the hydrophobic tail to escape from the aqueous environment while the hydrophilic head-group tends to remain immersed in the water [159]. One of the most important consequences of this behavior is self-assembly of surfactant monomers to form colloidal aggregates. Self-association of surfactants in the free aqueous solution may lead to a variety of structures, such as spherical and cylindrical micelles, bilayered vesicles, and hexagonal and lamellar phases, depending primarily on (1) the nature of the surfactant molecule (i.e., the *hydrophilic–lipophilic balance* HLB), (2) the total surfactant concentration, and (3) the composition and environment of the aqueous phase. The equilibrium curvature of a surfactant aggregate is a result of the balance between (1) the interactions of hydrophobic tails with water, (2) the interactions of tails with themselves, (3) the hydration of head-groups, and (4) the interaction between the hydrated head-groups mediated in the case of ionic groups by their ionic atmospheres. Account must also be taken of geometrical and packing factors that have to be satisfied, in addition to the energetic factors.

Any successful explanation of surfactant behavior at the solid–solution interface must include as a basic tenet the clustering of surfactant monomers to form periodic adsorbate structures, which are named the *interfacial aggregates*. The prevailing view is that the presence of the solid surface induces a decrease in the chemical (or electrochemical) potential of the surfactant solute in the system, and in consequence the surfactant tends to reproduce some aggregate structures encountered in free solution. Strong indications of the existence of surface-bound aggregates have been obtained using NMR [160], Raman [161], electron spin resonance [162,163], fluorescence decay [164–168], neutron reflection [169,170], surface force [171–175], atomic force microscopy [176–181], and calorimetric study [182–194]. Nevertheless, the precise aggregate structure and its evolution with increasing surface coverage remain controversial. The proposed interpretations of the experimental results reported on the apparently identical adsorption systems are sometimes contradictory, since insufficient attention has been given to the control of important physical factors, to the potentially of the experimental technique, or to the limitations of particular theoretical approaches.

The application of atomic force microscopy (AFM) to probe the aggregate structure has roused the interest of many researchers. Unfortunately, a serious disadvantage of this technique is that it cannot be used for powdered and porous solids, which are of great industrial importance. In contrast with the previously established models of more or less close-packed surfactant monolayers or bilayers, AFM images of interfacial aggregates formed on the atomically smooth surfaces at surfactant concentrations well above the critical micelle concentration (cmc) showed periodic structures with a high degree of curvature [179,180]. For quaternary ammonium cationics on amorphous silica, a random close-packed structure of spherical aggregates was revealed, with the observed nearest-neighbor distance between the aggregates varying inversely with the pH of the aqueous phase. In the case of a crystalline mica surface, surface-bound aggregates were organized into elongated cylindrical shapes. These adsorbate associations adopted the form of continuous wormlike stripes at high solution pH values. The resulting structures of surfactant aggregates on a crystalline hydrophobic surface of graphite were found to be consistent with parallel half-cylinders, in which the surfactant tails were in direct contact with the surface.

For fundamental reasons, a simple extension of the above patterns to other classes of adsorption system seems questionable. Surfactant adsorption at equilibrium concentrations below the cmc involves only surfactant monomers, whereas the direct adsorption of micelles on the surface has been detected above the cmc [195]. Nevertheless, the existence of isolated islands of hydrophobic surfactant aggregates at low bulk concentrations (well below the cmc) and low pH values has been corroborated from a combined AFM and the contact angle studies of the adsorption of cetyltrimethylammonium bromide (CTAB) onto mica [181].

The shape and size of the interfacial aggregate should be a compromise between the free curvature, as defined by the energetic, geometrical, and packing factors arising from the molecular structure of the surfactant in a given environment, and the constraints imposed by the direct solute–surface interactions and the porous structure of the adsorbent. When the presence of a flat solid surface hardly affects the self-assembly of surfactant molecules, the formation of micelle-like aggregates anchored to the surface is expected. Such aggregates will be characterized by a small number of solid–surfactant interaction contacts, and they will form at equilibrium concentrations only a little lower than the cmc. For example, the adsorption of some nonionic and zwitterionic surfactants on hydrophilic surfaces is consistent with this picture [167,168,192]. The extended interfacial aggregates with numerous surfactant–solid interaction contacts lie at the opposite end of the aggregate scale. This aggregation scheme can be envisaged for adsorption systems characterized by strong surfactant–solid interactions, if the underlying surface lattice imposes a high local density of directly adsorbed monomers paralleled by a strong cohesion between their hydrophobic tails. The adsorption of ionic surfactants on oppositely charged surfaces with a high charge density and a homogeneous charge distribution is a good example. Here interfacial aggregation may start at relatively low surface coverages [181]. According to Wängnerud and Jönsson [196], the local concentration of surfactant monomers in the solution region adjacent to a highly charged surface may exceed the cmc value, thereby inducing interfacial aggregation at equilibrium concentrations significantly lower than the cmc.

Interfacial aggregates formed in a real adsorption system may cover a wide spectrum of aggregate structures lying between the above two extremes. Given the diversity of solid surfaces and surfactant molecules, systematic studies are needed to explore the morphology of the adsorbed layer at different surface coverage ratios. This task requires the application of a variety of experimental techniques. Simultaneously, progress in the current understanding of the phenomenon depends on the correlation between theoretical and experimental studies.

The adsorption and interfacial aggregation of ionic surfactants on the charged surfaces of mineral oxides play a crucial role in many industrial applications. Although extensive fundamental studies have been performed on the subject over the years, there is still considerable scope for improving the current understanding of the adsorption mechanism. From a commercial standpoint, the cationic surfactant + negatively charged silica system is of great importance. Scientifically, this is the most extensively studied adsorption system. In this section the thermodynamic aspects of the adsorption of cationic surfactants on a negatively charged silica surface will be discussed. A theoretical model will be proposed to describe simultaneously the amount adsorbed and the differential molar enthalpy of displacement at different surface coverages.

B. Conceptualization of Ionic Surfactant Adsorption on Charged Solid Surfaces

The driving force for the adsorption of ionic surfactants on oppositely charged surfaces derives from two main categories of interaction: (1) adsorbate–surface interaction, allowing a direct adsorption of surfactant ions at the interface, and (2) hydrophobic bonding, responsible for the formation of interfacial aggregates.

The first category includes primarily coulombic attractions between ionic headgroups and charged surface sites inducing the formation of an electric double layer, as well as specific (i.e., nonelectrical) interactions of the surfactant head-groups and/or tails (CH_2 segments) with the surface. The direct adsorption of surfactant ions has a marked competitive character and is strongly affected by the charging behavior of the support [189–191,194,197–200]. The surface charge density of mineral oxides has been shown to be a function of the pH and electrolyte concentration in the bulk phase [e.g., 95,200]. The ionization of the surface groups also appears to increase as a result of surfactant adsorption [199]. A rise in the surface charge is accompanied by the uptake or release of protons and a consequent increase or decrease in pH of the aqueous solution [e.g., 189–191]. In consequence, the number of electrostatically bound head-groups in a given adsorption system may be different under different pH and ionic strength conditions. Specific attractions between the ionic head-groups and the charged surface sites (e.g., surface complexation model [95,200–202]) and/or hydrophobic association of the surfactant tail with the surface, especially for solids with low surface charge (at low salt concentrations and low or moderate pH values [187,190,199]), contributes to the improved screening of the surface charge by the surfactant ions. In the latter case, the mineral surface is locally dewetted and gains the appearance of being coated with a hydrocarbon. As a result, the effect of increasing hydrophobicity is expected.

The hydrophobic effect, which seeks to minimize the area of contact between the hydrophobic tails and water while maintaining maximum interaction between the hydrophilic moiety and water, significantly contributes to the adsorption phenomenon: it induces the formation of interfacial aggregates of varying size and structure. The beginning of interfacial aggregation is localized at different parts of the adsorption isotherm, depending on the nature of the adsorbing ions, the charging behavior of the solid surface in respect of the current environment of the aqueous phase, and the surface heterogeneity and its affinity for a given solvent [190,191,194,196,197,200,203–205]. The flocculation, contact angle, surface force, or calorimetric data evidence a hydrophilic character of many solid supports coated with the adsorbed surfactant layer in the neighborhood and above the cmc [171,173,175,189–191,206]. This means that the bilayer-like coverage represents complete saturation of the surface.

According to the four-region model, dating from the studies of Somasundaran and Fuerstenau [207], the adsorption isotherm can be divided into four distinct regions when plotted on a log–log scale. In region I, the individual surfactant ions adsorb with their head-groups in contact with the surface owing to the coulombic and specific interactions. A strong lateral attraction between the surfactant tails occurs in region II, where primary aggregates are formed. Based on the results of fluorescence, Raman, electron spin resonance, and contact angle studies for ionic surfactant adsorption on the oppositely charged surface of alumina, Somasundaran et al. [164–166,207] have concluded that the surfactant ions in the primary aggregates are oriented with their ionic head-groups toward the solid surface, thereby forming hydrophobic patches on the surface. As the adsorption progresses in region III, these hemimicellar aggregates grow in the direction perpendicular to the surface (the number of aggregates remains constant in this region). The oncoming ions can interpenetrate into the hemimicelles, leaving their head-groups in the bulk solution and rendering the solid surface hydrophilic again. The region II/region III transition is due to neutralization of the surface charge by the oppositely charged surfactant. In the adsorption plateau region IV, the chemical potential of surfactant monomers in the micellar equilibrium solution remains fairly constant and, consequently, their adsorption affinity levels off.

Scamehorn et al. [208] have constructed a theoretical model to describe interfacial aggregation for anionic surfactants onto alumina. The starting point of their approach was a modified Cases' concept of two-dimensional condensation [209–211]. Here bilayered aggregation was proposed to occur either subsequently to or simultaneously with the formation of a monolayered condensed phase. Later, the model of patchwise topography with a gaussian distribution of adsorption energy was introduced into theoretical considerations. It permitted to conclude that, even at low surface coverages, the adsorbed layer of surfactant consisted of aggregate patches having a bilayer character. Based on the pseudo–phase separation approach for the bilayered aggregates modeled on admicelles, Harwell et al. arrived at a similar conclusion [212]. In the next paper, Harwell and Yeskie [213] compared chemical (electrochemical) potentials of the surfactant monomer in a bilayered (*admicelle*) and a monolayered (*hemimicelle*) aggregates under certain conditions. Such an analysis enabled them to provide theoretical evidence for the preferential formation of admicelles on solid surfaces with a high surface charge density. On the contrary, low surface charge densities would favor an association pathway to hemimicelles. However, no attention was given to whether the "free" adsorbed monomers, which were neither hemimicelles nor admicelles, might have a lower chemical (electrochemical) potential under some conditions. In other words, the authors did not make any attempt to decide whether the admicelles were directly formed by the association of the adsorbed monomers or whether their formation proceeded via a hemimicelle pathway.

Gu et al. have propagated a "two-step" isotherm model for cationic surfactant adsorption on silica surfaces [214–216]. The first step on the theoretical curve corresponds to the compensation of the surface charge by the surfactant ions adsorbed "head-on" (i.e., ionic head-groups oriented towards the surface). These individually localized organic ions serve as "anchors" for the interfacial aggregates. In the second step, small surface micelles are formed owing to the hydrophobic association between the surfactant tails. Based on the mass-action treatment, estimates of the surface aggregation number could be made. An empirical approach, enabling calculation of the surface aggregation number from the first and second plateau on the isotherm, was also suggested.

An original approach to the problem has been proposed by Böhmer and Koopal [197,198,200]. To avoid any ad hoc assumptions about the structure of the adsorbed layer, they applied the self-consistent field lattice theory to calculate the segment density profile of the adsorbed surfactant ions as a function of the distance from a flat solid surface. This theory predicted successfully a model "two-step" isotherm for oxide surfaces with a constant surface charge at low salt concentrations and a model "four-region" isotherm for constant potential surfaces (metal oxides at a fixed pH). Unfortunately, the theoretical consideration did not go beyond the interpretation of adsorption isotherms. Moreover, this theory failed to take account of structure inhomogeneities parallel to the surface, necessary to describe the thoroughly interfacial aggregation of ionic surfactants.

The formation of interfacial aggregates was modeled by Li and Ruckenstein [217]. The model was based on the assumption that the adsorbed phase could be seen as a nonideal two-dimensional mixture of monomers, monolayered hemimicelles, bilayered admicelles, and "empty sites" on a solid surface with a uniform surface charge. The deviations from ideal mixing behavior were treated via a two-dimensional analogue of the Flory–Huggins equation for nonideal solutions. The formation of local aggregates of varying size and structure was proposed to be a compromise between the entropy, which tends to disperse the surfactant ions, and the interactions between them, which tend to provoke their association. The theoretical model was compared with the experimental adsorption isotherms for sodium alkylbenzenesulphonates on alumina and kaolinite, as well as for sodium dodecyl sulphate on alumina. The anionic surfactants were found to adsorb as monomers at low bulk concentrations. With increasing concentration, first monolayered and then bilayered aggregates dominated on the surface. The final adsorbed phase consisted mainly of bilayered admicelles. The theoretical adsorption isotherm was shown to be only slightly affected by the upper limit of the aggregate size, i_{max}, chosen for computation: the goodness-of-fit for $i_{max} = 10$ was equally good as that for $i_{max} = 1000$.

A somewhat different model of the surface phase was developed by Łajtar et al. [204,205]. The starting point for theoretical consideration was a mixture of oblate interfacial aggregates, interacting through "excluded volume" interactions. To explain the limited size of the surface aggregates, the energy of adsorption of a single monomer and the lateral interaction energy between two neighboring monomers in an aggregate were assumed to decrease with increasing aggregate size. The two assumptions were consistent with the two-dimensional condensation model for surface aggregation on a heterogeneous oxide surface with a "partially correlated" topography. The monomers were included among the monolayered hemimicelles by assuming their aggregation number equal to unity. Therefore a direct contact of the hydrophobic surfactant tail with the solid surface was neglected.

The proportion between the various adsorbed species and the size and the number of aggregates of a given type were allowed to undergo unrestricted changes along the adsorption isotherm. Theoretical expressions for the adsorption isotherms and isosteric heats of adsorption corresponding to the above model assumptions were developed. The resulting equations were subsequently fitted to the experimental adsorption and the calorimetric data for alkyltrimethylammonium bromides adsorbed on a hydrophilic silica surface at free pH and in the absence of extra salt [205]. A picture of the surface phase composed of individually adsorbed monomers, at low surfactant concentrations, and of small bilayered admicelles (the average aggregation number did not exceed 10), above a concentration of about one-fourth the cmc, was deduced from the best-fit parameter values. The formation of hemimicelles was essentially reduced. As there were some discrepancies observed between the experimental and theoretical differential heats of adsorption, especially in the region of very small surface coverages where individual adsorption dominated, it was evident that some details in the solid–adsorbate interaction had not been taken into account yet.

The above model has been refined to account for the effects of surface heterogeneity on the adsorption of surfactant monomers [218]. The insertion of the surface heterogeneity concept in the theoretical consideration markedly improved the agreement between theoretical and experimental enthalpy values for adsorption of benzyldimethyldodecylammonium bromide on a hydrophilic silica surface at free pH and in the absence of extra salt. Strong specific attraction between the cationic head-groups and the surface complexes SiO^- was indicated as the adsorbate–adsorbent interaction dominating the adsorption of surfactant cations as monomers in a region of low surface coverages. The changing conformation of the adsorbed surfactant tail, discussed previously on the basis of the multisite-occupancy adsorption model [219,220], was believed to have a secondary effect.

C. Theoretical Model for the Adsorption of Cationic Surfactants on a Hydrophilic Oxide Surface

Let us consider an adsorbed surfactant phase consisting of oblate-shaped aggregates built up of surfactant monomers and characterized by only two parameters: the cross-sectional area of the aggregate and the number of nearest-neighbor pairs in the aggregate. Suppose there exist two aggregate types: *monolayered* and *bilayered*. The monolayered aggregates comprise one layer of monomers oriented "head-on" towards the charged surface of a mineral oxide, with the surfactant tails forming a hydrophobic film in contact with the aqueous solution. Aggregates having such a structure are commonly referred to as hemimicelles. The bilayered aggregates, termed admicelles, are viewed as a classical bilayer of monomers. The orientation of surfactant molecules in the first layer is the same as that in a hemimicelle, and the molecules in the second layer are postulated to adsorb on those of the first layer with the polar head-groups outwardly disposed. The surface aggregates have an oblate, disklike shape. For an aggregate with a fixed number of monomers, the most probable structure will be the one characterized by the highest cohesive force (i.e., the largest number of nearest neighbor attractive interactions). The two assumed structures of surface aggregates are shown schematically in Fig. 21.

In our further considerations, two indices will be used to relate different quantities to a given surface disk. The subscript i will denote the number of monomers in a hemimicelle or in one layer of an admicelle. Thus it will determine the size of a disk (i.e., its radius R_i and its cross-sectional area S_i). The superscript s, ($s = h, a$), will refer a given quantity to either a hemimicelle or an admicelle.

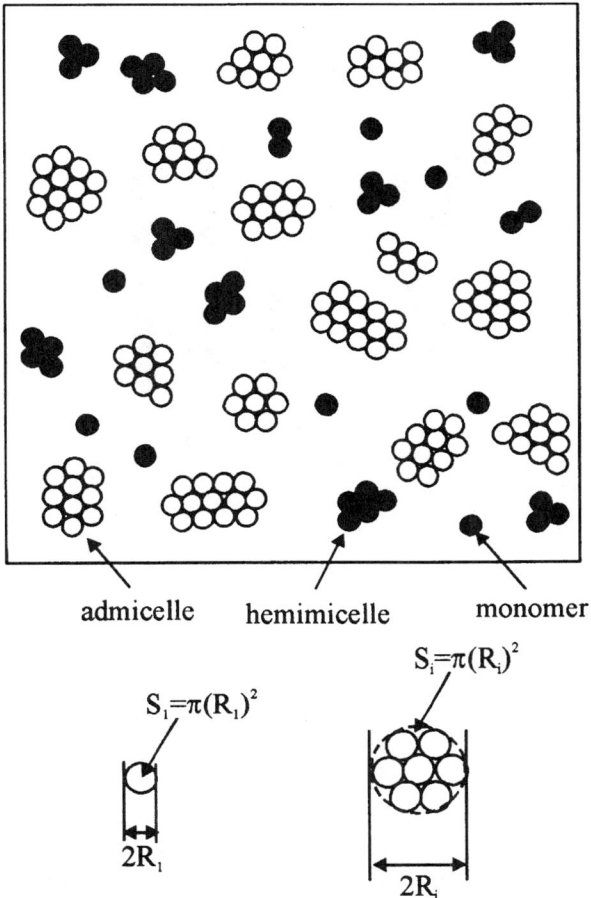

FIG. 21 Adsorbed phase seen from above. The black circles denote the monolayered aggregates (hemimicelles), whereas the white ones denote the bilayered aggregates (admicelles).

The total potential energy of one aggregate of size i is a sum of the energy of adsorption and the energy of lateral interaction between the monomers in the aggregate (Fig. 22). If we denote the energy of adsorption of a single monomer attached to the surface by its polar head-group by ε_1 and the energy of adsorption of a monomer in the second layer by ε_2, the total adsorption energy of aggregate i is

$$\varepsilon_i^s = i\varepsilon^s \qquad s = h, a \tag{82}$$

where $\varepsilon^a = \varepsilon^h + \varepsilon_2$.

Let l_i denote the number of the nearest-neighbor pairs in a hemimicelle composed of i monomers. Then the number of these pairs in an admicelle of size i will be $2l_i$. The energy of the lateral interaction in the aggregate can be written as

$$\omega_i^s = \delta^s l_i \omega \tag{83}$$

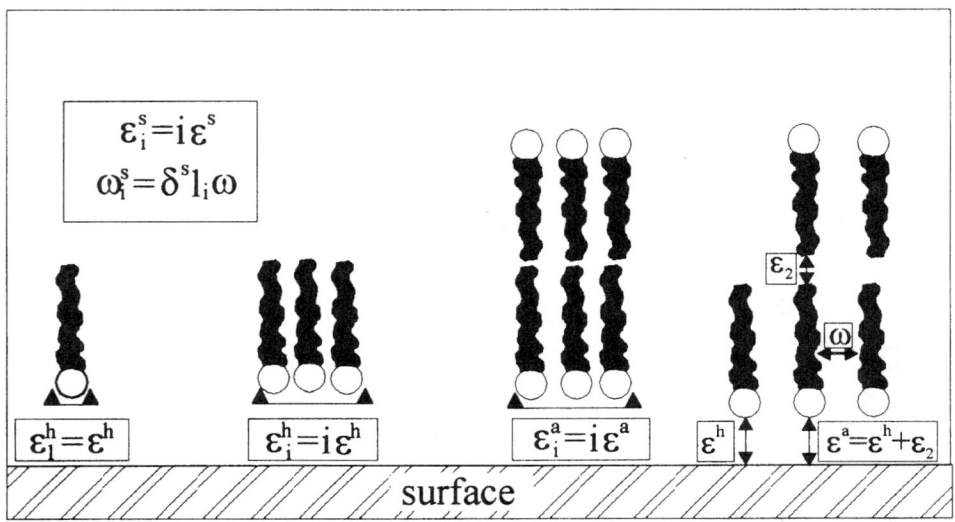

FIG. 22 The potential energy of surface aggregates.

where

$$\delta^s = \begin{cases} 1 & \text{for} \quad s = h \\ 2 & \text{for} \quad s = a \end{cases}$$

and ω is the energy of lateral interaction between two neighboring monomers in the aggregate. This pair interaction consists of the attraction between hydrophobic moieties of the neighboring surfactant molecules and of the electrostatic repulsion between their polar heads.

1. Isotherm of Adsorption

The total amount of surfactant, N, adsorbed on a solid surface is a function of the chemical potential of a monomer in the equilibrium bulk phase μ^b, namely,

$$\mu^b = \mu_0^b + kT \ln x \tag{84}$$

where x is the mole fraction of monomers in the bulk solution and μ_0^b is the standard chemical potential in the bulk phase.

The condition of thermodynamic equilibrium in the system has the form

$$\mu_i^s(\{N_i^s\}, N, T) = \delta^s i \mu^b(x, T) \tag{85}$$

where μ_i^s is the surface chemical potential of the aggregate of type (i, s), and $\{N_i^s\}$ is the distribution of monomers between aggregates of different types.

Suppose that the surface phase can be treated as a mixture of hard disks of different sizes formed on an *energetically homogeneous surface*. This means that we neglect all possible interactions between the surface aggregates other than the "excluded area" ones.

Under these circumstances, the canonical partition function for the adsorbed phase is

$$\ln Q(\{N_i^s\}, N, T) = \ln Z(\{N_i^s\}) + \sum_s \sum_{i=1} \left[N_i^s \ln f_i^s + N_i^s \left(\frac{i\varepsilon^s + \delta^s l_i \omega}{kT} \right) - \ln(N_i^s!) \right] \quad (86)$$

where

$$f_i^s = \delta^s i (q^s)^{\delta^s i} \Lambda^{-2} \quad (87)$$

In Eqs. (86) and (87), Z represents the configurational integral for the mixture of hard disks, q^s is the molecular internal partition function of a monomer in the surface aggregate, and Λ is the thermal wavelength of the monomer.

The scaled particle theory (STP) yields the following expression for the chemical potential of the surface aggregate of type (i, s)

$$\frac{\mu_i^s}{kT} = -\ln\left[f_i^s S_i \exp\left(\frac{i\varepsilon^s + \delta^s l_i \omega}{kT} \right) \right] + \ln\left(\frac{\Gamma_i^s}{1-\Theta} \right) + r_i \left(\frac{2A}{1-\Theta} \right)$$
$$+ s_i \left[\frac{\Gamma}{1-\Theta} + \left(\frac{A}{1-\Theta} \right)^2 \right] \quad (88)$$

where

$$\Gamma_i^s = \frac{N_i^s S_1}{S} \quad \text{and} \quad \Gamma = \sum_s \sum_i \Gamma_i^s \quad (89)$$

and

$$r_i = \frac{R_i}{R_1} \quad \text{and} \quad A = \sum_s \sum_i \Gamma_i^s r_i \quad (90)$$

and

$$s_i = \frac{S_i}{S_1} = \frac{\pi R_i^2}{\pi R_1^2} = r_i^2 \quad (91)$$

$$\Theta = \sum_s \sum_i \Gamma_i^s s_i = \sum_s \sum_i \Theta_i^s \quad (92)$$

where S_i is the cross sectional area of the ith disk, R_i is its radius, and S is the total area of the adsorbent surface. Different approximations can be used to calculate s_i. The following approximation has been validated in our previous publications [204,205]:

$$s_i = \frac{4i - 1}{3} \quad (93)$$

Equation (93) is correct for $i = 1, 7, 19, \ldots$, i.e., when the disks are fully symmetrical.

Combination of Eqs. (84,85,88–92) yields the equation system for the individual adsorption isotherms Γ_i^s and for the total adsorption isotherm Γ, that is,

$$\Gamma_i^s = (x)^{\delta^s i} K_i^s (1 - \Theta) \exp\left\{ -r_i \frac{2A}{1-\Theta} - s_i \left[\frac{\Gamma}{1-\Theta} + \left(\frac{A}{1-\Theta} \right)^2 \right] \right\} \quad \text{for } i \leq i_{\max} \quad (94)$$

where

$$K_i^s = [\Lambda^3(q)^{-1}]^{\delta^s i} f_i^s S_1 \exp\left(\frac{i\varepsilon^s + \delta^s l_i \omega}{kT}\right) \quad (95)$$

q is the internal molecular partition function of the surfactant molecule in the bulk phase and i_{max} is the maximum value of i.

The extension of the above theoretical considerations to a heterogeneous surface is not a trivial problem. From our previous numerical analysis of adsorption isotherms for cationic surfactants on oxide surfaces, it follows that the first step, observed in a wide range of initial bulk concentrations, corresponds to the adsorption of single monomers. Like simple cations, the cationic head-groups of surfactant molecules are adsorbed initially within the electrical double layer. In addition to the coulombic interactions, they also bond chemically to oxygen complexes SO$^-$ on the surface. It has been recently realized that such chemical bonds play a dominant role in the adsorption of simple cations within the electrical double layer [201,202]. Because of these strong interactions, surfactant monomers are highly immobilized on the active sites. It is thus reasonable to assume that the energetic heterogeneity of the surface affects only the adsorption of monomers [i.e., (1, h) type disks]. In their book, Rudziński and Everett [68] proposed a possible modification of Eq. (94) for monomer adsorption on a heterogeneous surface with random topography. According to this idea, Eq. (94) becomes, for $i=1$ and $s=h$,

$$\frac{\bar{\Gamma}_1^h}{1-\Theta} = \left\{x\bar{K}_1^h \exp\left\{-r_1 \frac{2A}{1-\Theta} - s_1\left[\frac{\Gamma}{1-\Theta} + \left(\frac{A}{1-\Theta}\right)^2\right]\right\}\right\}^{\alpha kT} \quad (96)$$

where $0 < \alpha kT < 1$ and α is the heterogeneity parameter characterizing the width of the energy distribution function. The constant $\bar{K}_1^h$ is defined as

$$\bar{K}_1^h = [\Lambda^3(q)^{-1}] f_1^h S_1 \exp\left(\frac{\bar{\varepsilon}^h}{kT}\right) \quad (97)$$

where $\bar{\varepsilon}^h$ has the meaning of the most probable value of the adsorption energy of a single monomer on a heterogeneous surface. Thus, taking into account the effects of surface heterogeneity on monomer adsorption, one has to replace the equation for Γ_1^h in the equation system Eq. (94) by $\bar{\Gamma}_1^h$ given in Eq. (96).

If the upper limit of the aggregate size is i_{max}, then we have to solve a system of at least $2i_{max}$ nonlinear equations. In the particular case of equation system Eq. (94), it is possible to reduce the numerical problem to solving a system of two nonlinear equations [204,205]. For this purpose, we introduce two new variables, ξ and φ, defined as

$$\xi = \frac{2A}{1-\Theta} \quad \text{and} \quad \varphi = \frac{\Gamma}{1-\Theta} + \left(\frac{A}{1-\Theta}\right)^2 \quad (98)$$

With this notation, the equation system Eq. (94) becomes

$$\Gamma_i^s = (1-\Theta)d_i^s \quad (99)$$

where, for disks of type (1, h),

$$d_1^h = [x\bar{K}_1^h \exp(-\xi r_1 - \varphi s_1)]^{\alpha kT} \quad (100)$$

and, for the others,

$$d_i^s = (x)^{\delta^s i} K_i^s \exp(-\xi r_i - \varphi s_i) \tag{101}$$

From Eqs. (98) and (99), it follows that

$$\frac{\Gamma}{1-\Theta} = \sum_s \sum_i d_i^s(\xi, \varphi) \tag{102}$$

$$\frac{A}{1-\Theta} = \sum_s \sum_i r_i d_i^s(\xi, \varphi) \tag{103}$$

Introducing the new variables on the left-hand side of Eqs. (102) and (103) yields the desired system of two nonlinear equations

$$F_1(\xi, \varphi) = \sum_s \sum_i r_i d_i^s(\xi, \varphi) - 0.5\xi = 0 \tag{104a}$$

$$F_2(\xi, \varphi) = \sum_s \sum_i d_i^s(\xi, \varphi) + 0.25\xi^2 - \varphi = 0 \tag{104b}$$

the solution of which gives ξ and φ. The last two variables determine unambiguously the distribution of adsorbed molecules among surface aggregates of various sizes and structures. Having evaluated ξ and φ, one can calculate the individual adsorption isotherms using the equation

$$\Gamma_i^s = \frac{d_i^s}{1 + \sum_s \sum_i d_i^s s_i} \quad \text{for} \quad 1 \leq i \leq i_{\max} \quad \text{and} \quad s = h, a \tag{105}$$

2. Differential Heat of Adsorption

Calorimetric measurements of the enthalpy changes associated with the surfactant adsorption occurring in a calorimetric cell at a constant pressure yield values of the differential molar heat of adsorption as a function of the surface coverage ratio [191,192,221]. For the adsorption systems studied, heats of adsorption reveal many important features, which can hardly be seen in the behavior of experimental adsorption isotherms [221,222]. It has been demonstrated [204,205,220] that a simultaneous analysis of the experimental adsorption isotherms and heats of adsorption allows for discrimination between different models of surfactant adsorption.

The differential molar heat accompanying the adsorption of surfactant molecules, Q_{st}, is given by

$$Q_{\mathrm{st}} = kT^2 \left[\frac{\partial \ln x}{\partial T}\right]_{N=\Sigma\Sigma\delta^s i N_i^s} = -kT^2 \frac{\sum_s \sum_i \delta^s i \Gamma_{iT}^{s'}}{\sum_s \sum_i \delta^s i \Gamma_{i\ln x}^{s'}} \tag{106}$$

The differentiation of Eq. (105) with respect to T and $\ln x$ gives derivatives $\Gamma_{iT}^{s'} = (\partial \Gamma_i^s / \partial T)$ and $\Gamma_{i\ln x}^{s'} = (\partial \Gamma_i^s / \partial \ln x)$ in the form

$$\Gamma_{iY}^{s'} = \frac{d_{iY}^{s'} - \Gamma_i^s \sum_s \sum_k d_{kY}^{s'} s_k}{1 + \sum_s \sum_k d_k^s s_k} \quad Y = T \text{ or } \ln x \tag{107}$$

The derivatives $d_{iY}^{s'} = \partial d_i^s / \partial Y$ can be calculated from Eqs. (100) and (101). For disks of type (1, h), we obtain

$$d_{1T}^{h'} = \alpha k T d_1^h \left[\frac{\partial \ln \bar{K}_1^h}{\partial T} - r_1 \xi_T' - s_1 \varphi_T' \right] + \frac{d_1^h}{T} \ln d_1^h \tag{108}$$

$$d_{1\ln x}^{h'} = \alpha k T d_1^h [1 - r_1 \xi_{\ln x}' - s_1 \varphi_{\ln x}'] \tag{109}$$

whereas, for the others, we have

$$d_{iT}^{s'} = d_i^s \left[\frac{\partial \ln K_i^s}{\partial T} - r_i \xi_T' - s_i \varphi_T' \right] \tag{110}$$

$$d_{i\ln x}^{s'} = d_i^s [\delta^s i - r_i \xi_{\ln x}' - s_i \varphi_{\ln x}'] \tag{111}$$

The derivatives ξ_T', $\xi_{\ln x}'$, φ_T', $\varphi_{\ln x}'$ can be calculated from equation system Eq. (104). The differentiation of this equation system with respect to T gives a system of two linear equations

$$d_1^h r_1 \left[(\alpha k T - 1) \frac{\partial \ln \bar{K}_1^h}{\partial T} + \frac{\ln d_1^h}{T} \right] + \sum_s \sum_k r_k d_k^s \frac{\partial \ln K_k^s}{\partial T}$$

$$- \left[s_1 d_1^h (\alpha k T - 1) + \sum_s \sum_k s_k d_k^s + 0.5 \right] \xi_T' \tag{112a}$$

$$- \left[r_1 s_1 d_1^h (\alpha k T - 1) + \sum_s \sum_i r_i s_i d_i^s \right] \varphi_T' = 0$$

and

$$d_1^h \left[(\alpha k T - 1) \frac{\partial \ln \bar{K}_1^h}{\partial T} + \frac{\ln d_1^h}{T} \right] + \sum_s \sum_k d_k^s \frac{\partial \ln K_k^s}{\partial T}$$

$$- (\alpha k T - 1) r_1 d_1^h \xi_T' - \left[s_1 d_1^h (\alpha k T - 1) + \frac{1}{1 - \Theta} \right] \varphi_T' = 0 \tag{112b}$$

the solution of which gives the derivatives ξ_T', φ_T'.

The differentiation of the equation system Eq. (104) with respect to $\ln x$ yields

$$\sum_s \sum_k \delta^s i r_k d_k^s + r_1 (\alpha k T - 1) d_1^h - \left[s_1 (\alpha k T - 1) d_1^h + \sum_s \sum_k s_k d_k^s + 0.5 \right] \ln \zeta_{\ln x}'$$

$$- \left[s_1 r_1 (\alpha k T - 1) d_1^h + \sum_s \sum_k s_k r_k d_k^s \right] \varphi_{\ln x}' = 0 \tag{113a}$$

and

$$\sum_s \sum_k \delta^s i d_k^s + (\alpha k T - 1) d_1^h - r_1 (\alpha k T - 1) d_1^h \xi_{\ln x}'$$

$$- \left[s_1 (\alpha k T - 1) d_1^h + \sum_s \sum_k s_k d_k^s + 1 \right] \varphi_{\ln x}' = 0 \tag{113b}$$

This is again a system of two linear equations from which the derivatives $\xi'_{\ln x}$ and $\varphi'_{\ln x}$ can be calculated.

Having evaluated the derivatives ξ'_Y, φ'_Y ($Y = T, \ln x$), the values of $\Gamma^{s'}_{iY}$ can be found from Eq. (107) and then the differential heat of adsorption is computed from Eq. (106).

Now, it is necessary to detail the quantity K_i^s defined in Eq. (95). Replacing f_i^s by Eq. (87) and assuming that the number of nearest-neighbor pairs in the aggregate is a linear function of aggregate size, $l_i = \beta_1 i + \beta_2$, the constant K_i^s can be rewritten as

$$K_i^s = \delta^s i S_1 \Lambda^{-2} (q^s \Lambda^3 (q)^{-1})^{\delta^s i} \exp\left\{\frac{i\varepsilon^s + \delta^s(\beta_1 i + \beta_2)\omega}{kT}\right\} \quad (114)$$

Numerous numerical exercises based on the above-discussed approach have shown that assuming a constant value of parameters ε and ω leads to a rapid two-dimensional condensation-like aggregation at a certain value of the surfactant concentration in the bulk phase. Formation of aggregates having finite dimensions occurred when the values of ε and/or ω were taken to decrease with increasing aggregate size. The decrease of ω with i may be caused by the same physical factors as those responsible for the finite size of micelles in free solution. In our opinion, however, it is the decrease of ε with i that is the main factor governing the formation of surface aggregates. The decrease of ε may be ascribed to the adsorption on a heterogeneous surface with a "partially correlated" topography [204,205]. In practice, both these parameters may be assumed to decrease linearly with increasing aggregate size. Therefore

$$\omega = \omega_0 - \Delta\omega i \quad (115a)$$
$$\varepsilon^s = \varepsilon_0^s - \Delta\varepsilon^s i \quad (115b)$$

Under such circumstances, the quantity K_i^s can be written in the compact form

$$K_i^s = ia^s \exp\{b^s i - c^s i^2\} \quad (116)$$

where

$$a^s = \delta^s S_1 \Lambda^{-2} \exp\left\{\frac{\delta^s \beta_2 \omega_0}{kT}\right\} \quad (117)$$

$$b^s = \ln[(q^s \Lambda^3 q^{-1})^{\delta^s}] + \frac{\varepsilon_0^s + \delta^s(\beta_1 \omega_0 - \beta_2 \Delta\omega)}{kT} \quad (118)$$

$$c^s = \frac{(\Delta\varepsilon^s + \delta^s \beta_1 \Delta\omega)}{kT} \quad (119)$$

For the purpose of further analysis, we shall assume that the values of the parameters a, b, and c characterizing the adsorption of a single monomer are the same as those valid for a monomer in a hemimicelle.

In order to compute the differential molar heat of adsorption, the values of the derivatives $\partial \ln K_i^s / \partial T$ have to be known. According to the approximation of Eq. (116), these derivatives are given by

$$\frac{\partial \ln K_i^s}{\partial T} = \frac{\partial \ln a^s}{\partial T} + i \frac{\partial b^s}{\partial T} - i^2 \frac{\partial c^s}{\partial T} \quad (120)$$

where

$$\frac{\partial \ln a^s}{\partial T} = \frac{1}{T}\left(1 - \ln\frac{a^s \Lambda^2}{\delta^s S_1}\right) \tag{121}$$

$$\frac{\partial b^s}{\partial T} = \frac{1}{T}\left[T\delta^s\left(\frac{\partial \ln q^s}{\partial T} - \frac{\partial \ln q}{\partial T}\right) - \frac{3\delta^s}{2} - \frac{\varepsilon_0^s + \delta^s(\beta_1\omega_0 - \beta_2\Delta\omega)}{kT}\right] \tag{122}$$

$$\frac{\partial c^s}{\partial T} = -\frac{c^s}{T} \tag{123}$$

The parameters a^s, b^s, c^s, and S_1 are found by fitting the theoretical equations to the experimental adsorption data. Then the derivatives $\partial a^s/\partial T$ and $\partial c^s/\partial T$ can be evaluated from Eqs. (121) and (123). To calculate the derivative $\partial b^s/\partial T$, the temperature dependence of the molecular partition functions q^s and q has to be known. Since such a relationship is unknown, we shall treat $\partial b^s/\partial T$ as an additional best-fit parameter, denoted by h^s (for s = h, a).

For the case of both hemimicelles and admicelles formed on the adsorbent surface, the relation between the theoretical quantities Γ_i^s and the experimentally measured adsorption Γ_t, usually expressed in moles per unit of surface area, takes the form

$$\Gamma_t = \frac{1}{N_A S}\sum_s\sum_i \delta^s i N_i^s = \frac{1}{N_A S_1}\sum_s\sum_i \delta^s i \Gamma_i^s \tag{124}$$

where N_A is the Avogadro number and S is the total surface area of the adsorbent.

The assumption that only monomers are adsorbed is consistent with the experimentally observed invariance of surfactant adsorption above the cmc. It can be derived from the requirement that the concentration of monomers in the bulk solution (chemical potential of surfactant) remains almost constant, when the cmc is reached. To calculate the monomer concentration corresponding to a given total surfactant concentration in the bulk solution, one can use, to a good approximation, the law of the mass action, that is,

$$K_n = \frac{x_n}{n(x)^n} \tag{125}$$

where K_n is the equilibrium constant for the formation of micelles of size n from the n single surfactant molecules, x_n is the mole fraction of the surfactant in the micelles, and x is the mole fraction of monomer. The equilibrium constant can be found from the relation

$$\ln K_n = \frac{-\Delta G_{\text{mic}}}{RT} = -n\ln\left(\frac{\text{cmc}}{55.5}\right) \tag{126}$$

where cmc is expressed in mol·dm^{-3}.

D. Comparison of the Model Predictions with Experimental Data

It is now commonly realized that the calorimetric effects accompanying the adsorption of surfactants at the oxide/aqueous solution interfaces carry very important information about the adsorption mechanism in these systems. In spite of this, the vast majority of published works on the theoretical approach to and modeling of surfactant adsorption has been devoted to the interpretation of adsorption isotherms which are not so much sensitive to the nature of the adsorption system as the enthalpies of adsorption. A reliable theory of the adsorption process should lead to a simultaneous good fit of both the experimental adsorption isotherms and the accompanying heats of adsorption. The proposed model brings, to our knowledge, a first successful attempt of this kind.

In order to check the ability of our model to predict quantitatively the adsorption behavior of ionic surfactants on oppositely charged oxide surfaces, we have chosen for analysis the experimental isotherms and differential enthalpy curves for benzyldimethyl dodecylammonium bromide (BDDAB) adsorbed on two different silica samples: (1) as-received nonporous, surface-hydroxylated amorphous silica obtained by precipitation from sodium silicate solution (referred to as SilNa), and (2) the SilNa sample subjected to a chemical pretreatment with 3N HCl and then subsequent water washing (referred to as SilH) [187,190]. It is important to recall that the adsorption experiments were carried out from deionized water at free pH and in the absence of background electrolyte. A steady increase in the concentration of sodium cations in the supernatant liquid was monitored during adsorption. It was ascribed to the release of certain mineral impurities from the silica surface by the adsorbing $BDDA^+$ ions. Since neither pH nor ionic strength were adjusted in the experiments, both the surface potential and the surface charge of the silica samples could vary to some extent as the adsorption progressed. The observed drop in the pH of the supernatant solution in a range of low and moderate bulk molalities was considered as evidence of a change in the surface charge in the course of adsorption. In the case of SilNa a marked ion exchange between Na^+ and $BDDA^+$ cations was also detected.

Negative silicas are classic model surfaces for studying the adsorption of cationic surfactants [e.g., 160,162,163,169, 179,187,190,194,196,199,205,206,214–216,218–220]. Due to ionizable silanol groups, the silica surface has a variable negative surface charge that increases with pH. The behavior of silica/electrolyte solution interface differs significantly from that of the metal hydroxides. For example, the surface charge vs. pH curves have a shape that is characteristic only of silica surface [199,223,224]. The hydroxylated surface of silica has a point of zero charge at about pH 2, but the density of negative charge remains low until pH 6 even at quite high electrolyte concentrations. The surface potential and total double-layer capacitance have been found to be much higher than the corresponding theoretical values calculated from the Nernst equation [198,223]. Moreover, the surface purity, hydrophobic/hydrophilic character, solubility, or charging behavior of silica in aqueous solutions can vary extensively according to the method used for its preparation [225].

There is strong experimental evidence that the surface of amorphous silica is strongly heterogeneous. The heterogeneity effect on adsorption on oxides is manifested mainly as the variation in the values of the "intrinsic" reaction constants of the surface complexation reactions defined by Eqs. (29a–29d). The dispersion of K_{a1}^{int} and K_{a2}^{int} values is responsible for the different values of the surface charge δ_0 on different surface areas, which affects the surfactant adsorption via columbic interactions [175,226]. The ability of a site to dissociate is dependent on the charging of neighboring sites, particularly if these sites are anchored to the same silicon atom. Then, the dispersion of the intrinsic constants K_C^{int} and K_A^{int} will lead to different values of the surface charge δ_β, also affecting the adsorption of surface aggregates. Further, in the case of surfactant adsorption also the dispersion of the energy of specific interactions between a surfactant molecule and the surface has to be taken into account.

That picture of surface heterogeneity is complicated still further by the effects arising from the nature of surface topography. The model of patchwise topography can hardly be accepted in the case of amorphous silica. Much more reliable seems to be the model of random topography assuming that the surface sites characterized by different values

TABLE 3 The Parameters Obtained from Fitting Best the Theoretical Adsorption Isotherms and Heats of Adsorption to Experimental Data

	S_1 (Å^2)	a^h	b^h	c^h	h^h (1/K)	a^a	b^a	c^a	h^a (1/K)	$kT\alpha$
SilNa	50	0.95	11.00	0.19	-5.6	$0.9598 \cdot 10^{-4}$	24.35	0.38	-14.5	0.6
SilH	50	0.95	5.30	0.17	-13.0	$0.9598 \cdot 10^{-4}$	20.55	0.20	-14.0	0.6

of adsorption parameters are distributed on a solid surface at random. That model was accepted by Rudziński et al. [79,80] in their studies of heterogeneity effects on adsorption of simple ions within the electrical double layer formed at water/oxide interfaces as well as by Narkiewicz-Michałek et al. [219,220] considering the adsorption of ionic surfactants at the water/oxide interface.

Nowadays, a general feeling is growing that the actual solid surfaces are characterized by a topography intermediate between patchwise and random. Such surfaces are called "partially correlated." As mentioned above, the concept of a partially correlated heterogeneous surface can be used to explain the decrease of the potential energy per monomer with the growing size of an aggregate, in our model of surface aggregation.

Although uncontrollable changes in the experimental conditions may affect to some extent the behavior of our systems, we decided to fit simultaneously the adsorption isotherms and enthalpy curves by our model equations. This strategy gave us the possibility of estimating the effect of surface heterogeneity on the adsorption mechanism in these systems, providing at the same time insight into the structure of the surface phase with an increasing surface coverage. It also allows estimating the values of the important energetic parameters of interest. In general, there are 10 parameters at the beginning to be fitted for each adsorption isotherm and the related enthalpy curve. However, the number of the adjustable parameters can be reduced by fitting simultaneously the theoretical equations for Γ_t and Q_{st} to experimental data obtained with the same surfactant molecule adsorbed on two silica samples. The parameters characterizing only a surfactant molecule, like S_1, a^h and a^a, should have the same values, irrespective of the adsorbent properties. Furthermore, the following relation between the parameters a^h and a^a can be deduced from Eq. (117):

$$a^s = \frac{2(a^h)^2}{S_1 \Lambda^{-2}} \tag{127}$$

so that only one of them is adjustable. The values of the bulk aggregation number, which is necessary to calculate the monomer concentration corresponding to a given total concentration of the surfactant in the bulk phase, were taken from the literature [227].

The agreement between theory and experiment is presented in Figs. 23–26, and the best-fit values of the model parameters are collected in Table 3. Figures 23 and 24 show the result of fitting our theoretical expressions to the experimental adsorption data for the two silica samples. The related fits to the experimental enthalpy values are plotted as functions of the quantity of surfactant adsorption in Figs. 25 and 26. The contributions of the monolayered and bilayered aggregates to the best-fit functions are represented as the dashed lines in these figures.

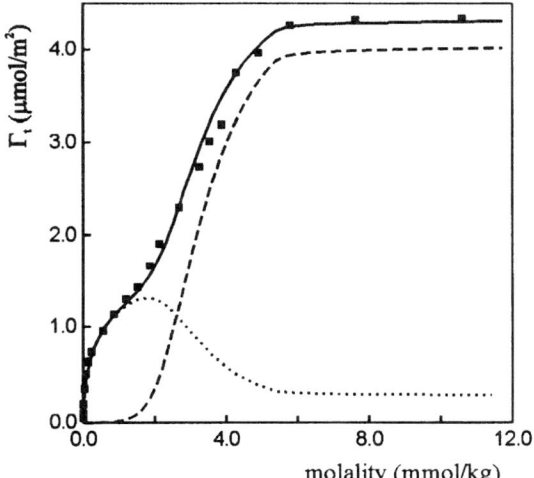

FIG. 23 Agreement between the experimental (■) and the theoretical (———) adsorption isotherms of BDDA$^+$ ions adsorbed on SilNa at 298 K and pH free. The theoretical curve is calculated using the parameters collected in Table 3. The dashed lines denote contributions to the total adsorption isotherm coming from hemimicelle (......) and admicelle (- - - - -) adsorption.

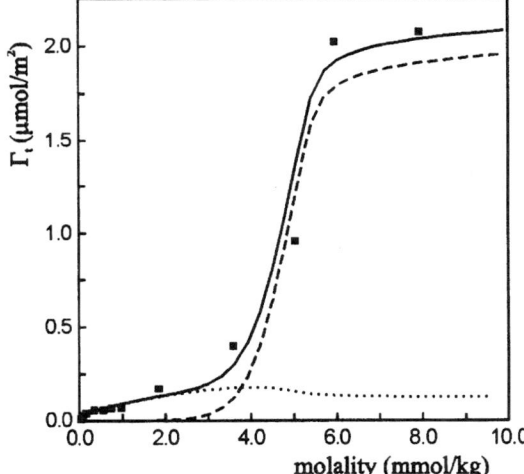

FIG. 24 Agreement between the experimental (■) and the theoretical (———) adsorption isotherms of BDDA$^+$ ions adsorbed on SilH at 298 K and pH free. The theoretical curve is calculated using the parameters collected in Table 3. The dashed lines denote contributions to the total adsorption isotherm coming from hemimicelle (......) and admicelle (- - - - -) adsorption.

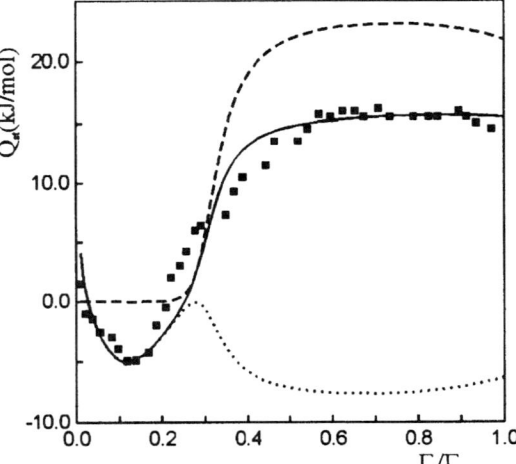

FIG. 25 Agreement between the experimental (■) and the theoretical (———) heat of adsorption of BDDA$^+$ ions adsorbed on SilNa at 298 K and pH free. The theoretical curve is calculated using the parameters collected in Table 3. The contributions coming from the hemimicelle (......) and admicelle (- - - - -) adsorption to the total differential heat of adsorption are also shown.

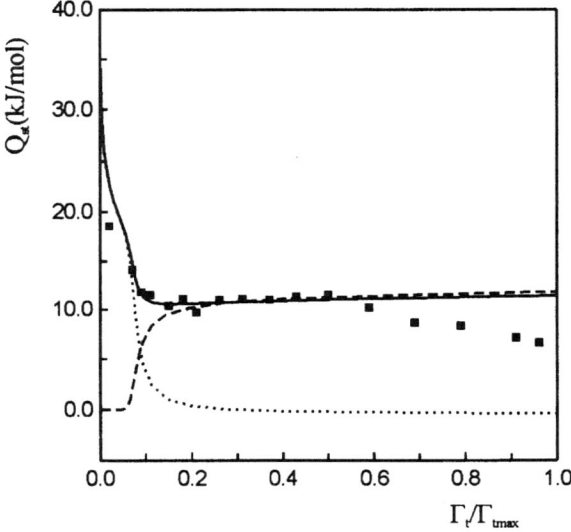

FIG. 26 Agreement between the experimental (■) and the theoretical (———) heat of adsorption of BDDA$^+$ ions adsorbed on SilH at 298 K and pH free. The theoretical curve is calculated using the parameters collected in Table 3. The contributions coming from the hemimicelle (......) and admicelle (- - - - -) adsorption to the total differential heat of adsorption are also shown.

It can be seen that the present theoretical model fits well the experimental adsorption isotherms and the accompanying calorimetric effects. Our previous model, neglecting the effect of surface heterogeneity on monomer adsorption, could not reproduce the decreasing part in the enthalpy curve at low surface coverages [205]. So, the present improvement is a strong argument for the role of the oxide surface heterogeneity in the mechanism of surfactant adsorption. The best-fit value of the "heterogeneity parameter" $kT\alpha = 0.6$ obtained for both adsorption systems indicates that the dispersion of the active cation–surface interactions are of the same type for both silica samples.

Looking to Figs. 23 and 24, one can see that both isotherms exhibit a two-step character. The first step corresponds to the adsorption of monomers and monolayered hemimicelles, whereas the second one is due to the bilayered admicelles forming at higher surfactant concentrations. This observation is in agreement with the quite common view that the "head-on" adsorption of surfactant ions dominates at low surfactant concentrations, while the net surface charge has an opposite sign [197]. The electrophoretic mobility measurements indicate that the isoelectric point is reached at an amount adsorbed of about 0.9 μmol m^{-2} for the SilNa system and of about 0.06 μmol m^{-2} for the SilH one [187,190]. This indicates that the negative charge density of the silica surface is much lower in the case of SilH. As a consequence, the average density of the "head-on" adsorbed molecules is too small to allow remarkable tail–tail interactions to exist and the subsequent formation of surface aggregates. This is manifested by a "hesitation" in adsorption at low bulk concentrations, leading to the appearance of a pseudoplateau in the isotherm of adsorption on the SilH surface. For the SilNa surface, exhibiting a higher charge density, this "hesitation" in adsorption is much less pronounced.

The curves of differential enthalpy of adsorption, plotted against the amount of BDDAB adsorbed on the two silica samples, had two distinct shapes, indicating a pronounced difference in the energetics of adsorption. The net energy of head-on adsorption is clearly included in the energetic parameters b^h and h^h: their best-fit values are much different for both the silica surfaces.

In order to find the source of the differences in adsorption behavior of our systems, we decided to analyze the evolution of the aggregate size with the surface coverage, predicted by our model. In Fig. 27 the theoretical mean aggregation numbers defined as

$$\bar{N}_{\text{agg.}} = \frac{\sum_s \sum_i \delta^s i N_i^s}{\sum_s \sum_i N_i^s} \tag{128}$$

are plotted against the surface coverage ratio $\Gamma_t/\Gamma_{\text{tmax}}$ for both the silica samples. It is clear that at low relative coverages the BDDA$^+$ ions adsorb as single monomers on both silica samples, but the further picture of the surface aggregation is much different for each of the silica samples. This can clearly be seen when one analyzes the distributions of the surface areas occupied by various aggregates among their aggregation numbers. This is presented in Figs. 28 and 29 for three values of the relative surface coverage ratio $\theta/\theta_{\text{max}}$ chosen at the beginning, in the middle, and at the end of the region where bilayered aggregates are formed. It is to be noted that, with increasing value of $\theta/\theta_{\text{max}}$ of SilNa, the heights of the distributions for admicelles increase but their widths do not show a marked change. The most probable aggregation number increases gradually, up to a value of 10 reached at $\theta/\theta_{\text{max}} = 0.99$. This means that in the region studied the admicelles grow both in size and in number. In contrast, the height of the distributions for hemimicelles decreases with increasing coverage ratio. For SilNa, the aggregates of this type are really

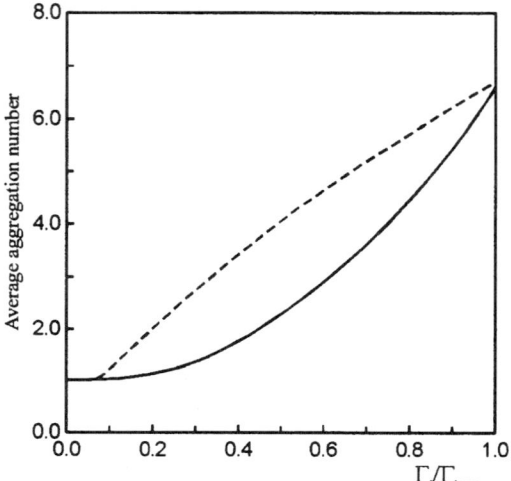

FIG. 27 Average aggregation number of the BDDA$^+$ surface aggregates formed on SilNa (———) and SilH (- - - - -) calculated using the parameters collected in Table 3 and plotted as a function of the relative surface coverage Γ_t/Γ_{tmax}.

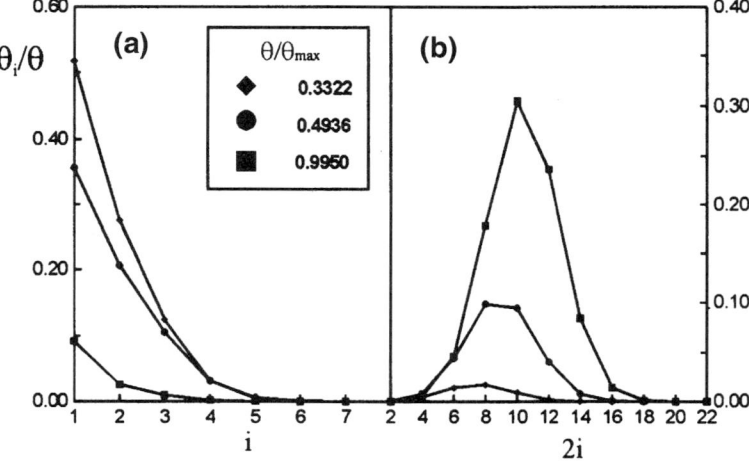

FIG. 28 Distributions of the surface areas occupied by the monolayered (a) and the bilayered surface aggregates (b) formed by BDDA$^+$ ions on the surface of SilNa at 298 K among their aggregation numbers calculated using the parameters from Table 3 for different values of the relative surface coverage θ/θ_{max}.

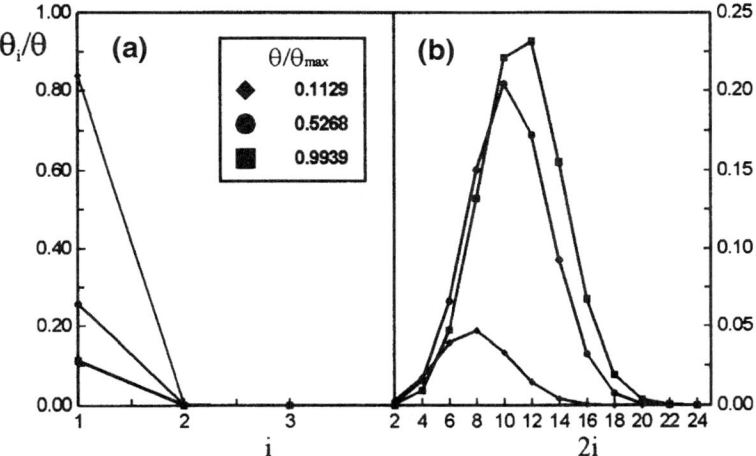

FIG. 29 Distributions of the surface areas occupied by the monolayered (a) and the bilayered surface aggregates (b) formed by $BDDA^+$ ions on the surface of SilH at 298 K among their aggregation numbers calculated using the parameters from Table 3 for different values of the relative surface coverage θ/θ_{max}.

very small and the surface area occupied by hemimicelles with an aggregation number greater than 4 is negligible. The most probable aggregation number corresponds to single monomers.

In the case of SilH, hemimicellar aggregates are absent on the surface, except of those with aggregation number equal to unity. As far as admicelles are concerned, the most probable aggregation number increases with increasing value of θ/θ_{max}, from a value of 6 at $\theta/\theta_{max} = 0.11$ to a value of 10–12 at $\theta/\theta_{max} = 0.99$. The distributions obtained for $\theta/\theta_{max} = 0.53$ and $\theta/\theta_{max} = 0.99$ are very similar, indicating thus that the ultimate structure of the adsorbed layer is reached already at moderate ratios of the surface coverage. Further increase in the amount adsorbed probably corresponds to a simple extension of the existing surface phase, without inducing remarkable modifications in its structure.

It may be concluded that the dominating admicellar aggregates on both the silica surfaces have the same size but differ much in number. SilNa is a more strongly charged support, and the maximum amount of BDDAB adsorbed on this surface is about twice that adsorbed on the SilH surface. The number of small surface aggregates existing at surface saturation should therefore be larger in the case of SilNa. This picture falls in line with images of spherical, isolated aggregates of tetradecyltrimethylammonium bromide on amorphous silica, recently obtained by atomic force microscopy [181]. With an increasing number of negative charges in the surface, the interfacial aggregates were found to keep the same structure, but the density of their packing was much higher.

The main difference in the aggregation mechanism on both silica surfaces lies in the hemimicellar region, and it is most clearly reflected in the behavior of the related enthalpies of adsorption. The very first points in the plot of Q_{st} against Γ in Figs. 25 and 26 show that there is a strongly exothermic contribution in the initial stage of adsorption, when almost only single monomers are present on the surface. This strong exothermicity may be ascribed to the adsorption on the high-affinity sites existing on the silica surface. Strong exothermic effects in the initial stages of adsorption have also been observed for the adsorption of zwitterionic and nonionic surfactants on silica [185,228], and for the anionic surfactants on alumina [228,229]. For the time being, this effect is most often attributed to the specific interactions of the surfactant head-groups with the most active adsorption centers on the surface. As the surface coverage increases, the differential heat of adsorption of BDDAB on SilNa decreases to endothermic values, whereas for SilH no endothermic effect is observed, and the adsorption process is exothermic in the whole range of adsorption. At the highest surface coverages, where admicellar aggregates are predicted to exist on the surface, the differences in the differential heats of adsorption disappear, and exothermic values are observed for both silica samples.

A significant endothermic contribution to the enthalpy of adsorption of BDDAB on SilNa, observed at low surface coverages, may be ascribed to desorption of the strongly structured interfacial water, due to the specific adsorption of a hydrophobic moiety and/or to the competitive character of the phenomenon, since a large part of $BDDA^+$ ions are adsorbed on the SilNa surface by cation exchange with Na^+ ions [187,190]. So far, however, neither the possibility of the alkyl chains associating with the adsorbent surface nor the competitive aspect of adsorption has been explicitly taken into account in our model.

Another explanation of the endothermic contribution to the enthalpy of adsorption of cationic surfactants on silica can be found in the paper by Wängnerud et al. [194]. They have suggested that a possible reason for the endothermic contribution observed at low adsorption densities may be the formation of small surfactant aggregates at the surface. Our model predicts the existence of such small aggregates (dimers, trimers) on the SilNa surface in the initial stage of adsorption. The molar enthalpy change accompanying the formation of such small aggregates in solution, and at the surface, will most likely differ from the molar enthalpy changes accompanying the formation of larger aggregates on the surface, or micelle formation in the bulk solution. The alkyl chains cannot be fully dehydrated, so a significant water contact will still be retained. The enthalpy of the formation of the small aggregates will probably have more of the character of pairwise (and higher-order) hydrophobic interaction than of bulk hydrophobic interaction as in micelle formation. These two types of hydrophobic interactions show distinctly different thermodynamic features [230–232]. For instance, the bulk interaction becomes weaker with increasing temperature, as shown by the increasing cmc and the decreasing aggregation numbers for micelle formation. On the contrary, the pairwise interaction becomes stronger with increasing temperature and is characterized by an endothermic enthalpy contribution.

In the seminal paper by Aniansson et al. [233] on the kinetic theory of micellar equilibria it follows that there is a large enthalpy difference between the transfer of a surfactant molecule from the aqueous solution to the interior of a micelle of optimum size, and the formation of a nucleus. Thus it is plausible that the formation of small hemimicellar aggregates in the initial stage of adsorption will have features in common with the formation of nuclei in solution. As the surface aggregates grow with increasing adsorption density, the process will more and more resemble micelle formation. Accordingly, the differential enthalpies of adsorption of BDDAB on SilNa can be expected to change from endothermic to increasingly more exothermic values with increasing surface coverage. In the case of the SilH system, for which only exothermic values of the enthalpy of adsorption are observed, our model does not predict the existence of small hemimicellar aggregates.

In conclusion, we can say that the observed calorimetric results are consistent with the interpretation of the adsorption process based on a two-dimensional micellization model proposed by us in the previous section. Taking into account the effect of energetic surface heterogeneity on monomer adsorption has led to an essential improvement of the agreement between theory and experiment, compared to our previous publication [205], where the heterogeneity effects were ignored. It is worth emphasizing once more that our model is the first in the world literature that allows us to fit at the same time the adsorption isotherms and the accompanying heats of adsorption and makes it possible to determine the size and the structure of the surface aggregates.

LIST OF SYMBOLS

A	variable defined in Eq. (90)
A_S	surface area of adsorbent
a	activity of ions in solution
a_H, a_C, a_A	bulk activities of proton, cation, and anion, respectively
a^s	parameter defined in Eq. (117)
B_s	constant [Eq. (33)]
b^s	parameter defined in Eq. (116)
c_+, c_-	electrical capacitance constants in the four-layer model [Eqs. (55ab)]
c_0, c_+, c_C, c_A	heterogeneity parameters
c_1	first integral capacitance in the triple-layer model
c_1^L, c_1^R	first integral capacitances for pH < PZC and pH > PZC, respectively [Eqs. (65)]
$c_1^{L,0}, c_1^{R,0}$	first integral capacitances for pH < PZC and pH > PZC in temperature T_0, respectively [Eqs. (65)]
c_i	heterogeneity parameters [Eq. (46)], $i = 0, +, A, C$
c^s	parameter defined in Eq. (118)
c_t	linear capacitance of the double electrical layer
D	heterogeneity parameter of distribution function [Eqs. (22a–22e)]
DEV	dimensionless standard deviation [Eqs. (28a–28e)]
d_1^h	quantity defined in Eq. (100)
d_i^s	quantity defined in Eq. (101)
E	energy of adsorption [Eq. (20)]
E_0	most probable value of E [Eqs. (22a–22c)]
E_1	lowest value of E [Eqs. (22d, 22e)]
ESP	dimensionless expectation of E [Eqs. (28a–28e)]
e	elementary charge
f_0, f_+, f_C, f_A	functions [Eqs. (36)]
f_i^s	canonical partition function of the surface aggregate of type (i, s)
$f_j^b, f_j^s, \langle f_j^s \rangle_T$	activity coefficient of component j
G	function [Eq. (73)]
H	activity of protons in solution
H_m^b	molar enthalpy of the liquid mixture

$H_{SL}^{\sigma(n)}, H_{S_0}^{\sigma(n)}$	reduced excess enthalpy per unit area for the solid–liquid and solid–vacuum interface
$H_{SV}^{\sigma(n)}, H_{LV}^{\sigma(n)}$	reduced excess enthalpy per unit area for the solid–vapor and liquid–vapor interface
i	number of monomers in hemimicelle or in one layer of an admicelle
i_{max}	maximum value of i
K_0	constant [Eq. (21a)]
K_0, K_+, K_C, K_A	equilibrium constants of reactions [Eqs. (35)]
$\bar{K}_1^h$	constant characterizing adsorption of the monomer (aggregate type 1, h)
$K_{a1}^{int}, K_{a2}^{int}, {}^*K_A^{int}, {}^*K_C^{int}$	equilibrium constants of reactions [Eqs. (29)]
K_n	constant characterizing micelle formation in the bulk phase
K_i^s	constant characterizing adsorption of hemimicelles (s = h) or admicells (s = a)
K_i	equilibrium constant of reaction i [Eqs. (34)], $i = 0, +, A, C$
K_i^0, K_i'	constants [Eqs. (51)]
K_j^0, K_j'	constants [Eqs. (48)]
k	Boltzmann's constant
l, m	fraction of the total nearest neighbors in the same lattice plane and in an adjacent lattice plane
1_i	number of the nearest-neighbor pair in the aggregate
M	number of surface groups in moles
M_S	surface area A_S or mass m_S of adsorbent
m_S	mass of adsorbent
M	total number of surfactant molecules in the adsorbed phase
N_A	Avogadro number
$\bar{N}_{agg.}$	average aggregation number in the adsorbed phase
N_i^s	number of the aggregates of type (i, s) in the surface phase
N_s	surface site density
n	parameter of distribution function [Eqs. (22d, 22e)] or maximum assumed size of the micelle [Eq. (125)]
n^0	initial number of moles of binary mixture in the immersion experiment [Eq. (4)]
n^a	number of moles of adsorption sites [Eq. (15a)]
n_C^P	initial number of moles of binary mixture in the flow cell [Eq. (6)]
n^s	number of moles of binary mixture in the adsorbed phase
n_T	final number of moles of binary mixture in the trap [Eq. (6)]
P^*	value defined in Eqs. (58cd)
PZC	point of zero charge
Q	canonical partition function for the adsorbed phase

Q_0, Q_+, Q_C, Q_A	molar differential heats of the formation of surface complexes [Eqs. (63)]
Q_{Av}	"average" heat of proton adsorption
$Q_{a1}, Q_{a2}, Q_{aC}, Q_{aA}$	nonconfigurational heats of appropriate reactions
$Q_i^{(c)}$	molar heats of adsorption in the model assuming small correlations
$Q_i^{(s)}$	molar heats of adsorption
Q_{ion}	nonconfigurational heat of adsorption of simple ions
Q_{pr}	heat effect of proton adsorption
Q_{st}	differential molar heat of adsorption
Q_w, Q_{wj}^0	enthalpy of immersion in a binary mixture and in pure liquid j
q	internal molecular partition function of a surfactant molecule in the bulk phase
q_j^b, q_j^s	molar partition function of component j
q^s	internal partition function of a surfactant molecule in a hemimicelle (s=h) or in an admicelle (s=a)
R_1	radius of an elementary disk
R_i	radius of the ith aggregate
r	adsorbate size ratio [Eq. (20)]
r_i	relative radius of a disk defined as R_i/R_1
$\check{r}$	constant [Eq. (71a)]
S	surface area of the adsorbent
S_1	surface area occupied by an elementary disk
S_i	surface covered by the ith disk
s_i	relative surface area of the ith aggregate
T	absolute temperature
W^p, W^v, W^b	dimensionless interaction parameters
w^p, w^v, w^b	interchange energy parameter for the adsorbed and the bulk lattice planes
X^*	value defined in Eqs. (58a)
x	mole fraction of a monomer in the bulk phase
x_2^b, x_2^s	mole fraction of component 2
x_2^0	initial composition of binary mixture in the immersion experiment [Eq. (4)]
x_2^P	initial composition of binary mixture in the flow cell [Eq. (6)]
x_n	mole fraction of the surfactant in the micelles
$Y^s, \langle Y^s \rangle_T$	local and global property of the adsorbed phase
Y^*	value defined in Eq. (58b)
Z	configurational integral for a mixture of hard disks
z	lattice coordination number
α	heterogeneity parameter
α_1^L, α_1^R	parameters [Eqs. (65)]
α_+, α_-	parameters [Eqs. (78)]
β	parameter [Eq. (38)]
β_1, β_2	parameters of the linear dependence $l_i = \beta_1 i + \beta_2$
γ	activity coefficient
Γ	variable defined in Eq. (89)

$\bar{\Gamma}_1^h$	mean value of Γ_1^h on a heterogeneous surface
Γ_i^s	variable defined in Eq. (89)
$\Gamma_2^{\sigma(n)}$	reduced adsorption of component 2
Γ_2^e	experimentally measured surface excess of component 2
Γ_t	total adsorbed amount in moles per square meter
λ	constant [Eq. (21b)]
$\Delta\omega$	parameter defined in Eq. (115a)
$\Delta\varepsilon^s$	parameter defined in Eq. (115b)
$\Delta_{dis}H$	enthalpy of displacement
$\Delta_{imm}H$, $\Delta_{imm}H_j^0$	enthalpy of immersion in a binary mixture and in pure liquid j
$\Delta_{mix}H_m^b$, $\Delta_{mix}H_m^s$, $\langle\Delta_{mix}H_m^s\rangle_T$	molar enthalpy of mixing of the two components to form a binary mixture
δ^*	charge defined in Eqs. (55c)
δ_0	surface charge
δ_β	charge at β plane
ε_0, ε_+, ε_C, ε_A	adsorption energy of surface complexes
ε_0^s	parameter defined in Eq. (115b)
ε^h	adsorption energy of a single monomer adsorbed head-on
ε_2	adsorption energy of a single monomer adsorbed in the second layer of an admicelle
ε_i	adsorption (binding) energy of the ith surface complex, $i = 0, +, A, C$
ε_i^0	the most probable adsorption energies, $i = 0, +, A, C$
$\bar{\varepsilon}^h$	mean adsorption energy of a single monomer on a heterogeneous surface
Θ	quantity defined in Eq. (92)
θ_0, θ_+, θ_C, θ_A, θ_-	surface coverages [Eqs. (31)]
θ_{it}, θ_{jt}	total surface coverages in the heterogeneous models [Eq. (50)]
Λ	thermal wavelength of a surfactant molecule
μ^b	chemical potential of the surfactant monomer in the bulk solution
μ_i^s	surface chemical potential of the aggregate of type (i, s)
$\mu_j^{0,b}$, $\mu_j^{0,s}$	standard chemical potential of component j
μ_j^b, μ_j^s	chemical potential of component j in solution
$\mu_{SO^-C^+}$, $\mu_{SOH_2^+A^-}$, μ_{SOH^0}, $\mu_{SOH_2^+}$	chemical potentials of surface complexes
ϕ_2^b	volume fraction of component 2 [Eq. (15b)]
ϕ_2^s, $<\phi_2^s>_T$	local and global surface fraction of component 2 [Eqs. 15a, 23b, 24a)]
ψ_0	surface potential
ψ_β	potential at the plane of specifically adsorbed counterions
χ	distribution of the adsorption energy [Eq. (22)]
Ω	set of allowed E-values [Eq. (22)]

ω	energy of lateral interactions between two neighboring monomers in an aggregate
ω_0	parameter defined in Eq. (115a)
ω_i^s	total energy of lateral interactions in the aggregate of type i, s

Superscripts

b	equilibrium bulk phase (bulk mixture)
s	adsorbed phase (surface mixture)

Subscripts

DI	Dirac δ distribution
GA	gamma distribution
GS	gaussian distribution
P	patchwise topography model
QG	gaussian-like distribution
R	random topography model
WB	Weibull distribution

REFERENCES

1. J. J. Kipling, *Adsorption from Solutions of Non-Electrolytes*, Academic Press, London, 1965.
2. D. H. Everett. Trans. Faraday Soc. *60*:1803 (1964); *61*:2478 (1965).
3. G. Schay and L. Gy. Nagy. J. Chim. Phys. *149* (1961).
4. G. Schay, L. Gy. Nagy, and T. Schekrenyesy. Periodica Polytechnica *4*:95 (1960); *6*:91 (1962)
5. R. Defay, I. Prigogine, A. Bellemans, and D. H. Everett, *Surface Tension and Adsorption* Longmans Green, London, 1966.
6. *Reporting Data on Adsorption from Solution at the Solid-Solution Interface*, prepared for publication by D. H. Everett. Pure Appl. Chem. *58*:967 (1986).
7. D. H. Everett and R. T. Podoll. J. Colloid Interface Sci. *82*:14 (1981).
8. H. E. Kern and G. H. Findenegg. J. Colloid Interface Sci. *75*:346 (1980).
9. P. Nikitas. J. Chem. Soc., Faraday Trans. 1 *80*:3315 (1984).
10. J. Torrent and F. Sanz. J. Electroanal. Chem. *303*:45 (1991).
11. R. Bennes. J. Electroanal. Chem. *105*:85 (1979).
12. J. E. Lane. Austr. J. Chem. *21*:827 (1968).
13. E. A. Guggenheim, *Mixtures*, Oxford University Press, London, 1952.
14. E. A. Guggenheim, *Applications of Statistical Mechanics*, Oxford University Press, London, 1966.
15. J. E. Lane, in *Adsorption from Solution at the Solid-Liquid Interface* (G. D. Parfitt and C. H. Rochester, eds.), Academic Press, 1983, Chap. 2.
16. S. G. Ash, D. H. Everett, and G. H. Findenegg. Trans. Faraday Soc. *64*:2639 (1968).
17. R. Altenberger and J. Stecki. Chem. Phys. Letters *5*:29 (1970).
18. I. Rusanov. Progr. Surf. Membr. Sci. *4*:57 (1971).
19. E. Brown, D. H. Everett, and C. J. Morgan. J. Chem. Soc., Faraday Trans. 1 *71*:883 (1975).
20. R. S. Hansen and U. H. Mai. J. Phys. Chem. *61*:573 (1957).
21. G. Delmas and D. Patterson. J. Phys. Chem. *64*:1827 (1960).
22. M. Siskova and E. Erdös. Collect. Chech. Chem. Commun. *25*:1729, 2599 (1960).
23. M. T. Coltharp and N. J. Hackerman. J. Colloid Interface Sci. *43*:176, 185 (1973).
24. W. Rudziński, J. Ościk, and A. Dąbrowski. Chem. Phys. Lett. *20*:5 (1973).

25. A. Dąbrowski, M. Jaroniec, and J. Toth. Acta Chim. Acad. Sci. Hung. *111*:311 (1982).
26. A. Dąbrowski, M. Jaroniec, and J. K. Garbacz. Thin Solid Films *103*:399 (1983).
27. A. Dąbrowski and M. Jaroniec. Chem. Eng. Sci. *40*:473 (1985).
28. M. Jaroniec, A. Patrykiejew, and M. Borówko. Progr. Surf. Membr. Sci. *14*:1 (1981).
29. M. Borówko and M. Jaroniec. Adv. Colloid Interface Sci. *19*:137 (1983).
30. M. Jaroniec and R. Madey, *Physical Adsorption on Heterogeneous Solids*, Elsevier, 1988.
31. J. Papenhuijzen and L. K. Koopal, in *Symposium on Adsorption from Solution*, Bristol, 1982, p. 210.
32. S. Sircar. J. Chem. Soc., Faraday Trans. 1 *79*:2085 (1983).
33. S. Sircar. Surface Sci. *148*:489 (1984).
34. S. Sircar. Langmuir *3*:369 (1987).
35. W. Rudziński, J. Narkiewicz-Michałek, and S. Partyka. J. Chem. Soc., Faraday Trans. 1 *78*:2361 (1982).
36. W. Rudziński, L. Łajtar, J. Zając, E. Wolfram, and I. Paszli. J. Colloid Interface Sci. *96*:339 (1983).
37. W. Rudziński, J. Zając, and C. C. Hsu. J. Colloid Interface Sci. *103*:528 (1985).
38. W. Rudziński and J. Zając. Acta Geod. Geoph. Montan. Hung. *20*:337 (1985).
39. W. Rudziński, J. Zając, I. Dekany, and F. Szanto. J. Colloid Interface Sci. *112*:473 (1986).
40. W. Rudziński, J. Zając, E. Wolfram, and I. Paszli. Colloids Surf. *22*:317 (1987).
41. W. Rudziński, J. Zając, and J. Michałek. Acta Chim. Hung. *125*:57 (1988).
42. J. Zając. Sci. Bull. Acad. Min. Metall. *19*:39 (1991); *21*:49 (1993).
43. Y. Ogino, S. Ozawa, W. Rudziński, and J. Zając. Z. Phys. Chem. *271*:759 (1990).
44. J. Michałek, J. Zając, and W. Rudziński. Langmuir *6*:1505 (1990).
45. C. Zettlemoyer and J. J. Chessick. Adv. Chem. Ser. No. *43*:88 (1964).
46. H. Everett, in *Proceedings of the Engineering Foundation Conference "Fundamentals of Adsorption"* (A. L. Myers and G. Belfort, eds.), New York, 1983, p. 1.
47. H. Everett. Prog. Coll. Polymer Sci. *65*:103 (1978).
48. *Proceedings of the BP Symposium on the Significance of the Heats of Adsorption at the Solid-Liquid Interface* (A. J. Groszek, ed.), BP Research Centre, Sunbury-on-Thames, 1971.
49. S. Sircar, J. Novosad, and A. L. Myers, I and EC Fundamentals *11*:246 (1972).
50. G. W. Woodbury, Jr. and L. A. Noll. Colloids Surf. *8*:1 (1983).
51. S. G. Ash, R. Bown, and D. H. Everett. J. Chem. Thermodyn. *5*:239 (1973).
52. Kurbanbekov, O. G. Larionov, K. V. Tchmutov, and M. D. Jubilevitch. Russ. J. Phys. Chem. *43*:1630 (1969).
53. H. Slaats, J. C. Kraak, W. J. T. Brugman, and H. Poppe, J. Chromatogr. *149*:255 (1978).
54. G. Schay, L. Gy. Nagy, and Gy. Racz. Acta Chim. Acad. Sci. Hung. *71*:23 (1972).
55. S. A. Busiev, S. I. Zvieriev, O. G. Larionov, and E. S. Jakubov. J. Chromatogr. *241*:287 (1982).
56. R. Denoyel, F. Rouquerol, and J. Rouquerol, in *Symposium on Adsorption from Solution*, Bristol, 1982, p. 224.
57. T. Allen and R. M. Patel, in *Proceedings of the BP Symposium on the Significance of the Heats of Adsorption at the Solid-Liquid Interface* (A. J. Groszek, ed.), BP Research Centre, Sunbury-on-Thames, 1971, p. 78.
58. M. W. Tideswell, in *Proceedings of the BP Symposium on the Significance of the Heats of Adsorption at the Solid-Liquid Interface* (A. J. Groszek, ed.), BP Research Centre, Sunbury-on-Thames, 1971, pp. 15 and 45.
59. E. H. M. Wright. Trans. Faraday Soc. *62*:1275 (1966); *63*:3026 (1967).
60. V. Smirnova, A. A. Kubasov, and K. V. Topichieva. Lect. Acad. Sci. USSR *139*:150 (1961).
61. S. Partyka and J. M. Douillard. J. Pet. Sci. Eng. *13*:95 (1995).
62. M. Douillard. J. Colloid Interface Sci. *182*:308 (1996).
63. G. J. Minkoff and R. H. E. Duffett. B. P. Magazine *13*:17 (1964).
64. F. E. Bartell and Y. Fu. J. Phys. Chem. *33*:1758 (1929).
65. E. Lane and T. H. Spurling. Austr. J. Chem. *29*:2103 (1976); *31*:933 (1978).
66. K. Snook and D. Henderson. J. Chem. Phys. *68*:2134 (1978).

67. K. Snook and W. van Megen. J. Chem. Phys. *70*:3099 (1979).
68. W. Rudziński and D. H. Everett, *Adsorption of Gases on Heterogeneous Surfaces*, Academic Press, New York, 1992.
69. R. C. Reid, J. M. Prausnitz, and T. K. Sherwood, *The Properties of Gases and Liquids*, McGraw-Hill, New York, 1977.
70. D. H. Everett, in *Colloid Science*, Specialist Periodical Reports, Vol. 1, p. 49, Chemical Society, London.
71. Gy. Nagy and G. Schay. Acta Chim. Acad. Sci. Hung. *39*:365 (1963).
72. S. Sircar and A. L. Myers. J. Phys. Chem. *74*:2828 (1970).
73. V. Kisielev and V. V. Khopina. Trans. Faraday Soc. *65*:1936 (1969).
74. F. Rodriguez-Reinoso and A. Linares-Solano, in *Chemistry and Physics of Carbon* (P. A. Thrower, ed.), Vol. 21, Marcel Dekker, New York, 1965, p. 1.
75. V. A. Bakaev. Surface Sci. *196*:571 (1988).
76. V. A. Bakaev and M. M. Dubinin. Dokl. Acad. Nauk SSSR *296*:369 (1987).
77. V. A. Bakaev and O. V. Chelnokova. Surface Sci. *215*:521 (1989).
78. G. Sposito. J. Colloid Interface Sci. *91*:329 (1983).
79. W. Rudziński, R. Charmas, S. Partyka, and A. Foissy. New J. Chem. *15*:327 (1991).
80. W. Rudziński, R. Charmas, S. Partyka, F. Thomas, and J. Y. Bottero. Langmuir *8*:1154 (1992).
81. J. Garcia-Miragaya and A. L. Page. Soil Sci. Soc. Am. J. *40*:658 (1976).
82. J. Garcia-Miragaya and A. L. Page. Water, Air, Soil Pollut. *9*:289 (1977).
83. J. J. Street. W. L. Lindsay, and B. R. Sabey. J. Environ. Qual., *6*:72 (1977).
84. M. M. Benjamin and J. O. Leckie. J. Colloid Interface Sci. *79*:209 (1981).
85. G. Sposito. Soil Sci. Soc. Am. J. *43*:197 (1979).
86. G. Sposito, in *Chemistry in the Soil Environment* (R. H. Dowdy, D. Baker, V. Volk, and J. Ryan, eds.), SSSA Spec. Pub., Soil Science Society of America, Madison, W. I. 1980.
87. M. M. Kinniburgh, J. A. Barkes, and M. Whitfield. J. Colloid Interface Sci. *95*:370 (1983).
88. W. H. Van Riemsdijk, G. H. Bolt, L. K. Koopal, and J. Blaakmeer. J. Colloid Interface Sci. *109*:219 (1986).
89. W. H. Van Riemsdijk, J. C. M. De Wit, L. K. Koopal, and G. H. Bolt. J. Colloid Interface Sci. *116*:511 (1987).
90. W. H. Van Riemsdijk, L. K. Koopal, and J. C. M. De Wit. J. Agric. Sci. *35*:241 (1987).
91. L. K. Koopal and W. H. Van Riemsdijk. J. Colloid Interface Sci. *128*:188 (1989).
92. A. W. M. Gibb and L. K. Koopal. J. Colloid Interface Sci. *134*:122 (1990).
93. W. Rudziński, R. Charmas, S. Partyka, and J. Y. Bottero. Langmuir *9*:2641 (1993).
94. D. E. Yates, S. Levine, and T. W. Hearly. J. Chem. Soc. Faraday Trans. I *70*:1807 (1974).
95. J. A. Davis, R. O. James, and J. O. Leckie. J. Colloid Interface Sci. *63*:480 (1978).
96. J. A. Davis and J. O. Leckie. J. Colloid Interface Sci. *67*:90 (1978).
97. K. F. Hayes and J. O. Leckie. J. Colloid Interface Sci. *115*:564 (1987).
98. K. F. Hayes, C. Papelis, and J. O. Leckie. J. Colloid Interface Sci. *125*:717 (1988).
99. A. P. Robertson and J. O. Leckie. J. Colloid Interface Sci. *118*:444 (1997).
100. D. A. Sverjensky and N. Sahai. Geochim. Cosmochim. Acta *60*:3773 (1996).
101. N. Sahai and D. A. Sverjensky. Geochim. Cosmochim. Acta *61*:2801 (1997).
102. N. Sahai and D. A. Sverjensky. Geochim. Cosmochim. Acta *61*:2827 (1997).
103. L. Pauling, in *The Nature of the Electrostatic Bond*, 3d ed., Cornell Univ. Press, Ithaca, NY, 1967.
104. G. H. Bolt and W. H. Van Riemsdijk, in *Soil Chemistry*, B. Physico-Chemical Models (G. H. Bolt, ed.), 2d ed., Elsevier, Amsterdam, 1982, p. 459.
105. T. Hiemstra, W. H. Van Riemsdijk, and G. H. Bolt. J. Colloid Interface Sci. *133*:91 (1989).
106. T. Hiemstra, J. C. W. De Witt, and W. H. Van Riemsdijk. J. Colloid Interface Sci. *133*:105 (1989).
107. T. Hiemstra and W. H. Van Riemsdijk. Colloids Surf. *59*:7 (1991).
108. T. Hiemstra and W. H. Van Riemsdijk. J. Colloid Interface Sci. *179*:488 (1996).
109. W. Rudziński, R. Charmas, and S. Partyka. Langmuir *7*:354 (1991).

110. W. Rudziński and R. Charmas. Adsorption 2:245 (1996).
111. W. Rudziński, R. Charmas, and T. Borowiecki, in *Adsorption on New and Modified Inorganic Sorbents* (A. Dąbrowski and V. A. Tertykh, eds.), Elsevier, Amsterdam, 1996, p. 357.
112. N. J. Barrow, J. W. Bowden, A. M. Posner, and J. P. Quirk. Aust. J. Soil Res. 19:309 (1981).
113. L. Bousse, N. F. De Rooij, and P. Bergveld. Surface Sci. 135:479 (1983).
114. R. Charmas, W. Piasecki, and W. Rudziński. Langmuir 11:3199 (1995).
115. R. Charmas and W. Piasecki. Langmuir 12:5458 (1996).
116. R. Charmas, in *Physical Adsorption: Experiment, Theory and Applications* (J. P. Fraissard and C. W. Conner, eds.), NATO-ASI-Series C, Vol. 491, Kluwer Academic Publishers, 1997, p. 609.
117. R. Charmas. Langmuir (in press).
118. L. Bousse, N. F. De Rooij, and P. Bergveld. IEEE Trans. Electron Devices 30:1263 (1983).
119. H. Van den Vlekkert, L. Bousse, and N. F. De Rooij. J. Colloid Interface Sci. 122:336 (1988).
120. N. J. Barrow, J. Gerth, and G. W. Brummer. J. Soil Sci. 40:437 (1989).
121. U. Rusch, M. Borkovec, J. Daicic, and W. H. Van Riemsdijk. J. Colloid Interface Sci. 191:247 (1997).
122. L. K. Koopal, W. H. Van Riemsdijk, and M. G. Roffey. J. Colloid Interface Sci. 118:117 (1987).
123. W. Rudziński, R. Charmas, and W. Piasecki. Adsorp. Sci. Technol. 14:25 (1996).
124. Y. G. Berube and P. L. de Bruyn. J. Colloid Interface Sci. 27:305 (1968).
125. P. H. Tewari and A. W. McLean. J. Colloid Interface Sci. 40:267 (1972).
126. P. H. Tewari and A. B. Campbell. J. Colloid Interface Sci. 55:531 (1976).
127. L. G. J. Fokkink, A. De Keizer, and J. Lyklema. J. Colloid Interface Sci. 127:116 (1989).
128. L. G. J. Fokkink, Ph.D. thesis, Agricultural University, Wageningen, 1987.
129. S. Subramanian, J. A. Schwarz, and Z. Hejase. J. Catal. 117:512 (1989).
130. J. F. Kuo and T. F. Yen. J. Colloid Interface Sci. 121:220 (1989).
131. K. Ch. Akratopulu, L. Vordonis, and A. Lycourghiotis. J. Chem. Soc. Faraday Trans. I 82:3697 (1986).
132. K. Ch. Akratopulu, C. Kordulis, and A. Lycourghiotis. J. Chem. Soc. Faraday Trans. 86:3437 (1990).
133. M. A. Blesa, A. J. G. Maroto, and A. E. Regazzoni. J. Colloid Interface Sci. 140:287 (1990).
134. Kosmulski, J. Matysiak, and J. Szczypa. J. Colloid Interface Sci. 164:280 (1994).
135. M. A. Blesa, N. M. Figliolia, A. J. G. Maroto, and A. E. Regazzoni. J. Colloid Interface Sci. 101:410 (1984).
136. A. E. Regazzoni, Ph.D. thesis, Universidad Nacional de Tucuman, Argentina, 1984.
137. P. V. Brady. Geochim. Cosmochim. Acta 56:2941 (1992).
138. M. Kosmulski. Colloids Surf. A 83:237 (1994).
139. M. L. Machesky, in *Chemical Modelling in Aqueous Systems II* (D. C. Melchior and R. L. Bassett, eds.), Amer. Chem. Soc. Sym. Ser. 416, Washington D.C., 1990.
140. D. A. Griffiths and D. W. Fuerstenau. J. Colloid Interface Sci. 80:271 (1981).
141. Foissy, Ph.D. thesis, Université de Franche-Comte Bescançon, 1985.
142. M. L. Machesky and M. A. Anderson. Langmuir 2:582 (1986).
143. S. R. Mehr, D. J. Eatough, L. D. Hansen, E. A. Lewis, and J. A. Davis. Thermochim. Acta 154:129 (1989).
144. A. De Keizer, L. G. J. Fokkink, and J. Lyklema. Colloids Surf. 49:149 (1990).
145. M. L. Machesky and P. F. Jacobs. Colloids Surf. 52:297 (1991).
146. M. L. Machesky and P. F. Jacobs. Colloids Surf. 53:315 (1991).
147. N. Kallay, S. Žalac, and G. Stefanić. Langmuir 9:3457 (1993).
148. W. H. Casay. J. Colloid Interface Sci. 163:407 (1994).
149. N. Kallay and S. Žalac. Croat. Chem. Acta 67:467 (1994).
150. S. Žalac and N. Kallay. Croat. Chem. Acta 69:119 (1996).
151. W. Rudziński, R. Charmas, J. M. Cases, M. François, F. Villieras, and L. J. Michot. Langmuir 13:483 (1997).

152. W. Rudziński, R. Charmas, W. Piasecki, J. M. Cases, M. François, F. Villieras, and L. J. Michot. Polish J. Chem. *71*:603 (1997).
153. W. Rudziński, R. Charmas, W. Piasecki, J. M. Cases, M. François, F. Villieras, and L. J. Michot. Colloids Surf. A. (in press).
154. W. Rudziński, R. Charmas, W. Piasecki, N. Kallay, J. M. Cases, M. François, F. Villieras, and L. J. Michot. Adsorption (in press).
155. W. Rudziński, R. Charmas, W. Piasecki, F. Thomas, F. Villieras, B. Prelot, and J. M. Cases. Langmuir (in press).
156. D. G. Hall. Langmuir *13*:91 (1997).
157. M. A. Blesa and N. Kallay. Adv. Colloid Interface Sci. *28*:111 (1988).
158. R. Sprycha. J. Colloid Interface Sci. *127*:12 (1989).
159. J. M. Rosen, *Surfactants and Interfacial Phenomena*, 2d ed., John Wiley, New York, 1989.
160. E. Söderlind and P. Stilbs. J. Colloid Interface Sci. *143*:586 (1991); Langmuir *9*:2024 (1993).
161. Somasundaran, J. T. Kunjappu, C. V. Kumar, N. J. Turro, and J. K. Barton. Langmuir *5*:215 (1989).
162. K. Esumi, T. Nagahama, and K. Meguro. Colloids Surf *57*:149 (1991).
163. E. Esumi, A. Sugimura, T. Yamada, and K. Meguro. Colloids Surf. *62*:249 (1992).
164. P. Chandar, P. Somasundaran, K. C. Waterman, and N. J. Turro. J. Phys. Chem. *91*:148 (1987).
165. P. Chandar, P. Somasundaran, and N. J. Turro. J. Colloid Interface Sci. *117*:31 (1987).
166. A. Fan, P. Somasundaran, and N. J. Turo. Langmuir *13*:506 (1997).
167. P. Levitz, A. El Miri, D. Keravis, and H. Van Damme. J. Colloid Interface Sci. *99*:484 (1984).
168. P. Levitz and H. Van Damme. J. Phys. Chem. *90*:1302 (1986).
169. R. Rennie, E. M. Lee, E. A. Simister, and R. K. Thomas. Langmuir *6*:1301 (1990).
170. D. C. McDermott, J. McCarney, R. K. Thomas, and A. R. Rennie. J. Colloid Interface Sci. *162*:304 (1994).
171. M. Pashley and J. N. Israelachvili. Colloids Surfaces *2*:169 (1981).
172. P. M. McGuiggan and R. M. Pashley. J. Colloid Interface Sci. *124*:560 (1988).
173. P. Kékicheff, H. K. Christenson, and B. W. Ninham. Colloids Surf. *40*:31 (1989).
174. J. L. Parker, V. Yaminsky, and P. M. Claesson. J. Phys. Chem. *97*:7706 (1993).
175. M. W. Rutland and J. L. Parker. Langmuir *10*:1110 (1994).
176. Y.-H. Tsao, S. X. Yang, and D. F. Evans. Langmuir *7*:3154 (1991).
177. M. W. Rutland and T. J. Senden. Langmuir *9*:412 (1993).
178. B. Johnson, C. J. Drummond, P. J. Scales, and S. Nishimura. Langmuir *11*:2367 (1995).
179. S. Manne and H. E. Gaub. Science *270*:1480 (1995).
180. S. Manne, T. E. Schäffer, Q. Huo, P. K. Hansma, D. E. Morse, G. D. Stucky, and I. A. Aksay. Langmuir *13*:6382 (1997).
181. G. Sharma, S. Basu, and M. M. Sharma. Langmuir *12*:6506 (1996).
182. G. W. Woodbury, Jr. and L. A. Noll. Colloids Surf. *33*:301 (1988).
183. R. Denoyel, F. Rouquerol, and J. Rouquerol. Colloids Surf. *37*:295 (1989).
184. G. H. Findenegg, B. Pasucha, and H. Strunk. Colloids Surf. *37*:223 (1989).
185. S. Partyka, M. Lindheimer, and B. Faucompre. Colloids Surf. *76*:267 (1993).
186. R. Denoyel, F. Giordano, and J. Rouquerol. Colloids Surf. *76*:141 (1993).
187. J. L. Trompette, J. Zając, E. Keh and S. Partyka. Langmuir *10*:812 (1994).
188. F. Giordano-Palmino, R. Denoyel, and J. Rouquerol. J. Colloid Interface Sci. *165*:82 (1994).
189. J. Zając, M. Lindheimer, and S. Partyka. Colloids Surfaces *98*:197 (1995).
190. J. Zając, J. L. Trompette, and S. Partyka. Langmuir *12*:1357 (1996).
191. J. Zając and S. Partyka, in *Adsorption on New and Modified Inorganic Sorbents* (A Dąbrowski and V. A. Tertykh, eds.), Elsevier, Amsterdam, 1996, Chapter III.6.
192. J. Zając, C. Chorro, M. Lindheimer, and S. Partyka. Langmuir *13*:1486 (1997).
193. Z. Kiraly, R. H. K. Börner, and G. H. Findenegg. Langmuir *13*:3308 (1997).
194. P. Wängnerud, D. Berling, and G. J. Olofsson. J. Colloid Interface Sci. *169*:365 (1995).
195. E. S. Pagac, D. C. Prieve, and R. D. Tilton. Langmuir *14*:2333 (1998).
196. P. Wängnerud and B. Jönsson. Langmuir *10*:3542 (1994).

197. M. R. Böhmer and L. K. Koopal. Langmuir 8:2649 (1992); 8:2660 (1992).
198. L. K. Koopal, in *Coagulation and Flocculation* (B. Dobias, ed.), Marcel Dekker, New York, 1993, Chapter IV.
199. T. Goloub, L. K. Koopal, B. H. Bijsterbosch, and M. P. Sidorova. Langmuir 12:3188 (1996).
200. L. K. Koopal, in *Adsorption on New and Modified Inorganic Sorbents* (A. Dąbrowski and V. A. Tertykh, eds.), Elsevier, Amsterdam, 1996, Chapter III.5.
201. W. Rudziński and R. Charmas. Langmuir 7:354 (1991).
202. W. Rudziński and R. Charmas. Langmuir 8:1154 (1992).
203. J. H. Harwell and M. A. Yeskie. J. Phys. Chem. 93:3372 (1989).
204. L. Łajtar, J. Narkiewicz-Michałek, W. Rudziński, and S. Partyka. Langmuir 9:3174 (1993).
205. L. Łajtar, J. Narkiewicz-Michałek, W. Rudziński, and S. Partyka. Langmuir 10:3754 (1994).
206. Z. M. Zorin, N. V. Churaev, N. E. Esipova, I. P. Sergeeva, V. D. Sobolev, and E. K. Gasanov. J. Colloid Interface Sci 152:170 (1992).
207. P. Somasundaran and D. W. Fuerstenau. J. Phys. Chem. 70:90 (1966).
208. J. F. Scamehorn, R. S. Schechter, and W. H. Wade. J. Colloid Interface Sci. 85:463 (1982).
209. J. M. Cases and F. Villieras. Langmuir 8:1251 (1992).
210. E. Rakotonarivo, J. Y. Bottero, and J. M. Cases. Colloids Surfaces 9:273 (1984).
211. E. Rakotonarivo, J. Y. Bottero, and J. M. Cases. Colloids Surfaces 16:153 (1985).
212. J. H. Harwell, J. C. Hoskins, and R. S. Schechter. Langmuir 1:251 (1985).
213. J. H. Harwell and M. A. Yeskie. J. Phys. Chem. 92:2346 (1988).
214. Y. Gao, J. Du, and T. Gu. J. Chem. Soc., Faraday Trans. I 83:2671 (1987).
215. H. Rupprecht and T. Gu. Colloid Polymer Sci. 269:506 (1991).
216. H. Rupprecht and J. Sigg. Polish J. Chem. 71:657 (1997).
217. B. Li and E. Ruckenstein. Langmuir 12:5052 (1996).
218. M. Drach, L. Łajtar, J. Narkiewicz-Michałek, W. Rudziński, and J. Zając. Colloids Surf. 145:243 (1998).
219. J. Narkiewicz-Michałek. J. Ber. Bunsen-Ges. Phys. Chem. 95:85 (1991).
220. J. Narkiewicz-Michałek, W. Rudziński, and S. Partyka. Langmuir 9:2630 (1993).
221. J. Zając. Colloids Surf. (in press).
222. J. Zając, J. L. Trompette, and S. Partyka. Polish J. Chem. 71:679 (1997).
223. G. H. Bolt. J. Phys. Chem. 61:166 (1957).
224. M. J. Meziani, J. Zając, D. J. Jones, J. Rozière, and S. Partyka. Langmuir 13:5409 (1997).
225. D. E. Yates and T. W. Healy. J. Colloid Interface Sci. 55:9 (1976).
226. Strazhesko, V. B. Strelko, V. N. Belyakov, and S. C. Rubank. J. Chromatogr. 102:191 (1974).
227. A. Malliaris, J. Lang, and R. Zana. J. Colloid Interface Sci. 110:237 (1986).
228. R. Denoyel and J. Rouquerol. J. Colloid Interface Sci. 143:555 (1991).
229. S. Partyka, W. Rudziński, B. Brun, and J. H. Clint. Langmuir 5:297 (1989).
230. I. R. Tasker and R. H. Wood. J. Phys. Chem. 86:4040 (1982).
231. D. G. Archer, H. J. Albert, D. E. White, and R. H. Wood. J. Colloid Interface Sci. 84:168 (1984).
232. D. G. Archer. J. Phys. Chem. 93:5272 (1989).
233. E. A. G. Aniansson, S. N. Wall, M. Almgren, H. Hoffmann, I. Kielmann, W. Ulbricht, R. Zana, J. Lang, and C. Tondre. J. Phys. Chem. 80905 (1976).

4
Surface Complexation Model for Solid–Liquid Interfaces

MAREK KOSMULSKI Institute of Catalysis, Polish Academy of Sciences, Krakow, and Department of Electrochemistry, Technical University of Lublin, Lublin, Poland

RYSZARD SPRYCHA Technical Center, Sun Chemical Corporation, Carlstadt, New Jersey

JERZY SZCZYPA[†] Department of Radiochemistry and Colloid Chemistry, Maria Curie-Skłodowska University, Lublin, Poland

I.	Introduction		164
II.	Mechanism of Surface Charge Creation at the Solid–Liquid Interface		165
	A.	Metal/electrolyte interface	165
	B.	Semiconductor/electrolyte interface	165
	C.	Ionic crystal (AgI)/electrolyte interface	165
	D.	Metal oxide (hydroxide)/electrolyte interface	165
III.	Electric Interfacial Layer Models Without Counterion Association		169
	A.	Constant capacitance	170
	B.	Diffuse layer	171
	C.	Stern model	174
	D.	1-pK model	174
	E.	1-pK multisite approach	177
	F.	MUSIC	178
IV.	Electric Interfacial Layer Models with Counterion Association		180
	A.	Empirical approach	180
	B.	Ion selectivity	182
	C.	Association	182
	D.	Surface complexation	183
V.	Determination of Surface Ionization and Complexation Constants		186
	A.	Surface charge	187
	B.	Electrokinetic data	191
	C.	Adsorption of counterions	196
VI.	Specific Adsorption		205
	A.	Methods	207
	B.	Proton stoichiometry	211

[†] Deceased.

C.	Models without electrostatics	211
D.	Models with single surface reaction	211
E.	Models with multiple surface reactions	213
F.	Adsorption enthalpy	215
G.	Adsorption competition between various multivalent ions	216
Symbols and Abbreviations		218
References		220

I. INTRODUCTION

Electrical charges at interfaces play a very important role in many practical applications and are crucial in understanding adsorption, stability, and other properties of colloidal systems. Detailed discussion of electrical aspects of interfacial phenomena can be found in a recent book by Lyklema [1]. Though each colloidal system is as a whole electroneutral, the properties of dispersed and continuous phases determine the charge distribution in the interfacial region. Due to electroneutrality conditions, charge at the solid surface is counterbalanced by an equal but opposite charge in the solution. Surface charge and countercharge at the solid–liquid interface form a so-called electrical interfacial layer (EIL). The structure of the electrical interfacial layer and the origin of the surface charge for different systems have been studied quite intensively in recent years, especially the EIL at the metal/electrolyte interface. The description of the electrical interfacial layer at the metal oxide (hydroxide)/electrolyte interface is much more complicated and less understood.

The purpose of this chapter is to discuss the properties of the electrical interfacial layer at the solid–liquid interface with special emphasis on an association of counterions at the oxide/electrolyte interface. In the second section, the mechanisms of surface charge creation at four different interfaces, metal/electrolyte, semiconductor/electrolyte, ionic crystal (AgI)/electrolyte, and metal oxide (hydroxide)/electrolyte, are generally discussed. The third section contains a description of the models of the electrical interfacial layer at the solid–liquid interface without counterion association. The models of the EIL that take into account the association of counterions are discussed in Section IV, where special emphasis is put on surface complexation models of the electrical interfacial layer.

Different models use different numbers of parameters. In Section V the methods of determination of surface ionization and complexation constants of the surface metal hydroxyl groups [− MOH] at the oxide/electrolyte interface are discussed. To evaluate those parameters the experimental surface charge, electrokinetic, or adsorption of counterions data can be used. This section focuses exclusively on the so-called "indifferent" electrolytes.

The effects of specific adsorption of ions on the properties of the electrical interfacial layer at the solid–liquid interface are discussed in Section VI. The content of this section covers the methods of specific adsorption determination, proton stoichiometry, EIL models without electrostatics, models with a single surface reaction, models with multiple surface reactions, enthalpy of adsorption, and adsorption competition between various multivalent ions.

As mentioned above, this chapter will focus mainly on oxide/electrolyte systems. For details on other solid–liquid systems not covered in this paper the reader is referred to the standard books and review papers on a specific subject.

II. MECHANISM OF SURFACE CHARGE CREATION AT THE SOLID–LIQUID INTERFACE

Solids can acquire net surface charge in many ways, such as external polarization (metals), preferential adsorption of certain ions, and isomorphic substitution (replacement of higher valence cations in the solid crystal structure by lower valence cations). The mechanism by which the solid surface is charged depends on the type of the solid and its environment.

A. Metal/Electrolyte Interface

Two extreme types of metal/electrolyte interfaces can be distinguished, perfectly polarizable and reversible. For a perfectly polarized electrode, there is no charge transfer across the interface (i.e., no current is passing). For reversible electrodes, a change in the electrode potential causes current flow (i.e., charge transfers across the interface) until the system reaches equilibrium.

A good example of a polarizable electrode is the mercury/electrolyte system. Because of a lack of charge transfer, such systems behave like condensers, and their capacitance can be measured. This measurement, along with independent measurements of surface tension of mercury polarized at different potentials, is extremely helpful in studies of the electrical interfacial layer at the metal/electrolyte interface. By external polarization of the mercury electrode, the surface of mercury acquires charge that can be positive or negative depending on the potential applied in a given system. Due to a very high concentration of free electrons in metals, the excess charge associated with metal/electrolyte is within an atomic distance from the metal surface on the metal side. On the other hand, the distribution of charge on the solution side of the EIL depends on the electrolyte type and its concentration and has a more or less diffusive character [2,3]. A tendency of different ions to adsorb at the mercury/electrolyte interface can be easily studied by measuring the surface tension of mercury as a function of applied potential and an electrolyte concentration.

Using the theory of electrocapillarity, the surface charge of mercury can be calculated from the Lippmann equation [2] as

$$\frac{\partial \gamma}{(\partial E)_{\mu,p,T}} = \sigma_0 \qquad (1)$$

When there is no net charge on the mercury surface (point of zero charge, pzc), according to Eq. (1) the surface tension of mercury reaches a maximum. The value of the potential difference at which the mercury surface acquires the pzc depends on the type and concentration of the electrolyte used. In the presence of specific adsorption (due to interactions other than electrostatic), the original value of the potential of the pzc shifts towards more negative potentials if anions are specifically adsorbed and towards more positive potentials for specifically adsorbed cations. A large number of papers dealing with the electrical interfacial layer at the mercury/electrolyte interface has been published; the reader is referred to the review papers and books [2–5].

There is no direct method of determining interfacial tension for solid metals, so Eq. (1) is useless for such systems. Also, standard electrokinetic methods usually fail for solid metals. Kallay et al. introduced an adhesion method to estimate the isoelectric point (iep) for conductive surfaces [6]; the rate of adhesion of latex particles (which are either positive or negative over the entire pH range of interest) to metal beads rapidly changes at the iep (the adhesion is fast when the signs of electrokinetic charge are opposite). The values of the iep obtained for particular metals coincide with those for their oxides. Since less noble metals in contact with aqueous solutions are covered with a layer of corresponding metal oxide (which defines the surface properties), this result is not very surprising.

B. Semiconductor/Electrolyte Interface

Contrary to metals, where free electrons are the only charge carriers, semiconductors possess two types of charge carriers, free electrons in the conduction band and holes in the valence band [7–9]. The numbers of charge carriers as well as the energy gap between conduction and valence bands are different for different semiconductors. For intrinsic semiconductors, for example silicon, the energy gap is small, and the number of electrons excited into the conduction band is equal to the number of holes created in the valence band. Some insulators may gain semiconductive properties when doped (impurity activated) with small amounts of other compounds. Such semiconductors are called nonintrinsic semiconductors. Charge carriers in nonintrinsic semiconductors can be free electrons (n type) when the electron donor is a dopant, or holes (p type) when the electron acceptor is used as a dopant. For doped semiconductors, the numbers of electrons and holes are not equal to each other.

Semiconductors can be charged in different ways. In intrinsic semiconductors, electrons and holes are not in equal numbers near the semiconductor surface. Due to a much lower number of charge carriers in semiconductors compared to metals, a so-called space charge layer exists, and its properties are dependent on the potential applied [7–10]. The distribution of charges in the space charge layer is analogous to the diffuse part of the electrical interfacial layer on the electrolyte side. The presence of ions in the solution can affect the bands in the space charge layer, and they may become bent up or down depending on the type of ions near the semiconductor surface. When no bending occurs, the space charge is equal to zero, and the potential is called a flat-band potential [7,11]. In other words, the flat-band potential is analogous to the potential of the point of zero charge.

Measurement of the flat-band potential is a common technique for studying the adsorption of ions at the semiconductor/electrolyte interface [7,11]. For nonintrinsic semiconductors (e.g., sulfides, oxides), their mechanism of surface charge creation is essentially the same as that for the appropriate sulfide or oxide (see Section II.D), because usually the amount of impurities (dopants) is low (only a fraction of 1%). More information on the electrical interfacial layer, adsorption, and charge transfer reactions at the semiconductor/electrolyte interface can be found elsewhere [7–12].

C. Ionic Crystal (AgI)/Electrolyte Interface

The mechanism of surface charge creation at the ionic crystal/electrolyte interface is essentially different from that described for metals and semiconductors. For example, silver iodide (an ionic crystal) will dissolve slightly when placed in water to satisfy the equilibrium required by the solubility product:

$$AgI \Leftrightarrow Ag^+ + I^- \qquad (2)$$

For AgI the $K_{sp} = 10^{-16}$, so one might expect that the equilibrium concentrations of Ag^+ and I^- ions are equal to each other and equal to 10^{-8} mol dm^{-3}. AgI crystal is electroneutral, and if during crystal dissolution equal amounts of Ag^+ and I^- ions pass into the solution the net charge on the AgI surface should remain unchanged and still equal to zero. Otherwise, AgI particles dispersed in water should attain their pzc and show no movement in the electric field.

It was found, however, that AgI particles dispersed in water are negatively charged, and this is due to the higher tendency of I^- ions to adsorb at the AgI surface (or the lower tendency of I^- ions to go into the solution). I^- (or Ag^+) ions adsorbed on the surface are indistinguishable from crystal structure ions. They are responsible for the charge and potential of AgI surface and are called potential determining ions (pdi).

Since the AgI surface in contact with water is negatively charged, its point of zero charge can be achieved by increasing the concentration of Ag^+ ions in the solution. It was found [13,14] that the surface of AgI crystals carries no net charge when $[Ag^+] = 10^{-5.5}$ mol dm^{-3}, i.e., pAg = 5.5 and simultaneously the pI = 10.5, which is required by the solubility product of AgI.

At equilibrium the electrochemical potentials of e.g. Ag^+ ions on the AgI surface and in the bulk of the solution are the same. Thus [14]

$$\mu_{Ag+s} + e\psi_0 = \mu_{Ag+b} + e\psi_b \tag{3}$$

and

$$\mu^0_{Ag+s} + kT \ln a_{Ag+s} + e\psi_0 = \mu^0_{Ag+b} + kT \ln a_{Ag+b} + e\psi_b \tag{4}$$

Assuming that the potential in the bulk of the solution $\psi_b = 0$, the surface potential is

$$\psi_0 - \psi_b = \psi_0 \tag{5}$$

From Eqs. (4) and (5) one obtains

$$e\psi_0 = [\mu^0_{Ag+b} - \mu^0_{Ag+s}] + kT \ln \frac{a_{Ag+b}}{a_{Ag+s}} \tag{6}$$

At the pzc ($\psi_0 = 0$), Eq. (6) takes the form

$$[\mu^0_{Ag+b} - \mu^0_{Ag+s}] + kT \ln \frac{a_{Ag+b(pzc)}}{a_{Ag+s(pzc)}} = 0 \tag{7}$$

As the amount of ions adsorbed on the surface is negligibly small compared to the number of ions in the ionic crystal structure, we can assume that the activities of Ag^+ ions at the charged and uncharged solid surfaces are the same:

$$a_{Ag+s} = a_{Ag+s(pzc)} \tag{8}$$

Taking into account Eq. (8) and subtracting Eq. (7) from Eq. (6), one obtains

$$\psi_0 = \frac{kT}{e} \ln \frac{a_{Ag+b}}{a_{Ag+b(pzc)}} \tag{9}$$

As seen from Eq. (9), the potential of AgI surface changes (according to the Nernst equation) by about 59.2 mV for every tenfold increase in the concentration of potential determining ions. In order to determine the relationship between the surface charge and the surface potential of AgI dispersion, the potentiometric titration technique can be used [15,16]. The surface charge density can be calculated as a difference between the amount of pdi added to the system and the amount of pdi remaining in the solution. To determine the concentration of pdi, e.g., Ag^+, in the solution, the ion selective Ag/AgI electrode can be used.

The measurements are performed at different concentrations of indifferent electrolyte to keep the diffuse part of the electrical interfacial layer the same while changing the concentrations of the potential determining ions. Details on potentiometric titration for the AgI/electrolyte interface and a discussion on the EIL in this system can be found in the literature [13–18].

Though the Nernst equation can be easily used to determine the surface potential of the AgI surface, this equation is not always valid for other systems. In many cases, e.g., for oxides, potential determining ions are not constituents of the bulk of the solid, and their activity depends on the surface charge—see Section II.D.

D. Metal Oxide (Hydroxide)/Electrolyte Interface

Pure metal oxide (in vacuum) consists of metal and oxygen atoms. In contact with water (water vapor or liquid water), the surface of metal oxide undergoes hydroxylation by dissociative adsorption of water molecules [19,20]. Finally, the oxide surface becomes covered with metal hydroxyl groups (−MOH). The fact that these groups exist only on the metal oxide surface and not in the bulk of the solid has a very important effect on the behavior of oxide/electrolyte systems, e.g., the nonapplicability of the Nernst equation to describe the surface potential of oxides.

Amphoteric surface hydroxyl groups in contact with electrolyte solution may accept or donate protons depending on the pH of the solution:

$$-MOH \Leftrightarrow -MO^- + H^+ \qquad (10)$$

$$-MOH_2^+ \Leftrightarrow -MOH + H^+ \qquad (11)$$

Reactions (10) and (11) are responsible for the creation of negatively charged $-MO^-$ and positively charged $-MOH_2^+$ surface groups, respectively. The pH at which the amount of negatively charged groups is equal to the number of positively charged groups, so that the net surface charge is equal to zero, is called (by analogy to the AgI/electrolyte system) the point of zero charge. Thus the H^+ and OH^- ions are considered potential determining (pd) ions for oxides.

At the pzc, the surface of oxide in water is mostly covered with neutral $-MOH$ groups and (possibly) small amounts of $-MO^-$ and $-MOH_2^+$ in equal concentrations. Since the charged groups are not constituents of the crystal lattice of the solid oxide, one can expect significant changes in their concentration when the pH of the electrolyte changes. Therefore we cannot assume, as we did for AgI [see Eq. (8)], that the activity of surface pd ions is independent of the surface charge. If so, one can expect that the Nernst equation is not obeyed in such systems, and the surface potential should change by less than 59.2 mV for each unit change in the pH. The deviation of the surface potential of oxides from the Nernst equation depends on the type of oxide and the extent of ionization of surface $-MOH$ groups. A number of researchers have studied this problem [21–24].

Recently, direct measurements of the surface potentials of oxides were attempted using the ion sensitive field effect transistor (ISFET) method. It was found that this technique could be helpful in differentiating between different mechanisms of oxide surface charging [25]. The surface charge of oxides can be determined by the potentiometric titration method [26]. In this method, a glass electrode is used to monitor the concentration of H$^+$ ions in the solution. In some cases, for sparingly soluble oxides, the H$^+$ or OH$^-$ ions can be consumed not only in reactions (10–11) but also in the process of oxide dissolution. For such systems, modified procedures of the potentiometric titration method are used to take this effect into account and calculate the real surface charge [27]. Interestingly, aggregation of solid particles near the pzc does not block a significant number of surface sites, i.e., the effective surface area in surface charging near and far from the pzc is practically the same [28].

III. ELECTRIC INTERFACIAL LAYER MODELS WITHOUT COUNTERION ASSOCIATION

Let us estimate the surface charge density of an oxide using reactions (10) and (11). The surface is assumed to be flat. The radii of curvature >0.1 μm are considered insignificant, but for nanoparticles, the surface dissociation depends on the particle size. Some experimental evidence has been found for silica [29].

The first, very rough, estimation is made taking into account the mass balance, i.e.,

$$[-\text{MOH}] + [-\text{MO}^-] + [-\text{MOH}_2^+] = N_s = \text{const} \tag{12}$$

neglecting the activity coefficients of the reagents in reactions (10) and (11), i.e., assuming that the thermodynamic equilibrium constants of reactions (10) and (11) can be expressed by the quotients of the concentrations of the reagents. This approach is normally used for amphoteric acid–base equilibria in solution when the ionic strength is low, and it leads to an approximately exponential increase in the absolute value of the surface charge density σ_0 as a function of |pH-pH$_{\text{pzc}}$| when N_s is much greater than the highest absolute value of experimental σ_0 (Fig. 1). When N_s is sufficiently high, the shape of the model curves obtained for the fitted values of the equilibrium constants of reactions (10) and (11) is practically independent of the choice of N_s. In order to obtain a model curve corresponding to experimental data, N_s must be assumed to be approximately equal to the highest absolute value of experimental σ_0 (Fig. 1). A single titration curve can be reasonably modeled using the above model with three adjustable parameters, i.e., the equilibrium constants of reactions (10) and (11), and N_s. However, the above approach is not acceptable for many reasons. First, it does not properly represent the effect of the ionic strength on the charging curves. Thus the calculated parameters are only valid for one ionic strength. The sigmoidal shape of the calculated charging curves (low N_s) is not reflected by experimental results. Moreover, the N_s obtained as an adjustable parameter in this model is much lower than the value estimated from crystallographic data, from IR spectra, or by tritium exchange.

An obvious shortcoming of the calculations shown in Fig. 1 is that they treat surface species in the same way as species in solution: the surface potential is neglected. Protons are known to be potential determining ions for oxides. Once the surface potential is defined, we can use the Boltzmann equation

$$c_s = c_b \exp\left(\frac{-ze\psi_0}{kT}\right) \tag{13}$$

to express the concentration of the ions near the surface. We still assume that the concentrations of ionic species are sufficiently low and that their activities can be represented by concentrations. The Nernstian surface potential with Eq. (10) and (11) leads to the

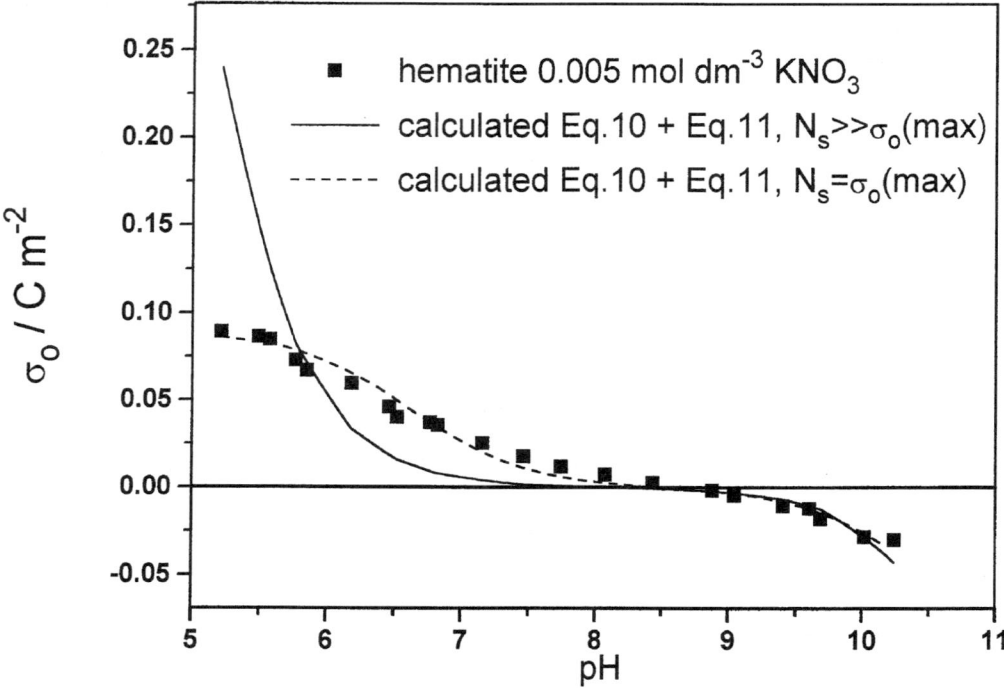

FIG. 1 Surface charge density of hematite in 0.005 mol dm^{-3} KNO$_3$. Symbols: experiment. (From Ref. 30.) Lines: calculated from Eqs. (10) and (11) neglecting the surface potential, pK = 11.59 and 4.72, 6 sites nm^{-2}, solid line; pK = 10.43 and 6.62, 0.53 sites nm^{-2}, dashed line.

constant ratios [−MO$^-$] : [−MOH] and [−MOH$_2^+$] : [−MOH], and this means that the surface charge density should be pH independent. This result is in clear discord with the experimental data.

A. Constant Capacitance

We can try to estimate the surface potential by assuming that the capacitance of the EIL is constant:

$$\frac{\sigma_0}{\psi_0} = C = \text{const} \tag{14}$$

Figure 2 shows the result of model calculations for the same experimental data as were used in Fig. 1. N_s is much greater than the highest absolute value of experimental σ_0. For the capacitance of 0.83 Fm^{-2}, the model reasonably reproduces the experimental data. Four adjustable parameters were used, namely the equilibrium constants of reactions (10) and (11), the electric interfacial layer capacitance, and N_s. When N_s is sufficiently high, the shape of the model curves obtained for the fitted values of the equilibrium constants and the electric interfacial layer capacitance is practically independent of N_s. The surface potentials calculated for the data set presented in Fig. 2 are lower than the Nernst potential by a factor of 2. Similar calculations can be carried

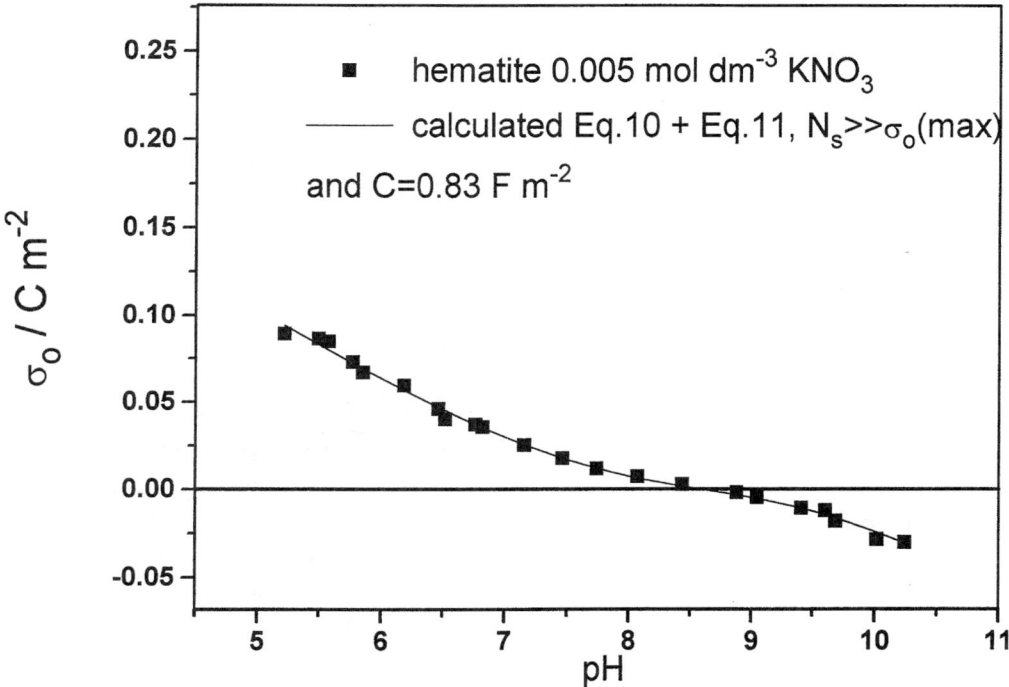

FIG. 2 Surface charge density of hematite in 0.005 mol dm^{-3} KNO$_3$. Symbols: experiment. Line: calculated from Eqs. (10) and (11), constant capacitance model, $C = 0.83$ F m^{-2}, pK = 11.14 and 6.08, 6 sites nm^{-2}.

out for charging data for higher ionic strengths, but they lead to different values of model parameters. The shortcoming of the constant capacitance model is that the model parameters (chiefly the capacitance) are variable, namely, they depend on the ionic strength. The capacitance of 0.83 Fm^{-2} is reasonable considering that the thickness of the electric interfacial layer is a fraction of 1 nm and the dielectric constant in the interfacial region is slightly lower than the bulk value for water.

B. Diffuse Layer

The diffuse layer model can predict (at least qualitatively) the ionic strength effect on the surface charging. It assumes that the surface charge is entirely balanced by the diffuse charge. For a flat surface,

$$\sigma_d = \frac{\varepsilon \kappa kT}{e}[\exp(e\psi_d/2kT) - \exp(-e\psi_d/2kT)] \tag{15}$$

Analogous equations (σ_d as a function of ψ_d) have been derived for other highly symmetric surfaces (e.g., spheres), but for typical Debye lengths in aqueous solutions, the surface curvature leads to significant deviations from Eq. (15) only for radii below 1 μm. The results obtained for N_s values much greater than the highest absolute value of experimental σ_0 are shown in Fig. 3. The diffuse layer model clearly overestimates the ionic strength effect on the charging curves at very low pH, but at pH above the pzc

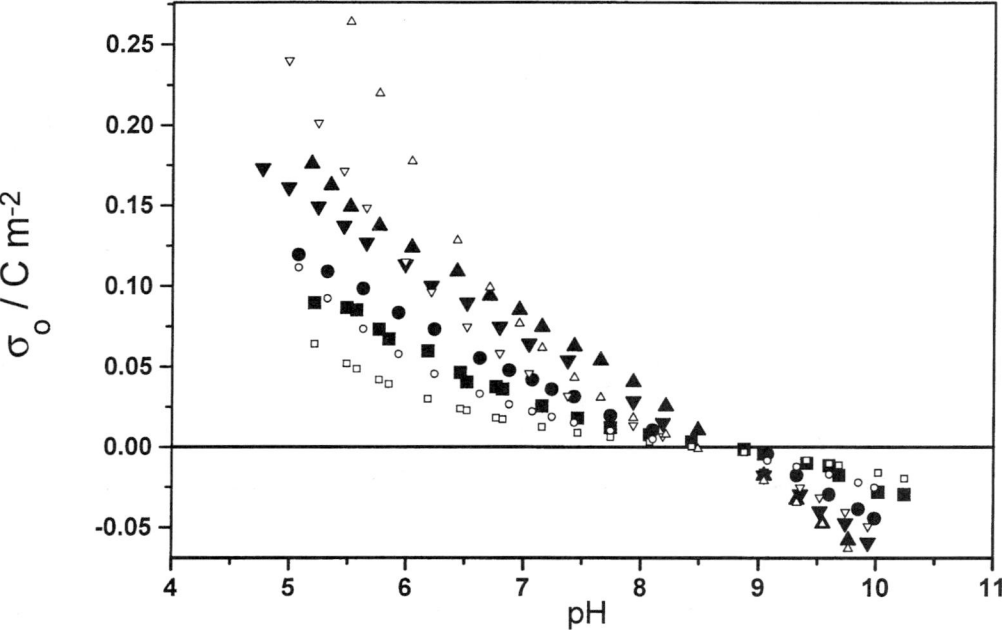

FIG. 3 Surface charge density of hematite in KNO_3: squares 0.005, circles 0.02, down triangles 0.2, up triangles 1 mol dm^{-3}. Full large symbols: experiment. Open small symbols: calculated from Eqs. (10) and (11), diffuse layer model, pK = 10.5 and 6.4, 6 sites nm^{-2}.

the model works reasonably. The quality of the fit can be improved when N_s equal to the highest experimentally observed σ_0 is assumed (Fig. 4). The diffuse layer model has two serious advantages over the constant capacitance model: it uses only three adjustable parameters: N_s and the equilibrium constants of reactions (10) and (11) (the constant capacitance model uses four parameters); and the same model parameters can be used at various ionic strengths. Figure 5 shows the surface potentials calculated for the same parameters of the diffuse layer model as the charging curves in Fig. 4. These potentials are nearly Nernstian at low ionic strengths near the pzc, and the absolute value of calculated surface potential at a given pH decreases when the ionic strength increases. The decrease in the surface potential when the ionic strength increases in reflected in the experimentally determined ζ potentials. Figure 6 shows the capacitances of the electric interfacial layer [Eq. (14)] calculated for the same parameters of the diffuse layer model as the charging curves in Fig. 4. The capacitance at a constant pH increases with the ionic strength, and for a given ionic strength it always shows a shallow minimum at the pzc.

Although the equilibrium constants of reactions (10) and (11) were fitted as two independent variables to obtain the model curves shown in Figs. 1–6, they are actually not independent, i.e., the sum of the pK must be equal to $2pH_{pzc}$ to obtain $\sigma_0 = 0$ at the pzc. Therefore actually we only have two adjustable parameters in the diffuse layer model, namely one equilibrium constant and the N_s.

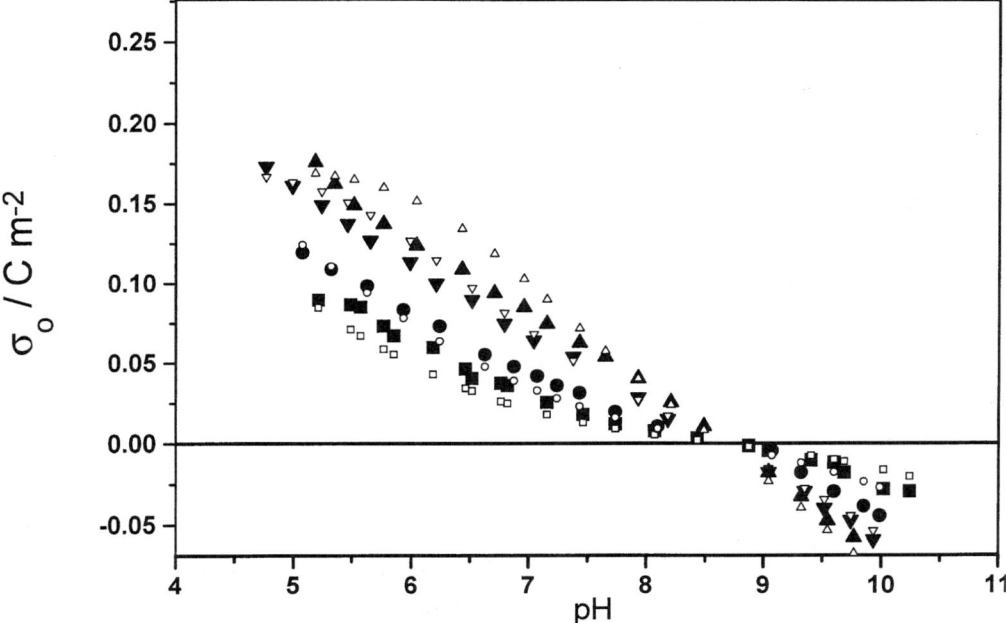

FIG. 4 As in Fig. 3, but pK = 9.45 and 7.81, 1.03 sites nm^{-2}.

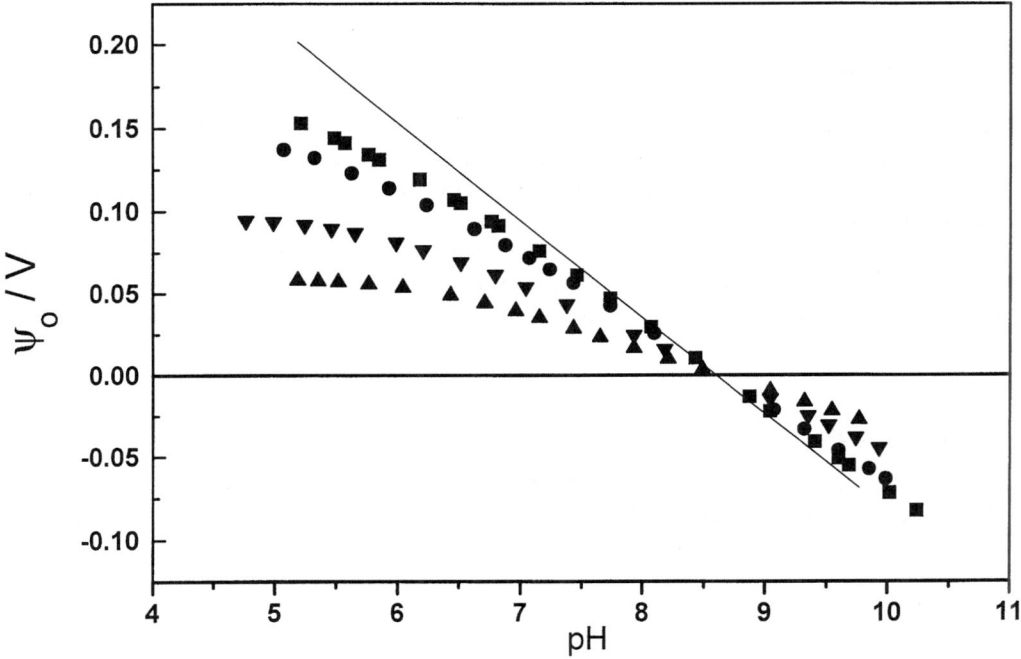

FIG. 5 Surface potential calculated for the same model as Fig. 4. Line: Nernst.

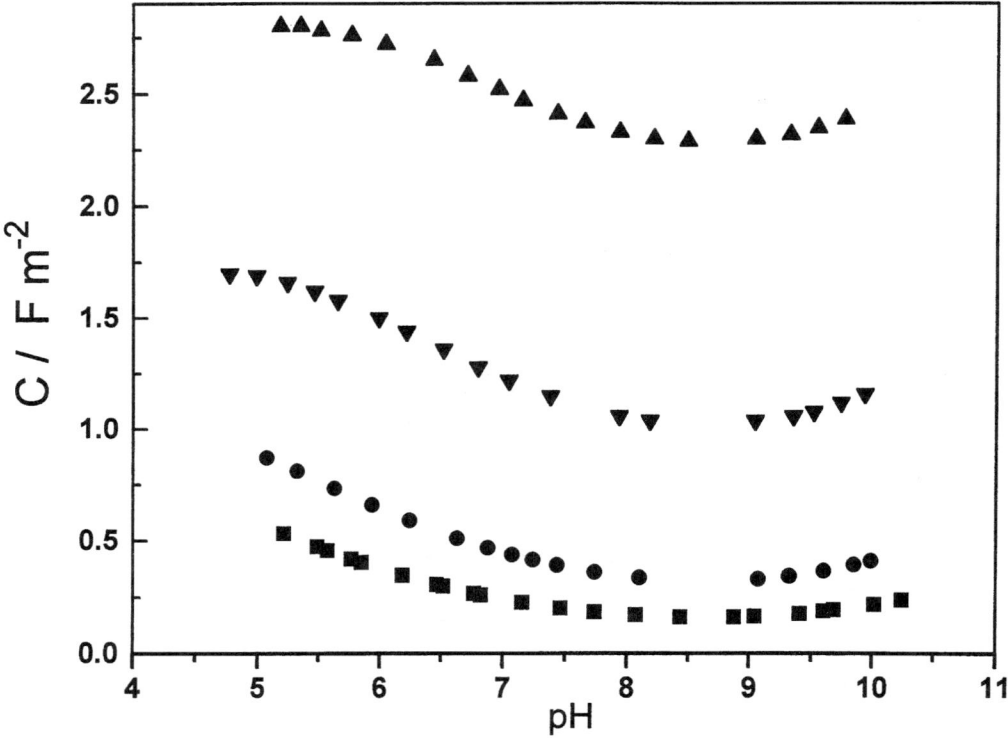

FIG. 6 Electric interfacial layer capacitance calculated for the same model as Fig. 4.

C. Stern Model

The ability of the diffuse layer model to model the charging curves can be substantially improved by combining it with the constant capacitance model. This leads to the Stern model with two capacitors in series: a part of the surface charge is neutralized by the Stern layer (constant capacitance) and the rest by a diffuse layer [Eq. (15)]. Figure 7 shows that the σ_0 calculated using the Stern model with a high N_s is in better agreement with the experimental data than that using diffuse layer model (Fig. 3). This was achieved by adding one adjustable parameter, namely, the Stern capacitance, to the model.

D. 1-pK Model

The model with reactions (10) and (11) involves three surface species, namely $-MO$, $-MOH_2^+$, and $-MOH$. In the 1-pK model [31], the neutral $-MOH$ groups are neglected and the surface charging is due to only one reaction,

$$-MOH^{1/2-} + H^+ = -MOH^{1/2+} \tag{16}$$

The logarithm of the equilibrium constant of this reaction must equal pH_{pzc} to obtain $\sigma_0 = 0$ at the pzc. The traditional approach using reactions (10) and (11) is called the 2-pK model since it uses two reactions whose equilibrium constants are used as model

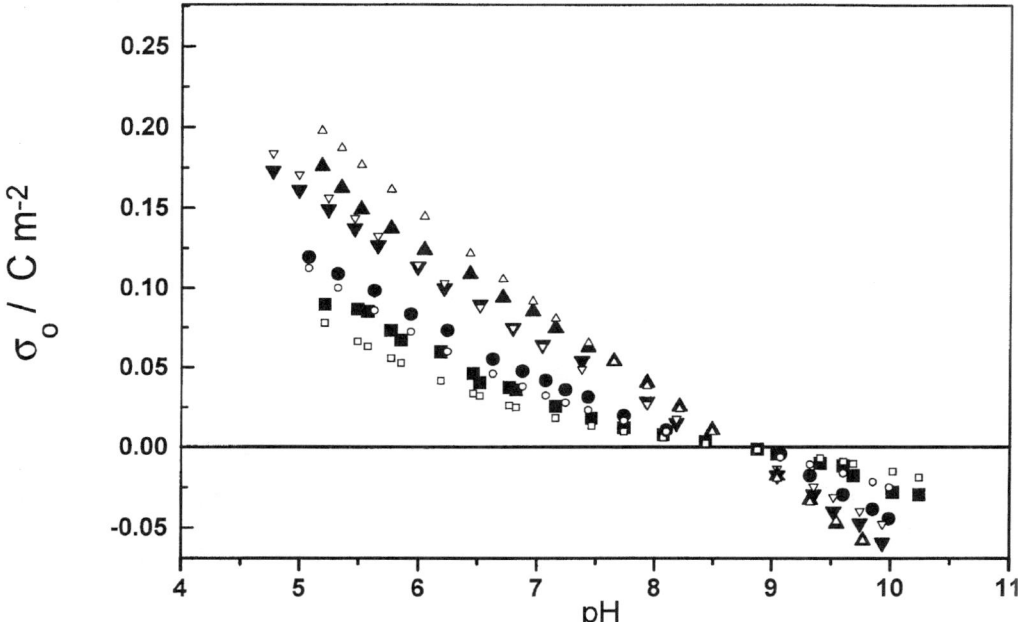

FIG. 7 Surface charge density of hematite: Stern model with Eqs. (10) and (11), pK = 9.21 and 8.12, $C = 1.5$ F m^{-2}, 6 sites nm^{-2}. Symbols have the same meanings as in Fig. 3.

parameters. According to the 1-pK model, a neutral surface has large and equal numbers of positively and negatively charged surface groups, while in the 2-pK model a neutral surface has chiefly neutral groups and only a few positively and negatively charged groups.

All the models discussed above in their 2-pK versions (constant capacitance, diffuse layer, Stern) can be converted into 1-pK versions. Let us discuss the 1-pK version of the diffuse layer model in more detail. Having the equilibrium constant of reaction (16) defined by the experimental pzc value, the only adjustable parameter in this model is the total number of ionizable surface groups. The results for the best-fit value are shown in Fig. 8. The experimental data are reasonably reproduced for low ionic strengths and for high ionic strengths only in the vicinity of the pzc. The fit is worse than that presented in Fig. 4, but an advantage of the 1-pK model is that the N_s and the pzc can be obtained not as fitting parameters but directly from independent experiments, so that the charging curves can be predicted without resorting to the titration data at all. Figure 9 shows the surface potentials calculated for a 1-pK version of the diffuse layer model. The potentials obtained for low ionic strengths are higher than for the fitted 2-pK-version parameters, but for high ionic strengths the deviation from the Nernst equation is more pronounced for the 1-pK model.

Equation (15) can be generalized as

$$-\text{MOH}^{x-} + \text{H}^+ = -\text{MOH}^{(1-x)+} \tag{17}$$

where $1 > x > 0$; Eq. (16) corresponds to $x = 1/2$. Also reactions (10) and (11) can be considered as special cases of Eq. (17) with $x = 1$ and 0, respectively.

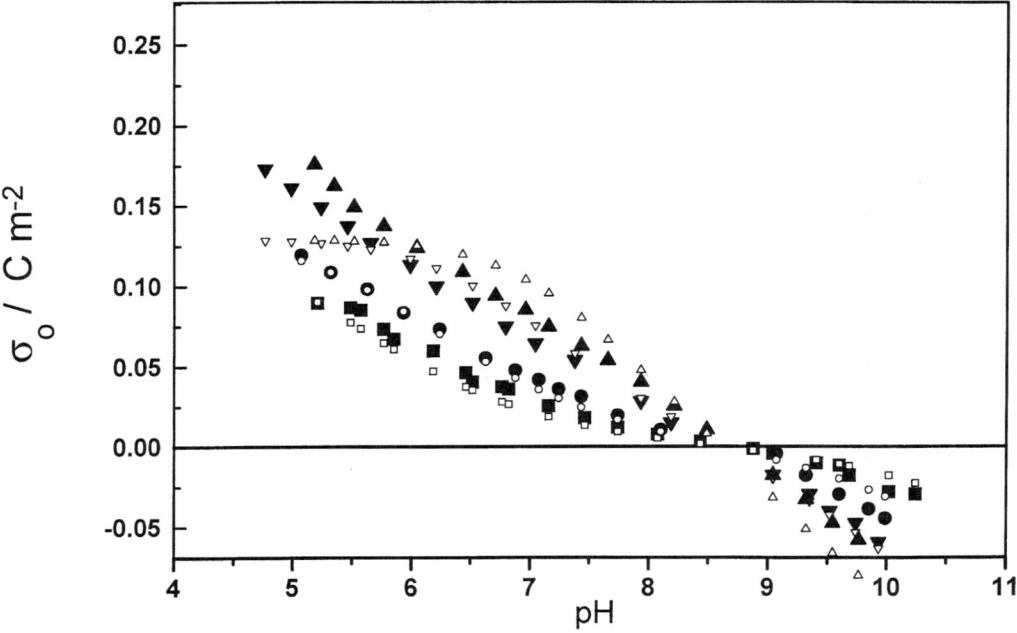

FIG. 8 Surface charge density of hematite: diffuse layer model with Eq. (16), 1.55 sites nm^{-2}. Symbols have the same meanings as in Fig. 3.

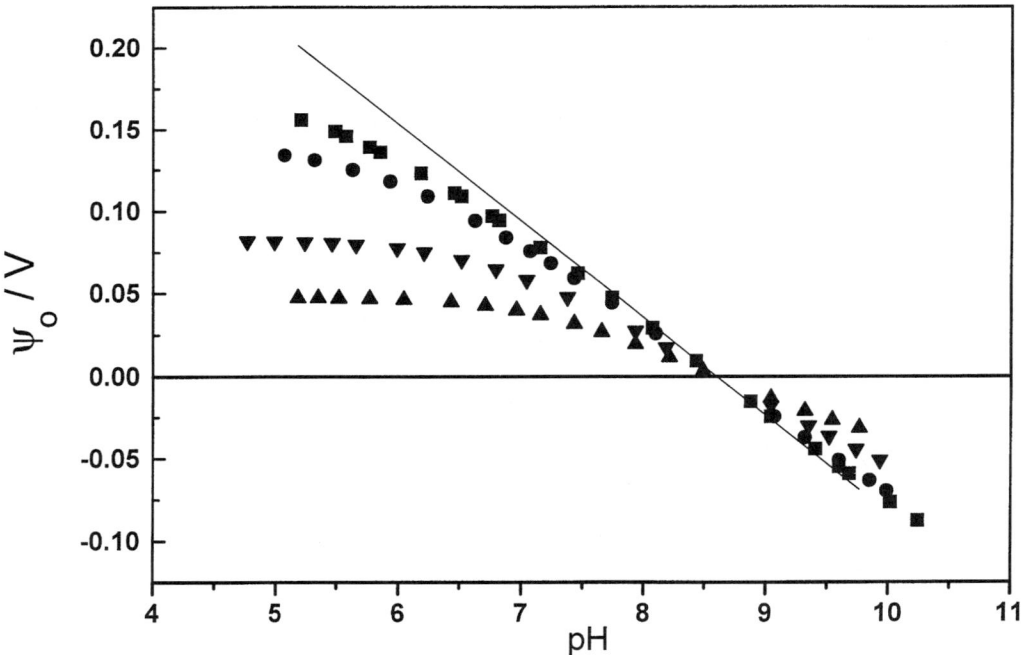

FIG. 9 Surface potential calculated for the same model as Fig. 8. Line: Nernst. Symbols have the same meanings as in Fig. 3.

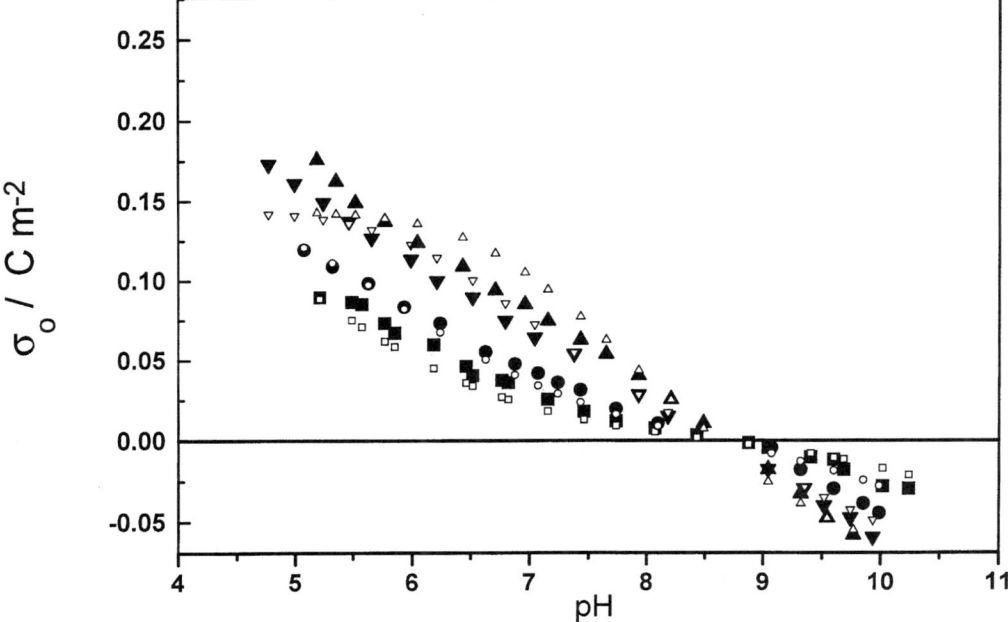

FIG. 10 Surface charge density of hematite diffuse layer model with Eq. (17), pK = 8.3, $x = 1/3$, 1.29 sites nm^{-2}. Symbols have the same meanings as in Fig. 3.

The equilibrium constant of Eq. (17) is unequivocally determined by the pzc, namely,

$$\log K = \mathrm{pH}_{\mathrm{pzc}} - \log\left(\frac{1-x}{x}\right) \tag{18}$$

Considering x as an adjustable parameter we find out that for the set of data used in this section to test various models of the EIL, $x = 1/3$ gives a better fit than $x = 1/2$. The model curves for the diffuse layer model with Eq. (17) and $x = 1/3$ are presented in Fig. 10.

E. 1-pK Multisite Approach

In the above discussion we assumed that one kind of surface site is responsible for surface charging. These sites can carry negative or positive charge when the protons are desorbed from or adsorbed on them. Each of these surface sites is an oxygen atom associated with one or more surface ions of metal in the oxide. These oxygen atoms are not identical: they have different coordination environments, and some of them are more acidic than others. Let us assume that there are two different kinds of surface sites: acidic sites, which can desorb protons and create negative surface charge [reaction (10)], and basic sites, which can adsorb protons and create positive surface charge [reaction (11)]. Let us combine this multisite approach (two sites in the present example) with the diffuse layer model. With two kinds of surface sites we have four adjustable parameters: concentrations of both kinds of surface sites and the equilibrium constants of reactions (10) and (11). In the multisite approach there is no simple relationship between the equilibrium constants of reactions (10) and (11) and the pzc. Figure 11 presents the results for the best fit parameters. The fit is much better than for any model presented before in this section.

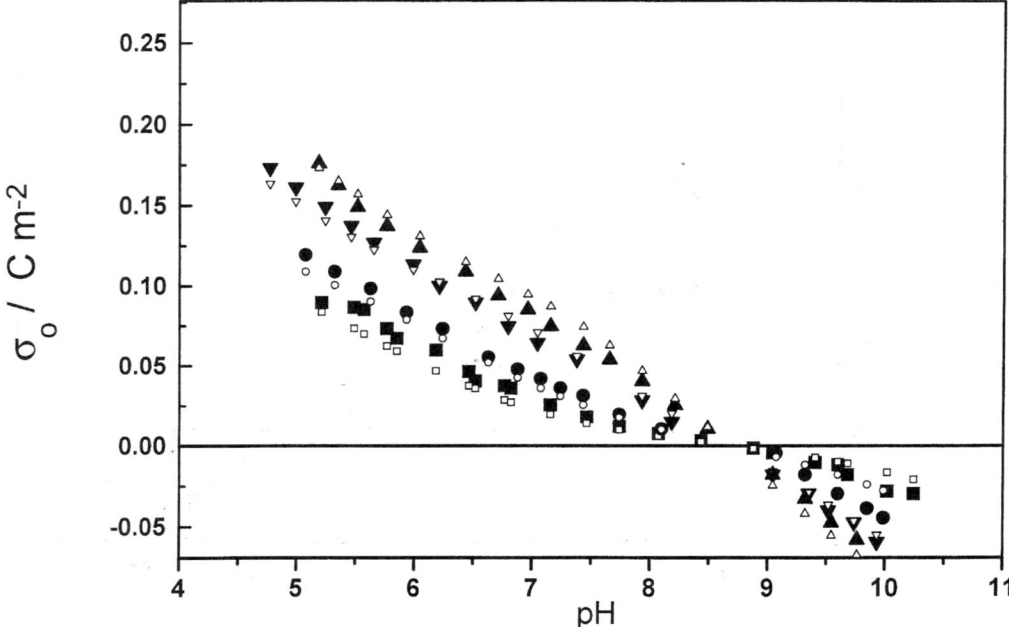

FIG. 11 Surface charge density of hematite diffuse layer model with two kinds of surface groups, acidic [Eq. (10)], pK = 6.46, 0.65 groups nm^{-2}, and basic [Eq. (11)], pK = 8.68, 1.26 groups nm^{-2}. Symbols have the same meanings as in Fig. 3.

Certainly the quality of the fit can still be improved, e.g., by considering more than two kinds of surface sites (and more adjustable parameters) or by combining the multisite approach with the Stern model rather than the diffuse layer model (one additional adjustable parameter).

F. MUSIC

Having found that the multisite approach is successful in modeling the surface charging of oxides in a series of curve fitting exercises, let us now discuss the physical grounds for the different acid–base behavior of different surface groups. The affinity of surface oxygen atoms to protons depends on their coordination environment. This effect is well known from the solution chemistry, e.g., the pK of silicic acid is a function of the degree of polymerization [32]. Careful analysis of the crystallographic structure of oxides led to the multisite complexation model (MUSIC) [33]. The original version of MUSIC was discussed and refined in many subsequent papers [34].

Let us define the bond valence v after Pauling as the charge of the cation divided by its coordination number:

$$v = \frac{Z}{CN} \tag{19}$$

In other words, the central metal ion is assumed to distribute uniformly its positive charge over all next-neighbor oxygen atoms. Using the concept of bond valence, we can define a formal charge of a surface $M_N OH$ group as $N \times v - 1$ rather than zero as it was in

Eq. (10) and Eq. (11). In a few special cases, when $N \times z = CN$, e.g., in Si–OH or Al$_2$OH surface groups, the bond valence concept leads to a zero formal charge of surface hydroxyl groups, but more often a fractional formal charge is obtained. For example, the formal charge of the Al$_3$O surface group equals $-1/2$, so its protonation can be written as

$$\equiv \text{Al}_3\text{O}^{-1/2} + \text{H}^+ = \ \equiv \text{Al}_3\text{OH}^{+1/2} \tag{20}$$

Thus the bond valence concept rationalizes the 1-pK model [Eq. (16)], or more generally Eq. (17); for example, considering the v for titanium in TiO$_2$, we obtain a formal charge of 1/3 for a hydroxyl group associated with two Ti atoms:

$$\equiv \text{Ti}_2\text{O}^{-2/3} + \text{H}^+ = \ \equiv \text{Ti}_2\text{OH}^{+1/3} \tag{21}$$

The nature and density of different surface species can be derived from the crystalline structure of the oxide, and they are different for different faces. The log K for reactions (20), (21), etc. can be estimated as

$$\log K = A - B \times \frac{Nv}{L} \tag{22}$$

where L is the metal–hydrogen distance (which can be found from crystallographic data), $B = 5.27$ nm, and A equals 18.4 for metals with 10 d electrons and 19.7 for central atoms of noble gas electronic structure; for hydroxocomplexes and for oxocomplexes, the corresponding A values are 32.2 and 34.5, respectively. The above A values apply for surface species, and they are somewhat higher than for ions in solution.

The log K values from Eq. (22) combined with the density of particular surface species derived from crystallographic data allow for a completely new approach: instead of treating site densities and log K values as adjustable parameters calculated from the charging data in MUSIC, we rather test whether the charging data comply with the model predictions. The Stern capacitance is the only adjustable parameter when we use the Stern model.

The above simplified approach to the bond valence of the oxygen can be refined by considering the contribution of hydrogen bonding to the neutralization of the valence of the oxygen atom. This neutralization $\sum s_j$ can be expressed as

$$\Sigma s_j = \Sigma s_{\text{Me}} + m s_{\text{H}} + n(1 - s_{\text{H}}) \tag{23}$$

where s_{Me} is the Pauling bond valence v [Eq. (19)] corrected for the actual metal–oxygen distance; the sum on the r.h.s. is taken over all metal atoms contributing to the neutralization (their contributions can be somewhat different when the interatomic distances are not identical); m is the number of oxygen s-orbitals occupied by hydrogen (0 for monomeric oxycomplexes and 1 for hydroxocomplexes); n is the number of unoccupied s-orbitals (3 for monomeric oxycomplexes and 2 for hydroxocomplexes in solution); and s_{H} is the bond valence of the H-donating bond and equals about 0.8 in water. The log K of protonation is a function of the undersaturation of the valence of oxygen V:

$$\log K = -A(\Sigma s_j - V) \tag{24}$$

where A is an empirical constant equal to 19.8. For surface oxygen atoms, $m + n = 2$ for steric reasons for oxygen atoms singly coordinated by metal, $m + n = 2$ or 1 for doubly coordinated and $m + n = 1$ for triply coordinated oxygen atoms. While in the older version of MUSIC the difference in formal charge of the surface group on protonation was 1, in the refined model [Eq. (23)] this difference is only 0.6 due to the hydrogen bonding. In the revised MUSIC, more than one type of surface groups having the same stoichiometry, e.g., =TiO, can be distinguished on the basis of different bond lengths. These groups have different formal charges and different log K [Eq. (24)].

The results of model calculations (MUSIC + Stern) confirm the experimentally observed trends. MUSIC leads to several interesting and important conclusions. For example, different crystallographic faces of goethite have different pzcs. Thus crystals of different morphologies can differ in their overall pzc. MUSIC explains also why the attempts to find a simple relationship between the ppzc of oxides and their properties, such as electronegativities or ionic radii, were not successful.

IV. ELECTRIC INTERFACIAL LAYER MODELS WITH COUNTERION ASSOCIATION

In the previous section, the role of the anions and cations of the supporting electrolyte in the formation of surface charge was not emphasized, although the ionic strength appears in the diffuse layer and Stern models as a parameter determining the magnitude of the diffuse charge. The diffuse layer model considers only electrostatic interactions of point charges, and ion specificity can only be modeled indirectly via the capacitance of the Stern layer. There is no simple relationship between the capacitance of the Stern layer and the ionic radius of the counterion or other well-defined parameters.

A. Empirical Approach

We can measure the magnitude of adsorption and describe the adsorption by means of adsorption isotherm equations. These results can be then analyzed to gain deeper insight into the mechanism of interactions between the adsorbent and the adsorbate, but even without sophisticated models and interpretation, such empirical knowledge is valuable in itself. Adsorption of ions on oxides is strongly pH dependent, but at fixed pH we can treat adsorption of ions in the same way as is usually done with uncharged species. Therefore experimental results are often presented in the form of adsorption isotherms, and the analysis of these isotherms gives clues about absorbent–adsorbate and adsorbate–adsorbate interactions. Similarities between dilute solutions and gases justify the application of adsorption isotherms derived for gas adsorption to the adsorption of ions from solution. In spite of a broad spectrum of equations for adsorption isotherms, whether empirical or theoretically derived, to account for various effects [35], most of the studies of ion adsorption on oxides use the Langmuir adsorption isotherm. This isotherm has been derived for localized monolayer adsorption on a homogeneous surface, without lateral interactions.

Zhang et al. [36] studied the adsorption of Na^+ on positively charged γ-alumina as a function of NaCl concentration. The pH was not adjusted, and it varied from 5.75 for the lowest to 6.25 for the highest electrolyte concentration. The experimentally determined Na^+ adsorption was negative, and the exclusion volume ($\Gamma_{Na}S/C_{Na}^0$) was plotted as function of $1/\kappa$, where S is the specific surface area. The negative adsorption of sodium was used to calculate diffuse layer potentials from the equation

$$\Gamma_{Na}S = \frac{2C_{Na}{}^0 S}{\kappa}[1 - \exp(-e\psi_d/2kT)] \qquad (25)$$

derived from the Boltzmann distribution. The calculated ψ_d decreases by about 30 mV per decade of ionic strength, but the ζ potentials measured by the same authors were rather insensitive to the ionic strength in the range 10^{-4}–10^{-2} mol dm^{-3} NaCl.

Adsorption isotherms of K$^+$ on negatively charged zirconia were studied at pH 7.5–10 and for initial concentrations 0.1–1 mol dm^{-3} [37]. The isotherms have Langmuirian shape without a distinct Henry range (no data for low initial concentrations) and with monolayer coverage (plateau of the Langmuir isotherm) ranging from 0.3 mol/kg (pH 7.5) to 0.7 mol/kg (pH 10). The level of the plateau increases linearly with pH. The authors did not use any specific equation of adsorption isotherm to interpret their results, but the existence of two different types of surface sites was postulated. At low pH values and low initial concentrations, the adsorption of potassium is higher than that of sodium or lithium, but at high initial concentrations and high pH values, this trend is reversed. This is interpreted in terms of activity coefficients of single ions: the activity coefficient of lithium in concentrated solutions of its salts is much higher than for other alkali metal cations.

Cerefolini and Boara [38] observed a decrease in concentration of alkali metal cations in water (no pH control) after addition of dried silica gel (except for 1.71×10^{-3} mol dm^{-3} NaCl, for which no depletion was observed). The results are interpreted in terms of segregation coefficients between bulk and absorbed water.

Static adsorption isotherms and the results of kinetic studies of sorption as well as dynamic adsorption curves of alkali metal cations on ultramicroporous silica glasses can be found in Ref. 39. Interestingly, these adsorbents show very low affinity to alkaline earth metal cations. Adsorption of alkali metal cations increases from Li to Cs.

B. Ion Selectivity

In some phenomena, the difference between various 1–1 salts is insignificant, but sometimes the difference between particular monovalent ions is striking. Some interesting and mysterious phenomena related to ion binding and ion specificity are discussed by Ninham and Yaminski [40].

The strong effect of the nature of alkali metal cations on the σ_0 of silica is well known. Tadros and Lyklema [41] have found that the negative σ_0 of silica is almost entirely balanced by the adsorption of counterions, so the sequence found for the σ_0, Cs>K>Na>Li, is also valid for the adsorption of cations.

Sonnefeld et al. [42] studied the cation specificity for silica (Aerosil 300) indirectly, i.e., they determined surface charge densities in the presence of various alkali chlorides over the concentration range 0.005 to 0.3 mol dm^{-3} and found the same sequence, Cs>Rb>K>Na>Li, for all concentrations studied.

The selectivity is often expressed as

$$K_{Na} = [Me_{ads}][Na_s][Na_{ads}]^{-1}[Me_s]^{-1} \tag{26}$$

The selectivity coefficient for Na is 1 by definition. The values 0.65 for Li, 1.8 for K, 2.4 for Rb, and 3.2 for Cs were reported by Tien [43].

Parida [44] reports the adsorption sequence K $\leq$ Na $\leq$ Li for β-FeOOH, with the cip of titration curves at pH 7.5 for alkali nitrates (0.001–0.1 mol dm^{-3}). In contrast, Li shows a significant interaction with the surfaces of hematite and γ-FeOOH.

Studies reporting adsorption isotherms of multivalent ions on oxides at constant pH are more numerous than for monovalent ions. Collections of literature references can be found in review papers on specific adsorption of ions [45]. Typically the adsorption isotherms cover many orders of magnitude of initial concentrations, and they are often plotted on a log–log scale. Linear log–log adsorption isotherms with a slope below one in a range of low initial concentrations were reported. This behavior has been interpreted in terms of surface heterogeneity. Applicability of discrete and continuous distributions of surface sites to interpret these adsorption isotherms has been hotly debated [46]. Adsorption of multivalent ions will be discussed in more detail in a separate section.

C. Association

The forces causing adsorption of ions can be classified as generic (due to attraction of oppositely charged ions, which can be treated as rigid spheres) and specific (with molecular orbital overlap). The term association will be used in the former and complexation in the latter case. Thus association is closely related to ion pairing in bulk solution (outer sphere complexes), and complexation is related to the formation of inner sphere complexes. The above definition is neither rigorous nor universally accepted, and there is no sharp border between these phenomena. For example, calculation of activities of metal ions in the presence of ligands in solution can be equally well performed when considering the complexes as separate species or using Pitzer formalism for metal–ligand interaction, i.e., treating these interactions as generic, when the metal–ligand complexes are not too strong.

The difference between ion pairing in solution and at the surface is due to the presence of the overall electric field of the surface in addition to the local field of the charged surface group of interest. Kallay and Tomić [47] used Bjerrum association theory with half of the space available for counterions. The shape of the association volume resembles a hallow sphere and its cross section resembles a half-moon. The central conception of this theory is the critical distance d_{crit}. Ions closer than d_{crit} are considered as associated; for aqueous solutions of 1–1 electrolytes, $d_{crit} = 0.35$ nm at 25°C. The second important parameter is the distance of closest approach (dca), which is the closest distance between the centers of interacting charges if the surface potential is zero, which is at the point of zero charge. However, by lowering or increasing the pH, the magnitude of the surface potential increases, which is followed by an increase in the critical distance. Once the potential is high enough to produce a critical distance larger than the distance of closest approach, the association of counterions takes place. As the surface potential increases, the counterion association constant also increases, which results in a higher degree of counterion association. The apparent association constant is defined as the ratio of the (average) concentration of ions in the "association volume" and in the whole system, and it is a function of the assumed critical distance and of the magnitude of surface potential, and also of the ionic strength. The distance of closest approach is different from ion to ion, and this is responsible for ion specificity. The association constant equals zero not only for the coions but even for the counterions when the surface potential is not high enough. For the distance of closest approach of 0.36 nm, the onset of the association of counterions at a surface potential of a few mV is predicted, but with a dca of 0.6 nm, the onset of association corresponds to surface potentials in excess of 60 mV (dependent on the ionic strength). In the range of low surface potential, the calculated association constant is very sensitive to the assumed critical distance, but at surface potentials in excess of 100 mV the association constants calculated for various critical distances converge. The calculated surface speciation suggests the absence of the "ionic pairs" formed by charged surface groups and ions of inert electrolyte at pH in the range $pH_{pzc} \pm 2$ pH units, but outside this range the contribution of ionic pairs to the total surface charge sharply increases, and beyond the range $pH_{pzc} \pm 3$ pH units, the ionic pairs are the dominating components of the surface charge. The model calculations predict a rather insignificant ion specificity. Kallay and Tomić derived their equation for the 2-pK model. Analogous calculations can be performed for models based on Eq. (17). For a fractional charge of the charged surface groups, the calculated degree of their association with the ions of inert electrolyte is even less pronounced than for the 2-pK approach.

D. Surface Complexation

It is well documented in the literature that the surface charge at the oxide electrolyte interface, determined by the potentiometric titration method, is considerably higher than the diffuse layer charge, determined by the use of microelectrophoresis, for example. To explain this phenomenon, the site-binding model of the electrical interfacial layer at the oxide/electrolyte interface was proposed [48,49]. The theory assumed that a considerable part of the surface charge at the oxide surface was counterbalanced by the adsorption of supporting electrolyte ions onto surface charged groups:

$$-MO^- + Y^+ \Leftrightarrow -MO^-Y^+ \tag{27}$$
$$-MOH_2^+ + X^- \Leftrightarrow -MOH_2^+X^- \tag{28}$$

Davis et al. [50,51] have suggested that it is more informative to write the reactions (27) and (28) as

$$-MOH_2^+X^- \Leftrightarrow -MOH + H_s^+ + X_s^- \tag{29}$$
$$-MOH + Y_s^+ \Leftrightarrow -MO^-Y^+ + H_s^+ \tag{30}$$

This formal change does not affect the basic idea of Yates et al. [48].

According to the site-binding model of the EIL, only potential determining ions H^+ and OH^- can be located in the innermost layer of the electrical interfacial layer. These ions contribute to the surface charge, σ_0, and experience the surface potential, ψ_0. Supporting electrolyte ions, which bind pairwise with oppositely charged surface groups, are located in the inner Helmholtz plane (IHP) (the same for cations and anions). Ions in the IHP are separated from the surface by a region of constant capacity C_1. They contribute to the charge σ_β and experience the potential ψ_β. The IHP is insulated from the outer Helmholtz plane (OHP) by a region of constant capacity C_2. The potential at the OHP (the onset of the diffuse layer) is ψ_d, and there is a corresponding diffuse layer charge σ_d.

In the surface complexation model (SCM) of the electrical interfacial layer at the oxide/electrolyte interface illustrated schematically in Fig. 12, the following four reactions are responsible for the creation of the surface charge (YX is a supporting electrolyte) [52]:

$$-MOH_2^+ \Leftrightarrow MOH + H_s^+ \tag{31}$$
$$-MOH \Leftrightarrow -MO^- + H_s^+ \tag{32}$$
$$-MOH_2^+X^- \Leftrightarrow -MOH + H_s^+ + X_\beta^- \tag{33}$$
$$-MOH + Y_\beta^+ \Leftrightarrow -MO^-Y^+ + H_s^+ \tag{34}$$

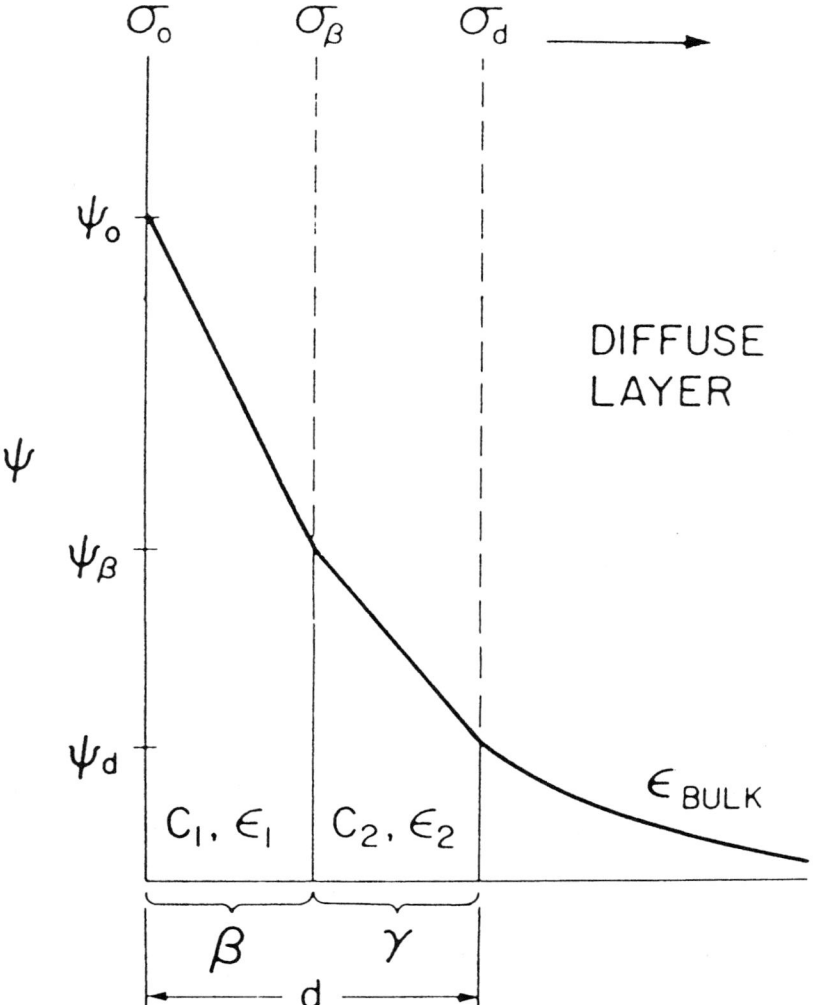

FIG. 12 Schematic representation of the charge distribution at an idealized planar surface and the potential decay away from the surface. (Reprinted from J. Colloid Interf. Sci., Ref. 50, © 1978 by Academic Press.)

According to Chan et al. [53], thermodynamic equilibrium constants of reactions (31–34) can be defined as

$$K_{a1}^{o\,int} = \frac{[-MOH][H_s^+]\gamma_0 \gamma_{Hs+}}{[-MOH_2^+]\gamma_+} \tag{35}$$

$$K_{a2}^{o\,int} = \frac{[-MO^-][H_s^+]\gamma_- \gamma_{Hs+}}{[-MOH]\gamma_0} \tag{36}$$

$$*K_{X-}^{o\,int} = \frac{[-MeOH][H_s^+][X_\beta^-]\gamma_0 \gamma_{Hs+} \gamma_{X\beta-}}{[-MOH_2^+X^-]\gamma_{+-}} \tag{37}$$

$$*K_{Y+}^{o\,int} = \frac{[-MO^-Y^+][H_s^+]\gamma_{+-} \gamma_{Hs+}}{[-MOH][Y_\beta^+]\gamma_0 \gamma_{Y\beta+}} \tag{38}$$

At equilibrium the following conditions are fulfilled for reactions (31,32):

$$\bar{\mu}_{MOH} + \bar{\mu}_{Hs+} = \bar{\mu}_{MOH2+} \tag{39}$$

and

$$\bar{\mu}_{MO-} + \bar{\mu}_{Hs+} = \bar{\mu}_{MOH} \tag{40}$$

where

$$\bar{\mu}_{Hs+} = \bar{\mu}_{Hb+} = \mu_{Hb+}^0 + kT\ln[H_s^+]\gamma_{Hs+} + e\psi_0 = \mu_{Hb+}^0 + kT\ln[H_b^+]\gamma_{Hb+} \tag{41}$$

Because both $[H_s^+]$ and $[H_b^+]$ are localized on the same solution phase,

$$[H_s^+]\gamma_{Hs+} = [H_b^+]\gamma_{Hb+}\exp\left(\frac{-e\psi_0}{kT}\right) \tag{42}$$

Taking into consideration arguments by Chan et al. [53] and assuming that $\gamma_0 = \gamma_+ = \gamma_-$, the ratios $\gamma_0/\gamma_+ = \gamma_-/\gamma_0 = 1$. A similar approach can be applied to reactions (33,34). As the result of the analysis by Smit and Holten [54], the ratios γ_{+-}/γ_0 and γ_0/γ_{+-} can be assumed to be constant. Eqs. (35–38) can be rearranged to obtain the so called effective intrinsic constants defined as

$$K_{a1}^{int} = \frac{[MOH][H_b^+]\gamma_{Hb+}}{[MOH_2^+]}\exp\left(\frac{-e\psi_0}{kT}\right) \tag{43}$$

$$K_{a2}^{int} = \frac{[-MO^-][H_b^+]\gamma_{Hb+}}{[-MOH]}\exp\left(\frac{-e\psi_0}{kT}\right) \tag{44}$$

$$*K_X^{int} = \frac{[-MOH][H_b^+][X_b^-]\gamma_{Hb+} \gamma_{Xb-}}{[MOH_2^+X^-]}\exp\frac{e(\psi_\beta - \psi_0)}{kT} \tag{45}$$

$$*K_Y^{int} = \frac{[-MO^-Y^+][H_b^+]\gamma_{Hb+}}{[-MOH][Y_b^+]\gamma_{Yb+}}\exp\frac{e(\psi_\beta - \psi_0)}{kT} \tag{46}$$

The charge localized in the σ_0 plane of the EIL is determined by the net amount of all charged surface groups:

$$\sigma_0 = B\{[-MOH_2^+] + [-MOH_2^+X^-] - [-MO^-] - [-MO^-Y^+]\} \tag{47}$$

The net charge localized in the σ_β plane due to specifically adsorbed ions is

$$\sigma_\beta = B\{[-MO^-Y^+] - [-MOH_2^+X^-]\} \tag{48}$$

The diffuse layer charge (determined from the Gouy–Chapman theory of the diffuse layer) can be expressed as

$$\sigma_d = -11.74(C)^{1/2}\sinh\frac{ze\psi_d}{2kT} \tag{49}$$

The entire system is electroneutral and thus

$$\sigma_0 + \sigma_\beta + \sigma_d = 0 \tag{50}$$

From the model presented schematically in Fig. 12, the constant C_1 and C_2 capacities are assumed and defined as

$$C_1 = \frac{\sigma_0}{\psi_0 - \psi_\beta} \tag{51}$$

and

$$C_2 = \frac{\sigma_d}{\psi_\beta - \psi_d} \tag{52}$$

The total amount of all kinds of surface groups per unit area is equal to

$$N_s = \{[-MOH_2^+] + [-MOH_2^+X^-] + [-MOH] + [-MO^-] + [-MO^-Y^+]\} \tag{53}$$

The entire set of Eqs. (43–53) can be solved numerically taking into account the determined earlier values of N_s, the surface ionization and complexation constants, and the C_1 and C_2 parameters. The potentials ψ_0 and ψ_β, and the concentrations of surface groups [–MOH] are independent variables [50]. Davis et al. [50,51] used the N_s values for oxides determined from crystallographic or tritium exchange data. The surface ionization and complexation constants for different oxides have been evaluated using the "double extrapolation technique" [55–57].

The SCM model of the electrical interfacial layer at oxide/electrolyte interface as proposed by Davis et al. [50] was modified by others to take into account the different sizes of anions and cations as well as the heterogeneity of the surface [58–60].

V. DETERMINATION OF SURFACE IONIZATION AND COMPLEXATION CONSTANTS

To model the adsorption of ions at the oxide/electrolyte interface, some parameters of the electrical interfacial layer have to be known. These parameters, e.g., N_s, can be determined by independent methods. Others, not measurable, like surface ionization and complexation constants, have to be determined using experimental data, e.g., potentiometric titration. Based on the potentiometric titration data for a given oxide, one can use graphical or numerical methods to determine the values of surface ionization and complexation constants. A comparison of both methods and their limitations for the oxide/electrolyte system was published by Koopal et al. [61]. A graphical double extrapolation method was proposed by James et al. [55–57], and the numerical optimizing method was proposed by Westall and Hohl [62–65]. Both methods have advantages and disadvantages.

To obtain reliable data using the graphical method, the densities of $[-MO^-]$ and $[-MOH_2^+]$ groups around the pzc should be small, i.e., the value of ΔpK_a should be large. This cannot be verified using potentiometric titration data. The extent of complexation is also important. For very weakly adsorbed ions even at higher ionic strengths the concentrations of e.g. $[-MO^-]$ and $[-MO^-Y^+]$ can be significant compared to those of $[-MOH_2^+X^-]$ and $[-MO^-Y^+]$.

For the numerical optimizing technique, basically nonlinear least square procedures, the above restrictions do not apply. However, for multiparameter optimization very complicated computer algorithms are required. The numerical method is limited by the error in the experimental data. According to Koopal et al. [61], with both the graphical and the numerical methods a good fit can be obtained, but none of the fits produces a distinctive set of parameters.

The description of different methods of surface ionization and complexation constant determination based on different experimental data, surface charge, electrokinetic potential, and supporting electrolyte ion adsorption densities, is presented below. This covers the double extrapolation method as proposed by James et al. [55–57] and its modifications. For more information on numerical optimization the reader is referred to the original papers [62–68].

A. Surface Charge

James et al. [55–57] used potentiometric titration data and the double extrapolation technique to evaluate the surface ionization and complexation constants. They assumed that reactions (31–34) were responsible for surface charge creation.

The model of the EIL proposed by Davis et al. [50] assumes a significant contribution of surface complexation reactions to the surface charge. Those authors showed that if complexation reactions at the oxide/electrolyte interface are neglected the values of "ionization constants" depend strongly on the electrolyte concentration [50]. It can be assumed that for pH > pH$_{pzc}$, $-\sigma_0 \approx B[-MO^-]$, and for pH < pH$_{pzc}$, $\sigma_0 \approx B[-MOH_2^+]$. By introducing $\alpha_- = -\sigma_0/N_s$ and $\alpha_+ = \sigma_0/N_s$, Eqs. (43,44) can be written in the logarithmic form as [50]

$$pK_{a1}^{int} = pH + \log \frac{\alpha_+}{1-\alpha_+} + \frac{e\psi_0}{2.3kT} \tag{54}$$

$$pK_{a2}^{int} = pH - \log \frac{\alpha_-}{1-\alpha_-} + \frac{e\psi_0}{2.3kT} \tag{55}$$

The acidity quotients can be defined as

$$pQ_{a1} = pH + \log \frac{\alpha_-}{1-\alpha_-} \tag{56}$$

$$pQ_{a2} = pH - \log \frac{\alpha_-}{1-\alpha_-} \tag{57}$$

From the plots of pQ_{a1} and pQ_{a2} as function of $\alpha_{+,-}$ one can determine the values of pK_{a1}^{int} and pK_{a2}^{int} by the extrapolation of these curves to $\alpha = 0$. The extrapolated constants determined in this way by Davis et al. [50] were only conditional constants useful for a specific

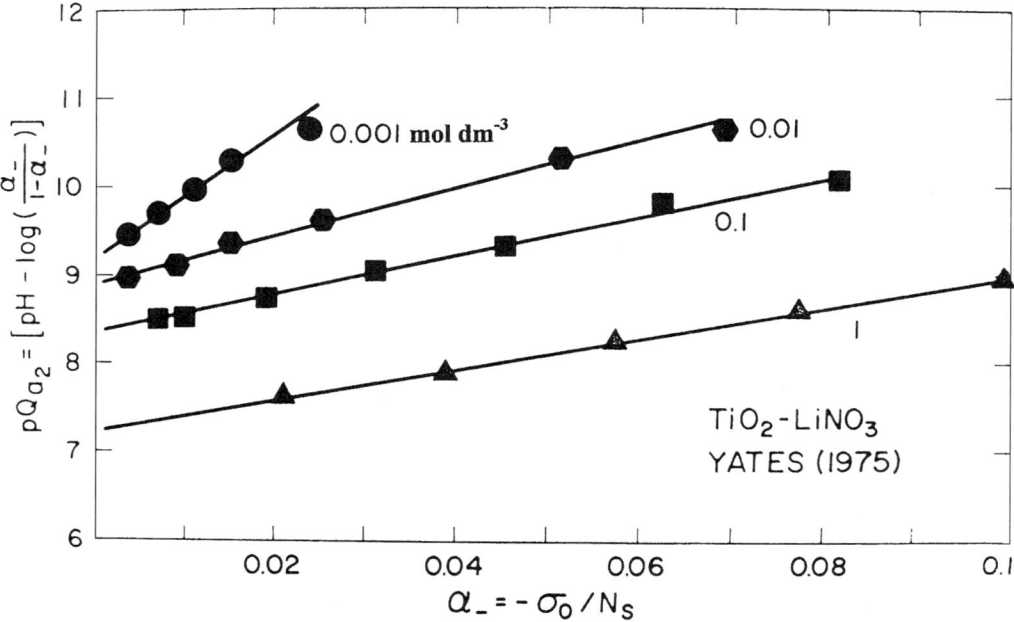

FIG. 13 p*Q_{a2} as a function of fractional surface ionization for TiO_2 in $LiNO_3$ solution from the data of Yates. (Reprinted from J. Colloid Interf. Sci., Ref. 50, © 1978 by Academic Press.)

electrolyte concentration. An example of such data for the $TiO_2/LiNO_3$ system is presented in Fig. 13. As observed, the values of conditional ionization constants depend strongly on electrolyte concentration.

To eliminate the effect of electrolyte concentration on the values of ionization constants, James et al. [55–57] proposed a double extrapolation technique to obtain pK_{a1}^{int} and pK_{a2}^{int}. In their method the acidity quotients pQ_{a2} and pQ_{a1} are plotted as functions of $\alpha_{+,-} + (C)^{1/2}$ and then extrapolated to $\alpha_{+,-} = 0$, i.e., to the pzc ($\psi_0 = 0$) and to $C = 0$. Different authors use different scales for the X axis, e.g., $\alpha_{+,-}$ [50], $\alpha_{+,-} + (C)^{1/2}$ [55], $10\alpha_{+,-} + (C)^{1/2}$ [52]. This has no effect on determination of the pK_{a1}^{int} and pK_{a2}^{int} constants. The only reason for adding $(C)^{1/2}$ is to separate, in a visible way, the curves for different ionic strengths. An example of double extrapolation is presented in Fig. 14.

At another extreme, when one neglects the contributions of $-MO^-$ and $-MOH_2^+$ groups, i.e., one assumes $-\sigma_0 \approx B[-MO^-Y^+]$ for pH > pH_{pzc} and $\sigma_0 \approx B[-MOH_2^+X^-]$ for pH < pH_{pzc}, the complexation constants for cation and anion can be expressed as [50]

$$p^*K_{Y+}^{int} = pH - \log\frac{\alpha_-}{1-\alpha_-} + \log[Y^+] + \frac{e(\psi_0 - \psi_\beta)}{2.3kT} \qquad (58)$$

$$p^*K_{X-}^{int} = pH + \log\frac{\alpha_+}{1-\alpha_+} - \log[X^-] + \frac{e(\psi_0 - \psi_\beta)}{2.3kT} \qquad (59)$$

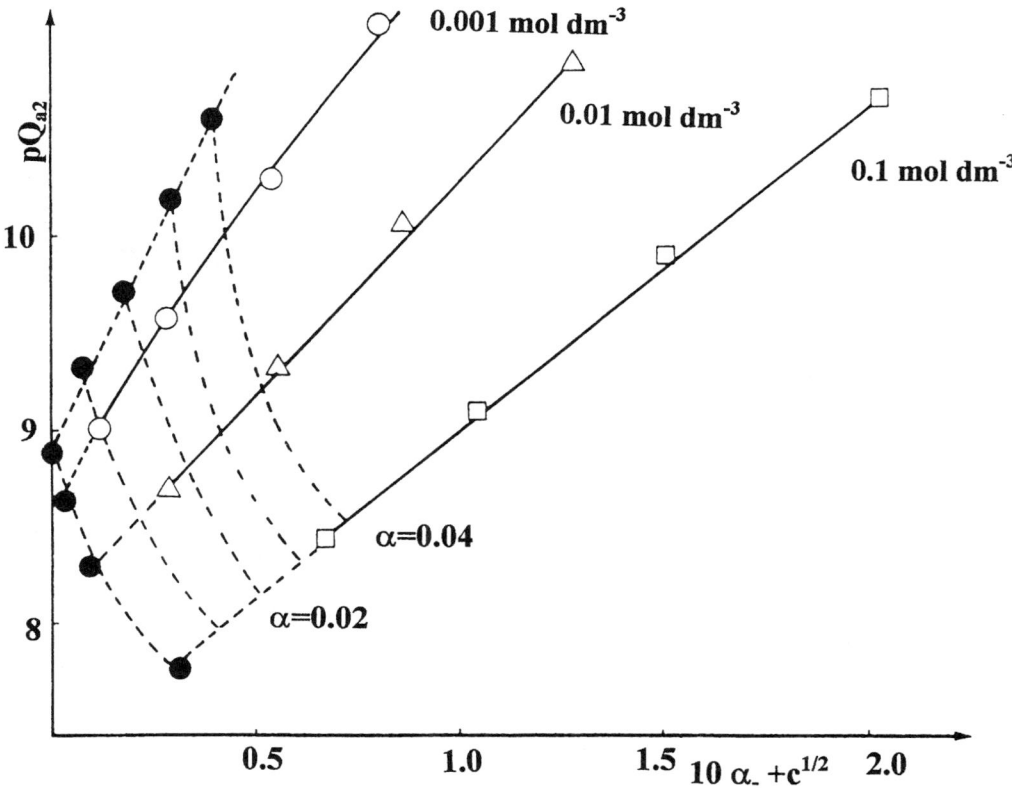

FIG. 14 Variation of pQ_{a2} values for anatase/NaCl solution system. The solid lines are experimental ones. The dashed lines and solid symbols are extrapolated ones. (Reprinted from J. Colloid Interf. Sci., Ref. 52, © 1984 by Academic Press.)

The appropriate complexation quotients are

$$p^*Q_{Y+} = pH - \log\frac{\alpha_-}{1-\alpha_-} + \log[Y^+] \qquad (60)$$

$$p^*Q_{X-} = pH + \log\frac{\alpha_+}{1-\alpha_+} - \log[X^-] \qquad (61)$$

By plotting the surface complexation quotients vs. α_+ or α_- and extrapolating the data to $\alpha = 0$ one can determine the value of the complexation constant. Example data for the $TiO_2/LiNO_3$ system are presented in Fig. 15. Contrary to Fig. 13, complexation quotients approach nearly the same value for different electrolyte concentrations. This means that surface complexes have a major contribution to the surface charge. However, for low electrolyte concentration and low values of α, significant deviations from linearity are observed. For low electrolyte, concentration, ionization and complexation processes may have comparable contributions to the surface charge.

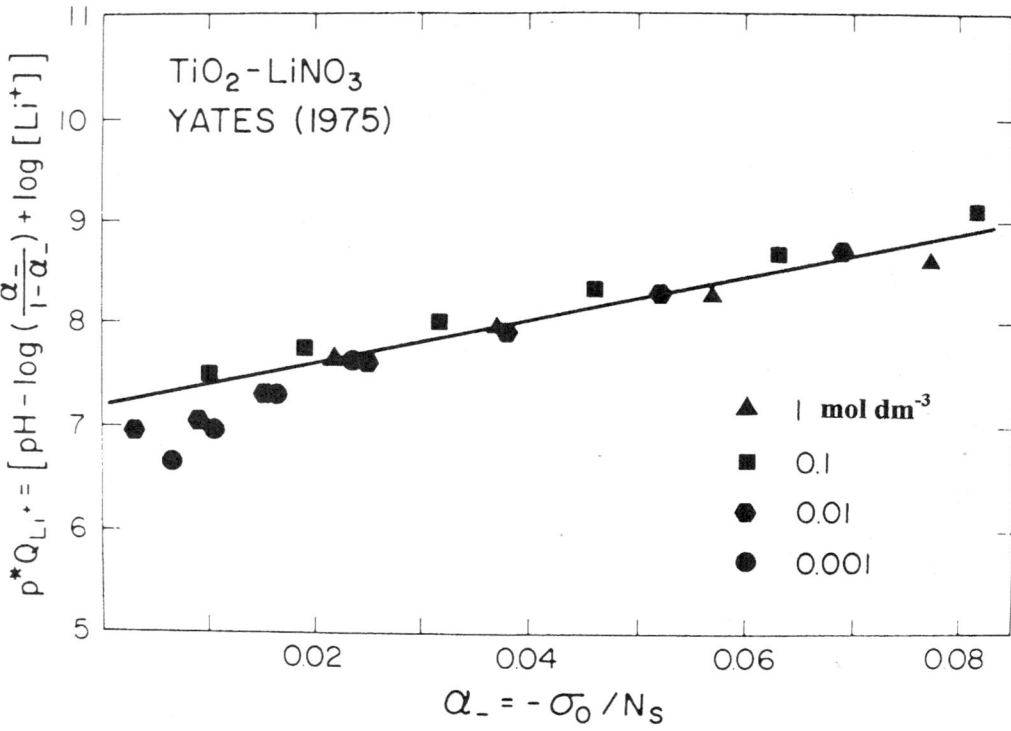

FIG. 15 p*Q_{Li+} as a function of fractional surface ionization for TiO_2 in $LiNO_3$ solution from the data of Yates. (Reprinted from J. Colloid Interf. Sci., Ref. 50, © 1978 by Academic Press.)

The surface complexation constant can be determined using the double extrapolation [55,57] method of complexation quotients—Eqs. (60,61)—to the pzc ($\psi_0 = \psi_\beta = 0$ and $\alpha = 0$) and to log $C = 0$ ($C = 1$). An example of such a procedure using potentiometric titration data for polystyrene latex [55] is presented in Fig. 16.

Though the method proposed by James et al. [55,57] is commonly used to determine the surface ionization and complexation constants for oxides, it suffers from some mathematical difficulties. According to Eqs. (56,57) and (60,61), the values of appropriate quotients tend to $+/-\infty$ if $\alpha \to 0$, i.e., when the system approaches the point of zero charge.

To overcome this difficulty, Janusz [68] proposed a new method of surface ionization and complexation constant determination by modifying the method by Schwarzenbach and Ackerman [69] that was used to determine the stability constants of complexes in the solution. Basic equations of this method are Eqs. (47) and (53) and (43–46). Janusz [68] introduced coefficients A_s (dependent on K_{a2} and *K_{Y+}) and B_s (dependent on K_{a1} *K_{X-}) and calculated them for different values of surface charge. By determining the values of A_s and B_s for different ionic strengths of the electrolyte, the surface ionization and complexation constants can be determined. Though Janusz did not use extrapolation but interpolation of his A_s and B_s coefficients to the pzc, his method is also based on some approximations regarding the A_s and B_s coefficients [68]. For oxides of low

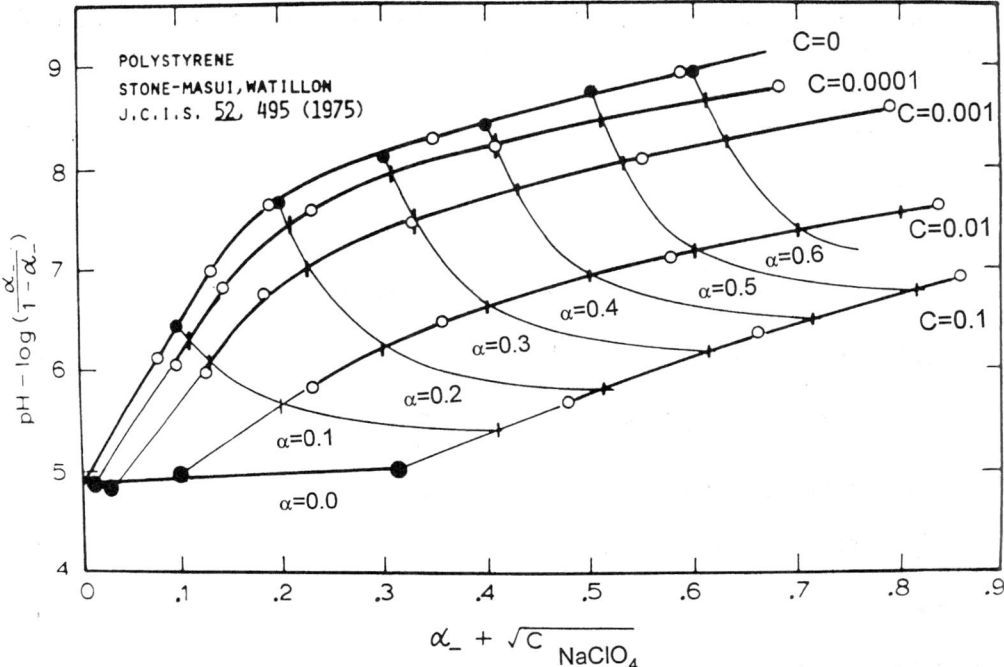

FIG. 16 The variation of the surface ionization activity quotient as a function of surface charge and concentration of supporting electrolyte. Lines are contours for constant electrolyte or constant surface charge. (Reprinted from J. Colloid Interf. Sci., Ref. 55, © Academic Press.)

solubility, e.g., TiO_2 and $FeOOH$, the agreement between the constants determined using Davis et al. [50,55] and Janusz [68] methods is good. Some discrepancies were observed for sparingly soluble oxides, e.g., ZnO [68].

B. Electrokinetic Data

In order to calculate the surface ionization constants of hydroxyl groups, $-MOH$, from Eqs. (43,44), the real $[-MO^-]$ and $[-MOH_2^+]$ components have to be known. These components determine the value of the diffuse layer charge and can be evaluated from electrokinetic data. Based on the charge balance equations—Eqs. (47–50)—one can write

$$-\sigma_d = B\{[-MOH_2^+] - [-MO^-]\} \tag{62}$$

For low ionic strengths, $C \leq 0.01$ M, it can be assumed that $\psi_d = \zeta$ [70] and the diffuse layer charge can be calculated from the diffuse layer theory using Eq. (49). Taking into account the above assumptions, a new method of surface ionization constant determination was proposed by Sprycha and Szczypa [71] using electrokinetic data for oxides. This method can be generalized to take advantage of electrokinetic data corresponding to higher ionic strengths, that is, we can calculate ψ_d from ζ [72] using Gouy–Chapman theory once the shear plane distance is known (or assumed). This approach is discussed in more detail in Chapter 6 of this book.

The concentrations of [–MO$^-$] and [–MOH$_2^+$] groups are comparable only in the vicinity of the pzc [19,50,73]. At a certain distance from the pzc one can assume that for pH>pH$_{pzc}$,

$$\sigma_d \approx -B[-\text{MO}^-] \tag{63}$$

and for pH<pH$_{pzc}$,

$$\sigma_d = B[-\text{MOH}_2^+] \tag{64}$$

Thus Eqs. (43,44) can be expressed in logarithmic form as

$$pK_{a1}^{int} = pH - \log[-\text{MOH}] - \log\frac{\sigma_{d+}}{B} + \frac{e\psi_0}{2.3kT} \tag{65}$$

$$pK_{a2}^{int} = pH + \log[-\text{MOH}] + \log\frac{|\sigma_{d-}|}{B} + \frac{e\psi_0}{2.3kT} \tag{66}$$

The respective acidity quotients are

$$pQ_{a1} = pH - \log[-\text{MOH}] + \log\frac{\sigma_{d+}}{B} \tag{67}$$

$$pQ_{a2} = pH + \log[-\text{MOH}] - \log\frac{|\sigma_{d-}|}{B} \tag{68}$$

Substituting Eqs. (63) and (64) into Eq. (53) one can write

$$[-\text{MOH}] \approx N_s - [-\text{MOH}_2^+\text{X}^-] - [-\text{MO}^-\text{Y}^+] - \frac{\sigma_{d+}}{B} \tag{69}$$

for pH<pH$_{pzc}$, and

$$[-\text{MOH}] \approx N_s - [-\text{MOH}_2^+\text{X}^-] - [-\text{MO}^-\text{Y}^+] - \frac{\sigma_{d-}}{B} \tag{70}$$

for pH>pH$_{pzc}$.

For low concentrations of the electrolyte one can use, as a first approximation, [–MOH]$\approx N_s$. Now the values of acidity quotients can be calculated for different pH values and electrolyte concentrations and plotted as a function of pH. To eliminate any effect of electrolyte concentration on the above procedure, the double extrapolation technique can be used (to pH = pH$_{pzc}$ and $C=0$). An example of the determination of the pK$_{a2}$ ionization constant for TiO$_2$ using the above procedure is presented in Fig. 17. The electrokinetic data were taken from the paper by Wiese and Healy [74].

The curves are extrapolated at first to pH$_{iep}$ + $\sqrt{C}$ for different ionic strengths. Then the straight line joining the extrapolated points is extrapolated to the pH$_{iep(pzc)}$ ($C=0$). The point of intersection of the extrapolated line and the vertical line pH$_{iep}$ determines the pK$_{a2}$ value. As was shown [71], the values of constants determined by the double extrapolation technique using electrokinetic data and potentiometric titration data were similar.

The slope of the curves in Fig. 17 can be used to evaluate the surface potential of the oxide. From Eqs. (65–68) one obtains

$$pK_{a1}^{int} - pQ_{a1} = \frac{e\psi_0}{2.3kT} \tag{71}$$

$$pK_{a2}^{int} - pQ_{a2} = \frac{e\psi_0}{2.3kT} \tag{72}$$

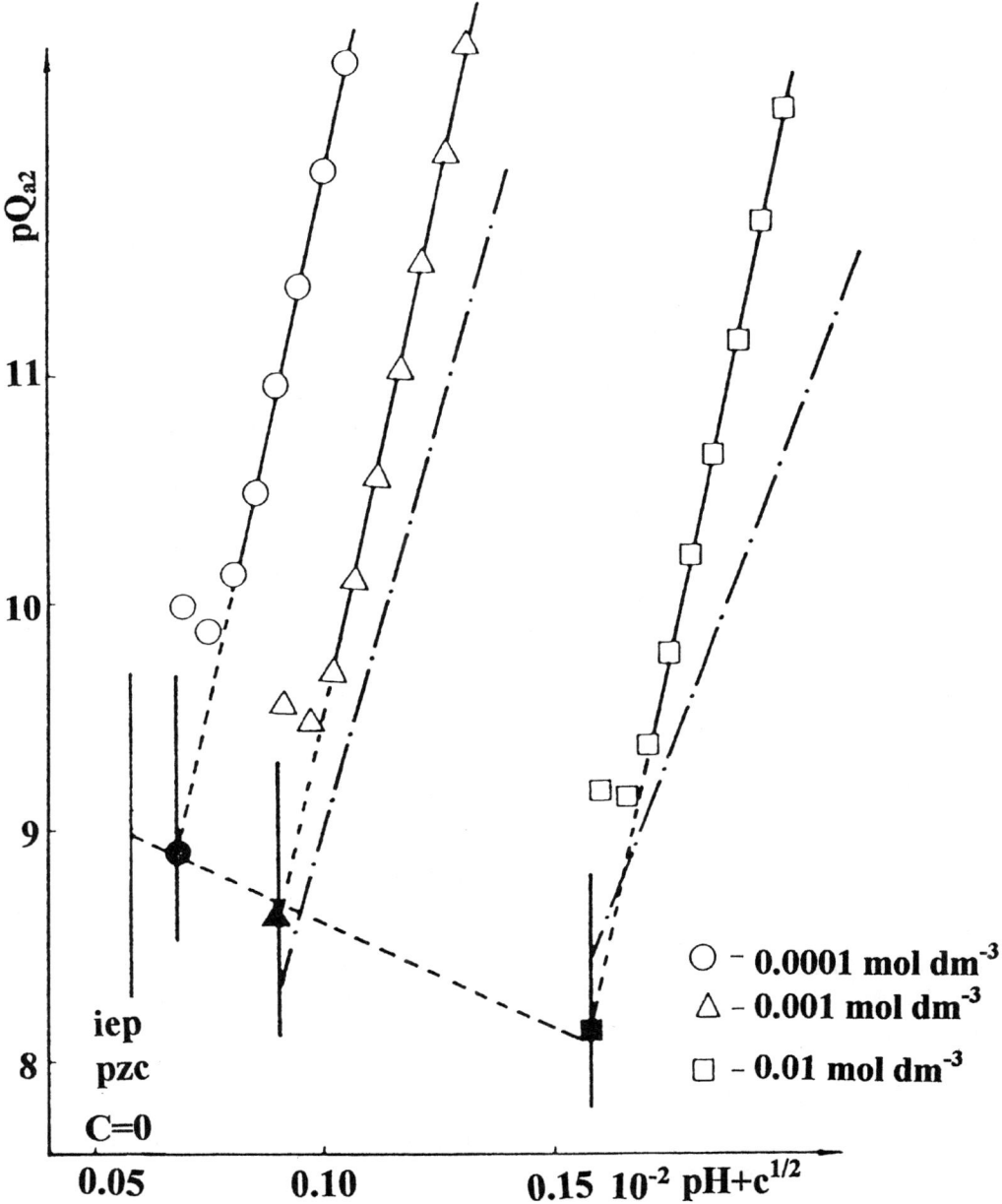

FIG. 17 pQ_{a2} vs. $0.01\ pH + c^{1/2}$ for TiO_2/KNO_3 system. The dashed lines are extrapolated ones. $(-\cdot-\cdot)$ Determined from potentiometric titration data. (Reprinted from J. Colloid Interf. Sci., Ref. 71, © 1987 Academic Press.)

or in the differential form,

$$-\frac{d\,pQ_{a1}}{d\,pH} = \frac{e}{2.3kT}\frac{d\psi_0}{d\,pH} \qquad (73)$$

$$-\frac{d\,pQ_{a2}}{d\,pH} = \frac{e}{2.3kT}\frac{d\psi_0}{d\,pH} \qquad (74)$$

The surface potential of oxides depends on electrolyte concentration. The largest differences between the different ionic strengths are observed in the vicinity of the pzc(iep) [56, 75]. As observed in Fig. 17, the slope of the extrapolated lines for different electrolyte concentrations is almost the same, which is in conflict with the theory. This disagreement is due to the invalidity of the assumption that the diffuse layer charge in the vicinity of the pzc is dominated by one type of charged group only. For accurate calculations, the contributions of both $-MO^-$ and $-MOH_2^+$ groups should be taken into account in determining acidity constants. However, there is no technique to measure independently the individual concentrations of these groups.

Sprycha and Szczypa [71] showed that for more accurate evaluation of the surface potential of the oxide from surface acidity quotients vs. pH plots, the straight line extrapolation to the iep should be replaced by the "curvilinear" extrapolation. The values of the surface ionization constants can be determined independently using surface charge [55–57] or electrokinetic data [71]. Having determined those constants, one can use the curvilinear interpolation of the acidity quotients calculated from Eqs. (67,68) to a predetermined value of pK_{a2}^{int} or pK_{a1}^{int} and then use those curves to evaluate the surface potential of a given oxide for different electrolyte concentrations. The surface potentials obtained in this way are more accurate in the vicinity of the pzc and consistent with the literature data [49,74].

An example of curvilinear interpolation of pQ_{a2} vs. pH curves for the Al_2O_3/KNO_3 system is presented in Fig. 18. This data was subsequently used to determine ψ_0 vs. pH relationships for this system—see Fig. 19. As observed, the slope of the curves near the iep increases as electrolyte concentration decreases, but far enough from the iep the curves have almost the same slope (about 56 mV per one pH unit). The data presented in Fig. 19 is in good agreement with model calculations by Wiese et al. [49] regarding the sequence and run of the curves ψ_0 vs. pH for different electrolyte concentrations.

The ψ_0 vs. pH relationships for oxides can be affected by the extent of ionization at the interface, i.e., the value of ΔpK_a,

$$\Delta pK_a = pK_{a2}^{int} - pK_{a1}^{int} \qquad (75)$$

By calculating $-MO^-$ and $-MOH_2^+$ components from Eqs. (43,44) and substituting those values into Eq. (62) one obtains

$$\sigma_d = \frac{K_{a2}^{int}[MOH]}{[H^+]\exp(-e\psi_0/kT)} - \frac{[-MOH][H^+]\exp(-e\psi_0/kT)}{K_{a1}^{int}} \qquad (76)$$

Eq. (53) can be rearranged to

$$[-MOH] + [-MO^-] + [-MOH_2^+] = N_s - [-MOH_2^+X^-] - [-MO^-Y^+] \qquad (77)$$

and from Eqs. (43,44) and (75) one obtains

$$\Delta pK_a = 2\log[-MOH] - \log[-MOH_2^+] - \log[-MO^-] \qquad (78)$$

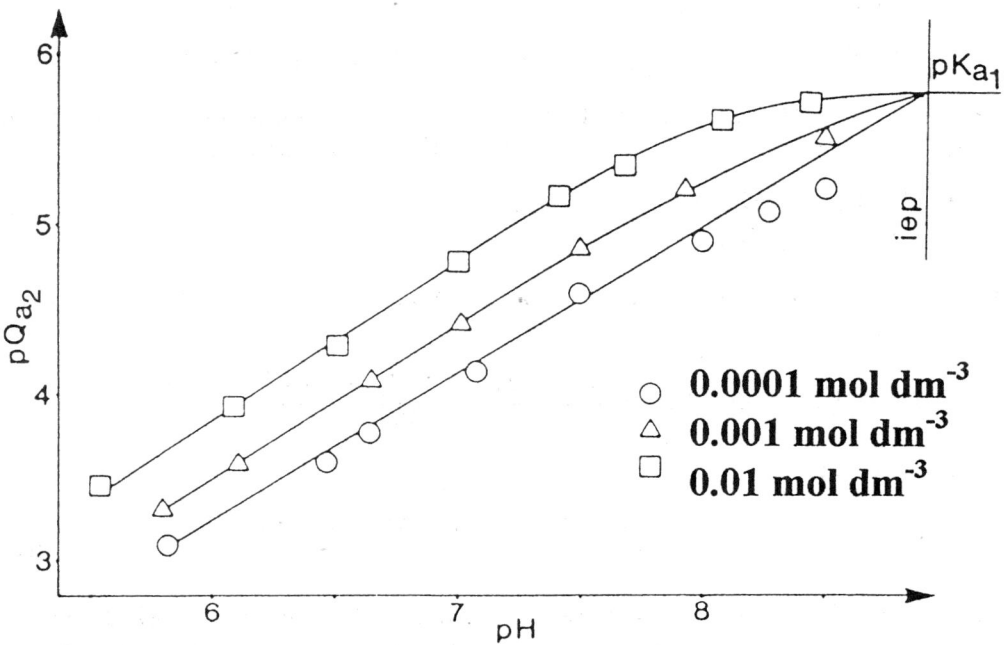

FIG. 18 pQ_{a2} vs. pH for Al_2O_3/KNO_3 system for different ionic strength obtained by curvilinear interpolation method. (Reprinted from J. Colloid Interf. Sci., Ref. 71, © 1987 by Academic Press.)

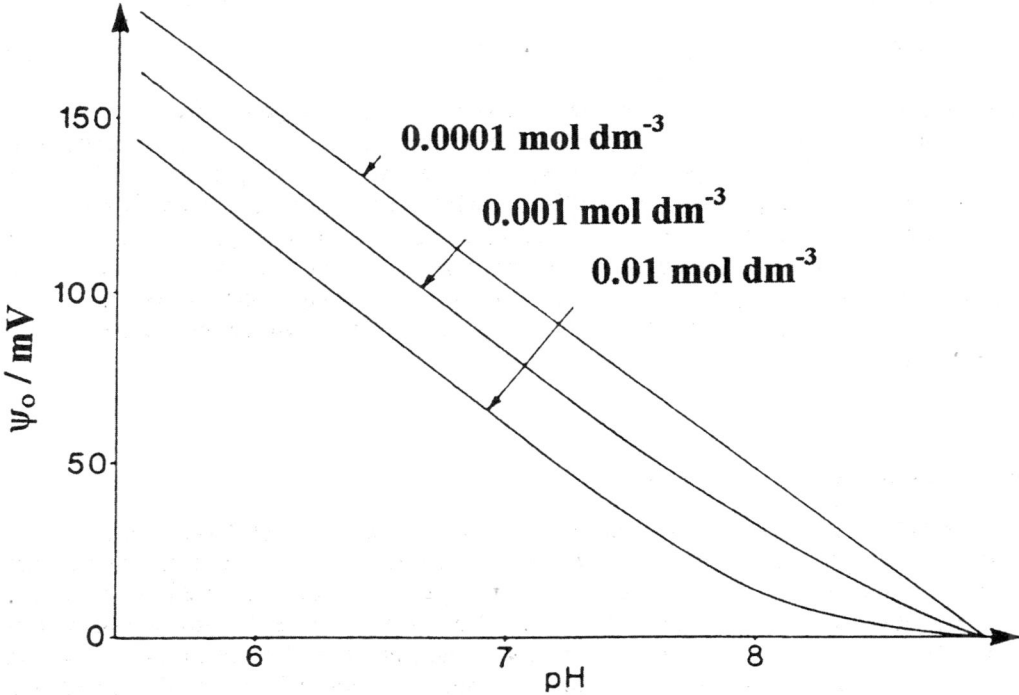

FIG. 19 Surface potential vs. pH for Al_2O_3 system for three different ionic strengths. (Reprinted from J. Colloid Interf. Sci., Ref. 71, © 1987 by Academic Press.)

Because at the pzc $[-MOH_2^+] = [-MO^-]$, Eq. (77) can be rewritten as

$$[-MeOH]_{pzc} + 2[-MeO^-]_{pzc} = N_s - [-MeOH_2^+X^-]_{pzc} - [-MeO^-Y^+]_{pzc} \tag{79}$$

The values of $[-MOH_2^+X^-]_{pzc}$ and $[-MO^-Y^+]_{pzc}$ can be determined by independent adsorption measurements [52]. From Eq. (78) at the pzc one obtains

$$\frac{[-MOH]_{pzc}}{[-MO^-]_{pzc}} = \frac{[-MOH]_{pzc}}{[-MOH_2^+]_{pzc}} = 10^{(\Delta pKa/2)} \tag{80}$$

By combining Eqs. (78) and (80) the contributions of ionization components at the pzc can be expressed as

$$[-MOH_2^+]_{pzc} = [-MO^-]_{pzc} = \frac{N_s - [-MOH_2^+X^-]_{pzc} - [-MO^-Y^+]_{pzc}}{2 + 10^{(\Delta pKa/2)}} \tag{81}$$

At the same time,

$$[-MOH]_{pzc} = \frac{\{N_s - [-MOH_2^+X^-]_{pzc} - [-MO^-Y^+]_{pzc}\} 10^{\Delta pKa/2}}{2 + 10^{(\Delta pKa/2)}} \tag{82}$$

The value of σ_d for oxides does not exceed, as a rule, ca. 0.02 C m^{-2} [50,76–81]. Hence the changes of $[-MO^-]$ and $[-MOH_2^+]$ vs. pH compared to $[-MOH]$ are rather small, and for the purpose of evaluation of $[-MOH]$ it can be assumed, as a first approximation, that $[-MO^-] \approx [-MOH_2^+] \approx [-MO^-]_{pzc} \approx [-MOH_2^+]_{pzc}$ in the entire pH range studied. As results from Eq. (82), $[-MOH]_{pzc}$ depends on ΔpK_a considerably when $\Delta pK_a \leq 2$. For $\Delta pK_a > 2$, the term $[10^{(\Delta pKa/2)}]/[2 + 10^{(\Delta pKa/2)}] \to 1$ and Eq. (82) can be simplified to

$$[-MOH] \approx N_s - [-MOH_2^+X^-] - [-MO^-Y^+] \tag{83}$$

and used with relatively good accuracy over the entire pH range studied.

For oxides with $\Delta pK_a > 2$, the ψ_0 vs. pH plots for different electrolyte concentrations can be constructed using Eqs. (76) and (82). Having known ψ_0 vs. pH relationships, the surface ionization components $[-MOH_2^+]$ and $[-MO^-]$ can be calculated from Eqs. (43,44). The surface ionization components, estimated by the above procedure, for the $Al_2O_3/NaCl$ system are presented in Fig. 20. The diffuse layer charge calculated from electrokinetic data using Eq. (49) is also presented for comparison. As observed, the assumptions expressed by Eqs. (63) and (64) are fulfilled better for higher electrolyte concentrations.

C. Adsorption of Counterions

As results from Eqs. (45,46), the concentrations of surface complexes have to be known to determine complexation constant values. In most cases, potentiometric titration data are only available for oxides, and therefore some approximations are used to determine the complexation constants as discussed in Section V.A. Recently, some researchers attempted to measure the extent of adsorption of indifferent electrolyte ions, as a function of their concentration and pH of the solution, onto the surface of some oxides [52,82–96]. By definition such electrolytes should show no adsorption when surface charge and oxide potential are equal to zero. This results directly from the models of the EIL assuming

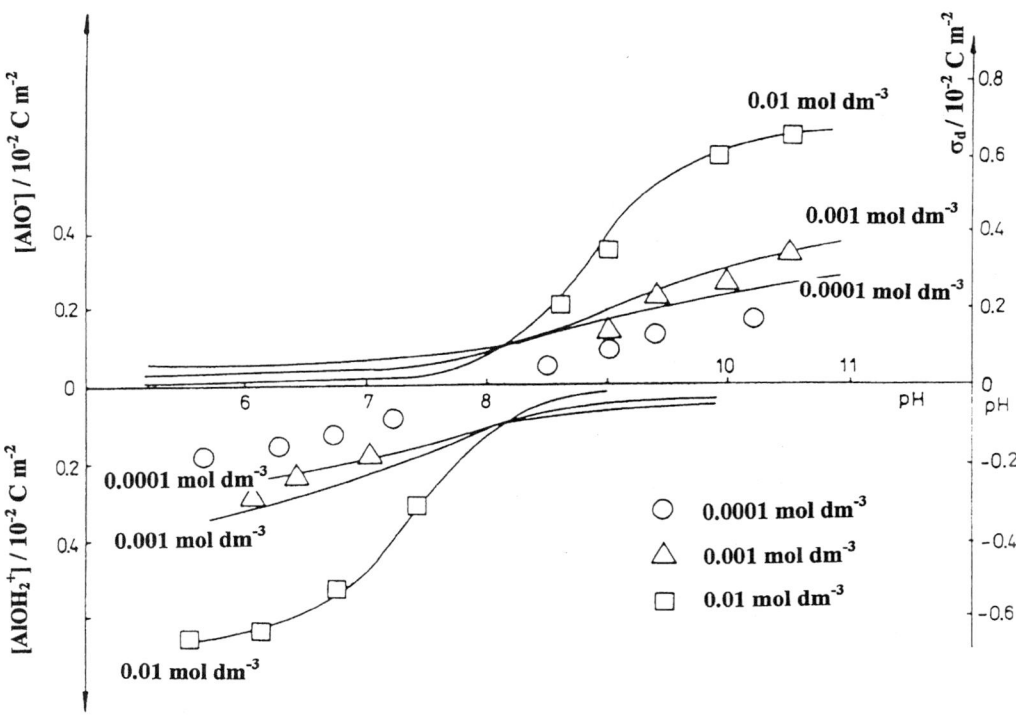

FIG. 20 Calculated [AlOH$_2^+$] and [AlO$^-$] components vs. pH of NaCl solution (lines) and diffuse layer charges (symbols). (Reprinted from J. Colloid Interf. Sci., Ref. 76, © 1989 by Academic Press.)

no complexation. It was found, however, that supporting electrolyte ions do adsorb at the pzc on the oxide surface. Due to relatively low affinity of these ions towards an oxide surface it is difficult to measure their adsorption.

Though adsorption experiments seem to confirm the validity of complexation models, it was observed in some cases that the extent of adsorption measured (using for example radiotracers) was higher than that predicted by the values of complexation constants determined employing potentiometric titration data [52]. If adsorption densities of anions and cations of the supporting electrolyte are known, one can determine the complexation constants using Eqs. (45,46). In logarithmic form one obtains

$$p^*K_{X-}^{int} = pH - \log[-MOH] - \log[X_b^-]\gamma_e + \log[-MOH_2^+X^-] + \frac{e(\psi_0 - \psi_\beta)}{2.3kT} \tag{84}$$

$$p^*K_{Y+}^{int} = pH + \log[-MOH] + \log[Y_b^+]\gamma_e - \log[-MO^-Y^+] + \frac{e(\psi_0 - \psi_\beta)}{2.3kT} \tag{85}$$

From the experimental data the values of the complexation quotients can only be determined

$$p^*Q_{X-}^{int} = pH - \log[-MOH] - \log[X_b^-]\gamma_e + \log[-MOH_2^+X^-] \tag{86}$$

$$p^*Q_{Y+}^{int} = pH + \log[-MOH] + \log[Y_b^+]\gamma_e - \log[-MO^-Y^+] \tag{87}$$

if Eqs. (69,70) are taken into account. The values of complexation constants can be determined directly from Eqs. (84,85) only at the pzc where $\psi_0 = \psi_\beta = 0$, because then the last terms in these equations disappear. The intersection of the curve $p^*Q_{X-}^{int}$ or $p^*Q_{Y+}^{int}$ vs. pH with the vertical line $pH = pH_{pzc}$ determines the value of complexation constant. Two examples of such curves for TiO_2 and Al_2O_3 are presented in Figs. 21 and 22, respectively.

The difference between the complexation constant and the complexation quotient from Eqs. (84,87) can be considered as a measure of $\psi_0 - \psi_\beta$ potential difference. For $pH < pH_{pzc}$,

$$p^*K_{X-}^{int} - p^*Q_{X-}^{int} = \frac{e(\psi_0 - \psi_\beta)}{2.3kT} \tag{88}$$

and for $pH > pH_{pzc}$,

$$p^*K_{Y+}^{int} - p^*Q_{Y+}^{int} = \frac{e(\psi_0 - \psi_\beta)}{2.3kT} \tag{89}$$

In the differential form one obtains

$$-\frac{d\, p*Q_{X-}}{d\, pH} = \frac{e}{2.3kT}\frac{d(\psi_0 - \psi_\beta)}{d\, pH} \tag{90}$$

and

$$-\frac{d\, p*Q_{Y+}}{d\, pH} = \frac{e}{2.3kT}\frac{d(\psi_0 - \psi_\beta)}{d\, pH} \tag{91}$$

respectively. The determined values of the $\psi_0 - \psi_\beta$ potential difference can be used along with potentiometric titration data to calculate the values of C_1 capacitance from Eq. (51).

From Eq. (51) one obtains

$$\sigma_0 = C_1(\psi_0 - \psi_\beta) \tag{92}$$

Examples of σ_0 vs. $(\psi_0 - \psi_\beta)$ curves, for different ionic strengths of the electrolyte, for the $Al_2O_3/NaCl$ system, are presented in Fig. 23. The slope of these curves is a measure of the capacitance C_1. As observed, the slope of the curves increases as electrolyte concentration increases. Moreover, the slopes in the anodic and cathodic ranges of pH are different. The complexation model assumes that the C_1 capacitance is constant and independent of the electrolyte concentration.

Having determined the $\psi_0 - \psi_\beta$ potential difference from adsorption data [Eqs. (86–89)], and the surface potential from electrokinetic data [Eqs. (71–74)], the potential distribution within the electrical interfacial layer can be estimated assuming that $\psi_d = \zeta$ [70]. An example of plots of ψ_0, ψ_β, and ψ_d potentials vs. pH for the $Al_2O_3/NaCl$ system is presented in Fig. 24.

The difference between complexation constants,

$$\Delta p^*K_{complex} = p^*K_{Y+}^{int} - p^*K_{X-}^{int} \tag{93}$$

is a measure of the extent of complexation in a given system. The higher the $\Delta p^*K_{complex}$, the lower the concentrations of surface complexes. Generally, by measuring the adsorption densities of supporting electrolyte ions, one can directly determine the surface charge components due to surface complex formation [52,82]. Another method of determination

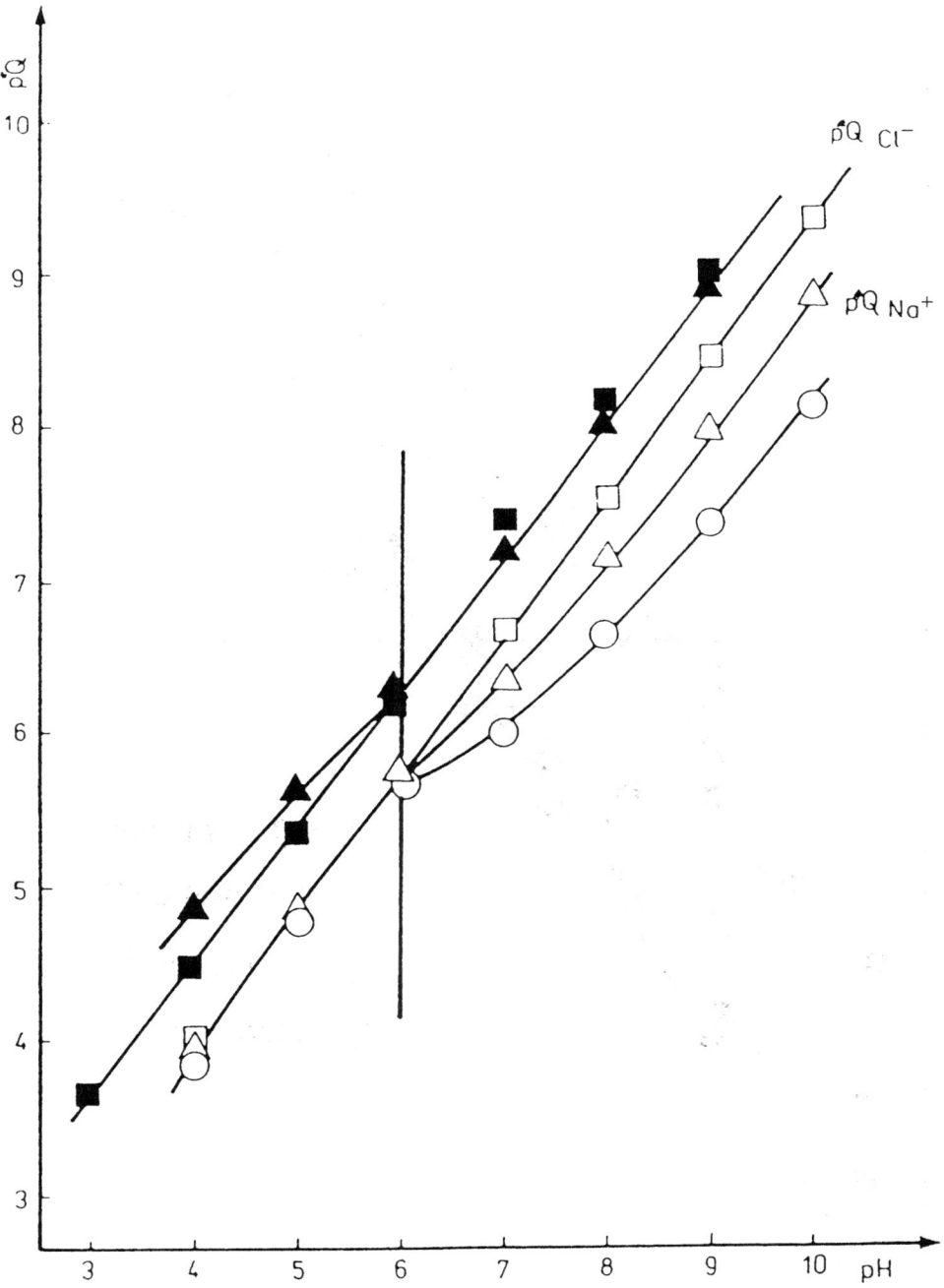

FIG. 21 Variation of p*Q_{Na+} (open symbols) and p*Q_{Cl-} (solid symbols) vs. pH of the anatase suspension and different ionic strengths: circles 0.001, triangles 0.01, squares 0.1. (Reprinted from J. Colloid Interf. Sci., Ref. 52, © 1984 by Academic Press.)

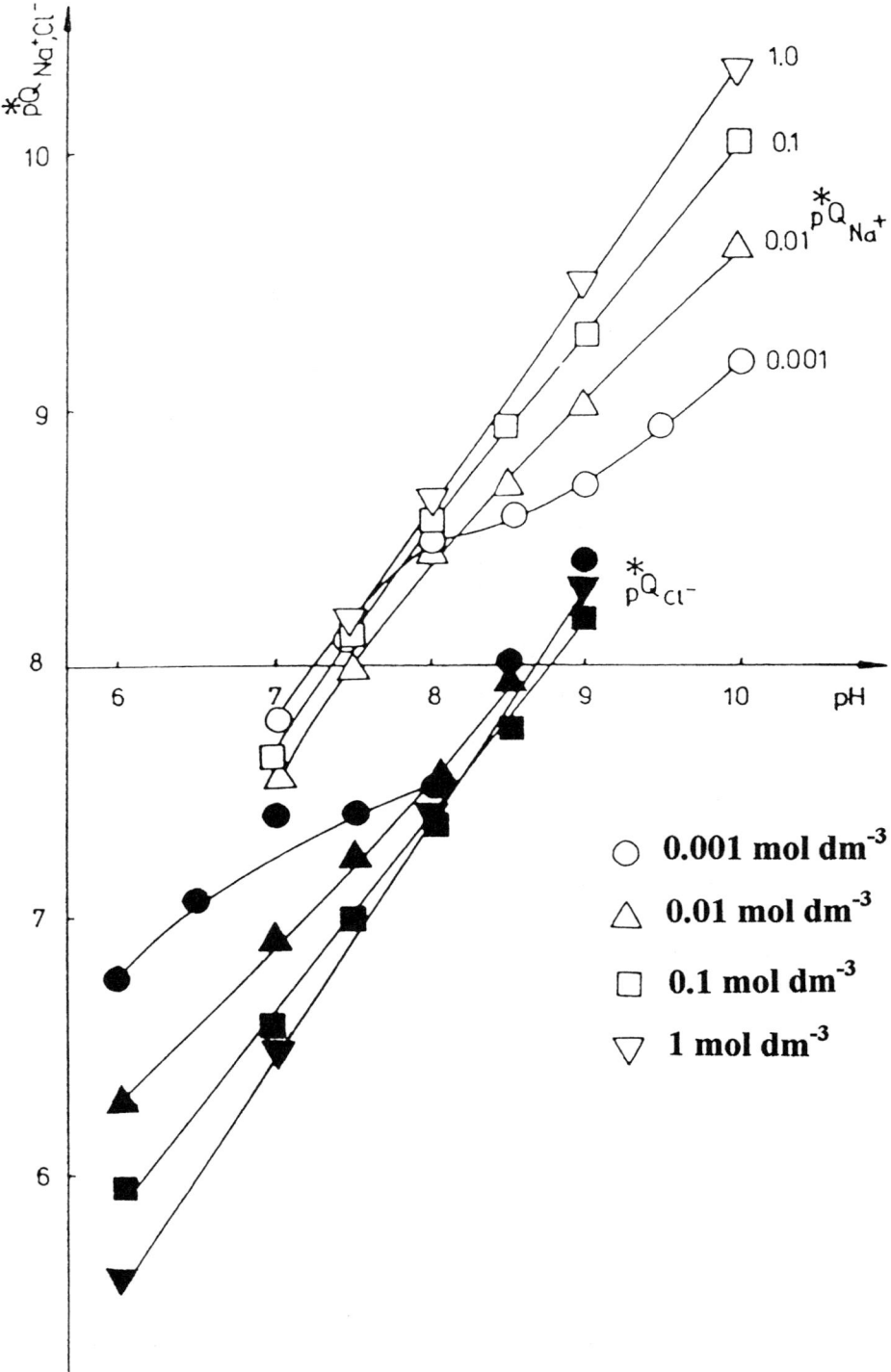

FIG. 22 Variation of p*Q_{Na+} (open symbols) and p*Q_{Cl-} (solid symbols) vs. pH of alumina suspension for different ionic strengths. (Reprinted from J. Colloid Interf. Sci., Ref. 82, © 1989 by Academic Press.)

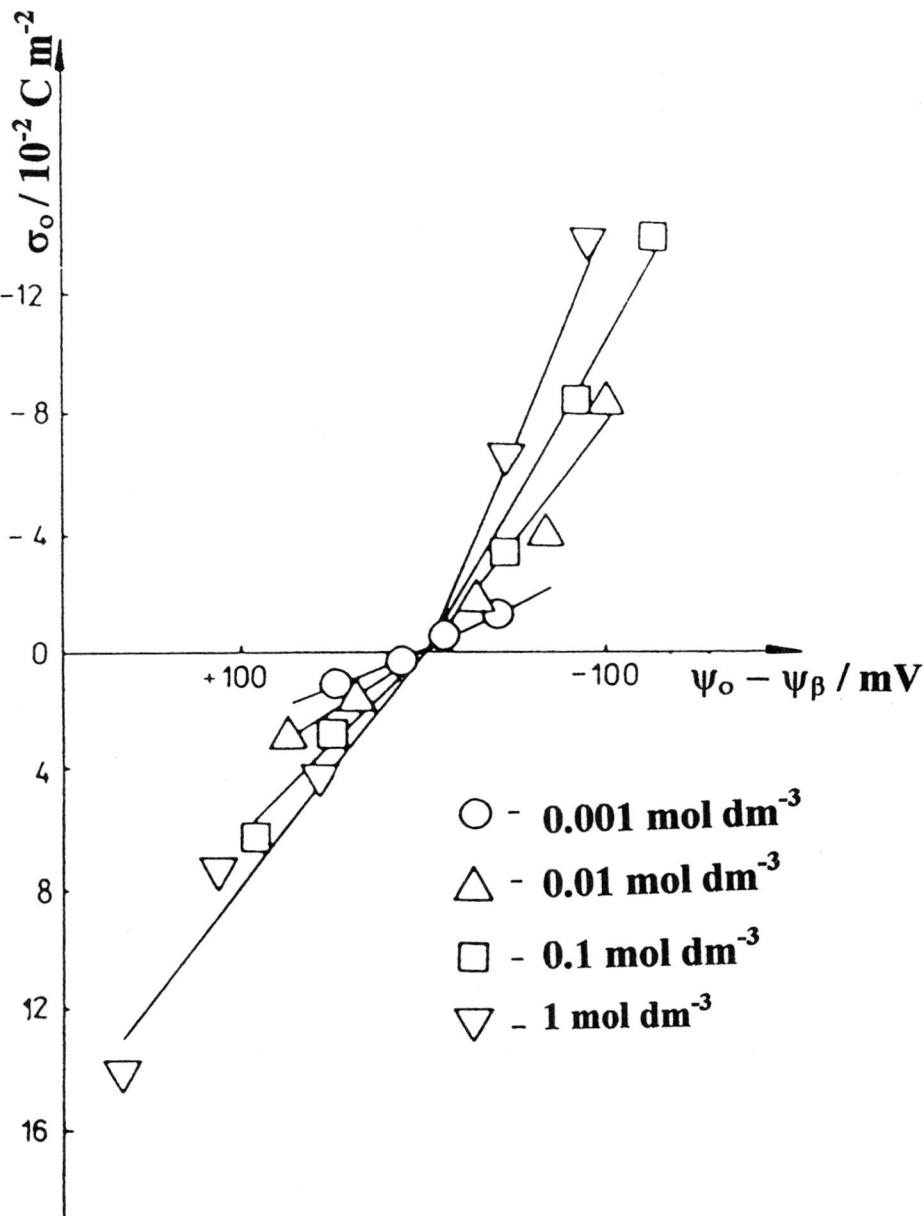

FIG. 23 Variation of surface charge vs. $(\psi_0 - \psi_\beta)$ for $Al_2O_3/NaCl$ system. (Reprinted from J. Colloid Interf. Sci., Ref. 82, © 1989 by Academic Press.)

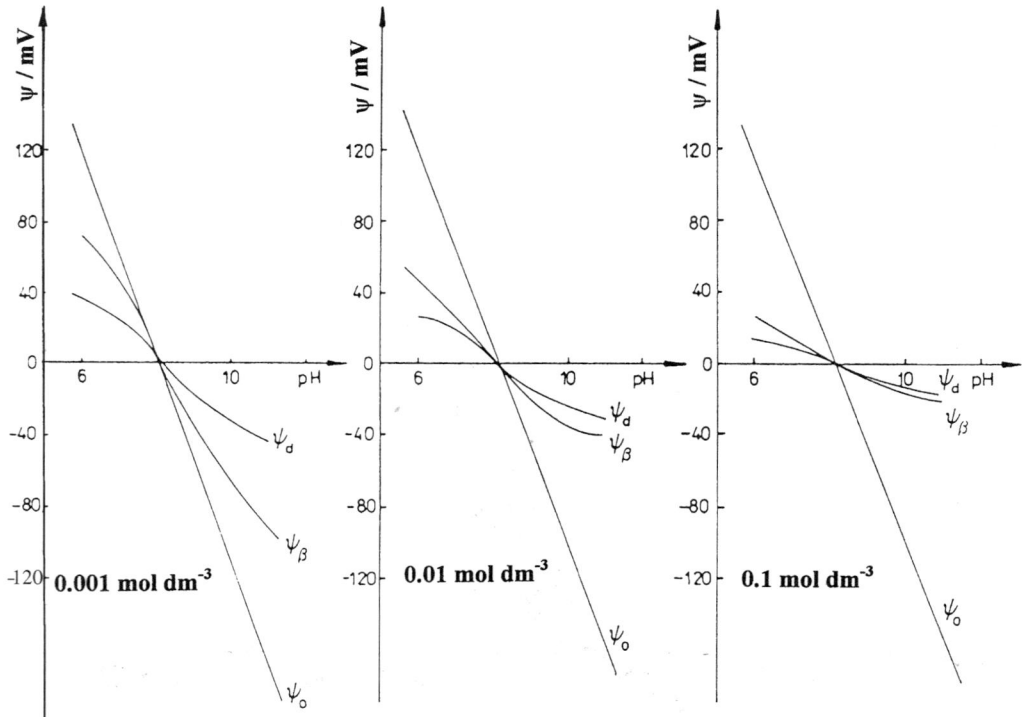

FIG. 24 Potential distribution within the electrical interfacial layer for the $Al_2O_3/NaCl$ system ($\Delta pK_a = 4.2$, $\Delta p^*K_{complex} = 1.1$). (Reprinted from J. Colloid Interf. Sci., Ref. 82, © 1989 by Academic Press.)

of the surface charge components, based on the calculation of the Esin–Markow coefficient (λ), can also be employed using potentiometric titration data [16,97–101]. The method was first applied for the Hg/electrolyte system [2]. The Esin–Markow coefficient measures the extent to which the applied potential must be changed to maintain the constant value of surface charge if the activity of the indifferent electrolyte is increased [2,16,99,100]:

$$\left(\frac{\partial E}{\partial \ln \gamma_{+-} C}\right)_{\sigma o} = \lambda \tag{94}$$

The Esin–Markow coefficient for colloidal systems was first applied by Lyklema [16,98–100]. For an AgI/electrolyte system the Esin–Markow coefficient can be defined as

$$\lambda = \left(\frac{d\text{pAg}}{d \log \gamma_{+-} C}\right)_{\sigma o} \tag{95}$$

By analogy, for oxides one obtains [98,99]

$$\lambda = \left(\frac{d\text{pH}}{d \log \gamma_{+-} C}\right)_{\sigma o} \tag{96}$$

From the analysis given by Lyklema [98,99], for 1 : 1 electrolytes, and using potentiometric titration data, the surface charge components can be expressed as

$$\sigma_+ = -\frac{\sigma_0}{2} + \frac{1}{2}\int_0^{\sigma_0} \lambda\, d\sigma_0 + \sigma_{+\mathrm{pzc}} \tag{97}$$

$$\sigma_- = -\frac{\sigma_0}{2} + \frac{1}{2}\int_0^{\sigma_0} \lambda\, d\sigma_0 + \sigma_{-\mathrm{pzc}} \tag{98}$$

In the absence of specific adsorption at the pzc, $\lambda = 0$ and

$$\frac{\partial \sigma_-}{\partial \sigma_0} = \frac{\partial \sigma_+}{\partial \sigma_0} = -0.5 \tag{99}$$

This means that for such systems anions and cations contribute equally to the compensation of the surface charge. It is, however, impossible to determine their contributions from Eqs. (97,98). An independent method is needed to do this. The values of the integration constants in Eqs. (97,98) have no meaning if one wants only to construct σ_0 vs. pH curves:

$$\sigma_0 = -(\sigma_+ = +\sigma_-) \tag{100}$$

because they have opposite signs and cancel. Knowledge of the $\sigma_{+\mathrm{pzc}} + \sigma_{-\mathrm{pzc}}$ constants is very important if one seeks to determine the real components of charge.

Considering surface charge components in terms of the surface complexation model of the electrical interfacial layer for a given system (1 : 1 electrolyte and pzc = iep) one obtains at the pzc $\sigma_{+\mathrm{pzc}} = -\sigma_{-\mathrm{pzc}} = f(\Delta p K_a, \Delta p^* K_{\mathrm{complex}})$. These integration constants can be calculated if the surface ionization and complexation constants for a given oxide are known [101]. Then using potentiometric titration data and Eqs. (97,98), one can determine the real components of the surface charge vs. pH. An example of the evaluation of σ_+ components of the surface charge for anatase, assuming constant value of $\Delta p K_a = 6$ and different values of $\Delta p^* K_{\mathrm{complex}}(0, 2, 4, \infty)$, is presented in Fig. 25. The obtained curves differ only by the constant. As seen from Fig. 25, if the adsorption of supporting electrolyte ions at the pzc is low enough, then the negative adsorption of e.g. cation for pH < pH$_{\mathrm{pzc}}$ should be observed. For low values of $\Delta p^* K_{\mathrm{complex}}$, e.g., zero, negative adsorption may not be observed experimentally. As shown above, one can determine the real integration constants in Eqs. (97,98), due to the formation of surface complexes, only by direct measurements of adsorption density of indifferent electrolyte ions.

It is well known from the literature on potentiometric titration that the surface charge at the silver iodide/electrolyte interface is significantly lower than the surface charge at the oxide/electrolyte interface [15,16,26,52,56]. On the other hand, the electrokinetic potential of silver iodide particles is considerably higher than the ζ potential of oxides in the same electrolyte solutions. This comparison shows that for some reasons "indifferent" electrolyte ions show higher affinity towards the oxide surface than towards the silver iodide surfaces. The above observations led Yates et al. [48,49] to develop the "site-binding" model of the electrical interfacial layer for oxides. Contrary to the models that assume zero adsorption of supporting electrolyte ions at the pzc ($\sigma_0 = 0$ and $\psi_0 = 0$), the surface complexation model allows some adsorption of these ions at the pzc. The extent of adsorption at the pzc depends on the values of complexation constants]Eqs. (37,38)], which are mostly determined from potentiometric titration data [50–57]. Though at the pzc, $\sigma_0 = 0$ and the surface macropotential $\psi_0 = 0$, the charged groups are still pre-

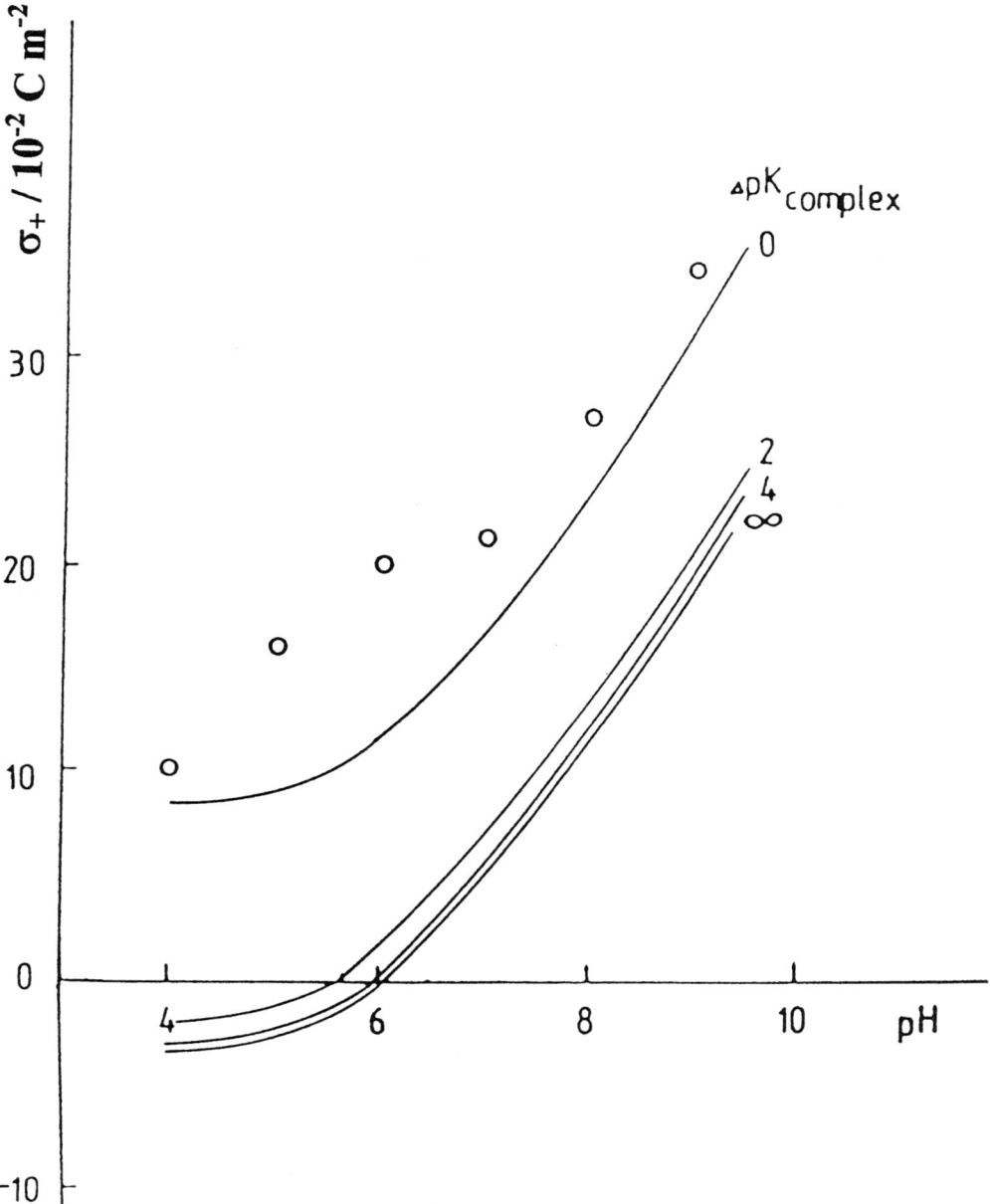

FIG. 25 $\sigma+$ ionic components of charge of anatase obtained for $\Delta pK = 6$ and different values of $\Delta pK_{complex}$. Experimental adsorption data (circles) are presented for comparison. (Reprinted from J. Colloid Interf. Sci., Ref. 101, © 1989 by Academic Press.)

sent on the oxide surface, and a local potential is different from zero. These local charged groups can interact with supporting electrolyte ions and be responsible for their adsorption at the pzc.

As was shown above, the real components of surface charge—Eq. (100)—can be determined only by direct adsorption measurements. Recently, a number of papers were published in which the authors measured positive adsorption of "indifferent" electrolyte ions (both anions and cations) at the pzc. The measurements were performed for different oxides such as SiO_2 [72,83,84,92], Al_2O_3 [54,82,90,91,92,95], ZrO_2 [87], Fe_2O_3 [93,94], and TiO_2 [52,85,86,87,92,94]. The examples of adsorption data are presented in Figs. 26–29. All researchers used the radiotracer technique to measure the adsorption density of the ions. The results obtained proved that complexation processes at the oxide/electrolyte interface indeed play an important role and are major contributors to the surface charge.

VI. SPECIFIC ADSORPTION

In the diffuse layer model the counterions of the supporting electrolyte neutralize the surface charge created by adsorption of potential determining ions. For multivalent counterions the drop of the absolute value of the potential in the interfacial region is steeper than for monovalent ions, and this is reflected in low absolute values of the ζ potential and low critical coagulation concentrations (ccc) (Schulze–Hardy rule). However, when the ccc is exceeded, the dispersion is often restabilized by the excess of multivalent counterions, while for monovalent counterions, the dispersion remains unstable up to very high electrolyte concentrations.

The sign of the ζ potential above the ccc is opposite to that below the ccc; thus the diffuse layer model is not sufficient to describe (even very roughly) the charging behavior of colloidal particles in the presence of multivalent ions. Therefore it is expedient to distinguish between adsorption of ions from inert electrolytes and specific adsorption. Specific adsorption is typical for multivalent ions, and electrostatic adsorption is typical for monovalent ions, but this rule is not categorical. For inert electrolytes the pzc and iep are independent of the electrolyte concentration, and for specific adsorption a systematic shift is observed: specifically adsorbed cations shift the pzc to lower pH and the iep to higher pH (and at high concentrations of some cations the ζ potential assumes only positive values, and there is no iep), and specific adsorption of anions shifts the pzc to higher pH and iep to lower pH.

The specific adsorption depends on the primary charging of the surface by the potential determining ions. For example, multivalent cations reverse the sign of the ζ potential of negatively charged oxides to positive, but a low pH values they do not adsorb, and their effect on the ζ potential is insignificant [102,103].

Chlorates (VII), nitrates (V), and halides of alkali metals at concentrations below 0.1 mol dm^{-3} usually do not affect the pzc or iep of (hydr)oxides. One clear exception from this rule is specific adsorption of halides on iron(III) (hydr)oxides, e.g., the cip of β-FeOOH titration curves shifts to higher pH values in the presence of chlorides (compared with nitrates, which are considered indifferent [44]). On the other hand, at sufficiently high ionic strength (a few tenths of 1 mol dm^{-3}) the iep of oxides shifts towards higher pH values (but the pzc is unaffected) [104].

The above phenomenological definition based on the shifts of pzc and iep does not explain the molecular mechanism of specific adsorption. Some modern methods offer a deeper insight into the mechanism of specific adsorption; a few examples are presented below.

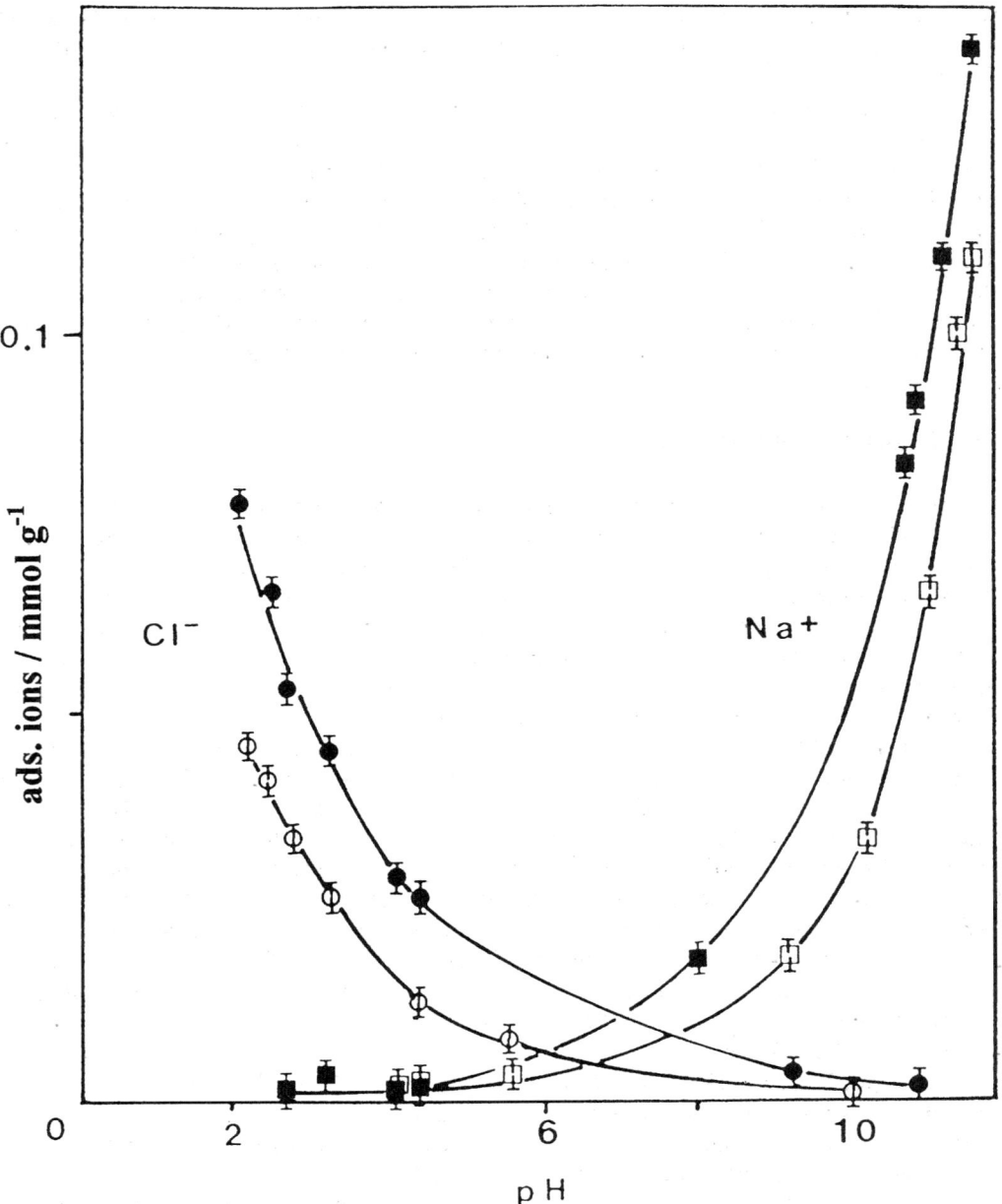

FIG. 26 Adsorption of Na$^+$ and Cl$^-$ counterions at various pH in 10^{-2} mol dm^{-3} NaCl for TiO$_2$ (open symbols) and for $w_{Pt} = 10\%$ Pt/TiO$_2$ sample (solid symbols). (Reprinted from J. Phys. Chem., Ref. 86, © 1986 by American Chemical Society.)

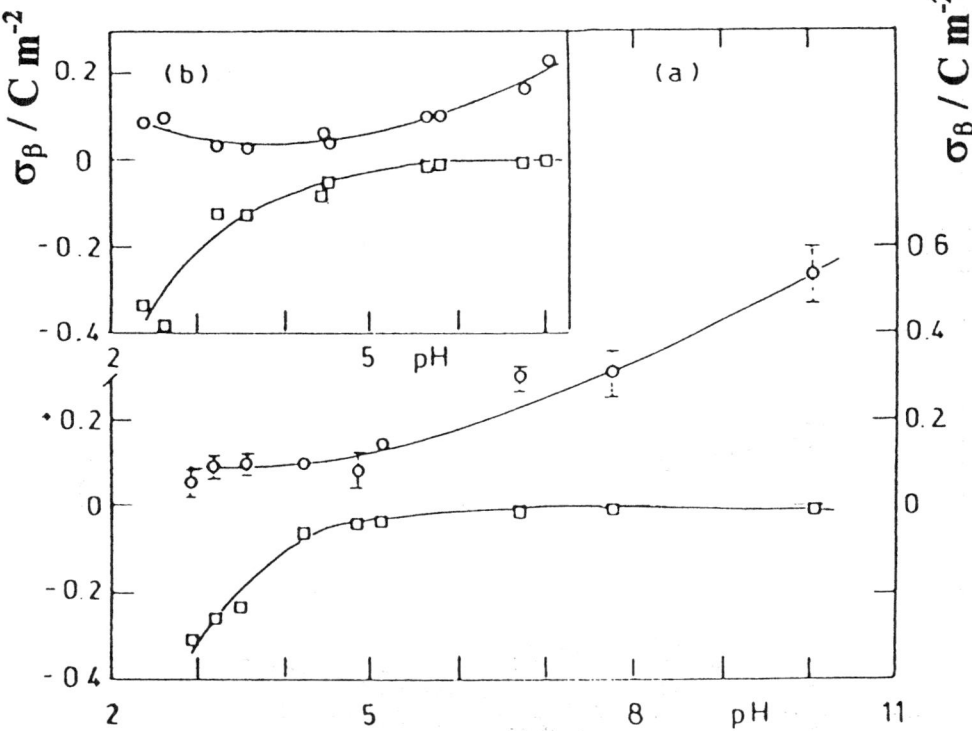

FIG. 27 Charge densities vs. pH curves of ions adsorbed in compact part of the electric interfacial layer of α-Al$_2$O$_3$ single crystals in about 0.05 mol dm^{-3} NaBr (a) and in about 0.01 mol dm^{-3} NaBr (b). Circles, Na; squares, Br. (Reprinted from J. Colloid Interf. Sci., Ref. 54, © 1980 by Academic Press.)

A. Methods

Atomic force microscopy (AFM) offers the possibility of measuring directly the force between two specimens immersed in solution and thus of verifying the theories of the electrical interfacial layer. Although commercial equipment is available on the market, the sample preparation and interpretation of measurements are far from trivial. So far we have not found any example of the application of AFM in the study of specific adsorption, but this method has been successful in the study of interactions in the presence of inert electrolytes. For example, the ζ potentials of silica derived from AFM measurements are in good agreement with those measured by electrophoresis and electroosmosis [105].

Time-resolved laser induced fluorescence spectroscopy (TRLFS) can be used in situ to study adsorption of cations, namely, trivalent actinides. In principle, the application of his method is not restricted to actinides, but lanthanides give only small chemical shifts, and for the d metal ions the absorption bands are too broad. The oxide must be transparent to laser light, and this limits the application of the method to silica. In TRLFS each species (in solution or adsorbed) shows a characteristic wavelength and fluorescence lifetime, which depend on the coordination environment of the central atom. Preliminary results are very promising [106]: in a dispersion of silica containing Cm(III), two

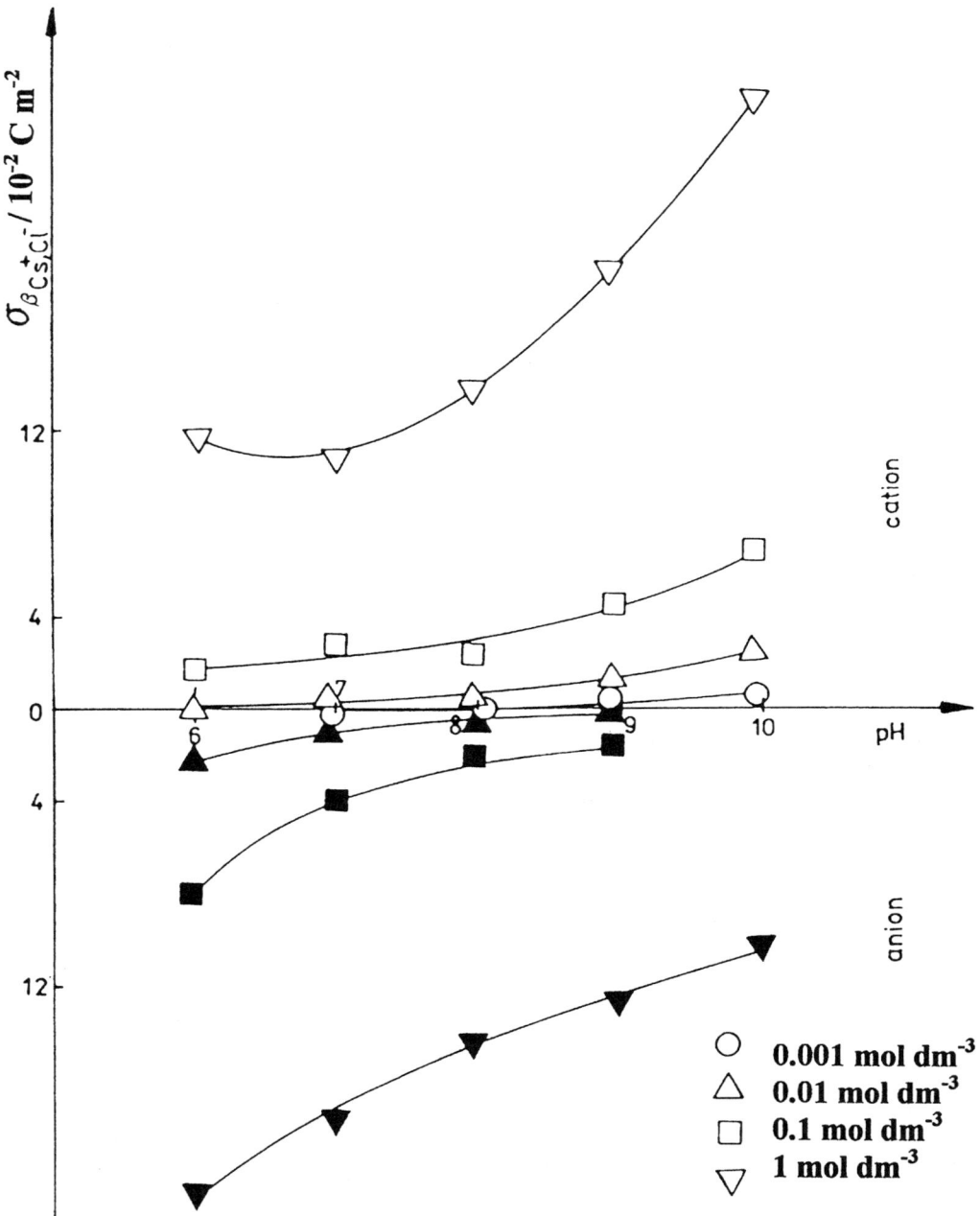

FIG. 28 Adsorption densities of Cs$^+$ (open symbols) and Cl$^-$ ions (solid symbols) vs. pH and electrolyte concentration. (Reprinted from J. Colloid Interf. Sci., Ref. 82, © 1989 by Academic Press.)

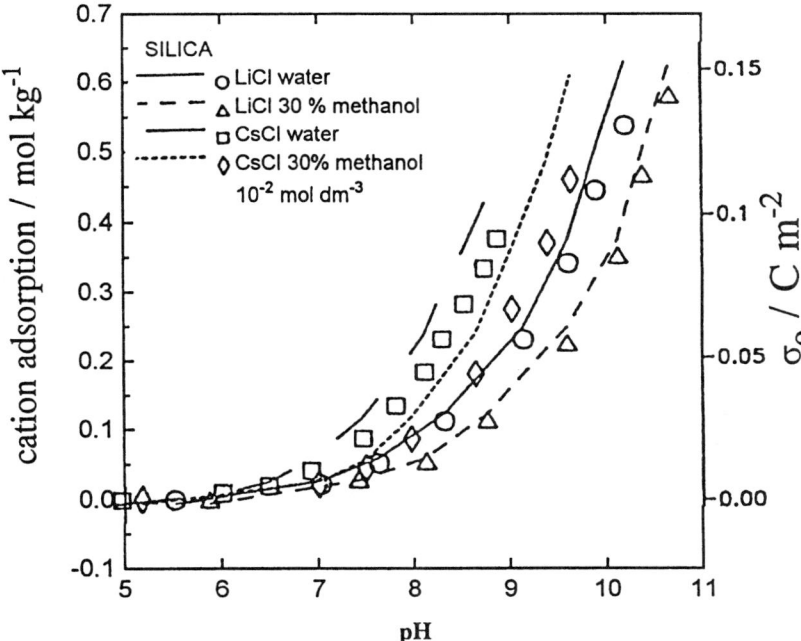

FIG. 29 Counterion adsorption and apparent σ_0 in the system silica 10^{-2} mol dm^{-3} LiCl and CsCl in water and aqueous methanol $w_{methanol} = 30\%$. (Reprinted from Pol. J. Chem., Ref. 92, © 1993 by Polish Chemical Society.)

additional peaks appear of hypothetical silica-Cm(III) surface complexes. The pH dependence on the intensities of these peaks (Fig. 30) is in line with the Cm(III) uptake curves. These peaks are shifted to longer wavelengths, and the emission decays more slowly (220 and 740 μs lifetimes) than for aqueous hydrolyzed species. The lifetime of the second peak suggests the incorporation of Cm(III) into the structure of silica.

X-ray absorption spectroscopy (XAS) also offers information about the coordination environment of the central atom of interest. Application of this method to the sorption of metal cations on oxides has been recently reviewed [107]. The advantage of XAS over TRLFS is that in the latter we can only estimate the degree of hydration of the central atom from the lifetime of fluorescence using a semiempirical formula, while XAS offers the possibility of quantitatively determining the number of the nearest (also second nearest, etc.) neighbors of the central atom and the interatomic distances when the atomic weights of particular elements are sufficiently different. Unfortunately, XAS requires rather high concentrations ($>10^{-4}$ mol dm^{-3}), while with TRLFS we can go down to micromolar or even nanomolar range.

Extended x-ray absorption fine structure (EXAFS) analysis shows that the coordination environment of Co on a silica surface is independent of the surface coverage, i.e., Co is octahedrally coordinated by six oxygen atoms at 0.21 nm and the second neigh-

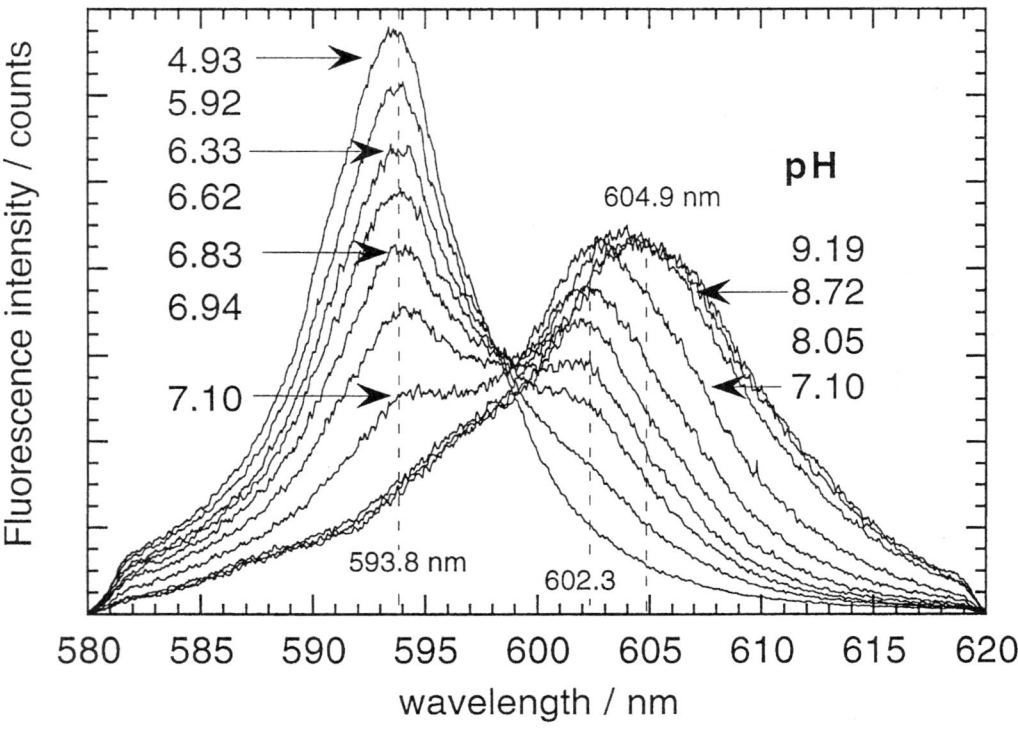

FIG. 30 Scaled fluorescence emission spectra of Cm(III) in silica dispersions (0.5 g dm^{-3}) at various pH. (Reprinted from Radiochim. Acta, Ref. 106, © R. Oldenbourg-Verlag.)

bors are Co atoms at 0.3 nm. Probably Co forms polynuclear complexes or disordered hydroxide-like precipitates [108]. Also Cu(II) forms clusters on silica even at very low surface coverage [109].

These and other similar results warn us against uncritical acceptance of simple models of adsorption with 1 : 1 or 1 : 2 metal–surface site complexes solely on the basis of the match between calculated and experimental adsorption curves. In many instances, however, surface precipitation or ion clustering can be excluded on the basis of spectroscopic measurements. For example, the same Cu(II) which clusters on silica resists any tendency to cluster on boehmite (AlOOH), as demonstrated in a study combining electron paramagnetic resonance (EPR) and XAS methods [110]. EPR gives us information about the state of adsorbed paramagnetic ions by comparison with the spectra of selected model compounds (e.g., metal hydroxide, metal ions in solution).

Spectroscopic methods indicate strong interactions between metal oxide surfaces and heavy metal ions (inner sphere surface complexation) and the absence of such strong interactions for alkali metal and alkaline earth metal cations. This result is not consistent with the definition of specific adsorption based on the shifts of pzc and iep discussed earlier in this section. The nature of the bond between oxide surfaces and alkaline earth metal cations was also studied using unconventional techniques. House [111] studied the dissolution rate of silica in the presence of salts. The model with and without surface complexation gave similar results for sodium salts, and for calcium the model with surface complexation works much better.

Spectroscopic studies of the adsorption of inorganic anions are sparse. The XAS techniques are not operative for elements of low atomic weight. Attenuated total reflection Fourier transformed infrared (ATR-FTIR) spectroscopy was developed in the 1980s, but nowadays this technique is rather forgotten. Tejedor-Tejedor and Anderson [112] determined the structure of the phosphate–goethite surface complex using this method.

B. Proton Stoichiometry

Adsorption of cations on oxides is accompanied by desorption of protons, and adsorption of anions is accompanied by adsorption of protons. The number of protons desorbed (or adsorbed) per one adsorbed cation (or anion) r is termed the proton stoichiometry parameter, and it is important in the verification of the adsorption models [113]. Experimental values of r can be obtained from potentiometric titrations when the number of adsorbed ions is sufficiently high, and these values are usually somewhat lower than the valence of the anion or cation [45]. For surface precipitation, r should be equal to the valence of the metal cation.

C. Models Without Electrostatics

In view of the above discussed coadsorption or release of protons, adsorption of ions on oxides is strongly pH dependent. The pH effect on the adsorption of ions is a combination of the mass action law and the electrostatic effect, and when r is lower than the valence of the ion, these two effects add up. Both lead to an increase of the adsorption of cations and a decrease of the adsorption of anions when the pH increases. Moreover, when the surface potential is a linear function of pH, both effects give the apparent adsorption constant as an exponential function of pH:

$$K_{\text{ads}} = \theta(1-\theta)^{-1} a^{-1} \, 10^{-r_1^* \text{pH}} \tag{101}$$

where θ is the surface coverage and r_1 is an empirical parameter (positive for cations and negative for anions) combining the proton stoichiometry parameter r and the Boltzmann factor. This equation has been successfully implemented to interpret the adsorption isotherms of metal cations at different pH values and at different temperatures [114]. In this approach, the number of surface sites is treated as an adjustable parameter, and it is determined for each pH value and each temperature separately from the experimental adsorption isotherm. Certainly, r_1 is another adjustable parameter in this model. We can try to reduce the number of adjustable parameters in the model assuming a constant number of surface sites.

D. Models with Single Surface Reaction

Equation (101) may reproduce the experimental adsorption data very well, but its parameters are purely empirical, and they do not have a physical sense. In order to obtain a real adsorption constant, we must first choose a electric interfacial layer model to account for the electrostatic contribution to the adsorption Gibbs energy and then combine this model with the reaction(s) responsible for the adsorption of anions or cations. Certainly, the value of the adsorption constant depends on the choice of the electric interfacial layer model and its parameters, so values reported in the literature that are calculated for different EIL models can be hardly compared. Dzombak and Morel [45]

TABLE 1 Adsorption Constants of Multivalent Cations [Reaction (102)]

Source	Oxide	Cations
K. M. Spark, B. B. Johnson, and J. D. Wells. Eur. J. Soil Sci. 46:621 (1995)	SiO_2	Zn -5.91 Cd -8.71 Cu(II) -5.12 Co -9.24
P. W. Schindler, B. Furst, R. Dick, and P. U. Wolf. J. Coll. Int. Sci. 55:469 (1976)	SiO_2	Cu(II) -7.00 Cd -9.20 Pb -6.65
S. E. Fendorf, G. M. Lamble, M. G. Stapleton, M. J. Kelley, and D. L. Sparks. Envir. Sci. Technol. 28:284 (1994)	SiO_2	Cr(III) 1.44
P. A. O'Day, C. J. Chisholm-Brause, S. N. Towle, G. A. Parks, and G. E. Brown. Geochim. Cosmochim. Acta 60:2515 (1996)	SiO_2	Cu(II) -8.19
M. Kosmulski. J. Coll. Int. Sci. 195:395 (1997)	SiO_2	Gd -6.00 Y -6.56
M. Kosmulski. J. Coll. Int. Sci. 190:212 (1997)	SiO_2	Ni -8.86
M. Kosmulski. Ber. Bunsenges. Phys. Chem. 98:1062 (1994)	SiO_2	Ca -7.32
M. Kosmulski. Coll. Surf 117:201 (1996)	SiO_2	Cd -9.15

For EIL parameters see text.

collected literature data for hydrous ferric oxide and calculated the adsorption constants for different anions and cations using one set of EIL parameters for all the data. For other (hydr)oxides such compilations are not available.

A surface complexation model (SCM) with a single surface species has a big advantage over models with multiple surface species (such models will be discussed later): the adsorption constant can be equivocally determined from experimental data. The values of equilibrium constants of multiple surface reactions obtained from limited data sets are ambiguous [115].

The reaction

$$-MOH + Me^{n+} \rightleftharpoons -MOMe^{(n-1)+} + H^+ \qquad (102)$$

has been very successful in modeling the adsorption of heavy metal cations on oxides. Summarized in Table 1 are the adsorption constants for an SCM with reaction (102) and the following TLM parameters: 6 sites nm^{-2}, $C_1 = 2.2$ Fm^{-2}, $C_2 = 0.2$ Fm^{-2}, log $K-5.71$, log $K_{cation}-7.24$, calculated from selected literature data.

The agreement of results from different sources obtained for Cd is satisfactory considering that different types of silica were used at different solid-to-liquid ratios and different initial concentrations of heavy metal ions and ionic strengths. On the other hand, the Cu(II) data show big discrepancies, probably because of clustering.

E. Models with Multiple Surface Reactions

1. Surface Heterogeneity

According to the MUSIC model, different surface sites of an oxide have different affinities to protons; thus the magnitude and often even the sign of the surface charge carried by different sites at a given pH can be different. This difference in electrostatic conditions certainly affects the adsorption of counterions, even when the chemical contribution to the Gibbs energy of adsorption is similar for different sites. To the best knowledge of the present authors, the MUSIC model has not been yet implemented for the interpretation of the adsorption of counterions. Dzombak and Morel [45] interpreted the effect of the initial concentration of heavy metal cations on their sorption on hydrous ferric oxide in terms of weak and strong surface sites. The strong and weak sites do not differ in their acid–base properties (a 2-pK model was used), but they differ by two or more orders of magnitude in their affinities to heavy metal cations. This model became very popular, although it does not explain the nature of the strong sites, and no spectroscopic proof for the existence of strong sites was found. In contrast, Rudziński et al. [46] used a continuous distribution of the Gibbs energies of adsorption of heavy metal cations to explain the nonlinearity of adsorption isotherms in the range of low concentrations of heavy metal cations.

2. Ternary Complexes

In the above discussions we have tacitly assumed that the surface species responsible for the adsorption of ions on oxides involve a surface site or sites and the ion of interest, for heavy metal cation adsorption of hydrolyzed species was also considered. The adsorbing species are not necessarily corresponding to the speciation in solution. For example, in the presence of 1 mol dm^{-3} of chlorides, Cd in solution occurs chiefly in the form of anionic and neutral chloride complexes, but the adsorption of cadmium on silica and alumina is cation-like, i.e., it increases when the pH increases [116]. A model with one surface complex, namely –SiOCd$^+$, reproduces the experimentally observed adsorption curves of Cd on silica at various concentrations of chlorides, but an analogous model underestimates the adsorption of Cd on alumina at high concentrations of chlorides. The agreement between the experimental and the calculated curves is much better when the formation of ternary complexes involving chloride is taken into account.

A model with one surface complex only is not capable of properly explaining the sorption of Hg(II) [117] from media containing chlorides. The formation of two surface complexes, namely –SiOHgOH and –SiOHgCl, was suggested. Formation of ternary complexes was invoked in many other papers [118]. They are very powerful in modeling extensive sets of data covering many orders of magnitude of ionic strength. The disadvantage of this approach is the absence of spectroscopic proof of the existence of such ternary complexes. Moreover, in models with many different complexes, the formation constants of these complexes cannot be unequivocally determined from the uptake curves, i.e., many sets of model parameters lead to practically identical model curves.

The concept of ternary surface complexes was recently used to explain the apparent surface heterogeneity of oxide surfaces [119]. Let us assume that the system contains an impurity whose concentration is in the micromolar range and that forms a very strong ternary complex with the metal ion of interest and the surface. For initial concentrations of heavy metal ions below the concentration of the impurity, the adsorption is governed primarily by the formation of the ternary complex, and the apparent adsorption constant is very high. However, when the initial concentration of the heavy metal ion cations increases above the concentration of the impurity, the contribution of the ternary complex to the total adsorption gradually decreases, and this leads to a decrease in the apparent adsorption constant.

3. Surface Complexation and Precipitation

The models neglecting the surface precipitation or formation of polynuclear surface complexes or formation of solid solution can often successfully reproduce the experimentally observed uptake curves. However, spectroscopic data directly show that in many systems simplified models considering only mononuclear surface species do not reflect the physical reality. Moreover, in many real systems, the percentage of uptake increases when the initial concentration increases at surface coverages on the order of a few percent, and this effect cannot be explained by surface complexation alone.

Katz and Hayes [120] considered three models to account for the contribution of Co polynuclear species to the uptake of Co by alumina. All these models involve one or more surface adsorption reactions and the reactions describing the solution chemistry of Co (or another heavy metal cation of interest).

(a) Solid Solution Model [121]. Implementation of this model requires the following additional reactions:

Surface precipitation

$$Co^{2+} + 2 H_2O = Co(OH)_2(s) + 2 H^+ \tag{103}$$

$$Al^{3+} + 3 H_2O = Al(OH)_3(s) + 3 H^+ \tag{104}$$

The mass action law expressions for the surface precipitation reactions involve a solid phase activity coefficient. In the simplest case of an ideal solid solution, the activities of the components are expressed as

$$\{Al(OH)_3(s)\} = [Al(OH)_3(s)]/([Al(OH)_3(s)] + [Co(OH)_2(s)]) \tag{105}$$

$$\{Co(OH)_2(s)\} = [Co(OH)_2(s)]/([Al(OH)_3(s)] + [Co(OH)_2(s)]) \tag{106}$$

Real solid solutions are often far from ideal, unfortunately, and to the best knowledge of the present authors there is no generally accepted theory by which to calculate the activities of the components of solid mixtures. Another difficulty in the implementation of the solid solution model is in the proper definition of the total amount of the component $Al(OH)_3(s)$ in the above reactions.

Solution chemistry of aluminum

The reactions are analogous to those reflecting the solution chemistry of cobalt, i.e.,

$$Al^{3+} + nH_2O = Al(OH)_n^{3-n} + nH^+ \quad n = 1, 2, 3, 4 \tag{107}$$

In the presence of anions forming stable complexes with aluminum, complexation reactions in solution involving aluminum must be also considered.

(b) Surface Polymer Model. By analogy to the multinuclear complexes of cobalt observed in solution, the surface polymers considered were $\mathrm{\equiv AlOCo_2(OH)_2^+}$ and $\mathrm{\equiv AlOCo_4(OH)_5^{2+}}$, and their stability constants are defined in the same way as for mononuclear complexes. Surface precipitation (reaction 103) is also considered, but in the surface polymer model and the activity of the solid cobalt hydroxide is set to 1.

(c) Continuum Model. The same surface species as in the previously described surface polymer model are considered. The activity of solid cobalt hydroxide is redefined as

$$\{Co(OH)_2(s)\} = \frac{[Co(OH)_2(s)]}{([\equiv AlOCo^+] + [\equiv AlOCo_2(OH)_2^+] + [\equiv AlOCo_4(OH)_5^{2+}] + [Co(OH)_2(s)]} \quad (108)$$

This definition of the activity of the cobalt hydroxide precipitate is much easier to implement than the definition used in the solid solution model, although both definitions express the same trend, namely the activity coefficient of the solid cobalt hydroxide varies with the pH and cobalt concentration. The continuum model has been successful in modeling the adsorption of Co on alumina over a broad range of initial concentrations of cobalt [120].

F. Adsorption Enthalpy

1. Calorimetry

Direct calorimetric measurements of surface charging of silica in the presence of 0.01 to 1 mol dm^{-3} of different electrolytes give different results suggesting that the adsorption of counterions significantly contributes to the measured overall heat [122]. The acid and base titrations in the presence of 1 mol dm^{-3} NaCl and KCl show a significant hysteresis which is not observed at lower ionic strengths. Not only calorimetric experiments are difficult but also their interpretation is not trivial [123]. Therefore, although some direct calorimetric values have been reported [124], most data on the enthalpy of adsorption of ions available in the literature have been obtained indirectly, from the temperature dependence of adsorption.

2. Temperature Effects on Stability Constants of Surface Complexes

The surface charge density at a given pH and ionic strength is more negative at higher temperatures. This creates more favorable electrostatic conditions for the association of cations and less favorable conditions for the association of anions. Therefore it is not surprising that the adsorption of cations increases and the adsorption of anions decreases with temperature. This applies to mono- and multivalent ions and to co- and counterions, as well.

The temperature effect on the sorption of multivalent ions can be quantified as a shift in the pH$_{50}$, the pH value at which 50% of the ions of interest are adsorbed. The position of the pH$_{50}$ is affected by the experimental conditions, e.g., the solid-to-liquid ratio, but the shift $[pH_{50}(T_2) - pH_{50}(T_1)]/(T_2 - T_1)$ is relatively insensitive to the solid-to-liquid ratio, ionic strength, and initial concentration of heavy metal cations [125]. This shift does not differ very much from the shift in the pzc of the oxide of interest. An advantage of this method of expressing the temperature effect on adsorption is that dpH_{50}/dT is obtained directly from experimental data without any model assumptions. With ions of inert electrolytes, the concept of pH$_{50}$ is not useful, as their uptake is very low even at high solid-to-liquid ratios and at favorable electrostatic conditions.

The most common way to express the temperature effect on the adsorption process is to calculate the enthalpy of adsorption. This approach requires certain model assumptions, and the results from different sources are not always compatible when the assumptions are different. The most rigorous approach is to define the surface reaction(s) responsible for the adsorption and to calculate the enthalpy from the temperature dependence of the equilibrium constant of this reaction,

$$\Delta H^\circ_{ads} = RT^2 \frac{d \ln K_{ads}}{dT} \tag{109}$$

The equilibrium constants of surface reactions depend on the model of the electrical interfacial layer, and the parameters of the model of the electrical interfacial layer (e.g., Stern layer capacitance) depend on the temperature. In sophisticated models of the electrical interfacial layer with many adjustable parameters, it is impossible to select unequivocally one best-fit set of parameters, because many different sets of parameters give almost identical model charging curves [126]. Thus these modes are not well suited to study the temperature effects on the binding constants of specifically adsorbed ions, and simple models of adsorption are preferred. The equilibrium constant of the reaction

$$\text{ion in solution} \leftrightarrow \text{ion adsorbed} \tag{110}$$

equals $c_{ads}/a_{solution}$, and it is certainly a function of experimental conditions. We can use this reaction to define the standard enthalpy of adsorption:

$$\Delta H^\circ_{ads} = \frac{R(a_2 - a_1)_{c_{ads},\ldots}}{1/T_2 - 1/T_1} \tag{111}$$

where a_i is the equilibrium activity in solution at the temperature T_i at constant c_{ads}. Hall [127] proposed a definition in which not only c_{ads} but the surface concentrations of all components are constant. This assumption, however, violates the condition of electroneutrality, i.e., in order to adjust concentrations of ions in the surface layer we have to change their concentrations in solution, and this cannot be done freely for all ions. Machesky [128] proposed a less rigorous but more practical definition, namely to keep c_{ads} and pH constant in Eq. (111). His definition gives endothermic enthalpies for cations (more favorable electrostatic conditions at higher temperatures) and exothermic enthalpies for anions (less favorable electrostatic conditions at higher temperatures). The effect of electrostatic conditions on the adsorption of ions can be avoided when c_{ads} and σ_0 are constant in Eq. (111). We can distinguish this quantity from the enthalpy defined according to Machesky by terming it enthalpy at constant σ_0, and Machesky's definition leads to enthalpy at constant pH. The enthalpies of adsorption of simple inorganic monovalent ions on oxides at constant σ_0 are close to zero. Some values taken directly from the literature are summarized in Table 2. Since different methods were used to determine these enthalpies, the values are not compatible. More data of this kind can be found in the review by Machesky [128].

G. Adsorption Competition Between Various Multivalent Ions

Adsorption of heavy metal cations is rather insensitive to ionic strength. However, relatively few systematic studies of ionic strength effects have been carried out, and most studies report results obtained at one ionic strength. On the other hand, there is a clear competition between alkali and alkaline earth metal cations for the surface sites of silica. The pH_{50} for Ca adsorption on silica increases by 2 pH units when the ionic strength increases from 0.001 to 0.1 mol dm^{-3} NaCl. Adsorption of Mg^{2+} on silica also decreases in the presence of alkali metal cations. Longer equilibration times lead to formation

TABLE 2 Standard Enthalpies of Adsorption of Cations on Oxides

Source	Oxide	Cations (kJ mol^{-1})
G. C. Bye, M. McEvoy, and M. A. Malati. J. Chem. Tech. Biotechn. *32*:781 (1982)	SiO$_2$	Cu(II) 23
M. A. Malati, M. McEvoy, and C. R. Harvey. Surf. Technol. *17*:165 (1982)	SiO$_2$	Cd 13
G. C. Bye, M. McEvoy, and M. A. Malati. J. Chem. Soc. Faraday. Trans. 1 *79*:2311 (1983)	SiO$_2$	Co 2 Ni 10
M. Kosmulski. Ber. Bunsenges. Phys. Chem. *98*:1062 (1994)	Al$_2$O$_3$ SiO$_2$	Eu 36 (at constant σ_0) Ca 0 (at constant σ_0) Ca 0 (at constant σ_0)
L. G. Fokkink, A. de Keizer, and J. Lyklema. J. Coll. Int. Sci. *135*:118 (1990)	Fe$_2$O$_3$ rutile	Cd -7 Cd 5
M. Kosmulski. Coll. Surf *117*:201 (1996)	Al$_2$O$_3$ SiO$_2$	Cd 0 (at constant σ_0) Cd 0 (at constant σ_0)
M. Kosmulski. Coll. Surf *83*:237 (1994)	SiO$_2$	Co 22 (at constant σ_0)

of magnesium hydroxysilicate (sepiolite) [129]. These results are in line with the XAS results and suggest the formation of inner sphere complexes by heavy metal cations and the absence of strong interaction with alkaline earth metal cations [130].

Many adsorption studies of heavy metal cations on oxides have been carried out in systems designed to mimic soil systems and containing magnesium and/or calcium [131]. We can ask ourselves if there is any significant difference between the adsorption behavior in the presence of these calcium and magnesium salts on the one hand and inert electrolytes on the other. Systematic studies of such effects are rare, and they show that there is no simple answer to this question: each system has to be treated individually. For example, the effect of alkaline earth metal cations on the adsorption of gadolinium on silica is insignificant, but the adsorption of nickel on silica considerably decreases in the presence of 10^{-2} mol dm^{-3} Ba^{2+} (but for Ba^{2+} concentrations $<10^{-3}$ mol dm^{-3} the effect is insignificant). The adsorption of Gd and Ni on alumina is insensitive to alkaline earth metal cations [125].

It would be also interesting to study the adsorption of heavy metal cations from solutions containing other heavy metal cations, especially in the systems where the individual adsorption isotherm suggests surface heterogeneity, e.g., to explain the nature of the "strong" surface sites. Are they the same for different heavy metal cations (this would result in adsorption competition) or are they different (no competition)? The apparent surface heterogeneity is observed in the concentration range where spectroscopic methods are ineffectual.

SYMBOLS AND ABBREVIATIONS

a	activity
A	empirical constant
AFM	atomic force microscopy
ATR-FTIR	attenuated total reflection Fourier transformed infrared
b (subscript)	bulk of the solution
B	conversion factor from mol dm^{-3} to C m^{-2}
B	empirical constant
C	concentration of the electrolyte
C_1, C_2	capacitance of the inner and outer regions of the compact layer, respectively
C	capacitance of electric interfacial layer
c	molarity of ionic species
c_{ads}	concentration of adsorbed species
ccc	critical coagulation concentration
cip	common intersection point
CN	coordination number
d_{crit}	critical distance in Bjerrum association theory
dca	distance of closest approach
e	proton charge
E	electric potential difference
EIL	electrical interfacial layer
EPR	electron paramagnetic resonance
EXAFS	extended x-ray absorption fine structure
iep	isoelectric point ($\zeta=0$)
IHP	inner Helmholtz plane
IR	infrared
k	Boltzmann constant
K	equilibrium constant
K_-	reciprocal equilibrium constant of reaction (10)
K_{a1}^{int}, K_{a2}^{int}	effective surface ionization constants
$K_{a1}^{o\,int}$, $K_{a2}^{o\,int}$	thermodynamic surface ionization constants
K_{ads}	equilibrium constant of surface reaction responsible for adsorption
K_{cation}	equilibrium constant of reaction (27)
K_{Na}	selectivity coefficient (with respect to sodium)
K_{sp}	solubility product
$*K_{Y+,X-}{}^{o\,int}$	thermodynamic surface complexation constants
$*K_{Y+,X-}{}^{int}$	effective surface complexation constants
L	metal–hydrogen distance
m	number of oxygen s-orbitals occupied by hydrogen
M	surface metal atom
MUSIC	multisite complexation model
n	number of unoccupied oxygen s-orbitals
N	number of next-neighbor metal ions
N_s	total number of surface sites available
OHP	outer Helmholtz plane
p	pressure
pdi	potential determining ion
ppzc	pristine point of zero charge

pzc	point of zero net surface charge ($\sigma_0 = 0$)
Q_{a1}, Q_{a2}	apparent surface ionization constants
$*Q_{Y+,X-}$	apparent surface complexation constants of cation and anion
R	gas constant
r	proton stoichiometry parameter
r_1	empirical constant combining the proton stoichiometry parameter and the Boltzmann factor
s_i	corrected bond valence
S	specific surface area
s (subscript)	surface
SCM	surface complexation model
T	temperature
TLM	triple layer model
TRLFS	time-resolved laser-induced fluorescence spectroscopy
v	bond valence
V	undersaturation of the valence of oxygen
x	formal charge of a surface oxygen
X^-, Y^+	anion and cation, respectively
XAS	x-ray absorption spectroscopy
Z	valence of ion
α_+, α_-	fraction of charged sites
β	plane of specifically adsorbed ions
γ	surface tension of liquid
γ_e	mean activity coefficient of the electrolyte
γ_0	activity coefficient of surface –MOH group
γ_-, γ_+	activity coefficients of surface –MO$^-$ and –MOH$_2^+$ groups, respectively
γ_{+-}, γ_{-+}	activity coefficients of surface –MOH$_2^+$X$^-$ and –MO$^-$Y$^+$ groups, respectively
$\gamma_{H+}, \gamma_{X-}, \gamma_{Y+}$	activity coefficients of hydrogen ion, anion and cation, respectively
Γ	surface excess
ΔH^0_{ads}	standard enthalpy of adsorption
ΔpK_a	$pK^{int}_{a2} - pK^{int}_{a1}$
$\Delta pK_{complex}$	$p*K^{int}_{Y+} - p*K^{int}_{X-}$
θ	surface coverage
κ	reciprocal Debye length
μ^0	standard chemical potential
μ	chemical potential
$\bar{\mu}$	electrochemical potential
ψ_b	electrical potential in the bulk of the solution
$\psi_0, \psi_\beta, \psi_d$	mean potentials in the plane of surface charge, adsorbed counterions, and at the start of the diffuse layer, respectively
$\sigma_0, \sigma_\beta, \sigma_d$	net charge densities at the surface, in the plane of adsorbed counterions and diffuse layer, respectively
σ_{d-}, σ_{d+}	diffuse layer charges in cathodic and anodic ranges, respectively
σ_-, σ_+	surface charge components
λ	Esin–Markow coefficient
ζ	electrokinetic potential (zeta potential)

REFERENCES

1. J. Lyklema, *Fundamentals of Interface and Colloid Science*, vol. II, Academic Press, New York, 1995.
2. D. C. Grahame. Chem. Rev. *41*:441 (1947). P. Delahay, *Double Layer and Electrode Kinetics*, Wiley-Interscience, New York, 1965.
3. R. Payne. J. Electroanal. Chem. *41*:277 (1973).
4. R. Payne, in *Progress in Surface and Membrane Science*, vol. 6, Academic Press, New York, 1973, pp. 51–123. J. R. MacDonald. J. Electroanal. Chem. *223*:1 (1987).
5. R. J. Hunter, *Foundations of Colloid Science*, vol. 1, Oxford University Press, Oxford, 1987.
6. N. Kallay, Ž. Torbić, E. Barouch, and J. Jednačak-Bišćan. J. Colloid Interface Sci. *118*:431 (1987). N. Kallay, Ž. Torbić, M. Golić, and E. Matijević, J. Phys. Chem. *95*:7028 (1991).
7. J. Leja, *Surface Chemistry of Froth Flotation*, Plenum Press, New York, London, 1982.
8. P. J. Boddy. J. Electroanal. Chem. *10*:199 (1965).
9. M. J. Sparnaay. Adv. Colloid Interface Sci. *1*:277 (1967).
10. I. Uchida, H. Akahoshi, and S. Toshima. J. Electroanal. Chem. *88*:79 (1978).
11. H. Gerischer. Electrochimica Acta *34*:1005 (1989).
12. H. Gerischer, in *Physical Chemistry*, vol. IXA, *Electrochemistry* (H. Eyring, ed.), Academic Press, New York, 1970, pp. 463–542.
13. B. H. Bijsterbosch and J. Lyklema. Adv. Colloid Interface Sci. *9*:147 (1978).
14. R. Hunter, *Zeta Potential in Colloid Science*, Academic Press, New York, 1981.
15. J. Lyklema and Th. G. Overbeek. J. Colloid Interface Sci. *16*:595 (1961).
16. J. Lyklema. J. Electroanal. Chem. *37*:53 (1972).
17. B. H. Bijsterbosch and J. Lyklema. J. Colloid Interface Sci. *28*:506 (1968).
18. J. Lyklema. Disc. Faraday. Soc. *42*:81 (1966).
19. H. P. Boehm. Disc. Faraday Soc. *52*:264 (1971).
20. G. A. Parks. Chem. Rev. *65*:177 (1965).
21. M. A. Blesa and N. Kallay. Adv. Colloid Interface Sci. *28*:111 (1988).
22. N. H. G. Penners, L. K. Koopal, and J. Lyklema. Colloids Surfaces *21*:459 (1986).
23. T. W. Healy, D. E. Yates, L. R. White, and D. Chan. J. Electroanal. Chem. *80*:57 (1977).
24. L. Bousse, N. F. de Roij, and P. Bergveld. Surface Sci. *135*:479 (1983).
25. R. E. G. van Hal, J. C. T. Eijkel, and P. Bergveld. Adv. Colloid Interface Sci. *69*:31 (1996).
26. G. A. Parks and P. L. de Bruyn. J. Phys. Chem. *66*:967 (1962). N. Kallay, D. Babić, and E. Matijević. Colloids Surfaces *19*:375 (1986). P. Hesleitner, N. Kallay, and E. Matijević. Langmuir *7*:178 (1991). N. Kallay, R. Sprycha, M. Tomić, S. Žalać, and Ž. Torbić. Croatica Chemica Acta *63*:467 (1990).
27. L. Blok and P. L. de Bruyn. J. Colloid Interface Sci. *32*:518 (1970). R. Sprycha, Colloids and Surfaces *5*:147 (1982).
28. P. Hesleitner, D. Babić, N. Kallay, and E. Matijević. Langmuir *3*:815 (1987).
29. J. Sonnefeld. J. Coll. Int. Sci. *155*:191 (1993). W. Janusz. Ads. Sci. Technol. *14*:151 (1996).
30. F. G. Fokkink, A. de Keizer, and J. Lyklema. J. Coll. Int. Sci. *127*:116 (1989).
31. W. H. van Riemsdijk, J. C. de Vit, L. K. Koopal, and G. H. Bolt. J. Coll. Int. Sci. *116*:511 (1987).
32. R. K. Iler, *The Chemistry of Silica*, John Wiley, New York, 1979.
33. T. Hiemstra, J. C. de Vit, and W. H. van Riemsdijk. J. Coll. Int. Sci. *133*:91 (1989). T. Hiemstra, W. H. van Riemsdijk, and G. H. Bolt. J. Coll. Int. Sci. *133*:105 (1989).
34. T. Hiemstra, P. Venema, and W. H. van Riemsdijk. J. Coll. Int. Sci. *184*:680 (1996). P. Venema, T. Hiemstra, P. G. Weidler, and W. H. van Riemsdijk. J. Coll. Int. Sci. *198*:282 (1998). L. K. Koopal. Electrochim. Acta *41*:2293 (1996). J. R. Rustad, A. R. Felmy, and B. P. Hay. Geochim. Cosmochim. Acta *60*, 1553, 1563 (1996).
35. W. Rudziński and D. H. Everett, *Adsorption of Gases on Heterogeneous Surfaces*, Academic Press, San Diego, 1991. M. Jaroniec and R. Madey, *Physical Adsorption on Heterogeneous Surfaces*, Elsevier, Amsterdam, 1988.
36. Z. Z. Zhang, D. L. Sparks, and N. C. Scrivner. J. Coll. Int. Sci. *162*:244 (1994).

37. S. K. Milonjić, Z. E. Ilić, and M. M. Kopecni. Coll. Surf. 6:167 (1983).
38. G. F. Cerefolini and G. Boara. J. Coll. Int. Sci. 161:232 (1993).
39. S. Kondo, T. Tamaki, and Y. Ozeki. Langmuir 3:349 (1987).
40. B. W. Ninham and V. Yaminski. Langmuir 13:2097 (1997).
41. T. F. Tadros and J. Lyklema. Electroanal. Chem. Int. Electrochem. 17:267 (1968).
42. J. Sonnefeld, A. Goebel, and W. Vogelsberg. Colloid Polym. Sci. 273:926 (1995). J. Sonnefeld. Colloid Polym. Sci. 273:932 (1995).
43. H. T. Tien. J. Phys. Chem. 69:350 (1965).
44. K. M. Parida. Ads. Sci. Technol. 3:89 (1986).
45. P. W. Schindler, in M. A. Anderson and A. J. Rubin, eds., *Adsorption of Inorganics at Solid-Liquid Interfaces*, Ann Arbor, 1981, Chapter 1. D. G. Kinnigurgh and M. L. Jackson, in *Adsorption of Inorganics at Solid-Liquid Interfaces* (M. A. Anderson and A. J. Rubin, eds.), Ann Arbor, 1981, p. 91. P. W. Schindler and W. Stumm, in *Aquatic Surface Chemistry* (W. Stumm, ed.), John Wiley, New York, 1987, p. 83. D. A. Dzombak and F. M. Morel, *Surface Complexation Modeling: Hydrous Ferric Oxide*, John Wiley, New York, 1990.
46. J. Lutzenkirchen and P. Behra. Langmuir 10:3916 (1994). W. Rudziński and R. Charmas. Langmuir 11:377 (1995).
47. N. Kallay and M. Tomić. Langmuir 4:559 (1988). M. Tomić and N. Kallay. Langmuir 4:565 (1988).
48. D. E. Yates, S. Levine, and T. W. Healy. J. Chem. Soc. Faraday Trans. I 70:1807 (1974).
49. G. R. Wiese, R. O. James, D. E. Yates, and T. W. Healy, in *International Review of Science*, Phys. Chem. Series 2., vol. 6 (J. O'M. Bockris, ed.), London, 1976.
50. J. A. Davis, R. O. James, and J. O. Leckie. J. Colloid Interface Sci. 63:480 (1978).
51. J. A. Davis and J. O. Leckie. J. Colloid Interface Sci. 67:90 (1978).
52. R. Sprycha. J. Colloid Interface Sci. 102:173 (1984).
53. D. Chan, J. W. Perram, L. R. White, and T. W. Healy. J. Chem. Soc. Faraday Trans. I 71:1046 (1975).
54. W. Smit and C. L. M. Holten. J. Colloid Interface Sci. 78:1 (1980).
55. R. O. James, J. A. Davis, and J. O. Leckie. J. Colloid Interface Sci. 65:331 (1978).
56. R. O. James, in *Adsorption of Inorganics at Solid–Liquid Interfaces* (M. A. Anderson and A. J. Rubin, eds.), Ann Arbor Sci. Ann Arbor, 1981, p. 219.
57. R. O. James and G. A. Parks, in *Surface and Colloid Science* (E. Matijevic, ed.), vol. 12, Wiley-Interscience, New York, 1982, pp. 119–216.
58. W. Smit. J. Colloid Interface Sci. 109:295;113:288 (1986).
59. W. Rudziński, R. Charmas, and S. Partyka. Langmuir 7:354 (1991).
60. W. Rudziński, R. Charmas, F. Thomas, and S. Partyka. Langmuir 8:1154 (1992).
61. L. K. Koopal, W. H. van Riemsdijk, and M. G. Roofey. J. Colloid Interface Sci. 118:117 (1987).
62. J. Westall and H. Hohl. Adv. Colloid Interface Sci. 12:265 (1980).
63. J. Westall, in *Particulates in Water* (M. C. Kavanagh and J. O. Leckie, eds.), Advances in Chemistry Series, No. 189, 1980, pp. 33–44.
64. J. Westall, in *Geochemical Processes at Mineral Surfaces* (J. A. Davis and K. F. Hayes, eds.), ACS Symposium Series 323, Washington, D.C., 1986, pp. 54–78.
65. J. Westall, in *Aquatic Surface Chemistry (W. Stumm, ed.)*, John Wiley, New York, 1987, pp. 3–32.
66. K. F. Hayes, G. Redden, W. Ela, and J. O. Leckie. J. Colloid Interface Sci. 142:448 (1991).
67. J. A. Davis and D. B. Kent, in *Review in Mineralogy*, vol. 23 (M. F. Hochella and A. F. White, eds.), 1990, pp. 177–260.
68. W. Janusz. Polish J. Chem. 65:799 (1991). W. Janusz, D.Sc. thesis, Lublin, 1994.
69. G. Schwarzenbach and H. Ackerman. Helvetica Chuimica Acta 31:1029 (1948).
70. H. C. Li and P. L. de Bruyn. Surface Sci. 5:203 (1966). R. J. Hunter. J. Colloid Interface Sci. 37:564 (1971).
71. R. Sprycha and J. Szczypa. J. Colloid Interface Sci. 102:288 (1984). R. Sprycha and J. Szczypa. J. Colloid Interface Sci. 115:590 (1987).

72. H. C. Li. Surface Sci. 5:203 (1966). R. J. Hunter and H. J. L. Wright. J. Colloid Interface Sci. 37:564 (1971).
73. W. Stumm, R. Kunnert, and L. Sigg. Croatica Chem. Acta 53:291 (1980).
74. G. R. Wiese and T. W. Healy. J. Colloid Interface Sci. 51:427 (1975).
75. T. W. Healy and L. R. White. Adv. Colloid Interface Sci. 9:303 (1978).
76. R. Sprycha. J. Colloid Interface Sci. 127:1 (1989).
77. R. O. James. Colloids and Surfaces 2:201 (1981).
78. R. E. Johnson. J. Colloid Interface Sci. 100:540 (1984).
79. R. J. Hunter, *Introduction to Modern Colloid Science*, Oxford Science, 1993.
80. P. Ney, Zeta Potentiale und Flotierbarkeit von Mineralen, Springer-Verlag, Vienna, New York, 1973.
81. H. L. Michael and D. J. A. Williams. J. Electroanal. Chem. 179:131 (1971).
82. R. Sprycha. J. Colloid Interface Sci. 127:12 (1989).
83. W. Smit, W. Holten, H. N. Stein, I. J. M. de Goeij, and H. M. J. Theelen. J. Colloid Interface Sci. 63:120 (1978).
84. W. Smith, W. Holten, H. N. Stein, I. J. M. de Goeij, and H. M. J. Theelen. J. Colloid Interface Sci. 67:397 (1978).
85. A. Foissy, A. M'Pandou, and J. M. Lamarche. Colloids Surfaces 5:363 (1982).
86. N. Jafferzic-Renault, P. Pichat, A. Foissy and R. Mercier. J. Phys. Chem. 90:2733 (1986).
87. W. Janusz. J. Radioanalytical Nucl. Chem. 125:393 (1988).
88. W. Janusz. J. Radioanalytical Nucl. Chem. 134:193 (1989).
89. W. Janusz. J. Colloid Interface Sci. 145:119 (1991).
90. S. Y. Shiao and R. E. Meyer. J. Inorg. Nucl. Chem. 43:3301 (1981).
91. S. Y. Shiao, Y. Egozy, and R. E. Meyer. J. Inorg. Nucl. Chem. 43:3309 (1981).
92. M. Kosmulski. J. Coll. Int. Sci. 135:590 (1990). M. Kosmulski. J. Coll. Int. Sci. 156:305 (1993). M. Kosmulski. Coll. Surf. 83:237 (1994). M. Kosmulski. Pol. J. Chem. 67:1831 (1993). M. Kosmulski. Bull. Pol. Acad. Sci. Chem. 41:325 (1994).
93. S. Chibowski and J. Szczypa. J. Colloid Interface Sci. 100:571 (1984). S. Chibowski. Materials Chem. Phys. 17:293 (1987).
94. S. Chibowski. Polish J. Chem. 59:1193 (1985).
95. W. Janusz. Materials Chem. Phys. 24:39 (1989).
96. R. Sprycha. J. Colloid Interface Sci. 96:551 (1983).
97. J. Lyklema. J. Colloid Interface Sci. 99:109 (1984).
98. A. Breeuwsma and J. Lyklema. J. Colloid Interface Sci. 99:109 (1984).
99. A. Breeuwsma and J. Lyklema. Disc. Faraday Soc. 52:324 (1971).
100. J. Lyklema. Trans. Faraday Soc. 59:189 (1963).
101. R. Sprycha, M. Kosmulski, and J. Szczypa. J. Colloid Interface Sci. 128:88 (1989).
102. R. O. James and T. H. Healy. J. Coll. Int. Sci. 40:53 (1972).
103. R. Herrera-Urbina and D. W. Fuerstenau. Coll. Surf. 98:25 (1995).
104. M. Kosmulski and J. B. Rosenholm. J. Phys. Chem. 100:11681 (1996).
105. P. G. Hartley, I. Larson, and P. J. Scales. Langmuir 13:2207 (1997).
106. K. H. Chung, R. Klenze, K. K. Park, P. Paviet-Hartmann, and J. I. Kim. Radiochim. Acta, 82:215 (1998).
107. K. F. Hayes and L. E. Katz, in *Physics and Chemistry of Mineral Surfaces* (P. V. Brady, ed.), CRC Press, Boca Raton, FL, 1996, p. 147.
108. P. O'Day, C. J. Chisholm-Brause. S. N. Towle, G. A. Parks, and G. E. Brown. Geochim. Cosmochim. Acta 60:2515 (1996).
109. K. Xia, A. Mehadi, R. W. Taylor, and W. F. Bleam. J. Coll. Int. Sci. 185:252 (1997).
110. F. J. Weesner and W. F. Bleam. J. Coll. Int. Sci. 196:79 (1997).
111. W. A. House. J. Coll. Int. Sci. 163:379 (1994).
112. M. I. Tejedor-Tejedor and M. A. Anderson. Langmuir 2:203 (1986).
113. P. Venema, T. Hiemstra, and W. H. van Riemsdijk. J. Coll. Int. Sci. 181:45 (1996).
114. D. P. Rodda, B. B. Johnson, and J. D. Wells. J. Coll. Int. Sci. 184:365, 564 (1996).
115. M. Kosmulski, paper in preparation.

116. M. Kosmulski. Coll. Surf. *117*:201 (1996).
117. C. Tiffreau, J. Lutzenkirchen, and P. Behra. J. Coll. Int. Sci. *172*:82 (1995).
118. N. Marmier, J. Dumonceau, J. Chupeau, and F. Fromage. C. R. Acad. Sci. Paris. Ser. *317*:311 (1993). L. Gunneriusson. J. Coll. Int. Sci. *163*:484 (1994).
119. M. Kosmulski et al., J. Coll. Int. Sci. *211*:410 (1999).
120. L. E. Katz and K. F. Hayes. J. Coll. Int. Sci. *170*:491 (1995).
121. K. J. Farley, D. A. Dzombak, and F. M. Morel. J. Coll. Int. Sci. *106*:226 (1985).
122. W. H. Casey. J. Coll. Int. Sci. *163*:407 (1994).
123. N. Kallay and S. Zalac. Croat. Chem. Acta *67*:467 (1994).
124. L. G. Fokkink. Ph.D. thesis. Agricultural University of Wageningen, 1987.
125. M. Kosmulski. J. Coll. Int. Sci. *192*:215 (1997).
126. L. E. Katz and K. F. Hayes. J. Coll. Int. Sci. *170*:477 (1995).
127. D. G. Hall. Langmuir *13*:91 (1997).
128. M. L. Machesky, in *Chemical Modeling in Aqueous Systems II* (D. C. Melchior and R. L. Bassett, eds.), ACS Symp. Ser. *416*:282 (1990).
129. D. B. Kent and M. Kastner. Geochim. Cosmochim. Acta *49*:1123 (1985).
130. K. F. Hayes and J. O. Leckie. J. Coll. Int. Sci. *115*:564 (1987).
131. G. W. Bruemmer, J. Gerth, and K. G. Tiller. J. Soil Sci. *39*:37 (1988).

5
Enthalpy of Surface Charging

NIKOLA KALLAY, TAJANA PREOČANIN, and SUZANA ŽALAC Faculty of Natural Sciences and Mathematics, University of Zagreb, Zagreb, Croatia

I.	Introduction	225
	A. Stoichiometry of chemical reactions	226
II.	Surface Reactions	227
	A. Mechanism of surface charging	228
	B. Counterion association	230
	C. Zero charge condition	232
III.	Enthalpy of Surface Charging Reactions	233
	A. Temperature dependency of the p.z.c.	233
	B. Calorimetry	236
	References	246

I. INTRODUCTION

The thermochemistry of surface reactions provides important information on the interactions of ionic species at an interface. The thermodynamics of interfacial reactions is rigorously treated in Chapter 3 of this book. This chapter is devoted to the interpretation of experimental data, which is not simple. The results are usually expressed in an empirical way as "enthalpy per adsorbed amount," and the mutual effects of adsorbed molecules are often neglected. In addition, an adsorption process may be accompanied by several others, either at the surface or in the bulk of the solution, so that "enthalpy per adsorbed amount" has no exact physical meaning. These problems are especially present in the case of surface–ionic interactions, since the binding of an ion to the surface influences all other equilibria at the interface through the electrostatic effects. For example, in the case of a metal oxide aqueous interface, several equilibria take place and influence each other. Thus the measured heat is not a consequence of one reaction only, so that the enthalpy of a particular surface reaction cannot be simply evaluated. Another problem is related to the separation of the electrostatic effects that is necessary in evaluation of the standard thermodynamic quantities.

Since the interpretation of the calorimetric data of ionic surface reactions requires a more rigorous (and tedious) treatment than that for solutions, the basic terms related to the reaction mechanism and the thermodynamics of chemical reactions will be reviewed, according to the IUPAC recommendations [1].

A. Stoichiometry of Chemical Reactions

The chemical process is presented by a reaction equation. Generally it can be written in the form

$$a\mathrm{A} + b\mathrm{B} + \cdots \rightarrow g\mathrm{G} + h\mathrm{H} + \cdots \qquad (1)$$

which means that a molecules of A and b molecules of B (and other reactant molecules) are transformed into g molecules of G and h molecules of H (and other products). A reaction equation is not an equation in the usual sense; it does not include only numbers, as mathematical equations do, or values of physical quantities, like physical equations. It is a "chemical" equation including species like molecules or ions. The term equation means that the number of atoms (free or bound) on the reactant side is equal to the number of atoms on the product side. Physical processes can also be presented by a reaction equation. For example, the melting of ice could be presented by a reaction equation denoting the transformation of one water molecule in the crystalline state to one water molecule in the liquid state. There are other less quantitative descriptions of chemical processes (not equations) indicating that a certain reactant is transformed into a certain product. In reaction equations the chemical symbols do not represent substances but chemical species, as molecules and ions. In a reaction equation the numbers of reactant and product species involved in one chemical transformation are denoted. In Eq. (1), a, b, g, and h are numbers of molecules. According to IUPAC, the stoichiometric number (v_i) is defined so that it is positive for products but negative for reactant species. For reaction (1),

$$v_\mathrm{A} = -a \qquad v_\mathrm{B} = -b \qquad v_\mathrm{G} = g \qquad v_\mathrm{H} = h \qquad (2)$$

The above definition of stoichiometric number enables us to write the reaction equation in a general way as

$$r = \sum_i v_i B_i = v_1 B_1 + v_2 B_2 + v_3 B_3 + \cdots \qquad (3)$$

where r is a general symbol for a chemical reaction representing one chemical transformation and B is a general symbol for a species involved in the reaction. The important issue is to consider a chemical reaction equation as a symbolic representation of a corresponding chemical transformation. Since transformations can be in principle counted, one can define the number of these transformations. Due to the reversibility one needs to distinguish between the number of forward, $N(r_\rightarrow)$, and backward, $N(r_\leftarrow)$, transformations, so that the number of net transformations, $N(r)$, which is of interest in thermochemistry, is given by

$$N(r) = N(r_\rightarrow) - N(r_\leftarrow) \tag{4}$$

Since the number of molecules in real systems, and consequently of chemical transformations, is high, it is a custom to divide it by the Avogadro constant, which results in the amount of molecules or amount of transformations. Previously these quantities did not have a specific name; they are simply called "number of moles." (By analogy one would use an expression like "number of ångstroms" instead of the bond length.) Now IUPAC recommends the term "amount" for this purpose, so that one can express the advancement of a chemical reaction by the amount of defined transformations, for which the term "extent of reaction" with symbol ξ was introduced [1].

For the same process one can write different reaction equations with respect to the mechanism, stoichiometric coefficients, and direction. Consequently, the values of physical quantities, such as reaction extent, enthalpy, and equilibrium constant, would depend on the way the reaction equation is written. The change in the amount of any component in the closed reaction system (Δn_B) is directly related to the extents ($\Delta \xi_i$) of all present reactions i. For species B,

$$\Delta n_B = \sum_i \nu_{B,i} \Delta \xi_i \tag{5}$$

The change of the enthalpy of a reaction system (ΔH) depends on the extents of reactions in a similar way:

$$\Delta H = \sum_i \Delta_r H_i \Delta \xi_i \tag{6}$$

where $\Delta_r H_i$ is (molar) reaction enthalpy of the ith reaction. Generally, (molar) reaction quantities are defined as a change of respective quantity for the reaction system "per amount of transformations." Accordingly, the (molar) reaction enthalpy would be defined as

$$\Delta_r H = \frac{dH}{d\xi} = \frac{H}{\Delta \xi} \tag{7}$$

Note that the above definition is valid in the case of only one reaction taking place in the system. In the case of more than one simultaneous or consecutive reaction, the use of the above definition will result in the sum of reaction enthalpies of all present reaction, provided that their extents are equivalent.

II. SURFACE REACTIONS

There are several proposed mechanisms of surface charging of solids as i.e. metal oxides in an aqueous environment. All of them describe interactions of (active) surface sites with potential determining ions. In this section the main approaches will be reviewed. They can be described as (1) binding of potential determining ions [2,3], (2) protonation and deprotonation of amphoteric surface sites [4,5], (3) two-step protonation [6], and (4) the 1-pK model [7]. The first three approaches can be considered as different modifications of the 2-pK model. Note that the meaning of the terms 1- and 2-pK model is related to one or two surface reactions and consequently equilibrium constants being responsible for charging of the surface [8,9].

A. Mechanism of Surface Charging

The mechanism of surface charging will be demonstrated by the example of metal oxides.

1. Binding of Potential Determining Ions

Adsorption of potential determining ions for the case of metal oxides can be represented by the binding of H^+ and OH^- ions to surface groups $\equiv S$. Binding of protons (reaction bp) is described by

$$\equiv S + H^+ \rightarrow \equiv SH^+ \qquad K_{bp}, \Delta_{bp}H, \Delta\xi_{bp} \tag{8}$$

while for binding of hydroxide ions (reaction bh) the following reaction equation is used:

$$\equiv S + OH^- \rightarrow \equiv SOH^- \qquad K_{bh}, \Delta_{bh}H, \Delta\xi_{bh} \tag{9}$$

Each reaction is characterized by the change in reaction extent ($\Delta\xi$), enthalpy (ΔH), and the equilibrium constant (K). The above approach is rarely used for metal oxides but is representative for other salts like silver iodide for which potential determining ions are Ag^+ and I^- ions.

2. Protonation and Deprotonation of Amphoteric Surface Sites

It is assumed that hydration of the surface results in amphoteric $\equiv MOH$ groups (M denotes a metal atom). Surface protonation (reaction p) is

$$\equiv MOH + H^+ \rightarrow \equiv MOH_2^+ \qquad K_p, \Delta_p H, \Delta\xi_p \tag{10}$$

while surface deprotonation (reaction d) is

$$\equiv MOH \rightarrow \equiv MO^- + H^+ \qquad K_d, \Delta_d H, \Delta\xi_d \tag{11}$$

3. Two-Step Protonation

This approach is similar to the latter one and differs only in the direction of one of the processes. The first protonation step (reaction p1),

$$\equiv MO + H^+ \rightarrow \equiv MOH \qquad K_{p1}, \Delta_{p1}H, \Delta\xi_{p1} \tag{12}$$

is opposite to reaction d, Eq. (11), while the second protonation step (reaction p2),

$$\equiv \text{MOH} + \text{H}^+ \rightarrow \equiv \text{MOH}_2^+ \qquad K_{p2}, \Delta_{p2}H, \Delta\xi_{p2} \qquad (13)$$

is identical to reaction p, Eq. (10).

All the above three stoichiometries can be used for the description of the same process, and the relationships between corresponding reaction quantities are simple:

$$\Delta_p H = \Delta_{bp} H = \Delta_{p2} H \qquad (14)$$
$$\Delta_d H = \Delta_{bh} H - \Delta_n H = -\Delta_{p1} H \qquad (15)$$

(where $\Delta_n H$ is the enthalpy of bulk neutralization)

$$\text{OH}^- + \text{H}^+ \rightarrow \text{H}_2\text{O} \qquad K_n(=1/K_w), \Delta_n H, \Delta\xi_n \qquad (16)$$

which should be considered when deprotonation, Eq. (11), is compared to the binding of OH^- ions, Eq. (9). According to the surface complexation model (SCM) [10], the electrostatic effect on the surface equilibrium is considered through an exponential term including electrostatic potential at the 0-plane (ϕ_0), in which charged surface groups, $\equiv\text{MOH}_2^+$ and $\equiv\text{MO}^-$, are located. For protonation and deprotonation, the equilibrium constants are defined in terms of bulk activities (a) and surface concentrations (Γ) as

$$K_p = \exp\left(\frac{F\phi_0}{RT}\right) \frac{\Gamma(\text{MOH}_2^+)}{a(\text{H}^+)\,\Gamma(\text{MOH})} \qquad (17)$$

and

$$K_d = \exp\left(\frac{F\phi_0}{RT}\right) \frac{\Gamma(\text{MO}^-)\,a(\text{H}^+)}{\Gamma(\text{MOH})} \qquad (18)$$

These constants are directly related to other constants corresponding to the stoichiometries written in different ways:

$$K_p = K_{p2} = K_{bp} \qquad (19)$$
$$K_d = \frac{1}{K_p} = \frac{K_{bh}}{K_n} (= K_{bh} K_w) \qquad (20)$$

4. 1-pK Model

A quite different mechanism of surface charging is described by the so-called 1-pK model. This approach does not assume electrically neutral amphoteric surface groups, so that the surface equilibrium is described by

$$\equiv M(OH)_n^{x/y} + H^+ \rightarrow \equiv M(OH)_n H^{1+x/y} \qquad K_{x/y}, \Delta_{x/y}H, \Delta\xi_{x/y} \qquad (21)$$

$$K_{x/y} = \exp\left(\frac{F\phi_0}{RT}\right) \frac{\Gamma(M(OH)_n H^{1+x/y})}{a(H^+)\,\Gamma(M(OH)_n H^{x/y})} \qquad (22)$$

Charge numbers of surface groups are x/y and $(1 - x/y)$ and depend on the coordination of metal atoms in metal oxides.

In the case of hematite, $x/y = 1/2$, so that the surface equilibrium is described by

$$\equiv Fe(OH)^{-1/2} + H^+ \rightarrow \equiv Fe(OH)H^{+1/2} \qquad K_{1/2}, \Delta_{1/2}H, \Delta\xi_{1/2} \qquad (23)$$

$$K_{1/2} = \exp\left(\frac{F\phi_0}{RT}\right) \frac{\Gamma(Fe(OH)^{1/2})}{a(H^+)\,\Gamma(Fe(OH)H^{-1/2})} \qquad (24)$$

B. Counterion Association

Even in the most simple systems, like metal oxides dispersed in aqueous solution of strong (neutral) electrolytes, the counterions may get associated with the oppositely charged surface groups [11]. In the case of so-called neutral electrolytes, the bond is electrostatic, and the association is enabled by attractive forces between the two interacting charges and influenced by overall electrostatic potential. If the protonation–deprotonation mechanism of charging is assumed, the association of negative sites with counterions C^+ (reaction c) is represented by

$$\equiv MO^- + C^+ \rightarrow \equiv MO^- \cdot C^+ \qquad K_c, \Delta_c H, \Delta\xi_c \qquad (25)$$

while the association of positive sites with counterions A^- (reaction a) is

$$\equiv MOH_2^+ + A^- \rightarrow \equiv MOH_2^+ \cdot A^- \qquad K_a, \Delta_c H, \Delta\xi_a \qquad (26)$$

The equilibrium constants of these reactions can be defined according to the SCM as

$$K_c = \exp\left(\frac{F\phi_\beta}{RT}\right) \frac{\Gamma(MO^- \cdot C^+)}{a(C^+)\,\Gamma(MO^-)} \qquad (27)$$

and

$$K_a = \exp\left(\frac{-F\phi_\beta}{RT}\right) \frac{\Gamma(MOH_2^+ \cdot A^-)}{\Gamma(MOH_2^+)\,a(A^-)} \qquad (28)$$

where ϕ_β denotes the electrostatic potential affecting the state of the associated counterions, i.e., of the β-plane in which the centers of bound counterions are located. The relationship between ϕ_0 and ϕ_β involves capacity depending on the kind of counterion.

Enthalpy of Surface Charging

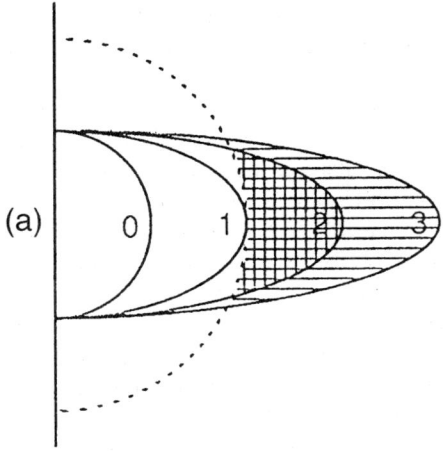

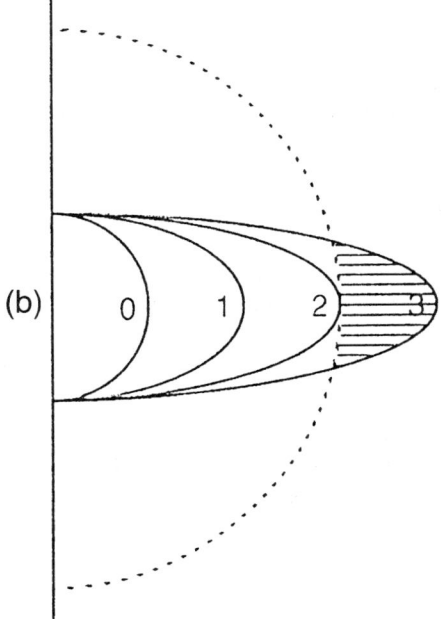

FIG. 1 Schematic presentation of critical boundaries of counterion association space around a fixed central surface charged group for gradually increasing surface potentials [13,14]. Dashed lines are minimum distances between centers of central ion and counterion for relatively small (a) and large (b) counterions. The numbers represent the values of the surface potential on the arbitrary scale. The shadowed area in the association space: for relative potential 2, vertical shading, and for relative potential 3, horizontal shading.

The above treatment indicates that the association of counterions is present at any condition, i.e., at any pH, but is noticeable when the potential is high enough to attract the counterions from the solution.

Another treatment is based on the Boltzmann statistics and is similar to that applied by Bjerrum [12] for ionic association in the bulk of the electrolyte solution [13,14]. In the case of surface association, half of the space should be disregarded, since counterions cannot be distributed within the solid phase. Also, in addition to the coulombic potential between two interacting charges, the effect of the overall potential in the electrical interfacial layer should be considered. This approach indicates that association may take place when the surface potential is high enough, which depends on the closest distance of interacting charges. Accordingly, the theory predicts the absence of association in the pH region around the zero charge condition, and pronounced association for effectively smaller counterions (lyotropic effect). Above the critical surface potential (pH) the association is promoted by further increase of potential (pH), as in the case of the SCM. The schematic presentation of the association space for small and large counterions is given in Fig. 1.

C. Zero Charge Condition

The zero charge condition is expressed by two quantities, the point of zero charge (p.z.c.) and the isoelectric point (i.e.p.). According to the protonation–deprotonation mechanism, the p.z.c. is defined as

$$\Gamma(\text{MO}^-) + \Gamma(\text{MOH}_2^+ \cdot \text{A}^-) = \Gamma(\text{MO}^- \cdot \text{C}^+) + \Gamma(\text{MOH}_2^+) \tag{29}$$

At low ionic strength and also at low surface potentials the association of counterions is negligible, so that Eq. (29) is reduced, and also the potential at the surface becomes zero:

$$\Gamma(\text{MO}^-) = \Gamma(\text{MOH}_2^+) \qquad \phi_0 = 0 \tag{30}$$

The latter condition ($\phi_0 = 0$) requires the absence of specific adsorption of other ions, but it may be applied for the case of counterion association assuming that $\Gamma(\text{MOH}_2^+ \cdot \text{A}^-) \approx \Gamma(\text{MO}^- \cdot \text{C}^+)$. In such a case the p.z.c. coincides with the electrokinetic i.e.p.

According to relationships (17) and (18), the point of zero charge is given by the protonation and deprotonation equilibrium constants

$$\text{pH}_{\text{pzc}} = \frac{1}{2} \log \frac{K_p}{K_d} \tag{31}$$

By using relationships (19) and (20), one can simply express pH_{pzc} by the equilibrium constants corresponding to other 2-pK stoichiometries. Somehow a different situation exists when the 1-pK model is applied. For the general case of the 1-pK mechanism, Eq. (21),

$$\text{pH}_{\text{pzc}} = \log K_{x/y} - \log \frac{x}{x-y} \tag{32}$$

In the case of hematite, $x/y = -1/2$ so that $\text{pH}_{\text{pzc}} = \log K_{1/2}$.

The above treatment enables us to relate equilibrium constants for two different assumed mechanisms of the charging process, i.e., for 1-pK and 2-pK models:

$$K_{x/y} = \frac{1}{2}\log\frac{K_p}{K_d} + \log\frac{x}{x-y} \tag{33}$$

III. ENTHALPY OF SURFACE CHARGING REACTIONS

In principle, the enthalpy of any chemical reaction could be obtained directly by measuring the heat transferred from the reaction system to the surroundings at constant temperature and pressure. Reaction enthalpy could be also obtained by measuring the temperature dependency of an equilibrium parameter such as the equilibrium constant of a given reaction. As will be shown later, measurement of the temperature dependency of the point of zero charge provides standard reaction enthalpy and entropy, the values of which depend on how the reaction stoichiometry is expressed. The calorimetric experiments for surface reactions associated with charging of the interface cannot be simply interpreted. The first problem is to distinguish between contributions of several reactions taking place at the interface and also in the bulk of the solution, while the second is related to the electrostatic effects on the enthalpy. In the forthcoming sections the solutions of these problems will be presented.

A. Temperature Dependency of the p.z.c.

The interpretation of measurements of the temperature dependency of the point of zero charge is relatively simple [15–24]. In the absence of ions that would specifically adsorb, the p.z.c. coincides with the i.e.p., enabling application of both potentiometric and electrokinetic methods.

There are several techniques developed for the determination of the isoelectric point and of the point of zero charge. The common method for i.e.p. determination is electrophoresis. The problem associated with this method is shear plane adjustment, and the fact that most of the devices do not ensure accurate regulation of the temperature. Potentiometric acid–base titration of metal oxide suspensions yields the p.z.c. as the intercept of the surface charge function at different ionic strengths. The problem with this method lies in that the intersection of two functions that are not steep cannot be determined with high accuracy. The pH shift method is based on the same phenomenon: the pH of a suspension that does not change by addition of electrolyte is taken as the p.z.c. Both experimental approaches cannot avoid comparison of pH values at low and high neutral electrolyte concentrations, which causes two problems. First, counterion association cannot be simply neglected at high ionic strength, and second, which is even more important, the potential of the reference electrode depends slightly on the ionic strength, thus influencing the pH readings. Accordingly, the same electromotive force at low and high ionic strength does not correspond to the same activity of H^+ ions.

The mass titration method [25], introduced recently, avoids the above problems. The basis of this method lies in that the pH of the suspension approaches the point of zero charge as the concentration of the solid phase increases. This is true only for pure samples that do not contain any acid or base. The advantage of the mass titration method is that one may perform experiments at constant and very low ionic strength, so that problems

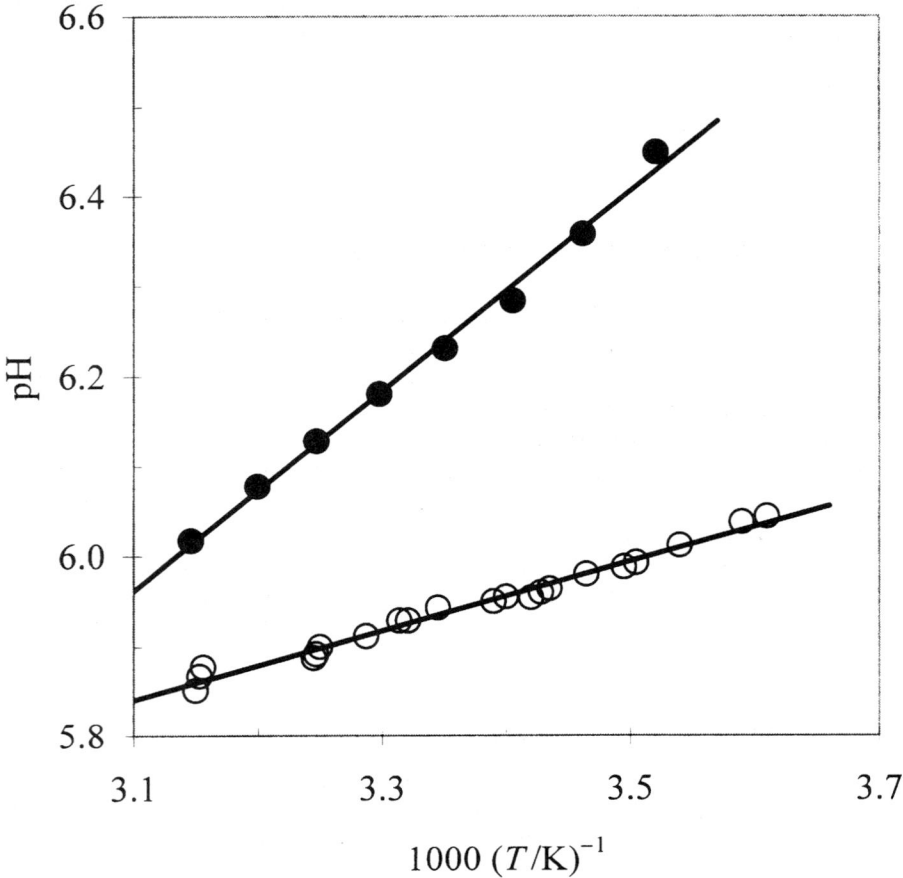

FIG. 2 Temperature dependency of the point zero charge of hematite (●) and anatase (○). Plot according to Eq. (35). For hematite [38]: $\Delta_{ch}H° = 42.5$ kJ mol^{-1}; $\Delta_{ch}S° = 96.4$ kJ mol^{-1} K^{-1}. For anatase [34]: $\Delta_{ch}H° = 14.6$ kJ mol^{-1} $\Delta_{ch}S° = 178.2$ kJ mol^{-1} K^{-1}.

of counterion association and the reference electrode disappear. Measurement of the temperature dependency of the p.z.c. is also simple [22–24], one prepares a concentrated suspension of a purified metal oxide. The purity could be also checked on the basis of the mass titration method [26] (acid–base titration of a concentrated suspension). Ionic strength could be kept as low as e.g., 10^{-5} mol dm^{-3}, depending on the pH$_{pzc}$. The system should be kept under an inert atmosphere and thermostatted. One changes the temperature of the same sample and measures the pH. In doing so one should calibrate electrodes at each temperature. The equilibration time can be reduced by the application of ultrasound.

As a result, one obtains the pH$_{pzc}$ values at several temperatures. The interpretation of data is based on a general expression including standard thermodynamic quantities:

$$-RT \ln K_r° = \Delta_r G° = \Delta_r H° - T\Delta_r S° \tag{34}$$

Enthalpy of Surface Charging

Combining the above expression with the relationship between the pH_{pzc} and the equilibrium constants, Eq. (31), of protonation, Eq. (10), and deprotonation, Eq. (11), reactions one obtains

$$pH_{pzc} = \frac{\Delta_d H^\circ - \Delta_p H^\circ}{2RT \ln 10} - \frac{\Delta_d S^\circ - \Delta_p S^\circ}{2R \ln 10} \tag{35}$$

so that the linear plot of pH_{pzc} vs. $1/T$ provides $(\Delta_d H^\circ - \Delta_p H^\circ)$ from the slope and $(\Delta_d S^\circ - \Delta_p S^\circ)$ from the intercept. It is clear that the temperature dependency of the point of zero charge cannot provide separate enthalpy values for protonation and deprotonation reactions, but rather their difference. If another stoichiometry for the same process is assumed, when expressing the results one needs to take into account that

$$\Delta_d H^\circ - \Delta_p H^\circ = -\Delta_{p1} H^\circ - \Delta_{p2} H^\circ = \Delta_{bh} H^\circ - \Delta_n H^\circ - \Delta_{bp} H^\circ \tag{36}$$

With respect to the above approach, the 1-pK model considers a quite different mechanism. It is one-step protonation of negatively charged groups. Also, this model assumes the absence of neutral surface groups; they are either negatively or positively charged. According to Eq. (32), the temperature dependency of the p.z.c. is given by

$$pH_{pzc} = \frac{-\Delta_{xy} H^\circ}{RT \ln 10} + \frac{\Delta_{xy} S^\circ}{RT \ln 10} - \log \frac{x}{x - y} \tag{37}$$

The slope in presentation pH_{pzc} vs. $1/T$ yields the (molar) enthalpy for the assumed 1-pK mechanism, which is related to the enthalpy for the 2-pK mechanism by

$$\Delta_d H^\circ - \Delta_p H^\circ = -2\Delta_{x/y} H^\circ \tag{38}$$

As a contrast to enthalpy, the evaluation of entropy associated with the 1-pK mechanism requires knowledge of the charge numbers of the surface groups, i.e. of the x and y values. Accordingly, the following relationship holds:

$$\Delta_d S^\circ - \Delta_p S^\circ = -2\Delta_{x/y} S^\circ + 2R \ln \frac{x}{x - y} \tag{39}$$

Figure 2 represents the temperature dependency of the point of zero charge for hematite and anatase as obtained by the mass titration method. The temperature dependency of the point of zero charge of the metal oxide–aqueous interface and other solid/liquid interfaces provides standard thermodynamic quantities such as standard reaction enthalpy and standard reaction entropy. In the case of enthalpy the term "standard" means the ideal state, namely the absence of electrostatic interactions at the surface, while for entropy the term standard is also related to the choice of standard states such as the hypothetical "ideal one-molar solution." For described mechanisms of surface reactions, the question of standard states of interfacial species disappears, since the chosen values or reference surface concentrations cancel. However, it implies the absence of electrostatic effects, which is by definition achieved at the point of zero charge. The temperature dependency of the point of zero charge did not enable us to distinguish between deprotonation and protonation processes; only the difference could be obtained.

B. Calorimetry

Calorimetry is considered as a direct method for the determination of reaction enthalpies. If the experiment is performed at constant pressure, and if the temperature of the reaction system is the same before and after completion of the chemical process, the heat exchanged with the surroundings corresponds to the change of the system enthalpy. Commercial calorimeters do not operate in such a mode, but the change of the enthalpy of the system can still be evaluated from the temperature change in the system. The interpretation of calorimetric data requires some corrections such as for dilution heats. Another problem is to derive an intensive quantity from the extensive one, i.e., to obtain the reaction enthalpy from the change of the enthalpy of the reaction system. For reactions in a homogeneous phase, the procedure is usually uniform and clear. Authors would follow the practice of expressing the enthalpy "per mole of a product" and more recently, according to IUPAC recommendations, "per mole of reaction," which means per extent of reaction, as described by the reaction equation [1]. These two procedure are consistent provided that only one reaction takes place in the calorimeter and also that consecutive reactions proceed by the same extent. In the area of surface chemistry, the situation is more complicated. If one expresses the enthalpy "per mole of adsorbed species," the physical meaning of such an expression may be not clear. For example, in the system where adsorption takes place, usually some other reactions (either at the interface or in the bulk of the solution) may have changed their equilibrium state with extents different from that of the adsorption process. Even in the case of equal extents, one should not simply attribute the enthalpy to the adsorption only. These problems are especially noticeable in the case of reactions at the metal oxide–aqueous interface where several surface reactions change their equilibrium, due to the addition of acid or base to the suspension. The reason for the troubles is that surface reactions do not proceed quantitatively but rather with different extents depending on their equilibrium constants and the bulk concentrations of interacting species. For example, if the system undergoes the 2-pK mechanism, one needs to consider protonation, Eq. (10), deprotonation, Eq. (11), association of cations, Eq. (25), and of anions, Eq. (26). In addition, in the bulk of the solution the change in the water ionization equilibrium takes place. Therefore, if one expresses the enthalpy "per amount of consumed H^+ ions" or "per surface charge," the result cannot be simply related to the reaction quantities characterizing a particular surface process. The additional task is to deal with the electrostatic effects on the enthalpy of surface reactions. Therefore it is not surprising that, despite its importance, only a few reports on the enthalpy of interfacial reactions have been published in the literature [27–38,50]. In the forthcoming section, the design of a calorimetric experiment that avoids the problems associated with counterion association and electrostatic effects will be presented. For these experiments, the term "symmetric calorimetric experiment" will be used [34]. Also, calorimetric experiments (the results of which are influenced by electrostatic effects) will be described in detail, together with the interpretation of the data [38].

1. Surface Potential

Usually, the meaning of the term "surface potential" depends on the context. In the case of colloid stability, when electrostatic interaction energy between two approaching surfaces is calculated, by "surface potential" one understands the electrostatic potential at the onset of the diffuse layer ϕ_d. However, when the electrostatic effect on the enthalpy of surface reactions in the 0-plane is considered, the term "surface potential" applies

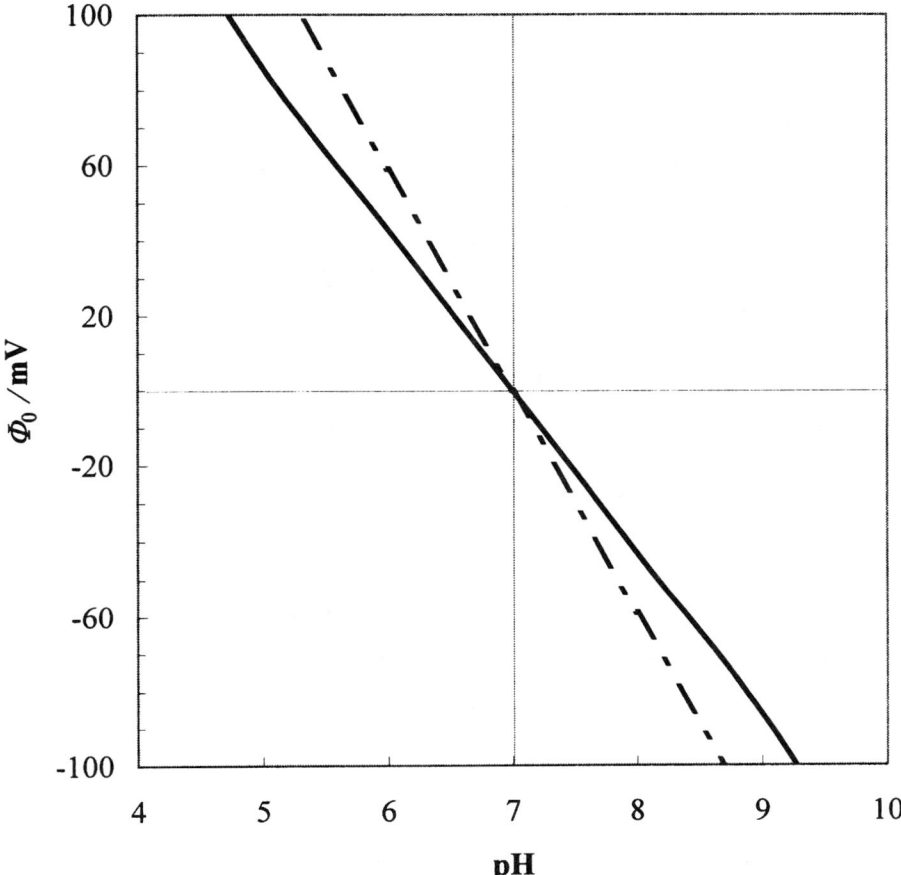

FIG. 3 Dependency of potential ϕ_0 on pH. The solid line is obtained by using the surface complexation model [11]. The slope was obtained as -43 mV, i.e. $\alpha = 0.73$. In calculations the following values of parameters were used: $K_p = 10^7$, $K_d = 10^{-9}$ (pH$_{pzc} = 7$), $\Gamma_{tot} = 10^{-5}$ mol m^{-2}, $T = 298$K, $\epsilon_r = 78.5$, $I_c = 5 \times 10^{-3}$ mol dm^{-3}. The capacitances of the first and second layer were 1.5 F m^{-2} and 0.2 F m^{-2}, respectively (see Chapter 6.). For comparison the dashed line represents Nernstian behavior with a slope of $-RT \ln 10/F$.

to ϕ_0. It is commonly accepted that ϕ_d follows the electrokinetic ζ-potential, being somehow higher in magnitude, and also that ϕ_0 is close to the Nernst potential. In this section the ϕ_0(pH) function will be analyzed on the basis of the SCM [39].

According to Eqs. (17), (18), and (31), the pH dependence of the surface potential is given by

$$\phi_0 = \frac{RT \ln 10}{zF}(\text{pH}_{pzc} - \text{pH}) - \frac{RT \ln 10}{zF} \log \frac{\Gamma(\text{MOH}_2^+)}{\Gamma(\text{MO}^-)} \tag{40}$$

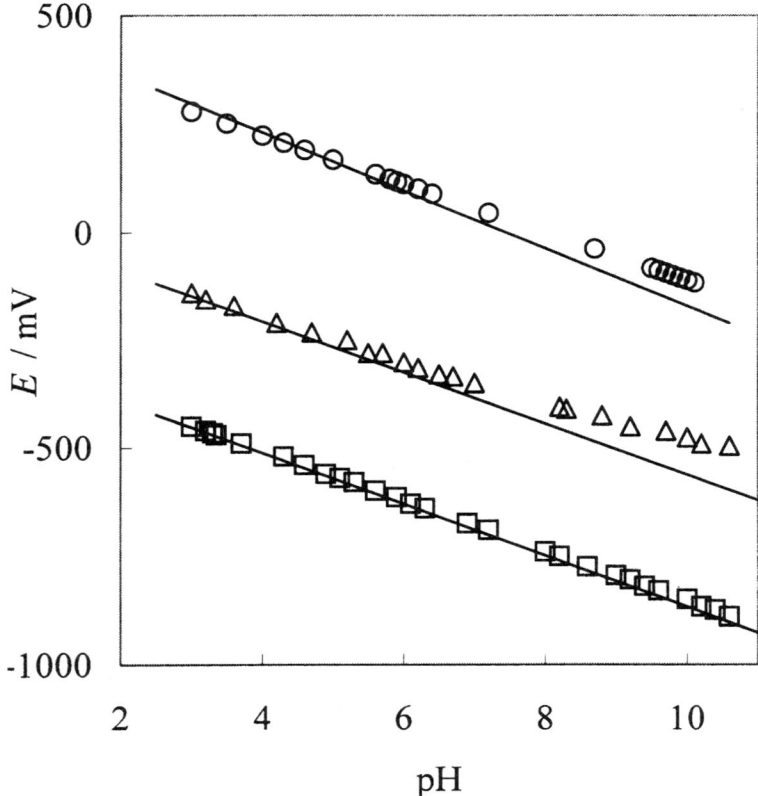

FIG. 4 Dependency of electrode potential on pH at 25°C and ionic strength of 0.1 mol dm^{-3} for different electrodes [11]: commercial glass electrode filled with mercury (□), electrode made from ordinary laboratory glass filled with mercury (○); electrode made from ordinary laboratory glass filled with HCl with internal AgCl/Ag electrode (Δ); solid lines represent the Nernstian slope of -59.2 mV.

If surface concentrations of MOH_2^+ and MO^- are very high with respect to the their difference, i.e. with respect to the surface charge density, the ratio $\Gamma(MOH_2^+) / \Gamma(MO^-)$ will be close to 1 in the whole pH region, so that ϕ_0 corresponds to the Nernstian potential [10,39–41]. Such conditions apply to oxides with high K_p and K_d values, and are characterized by high concentrations of active surface sites. Accordingly the surface potential can be approximated by

$$\phi_0 = \frac{RT \ln 10}{zF}(pH_{pzc} - pH)\alpha \tag{41}$$

The coefficient α describes the deviation from the ideal Nernstian behavior and is equal to 1 in the ideal case. For real systems, the expected values of α are below 1 and depend on the kind of oxide. For example, adsorption measurements on rutile [42] resulted in $\alpha = 0.69$, which was later confirmed by direct measurement with a Ti/TiO$_2$ electrode [43]. For hematite, the value of α, obtained by interpretation of adsorption measurements [44], was found to be 0.8 and 0.75 (see Chapter 6). Potentiometric measurements with a platinum electrode covered by hematite [45] suggest "ideal" behavior of the hematite surface, i.e., the slope of -59 mV/pH at 20°C. This finding can be explained if one considers the porosity of the hematite layer enabling direct contact of solution with metallic surface. Another possible explanation lies in the finding of Lyklema who suggested [46,47] a "porous double layer concept" for hematite, which means that surface charge is not created at a plane but rather within the porous solid layer. If so, the density of active surface sites becomes markedly larger, so that the interface approaches the "ideal" Nernstian behavior. The problem of the surface potential at the 0-plane was extensively discussed by Lyklema [48]. Figure 3 demonstrates that the surface complexation model leads to ϕ_0 dependency on pH with the slope lower than would be expected on the basis of the Nernst equation.

As mentioned above, direct measurements of surface potential require a nonporous layer of an oxide being in contact with a conductor on one side and with the electrolyte solution on the other side. Such a condition was achieved with glass by pouring mercury into the glass bulb. The results with different kinds of glasses are presented in Fig. 4. It can be seen that the glass surface exhibits properties close to the Nerstian, especially in the case of glass used for the construction of pH-electrodes. Such a finding was expected, since glass in contact with water forms a kind of gel layer with an extremely large number of active sites. In addition, preliminary experiments [49] with an "ice electrode" also showed behavior close to the Nernstian.

2. Extent of Surface Reactions

The change in amounts (moles) of H$^+$ ions, $\Delta n(\text{H}^+)$, due to the addition of small portions of acid or base to the suspension in the calorimeter, is related to the extents of protonation, deprotonation, and bulk neutralization and could be obtained from initial and final pH values pH$_1$ and pH$_2$ ($\Delta n = n_2 - n_1$):

$$\Delta n(\text{H}^+) = \Delta n(\text{HNO}_3) - \Delta \xi_p + \Delta \xi_d - \Delta \xi_n = c^o \left(\frac{V_2 10^{-\text{pH}_2}}{y_2} - \frac{V_1 10^{-\text{pH}_1}}{y_1} \right) \quad (42)$$

where $\Delta n(\text{HNO}_3)$ is the change in the amount of added acid (if added), V is the volume of suspension after each addition, and y is the activity coefficient calculated by the Debye–Hückel law of electrolyte solution.

Change in amounts (moles) of OH$^-$ ions in the bulk of the solution $\Delta n(\text{OH}^-)$ could be obtained from pH measurements and is related to the extent of bulk neutralization by

$$\Delta n(\text{OH}^-) = K_w^o c^o \left(\frac{V_2 10^{-\text{pH}_2}}{y_2} - \frac{V_1 10^{-\text{pH}_1}}{y_1} \right) = \Delta n(\text{KOH}) - \Delta \xi_n \quad (43)$$

where $\Delta n(\text{KOH})$ is the amount of added base (if added).

From Eq. (43) one could calculate the extent of bulk neutralization (Eq. 16). The ratio between extents of protonation and deprotonation, D, is

$$D = \frac{\Delta \xi_p}{\Delta \xi_d} = \frac{\Delta \Gamma(\text{MOH}_2^+)}{\Delta \Gamma(\text{MO}^-)} \quad (44)$$

which is according to Eqs. (17) and (18) equal to

$$D = \frac{K_p[\Gamma_2(MOH)a_2(H^+)\exp(-\phi_{0_2}/RT) - \Gamma_1(MOH)a_1(H^+)\exp(-\phi_{0_1}/RT)]}{K_d[\Gamma_2(MOH)a_2(H^+)^{-1}\exp(\phi_{0_2}/RT) - \Gamma_1(MOH)a_1(H^+)^{-1}\exp(\phi_{0_1}/RT)]} \quad (45)$$

Since a small portion of acid or base is added to the suspension, the amount of neutral sites does not change significantly, so that the combination of Eqs. (31), (49), and (45) yields

$$D = -10^{2(1-\alpha)(\mathrm{pHpzc}-\overline{\mathrm{pH}})} \quad (46)$$

where $\overline{\mathrm{pH}}$ is the mean value of initial and final pH [$\overline{\mathrm{pH}} = (\mathrm{pH}_1 + \mathrm{pH}_2)/2$].

The difference between extents of protonation and deprotonation, B, is according Eqs. (42) and (43) equal to

$$B = \Delta\xi_p - \Delta\xi_d = \Delta n(\mathrm{HNO_3})$$
$$- \Delta n(\mathrm{KOH}) - c^\circ \left(\frac{V_2(10^{-\mathrm{pH}_2} - K_w^\circ 10^{\mathrm{pH}_2})}{y_2} - \frac{V_1(10^{-\mathrm{pH}_1} - K_w^\circ 10^{\mathrm{pH}_1})}{y_1} \right) \quad (47)$$

Equations (43), (45), and (46) enable the separation of extents of protonation and deprotonation by using

$$\Delta\xi_p = \frac{BD}{D-1} \quad (48)$$
$$\Delta\xi_d = \frac{B}{D-1} \quad (49)$$

3. Electrostatic Effect on Thermodynamic Quantities

Generally, the effects due to charging of the interface can be separated into chemical and electrostatic parts, so that the Gibbs energy of a particular surface reaction is the sum of a "chemical" (standard, $\Delta_r G^\circ$) and an electrostatic ($\Delta_r G_{el}$) contribution:

$$\Delta_r G = \Delta_r G^\circ + \Delta_r G_{el} \quad (50)$$

where the electrostatic contribution to the Gibbs energy of a surface reaction taking place at the 0-plane is related to surface potential by

$$\Delta_r G_{el} = -zF\phi_0 \quad (51)$$

Enthalpy of Surface Charging

In the above equation, z is the charge number of interacting surface groups. Accordingly, the enthalpy of surface charging is a sum of chemical (standard) and electrostatic contributions:

$$\Delta_r H = \Delta_r H° + \Delta_r H_{el} \tag{52}$$

The electrostatic contribution to the enthalpy is related to the electrostatic contribution to the Gibbs energy by

$$\Delta_r H_{el} = \Delta_r G_{el} + T \Delta_r S_{el} \tag{53}$$

and is not directly related to the surface potential.

4. Standard (Chemical) Enthalpy—Symmetric Calorimetric Experiment

It is possible to avoid electrostatic effects to the enthalpy of surface charging by performing the "symmetric" calorimetric experiment in which the point of zero charge is exactly the mean value of initial and final pH [34].

The interpretation of calorimetric data requires knowledge of the heat of neutralization so that the measured heat, Q, according to the 2-pK model, is related to the enthalpy of protonation and deprotonation reactions by

$$Q - \Delta_n H \Delta \xi_n = (\Delta_p H° + \Delta_p H_{el}) \Delta \xi_p + (\Delta_d H° + \Delta_d H_{el}) \Delta \xi_d \tag{54}$$

If addition of base (or acid) is chosen so that the mean of the initial and final pH values equals the p.z.c. ($\overline{pH} = pH_{pzc}$) or in the case of ideal Nernstian behavior $\alpha = 1$, according to Eqs. (44) and (46) the extents of protonation and deprotonation will be equal in magnitude but opposite in sign: $\Delta \xi_p = -\Delta \xi_d$. It can be assumed that for such a case the initial and final surface potential ($\phi_{0,1}$ and $\phi_{0,2}$) are equal in magnitude but opposite in sign, so that the electrostatic contributions to the enthalpy of protonation and deprotonation are related as

$$\Delta_p H_{el} = -\Delta_d H_{el} \tag{55}$$

Accordingly, the measured heat is equal to

$$Q - \Delta_n H \Delta \xi_n = (\Delta_d H° - \Delta_p H°) \Delta \xi_d = \Delta_{ch} H° \Delta \xi_d \tag{56}$$

According to the above relationship, by performing symmetric calorimetric experiments it is possible to obtain the difference between standard enthalpies of protonation and deprotonation, denoted as $\Delta_{ch} H°$, but distinguishing between these two quantities cannot be achieved. Symmetric calorimetric experiments were performed with anatase and silica and resulted in $\Delta_{ch} H° = 14.7$ kJ mol^{-1} (anatase) [34] and 34.4 kJ mol^{-1} (silica) [22]. These results were confirmed by measurements of the pH$_{pzc}$ dependence of temperature (14.6 kJ mol^{-1} and 32.5 kJ mol^{-1}, respectively) ref. Relations between values corresponding to different stoichiometries are described by Eqs. (14), (15), and (38).

5. Electrostatic Contribution to the Enthalpy—Calorimetric Titration

Outside the point of zero charge region, the electrostatic contribution [Eq. (41)] to the enthalpy of the surface reaction is expected to be significant. The calorimetric experiments in these regions are performed so that one titrates a suspension by adding subsequent

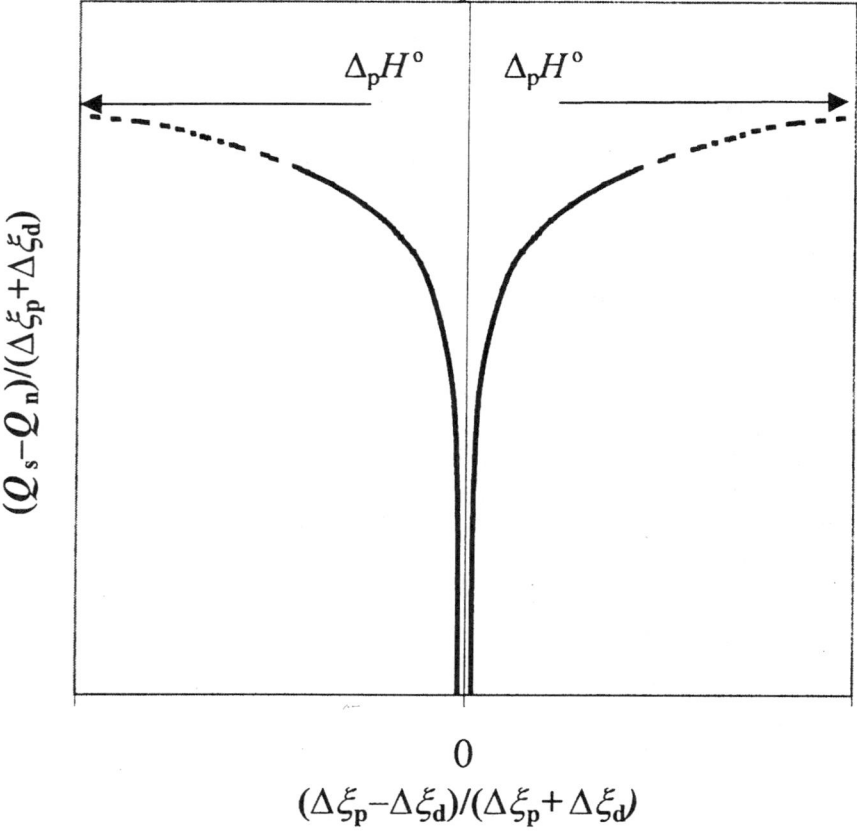

FIG. 5 Schematic presentation of the interpretation of calorimetric titration data according to Eq. (57).

portions of acid or base and measures corresponding enthalpy and pH change. In order to avoid coagulation in the isoelectric region one should perform two titration runs, one above and the other below the i.e.p.

The interpretation of the data requires knowledge of the heat of neutralization due to the reaction of H^+ and OH^- ions in the bulk of the solution, which can be obtained by separate calorimetric titration experiments in the absence of the solid phase, or it can be calculated from the extent of neutralization [Eq. (43)] using the known value of (molar) enthalpy of neutralization. To obtain the extents of surface reactions [Eqs. (42) and (43)] one determines the change in the amounts (moles) of H^+ ions in the bulk of the solution by pH measurements.

According to Eqs. (48), (49), and (54), the interpretation of the data could be based on the relationship

$$\frac{Q_s - \Delta_{ch}H^\circ \Delta \xi_d}{\Delta \xi_p + \Delta \xi_d} = \Delta_p H^\circ + \frac{\Delta \xi_p - \Delta \xi_d}{\Delta \xi_p + \Delta \xi_d} \Delta_p H_{el} \qquad (57)$$

Enthalpy of Surface Charging

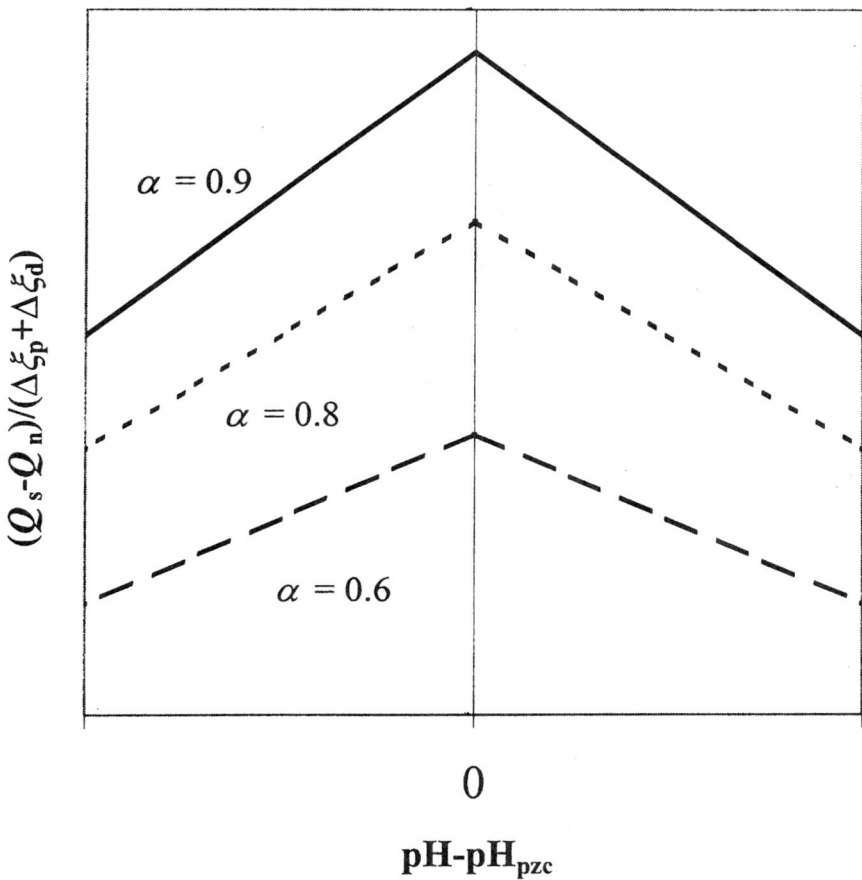

FIG. 6 Schematic presentation of the interpretation of calorimetric titration data. The linearization is based on Eq. (57). The effect of the α value on the intercept yielding the $\Delta_p H^\circ$ value.

The quantity on the left hand side (LHS) of Eq. (57) can be calculated assuming that the value of $\Delta_{ch} H^\circ$ is obtained from symmetric calorimetric experiments or from the temperature dependency of pH_{pzc}. It can be plotted as a function of $(\Delta \xi_p - \Delta \xi_d)/(\Delta \xi_p + \Delta \xi_d)$, and the slope of the tangent at a certain pH would correspond to $\Delta_p H_{el}$. As the condition approaches the zero charge condition ($\Delta_p H_{el} = 0$), the function becomes flat, reaching its constant value equal to $\Delta_p H^\circ$. The experiments in the basic and acidic region should produce the same limiting value of the function, i.e., the same $\Delta_p H^\circ$, which is the test for the proper choice of the $\Delta_{ch} H^\circ$ value. The problem of the above procedure is related to experiments in which no data points exist close enough to the p.z.c. To solve this problem one can plot the LHS of Eq. (57) as a function of $pH - pH_{pzc}$, and since this function is approximately linear, the extrapolation to pH_{pzc} yields $\Delta_p H^\circ$. Again, the proper value of $\Delta_{ch} H^\circ$ yields the same $\Delta_p H^\circ$ value for data in the acidic and in the basic region.

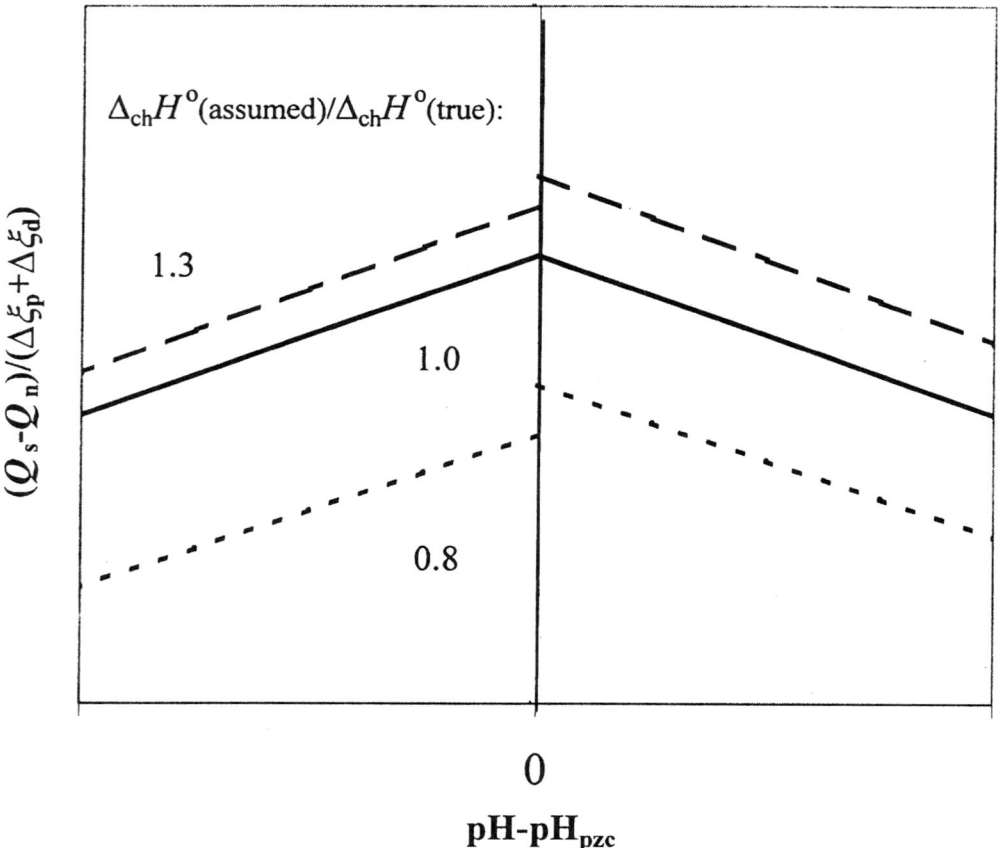

FIG. 7 Schematic presentation of the interpretation of calorimetric titration data. The linearization is based on Eq. (57). The effect of the $\Delta_{ch}H^\circ$ value on the intercept yielding the $\Delta_p H^\circ$ value.

Once the $\Delta_p H^\circ$ value is known, $\Delta_d H^\circ$ can be obtained from $\Delta_{ch}H^\circ$ Eq. (56). Also, $\Delta_p H_{el}$ and $\Delta_d H_{el}$ can be calculated by Eq. (56) and analyzed as a function of pH.

Figure 5 is a schematic presentation of a calorimetric titration of a metal oxide suspension as interpreted on the basis of Eq. (57). Both branches of the function start from $-\infty$ at the zero value on the x-axis, which corresponds to pH$\gg$pH$_{pzc}$ and also to pH$\ll$pH$_{pzc}$. The functions in basic and in acidic regions approach a constant value at pH$\rightarrow$pH$_{pzc}$. This limiting value is equal to the standard enthalpy of protonation ($\Delta_p H^\circ$). The plot shown above is not suitable for the interpretation of experimental results because the data points in the pH region close to the p.z.c., necessary for $\Delta_p H_{el}$ evaluation, are not reliable due to the coagulation of the suspended particles. The plot in Fig. 6 is more suitable, since it is approximately linear, which enables extrapolation to the p.z.c. condition.

According to Eq. (57), the intercept of both branches should be at p.z.c. (pH $-$ pH$_{pzc}$ = 0). Figure 6 also demonstrates that the proper choice of α is essential in the interpretation of calorimetric data. The interpretation also requires knowledge of the $\Delta_{ch}H^\circ$ value. This value can be determined independently by "symmetric"

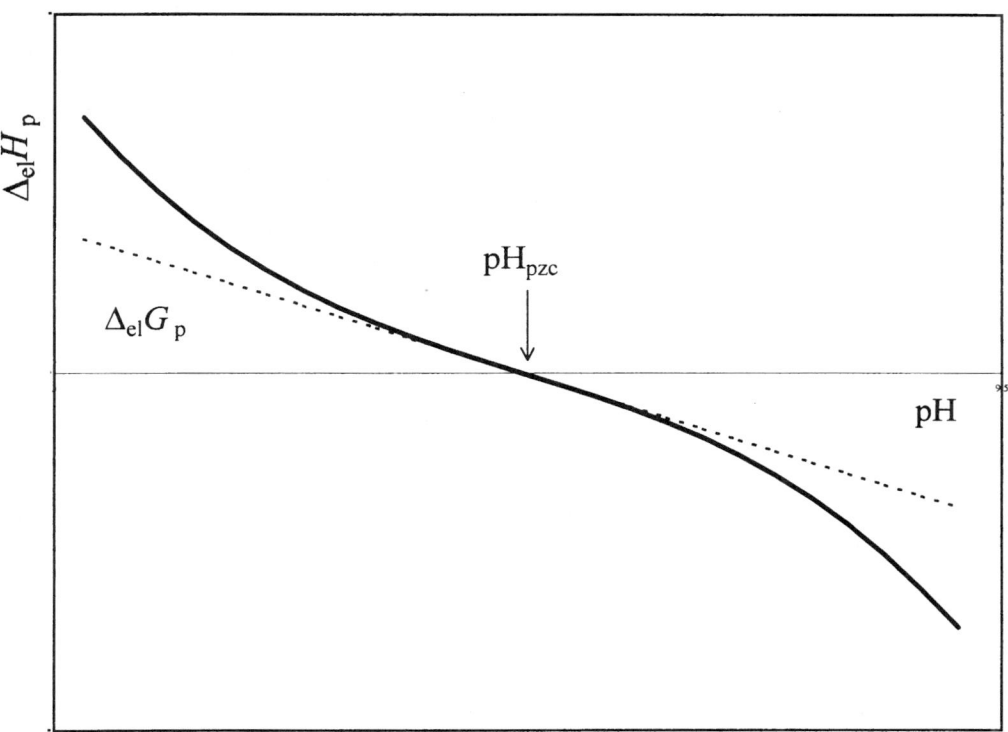

FIG. 8 Electrostatic contribution to the enthalpy of protonation according to Eq. (57), as obtained for hematite [38] ($\Delta_d H^\circ = 74.5$ kJ mol^{-1}, $\Delta_p H^\circ = 32$ kJ mol^{-1}, and $\alpha = 0.8$). The dashed line represents ΔG_{el} calculated from Eq. (51).

calorimetric experiments or more simply by measuring the p.z.c. dependency on temperature. The effect of the assumed $\Delta_{ch} H^\circ$ value is demonstrated in Fig. 7. It is obvious that the intercept of the function in basic and acidic regions is at the p.z.c. only if the proper value of $\Delta_{ch} H^\circ$ is used.

The interpretation of the data, as presented in this chapter, was used for the evaluation of the standard enthalpies of protonation and deprotonation reactions at the hematite–aqueous interface [38]. The interpretation of calorimetric titration for a hematite–aqueous suspension enabled us to distinguish between protonation and deprotonation reactions, so that the following standard enthalpies were obtained: $\Delta_p H^\circ = 32.0$ kJ mol^{-1} and $\Delta_d H^\circ = 74.5$ kJ mol^{-1}. Once these values are known, the electrostatic effect on the enthalpy could be evaluated by means of Eq. (57). The results for the hematite–aqueous interface are presented in Fig. 8. As expected, due to the repulsion, the electrostatic part of the protonation enthalpy is positive for the positively charged interface and negative in the pH region where the surface bears a negative charge. As expected, it is zero at the point of zero charge. The electrostatic part of the Gibbs energy of protonation can be calculated *via* Eq. (51) using the values of the surface potential obtained from Eq. (41). As shown in Fig. 8, in the region around the p.z.c. the values of electrostatic contributions to the Gibbs energy and enthalpy are similar, indicating the absence of electrostatic effects on the entropy term. However, this effect becomes significant as the positive or negative charge of the surface increases.

The above interpretation is based on the 2-pK model. Similar treatment can be applied using the 1-pK model. In such a case the interpretation is simpler, since only one reaction is assumed to take place at the interface. This chapter shows the complexity of the rigorous interpretation of calorimetric data even in the most simple case of a metal oxide suspended in diluted solution of a neutral electrolyte. In the more complicated situation, when specific adsorption takes place, one can either express the results on the empirical scale or apply a more sophisticated treatment that may include assumptions that cannot be easily verified.

REFERENCES

1. I. Mills, T. Cvitaš, N. Kallay, K. Homann, and K. Kochitsu, *Quantities, Unities and Symbols in Physical Chemistry*, 2d ed., Blackwell, Oxford, 1993.
2. W. Stumm, C. P. Huang, and S. R. Jenkins. Croat. Chem. Acta 42:223 (1970).
3. A. Breeuwsma and J. Lyklema. J. Colloid Interface Sci. 43:437 (1973).
4. D. E. Yates, S. Levine, and T. W. Healy. J. Chem. Soc. Faraday Trans. I 70:1807 (1974).
5. H. Hohl and W. Stumm. J. Colloid Interface Sci. 55:281 (1976).
6. J. Westall and H. Hohl. Adv. Colloid Interface Sci. 12:265 (1980).
7. V. H. van Riemsdijk, G. H. Bolt, L. K. Koopal, and J. Blaakmeer. J. Colloid Interface Sci. 109:219 (1986).
8. M. Borkovec. Langmuir 13:2608 (1997).
9. J. Luetzenkirchen. Environ. Sci. Technol. 32:3149 (1998).
10. L. K. Koopal, W. H. van Riemsijk, and M. G. Roffey. J Colloid Interface Sci. 118:117 (1987).
11. N. Kallay, R. Sprycha, M. Tomić, S. Žalac, and Ž. Torbić. Croat. Chem. Acta 63:467 (1990).
12. J. O. M. Bockris and A. K. N. Reddy, in *Modern Electrochemistry*, Vol. 1, Plenum Press, New York, 1970.
13. N. Kallay and M. Tomić. Langmuir 4:559 (1988).
14. M. Tomić and N. Kallay. Langmuir 4:565 (1988).
15. Y. G. Berube and P. L. de Bruyn. J. Colloid Interface Sci. 27:305 (1968).
16. P. H. Tewari and A. B. Campbell. J. Colloid Interface Sci. 55:531 (1976).
17. M. A. Blesa, N. M. Figliolia, A. J. G. Maroto, and A. E. Regazzoni. J. Colloid Interface Sci. 101:410 (1984).
18. L. G. H. Fokkink, A. de Keizer, and J. Lyklema. J. Colloid Interface Sci. 127:116 (1989).
19. L. G. H. Fokkink, A. de Keizer, and J. Lykleman. J. Colloid Interface Sci. 135:118 (1990).
20. M. A. Blesa. J. Colloid Interface Sci. 140:287 (1990).
21. M. L. Machesky, D. J. Wesolowski, D. A. Palmer, and K. Ichiro-Hayashi. J. Colloid Interface Sci. 200:298 (1998).
22. N. Kallay and S. Žalac. Croat. Chem. Acta. 67:467 (1994).
23. N. Kallay, S. Žalac, J. Ćulin, U. Bieger, A. Pohlmeier, and H. D. Narres. Progr. Colloid Polym. Sci. 95:108 (1994).
24. N. Kallay, S. Žalac, and I. Kobal, in *Problems in Modeling the Electrical Interfacial Layer in Metal/Oxide Aqueous System*, in *Adsorption on New and Modified Inorganic Sorbents* (A. Dabrowski and V. A. Tertykh, eds.), Elsevier Science B.V., Amsterdam, 1996.
25. J. S. Noh and J. A. Schwarz. J. Colloid Interface Sci. 130:157 (1989).
26. S. Žalac and N. Kallay. J. Colloid Interface Sci. 149:233 (1992).
27. M. L. Machesky and M. A. Anderson. Langmuir 2:582 (1986).
28. M. L. Machesky, B. L. Bischoff, and M. A. Anderson. Environ. Sci. Technol. 23:580 (1989).
29. K. A. Wierer and B. Dobiaš. J. Colloid Interface Sci. 122:171 (1988).

30. S. R. Mehr, D. J. Eatough, L. D. Hansen, E. A. Lewis, and J. A. Davis. Thermochim. Acta *154*:129 (1989).
31. A. de Keizer, L. G. J. Fokkink, and J. Lyklema. Colloids Surf. *49*:149 (1990).
32. M. L. Machesky and P. F. Jacobs. Colloids Surf. *53*:297 (1991).
33. W. H. Casey. J. Colloid Interface Sci. *163*:407 (1994).
34. N. Kallay, S. Žalac, and G. Štefanić. Langmuir *9*:3457 (1993).
35. W. Rudziński, R. Charmas, J. M. Cases, M. François, F. Villieras, and L. J. Michot. Langmuir *13*:483 (1997).
36. W. Rudziński, R. Charmas, W. Piasecki, J. M. Cases, M. François, F. Villieras, and L. J. Michot. Colloids Surf. *137*:57 (1998).
37. R. Charmas. Langmuir *14*:6179 (1998).
38. N. Kallay, T. Preočanin, S. Žalac, H. Lewandowski, and H. D. Narres. J. Colloid Interface Sci., *211*:401 (1999).
39. M. A. Blesa and N. Kallay. Adv. Colloid Interface Sci. *28*:111 (1988).
40. Y. G. Berube and L. de Bruyn. J. Colloid Interface Sci. *27*:305 (1968).
41. T. W. Healy, D. E. Yates, L. R. White, and D. Chan. J. Electroanal. Chem. *80*:57 (1977).
42. N. Kallay, D. Babić, and E. Matijević. Colloids Surf. *19*:375 (1986).
43. M. J. Avena, O. R. Camara, and C. P. De Pauli. Colloids Surf. *69*:217 (1993).
44. M. Čolić, D. W. Fuerstenau, N. Kallay, and E. Matijević. Colloids Surf. *59*:169 (1991).
45. N. H. G. Penners, L. K. Koopal, and J. Lyklema. Colloids Surf. *21*:457 (1986).
46. J. Lyklema. J. Electroanal. Chem. *18*:341 (1968).
47. J. Lyklema. Croat. Chem. Acta *43*:249 (1971).
48. J. Lyklema, in *Fundamentals of Interface and Colloid Science*, Vol. II, Academic Press, London, 1995.
49. N. Kallay, D. Čakara, and T. Preočanin. To be published.
50. W. Rudziński, R. Charmas, W. Piasecki, N. Kallay, J. M. Cases, M. François, F. Villieras, and L. J. Michot. Adsorption J. *4*:287 (1998).

6
Interpretation of Interfacial Equilibria on the Basis of Adsorption and Electrokinetic Data

NIKOLA KALLAY, DAVOR KOVAČEVIĆ, and ANA ČOP Faculty of Natural Sciences and Mathematics, University of Zagreb, Zagreb, Croatia

I.	Introduction	249
II.	Electrical Interfacial Layer	250
III.	Langmuir Isotherm	253
	A. Empirical approach	254
	B. Extended Langmuir isotherm	255
IV.	Surface Complexation Model	257
	A. Mechanism of surface reactions	258
	B. Surface charging	258
	C. Adsorption of (large) organic ions	259
	D. Adsorption of heavy metals on metal oxides	267
	References	271

I. INTRODUCTION

Numerous phenomena in colloidal systems are influenced by interfacial properties. This chapter considers solid–liquid interfaces and will be devoted to metal oxides. One of the most important phenomena in interfacial chemistry is the electrostatic charging of the surface due to interactions with ionic species from the bulk of the solution. In considering this problem one should distinguish the conductive metallic surfaces from other solids such as ionic crystals. In past decades, equilibria at metal oxide interfaces have been interpreted on the basis of the surface complexation model (SCM) [1] which is sometimes called the site binding model. The main feature of this approach is based on interactions of bulk ions with specified surface groups. The resulting interfacial species are called either surface complexes or associates depending on the nature of the chemical bond. Several reactions may take place at the interface, and this problem is discussed in details in Chapters 4 and 5 of this book.

The aim of this chapter is to demonstrate the application of electrokinetic measurements in the interpretation of equilibria at interfaces. The electrokinetic measurements by themselves may be used to elucidate processes taking place at the surface [2]. Also, they may be used as criteria for the verification of a certain theoretical model in order to verify its applicability. For example, one can interpret the experimental surface charge vs. pH function by a certain model and obtain the values of model parameters. Using these values one can calculate the pH dependency of the potential at the onset of the diffuse layer and compare these values with measured electrokinetic potentials. If the results are in agreement, i.e. if electrokinetic potentials are equal or slightly lower than the calculated values of the potentials at the onset of the diffuse layer, one may conclude that the applied theoretical model does not disagree markedly with reality.

Quite another approach is to interpret the adsorption and electrokinetic data simultaneously on the basis of a certain theoretical model. In this chapter, one of the possible models will be used to demonstrate how powerful this approach can be. The advantage of such an interpretation lies in that one may be able to distinguish between different theoretical assumptions. If one interprets the adsorption results only, it may happen that data could be successfully fitted by different theoretical models. The consequence is that one is not able to determine the mechanism of the process, and also, the values of the parameters obtained by such a procedure have doubtful physical meaning. The problem can be solved by improving the accuracy of measurements, by examining the systems at different conditions, or by applying some other independent experimental methods. Spectroscopic data are often essential in postulating the theoretical model needed for the interpretation of the data. In the case of additional electrokinetic measurements, one introduces another set of data to which the theoretical function should be fitted. Consequently, the restrictions are more narrow, and those theoretical models that do not describe properly the real situation will fail.

In the next sections the theoretical models will be described, and some results of the simultaneous interpretation of adsorption and electrokinetic measurements will be demonstrated.

II. ELECTRICAL INTERFACIAL LAYER

In the electrical interfacial layer, the electrostatic potential changes gradually from the surface towards the bulk of the solution. It is commonly accepted that the bulk of the solution is the reference point with the (average) potential being zero. Charged species in the interfacial layer are exposed to the effect of the electrostatic potentials, so that the solution of any equilibrium state requires knowledge of the relationships between surface charge densities and potentials, which enables the quantitative interpretation of the experimental data. The phenomenological approach described in this chapter is based on the "primitive" models of the interfacial layer. The term primitive is related to at least two approximations. At first, the permittivity of the interfacial layer is assumed to be constant, and equal to the one in the bulk of the solution. Secondly, the assumption of homogeneous charge distribution is used, which requires a constant potential of a certain plane parallel to the surface. This assumption approaches reality when the distances between the ions are on the same scale, i.e., if the ions distributed in the interfacial layer are separated from the surface by a distance larger than the mutual distances between surface charged groups [3]. Figure 1 is a schematic presentation of the electrical interfacial layer (EIL) for the simple case of a metal oxide in the "neutral" electrolyte aqueous solution. An electrolyte is neutral if the ions do not form chemical bonds with the surface sites. The scheme corresponds to the positive net surface charge, i.e., to the interface with positively charged groups prevailing over those that are negatively charged. Such a situation occurs in the pH region below the point of zero charge. It is worth noting that in reality most of the amphoteric groups remain uncharged.

Interpretation of Interfacial Equilibria

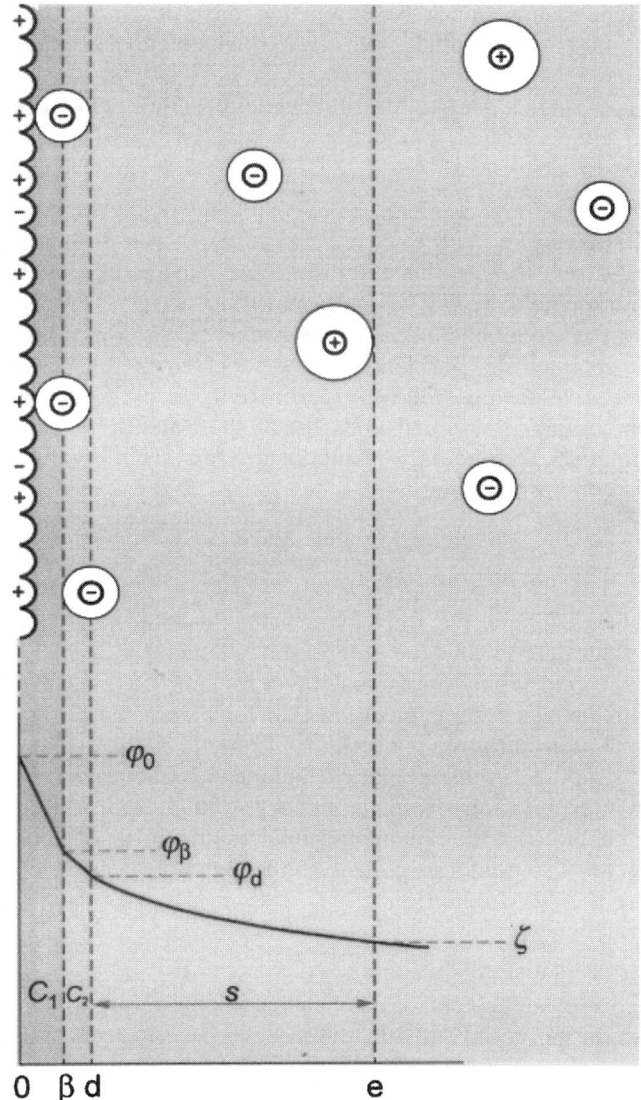

FIG. 1 Schematic presentation of the electrical interfacial layer (EIL) for the simple case of a metal oxide in a "neutral" electrolyte aqueous solution.

The scheme of Fig. 1 is in accordance with a general model introduced by Leckie et al. [4], which is sometimes called the triple layer model (TLM). The first plane is at the solid surface and is denoted the 0-plane. In this plane the centers of surface charges, created by interactions with potential determining ions, are located. For metal oxides, potential determining ions are H^+ and OH^- ions. The electrostatic potential affecting the state of charged groups in the 0-plane is denoted by ϕ_0. The second plane is the β-plane in which the centers of associated counterions are located and is characterized by the potential ϕ_β. The separation b_1 of this plane from the 0-plane is determined by ionic dimensions including the hydration shell. In the region between the 0- and β-planes a linear potential drop is assumed. This assumption is based on the constant capacitance concept, since 0- and β-planes are taken as parallel, with a dielectricum in the space between them.

Accordingly, the capacitance (per surface area) of the first capacitor, C_1, is defined as

$$C_1 = \frac{\varepsilon_1}{b_1} \tag{1}$$

where ε_1 is the permittivity of the first layer, i.e., of the space between the 0- and β-planes. In the layer next to the solid surface, structuring of the water molecules may be expected, so that one needs to take into account the lowering of the permittivity with respect to the bulk phase. The problem is that the ε_1 value is unknown. However, C_1 is commonly taken as an adjustable parameter, so that this problem may be reflected in the value of b_1, which is of course on the relative scale if the bulk permittivity is used. In the interpretation of the surface charge data the assumed linearity of the potential within the first layer does not affect the final result. The only results of interest are the values of ϕ_0 and ϕ_β. Their relationship is commonly given as

$$\phi_0 = \frac{\sigma_0}{C_1} + \phi_\beta \tag{2}$$

where σ_0 is surface charge density of the 0-plane. The third plane is the onset of the diffuse layer, denoted as the d-plane, and is characterized by the potential ϕ_d. Generally, it can be assumed that β- and d-planes form a second capacitor of capacitance (per surface area) C_2. The existence of the second capacitor is based on assumption that ions from the diffuse layer cannot approach the β-plane. This assumption is necessary if one takes the d-plane as identical to the electrokinetic slipping or shear plane (e-plane) characterized by ζ-potential, i.e., $\phi_d = \zeta$, since the interpretation of surface charge data requires that $|\phi_\beta| > |\zeta|$. The problem is solved by introducing the e-plane as separated from the d-plane by a distance s, which makes $|\phi_d| > |\zeta|$, so that the plausible approximation $\phi_\beta \approx \phi_d$ can be used. Therefore, when using the TLM equation

$$\phi_\beta = \frac{\sigma_\beta}{C_2} + \phi_d \tag{3}$$

one can introduce $C_2 \to \infty$, so that the potential drop in the layer between the β- and d-planes disappears.

The onset of the diffuse layer is the d-plane, and theoretically this layer is extended to infinity, because all the charge directly bound to the surface (σ_s) should be compensated by a counter charge in the diffuse layer (σ_d). For the diffuse layer, the Gouy–Chapman theory is used, introducing the reciprocal thickness of the diffuse layer κ as a measure of its extension. It is a function of ionic strength (I_c), permittivity (ε), and temperature (T) according to

$$\kappa = \sqrt{\frac{2F^2 I_c}{\varepsilon RT}} \tag{4}$$

The Gouy–Chapman theory provides the relationship between surface charge density and the potential at the onset of the diffuse layer:

$$\sigma_s = -\sigma_d = -\sqrt{8RT\varepsilon I_c} \sinh\left(\frac{\phi_d F}{2RT}\right) = -\frac{4FI_c}{\kappa} \sinh\left(\frac{\phi_d F}{2RT}\right) \tag{5}$$

where σ_s corresponds to the charges fixed to the surface and σ_d to the charge distributed in the diffuse layer. Surface charge density and potential in the diffuse layer decrease from the d-plane in the direction of the bulk of the solution. The Gouy–Chapman theory provides also the relationship between the potentials at different separations from the surface, so that the potential at the onset of the diffuse layer could be calculated from the electrokinetic ζ-potential if the concept of an electrokinetic slipping plane separation (s) is introduced:

$$\phi_d = \frac{2RT}{F} \ln \frac{\exp(-s\kappa) + \tanh(F\zeta/4RT)}{\exp(-s\kappa) - \tanh(F\zeta/4RT)} \quad (6)$$

The interpretation of interfacial equilibrium data results [5–13] in the slipping plane separation of about 15 Å when the $\phi_\beta \approx \phi_d$ approximation is used. The concept is useful in the theoretical treatment of experimental data, since the measured electrokinetic potential may be used as additional information on the behavior of the interface. The idea of a slipping plane separation was suggested by Eversole and coworkers [14,15] almost sixty years ago. They have interpreted the data on the ionic strength dependency of the ζ-potential using the Gouy–Chapman theory. The Eq. (6) could be transformed into

$$\ln \tanh\left(\frac{F\zeta}{4RT}\right) = \ln \tanh\left(\frac{F\phi_d}{4RT}\right) - \kappa s \quad (7)$$

which suggests the linearity of $\ln \tanh(F\zeta/4RT)$ with respect to κ, with slope equal to the slipping plane separation. In Fig. 2 the electrokinetic data of Gaudin and Fuerstenau[16] for quartz are displayed.

It is obvious that no linearity was obtained, and this finding resulted in the abundance of the slipping plane separation concept. However, the reason for nonlinearity lies in that ϕ_d is not a constant but rather decreases with the increase of electrolyte concentration [17] so that the slope of the tangent is always higher than the value of the slipping plane separation. Accordingly, the nonlinearity of the presentation in Fig. 2 does not suggest that the slipping plane separation concept is unrealistic. On the contrary, as will be shown in this chapter, this concept is useful in the interpretation of interfacial equilibria.

III. LANGMUIR ISOTHERM

The Langmuir isotherm is often used for the interpretation of adsorption data, especially for the specific adsorption of molecules or ions that are chemically bound to the surface. In many cases this isotherm is used for the interpretation of adsorption data without considering the model on which it is based. For ionic adsorption, at least one requirement is violated—the ions at the interface are not independent but rather they exhibit pronounced mutual electrostatic interactions. Since the Langmuir isotherm is a popular tool, widely used in treating the adsorption data, some consequences of its use will be discussed here, as well as its refinement in solving the problems due to electrostatic interactions and the association–dissociation equilibria in the bulk of the solution.

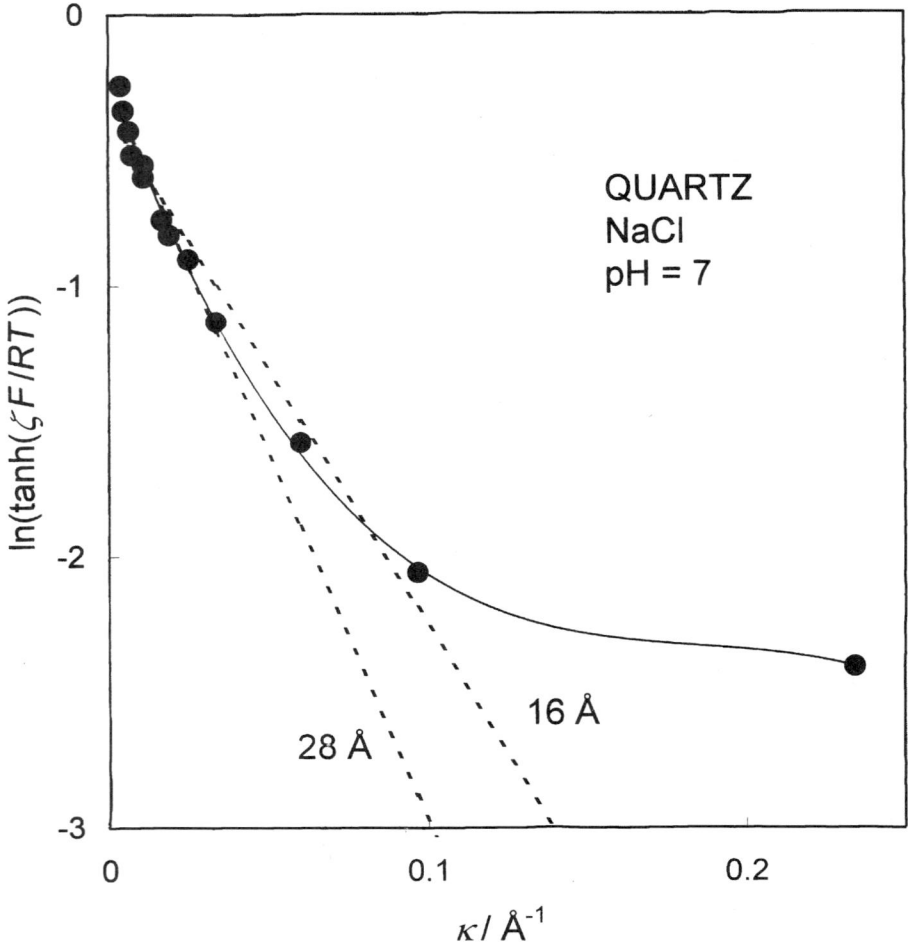

FIG. 2 The determination of the slipping plane separation (s) for quartz in aqueous solution of NaCl as suggested by Eversole and coworkers [14]. The plot is according to Eq. (7). Dashed lines correspond to different values of s as denoted in the figure. The experimental data were taken from Ref. 16. (From Ref. 37.)

A. Empirical Approach

Adsorption of ionic species could be interpreted on the basis of the Langmuir isotherm although charged ionic species mutually interact at the interface. The shape of the adsorption curve in most cases exhibits a plateau, so that the common plot for the equilibrium state, reciprocal surface concentration vs. reciprocal bulk concentration, can be linear according to

$$\frac{1}{\Gamma} = \frac{1}{\Gamma_{max}} + \frac{1}{\Gamma_{max} K_C} \tag{8}$$

where c is the bulk equilibrium concentration of the adsorbing species, Γ is the surface concentration of the adsorbed species (amount of species per surface area), and Γ_{max} is related to the occupied surface area per adsorbed molecule a_s by

$$\Gamma_{max} = \frac{1}{L \cdot a_s} \tag{9}$$

where is L the Avogadro constant. Formally, using the above equation one can obtain [18] the adsorption equilibrium constant K and the area occupied by one adsorbed ion a_s. The problem is that the resulting value of a_s depends on the pH and that the value of K is either constant or affected by the pH in an unexpected manner. For such a finding one may say that the a_s value corresponds to the area physically occupied by an adsorbed ion plus the area protected by electrostatic repulsion, but such a rationalization still does not have an exact physical meaning.

B. Extended Langmuir Isotherm

By analyzing the requirements connected to the Langmuir isotherm one can easily find the way to refine the empirical approach by introducing an electrostatic interaction term that may be considered either as a correction of the equilibrium constant or as the activity coefficient of the adsorbed charged species. Also, in the case of the association–dissociation equilibrium in the bulk of the solution one should use the equilibrium concentration or activity of those species that actually adsorb. Accordingly, the refined adsorption isotherm of the Langmuir type for ionic species i of the charge number z would read

$$\frac{1}{\Gamma_i} = \frac{1}{\Gamma_{max}} + \frac{1}{\Gamma_{max} K_i \exp(-z_i F \phi_a / RT) a_i} \tag{10}$$

where a_i is the bulk activity of adsorbing species i, z_i their charge number, Γ_i the surface concentration of the adsorbed species i, and ϕ_a the surface potential affecting the state of the adsorbed species i. In the interpretation of adsorption experiments one uses electrokinetic data to evaluate ϕ_a by using the approximation $\phi_a \approx \phi_d$. In doing so one calculates ϕ_d from the ζ-potential by Eq. (6), using the slipping plane separation as an adjustable parameter. The above treatment was successfully applied to the adsorption of citric, oxalic, and iminodiacetic on hematite [19,20] and will be demonstrated in the forthcoming paragraph using data on the adsorption of salicylic acid on hematite and titania. The refined Langmuir isotherm does not provide information on the molecular level, i.e., it does not consider an active surface group as reactant, but rather the surface itself. However, as will be demonstrated, it does result in the charge number of species that actually adsorb at the surface, thus indicating the mechanism of the surface binding process. The area occupied by one ion at the surface can be also used as information about the orientation of adsorbed ionic species at the surface, while the equilibrium constant is useful for comparison of adsorption processes in different systems.

1. Adsorption of Salicylic Acid on Hematite and Titania

The modified Langmuir isotherm (Eq. 10) was applied to the adsorption of salicylic acid on hematite and titania [12,13]. The adsorption isotherm and the corresponding electrokinetic mobilities of hematite particles were measured as a function of pH (Fig. 3).

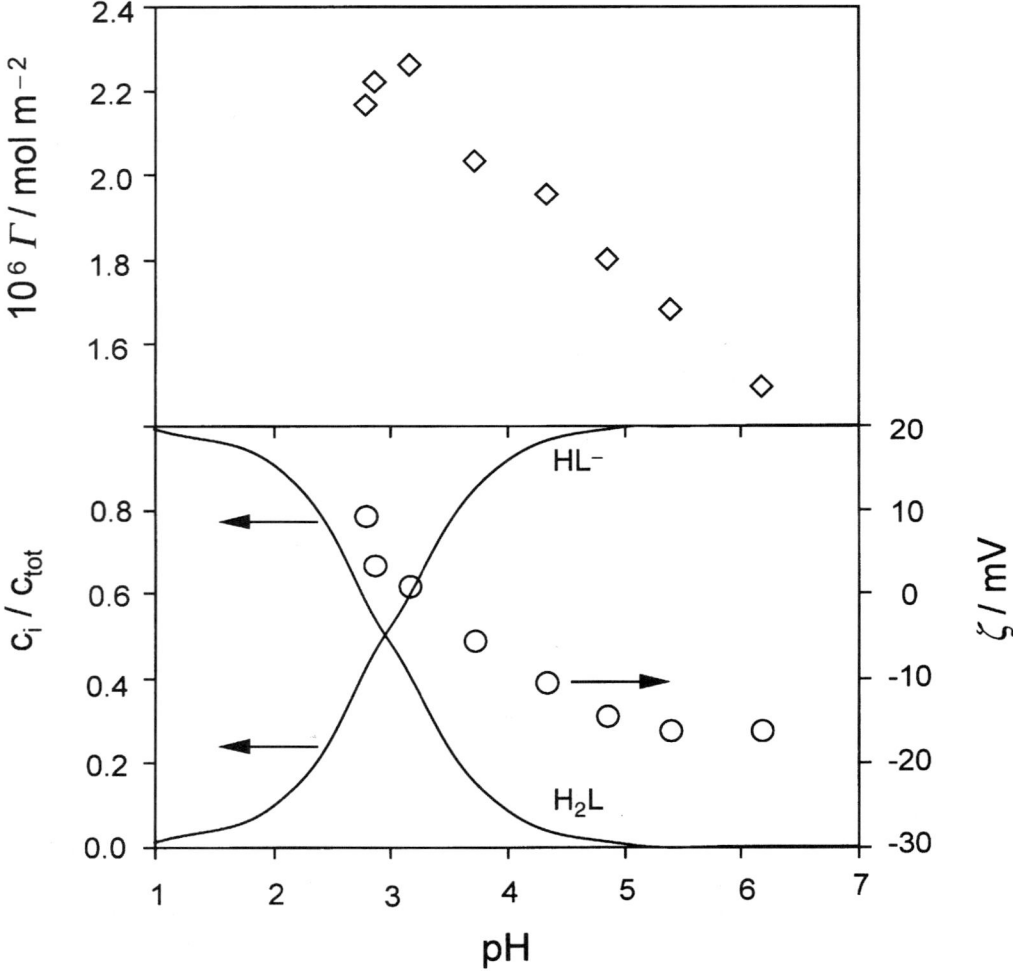

FIG. 3 Adsorption of salicylic acid on hematite. The effect of pH on the surface concentration of salicylic acid on hematite ($\diamond$) and on the electrokinetic potential of hematite particles ($\bigcirc$) compared with the "speciation" diagram of salicylic acid (———). Mass concentration of hematite was 200 g dm^{-3}, initial concentration of salicylic acid was $7.21 \cdot 10^{-3}$ mol dm^{-3}, and temperature was 20°C. The ionic strength of $I_c = 1.6 \cdot 10^{-2}$ mol dm^{-3} was due to the initial concentration of salicylic acid and the addition of NaOH or HNO$_3$, used for pH adjustment. (Data taken from Ref. 12.) (Note that for electrokinetic experiments a low mass concentration of particles was used.)

Salicylic acid dissociates in two steps, so that in the solution neutral H_2L, singly charged HL^- and doubly charged L^{2-} species are present. Their amount depends on pH and could be calculated using corresponding equilibrium constants. In the lower part of Fig. 3 the so called speciation diagram of salicylic acid is demonstrated, showing the pH dependency of the fraction of different salicylic species.

The adsorption equilibrium was interpreted [12] considering the dissociation of the acid in the bulk of the solution and the electrostatic interactions between the surface and the adsorbed ions. Several different possibilities (adsorption of neutral, singly, and doubly charged species) were tested on the basis of Eq. (10) using the equilibrium concentrations of species which were assumed to adsorb at the interface. The potential affecting the state of adsorbed salicylate species was assumed to be equal to the potential at the onset of the diffuse layer, which was calculated from the ζ-potential by Eq. (6). In the calculations, different values of slipping plane separation s were used. Linearity was obtained only when the adsorption of singly charged HL^- ions was assumed, and the best straight line was obtained for $s = 15$ Å The area occupied by one adsorbed HL^- ion was found to be 87 ± 15 Å^2, and the adsorption equilibrium constant was obtained as $\log K = 3.5 \pm 0.1$.

The same procedure was applied for the adsorption of salicylic acid on titania [13], but none of the above assumptions regarding adsorbed species was found appropriate. Therefore, a more complex assumption was introduced, i.e., the simultaneous adsorption of different ionic species. It was concluded that both neutral and singly charged species are adsorbable, and adsorption equilibrium constants were determined as $\log K_0 = 2.15 \pm 0.15$ and $\log K_1 = 1.77 \pm 0.07$.

Adsorption and electrokinetic data could be analyzed also in a semiquantitative way by comparing them with a speciation diagram. For example, one may assume adsorption of neutral H_2L species. No electrostatic effects are expected, so that adsorption density should follow the pH dependency of the fraction of H_2L species (Fig. 3), which is not the case. Therefore this assumption must be excluded. Also, adsorption of doubly charged L^{2-} species cannot be expected, since their equilibrium concentration in the pH range from 2 to 7 is negligible. Adsorption of singly charged HL^- species is a plausible assumption. Their fraction increases from pH 2 to 5. In this region the positive charge of the surface decreases up to the i.e.p. at $pH = 3.5$, and a further increase in pH is followed by an increase of negative potential (Fig. 3). These two effects practically compensate; as the pH increases, the fraction of adsorbable species increases, which promotes adsorption, but at the same time the pH has also the opposite effect, since at higher pH values the more negative surface will repel negatively charged salicylate ions.

IV. SURFACE COMPLEXATION MODEL

In contrast to the treatment of adsorption data by an adsorption isotherm, the surface complexation model (SCM) [1] considers the interactions of defined surface species with adsorbable molecules or ions. The advantage of such an approach lies in consideration of the mechanism of the adsorption process on the molecular scale, so that information regarding the structure of surface species obtained by e.g. spectroscopy can be related to the adsorption results. In the forthcoming sections, the interpretation of adsorption and electrokinetic data based on the SCM will be presented.

A. Mechanism of Surface Reactions

According to the surface complexation model (2-pK concept) [1] the surface charging of metal oxides is due to protonation (p) and deprotonation (d) of amphoteric surface $\equiv$MOH groups (see Chapters 4 and 5). For metal oxides, the potential determining ions are H$^+$ and OH$^-$ ions, and their interactions in the 0-plane are described by

$$\equiv \text{MOH} + \text{H}^+ \rightarrow \equiv \text{MOH}_2^+ \qquad K_p = \exp\left(\frac{F\phi_0}{RT}\right) \frac{\Gamma(\text{MOH}_2^+)}{\Gamma(\text{MOH})a(\text{H}^+)} \qquad (11)$$

$$\equiv \text{MOH} \rightarrow \equiv \text{MO}^- + \text{H}^+ \qquad K_d = \exp\left(\frac{-F\phi_0}{RT}\right) \frac{\Gamma(\text{MO}^-)a(\text{H}^+)}{\Gamma(\text{MOH})} \qquad (12)$$

where K_p and K_d are the corresponding equilibrium constants. Note that in the literature the term "intrinsic equilibrium constant" is often used for K_p and K_d. The reason for such a practice is historical. The "intrinsic approach" considers the binding of e.g. H$^+$ ion to a surface site as a two-step process. The first step was taken to be the transfer of H$^+$ ion from the bulk phase to the interface (intrinsic state characterized by potential ϕ_0), and this equilibrium was treated by the Boltzmann statistics. The second process is the binding of H$^+$ ion in the "intrinsic state" to the surface site, and this step is characterized by the "intrinsic" equilibrium constant. This approach is correct but could be replaced by a more direct one introducing the activity coefficients of interfacial species. However, the resulting equations are the same.

Ions of the charge opposite with respect to the surface (counterions) may form associates with the charged surface groups. The equilibria are commonly described by

$$\equiv \text{MOH}_2^+ + \text{A}^- \rightarrow \equiv \text{MOH}_2^+ \cdot \text{A} \qquad K_a = \exp\left(\frac{-F\phi_\beta}{RT}\right) \frac{\Gamma(\text{MOH}_2^+ \cdot \text{A}^-)}{a(\text{A}^-)\Gamma(\text{MOH}_2^+)} \qquad (13)$$

$$\equiv \text{MO}^- + \text{C}^+ \rightarrow \equiv \text{MO}^- \cdot \text{C}^+ \qquad K_c = \exp\left(\frac{F\phi_\beta}{RT}\right) \frac{\Gamma(\text{MO}^- \cdot \text{C}^+)}{\Gamma(\text{MO}^-)a(\text{C}^+)} \qquad (14)$$

where C$^+$ and A$^-$ denote cations and anions, respectively.

In the case of purely electrostatic interactions, the statistic approach to the equilibrium of counterion association was also introduced. This approach, based on the Bjerrum model, is described in Chapters 4 and 5. Both approaches result in similar findings: the association of counterions is more pronounced at higher ionic strengths, and in practice it takes place when the surface potential is high enough and opposite in sign with respect to the ion that undergoes association. Counterion association will be more pronounced at higher surface potentials for both theoretical concepts.

B. Surface Charging

Charging of a metal oxide aqueous interface in the neutral electrolyte system is described by the equilibria (11–14). Accordingly, surface charge densities in 0- and β-planes are related to surface concentrations of relevant ionic species by

$$\sigma_0 = F(\Gamma(\text{MOH}_2^+) + \Gamma(\text{MOH}_2^+ \cdot \text{A}^-) - \Gamma(\text{MO}^-) - \Gamma(\text{MO}^- \cdot \text{C}^+)) \qquad (15)$$

$$\sigma_\beta = F(\Gamma(\text{MO}^- \cdot \text{C}^+) - \Gamma(\text{MOH}_2^+ \cdot \text{A}^-)) \qquad (16)$$

The net surface charge density (σ_s), opposite in sign to that in the diffuse layer (σ_d), is given by

$$\sigma_s = -\sigma_d = F(\Gamma(\text{MOH}_2^+) - \Gamma(\text{MO}^-)) \qquad (17)$$

while the total concentration of surface sites, Γ_{tot}, is equal to

$$\Gamma_{tot} = \Gamma(\text{MOH}) + \Gamma(\text{MOH}_2^+) + \Gamma(\text{MO}^-) + \Gamma(\text{MO}^- \cdot \text{C}^+) + \Gamma(\text{MOH}_2^+ \cdot \text{A}^-) \qquad (18)$$

In solving the equilibrium state at the interface for a "simple" case of a metal oxide in an aqueous electrolyte solution one deals with numerous parameters: the total number of surface sites [Eq. (18)], four equilibrium constants [Eqs. (11–14)], three equations defining the surface charge densities [Eqs. (15–17)], and several equations relating surface potentials with corresponding surface charge densities [Eqs. (2,3,5)]. When interpreting the measured σ_0(pH) function, as obtained by potentiometric titration of a suspension, one needs to adjust the values of several equilibrium parameters, which is not a reliable approach to the problem. As expected, the experimental data can be fitted successfully by different modifications of the surface complexation mode [21], so that the physical meaning of evaluated parameters remains questionable. By introducing the electrokinetic data, in addition to the surface charge data, and using Eq. (6) for evaluation of ϕ_d, the results become more accurate. For example, it was shown that the application of such an approach explains the lyotropic effect [22,23]. The ϕ_d potential was calculated from equilibrium parameters, and the coagulation behavior of the system was also explained. It was shown that coagulation kinetics is mainly determined by the association of counterions with the charged surface sites reducing potential at the onset of the diffuse layer. Specificities of different counterions of the same charge were explained in terms of their effective sizes influencing the values of the association equilibrium constants and the capacitance in the first layer. Also, in the pH region close to the zero surface charge, where counterion association is negligible, the compression of the diffuse layer is responsible for coagulation, so that the specificity of the counterions disappears.

C. Adsorption of (Large) Organic Ions

In the literature one can find different variations of the surface complexation model applied to the binding of organic species on metal oxides [24–27]. The differences are reflected in the assumed mechanisms of surface reactions and the postulated structures of the electrical interfacial layer. Figure 4 is a schematic presentation of the electrical interfacial layer (EIL) at a metal oxide surface with adsorbed monovalent organic anions. This figure is representative for salicylic acid adsorption according to the analysis presented in this section. It corresponds to pH = 4.5 where positive surface charged groups $\equiv \text{MOH}_2^+$ dominate over the negative $\equiv \text{MO}^-$ surface species, the amount of which was found to be negligible. Due to strong adsorption of salicylic singly charged ionic species, a reversal of the original positive charge occurs, so that the net surface charge is negative and is compensated by a positive charge in the diffuse layer. The large ionic salicylic species are located in the a-plane covering several $\equiv$MOH surface groups.

Adsorption of salicylic acid on hematite is often used as a model system for studying interfacial phenomena of metal oxides in the presence of organic ions [28–30]. The interpretation based on the extended Langmuir isotherm did not consider the mechanism of binding of salicylic acid onto hematite; adsorption is just treated as the accumulation of molecules at the surface. On the other hand, the interpretation based on the surface complexation model requires a definition of surface reactions. The starting point in

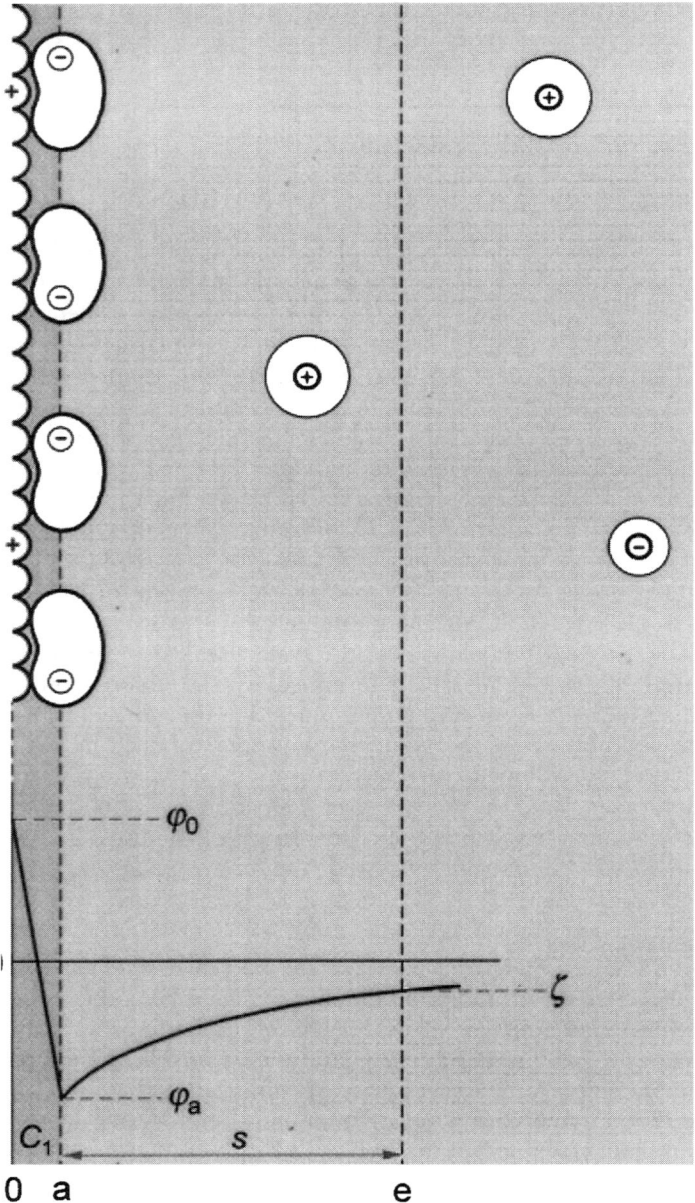

FIG. 4 Schematic presentation of the electrical interfacial layer (EIL) at a metal oxide surface with adsorbed organic anions.

the determination of surface reactions is spectroscopy data, which suggest several possible different structures of surface complexes in the salicylic acid/hematite system [28–30]. The choice of the surface complex structure, i.e., of the appropriate surface reaction, can be made by considering the results of adsorption and electrokinetic data [11]. Since the interpretation based on the extended Langmuir isotherm showed that singly charged salicylate ions are bound to the surface, one can conclude that the following surface reaction takes place:

$$\equiv MOH + HL^- \rightarrow \equiv ML^- + H_2O$$
$$K(HL^-) = \exp\left(\frac{-F\phi_a}{RT}\right) \frac{\Gamma(ML^-)}{\Gamma(MOH)a(HL^-)} \tag{19}$$

The same process can be expressed by differently written reaction equations. For example, the equation

$$\equiv MOH + H^+ + HL^- \rightarrow \equiv MOH_2^+ HL^- \tag{20}$$

is equivalent to Eq. (19) and results in the same values of the equilibrium parameters. The only difference is the state of the water molecule at the interface. According to Eq. (19), the water molecule is released from the surface complex, and since the process takes place in an aqueous environment, the formulation of the equilibrium conditions remains the same.

For the case of the adsorption of relatively large organic species one should also consider that a bound organic molecule does not cover just one surface site, the one to which it is bound, but also several adjacent surface groups. This so-called "umbrella effect" should be significant in the case of salicylate ions and is reflected in the summation of surface sites. The following equation takes into account that one adsorbed salicylate ion, forming an ML^- surface complex, excludes from the binding process f neutral $\equiv MOH$ species, one of which is bound and the others covered, preventing them from being active in the adsorption process:

$$\Gamma_{tot} = \Gamma(MOH) + \Gamma(MOH_2^+) + \Gamma(MO^-) + f\,\Gamma(ML^-) \tag{21}$$

where Γ_{tot} denotes the total surface concentration of the active sites.

Generally, the surface charge density in the inner part of the electrical interfacial layer (0-plane) is given by

$$\sigma_0 = F(\Gamma(MOH_2^+) - \Gamma(MO^-)) \tag{22}$$

while the surface charge density in the adsorption plane (a-plane) is

$$\sigma_a = -F(\Gamma(HL^-)) \tag{23}$$

As mentioned before, the effective (net) surface charge density (σ_s) is equal in magnitude, but opposite in sign, to the surface charge density in the diffuse layer (σ_d). At low ionic strength, taking into account the contribution of specifically adsorbed salicylate ions, the following relationship holds:

$$\sigma_s = -\sigma_d = F(\Gamma(\text{MOH}_2^+) - \Gamma(\text{MO}^-) - \Gamma(\text{HL}^-)) \tag{24}$$

The value of σ_d could be obtained via Eq. (3) from the electrostatic potential at the onset of the diffuse layer ϕ_d, which could be calculated from the measured values of the ζ-potential [Eq. (6)]. According to Eqs. (22,24), the surface charge density on the 0-plane could be determined by

$$\sigma_0 = \sigma_s - F\Gamma(\text{HL}^-) \tag{25}$$

The above equation is useful for calculation of surface charge density in the 0-plane using the value of σ_s, as obtained from electrokinetic measurements [Eqs. (5) and (6)] and measured adsorption density $\Gamma(\text{HL}^-)$. The potential drop between the 0-plane and the a-plane in which adsorbed HL^- species are located can be calculated via Eq. (2) using the approximation $\beta = d = a$, i.e., $\phi_\beta = \phi_d = \phi_a$. Once ϕ_0 is known, the surface concentrations of positive and negative surface sites can be calculated as follows. At the point of zero charge (pH_{pzc}) in the absence of specific adsorption,

$$\text{pH}_{\text{pzc}} = 0.5 \log\left(\frac{K_p}{K_d}\right) \tag{26}$$

According to Eqs. (11,12,26), the ratio $\Gamma(\text{MOH}_2^+)/\Gamma(\text{MO}^-)$ can be obtained from the values of the surface potential ϕ_0 for any pH by

$$\frac{\Gamma(\text{MOH}_2^+)}{\Gamma(\text{MO}^-)} = \frac{1}{a_{\text{pzc}}^2(\text{H}^+)} \exp\left(\frac{-2\phi_0}{RT}\right) a^2(\text{H}^+) \tag{27}$$

Once this ratio is known, one can simply calculate the charge density of negative sites by the following relationship, obtained by rearranging Eq. (15), for the case of negligible counterion association:

$$\Gamma(\text{MO}^-) = \frac{(\sigma_0/F)}{(\Gamma(\text{MOH}_2^+)/\Gamma(\text{MO}^-)) - 1} \tag{28}$$

The density of positive surface sites is then given by

$$\Gamma(\text{MOH}_2^+) = \frac{\sigma_0}{F} - \Gamma(\text{MO}^-) \tag{29}$$

Interpretation of Interfacial Equilibria

By dividing the equilibrium expressions for the binding of protons (11) and of salicylate ions (19), and by introducing the capacitance (2), one obtains

$$\ln\frac{\Gamma(MOH_2^+)a(HL^-)}{\Gamma(ML^-)a(H^+)} + \frac{2\phi_d F}{RT} = \ln\frac{K_p}{K(HL^-)} - \frac{(\Gamma(MOH_2^+) - \Gamma(MO^-))F^2}{C_1 RT} \quad (30)$$

By plotting the LHS of Eq. (30) as the function of $(\Gamma(MOH_2^+) - \Gamma(MO^-))$, one should get the straight line yielding the ratio of $K_p/K(HL^-)$ (intercept), while the slope provides the value of the capacitance C_1. This approach requires the application of an iterative procedure, since the necessary data on $\Gamma(MOH_2^+)$ and $\Gamma(MO^-)$ depend on the chosen value of C_1. From Eqs. (11) and (21) one can derive the expression

$$\frac{\exp(-\phi_0 F/RT)}{\Gamma(MOH_2^+)}(\Gamma_{tot} - \Gamma(MOH_2^+)) = \frac{1}{K_p} + f\frac{\Gamma(HL^-)\exp(-\phi_0 F/RT)a(H^+)}{\Gamma(MOH_2^+)} \quad (31)$$

which could be used for evaluation of K_p and f values. One plots

$$\frac{\exp(-\phi_0 F/RT)}{\Gamma(MOH_2^+)}(\Gamma_{tot} - \Gamma(MOH_2^+))$$

vs.

$$\frac{\Gamma(HL^-)\exp(-\phi_0 F/RT)a(H^+)}{\Gamma(MOH_2^+)}$$

and the slope should correspond to a number (f) of neutral $\equiv$MOH species covered and deactivated by one adsorbed salicylate ion. In addition, the intercept provides the K_p value, which can be then used to obtain the value of the deprotonation equilibrium constant (K_d), using Eq. (26). Once the concentration of all the surface species, and also the electrostatic potential affecting the state of adsorbed ions, is known, one can readily calculate the equilibrium constant for the binding of salicylate ions, Eq. (19). However, in doing so one needs to use certain values of C_1, Γ_{tot}, and the umbrella coefficient f.

For interpretation of the equilibrium state in the salicylic acid/hematite system, the adsorption and electrokinetic data, shown in Fig. 3, were used, as well as corresponding values of $a(HL^-)$. The point of zero charge of hematite in the absence of specific adsorption is also necessary: $pH_{pzc} = 6.5$. In the first step, the effective (net) surface charge density (σ_s) was calculated via Eq. (24) from the potential at the onset of the diffuse layer φ_d, which was obtained from the measured ζ-potential, Eq. (6). According to Eq. (25), the surface charge density in the 0-plane was then calculated using the values of the surface concentration of adsorbed HL^- species. The next step was the calculation of the surface potential ϕ_0 via Eq. (2). For this purpose, the value of the (constant) capacitance C_1 is needed. Since the dependency of ϕ_0 on pH should show a quasi-Nernstian behavior [10,31–37], the limiting values of C_1 are obtained as 1.4 and 1.7 F m^{-2} yielding the slopes of -46 and -42 mV/pH, respectively (Fig. 5).

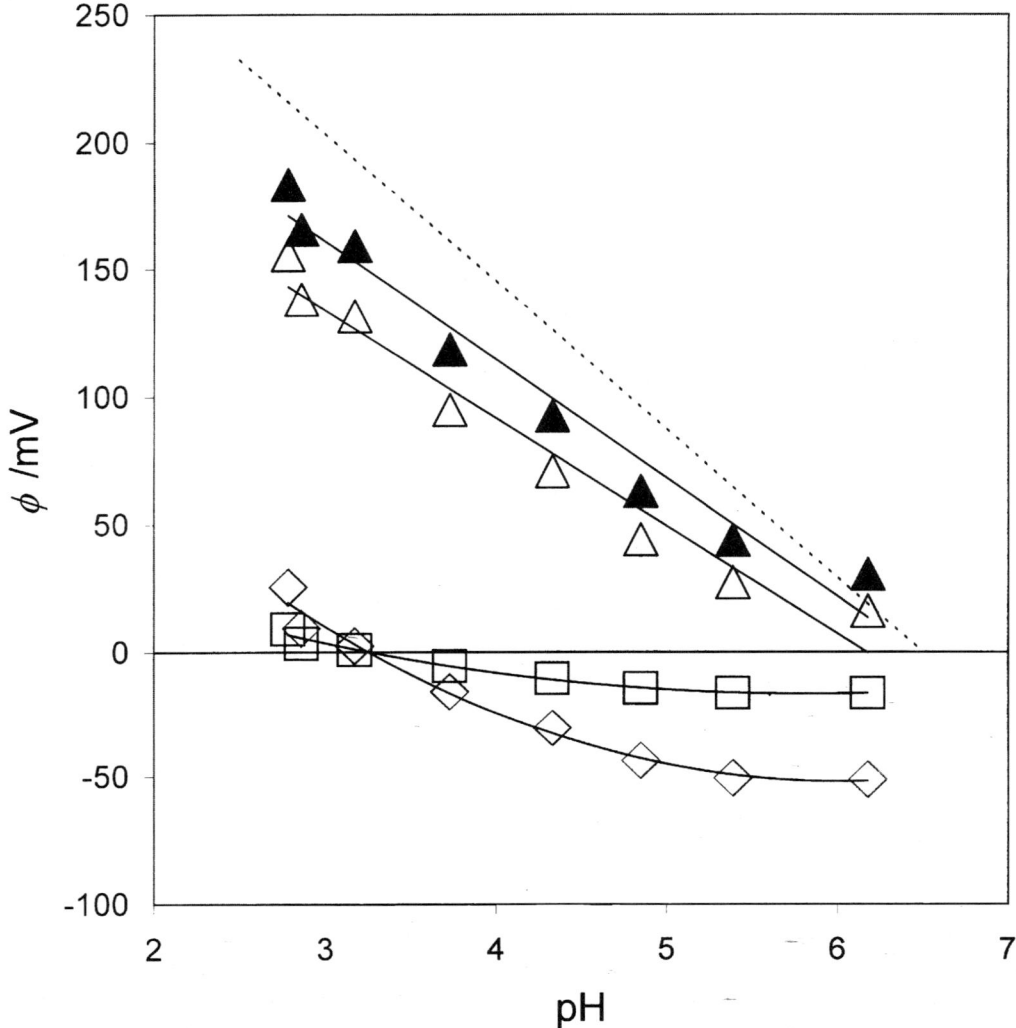

FIG. 5 Adsorption of salicylic acid on hematite. Data are taken from the experiment shown in Fig. 3. Potentials in the interfacial layer as a function of pH ($s = 15$ Å); measured ζ-potentials ($\square$), calculated ϕ_d ($\diamond$), and ϕ_0 ($\blacktriangle$,$\triangle$). The dashed line presents the theoretical Nernstian slope at 20°C of -58.2 mV. In calculations of f, two limiting values of C_1 were used: $C_1 = 1.4$ F m^{-2} ($\blacktriangle$) and $C_1 = 1.7$ F m^{-2} ($\triangle$). (From Ref. 11.)

Concentrations of positive and negative surface species were obtained from Eqs. (27–29) using ϕ_0 values as obtained for two limiting values of C_1 (Fig. 6). The concentrations of surface species obtained by the above procedure were used in linear plots according to Eq. (30). Since the slope yields the value of C_1, the used values of this parameter (first part of interpretation) could be tested. On the basis of the ϕ_0(pH) function (Fig. 5), the value of C_1 was estimated as 1.4 to 1.7 F m^{-2}. Since the slope of the linear presentation according to Eq. (30) agreed with this estimation, it may be concluded that

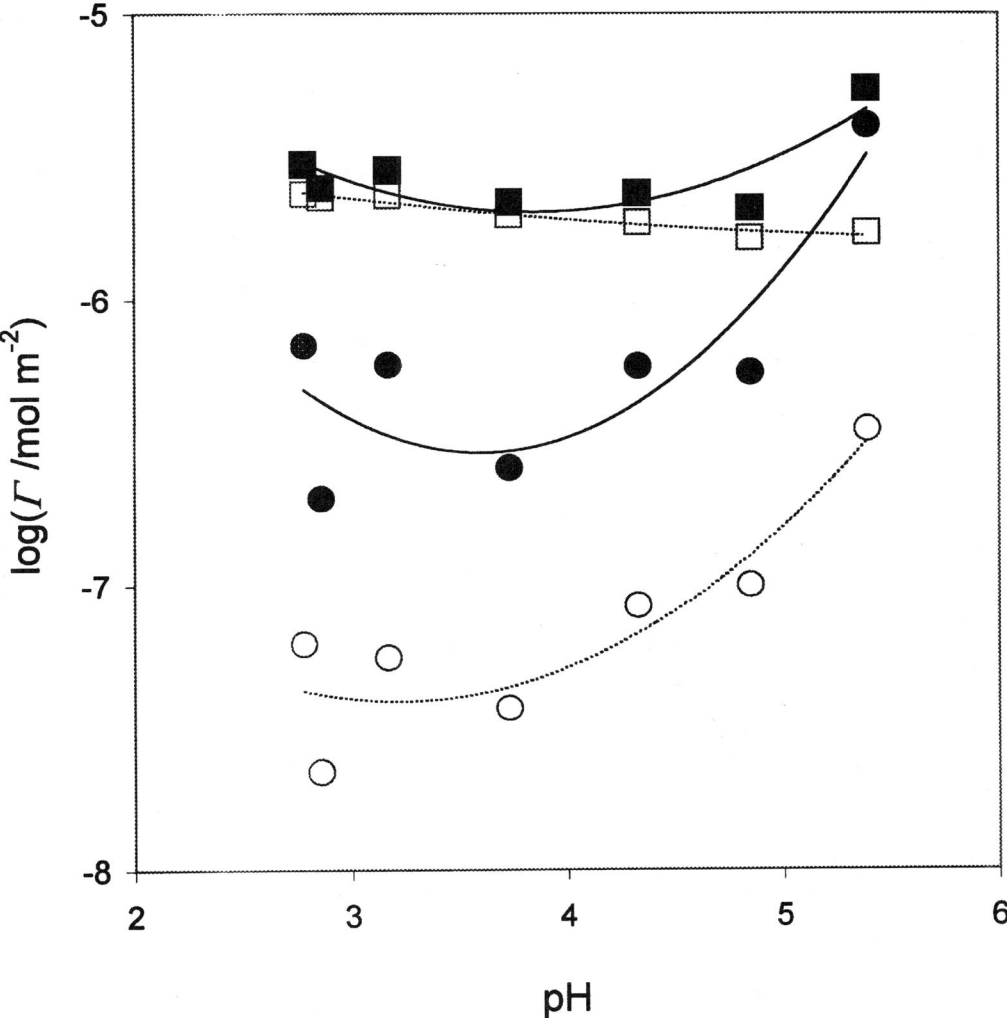

FIG. 6 Adsorption of salicylic acid on hematite. Data are taken from the experiment shown in Fig. 3. Surface concentrations of positive MOH_2^+ (■,□) and negative MO^- (●,○) surface sites. In calculation, two limiting values of C_1 were used: $C_1 = 1.4$ F m^{-2} (■,●) and $C_1 = 1.7$ F m^{-2} (□,○). (From Ref. 11.)

the assumed range of the capacitance is representative for the interfacial layer in the presence of adsorbed salicylate ions. The value of $\ln(K_p/K(HL^-))$ was roughly estimated as between 5 and 7 due to the scattering of data points.

The number of surface sites covered with a salicylate ion was obtained from the slope in the linear plot according to Eq. (31), which is presented in Fig. 7. Both limiting values of the capacitance produced similar slopes, so that the umbrella coefficient is obtained as $f = 6 \pm 1$. This result slightly depends on the assumed value of the total concentration of surface sites. However, it may be concluded that the salicylate ion is bound to one

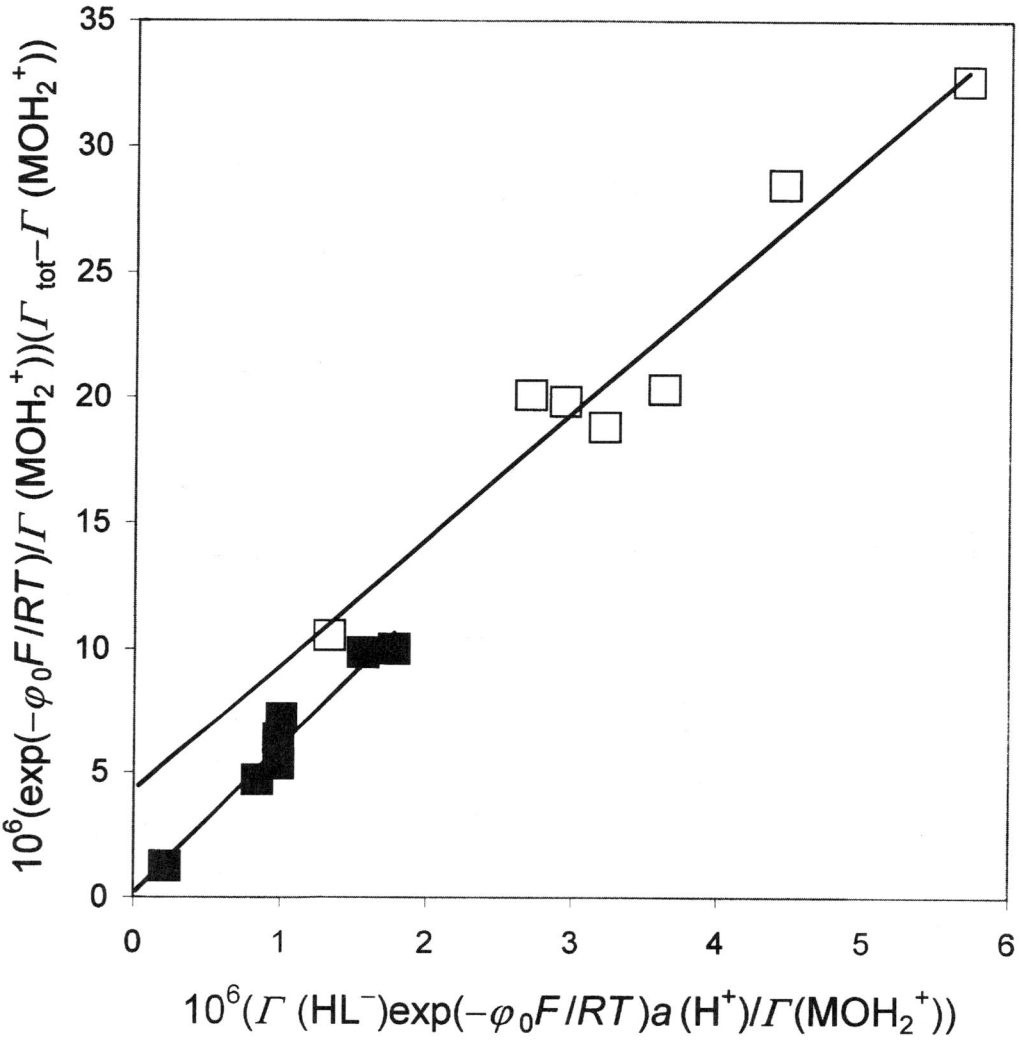

FIG. 7 Adsorption of salicylic acid on hematite. Data are taken from the experiment shown in Fig. 3 and are interpreted by Eq. (31). In calculation, two limiting values of C_1 were used: $C_1 = 1.4$ F m^{-2} (■) and $C_1 = 1.7$ F m^{-2} (□); $\Gamma_{tot} = 1.5 \cdot 10^{-5}$ mol m^{-2}. (From Ref. 11.)

neutral ≡MOH group but covers approximately four to six others, which is reasonable considering the size of the salicylate ion. It is also in accordance with the interpretation based on the extended Langmuir isotherm [Eq. (10)] yielding the area per absorbed ion of about 90 Å^2. The equilibrium constant of protonation K_p is obtained from the intercept and lies in the range between $2 \cdot 10^5$ and $7 \cdot 10^6$. This result agrees with the values obtained by the potentiometric measurements in the absence of specific adsorption [7], i.e., by interpretation of the σ_0(pH) function. The equilibrium constant of deprotonation K_d can be obtained from K_p and pH$_{pzc}$ (= 6.5), via Eq. (26), and is between

Interpretation of Interfacial Equilibria

$7 \cdot 10^{-7}$ and $2 \cdot 10^{-8}$. Once the concentration of all surface species, and also the electrostatic potential affecting the state of adsorbed ions, are known, it is possible to calculate the equilibrium constant for the binding of salicylate ions via Eq. (19). The obtained value of $K(\text{HL}^-)$ is estimated to be in the range from 100 to 1000.

D. Adsorption of Heavy Metals on Metal Oxides

Interfacial equilibria in the system metal oxide/heavy metal electrolyte solution can also be investigated on the basis of adsorption and electrokinetic data [38]. Heavy metal ions exhibit pronounced affinity for adsorption at metal oxide interfaces. Figure 8 is a schematic presentation of the electrical interfacial layer for such a case. It applies to the adsorption of hydrolyzed cadmium ions on a goethite surface at pH ≈ 6, at which the net surface charge in the 0-plane is negative and the CdOH^+ species are bound to negative MO^- groups. A reversal of charge at the interface occurs and is demonstrated by potentials at different planes. As an example, the adsorption of cadmium on goethite, which is often used as a model system for the adsorption of heavy metals on metal oxides [39–42], will be described. Several possibilities of interactions with surface MO^- groups exist for the binding of Cd-species:

$$\text{MO}^- + \text{Cd}^{2+} \rightarrow \text{MO} \cdot \text{Cd}^+ \qquad K_{a(1.2)}\exp\left(\frac{-2\phi_a F}{RT}\right) = \frac{\Gamma(\text{MO} \cdot \text{Cd}^+)}{\Gamma(\text{MO}^-)a(\text{Cd}^{2+})} \tag{32}$$

$$(\text{MO}^-)_2 + \text{Cd}^{2+} \rightarrow (\text{MO})_2 \cdot \text{Cd} \qquad K_{a(2.2)}\exp\left(\frac{-2\phi_a F}{RT}\right) = \frac{\Gamma(\text{MO})_2 \cdot \text{Cd})}{\Gamma(\text{MO}^-)a(\text{Cd}^{2+})} \tag{33}$$

$$\text{MO}^- + \text{CdOH}^+ \rightarrow \text{MO} \cdot \text{CdOH} \qquad K_{a(1.1)}\exp\left(\frac{-\phi_a F}{RT}\right) = \frac{\Gamma(\text{MO} \cdot \text{CdOH})}{\Gamma(\text{MO}^-)a(\text{CdOH}^+)} \tag{34}$$

$$\text{MO}^- + \text{Cd}(\text{OH})_2 \rightarrow \text{MO} \cdot \text{Cd}(\text{OH})_2^- \qquad K_{a(1.0)} = \frac{\Gamma(\text{MO} \cdot \text{Cd}(\text{OH})_2^-)}{\Gamma(\text{MO}^-)a(\text{Cd}(\text{OH})_2)} \tag{35}$$

where ϕ_a is the electrostatic potential of the plane a in which the adsorbed Cd species are located.

The total concentration of active surface sites in the interfacial layer is generally defined as

$$\begin{aligned}\Gamma_{\text{tot}} =& \Gamma(\text{MOH}) + \Gamma(\text{MOH}_2^+) + \Gamma(\text{MO}^-) + \Gamma(\text{MO} \cdot \text{Cd}^+) + \Gamma(\text{MO} \cdot \text{CdOH}) \\ & + 2\Gamma(\text{MO} \cdot \text{Cd}(\text{OH})_2^-) + \Gamma((\text{MO})_2 \cdot \text{Cd})\end{aligned} \tag{36}$$

The surface charge densities in the 0-plane and the a-plane are given by

$$\begin{aligned}\sigma_0 =& F(\Gamma(\text{MOH}_2^+) - \Gamma(\text{MO}^-) - \Gamma(\text{MO} \cdot \text{Cd}^+) - \Gamma(\text{MO} \cdot \text{CdOH}) \\ & - \Gamma(\text{MO} \cdot \text{Cd}(\text{OH})_2^-) - 2\Gamma((\text{MO})_2 \cdot \text{Cd}))\end{aligned} \tag{37}$$

$$\sigma_a = F(2\Gamma(\text{MO} \cdot \text{Cd}^+) + \Gamma(\text{MO} \cdot \text{CdOH}) + 2\Gamma((\text{MO})_2 \cdot \text{Cd})) \tag{38}$$

Again, the surface charge density in the diffuse layer (σ_d) is equal in magnitude, but opposite in sign, to the net charge bound to the surface (σ_s):

$$\begin{aligned}\sigma_s = -\sigma_d = \sigma_0 + \sigma_a =& F(\Gamma(\text{MOH}_2^+) - \Gamma(\text{MO}^-) + \Gamma(\text{MO} \cdot \text{Cd}^+) \\ & - \Gamma(\text{MO} \cdot \text{Cd}(\text{OH})_2^-)\end{aligned} \tag{39}$$

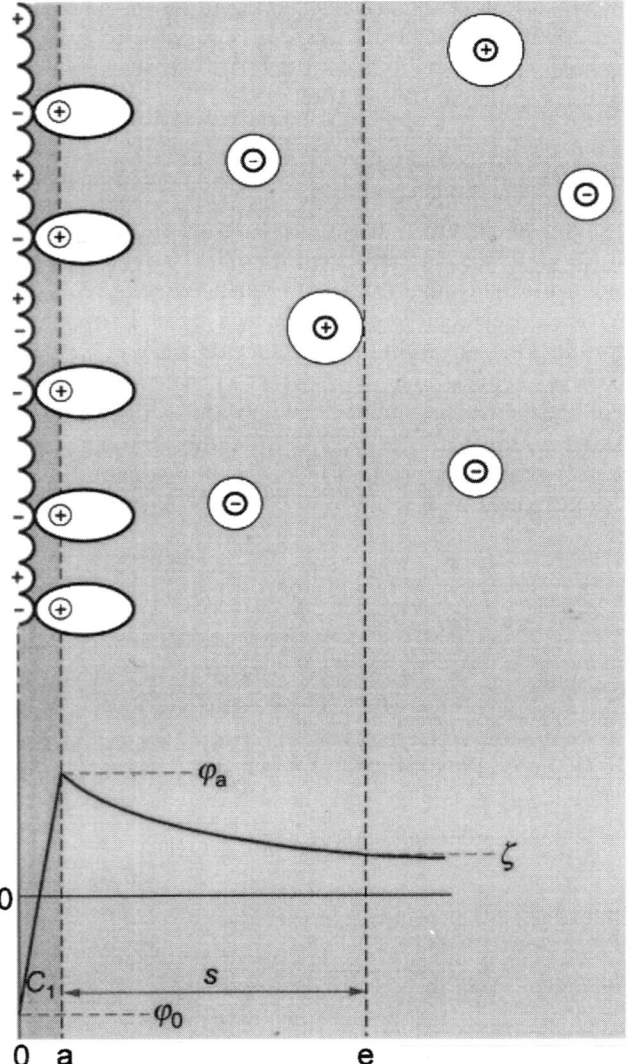

FIG. 8 Schematic presentation of the electrical interfacial layer (EIL) at a metal oxide surface with adsorbed metal ions.

Combining Eqs. (11,12,32–35,39), the expression for the surface concentration of negatively charged surface species $\Gamma(\mathrm{MO}^-)$ is obtained as

$$\Gamma(\mathrm{MO}^-) = \frac{\sigma_s/F - (z-n)\Gamma(\mathrm{Cd})}{(K_p/K_d)\exp(-2\phi_0 F/RT)a(\mathrm{H}^+)^2 - 1} \tag{40}$$

where $\Gamma(\mathrm{Cd})$ denotes the surface concentration of adsorbed cadmium species with the charge number z and n is the number of MO^- groups that build a surface complex. Equation (40) is valid for one of the possible mechanisms of the binding of Cd species. For the calculation of ϕ_0 and σ_s, Eqs. (2,5,6) are used.

In the interpretation of experimental data the following procedure was used:

$$\zeta \xrightarrow{s=15\text{Å}} \phi_d(\approx \phi_a) \xrightarrow{\text{G.C.}} \sigma_s(=-\sigma_d) \xrightarrow{-\sigma_a} \sigma_0 \xrightarrow{C_1} \phi_0 \tag{41}$$

For the calculation of ϕ_d ($\approx \phi_a$) from the measured ζ-potentials [Eq. (6)], the value of the electrokinetic slipping plane separation (s) should be assumed. Values of s from 5 to 25 Å were used, and no significant effect on the final results was found. Therefore $s = 15$ Å was used, since it was found to be representative for the metal oxide/aqueous interface [5–10,12,13]. In the second step, the Gouy–Chapman theory was used, enabling the calculation of σ_d and consequently the net surface charge density σ_s [Eq. (5)].

The next step is the evaluation of the surface charge density at the 0-plane; the surface charge density in the a-layer should be subtracted from σ_s. For this purpose, the value of σ_a should be calculated on the basis of Eq. (23). The problem could be solved assuming that only one kind of Cd species is adsorbable. All possible assumptions (binding of neutral, singly, and doubly charged species) were tested. The results obtained for the adsorption of singly charged species [Eq. (34)] satisfied all requirements, while the assumptions of binding of neutral and doubly charged species resulted in disagreement with theoretical requirements.

For evaluation of ϕ_0 from σ_0 [Eq. (2)], the value of the capacitance of the first interfacial layer C_1 is needed. The proper choice of C_1 value can be verified by considering the final results of the interpretation, as will be demonstrated. Once ϕ_0 is calculated, the surface concentrations of MO^- species and, consequently, adsorption equilibrium constants can be obtained. In doing so, the value of $K_p/K_d = 10^{17}$ was used, as obtained from the point of zero charge of goethite in the absence of cadmium [Eq. (26)].

The choice of a proper assumption regarding the mechanism of adsorption and of the representative value of C_1 is based on several requirements, which are

1. The calculated values of $\Gamma(MO^-)$ and K_a should be positive in the entire pH-region.
2. The slope of ϕ_0 vs. pH must be negative and (slightly) lower in magnitude than 58 mV/pH.
3. All data points should produce the same value of K_a.

The "Nernstian slope" $d\phi_0/d\text{pH}$ enabled the detection of the lower limit of capacitance, resulting in the conclusion that C_1 should be higher than 1.95 F m^{-2}. This finding agrees with the criterion based on the minimum of the standard deviation of K_a, obtained for $C_1 = 2$ F m^{-2}. The minimum is shallow, but one can still conclude that C_1 should be lower than 2.5 F m^{-2}. Accordingly, it was concluded that C_1 is approximately 2.2 F m^{-2}. Consequently, the limiting value of log $K_{a(1,1)}$ is 13.3 with the corresponding "Nernstian" slope of -51 mV/pH.

Figure 9 displays the adsorption data together with the calculated ϕ_0 and ϕ_d potentials in the interfacial layer of the goethite aqueous system in the presence of adsorbed cadmium species, as well as the measured ζ-potentials. Such a presentation can be used for a semiquantitative analysis. In the examined region, the potential affecting the state of dissolved Cd species is approximately constant. Thus the adsorption isotherm should follow the bulk equilibrium concentration of those species that actually adsorb. According to the equilibrium constants describing the hydrolysis of Cd species in the bulk of the solution, one can conclude that $CdOH^+$ species are bound to the surface, which indicates the mechanism of the adsorption process.

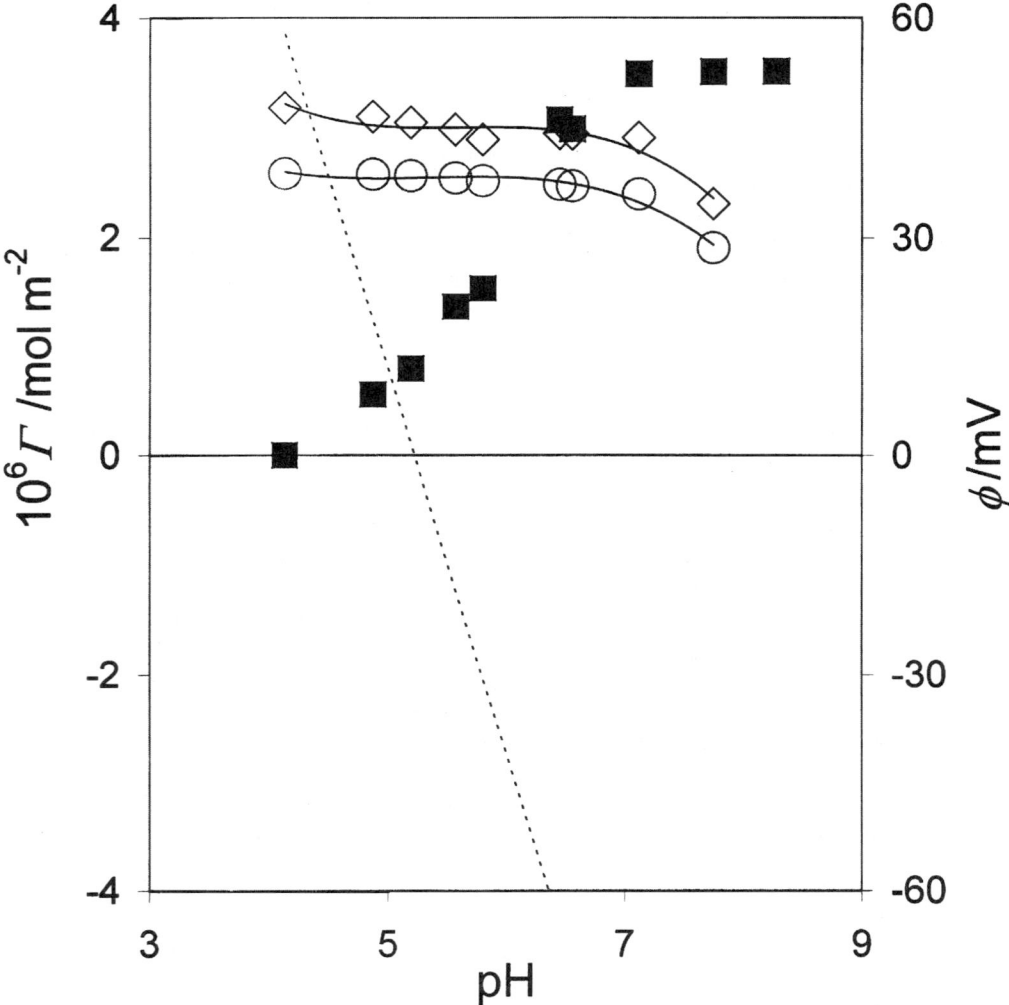

FIG. 9 Adsorption of cadmium at the goethite aqueous interface. Dependency of the surface concentration of Cd species (■), measured ζ-potential (○), calculated ϕ_d (◇), and ϕ_0 (- - -) potentials on pH at 20°C and $I_c = 1 \cdot 10^{-2}$ mol dm^{-3}. In adsorption measurements, the mass concentration of goethite was 10 g dm^{-3}, and the initial concentration of Cd was $2.5 \cdot 10^{-3}$ mol dm^{-3}. Potentials at the inner plane of the Helmholtz layer (ϕ_0) were calculated using the capacitance value $C_1 = 2.2$ F m^{-2}.

In this chapter, the application of electrokinetic measurements in the interpretation of equilibrium in the electrical interfacial layer was demonstrated. The problem that remains unsolved is the evaluation of the ζ-potential from the electrokinetic data. However, in any case the calculated ζ-potential would follow the ϕ_d potential, so that any problem in its evaluation will be reflected in the fitted value of the corresponding slipping plane separation. The value of this parameter is anyway apparent due to the common usage of the bulk permittivity.

REFERENCES

1. D. E. Yates, S. Levine, and T. W. Healy. J. Chem. Soc. Faraday Trans. I *70*:1807 (1974).
2. R. Sprycha. J. Colloid Interface Sci. *110*:278 (1986).
3. M. Mirnik. Croat. Chem. Acta *42*:161 (1970).
4. J. A. Davis, R. O. James, and J. O. Leckie. J. Colloid Interface Sci. *63*:480 (1978).
5. R. Torres, N. Kallay, and E. Matijević. Langmuir *4*:706 (1988).
6. P. Hesleitner, D. Babić, N. Kallay, and E. Matijević. Langmuir *3*:178 (1991).
7. P. Hesleitner, D. Babić, N. Kallay, and E. Matijević. Langmuir *3*:1554 (1991).
8. R. Sprycha and J. Szczypa. J. Colloid Interface Sci. *102*:288 (1984).
9. T. W. Healy and L. R. White. Adv. Colloid Interface Sci. *9*:303 (1978).
10. R. Sprycha and E. Matijević. Langmuir *5*:479 (1989).
11. D. Kovačević, I. Kobal, and N. Kallay. Croat. Chem. Acta *71*:1139 (1998).
12. D. Kovačević, N. Kallay, I. Antol, A. Pohlmeier, H. Lewandowski, and H. D. Narres. Colloids Surf. *140*:261 (1998).
13. N. Kallay, A. Čop, D. Kovačević, and A. Pohlmeier. Progr. Colloid Polymer Sci. *109*:221 (1998).
14. W. G. Eversole and P. H. Lahr. J. Chem. Phys. *9*:530 (1941).
15. W. G. Eversole and W. W. Boardman. J. Chem. Phys. *9*:798 (1941).
16. A. M. Gaudin and D. W. Fuerstenau. Trans. AIME *202*:66 (1955).
17. M. A. Blesa and N. Kallay. Adv. Colloid Interface Sci. *28*:111 (1988).
18. P. H. Tewari, A. B. Campbell, and W. Lee. Can. J. Chem. *50*:1642 (1972).
19. N. Kallay and E. Matijević. Langmuir *1*:195 (1985).
20. Y. Zhang, N. Kallay, and E. Matijević. Langmuir *1*:201 (1985).
21. J. Westall and H. Hohl. Adv. Colloid Interface Sci. *12*:265 (1980).
22. M. Čolić, D. W. Fuerstenau, N. Kallay, and E. Matijević. Colloids Surfaces *59*:169 (1991).
23. N. Kallay, M. Čolić, D. W. Fuerstenau, H. M. Jang, and E. Matijević. Colloid Polym. Sci. *272*:554 (1994).
24. N. Nilsson, P. Persson, L. Lövgren, and S. Sjöberg. Geochim. Cosmochim. Acta *60*:4385 (1996).
25. R. Rodriguez, M. A. Blesa, and A. E. Regazzoni. J. Colloid Interface Sci. *177*:122 (1996).
26. R. Kummert and W. Stumm. J. Colloid Interface Sci. *75*:373 (1980).
27. A. E. Regazzoni, P. Mandelbaum, M. Matsuyoshi, S. Schiller, S. A. Bilmes, and M. A. Blesa. Langmuir *14*:868 (1998).
28. E. C. Yost, M. I. Tejedor-Tejedor, and M. A. Anderson. Environ. Sci. Technol. *24*:822 (1990).
29. M. V. Biber and W. Stumm. Environ. Sci. Technol. *28*:763 (1994).
30. J. D. Kubicki, M. J. Itoh, L. M. Schroeter, and S. E. Apitz. Environ. Sci. Technol. *31*:1151 (1997).
31. N. H. G. Penners, L. K. Koopal, and J. Lyklema. Colloids Surf. *21*:457 (1986).
32. L. Bousse, N. F. De Roij, and P. Bergveld. Surf. Sci. *135*:479 (1983).
33. S. Ardizzone, P. Siviglia, and S. Trasatti. J. Electroanal. Chem. *122*:395 (1981).
34. N. Kallay, D. Babić, and E. Matijević. Colloids Surf. *19*:375 (1986).
35. H. C. Chang, T. W. Healy, and E. Matijević. J. Colloid Interface Sci. *92*:469 (1983).
36. M. J. Avena, O. R. Camara, and C. P. De Pauli. Colloids Surf. *69*:217 (1993).
37. N. Kallay, R. Sprycha, M. Tomić, S. Žalac, and Ž. Torbić. Croat. Chem. Acta *63*:467 (1990).
38. D. Kovačević, A. Pohlmeier, G. Özbas, H. D. Narres, and N. Kallay. Progr. Colloid Polymer Sci. *112*:183 (1999).
39. L. Gunneriusson. J. Colloid Interface Sci. *163*:484 (1994).
40. B. B. Johnson. Environ. Sci. Technol. *24*:112 (1990).
41. K. F. Hayes and J. O. Leckie. J. Colloid Interface Sci. *115*:564 (1987).
42. P. Venema, T. Hiemstra, and W. H. van Riemsdijk. J. Colloid Interface Sci. *183*:515 (1996).
43. M. Djafer, I. Lamy, and M. Terce. Progr. Colloid Polymer Sci. *79*:150 (1989).

7

Electrical Interfacial Layer in Nonaqueous Solvents

MAREK KOSMULSKI Institute of Catalysis, Polish Academy of Sciences, Krakow, and Department of Electrochemistry, Technical University of Lublin, Lublin, Poland

I.	Introduction	273
II.	Mechanism of Surface Charging in Nonaqueous Media	274
III.	Experimental Methods	278
IV.	Ions in Solution	280
	A. pH scales	282
V.	Empirical Solvent Scales	285
VI.	Surface Charge	289
	A. Silica	290
	B. Rutile	296
	C. Anatase	296
	D. Alumina	297
	E. Hematite	298
VII.	ζ-Potentials	299
	A. ζ-potentials in mixed solvents rich in water	299
	B. ζ-potentials in nonaqueous or almost nonaqueous solvents	301
VIII.	Stability	307
	A. Silica	308
	B. Goethite	308
	C. Alumina	308
	Note Added in Proof	308
	Symbols and Abbreviations	308
	References	310

I. INTRODUCTION

There are applications of chemistry in which "solution" and "aqueous solution" or "solvent" and "water" are almost pairs of synonyms. Although theoretically water can be replaced by another solvent, the abundance and low cost of water and its excellent physical, chemical, and physiological properties make it hard to imagine that any other solvent could replace water in certain fields; and in view of the growing ecological awareness, the "environment-friendly" technologies based on water have a much better prospect than those in which nonaqueous solvents are involved, so water is probably also the solvent of the future.

The solvent water is the one in which more fundamental studies have been carried out than in any other solvent, and the literature data on the physical properties of aqueous solutions are readily available. In contrast, there are rather few sources where extensive collections of data on solutions in mixed and nonaqueous solvents can be found [1].

Water is an exceptional solvent, and the results obtained in aqueous solutions cannot necessarily be generalized for other solvents. The unusually high freezing and boiling points compared with other compounds of similar molecular weight, the expansion on freezing, the maximum of density of liquid water, a high relative permittivity of liquid water, an even higher one for ice, and a high ionic electric conductance for hydrated protons and hydroxyl anions can be given as examples of the outstanding properties of water.

It is believed that the bulk properties of water do not reflect the properties of thin layers of water in the interfacial region, especially near a charged solid surface. For example, the relative permittivity of partially organized dipoles of water at such surfaces is considerably lower than in bulk water, and values of ϵ_r^* down to 6 were proposed. In early 1970s, the existence of a separate phase called "polywater" or "anomalous water" near the surface was postulated, but this idea was finally abandoned [2].

The idea of an electrical double layer was first introduced for the mercury–aqueous solution interface, and studies of electrocapillarity in mixed solvents followed very soon. Admixtures of organic solvents lead to a decrease of the interfacial tension in the vicinity of the electrocapillary maximum, but far from the electrocapillary maximum the solvent effects on the electrocapillary curves of mercury are negligible. Details can be found in handbooks of electrochemistry and they will not be discussed here. Later, strong solvent effects on the electrical double layer at AgI were found. The point of zero charge and isoelectric point are shifted by more than one pAg unit at mole fractions of organic solvents on the order of 1%. These studies were recently summarized by Lyklema [3]. The present review is focused on electrical double layers in systems of oxide in solutions of simple inorganic (usually (1–1) electrolyte. Many surface charging studies of oxides in nonaqueous media were carried out in the presence of strongly species like surfactants and polymers, but these results will not be discussed here.

The surface charging in nonaqueous and mixed solvents has been reviewed by Lykema [4], Parfitt and Peacock [5], and Kitahara [6], and more recently by Morrison [7] and Kosmulski [8]. The present review reports the recent progress in the field and discusses some aspects that have not been considered in earlier reviews.

II. MECHANISM OF SURFACE CHARGING IN NONAQUEOUS MEDIA

The surface charge at an insulator–solution interface may be a result of preferential adsorption of ions from the solution or of the dissociation of surface groups. The bulk solution remains electrically neutral, and the surface charge must be balanced by a

---

* For the meanings of symbols and abbreviations see at the end of this chapter.

counterchange (layer of counterions) in the closest vicinity of the surface. Therefore the surface charging phenomena are often referred to as the formation of an electrical double layer.

The surface charging process of oxides in aqueous systems is rather well understood. Unless subjected to drying at extreme conditions, the surface of an oxide is covered by chemisorbed water, which forms surface hydroxyl groups SOH where S is a surface metal ion. The presence of such groups has been confirmed spectroscopically. IR spectroscopy is also one of the methods of determining the surface concentration of the hydroxyl groups. In contact with aqueous solution, a surface hydroxyl group can gain a proton from the solution to form a positive surface charge (which is balanced by counterions, in this case anions of the supporting electrolyte in the adherent solution) or dissociate a proton to form a negative surface charge balanced by cations of the supporting electrolyte.

$$SO^- \xrightleftharpoons[\text{low activity of protons in solution}]{} SOH \xrightleftharpoons[]{\text{high activity of protons in solution}} SOH_2^+ \qquad (1)$$

Reaction (1.1) is characterized by two surface equilibrium constants. Unfortunately, there is no direct method to check if an electrically neutral surface is predominantly covered by neutral SOH groups or by large (and equal) concentrations of SO^- and SOH_2^+ groups. In the more recent 1-pK model [9], the neutral SOH groups are not considered at all, and an electrically neutral surface is covered by an equal number of negatively charged $SO^{1/2-}$ and positively charged $SOH^{1/2+}$ surface groups.

$$SO^{1/2-} \xrightleftharpoons[\text{low activity of protons in solution}]{\text{high activity of protons in solution}} SOH^{1/2+} \qquad (2)$$

Protons and hydroxyl anions are potential determining ions for oxides. Application of the mass law to reactions (1) and (2) leads to the conclusion that when pH is greater than a characteristic value for a given oxide the oxide is negatively charged, and for pH below this value the oxide is positively charged. This characteristic pH value is called the point of zero charge (pzc). The pzc value is sensitive to specific adsorption, and the values obtained in the absence of specific adsorption are referred to as the pristine point of zero charge (ppzc). Table 1 presents ppzc values for some oxides. For more extensive collections of ppzc values, cf. Ref. 3 and 10.

The pzc values in Table 1 represent well-established values confirmed in many laboratories, using various experimental methods. These values are likely to be representative, i.e., any sample of a given oxide should have a ppzc equal or close to the value reported in Table 1. Significantly different values can be explained in terms of specific adsorption or experimental errors. Some common oxides, e.g., SiO_2, MnO_2, and SnO_2, are omitted in Table 1. The reason is that for these oxides, one sharp value can hardly be proposed. For example, the ppzc of SiO_2 is lower than that of any oxide listed in Table 1, but the literature data are scattered, and values from below 2 to above 3 are quoted.

In most studies of surface charging, oxides are assumed to be completely insoluble. Titania and hematite are examples of rather insoluble oxides, and for practical purposes the effects related to their solubility can be neglected. However, many common oxides,

TABLE 1 Pristine Points of Zero Charge of Some Oxides at 25°C. Less Reliable Values Are Given in Parentheses

Oxide	ppzc
Nb_2O_5	4.1
RuO_2	5.4
TiO_2	5.8
ZrO_2	7.8
CeO_2	8.1
Fe_2O_3	9.0
$\alpha\text{-}Al_2O_3$	9.1
Y_2O_3	(9.1)
ZnO	(9.2)
MgO	(12.0)

Source: Adapted from Ref. 11.

e.g., silica and alumina, show a significant solubility. At high pH values, the dissolution of amphoteric oxides leads to anions in solution, and a positive charge is left behind on the surface. The opposite effect occurs at low pH values.

Reactions (1) and (2) may be considered also as adsorption and/or desorption of potential determining ions being the main charging mechanism in the absence of specific adsorption, i.e., in solutions of inert electrolytes (alkali perchlorates and nitrates, sometimes also halides). Specific adsorption of ions is very common and occurs in most systems of practical importance. For example, multivalent ions and ionic surfactants adsorb specifically on oxides and affect the surface charge and the location of the pzc. The mechanism of surface charging as a result of the adsorption of multivalent ions of oxides is discussed in more detail in the chapter on counterion association in this book.

According to Eqs. (1) and (2), the surface charge density of oxides in the absence of specific adsorption can be determined from the uptake of H^+ or OH^- ions from the solution. This method is based on our knowledge that H^+ and OH^- ions are potential determining ions, and that their adsorption/desorption governs the surface charge. Adsorption of the potential determining ions is accompanied by adsorption of an approximately equal amount of counterions, and the total charge of adsorbed ions is close to zero, but we know that the potential determining ions on the one hand, and counterions on the other, are adsorbed at various distances from the surface, and the charge of counterions does not contribute to the surface charge. In water–organic mixtures, the degree of association of counterions is higher than in water: the "critical distance" for ion pairing is a function of the permittivity [12]. Counterion association is discussed in more detail in a separate chapter in this book.

However, in the presence of specific adsorption of ions, the uptake of H^+ or OH^- ions from solution cannot be identified with the surface charge. In the presence of specific adsorption, the surface charge is balanced by counterions in the adherent solution in a similar manner as in inert electrolytes. As a compromise between the attraction of mobile counterions by the fixed surface charge and the mutual repulsion of counterions, the diffuse layer is formed, that is, counterions balancing the surface charge are found at various distances from the surface. When a charged particle moves with respect to the surrounding medium (as a result of gravity, an electric field, etc.), most of the counterions move with the particle, but some fraction of counterions is left behind with the bulk phase. At a sufficiently low concentration of the supporting electrolyte, this fraction is significant enough to give rise to electrokinetic phenomena. The distribution of ions in the diffuse layer can be estimated from the Boltzmann equation

$$c = c_0 \exp\left(\frac{-ze\phi}{kT}\right) \qquad (3)$$

where c_0 is the bulk concentration of ions carrying z elementary charges (z is positive for cations and negative for anions) and ϕ is the local electrostatic potential. For high surface potentials and high ionic strengths, Eq. (3) leads to concentrations that do not have physical sense. Therefore more sophisticated models, e.g., a triple layer model, were introduced to describe the distribution of ions in the vicinity of a charged surface. A common feature of these models is that the outermost "layer" is a diffuse layer containing an excess of counterions whose distribution is governed by the Boltzmann equation. The electrokinetic charge can be measured when a significant excess of counterions is found beyond the shear plane.

The above charging mechanism [reaction (1) or (2)], originally proposed for water, applies also for mixed solvents containing water, and for water-free protic solvents, when the relative permittivity is high enough. These solvents can donate or absorb protons [reaction (1) or (2)], and they can dissolve salts, which build the diffuse layer. Aprotic solvents have a limited ability to donate protons, but some of them are able to bind protons dissociated by the solid surface. Thus an aprotic character of the solvent is not an obstacle in the formation of a negative surface charge, when the electron pair donation ability of the solvent is high enough. The concentrations of ions in the presence of strong electric fields can be enhanced due to the Debye–Falkenhagen effect [13].

Electron transfer has been considered as a possible charging mechanism in media where ions are practically absent. Such a transfer is possible for solvent molecules directly interacting with the solid surface, but a diffuse layer consisting of ionized solvent molecules is hard to imagine. A more plausible charging mechanism in nonaqueous solvents is by ionic surfactants or related molecules capable of forming ions that are large and thus relatively stable against coalescence, while ions of opposite charge are adsorbed by solid particles. These ionophores can be introduced on purpose or present as impurities. In the latter case, the electrophoretic mobilities will be irreproducible, while ionic surfactants introduced in concentrations considerably exceeding the concentration of impurities lead to reproducible and easy-to-control mobilities. In contrast with aqueous systems, for which anionic surfactants give negative ζ-potentials as a result of the adsorption of high molecular mass ions, in nonaqueous systems anionic surfactants will rather lead to positive ζ-potentials with an excess of high molecular weight ions in solution and small inorganic cations adsorbed on the surface. Certain nonionic compounds enhance the conductivity of inorganic salts in nonaqueous solvents by many orders of magnitude. This synergism is due to the wrapping of molecules around inorganic ions, thus protecting them against coalescence. Again, these substances can be either added on purpose or present as impurities, an in the later case the electrokinetic properties remain beyond control.

In systems of high electrolyte concentration, the σ_0 significantly exceeds the electrokinetic charge, but in nonaqueous systems whose ionic strength is very low, the σ_0 and the electrokinetic charge are practically equal [4–8] and

$$\sigma_0 = \frac{\zeta\varepsilon}{a} \tag{4}$$

where a is the particle radius. This is because there are only a few counterions between the surface and the shear plane.

III. EXPERIMENTAL METHODS

In the absence of specific adsorption, the surface charge density of oxides in aqueous systems and in mixed solvents rich in water is obtained from potentiometric titration. To avoid change in the composition of a mixed solvent in the course of titration, the titrant (acid or base) should be a solution in a given solvent rather than a purely aqueous solution. This is specially important when the volume of the titrant is large compared with the initial volume of the solution. Titrations are usually carried out in an inert gas atmosphere to avoid the absorption of carbon dioxide. In practice, an inert gas is bubbled through or over the solution. Minor variations in solvent composition in the course of titration can be caused by different volatilities of the components of the solvent mixture. This effect can be avoided when the inert gas is first bubbled through a "washer" containing the same solvent mixture as that in the reaction vessel. In the presence of specific adsorption, the contribution of specifically adsorbed ions can be accounted for by a properly designed adsorption experiment, e.g., measurement of the uptake of given ions from the solution. One must be aware of a semantic problem here, i.e., some authors define the surface charge as the primary surface charge, i.e., the contribution of the net proton balance, and they neglect the contribution of specifically adsorbed ions, e.g., heavy metal cations. The problems with pH measurements in nonaqueous systems discussed in the next paragraph limit the application of potentiometric titration to mixed solvents rich in water.

Inert electrolyte titration can be used to find the pzc, when the values of σ_0 beyond the pzc are not required. This method leads to very accurate values of the common intersection point(cip) of titration curves, which in case of an inert electrolyte is identical with the pzc. Addition of small amounts of a concentrated solution of inert electrolyte causes a shift of pH toward the pzc, and at the pzc no shift is observed. This method can be used in mixed solvents, but the acquired information is useful only when the pH-meter reading has a well established physical sense. When a solution of inert electrolyte in a given solvent is used as a titrant rather than an aqueous solution, errors related to variations in solvent composition during the titration can be avoided. It should be emphasized, however, that the cip obtained in a mixed solvent is not necessarily equivalent to the pzc, even when cip = pzc for the same oxide and the same salt in a purely aqueous system. Either a cation or an anion can be specifically adsorbed from a solvent mixture.

Long ago, methods based on electro-osmosis were used to study electrokinetic properties of particles in nonaqueous media. More recently, electrophoresis (or microelectrophoresis) became the dominating method to determine the ζ-potential (electrokinetic potentials), also in nonaqueous media. This is mainly due to the availability of modern instruments with automatic determination of the velocity of particles, thus eliminating the problem of rather low mobilities of particles in media of low ϵ. New designs of electrophoretic cells make it possible to eliminate the effects of electro-osmosis in electrokinetic measurements.

Electrophoretic cells specially designed for measurements in nonaqueous solvents are available. Large flat electrodes with a narrow gap between them ensure a uniform electric field in the scattering volume. Nevertheless, most ζ-potential data for nonaqueous solvents reported in the literature, even quite recent, are based on measurements carried out in "ordinary" electrophoretic cells designed for aqueous systems, i.e., with relatively small electrodes and a broad distance between them. With nonpolar solvents the electric fields in such cells are not uniform, and in extreme cases the electric field is present outside the cell, so that even highly charged particles do not show any mobility.

The electroacoustic method can also be used in nonaqueous and mixed solvents when the ionic strength and thus the conductivity is high enough. In the case of the Acoustosizer, a large amount of solid is necessary to fill the cell. Therefore the instrument is suitable for carrying out titrations and collecting many data points using the same sample of solid. It is difficult to insulate the suspension from the atmosphere. In addition to problems with carbon dioxide, which are also faced in purely aqueous systems, the instrument is not suitable for work with very toxic and flammable organic solvents, and also the composition of solvent mixtures may change in the course of the titration. Some commercial zetameters display ζ-potential values calculated from the Smoluchowski equation:

$$\text{Electrophoretic mobility} = \frac{\zeta \varepsilon}{\eta}$$

assuming that the measurement has been carried out in water at 25°C. These values have to be recalculated when the measurement is carried out in a nonaqueous or mixed solvent. When $\kappa a << 1$ (Hückel limit) or $\kappa a > 100$ (Smoluchowski limit), where κ is the reciprocal Debye length and a is the particle radius, it is enough to know the viscosity and relative permittivity of the medium to calculate the actual ζ-potential in a solvent.

$$\zeta = 1.5 \zeta_{\text{displayed}} \times \frac{\varepsilon_{\text{water}}}{\varepsilon_{\text{solvent}}} \times \frac{\eta_{\text{solvent}}}{\eta_{\text{water}}} \qquad \kappa a << 1 \qquad (5)$$

$$\zeta = \zeta_{\text{displayed}} \times \frac{\varepsilon_{\text{water}}}{\varepsilon_{\text{solvent}}} \times \frac{\eta_{\text{solvent}}}{\eta_{\text{water}}} \qquad \kappa a > 100 \qquad (6)$$

Otherwise, the conversion procedure is much more complex. First, the electrophoretic mobility must be determined from the Smoluchowski equation using ϵ and η of water (unless the zetameter can be switched to display the mobility). Then a commercial computer program [14] can be used to calculate the electrophoretic mobility from the ζ-potential using ϵ and η of the solvent. Since the ζ-potentials are the input data in this program, the practical procedure is to prepare a kind of calibration curve covering the range of electrophoretic mobilities of interest. The concentrations of particular ionic species due to association are often lower than the analytical concentrations of corresponding salts. Their mobilities in the organic medium must also be known. Availability of relevant data (dissociation constants of salts, mobilities of ions) from the literature is limited. On the other hand, only very high errors (by an order of magnitude) in the concentrations of particular ionic species and their mobilities used in the calculations lead to significant errors in the calculated ζ-potentials. Therefore approximate methods, e.g., the Walden rule, are often used to estimate the missing data. It should be emphasized that simple inorganic salts in principle do not obey the Walden rule, so the calculated mobilities of ions reflect only the order of magnitude. A number of approximate formulae to calculate ζ-potentials, valid for $\kappa a > 10$, have been proposed [3]. They offer an attractive alternative to exact calculations, which are very tedious.

The above procedures apply only to zetameters based on microelectrophoresis. With the Acoustosizer, the corresponding correction equation is

$$\zeta = \zeta_{\text{displayed}} \times \frac{\eta_{\text{solvent}}}{\eta_{\text{water}}} \times \frac{\rho_{\text{solvent}}}{\rho_{\text{water}}} \times \frac{\rho_{\text{solid}} - \rho_{\text{water}}}{\rho_{\text{solid}} - \rho_{\text{solvent}}} \qquad \kappa a > 100 \tag{7}$$

The iep can also be estimated using indirect methods, i.e., from colloid stability or adhesion.

IV. IONS IN SOLUTION

Even in a pure nonaqueous solvent SH, where S is the part of the solvent molecule dissociable as an anion, ions exist as a result of autoprotolysis:

$$2\text{SH} = \text{SH}_2^+ + \text{S}^- \tag{8}$$

The concentrations of lyonium and lyate ions in a pure solvent are equal to the square root of the autoprotolysis constant.

The autoprotolysis in mixed solvents is more complicated. Reaction (8) in a binary water + ZH mixture symbolizes a combination of the following reactions:

$$2\text{H}_2\text{O} = \text{H}_3\text{O}^+ + \text{OH}^- \tag{9}$$
$$\text{H}_2\text{O} + \text{ZH} = \text{H}_3\text{O}^+ + \text{Z}^- \tag{10}$$
$$2\text{ZH} = \text{ZH}_2^+ + \text{Z}^- \tag{11}$$
$$\text{ZH} + \text{H}_2\text{O} = \text{ZH}_2^+ + \text{OH}^- \tag{12}$$

When ZH is completely inert, reaction (9) is the only significant process. However, absolutely inert solvents do not exist. Table 2 shows that also aprotic solvents exhibit some degree of autoprotolysis, and when the concentration of water is low enough, the contributions of reactions (10)–(12) might be significant even in relatively inert solvents.

The EMF of the cell,

$$\text{Pt} \mid \text{H}_2(101325\text{Pa}) \mid \text{KS} + \text{KCl in SH} \mid \text{AgCl} \mid \text{Ag} \mid \text{Pt} \tag{13}$$

where KS is potassium lyate equals

$$E = E^0 - \log\left(\frac{a_{\text{SH2+}} a_{\text{Cl-}}}{a_{\text{SH}}}\right) RT \frac{\ln 10}{F} \tag{14}$$

where the superscript [0] denotes the standard state. Extrapolation of the expression

$$\frac{(E - E^0)F}{RT \ln 10} + \log \frac{m_{\text{Cl-}}}{m_{\text{S-}}} \tag{15}$$

to zero ionic strength gives a molal pK_{ap}, where $m_{\text{Cl-}}$ and $m_{\text{S-}}$ represent molalities of KCl

Electrical Interfacial Layer

TABLE 2 Autoprotolysis Constants of Solvents at 25°C (Molal Scale)

Solvent	pK_{ap}	Ref.
Ethanolamine	5.2	16
Formic acid	6.2	17
Hydrogen peroxide	12.7	18
Hydrazine	13	19
Water	13.997	20
Acetic acid	14.5	17
Deuterium oxide	14.955	21
Ethylene diamine	15.2	22
Ethanediol	15.84	23
Methanol	16.45	24
Propanediol	17.21	23
Diethylene glycol	17.4	25
Ethanol	18.67	26
1-Propanol	19.24	26
Ethyl cellosolve	19.3	25
1-Octanol	19.44	27
1-Hexanol	19.74	27
2-Propanol	20.58	27
1-Pentanol	20.65	27
1-Butanol	21.56	27
2-Methyl 2-propanol	28.5	28
Ethyl-methylketone	31	29
DMSO	33.4	30

Source: Adapted from Ref. 15.

and KS, respectively. It can be converted into molar pK_{ap} using the formula

$$pK_{ap}(\text{molar}) = pK_{ap}(\text{molal}) - 2\log\frac{\rho}{\rho_0} \qquad (16)$$

where ρ denotes density and ρ_0 is 1 kg dm^{-3}.

Most salts are strong electrolytes in water, but this is not necessarily the case in a nonaqueous solvent. The following equation can be used to estimate the association constant of a 1–1 electrolyte [31]:

$$K_A = \frac{4000\pi N_A r_i^3}{3} \times \exp\frac{e^2}{(4\pi\varepsilon r_i kT)} \qquad (17)$$

The results are very sensitive to the ion size parameter r_i, which unfortunately cannot be directly measured and may assume different values in different solvents.

Some salts show significant solubilities in nonaqueous solvents, but usually these solubilities are lower than in water (e.g., Table 3). Since the solubility product (in terms of activities) of a salt is solvent independent, this can be interpreted as the enhanced activity of ions in nonaqueous media. Inert solvents do not stabilize ions, but dissolution of ionic pairs or larger aggregates is an entropically favorable process. Thus the concen-

TABLE 3 Solubility of NaClO$_4$ in Water and Organic Solvents at 25°C, Expressed as Mass Fraction of NaClO$_4$

Solvent	Solubility (%)
Water	73.21
Acetone	34.1
Methanol	33.93
Ethanol	12.82
Ethyl acetate	8.8
1-Propanol	4.66
1-Butanol	1.83
2-Methyl 1-propanol	0.78

Source: Ref. 33.

TABLE 4 Solubility of NaCl in Organic Solvents at 25°C, Expressed as Mass Fraction of NaCl

Solvent	Solubility (%)	Ref.
Water	26.43	
Formic acid	5.21	34
Methanol	1.29	35
Ethanol	0.065	35
1-Propanol	0.012	35
2-Butanol	0.00047	36
Acetonitrile	0.00025	34
Acetone	0.00004	37

trations of simple ions in nonaqueous solvents are in principle much lower than the solubilities listed in Tables 3 and 4. Activities of single ions can be determined using extrathermodynamic assumptions [32].

The ability of a solvent to dissolve salts cannot be explained in terms of a single property of a solvent. For example, the solubility of NaClO$_4$ in acetone is only slightly lower than in water, while NaCl (Table 4) is practically insoluble in acetone. Detailed information about solubilities of particular salts in different organic solvents can be found in *Gmelin's Handbook* [38].

A. pH Scales

The mechanism of electrical charging of oxide–solution interfaces in polar nonaqueous and partly aqueous solvents is similar to that in aqueous solutions. Since in all these cases the process of surface charging is very pH sensitive, the questions of correctly defining, measuring, and interpreting pH are crucially important. However, even for aqueous solutions, these tasks are anything but trivial, let alone for partly aqueous and nonaqueous media.

As the basic theory and practice of pH determination are discussed in sufficient detail in Chapter 1 (Section III.C), only a few of the most important formulae will be cited here, for easier readability.

Under the so-called operational definition [39], the pH scale is defined by means of five standard aqueous buffer solutions whose values, denoted as pH(S), have been deduced from potentiometric measurements on cells without transference, in a relatively broad temperature range. The pH response of the glass electrode (GE) cell,

$$\text{GE} \mid \text{H}^+(\text{aq}) + \text{other ions} \parallel \text{KCl(aq, sat.)} \mid \text{RE} \tag{18}$$

RE being the reference electrode—nowadays the GE cell is commonly used for pH measurement—is approximately linear. The value of the slope, $|\partial E/\partial \text{pH}|$, is usually a few percent lower than the Nernstian value, $(RT \ln 10)/F$. Unfortunately, the EMF of cell (18) contains a component, called the asymmetry potential, that slowly fluctuates in an unpredictable way. Therefore, *at least* two standard buffer solutions are required for calibrating the glass electrode cell, and the calibration needs to be repeated every several hours. The calibration formula for the two-buffer case is

$$\text{pH} = \text{pH}(S_1) + \frac{\text{pH}(S_2) - \text{pH}(S_1)}{E_2 - E_1} \cdot (E - E_1). \tag{19}$$

Here E_1 and E_2 denote the e.m.f.'s measured with buffer solutions S_1 and S_2, respectively, while the unsubscripted symbols refer to the solution under study.

The linearity of Eq. (19) can be checked by measuring pH value(s) of one or more additional buffers with pH(S) value(s) lying between pH(S_1) and pH(S_2). A linear response means that the liquid junction potential is reasonably constant in the range pH(S_1) $\leq$ pH $\leq$ pH(S_2). This will usually hold for the solution under study as well, provided its ionic strength does not differ too much from those of the calibration buffers.

The value of pH measured in accordance with the operational definition can be interpreted as a negative decadic logarithm of the "activity" of H$^+$, i.e.,

$$\text{pH} = -\lg\left(\frac{[\text{H}^+]\bar{\gamma}_\pm}{\text{mol dm}^{-3}}\right) \tag{20}$$

provided the following three conditions are all met:
The liquid junction potential remains approximately constant throughout.
The solution under study is neither too acid nor too alkaline (2 < pH < 12, say).
Its ionic strength is low or moderate (less than, approximately, 0.1 mol dm^{-3}).

Although Eq. (19) has been originally introduced for water, the potential of hydrogen or the glass electrode can also be used to evaluate the pH in these nonaqueous or mixed solvent media whose relative permittivity is sufficiently high [40]. Electrometrical methods require a sufficiently high conductivity of the solution.

The suitability of the glass electrode as a proton sensing electrode, not only in pure water, is generally accepted, although it has also been criticized [41]. An application of Eq. (19) for a nonaqueous (or mixed) solvent does not pose any additional difficulties in comparison with pure water. When SCE is used as the reference electrode, the assumption that the liquid junction potential does not depend on the pH must be also made. An application of Eq. (19) for nonaqueous systems requires buffer solutions. One possibility is that the pH electrode is calibrated using pH buffers in the same solvent as the unknown solution. Such pH buffers can be defined using a similar convention as that previously described for water. The available recipes for pH buffers in nonaqueous and mixed solvents cover only a limited range of solvents and solvent mixtures (mostly water–organic). For some mixed solvents the problem of limited availability of pH buffers can be solved by interpolation or extrapolation. Empirical formulas to calculate a pH value of a buffer solution in various solvent mixtures involving various organic cosolvents have been also proposed [39]. Certainly, the standard state is now defined as a solution whose concentration is 1 mol dm^{-3} with respect to protons, and the properties are as at an infinite dilution in this solvent. This standard state is hypothetical, but it is not more hypothetical than the corresponding standard state in water. The pH scale defined for a solvent by means of Eq. (20) is called a solvent scale. The pH values on the solvent scale and the pH values on the aqueous scale (or on a solvent scale defined for another solvent) are unrelated. This is because these pH values refer to different standard states. For example, pH = 2 means that the activity of protons is 1/100 of that in the standard state (in a given solvent), but the relationship between the proton activities in standard states in different solvents remains unknown. Since the relative permittivity of most organic and mixed solvents is lower than that of water, enhanced coulombic interactions, and thus high activity coefficients of ions in nonaqueous systems compared with water, are expected. This phenomenon is often termed the primary solvent effect, and it is manifested in the (typically) low solubility of salts in nonaqueous systems.

The relationship between the solvent and the aqueous pH scale is expressed by the equation

$$\text{pH (aqueous scale)} = \text{pH (solvent scale)} - \log \gamma_t \qquad (21)$$

$$\gamma_t = \exp\left(\frac{\Delta_t G_H^0}{RT}\right) \qquad (22)$$

where $\Delta_t G_H^0$ is the standard molar Gibbs energy of proton transfer from water into a given solvent. Gibbs energies of transfer can be directly determined for salts, but some convention is needed to split such a Gibbs energy into contributions of single ions. The widely accepted convention is that two large ions of opposite sign but similar structure have the same $\Delta_t G_{ion}^0$. The validity of this assumption for tetraphenyloborate and tetraphenyloarsonium was studied and confirmed by Kim [42]. A few values of $\Delta_t G_H^0$ are summarized in Table 5.

The pH scale in a given solvent spans from 1 molal lyonium to 1 molal lyate in this solvent, i.e., from zero to pK$_{ap}$ in solvent scale. In aqueous scale, the last column in Table 5 is the low point of the pH scale in a solvent, according to Eq. (21) and the high point is $-\log(e) \Delta_t G_H^0 / RT + \text{pK}_{ap}$, respectively.

TABLE 5 Standard Gibbs Energies (in kJ mol^{-1}) of Transfer of Proton from Water to Nonaqueous Solvents

Solvent	$\Delta_t G_H^0$	$-\log(e) \Delta_t G_H^0 / RT$
Propylene carbonate	50	-8.76
Acetonitrile	46.4	-8.13
Nitrobenzene	33	-5.78
Ethanol	11.1	-1.95
Methanol	10.4	-1.82
1-Propanol	9	-1.58
Ethanediol	5	-0.88
Water	0	0
DMF	-18	$+3.15$
DMSO	-19.4	$+3.4$

Source: Ref. 43.

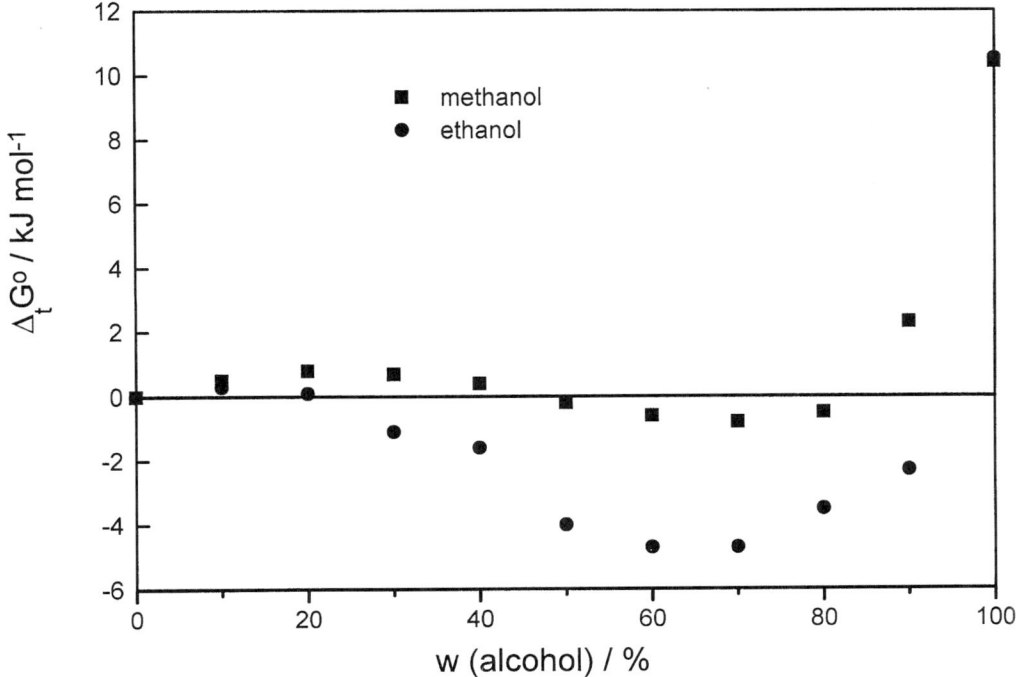

FIG. 1 Standard Gibbs energies of transfer of proton from water to aqueous alcohols.

The $\Delta_t G_H^0$ values for organic–water solvent mixtures are also available. Not necessarily are they linear functions of the solvent composition. Two examples, namely ethanol–water and methanol–water mixtures, are shown in Fig. 1.

Instead of using Eq. (21) we can apply standard buffers in given solvent whose pH values in aqueous scale are known. Recipes for such buffers are available in the literature [44]. Certainly, the extrapolation to zero $[Cl^-]$ cannot be used to assign the pH values of these buffers. The other method to measure the pH in mixed or nonaqueous solvents is to calibrate the electrode by means of aqueous buffers and introduce a correction ΔE_j for a difference in liquid junction potentials between the KCl salt bridge of the calomel electrode and water on the one hand and the given solvent on the other hand. Interesting enough, for dilute (mass fraction up to about 30%) aqueous ethanol and methanol, this correction is negligibly small (less than 10 mV or 0.2 pH unit), so with a pH electrode calibrated in aqueous buffers the pH meter display in these solvents gives a good estimate of the pH in aqueous scale (Fig. 2, data from Ref. 45). For nonaqueous solvents, the ΔE_j is larger, e.g., -35 mV for methanol [46], -71 mV for ethanol [47], and $+250$ mV for acetonitrile [48]. The ΔE_j data are only available for a few solvents or solvent mixtures.

V. EMPIRICAL SOLVENT SCALES

Let us consider a solvent-dependent quantity, e.g., an equilibrium constant or rate constant of a chemical reaction. Such a quantity can be a basis for a solvent scale with each solvent characterized by a number. In this manner an infinite number of solvent scales

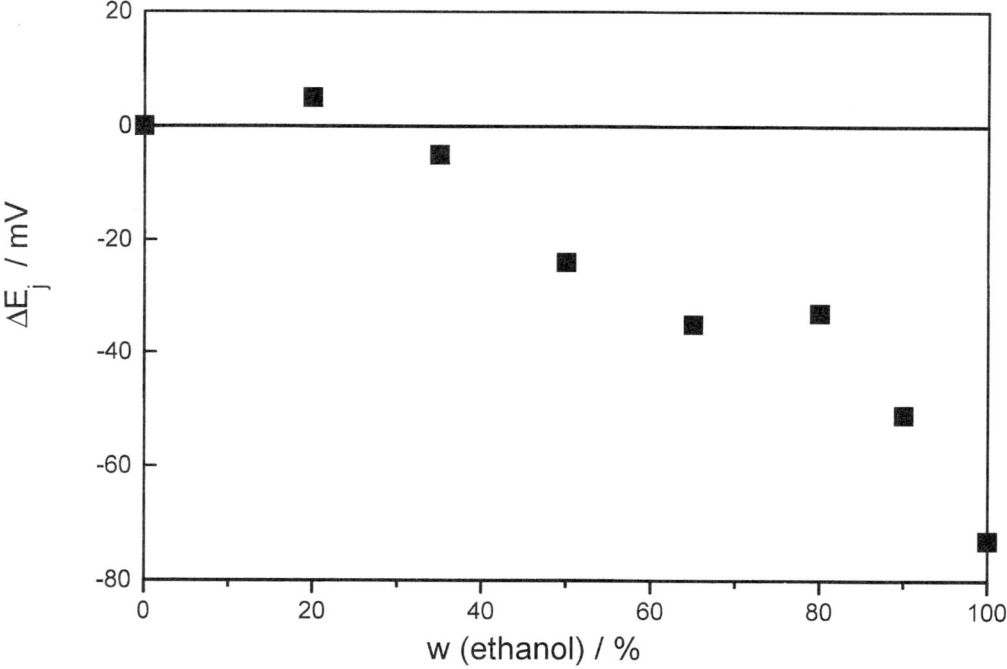

FIG. 2 Difference in liquid junction potentials between KCl salt bridge and water and aqueous ethanol.

can be built. It is not a surprise that these solvent scales are to some extent interrelated. On the other hand, the solvent scales can hardly be explained in terms of a single physical property of a solvent, e.g., relative permittivity, refractive index, etc., or a combination thereof. This led to the idea of empirical solvent scales. Standard Gibbs energies and activation energies of chemical reactions depend on the solvating power of the solvent in which the reaction is performed. Certainly, the solvation power depends on the nature of the solute (e.g., anions vs. cations). Therefore it would be expedient to find possibly a few model processes representing the solvation power of a solvent toward different types of solutes.

Kamlet et al. [49] have proposed the expression

$$XYZ = XYZ_0 + a\alpha + b\beta + s\pi^* \tag{23}$$

where XYZ is a quantity characterizing a solvent-dependent process, XYZ_0, a, b, and s are solvent-independent parameters characterizing this process, and the parameters α, β, and π^* characterize a solvent but are independent of the process. These three parameters are derived from "solvatochromic" properties, i.e., energies of the longest wavelength absorption peaks of specially selected probe molecules; α characterizes the hydrogen bond donation ability, β characterizes the hydrogen bond acceptance ability, and π^* is the polarity/polarizability parameter. These three parameters are dimensionless; their values are so normalized that for most solvents they fall between 0 and 1. Equation (23) can be expanded by the addition of (a) further term(s) to the r.h.s.

In Eq. (23), the solvent effect is split into three kinds of interactions, each characterized by one parameter. Also one-parameter approaches based on solvatochromism have been very successful. In these approaches,

$$XYZ = XYZ_0 + aX \tag{24}$$

where a is a solvent-independent parameter characterizing the process of interest and $X =$ Kosower's Z parameter or Reichardt's $E_T(30)$ parameter. Both parameters can be obtained as

$$E_T = \frac{hcN_A}{\lambda_{max}} \tag{25}$$

where λ_{max} is the wavelength of the maximum of the longest wavelengths in the absorption spectrum of a dilute solution of the corresponding solvatochromic dye in a given solvent, measured at 25°C, at atmospheric pressure. The $E_T(30)$ parameter can also be normalized to obtain a dimensionless E_T^N parameter, whose values range from 0 to 1. Usually it is a problem to find a solvatochromic dye sufficiently soluble in a polar solvent on the one hand and in apolar solvents on the other. This difficulty can be circumvented by using two probe molecules with analogous structures, whose solubilities are adjusted by suitable substituents, and whose E_T are linearly correlated. A successful synthesis of such probe molecules contributed to the success of the $E_T(30)$ scale.

The Kamlet and Taft Z and $E_T(30)$ are the most popular but not the only solvent scales based on solvatochromic dyes. Other examples of empirical solvent scales based on solvatochromism can be found, e.g., in Ref. 50.

In addition to solvatochromism, other solvent-dependent quantities were employed to build solvent scales.

$$XYZ = XYZ_0 + aAN + bDN \tag{26}$$
$$XYZ = XYZ_0 + aAcity + bBasity \tag{27}$$

The acceptor number AN is a dimensionless parameter ranging from 0 to 100 and representing the NMR chemical shift of ^{31}P in triethyloxophosphine. The donor number DN represents the heat of dissolution of $SbCl_5$ (a strong Lewis acid) in given solvent. Acity and Basity (sic! not acidity and basicity) are dimensionless parameters expressing the ability to solvate anions and cations, respectively. They were obtained by statistical evaluation of many solvent-dependent quantities taken from the literature and normalized to range from 0 to 1 for most solvents.

The values of solvent parameters are listed in Table 6. The solvents are listed in order of increasing polarity, expressed by the $E_T(30)$ parameter. Some of the solvent scales whose values are listed in Table 6 are interrelated, e.g., α, AN, and Acity of $E_T(30)$ and Z. These linear correlations suggest that in spite of different methods of measurement these solvent scales express in principle the same property of the solvent. This property of solvent scales makes it possible to estimate the values of parameters that cannot be measured directly, e.g., because of insufficient solubility of the probe molecules in the solvent of interest. Even better correlations are observed when one solvent parameter is expressed as a linear combination of two other parameters. For example, if we assume that α is really the measure of the hydrogen bond donation ability, and π^* of the polarity, and express other solvent parameters as linear combinations of α and π^*, it appears that

TABLE 6 Empirical Solvent Scales and Relative Permittivities

Solvent	α	β	π^*	$E_T(30)$	DN[†]	AN	Acity	Basity	Z	δ [52]	$\varepsilon_r^{\ddagger}$ [53]
n-Hexane	0	0	−0.04	31	(0)	0	0.01	−0.01		14.9	2.023
Benzene	0	0.10	0.59	34.3	0.1	8.2	0.15	0.59	54.0	18.8	2.275
Dioxane	0	0.37	0.55	36.0	14.3	10.3	0.19	0.67	64.5	20.5	2.209
THF	0	0.55	0.58	37.4	20.0	8	0.17	0.67	58.8	18.6	7.58
Ethyl acetate	0	0.45	0.55	38.1	17.1	9.3	0.21	0.59	64.0	18.6	6.02
Chloroform	0.20	0.10	0.58	39.1	(4)	23.1	0.42	0.73	63.2	19.0	4.9
Pyridine	0	0.64	0.87	40.5	33.1	14.2	0.24	0.96	64.0	21.9	12.3
Benzonitrile	0	0.37	0.90	41.5	11.9	15.5	0.30	0.87	65.0	17.2	25.2
Nitrobenzene	0	0.30	1.01	(41.5)[§]	4.4	14.8	0.29	0.86		20.5	34.82
Acetone	0.08	0.43	0.71	42.2	17.0	12.5	0.25	0.81	65.7	20.2	20.7
DMF	0	0.69	0.88	(43.8)	26.6	16	0.30	0.93	68.4	24.8	36.71
DMSO	0	0.76	1.00	45.1	29.8	19.3	0.34	1.08	70.2	24.5	46.68
Acetonitrile	0.19	0.40	0.75	45.6	14.1	18.9	0.37	0.86	71.3	24.3	37.5
Nitromethane	0.22	0.06	0.85	46.3	2.7	20.5	0.39	0.92	71.2	26.0	35.8
Propylene carbonate	0	0.40	0.83	(46.6)	15.1	18.3			72.4	27.2	66.1
2-Propanol	0.76	0.84	0.48	(49.2)	(36)	33.5	0.59	0.44	76.3	23.5	19.92
1-Butanol	0.84	0.84	0.47	(50.2)	(29)	36.8	0.61	0.43	77.7	23.3	17.51
1-Propanol	0.84	0.90	0.52	50.7		37.3	0.63	0.44	78.3	24.3	20.33
Ethanol	0.86	0.75	0.54	51.9	(32)	37.1	0.66	0.45	79.6	26.0	24.55
Methanol	0.98	0.66	0.60	55.4	(30)	41.3	0.75	0.50	83.6	29.6	32.70
Ethanediol	0.90	0.52	0.92	56.3	(20)		0.78	0.84	85.1	29.9	37.7
Formamide	0.71	0.48	0.97	(56.6)	(24)?	39.8	0.66	1.00	73.3	39.3	111.0
Propanetriol	1.21	0.51	0.62	57	(19)				82.7	33.7	42.5
Formic acid	1.23	0.38	0.65	(57.7)	(19)	83.6	1.18	0.51		24.7	58.5
Water	1.17	0.47	1.09	63.1	(18)	54.8	1	1	94.6	47.9	78.39

[†]The values in parentheses were calculated from correlations between solvent scales.
[‡]Most values are for 25°C, but 20 and even 15°C data are used for some solvents.
[§]The $E_T(30)$ values in parentheses: different values are quoted in Ref. 50.
Source: Ref. 51.

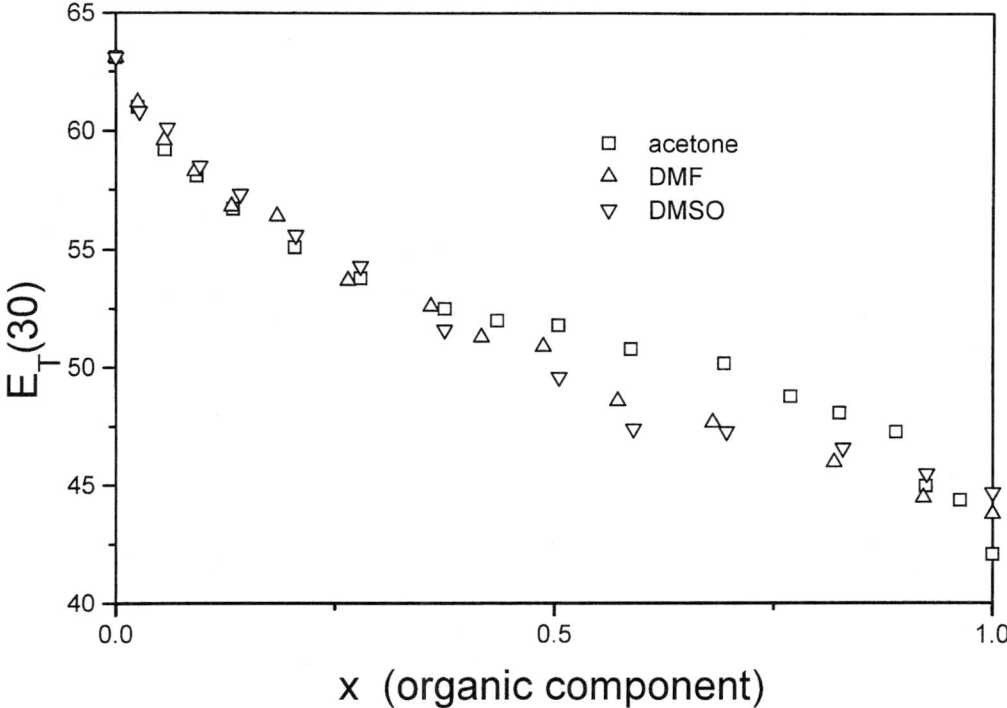

FIG. 3 $E_T(30)$ in mixed solvents. (From Ref. 54.)

$E_T(30)$ and Z are acidity scales with small contributions of polarity. Also solvent scales not discussed in the present chapter are linearly correlated with one of the above solvent scales or a combination thereof.

Technically, it is no problem to measure solvatochromic shifts (and other quantities defining solvent scales) in mixtures of solvents, and literature data are abundant, but the interpretation of such results is not obvious. The nonlinearity in the dependence of $E_T(30)$ on the solvent composition is most likely due to preferential solvation of the probe molecule by the components of the mixture. Typical courses of $E_T(30)$ as a function of the mole fraction are plotted in Fig. 3.

VI. SURFACE CHARGE

As discussed above, in view of the difficulties related to pH measurement and the interpretation of the results, the surface charging data are available only for mixed solvents rich in water. The values of σ_0 in nonaqueous or almost nonaqueous systems can be calculated from electrokinetic results, but since the pH cannot be measured, the results refer to some "natural" pH value dependent on the pretreatment of the solid sample, which is not necessarily the same for different samples of the same oxide, and which is very sensitive to impurities of the solvent. Comparison of quantities characterizing solvent mixtures involves difficulties that are not encountered when pure organic solvents are considered.

In order to compare the effects of two different organic cosolvents on σ_0, their solutions of equal ionic concentrations should be considered. The term "equal concentration" depends on the units used to measure the concentration. When two solvents that are compared have similar molecular weight, these units do not play a very significant role, but for example in equimolal aqueous solutions of methanol and DMSO in water, the mass concentration of the latter is over two times higher. In other words, the effects of low molecular weight compounds will be high when the effect is related to the mass concentration or fraction, and much lower (compared with other solvents) when the effect is related to the molal concentration. A similar problem is encountered with almost nonaqueous solvents: a certain amount of water in a low molecular weight nonaqueous solvent can be a much lower mole fraction than the same weight concentration of water in a high molecular weight nonaqueous solvent.

In addition to the most common ways of expressing the concentration, i.e., by weight on the one hand and by molality, molarity, and mole fraction on the other, other ideas can also be proposed, e.g., in view of the effect of the relative permittivity on the dissociation of electrolytes and the capacity of the double layer, it may be expedient to compare "isodielectric" mixed solvents, i.e., mixtures having the same relative permittivity, in studies of the solvent effects on the surface charging.

Another question is how to treat the ionic strength in mixed solvents. Certainly, in equimolar solutions of salts in various solvents, the activities of electrolytes and of particular ions can be different (cf., e.g., solubilities). Therefore it is questionable whether the experiments carried out at the same analytical concentration of an electrolyte in different solvents can be compared. Perhaps one should rather compare experiments carried out at equal activities of this electrolyte, or perhaps at a constant activity of counterions.

A. Silica

The most extensive studies of surface charging in mixed solvents were carried out for silica [55] over the pH range 6–8. At lower pH values, the surface charge density of silica is very low. At higher pH values, the solubility of silica becomes significant compared with σ_0.

In the presence of mixed solvents, the absolute value of the negative σ_0 of silica is lower than in water, with a few exceptions [55], e.g., heavy water and glycerol, whose admixture does not affect the surface charge density of silica. For a few organic cosolvents (methanol, DMSO), the solvent effect (relative decrease of σ_0) is proportional to the concentration of the organic cosolvent. The dependence of the σ_0 of silica on methanol concentration is plotted in Fig. 4. For other cosolvents, their small amounts lead to a relatively significant decrease in the absolute value of σ_0, while the effect of a further increase of the organic cosolvent concentration is relatively insignificant. Two examples of such a behavior, i.e., the dependence of the σ_0 of silica on DMF and 2-propanol concentration, are plotted in Figs. 5 and 6. The solvent effect on the σ_0 of silica is rather insensitive to the pH and the nature and concentration of the supporting electrolyte, but a very accurate comparison indicates that the solvent effect (relative decrease in σ_0) is as follows.

> It is somewhat more pronounced at higher ionic strengths.
> It becomes somewhat more pronounced in the series KCl, NaCl, LiCl, CsCl.

Some qualitative trends in the effect of the nature of the solvent are apparent: the solvent effect (relative decrease in σ_0) is as follows.

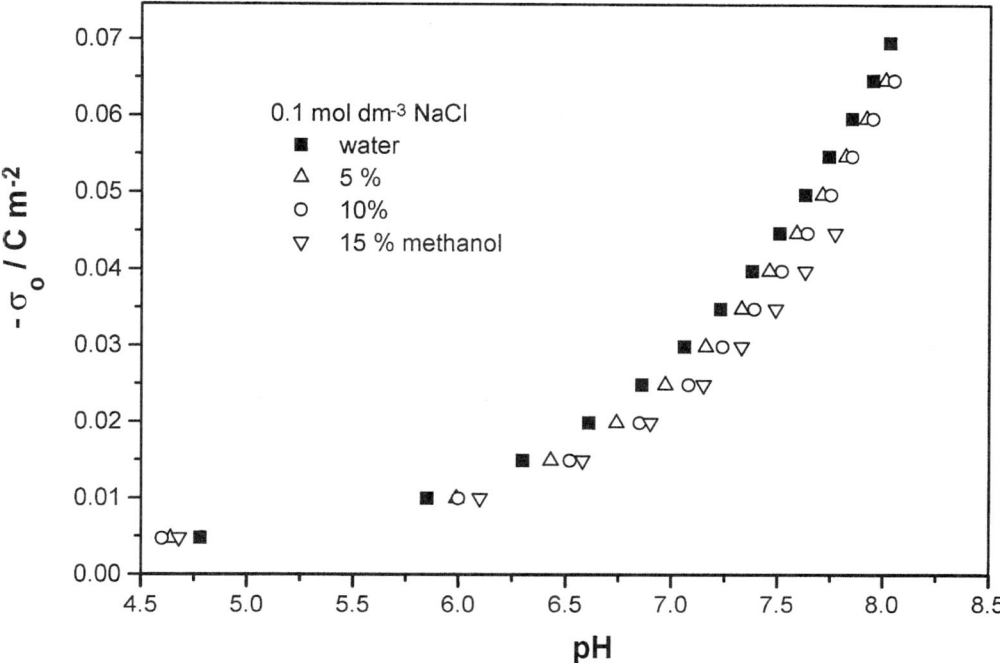

FIG. 4 Surface charge density of silica in aqueous methanol, 0.1 mol dm^{-3} KCl, 25°C. Concentration of the organic component is expressed as the mass fraction.

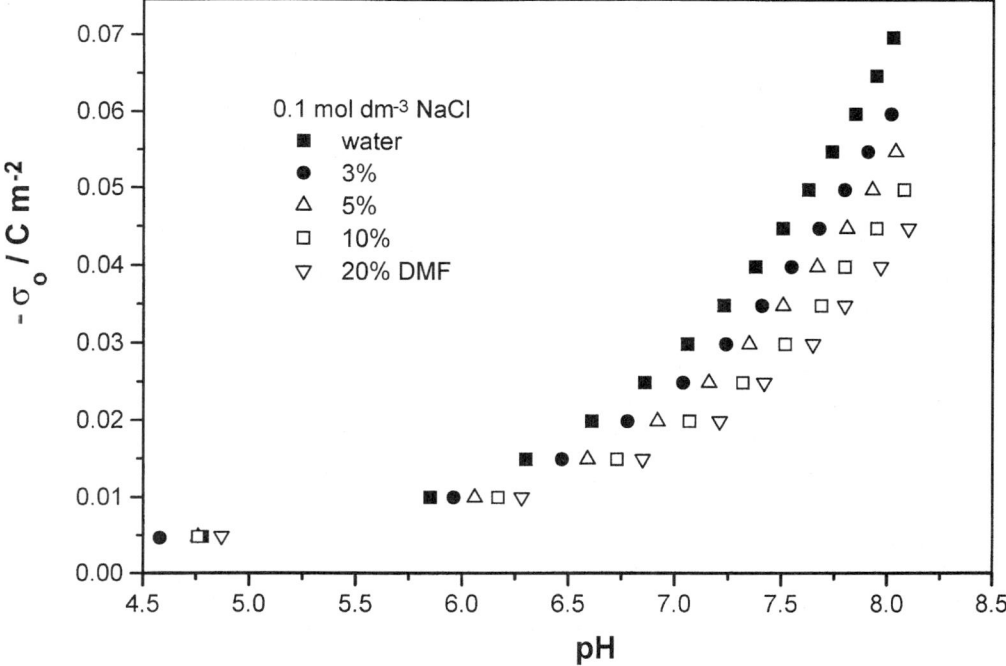

FIG. 5 Surface charge density of silica in aqueous dimethyl formamide, 0.1 mol dm^{-3} KCl, 25°C. Concentration of the organic component is expressed as the mass fraction.

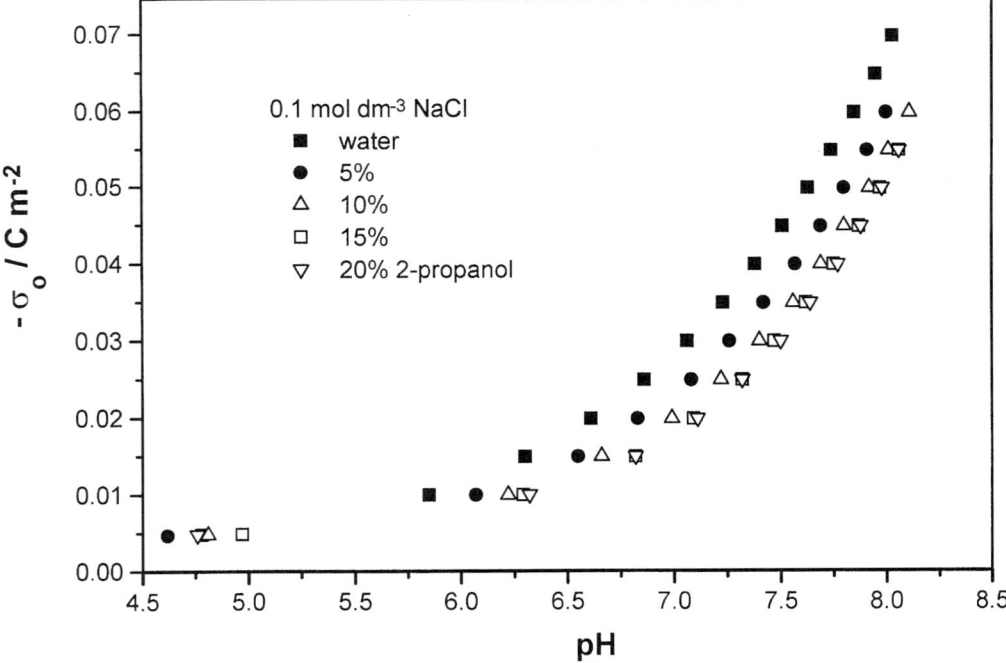

FIG. 6 Surface charge density of silica in aqueous 2-propanol, 0.1 mol dm^{-3} KCl, 25°C. Concentration of the organic component is expressed as the mass fraction.

It is more pronounced when the length of the hydrocarbon chain increases in a series of homologues, e.g., diols, Fig. 7. The solvent effect for pentanediol is equal to that of butanediol, at equal weight concentrations of these organic cosolvents. However, the molar concentration of pentanediol in Fig. 7 is considerably lower than that of butanediol. A similar trend is observed for monoalcohols.

It is similar for isomers. Figure 8 shows equal solvent effects for two isomeric butanediols and a slightly more pronounced effect of ethoxyethanol. Other pairs of isomers, e.g., 1- and 2-proponol, also exert similar effects.

It is less pronounced when the number of polar groups increases at constant length of the hydrocarbon chain. Figure 9 shows that the relative decrease in σ_0 is significant for ethanol but rather insignificant for ethanediol. Propanol, propanediol, and propanetriol and C_4 alcohols with up to four hydroxyl groups are another example for this rule of thumb.

Heavy water (mass fraction up to 50%) does not affect the σ_0 (Fig. 10).

The above trends qualitatively agree with solvent scales expressing the solvent polarity and with the relative permittivity of pure organic solvents. It may be tempting to relate the solvent effects to the relative permittivity of the mixture. However, the experimental data show that the relative permittivity of the medium is not the factor governing the surface charge of silica in organic–water mixtures. Mixtures of water with DMSO, $w_{H_2O} > 60\%$, have a relative permittivity close to that of pure water (Fig. 11), while the presence of glycerol leads to a significant decrease in the relative permittivity at relatively low concentrations of the organic solvent. The trend shown in Fig. 11 for glycerol is typical for organic solvents. However, DMSO belongs to the organic

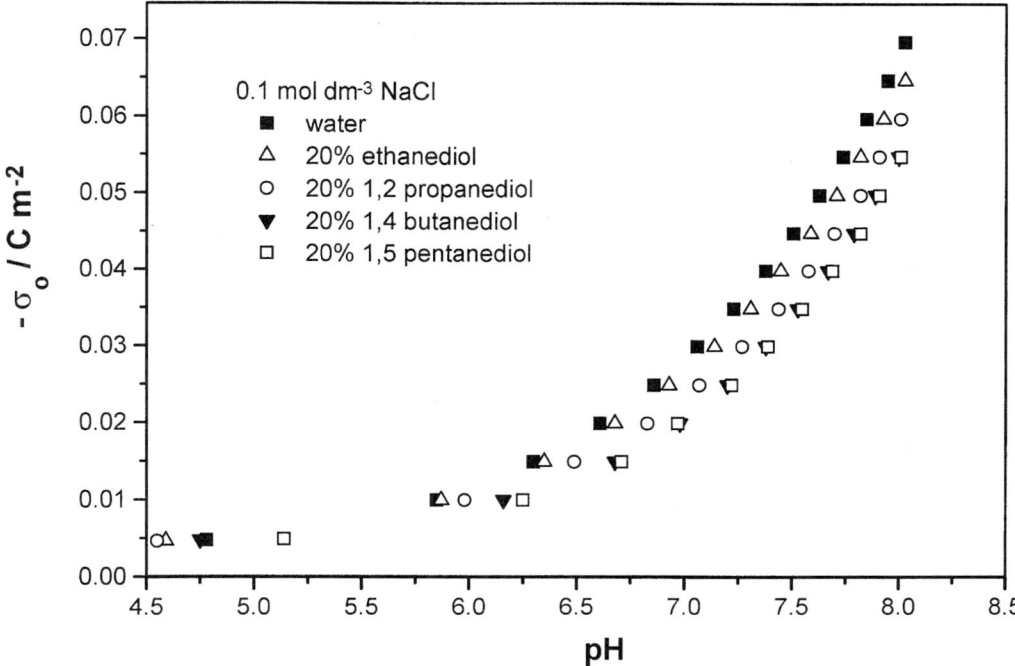

FIG. 7 Surface charge density of silica in aqueous diols, $w_{diol} = 20\%$, 0.1 mol dm^{-3} KCl, 25°C.

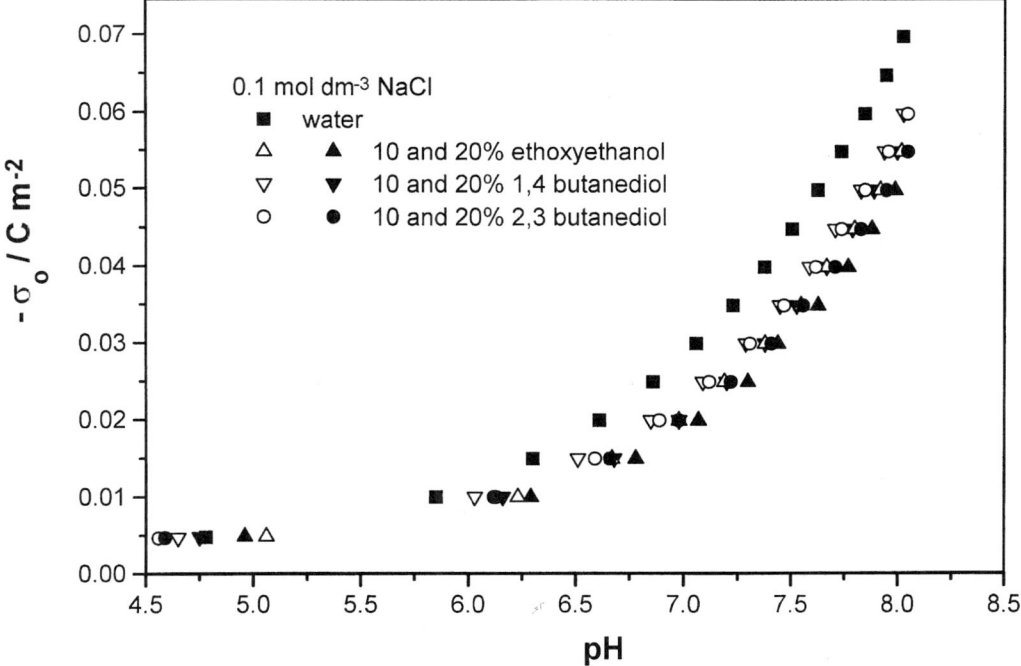

FIG. 8 Surface charge density of silica in 10% and 20% aqueous ethoxyethanol and butanediols, 0.1 mol dm^{-3} KCl, 25°C. Concentration of the organic component is expressed as the mass fraction.

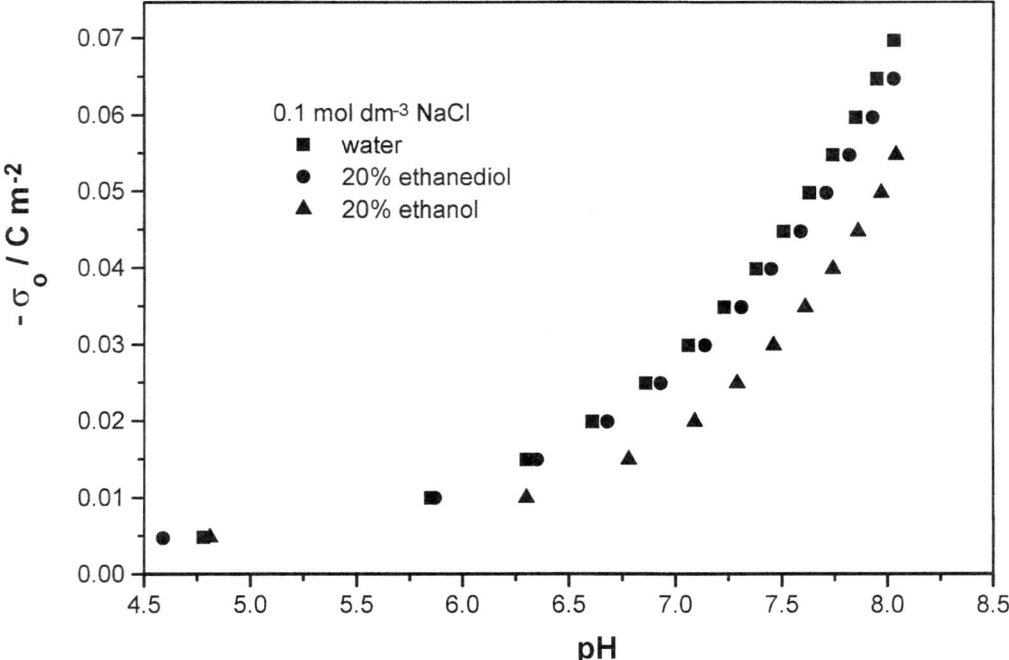

FIG. 9 Surface charge density of silica in aqueous ethanol and ethanediol, $w_{\text{alcohol}} = 20\%$, 0.1 mol dm^{-3} KCl, 25°C.

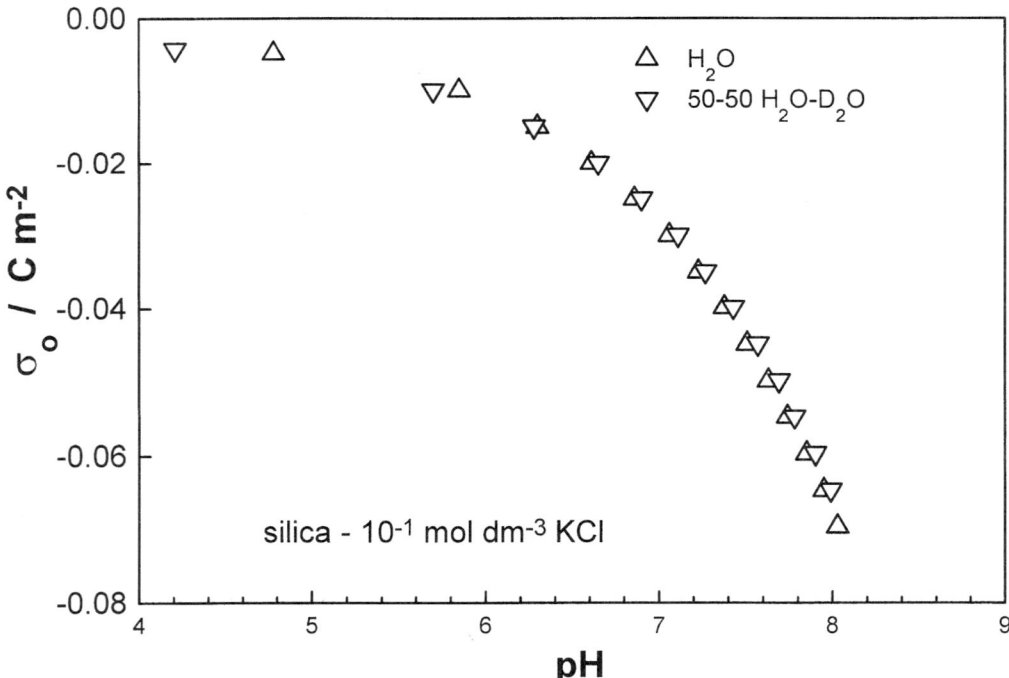

FIG. 10 Surface charge density of silica in 50–50 H$_2$O-D$_2$O (mass fraction), 0.1 mol dm^{-3} KCl, 25°C.

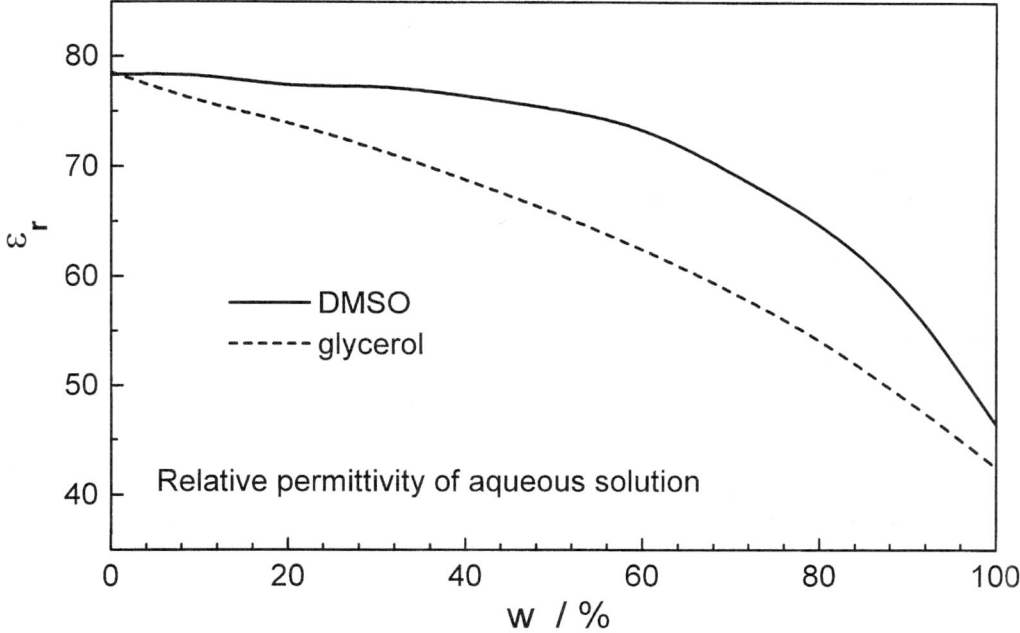

FIG. 11 Relative permittivity of aqueous glycerol and aqueous DMSO.

solvents, whose effect on the σ_0 of silica is significant, while the effect of glycerol is negligible; at a sufficiently low ionic strength, the σ_0 of silica increases in the presence of glycerol (this trend is opposite to that of most other solvents).

The solvent composition near the surface of silica is not necessarily the same as the bulk composition. The results obtained in alcohol–water mixtures suggest that the adsorption of organic solvents on silica from their dilute aqueous solutions is positive [56,57]. This explains the strong effect of many organic solvents at low concentrations.

The $E_T(30)$ parameter (values for pure organic solvents) is better correlated with the σ_0 of silica in organic–water mixtures than any other solvent scale (correlation with the Z scale is only slightly worse). The σ_0 is high for solvent having a high $E_T(30)$. Thus the solvent polarity parameter $E_T(30)$ seems to represent the ability of silica to dissociate protons into solution better than the relative permittivity. This is probably because the solvent polarity scales characterize the ability of the solvent to solvate particular ionic species. Even better correlation can be achieved in a two-parameter approach, when the solvent is characterized by a linear combination of $E_T(30)$ or Z solvent scale on the one hand, and β hydrogen bond acceptance ability or DN donor number on the other hand. The contribution of either of the latter two solvent parameters in these linear combinations is low as compared with that of $E_T(30)$ or Z, and high values of β or DN lead to high negative σ_0 of silica.

The adsorption of monovalent cations of inert electrolyte on silica is correlated with the negative σ_0. It has been shown (for a limited number of solvents in the range of low concentrations, $w_{\text{organic component}}$ up to 30%) that the ratio of the σ_0 of silica in mixed solvent and water is approximately proportional to the corresponding ratio of adsorption of different alkali metals (Li, Na, Cs) [58] and for different temperatures [59]. These results suggest that adsorption of alkali metal ions on silica can be used to estimate the σ_0 and thus to study the solvent effects on the σ_0 of silica for solvents in which the pH cannot be measured or the σ_0 data are not available. For example, the decrease of adsorption of Li^+ at pH 9 in the presence of 0.2 mol dm^{-3} urea, 2 mol dm^{-3} N-methyl acetamide, and 2 mol dm^{-3} methylurea [60] can be interpreted as a decrease of σ_0 in mixed solvents. On the other hand, adsorption of alkali metal cations from aqueous methanol, $w_{methanol} = 50\%$, is approximately equal to that from water, while the surface charge density in the mixed solvent is considerably lower. Thus the negative surface charge is neutralized to higher degree by counterions in the mixed solvent, and the negative diffuse charge decreases, and this leads to lower ζ-potentials.

The surface of controlled pore glass CPG is similar to that of silica, and similar solvent effects are observed [61]:

The presence of ethanol, 1-propanol, and 1-butanol leads to a decrease of the absolute value of the negative surface charge.

This effect is observed at various pH values and ionic strengths and for various electrolytes.

The relative decrease of σ_0 of CPG at equal weight concentrations of the alcohols is rather insensitive to the chain length. However, the effect related to molal concentration of alcohols (in bulk solution) is higher when the chain length increases.

B. Rutile

A qualitatively similar decrease of the absolute value of the σ_0 in mixed solvents as discussed above for silica was also observed for rutile [62]. The data for the two organic cosolvents studied (ethanol and 1-butanol) and one electrolyte (NaCl) show the following trends.

The relative decrease of σ_0 of rutile in the presence of mixed solvents is rather insensitive to pH and ionic strength.
The solvent effect is more pronounced for longer chains.
The pzc is not shifted in the presence of organic solvents.
The latter conclusion was also confirmed for dioxane ($w_{dioxane}$ up to 40%) [63].

C. Anatase

In view of the results obtained for rutile, it is rather unexpected that the σ_0 of another crystalline modification of titania, namely anatase, is insensitive to an admixture of methanol and ethanol [64] ($w_{alcohol}$ up to 30%) over a broad range of pH and ionic strengths (10^{-3} to 1 mol dm^{-3} KCl). Ethanol belongs to the solvents exerting a relatively significant effect on the charging curves of silica and rutile, and it is difficult to explain why the same cosolvent has a negligible effect on the σ_0 of anatase.

Janusz et al. [65] have confirmed that even for $w_{ethanol} = 50\%$, the σ_0 of anatase in 0.01 mol dm^{-3} NaCl is identical to that in water. On the other hand, the σ_0 in the mixed solvent measured by these authors was considerably higher than in water for 0.1 mol dm^{-3} NaCl but considerably lower for 0.001 mol dm^{-3} NaCl, without a significant shift of the pzc. These differences between water and the mixed solvent are only observed for the negative branch of the charging curves, while the positive branches are practically identical.

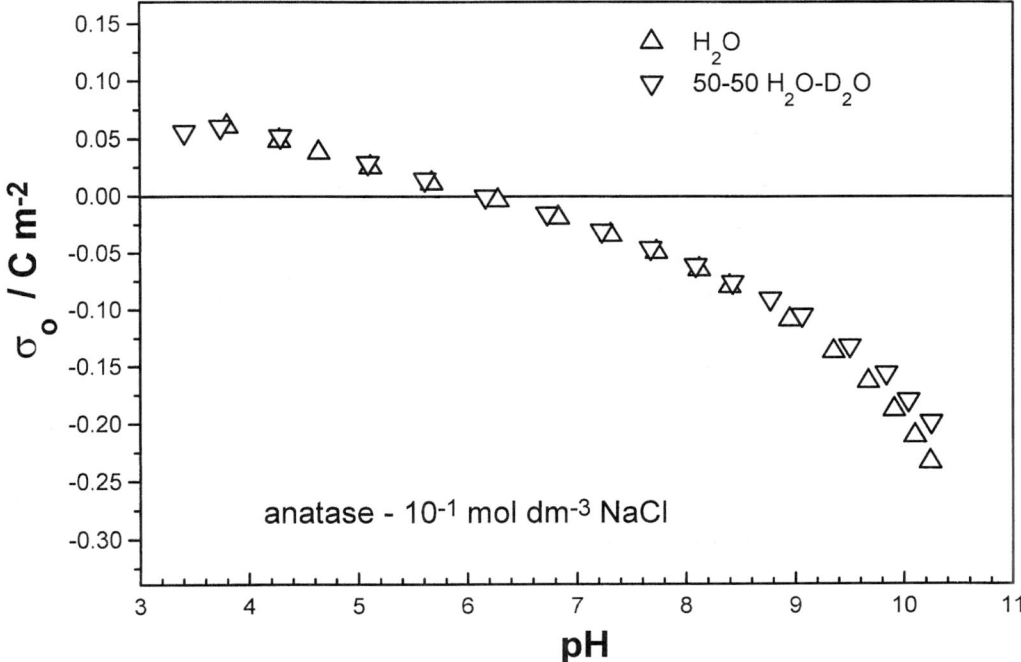

FIG. 12 Surface charge density of anatase in 50–50 H_2O-D_2O (mass fraction), 0.1 mol dm^{-3} NaCl, 25°C.

More recently, surface charging of anatase in a number of mixed solvents was studied [66]. Only heavy water (w_{D2O} up to 50%, Fig. 12) does not affect the σ_0 of anatase. For organic solvents, the negative branch of the charging curves is also unaffected with the one exception of DMSO, which causes a decrease in the absolute value of the negative surface charge, though this is less significant than the effect of the same solvent on the σ_0 of silica (Fig. 13). The DMSO effect observed for anatase is rather independent of the nature and concentration of the supporting electrolyte.

On the other hand, the positive surface charge of anatase (below the pzc) increases in the presence of DMSO, and the pzc is shifted to higher pH values than the pzc in water. Not only DMSO but also other organic solvents (dioxane, THF, glycerol) cause an increase of the positive surface charge of anatase, but with the latter solvents no significant shift of the pzc is observed.

D. Alumina

A series of experiments performed with *n*-alcohols [67] did not indicate a significant effect of these solvents on the σ_0 of, and sorption of the ions of supporting electrolyte (NaCl) on, alumina. Accordingly, the pzc is also insensitive to the presence of alcohols. On the other hand, the common intersection point (cip) of the charging curves of alumina clearly shifts to higher pH values in the presence of dioxane [63]. This result was erroneously interpreted as a shift of the pzc to higher pH values. A later study indicated that cip = pzc at the presence of inert electrolytes in water, but for the same electrolytes the cip does not necessarily coincide with the pzc in mixed solvents [66]. In the presence of dioxane, the values of

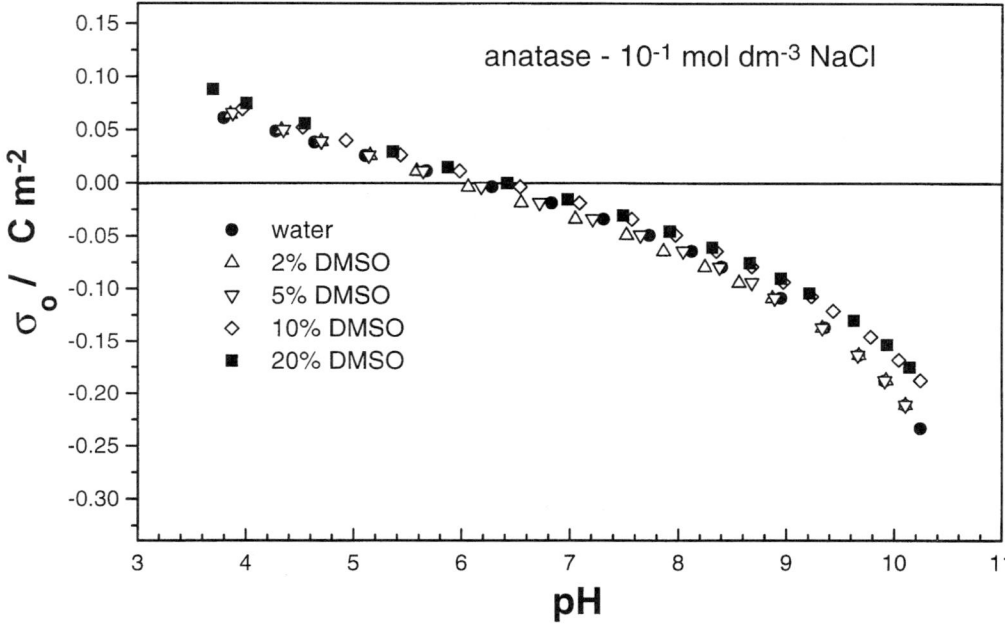

FIG. 13 Surface charge density of anatase in aqueous DMSO, 0.1 mol dm^{-3} NaCl, 25°C. Concentration of the organic component is expressed as the mass fraction.

σ_0 are equal to those observed in water when the surface carries a high positive or high negative charge. At pH values near pzc, the presence of dioxane leads to more negative values of σ_0 and the pzc is shifted to lower pH values. This effect is more significant at low ionic strengths (Fig. 14).

On the other hand, the pzc of alumina is shifted to higher pH values in the presence of DMSO as a result of a decrease in the negative surface charge in the mixed solvent, analogous to that observed for silica and anatase.

E. Hematite

Hesleitner et al. [68] have found that the effect of ethanol and methanol ($w_{alcohol}$ up to 50% by weight) on the surface charge density of hematite in 0.001 and 0.01 mol dm^{-3} NaNO$_3$ is insignificant.

Sworska et al. [69] have chosen NaCl for studies of the solvent effect on the surface charging of hematite. NaCl is not an inert electrolyte in this system; the chloride anions are specifically adsorbed. The result is that the common intersection point of the charging curves does not coincide with the pzc, which in turn is a function of the ionic strength. They have confirmed the absence of a significant difference between the charging curves of hematite in water on the one hand and aqueous ethanol $w_{ethanol} = 50\%$ on the other.

Szczypa et al. [70] measured the adsorption of Na$^+$ and Cl$^-$ ions on hematite from DMSO. Adsorption of the cations was significantly higher than that of the anions. Adsorption of both anions and cations from DMSO was more than 10 times higher than from water.

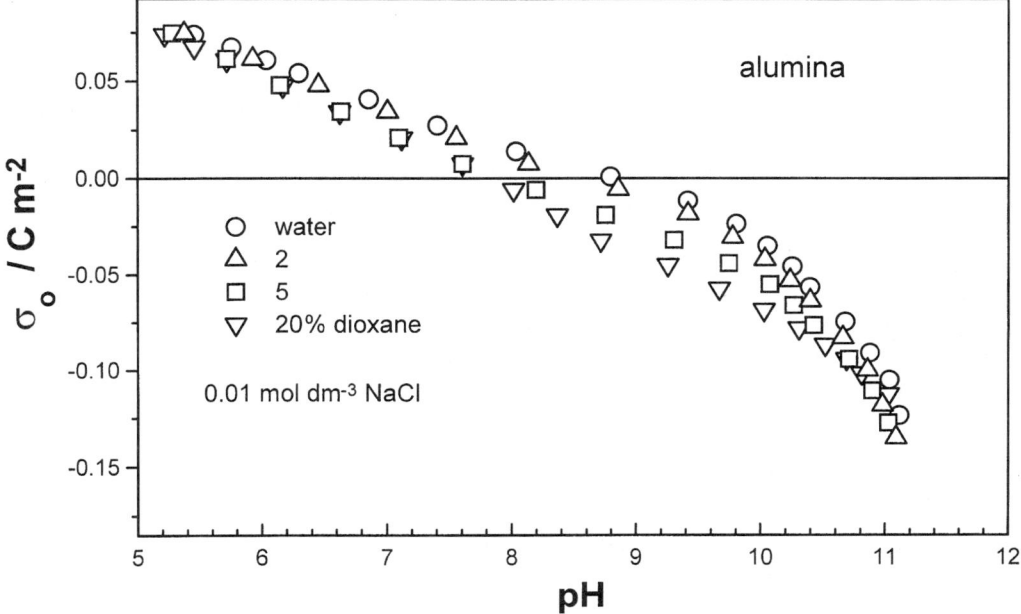

FIG. 14 Surface charge density of alumina in aqueous dioxane, 0.1 mol dm^{-3} NaCl, 25°C. Concentration of the organic component is expressed as the mass fraction.

The common intersection point of charging curves of hematite in KCl (from 1 : 1 electrolyte titration) shifts to lower pH values when water is replaced by water–dioxane mixtures. KCl is not an inert electrolyte in this system, since the chloride anions are specifically adsorbed. The cip of hematite in the presence of KNO$_3$ also shifts to lower pH values in dioxane–water mixtures ($w_{dioxane}$ up to 40%), but the further increase of dioxane concentration leads to a higher cip [71].

VII. ζ-POTENTIALS

A. ζ-Potentials in Mixed Solvents Rich in Water

In purely aqueous systems, the iep of oxides in 1 : 1 electrolytes is rather insensitive to the nature and concentration of these electrolytes. However, recent studies suggest that at a sufficiently high ionic strength (a few tenths of 1 mol dm^{-3}) the iep of oxides shifts toward higher pH values [72]. For some lithium and sodium salts, above a critical concentration, characteristic for a given salt, only positive ζ-potentials of anatase are observed (no iep) [72]. In this respect, the electrokinetic properties of oxides are different from the surface charging, since the pzc remains constant even at extremely high ionic strengths. At very high ionic strengths the physical properties of water, e.g., the relative permittivity, are significantly different from those of pure water. Most of the water molecules in such systems are engaged in solvation sheaths of ions (chiefly cations). Thus the solvent water at high ionic strengths is in some sense different from water at low ionic strengths. In mixed solvents rich in water, a similar shift of the iep is observed at much lower ionic strengths than in water.

1. Silica

At low ionic strengths and concentrations of organic cosolvents, the electrophoretic mobilities of silica in aqueous ethanol, methanol, and dioxane are considerably lower than in water. This is a result of a high viscosity and a low relative permittivity. When, however, ζ-potentials are calculated, the results obtained in water and mixed solvents match surprisingly well when the ionic strength and concentration of organic cosolvent are low enough [73]. At higher ionic strengths and higher concentrations of the organic component, the ζ-potentials in mixed solvents do not differ from those measured in water at pH < 7, but at higher pH values the negative ζ-potentials clearly decrease when the concentration of the organic component increases. Finally, at 0.1 mol dm^{-3} KCl and $w_{methanol} = 30\%$ or $w_{ethanol} = 30\%$, not only the negative ζ-potentials decrease but also the iep shifts to higher pH values. One possible explanation is in terms of the activity of the counterions, which is significantly higher in mixed solvents than in water. When instead of solutions of equal molar concentrations, solutions of equal activity of the K$^+$ cation in various solvents are compared, the ζ-potentials are rather insensitive to the composition of the mixed solvent.

Schwer and Kenndler [74] studied electro-osmotic velocities in fused silica capillaries. They used an electrolyte consisting of 10^{-2} mol dm^{-3} KCl and 10^{-3} mol dm^{-3} H$_3$PO$_4$ and adjusted the pH. The calculated ζ-potentials at high pH values decrease linearly with the mole fraction of the organic component for n-alcohols, DMSO, and acetonitrile, but the addition of small amounts of acetone leads to ζ-potentials higher than in water. However, the negative ζ-potentials in 50–50 (mole fraction) acetone–water mixture is half of that in aqueous systems. No reversal of sign of the ζ-potentials was directly observed, but an extrapolation of the electrokinetic curves suggests that the iep is higher in organic–water mixtures than in pure water.

2. Titania

The admixture of organic solvents exerts a more significant but qualitatively similar effect on the electrokinetic behavior of anatase, as was discussed above for silica, i.e., at ionic strengths > 10^{-2} mol dm^{-3} and $w_{organic\ component} > 20\%$ (lower n-alcohols), the negative ζ-potentials decrease and the iep shifts to higher pH values, and at even higher ionic strengths there is no iep at all (only positive ζ-potentials). It has been shown that the nature of the supporting electrolyte (e.g., LiCl vs. CsCl, and KCl vs. KI) does not play a significant role in this phenomenon [64]. At high methanol and ethanol concentrations, the decrease of the negative ζ-potentials was accompanied by a significant increase of positive ζ-potentials. This electrokinetic behavior of anatase in mixed solvents was recently confirmed using an electroacoustic method [75]. The result indicates that the difference between the surface charging (no shift of pzc) and electrokinetic (clear shift of iep) behavior of anatase in mixed solvents is not due to the difference in the solid-to-liquid ratio used in a potentiometric titration on the one hand and electrophoresis on the other hand. The electroacoustic measurements are carried out at a solid-to-liquid ratio similar to that used in titrations.

The electrokinetic results obtained by Janusz et al. [65] in aqueous ethanol, $w_{ethanol} = 50\%$, qualitatively confirm the trends observed at lower concentrations of the organic cosolvent. The iep is insensitive to the composition of the solvent at low ionic strengths, but at high ionic strengths it clearly shifts toward higher pH values, and the absolute values of the negative ζ-potentials in aqueous ethanol, $w_{ethanol} = 50\%$, are lower by a factor of 2 for 0.01 mol dm^{-3} NaCl and by a factor greater than 5 for 0.1 mol dm^{-3} NaCl than in water, while at lower ionic strengths they are relatively insensitive to the nature of the solvent.

3. Alumina

The iep is shifted to higher pH values and the absolute values of negative ζ-potentials decrease by a factor higher than 5 in 10^{-1} mol dm^{-3} NaCl when water is replaced by aqueous ethanol, $w_{\text{ethanol}} = 30\%$, while the positive branch of the electrokinetic curves is only slightly affected [67].

4. Hematite

Hesleitner et al. [68] have observed a decrease of the negative ζ-potentials of hematite in aqueous ethanol or methanol, $w_{\text{alcohol}} = 50\%$ at 10^{-2} mol dm^{-3} NaNO$_3$. For 10^{-3} mol dm^{-3} NaNO$_3$ this effect is less significant. Practically no shift of the iep was observed, but results from the same laboratory [76] suggest that at 10^{-1} mol dm^{-3} NaNO$_3$ the iep is shifted to higher pH values. Electrophoretic mobilities of hematite in aqueous ethanol, $w_{\text{ethanol}} = 50\%$, were also measured by Sworska et al. [69]. These authors have also found a shift of the iep toward higher pH values, although the iep in the mixed solvent was relatively insensitive to the ionic strength (0.001 to 0.1 mol dm^{-3} NaCl).

5. Yttria

At 10^{-1} mol dm^{-3} KCl, already in pure water the iep is higher than the pristine value. The presence of ethanol, $w_{\text{ethanol}} = 10\%$, causes a further shift of the iep by 1 pH unit and at $w_{\text{ethanol}} = 20\%$ there is no iep at all (positive ζ-potentials over the entire pH range) [64].

B. ζ-Potentials in Nonaqueous or Almost Nonaqueous Solvents

This section reports the results obtained in media in which the pH cannot be measured, and where the ζ-potentials can only be related to analytical concentrations of electrolytes.

1. Silica

Silica is negatively charged in nonaqueous or almost nonaqueous solvents. Labib and Williams [77] studied the effect of traces of moisture on the ζ-potential of glass spheres, and Spange et al. [78] studied the moisture effect on the ζ-potential of Aerosil in a number of aprotic solvents. In the both studies, the presence of water induces a decrease of the absolute value of the negative ζ-potential of silica. The ζ-potential of silica measured 5 min after the addition of water to a dry system is practically equal to that in a dry system, so the process governing the decrease of the negative ζ-potential in the presence of moisture is rather slow. Labib and Williams and Spange et al. demonstrate rather weak correlations between their ζ-potentials and the DN of the solvent. On the other hand, Spange et al. have found a good correlation between the maximum of absorption of a solvatochromic dye, dicyano-bis-(1,10-fenantroline)-ferric II adsorbed on Aerosil and the ζ-potential measured in a given solvent for solvents showing moderate donor and weak acceptor properties. Spange et al. compare also the ζ-potentials obtained from electrophoretic measurements on the one hand and electro-osmotic measurements on the other. The absolute value of the latter is lower by a factor of 4 to 10, but the sign is consistent.

The electrokinetic charge of quartz in DMSO becomes more negative when the concentration of 1–1 electrolytes (LiBr, NaBr) increases up to 10^{-2} mol dm^{-3}, but the absolute value of the ζ-potential decreases. A similar trend was observed for KBr solutions in DMSO containing different amounts of water, w_{water} ranging from 0.5 to 3.5% [79]. On the other hand, the negative electrokinetic charge of silica in solutions of tetraethylammonium and tetramethylammonium bromides has a maximum at about 10^{-3} mol dm^{-3} of these salts, and the further increase of the ionic strength leads to a decrease in the negative electrokinetic charge, but the charge reversal is not reached. The adsorption of bromide anions on silica from DMSO is higher than the adsorption of Li$^+$ cations. This result suggests that the negative electrokinetic charge of silica in solutions of LiBr in DMSO is due to an excess of Br$^-$ anions beyond the slipping plane and an excess of Li$^+$ cations in the diffuse layer. On the other hand, for NaBr and KBr solutions in DMSO, adsorption of the cations prevails. This adsorption behavior is similar to that in purely aqueous systems, and this may suggest similar charging mechanisms. Cations of the supporting electrolyte (whose adsorption is higher than that of anions) clearly play the role of counterions, and the negative surface charge must be due to even higher adsorption of potential determining ions. In systems containing significant amounts of water, protons and hydroxyl ions play the role of pdi as in purely aqueous systems. Even in an aprotic nonaqueous solvent like DMSO, a negative surface charge of silica can be a result of the dissociation of protons from the surface, which then react with solvent molecules to form lyonium ions.

Two different behaviors were observed on addition of CsCl to suspensions of silica in organic solvents containing water, $w_{water} = 1\%$. In organic solvents of low relative permittivity ($\epsilon_r < 25$), the sign of the ζ-potential is reversed from negative to positive at a sufficiently high ionic strength [80]. A similar reversal of sign was observed in solvent mixtures containing more water, $w_{water} > 1\%$, provided that the relative permittivity of the mixture was low enough. For water–dioxane mixtures, the reversal was observed for $w_{dioxane} > 70\%$ ($\epsilon_r < 19$); thus this "critical" value of the relative permittivity is somewhat lower than that mentioned above for mixtures of organic solvents at $w_{water} = 1\%$. On the other hand, at a high relative permittivity, the reversal of sign is not observed, but the ζ-potential asymptotically tends to zero. This "asymptotic" behavior is also observed in aqueous systems. Examples of asymptotic and sign reversal behavior are given in Fig. 15. Attempts to find a correlation between the CsCl concentrations at which the sign of the ζ-potential of silica in a given solvent is reversed and empirical solvent scales were not successful.

The role of the relative permittivity and "asymptotic" or "sign reversing" behavior discussed above is confirmed by results published in earlier studies. Asymptotic behavior was found for DMSO ($\epsilon_r = 46.6$), but a clear sign reversal was observed in ethanolic solution ($\epsilon_r = 24.3$) in the same laboratory [81]. The salt concentration causing the sign reversal strongly depends on the nature of the alkali metal cation; the corresponding concentration of LiBr is about 30 times higher than the sign reversing concentration of KBr. This lyotropic series of cations coincides with directly measured adsorption isotherms of cations: the adsorption of potassium on silica from ethanolic solution is much higher than the adsorption of sodium, which in turn is much higher than the adsorption of bromide. This result shows that the reversal of the sign of the electrokinetic potential is due to the adsorption of alkali metal cations beyond the shear plane. The reversal of the sign of the ζ-potential of silica in aqueous-dioxane, $w_{dioxane} = 90$ and 95%, was observed at sufficiently high KCl and CsCl concentrations [82]. The lyotropic series of alkali metal cations discussed above for ethanol is also observed in aqueous dioxane. This is a

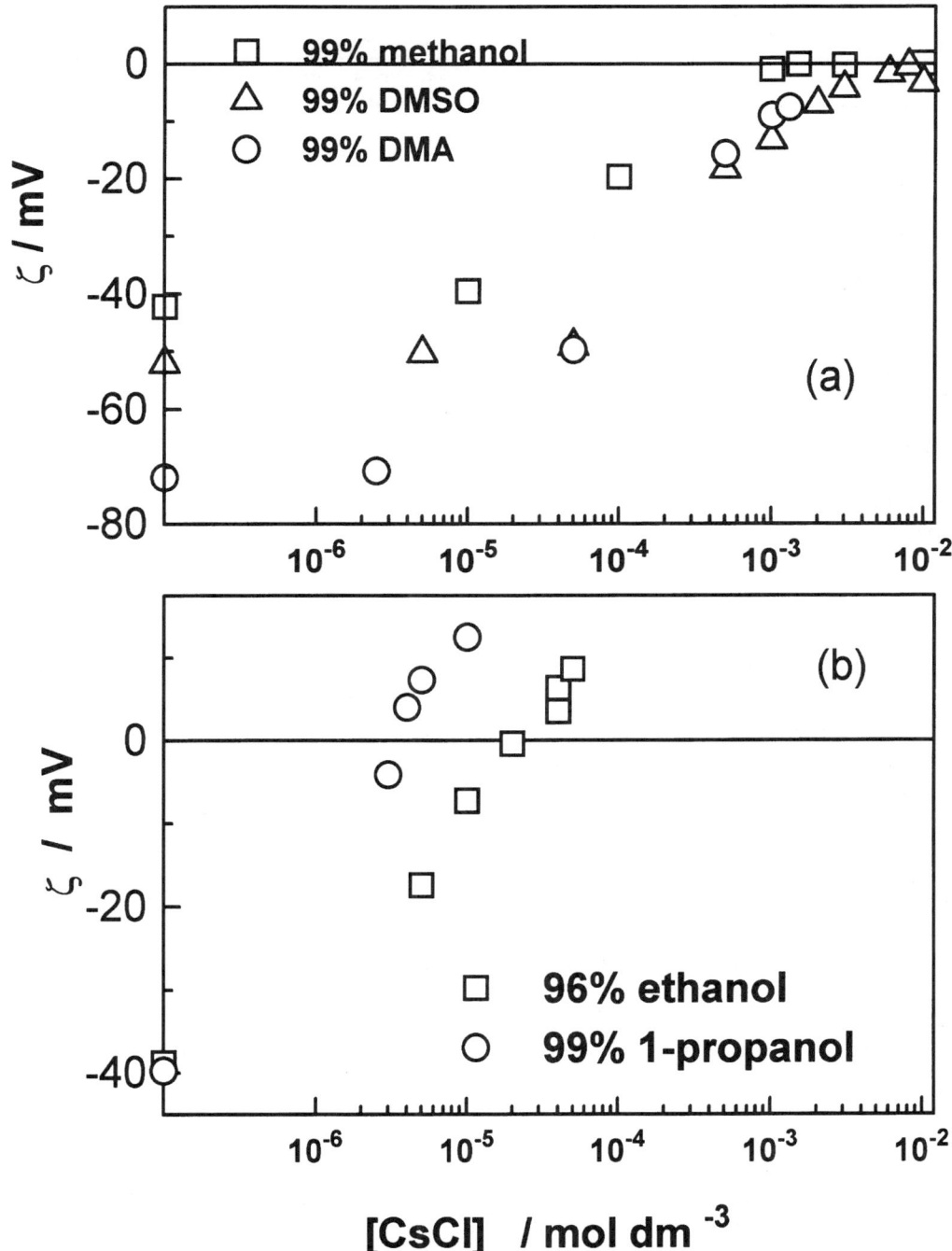

FIG. 15 Asymptotic behavior (a) and sign reversal (b) of silica in CsCl solutions in organic solvents. Concentration of the organic component is expressed as the mass fraction.

TABLE 7 The ζ-Potentials (in mV) of Titania in Nonaqueous or Almost Nonaqueous Solvents, No Salt Added. All Concentrations Are Expressed as $w_{\text{organic component}}$

Solvent	Anatase [85]	Rutile [86]	Rutile [79]	Rutile [87]	Rutile [77]
95% dioxane	−40				
Methanol	−18 (99%)	18			
Ethanol	−11 (96%)	−28		−14	
1-Propanol	−48 (99%)	26			
2-Propanol	−34 (99%)				
1-Butanol		−14	−18		
1,2-Dichloroethane					48
Acetic anhydride					−77
Ethyl acetate					−44
Dimethyl acetamide					−10
Nitromethane					−9
DMSO	−20 (99%)				−34
Acetone	−89 (99%)			−25	
Acetonitrile				−3	
Formamide				−51	
Diethylamine				−163	

significant difference between systems rich in water, for which the nature of the 1–1 electrolyte does not play a significant role, and nonaqueous or almost nonaqueous systems, in which at least the nature of the cation of a 1–1 electrolyte is important.

While the values of the ζ-potentials obtained without the addition of salt in Refs. 80 and 82 are considerably different, the CsCl concentrations necessary to reverse the sign of the ζ-potential are consistent.

Ketelson et al. [83] studied the effect of salts on the electrophoretic mobility of silica in acetone containing traces of water, $w_{\text{acetone}} = 99.7\%$. CaI_2 induced a reversal of sign from negative to positive at 7×10^{-4} mol dm^{-3} CaI_2. This salt (up to 10^{-1} mol dm^{-3}) did not induce reversal of sign in water at pH 8. On the other hand, NaI (up to 10^{-2} mol dm^{-3}) did not induce reversal of sign in acetone, $w_{\text{acetone}} = 99.7\%$.

The same authors [84] report electrophoretic mobilities of silica in 10^{-3} mol dm^{-3} CaI_2 in acetone–water mixtures. The sign is positive in mixtures rich in water, but at $\epsilon_r = 25$ the sign is reversed to positive. The ζ-potential of silica in 1.5×10^{-3} mol dm^{-3} NaI and with no salt added is rather insensitive to the composition of the acetone–water mixture.

2. Titania

Titania is negatively charged in nonaqueous or almost nonaqueous solvents. Typical values of the ζ-potential are given in Table 7.

A reversal of the sign of the ζ-potential of titania in butanol at 3×10^{-4} mol dm^{-3} NaBr was observed [86]. The electrokinetic potential is negative at low electrolyte concentrations; it reaches a maximum and then decreases. The sign of the ζ-potential of anatase is reversed from negative to positive at $w_{\text{dioxane}} = 95\%$, $w_{\text{methanol}} = 99\%$, $w_{\text{ethanol}} = 96\%$, $w_{\text{propanol}} = 99\%$ (1- and 2-propanol), and $w_{\text{DMSO}} = 99\%$ at a sufficiently high concentration of CsCl and at $w_{\text{acetone}} = 99\%$ [85]. CsCl is not sufficiently soluble in the latter solvent, and it must be replaced by perchlorate or iodide.

Reversal of the sign of the ζ-potential of monodispersed spherical anatase particles at 5×10^{-3} mol dm^{-3} NaCl was observed in commercial anhydrous ethanol [65]. This concentration is considerably higher than the c_{rev} of CsCl (the concentration at which the sign of the ζ-potential is reversed to positive) observed for commercial anatase at $w_{ethanol} = 96\%$, according to the general rule that the c_{rev} is lower for CsCl than for NaCl. Interesting enough is that at low ionic strengths the adsorption of Na$^+$ is higher than that of Cl$^-$ ions, and at ionic strengths higher than 10^{-3} mol dm^{-3}, adsorption of the anions of the supporting electrolyte prevails. This suggests that the negative ζ-potential of anatase in dehydrated ethanol at low ionic strengths is a result of adsorption of OH$^-$ (or lyate) ions at or dissociation of protons from the surface, which is partially balanced by a positive adsorption of cations of the supporting electrolyte, and that the positive ζ-potential of anatase in dehydrated ethanol at high ionic strengths is a result of adsorption of H$^+$ (or lyonium) ions at the surface, which is partially balanced by a positive adsorption of anions of the supporting electrolyte.

On the other hand, in aqueous ethanol, $w_{ethanol} = 50\%$, no reversal of the sign of the ζ-potential was observed up to 10^{-2} mol dm^{-3} NaCl.

3. Alumina

Labib and Williams [77] report values of the ζ-potential of alumina (90% in the α-form and 10% in the γ-form) in a number of nonaqueous solvents. The values obtained in ethylenediamine, DMSO, THF, ethyl acetate, acetonitrile, and acetic anhydride were negative and rather insensitive to traces of moisture. For nitromethane, nitrobenzene, and 1,2-dichloroethane, the ζ-potential of alumina was negative in the dry state, but traces of moisture caused a reversal of sign of the ζ-potential. The ζ-potential of alumina in dichloroethane was positive in the presence and absence of moisture. This result was interpreted in terms of the DN solvent scale: solvents that have DN > 10 give negative ζ-potentials irrespective of the presence of water, and those with 0 < DN < 10 give positive ζ-potentials in the dry state, but negative ζ-potentials in the wet state. Finally, the ζ-potential of alumina in solvents with negative DN values is positive in the presence and absence of water. Only one example of such a solvent, namely dichloromethane, was found by Labib and Williams, who assumed a significantly negative DN value for this solvent. However, the correlation between empirical solvent scales leads to DN = 1 for dichloromethane. Labib and Williams claim that the DN of dry alumina is about 10 and the DN of wet alumina is about zero. Within the group of solvents whose DN > 10 there is no correlation between the value of the DN and the value of the ζ-potential.

The ζ-potentials of alumina determined by other authors do not confirm the above "pzc" of alumina on the DN scale. A clearly positive ζ-potential in 1-butanol was found. The DN of butanol estimated from correlations between solvent scales is 29, i.e., more than for most solvents considered in Ref. 77. The DN values estimated for other lower n-alcohols are also high, and this should lead to negative ζ-potentials in the dry and wet states. The experimental ζ-potentials [88] are negative, indeed, in pure alcohols, but in the presence of water the sign is reversed (Table 8). The same author reports positive ζ-potentials of aluminum hydroxide in the same solvents in the dry and wet states.

The addition of NaBr leads to a decrease of the positive ζ-potentials of alumina without charge reversal [89].

Wang et al. [90] report ζ-potentials of α-alumina in ethanol containing water, $w_{water} = 0.2\%$, as a function of pH. The pattern of the electrokinetic curves is similar to that obtained in water, i.e., positive ζ-potentials are obtained below the iep and negative ζ-potentials above the iep, which falls at pa_H 8.33 (aqueous scale). When the measure-

TABLE 8 The ζ-Potentials (in mV) of Alumina in Nonaqueous or Almost Nonaqueous Solvents, No Salt Added

Solvent	α-form [87]	80% α-form [89]	90% α-form wet [77]	90% α-form dry [77]	α-form dry [88]	α-form wet [88]
Acetonitrile	−36		−58	−70		
Acetone	−29					
Formamide	−12					
Diethylamine	−49					
Nitrobenzene	±3		30			
Ethylenediamine	−43		−40	−42		
Pyridine	−22					
1-Butanol		25			−27	13
Methylene chloride			42	55		
Dichloroethylene			65	−90		
Nitromethane			24	−18		
Acetic anhydride			−50	−72		
Ethylene acetate			−38			
DMSO			−58	−60		
THF			−46			
$C_3 - C_5$ n- and iso-alcohols					−26..−28	13..15

ments are carried out in 10^{-3} mol dm^{-3} LiCl, the absolute value of the ζ-potentials is lower than without salt addition, but the iep is not affected. The above iep is similar to the values typically obtained for α-alumina in an aqueous system. Unfortunately, the authors do not report results for purely aqueous systems for comparison.

4. Hematite

Positive ζ-potentials were observed in DMSO [70] at 10^{-4} to 10^{-3} mol dm^{-3} NaCl, but in aqueous DMSO, $w_{DMSO} = 50\%$, the sign was negative at 10^{-5} to 10^{-2} mol dm^{-3} NaCl.

5. Magnesia

The ζ-potentials obtained by Labib and Williams [77] were rather insensitive to the presence of moisture. Positive ζ-potentials are reported for solvents whose DN < 10 and negative ζ-potentials for solvents whose DN > 30. For 10 < DN < 30, a random mixture of positive and negative ζ-potentials is obtained. Hamieh and Schultz [91] claim DN = 20 for MgO. The results from both studies are summarized in Table 9.

6. Tantalia

The isoelectric point of monodispersed, spherical Ta_2O_5 particles at pH 5.1 in ethanol (doubly dehydrated by means of a 3 Å molecular sieve) was reported [92]. The same sample had an iep at pH 4.1 in water. The ionic strength was not specified but was probably low ("pH adjusted with hydrochloric acid and sodium hydroxide"). The absolute values of the ζ-potential in ethanol are higher than in water for both the positive and the negative branch of the electrokinetic curves. This is a rare example of an electrokinetic study

TABLE 9 The ζ-Potentials (in mV) of Magnesia in Nonaqueous Solvents, No Salt Added

Solvent	Ref. 77 (dry)	Ref. 91
Nitrobenzene	42	
1,2-Dichloroethane	40	
Dichloromethane		27
Acetic anhydride	18	
Chloroform		14
Nitromethane	20	−32
THF		−20
Ethylene diamine	−25	
Ether		−26
DMSO	−26	
Acetonitrile	−26	−65
Ethyl acetate	−46	
Acetone		−53

in a completely nonaqueous solvent where the pH values are reported. No details about the pH measurements in nonaqueous ethanol are included. Most likely, the pH values in ethanol are just displays on the pH-meter with a combination electrode standardized by means of aqueous buffers, and the problems related to the liquid junction potential was not considered.

7. Zinc Oxide

Logtenberg and Stein [93] report negative ζ-potentials of zinc oxide in lower alcohols. The presence of water $w_{water} = 0.1\%$, leads to the reversal of the sign of the ζ-potentials in ethanol and propanol, but not for methanol.

The same authors studied the effects of the addition of HCl and NaOH on the ζ-potentials of zinc oxide in alcoholic solutions. No clear trend was observed.

Hamieh and Schultz [91] report a positive ζ-potential of zinc oxide in dichloromethane, zero (±10 mV) for chloroform, and negative values for THF, ether, nitromethane, acetone, and acetonitrile. Their ζ-potentials of zinc oxide are very well correlated with the values obtained form magnesia, and the authors claim that DN = 18 for ZnO.

VIII. STABILITY

Many physical properties of aqueous dispersions, e.g., adhesion and rheological properties, are correlated with the ζ-potential, and they can be used to estimate the iep. Probably similar correlations exist in mixed solvents and in nonaqueous polar solvents. The correlation between the minimum of stability and the iep is well known for aqueous colloids, and a few attempts to find a similar correlation in organic media are presented in this section.

Stability in nonpolar media was recently discussed by van der Hoeven and Lyklema [94]. The coagulation concentration roughly derived from the DLVO theory is proportional to ϵ^3, so the stability should decrease when water is replaced by a mixed solvent.

A. Silica

Silica shows an anomalous stability behavior in water; it is stable in the vicinity of the iep. Therefore one can hardly expect the minimum of stability would coincide with the iep in other solvent systems. Nevertheless, a correlation between stability and ζ-potential of silica in a nearly nonaqueous system has been reported: two critical coagulation concentrations (ccc) of CaI_2 in acetone ($w_{water} = 0.3\%$) were observed. A negative sol is stable below the first ccc, and a positive sol is stable below the second ccc [84].

B. Goethite

Positive ζ-potentials of goethite were observed in methanol, ethanol, propanol, and acetone and their mixtures with water [95]. These values can hardly be compared with the ζ-potentials in "pure" organic solvents reported in the previous section, for the dispersions were obtained by the dilution of an acidic stock solution. Thus these dispersions contain about 2.3×10^{-4} mol dm^{-3} HNO_3. Kosmulski and Matijevic [82] showed that the sign of the ζ-potential of silica in dioxane–water mixtures can be reversed from negative to positive in the presence of about 10^{-5} mol dm^{-3} acid, and probably the ζ-potentials of other oxides in mixed or nonaqueous solvents are also positive at a sufficiently high concentration of acid.

The coagulation concentration of $LiNO_3$ is 60×10^{-3} mol dm^{-3} for water, 18×10^{-3} mol dm^{-3} for 99.5% methanol ($w_{water} = 0.5\%$), 3×10^{-3} mol dm^{-3} for ethanol ($w_{water} = 0.5\%$), 0.1×10^{-3} mol dm^{-3} for 2-propanol ($w_{water} = 0.5\%$), and 0.001×10^{-3} mol dm^{-3} for acetone ($w_{water} = 0.5\%$). The ratio of the coagulation concentration to ϵ^3 is approximately constant for aqueous ethanol and aqueous acetone ($w_{organic\ component}$ up to 50%), but the coagulation concentrations of 2-propanol and acetone–water mixtures rich in dioxane are much lower than those derived from ϵ^3.

C. Alumina

The stability ratio as a function of the pH of an ethanolic solution (no salt added) displays a broad minimum in the vicinity of the pzc. No clear minimum was observed for 10^{-3} mol dm^{-3} LiCl in ethanol (low stability over the entire pH range) [90].

It is still not clear if the electrokinetic potential of oxides in mixed solvents (a shift of the iep to higher pH values at high ionic strength) is accompanied by a shift of the pH range of low stability of colloidal particles. Unfortunately, this pH range is relatively broad at the ionic strength of interest, even in water. It would be also interesting to study the effect of the shift of the iep in mixed solvents on the rheological properties of dense dispersions.

NOTE ADDED IN PROOF

A completely new approach to surface charging in nonaqueous media was recently presented [96].

SYMBOLS AND ABBREVIATIONS

1-pK   model using reaction (2) for surface charging
2-pK   model using reaction (1) for surface charging

a	particle radius
a	solvent-independent parameter characterizing a solvent-dependent process
A	coefficient in Debye–Hückel and related equations
a_i	activity of the species i
AN	acceptor number
b	solvent-independent parameter characterizing a solvent-dependent process
c	velocity of light in vacuum
c_i	molarity
cip	common intersection point
c_{rev}	salt concentration necessary to reverse the sign of electrokinetic potential
DMF	N,N,-dimethyl formamide
DMSO	dimethyl sulfoxide
DN	donor number
e	charge of proton
e	basis of natural logarithms
E	electromotoric force
EMF	electromotoric force
E_T^N	dimensionless polarity/polarizability parameter (Reichardt)
$E_T(30)$	polarity/polarizability parameter (Reichardt)
F	Faraday constant
GE	glass electrode
h	Planck constant
I	ionic strength
iep	isoelectric point
IR	infrared
k	Boltzmann constant
K_A	association constant
K_{ap}	autoprotolysis constant
m_i	molality of the component i
N_A	Avogadro constant
NMR	nuclear magnetic resonance
p	$-\log$
pdi	potential determining ion
ppzc	pristine point of zero charge
pzc	point of zero charge
R	gas constant
r_i	ion size parameter
s	solvent-independent parameter characterizing a solvent-dependent process
S	surface metal ion
SCE	saturated calomel electrode
SH	a molecule of organic solvent
SOH	surface hydroxyl group
T	temperature
THF	tetrahydrofuran
w_i	mass fraction of the species i
x_i	mole fraction of the species i
X	solvent scale
XYZ	quantity characterizing a solvent-dependent process
XYZ_0	solvent-independent parameter characterizing a solvent-dependent process
y_i	molar activity coefficient of the species i

z_i	valence of the ion i (positive for cations, negative for anions)
Z	polarity/polarizability parameter (Kosower)
ZH	a molecule of organic solvent
α	hydrogen bond donation ability
β	hydrogen bond acceptance ability
γ_i	molal activity coefficient of the species i
γ_t	transfer activity coefficient of proton
δ	Hildebrand solubility parameter
$\Delta_t G_H^0$	standard molar Gibbs energy of proton transfer from water into solvent
ΔE_j	difference in liquid junction potentials
ϵ	permittivity
ϵ_r	relative permittivity
ζ	electrokinetic potential
η	viscosity
κ	reciprocal Debye length
λ	wavelength
π^*	polarity/polarizability parameter
ρ	specific density
σ_0	surface charge density
ϕ	local electrostatic potential

REFERENCES

1. G. J. Janz and R. P. T. Tomkins, *Nonaqueous Electrolytes Handbook*, Academic Press, New York, Vol. I, 1972, vol. III, 1973. *Landolt-Börnstein's Handbook*, n.s. Springer-Verlag, Berlin.
2. A. W. Adamson, *Physical Chemistry of Surfaces*, 3d ed., John Wiley, New York, 1990, p. 283.
3. J. Lyklema, *Fundamentals of Interface and Colloid Science*, Vol. 2, Academic Press, New York, 1995.
4. J. Lyklema. Adv. Colloid Int. Sci. *2*:65 (1968).
5. G. D. Parfitt and J. Peacock. Surface and Colloid Science *10*:163 (1978).
6. A. Kitahara, in *Electrical Phenomena at Interfaces* (A. Kitahara and A. Watanabe, eds.), Marcel Dekker, New York, 1984, p. 119; Adv. Colloid Int. Sci. *38*:1 (1992).
7. I. D. Morrison. Colloids Surf. *71*:1 (1993).
8. M. Kosmulski. Colloids Surf. *95*:81 (1995).
9. W. H. van Riemsdijk, J. C. M. de Vit, L. K. Koopal, and G. H. Bolt. J. Colloid Int. Sci. *116*:511 (1987).
10. G. A. Parks. Chem. Rev. *65*:177 (1965). A. Daghetti, G. Lodi, and S. Trasatti. Mater. Chem. Phys. *8*:1 (1983). S. Adrizzone and S. Trasatti. Adv. Colloid Int. Sci. *64*:173 (1996).
11. M. Kosmulski, Langmuir *13*:6315 (1997).
12. N. Kallay and M. Tomic. Langmuir *4*:559 (1988). M. Tomic and N. Kallay. Langmuir *4*:565 (1988).
13. P. Debye and H. Falkenhagen. Phys. Z. *29*:121; 401 (1928).
14. Winmobil ver. 2. 10, by Ch. Mangelsdorf, D. Chan, and L. White, University of Melbourne.
15. S. Rondinini, P. Longhi, P. R. Mussini, and T. Mussini. Pure Appl. Chem. *59*:1693 (1987).
16. F. Masure and R. Schaal. Bull. Soc. Chim. France, 1372 (1956).
17. I. M. Kolthoff and S. Bruckenstein in *Treatise on Analytical Chemistry* (I. M. Kolthoff and P. J. Elving eds.), Part 1, Vol. 1, Interscience, New York, 1959, Chapter 13.
18. A. G. Mitchell and W. F. K. Wynne-Jones. Trans. Faraday Soc. *52*:824 (1956).
19. L. J. Vieland and R. P. Sweard. J. Phys. Chem. *59*:466 (1955).

20. H. S. Harned and B. B. Owen, *The Physical Chemistry of Electrolytic Solutions*, 3d ed., Reinhold, New York, 1958, pp. 644–645. H. S. Harned and R. A. Robinson, Trans. Faraday Soc. 36:973 (1940). R. A. Robinson and R. H. Stokes, *Electrolyte Solutions*, 2d rev. ed., Butterworths, London, 1965, pp. 363, 544.
21. A. K. Covington, R. A. Robinson, and R. G. Bates, J. Phys. Chem. 70:3820 (1966).
22. W. B. Schaap, R. E. Bayer, J. R. Siefker, J. L. Kim, P. W. Brewster, and F. C. Schmidt. Rec. Chem. Prog. 22:197 (1961).
23. K. K. Kundu, P. K. Chattopadhyay, D. Dana, and M. N. Das, J. Phys. Chem. 74:2633 (1970).
24. G. Briere, B. Crochon, and N. Felici. Compt. Rend. 254:4458 (1962).
25. M. L. London. J. Chim. Phys. 48:C34 (1951).
26. A. Teze and R. Schaal. Compt. Rend. 253:114 (1961). R. Schaal and A. Teze. Bull. Soc. Chim. France 1783 (1961).
27. A. R. Kreshkov, N. Sh. Aldarova, and N. T. Smolova. Zh. Fiz. Khim. 43:2846 (1969).
28. I. M. Kolthoff and M. K. Chantooni. Anal Chem. 51:1301 (1979). Y. Marcus, Pure Appl. Chem. 58:1411 (1986).
28. L. N. Bykova and S. I. Petrov. Russian Chem. Rev. 39:766 (1970); 41:975 (1972).
30. J. Courtot-Coupez and M. Le Demezet. Bull. Soc. Chim. France 1033 (1969), Compt. Rend. C 266:1438 (1968).
31. R. M. Fuoss. J. Am. Chem. Soc. 80:5059 (1958).
32. Y. Marcus. Pure Appl. Chem. 62:899 (1990).
33. H. H. Willard and G. F. Smith. J. Am. Chem. Soc. 45:286 (1923).
34. T. Pavlopoulos and H. Strehlow. Z. physik. Chem. 202:474 (1954).
35. W. E. S. Turner and C. C. Bissett. J. Chem. Soc. 103:1904 (1913).
36. R. G. Larson and H. Hunt. J. Phys. Chem. 43:417 (1939).
37. Interpolation from A. Laanung. Z. physik. Chem. (A) 161:255 (1932).
38. *Gmelin's Handbook of Inorganic Chemistry*, Verlag Chemie, Weinheim.
39. R. G. Bates. Pure Appl. Chem. 18:421 (1969). P. Longhi, T. Mussini, and S. Rondinini. Anal. Chem. 86:2290 (1986).
40. T. Mussini, A. K. Covington, P. Longhi, and S. Rondinini. Pure Appl. Chem. 57:865 (1985).
41. K. L. Cheng. Microchem. J. 42:5 (1990).
42. J. I. Kim. J. Phys. Chem. 82:191 (1978).
43. Y. Marcus. Pure Appl. Chem. 53:955 (1983).
44. H. Galster, *pH Measurement*, VCH, Weinheim, 1991.
45. O. Popowych and R. P. T. Tomkins, *Nonaqueous Solution Chemistry*, John Wiley, New York, 1981.
46. C. L. de Ligny and M. Rehbach. Rec. Chim. 79:727 (1960).
47. R. G. Bates, M. Paabo, and R. A. Robinson. J. Phys. Chem. 67:1833 (1963).
48. I. M. Kolthoff and F. G. Thomas. J. Phys. Chem. 69:3049 (1965).
49. M. L. Kamlet, L. M. Abboud, M. H. Abraha, and R. W. Taft. J. Org. Chem. 48:2877 (1983).
50. Ch. Reichardt. Chem. Rev. 94:2319 (1994).
51. Y. Marcus. Chemical Society Reviews 22:409 (1993).
52. A. F. M. Barton, *CRC Handbook of Solubility Parameters and Others Cohesion Parameters*, CRC Press, FL, Boca Raton, 1983, pp. 142–149.
53. Y. Marcus, *Ion Solvation*, John Wiley, New York, 1985.
54. R. D. Skwierczynski and K. A. Connors. J. Chem. Soc. Perkin Trans. 2:467 (1994).
55. M. Kosmulski. J. Colloid Interface Sci. 179:128 (1996).
56. M. Kosmulski. J. Colloid Interface Sci. 156:305 (1993).
57. M. Kosmulski. Bull. Pol. Acad. Sci. 41:325 (1994).
58. M. Kosmulski. Polish J. Chem. 67:1831 (1993).
59. M. Kosmulski. Colloids Surf. A. 83:237 (1994).
60. G. Peschel and P. Ludwig. Ber. Bunsenges. Phys. Chem. 91:536 (1987).
61. J. Szczypa, I. Kajdewicz, and M. Kosmulski. J. Colloid Interface Sci. 137:157 (1990).
62. J. Szczypa, L. Wasowska, and M. Kosmulski. J. Colloid Interface Sci. 126:592 (1988).
63. M. Kosmulski, J. Matysiak, and J. Szczypa. J. Colloid Interface Sci. 164:280 (1994).

64. M. Kosmulski and E. Matijevic. Colloids Surf. *64*:57 (1992).
65. W. Janusz, A. Sworska, and J. Szczypa, ACS, 70th Colloid and Surface Science Symposium, Potsdam, NY, June 16–19, 1996, paper 357.
66. M. Kosmulski and A. Plak. Colloids Surf A. *149*:409 (1999).
67. M. Kosmulski. J. Colloid Interface Sci. *135*:590 (1990).
68. P. Hesleitner, N. Kallay, and E. Matijevic. Langmuir *7*:178, 1554 (1991).
69. A. Sworska, J. Szczypa, and W. Janusz, Colloids Surf A. *149*:421 (1999).
70. J. Szczypa, W. Staszczuk, and W. Janusz, 8th International Conference on Surface and Colloid Science, Adelaide 1994, paper P186.
71. M. Kosmulski, J. Matysiak, and J. Szczypa. Bull. Pol. Acad. Sci. *41*:333 (1994).
72. M. Kosmulski and J. B. Rosenholm. J. Phys. Chem. *100*:11681 (1996).
73. M. Kosmulski and E. Matijevic. Langmuir *8*:1060 (1992).
74. Ch. Schwer and E. Kenndler. Anal. Chem. *63*:1801 (1991).
75. M. Kosmulski, J. Gustafsson, S. Durand-Vidal, and J. B. Rosenholm. Colloids Surf. A *157*:245–259 (1999).
76. N. Ryde, H. Kihira, and E. Matijevic. J. Colloid Int. Sci. *151*:421 (1992).
77. M. E. Labib and R. Williams. J. Colloid Interface Sci. *97*:356 (1984). Colloid Polym. Sci. *264*:533 (1986).
78. S. Spange, F. Simon, G. Heublein, H.-J. Jacobasch, and M. Boerner, Colloid Polym. Sci. *269*:173 (1991).
79. V. I. Varzhel, A. N. Zhukov, L. G. Levashova, and S. V. Uspenskaya. Vestnik LGU *4*:99 (1990).
80. M. Kosmulski, P. Eriksson, C. Brancewicz, and J. B. Rosenholm, Colloids Surf A, in press.
81. A. N. Zhukov and V. I. Varzhel. Kolloid. Zh. *52*:781 (1990).
82. M. Kosmulski and E. Matijevic. Langmuir *7*:2066 (1991).
83. H. A. Ketelson, R. Pelton, and M. A. Brook. Langmuir *12*:1134 (1996).
84. H. A. Ketelson, R. Pelton, and M. A. Brook. J. Colloid Int. Sci. *179*:600 (1996).
85. M. Kosmulski, in *Fine Particles Science and Technology*, (E. Pelizzetti, ed.), Kluwer, Dordrecht, 1996, p. 185.
86. P. Jackson and G. D. Parfitt. Kolloid Z. Z. Polymer *244*:240 (1971).
87. B. Siffert, J. Eleli-Letsango, A. Jada, and E. Papirer. Colloids Surf. *92*:107 (1994).
88. L. A. Romo. Disc. Faraday Soc. *42*:322 (1966).
89. A. N. Zhukov, L. G. Levashova, and A. Yu Menschikova. Vestnik LGU *3*:61 (1989).
90. G. Wang, P. Sarkar, and P. S. Nicholson. J. A. Ceram. Soc. *80*:965 (1997).
91. T. Hamieh and J. Schultz. C. R. Acad. Sci. Paris. Ser. II. *322*:691 (1996).
92. K. Nakanishi, Y. Takamiya, and T. Shimohira. Yogyo Kyokaishi *94*:1023 (1986).
93. E. H. P. Logtenberg and H. N. Stein. Colloids Surf. *17*:305 (1986).
94. Ph. C. van der Hoeven and J. Lyklema. Adv. Colloid Int. Sci. *42*:205 (1992).
95. N. de Rooy, P. L. de Bruyn, and J. T. Overbeek. J. Colloid Int. Sci. *75*:542 (1980).
96. M. Kosmulski, P. Eriksson, and J. B. Rosenholm. Anal. Chem. *71*:2518 (1999).

8
Immersion Phenomena

JEAN MARC DOUILLARD LAMMI, Montpellier University of Sciences, Montpellier, France

I.	Introduction	313
II.	Immersion and Wetting Calorimetry: General View	314
	A. Surface thermodynamic quantities	314
	B. Wetting enthalpy and surface state of the solid	315
	C. Enthalpy of immersion	317
	D. Experimental	318
III.	Surface Energies of Solids	320
	A. Classical definitions	320
	B. Surface energy	320
IV.	Enthalpy of Immersion: Experimental Results	324
	A. Case of water	325
	B. Case of alkanes	328
	C. Case of polar liquids	329
V.	Analysis of Results	331
	A. Introduction	331
	B. The model	332
	C. Determination of solid surface tensions	337
VI.	Surface Energy of High-Energy Solids	339
	A. Surface enthalpy	339
	B. Surface tension	343
	Conclusion	345
	References	346

I. INTRODUCTION

The drops forming on the upper part of a glass containing a wine rich in alcohol illustrate the magnitude of the physical and chemical effects of wetting. The thermal effect linked to wetting phenomena is so great that it was quantified as early as 1822 by Pouillet [1] in Marseille (France). Wetting phenomena are currently of great importance in industrial areas such as lubrication [2], adsorption [3], flotation, extraction, chromatography, and membrane processes [4]. They also have an importance in graphic arts and technology.

The key point is the difference in structure between the three interfacial layers (two solid–fluid interfaces and one fluid–fluid). These cause large molecular reorganization at the triple line of contact [5]. For instance, some of these structures in the liquid phase at the solid–liquid interface can develop on a very large scale in liquid phases [6], causing a gel structure in some cases [7]. At first sight, one is left to distinguish at the microscopic scale between these effects by a correlation length, or alternatively, if one is using thermodynamic reasoning, by the entropic character. At a preliminary level of understanding, many chemists prefer to build scales of interactions. For example, to distinguish between repulsive and attractive terms, they use interaction concepts such as "hydrophilic" and "hydrophobic," even though these words are often used with different intentions depending on the field of science concerned [8,9].

In this context, surveying heats of wetting literature gives rise to some surprising points. Classically, a thermodynamic analysis is the comparison of mechanical and thermal effects (G and H, say). But in interfacial science, the two effects have often been analyzed separately, the mechanical effects being grouped under the term "wetting phenomena" and the thermal effects under the terms "adsorption" and "immersion." Therefore authors concerned with thermal effects [3,10] were not interested in mechanical effects, and vice versa [11]. One very typical example is the famous book of Defay, Prigogine, Bellemans, and Everett, *Surface Tension and Adsorption* [12], where the words "immersion" and "heat of wetting" are absent from the index.

In fact, adsorption and wetting are two faces of the same coin, the interactions between solids and fluids. The immersion of a solid in a liquid is the adsorption of a vapor followed by the wetting of the liquid on the vapor layers. Of course, in order to interpret, it could seem easier to treat each phenomenon separately. But to us it seems more fruitful to consider the problem as a whole.

In this chapter we will try to present the thermodynamics of immersion and recall the relationship between immersion, adsorption, and wetting in the case of pure phases. After a summary of the different experimental trends observed in the literature and in the experiments performed in this laboratory, we will propose an interpretation of immersion using a semiempirical model.

Using this model, it is possible to see a logical trend in the different values of heats of adsorption, heats of immersion and contact angles reported in the literature. A classification of the surface energy of divided solids, useful in chemical engineering, is a result of this work.

II. IMMERSION AND WETTING CALORIMETRY: GENERAL VIEW

A. Surface Thermodynamic Quantities

The interfaces at a solid phase can be treated with the same intensive and extensive parameters as other interfaces as long as the solid is not submitted to a stress [12–14]. Therefore, at equilibrium, an interface near a solid phase can be considered as an autonomous phase, characterized by the thermodynamic parameters T, S, p, V, n_i, μ_i, γ, and A (temperature, entropy, pressure, volume, number of molecules of different species, chemical potential, surface tension, and surface area, respectively).

A total surface energy can be defined. This has to be expressed differently if one deals with a surface between pure phases or between mixtures.

As a good approximation, for condensed phases [15], the surface energy E^S and the surface enthalpy H^S are not distinguished. The surface enthalpy is given as

$$H^S = TS^S + G^S \tag{1}$$

where G^S is the surface Gibbs energy, or alternatively,

$$H^S = TS^S + \sum_i n_i^S \mu_i^S + \gamma A \tag{2}$$

where the n_i^S are the number of molecules in the interface, and the μ_i^S are the chemical potentials of the i species in the interface.

For pure phases, or interfaces where only one species is present, or for processes where the chemical potentials or the number of molecules are constant, the classical form results:

$$H^S = TS^S + \gamma A + n_i \mu_i^o \tag{3}$$

The terms per unit area can easily be defined as

$$H^{S(A)} = \left(\frac{\partial H^S}{\partial A}\right)_{T,p,n_i} \quad \text{and} \quad G^{S(A)} = \left(\frac{\partial G^S}{\partial A}\right)_{T,p,n_i} = \gamma \tag{4}$$

B. Wetting Enthalpy and Surface State of the Solid

The immersion experiment is very simple. One has to immerse a solid in a liquid, at constant temperature and pressure. Generally these experiments are performed at pressures near to ambient. The influence of the pressure is negligible [16].

There is an ambiguity in the above description. This is the initial surface state of the solid. If the solid is surrounded by a pure vapor phase, then the experiment performed is one of wetting.

Indeed, classically the wetting is studied by forming a drop of liquid [17,18] on a solid plate (see Fig. 1). But this is only one of the possible experiments. For example, we can quote the "capillary rise" [18], or the Washburn method [19], using the rise of a liquid in a powder bed. The key point is, what defines wetting is the contact of three mutually insoluble phases.

The physics of wetting is described by the Young equation [20]

$$\gamma_{LV} \cos \theta = \gamma_{SV} - \gamma_{SL} \tag{5}$$

where γ_{ij} is the interfacial tension between two phases i and j and θ the contact angle, conventionally taken in the denser phase. The subscripts L, V, and S refer respectively to the liquid, vapor, and solid phases. The contact angle is conventionally taken in the denser fluid phase.

This equation is simply a mechanical balance between the three interfacial tensions in contact at the triple line. The practical problem in this equation is that it is not possible to measure the two solid–fluid surface tensions γ_{SV} and γ_{SL} [21].

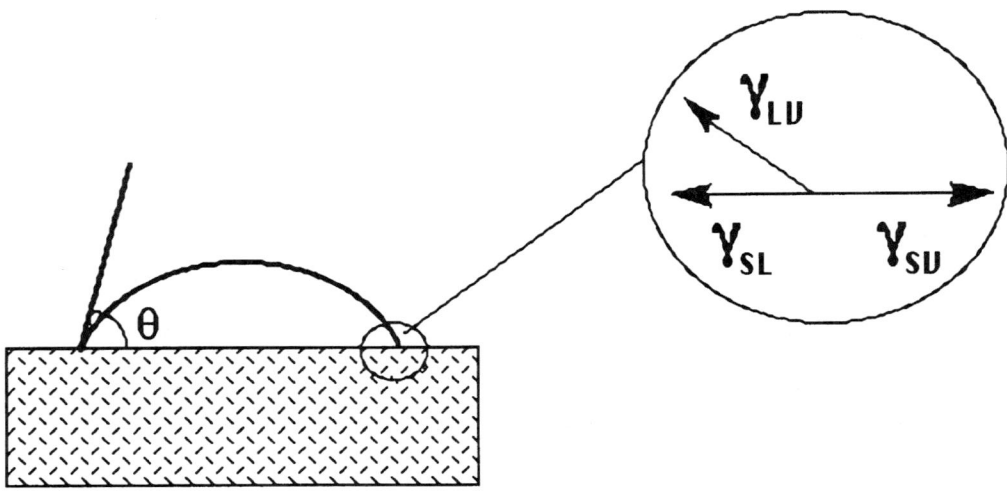

FIG. 1 A schematic view of a liquid drop on a solid plate.

The link between the mechanical and the thermal effects is very simple if one deals with pure phases. From Eq. (3), at constant pressure and temperature, the Gibbs energy variation dG is the variation of the surface tension concerned:

$$dG = A\, d\gamma \tag{6}$$

with A being the surface area involved.

In the same conditions, if the experiment is performed in a calorimeter, one can use the general relationship [22] between enthalpy and Gibbs energy:

$$\Delta H = \Delta G - T\left(\frac{\partial \Delta G}{\partial T}\right)_p \tag{7}$$

Therefore one can write the two following relationships defining a Gibbs energy and an enthalpy of wetting, noted below by surface area unit:

$$\Delta_W G^{(A)} = (\gamma_{SL} - \gamma_{SV}) \tag{8}$$

and

$$\Delta_W H^{(A)} = (\gamma_{SL} - \gamma_{SV}) - T\left(\frac{\partial(\gamma_{SL} - \gamma_{SV})}{\partial T}\right)_p \tag{9}$$

These relationships suggest that the heat effects involved are exothermic, the surface tension γ_{SV} being greater than γ_{SL}.

Putting the Young equation in (9), one obtains

$$-\Delta_W H^{(A)} = \gamma_{LV}\cos\theta - T\left(\frac{\partial(\gamma_{LV}\cos\theta)}{\partial T}\right)_p \qquad (10)$$

Hence in the case of perfect wetting one has

$$-\Delta_W H^{(A)}_{\theta=0} = \gamma_{LV} - T\left(\frac{\partial(\gamma_{LV})}{\partial T}\right)_p = H^{(A)}_{LV} \qquad (11)$$

$H^{(A)}_{LV}$ is frequently denoted H_L and called the surface enthalpy of the liquid. It is determined by surface tension measurements. Rigorously speaking, it is a reduced surface enthalpy, i.e., a derivative of the actual surface enthalpy, which does not take into account the actual number of molecules in the studied interface.

These relations define a very simple experiment that has to be called "wetting calorimetry" and that has been used by Jura and Harkins [23] to determine the specific surface area of powders. Unfortunately this experiment is often confused with "immersion calorimetry" [24], because the solid–vapor interface (and therefore the related tension) depends on the partial pressure of the surrounding vapor [3,10]. For instance, under a vacuum the solid–vapor tension vanishes, being replaced by the tension of the solid in equilibrium with its own vapor without stress, γ_S^o, well defined theoretically [13,14].

In fact, it is possible to define a certain number of surface states of the solid corresponding to the different values of the vapor partial pressure. These surface states lie between two extremes, the solid under vacuum, and the solid with a vapor adsorbed, near liquefaction.

C. Enthalpy of Immersion

Using an adapted calorimetric device, it is possible to immerse a solid [24], maintained under a vacuum, directly into a liquid phase. The solid is maintained under vacuum in a closed glass bulb, which is immersed in the liquid, and then the bulb in broken.

The different thermal effects will be listed in the following section. The main effect is due to the disappearance of the solid–vacuum interface and to the creation of a solid–liquid interface [25]. If the liquid is in a sufficient quantity, electric charge effects are negligible, provided that the liquid phase is pure [26].

The equations describing the exact immersion process are very near to the above ones. The solid in equilibrium with its own vapor without strain is characterized by the tension noted γ_S^o.

Therefore the Gibbs energy variation by surface area is

$$\Delta_{imm} G^{(A)} = \gamma_{SL} - \gamma_S^o = \gamma_{SL} - \gamma_S^o + \gamma_{SV} = -\pi_e - \gamma_{LV}\cos\theta \qquad (12)$$

$\Delta_{imm} G^{(A)}$ is the immersion Gibbs energy per unit surface area. π_e is the maximum pressure of the film due to the vapor of the liquid involved (i.e., $\pi_e = \gamma_S^o - \gamma_{SV}^{lim}$; the film pressure which varies with the surrounding pressure is denoted π_s; π_s tends to π_e [27] when p tends to p^o). Equation (12) does not correspond to a unique experiment; it is classical in thermodynamics. However, it is possible to compute the film pressure using a vapor adsorption experiment:

$$\pi_s = RT \int_{p=0}^{p} \Gamma \, d\ln p \qquad (13)$$

Deduced from the above equations, the immersion enthalpy per area is equal to

$$\Delta_{\text{imm}} H^{(A)} = -\pi_e - \gamma_{\text{LV}}(\cos\theta) - T\left(\frac{\partial[-\pi_e - \gamma_{\text{LV}}(\cos\theta)]}{\partial T}\right)_P \tag{14}$$

In the case of perfect wetting, one has

$$\Delta_{\text{imm}} H^{(A)} = -\gamma_{\text{SL}} - \gamma_S^o - T\left(\frac{\partial[-\gamma_{\text{SL}} - \gamma_S^o]}{\partial T}\right)_p \tag{15}$$

In an equivalent form,

$$-\Delta_{\text{imm}} H^{(A)} = \gamma_S^o - \gamma_{\text{SL}} - T\left(\frac{\partial[\gamma_S^o - \gamma_{\text{SL}}]}{\partial T}\right)_p \tag{16}$$

It is useful to remark here that generally, the surface enthalpy H^S can be written as the product of an intensity factor and a capacity factor.

$$H^S = H^{S(A)}(mJ\, m^{-2}) \cdot A(m^2) \tag{17}$$

with H^S defined by Eq. (2).

This allows the definition of a reduced surface enthalpy per unit area, $(\partial H_i^S / \partial A)_{T,p,\Sigma n_i \mu_i}$, where H_i^S is the total enthalpy of the considered surface of the species i. This corresponds [12] to the enthalpy necessary to extend a surface area unit of the surface, the chemical potential term $\Sigma n_i \mu_i$ being kept constant. This is the thermal effect linked to the surface tension variation at constant temperature and pressure:

$$\left(\frac{\partial H_i^S}{\partial A}\right)_{\Sigma n_i \mu_i} = \left[\partial\gamma_i - T\left(\frac{\partial\gamma_i}{\partial T}\right)_p\right] = H'{}_i^{S\,(A)} \tag{18}$$

When dealing with pure phases, or keeping n_i or μ_i constant, the variations of the reduced surface enthalpy per unit area are equal to the variations of the total surface enthalpy per unit area. With such a definition, the enthalpy of immersion per unit surface area is

$$-\Delta_{\text{imm}} H^{(A)} = H_S^{o(A)} - H_{\text{SL}}^{(A)} \tag{19}$$

which is equivalent to $-\Delta_{\text{imm}} H^{(A)} = H_S^{\prime o(A)} - H_{\text{SL}}^{\prime(A)}$. The enthalpy of wetting is

$$-\Delta_W H^{(A)} = H_{\text{SV}}^{(A)} - H_{\text{SL}}^{(A)} \tag{20}$$

which is equivalent to $-\Delta_W H^{(A)} = H_{\text{SV}}^{\prime(A)} - H_{\text{SL}}^{\prime(A)}$.

D. Experimental

In many experimental cases reported in the literature, a differential conduction microcalorimeter is used [28], due to the high precision and sensitivity obtained with this type of device. We used a Tian–Calvet microcalorimeter of this type [24]. The half part of the calorimeter is sketched in Fig. 2 with a typical immersion setup. The powdered solid sample is located in a glass ampoule. The glass ampoule is brittle and is sealed under a controlled pressure before being fixed to the end of a glass rod. The glass ampoule is placed inside the calorimetric cell, to be broken when the experimenter pushes down the glass rod. During this operation, heat due to the friction of the rod on a ring placed

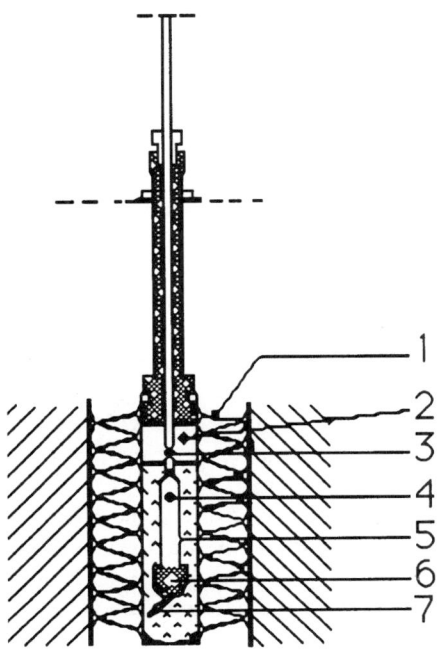

FIG. 2 Calorimetric cell designed for immersion studies. 1: thermopiles; 2: dead volume of the cell; 3: glass rod; 4: dead volume of the bulb; 5: bulb; 6: powder; 7: liquid.

at the upper part of the cell is probably produced (the ring is used to maintain a seal). This heat is not recorded due to the location of the O-ring about 8 cm above the thermopile. When the ampoule breaks, a volume v of the glass rod enters the liquid. Due to the equivalent rise of the liquid level, a heat is produced that is equal to the compression of the gas phase. Another heat is due to the breaking of the ampoule. These last two terms are small, with an order of 1 and 5 mJ, respectively.

The experimental heat can contain some artefactual heats (swelling dissolution, charge effects ...). The main contributions that can be recorded are the addition of the enthalpy of liquefaction of the vapor previously filling the bulb (corresponding volume v' and pressure p') and the enthalpy of vaporization of the vapor that fills the volume created by the lowering of the liquid level.

These effects can be written $[\Delta_{liq}H(p' - p°)v'/RT]$. They vary with the filling of the ampoule and of the calorimetric cell. The best way to determine and remove this term from the calculation is the use of a set of blank experiments.

The reproducibility is about 30 mJ. The values given here are the average values of three experiments. The calorimetric cell is introduced inside the calorimeter some hours (minimum 2 hours) before the actual experiment to obtain thermal equilibrium. Before the experiment, the solids are kept under vacuum lower than 10^{-3} torr at 150°C. (In the case of organic solids, we used lower temperatures.)

In order to obtain the enthalpy per unit area, it is useful to correspond the effective surface area to the specific surface area (in $m^2 \, g^{-1}$). The specific surface area is measured by classical methods, such as nitrogen adsorption [10,29,30] or the Harkins–Jura method [23,24]. The weight of solid in the ampoule is determined with a balance.

III. SURFACE ENERGIES OF SOLIDS

A. Classical Definitions

Each state of matter can be accurately described by a certain thermodynamic variable. For the vapor phase, the pV term is the major one; for mixtures, it is the $\Sigma n_i \mu_i$ term; and for a surface phase, the γA term.

Concerning solids, the main research effort during the last decades has been expended on the description of the surface area A (value, fractality, precision, ...) with focus on [10,29] the specific surface area A_s (e.g., per mass of solid). But generally speaking, interaction at the surface of a solid can take place because of the forces residing in the surface. If the solid were inert, it would not matter whether the solid had a large or a small specific surface area. In fact, the tension γ being an intensive parameter, it is the dominant term in a surface from the qualitative interpretation viewpoint. As shown in Section 1, the tension γ is linked to the surface enthalpy by a derivative with respect to the surface area. Hence the surface enthalpy is also a qualitative parameter and is essential to be known for the interpretion of observed phenomena.

Therefore, using immersion calorimetry, we can define and measure some effective and efficient values. Firstly, the reduced surface enthalpy per unit area of the solid, without strain, in equilibrium with its own vapor, is denoted here as $H_S^{\prime o(A)}$. Its value depends only on the nature and arrangement of atoms in the surface and on the bonds between them. For a given solid substance, $H_S^{\prime o(A)}$ depends on the considered crystal face and on the history of the sample, surfaces being very reactive and sensitive to stresses, strain, and pollution. $H_S^{\prime o(A)}$ can be considered as a constant for a sample prepared in controlled conditions.

Secondly, as we have stated before, it is possible to link this part of the intensity of surface enthalpy to the surface tension [same equation as the general form (17)], for instance for a solid in equilibrium with its own vapor,

$$H_S^{\prime o(A)} = \gamma_S^o - T\left(\frac{\partial \gamma_S^o}{\partial T}\right)_p \tag{21}$$

Generally, in this viewpoint, with physical processes concerning interfaces between pure substances, an enthalpy variation corresponds to a surface tension variation. These coupled terms are listed in Table 1.

Obviously, following these definitions, one has the relation

$$\Delta_{imm} H^{(A)} = \Delta_W H^{(A)} + \Delta_{ads} H^{(A)} \tag{22}$$

expressing that the immersion enthalpy is the sum of two different enthalpies due respectively to vapor adsorption and to wetting of the liquid phase on the vapor adsorbed.

B. Surface Energy

Considering that the main problems encountered with solids are wettability (in petroleum wells for instance [31]) and adhesion or lubrication, it could seem logical that researchers have focused on the "efficient" thermodynamic parameter, surface tension, linked to the Gibbs energy. But in fact there are only a few values of surface tension of solids (or of surface energy) in the literature, although Schrader [32] proposed to define the solids as belonging to three different groups: low energy (0 mN m^{-1} < γ < 72 mN m^{-1}), middle energy (72 mN m^{-1} < γ < 200 mN m^{-1}), and high energy (>200 mN m^{-1}). The surface

TABLE 1 Nomenclature of Tension Variation Effects and Corresponding Enthalpy Variations

Tension variation	Corresponding variation of enthalpy per area
Vapor adsorption: $\gamma_S^o - \gamma_{SV} = \pi_e$ π_s: film pressure	Integral enthalpy of vapor adsorption: $\Delta_{ads}H^{(A)} = \pi_e - T\left(\dfrac{\partial \pi_e}{\partial T}\right)_p$
Differential vapor adsorption: $\dfrac{\partial \pi_e}{\partial n_i}$	Differential enthalpy of vapor adsorption: $\dfrac{\partial \Delta_{ads}H^{(A)}}{\partial n_i}$
Immersion: $\gamma_S^o - \gamma_{SL}$	Enthalpy of immersion: $\Delta_{imm}H^{(A)} = \left[(\gamma_S^o - \gamma_{SL}) - T\left(\dfrac{\partial(\gamma_S^o - \gamma_{SL})}{\partial T}\right)_p\right]$
Wetting: $\gamma_S^o - \gamma_{SL} = \gamma_{LV}\cos\theta$	Enthalpy of wetting: $\Delta_W H^{(A)} = \left[(\gamma_{LV}\cos\theta) - T\left(\dfrac{\partial(\gamma_{LV}\cos\theta)}{\partial T}\right)_p\right]$

tension of solids is mainly obtained by studying the influence of the surface area of the sample on the energy of crystallization or of dissolution. Some other experimental values have been obtained by cleavage experiments or by applying stress. They are reviewed in the book of Adamson [3]. An analysis of the surface behavior of some iron oxides (goethite and hematite) not quoted by Adamson was performed by Langmuir and has been recalculated by Diakonov et al. [33]. These values are reported in Table 2.

Two other general techniques have been used with a view to characterization of the surface tension in solids: contact angle and inverse gas chromatography (IGC). These two techniques are carried out in the presence of the liquid phase or of a vapor phase, implying that they are only able to give solid–vapor surface tension values.

1. Contact Angle

The Young equation describing the static of contact angle has been presented in Section I. Phenomenologically, this equation can be seen as a competition between two surface tensions: a solid–fluid surface tension (that we will denote γ_{SF}^*) and a liquid–vapor surface tension.

$$\cos\theta = \frac{\gamma_{SV} - \gamma_{SL}}{\gamma_{LV}} = \frac{\gamma_{SF}^*}{\gamma_{LV}} \tag{23}$$

γ_{SF}^* can be assumed to be the major fraction of γ_{SV}, γ_{SL} being generally considered as low.

TABLE 2 Literature Results of Surface Tension of Solids and of Reduced Surface Enthalpy

Solid	Surface energy (mJ m^{-2})
CaO	$H_S^{\prime o(A)}$: 1310
Ca(OH)$_2$	$H_S^{\prime o(A)}$: 1180
Copper	γ_S^o: 1370
Gold	γ_S^o: 1300–1700
Solid paraffin	γ_S^o: 50
Sodium chloride	$H_S^{\prime o(A)}$: 276
Mica in air	γ_{SV}: 375
Mica in vacuum	γ_S^o: 4500
Tobermite	$H_S^{\prime o(A)}$: 386
Magnesium oxide	$H_S^{\prime o(A)}$: 1040
Goethite	γ_S^o: 1400
Hematite	γ_S^o: 950

In the case where the two fluid phases are liquid, one has an equivalent relationship, i.e.,

$$\cos \theta = \frac{\gamma_{SL_1} - \gamma_{SL_2}}{\gamma_{L_1 L_2}} \simeq \frac{\gamma_{SL}^*}{\gamma_{LV}} \tag{24}$$

In this case, one solid–liquid interfacial tension is much lower than the other and is often neglected.

On this basis, at the solid–liquid–vapor triple line, the behavior of a solid versus a certain liquid allows the phenomenological behavior to be considered. If the liquid wets the solid, the solid has a surface tension of γ_{SL}^* that is higher than that of the liquid. If the liquid does not spread on the solid, the solid surface tension is lower than the liquid one.

When water is chosen as a reference liquid, the wetted solids are called "high-energy" solids, and the solids presenting water drops are called "low-energy" solids.

This route of analysis was chosen by Zisman and co-workers (34–39). They were contact angle values to determine a "critical" surface tension. The lowest surface tension of the liquids spreading on the studied solid is considered as equal to the surface tension of the solid. This "solid surface tension" is called the "critical surface tension of the solid," γ_c. In this viewpoint, it is possible to write Eq. (23) in the form

$$\cos \theta \simeq \frac{\gamma_{SF}^*}{\gamma_{LV}} = \frac{\gamma_c}{\gamma_{LV}} - i \tag{25}$$

i being an interaction term between the solid and the considered fluid, causing the variation of the contact angle. The limiting case is $\cos \theta = 1$, where the contact angle is nil and i is at a minimum (Zisman assumes that $i = 0$ and this implies $\gamma_c = \gamma_{LV}$).

With this assumption, polymer solids can be classified by the value of their critical tension. The same trend can be observed for different surface groups. In this framework, fluor groups are "very low energy" solids (<10 mN m^{-1}) and methyl groups have a surface tension about 25 mN m^{-1}. A general order of magnitude for different solids is reported in Table 3.

TABLE 3 Some Typical Critical Surface Tension of Polymeric Materials

Solid	γ_c (mN m^{-1})
Nylon 6-6	45
Polyethylene	32
PTFE	19
Polyhexafluoropropene	16

Source: Ref. 36.

The values of Table 3 are given to ±5 mN m^{-1}. We have to stress that in many following papers, the critical tension is confused with an ideal solid–vapor surface tension [40]. But this idea rests on two controversial assumptions: (1) the solid–liquid surface tension (present in the term i) being considered as nil, and (2) the vapor is considered as inert, i.e., the vapor phase is the air, and is not modified by the presence of the other fluid.

In fact, contact angle techniques only give a solid–vapor surface tension [41]. Indeed, the experiments are always performed in the presence of the liquid phase, i.e., at saturation pressure. Moreover, they solely give values for "low energy" solids, the experiment only being possible when the liquid does not spread on the solid.

The approximated form of the Zisman definition [Eq. (25)] can be compared with the definition of enthalpy of immersion [Eq. (20)]. Both equations appear as differences between a constant term characteristic of the solid and a varying interaction term. Only Eq. (20) is thermodynamically correct, both critical tension and i being terms that regroup different thermodynamic parameters.

2. Inverse Gas Chromatography

The inverse gas chromatography (IGC) technique has been widely used [42–52] to study the surface properties of organic and inorganic materials. IGC is based on the adsorption by a solid surface of selected probes at different concentrations. The net specific retention volume, per adsorbent gram at a certain temperature, can be obtained by determination of the adsorbent quantity packed in the chromatographic column, the column temperature, the flow rate of the carrier gas, and the retention time of the probes. The net specific retention volume per mass (V_g) is related to the enthalpy of vapor adsorption by an equation of the type

$$-\Delta_{\text{ads}} H = R \frac{d[\ln V_g]}{d(1/T)} \tag{26}$$

Assuming a certain ratio between enthalpy and Gibbs energy, the Gibbs energy of adsorption is also related to the specific retention volume V_g:

$$-\Delta_{\text{ads}} G = RT \ln V_g + C \tag{27}$$

where C is a constant depending on the reference state. The surface tension is computed using a semiempirical equation:

$$\gamma_{SV} = \frac{1}{\gamma(-X-)} \left[\frac{\Delta G(-X-)}{2N_a a} \right]^2 \quad (28)$$

where $\gamma(-X-)$ is the surface tension of the concerned chemical group and $\Delta G(-X-)$ is the free energy of the same group. This last value is obtained from the slope of $\ln V_g$ versus the number of atoms in a series. a is the surface area occupied by the group.

On this basis, the surface tension of several high-energy solids has been obtained. But this parameter has to be questioned, because it is a solid–vapor tension at a certain coverage, obtained by an extrapolation assumed at zero coverage. It is in fact difficult to state if it is a solid–vector phase surface tension or a solid–probe vapor tension.

Frequently, when using this technique, the surface tension is separated into a polar part and a dispersive part, the latter one being determined using alkanes as probes. The values obtained [52] show a scale of tension where silicas are in the range 60–80 mN m^{-1} and clays about 150–200 mN m^{-1}. An influence of the lamellarity has been also observed in the case of silicas, because lamellar silicas have an energy greater than amorphous ones, i.e., 200 mN m^{-1}.

The general trend we can deduce from these results is that the surface tension of solids increases from polymers to lamellar crystals, i.e., with crystallinity. The chemical nature of the external groups (chemical groups bounded to the surface and pointing out into the fluid phase) is a second-order parameter. The fluor groups tend to lower surface tension and the oxygenated groups tend to increase surface tension. All these points can obviously be related to the general chemical concepts of interactions.

3. Symbiosis

On the basis of the above results and considerations, a rough picture of the scale of the surface energy of solids can be drawn. The different values of surface tension for a solid in equilibrium with its own vapor and of solid–vapor surface tension listed above are in agreement with the general view that the film pressure is proportional to the solid surface tension, at a first level of approximation. Therefore we can assume an order and a range of magnitude for the assumed film pressures. These values are conceived as a rough guideline for preview. They are listed in Table 4. A clear evolution of γ can be easily seen. This depends on the type of interaction brought about by the solid, i.e., apolar, acid–base, metallic bond, etc. This scale will be analyzed more quantitatively in the next parts of this chapter. We have to stress that this order is more or less the same as the order in Moh hardness values. Pearson has recently reviewed [53] the links between Moh hardness, physical hardness (related to compressibility of the solid), and chemical hardness. In fact, the rough picture drawn here is in agreement with Pearson's view.

IV. ENTHALPY OF IMMERSION: EXPERIMENTAL RESULTS

The simple immersion experiment described in Section II gives a very complex value involving parameters that can be difficult to interpret. As a result, this experiment has been developed in only a few laboratories [23,24,52,54–92] with the main attention on immersion in water. The trends observed in surface energy evaluations are confirmed by the immersion experiments. Therefore the results from the literature can be grouped by the characteristic of the solid. For example, solids wetted by water and solids

TABLE 4 Measured and Assumed (in Italics) Surface Tensions of Typical Solids Versus Vacuum and Versus Air

Solid	γ_S^0 (mN m^{-1})	γ_{SV} (mN m^{-1})
Mica	4500	375
LiF (100)	340	≈ *170*
MgO (100)	1200	≈ *500*
CaCO$_3$	230	≈ *100*
NaCl	110	?
Pt	1700	≈ *700*
Fe	1800	≈ *800*
W	2300	≈ *900*
Polystyrene	≈ *80*	40
Nylon	≈ *70*	33
Paraffin	≈ *50*	23
PTFE	≈ *27*	19.5

TABLE 5 Typical Literature Values for Some Solids Immersed in Water at Ambient Temperature (mJ m^{-2})

Solid	Enthalpy of immersion (average value of the literature)
Teflon (FEP)	−6
Graphon	−32
Graphite	−50
Silica	−100 −260
Rutile	−400 −700
Aluminoxid	−430
Zinc oxide	−475
Quartz	−450 −850
Calcite	−480
Zeolite NaX	−500
Anatase	−510 −850
Cristobalite	−670
Copper	−725
Alumina	−900
Thorium oxide	−1360

nonwetted, or alternatively, low energy, middle energy, and high energy. We will see below that some observations with certain liquids cannot be generalized. Therefore we cannot restrict the view of experimental results to differences due to the solids. Moreover, some peculiar effects, such as the enhancement of the immersion enthalpy due to micropores, also have to be mentioned. In the following part, we will present the results obtained with water. The influence of liquids other than water will be examined in the second and third parts of this section.

A. Case of Water

Due to the nature of covalent and ionic bonds, "high surface energy" solids are always inorganic solids. This group is composed mainly of silicates and alumino-silicates, such as silicas, quartz, clays, zeolites, and micas. Concerning immersion, the most studied solid

TABLE 6 Surface Area and Immersion Enthalpies in Water of the Different Solid Samples Studied in the Laboratory

Solid	Origin	Specific surface area ($m^2\ g^{-1}$)	Enthalpy of immersion in water ($mJ\ m^{-2}$)
Dolomite ($CaMg(CO_3)_2$)	Réference IFP	0.43	-1253
Chlorite	Luzenac	2.5	-868
Alumina	α Union Carbide	9.2	-704
$MgCO_3$	Réference Luzenac Europe	11.2	-754
Calcite ($CaCO_3$)	Gregory, Boottley and Lloyd (GB)	2.45	-535
Quartz	Sifraco C600	4.3	-510
Quartz	Sifraco C800	5.2	-383
Kaolinite	Ploemeur	16	-287
Kaolinite	Charentes	20	-308
Kaolinite	Provins	47	-330
Silica	XOB 015 (RP)	29	-237
Silica	Cabosil (Cabot)	207	-170
Silica	SiNa (RP)	40	-1332
Illite	Hungary	127	-364
Illite	Brives-Charensac	37	-212
Illite	Vosges	30	-351
Talc 1	Luzenac	3.0	-358
Talc 2	1 + grinding	6.0	-202
Talc 3	2 + grinding	9.2	-183
Talc	Spain	6.5	-356
Talc	Luzenac	1.9	-321
Talc	Italy	3.0	-311
Graphon	Cabot	82	-29
Teflon		0.5	-6
Sulphur 1	Soufres Réunis	30.4	-6
Sulphur 2	Soufres Réunis	19.7	-20
Sulphur 3	Soufres Réunis	12.4	-21

is quartz [23,24,61,63,72,74–76,84]. The results depend on the grinding and the washing of the sample. In Table 5 we list the orders of magnitude observed in the literature, and in Table 6, we report the results obtained in this laboratory. All the solids reported in Table 6 are described in detail elsewhere [26,93–109]. We have focused our present list on silicates, clays, and sulphur. These samples are of industrial interest. Our results show the great influence of preparation on the surface properties of chemical compounds that are otherwise identical. Generally, industrial compounds are more sensitive to this type of problem than solids obtained with more qualitative procedures. This can be clearly seen in Table 5, when one observes the variations of the enthalpy for different solids. For solids of industrial interest, such as quartz and silicas, the variation is large.

The case of clays [60,87] is different. Clays are compounds of the same chemical family, but with very different chemical composition, depending on the type of clay (illite, kaolinite, etc.) and on the origin of the sample. We will see later that the quantity of aluminum present in the sample is a parameter of greatest importance to surface energy.

We also report the results obtained with a talc, ground to obtain different characteristics, and of three very pure talcs extracted from three different locations, with very different lamellarity. We finally report a result obtained with a powder of Teflon (see also [62]). One also has to note here that the two quartzes reported in Table 6 are ground and probably amorphous on the surface.

In the immersion experiments, the defects present at the surface of the sample have a very strong influence in two opposite directions. Impurities adsorbed, such as surfactants and greases, tend to decrease the solid surface tension and consequently the immersion enthalpy. Different methods of preparation for the same material have a pronounced effect on the immersion enthalpy.

Some metals prepared at high temperature are annealed and their surface energy is lowered (a typical example is copper [58] where the enthalpy can increase from -725 to -235 mJ m^{-2} depending on the temperature of preparation).

However, substitution of surface atoms and groups by atoms or groups with a greater reactivity tend to increase the immersion enthalpy. In this trend, alumino-silicates can be viewed as intermediate products between pure silica [54,55] and pure alumina [87]. The observed immersion enthalpies reflect this viewpoint, the obtained enthalpy of immersion being intermediate between the average values of silica and alumina. This indicates that the experimental enthalpy can be interpreted as the sum of contributions of individual enthalpies of interaction between the different types of groups of the solid and the fluid molecules. Thermal treatment also modifies external chemical bonds, and can also modify the surface network and therefore the surface energy. Silica has been extensively studied [23,24,52,63], and the classical interpretation of the variation of immersion enthalpy versus the treatment temperature is a result of the modification of the types of silanol (individual or grouped) and of their surface densities. Some studies using different activation conditions have been published. The silanol groups can also be dehydroxylated by surface treatment. An order of energy between the different types of oxides of silicon has been proposed following this idea [110].

Grinding obviously has an influence, combining the different effects listed above. Grinding can modify the ratio between different crystallographic planes at the surface, can introduce some impurities (the metal of the grinder, mainly) and can amorphize the surface by mechanical and thermal energy transfers. This point has not been treated in detail; a complex work of comparison between immersion and a surface spectroscopic method is needed.

Another point important to emphasize is that talc is very often considered as a "hydrophobic" solid, a "solid oil." For instance, it has been shown from wetting enthalpy measurements that water gives large contact angle values on talc [100,108]. However, the immersion enthalpies of talc in water are of the same order of magnitude as the hydrophilic solids: quartz, hydrophilic silica, and clay minerals. This indicates great differences of behavior between the solid–vacuum interface and the solid–atmosphere interface.

Organic solids, meaning the solids formed by bonds between carbon, fluor, or sulphur groups, give very low enthalpy of immersion with water. This is classically interpreted [8,78,80] in terms of repulsion between these molecules and the structure of water. But this point is controversial in the sense that the immersion enthalpy of organic solids in water is weak but exothermic. This indicates in fact the same type of interaction as that between minerals and water, but at a lower value.

Some very high values have been reported in the peculiar case of swelling materials [85,109]. For instance, a value of about -540 mJ g^{-1} is frequently reported for the immersion of cellulose in water. Knowing that the specific surface area of this type of powder is generally estimated at around 1 m^2 g^{-1}, one obtains an immersion enthalpy of about -540 mJ m^{-2}, a high value compared with other products, especially if one considers that cellulose is a polymer with a low-energy surface. The same trend is observed with swelling clays, such as montmorillonites. Values of about -70 J g^{-1} are reported [111] for samples with a specific surface area of 50 m^2 g^{-1}, i.e., immersion enthalpies of about -1400 mJ m^{-2}, to be compared with the -300 mJ m^{-2} typical of kaolinites, which are very similar clays, although nonswelling. Both cases are interpreted in the same way. In the experiments reported, the heat of swelling is not subtracted from the experimental heat, because it is difficult to determine this complex parameter by an independent experiment. On the basis of the trends of results concerning the other chemical compounds reported here, we can estimate the heat of swelling at about -1100 mJ m^{-2} for montmorillonite. The case of cellulose is more complex. Considering that estimation of surface tension gives a value around -50 mN m^{-1}, we can estimate the surface energy to be around -100 mJ m^{-2}. On this basis, the swelling heat of cellulose is expected around -400 mJ g^{-1}. Comparison of immersion results of the same solid but in a nonswelling solvent confirms this analysis.

B. Case of Alkanes

The results of the immersion of different solids in water can be considered as an estimation of the hydrophilicity of these solids. In this viewpoint of hydrophilic/hydrophobic interactions, it is interesting to report the results obtained for the immersion in n-alkanes [59,84]. Considerable care has to be taken to control the presence of trace water in the alkane. This presence has an influence on the quality of the calorimetric results. Generally, samples have to be dried with molecular sieves and analyzed for water content.

For comparison, when reporting our results in Table 7, we have kept the same order of appearance as in Table 6. Surprisingly, the alumino-silicates appear quite in the same order, meaning that at first sight the more hydrophilic solids are also the more hydrophobic, if one considers the affinity for apolar compounds as a scale of hydrophobicity. The trend observed shows that minerals give enthalpies lower than -100 mJ m^{-2}, and organic solids give enthalpies higher than -100 mJ m^{-2}. Activated carbons, which are both hydrophobic (the carbon backbone) and hydrophilic (the polar surface groups) are at the crossover [90,91]. The conclusion from these results is the idea that it is not possible to order the solids on such a unique parameter as an immersion enthalpy in only one fluid. As shown in the thermodynamic definition of this parameter, the solid surface enthalpy has paramount importance in the final value, which cannot simply be restricted to an interaction force value.

The results presented here are in good agreement with literature results. Some values have been given [90] for carbons in cyclohexane (-103 mJ m^{-2}) or [91] in n-dodecane (-100 mJ m^{-2}), heptane, or hexadecane (-120 mJ m^{-2}), and for teflon in n-heptane [62]. Silica has been immersed in iso-octane [57] (-55 mJ m^{-2}), heptane, or hexadecane (-50 mJ m^{-2}). A peculiar case is the immersion of metals in alkanes. In this case strong interactions take place. For instance, the immersion enthalpy of copper in cyclohexane [58] is -740 mJ m^{-2}, the same as in water. This point is confirmed by the immersion of copper in benzene or in methanol (respectively -880 and -950 mJ m^{-2}).

TABLE 7 Immersion Enthalpies of the Solid Samples in Decane or in *n*-Heptane at Ambient Conditions (mJ m^{-2})

Solid		Immersion in decane	Immersion in heptane
Dolomite			−316
Chlorite			−202
Alumina			−162
MgCO$_3$			−126
Calcite		−206	−236
Quartz	C 800		−104
Kaolinite	Ploemeur		−96
Kaolinite	Charentes		−106
Kaolinite	Provins		−86
Silica	XOB O15		−140
Silica	Cabosil		−85
Illite	Hungary		−103
Illite	Brives-C		−83
Illite	Vosges		−90
Talc	1		−119
Talc	2		−139
Talc	3		−152
Talc	Spain		−84
Talc	Luzenac		−109
Talc	Italy		−151
Graphon			−120
Teflon			−58
Sulphur	1	−73	
Sulphur	2	−106	
Sulphur	3	−44	

C. Case of Polar Liquids

In this group, we also consider the case of nonpolar but polarizable liquids, such as benzene. Indeed, this is the key point in immersion enthalpy. We observe the same general order as for alkanes and for water. In fact, this order is able to summarize the results, being an order of the energy between "high-energy" solids and "low-energy" solids, and confirms the link between contact angle phenomena and immersion phenomena. But as in the preceding case, the fact that immersion enthalpy is the result of a combination of two parameters, the solid surface enthalpy and the enthalpy of interaction, explains the complexity of experimental trends. Firstly, the solid surface enthalpy is greater (at first approximation, the enthalpy of fusion of the solid), as is the immersion enthalpy. Secondly, both the capacity of interaction between the solid and the fluid, and the immersion enthalpy, are greater. Thus the visible order between the solids is explained. The order between the fluids is also explained, formamide between more reactive than pyridine and benzene. Exceptions are observed. Some of them are probably due to epitaxy phenomena between the solid and the fluid. For instance, the bond distance between atoms of the sheets of talcs are the same as in the benzene molecule, which can cause the high energy observed for this couple.

TABLE 8 Immersion Enthalpies of the Solid Samples in Some Polar or Polarizable Liquids at Ambient Conditions (mJ m^{-2})

Solid		Immersion in benzene	Immersion in pyridine	Immersion in formamide
Dolomite		−452		
Chlorite				−1006
Alumina				−585
MgCO$_3$				−674
Calcite		−289	−464	
Quartz	C800	−157	−317	
Kaolinite	Ploemeur	−156	−382	
Kaolinite	Charentes	−171	−410	
Kaolinite	Provins	−148	−404	
Silica	XOB O15	183		
Silica	Cabosil			−289
Illite	Hungary	−151	−449	
Illite	Brives-C			−331
Illite	Vosges	−161	−387	
Talc	1			−429
Talc	2			−284
Talc	3			−275
Talc	Spain			−531
Talc	Luzenac	−281		−416
Talc	Italy			−345
Graphon		−114		

Here too, these results are in good agreement with literature values. Robert [79] reported 110 mJ m^{-2} for the immersion of carbon in benzene. However, some high values have been reported by Denoyel et al. [91] for microporous charcoals. They reported values in benzene between −77 and −194 J g^{-1} for samples of different specific surface areas. Starting from the BET specific surface area (from 563 to 1900 m^2 g^{-1} in these cases), this give immersion enthalpies per unit surface area of about −140 mJ m^{-2}, decreasing for samples of very high specific surface area. Compared to the chemical nature of these active carbons, this value appears too high. Therefore Denoyel et al. attributed these high values to the fact that immersion enthalpy is sensitive to the whole surface area, and to the presence of a micropore volume. This presence leads to a large underestimation of the actual surface area, by the BET/vapor adsorption method. Thus Denoyel et al. compute a new specific surface area using their immersion results and an equivalent form of our Eq. (17). In fact, they assume a value of the surface energy, considering that this energy is constant for a given chemical family, such as carbons. The average value they finally obtain for $\Delta_{imm}H$ is around −110 mJ m^{-2}, very near that of Robert.

This analysis compared to the generally reported results shows that the immersion enthalpy per unit area is a constant term, characteristic of the surface of a chemical species and varying only at the level of 10%, say, with the surface state. Therefore, at a first level of interpretation, the results we report can be used to determine the specific surface area of other solid samples, by dividing the experimental heat of immersion by $\Delta_{imm}H$ expressed in mJ m^{-2}. This is in agreement with an idea we will develop in the next sections: that the interaction term between a solid and a fluid is mainly due to the solid characteristics.

V. ANALYSIS OF RESULTS

A. Introduction

As seen above, the enthalpy of immersion can be viewed as a difference between a value characteristic of the solid and an intermolecular enthalpy:

$$-\Delta_{imm}H = H_S^{o(A)} - H_{SL}^{(A)} \tag{19}$$

The first term depends only on the solid. The second term depends both on the liquid and on the solid.

In the thermodynamic conditions described above (pure phases, constant temperature and pressure, no swelling, no dissolution), these enthalpies can be replaced by a term relative to surface tension:

$$-\Delta_{imm}H = \left[\gamma_S^o - T\left(\frac{\partial \gamma_S^o}{\partial T}\right)_P\right] - \left[\gamma_{SL} - T\left(\frac{\partial \gamma_{SL}}{\partial T}\right)_P\right] \tag{28a}$$

The first term has been precisely defined and discussed [13]. It is clear that it is strongly correlated to the average bond energy of the crystal, with some modification being due to rearrangements of the electronic density of the surface and to the differences of the average number of neighboring atoms. Moreover, solid surfaces are not homogeneous planes but present steps, kinks, and all sorts of physical defects. A very simple way to estimate this tension and the corresponding enthalpy consists of considering this enthalpy as a fraction (say 80%) of the enthalpy of vaporization of the considered solid [17]. The interesting point here is that this value does not change with different immersing liquids.

However, when changing the liquid, the solid–liquid enthalpy is expected to change. The interpretation of these observed variations belongs to the general problem of intermolecular forces [9].

It is possible to consider the solid as a heterogeneous phase composed of individual atoms, and to obtain the total interaction energy by summing the different interactions between each atom and the liquid. This summation could be averaged by a term [112] (the "heterogeneity") reflecting the different orientations of the atoms in different locations, such as on the faces and edges of the solid grain. Alternatively, the total immersion enthalpy can be divided into energetic terms taking into account the actual differences between adsorption sites, i.e. due to the location of crystallographic planes, to the presence of compensation cations, chemical groups, and so on. A general equation of the following type can be given:

$$\Delta_{imm}H = \sum_i \theta_i S_i \tag{29}$$

where S_i is the strength of interaction of the site i and θ_i the number of sites i per surface area.

But this approach could be very difficult, because the different locations and orientations of actual solids are generally unknown. Moreover, the precise number of surface atoms and groups, e.g., the real thermodynamics of the surface compared to the bulk, is unknown [3].

Therefore, a more phenomenological approach to calculate intermolecular energies is preferable. In this sense, we propose below a model in order to interpret immersion, adsorption, and wetting.

B. The Model

1. General Case

For a long time (for example see Bondi and Simkin [113]), it appeared that the intermolecular energy between two species could be attributed to different additive terms, each characteristic of one specific type of interaction. Unfortunately, each type of interaction being characterized by a length, the different types of interaction do not act in the same way, and the final equations describing an intermolecular energy can be complicated.

Following Kitaura and Morokuma, the intermolecular energy between species 1 and 2 can be sketchily written as [114]:

$$\Delta E_{12} = \Delta E^{ES} + \Delta E^{PL} + \Delta E^{EX} + \Delta E^{CT} + \Delta E^{MIX} \tag{30}$$

where ES refers to the electrostatic component of the energy, PL to polarization, EX to exchange repulsion, CT to charge transfer, and MIX to coupling.

More generally, one can write

$$\Delta E_{12} = \Sigma \Delta E^i \tag{31}$$

where

$$\Delta E^i = E^i_{12} - E^i_1 - E^i_2$$

and

$$E^i_{12} = f_i(E^i_1 + E^i_2)$$

f_i being a characteristic function of each interaction i.

In our particular case, where we study interactions between condensed phases, it is possible to assume the following equality between internal energy and enthalpy [15,22]:

$$\Delta E \approx \Delta H$$

and consequently to write the intermolecular interaction enthalpy in the form

$$\Delta H_{12} = \Delta H^{ES} + \Delta H^{PL} + \Delta H^{EX} + \Delta H^{CT} + \Delta H^{MIX} \tag{32}$$

The question arising here is the difficulty of linking the theoretical interaction terms to the actual experiments. An interaction is not a molecule, and a molecule deals with more than one interaction. This point led great number of researchers to propose semiempirical equations-linking the theoretical interactions to parameters that can be estimated in an experimental way. The most useful experimental term is the interaction enthalpy. Many semiempirical models have been proposed. We will recall here the one due to Drago and coworkers [115]. It is very simple equation, linking the acid–base interaction enthalpy to some properties of frontier orbitals:

$$\Delta H_{AB} = E_A E_B + C_A C_B \tag{33}$$

where E_I is the susceptibility of the species I to take part in an electrostatic interaction, and C_I the susceptibility of the Ith species to take part in a covalent bond.

This last equation is only one of several possible useful equations, but it is the most used. We recall it here as an example, helping to keep in mind that such equations are experimental adaptations of exact theories. The parameters or "components" used do not directly correspond to the theoretical components. They are functions of them.

2. General Application to the Surface

In actual cases, the exact forms of the equations given above depend on the phase density. Hence, in the surface, which can be considered as an autonomous phase [12] if at chemical equilibrium, we can consider that the surface has the possibility to be more or less dense than the adjacent phases [116]. It is then possible to use the above equations to describe the interfacial interactions and predict the behavior of the system.

Therefore, in a liquid–solid or liquid–liquid surface (denoted σ), between two pure species, denoted 1 and 2, we can write

$$\Delta H_{12}^\sigma = (\Delta H^{ES})^\sigma + (\Delta H^{PL})^\sigma + (\Delta H^{EX})^\sigma + (\Delta H^{CT})^\sigma + (\Delta H^{MIX})^\sigma \tag{34}$$

Consequently, the interfacial interaction enthalpy between a solid species A and a liquid species B can be deduced using an equation of the Drago form:

$$\Delta H_{AB}^\sigma = k^\sigma (E_A E_B + C_A C_B) \tag{35}$$

k^σ being a density factor, written here to keep in mind that it is an equation used in the surface phase. In this form it is not necessary to define the terms E_I^σ and C_I^σ, characteristic of the interface, which could be confusing.

Surprisingly, such an equation can be also used to calculate the energy of a liquid–liquid interface, assuming that the interaction enthalpy ΔH_{AB}^σ is similar to the interfacial enthalpy ΔH_{12}^σ. Indeed, the late Fowkes [117] used it to calculate an interaction term between water and benzene, using Drago's values, and a density factor of 1. He obtained $\Delta H_{12}^\sigma = 5$ kJ mol^{-1}. This choice of k^σ confuses an interfacial interaction and a bulk interaction, but we will admit heuristically that this point is not so important. Starting from a molar energy [17], it is quite easy to compute the energy related to a surface area. Using the most simple calculation, established from the density and mass of the molecules, and considering that the surface is an equimolar mixture of water and benzene, we obtain a ratio of

$$k^\sigma N_A \left(\frac{m}{d}\right)^{2/3} = 1.16 \, 10^5 \, \text{m}^2 \, \text{mol}^{-1} \tag{36}$$

with N_A being Avogadro's number, $k^\sigma = 1$, and m the average molecular mass: $(78 + 18)/2 N_A$.

We then obtain an interfacial enthalpy of 43.1 mJ m^{-2}, which compares very satisfactorily to the surface tension (35 mN m^{-1}) and justifies the heuristic approximation used above.

Although this approach is very fruitful, this type of value is the only one that can be obtained in a rigorous way. These terms can be also obtained by microcalorimetry. As we have seen above, the interaction between a solid and a vapor is linked with an adsorption enthalpy, and the interaction between a solid and a liquid is linked with an immersion enthalpy or a wetting enthalpy, depending on the experimental starting pressure.

Following the recent evolutions of the type of parametrization used by Drago [9], and also some developments from the recent literature on colloid science [118], we will consider here that it is necessary and sufficient to characterize a molecular species by a set of three terms. Starting from calorimetric experiments, this allows intermolecular interactions to be computed. Therefore we propose a general analysis of immersion, adsorption, and wettability in terms of three surface components, namely dispersive, acid, and basic, generalizing the most powerful experimental method, the Good–Van Oss–Chaudhury (GVC) method [119–124]. But one has to emphasize that, generally speaking, the point concerning the real number of interactions acting in the surface is not so important, mainly because experiments are performed with real molecules, and real molecules are representative of more than one type of interaction. Therefore the number of interactions deduced from trends of experimental results is not the actual number of theoretical interactions defined in a theoretical approach.

One of the main points of the model we propose is the possibility of interpreting experimental enthalpies related to the surface in terms of "surface enthalpy components." Another point is that this approach is also valid for "high-energy" solids, if there is no chemisorption, but only physical adsorption, in the studied system.

3. Basic Equations

Basic ideas of the general model we propose are the same as in the GVC model. Repulsive interactions are fully additive, although attractive interactions add following some exclusion rules. In this view, polar interactions are only additive; they cannot interact with dispersive terms. These points can be accepted from the point of view of theoretical chemistry but are controversial. For instance, from the viewpoint of chemical reactivity science, alkanes can be considered as bases.

The basic equation is the definition of the surface enthalpy of interaction between a molecule 1 and a molecule 2:

$$H_{12}^\sigma = H_1^\sigma + H_2^\sigma - 2((H_1^{\sigma\text{LW}} \cdot H_2^{\sigma\text{LW}})^{1/2} + (H_1^{\sigma+} \cdot H_2^{\sigma-})^{1/2} + (H_1^{\sigma-} \cdot H_2^{\sigma+})^{1/2} \tag{37}$$

with

$$H_i^\sigma = H_i^{\sigma\text{LW}} + 2(H_i^{\sigma+} \cdot H_i^{\sigma-})^{1/2} \tag{38}$$

The term LW represents the apolar contribution (the letters L and W correspond to the initials of Lifschitz and van der Waals). The acid and basic contributions are respectively denoted as plus and minus. The modern view of acidity [9] implies that these terms are only parameters related to the theoretical acid strength S_A [53], without knowing the relationship in details.

The varying term in the measure of the enthalpy of immersion is the "intermolecular" solid–liquid enthalpy per unit of surface area. We can use the "reduced" enthalpies per unit of surface area and the enthalpies interchangeably, because we are in

Immersion Phenomena

unary phases with T and p constant: the variations are the same. Thus we can write

$$H_{SL}^\sigma = H_S^\sigma + H_L^\sigma - 2((H_S^{\sigma LW} \cdot H_L^{\sigma LW})^{1/2} \\ + (H_S^{\sigma+} \cdot H_L^{\sigma-})^{1/2} + (H_S^{\sigma-} \cdot H_L^{\sigma+})^{1/2} \quad (39)$$

where we can substitute: $H_S^\sigma = H_S^{\circ(A)}$ and $H_L^\sigma = H_{LV}^{(A)}$. From this, we can deduce the following equation, expressed per unit area, without using the notation (A) for sake of simplicity:

$$-\Delta_{imm}H = -H_{LV} + 2((H_S^{\circ LW} \cdot H_{LV}^{LW})^{1/2}(H_S^{o+} \cdot H_{LV}^-)^{1/2} + (H_S^{o-} \cdot H_{LV}^+)^{1/2} \quad (40)$$

If one determines the enthalpy of immersion in a liquid without apolar interactions, the equation is greatly simplified, and the measurement allows the determination of one unknown term, the surface apolar enthalpy:

$$-\Delta_{imm}H = -H_{LV} + 2((H_S^{\circ LW} \cdot H_{LV}^{LW})^{1/2}) \quad (41)$$

Proceeding in this way with different probe liquids, carrying the three different types of interaction with different percentages (their reduced surface enthalpy can be determined by tensiometry measurements at different temperatures), it is possible to determine the unknown parameters.

This three-parameters model has been chosen after a phenomenological analysis of our results. We will also see below that immersion results are in agreement with the trends of vapor adsorption isotherms. This indicates a general behavior of surface intermolecular forces, more or less in agreement with a classical analysis of solvation phenomena [9,53,115,118,125]. In fact, very generally, a scale of interactions could be drawn that is valid in most phases. We can represent (Fig. 3) the intermolecular potential interaction as dependent on some parameters, ordered as dimensions of space. The first dimension reflects the influence and respective proportion of van der Waals interactions and of polar interaction (this last term is mainly the value of the dielectric constant with the particular case of the possibility of a self-dissociation of the molecule). The second dimension reflects the "hard–soft" parameters, which is second order, but which is very important. The "poor solvents" are at the bottom of the scale and the "good solvents" at the top. The mutual solubility of two different molecules depends on their distance of separation on this scale. It the two molecules are very near on the picture, they are mutually soluble. If they are condensed in different mutually insoluble phases (a solid and a liquid for instance) the interaction (i.e., the adsorption) decreases as the distance increases. An interfacial tension can only occur between mutually insoluble phases. Therefore the interfacial tension is dependent on the separation between molecules on this simple scale.

The model we propose is simply an operative and quantitative way to use this scale, which has the ability to explain quantitatively results as different as wetting (three phases) and adsorption (two phases).

4. Numerical Application

The point now is the choice of probe fluid molecules, and the determination of values of corresponding reduced surface enthalpies, for the probe fluid molecules. There is an abundant literature on the choice of surface tension components [126], even though there is not a general agreement on the number of components. Many scientists divide the surface tension into a polar and an apolar term, the word polar referring in fact to interactions "like an alkane." Users of the GVC model confound the value of the apolar component

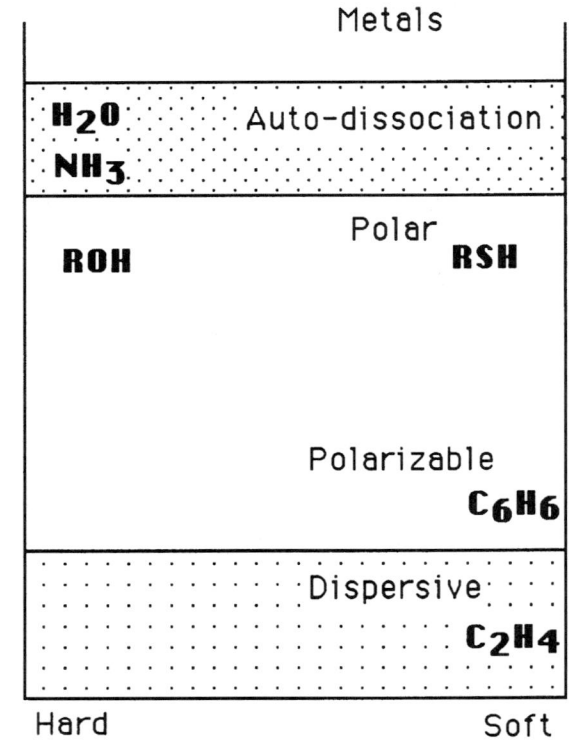

FIG. 3 A simple view of the interactions between molecules depending on two parameters, hardness and dipole moment. Some typical molecules are located in the corresponding domain. The strength of intermolecular forces and the interfacial tension depend on the distance between molecules in this map.

TABLE 9 Values of the Enthalpy Components used for the Different Liquid Probes (mJ m^{-2})

Liquid	H_{LV}	H^{LW}	H^+	H^-
Heptane	48.9	48.9	—	—
Decane	50.8	50.8	—	—
Water	118	36	41	41
Benzene	66.7	66.7	—	4.56
Pyridine	75.5	20.5	4.1	5.3
Formamide	82.1	55.2	3.23	56

with the "Lifshitz–Van der Waals" term, meaning that both terms describe the same type of interaction. Their acid–base scale assumes that water is amphoteric, i.e., the acid and basic components for water are equal. This idea is controversial [127], other fields of chemistry thinking that water is a hard base [53]. But this choice is not so important, because such models are not describing actual interactions, as studied by theoretical chemistry and physics. They are only scales, and each scale needs a starting point. It is easy to go from one well-defined scale to another. With this basic idea, we have followed the GVC

Immersion Phenomena

model, and we use this scale to establish the reduced surface enthalpy of the probe liquids we use, namely heptane, decane, benzene, water, formamide, chloroform. Pyridine is a special case, discussed elsewhere [106]. The relative importance of the basic and acid component we obtain in the calculation is only a relative trend, the signification of which depends on the GVC assumptions. Therefore we do not focus on this point but only on the apolar/polar ratio, and on the total surface enthalpy we obtain for each solid studied, which is a very rare and useful information in view of wettability understanding. The values used are reported in Table 9.

C. Determination of Solid Surface Tensions

It is possible to deduce from surface enthalpies the solid surface tension. We will write the well-known relationships [12] in the following form (at constant surface area and constant temperature and pressure):

$$dH^\sigma = T\,dS^\sigma + \Sigma n_i^\sigma\,d\mu_i + A\,d\gamma$$
$$dG^\sigma = \Sigma n_i^\sigma\,d\mu_i + A\,d\gamma \qquad (42)$$

where A is the surface area and γ the tension. At chemical equilibrium between two insoluble phases, such as a solid and a liquid, we have

$$dH^\sigma = T\,dS^\sigma + A\,d\gamma$$
$$dG^\sigma = A\,d\gamma \qquad (43)$$

It is then possible to write the following relation between enthalpy and free enthalpy at constant temperatures:

$$\Delta G^\sigma = f\,\Delta H^\sigma \qquad (44)$$

This equation imposes the following definition of the term f:

$$f = 1 - T\frac{\Delta S^\sigma}{\Delta H^\sigma} \qquad (45)$$

In our case, we can write

$$f = 1 - T\frac{\Delta_{\text{imm}} S}{\Delta_{\text{imm}} H}$$

A priori, f cannot be independent of the interfacial system studied, and obviously f depends on the temperature. However, using immersion calorimetry and adsorption isotherms, f can be calculated. We have carried out some experiments of vapor adsorption to determine the adsorption isotherms. An example is given in Fig. 4. This shows a remarkable trend, e.g., when comparing isotherms for the same solid. An increase of the quantity adsorbed is due to the polarity of the adsorbing vapor. This trend is obvious in literature results [3,128,129] even though it has never been highlighted. It is in clear agreement with the view developed above, of a hierarchy of the interaction types, going from simple van der Waals interactions to polar effects. Starting from our experimental results, it can be remarked that f is quite constant, with a value near of 0.45 (the experiments were performed with silicates and the usual solvents: some alkanes, aromatics, water). The assumption that f is constant is often made, and more frequently in theory [130]. To the best of our knowledge, in surface science, Fowkes was the first to make this

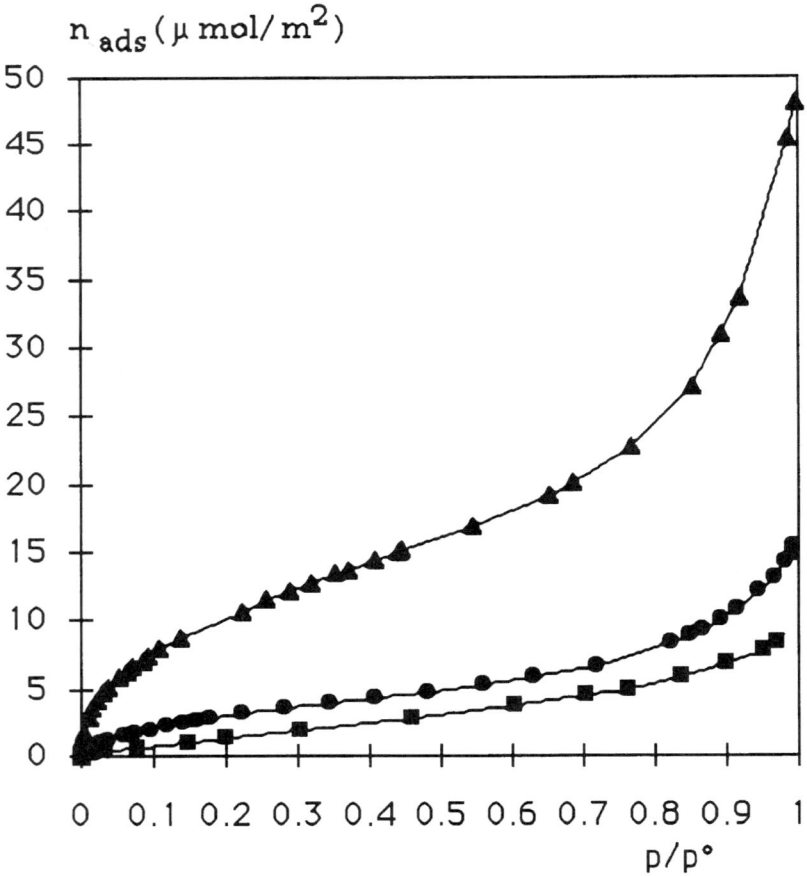

FIG. 4 Adsorption isotherms at 35°C of heptane (squares), benzene (circles), and water (triangles) onto the C800 quartz sample. Experimental conditions and procedures are described in Ref. 104.

assumption, even if Fowkes in fact wrote this parameter in a different manner. He introduced another parameter called "interfacial concentration of acid–base interactions," allowing him to obtain an f parameter of 0.38.

Experimentally, f can be computed starting from results obtained by Mortier and coworkers [131]. f varies between 0.3 and 0.4 for some hexane/zeolite and benzene/zeolites systems, at ambient temperature. This small difference with our results suggests a similarity of structure between the different mineral solids studied (silicates and alumino-silicates), for the low range of temperature concerned (near ambient). We have fixed a value of $f = 0.41$ for the Lifshitz–van der Waals component and of 0.48 for the acid component obtained using results with benzene. Finally, the values obtained using results with water were analyzed with $f = 0.5$.

To assume f constant allows the following equations to be written:

$$fH_{12}^\sigma = f(H_1^\sigma + H_2^\sigma - 2((H_1^{\sigma LW} \cdot H_2^{\sigma LW})^{1/2} + (H_1^{\sigma +} \cdot H_2^{\sigma -})^{1/2} + (H_1^{\sigma -} \cdot H_2^{\sigma +})^{1.2})) \quad (46)$$

which is strictly equivalent to

$$G^\sigma_{12} = G^\sigma_1 + G^\sigma_2 - 2((G^{\sigma\,LW}_1 \cdot G^{\sigma\,LW}_2)^{1/2} + (G^{\sigma+}_1 \cdot G^{\sigma-}_2)^{1/2} + (G^{\sigma-}_1 \cdot G^{\sigma+}_2)^{1/2}) \qquad (47)$$

where the G^σ_{ij} are free enthalpies per surface area, and to

$$\gamma_{12} = \gamma_1 + \gamma_2 - 2((\gamma^{LW}_1 \cdot \gamma^{LW}_2)^{1/2} + (\gamma^+_1 \cdot \gamma^-_2)^{1/2} + (\gamma^-_1 \cdot \gamma^+_2)^{1/2}) \qquad (48)$$

which is the basic equation of the Good–Van Oss–Chaudhury model.

This indicates clearly that the Good–Van Oss–Chaudhury model is simply a first approximation of the more general model developed above. It assumes that the f coefficient, i.e., the variation of entropy during the contact of the three phases at the triple line, is the same for all the molecules in contact. The low variation of f observed is in quite good agreement with this idea, even if one expects that the small variation we observe can have great consequence in the details of the results.

VI. SURFACE ENERGY OF HIGH-ENERGY SOLIDS

A. Surface Enthalpy

1. Results

Using the model defined above, it is easy to compute the components of the surface enthalpy per unit area for each solid sample. The values are listed in Table 10. Results for $H^{0(A)}_S$ shows clearly an order, respecting mainly the order of the enthalpy of fusion of the selected solids. For instance, it can be recalled that the enthalpy of fusion of quartz and alumina are respectively about 14 kJ mol^{-1} (1470°C) and 109 kJ mol^{-1} (2045°C), to be compared for instance with an enthalpy of fusion of 146 joules per gram for paraffin. More precisely, this order of surface bond energy also reflects an order of surface states, influenced on the one hand by the details of the bulk solid network (bond energy, crystallographic structure) and on the other by the details of the surface bonds. There is some influence by the repartition of chemical groups present at the surface, and some aspects of its history (formation, mechanical strain). The low polar values obtained for Teflon and sulphurs 1 and 3 are probably not significant, and due to impurities. These types of samples can be considered as apolar, and the polar values we give are a good estimation of the precision and signification of our results.

Moreover, the balance between acidic and basic energy is totally dependent on the choice of assessment we have made for water and consequently reflects a ratio due to a certain reference more than a reality. This point suggests that we consider the ratio between polar and apolar energy as the more representative value. Even in this type of experimental trend, the differences between same-family solids, such as the group of illites or the group of kaolinites, are clearly marked. The difference between high-energy solids and low-energy solids is also very apparent. Silicas, talcs, kaolinites, and illites are middle-energy solids [32].

The values of the apolar energy do not vary in the same proportion as the polar energy. They are more dependent on the nature of the phase than on the chemical nature of the product. The same point is obtained for liquids where the apolar surface tension approximately varies between 20 and 30 mN m^{-1} for the majority of liquids [3] at ambient temperature (metal liquids excepted).

TABLE 10 Components of the Reduced Surface Enthalpy and Total Reduced Surface Enthalpy at Ambient Temperature Using our Model and Immersion Results (mJ m^{-2})

Solids		H_S^{LW}	H_S^+	H_S^-	H_S^o
Dolomite		681	469	3705	3318
Chlorite		322	2832	47	1052
Alumina		227	531	728	1470
MgCO$_3$		156	1174	251	1241
Calcite		415	29	701	700
Quartz	C800	119	199	218	535
Kaolinite	Ploemeur	107	158	87	341
Kaolinite	Charentes	122	180	89	376
Kaolinite	Provins	93	179	157	428
Silica	XOB O15	182	47	67	294
Silica	Cabosil	91	124	5.7	144
Illite	Hungary	118	89	324	458
Illite	Brives-C	89	229	3	141
Illite	Vosges	98	239	141	465
Talc	1	144	444	23	346
Talc	2	180	115	2.7	215
Talc	3	206	90	0.3	216
Talc	Spain	92	550	21	305
Talc	Luzenac	121	477	4	206
Talc	Italy	134	233	54	357
Graphon		146	—	—	146
Teflon		58	1.6	1.6	61
Sulphur	1	75	0.6	0.6	76
Sulphur	2	121	—	—	121
Sulphur	3	44	5	5	54

2. Discussion

As seen before, the solid surface energy is only one of two parameters governing the enthalpies of immersion. In this sense the energetic order of the immersion enthalpies is not strictly respected by the solid surface enthalpies. These last parameters are, strictly speaking, characteristic of the solid as a pure phase. Therefore, it is logical that the solid surface enthalpies follow the order of the bonding energies of the different solids, a term which can be related to the electronegativity of the atoms constituting the solid [9,53]. The difference observed in our results between carbon, silicates, and alumino-silicates is apparent. Moreover, the influence of the quantity of aluminum atoms in the framework is of paramount importance (see Table 11). These results are in perfect agreement with the orders of reactivity as determined in catalysis science [132].

We can also observe that for the same type of solid, specifically talc and chlorite, the nature of the bond involved in the surface phase by a certain atom is of great importance. Chlorites are characterized by the presence of oxygen molecules bound to hydrogen (i.e., silanol groups) on one of the external sheets of their structure. Talcs present only dehydroxylated oxygens on their external faces [133]. The result of this is very clear

TABLE 11 Quantity of Aluminum in the Bulk Structure and Lifshitz–van der Waals Term of the Reduced Surface Enthalpy for Two Groups of Clays

Solids		Weight % of Al_2O_3	H_S^{LW} (mJ m^{-2})
Kaolinite	Charentes	38	122
Kaolinite	Ploemeur	35.4	107
Kaolinite	Provins	33.3	93
Illite	Hungary	24.4	118
Illite	Vosges	18.7	98
Illite	Brives C.	17.4	89

in the respective enthalpy values, because silanol groups are very reactive. These results show that talc is highly polar, essentially because of its strong surface acidity. This point is not a surprise when compared with the strong surface acidity of the clay minerals like kaolinite [134], attapulgite [134,135] and montmorillonite [134]. It appears from other results on the adsorption of strong bases [136–138] that alumino-silicates are more acidic than magnesia-silicates. Kaolinite and illite, which are alumino-silicate clays, have a higher Lewis acid component, when determined by our method, than talc, which is a magnesium-silicate clay. Generally, the surface acidity is attributed to the hydrogen of the surface hydroxyl groups [139] of AlOH, MgOH, and SiOH. Obviously, the surface acidity of these talcs increases with their surface hydroxyl group content: i.e., when the lateral surface area percentage increases.

The total picture can be described as follows. With the exception of the massive crystals found in rocks, solids usually exist as colloidal particles, with a maximum dimension of less than 1 micron. This is the case in our study. Such particles have a high surface-to-volume ratio, hence a significant proportion of the atoms located at the particle surface. The surface hydroxyl groups are the principal functional groups of the solids studied. They can arise from irreversible adsorption of water or from structural OH. If one considers adsorption and wetting as a chemical reaction, they are the reactive chemical entities at the surface of the solid. They possess a double pair of electrons together with a dissociable hydrogen atom that enables them to react with both acids and bases. Due to this fact the majority of the solids we study are amphoteric. Bulk structure considerations indicate that the surface hydroxyl groups can be coordinated to different atoms (Fe, Al, Si, C, Mg, etc.) Moreover, they can be coordinated to one, two, or three underlying atoms. Consequently, some different types of hydroxyls may appear: simple, geminal (i.e., doubly bonded to the same atom), doubly, or triply coordinated. The overall density of these groups depends both on the crystal structure and on the extent of development of the different crystal faces. Due to the difference between the numbers of underlying atoms that are coordinated to the surface functional groups and to the distance between the chemical groups, the acidity and the reactivity of the different types of surface groups vary. On this basis, the surface of a solid can be viewed as a patchwork composed of a certain number of chemical groups, separated by a certain distance. This distance can be quantified by a "surface density" of the groups. One can therefore distinguish between an attraction for the fluid molecule that is related to the influence of isolated groups and an attraction due to the interaction between groups. (The separation between the two terms is not so clear but is useful in the reasoning.) The first term has an influence on the surface enthalpy due to the multiplication by the net number of surface groups and is clearly related to the composition of the solid (i.e., the stoichiometry of the different groups). The second term is more global and is related to the actual morphology of the crystal. In fact, if the crystals were perfect, the ratio between different groups located at different places such as "basal," "lateral," and "edge" should be known; and the relation between the first and the second term could be established. Generally, the actual solid grains are broken, fractured, and the proportions between the different types of bonds are difficult to estimate.

Concerning the first term, the solid attraction for the fluid molecule due to the addition of the different attractions of surface groups is also influenced by the strength of the bonds in the framework. This "hardness" of the solid influences the electronegativity of each group. The result can be viewed as a "collective attraction" to be narrowed with the "intrinsic framework electronegativity," a concept developed by Mortier and coworkers [131]. As our results are in good agreement with the trends observed by Mortier, we can recall some of the remarks he developed: the lowest electronegativity calculated for solids corresponds with the densest framework, and the highest electronegativity corresponds with the most open framework. The most attractive chemical group for the external fluid molecule is the hydroxyl group. The electronegativity of this group depends on the atom to which it is bound. Clearly, from our results, hydroxyls are more attractive when bound to networks rich in aluminum. This raises the question whether it is the influence of the atom on the surface group or on the network that explains the result.

3. Influence of the Shape of the Grain

The immersion enthalpies in n-heptane, water, and formamide of the three pure talc samples of different lamellarity are reported in Table 6. These three samples are well described in Ref. 108.

In the apolar solvent (n-heptane), one observes an increase in the immersion enthalpies with sample lamellarity. The order of this gradation is reversed in the case of the polar solvents, water and formamide.

When the lamellarity parameter changes from one sample to another, the ratio between basal and lateral surface areas changes. It is an important effect, influencing the immersion enthalpy results. This modifies the partition of polar groups on the surface of the grain more than the force of the chemical bond.

Our results in the case of the n-heptane, which is apolar, are difficult to explain. Generally there is a weak influence by the polarity of the solid on the immersion enthalpies in apolar solvents. Nevertheless, one can observe an increase in the immersion enthalpies in n-heptane with the sample lamellarity, an effect that may come from the orientation of the solvent molecules on the solid surface, the talc samples being characterised by an extraordinary planarity. Accordingly, one can suppose that n-heptane molecules are more organized on the regular, large, and plane basal surfaces of Italian talc than on the disorganized and short basal surfaces of Spanish talc. To verify this orientation effect on the solid liquid interactions, the immersion enthalpies of our talc samples in a very branched alkane : iso-octane have been determined.

The experimental results are reported in Table 12. The three talc samples give similar immersion values in isooctane. The gap between the immersion enthalpies in n-heptane and in iso-octane is nil for Spanish talc, and it increases strongly when the lamellarity of the studied samples increases. These results show a large influence of the surface geometry of the adsorbed molecule, imposed by the crystal morphology, on the interaction energy.

TABLE 12 Enthalpies of Immersion of Three Different Talcs in Iso-octane at 25°C

Talc sample	$\Delta_{imm}H$ (mJ m^{-2})
Spain	−84
Luzenac	−103
Italy	−111

The lamellae of Spanish talc have the smallest sizes, so the ratio of basal surface to lateral surface is the lowest. Considering that lateral surfaces are composed by SiOH and MgOH, they can interact strongly with polar solvents through strong hydrogen bonds. This is the reason that this sample gives the largest immersion enthalpies in water and formamide. In contrast, the Italian talc sample, which has the highest lamellae sizes and consequently the lowest lateral surface percentage, gives the lowest immersion enthalpies in water and formamide. Effectively, the basal surfaces are constituted only by siloxane groups, which cannot form strong hydrogen bonds. Yariv [140] has more precisely identified the origin of these hydrophilic/hydrophobic properties of the lateral and basal surfaces for minerals. From these results, it appears that the talcs and the other mineral solids are made up of surface patches with opposing properties. Some are high-energy surfaces and others are low-energy surfaces. Lateral surfaces with sites that can form strong hydrogen bonds are the more hydrophilic; basal surfaces are hydrophobic, especially in the case of talc, because talc has no substitution of silica by aluminum or iron. This view can be generalized to other solid powders. The surfaces of all of them are patchworks of faces with different energies. Each mechanical or chemical modification of the initial patchwork, due to cleaning or grinding, modifies the partition of interaction present at the surface.

B. Surface Tension

1. General Phenomenology

Starting from Eqs. (48) and (14), it appears that it is possible to compute the surface tension of solids if experiments on the adsorption of probe vapors are carried out. This has been done for some samples of our set. We have observed a general trend that is in agreement with literature results. As we have seen, high-energy solids are characterized by the existence of three types of intermolecular potential energy. Therefore they possess an attractive energy for dispersive fluids rather than for polar fluids. Obviously, the interaction energy, which influences the number of molecules adsorbed per unit area, is greater for polar molecules than for apolar molecules, but this last term is never negligible. Hence in each case we expect to obtain adsorption isotherms per unit area where the quantity of alkane adsorbed is always lower than polar (or polarizable) molecules. Also, molecules with low dielectric constants are expected to be less adsorbed than molecules of high-energy constants. Generally, water is the molecule adsorbed in the greatest quantity. In our case, the isotherms obtained were of Type II [10]. However, the energetic model we use is not dependent on the porosity of the sample (which can give a type IV or type I isotherm in IUPAC classification).

TABLE 13 Surface Tension Components and Total Surface Tension of the Solid in Equilibrium with its Own Vapor Using Our Model (mN m^{-1})

Solids		γ_S^{LW}	γ_S^+	γ_S^-	γ_S^o
Dolomite		306	234	1852	1623
Chlorite		145	1416	23	506
Alumina		102	265	364	723
MgCO$_3$		70	587	125	612
Calcite		187	14	350	240
Quartz	C800	53	99	109	261
Kaolinite	Ploemeur	48	79	43	164
Kaolinite	Charentes	55	90	44	181
Kaolinite	Provins	42	89	78	209
Silica	XOB O15	82	23	33	137
Silica	Cabosil	41	62	3	55
Illite	Hungary	53	44	162	222
Illite	Brives-C	40	114	1.5	66
Illite	Vosges	44	119	70	226
Talc	1	65	222	11	164
Talc	2	81	57	1.3	98
Talc	3	93	45	0.1	97
Talc	Spain	40	275	10	148
Talc	Luzenac	53	238	2	97
Talc	Italy	59	116	27	172
Graphon		66	—	—	66
Teflon		26	0.8	0.8	28
Sulphur	1	34	0.3	0.3	35
Sulphur	2	54	—	—	54
Sulphur	3	20	5	5	30

The results of the assumption discussed above, considering that enthalpy and free enthalpy are simply linked at constant temperature, are listed in Table 13. They are not surprising, following the order obtained for the solid surface reduced enthalpy; these values can be compared with the literature results as reported by Adamson [3]. The approximated results we obtain are in good agreement with the exact results of the literature, and they make sense. It is clear that the lamellarity of the sample has a great influence on the surface tension of the solid, mica giving the upper value of all the solids studied. Crystals partially soluble in water do not give very high values of surface tension. The other trends have been discussed before. From this table, it is difficult to predict the behavior with mixtures, solute molecules, or impurities and pollutants in actual conditions. The three parameters of the tension that have an influence are apolarity, acidity, and basicity. A weak base will adsorb preferentially on a solid of a strong acid character, although strong bases should make no distinction between acid solids. Macromolecules, even polar, would tend to adsorb preferentially on solids with a strong apolar character, but the distance between surface sites has to be known to predict the outcome.

2. Consequences of the Usual Theories

One of the most important results of our work is the range of values obtained for the solid–liquid surface tension. Even though this value varies from low-energy to high-energy solids, it is clear that a solid–liquid surface tension is always positive and greater than

20 mN m^{-1}. This point has an important consequence for the classical theories of contact angle. Following Zisman and Good, the majority of researchers have considered that the solid–liquid surface tension is negligible. This assumption has allowed scientists to consider that the minimal surface tension in a family of molecules (for instance alkanes) that are able to spread on a solid surface is the surface tension of the solid. Our results clearly demonstrate that this minimal surface tension is not the solid–vapor surface tension. More generally, all the theories considering that the solid–liquid surface tension is nil cannot give a solid–vapor surface tension of the solid. They only give a value in the correct order of magnitude, but lower than the exact one. This point casts a shadow on the theory of Neumann and coworkers [141], called the equation-of-state approach. But in fact, the shadow is not very important. It simply means that such theories give a value of the difference $\gamma_{SV} - \gamma_{SL}$. One has to emphasise that our results demonstrate that the value of the solid–liquid surface tension depends strongly on the magnitude of the solid energy, and hence that the solid–liquid surface tension must have low values for polymers. The assumption quoted above has been made in this field and therefore is not so great.

In fact, this assumption was surprising, because surface tension is linked to the mutual solubility of phases. For example, one can consider a nil surface tension in the case of the mixture of vapor phases, or for a critical interface at the critical point. But these cases are limiting cases, and one generally expects a positive value of the surface tension for nonmixing phases. Thus the solid–liquid surface tension is probably related to the solubility product of the solid concerned in the chosen fluid. The more soluble the solid, the lower must be the surface tension. But this question of the solubility of solids is a very complex one. For ionic solids, for example, it is the result of a balance between the Madelung energy of the crystal, which depends on the radius of the ions, and the hydration energy of the ions, which also depends on the radius of each ion, but following a different equation. The key point here is that what is obtained using contact angles is a tension related to the solid–vapor surface tension.

CONCLUSION

Some points can be deduced from the experimental results and numerical trends we obtain.

1. Schrader's ideas, which divide solids into three classes, namely, low energy, middle energy, and high energy, are valid for powders.
2. The pertinent parameter for describing solids is the surface energy (either enthalpy or free enthalpy) of the solid in equilibrium with its own vapor.
3. The heat of fusion value is a good approximation allowing the surface enthalpy of powders to be estimated.
4. Many of the values of γ_{SV} published in the literature have been underestimated, due to the assumption that π_e is negligible. The critical surface tension γ_c is clearly a minimal value, lower than the actual solid–vapor tension. The difference is simply the film pressure of the vapor concerned, or if in air, the film pressure of water. That the assumption of a negligible π_e is false was actually very clear in the drawing of IUPAC isotherms classification. Low-energy solids correspond to type III and V isotherms. For such isotherms, the adsorbed quantity is never nil. It increases drastically for high values of the partial pressure, i.e., for great activities of the vapor.

5. Immersion enthalpy results can be simply interpreted on the basis of a scale of surface energy of the solid powders, and by using the idea that the solid–vapor or solid–liquid strength of interaction is dependent on the surface energy of the solid. This point was controversial and is for example denied by Neumann and coworkers. We think that our results are important in this debate.

REFERENCES

1. M. C. S. Pouillet. Ann. Chim. Phys. 20:141 (1822).
2. A. J. Groszek, in *Proc. NASA Sponsored Symp. on Interdisciplinary Approach to Liquid Lubricant Technol.*, Cleveland, Ohio, January 11–13, 1972, p. 477.
3. A. W. Adamson, *Physical Chemistry of Surfaces*, John Wiley, New York, 1976.
4. *Handbook of Separation Process Technology* (R. W. Rousseau, ed.), John Wiley, New York, 1987.
5. P. G. De Gennes. Rev. Mod. Phys. 57:827 (1985).
6. J. N. Israelachvili and R. M. Pashley. Nature 306:249 (1983).
7. J. Fripiat, M. Cases, M. Francois, and M. Letellier. J. Colloid Interface Sci. 89:378 (1982).
8. C. Tanford, *The Hydrophobic Effect: Formation of Micelles and Biological Membranes*, John Wiley, New York, 1973.
9. J. E. Huheey, E. A. Keiter, and R. L. Keiter, *Inorganic Chemistry: Principles of Structure and Reactivity*, Harper Collins, New York, 1993.
10. S. J. Gregg and K. S. W. Sing, *Adsorption, Surface Area and Porosity*, Academic Press, London, 1982.
11. J. Guastalla. J. Colloid Interface Sci. 11:623 (1956).
12. R. Defay, I. Prigogine, A. Bellemans, and D. H. Everett, *Surface Tension and Adsorption* Longmans, London, 1966.
13. R. G. Linford, Chem. Rev. 78:81 (1978).
14. D. M. Mohilner and T. R. Beck, J. Phys. Chem. 83:1160 (1979).
15. J. A. V. Butler, *Chemical Thermodynamics*, Macmillan, London, 1962.
16. J. C. Eriksonn. Acta Chem. Scand. 16:2199 (1962).
17. J. T. Davies and E. K. Rideal, *Interfacial Phenomena*, Academic Press, London, 1961.
18. A. W. Neumann. Adv. Colloid Interface Sci. 4:1 (1974).
19. E. W. Washburn. Phys. Rev., Ser. 2 17:273 (1921).
20. T. Young, *Miscellaneous Works* (G. Peacock, ed.), Murray, London, Vol. 1, 1855.
21. D. H. Everett, *Basic Principles of Colloid Science*, Royal Society of Chemistry, 1988.
22. I. Prigogine and R. Defay, *Chemical Thermodynamics*, Longmans, London, 1954.
23. G. Jura and W. D. Harkins, J.A.C.S. 66:1356 (1944).
24. S. Partyka, F. Rouquerol, and J. Rouquerol. J. Colloid Interface Sci. 68:30 (1979).
25. G. Jura and T. L. Hill. J.A.C.S. 74:1598 (1952).
26. V. Médout-Marère, H. Malandrini, T. Zoungrana, J. M. Douillard, and S. Partyka. J. Pet. Sci. and Eng. 20:223 (1998).
27. P. C. Hiementz, *Principles of Colloid and Surface Chemistry*, Marcel Dekker, New York, 1986.
28. E. Calvet and H. Prat, *Microcalorimétrie*, Masson, Paris 1956.
29. A. Lecloux, and J. P. Pirard. J. Colloid Interface Sci. 70:265 (1979).
30. *Physical Adsorption: Experiment, Theory and Applications* (J. Fraissard and C. W. Conner, eds.), Nato ASI Series, Kluwer AC, Dordrecht, 1997.
31. L. Cuiec, in *Interfacial Phenomena in Oil Recovery* (N. R. Morrow, ed.), Marcel Dekker, New York, 1990, p. 319.
32. M. E. Schrader, in *Modern Approaches to Wettability* (M. E. Schrader and I. Loeb, eds.), Plenum Press, New York, 1992, p. 101.
33. I. Diakonov, I. Khodakovsky, J. Schott, and E. Sergeeva. Eur. J. Min. 6:967 (1994).

34. H. W. Fox and W. A. Zisman. J. Colloid Sci. *5*:514 (1950).
35. M. K. Bernett and W. A. Zisman. J. Phys. Chem. *62*:1241 (1959).
36. E. G. Shafrin and W. A. Zisman. J. Phys. Chem. *63*:519 (1960).
37. W. A. Zisman. Adv. Chem., Ser. 43 (1964).
38. M. K. Bernett and W. A. Zisman. J. Colloid Interface Sci. *29*:413 (1969).
39. W. A. Zisman, *Adhesion Science Technology*, 9, A, Plenum Press, New York, 1975.
40. M. Zaborski, A. Vidal, E. Papirer, and J. C. Morawski. Makromol. Chem. Macromol. Symp. *23*:307 (1989).
41. J. Cognard. J. Chim. Phys. *84*:357 (1987).
42. C. Saint-Flour and E. Papirer. Ind. Eng. Chem. Prod. Res. Rev. *21*:337 (1982).
43. H. P. Schreiber, M. R. E. Wertheimer, and M. Lambla. J. Appl. Polym Sci. *27*:2269 (1982).
44. C. Saint-Flour and E. Papirer. J. Colloid Interface Sci. *91*:69 (1983).
45. E. Papirer, P. Roland, M. Nardin, and H. Balard. J. Colloid Interface Sci. *113*:62 (1986).
46. A. Vidal, E. Papirer, W. Meng Jiao, and J. B. Donnet. Chromatographia *23*:121 (1987).
47. M. Ciganek and M. Dressler. Polym. Composites *10*:194 (1989).
48. D. Williams, in *Controlled Interphases in Composite Materials* (H. Ishida, ed.), Elsevier New York, 1990, p. 219.
49. I. Tijburg, J. Jagiello, A. Vidal, and E. Papirer. Langmuir *7*:2243 (1991).
50. F. J. Lopez-Garzon, M. Pyda, and M. Domingo-Garcia. Langmuir *9*:531 (1993).
51. D. P. Kamdem, S. K. Bose, and P. Luner. Langmuir *9*:3039 (1993).
52. A. P. Legrand, H. Hommel, A. Tuel, A. Vidal, H. Balard, E. Papirer, P. Levitz, M. Czernichwi, R. Erre, H. Van Damme, J. P. Gallas, J. F. Hemidy, J. C. Lavaley, O. Barres, A. Burneau, and Y. Grillet. Adv. Colloid Interface Sci. *33*:91 (1990).
53. R. G. Pearson, *Chemical Hardness*, Wiley-VCH, New York, 1997.
54. W. D. Harkins and G. E. Boyd. J.A.C.S. *64*:1190 (1942).
55. A. C. Zettlemoyer, G. J. Young, J. J. Chessick, and T. H. Healey. J. Phys. Chem. *57* 649 (1953).
56. G. J. Young, J. J. Chessick, T. H. Healey, A. C. Zettlemoyer, and G. J. Young. J. Phys. Chem. *58*:313 (1954).
57. T. H. Healey, J. J. Chessick, A. C. Zettlemoyer, and G. J. Young. J. Phys. Chem. *58*:887 (1954).
58. F. E. Bartell and R. Murray Suggitt. J. Phys. Chem. *58*:36 (1954).
59. J. J. Chessick, A. C. Zettlemoyer, T. H. Healey, and G. J. Young. Can J. Chem. *33*:251 (1955).
60. W. H. Slabaugh. J. Phys. Chem. *59*:1022 (1955).
61. N. Hackerman and A. C. Hall. J. Phys. Chem. *62*:1212 (1958).
62. R. J. Good, L. A. Girifalco, and G. Kraus. J. Phys. Chem. *62*:1418 (1958).
63. A. C. Makrides and N. Hackerman. J. Phys. Chem. *63*:594 (1959).
64. W. H. Wade and N. Hackerman. J. Phys. Chem. *63*:1639 (1959).
65. J. J. Chessick and A. C. Zettlemoyer. Adv. Catalysis *11*:263 (1959).
66. W. H. Slabaugh. J. Phys. Chem. *63*:1333 (1959).
67. W. H. Wade, R. L. Every, and N. Hackerman. J. Phys. Chem. *64*:355 (1960).
68. W. H. Wade and N. Hackerman. J. Phys. Chem. *64*:1196 (1960).
69. W. H. Wade and N. Hackerman. J. Phys. Chem. *65*:1681 (1961).
70. C. M. Hollabaugh and J. J. Chessick. J. Phys. Chem. *65*:109 (1961).
71. J. W. Whalen, *Advances in Chemistry Series*, 33, ACS, Washington, 1961, p. 281.
72. J. W. Whalen. J. Phys. Chem. *65*:1676 (1961).
73. R. L. Every, W. H. Wade, and N. Hackerman. J. Phys. Chem. *65*:937 (1961).
74. J. W. Whalen. J. Phys. Chem. *66*:511 (1962).
75. J. W. Whalen. J. Phys. Chem. *67*:2114 (1963).
76. W. H. Wade. J. Phys. Chem. *68*:1029 (1964).
77. A. C. Zettlemoyer and J. J. Chessick. Adv. Chem. Ser. *43*:88 (1964).
78. H. F. Holmes, E. L. Fuller, and C. H. Secoy. J. Phys. Chem. *70*:436 (1966).
79. L. Robert. Mém. Soc. Chem. France *5/412*:147 (1967).
80. P. Roy and D. W. Fuerstenau. J. Colloid Interface Sci. *26*:102 (1968).

81. H. F. Holmes, E. L. Fuller, and C. H. Secoy. J. Phys. Chem. *72*:2095 (1968).
82. B. Coughlan and W. M. Caroll. J. Chem. Soc. Faraday Trans. *72*:2016 (1976).
83. G. Goujon and B. Mutaftschiev. J. Chim. Phys. *73*:351 (1976).
84. J. W. Whalen and P. C. Hu. J. Colloid Interface Sci. *65*:460 (1978).
85. R. G. Hollenbeck, G. E. Peck, and D. O. Kildsig. J. Pharm. Sci. *67*:1599 (1978).
86. D. A. Griffiths and D. W. Fuerstenau. J. Colloid Interface Sci. *80*:271 (1981).
87. I. Dekany, F. Szanto, and L. G. Nagy. J. Colloid Interface Sci. *109*:376 (1986).
88. E. M. Arnett and K. F. Cassidy. Rev. Chem. Intermediates *9*:27 (1988).
89. N. D. Sanders and C. F. Keweshan. J. Colloid Interface Sci. *124*:606 (1988).
90. M. L. Gonzalez-Martin, B. Janczuk, L. Labajos-Broncano, and J. M. Bruque. Langmuir *13*:5991 (1997).
91. R. Denoyel, J. Fernandez-Colinas, Y. Grillet, and J. Rouquerol. Langmuir *9*:515 (1993).
92. M. T. Gonzalez, A. Sepulveda-Escribano, M. Molina-Sabio, and F. Rodriguez-Reinoso. Langmuir *11*:2151 (1995).
93. J. M. Douillard, A. Berrada, T. Zoungrana, S. Partyka, and H. Toulhoat, in *Physical Chemistry of Colloids and Interfaces in Oil Production* (H. Toulhoat and J. Lecourtier, eds.), Technip, Paris, 1992.
94. J. M. Douillard, S. Pougnet, B. Faucompré, and S. Partyka. J. Colloid Interface Sci. *154*:113 (1992).
95. S. Lagerge, P. Rousset, J. M. Douillard, and S. Partyka. Colloids and Surfaces A *80*:261 (1993).
96. J. M. Douillard, M. Elwafir, and S. Partyka. J. Colloid Interface Sci. *164*:238 (1994).
97. J. M. Douillard, H. Malandrini, T. Zoungrana, F. Clauss, and S. Partyka. J. Therm. Anal. *41*:1325 (1994).
98. T. Zoungrana, J. M. Douillard, and S. Partyka. J. Therm. Anal. *41*:1287 (1994).
99. T. Zoungrana, Ph.D. thesis, Montpellier II, France (1994).
100. H. Malandrini, Ph.D. thesis, Montpellier II, France (1995).
101. J. M. Douillard, T. Zoungrana, and S. Partyka. J. Pet. Sci. Eng. *14*:51 (1995).
102. T. Zoungrana, A. Berrada, J. M. Douillard, and S. Partyka. Langmuir *11*:1760 (1995).
103. S. Partyka, and J. M. Douillard. J. Pet. Sci. Eng. *13*:232 (1995).
104. J. M. Douillard, H. Malandrini, T. Zoungrana, and S. Partyka, in *Proceedings of the 3rd International Symposium on Evaluation of Reservoir Wettability and its Effect on Oil Recovery*, (N. R. Morrow, ed.), Laramie, NY (USA), 1996.
105. J. M. Douillard. J. Colloid Interface Sci. *182*:308 (1996).
106. H. Malandrini, R. Sarraf, B. Faucompré, S. Partyka, and J. M. Douillard. Langmuir *13*:1337 (1997).
107. R. Sarraf, B. Faucompré, J. M. Douillard, and S. Partyka. Langmuir *13*:1274 (1997).
108. H. Malandrini, F. Clauss, S. Partyka, and J. M. Douillard. J. Colloid Interface Sci. *194*:183 (1997).
109. V. Médout-Marère. H. Belarbi, P. Thomas, F. Morato, J. C. Giuntini, and J. M. Douillard. J. Colloid Interface Sci. *202*:139 (1998).
110. D. Nyfeler and T. Armbruster. Am. Mineral. *83*:119 (1998).
111. J. M. Cases, I. Bérend, G. Besson, M. François, J. P. Uriot, F. Thomas, and J. E. Poirier. Langmuir *8*:2730 (1992).
112. W. Rudzinski and D. H. Everett, *Adsorption of Gases in Heterogeneous Surfaces*, Academic Press, London, 1992.
113. A. Bondi and D. J. Simkin. AIChe J. *3*:473 (1957).
114. K. Kitaura and K. Morokuma. Int. J. Quantum Chem. *10*:325 (1976).
115. R. S. Drago, G. C. Vogel, and T. E. Needham. J.A.C.S. *93*:60 (1971).
116. E. Tronel-Peyroz, J. M. Douillard, L. Ténebre, R. Bennes, and M. Privat. Langmuir *3*:1027 (1987).
117. F. M. Fowkes. J. Adh. Sci. Technol. *4*:669 (1990).
118. B. W. Ninham and V. Yaminsky. Langmuir *13*:2097 (1997).
119. C. J. Van Oss, R. J. Good, and M. K. Chaudhury. J. Colloid Interface Sci. *111*:378 (1986).

120. C. J. Van Oss, R. J. Good, and M. K. Chaudhury. Langmuir *4*:884 (1988).
121. C. J. Van Oss, R. F. Giese, and R. J. Good. Langmuir *6*:1711 (1990).
122. C. J. Van Oss. J. Dispersion and Technology *12*:201 (1991).
123. R. J. Good, N. R. Srivatsa, M. Islam, H. L. T. Huang, and C. J. Van Oss, in *Acid-Base Interactions* (K. L. Mittal and H. R. Anderson, eds.), VSP, Utrecht, 1991.
124. C. J. Van Oss. Colloids and Surfaces *78*:1 (1993).
125. V. Gutmann, *The Donor-Acceptor Approach to Molecular Interactions*, Plenum Press, New York, 1978.
126. A. Holländer. J. Colloid Interface Sci. *169*:493 (1995).
127. L. H. Lee, in *Contact Angle, Wettability and Adhesion* (K. L. Mittal, ed.), VSP, Utrecht, 1993.
128. C. S. Brooks. J. Phys. Chem. *64*:532 (1960).
129. J. J. Jurinak and D. H. Volman. J. Phys. Chem. *65*:150 (1961).
130. M. D. Vrbanac and J. C. Berg, in *Acid-Base Interactions: Relevance to Adhesion Science Technology* (K. L. Mittal and H. Anderson, Jr., eds.), VSP, Utrecht, 1991, p. 67
131. J. Jänchen, H. Stach, L. Uytterhoeven, and W. J. Mortier. J. Phys. Chem. *100*:12489 (1996).
132. *The Chemical Physics of Solid Surfaces and Heterogeneous Catalysis* (D. A. King and D. P. Woodruff, eds.), Elsevier, Amsterdam, 1982.
133. J. H. Rayner and G. Brown. Clays and Clay Minerals *21*:103 (1973).
134. H. A. Benesi. J. Phys. Chem. *61*:970 (1957).
135. D. H. Salomon, J. D. Swift, and A. J. Murphy. J. Macromol. Sci. *A5*:441 (1971).
136. M. Deeba and W. K. Hall. J. Catalysis *60*:417 (1979).
137. G. I. Kapustin, T. R. Brueva, A. L. Klyachko, S. Beran, and B. Wichterlova. Appl. Catal. *42*:239 (1988).
138. M. Kojima, J. C. Q. Fletcher, and C. T. O'Connor. Appl. Catal. *28*:169 (1986).
139. A. M. Youssef, Surface Technol. *9*:187 (1979).
140. S. Yariv, in *Modern Approach to Wettability* (M. E. Schrader and G. I. Loeb, eds.), Plenum Press, London, 1992.
141. J. K. Spelt, D. Li, and A. W. Neumann, in *Modern Approaches to Wettability* (M. E. Schrader and G. I. Loeb, eds.), Plenum Press, London, 1992, p. 101.

9
Ionic Adsorption/Desorption Kinetics

KAZUAKI HACHIYA Department of Mechanical Engineering, Okayama University of Science, Okayama, Japan

YOSHIKO MORIYAMA and KUNIO TAKEDA Department of Applied Chemistry, Okayama University of Science, Okayama, Japan

I.	Introduction	351
II.	Theoretical Treatment	352
	A. Single-step adsorption/desorption models	352
	B. Multistep adsorption/desorption models	357
III.	Experimental Methods	363
	A. Pressure-jump method	363
	B. Electric field-jump method	368
	C. Concentration-jump method	370
IV.	Adsorption/Desorption of Cations and Anions on Metal Oxides	371
	A. Adsorption/desorption of potential determining ions	371
	B. Adsorption/desorption of anions	372
	C. Adsorption/desorption of metal ions	380
	D. Intercalation of ions into the interlamellar layer	385
V.	Adsorption/Desorption of Metal Ions, Amino Acids, and Charged Proteins on Ion-Exchange Sephadex	386
	A. Proton association/dissociation of functional groups on ion-exchange sephadex	388
	B. Adsorption/desorption of alkali metal ions	391
	C. Adsorption/desorption of amino acids	394
	D. Adsorption/desorption of charged proteins	395
VI.	Conclusion and Further Problems	399
	References	400

I. INTRODUCTION

The rate of adsorption of a substance to a particle surface is usually measured by a periodical batch experiment. After a constant amount of sample is withdrawn from a large volume of suspension each time, the amount of the substance adsorbed on the particle is determined [1,2]. Since it takes, at least, several minutes for a series of the above procedures, a fast adsorption process cannot be followed by such a batch method. Fortunately, particles such as metal oxide have functional groups like the amphoteric surface hydroxyl groups [3]. When the particle surface becomes positive or negative because of an association/dissociation of the functional groups, the adsorptions of various substances onto particles often occur as ionic reactions. Then a concentration change of ion, which is caused by the adsorption, can be immediately followed by a change in the conductivity. Even a small conductivity change can be measured sensitively with a Wheatstone bridge circuit. Fast reactions in the suspensions are now observed by the pressure-jump, electric field-jump, and concentration-jump methods with conductometric detection. These methods have been developed on the basis of the principle of chemical relaxation [4]: a small perturbation, applied to a certain chemical equilibrium by a change of pressure, temperature, or concentration, causes a shift to a new equilibrium state. This shift can be measured as a chemical relaxation with suitable detections. In Section III, the relaxation methods will be explained. These methods are applied to measure the adsorption/desorptions of various ions on metal oxides and ion-exchange Sephadexes. The reaction mechanisms are introduced in Sections IV and V.

The rate of a reaction can be analyzed by solving the rate equation. It is not easy to solve rate equations of multistep reactions [5]. In fact, ion adsorption/desorptions occurring on the particle surface are heterogeneous reactions that often include several elementary steps. However, the relaxation method can easily give solutions of rate equations for complicated multistep mechanisms as well as simple one- and two-step mechanisms [6]. The rate equations are usually expressed as functions of very small concentration changes from the equilibrium concentrations. The theoretical treatments of rate equations will be explained for ion adsorption/desorptions in Section II.

II. THEORETICAL TREATMENT

A. Single-Step Adsorption/Desorption Models

Various ions are adsorbed/desorbed on particles such as metal oxides, and these processes have been represented by many reaction mechanisms. Among the mechanisms, a single step adsorption/desorption mechanism is the simplest. This is further classified into Langmuir-type and ion exchange-type adsorption/desorption mechanisms. In the following sections, the equations of the relaxation times will be derived for the two types of mechanisms, and the role of the potential terms in the rate constants will be explained in the description of the electrostatic repulsion/attraction in the adsorption/desorption process.

1. Langmuir-Type Adsorption/Desorption

Let us consider the following adsorption/desorption mechanism:

$$-S^- + A^+ \underset{k_{-1}}{\overset{k_1}{\rightleftharpoons}} -S^-A^+ \tag{1}$$

where $-S^-$, A^+, $-S^-A^+$, and k_i ($i = \pm 1$) denote the negatively charged site, the cation, the state of A^+ adsorbed on the site, and the rate constant, respectively. In the ion adsorption/desorption, the sites are postulated to be uniformly located on the particle surface. The equilibrium constant, K_1, in scheme (1) is represented by

$$K_1 = \frac{k_1}{k_{-1}} = \frac{[S^-A^+]}{[-S^-][A^+]} = \frac{[-S^-A^+]}{([-S^-]_T - [-S^-A^+])[A^+]} \tag{2}$$

where [ ] denotes the concentration and $[-S^-]_T$ is the total concentration of sites. From

Ionic Adsorption/Desorption Kinetics

Eq. (2), the amount of A^+ adsorbed on the particle is given by

$$[-S^-A^+] = \frac{K_1[-S^-]_T[A^+]}{1 + K_1[A^+]} \tag{3}$$

Equation (3) is the Langmuir adsorption isotherm, and then the reaction represented by scheme (1) is called Langmuir-type adsorption/desorption.

The rate equation for the reaction in scheme (1) is represented as [6]

$$-\frac{d[A^+]}{dt} = k_1[-S^-][A^+] - k_{-1}[-S^-A^+] \tag{4}$$

At equilibrium, Eq. (4) becomes

$$-\frac{d[A^+]}{dt} = k_1\overline{[-S^-]}\,\overline{[A^+]} - k_{-1}\overline{[-S^-A^+]} = 0 \tag{5}$$

where [ ] denotes the equilibrium concentration. The bulk and surface concentrations at the equilibrium state are changed by a perturbation such as a rapid pressure change:

$$[A^+] = \overline{[A^+]} + \Delta[A^+] \tag{6a}$$
$$[-S^-] = \overline{[-S^-]} + \Delta[-S^-] \tag{6b}$$
$$[-S^-A^+] = \overline{[-S^-A^+]} + \Delta[-S^-A^+] \tag{6c}$$

where $\Delta[A^+]$ is a small deviation from the equilibrium concentration, $\overline{[A^+]}$. Since the square term, $\Delta[A^+]\Delta[-S^-]$, is assumed to be negligibly small compared to other terms such as $\overline{[-S^-]}\Delta[-S^-]$, Eq. (4) is rewritten by using Eqs. (5) and (6) as

$$\begin{aligned}-\frac{d\Delta[A^+]}{dt} &= \{k_1(\overline{[-S^-]} + \overline{[A^+]}) + k_{-1}\}\Delta[A^+] \\ &= \tau^{-1}\Delta[A^+]\end{aligned} \tag{7}$$

and then we obtain

$$\tau^{-1} = k_1(\overline{[-S^-]} + \overline{[A^+]}) + k_{-1} \tag{8}$$

where τ^{-1} denotes the reciprocal relaxation time, and τ^{-1} corresponds to an apparent first-order rate constant in the usual kinetics [7,8]. Hereafter, the bar over the concentration symbol will be omitted for brevity.

The adsorption of ions on a charged surface is affected by an electrostatic interaction. For example, the concentration of A^+ ion, $[A^+]_s$, at some surface, s, in the electrical double layer is related to the bulk concentration, $[A^+]$, by the Boltzmann distribution [3],

$$[A^+]_s = [A^+]\exp\left(-\frac{e\psi_s}{k_B T}\right) \tag{9}$$

where e, ψ_s, k_B, and T denote the elementary charge, the potential at the s plane, the Boltzmann constant, and the absolute temperature, respectively. As is illustrated in Fig. 1, a potential determining ion like H^+ is located at the 0-plane. Then the potential ψ_s for the potential determining ion is equal to that at the 0-plane, ψ_0. On the other hand, a counterion such as Na^+ forms an ion pair with a surface site at the β-plane, the potential

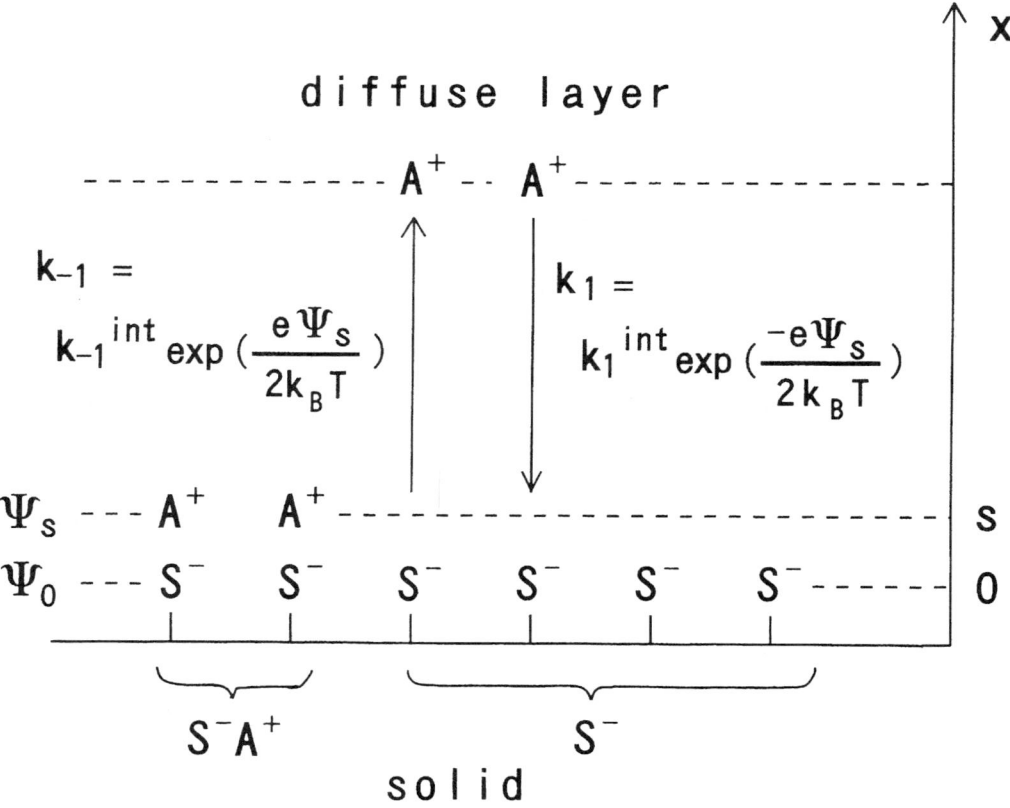

FIG. 1 Schematic representation of the electrical double layer and the rate constants.

at which, ψ_β, is equal to ψ_s. Considering the surface potential, the apparent equilibrium constant, $K_{1,\text{app}}$, for the reaction in scheme (1) is represented as [3]

$$K_1 = K_{1,\text{app}} = \frac{[-S^-A^+]}{[-S^-][A^+]} = \frac{[S^-A^+]}{[-S^-][A^+]_s}\exp\left(-\frac{e\psi_s}{k_BT}\right)$$
$$= K_1^{\text{int}}\exp\left(-\frac{e\psi_s}{k_BT}\right) \qquad (10)$$

where the superscript int indicates intrinsic. This equilibrium constant is related to the rate constants as [9,10]

$$k_1 = \frac{k_1}{k_{-1}} = \frac{k_1^{\text{int}}\exp\left(-\frac{e\psi_1^\ddagger}{k_BT}\right)}{k_{-1}^{\text{int}}\exp\left(\frac{e\psi_{-1}^\ddagger}{k_BT}\right)} = K_1^{\text{int}}\exp\left(-\frac{e(\psi_1^\ddagger + \psi_{-1}^\ddagger)}{k_BT}\right) \qquad (11)$$

with

$$\psi_1^\ddagger = \psi_{-1}^\ddagger = \frac{\psi_s}{2} \qquad (12)$$

where $\psi_1^\ddagger$ and $\psi_{-1}^\ddagger$ are the electric activation potentials of adsorption and of desorption processes, respectively. Equation (12) can be derived on the basis of two concepts. One concept is that the adsorption rate is accelerated but the desorption rate is decelerated in the ion adsorption/desorption on the surface site with the opposite charge. The other is that an ion adsorbed at a place in the electrical double layer is physically affected by the same activation potential as an ion desorbed at the same place (Fig. 1). Thus the rate equation in Eq. (4) and the reciprocal relaxation time in Eq. (8) are rewritten as

$$-\frac{d[A^+]}{dt} = k_1^{int}\exp\left(-\frac{e\psi_s}{2k_BT}\right)[-S^-][A^+] - k_{-1}^{int}\exp\left(\frac{e\psi_s}{2k_BT}\right)[-S^-A^+] \qquad (13)$$

and

$$\tau^{-1} = k_1^{int}\exp\left(-\frac{e\psi_s}{2k_BT}\right)([-S^-]+[A^+]) + k_{-1}^{int}\exp\left(\frac{e\psi_s}{2k_BT}\right) \qquad (14)$$

In order to obtain the intrinsic rate constants, Eq. (14) is transformed to

$$\tau^{-1}\exp\left(-\frac{e\psi_s}{2k_BT}\right) = k_1^{int}\exp\left(-\frac{e\psi_s}{k_BT}\right)([-S^-]+[A^+]) + k_{-1}^{int} \qquad (15)$$

As is evident from Eq. (15), a linear relationship is obtained in a plot of $\tau^{-1}\exp(-e\psi_s/2k_BT)$ against $\exp(-e\psi_s/k_BT)([-S^-]+[A^+])$ as shown in Fig. 2. The k_1^{int} and k_{-1}^{int} can be obtained from the slope and the intercept of the linear plot, respectively.

2. Ion Exchange-Type Adsorption/Desorption

As ion-exchange reaction between a cation, A^+, on the site, $-S^-$, and another cation, B^+, is represented as

$$-S^-A^+ + B^+ \underset{k_{-1}}{\overset{k_1}{\rightleftharpoons}} -S^-B^+ + A^+ \qquad (16)$$

As described in the previous section, the reciprocal relaxation time for the reaction in scheme (16) is given by the equation [6]

$$\tau^{-1} = k_1([-S^-A^+]+[B^+]) + k_{-1}([-S^-B^+]+[A^+]) \qquad (17)$$

When the rate constants are defined in a similar manner to Eqs. (10)–(12), they are related to K_1 for scheme (16) as

$$K_1 = \frac{[-S^-B^+][A^+]}{[-S^-A^+][B^+]} = K_1^{int} = \frac{k_1}{k_{-1}} = \frac{k_1^{int}}{k_{-1}^{int}} \qquad (18)$$

The adsorption and desorption rate constants of A^+ (or B^+) in Eq. (18) seem to be independent of the surface potential. This equation is valid only when the adsorption of A^+ and the desorption of B^+ occur at the same time on the surface site or vice versa, since the activation potentials of both adsorption and desorption are compensated with each other in the case of simultaneous adsorption and desorption.

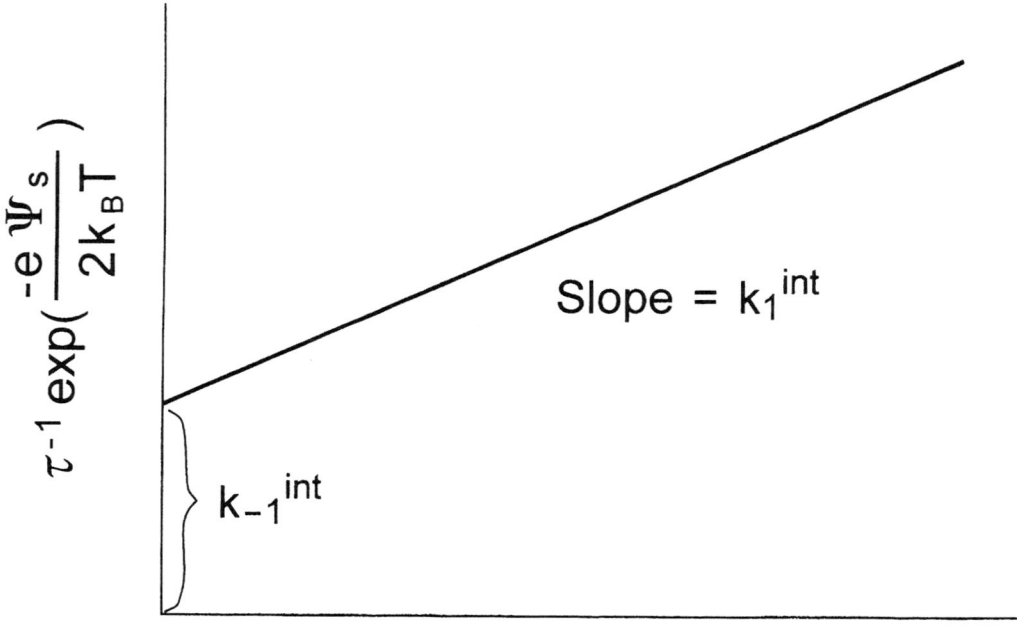

FIG. 2 Determination of rate constants from the plot of $\tau^{-1}\exp(-e\psi_s/2k_BT)$ against $\exp(-e\psi_s/k_BT)\,([-S^-]+[A^+])$ for the Langmuir-type adsorption/desorption.

In general, there are empty charged sites to which cations are not adsorbed. On the other hand, the surface charge cannot be completely neutralized by the formation of both $-S^-A^+$ and $-S^-B^+$, since the hydrated ions of A^+ and B^+ are not directly adsorbed on the plane of $-S^-$ (Fig. 1) [11,12]. As a result, the adsorption/desorption of A^+ and B^+ in scheme (16) is affected by the surface potential. In order to consider the potential terms in the rate constants in the above system, mechanism (16) is divided into two elementary steps:

$$-S^-A^+ + B^+ \underset{k_{-11}}{\overset{k_{11}}{\rightleftharpoons}} -S^- \cdots \begin{matrix} A^+ \\ \\ B^+ \end{matrix} \underset{k_{-12}}{\overset{k_{12}}{\rightleftharpoons}} -S^-B^+ + A^+ \qquad (18a)$$

$$\text{(step 1, } K_{11}) \qquad \text{(step 2, } K_{12})$$

where the intermediate species in the middle of scheme (18a) is an encounter complex. If this encounter complex is assumed to be in a steady state, we can derive the equation of the reciprocal relaxation time, which is similar to Eq. (17), as [13]

$$\tau^{-1} = \frac{k_{11}k_{12}}{k_{-11}+k_{12}}([-S^-A^+]+[B^+]) + \frac{k_{-11}k_{-12}}{k_{-11}+k_{12}}([-S^-B^+]+[A^+]) \qquad (19)$$

The equilibrium and rate constants in steps 1 and 2 of mechanism (18a) are given by

$$K_{11} = \frac{[-S^- - A^+ \cdots B^+]}{[-S^-A^+][B^+]} = \frac{[-S^- - A^+ \cdots B^+]}{[-S^-A^+][B^+]_s} \exp\left(-\frac{e\psi_s}{k_B T}\right)$$

$$= K_{11}^{int} \exp\left(-\frac{e\psi_s}{k_B T}\right) = \frac{k_{11}}{k_{-11}} = \frac{k_{11}^{int} \exp\left(-\frac{e\psi_s}{2k_B T}\right)}{k_{-11}^{int} \exp\left(\frac{e\psi_s}{2k_B T}\right)} \quad (20)$$

and

$$K_{12} = \frac{[-S^-B^+][A^+]}{[-S^- - A^+ \ldots B^+]} = K_{12}^{int} \exp\left(\frac{e\psi_s}{k_B T}\right) = \frac{k_{12}}{k_{-12}} = \frac{k_{12}^{int} \exp\left(\frac{e\psi_s}{2k_B T}\right)}{k_{-12}^{int} \exp\left(-\frac{e\psi_s}{2k_B T}\right)} \quad (21)$$

If we assume that step 1 equilibrates much faster than step 2 (i.e., $k_{-11} \gg k_{12}$) in mechanism (18), Eq. (19) is rewritten as

$$\tau^{-1} = k_{12} K_{11}([-S^-A^+] + [B^+]) + k_{-12}([-S^-B^+] + [A^+]) \quad (22)$$

Substituting Eqs. (20) and (21) into Eq. (22) gives

$$\tau^{-1} = k_1^{int} \exp\left(-\frac{e\psi_s}{2k_B T}\right)([-S^-A^+] + [B^+]) + k_{-1}^{int} \exp\left(-\frac{e\psi_s}{2k_B T}\right) \\ ([-S^-B^+] + [A^+]) \quad (23)$$

with

$$k_1^{int} = k_{12}^{int} K_{11}^{int} \quad (24a)$$
$$k_{-1}^{int} = k_{-12}^{int} \quad (24b)$$

In the case that $k_{-11} \ll k_{12}$, Eq. (23) is also derived, but the rate constants are different from those in Eq. (24). The rate constants in Eq. (23) are related to those in Eq. (25) by

$$k_1^{int} = k_{11}^{int} \quad (25a)$$
$$k_{-1}^{int} = k_{-11}^{int} K_{12}^{int} \quad (25b)$$

Equation (23) indicates that the ion-exchange reaction is affected by the electrostatic potential.

B. Multistep Adsorption/Desorption Models

A solution of the rate equation for a multistep reaction can be obtained by solving complicated integral equations in the ordinary chemical kinetics [5,7,8]. On the basis of the chemical relaxation method, however, linearized rate equations can be easily obtained for various reactions [6]. In this section, a two-step reaction mechanism of an ion-pair formation accompanying conformational change of adsorbent is introduced, and then complicated processes such as metal complex formation and protein adsorption/desorption are introduced.

1. Ion-Pair Formation Accompanying Conformational Change of a Particle

When ions are adsorbed on a particle with a soft structure, the volume and shape of the particle are affected by the electrostatic interaction such as a neutralization of the surface charge. Reaction scheme (26) describes the ion-pair formation accompanying a conformational change of such a particle [14]:

$$-S^- + A^+ \underset{k_{-1}}{\overset{k_1}{\rightleftharpoons}} (-S^-A^+)_I \underset{k_{-2}}{\overset{k_2}{\rightleftharpoons}} (-S^-A^+)_{II} \qquad (26)$$

(step 1, K_1) (step 2, K_2)

where $(-S^-A^+)_I$ and $(-S^-A^+)_{II}$ denote the two types of ion pairs, and the process of step 2 represents the conformational change induced by the A^+ adsorption/desorption. The linearized rate equations in steps 1 and 2 for small concentration deviations are given by [4]

$$-\frac{d\Delta[A^+]}{dt} = \{k_1([-S^-]+[A^+])+k_{-1}\}\Delta[A^+]+k_{-1}\Delta[(-S^-A^+)_{II}] \qquad (27)$$

and

$$-\frac{d\Delta[-S^-A^+]}{dt} = k_2\Delta[A^+]+(k_2+k_{-2})\Delta[(-S^-A^+)_{II}] \qquad (28)$$

with the following stoichiometric relations:

$$\Delta[S^-] = \Delta[A^+]$$

and

$$\Delta[S^-] + \Delta[(-S^-A^+)_I] + \Delta[(-S^-A^+)_{II}] = 0$$

Equations (27) and (28) are represented in the general forms

$$-\frac{d\Delta[A^+]}{dt} = a_{11}\Delta[A^+]+a_{12}\Delta[(-S^-A^+)_{II}] \qquad (29)$$

$$-\frac{d\Delta[-S^-A^+]}{dt} = a_{21}\Delta[A^+]+a_{22}\Delta[(-S^-A^+)_{II}] \qquad (30)$$

with

$$a_{11} = k_1([-S^-]+[A^+])+k_{-1} \tag{31a}$$
$$a_{12} = k_{-1} \tag{31b}$$
$$a_{21} = k_2 \tag{31c}$$
$$a_{22} = k_2 + k_{-2} \tag{31d}$$

The reciprocal relaxation times are given by the eigenvalues of the determinantal equation

$$\begin{vmatrix} a_{11} - \tau^{-1} & a_{12} \\ a_{21} & a_{22} - \tau^{-1} \end{vmatrix} = 0 \tag{32}$$

The two solutions of Eq. (32) are represented as

$$\tau^{-1} = \frac{1}{2}\left(\Sigma \pm \sqrt{\Sigma^2 - 4\Pi}\right) \tag{33}$$

with

$$\Sigma = a_{11} + a_{22}$$
$$\Pi = a_{11}a_{22} - a_{12}a_{21}$$

When the conformational change of the particle is much slower than the ion-pair formation (i.e., $a_{11} \gg a_{22}$), the reciprocal fast and slow relaxation times, τ_f^{-1} and τ_s^{-1}, are derived from Eq. (33) by

$$\tau_f^{-1} = a_{11} = k_1([-S^-]+[A^+]) + k_{-1} \tag{34}$$

and

$$\tau_s^{-1} = a_{22} - \frac{a_{12}a_{21}}{a_{11}}$$
$$= k_2 \frac{k_1([-S^-]+[A^+])}{k_1([-S^-]+[A^+]) + k_{-1}} + k_{-2} \tag{35}$$

The equilibrium and rate constants in steps 1 and 2 of mechanism (26) are represented as

$$K_1 = \frac{[-S^-A^+]}{[-S^-][A^+]} = K_1^{int}\exp\left(-\frac{e\psi_s}{k_B T}\right) = \frac{k_1}{k_{-1}} = \frac{k_1^{int}\exp\left(-\dfrac{e\psi_s}{2k_B T}\right)}{k_{-1}^{int}\exp\left(\dfrac{e\psi_s}{2k_B T}\right)} \tag{36}$$

and

$$K_2 = \frac{[(-S^-A^+)_{II}]}{[(-S^-A^+)_{I}]} = K_2^{int} = \frac{k_2}{k_{-2}} = \frac{k_2^{int}}{k_{-2}^{int}} \tag{37}$$

It is interesting that the rate constants, k_2 and k_{-2}, in Eq. (37) are independent of surface potential, since there is no migrating process of ions from the bulk phase to the particle surface in step 2 of mechanism (26). The substitution of Eqs. (36) and (37) into Eq. (35) leads to the following equation of the reciprocal relaxation time for the rate determining step:

$$\tau_s^{-1} = k_1^{int} \frac{A}{A+1} + k_{-1}^{int} \tag{38}$$

with

$$A = K_1^{int} \exp\left(-\frac{e\psi_s}{k_B T}\right)([-S^-] + [A^+])$$

2. Adsorption/Desorption of a Divalent Metal Ion Accompanying Complex Formation

The hydrated divalent metal ion, $M(H_2O)_n^{2+}$, which is adsorbed on metal oxide, forms several types of complexes with a surface hydroxyl group, $-SH$. The surface complex formation of the metal ion is described by [9]

$$
\begin{array}{ccc}
& k_{11} & k_{12} \\
-SH & \rightleftarrows & -S\!\!<\!\!\genfrac{}{}{0pt}{}{H}{M(H_2O)_{n-1}^{2+}} \rightleftarrows -SM(H_2O)_{n-1}^+ \\
& k_{-11} & k_{-12} \\
M(H_2O)_n^{2+} & H_2O & H^+ \\
(\text{step } 11, K_{11}) & & (\text{step } 12, K_{12})
\end{array}
\tag{39}
$$

where $-S-H-M(H_2O)_{n-1}^+$ and $-SM(H_2O)_{n-1}^+$ denote the encounter complex and the surface complex, respectively. When a water molecule is released from $M(H_2O)_n^{2+}$, the metal ion can be covalently adsorbed to a surface hydroxyl group in the process of step 11. The equilibrium and rate constants in steps 11 and 12 are given by

$$K_{11} = \frac{[-S-H-M(H_2O)_{n-1}^{2+}]}{[-SH][M(H_2O)_n^{2+}]} = \frac{[-S-H-M(H_2O)_{n-1}^{2+}]}{[-SH][M(H_2O)_n^{2+}]_s} \exp\left(-\frac{2e\psi_s}{k_B T}\right)$$

$$= K_{11}^{int} \exp\left(-\frac{2e\psi_s}{k_B T}\right) = \frac{k_{11}}{k_{-11}} = \frac{k_{11}^{int} \exp\left(-\frac{e\psi_s}{k_B T}\right)}{k_{-11}^{int} \exp\left(\frac{e\psi_s}{k_B T}\right)} \tag{40}$$

and

$$K_{12} = \frac{[-SM(H_2O)_{n-1}^+][H^+]}{[-S-H-M(H_2O)_{n-1}^{2+}]} = K_{12}^{int} \exp\left(\frac{e\psi_s}{k_B T}\right) = \frac{k_{12}}{k_{-12}} = \frac{k_{12}^{int} \exp\left(\frac{e\psi_s}{2k_B T}\right)}{k_{-12}^{int} \exp\left(-\frac{e\psi_s}{2k_B T}\right)} \tag{41}$$

with

$$[M(H_2O)_n^{2+}] = [M(H_2O)_n^{2+}]_s \exp\left(\frac{2e\psi_s}{k_B T}\right) \tag{42}$$

If $-S-H-M(H_2O)_{n-1}^+$ is assumed to be a complex in a steady state, the reciprocal relaxation time is derived by [13]

$$\tau^{-1} = \frac{k_{11}^{int}\exp\left(-\frac{e\psi_s}{k_B T}\right)k_{12}^{int}\exp\left(\frac{e\psi_s}{2k_B T}\right)}{k_{-11}^{int}\exp\left(\frac{e\psi_s}{k_B T}\right) + k_{12}^{int}\exp\left(\frac{e\psi_s}{2k_B T}\right)}([-SH] + [M(H_2O)_n^{2+}])$$
$$+ \frac{k_{-11}^{int}\exp\left(\frac{e\psi_s}{k_B T}\right)k_{-12}^{int}\exp\left(-\frac{e\psi_s}{2k_B T}\right)}{k_{-11}^{int}\exp\left(\frac{e\psi_s}{k_B T}\right) + k_{12}^{int}\exp\left(\frac{e\psi_s}{2k_B T}\right)}([-SM(H_2O)_{n-1}] + [H^+]) \tag{43}$$

If the release process of one water molecule in step 11 of mechanism (39) is a rate determining step (i.e., $k_{12} \gg k_{-11}$), Eq. (43) becomes

$$\tau_{-1} = k_{-1}^{int}\{[-SM(H_2O)_{n-1}^+] + [H^+] + K_1^{int}\exp\left(-\frac{e\psi_s}{k_B T}\right)([-SH] + M(H_2O)_n^{2+}]\} \tag{44}$$

with

$$k_1^{int} = k_{11}^{int} \tag{45a}$$

$$k_{-1}^{int} = \frac{k_{-11}^{int} k_{-12}^{int}}{k_{12}^{int}} = k_{-11}^{int}(K_{12}^{int})^{-1} \tag{45b}$$

3. Adsorption/Desorption of Charged Protein or Multivalent Polyelectrolyte

Let us consider protein adsorption/desorption on particles such as ion-exchange Sephadex. Proteins or polyelectrolytes are made up of large molecules that have many charges resulting from functional groups like amino acid residues. They are exchanged with several ions located on a particle, as is represented by the mechanism [15]

$$n(-S^-A^+) + P \rightleftharpoons (-S^-)_n P + nA^+ \tag{46}$$

where P denotes the protein molecule. This is the overall model of protein adsorption/desorption. A protein molecule has several or many functional groups that can interact with surface sites on a particle [16,17]. When a protein molecule is adsorbed on a particle, only a particular number of functional groups are assumed to be bound consecutively to the surface sites. This means that the binding of these functional groups to the surface sites does not occur simultaneously, as is shown in mechanism (46). Thus the protein adsorption/desorption is represented as the consecutive reaction mechanism [15]

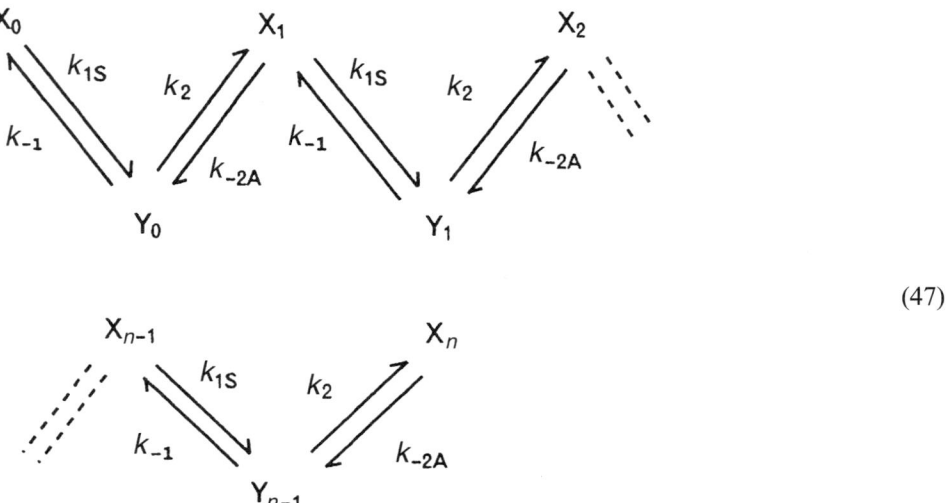

(47)

where X_i ($i=0$ to n) denotes the protein molecule that is bound to i charged surface sites $(-S^-)_i$, and Y_i denotes the protein molecule that is bound to one surface ion-pair $(-S^-A^+)$ in addition to i charged sites. The ith adsorption process in scheme (47) might be described in the scheme

$$P(-S^-)_i + -S^-A^+ \rightleftharpoons P(-S^-)_i(-S^-A^+) \rightleftharpoons P(-S^-)_{i+1} + A^+$$
$$\quad X_i \qquad\qquad\qquad\qquad Y_i \qquad\qquad\qquad X_{i+1}$$

(48)

The process in scheme (48) includes two elementary steps: one is the adsorption step of one more functional group of $P(-S^-)_i$ to a surface ion-pair on the particle ($X_i \rightleftharpoons Y_i$), and the other is the release step of an A^+ ion ($Y_i \rightleftharpoons X_{i+1}$).

Under the condition that the concentrations of X_i and Y_i are much smaller than those of $-S^-A^+$ and A^+, the rate constants in mechanism (47) can be derived through the pseudo–first-order approximation [18]

$$k_{1s} = k_1^{int}\exp\left(-\frac{e\psi_s}{2k_BT}\right)[-S^-A^+] \tag{49a}$$

$$k_{-1} = k_{-1}^{int}\exp\left(\frac{e\psi_s}{2k_BT}\right) \tag{49b}$$

$$k_2 = k_2^{int}\exp\left(\frac{e\psi_s}{2k_BT}\right) \tag{49c}$$

$$k_{-2A} = k_{-2}^{int}\exp\left(-\frac{e\psi_s}{2k_BT}\right)[A^+] \tag{49d}$$

A linearized rate equation can be given by the summation over all the rate equations for ΔX_i in scheme (47) [15]:

$$-\frac{d}{dt}\sum_{i=0}^{n}\Delta X_i = \sum_{i=0}^{n-1}k_{1s}\Delta X_i - \sum_{l=0}^{n-1}k_{-1}\Delta Y_l + \sum_{i=1}^{n}k_{-2A}\Delta X_i - \sum_{l=1}^{n}k_2\Delta Y_{l-1} \qquad (50)$$

If new equilibria of the release steps of A^+ are attained more rapidly than those of the adsorption steps, Eq. (50) becomes

$$-\frac{d}{dt}\sum_{i=0}^{n}\Delta X_i = \sum_{i=0}^{n-1}k_{1s}\Delta X_i - \sum_{l=0}^{n-1}k_{-1}\Delta Y_l \qquad (51)$$

When n is large in scheme (47), the following stoichiometric relation can be obtained:

$$\sum_{i=0}^{n}\Delta X_i \doteqdot \sum_{i=0}^{n-1}\Delta X_i \doteqdot -\sum_{l=0}^{n-1}\Delta Y_l \qquad (52)$$

Substituting Eq. (52) into Eq. (51) gives

$$-\frac{d}{dt}\sum_{i=0}^{n-1}\Delta X_i = (k_{1s} + k_{-1})\sum_{i=0}^{n-1}\Delta X_i$$
$$= \tau^{-1}\sum_{i=0}^{n-1}\Delta X_i \qquad (53)$$

The equation of τ^{-1} in Eq. (53) can be rewritten by using Eqs. (49a) and (49b) as

$$\tau^{-1}\exp\left(-\frac{e\psi_s}{2k_BT}\right) = k_1^{\text{int}}\exp\left(-\frac{e\psi_s}{k_BT}\right)[-S^-A^+] + k_{-1}^{\text{int}} \qquad (54)$$

III. EXPERIMENTAL METHODS

A. Pressure-Jump Method

When hydrated ions are adsorbed on a particle surface in suspension, a part of the water molecules are released from the ions. The release of water molecules causes a small increase in the volume of the suspension, because the amount of free water molecules increases in the bulk phase. When pressure is applied to a suspension containing such hydrated ions, the adsorption/desorption equilibrium of the ions, which necessarily includes the hydration or dehydration steps, can be shifted according to Le Chatelier's principle. The relation of the equilibrium constant, K, with the standard molar volume change of reaction, ΔV, is expressed as [4,19]

$$\left(\frac{\partial \ln K}{\partial p}\right)_\tau = -\frac{\Delta V}{RT} \qquad (55)$$

where p and R are the pressure and the gas constant, respectively. Under the condition that the application of pressure does not affect the volume and shape of the particle, but contracts the volume of the suspension itself, the conductivity of the suspension increases with increasing pressure. This increase in the conductivity results from the volume con-

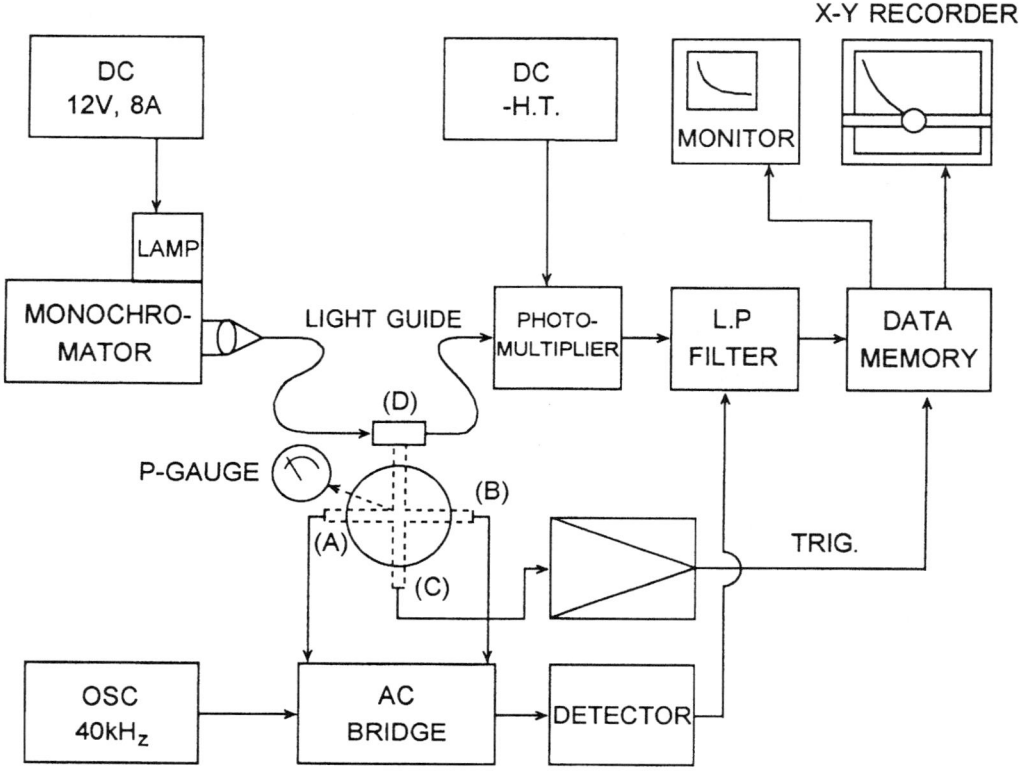

FIG. 3 Schematic diagram of the pressure-jump apparatus with conductivity and absorbance detections: (A) and (B) conductivity sample and reference cells; (C) $BaTiO_3$ piezoelectric element; (D) spectrophotometric sample cell. (From Ref. 22; reproduced by permission of the American Chemical Society, Washington, D.C.)

traction of the suspension as well as the release of ions from the particles. However, the velocity of the volume contraction is as fast as the velocity of sound. If the rate of ion adsorption/desorption is slower than that of the change in volume, the adsorption/desorption reaction can be observed by detecting the conductivity change. Therefore perturbation by a rapid pressure change is an effective way to follow the ion adsorption/desorption process.

The pressure-jump method has so far been reviewed by several authors [4,10,19–21]. A schematic diagram of the pressure-jump apparatus is shown in Fig. 3 [22]. The ion adsorption/desorption is followed mainly by observing the change in the conductivity. The conductivity cell is shown on a large scale in Fig. 4. This is an improved version of the cell described by Knoche [19,23]. Two cells with almost the same cell constants are filled with about 100 μL of sample suspension and reference solution and are mounted in the autoclave, whose temperature is controlled by circulating water from a thermostat, as shown in Fig. 5. Pressure is applied to the autoclave by a pressure pump, and circulating water has the role of a pressure transducing medium. The autoclave is closed with a bronze or phosphor bronze diaphragm 0.05 mm thick, and then the diaphragm is spontaneously ruptured at about 100 atm ($\approx$ 10 MPa). Then the pressure falls down from 100 atm to the atmospheric pressure within 80 μs. This is called the "pressure jump." The bursting

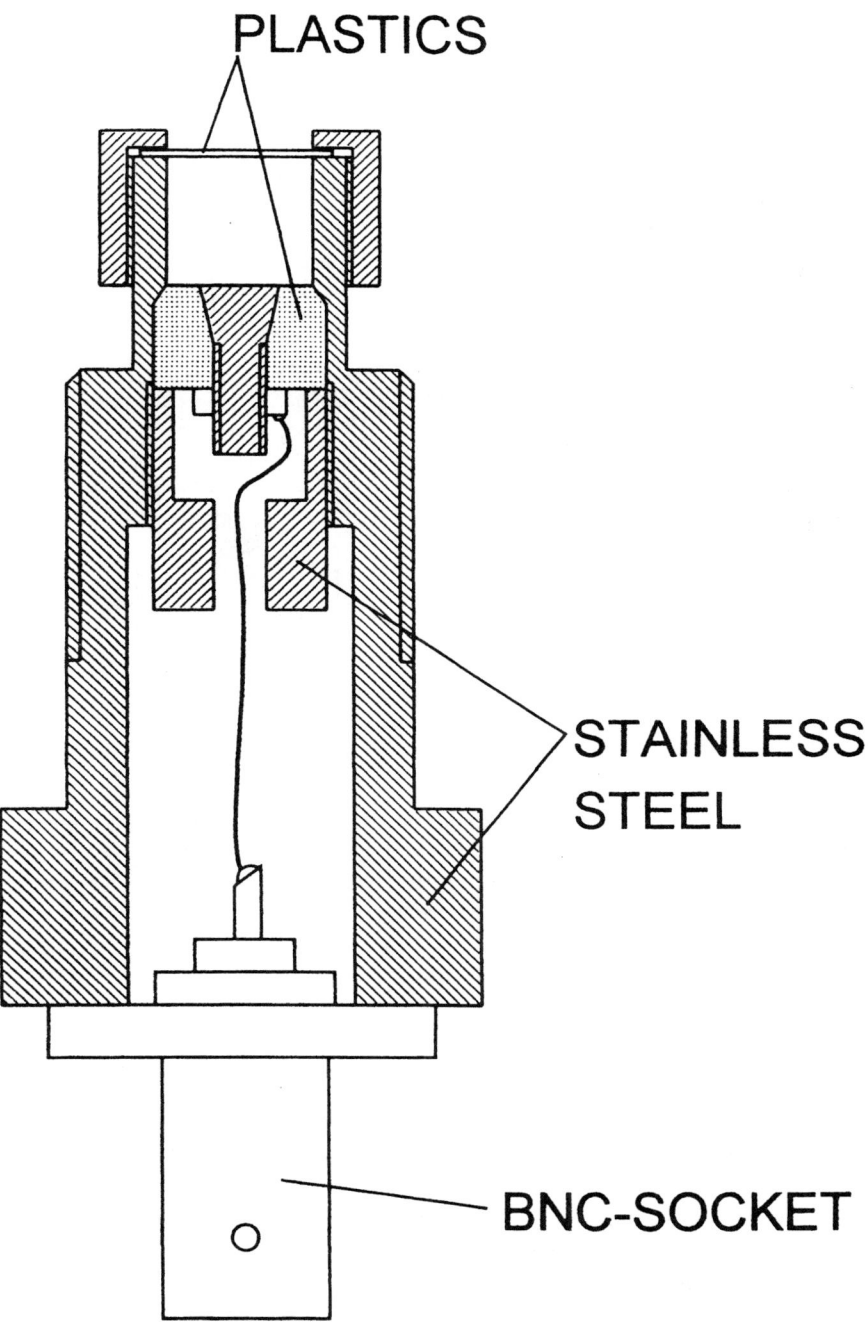

FIG. 4 Sectional view of the conductivity cell.

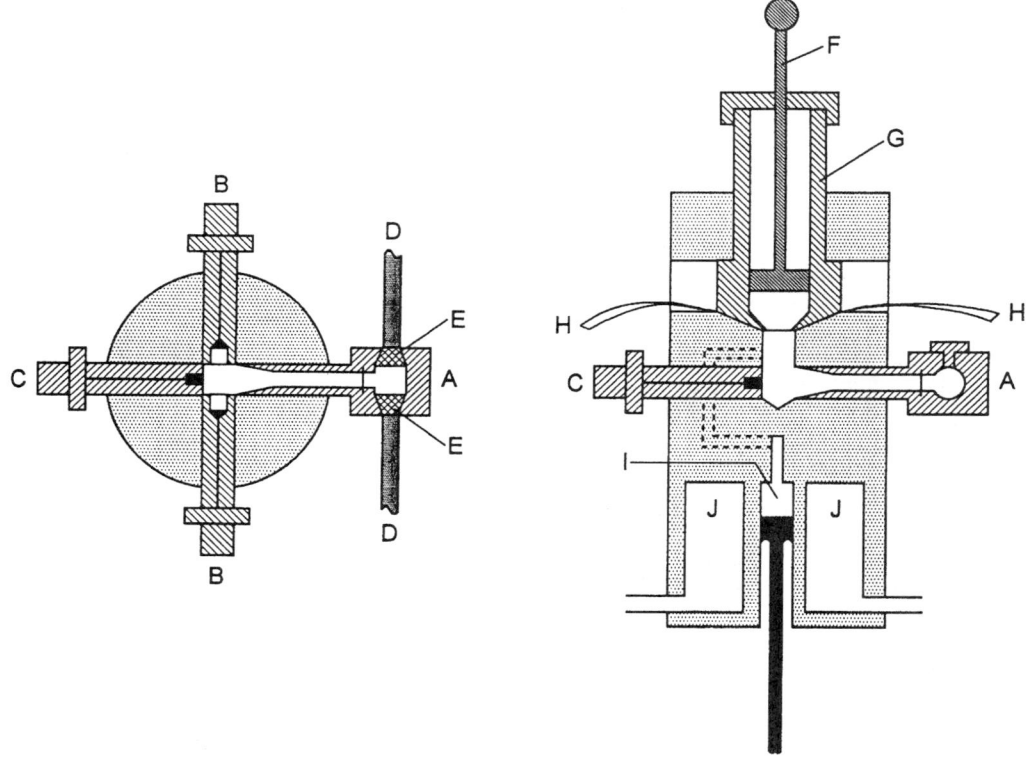

FIG. 5 Sectional views of the autoclave: (A) spectrophotometric cell, (B) conductivity sample and reference cells, (C) BaTiO$_3$ piezoelectric element, (D) light guides, (E) conical lenses, (F) vacuum pump, (G) bayonet socket, (H) burst diaphragm, (I) pressure pump, and (J) heat exchanger. (From Ref. 22; reproduced by permission of the American Chemical Society, Washington, D.C.)

pressure can be controlled by changing the thickness of the diaphragm. The bursting shock wave produces a triggering signal with the BaTiO$_3$ piezoelectric element. Without changing the sample and reference solutions in the conductivity cells, the measurement can be easily repeated by loosening the bayonet socket by a quarter turn and shifting the strip of the diaphragm.

A small difference between the conductances of the sample and reference cells, which is induced by the pressure jump, is measured with time by the Wheatstone bridge circuit with a frequency of 40 kHz or higher (Fig. 3). The AC signal of the conductance change is detected to the DC signal with the demodulation circuit in Fig. 6. By using this circuit, the direction of the increase or decrease in the conductance can also be reproduced exactly as the direction of the change in the DC voltage. The detected signal is amplified by as much as about 10^4 times and then is monitored with a digital storage oscilloscope. The relaxation signal is transferred from the oscilloscope to a computer in order to calculate the relaxation time. In the pressure-jump measurement, the relaxation amplitude of the ion adsorption/desorption on the particle surface is comparatively small. When the pressure applied to the sample is increased from 100 to 200 atm or when the relaxation

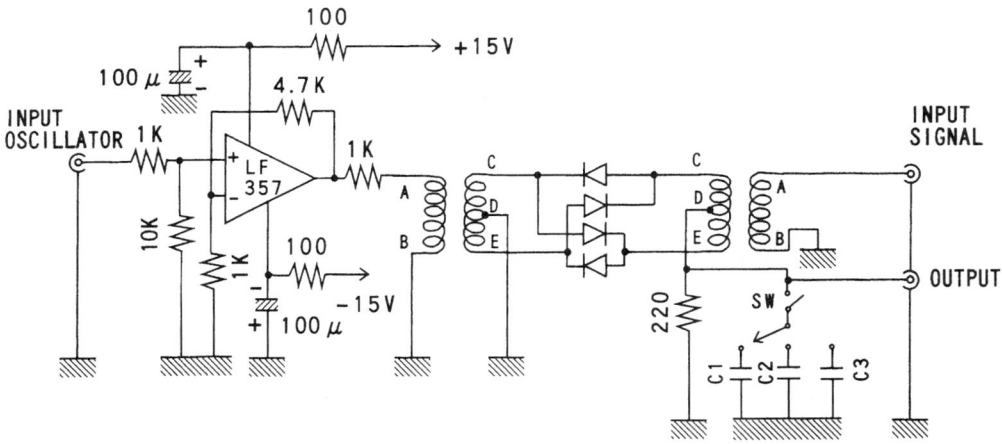

FIG. 6 Circuit diagram of the demodulation circuit. The capacitances of C1 to C3 are from 0.047 to 1 μF. The inductances between A and B, C and D, and D and E are adjusted to be equal.

signal is arithmetically averaged several times with the digital storage oscilloscope, a noisy and small signal of relaxation becomes a clear one. In order to improve the signal-to-noise ratio, a repetitive pressure-jump measurement is also recommended [24].

The relaxation can be measured spectrophotometrically by the pressure-jump method. Figure 5 also shows a spectrophotometric sample cell, which is located outside the autoclave and connected with a monochrometer and a photomultiplier by flexible light guides (Fig. 3) [22]. A change in turbidity of the suspension is measured by this detection. The turbidity measurement can be carried out simultaneously with the conductivity measurement. Simultaneous detection is effective to judge whether the observed relaxation is caused by a chemical reaction or by physical phenomena such as aggregation and expansion of particles.

A conductivity change of a suspension occurs when charged particles are sedimenting or when sedimented particles are slightly stirred by the shock of the pressure jump. Since the conductivity change caused by the sedimentation is larger than that caused by a chemical relaxation, the particle size must be kept smaller than 0.1 μm in order to prevent sedimentation during the kinetic measurement. Colloidal γ-Al_2O_3 particles have been used for the measurement of the adsorption/desorption of metal ions [22], while zeolite and ion-exchange Sephadex, which have comparatively large particle sizes, have been ground in a mill, and then their particles with uniform size have been selected [25]. The particles such as TiO_2 or Al_2O_3 are attached electrostatically to the electrodes and the wall of the conductivity cell. In an accidental case, the sample, containing such particles, is dried, and then is coagulated at the bottom of the cell. Since the above particles, attached to the cell, cannot be removed at all by washing with detergent solutions or strong acids or bases, they must be wiped out with a soft tissue or must be mechanically removed by an ultrasonic cleaning bath after each measurement. If a slight amount of the particles is left in the cell, the reaction for the previous sample is often observed as a small relaxation. On the other hand, the conductivity cells might be sometimes broken, since a pressure of 100 or 200 atm is repeatedly applied to them. At every measurement, therefore, the cell must be checked with NaCl solution, in which no relaxation is observed, and with some standard solution, in which a typical relaxation can be observed. The present authors have used a 10 mM sodium dodecyl sulfate solution, in which we can observe a relaxation with a relaxation time of about 2 ms [26]. The cell must be repaired if broken.

B. Electric Field-Jump Method

When ions are adsorbed to charged functional groups on the particle surface, ion pairs are formed between them. If a high electric field of 10–100 kV/cm is applied to the suspension containing these particles, the ion pairs are broken according to the dissociation field effect (i.e., the second Wien effect) [4,20]. Then the chemical equilibrium is shifted to a new equilibrium, whose equilibrium constant, K, is expressed by the equation [4]

$$\left(\frac{\partial \ln K}{\partial |E|}\right)_{p,T} = \frac{z_A u_A - z_B u_B}{u_A + u_B} \frac{|z_A z_B| e^3}{8\pi\varepsilon\varepsilon_0 (k_B T)^2} \tag{56}$$

where E is the electric field, z the valence of the ion, u the mobility of the ion, and ϵ and ϵ_0 the permittivities of solution and vacuum, respectively. The dissociation process can be measured by following a change in the conductivity of the suspension. The electric field-jump method is applicable to reactions with relaxation times of the order of 0.1–20 μs, which are much faster than those measured by the pressure-jump method.

The electric field-jump apparatus, in which a coaxial delay cable was used for discharging the high voltage instead of a capacitor, was originally constructed by Ilgenfritz [27]. The other types of electric field-jump methods have been described elsewhere [4,20,28]. The block diagram of the apparatus used by the authors is shown in Fig. 7 [29]. The sample cell with the electrode distance of 0.3 cm and the reference cell with the variable electrode distance are built into a high-field Wheatstone bridge circuit. The reference cell contains an NaCl solution of approximately the same resistance as the sample solution. The resistances of both cells are precisely matched by adjusting the distance between the two electrodes in the reference cell. A high voltage, charged at the coaxial cable, is abruptly discharged and applied to the cells, when it attains to a given voltage level, which can be adjusted by changing the distance of the spark gap. The discharge from the coaxial cable can produce a rectangular electric-field pulse with a duration of 10 or 20 μs and with rise and decay times of less than 0.1 μs. The voltage difference between points A and B in Fig. 7 is detected with the differential amplifier of an oscilloscope. This voltage difference is generated by the shift of a chemical equilibrium due to the dissociation-field effect. When the values of the two resistances on the lower branches of the Wheatstone bridge circuit in Fig. 7 are adjusted to be equally small, the output signals can be controlled within a range of voltage less than 10 V. This prevents destroying the differential amplifier.

When a relaxation is observed by the kinetic method with conductivity detection, we must always distinguish experimentally whether the conductivity change arises from a chemical reaction or from a physical phenomenon. Relaxation phenomena induced by the applied electric field might be due to a migration or orientation of charged particles, and a change in the electrical double layer around the particles as well as the ionic adsorption/desorption reaction. If the relaxation arises merely from the

Ionic Adsorption/Desorption Kinetics

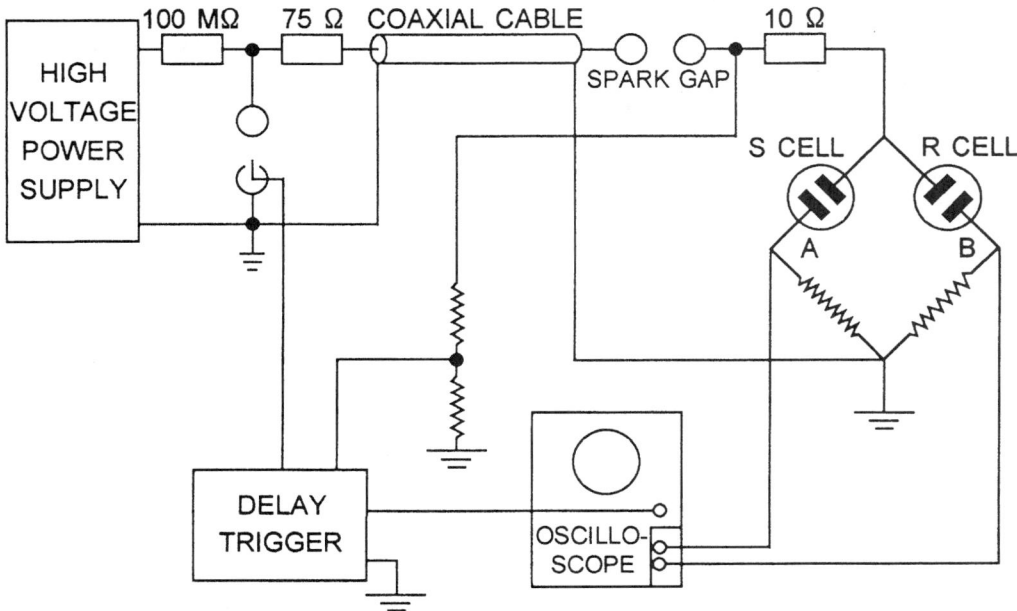

FIG. 7 Schematic diagram of the electric field-jump apparatus. (From Ref. 29; reproduced by permission of the American Chemical Society, Washington, D.C.)

adsorption/desorption of ions on the particle surface, the relaxation time is independent of the applied electric field, E, while the change in the conductivity, $\Delta\kappa$ (i.e., the relaxation amplitude), is proportional to E as [30–33]

$$\frac{\Delta\kappa}{\kappa} = \frac{1-\alpha}{2-\alpha} \left| \frac{z_A u_A - z_B u_B}{u_A + u_B} \right| \frac{|z_A z_B| e^3}{8\pi\varepsilon\varepsilon_0 (k_B T)^2} E \tag{57}$$

where α is the degree of dissociation. A particle with an electric dipole moment is oriented in the direction of the applied electric field. This phenomenon is called the Karr effect. The amplitude of the orientation is proportional to E^2 [34]. On the other hand, the velocity of particle flow is represented by the equation [35]

$$v = \mu E \tag{58}$$

where μ is the electrophoretic mobility of a particle. If an electric field of 10^4 V/cm is applied to a particle such as γ-Al_2O_3 with a mobility of 2×10^{-4} cm^2/sV, we can calculate the v value to be 2 cm/s from Eq. (58). When the electrode distance is 0.3 cm, it takes about 0.1 s for the particle to migrate from one end to the other. However, the particle migration can be ignored in the electric field-jump measurement, since the relaxation time observed by this method is of the order of 1 μs. In a manner similar to the pressure-jump method, we have to confirm that turbidity does not change in the electric field-jump measurement. This is because a turbidity change would be caused also by one of the physical phenomena, a rapid polarization of the electrical double layer, which is followed by ion adsorption [32]. The check of turbidity must be made at a low turbidity by using a suspension of low particle concentration. Every observed relaxation must be examined from various angles although the usual spectrophotometric measurement is impossible for the turbid system.

C. Concentration-Jump Method

The electrostatic binding of Sephadex with a multivalent macromolecule such as a charged protein is much stronger than that with a small monovalent or divalent ion. Thus the adsorption equilibrium of the macromolecule would not be perturbed easily by the pressure-jump and electric field-jump methods. Change of the pressure of 100 atm or of the electric field of 10 kV/cm is not enough to cause an appreciable shift of such an equilibrium [4]. However, by mixing the protein solution rapidly with the suspension containing particles, the protein adsorption process can be detected by following the conductivity change of the suspension. This mixing is made in the conductivity cell, which is built in the Wheatstone bridge circuit as used in the above pressure-jump apparatus. Then a fast adsorption process can be detected with time as a conductivity change. This is the concentration-jump method with conductivity detection. One of the concentration-jump methods is a conductance stopped-flow method [36–39]. Two reactant solutions in syringes are pushed simultaneously into a mixing chamber by a gas like N_2 at high pressure, and the flow is instantaneously stopped when the plungers of the syringes reach a mechanical stop. In the stopped-flow method, the mixing or dead time is about 1 ms [36]. The present concentration-jump measurement can be carried out simply by injecting two reactant solutions with pipets into a conventional conductivity cell. Particular attention should be paid to keep the cell and the reactant solutions at the same temperature. The dead time of this mixing method is within 1 s [15]. Instead of using the Wheatstone bridge circuit, the conductivity change can be directly measured using an electric conductivity meter. The signal is also saved with a digital storage oscilloscope connected to a computer [40,41]. Since the concentration-jump method is used in almost the same time range as the pressure-jump method, both methods are applicable to the same system [40].

The concentration-jump apparatus with conductivity detection is an effective tool for the study of ion or protein adsorption/desorption, but it has a few problems. When two reactants are mixed in the conductivity cell, a conductivity change might be induced also by the heat of reaction or by the heat of dissolution that is given merely by the mixing of two solutions. When the rate of a chemical reaction is the same as that of an exothermic or endothermic process of the corresponding reaction, the chemical reaction can be measured as a conductivity change. However, if the temperature change in the suspension, resulting from the exothermic or endothermic process, occurs more slowly than a chemical reaction, it is difficult to distinguish a conductivity change due to a chemical process and that due to a physical one. In this case, we have to investigate whether there is concentration and/or temperature dependence of the relaxation time that would be adequate for a real chemical reaction [6,42]. One more big problem is as follows. Particles are adhered on the walls of the conductivity cell and on the inside of the narrow pipes used in the concentration-jump apparatus. These remaining particles cannot be removed easily

IV. ADSORPTION/DESORPTION OF CATIONS AND ANIONS ON METAL OXIDES

A. Adsorption/Desorption of Potential Determining Ions

A metal oxide such as TiO_2 has amphoteric surface groups, $-SOH$. The surface group is $-TiOH$ for TiO_2, $-AlOH$ for $\gamma-Al_2O_3$, and $-FeOH$ for hematite and magnetite. Below the pH of the zero point of charge, pH_{zpc}, the metal oxide surface becomes positive, since H^+ is adsorbed to the surface functional group [3,10,43]:

$$-SOH + H^+ \underset{k_{-1}}{\overset{k_1}{\rightleftharpoons}} -SOH_2^+ \tag{59}$$

The acid dissociation constant of the surface hydroxyl group, K_a, is related to the equilibrium constant, K_1, for mechanism (59) [10,43]:

$$K_a = \frac{[SOH][H^+]}{[-SOH_2^+]} = \frac{k_{-1}}{k_1} = K_1^{-1} \tag{60}$$

At a pH above pH_{zpc}, the surface becomes negative owing to the following reactions of the functional groups:

$$-SOH \rightleftharpoons -SO^- + H^+ \tag{61}$$
$$-SOH + OH^- \rightleftharpoons SO^- + H_2O \tag{62}$$

Schemes (59), (61), and (62) indicate the fundamental reactions of the potential determining ions, H^+ and OH^-, on the particle surface. When a change in the volume of the suspension is caused by the adsorption/desorption of these ions, a relaxation is observed as a change in the conductivity by the pressure-jump method [44–48]. Since the reactions in schemes (59), (61), and (62) are Langmuir-type adsorption/desorptions, the observed relaxations can been analyzed by Eq. (15) in Theoretical Treatment (Section II). If counterions such as Na^+ or Cl^- are adsorbed to the negative or positive surface groups to form ion pairs, schemes (59) and (62) are rewritten more strictly [46] as

$$-SOH + H^+ \underset{k_{-1}}{\overset{k_1}{\rightleftharpoons}} -SOH_2^+ \underset{}{\overset{A^-}{\rightleftharpoons}} -SOH_2^+ \cdots A^- \tag{63}$$

and

$$-SOH + OH^- \underset{H_2O}{\rightleftharpoons} -SO^- \underset{}{\overset{M^+}{\rightleftharpoons}} -SO^- \cdots M^+ \tag{64}$$

where A^- and M^+ denote the negative and positive counterions, respectively. In these reactions, the rate determining steps are the acid dissociation reactions of the hydroxyl groups, since the adsorption/desorption of the counterions is much faster than that of the potential determining ions. The relaxation time for the acid dissociation/association is scheme (63) can be given [46] by

$$\tau^{-1} = k_1^{int}\left(\frac{-e\psi_0}{2k_BT}\right)\left([-SOH] + [H^+] + K_1^{-1}\frac{K_{anion}^{-1} + [-SOH_2^+]}{K_{anion}^{-1} + [-SOH_2^+] + [A^-]}\right) \quad (65)$$

where the equilibrium constant of the ion-pair formation, K_{anion}, is

$$K_{anion} = \frac{[-SOH_2^+ \cdots A^-]}{[-SOH_2^+][A^-]} = K_{anion}^{int}\exp\left(\frac{e\psi_\beta}{k_BT}\right) \quad (66)$$

The electrostatic potential at the surface of the potential determining ion, ψ_0, can be calculated from the pH dependence of the acid dissociation constant, which is obtained by the titration of the suspension containing metal oxide particles [3,46]:

$$\psi_0 = \frac{2.303k_BT}{e}(pK_a^{int} - pK_a) \quad (67)$$

In the H^+ adsorption/desorption on hematite and magnetite, the relaxation times vary as a function of pH as shown in Fig. 8. A plot of τ^{-1} against the concentration term in Eq. (65) will yield a straight line (Fig. 9). The rate constant, k_1^{int}, can be obtained from the straight line, while k_{-1}^{int} is calculated by using k_1^{int}, K_a^{int}, and K_{anion}^{int} [46].

The rate constants of H^+ adsorption/desorption on the various metal oxides are compared in Fig. 10: log k_1^{int}, and log k_{-1}^{int}, are plotted against log K_a^{int} [48]. The k_1^{int} value is constant on the order of 10^5 mol^{-1} dm^3 s^{-1}. This indicates that the H^+ adsorption process is a diffusion controlled one, and then the k_1^{int} value is independent of the acid property of the surface site [43,44,49]. On the other hand, log k_{-1}^{int} is proportional to log K_a^{int}, which means that the rate of the H^+ desorption becomes slower as the binding energy between the metal oxide and H^+ becomes higher (i.e., the pK_a value becomes larger) [43]. In comparing the above adsorption and desorption rate constants, we shall find that the difference in the acid dissociation constants of the metal oxides, obtained from many titration data, seems to result from the difference in the rate constants of the H^+ desorption process.

B. Adsorption/Desorption of Anions

When protons are adsorbed onto a metal oxide such as γ-Al_2O_3, the surface becomes positive, as is described in the previous section. Anions such as phosphate ion are electrostatically adsorbed on the positive surface. Figure 11 shows that the amount of phosphate ion adsorbed on the γ-Al_2O_3 surface depends on the pH of the suspension [50]. In the pH region where the amount of adsorbed phosphate ion varies, a double relaxation is observed as a conductivity change by the pressure-jump method, as shown in Fig. 12. The reciprocal fast relaxation time increases with increasing pH, while the slow one decreases with increasing pH. Since the pH_{zpc} of the suspension containing the phosphate ion is lower than that in the absence of the ion (Fig. 11), the adsorption is caused by the interactions of $H_2PO_4^-$ and HPO_4^{2-} with the portonated surface hydroxyl group,

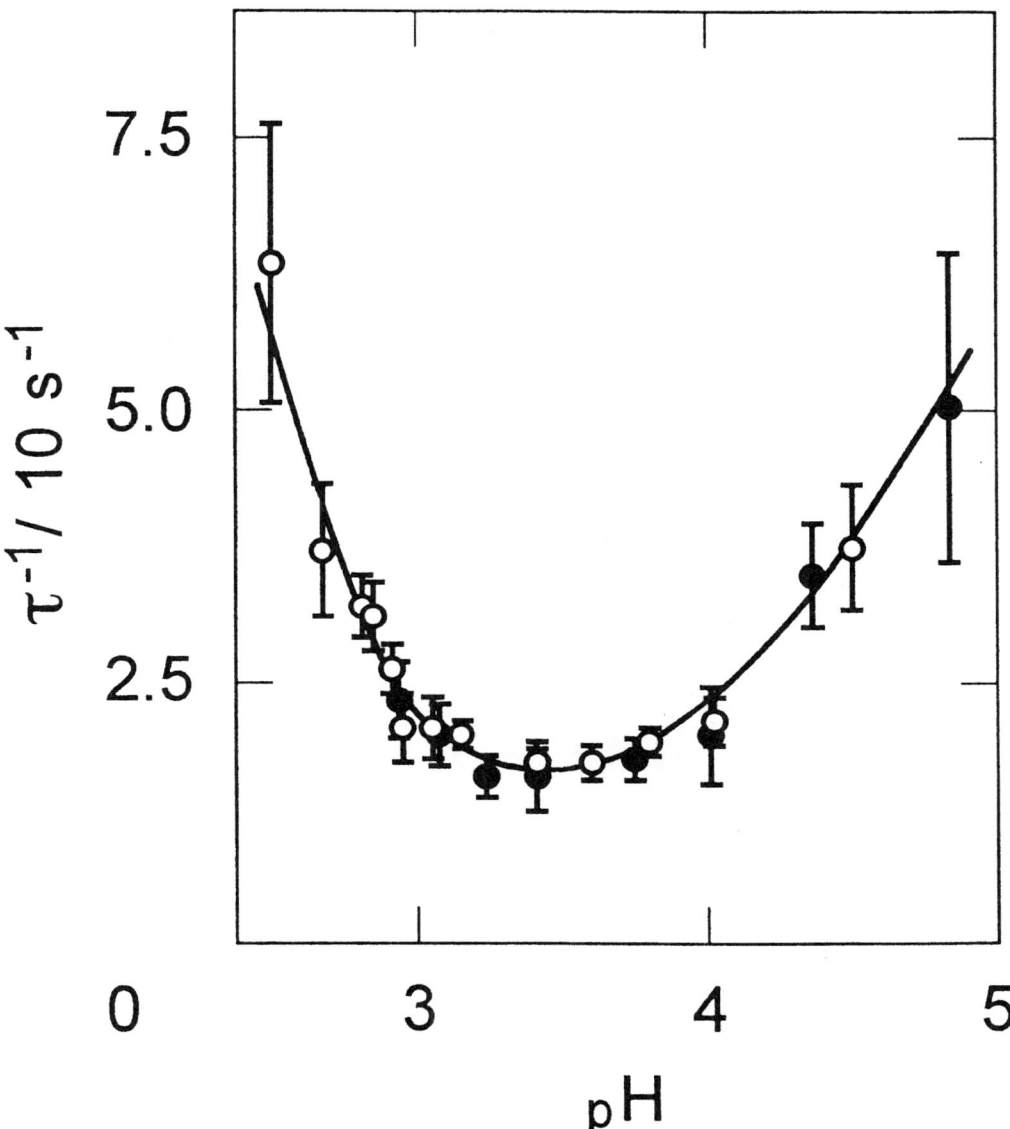

FIG. 8 pH dependence of the reciprocal relaxation times obtained in the acidic hematite (○) and magnetite (●) suspensions at ionic strength $I = 2 \times 10^{-3}$ M and 25°C. (From Ref. 46; reproduced by permission of the American Chemical Society, Washington, D.C.)

$-AlOH_2^+$. Thus the relaxations are assigned to the following adsorption/desorption reactions [50]:

$$-AlOH_2^+ + H_2PO_4^- \underset{k_{-1}}{\overset{k_1}{\rightleftharpoons}} -AlOH_2^+ \ldots H_2PO_4^- \qquad (68)$$

$$-AlOH_2^+ + HPO_4^{2-} \underset{k_{-2}}{\overset{k_2}{\rightleftharpoons}} -AlOH_2^+ \ldots HPO_4^{2-} \qquad (69)$$

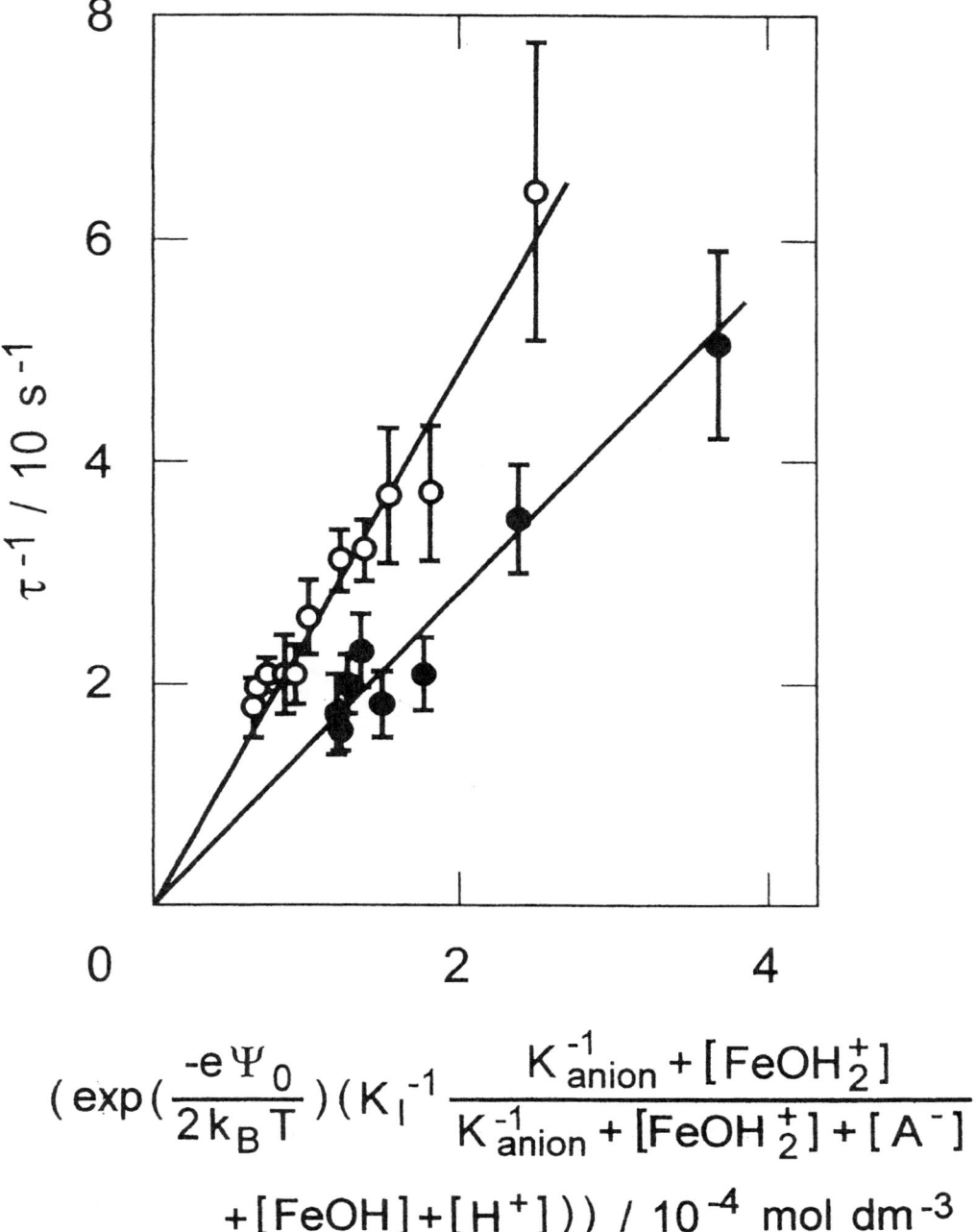

FIG. 9 τ^{-1} vs. the concentration term in Eq. (65) at 25°C: hematite (○) and magnetite (●). (From Ref. 46; reproduced by permission of the American Chemical Society, Washington, D.C.)

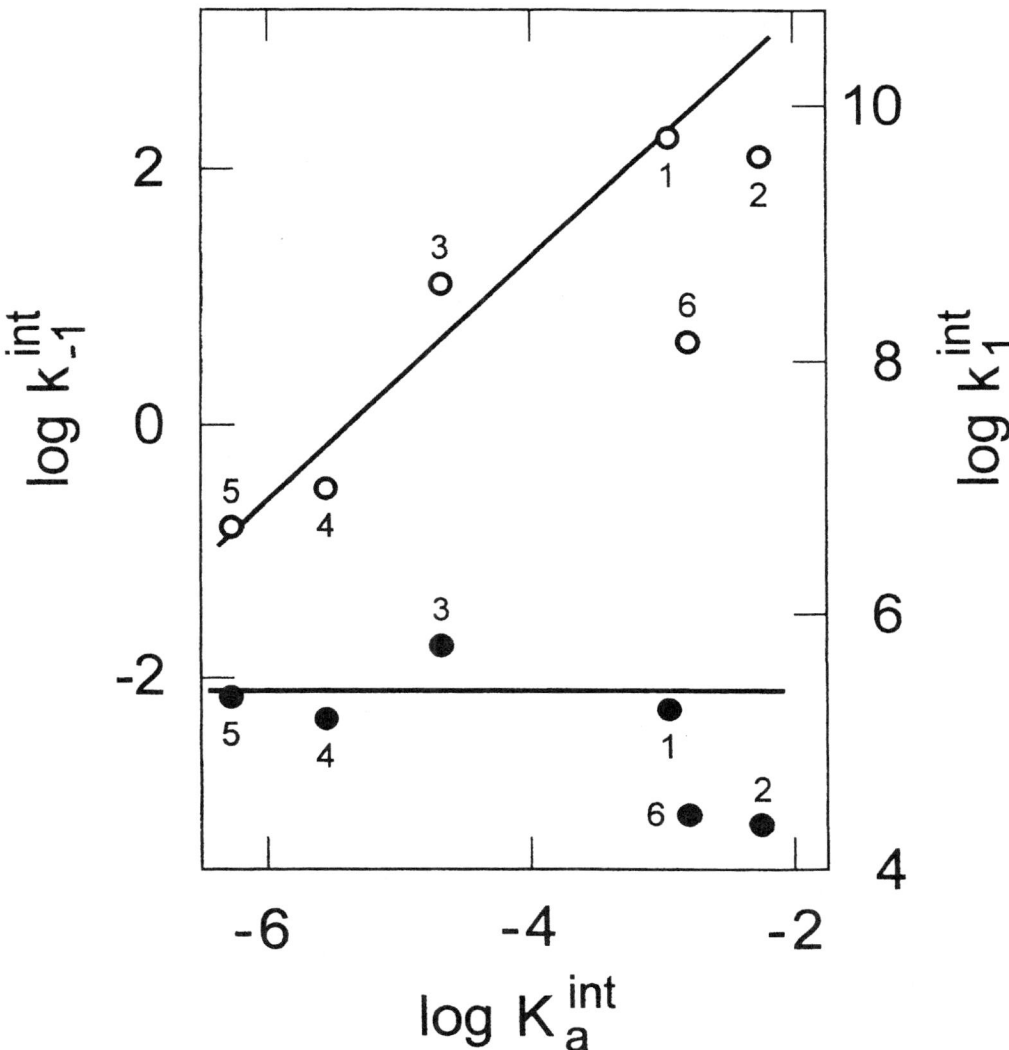

FIG. 10 Relationships between k_1^{int} (●), k_{-1}^{int} (○), and acidity constant K_a^{int}: (1) α-zirconium phosphate [48]; (2) γ-zirconium phosphate [48]; (3) TiO$_2$ [44]; (4) Fe$_3$O$_4$ [46]; (5) Fe$_2$O$_3$ [46]; (6) silica-alumina [47]. (From Ref. 48; reproduced by permission of the American Chemical Society, Washington, D.C.)

where $-AlOH_2^+ \ldots H_2PO_4^-$ and $-AlOH_2^+ \ldots HPO_4^{2-}$ are the adsorbed states of the corresponding phosphate ions. Here it is assumed that the rate of the adsorption/desorption of HPO_4^{2-} is faster than that of $H_2PO_4^-$. The pH dependence of the adsorbed amount of phosphate ion can be also simulated on the basis of the mechanisms (68) and (69) as is illustrated in Fig. 11. The adsorption/desorption of chromate ions on the γ-Al$_2$O$_3$ surface can be also analyzed by using mechanisms similar to (68) and (69) [51]. The rate constants for the phosphate and chromate adsorption/desorptions are listed in Table 1. The adsorption rate constants, k_1^{int} and k_2^{int}, for $HCrO_4^-$ and CrO_4^{2-} are one and

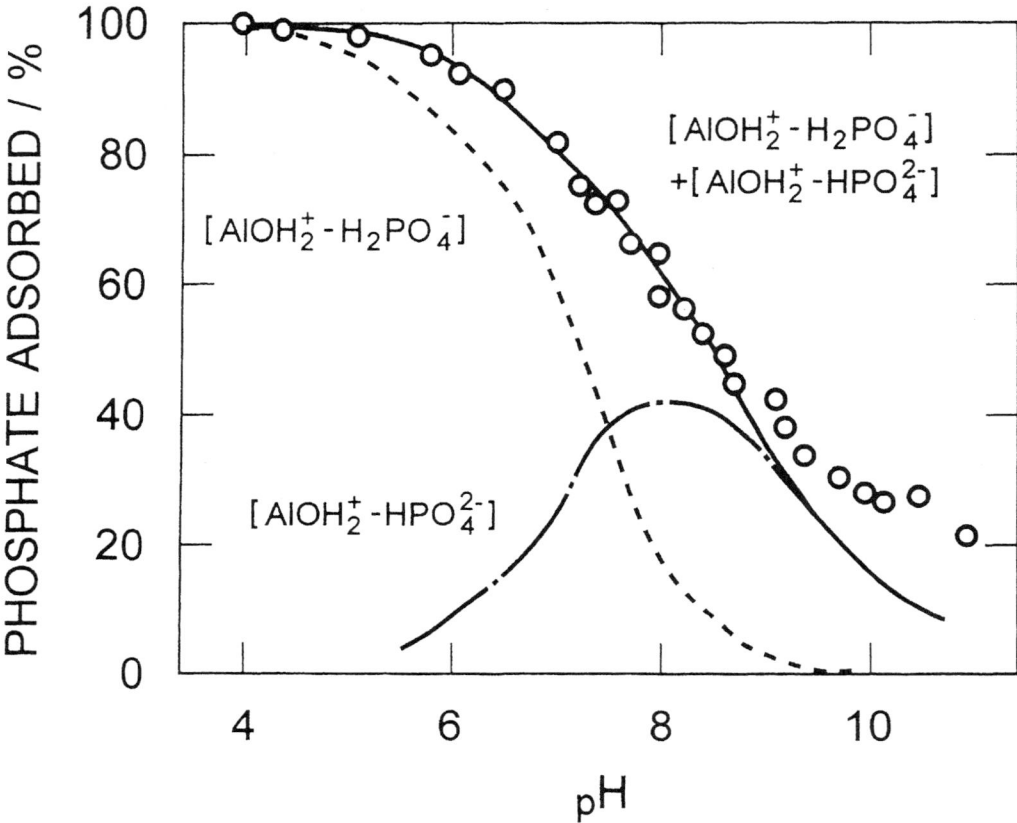

FIG. 11 pH dependence of the amount of phosphate ion adsorbed onto γ-Al$_2$O$_3$ surface (○) at particle concentration, $C_p = 30$ g dm^{-3}, phosphate concentration = 8 mM, $I = 1.5 \times 10^{-2}$ M, and 25°C. The solid line indicates the sum of the calculated amounts of adsorbed H$_2$PO$_4^-$ (- - -) and HPO$_4^{2-}$ (- · -). (From Ref. 50; reproduced by permission of the American Chemical Society, Washington, D.C.)

two orders of magnitude smaller than those for H$_2$PO$_4^-$ and HPO$_4^{2-}$, respectively, while the desorption rate constants, k_{-1}^{int} and k_{-2}^{int}, for the former are one order of magnitude larger than those for the latter. The difference in the rate constants means that the interaction of –AlOH$_2^+$ with the phosphate ions is stronger than that with the chromate ions [43,51].

In the pH region where the H$^+$ adsorption/desorption as expressed in scheme (59) does not occur, a single relaxation is observed in the TiO$_2$ suspension containing KIO$_3$ by the pressure-jump method [52]. When the KIO$_3$ solutions (pH 6.0) of 5 to 20 mM are added to the TiO$_2$ suspension (pH 5.8), the pH of the suspension increases by 1.2 to 1.4. The increment in pH indicates that OH$^-$ is released from the TiO$_2$ surface upon the IO$_3^-$ adsorption, and that this reaction would be different from the ion-pair formation

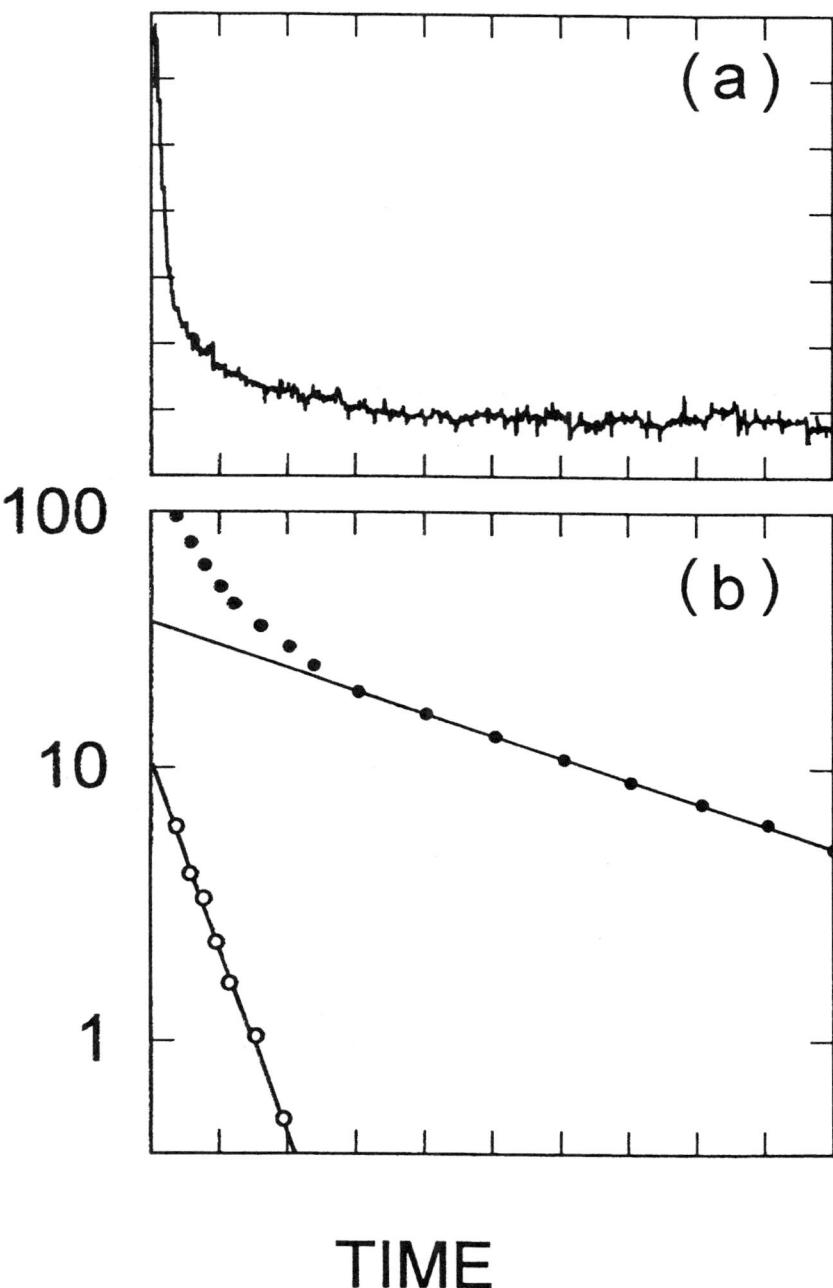

FIG. 12 (a) Typical double relaxation curve in the γ-Al_2O_3-phosphate system observed by the pressure-jump method with conductivity detection: $C_p = 30$ g dm^{-3}, phosphate concentration = 8 mM, $I = 1.5 \times 10^{-2}$ M at 25°C; sweep, 20 ms/division. (b) The semilogarithmic plot of the relaxation curve; sweep, 10 ms/division. (From Ref. 50; reproduced by permission of the American Chemical Society, Washington, D.C.)

TABLE 1 Rate Constants of Adsorption/Desorption of Phosphate and Chromate Ions

Anion	k_1^{int} (mol^{-1}dm^{3}s^{-1})	k_{-1}^{int} (s^{-1})	k_2^{int} (mol^{-1}dm^{3}s^{-1})	k_{-2}^{int} (s^{-1})
Phosphate	4.1×10^5	2.3	1.1×10^7	2.7
Chromate	5.3×10^4	1.9×10	9.9×10^4	5.2×10

Source: Ref. 51. (Reproduced by permission of the American Chemical Society, Washington, D.C.)

of anion with the surface hydroxyl group in schemes (68) and (69). Then the single relaxation can be attributed to the ion-exchange reaction of IO_3^- with OH^- of the surface hydroxyl group as follows [52]:

$$-TiO^- \underset{H_2O}{\rightleftarrows} -TiOH \underset{OH^-\ IO_3^-}{\rightleftarrows} -TiOH \cdots IO_3^- \underset{OH^-}{\rightleftarrows} -TiIO_3 \quad (70)$$

$$\text{(step 1)} \qquad \text{(step 2)}$$

where $-TiOH \cdots IO_3^-$ and $-TiIO_3$ represent the stable complex formed by the adsorption of IO_3^- and the surface complex, respectively. The pressure-jump method has been applied to follow the adsorptions of oxyanions, molybdate [53], sulfate [54], selenite [55], arsenate [56], and chromate [56] on goethite. In the kinetic analysis of the oxyanion adsorptions, the processes in steps 1 and 2 of scheme (70) are interpreted as the formations of outer-sphere and inner-sphere complexes, respectively [54,55].

A different type of adsorption/desorption mechanism has been proposed for the ion-pair formation studied by the electric field-jump method [32]. This reaction occurs in a very fast time range of 10 μs order, compared with those observed by the pressure-jump method. When an anion, A^-, is adsorbed to a surface group of $-SOH_2^+$, the following two-step reaction mechanism has been proposed:

$$A^- \underset{k_{-1}}{\overset{k_1}{\rightleftharpoons}} A_\beta^- \quad (71)$$

$$-SOH_2^+ + A_\beta^- \underset{k_{-2}}{\overset{k_2}{\rightleftharpoons}} -SOH_2^+ \cdots A_\beta^- \quad (72)$$

where A_β^- is the anion located at the β-plane, and schemes (71) and (72) denote the diffusion process of A^- from the bulk phase to the β-plane and the recombination/dissociation process on the metal oxide surface, respectively. The apparent overall equilibrium constant of the ion-pair formation, K_{anion}, in Eq. (66) is rewritten for schemes (71) and (72) as

$$K_{\text{anion}} = \frac{[-SOH_2^+ \cdots A_\beta^-]}{[-SOH_2^+][A^-]} = \frac{[-SOH_2^+ \cdots A_\beta^-]}{[-SOH_2^+][A^-]_\beta} \exp\left(\frac{e\psi_\beta}{k_B T}\right) \quad (73)$$

The A^- concentration in the bulk phase is related to that in the β-plane by

$$[A^-] = [A^-]_\beta \exp\left(-\frac{e\psi_\beta}{k_B T}\right) \tag{74}$$

By substituting Eq. (74) into Eq. (73), we obtain [32]

$$K_{\text{anion}} = \frac{[-SOH_2^+ \cdots A^-]}{[-SOH_2^+][A^-]_\beta} \times \frac{[A^-]_\beta}{[A^-]} \tag{75}$$

The first and second terms on the right-hand side of Eq. (75) represent the equilibrium constants of the recombination/dissociation process in scheme (72) and of the ion-diffusing process in scheme (71), respectively.

If the diffusion process is much faster than the recombination/dissociation process, the reciprocal relaxation time of the rate determining step is represented [32] by

$$\tau^{-1} = k_2^{\text{int}}\left(\frac{1}{1+K_d^{-1}}[-SOH_2^+] + [A^-]_\beta\right) + k_{-2}^{\text{int}} \tag{76}$$

where

$$K_d = \frac{[A^-]_\beta}{[A^-]} = \exp\left(\frac{e\psi_\beta}{k_B T}\right) \tag{77}$$

The derivation of the τ^{-1} equation is quite different from that in Theoretical Treatment (Section II). Whether the above derivation is a general procedure for the analysis of ion adsorption/desorption or not must be discussed in future. For example, when the ion-pair formation process is not divided into the two processes in schemes (71) and (72), the equation of τ^{-1} is given by the following equation similar to Eq. (14) in Theoretical Treatment (it should be noted that the signs of charges of counterion and surface site and the sign of the potential are opposite) [32]:

$$\tau^{-1} = k_1^{\text{int}}\exp\left(\frac{e\psi_\beta}{2k_B T}\right)([-SOH_2^+] + [A^-]) + k_{-1}^{\text{int}}\exp\left(-\frac{e\psi_\beta}{2k_B T}\right) \tag{78}$$

Both Eq. (76) and Eq. (78) represent the τ^{-1} equations for the rate determining steps of the ion-pair formations. However, the adsorption rate constant in Eq. (76) is $k_2^{\text{int}}/(1+K_d^{-1})$, while that in Eq. (78) is $k_1^{\text{int}}\exp(e\psi_\beta/2k_B T)$. The former rate constant is not dependent so much on the surface potential as the latter. The difference between the adsorption rate constants might result from the introduction of an equilibrium of the ion-diffusing proces (71). In other words, the potential term, $\exp(e\psi_\beta/2k_B T)$, used frequently in the electrical double layer model [3], suggests an equilibrium constant according to the present model.

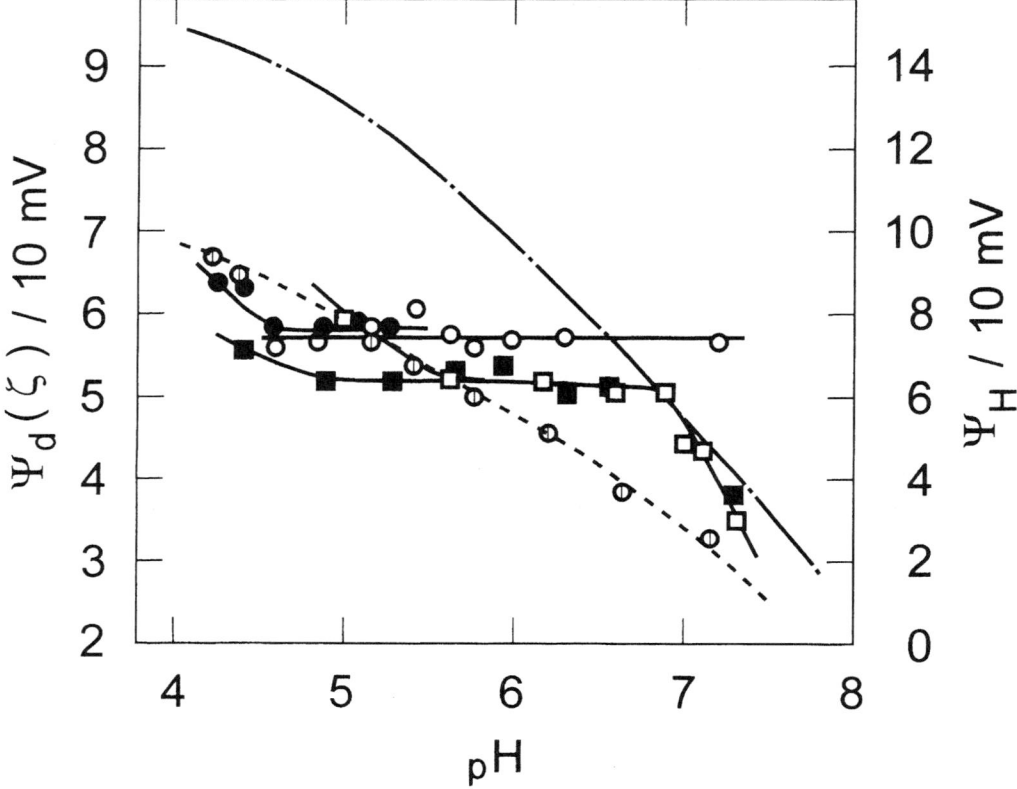

FIG. 13 pH dependence of ζ-potentials (———) in the γ-Al$_2$O$_3$-M (NO$_3$)$_2$ systems and those of ζ-potential (- - - -) and ψ_H (— · —) in the γ-Al$_2$O$_3$-HNO$_3$ system of $C_p = 30$ g dm^{-3}, and $I = 1.5 \times 10^{-2}$ M at 25°C: (○) Pb^{2+}, (●) Cu^{2+}, (□) Co^{2+}, (■) Zn^{2+}, (◐) H$^+$. (From Ref. 57; reproduced by permission of the American Society, Washington, D.C.)

C. Adsorption/Desorption of Metal Ions

Divalent metal ions, Cu^{2+}, Mn^{2+}, Zn^{2+}, Co^{2+}, and Pb^{2+} are specifically adsorbed on the surface of a metal oxide like γ-Al$_2$O$_3$ and are exchanged with H$^+$ of the surface hydroxyl groups [57–59]. The adsorption of the specifically adsorbed ions causes a charge reversal of the ζ-potential: the potential of a particle becomes constant as shown in Fig. 13 [57] or increases with increasing pH of the suspension [60]. Then the pH$_{ZPC}$ of γ-Al$_2$O$_3$ is shifted to a higher pH. In the suspension containing monovalent cations such as Na$^+$ which are not specifically adsorbed, however, the ζ-potential monotonically decreases with increasing pH of the suspension (Fig. 13). In earlier studies of metal ion adsorption onto a metal oxide, the amount of divalent metal ion, M^{2+}, added in the suspension was of the comparatively small order of 10^{-6} to 10^{-4} M [58,59,61]. The M^{2+} adsorption was analyzed by the model that M^{2+} as well as a monovalent cation was bound electrostatically to the negatively chargd surface hydroxyl group [58]. However, such a model cannot be applied to analyze the adsorption/desorption occurring in a higher concentration region of more than 10^{-3} M, which is required for the pressure-jump experiment [57].

The binding of M^{2+} to the oxygen atom of the surface hydroxyl group has been found to be covalent on the basis of the infrared measurement [62]. This indicates that a part of the water molecules are released from the hydrated M^{2+} ion before the direct binding of M^{2+} to the surface hydroxyl group. Then, the M^{2+} forms a complex with the surface hydroxyl group according to the following coordinate binding model [57]:

$$-\text{AlOH}_2^+ \underset{}{\overset{}{\rightleftarrows}} -\text{AlOH} \underset{k_{-1}}{\overset{k_1}{\rightleftarrows}} -\text{AlOM}(\text{H}_2\text{O})_{n-1}^+$$

$$\text{H}^+ \quad \text{M}(\text{H}_2\text{O})_n^{2+} \qquad \text{H}^+ + \text{H}_2\text{O}$$

$$(K_{a1}) \qquad\qquad (\text{step 1, } K_1) \tag{79}$$

$$\underset{k_{-2}}{\overset{k_2}{\rightleftarrows}} -\text{AlOMOH}(\text{H}_2\text{O})_{n-2}$$

$$\text{H}^+$$

$$(\text{step 2, } K_2)$$

where $\text{AlOM}(\text{H}_2\text{O})_{n-1}^+$ and $\text{AlOMOH}(\text{H}_2\text{O})_{n-2}$ denote two kinds of surface complexes formed by the adsorption of hydrated metal ion, $\text{M}(\text{H}_2\text{O})_n^{2+}$. In step 1 of mechanism (79), a water molecule is released from $\text{M}(\text{H}_2\text{O})_n^{2+}$ by the coordinate binding of M^{2+} to the oxygen atom. This means that M^{2+} is not adsorbed on the β-plane of the ion-pair formation but on the same surface plane as H^+. The surface potential, ψ_0, and surface charge density, σ_0, can be given [57] by

$$\psi_0 = \psi_M + \psi_H \tag{80}$$

$$\sigma_0 = \sigma_M + \sigma_H$$
$$= \frac{F}{[P]S}([-\text{AlOH}_2^+] + [-\text{AlOM}(\text{H}_2\text{O})_{n-1}^+]) \tag{81}$$

where ψ_M and ψ_H are the surface potentials created by adsorbed M^{2+} and H^+, respectively, σ_M and σ_H are the corresponding surface charge densities, and F, $[P]$, and S denote the Faraday constant, the particle concentration, and the specific surface area, respectively. When the pH of the γ-Al_2O_3 suspension increases on the addition of base, the amount of AlOH_2^+ decreases, while the amount of $\text{AlOM}(\text{H}_2\text{O})_{n-1}^+$ increases. Since the sum of the surface charge densities of σ_M and σ_H is kept to be nearly constant in Eq. (81), the charge reversal in Fig. 13 can be explained qualitatively. The equilibrium constants

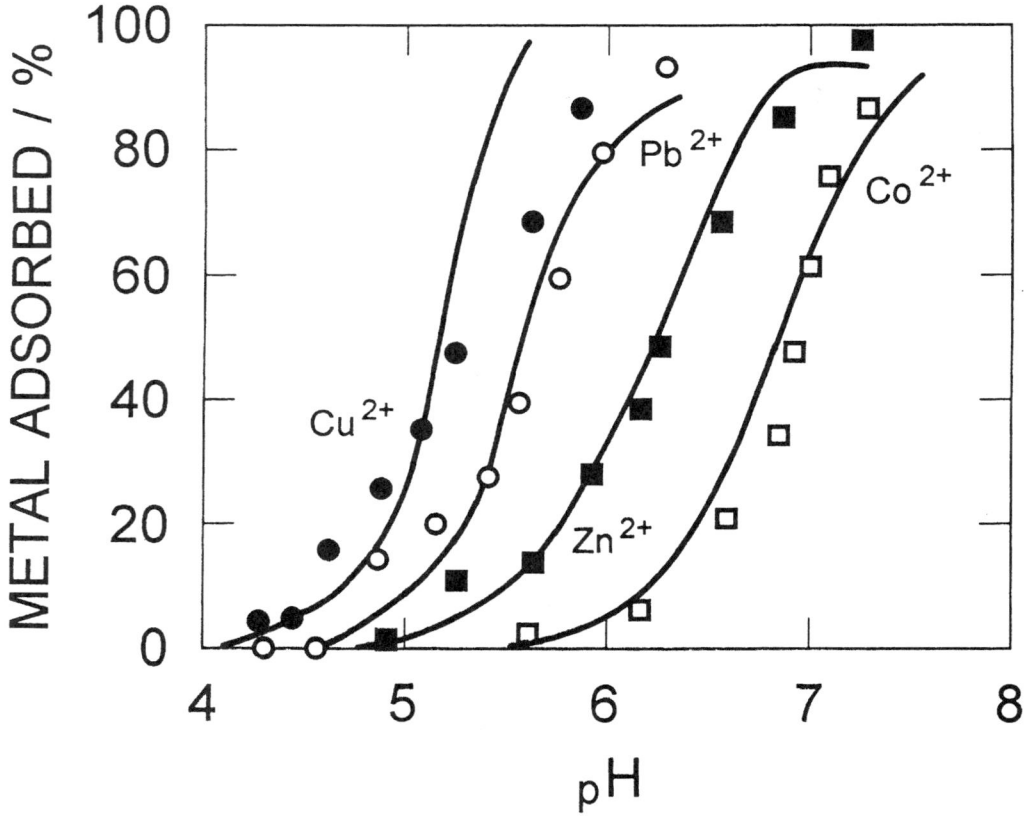

FIG. 14 pH dependence of divalent metal ions adsorbed onto γ-Al_2O_3 surface at $C_p = 30$ g dm^{-3}, total metal ion concentration = 3 mM, $I = 7.5 \times 10^{-3}$ M, and 25°C; (○) Pb^{2+}, (●) Cu^{2+}, (□) Co^{2+}, (■) Zn^{2+}. (From Ref. 57; reproduced by permission of the American Chemical Society, Washington, D.C.)

for mechanism (79) are given by

$$K_1 = \frac{[-\text{AlOM}(H_2O)_{n-1}^+][H^+]}{[-\text{AlOH}][M(H_2O)_n^{2+}]} = K_1^{\text{int}} \exp\left(-\frac{e\psi_o}{k_B T}\right) \quad (82)$$

$$K_2 = \frac{[-\text{AlOMOH}(H_2O)_{n-2}][H^+]}{[-\text{AlOM}(H_2O)_{n-1}^+]} = K_2^{\text{int}} \exp\left(\frac{e\psi_o}{k_B T}\right) \quad (83)$$

The pH dependene of the amount of M^{2+} adsorbed on the metal oxide in Fig. 14 can be analyzed numerically by using the above equilibrium constants [57]. The values of K_1^{int} and $K_1^{\text{int}} K_2^{\text{int}}$ are plotted against the hydrolysis constants, K_h, of the metal ions in Fig. 15. The correlation between these constants indicates that surface complexes, $\text{AlOM}(H_2O)_{n-1}^+$ and $\text{AlOMOH}(H_2O)_{n-2}$, are formed by the hydrolysis of the adsorbed metal ion, i.e., a surface hydrolysis [57].

A relaxation is observed in the γ-Al_2O_3 suspension containing Cu^{2+}, Mn^{2+}, Zn^{2+}, Co^{2+}, or Pb^{2+} by the pressure-jump method. The reciprocal relaxation time increases with increasing M^{2+} concentration and with increasing pH [9]. The dependences of τ^{-1} on M^{2+}

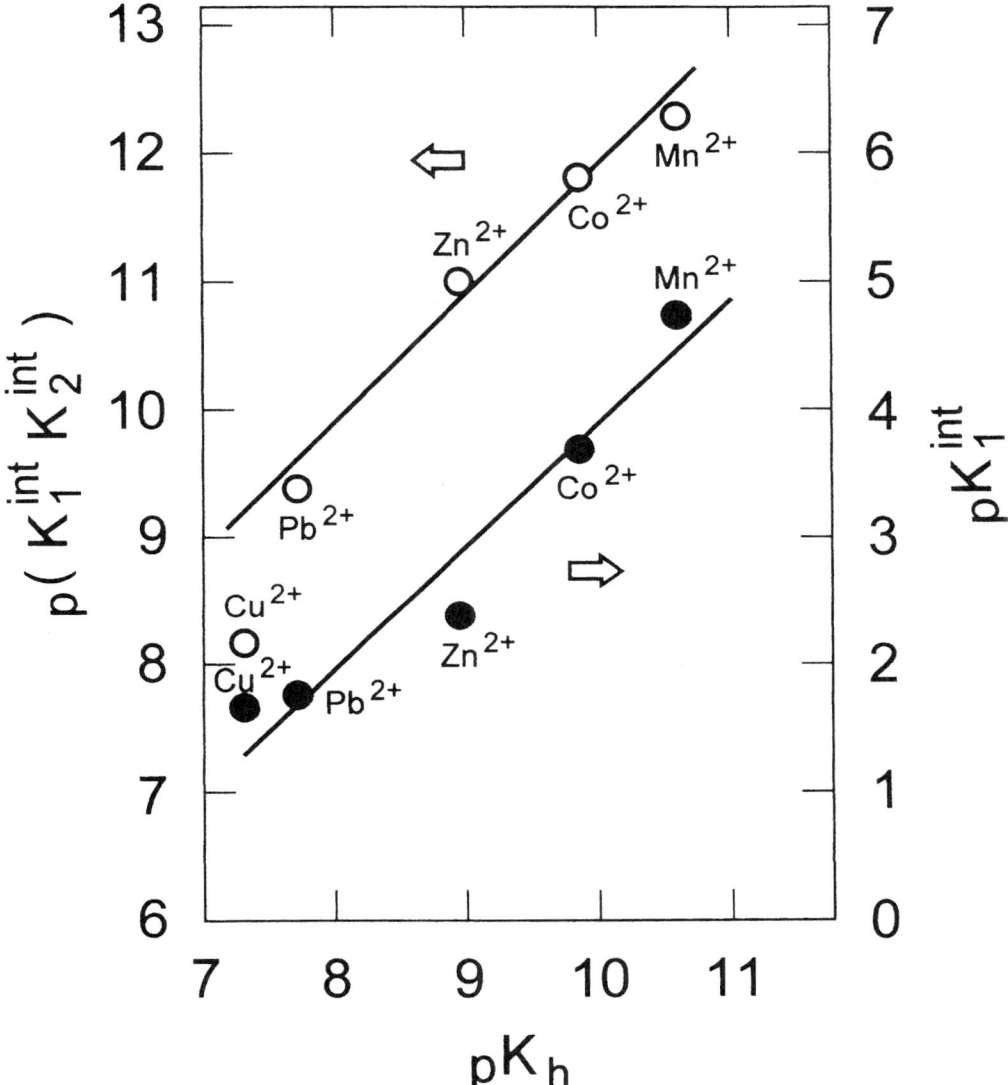

FIG. 15 Logarithmic plot of $K_1^{int}K_2^{int}$ (○) and K_1^{int} (●) vs. the hydrolysis constants of the metal ions. (From Ref. 57; reproduced by permission of the American Chemical Society, Washington, D.C.)

concentration and on pH cannot be accounted for by the adsorption/desorption mechanism (79), by which the pH dependence of the M^{2+} adsorption in Fig. 14 can be analyzed successfully. This hints that several steps in mechanism (79) would consist of more elementary steps. The step 1 of this mechanism is further divided into two elementary steps as represented in mechanism (39) in Theoretical Treatment (Section II). If so, the equation of τ^{-1} is given by Eq. (44). Indeed, the M^{2+} concentration and pH dependences of τ^{-1} can be explained by Eq. (44) with the aid of the surface potential calculated from the ζ-potential. The rate constants are determined from a slope of a linear plot of τ^{-1} against the concentration term of Eq. (44). Figure 16 indicates that the adsorption rate

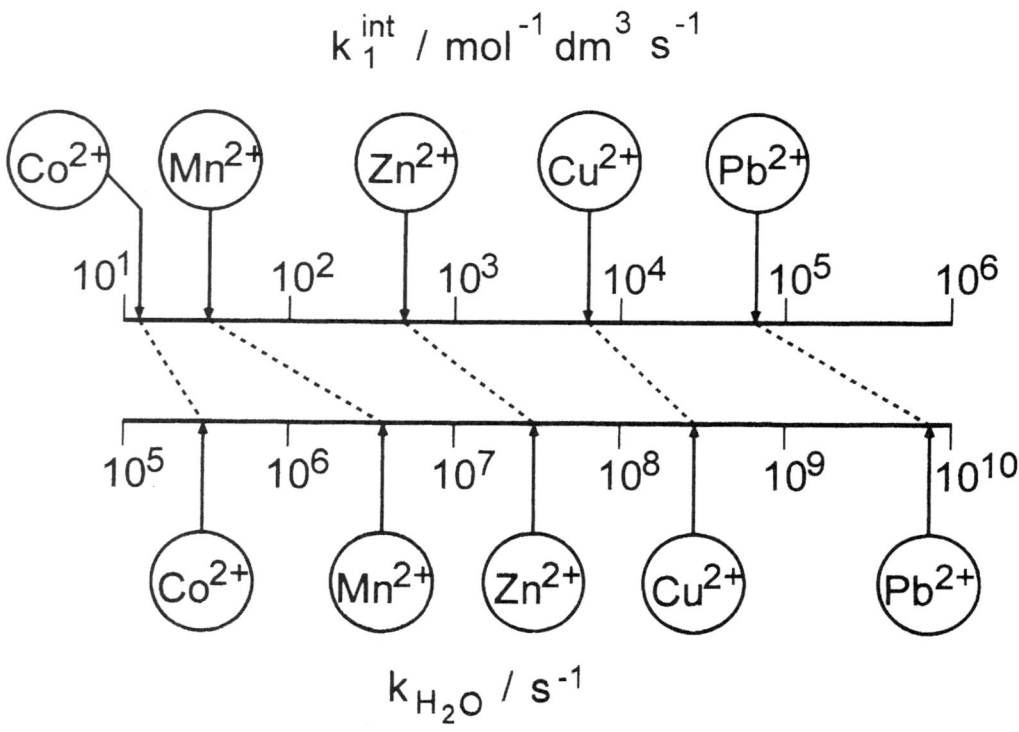

FIG. 16 Relationship between the adsorption rate constants, k_1^{int}, and the rate constants for the release of a water molecule from hydrated metal ion in the metal complex formation, k_{H_2O}, for the divalent metal ions. (From Ref. 9; reproduced by permission of the American Chemical Society, Washington, D.C.)

constant, k_1^{int} ($= k_{11}^{int}$), correlates with the rate constant for the release of a water molecule from a hydrated metal ion, k_{H_2O}, in the metal complex formation in homogeneous systems [9,63]. This correlation between the k_1^{int} and k_{H_2O} values would support the validity of the coordinate adsorption/desorption model of M^{2+} as described in mechanism (39), in which M^{2+} is assumed to be adsorbed on the same surface plane as the potential determining ion [9]. We can discriminate the adsorption/desorption of M^{2+} with that of a monovalent cation such as Na^+ on the metal oxide: a surface complex is formed by the coordinate binding of M^{2+} to the surface hydroxyl group (i.e., inner-sphere surface complex), and an ion pair is formed by the electrostatic binding of the monovalent cation to the surface hydroxyl group (i.e., outer-sphere surface complex) [9,57]. Through the discussion on these surface-complex and ion-pair formations, the adsorption/desorption model of the divalent metal ion has been developed to the three-layer model (TLM) [64,65].

In kinetic studies on metal ion adsorption/desorptions, the suspension of γ-Al_2O_3 has mostly been used as a counterpart. This suspension is adequate for pressure-jump experiments. In this suspension, a relaxation due to M^{2+} adsorption/desorption is observed in the time range of this method, while a relaxation due to proton association/dissociation is not observed. This indicates that the latter relaxation does not interfere with the observation of the former relaxation. Relaxations have been observed in the γ-Al_2O_3 suspensions containing uranyl ions [66] or trivalent cations, Cr(III) [67], Ga(III) [68], and In(III) [68]. The relaxation for the suspension containing uranyl ions is attributed to the adsorption/desorption of $(UO_2)_3(OH)_5^+$ on the surface hydroxyl group [66]. On the other hand, the relaxation for the suspension containing trivalent cations, M^{3+}, is interpreted as proton release from surface hydroxyl groups followed by the attachments of M^{3+} and MOH^{2+} which form bidentate, $(-AlO)_2M^+$, and monodentate, $-AlOMOH^+$, respectively [67,68].

D. Intercalation of Ions into the Interlamellar Layer

The intercalation of ions into the interlamellar layer region is represented by the scheme [43,69]:

$$\begin{array}{cccc} S(A) & S(A)M & S(M)A & S(M) \\ \text{////// } \underset{k_{-1}}{\overset{k_1}{\rightleftharpoons}} \text{////// } \underset{k_{-2}}{\overset{k_2}{\rightleftharpoons}} \text{////// } \underset{k_{-3}}{\overset{k_3}{\rightleftharpoons}} \text{////// } \\ A \quad \quad M \quad \quad M \quad \quad A \quad \quad M \\ \text{//////} \quad \text{//////} \quad \text{//////} \quad \text{//////} \\ M & & A & \end{array} \qquad (84)$$

step 1   step 2   step 3

where S, M, and A denote the site in the interlayer, the intercalating ion, and the exchangeable ion which is initially bound to the site, respectively. After M is adsorbed at the entrance of the interlayer in step 1, it deeply intercalates by intracrystalline exchange in step 2, and then A is released from the interlayer in step 3. S(A), S(A)M, S(M)A, and S(M) represent the states of ions bound in the interlayer. The ion expressed by A corresponds to H^+ for α-zirconium phosphate (α-ZrP) [70,71] and zeolite H-ZSM-5 [72,73], to Na^+ for zeolite 4A (Z-4A) [74–76] and montmorillonite [69], and to Cl^- for hydrotalcite-like compound [77], while the interlayer is initially vacant for TiS_2 [78–80]. Although it takes 24–72 hours for the equilibration in the process of step 3 in a typical intercalation, two chemical relaxations due to the reactions in steps 1 and 2 have been observed on the orders of milliseconds by the pressure-jump and the stopped-flow methods [69–80]. If step 1 is faster than step 2 (and that step 3 is much slower than steps 1 and 2), the fast and slow relaxation times can be represented [70] by

$$\tau_1^{-1} = k_1([S(A)] + [M]) + k_{-1} \qquad (85)$$

$$\tau_2^{-1} = k_2 \frac{[S(A)] + [M]}{[S(A)] + [M] + K_1^{-1}} + k_{-2} \qquad (86)$$

Since a large change in the surface potential is not caused by the exchange of intercalating ions with exchangeable ions bound on the host lattice in the interlayer, the reaction can be analyzed by Eqs. (85) and (86) [69,72–76]. In the cases of ZrP and TiS_2, however, the potential terms are introduced into these equations, since large changes in the potentials are observed for the ion intercalations into the particles [70,71,77–80].

The size of a guest ion and the interlayer distance result in a stereoselectivity for the intercalation of ions. Figure 17 shows the plots of the rate constants against the interlayer distance in the intercalation of alkali metal ions into α-ZrP. The size of the alkali metal ion determines the rates of steps 1 and 2 in mechanism (84) [70]. However, the intercalations of these ions into TiS_2 and Z-4A are not influenced by the size of the intercalating ion, since the layered structure of TiS_2 and the cage structure of Z-4A are rigid [75,80].

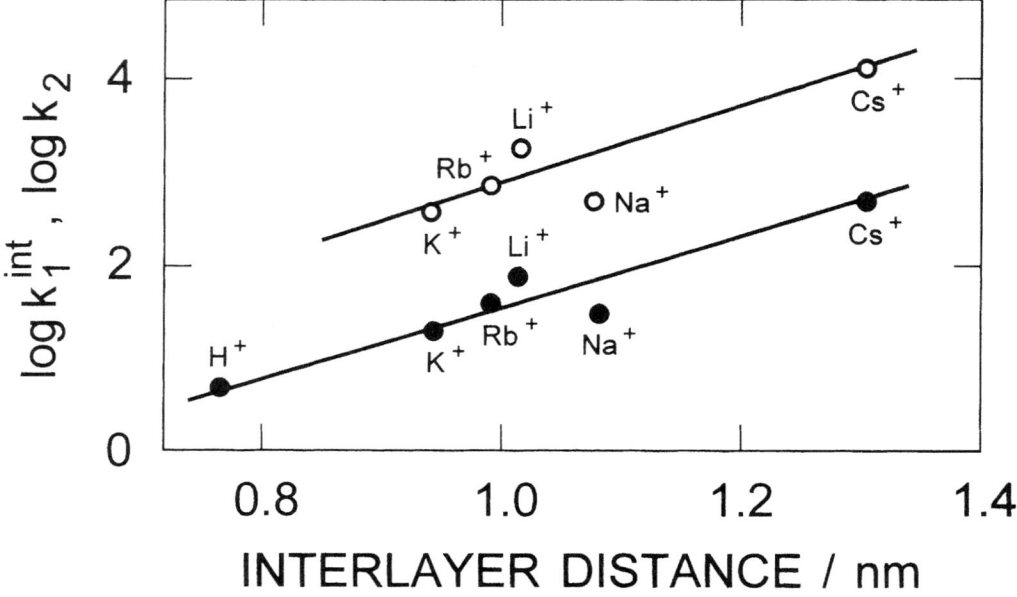

FIG. 17 Plots of log k_1^{int} (○) and log k_2 (●) vs. interlayer distance. (From Ref. 70; reproduced by permission of the American Chemical Society, Washington, D.C.)

The ion sizes of ammonium and amine ions influence the intercalation rates into Z-4A particles. The equilibrium and rate constants are plotted against the ion size in Figs. 18 and 19 [74]. Since the intercalation rate constants, k_1 and k_2, decrease with an increase in the cation volume, smaller ions would be able to intercalate more favorably into the solid phase: this corresponds to an ion-size effect like an ion sieve [74]. In fact, neither relaxation nor ion exchange has been observed for i-$C_3H_7NH_3^+$, $(CH_3)_3NH^+$, or $(CH_3)_4N^+$ with volumes greater than 0.12 nm^3 (i.e., 120 Å^3) (Figs. 18 and 19). The deintercalation rate constant, k_{-1}, decreases with an increase in the pK_a of ammonium and amines as shown in Fig. 20 [74]. This tendency is similar to that of H$^+$ desorption from the metal oxide as is illustrated in Fig. 10. If the amine cation is charged more positively, the binding energy of the cation to the dissociated surface site would increase. As a result, the deintercalation rate of amine seems to decrease with an increase in the binding energy.

V. ADSORPTION/DESORPTION OF METAL IONS, AMINO ACIDS, AND CHARGED PROTEINS ON ION-EXCHANGE SEPHADEX

Sephadex is the name of a commercially produced particle with a network structure, which consists of dextran chains linked three-dimensionally [81,82]. Various amino acids and proteins have been separated or purified by chromatography using this ion-exchange Sephadex [83,84], on which functional groups such as carboxymethyl (CM) and diethylaminoethyl (DEAE) groups are located. In this section, we introduce kinetic studies of the fundamental reactions of the functional groups on the Sephadexes, counterion adsorption/desorptions, and ion-exchange reactions.

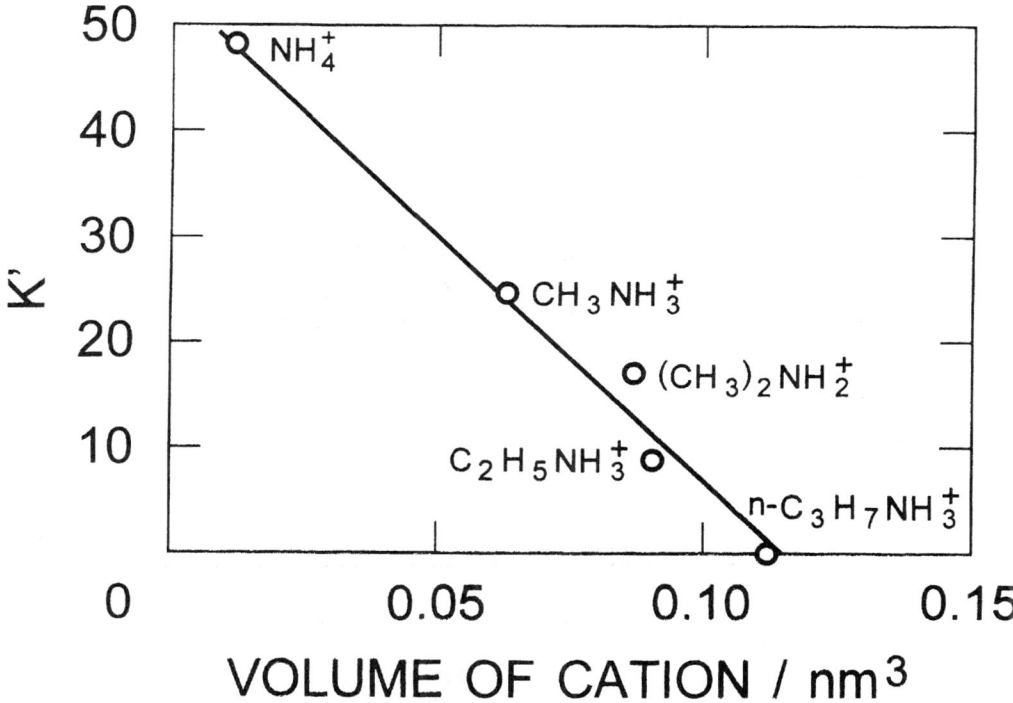

FIG. 18 Plot of the ion exchange constant, K', vs. volume of cation. K' is defined as $[M_{ads}][Na^+]/[S(Na)][M]$. (From Ref. 74; reproduced by permission of the American Chemical Society, Washington, D.C.)

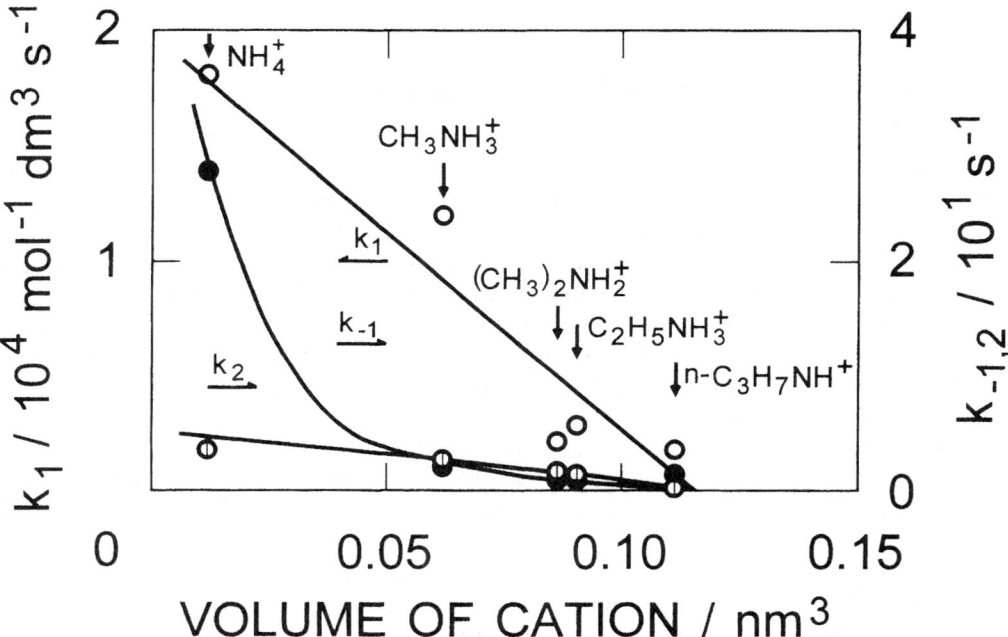

FIG. 19 Plot of k_1 (○), k_{-1} (●), and k_2 (⦶) vs. volume of cation. (From Ref. 74; reproduced by permission of the American Chemical Society, Washington, D.C.)

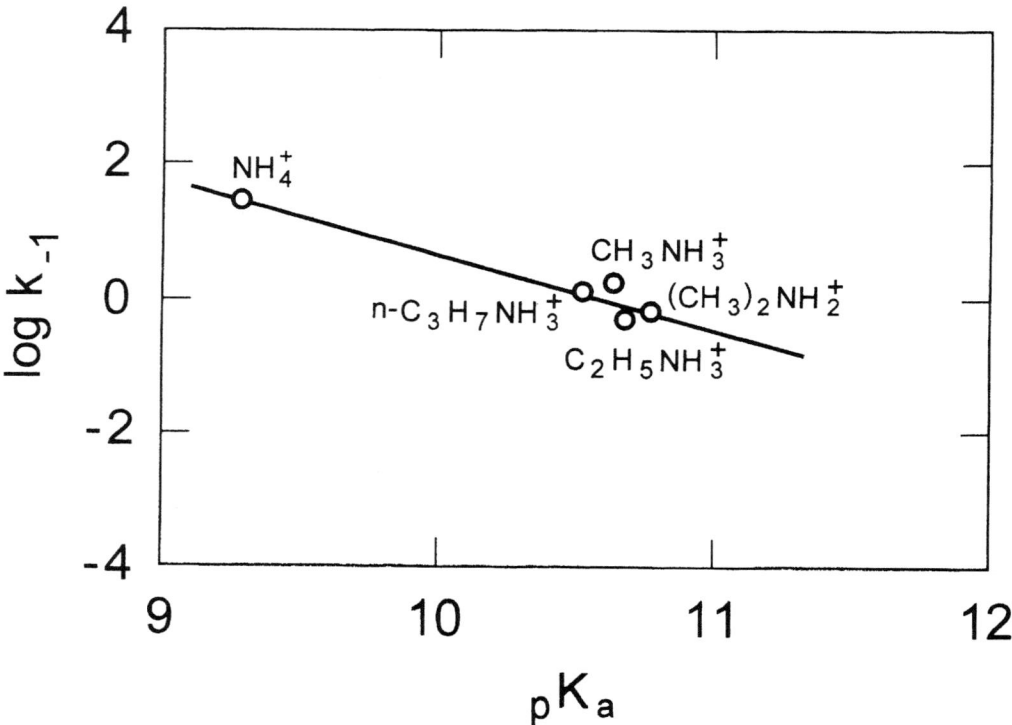

FIG. 20 Plot of log k_{-1} vs. pK_a. (From Ref. 74; reproduced by permission of the American Chemical Society, Washington, D.C.)

A. Proton Association/Dissociation of Functional Groups on Ion-Exchange Sephadex

On the addition of base, the CM groups on the Sephadex surface are deprotonated to become negative. The surface of Sephadex is different from that of metal oxide with an amphoteric property. The amphoteric property is defined as the character that the surface charge widely changes from positive to negative with pH change, as is indicated in schemes (59), (61), and (62). When the CM Sephadex suspension is titrated with NaOH, the pH of the suspension increases from pH 3.2 to pH 4.1. However, the Na^+ concentration is kept almost constant, even though the added amount of NaOH increases (Fig. 21). This indicates that almost all of the Na^+ ions, which are added as NaOH in the bulk phase, are strongly adsorbed to the dissociated CM groups. The ion-pair formation of the counterion in a Sephadex suspension is very different from that in a metal oxide suspension. In other words, the interaction of the counterion with the CM group is much stronger than that with the surface hydroxyl group of a metal oxide, as is described in Section IV.A.

In the NaOH concentration range, where 0 to 10% of the CM groups dissociate in Fig. 21, a single relaxation has been observed by the pressure-jump method with electric conductivity detection: the conductivity of the suspension decreases with time after the pressure of 100 atm is suddenly dropped to atmospheric pressure [25]. This is the same tendency in the conductivity change as has been observed in the metal oxide suspensions. If

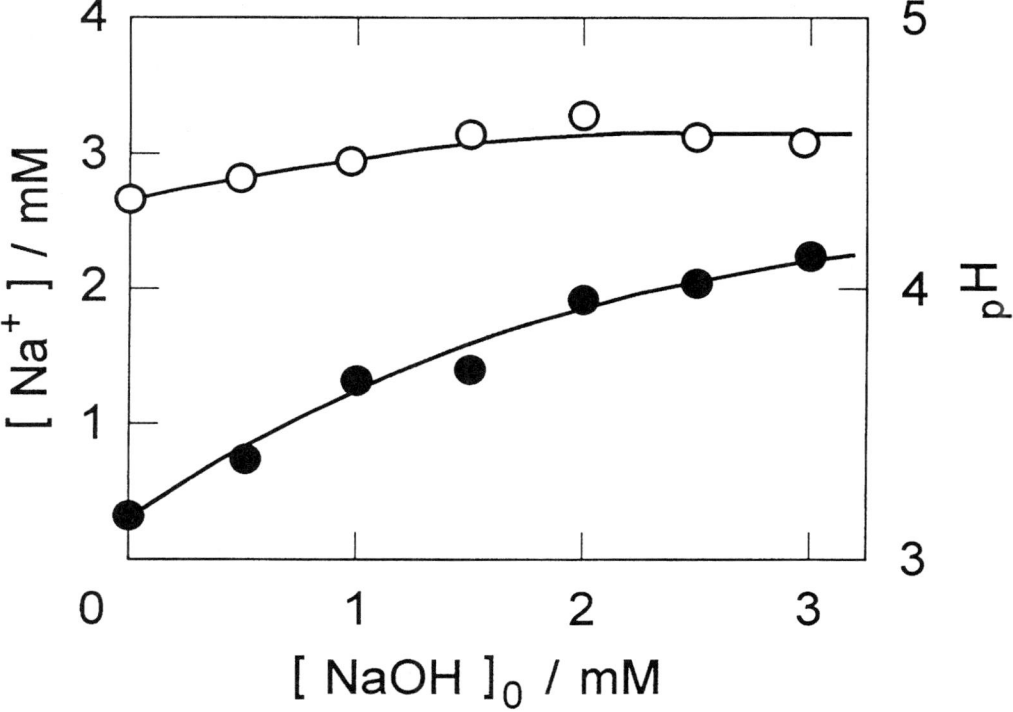

FIG. 21 Dependence of pH (●) and free Na^+ concentration (○) on the amount of NaOH added to Sephadex C-25 suspension (6 g dm^{-3}) containing 3 mM NaCl. (From Ref. 25; reproduced by permission of Academic Press, New York.)

the CM group, as well as the surface hydroxyl group on the metal oxide, does not strongly form an ion pair with a counterion, the proton association/dissociation of the CM group could be represented by the following modified mechanism of scheme (64):

$$-COOH \rightleftharpoons -COO^- \rightleftharpoons -COO^- \cdots M^+$$
$$H^+ \qquad M^+$$

(64′)

However, since the ion pairs are formed on most of the dissociated functional groups, the free dissociated CM groups, $-COO^-$, scarcely exist. Thus the protonation/deprotonation of the CM groups should be written as [25]

$$\begin{array}{c} ① \\ -COOH \end{array} \underset{k_{-1}}{\overset{k_1}{\rightleftharpoons}} \begin{array}{c} ② \\ -COO^- \end{array}$$

$Na^+ \searrow (K_{Na}') \qquad H^+ \qquad Na^+ \searrow (K_{Na})$

$$\begin{array}{c} -COO-H\cdots Na^+ \\ ③ \end{array} \underset{k_{-2}}{\overset{k_2}{\rightleftharpoons}} \begin{array}{c} -COO^-\cdots Na^+ \\ ④ \end{array}$$

(87)

where $-COO^-\cdots Na^+$ and $-COO\text{-}H\cdots Na^+$ are the ion pairs between the dissociated functional group and Na^+ and the encounter complex, respectively. Now, we assume that the processes of ion-pair formation, ① ⇌ ③ and ② ⇌ ④, are faster than those of the proton association/dissociations, ① ⇌ ② and ③ ⇌ ④. This is different from the assumption that we have made previously [25], since the counterion adsorption/desorption has so far been studied in more detail as is described in the next section [14]. The linearized rate equation of proton concentration in mechanism (87) is given by using the total concentration of the dissociated CM groups $[-COO^-]_t$, which is the sum of $[-COO^-]$ and $[-COO^-\cdots Na^+]$ [85]:

$$-\frac{d\Delta[H^+]}{dt} = \left\{\frac{1}{D}(k_{-1} + k_{-2}K_{Na}[Na^+])([H^+] + [-COO^-]_t) + \frac{1}{E}(k_1 + k_2 K'_{Na}[NA^+])\right\}\Delta[H^+] \tag{88}$$

where

$$D = 1 + K_{Na}[Na^+]$$

and

$$E = 1 + K'_{Na}[Na^+]$$

Under the condition that the concentrations of $-COO^-$ and $-COO\text{-}H\cdots Na^+$ are much smaller than those of $-COOH$ and $-COO^-\cdots Na^+$ (i.e., $K_{Na}[Na^+] = [-COO^-\cdots Na^+]/[-COO^-] \gg 1$ and $K'_{Na}[Na^+] = [-COO-H\cdots Na^+]/[-COOH] \ll 1$), Eq. (88) can be rewritten as

$$-\frac{d\Delta[H^+]}{dt} = \{k_{-2}([H^+] + [-COO^-]_t) + k_1\}\Delta[H^+]\} \tag{89}$$
$$= \frac{1}{\tau}\Delta[H^+]$$

Judging from the form of the reciprocal relaxation time in Eq. (89), the whole processes in mechanism (87) are approximated to the following proton association/dissociation process of the CM group, in which ion-pair formation is ignored:

$$-COOH \underset{k_{-1}}{\overset{k_1}{\rightleftharpoons}} -COO^- + H^+ \tag{90}$$

The relaxation time is given by the same equation as is obtained from a simple Langmuir type analysis in Theoretical Treatment, Section A.1. In this analysis, the concepts of the surface potentials and the electrical double layer model are introduced in a manner similar to the analysis of the reactions on metal oxides in the previous section. The rate and equilibrium constants are listed in Table 2 [25].

The proton association/dissociation on DEAE Sephadex is represented as [86]

$$-(CH_2)_2N^+H(C_2H_5)_2 \underset{k_{-1}}{\overset{k_1}{\rightleftharpoons}} -(CH_2)_2N(C_2H_5)_2 + H^+ \tag{91}$$

where $-(CH_2)_2N(C_2H_5)_2$ denotes the diethylaminoethyl group, which is a weak base. A single relaxation of the order of s, observed by the pressure-jump method, is attributed to reaction (91). The rate and equilibrium constants are listed in Table 2.

TABLE 2 Rate Constants and Equilibrium Constants for Proton Association/Dissociations of Functional Groups of Sephadex

Sephadex	k_1^{int} (s^{-1})	k_{-1}^{int} (mol^{-1}dm^{3}s^{-1})	$(k_1^{int})^a$ (mol dm^{-3})	$(k_1^{int})^b$ (mol dm^{-3})
CM[c]	1.7×10	3.5×10^3	4.7×10^{-3}	1.0×10^{-3}
DEAE[d]	1.8×10^{-1}	1.7×10^4	11×10^{-6}	2×10^{-6}

[a] The equilibrium constant is calculated from the k_i^{int} and k_{-1}^{int} values.
[b] This constant is determined by the titration.
[c] *Source*: Ref. 25. (Reproduced by permission of Academic Press, New York.)
[d] *Source*: Ref. 86. (Reproduced by permission of the American Chemical Society, Washington, D.C.)

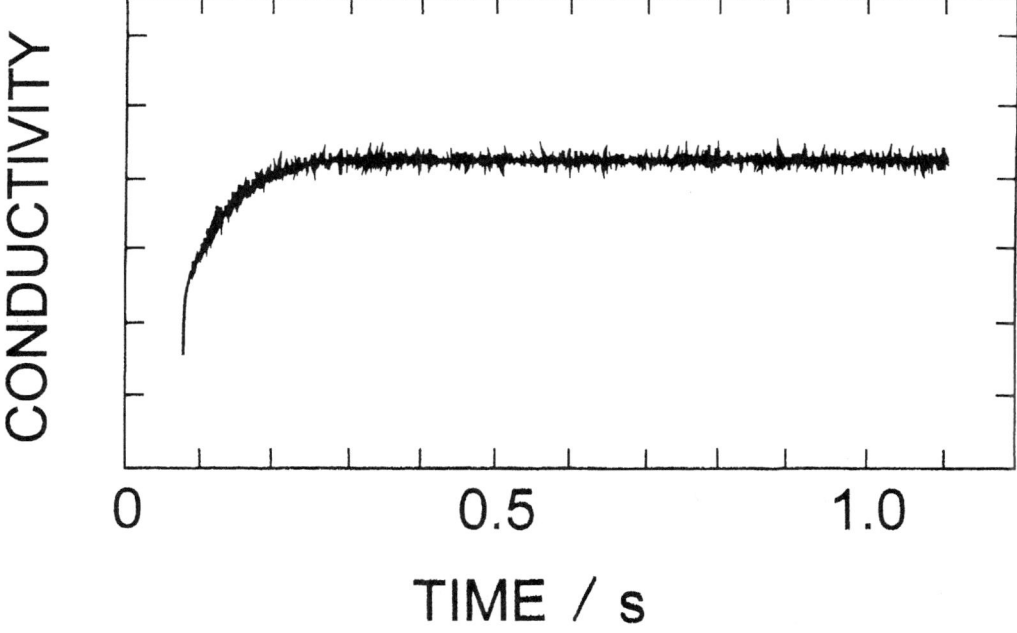

FIG. 22 Typical relaxation curve observed in the Sephadex C-25 suspension (10 g dm^{-3}) containing 20 mM NaOH and 0.7 mM NaCl at 25°C. (From Ref. 14; reproduced by permission of Academic Press, New York.)

B. Adsorption/Desorption of Alkali Metal Ions

The relaxation observed in the CM Sephadex suspension is attributed to the proton adsorption/desorption of the CM groups in the previous section. However, the relaxation disappears with increasing pH on the addition of bases, LiOH, NaOH, and KOH. Instead, a new relaxation of 0.1 s order is observed by the pressure-jump method as shown in Fig. 22 [14]. The dissociation degree of the functional group is 10–100% at the pH region where the relaxation is observed. Since most of the added alkali metal ions form ion pairs with the dissociated CM groups on the Sephadex, free cations are scarcely left in the bulk phase. Thus the kinetic measurements have been carried out at ionic strength of 1×10^{-3} M,

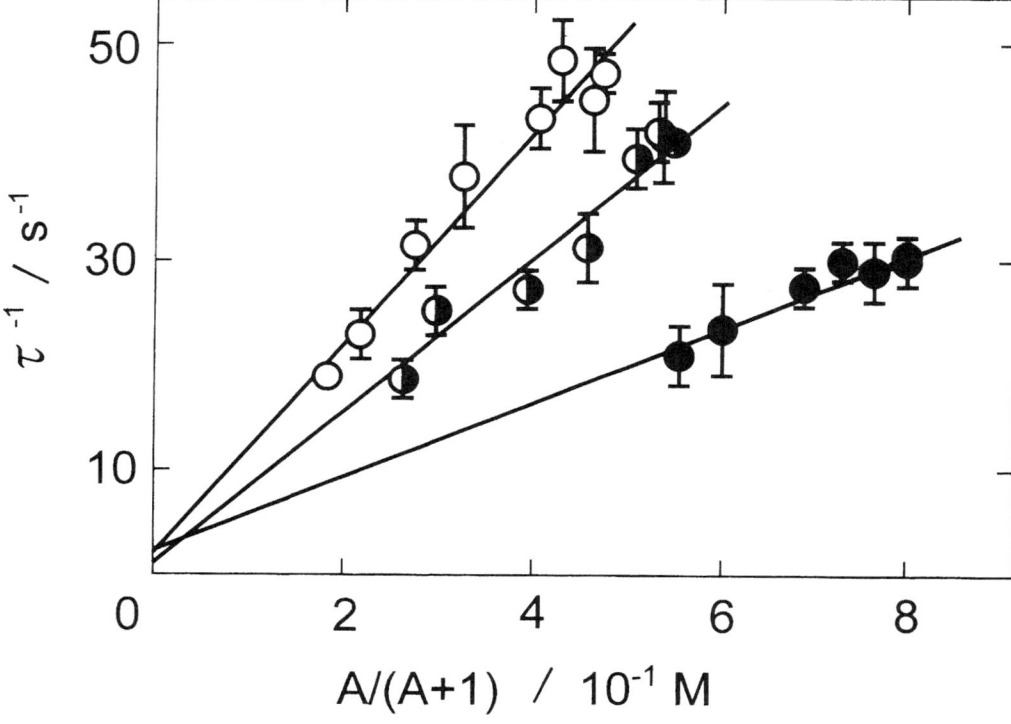

FIG. 23 Plot of τ^{-1} against the concentration term in Eq. (38): (○) Li$^+$, (◐) Na$^+$, and (●) K$^+$. (From Ref. 14; reproduced by permission of Academic Press, New York.)

which is controlled by the addition of alkali chlorides. The conductivity of the suspension in Fig. 22 increases with time after the pressure of 100 atm is suddenly dropped to atmospheric pressure. The direction of the conductivity change is opposite to that observed in the metal oxide suspensions in Section IV. The difference in the direction of the conductivity change seems to be related to the structures of the particles: metal oxides such as TiO_2 and γ-Al_2O_3 have hard structures that cannot be deformed by high pressure, while the Sephadex particle has a soft spongy structure consisting of flexible dextran chains [81]. A conformational change might be induced in the Sephadex particle with a soft and flexible structure. A linear relation in Fig. 23 can be interpreted by the conformational change model of the Sephadex particles accompanying the ion-pair formation between alkali metal ions and the dissociated CM groups, as is described in scheme (26) of Theoretical Treatment, Section II [14]. The adsorption/desorption of NH_4^+ on the CM Sephadex can be also analyzed by the same mechanism as that of alkali metal ions [87]. In the adsorption/desorption of alkali metal ions, however, only the conformational change process of Sephadex in scheme (26) can be observed as a relaxation by the pressure-jump method. The ion-pair formation process between alkali metal ions and the dissociated CM groups is probably too fast to be measured by the above method.

The rate constants can be evaluated from a linear plot of the reciprocal relaxation time against the concentration term of Eq. (38) as is explained in the previous sections. The rate constant of the conformational change, k_2^{int}, is inversely proportional to the

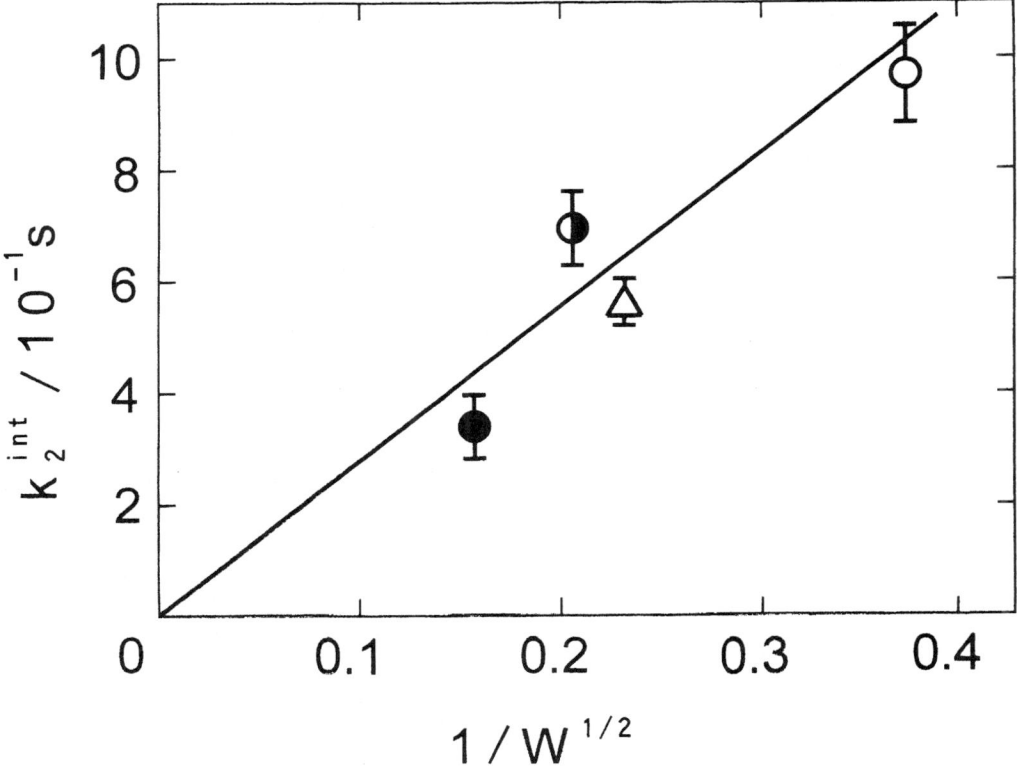

FIG. 24 Plot of k_2^{int} against $1/W^{1/2}$: (○) Li$^+$ [14], (◐) Na$^+$ [14], (●) K$^+$ [14], and (△) NH$_4^+$ [87]. (From Ref. 87; reproduced by permission of Academic Press, New York.)

square root of the atomic weight, W, of the alkali metal ion, as shown in Fig. 24 [14]:

$$k_2^{int} = \frac{C}{W^{1/2}} \tag{92}$$

where C is a constant. This has no relation to the selectivity sequence of alkali metal ions on the CM Sephadex, Li$^+$ < K$^+$ < Na$^+$, as determined by the equilibrium analysis. The Sephadex particle consists of flexible dextran chains. A movement of the dextran chains, on which alkali metal ions are adsorbed, would cause a conformational change of the particle. Here, we assume that the movement of a dextran chain is analogous to the vibration of a massless string of tension T and total length $2L$ with a single object of mass W, which is spaced at equal intervals. The frequency of vibration for the fundamental mode, v_0, can be given by the equation [88,89]

$$v_0 = \frac{1}{\pi}\left(\frac{T}{2WL}\right)^{1/2} = \frac{C'}{W^{1/2}} \tag{93}$$

In this equation, v_0 is inversely proportional to the square root of W. The conformational change of the Sephadex particle might result from such a movement of dextran chains [14].

Since an ion is assumed to be located on a string in the derivation of Eq. (93), alkali metal ions might be directly bound to the dissociated CM groups on the dextran chain [14] but not be captured uniformly in the electrical double layer inside the Sephadex particle [90].

Cl^- adsorption/desorption on DEAE Sephadex has been observed as a single relaxation by the pressure-jump method [86]. The conductivity of the suspension changes with time after the pressure jump in a manner similar to the case of the relaxation observed in the CM Sephadex suspension. The direction of the conductivity change is opposite to that observed in metal oxide suspensions. This tendency for conductivity change seems to be characteristic of reactions occurring in Sephadex suspensions.

C. Adsorption/Desorption of Amino Acids

Basic amino acids, lysine, arginine, and histidine, are strong bases, and more than 85% of these amino acids can be adsorbed on CM Sephadex. The properties of the basic amino acids are very similar to those of bases such as NaOH and KOH. The relaxations due to the conformational change of Sephadex are measured in the suspensions containing the bases of alkali metal ions as described in the previous section. In the Sephadex suspension containing lysine or arginine, a single relaxation of 0.1 s order is also observed. However, no relaxation is found in the suspension containing histidine [91]. The directions of the conductivity changes of the relaxations, observed in the suspensions containing the two amino acids, agree with those observed in the CM Sephadex suspensions containing the bases of the alkali metal ions. Thus the relaxations can be similarly analyzed by the mechanism of the conformational change of a Sephadex particle accompanying ion-pair formation, where M^+ is replaced by an amino acid cation in mechanism (26) [91].

Kinetic study of the adsorption/desorption of these amino acids will give us useful information about the conformational change of the Sephadex particle in addition to the information obtained in the case of alkali metal ions. Since the three amino acids have different dissociation constants of side chains, the occurrence of the relaxation must be related to the charged states of the amino acids. The relaxation is observed in the pH range of 4.5 to 7. In this pH range, most of the lysine or arginine in the bulk phase exists as an apparently univalent cation, Lys^+ or Arg^+, which has two protonated amino groups and one dissociated carboxyl group. When these cations are adsorbed on the negatively charged surface of Sephadex with a flexible structure, the volume of the particle decreases. Upon the pressure change of 100 to 1 atm in suspension containing the above particles, the conductivity increases with time. This phenomenon indicates that the amino acid cations are released from the particle surface as the volume of the particle changes. On the other hand, histidine exists as an apparently neutral zwitterion, His^0, or a univalent cation, His^+, and the fraction of His^0 increases from 10 to 80% with increasing pH [92]. Thus a considerable amount of His^0 is expected to form an ion pair with a dissociated CM group in the form of $-COO^-\ldots His^0$. The negative charges of the dissociated CM groups cannot be neutralized by this type of ion-pair formation. In the pH range below 4, no relaxation is observed, since most of the histidine exists as the univalent cation, while the CM group is undissociated [91].

In the pH range of 4.5–7, acidic and neutral amino acids exist mostly as negative and neutral zwitterions, respectively. These amino acids cannot compensate for the surface negative charges, and thus the volume of the Sephadex is not contracted by the adsorptions of these acidic and neutral amino acids. When salts such as halides of amino acids or alkali metal ions are added to the CM Sephadex suspension, CM groups would be mostly undissociated. This is because the pH of the suspension does not increase on the addition of these salts. In these suspensions no relaxation has been found. Therefore the relaxation arising from the conformational change of the particle can be observed only in the case that both the dissociated CM groups and the apparently univalent amino acid cations exist sufficiently in the suspension [91].

DEAE Sephadex has a positively charged surface, and so negative ions such as Cl^- form ion pairs with DEAE groups. The acidic amino acids like aspartic acid and glutamic acid are exchanged with Cl^- on the DEAE Sephadex surface. This process has been observed as a relaxation by the pressure-jump method. However, no relaxation has been found in the suspension containing basic and neutral amino acids [93]. Since the direction of the conductivity change in the relaxation observed in the DEAE Sephadex suspension is the same as that observed in the metal oxide suspension, the relaxation might not be due to a conformational change but to an ion-exchange process of the amino acid with the counterion [93].

D. Adsorption/Desorption of Charged Proteins

Ion-exchange Sephadexes have been used for the chromatography of proteins. In this process, proteins are adsorbed on the ion exchangers and are exchanged with ions such as alkali metal ions on the exchanger surfaces. At the beginning of the studies, the pressure-jump method was applied to the CM Sephadex suspension containing various proteins, but no relaxation was observed. Since both ion-exchange Sephadex and protein are macromolecules with many charges (i.e., polyelectrolytes), the interaction between these molecules would be much stronger and more complicated than those between the Sephadex and small ions such as alkali metal ions. Thus, the equilibrium of the protein adsorption/desorption cannot presumably be perturbed by the pressure change of 100 atm order. However, the adsorption/desorption process of charged protein such as lysozyme on the CM Sephadex can be followed as a function of time by a rapid mixing method with conductivity detection [15]. In this method, the protein solution is rapidly mixed with the Sephadex suspension in the conductivity cell. Prior to the mixing, the pH values of both the solution and the suspension are adjusted to be exactly equal in the thermostat. Figure 25 shows that the conductivity increases exponentially with time as the protein is adsorbed on the Sephadex. In this process, the release of protons from the Sephadex is observed in the same time scale by the pH measurement (Fig. 25). From the pH increment of the suspension, the number of released protons, n, has been determined to be 5.6 per adsorbed lysozyme molecule [15]. The lysozyme molecule has 15–18 positive amino acid residues at pH 3–4 [16]. Since the number n is smaller than that of such amino acid residues, only a part of the positive sites on the protein molecule are used for the protein adsorption. As a result, the charge reversal of the ζ-potential is caused by the protein adsorption on the Sephadex particle as shown in Fig. 26. This is a typical phenomenon in the protein adsorption on the Sephadex particle.

When several protons are released by lysozyme adsorption on CM Sephadex, the overall reaction can be represented by mechanism (46) in Theoretical Treatment, Section II. In an actual process, the n sites on the protein molecule would not be bound simultaneously to CM groups but consecutively to the groups as is given by the reaction mechanism (47). The equation of τ^{-1} can be derived as Eq. (54). A plot of τ^{-1} versus the concentration term will yield a straight line from which the rate constants, k_1^{int} and k_{-1}^{int}, can be obtained, as shown in Fig. 27 [15]. The adsorption/desorption of myoglobin on the CM Sephadex can also be analyzed by the same mechanism, and the rate constants are listed in Table 3. The adsorption rate constant for myoglobin is smaller than that for lysozyme, although the number of the positive charges of myoglobin is 30 per molecule under the experimental condition, being larger than that of lysozyme. However, lysozyme

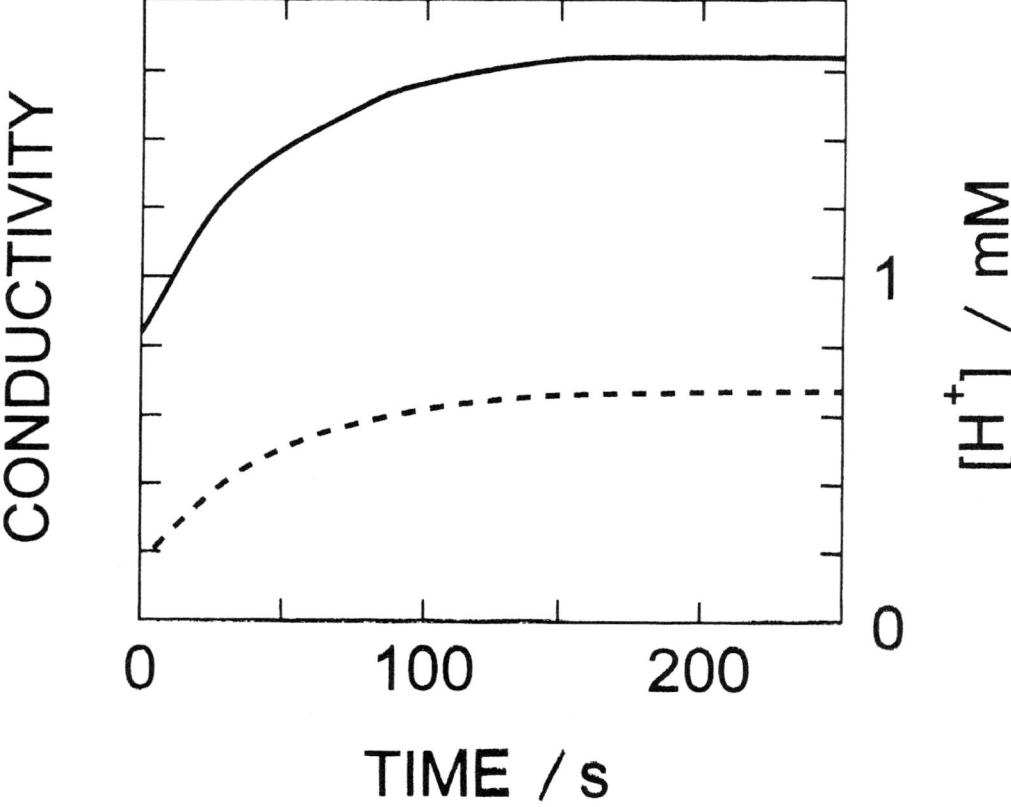

FIG. 25 Representative relaxation curve detected by change of conductivity (——) and proton concentration (- - - -) after mixing lysozyme solution with Sephadex C-25 suspension at 25°C. Final concentrations of lysozyme and Sephadex are 2×10^{-4} M and 2 g dm^{-3}, respectively. (From Ref. 15; reproduced by permission of Steinkopff, Darmstadt, Germany.)

structure becomes compact with decreasing pH, while myoglobin has a loosed structure even at pH 3. It is characteristic in protein adsorption that the rate of the reaction would be influenced by not only the number of charges but also the structure of the molecule. Therefore the model of protein adsorption/desorption might be essentially different from that of small ion adsorption/desorption as is described in the previous sections.

Ion-exchange Sephadex is constituted by dextran chains with functional groups. In a detailed analysis of protein adsorption/desorption, the complex formation of protein with the linear dextran chain would play an important role in the ion-exchange reaction. In the field of colloid titration, the complexations between linear polyelectrolytes and proteins have been widely studied by the precipitation technique [94,95] or by dynamic and electrophoretic light scattering methods [96,97]. Sodium dextran sulphate consists of a linear dextran with sulphate groups. By the electrophoretic light scattering method, a charge reversal of the electrophoretic mobility of the dextran has been observed upon the addition of lysozyme as shown in Fig. 28. The number of released Na$^+$ per lysozyme bound to the dextran chain is 4.5 [98], which is a number similar to that of released protons per lysozyme bound to the CM Sephadex [15]. There is a resemblance between the ion-exchange Sephadex and the linear ion-exchange dextran. However, no relaxation

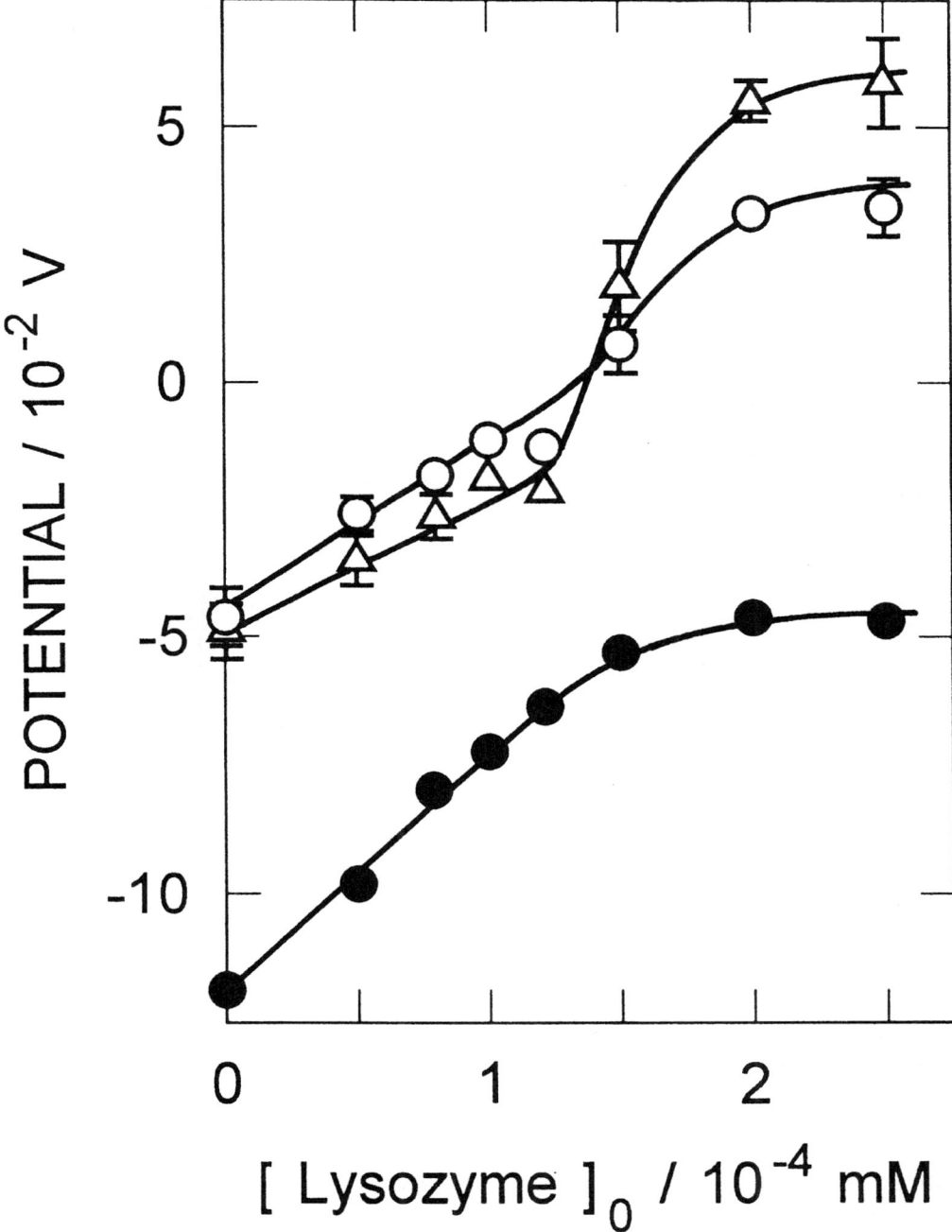

FIG. 26 Dependence of ζ-potential ($\bigcirc$), ψ_β ($\triangle$), and ψ_0 ($\bullet$) on concentration of lysozyme added to Sephadex C-25 suspension (2 g dm^{-3}). (From Ref. 15; reproduced by permission of Steinkopff, Darmstadt, Germany.)

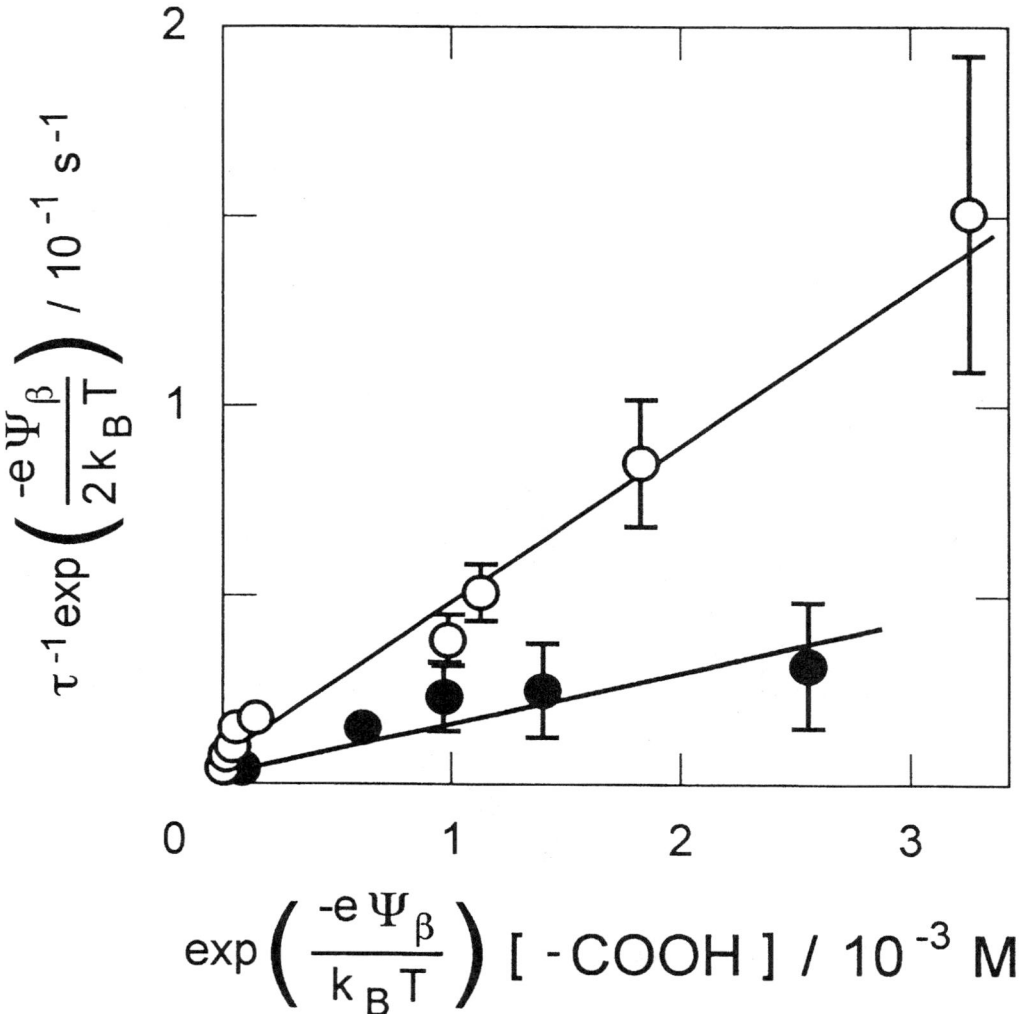

FIG. 27 Plot of the left term containing τ^{-1} in Eq. (54) against the concentration term in the same equation; (○) lysozyme and (●) myoglobin. (From Ref. 15; reproduced by permission of Steinkopff, Darmstadt, Germany.)

TABLE 3 Rate Constants and Equilibrium Constants for Protein Adsorption/Desorptions on CM Sephadex

Protein	k_1^{int} (mol^{-1} dm^{3}s^{-1})	k_{-1}^{int} (s^{-1})	$(K_1^{\text{int}})^a$ (mol^{-1} dm^3)
Lysozyme[b]	4.3×10	6.6×10^{-3}	6.4×10^3
Myoglobin	1.2×10	5.4×10^{-3}	2.2×10^3

[a] The equilibrium constant is calculated from the k_1^{int} and k_{-1}^{int} values.
[b] *Source*: Ref. 15. (Reproduced by permission of Steinkopff, Darmstadt, Germany.)

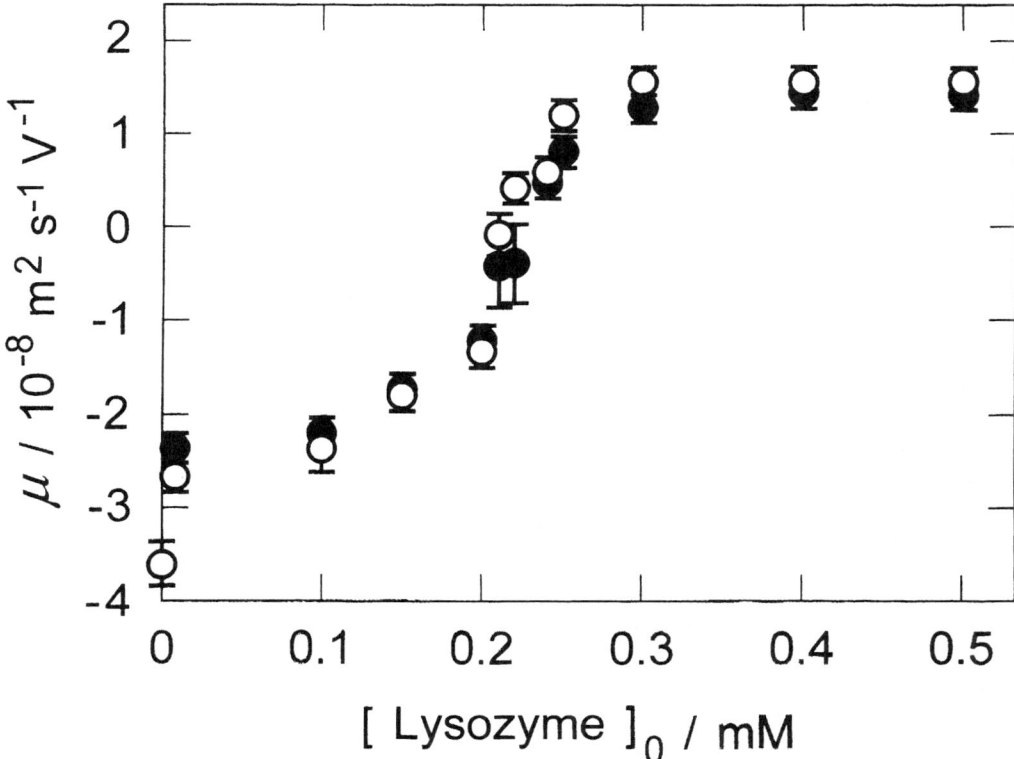

FIG. 28 Electrophoretic mobilities of sodium dextran sulfates as a function of the added lysozyme concentration at 25°C; (○) molecular weight (Mw) = 5.0×10^5, lysozyme concentration = 1.0 μM. (●) Mw = 2.5×10^4, lysozyme concentration = 20 μM. The ionic strength of each solution is kept almost constant by the addition of 10 mM NaCl. (From Ref. 98; reproduced by permission of Academic Press, New York.)

has been observed yet in the sodium dextran sulfate solution containing lysozyme. Kinetic and equilibrium studies of charged protein adsorption/desorption on ion-exchange dextrans might give us a key for analyzing a detail of the adsorption/desorption on the Sephadex surface as well.

VI. CONCLUSION AND FURTHER PROBLEMS

The kinetic studies of ion adsorption/desorptions on metal oxides and ion-exchange Sephadexes have been investigated mainly by the pressure-jump method. Many reaction mechanisms for these processes have been proposed. However, only one or two processes in the proposed multistep reactions have been observed experimentally as a single or double relaxation. The other steps are simply assumed to be very fast or very slow. For example, proton association/dissociation reactions of the surface hydroxyl groups on γ-Al_2O_3 and SiO_2 have not been measured yet (Sections IV.A and C). Ion-pair formation has only been assumed to be much faster in the adsorption/desorption of the alkali metal ions on CM Sephadex (Section V.A).

There must be faster adsorption/desorptions than those observed by the electric field-jump method, which provides the fastest time range, 0.1–20 µs, for these kinetic studies. Fast reactions in solutions can now be followed of the order of nanoseconds or picoseconds by spectrophotometric detections with laser light [99]. However, spectrophotometric methods cannot be applied to reactions in suspensions, since laser light, which is scattered by particles in a suspension, does not penetrate through a suspension. In this text, the conductometric detection is demonstrated as a convenient method of following a fast reaction in suspension. However, this detection has little usefulness in the discrimination of plural reactions occurring simultaneously in a similar time range. For example, it is difficult to distinguish the adsorption/desorption of divalent metal ions on TiO_2 and the proton association/dissociation of the surface hydroxyl group on TiO_2 (Section IV.A), both of which are expected to be measured in the time range of the pressure-jump measurement. Therefore a newly designed apparatus has been expected in order to observe and characterize many processes occurring in suspension.

Equilibrium and kinetic studies of the ion adsorption/desorptions on metal oxides have been widely investigated as introduced in this chapter. The electrical double layer model for the metal oxide system seems to be appreciably developed [64,65]. In comparison with this system, there remain several problems that have not been solved for ion adsorption/desorptions on ion-exchange Sephadexes. One of them is the bound state of counterions to ion-exchange Sephadex. Miyajima et al. suggest that counterions are distributed uniformly inside the gel phase [90,100]. In Section V.B, on the other hand, these ions are explained to be strongly bound to the functional groups on the dextran chain by the present authors. The latter model has been proposed on the basis of kinetic experiments only for the system of CM Sephadex and alkali metal ions. Thus the bound state of various counterions to ion-exchange Sephadex is expected to be clarified by further kinetic and equilibrium studies.

Ion-exchange Sephadex is a soft particle in which dextran chains are linked three-dimensionally. Ions can penetrate freely inside the Sephadex particle. Thus the structure of Sephadex is apparently different from that of metal oxide. This means that the potentials in the electrical double layer model, used for analyses of ion adsorptions on metal oxides, cannot be strictly applied to the Sephadex system. The electrophoretic mobility and potential of such a soft particle have been widely studied by Ohshima et al. [101–105]. In their model, however, the relation between the potential and the acid dissociation constant of the functional group is not defined. The bindings of the potential determining ion and the counterion with the surface functional groups have not been discriminated in a similar manner to the three-layer model for the metal oxide system (Section IV.C) [64,65]. Therefore the electrical double layer model must also be established in ion adsorption/desorption on Sephadex.

REFERENCES

1. E. E. Graham and C. F. Fook. AIChE J. *28*:245 (1982).
2. H. S. Tsou and E. E. Graham. AIChE J. *31*:1959 (1985).
3. J. A. Davis, R. O. James, and J. O. Leckie. J. Colloid Interface Sci. *63*:480 (1978).
4. C. F. Bernasconi, in *Relaxation Kinetics*, Academic Press, New York, 1976, pp. 178–243.
5. I. H. Segel, in *Enzyme Kinetics*, John Wiley, New York, 1993, pp. 18–64.
6. C. F. Bernasconi, in *Relaxation Kinetics*, Academic Press, New York, 1976, pp. 3–75.
7. J. H. Espenson, in *Chemical Kinetics and Reaction Mechanisms*, McGraw-Hill, New York, 1981, Chapters 3 and 10.

8. K. J. Laidler, in *Chemical Kinetics*, Harper and Row, New York, 1987, Chapter 2.
9. K. Hachiya, M. Sasaki, T. Ikeda, N. Mikami, and T. Yasunaga. J. Phys. Chem. *88*:27 (1984).
10. K. Hachiya, K. Takeda, and T. Yasunaga. Adsorpt. Sci. Technol. *4*:25 (1987).
11. G. R. Wiese, R. O. James, and T. W. Healy. Discuss. Faraday Soc. *52*:302 (1971).
12. S. Usui, in *Electrical Phenomena at Interfaces* (A. Kitahara and A Watanabe, eds.), Surfactant Science Series Vol. 15, Marcel Dekker, New York, 1984, pp. 30–42.
13. C. F. Bernasconi, in *Relaxation Kinetics*, Academic Press, New York, 1976, pp. 30–31.
14. K. Hachiya and K. Yamaguchi. J. Colloid Interface Sci. *162*:189 (1994).
15. K. Yamamoto, K. Hachiya, and K. Takeda. Colloid Polym. Sci. *270*:878 (1992).
16. J. J. Bergers, M. H. Vingerhoeds, L. v. Bloois, J. N. Herron, L. H. M. Janssen, M. J. E. Fischer, and D. J. A. Crommelin. Biochemistry *32*:4641 (1993).
17. E. E. Conn and P. K. Stumpf, in *Outlines of Biochemistry*, John Wiley, New York, 1976, Chapter 4.
18. C. F. Bernasconi, in *Relaxation Kinetics*, Academic Press, New York, 1976, pp. 11–12.
19. W. Knoche, in *Techniques of Chemistry* (G. G. Hammes, ed.), Vol. 6, Part 2, Wiley-Interscience, New York, 1973, pp. 187–210.
20. H. Strehlow and W. Knoche, in *Monographs in Modern Chemistry* (H. F. Ebel, ed.), Vol. 10, Verlag Chemie, New York, 1977, pp. 5–45.
21. W. Knoche and G. Wiese. Chem. Instrum. *5*:91 (1973–1974).
22. K. Hachiya, M. Ashida, M. Sasaki, H. Kan, T. Inoue, and T. Yasunaga. J. Phys. Chem. *83*:1866 (1979).
23. W. Knoche, in *Chemical and Biological Applications of Relaxation Spectrometry* (W. Jones, ed.), D. Reidel Publishing, Dordrecht, Holland, 1975, pp. 91–102.
24. T. Yasunaga, N. Tatsumoto, S. Harada, and M. Hiraishi. Rev. Sci. Instrum. *49*:1747 (1978).
25. K. Hachiya, K. Yamamoto, T. Inoue, and K. Takeda. J. Colloid Interface Sci. *150*:270 (1992).
26. K. Takeda and T. Yasunaga. J. Colloid Interface Sci. *40*:127 (1972).
27. G. Ilgenfritz, Ph.D. thesis, George August University, Goettingen, 1966.
28. L. D. Maeyer and A. Persoons, in *Techniques of Chemistry* (G. G. Hammes, ed.), Vol. 6, Part 2, Wiley-Interscience, New York, 1973, pp. 211–235.
29. Y. Tsuji, T. Yasunaga, T. Sano, and H. Ushio. J. Am. Chem. Soc. *98*:813 (1976).
30. L. Onsager. J. Chem. Phys. *2*:599 (1934).
31. M. Eigen and G. Schwartz. J. Colloid Sci. *12*:181 (1957).
32. M. Sasaki, M. Moriya, T. Yasunaga, and R. D. Astumian. J. Phys. Chem. *87*:1449 (1983).
33. M. Sasaki, T. Inoue, T. Yasunaga, and Z. A. Schelly. J. Phys. Chem. *91*:4827 (1987).
34. I. Tinoco, Jr., and K. Yamaoka. J. Chem. Phys. *63*:423 (1959).
35. A. W. Adamson, in *Physical Chemistry of Surfaces*, John Wiley, New York, 1976, pp 208–216.
36. T. Okubo, H. Kitano, T. Ishiwatari, and N. Ise. Proc. Royal Soc. Lond. Ser. A. *366*:81 (1979).
37. J. A. Sirs. Trans. Faraday Soc. *54*:201 (1958).
38. R. H. Prince. Trans. Faraday Soc. *54*:838 (1958).
39. T. Ikeda, J. Nakahara, M. Sasaki, and T. Yasunaga. J. Colloid Interface Sci. *97*:279 (1984).
40. S. Miyamoto and M. Tagawa. Colloid Polym. Sci. *263*:597 (1985).
41. S. Miyamoto. J. Chem. Soc. Jpn. *167* (1998).
42. J. H. Espenson, in *Chemical Kinetics and Reaction Mechanisms*, McGraw-Hill, New York, 1981, Chapter 6.
43. T. Yasunaga and T. Ikeda, in *Geochemical Processes at Mineral Surface* (J. A. Davis and K. F. Hayes, ed.), ACS Symposium Series, Vol. 323, 1986, Chapter 12.
44. M. Ashida, M. Sasaki, H. Kan, T. Yasunaga, K. Hachiya, and T. Inoue. J. Colloid Interface Sci. *67*:19 (1978).
45. M. Ashida, M. Sasaki, K. Hachiya, and T. Yasunaga. J. Colloid Interface Sci. *74*:572 (1980).
46. R. D. Astumian, M. Sasaki, T. Yasunaga, and Z. A. Schelly. J. Phys. Chem. *85*:3832 (1981).
47. T. Ikeda, M. Sasaki, K. Hachiya, R. D. Astumian, T. Yasunaga, and Z. A. Schelly. J. Phys. Chem. *86*:3861 (1982).

48. M. Sasaki, N. Mikami, T. Ikeda, K. Hachiya, and T. Yasunaga. J. Phys. Chem. 86:5230 (1982).
49. R. D. Astumian and Z. A. Schelly. J. Am. Chem. Soc. 106:304 (1984).
50. N. Mikami, M. Sasaki, K. Hachiya, R. D. Astumian, T. Ikeda, and T. Yasunaga. J. Phys. Chem. 87:1454 (1983).
51. N. Mikami, M. Sasaki, T. Kikuchi, and T. Yasunaga. J. Phys. Chem. 87:5245 (1983).
52. K. Hachiya, M. Ashida, M. Sasaki, M. Karasuda, and T. Yasunaga. J. Phys. Chem. 84:2292 (1980).
53. P. C. Zhang and D. L. Sparks. Soil Soc. Am. J. 53:1028 (1989).
54. P. C. Zhang and D. L. Sparks. Soil Sci. Soc. Am. J. 54:1266 (1990).
55. P. C. Zhang and D. L. Sparks. Environ. Sci. Technol. 24:1848 (1990).
56. P. R. Grossl, M. Eick, D. L. Sparks, S. Goldberg, and C. C. Ainsworth. Environ. Sci. Technol. 31:321 (1997).
57. K. Hachiya, M. Sasaki, Y. Saruta, N. Mikami, and T. Yasunaga. J. Phys. Chem. 88:23 (1984).
58. J. A. Davis and J. O. Leckie. J. Colloid Interface Sci. 67:90 (1978).
59. H. Hohl and W. Stumm. J. Colloid Interface Sci. 55:281 (1976).
60. R. O. James and T. W. Healy. J. Colloid Interface Sci. 40:53 (1972).
61. R. O. James, P. T. Stiglich, and T. W. Healy. Faraday Discuss. Chem. Soc. 59:142 (1975).
62. M. B. McBride. Soil Sci. Soc. Am. J. 42:27 (1978).
63. H. Diebler, M. Eigen, G. Ilgenfritz, G. Maass, and R. Winkler. Pure Appl. Chem. 20:93 (1963).
64. K. F. Hayes and J. O. Leckie, J. Colloid Interface Sci. 115:564 (1987).
65. K. F. Hayes, G. Redden, W. Ela, and J. O. Leckie. J. Colloid Interface Sci. 142:448 (1991).
66. N. Mikami, M. Sasaki, K. Hachiya, and T. Yasunaga. J. Phys. Chem. 87:5478 (1983).
67. K.-S. Chang, C.-F. Lin, D.-Y. Lee, S.-L. Lo, and T. Yasunaga. J. Colloid Interface Sci. 165:169 (1994).
68. C.-F. Lin, K.-S. Chang, C.-W. Tsay, D.-Y. Lee, S.-L. Lo, and T. Yasunaga. J. Colloid Interface Sci. 188:201 (1997).
69. T. Ikeda and T. Yasunaga. J. Phys. Chem. 88:1253 (1984).
70. N. Mikami, M. Sasaki, T. Yasunaga, and K. F. Hayes. J. Phys. Chem. 88:3229 (1984).
71. N. Mikami, M. Sasaki, N. Kawamura, K. F. Hayes, and T. Yasunaga. J. Phys. Chem. 90:2757 (1986).
72. T. Ikeda, M. Sasaki, and T. Yasunaga. J. Colloid Interface Sci. 98:192 (1984).
73. T. Ikeda and T. Yasunaga. J. Colloid Interface Sci. 99:183 (1984).
74. T. Ikeda, M. Sasaki, and T. Yasunaga. J. Phys. Chem. 87:745 (1983).
75. T. Ikeda, J. Nakahara, M. Sasaki, and T. Yasunaga. J. Colloid Interface Sci. 97:278 (1984).
76. T. Ikeda, M. Sasaki, and T. Yasunaga. Faraday Discuss. Chem. Soc. 77:223 (1984).
77. T. Ikeda, M. Sasaki, and T. Yasunaga. J. Am. Chem. Soc. 106:5772 (1984).
78. H. Negishi, M. Sasaki, T. Yasunaga, and M. Inoue. J. Phys. Chem. 88:1455 (1984).
79. M. Sasaki, H. Negishi, M. Inoue, and T. Yasunaga. J. Phys. Chem. 88:3082 (1984).
80. M. Sasaki, H. Negishi, H. Ohuchi, M. Inoue, and T. Yasunaga. J. Phys. Chem. 89:1970 (1984).
81. G. M. Flodin and B. G. Ingelman, U.S. Patent 3,042,667 (1959).
82. G. M. Flodin, U.S. Patent 3,208,994 (1962).
83. J. Novotny, FEBS Lett. 14:7 (1971).
84. H. Oe, E. Doi, and M. Hirose. J. Biochem. 103:1066 (1988).
85. C. F. Bernasconi, in *Relaxation Kinetics*, Academic Press, New York, 1976, pp. 63–67.
86. K. Hachiya, M. Sasaki, Y. Nabeshima, N. Mikami, and T. Yasunaga. J. Phys. Chem. 88:1623 (1984).
87. K. Hachiya, K. Yamaguchi and K. Takeda. J. Colloid Interface Sci. 178:374 (1996).
88. J. Orear, in *Physics*, Macmillan, New York, 1979, pp. 227–228.
89. A. P. French, in *Vibrations and Waves*, Norton, New York, 1971, pp. 136–141.
90. J. A. Marinsky, T. Miyajima, E. Högfeldt, and M. Muhammed. React. Polym. 11:279 (1989).
91. K. Hachiya, K. Yamaguchi, and K. Takeda. J. Colloid Interface Sci. 170:249 (1995).

92. A. L. Lehninger, in *Principles of Biochemistry*, Worth Publishers, New York, 1982.
93. M. Sasaki, K. Hachiya, T. Ikeda, and T. Yasunaga. J. Phys. Chem. *88*:1627 (1984).
94. E. Kokufuta, H. Shimizu, and I. Nakamura. Macromolecules *15*:1618 (1982).
95. J. M. Park, B. B. Muhoberac, P. L. Dubin, and J. Xia. Macromolecules *25*:290 (1992).
96. J. Xia, P. L. Dubin, and E. Kokufuta. Macromolecules *26*:6688 (1993).
97. J. Xia, P. L. Dubin, Y. Kim, B. B. Muhoberac, and V. J. Klimkowski, J. Phys. Chem. *97*:4528 (1993).
98. K. Yamaguchi, K. Hachiya, and K. Takeda. J. Colloid Interface Sci. *179*:249 (1996).
99. J. F. Riordan and B. L. Vallee (eds.), in *Metallobiochemistry*, Methods in Enzymology Vol. 226, Academic Press, New York, 1993, pp. 119–198.
100. T. Miyajima, in *Ion Exchange and Solvent Extraction* (J. A. Marinsky and Y. Marcus, eds.), A Series of Advances Vol. 12, Marcel Dekker, New York, 1995, Chapter 7.
101. H. Ohshima and T. Kondou. J. Colloid Interface Sci. *130*:281 (1989).
102. H. Ohshima and T. Kondou. Biophys. Chem. *39*:191 (1991).
103. H. Ohshima and T. Kondou. Colloid Polym. Sci. *271*:1191 (1993).
104. H. Ohshima. J. Colloid Interface Sci. *163*:474 (1994).
105. E. Makino, K. Suzuki, Y. Sakurai, T. Okano, and H. Ohshima. J. Colloid Interface Sci. *174*:400 (1995).

10
Quantitative Analysis of Protein Adsorption Kinetics

VLADIMIR HLADY, C.-H. HO, and D. W. BRITT Department of Bioengineering, University of Utah, Salt Lake City, Utah

I.	Protein–Surface Interactions	405
II.	Methods to Study Protein Adsorption	406
III.	Protein Adsorption Kinetics	407
	A. Quantitative analysis of protein adsorption	407
	B. Computation of the flux of protein molecules to the adsorbent surface	407
	C. Modeling of the kinetics of lipoprotein adsorption	410
IV.	Conclusions	415
	List of Symbols	417
	References	417

I. PROTEIN–SURFACE INTERACTIONS

Protein adsorption from aqueous solutions is determined by the properties at two interfaces, one between the protein and the aqueous solution and the other between the adsorbent surface and the solution, and how they "match" each other. The protein–solution interface is defined by a subtle interplay between polar and nonpolar interactions [1]. These two types of interactions govern protein stability and will also play a decisive role in protein interactions at the adsorbent surface. Other factors include the adsorbent's surface energetics, charge, surface rugosity, and the structure of water at both interfaces, i.e., their respective hydrophilicity and interfacial hydration layers [2,3].

Most surfaces acquire charge when exposed to ionic solution [4]. Because of the long-range nature of the electrostatic interactions, a protein may be guided into a unique orientation when approaching an oppositely charged surface. Although a charged protein is expected to prefer adsorption onto an oppositely charged surface, the osmotic pressure of counterions, desolvation of charged groups, and burying of charges into a nonpolar environment may counteract electrostatic attraction. When considered alone, electrostatic interactions may not fully account for the adsorption of protein to a charged interface. In the case of like charges between the surface and a protein, an energy barrier to adsorption may develop affecting the adsorption rate. If the overall charge of a protein is zero, i.e., the protein adsorbs at its isoelectric point, the electrostatic interaction between the protein and the surface will still contribute, because the distribution of charges on the protein surface is, as a rule, not uniform.

Many proteins, especially small globular proteins such as lysozyme, α-lactalbumin, and others, have some of their hydrophobic residues exposed and in contact with aqueous solution [5]. This contact is energetically costly; the enthalpy of these interactions is generally small and negative, but water molecules in the vicinity of protein hydrophobic residues lose some of their entropy relative to the bulk water. Despite these exposed hydrophobic residues a protein molecule will still dissolve in water if it has enough polar interactions to overcome these entropic losses. The adsorbent surface may also have nonpolar, hydrophobic character. The free energy of the system can further decrease when a hydrophobic protein residue adsorbs onto the hydrophobic surface. Dehydration of a hydrophobic domain of protein and a hydrophobic surface releases bound water, thereby resulting in a large positive entropy change and favorable binding [6]. Protein adsorption is further complicated since hydrophobic, polar, and electrostatic interactions may act simultaneously in a nonadditive fashion. In laboratory experiments this complexity can be simplified by using model adsorbents with controlled surface chemical and physical properties.

II. METHODS TO STUDY PROTEIN ADSORPTION

In order to characterize and predict protein adsorption, one seeks information about adsorption isotherms, adsorption kinetics (preferably measured in situ), conformation of adsorbed proteins, number and character of surface-bound protein segments, and the physical parameters describing the adsorbed protein layer [3,7]. When combined, such information can answer questions about the mechanism of protein adsorption and desorption from surfaces. Many different techniques can be applied to study protein adsorption at interfaces [8]. The most powerful techniques include optical and spectroscopic methods, because they can provide insight into both protein conformation and dynamics at interfaces [9]. In this chapter we will show an example of a protein adsorption study that utilizes total internal reflection fluorescence spectroscopy (TIRF), a method that offers great versatility and sensitivity [10]. TIRF allows selective excitation of fluorescent protein molecules in close (approx. 10^{-7} m) proximity to the adsorbent surface and can be used to measure the interfacial concentration of protein fluorophores in situ and in real time. Spatially resolved TIRF is a spectroscopic imaging variant of the same technique capable of measuring protein adsorption and desorption processes along one surface dimension [11]. In the subsequent sections of this chapter, we will show how the analysis of an experimental TIRF low density lipoprotein (LDL) adsorption onto the one-dimensional surface density gradient of octadecyldimethylsilyl chains on fused silica (C18-silica gradient) can be used to characterize protein–surface interactions.

The reason to focus on the adsorption kinetics rather than on the adsorption isotherm is largely pragmatic. In order to have rigorous thermodynamic meaning, an adsorption isotherm, a function which relates the surface concentration of protein, Γ_p, to the solution concentration of protein, c_p, has to be measured at equilibrium. Typically, Γ_p increases sharply at low solution concentration of protein, tapers off at higher protein concentrations, and often approaches a limiting value, indicating that the adsorbing surface is saturated with protein molecules. The molecular shape of a protein molecule and the heterogeneity of its surface bring a number of energetically different configurations of adsorbed molecules. For example, a nonspherical protein may display discontinuities in the adsorption isotherms, which may indicate different concentration-dependent orientations of adsorbed proteins at the interface [12]. Other processes, like conformational change or lateral rearrangements in the adsorbed layer, may run concurrently with the adsorption process. Most experimentally measured protein adsorption isotherms may not be the true thermodynamic isotherms.

III. PROTEIN ADSORPTION KINETICS

A. Quantitative Analysis of Protein Adsorption

It is generally accepted that the process of protein adsorption has the following steps:
1. Transport towards the interface
2. Attachment at the interface
3. Eventual structural rearrangements in the adsorbed state
4. Detachment from the interface
5. Transport away from the interface

In principle, each of these steps can determine the overall rate of the adsorption process. Quantitative analysis of the protein adsorption kinetics requires that the protein surface concentration be known as a function of time. In the case of C18-silica gradient surfaces, it is possible to design the experiment so that the adsorption to one region of the gradient surface is initially rate-limited by the process of transport to the interface, while in the other region of the gradient surface the rate-limiting process may be different. Since the experimentally measured protein adsorption kinetics cannot proceed faster than the rate of the supply of protein molecules to the adsorbing surface, we have often utilized the mass-transport regime as a way to calibrate the TIRF-measured fluorescence intensity in terms of protein surface concentration [13,14].

B. Computation of the Flux of Protein Molecules to the Adsorbent Surface

A mass transport of protein molecules in a solution that flows in a rectangular flow channel is fully described elsewhere [15–17]. We start the analysis from an analytical solution to the equivalent heat-transfer model derived by Leveque [18]. The analysis assumes that the walls of the flow channel act as sinks for all arriving molecules so that the flux of protein molecules to the surface, J_p (kg m^{-2} s^{-1}), is*

$$J_p = \left[\Gamma\left(\frac{4}{3}\right)\right]^{-1} 9^{-\frac{1}{3}} \left(\frac{6q}{b^2 w l D_p}\right)^{\frac{1}{3}} D_p c_p \tag{1}$$

where D_p (m^2 s^{-1}) is the diffusion coefficient of the protein, c_p (kg m^{-3}) is protein solution mass concentration, $\Gamma(4/3)$ is the gamma function of $4/3$, l (m) is the distance along the rectangular flow channel measured from its entrance, b (m) is the thickness of the flow channel, q (m^3 s^{-1}) is the volume flow rate of protein solution, and w (m) is the

---

*Throughout this chapter we will use the mass concentration SI units instead of the molar (or molecular) units. Accordingly, Γ_p and c_p will have the units of mass area^{-1} and mass volume^{-1}, respectively. The conversion of mass concentration units into molar (or molecular) units requires that the molecular mass of protein, M_p, be known, which may not always be the case.

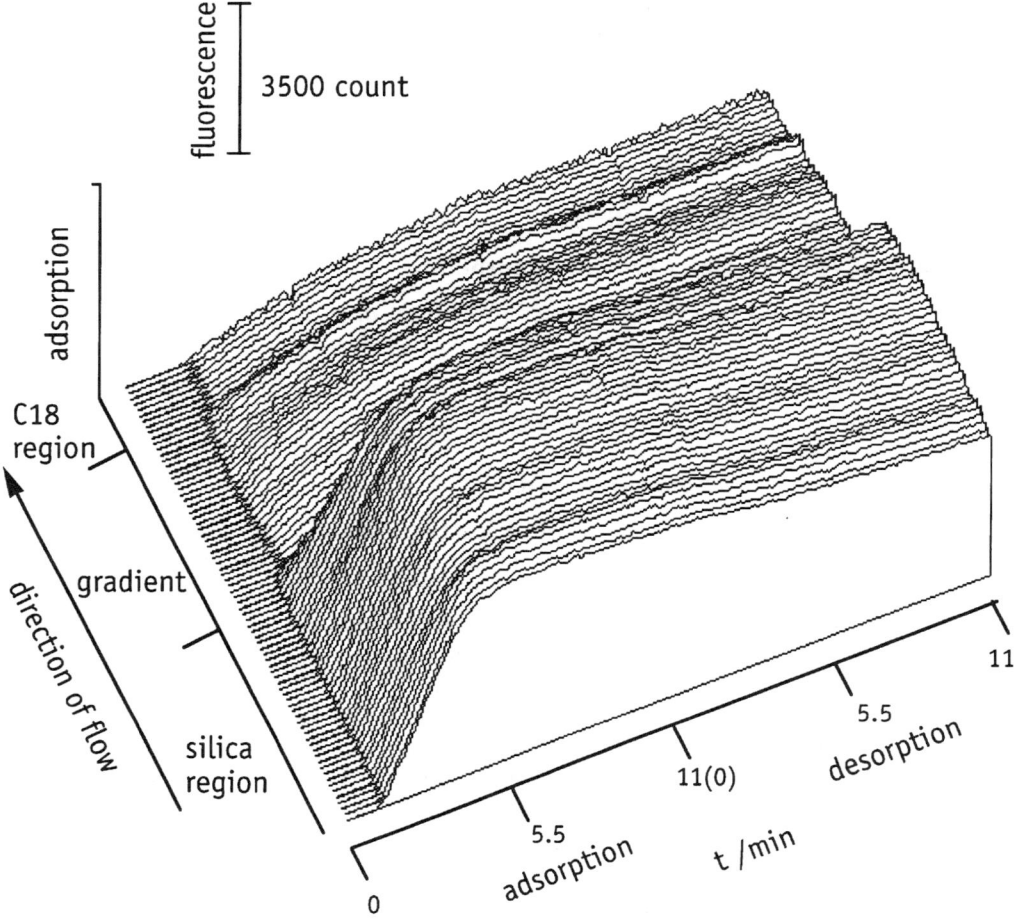

FIG. 1a Composite 3-D plot of FITC-LDL adsorption kinetics measured along the C18-silica gradient surface using spatially resolved TIRF. The TIRF experiment consisted of two 11 min segments: the adsorption segment, during which the FITC-LDL solution, $c_{LDL} = 4 \times 10^{-2}$ kg m^{-3}, was flowed through the TIRF cell, and the desorption segment, during which the flow was switched to the buffer solution.

width of the flow channel. In the case of pure transport-limited adsorption and no desorption, J_p is equal to the rate of protein adsorption to the surface, $d\Gamma_p(t)/dt$ (kg m^{-2} s^{-1}).

The example used below is the adsorption kinetics of fluorescein-labeled low-density lipoprotein (FITC-LDL) onto the C18-silica gradient surface. The FITC-LDL adsorption was measured using the spatially resolved total internal reflection fluorescence (TIRF) technique [19]. Figure 1a shows the result of the FITC-LDL TIRF adsorption experiment: fluorescence intensity from the adsorbed FITC-LDL molecules is measured every second along the C18-silica gradient surface and combined in a composite fluorescence intensity vs. time vs. surface gradient position plot. The TIRF-measured protein adsorption was quantified using an independent experiment in which the adsorption of ^{125}I-labeled LDL was carried out under experimental conditions identical to the TIRF experiments,

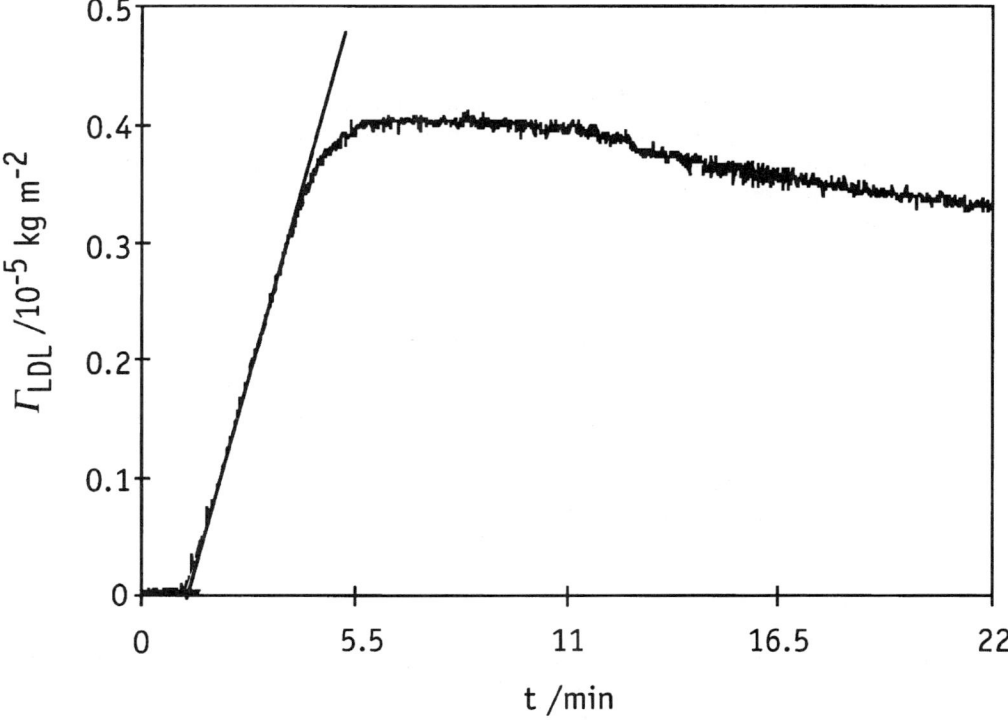

FIG. 1b Calibrated FITC-LDL adsorption and desorption kinetics on the hydrophilic region of the C18-silica gradient surface. The "transport-limited adsorption" slope, shown as a straight line, was computed from J_p (Eq. 1) using $q = 1.4 \times 10^{-8}$ m^3 s^{-1}, $b = 5 \times 10^{-4}$ m, $w = 5 \times 10^{-3}$ m, $l = 2.8 \times 10^{-2}$ m, $D_{LDL} = 1.8 \times 10^{-11}$ m^2 s^{-1}, and $c_{LDL} = 4 \times 10^{-2}$ kg m^{-3}.

and the surface mass concentration of LDL along the C18-silica gradient was determined at the end of the adsorption/desorption cycle using quantitative autoradiography [20]. The TIRF-measured FITC-LDL fluorescence intensity was then converted into the LDL surface mass concentration using the autoradiography results. The calibrated FITC-LDL adsorption and desorption kinetics onto the hydrophilic silica end of the gradient surface are shown in Fig. 1b.

In the case of transport-limited protein adsorption and very slow or no desorption, the Leveque equation (Eq. 1) predicts that the protein surface concentration increases linearly with time. The rate of transport-limited adsorption is proportional to $D_p^{2/3}$, c_p, and the wall shear rate, $\gamma^{1/3}$. The wall shear rate is defined as

$$\gamma = \frac{6q}{b^2 w} \qquad (2)$$

Since the desorption rate of FITC-LDL was found to be very slow [19], the FITC-LDL surface mass concentration, Γ_{LDL}, could have been calibrated directly from the initial linear slope of TIRF-measured fluorescence increase with time, i.e., by equating initial $d(fluorescence)/dt$ to J_p (Eq. 1). This so-called "transport-limited adsorption" slope, computed using $q = 1.4 \times 10^{-8}$ m³ s⁻¹, $b = 5 \times 10^{-4}$ m, $w = 5 \times 10^{-3}$ m, $l = 2.8 \times 10^{-2}$ m, $D_{LDL} = 1.8 \times 10^{-11}$ m² s⁻¹ [21], and $c_{LDL} = 4 \times 10^{-2}$ kg m⁻³ and shown as a straight line in Fig. 1b, was found to be in good agreement with the TIRF results calibrated through autoradiography [19]. The flux of FITC-LDL to the other regions of the C18-silica gradient surface differed only by the factor $l^{-1/3}$. Note that in both TIRF quantification schemes it was also assumed that the quantum yield of adsorbed FITC-LDL fluorescence was constant during the process of adsorption. In the subsequent analysis, the quantum yield of fluorescence of adsorbed FITC-LDL was also assumed to be identical along the C18-silica gradient surface.

C. Modeling of the Kinetics of Lipoprotein Adsorption

A simple protein adsorption model, to which the experimental FITC-LDL adsorption data were to be fitted, comprised only two processes: adsorption and desorption. The model incorporates the rate of diffusion through a concentration boundary layer (often called the "unstirred" layer) of constant thickness by solving the Fick's law. The order of magnitude thickness of this layer, δ, is:

$$\gamma = O\left(\frac{D_p l}{\gamma}\right)^{\frac{1}{3}} \tag{3}$$

It has been shown elsewhere that a linear protein concentration profile across the "unstirred" layer develops rather rapidly [22]. The model, which closely follows the arguments given in Ref. 22, starts by balancing the rates of protein adsorption and desorption to find the overall adsorption rate, $d\Gamma_{LDL}(t)/dt$.

$$\frac{d\Gamma_{LDL}(t)}{dt} = k_{on}[\Gamma_{LDL}^{max} - \Gamma_{LDL}(t)]c_{LDL}(0, t) - k_{off}\Gamma_{LDL}(t) \tag{4}$$

where k_{on} (in units of concentration⁻¹ time⁻¹) and k_{off} (in units of time⁻¹) are the intrinsic adsorption and desorption rate constants, respectively, Γ_{LDL}^{max} is the maximum lipoprotein surface mass concentration, $\Gamma_{LDL}(t)$ is the lipoprotein surface mass concentration at time t, and $c_{LDL}(0, t)$ is a function that describes the LDL concentration at distance $y = 0$ from the adsorbent surface at time t, where y is measured normal to the surface. Equation 4 is solved with the following boundary conditions [22]:

(a) the rate of adsorption is linearly dependent to the concentration gradient that exists right next to the surface, $\partial c_{LDL}(0, t)/\partial y$:

$$\frac{d\Gamma_{LDL}(t)}{dt} = D_{LDL}\frac{\partial}{\partial y}c_{LDL}(0, t) \tag{5}$$

(b) lipoprotein concentration at distances δ or greater from the surface is equal to the lipoprotein bulk solution concentration:

$$c_{LDL}(\gamma, t) = c_{LDL} \tag{6}$$

The adsorption model is based on the assumption that a linear protein concentration gradient in the "unstirred" layer exists at all times [22]. In such a case the boundary condition (a) (Eq. 5) can be recast in the form

$$\frac{d\Gamma_{LDL}(t)}{dt} = D_{LDL}\frac{\partial}{\partial y}c_{LDL}(0, t) = D_{LDL}\left[\frac{c_{LDL} - c_{LDL}(0, t)}{\delta}\right] \tag{7}$$

or

$$c_{LDL}(0, t) = c_{LDL} - \frac{\delta}{D_{LDL}} \frac{d\Gamma_{LDL}(t)}{dt} \quad (8)$$

Substituting Eq. 8 into Eq. 4, the same can be rearranged as

$$\frac{d\Gamma_{LDL}(t)}{ft} = k_{on}^{app}[\Gamma_{LDL}^{max} - \Gamma_{LDL}(t)]c_{LDL} - k_{off}^{app}\Gamma_{LDL}(t) \quad (9)$$

where k_{on}^{app} and k_{off}^{app} are the apparent adsorption rate constant and desorption rate constant, defined as

$$k_{on}^{app} = \frac{k_{on}D_{LDL}}{D_{LDL} + \delta k_{on}[\Gamma_{LDL}^{max} - \Gamma_{LDL}(t)]} \quad (10)$$

$$k_{off}^{app} = \frac{k_{off}D_{LDL}}{D_{LDL} + \delta k_{on}[\Gamma_{LDL}^{max} - \Gamma_{LDL}(t)]} \quad (11)$$

The adsorption model allows the intrinsic adsorption and desorption rate constants, k_{on} and k_{off}, to be defined as empirical exponential functions of protein surface concentration, Γ:

$$k_{on} = k_1 e^{-\alpha\Gamma} \quad (12)$$

$$k_{off} = k_{-1} e^{\beta\Gamma} \quad (13)$$

where k_1 is the initial intrinsic adsorption rate constant, k_{-1} is the initial intrinsic desorption rate constant, and α and β are the adsorption and desorption "cooperativity" constants [22].

A simple computer program was used to solve the differential Eq. 9 numerically to simulate the adsorption kinetics of protein using a set of the known parameters: Γ_{LDL}^{max}, D_{LDL}, δ, k_1, k_{-1}, α, β, and c_{LDL} [19,23].

1. Finding the Desorption Rate Constant

The first step in fitting the experimental LDL adsorption results is to find k_{-1} from the desorption segment of the TIRF experiments ($t > 11$ minutes, Fig. 1). Since the protein desorption experiments were carried out with a buffer solution ($c_{LDL} = 0$), Eq. 9 can be written as

$$\frac{d\Gamma_{LDL}(t)}{dt} = -k_{off}^{app}\Gamma_{LDL}(t) \quad (14)$$

At reasonably high protein surface coverage there will be no free binding sites available for readsorption of desorbed molecules, i.e., $[\Gamma_{LDL}^{max} - \Gamma_{LDL}(t)] \to 0$, so that

$$k_{off}^{app} \approx k_{off} \quad (15)$$

In this case, the desorption segment of the LDL adsorption experiment can be modeled as

$$\frac{d\Gamma_{\text{LDL}}(t)}{dt} = -k_{\text{off}}\Gamma_{\text{LDL}}(t) \qquad (16)$$

or in an integral form as

$$\Gamma_{\text{LDL}}(t) = \Gamma_{\text{LDL}}(t_{\text{d-beg}})e^{-tk_{\text{off}}} \qquad (17)$$

where $\Gamma_{\text{LDL}}(t_{\text{d-beg}})$ and $\Gamma_{\text{LDL}}(t)$ are the LDL surface mass concentrations at the beginning of the desorption segment ($t_{\text{d-beg}}$) and at the time t, respectively. The fitting of the experimental FITC-LDL desorption results was carried out for the two ends of the C18-silica gradient surface, i.e., for the hydrophilic silica where the fraction of the surface covered with C18 chains, θ_{C18}, was zero, and on the hydrophobic end-silica surface where the average fraction of the surface covered with C18 chains,* θ_{C18}, was 0.9.

2. Modeling of the Adsorption Process

Once the desorption rate has been found, one proceeds with the modeling of the adsorption segment of the experiment. First, the approximate thickness of the "unstirred" layer, δ, is calculated using Eq. 3. The magnitude of δ depends on the actual flow cell dimensions and on the flow rate used in the experiments. In the FITC-LDL adsorption experiments, δ amounted to 3.1×10^{-5} m at the hydrophilic region and 3.6×10^{-5} m at the hydrophobic region of the C18-silica gradient surface, respectively. The difference between the two was due to the distance between these two regions in the TIRF flow cell. The other parameter needed is $\Gamma_{\text{LDL}}^{\max}$. In principle the fitting program can find its "best" $\Gamma_{\text{LDL}}^{\max}$ value. Alternatively, one can measure $\Gamma_{\text{LDL}}^{\max}$ directly. The potential problem with any measured $\Gamma_{\text{p}}^{\max}$ parameter is related to the degree of uncertainty about the conformational state of adsorbed protein molecules. Namely, if the adsorbed protein undergoes a conformational change, the measured $\Gamma_{\text{p}}^{\max}$ may represent the adsorption saturation of denatured protein whose adsorption kinetics may not be measured by the experiment. In the present example the $\Gamma_{\text{LDL}}^{\max}$ have been set to 4×10^{-6} and 3×10^{-6} kg LDL m^{-2} for the hydrophilic and hydrophobic regions of the C18-silica gradient surface, respectively. These two $\Gamma_{\text{LDL}}^{\max}$ values agreed very well with the previous study of LDL adsorption onto the hydrophobic and hydrophilic silica [21]. The fitting of the experimental data was carried out with k_1 and α as the only two adjustable parameters (recall that the parameters k_{off}, k_{-1}, and β were found from the desorption segment of the experiment as described in Section III.C.1). A comparison between the modeled adsorption (solid line) and the experimental results (symbols) for the two regions of the C18-silica gradient surface is shown in Fig. 2. The parameters used to achieve the two fits are listed in Table 1. Note that the modeling accurately reproduced the linear increase of FITC-LDL adsorption onto the hydrophilic silica region due to the transport-limited adsorption rate.

---

* The surface coverage of C18 chains was calculated from the advancing water contact angles, θ_{adv}, using the Cassie equation [13,19]. The C18-region of the gradient surface displayed $\theta_{\text{adv}} = 104°$.

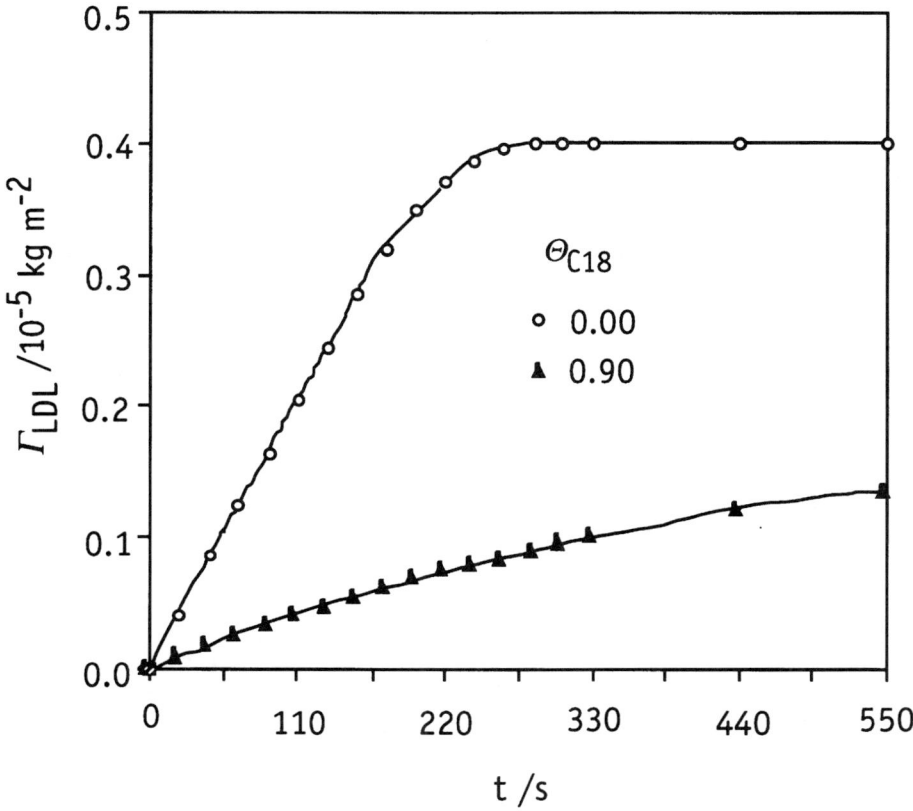

FIG. 2 Comparison between the modeled adsorption (solid lines) and the experimental results (symbols) of FITC-LDL adsorption onto the hydrophilic and hydrophobic regions of the C18-silica gradient surface.

3. Interpretation of the Adsorption and Desorption Rate Constants

How far can one go with the interpretation of the model parameters? Clearly, one would need to vary at least some experimental variables and repeat the experiment→quantification→fitting procedure. The easiest variables to vary are protein concentration, pH, and temperature, T. The effect of protein concentration change is straightforward, see Eqs. (1) and (4), and it can be used to test the quantification and the fitting routine. Protein adsorption from solutions of higher concentration may not be transport-limited any more. The variation of pH may be a useful experimental variable if the charge of adsorbing surface changes with pH. In such a case, the difference between the model parameters will reflect change in electrostatic interactions between the protein and the surface. Experimentally accessible temperature range may be somewhat restricted due to the limited protein stability at high and low T. In principle, by varying T one can experimentally establish the functions $k_1(T)$ and $k_{-1}(T)$, which may be subsequently used to calculate the adsorption energy.

TABLE 1 Adsorption Parameters Obtained by Fitting the Experimental LDL Adsorption (Fig. 1) to the Model Given by Eqs. (9–13). The Fitting Parameters for an Ideal C18 Monolayer Have Been Obtained by Linear Extrapolation [19]

	Hydrophilic silica	C18-covered silica	Ideal C18 monolayer
$\theta_{adv}/°$	0	104	112
Θ_{C18}	0	0.9	1.0
γ/m	3.1×10^{-5}	3.6×10^{-5}	3.6×10^{-5}
$\alpha/\text{kg}^{-1}\ \text{m}^2$	-2.0×10^5	2.5×10^5	2.0×10^5
$\beta/\text{kg}^{-1}\ \text{m}^2$	0	0	0
Γ_{LDL}^{max}, kg m^{-2}	4.0×10^{-6}	3.0×10^{-6}	2.6×10^{-6}
$k_1/\text{kg}^{-1}\ \text{m}^3\ \text{s}^{-1}$	7.5×10^{-1}	4.5×10^{-2}	2.2×10^{-2}
k_{-1}, s^{-1}	2.9×10^4	2.3×10^4	2.3×10^4
$K/\text{kg}^{-1}\ \text{m}^3$	2.6×10^3	2.0×10^2	1.0×10^2
$(K/\text{M}^{-1})^*$	(6.5×10^9)	(4.9×10^8)	(2.4×10^8)

* The K values in units of M^{-1} were calculated by assuming $M_{LDL} = 4.35 \times 10^{-21}$ kg [21].

In a more limited study, such as the one shown here, one can estimate the apparent LDL adsorption equilibrium constant, K_{LDL}, from the k_1/k_{-1} ratio.* In the case of FITC-LDL adsorption, the adsorption equilibrium constant for the negatively charged hydrophilic silica surface, $K_{LDL}^{silica} = 2.6 \times 10^3$ m^3 kg^{-1} (or 6.5×10^9 M^{-1}), was by an order of magnitude larger than the affinity for the hydrophobic C18-silica surface, $K_{LDL}^{C18} = 2.0 \times 10^2$ m^3 kg^{-1} (or 4.9×10^8 M^{-1}). Since the two respective desorption rate constants were only slightly different (see Table 1), one can conclude that the difference in K_{LDL}s between the different regions of the C18-silica gradient surface is primarily due to the difference in the intrinsic adsorption rate constants. LDL is known to bind strongly to negatively charged adsorbents; in fact, some of the best commercially available LDL-apheresis devices use negatively charged dextran sulfate adsorbent [24,25]. Thus the electrostatic interactions may be responsible for the higher adsorption affinity of LDL of the negatively charged silica surface.

The mechanism of LDL adsorption onto the hydrophobic region of the C18-silica gradient surface is less obvious. The adsorption kinetics suggests an energy barrier to adsorption. The origin of this energy barrier may be in the improper orientation of the LDL particle during some collisions with the hydrophobic surface. When only a small fraction of all possible orientations leads to adsorption, the experimental adsorption rate will decrease and the adsorption will not be transport limited.

The adsorption onto the intermediary region of the C18-silica gradient surface is also interesting. It has been shown elsewhere [19] that a simple additive two-sites model, which accounts for the fractional surface coverage of silica and C18 chains in the gradient region,

---

* In order to estimate K_{LDL} in M^{-1} units, one has to know the molecular mass of the LDL molecule, M_{LDL}, and convert k_1 from (mass concentration)$^{-1}$ time^{-1} units into (molar concentration)$^{-1}$ time^{-1} units. While the exact molecular mass of the LDL particle is often not known, this conversion has been performed here by assuming M_{LDL} to be 4.35×10^{-21} kg [21].

is sufficient to fit the experimental adsorption data. It appears that the adsorption processes on these two types of sites do not affect each other. This finding may indicate that the size of C18 islands on a silica surface is larger than the size of the LDL molecule.

4. Comments on the Adsorption "Cooperativity"

In order to obtain the best fit between the model and the experiment, an adsorption "cooperativity" constant ($\alpha = 2.5 \times 10^5$ m^2 kg^{-1}, Table 1) had to be introduced in fitting the FITC-LDL adsorption onto the hydrophobic region of the C18-silica gradient surface. One interpretation of this negative adsorption "cooperativity" is that a repulsive potential exists around each adsorbed protein molecule. Such a potential makes the subsequent attachment of molecules in the neighborhood of adsorbed molecules more difficult. The repulsive potential around the LDL molecules adsorbed onto the hydrophobic surface probably originates from the uncompensated positive charge of LDL. It was quite puzzling to find that a small positive adsorption "cooperativity" ($\alpha = -2 \times 10^5$ m^2 kg^{-1}, Table 1; note that the sign of the "cooperativity" constant is opposite to the cooperativity itself) had to be introduced in order to fit the experimental FITC-LDL adsorption onto the hydrophilic silica. This positive "cooperativity" implies that the adsorbed molecules like to "stick" next to each other more than to be adsorbed isolated on the surface. The molecular origin of the attractive potential that drives this positive "cooperativity" is not known. One corroborating piece of evidence that such an attractive potential was not an artifact introduced by the fitting routine was found in scanning force microscopy (SFM) LDL adsorption experiments [26]. Figure 3 shows a 5×5 μm^2 area with the LDL molecules adsorbed on a negatively charged mica surface imaged by SFM. Many adsorbed LDL molecules are found aggregated in "L"-shaped (indicated by white arrows) and "clover"-shaped (indicated by black arrows) tetrameric patterns as well as in dimer and trimer clusters. The reason the aggregation did not result in aggregates larger than tetramers is not known. The in situ SFM imaging of LDL adsorption on mica indicated that the aggregation occurred during the adsorption processes [26]. One can speculate that the adsorption-induced aggregation of LDL could have a physiological relation to the pathological formation of lipoprotein deposits in the cardiovascular system.

IV. CONCLUSIONS

In our opinion the ultimate purpose of any protein adsorption experiment is the extraction of physicochemical parameters, such as the adsorption and desorption rates and the cooperativity constants. These parameters can provide a condensed description of the observed process, and when combined with parameters obtained from other adsorption experiments can be used to predict interfacial behavior of a particular protein at a given interface. In the analysis of the experimental adsorption data, it is often necessary to introduce a model of the adsorption process. Deciding which model to use can be rather difficult. While some simple models will overlook particular details of protein adsorption, other more complex models with many fittable parameters may indiscriminately fit almost any experimentally observed behaviors. The quality of the experimental results (i.e., signal-to-noise ratio, experimental errors, etc.) is often the limiting factor in determining how complex the adsorption model can be.

In this chapter, a simple model of protein adsorption was chosen to fit the experimental low-density lipoprotein adsorption kinetics onto the hydrophilic and hydrophobic silica surfaces. The use of a single C18-silica gradient surface simplified

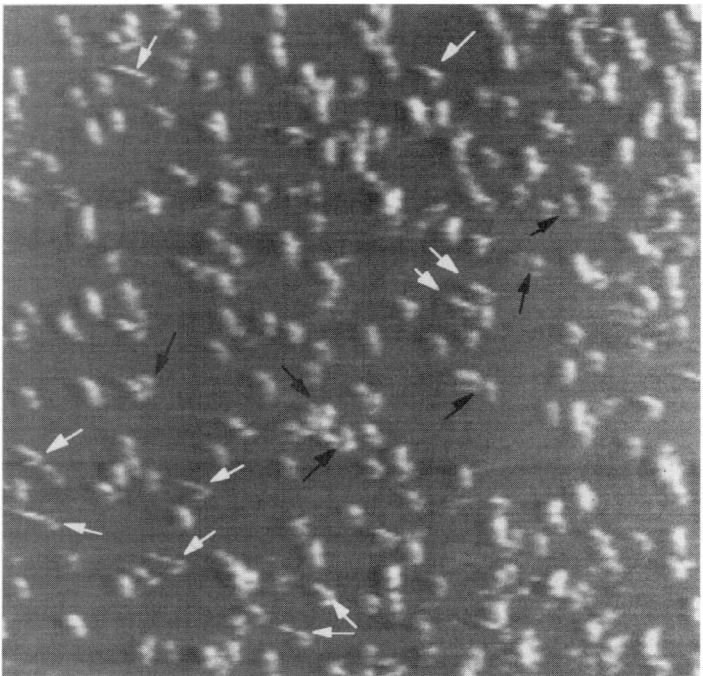

FIG. 3 SFM image (5×5 μm² area) of LDL adsorbed on freshly cleaved mica. Adsorption was allowed to take place for 15 min, and the mica was then rinsed and imaged by SFM in air. Note the "L"-shaped (indicated by white arrows) and "clover"-shaped (indicated by black arrow) LDL tetramer patterns. Dimers and trimers of LDL are also apparent, but to a lesser extent than tetramers.

the comparison between the LDL adsorption onto the two surfaces, since the mass transfer, adsorption, and detection of adsorbed protein fluorescence occurred under almost identical experimental conditions. The experimental TIRF results were first calibrated with an independent method in order to obtain rigorously quantitative adsorption data. The adsorption model used here allowed only two processes: adsorption and desorption. By accurately computing the concentration of protein at the interface as a function of time and empirically correcting for adsorption "cooperativity," the model was found to fit the experiments very well. An additional experimental support for the positive "cooperativity" of LDL adsorption to negatively charged silica was found in the SFM images of LDL adsorbed on the similarly charged surface of mica. The adsorbed LDL molecules on mica were found aggregated in tetramers with a small fraction of dimers and trimers. The molecular origin of this positive cooperativity is presently unknown and remains to be further studied and elucidated.

ACKNOWLEDGMENTS

The authors thank H. P. Jennissen for helpful discussions about the role of cooperativity in protein adsorption. This work was financially supported by the NIH grant R01 HL-44538 and by the Center for Biopolymers at Interfaces, University of Utah.

LIST OF SYMBOLS

Variables and Constants

b	thickness of rectangular flow channel, m
c	solution mass concentration, kg m^{-3}
$c(0, t)$	solution mass concentration next to the adsorbent surface at time t, kg m^{-3}
D	diffusion coefficient, m^2 s^{-1}
J	flux of molecules to the surface, kg m^{-2} s^{-1}
k_{on}	intrinsic adsorption rate constants, kg^{-1} m^3 s^{-1}
k_{off}	intrinsic desorption rate constants, s^{-1}
k_1	initial intrinsic adsorption rate constant, kg^{-1} m^3 s^{-1}
k_{-1}	initial intrinsic desorption rate constant, s^{-1}
K	adsorption equilibrium constant, kg^{-1} m^3
l	distance along rectangular flow channel measured from its entrance, m
M	molecular mass, kg
$O(X)$	of the order of X
q	volume flow rate of solution, m^3 s^{-1}
t	time, s
T	temperature, K
w	width of rectangular flow channel, m
y	distance normal to the adsorbent surface, m
α	adsorption "cooperativity" constant, kg^{-1} m^2
β	desorption "cooperativity" constant, kg^{-1} m^2
γ	wall shear rate, s^{-1}
Γ	surface mass concentration of protein, kg m^{-2}
$\Gamma(4/3)$	gamma function of 4/3
δ	thickness of concentration boundary (i.e. "unstirred") layer, m
θ	contact angle, °
Θ	fraction of surface covered with C18 chains

Superscripts and Subscripts

adv	advancing
app	apparent
C18	octadecyldimethylsilyl (C18-)
d-beg	beginning of the desorption experiment segment
LDL	low density lipoprotein
p	protein
max	maximum
sil	silica

REFERENCES

1. C. Branden and J. Tooze, *Introduction to Protein Structure*, Garland, New York, 1991.
2. J. D. Andrade, in *Surface and Interfacial Aspects of Biomedical Polymers, Vol. 1. Surface Chemistry and Physics* (J. D. Andrade, ed.), Plenum Press, New York, 1985, pp. 1–13.
3. J. D. Andrade, in *Surface and Interfacial Aspects of Biomedical Polymers, Vol. 2. Protein Adsorption* (J. D. Andrade, ed.), Plenum Press, New York, 1985, pp. 1–80.
4. J. Israelachvili, *Intermolecular and Surface Forces*, Academic Press, London, 1992.

5. C. A. Haynes and W. Norde, J. Colloid Interface Sci. *169*:313 (1995).
6. W. Norde, in *Biopolymers at Interfaces* (M. Malmsten, ed.), Marcel Dekker, New York, 1998, pp. 27–54.
7. W. Norde. Adv. Colloid Interface Sci. *25*:267 (1986).
8. J. J. Ramsden. Quart. Rev. Biophys. *27*:41 (1993).
9. V. Hlady and J. Buijs, in *Biopolymers at Interfaces* (M. Malmsten, ed.), Marcel Dekker, New York, 1998, pp. 181–220.
10. V. Hlady, R. A. Wagenen, and J. D. Andrade, in *Surface and Interfacial Aspects of Biomedical Polymers, Vol. 2. Protein Adsorption* (J. D. Andrade, ed.), Plenum Press, New York, 1985, pp. 81–118.
11. V. Hlady. Appl. Spectroscopy *45*:246 (1991).
12. C. S. Lee and G. Beltfort. Proc. Natl. Acad. Sci. USA *86*:8392 (1989).
13. Y. S. Lin and V. Hlady. Colloids Surf. B. Biointerfaces *2*:482 (1994).
14. Y. S. Lin and V. Hlady. Colloids Surf. B. Biointerfaces *4*:65 (1995).
15. B. K. Lok, Y.-L. Cheng, and C. R. Robertson. J. Colloid Interface Sci. *9*:104 (1983).
16. B. K. Lok, Y.-L. Cheng, and C. R. Robertson. J. Colloid Interface Sci. *9*:87 (1983).
17. D. Kim, W. Chao, and R. L. Bessinger. J. Colloid Interface Sci. *159*:1 (1993).
18. M. Leveque. Ann. Miner. *13*:284 (1928).
19. C.-H. Ho and V. Hlady, in *Protein at Interfaces II* (T. Horbett and J. Brash, eds.), ACS Symp. Ser. *602*:371 (1995).
20. Y.-S. Lin, V. Hlady, and J. Janatova. Biomaterials *13*:61 (1992).
21. V. Hlady, J. Rickel, and J. D. Andrade. Colloids Surf. *34*:171 (1988/89).
22. J. W. Corsel, G. M. Willems, J. M. M. Kop, P. A. Cuypers, and W. T. Hermens. J. Colloid Interface Sci. *111*:544 (1986).
23. D. G. Horsley, M. S. thesis, University of Utah, Salt Lake City, 1988.
24. S. Yokoyama, R. Hayshi, T. Kikawa, N. Tani, S. Takada, K. Hatanaka, and A. Yamamoto. Arteriosclerosis *4*:276 (1984).
25. V. Ikonomov, W. Samtleben, B. Schmidt, M. Blumenstein, and H. J. Gurland. Int. J. Artif. Organs *15*:312 (1992).
26. C.-H. Ho, D. W. Britt, and V. Hlady. J. Mol. Recogn. *9*:444 (1996).

11
Microcalorimetric Studies of Biomaterials at Interface

WEN-YIH CHEN and FU-YUNG LIN Department of Chemical Engineering, National Central University, Chung-Li, Taiwan

CHING-FA WU Project Management Office, Sintong Chemical Industrial Company Ltd., Taoyuan, Taiwan

I.	Introduction	419
II.	Behavior of Biomaterial at Interface	420
	A. Liquid–liquid interface	420
	B. Solid–liquid interface	421
III.	Isothermal Titration Microcalorimetry—Direct Measurement of the Adsorption Enthalpy	422
IV.	Applications of ITC on the Biomaterial at Interface	423
	A. Ion-exchange system	423
	B. Immobilized metal ion affinity system	424
	C. Salt concentration effects on the enthalpy of adsorption	427
V.	Conclusion	430
	References	431

I. INTRODUCTION

Biomaterials such as proteins, peptides, polynucleotides, and cells exhibit different behaviors at solid–liquid, liquid–liquid, and liquid–air interfaces. The characteristics associated with the behaviors are attributed to properties of the biomaterial, of the interface, and of the microenvironment of the interface. Nevertheless, bioindustrial applications involving interfaces provide the impetus for understanding the behavior of biomaterials at interface. Examples of industrial applications involving the characteristics of biomaterials at interface are numerous. For instance, the biocomparability of a material heavily relies on the extent of blood proteins, such as fibrinogen and albumin, as well as adsorption on the material surface. The resolution of the protein purification by chromatography is ascribed to the adsorption/desorption of the proteins at the interface. Other applications involving enzyme or cell immobilization for reactions or water treatment contain biomaterial at interface. Therefore further understanding of the interaction between the biomaterial and the interface is highly desired.

Biomaterial behavior at any interface is attributed to all the interacting forces between the biomaterial and the interface. These forces include electrostatic, van der Waals force, hydrophobicity, and even bond formation. For biomaterials, which are mostly macromolecules in an aqueous or buffer solution, the "exposed" surface properties and the conformational change of the biomaterial and the hydration/dehydration process of biomaterial and interface should all be considered when discussing the interactions. The above phenomena involving the biomaterial and interface, however, will become evident in thermodynamics parameters such as enthalpy and entropy change of the interactions. Thus measuring the heat and equilibrium data of the interaction with various controlled parameters of the experiments reveals fundamental information on the interaction mechanism.

This chapter highlights the applications of an isothermal titration microcalorimetry (ITC) to measure the binding interaction enthalpy and entropy and also the binding equilibrium. The enthalpy and entropy data are valuable information for describing biomaterial at interface. Notably, the discussions assume a reversible binding between the biomaterial and the interface. For an irreversible interaction, the ability to derive the free energy from the binding equilibrium data is doubtful. Nevertheless, direct measurement of the binding enthalpy by ITC would facilitate discussions of the binding heat for various binding conditions and binding amounts.

Binding or the biomaterial's behavior at the interface should be examined by thermodynamics and should also be characterized by various environmental factors, such as the pH value, salt concentration, and temperature, and by properties of the biomaterial itself, such as conformation. The influence of these factors reveals variations in binding ability and capability. To gain further insight, variations in binding ability and capability are attributed to the alternations of molecular interactions between biomaterials and interface and between biomaterials. Nevertheless, understanding the interaction mechanism would advance industrial applications of biomaterial at interface. Such a motivation would also facilitate the measurement of thermodynamics information of the interaction between biomaterial and interface.

II. BEHAVIOR OF BIOMATERIAL AT INTERFACE

The behavior of biomaterial at interface is a complex system, for different properties of interface provided and for the macromolecular structure intricacy of the biomaterial. However, previous literature suggests that [1–15] with a kinetics or thermodynamics approach and with the assistance of molecular modeling, the biomaterial behavior at interface can be analyzed or predicted. In the following we analyze the biomaterial behavior at liquid–liquid and solid–liquid interfaces from the perspective of thermodynamics and the molecular level. More importantly, the subsequent behaviors of biomaterial at interface is divided into several processes. Also, discussed herein is how the individual process influences biomaterial behavior at interface in terms of thermodynamics.

A. Liquid–Liquid Interface

Partition is always used to describe biomaterial at liquid–liquid interface. The most widely used industrial applications of partitioning biomaterial is liquid–liquid extraction such as aqueous two-phases [16–27] or reverse micelle systems [28–30]. The imprint theory of these system is the partition driving force of hydrophobicity, electrostatic, and/or special functional group interactions between the biomaterial and the phases of the system. Analyzing the partition behavior allows us to classify the partition mechanism into three subprocesses:

1. Dehydration of biomaterials: Most of the exposed surfaces of biomaterials are hydrophilic residues. For the partition to occur, the water molecules around the exposed surface must be removed or rearranged. For the dehydration process, enthalpy is necessary, and the process would have entropic gain.
2. Diffusion of biomaterial to interface: Diffusion of biomaterial to interface is generally accompanied by a lowering of the potential energy with the expense of enthalpy. For entropy change with the process, the feasibility of rearranging the dehydration water molecules, the solvent molecules, and even conformation change of the biomaterial should be considered.
3. Redistribution of biomaterials at interface: Redistribution of biomaterials is motivated by lowering the system's cumulative free energy for a stable thermodynamic state of the system, including the biomaterial. In terms of the biomaterial properties in phases and the biomaterial size, Albertsson [31] elucidated the relationship between the partition coefficient and the free energy of the biomaterial in the phases.

B. Solid–Liquid Interface

Adsorption is the major interaction mechanism between biomaterial and solid interface. In addition, the adsorption interaction has attracted significant attention owing to the comprehensive applications of adsorption on solid interface such as protein purification, gene therapy, and cell immobilization. Studies involving the interactions between biomaterials and solid interface are multiplicative by the surface modification of the solid surface. However, modification of a solid surface provides not only variety in the adsorption for any specific purpose but also a good model system to study the adsorption mechanism. For instance, immobilization of metal ion as an affinity ligand for protein purification [32–38] or bifunctional groups modification of a solid surface with a more conventional solid surface with hydrophobic and electrostatic functional groups [39–42] are good examples of how to provide an adsorption scheme for purification or immobilization.

However, the adsorption scheme, in general, has to adhere to the following steps as discussed in the liquid–liquid interface:

1. Dehydration of biomaterials: As is generally known, the adsorbed or bound water molecules or the counterions on the biomaterial surface must be removed before adsorption. Removal of the molecules comes at the expense of both the enthalpy of the system and the entropy gain with respect to the water and ion molecules. In light of the above discussions, the pH and ionic strength of the environment and even the conformational change of the biomaterial must be carefully considered.
2. Dehydration of adsorbents: As for biomaterials, discharging the bound water or ion molecules from the solid surface is necessary to adsorb the biomaterial. Facilitating the removal of molecules is a function of solid surface properties such as hydrophobicity, electrostatic state, and functional group for possible chemical bonding. In addition, removing the water molecules or the counterions from the surface is an energy consuming process, in which the system enthalpy ultimately decreases and the entropy increases with respect to the bound molecules.

3. Interaction between biomaterial and solid surface: The adsorption process of the biomaterial into solid surface is the result of the interactions between the biomaterial and the solid surface. The interaction is promoted by the van der Waals force, the electrostatic attractive force, and/or bond forming such as hydrogen bonding or the formation of a coordination compound. The interaction is usually accompanied by the system enthalpy release and entropy loss in respect to the biomaterial molecules. Therefore the enthalpy change of the interaction plays an important role for the system free energy for the adsorption process to happen.
4. Conformational rearrangement and reorientation of biomaterials: The molecular interactions between the adsorbed biomaterials must be considered in the adsorption mechanism. The interactions between biomaterials, e.g., the steric hinderance, electrostatic repulsive force, and interactions, must be balanced by the molecular conformational rearrangement or the reorientation of the adsorption. The conformational change of the biomaterials is primarily the enthalpy needed from the environment and is a function of the conformational rigidity of the biomaterial.

III. ISOTHERMAL TITRATION MICROCALORIMETRY—DIRECT MEASUREMENT OF THE ADSORPTION ENTHALPY

To elucidate the biomaterial adsorption process by microcalorimetry, the biomaterial adsorption process was divided into several substeps as mentioned earlier. In this section, the enthalpy of the adsorption is directly measured.

Generally, the feasibility of implementing a reaction process at constant pressure and temperature is determined by the Gibbs free energy of the reaction (ΔG). For a spontaneous reaction, the Gibbs free energy of the system for the process is negative. Assuming that the adsorption is a reversible process, the thermodynamics theory of the free energy with enthalpy and entropy can be applied to the adsorption process. In addition, directly measured enthalpy and the adsorption substeps allow us qualitatively or even quantitatively to analyze the thermodynamics information of the individual substeps of the adsorption.

The standard Gibbs free energy change of the system is

$$\Delta G^0 = -RT \ln K \tag{1}$$

and

$$\Delta G^0 = \Delta H^0 - T\Delta S^0 \tag{2}$$

For a reversible process,

$$\frac{\Delta H^0_{ads}}{T^2} = -\frac{\partial(\Delta G^0_{ads}/T)}{\partial T} \tag{3}$$

$$\Delta S^0_{ads} = \frac{-\partial(\Delta G^0_{ads})}{\partial T} \tag{4}$$

Owing to the conformational change of biomaterial and the rearrangement of the solvent molecules, the system's entropy constantly changes as the adsorption process proceeds.

In addition, the entropy cannot be measured, although previous attempts have been made to predict the translational entropy loss [43]. Moreover, direct measurement of the enthalpy variation of the process by microcalorimetry is a convenient tool for obtaining thermodynamics information.

At a constant temperature and pressure of a liquid system, the exchanged heat equals the enthalpy change of the system:

$$Q = H \qquad (5)$$

With the above equations, the enthalpy change could be resolved with the individual steps with the various adsorption conditions.

IV. APPLICATIONS OF ITC ON THE BIOMATERIAL AT INTERFACE

To gain further insight into the applications of ITC on the studies of biomaterial at interface, a survey is performed of the ion-exchange system and immobilized metal ion system as our models. Also discussed herein are details on measurements of the heat change of the protein adsorption onto the systems. The dominant interaction between protein and ion-exchange resin is the electrostatic force. Meanwhile, the immobilized metal ion system forms the coordination compound. These two types of interaction mechanism furnish not only the major interaction for the applications but also the interaction in a sensible heat change.

A. Ion-Exchange System

Ion exchange is the mostly widely used bioseparation system in industry. The ion-exchange resin surface is either hydrophilic cellulosic or a polymer material; the protein adsorption onto the resin surface is driven primarily by an electrostatic attractive force and by hydrophobicity interaction. These driving forces are functions of factors such as the protein structure, the pH value of the solution, and the salt concentration. Norde et al. [44–49] have examined the thermodynamic approach to the complex system. With the isotherm of the protein adsorption onto the ion-exchange resin and the information on the heat of adsorption measured by microcalorimetry, Norde et al. [50–54] discussed the adsorption behavior in a systematic manner on the molecular level. For instance [51], for the adsorption of human plasma albumin and ribonuclease onto a hematite (α-Fe_2O_3) surface, they used LKB batch microcalorimetry to measure the adsorption enthalpy isothermally. Under constant temperature and ionic strength, those investigators also varied the charge of the protein and the adsorbent by altering the solution pH value. By doing so, they obtained the enthalpy of adsorption per amount of protein adsorbed. In all the examined pH region, the positive adsorption enthalpy indicates an entropy-driven binding process. Analysis of the variance of the ΔH_{ads} with the amount of protein adsorbed reveals that ΔH_{ads} increases with the protein adsorbed and with the net charge of the protein. Examination of the relationship between the amount of protein and ΔH_{ads} indicates that the protein interaction at the resin surface heavily influences the protein adsorption. Later on, Norde et al. [55] summarized the data of lysozyme adsorption enthalpy information by ITC.

Bowen and Hughes [56] selected two types of ion-exchange resins to study bovine serum albumin (BSA) adsorption by ITC. According to their results, positive ΔH_{ads} was obtained with BSA adsorption on Whatman QA52—a cellulosic, quaternary amine-bearing, strong anion exchanger, and also, for PL-SAX1000 adsorbate, a strong anion and fully quaternized polyethyleneimine coupled to a synthetic polymer matrix, negative ΔH_{ads} was acquired. These results strongly suggest that the entropy contribution by the solvent molecule rearrangements significantly influences the adsorption process. Bowen and Hughes [56] also discussed the enthalpy change from negative to positive values during the protein binding process. That investigation postulated that in the first step of binding, the adsorption is driven by the electrostatic attraction, which provides a heat release (negative ΔH_{ads}). In addition, as the amount of adsorbed protein increases, the interaction between the protein molecules increases, and the interaction forces the protein molecule's structure rearrangement or reorientation for adsorption. These interactions must be compensated for by decreasing the entropy of the protein molecules.

B. Immobilized Metal Ion Affinity System

Porath et al. [57] proposed immobilized metal ion affinity chromatography (IMAC) in 1975. IMAC is a separation method based on interfacial interaction between biomaterials and metal ions immobilized to a solid support in a solution. The specific interaction of a protein with immobilized metal ions frequently varies with the topography of the protein surface and the chemistry and physical conditions of the interaction [58–63]. Although previous literature [64–70] has established the factors that contribute toward the interaction, relatively few studies apply a thermodynamics approach to the IMA method. The thermodynamic characteristic functions provide valuable information to understand thoroughly protein adsorption at the IMA surface.

A detailed presentation is made in the following of the ITC method of measuring adsorption heat of protein with immobilized metal. Examined as well are instruments, effects of the binding conditions, amount of protein adsorbed, and different immobilized ions on the adsorption enthalpy. The molecular level discussion and phenomena presented in the following for protein should be applicable to other systems involving protein behavior at interface.

The microcalorimeter used is a Thermal Activity Monitor (Thermometric AB, Sweden) controlled by Digitam software. The microreaction system is a titration mode with a 4 mL stainless steel ampoule. Continuous heat leakage measurements are taken in an isothermal system. The heat of adsorption flows through high-sensitivity thermopiles surrounded by a heat sink stabilized at $\pm 2 \times 10^{-4} \,^{\circ}\text{C}$. The magnitude of heat exchange of a thermopile with a heat sink is proportional to the time interval of the voltage signal. The electrical calibration performance allows us to quantify the results. During the measurement, the gel (CS-IDA-Cu(II)) was suspended in the equilibrium buffer solution and was placed in the ampoule and stirred at 120 rpm. When thermal equilibrium between the ampoule and the heat sink was reached, a 20 μL protein solution prepared in an equilibrium buffer solution was titrated into the dispersion gel suspension with a Hamilton microliter syringe at 25 min intervals.

Before calculating the enthalpy of adsorption from titration to obtain the net heat of interaction between protein and immobilized metal ion, the heat of titration should be corrected by the following effects: (1) heat of dilution of protein, (2) heat of dilution of gel, and (3) heat of interaction between protein and gel without immobilized ions (e.g., CS-IDA). With the above heat data and the adsorption isotherm, the adsorption behavior of protein onto the gel can be interpreted from a thermodynamic perspective.

The interaction mechanism of a protein with immobilized ions is the formation of a coordination compound by the electron pair donor from the exposed amino acid residue of protein with a transition metal ion such as Cu(II), Fe(III), and Ni(II). The electron donor groups of protein include imidazole side chains of histidine, sulfhydryl groups of cysteine, or residues of phosphorylated serine. Previous literature suggests that the energy of formation of a coordination compound is in the order of ~ 5 kcal/mole [71] and is higher than the energy of electrostatic or hydrophobic interactions. In the 1990s, Yip and Hutchens [72], Arnold et al. [70], and Hearn et al. [73] have investigated IMAC binding from a thermodynamic perspective. The thermodynamic information they obtained, however, were derived from isotherm data and thermodynamic relationships for reversible binding. Chen et al. [74,75] performed direct enthalpy measurements of protein adsorption of the IMAC system with discussions of the effects of salt concentration and pH value.

The heterogeneity of the adsorption enthalpy with the adsorbed amount of protein can be predicted according to the accessibility of each protein electron donor group and the affinity of the interaction between the donor group and the immobilized metal ion. Moreover, the effects of adsorbed protein molecule interaction on the binding are also reflected in the enthalpy of the adsorption with different amounts of protein adsorbed. According to Arnold et al. [70], two binding mechanisms can influence the binding enthalpy, i.e., (1) the interaction between the adsorbed protein molecules and (2) the arrangement of metal ion ligands on the surface. Therefore direct measurement of the enthalpy of the adsorption with various binding conditions facilitates a discussion of the binding mechanism.

The following are examples of the binding enthalpy measurement for the IMAC system of lysozyme with immobilized Cu(II) and Fe(III) of CS-IDA with various binding conditions. For the pH effect, the titration of lysozyme in a CS-IDA-Cu(II) suspension was performed at various pH values and in a 0.1 M NaCl, 20 mM phosphate solution, at a temperature of 298 K. Figure 1 summarizes the results obtained by ΔH_{ads} versus the extent of adsorption of lysozyme A_n. The enthalpy of adsorption ΔH_{ads} is calculated by

$$Q_{ads} = V \times A_n \times H_{ads} \tag{6}$$

where V represents the volume of CS-IDA-Cu(II) gel in the ampoule, A_n is the adsorbed amount of lysozyme on the gel, and ΔH_{ads} is the enthalpy of adsorption. A_n can be obtained from the isotherm. Because the pK value of the side chain of histidine residues on lysozyme is 5.71, the experimental pH values are designed to cover the deprotonation and protonation states of the imidazole group. The pH value presents an obvious effect on the enthalpy of adsorption (ΔH_{ads}) as illustrated in Fig. 1. At the pH value of the histidine deprotonation state, the formation of a coordination compound between immobilized ion and protein is the major binding force, and the binding heat is exothermic, while at a pH value lower than the pK of a histidine side chain, the binding enthalpy becomes endothermic. This result is the same as for the adsorption of imidazole with CS-IDA-Cu(II) [74]. Therefore the enthalpy change provides an indicator of the strength at various pH values.

Interestingly, in Fig. 1, endothermic enthalpy of adsorption as high as 10 kJ/mol (on the basis of the adsorbed lysozyme) is present at pH = 5.5. A negative value of ΔG^0_{ads} is required for the adsorption to occur spontaneously. Therefore the adsorption of lysozyme onto CS-IDA-Cu(II) is concluded to be an entropy-driven phenomenon. A more thorough discussion of lysozyme adsorbed onto the gel necessitated considering the following processes as discussed above: (a) water or ion molecules surrounding imidazole are excluded, i.e., the dehydration or "deionization" (removing of the electrical double layer) process of lysozyme; (b) water molecules surrounding the immobilized metal ion are excluded, i.e., the dehydration or decoordination process of immobilized metal ion; (c) coordination

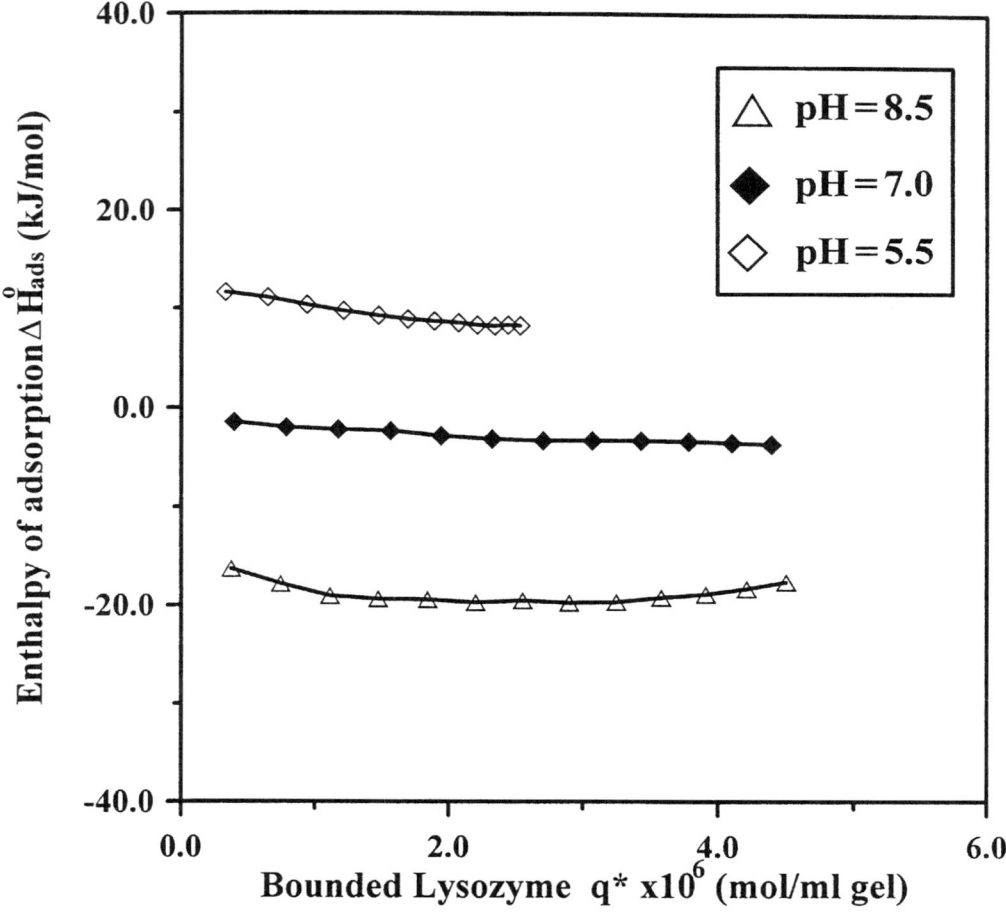

FIG. 1 Enthalpy of adsorption (ΔH_{ads}) of lysozyme onto CS-IDA-Cu(II) gel at 0.1 M NaCl, and 20 mM phosphate and various pH values.

interaction and/or nonspecific interaction is formed between lysozyme and immobilized metal ions; (d) the rearrangement of the conformation of lysozyme; and (e) rearranging the excluded water or ion molecules in a bulk solution, contributing toward an increment in entropy. Both dehydration and deionization processes, i.e., (a) and (b), are endothermic processes. In addition, lysozyme is a macromolecule (molecular weight 14,000 Da, dimensions $4.6 \times 3.0 \times 3.0$ nm^3), and more than one mole of water or ion molecules is removed or dehydrated when a mole of lysozyme coordinates with an immobilized copper ion. The exothermic process (c) is minimized at pH = 5.5 in the range of the pH values of this study, since the nitrogen atom of the imidazole group is at a protonated state, so the formation of a coordination compound is unlikely. Quantifying the heat required for removing of the counterion of the electrical double layer (EDL) from the neighborhood of the histidine residue, i.e., process (a), is difficult. It is, however, an endothermic process, since the primary structure of the lysozyme molecule around the histidine residue is given as ··· -Ala-Met-Lys-Arg-His-Gly-Leu-···, where lysine and arginine are adjacent to the histidine residue that has a positive charge side chain at pH 5.5–8.5, and the EDL is formed mainly by Cl^- around the side chains. Moreover, by estimating the exposed surface area of the histidine residue of lysozyme [76], it may be concluded that the accessibility of the exposed histidine residue on lysozyme surface is low. In other words, this exposed histidine residue may locate at a hydrophobic region of lysozyme. Thus lysozyme would have to rearrange the conformation, such as by breaking the hydrogen bond or salt bridge, in order to reduce the contact surface tension between the hydrophobic region around the histidine residue and the hydrophilic surface of the CS-IDA gel. However, a total rearrangement of the stable lysozyme conformation of the process (d) is doubtful if one considers the change of pH from 5.0 to 8.5 and the addition of 0.1–1 M NaCl. To gain an idea of the relative amount of enthalpy for removing the EDL and the structural rearrangement of protein, Norde [45] assigned a value for the enthalpy of the structural rearrangement and the deionization processes of lysozyme and αlactalbumin adsorbed onto polystyrene with negative surface as high as 50 kJ/mol and 150 kJ/mol, respectively. The experimental results of this investigation reveal that the enthalpy of adsorption is 20 kJ/mol (based on adsorbed lysozyme) at pH = 5.5. Therefore the contribution of the entropy, in order to make the binding free energy less than zero, is the impetus for the adsorption at pH = 5.5.

C. Salt Concentration Effects on the Enthalpy of Adsorption

The previous literature has discussed the effects of salt concentration on the binding capability and capacity of imidazole and proteins with immobilized metal ions [76,77]. From the perspective of the binding mechanism, the enthalpy of adsorption would be affected not only by the pH value on the electron-donating capability but also by the salt concentration on the hydrophobic and electrostatic interaction between lysozyme and immobilized metal ions. In this investigation, experiments were performed by varying the NaCl concentrations from zero to 1.0 M at pH values higher and lower than the pKa value of the imidazole group of exposed histidine residue on lysozyme surface, and at 20 mM phosphate, at the temperature of 298 K. Figures 2, 3, and 4 summarize the results obtained from the enthalpy change in the adsorption of lysozyme with CS-IDA-Cu(II) at various NaCl at pH = 5.5, 7.0, and 8.5, respectively. Lysozyme possesses low charge capacity on its surface at pH = 8.5 due to pI = 11.2. When the NaCl concentration is as high as 0.5 M and 1.0 M in a buffer solution, lysozyme aggregated to precipitation. Therefore the titration experiment cannot be performed.

It is interesting that the adsorption process was endothermic in the absence of electrolyte when the pH value of the environment was lower (pH = 5.5) and higher (pH = 7.0 and 8.5) than the pK value of the imidazole group of the histidine residue. Restated, the degree of the endothermic amount of both the dehydration and the structural rearrangement processes is large enough in the absence of NaCl to overcome the degree of the exothermic amount of coordination interaction even at pH = 8.5. The adsorption enthalpy changes from positive to negative value at a higher NaCl concentration, as illustrated in Figs. 2, 3, and 4. Possible explanations are (1) a reduction in the degree of the endothermic amount of the dehydration and the removed of EDL processes [i.e., (a) and (b) processes] at a higher salt buffer concentration, (2) an increment in the degree of exothermic amount of the coordination interaction [(c) process], and the nonspecific binding between the immobilized metal ion and protein. The dehydration process is not a function of salt concentration, as the binding mechanism between water molecules with protein is of the nature of hydrophilic interaction or hydrogen bonding of the induced dipole moment. On the other hand, the enthalpy required for the removing of EDL is

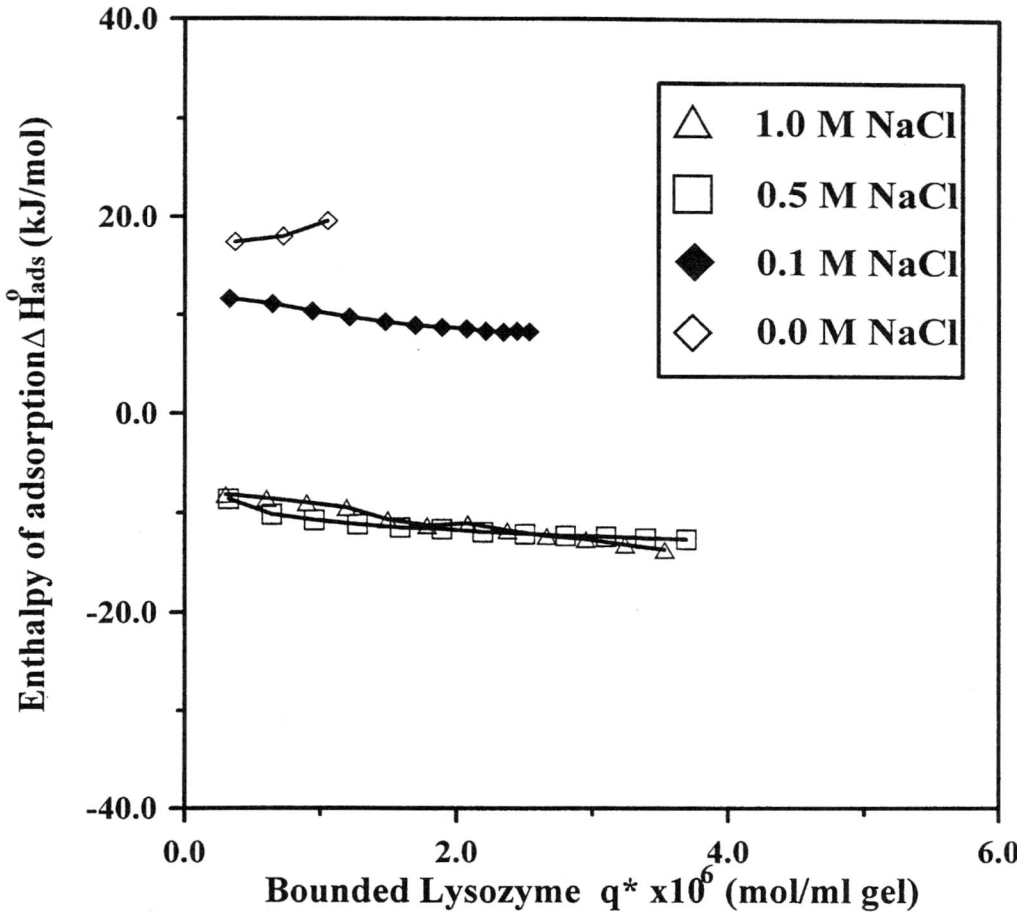

FIG. 2 Enthalpy of adsorption (ΔH_{ads}) of lysozyme onto CS-IDA-Cu(II) gel at pH 5.5, and 20 mM phosphate and various NaCl concentrations.

confirmed in that ΔH_{ads} values at 0.5 M and 1.0 M ($I = 0.5$ and 1.0) NaCl concentration solution are almost same and show the same concentration dependence at pH 5.5 and 7.0 in Figs. 4 and 5. Since the Cl^- effect becomes zero at $I = 0.0$ M, ΔH_{ads} would be simply positive at pH 5.0, 7.0, and 8.5 in Figs. 2–4 and show almost the same ΔH_{ads} values at the low bound protein concentrations. However, ΔH_{ads} becomes negative at $I = 0.5$ M and 1.0 M because lysozyme is adsorbed on the IDA-Cu(II) group after the exclusion of EDL. At $I = 0.1$ M, the sign of ΔH_{ads} changes from positive to negative with the increase in pH. This indicates that the EDL formation is influenced by the change in the charge of the side chain of the histidine residue. Therefore the salt concentration contributes in a major way to the heat released of the nonspecific binding and the heat requirement of the removing EDL of histidine and neighboring residues. the 10–20 kJ/mol (on the basis of adsorbed lysozyme) enthalpy difference of salt concentration from zero to

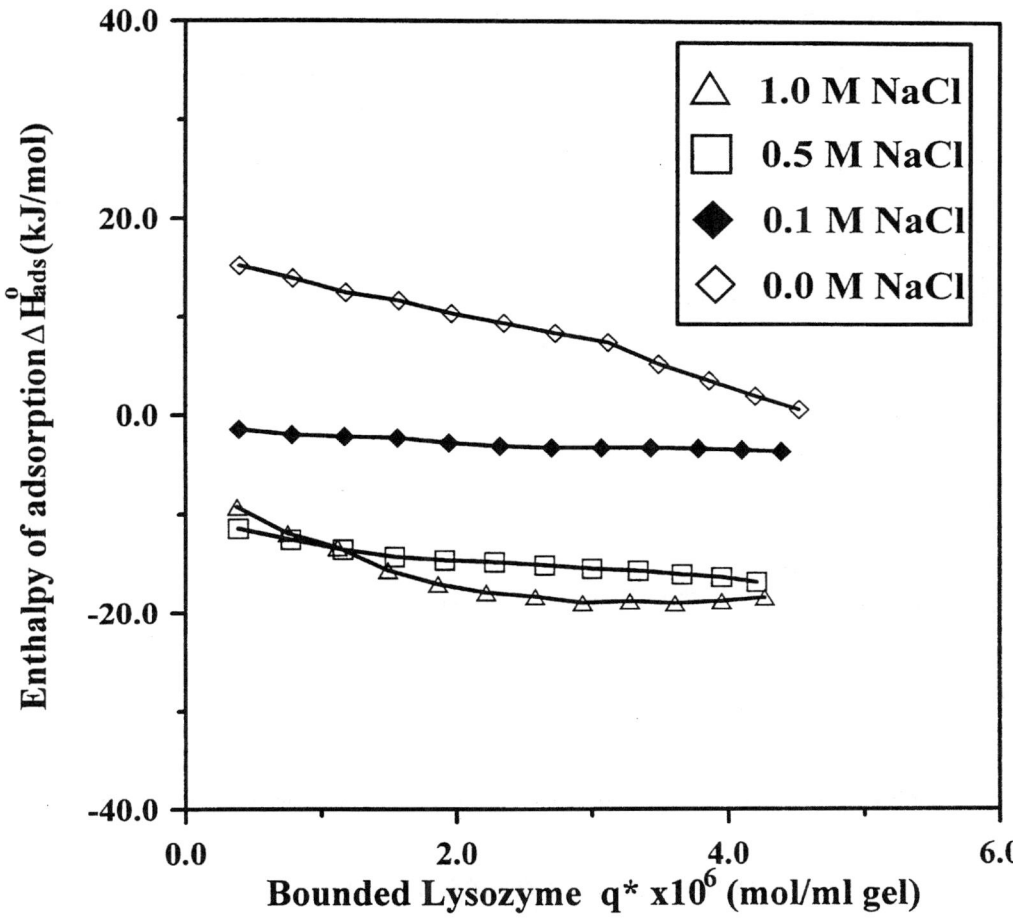

FIG. 3 Enthalpy of adsorption (ΔH_{ads}) of lysozyme onto CS-IDA-Cu(II) gel at pH 7.0, and 20 mM phosphate and various NaCl concentrations.

about 0.1 M is reasonable compared to the adsorption enthalpy difference of 60–100 kJ/mol of bovine serum albumin onto PL-SAL ion-exchanger resin of salt concentration between 0.0 M and 0.1 M NaCl [56].

Finally, the ΔH_{ads} of lysozyme adsorption onto different immobilized metal ions at pH 8.5 is presented in Fig. 5. At the low concentrations of lysozyme at solution phase, the ΔH_{ads} with Cu(II) and Fe(III) does not vary with the concentration of lysozyme, while, as the adsorption amount reaches a higher concentration, the ΔH_{ads} of Cu(II) drops and the heat releases from the system is larger than that from the Fe(III) system. This phenomenon suggests that the binding force of lysozyme with Cu(II) is higher than with Fe(III), and the results can also be confirmed by the theory of hard and soft acids and bases by Pearson [78–80]. The figure also gives information on the relative value of the enthalpy of each of the steps during the binding processes that were proposed above.

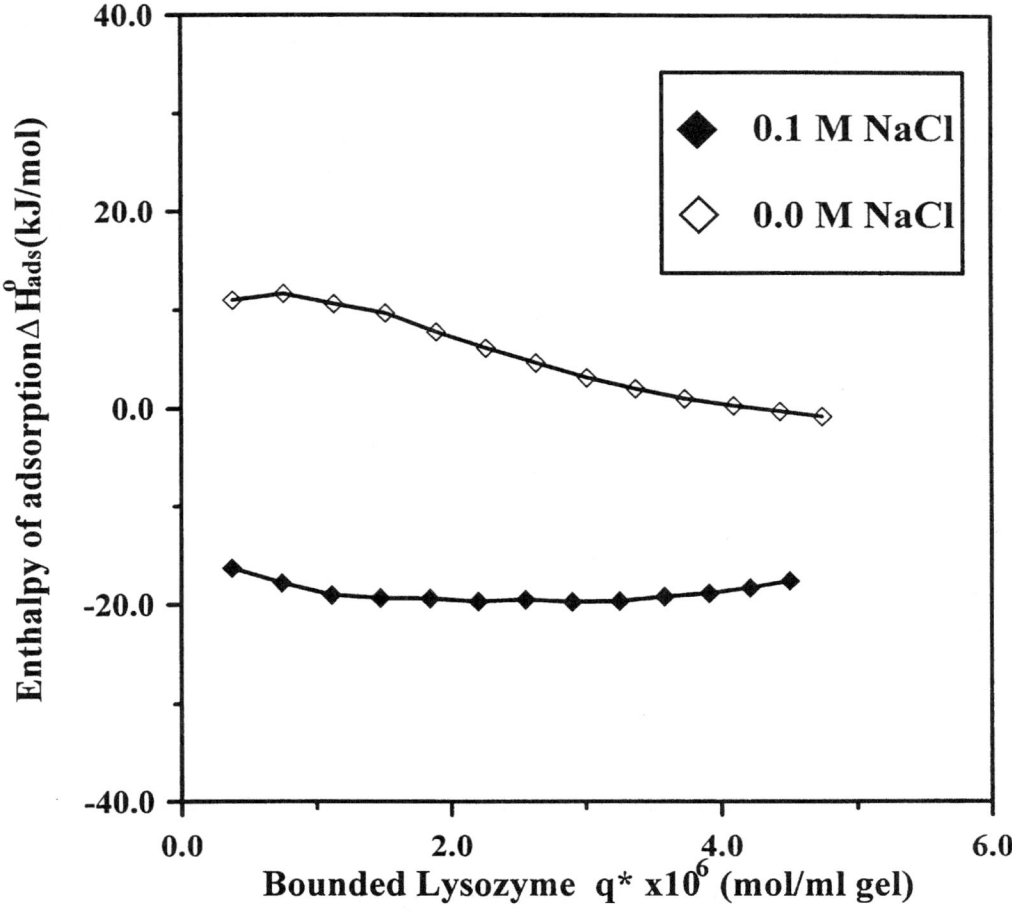

FIG. 4 Enthalpy of adsorption (ΔH_{ads}) of lysozyme onto CS-IDA-Cu(II) gel at pH 8.5, and 20 mM phosphate and various NaCl concentrations.

V. CONCLUSION

In this chapter, we pointed out applications of microcalorimetry in the study of biomaterials at interface. Analysis of the heat measured by microcalorimetry of biomaterial at interface reveals directly not only the adsorption enthalpy but also, and more importantly, the binding mechanism and the behavior of biomaterial at interface. For the sake of discussions of the information from microcalorimetry for biomaterial at interface, we divided and pictured the binding process into five substeps, and the enthalpy and entropy contributions to the binding free energy were qualitatively discussed. Thermodynamic data from microcalorimetry for the ion-exchange system and in the IMAC system were discussed in detail. We conclude that microcalorimetry is a powerful tool for direct measurement of enthalpy in a study of interactions between biomaterial and surface, and the results obtained provide valuable thermodynamic information about binding mechanisms.

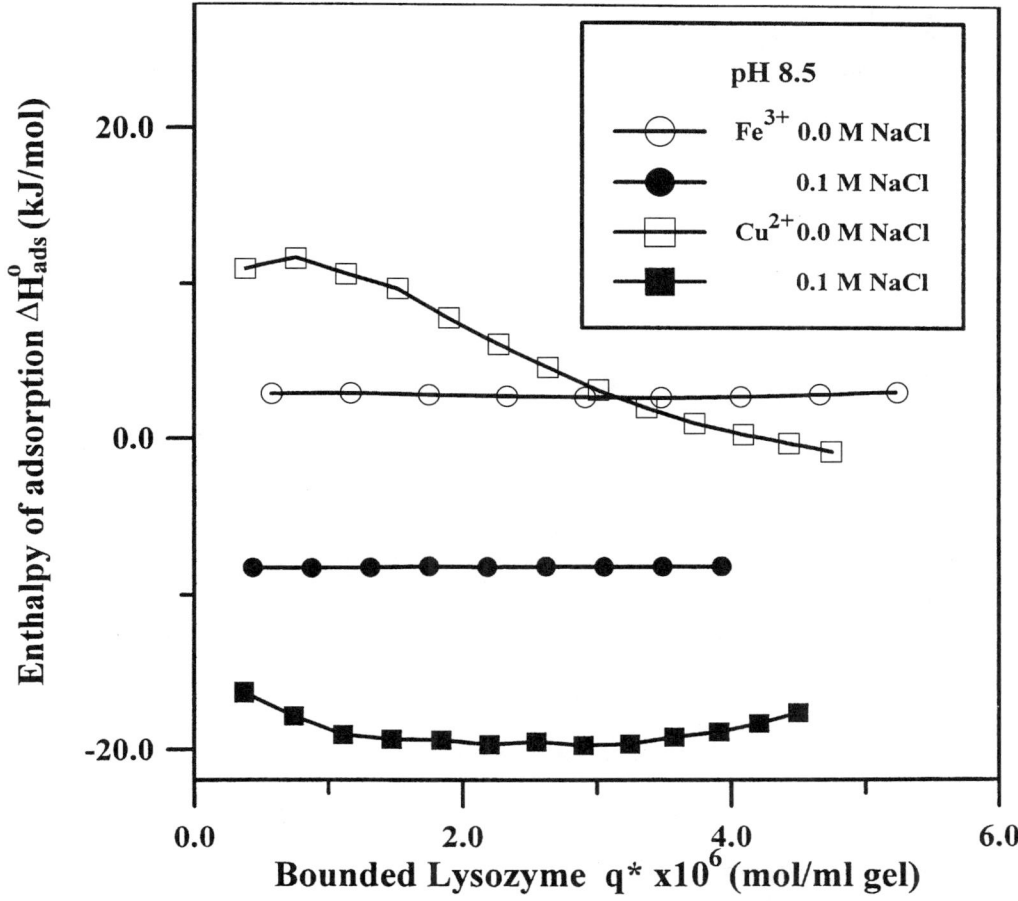

FIG. 5 Enthalpy of adsorption (ΔH_{ads}) of lysozyme onto CS-IDA-Cu(II) or CS-IDA-Fe(III) gel at pH 8.5, and 20 mM phosphate and various NaCl concentrations.

REFERENCES

1. S. Vunnum, S. R. Gallant, Y. J. Kim, and S. M. Cramer. Chem. Eng. Sci. *50*:1785 (1995).
2. S. R. Gallant, A. Kundu, and S. M. Cramer. J. Chromatogr. A *702*:125 (1995).
3. R. J. Yon. J. Chromatogr. *457*:13 (1988).
4. F. H. Arnold, S. A. Schofield, and H. W. Blanch. J. Chromatogr. *335*:1 (1986).
5. H. P. Jennissen. J. Chromatogr. *215*:73 (1981).
6. A. A. Gorbunov, A. Y. Lukyanov, V. A. Pasechnik, and A. V. Vakhrushev. J. Chromatogr. *365*:205 (1986).
7. A. Barroug, P. G. Rouxhet, and J. Lemaitre. Colloids Surfaces. *37*:339 (1989).
8. R. Blanco, A. Arai, N. Grinberg, D. M. Yarmush, and B. L. Karger. J. Chromatogr. *482*:1 (1989).
9. C. F. Lu, A. Nadarajah, and K. K. Chittur, J. Colloid Interface Sci. *168*:152 (1994).
10. M. M. Habbaba and K. O. Ulgen. J. Chem. Technol. Biotechnol. *69*:405 (1997).
11. C. T. Shibata and A. M. Lenhoff. J. Colloid Interface Sci. *148*:485 (1992).
12. G. A. Bornzin and I. F. Miller. J. Colloid Interface Sci. *86*:539 (1982).

13. V. B. Fainerman, R. Miller, R. Wustneck. J. Colloid Interface Sci. *183*:26 (1996).
14. S. W. Lin, R. Blanco, and B. L. Karger. J. Chromatogr. *557*:369 (1991).
15. V. Noinville, C. Vidal-Madjar, and B. Sebille. J. Phys. Chem. *99*:1516 (1995).
16. K. Berggren, H.-O. Johansson, and F. Tjerneld. J. Chromatogr. A *718*:67 (1995).
17. K. D. Cole, T. K. Lee, and H. Lubon. Appl. Biochem. Biotechnol. *67*:97 (1997).
18. M. Carlsson, K. Berggren, P. Linse, A. Veide, and F. Tjerneld. J. Chromatogr. A *756*:107 (1996).
19. W.-Y. Chen, C.-G. Shu, J.-Y. Chen, and J.-F. Lee, J. Chem. Eng. Jpn. *27*:688 (1994).
20. G. Birkenmeier, M. A. Vijayalakshmi, T. Stigbrand, and G. Kopperschlager. J. Chromatogr. *539*:267 (1991).
21. G. E. Wuenschell, E. Naranjo, and F. H. Arnold. Bioprocess Eng. *5*:199 (1990).
22. H. Walter and K. E. Widen. J. Chromatogr. *641*:279 (1993).
23. B. H. Chung and F. H. Arnold. Biotechnol. Lett. *13*:615 (1991).
24. P. P. Godbole, R. Tsai, and W. M. Clark. Biotechnol. Bioeng. *38*:535 (1991).
25. M. E. Van Dam, G. E. Wuenschell, and F. H. Arnold. Biotechnol. Appl. Biochem. *11*:492 (1989).
26. H. G. Botros, G. Birkenmeier, A. Otto, G. Kopperschlager, and M. A. Vijayalakshmi. Biochim. Biophys. Acta *1074*:69 (1991).
27. J. Carlsson, K. Mosbach, and L. Bulow. Biotechnol. Bioeng. *221*:51 (1996).
28. G. J. Lye, J. A. Asenjo, and D. L. Pyle. Biotechnol. Bioeng. *47*:509 (1995).
29. Z.-Y. Hu and E. Gulari. J. Chem. Technol. Biotechnol. *65*:45 (1996).
30. Z.-Y. Hu and E. Gulari. Biotechnol. Bioeng. *50*:203 (1996).
31. P. A. Albertsson, in *Partition of Cell Particles and Macromolecules*, 3d ed., John Wiley, New York, 1986, pp. 1–20.
32. J. Porath. Trend Anal. Chem. *7*:254 (1988).
33. J. Porath. Protein Express. Purif. *3*:263 (1992).
34. F. H. Arnold. Biotechnol. *9*:151 (1991).
35. J. W. Wong, R. L. Albright, and N.-H. L. Wang. Sep. Purif. Method. *20*:49 (1991).
36. L. D. Holmes and M. R. Schiller. J. Liq. Chrom. Rel. Technol. *20*:123 (1997).
37. E. Sulkowski. Trend Biotechnol. *3*:1 (1985).
38. E. Sulkowski. BioEssays. *10*:170 (1989).
39. L. Sun and P. W. Carr. Anal. Chem. *67*:2517 (1995).
40. L. Sun and P. W. Carr. Anal. Chem. *67*:3717 (1995).
41. W. A. Schafer and P. W. Carr. J. Chromatogr. *587*149 (1991).
42. A. M. Clausen and P. W. Carr. Anal. Chem. *70*:378 (1998).
43. K. P. Murphy, D. Xie, K. S. Thompson, L. M. Amzel, and E. Freire. Protein Struct. Funct. Genet. *18*:63 (1994).
44. W. Norde. J. Disper. Sci. Technol. *13*:363 (1992).
45. W. Norde. Pure Appl. Chem. *66*:491 (1994).
46. W. Norde and J. Lyklema. J. Colloid Interface Sci. *66*:295 (1978).
47. W. Norde and J. Lyklema. Colloids Surfaces. *38*:1 (1989).
48. T. Arai and W. Norde. Colloids Surfaces. *51*:1 (1990).
49. W. Norde and C. A. Haynes, in *Protein at Interfaces II* (T. A. Horbeet and J. L. Brash, eds.), ACS Symp. Ser. 602, American Chemical Society, Washington, D.C., 1995, pp. 26–40.
50. W. Norde and J. Lyklema. J. Colloid Interface Sci. *66*:257 (1978).
51. P. G. Koutsoukos, W. Norde, and J. Lykleman. J. Colloid Interface Sci. *95*:385 (1983).
52. W. Norde and J. Lyklema. J. Colloid Interface Sci. *66*:285 (1978).
53. W. Norde and J. Lyklema. J. Colloid Interface Sci. *66*:266 (1978).
54. P. G. Koutsoukos, C. A. Mumme-Young, W. Norde, and J. Lyklema. Colloids Surfaces. *5*:93 (1982).
55. C. A. Haynes, E. Sliwinsky, and W. Norde. J. Colloid Interface Sci. *164*:394 (1994).
56. W. R. Bowen and D. T. Hughes. J. Colloid Interface Sci. *158*:395 (1993).
57. J. Porath, J. Carlsson, I. Olsson, and G. Belfrage. Nature *258*:598 (1975).
58. R. D. Johnson and F. H. Arnold. Biotechnol. Bioeng. *48*:437 (1995).

59. M. Zachariou, I. Traverso, L. Spiccia, and T. W. Hearn. Anal. Chem. *69*:813 (1997).
60. M. Zachariou and M. T. W. Hearn. J. Protein Chem. *14*:419 (1995).
61. P. Chakrabarti. Protein Eng. *4*:57 (1990).
62. Z. E. Rassi and C. Horváth. J. Chromatogr. *359*:241 (1986).
63. R. J. Todd, R. D. Johnson, and F. H. Arnold. J. Chromatogr. A *662*:13 (1994).
64. E. S. Hemdan, Y.-J. Zhao, E. Sulkowski, and J. Porath. Proc. Natl. Acad. Sci. U.S.A. *86*:1811 (1989).
65. R. J. Todd, M. E. Van Dam, D. Casimiro, B. L. Haymore, and F. H. Arnold. Protein Struct. Funct. Genet. *10*:156 (1991).
66. T. W. Hutchens and T.-T. Yip. J. Chromatogr. *536*:1 (1991).
67. T. W. Hutchens and T.-T. Yip. J. Chromatogr. *500*:531 (1990).
68. W.-Y. Chen, C.-F. Wu, and C.-C. Liu. J. Colloid Interface Sci. *180*:135 (1996).
69. F. B. Anspach. J. Chromatogr. A *676*:249 (1994).
70. R. D. Johnson, Z.-G. Wang, and F. H. Arnold. J. Phys. Chem. *100*:5134 (1996).
71. S. Mallik, S. Plunkett, P. K. Dhal, R. D. Johnson, D. Pack, D. Shnek, and F. H. Arnold. New J. Chem. *18*:299 (1994).
72. T. W. Hutchens and T.-T. Yip. J. Inorg. Biochem. *42*:105 (1991).
73. G. M. S. Finette, Q.-M. Mao, and M. T. W. Hearn. J. Chromatogr. A *763*:71 (1997).
74. C.-F. Wu, W.-Y. Chen, and J.-F, Lee, J. Colloid Interface Sci. *28*:419 (1995).
75. W.-Y. Chen, J.-F. Lee, C.-F. Wu, and H.-K. Tsao. J. Colloid Interface Sci. *190*:49 (1997).
76. C.-F. Wu, W.-Y. Chen, and H.-S. Liu. J. Chem. Eng. Jpn. *28*:419 (1995).
77. E. S. Hemdan and J. Porath. J. Chromatogr. *323*:265 (1985).
78. R. G. Pearson. J. Chem. Educ. *45*:643 (1968).
79. R. G. Pearson. J. Chem. Educ. *45*:581 (1968).
80. R. G. Pearson. Coor. Chem. Rev. *100*:403 (1990).

12

Kinetics and Mechanisms of Crystal Growth in Aqueous Systems

LJERKA BREČEVIĆ and DAMIR KRALJ Department of Materials Chemistry, Ruđer Bošković Institute, Zagreb, Croatia

I.	Introduction	435
II.	Solution Stability	436
III.	Nucleation	439
	A. Homogeneous nucleation	440
	B. Heterogeneous nucleation	443
	C. Secondary nucleation	444
IV.	Crystal Growth	444
	A. Transport controlled growth	446
	B. Surface reaction controlled growth	449
	C. Growth controlled by different mechanisms	457
V.	Crystal Dissolution	458
VI.	Influence of Impurities	459
VII.	Aging	462
	A. Flocculation	462
	B. Ostwald ripening	463
	C. Transformation of metastable phases	463
VIII.	Experimental Approach	464
	Symbols	469
	References	471

I. INTRODUCTION

Considerable interest in the chemistry and physics of solids has led to research on crystal growth. This has also given rise to the development of the scientific basis of crystallization. However, most publications on the theory of crystal growth have elaborated the physical characteristics of the processes involved rather than the chemical phenomena occurring. Besides, most measurements of the rates of crystal growth from solution have been performed with rather soluble substances [1,2], so that the present state of knowledge of the crystallization of readily soluble substances is better than that of sparingly soluble ones. A relatively high supersaturated solution is requested for particles of sparingly soluble substances to be formed, so most processes involved often occur simultaneously and proceed rapidly. Therefore it is difficult to study these processes independently of each other. Still, with less soluble substances, a larger range of supersaturations can be achieved, which makes it possible to investigate the kinetics of crystallization processes separately.

The crystallization of sparingly soluble substances from solutions, also called precipitation, is one of the most important operations in chemical, pharmaceutical, and many other process industries. In medicine and biology, precipitation plays an important role in normal and pathological mineralization (formation of bones, teeth, shells, calculis, etc.). Precipitation has to be considered also in geology and oceanology, as it is responsible for the formation of many sedimentary rocks. Traditionally, precipitation is connected with analytical chemistry, where it is used for assaying of elements and compounds and their identification, isolation, purification, and separation. Hence it follows that precipitation processes are of interest in many diverse areas of science, and that these processes have also been studied for different purposes.

Nowadays, a generally accepted scheme of precipitation processes (Fig. 1) is that proposed by Nielsen [3]. According to this scheme, precipitation starts with the formation of nuclei in a supersaturated solution. Nucleation occurs either on impurity particles already present in the solution (heteronuclei) or by the formation of associates of ions and molecules in metastable equilibrium with the solution (embryos, homogeneous nuclei). The process continues by growing nuclei into crystallites, and sometimes also by the formation of secondary nuclei, and terminates by aging of solid particles to a size that causes their sedimentation. Theoretically, the aging of crystals results in the formation of one crystal in equilibrium with the saturated solution.

This chapter is devoted mainly to the current status of knowledge in the area of the kinetics and mechanisms of processes responsible for the formation of crystals of sparingly soluble salts from supersaturated aqueous solutions. Although we have selected a limited area, we give only an outline of the subject necessary for a better understanding of these phenomena, mostly its theoretical aspect, since the literature in this field is very extensive.

II. SOLUTION STABILITY

To understand the kinetics and mechanisms of crystal formation, some knowledge of the thermodynamics of solution and of its properties is needed.

An electrolyte solution can be either stable (undersaturated and saturated) or unstable (supersaturated) with respect to the formation of a solid electrolyte $A_\alpha B_\beta$. The stability of the solution depends on the relation between the electrolyte concentration in solution, which may be expressed as the product of ionic activities,

$$\Pi \equiv a_A^\alpha a_B^\beta \tag{1}$$

and the electrolyte concentration in equilibrium, expressed by analogy as the thermodynamic equilibrium constant of dissolution, often called the "thermodynamic solubility product," K_{sp}. If $\Pi \leq K_{sp}$, the solution by itself is stable, and the formation of a new solid phase is not possible. At $\Pi < K_{sp}$, the crystals, if present in the solution, will dissolve, whereas at $\Pi = K_{sp}$, the system is stable. The formation of one or more solid phases is possible only at $\Pi > K_{sp}$. This excess of the solute concentration above the

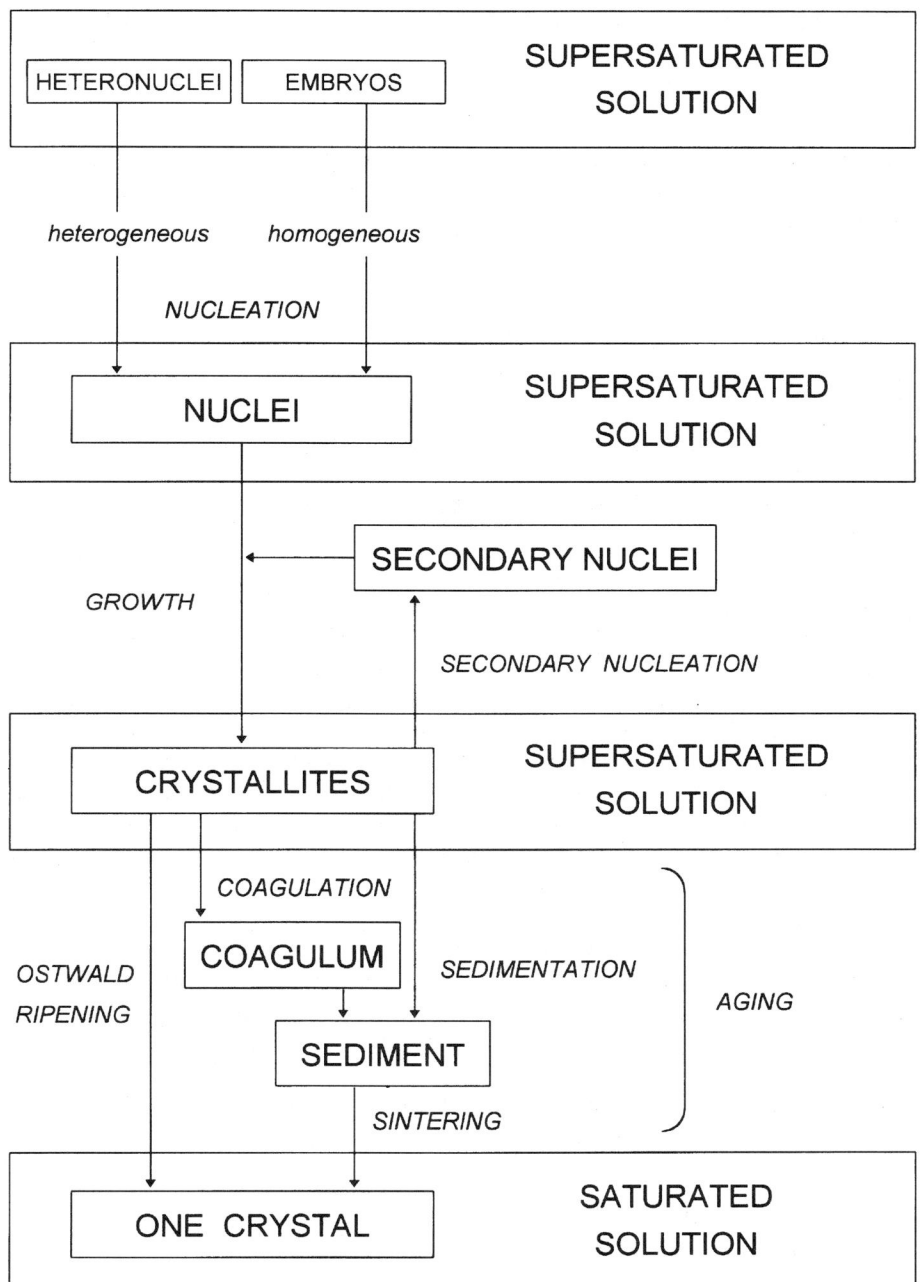

FIG. 1 Schematic representation of pathways and stages of precipitation processes. (From Ref. 3.)

solubility, i.e., supersaturation, can be defined in terms of chemical potentials as

$$\ln S = \frac{\Delta \mu}{RT} \tag{2}$$

where R is the gas constant, T is the thermodynamic temperature, and $\Delta \mu$ is the difference in chemical potential between the solute in a supersaturated solution, μ_2, and the solute in equilibrium with the crystal phase, μ_1:

$$\Delta \mu = \mu_1 - \mu_2 \tag{3}$$

In general, the chemical potential is expressed by

$$\mu = \mu^\circ + RT \ln a \tag{4}$$

where μ° represents the chemical potential in the standard state and a represents the activity of the solute.

Combination of Eqs. (2), (3), and (4) provides the definition for supersaturation:

$$S = \frac{a}{a_s} \tag{5}$$

which is dimensionless and where a and a_s are the activities of the solute in a supersaturated solution and in equilibrium, respectively. The activity is expressed as $a = \gamma \, c/c^\circ$, γ being the activity coefficient, c representing the concentration of the solute, and c° the standard value of concentration, usually $c^\circ = 1$ mol dm^{-3}. The activity coefficient denotes the deviation of a real solution from an ideal solution, in which constituent ions do not interact with each other and where the value of γ equals 1. Depending on the solution concentration and the solution properties, the deviations from ideality can be larger or smaller. Thus, for example, the dilute solutions have activity coefficients closer to unity with respect to the more concentrated solutions. Still, in practical use, especially for sparingly soluble salts, simplifications, such as $a/a_s \approx c/c_s$, are often introduced into calculations. The determination of activity coefficients, particularly in strongly supersaturated solutions, is not always an easy task. As such solutions are unstable, the experimental determination of activity coefficients is not reliable. Only estimations using theoretical statements and appropriate correlations are applicable. For this purpose, the Debye–Hückel theory is commonly used.

The common ways to define supersaturation are as follows:

The quotient of ionic products, Π/K_{sp}
The saturation ratio, $S = a/a_s$ or $S = (\Pi/K_{sp})^{1/\nu}$, if defined for formula units consisting of $\nu = \alpha + \beta$ ions
The reaction affinity, $\phi = RT \ln (\Pi/K_{sp}) = \nu RT \ln (a/a_s)$
The absolute supersaturation, $a - a_s$
The relative supersaturation, $\sigma = S - 1 = (a - a_s)/a_s$

The selection of a particular definition depends on the properties of the electrolyte solution and also often on the data available.

In practice, a supersaturated solution can be prepared from a saturated or an undersaturated solution by changing the temperature or the pH, by adding a poor solvent to a solution with a good solvent, or by a chemical reaction. The chemical reaction is frequently used and often involves reaction between two solutions, but a chemical reaction between a solution and a gas or between a solution and a solid can also be applied. In order to achieve a needed supersaturation upon mixing reactants and also to be able to determine the driving force for the formation of precipitate, the solubilities of both the reactants and the resulting solid phase at a given temperature are to be known.

The level of supersaturation influences the rates of precipitation processes. It also affects the properties of the electrolyte solution and of the resulting solid phase(s).

III. NUCLEATION

In the formation of a new solid phase in a homogeneous solution, nucleation is the first and, energetically, often the most difficult in the sequence of events (nucleation, crystal growth, and aging processes). This step will start only if the energy barrier associated with the extra surface energy of the formation of small clusters (nuclei) is overcome. In order to provide this driving force, a supersaturated solution is needed. When nucleation takes place in a closed system in which there is no exchange of matter with the surroundings, a gradual change of supersaturation will accompany the sequence of precipitation processes. However, although the thermodynamic prerequisite for precipitate formation is made by establishing the supersaturated solution, there is always a period of time when no changes in a physical property of the solution, and consequently no precipitate, can be detected. This period of time, elapsing between the creation of supersaturation and the very first detectable changes in the system due to the formation of solid particles, is called the induction period. The length of the induction period may vary from a fraction of a second to a few years, which is mostly the consequence of different nucleation mechanisms at various supersaturations. In addition, the sensitivity of the detection method being applied [4–6] and the various possibilities of defining a measure may also to some extent affect the length of the induction period [7].

Walton [8] defined the induction period, t_I, as a sum of the time intervals required for the achievement of the steady state of the nucleus distribution, t_d, [9,10], the formation of the critical nucleus, t_n, and the growth of the critical nucleus to detectable size, t_g:

$$t_I = t_d + t_n + t_g \tag{6}$$

For additional information on the definition of the measure of the induction period, the reader is referred to Refs. 7 and 11.

Contemplated in another manner, the maximum stable supersaturation is a function of time and method of observation, as well as of the nature of the solid phase formed and the presence of impurities in the system. This limiting supersaturation for a given system can be determined either by gradually increasing reactant concentrations under isothermal conditions, or by supercooling at constant composition. The terms in use for such supersaturation are the metastable limit, S_m, or the critical supersaturation, S^*. The first is the characteristic value of supersaturation at which nucleation takes place on impurity particles (heterogeneous nucleation). The second is the value of supersaturation at which a significant increase in the number of particles becomes detectable, as a consequence of aggregation of constituent units in the formation of nuclei (homogeneous nucleation). Since the formation of the solid phase, after exceeding the metastable limit, is initiated by impurities, the number of precipitated particles (nuclei) is limited to the number of impurities present in the system and their efficiency in catalyzing heterogeneous nucleation. Therefore the number of particles as well as the limiting supersaturation depend strongly upon the preparation and cleaning of the solution. The number density of particles in such solutions usually ranges from 10^6 to 10^7 particles per cm^3. At higher

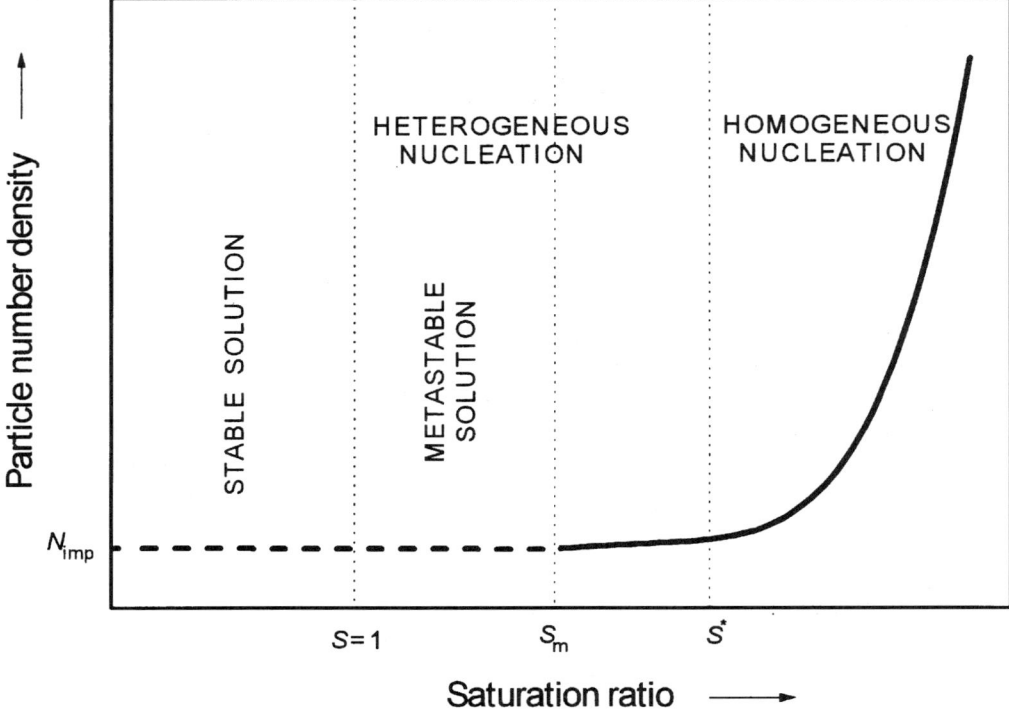

FIG. 2 Number density of precipitate particles as a function of saturation ratio. The curve corresponds to the system in which the formation of the precipitate is initiated by N_{imp} impurity particles, heteronuclei. S_m and S^* are, respectively, the metastability limit and the critical supersaturation for homogeneous nucleation.

supersaturations, heterogeneous nucleation is joined with homogeneous nucleation. The latter becomes the dominant mechanism responsible for the solid phase formation and manifests itself in a significant increase in the rate of nucleation, and consequently in an increase in the number of nuclei (up to several orders of magnitude). Both these mechanisms of nucleation are called primary nucleation, in distinction to the so-called secondary nucleation, in which the formation of a new phase is influenced by the presence of the solid phase of the precipitating material. Figure 2 shows a schematic diagram related to the number of particles produced in solutions of various supersaturations, and the types of nucleation mechanisms.

A. Homogeneous Nucleation

The onset of homogeneous nucleation occurs after the critical supersaturation for homogeneous nucleation has been overcome. The nuclei are formed spontaneously as a consequence of collisions between ions and molecules, by gradual building of ionic or molecular associates, embryos, and nuclei. The kinetic mechanism describing the supersaturated solution is

$$X + X \leftrightarrow X_2$$
$$X_2 + X \leftrightarrow X_3$$
$$\vdots$$
$$X_{i-1} \xrightleftharpoons{\pm X} X_i \xrightleftharpoons{\pm X} X_{i+1}$$
$$\vdots$$

where a single ion or molecule, X, forms an associate that contains i ions or molecules, X_i, and which again by capturing or losing a single unit becomes an X_{i+1} or X_{i-1} aggregate of single units, finally forming a nucleus. This description of the formation of a solid phase nucleus in a supersaturated solution is comparable with the classical nucleation theory model made for homogeneous nucleation in vapor phase [10,12,13]. The classical theory conceives a homogeneous nucleus as an aggregate of critical size in dynamic equilibrium with solution. The change of the Gibbs energy, ΔG, caused by the isothermal formation of the nucleus having volume, V, and surface area, A, can be expressed as

$$\Delta G = V \Delta \mu_v + A \sigma \tag{7}$$

where $\Delta \mu_v = \Delta \mu / V_m$ is the change of the Gibbs energy per molar volume of the new phase and σ is the interfacial tension. For a supersaturated solution $\Delta \mu_v < 0$ since $\mu_2 > \mu_1$ [Eq. (3)], so that the first term of Eq. (7) is always negative. The contribution of this term to the Gibbs energy depends on the nucleus size, which is small for a small volume of nuclei and vice versa. The energetic contribution of the surface of the new phase, expressed in the second term, becomes important for small nuclei, whereas it may be neglected for large ones [7]. By introducing the assumption that all particles formed are of the same shape, so that $A = f V^{2/3}$ and the geometric factor $f = 4A^3/27V^2$ is a constant value, Eq. (7) adopts the form

$$\Delta G = V \Delta \mu_v + f V^{2/3} \sigma \tag{8}$$

For a spherical nucleus of radius r it becomes

$$\Delta G = \frac{4 r^3 \pi}{3} \Delta \mu_v + 4 r^2 \pi \sigma \tag{9}$$

The plot of Eq. (9) for constant values of $\Delta \mu_v$ and σ is shown in Fig. 3. The function resulting from the two terms goes through a maximum at $r = r^*$. At this maximum the Gibbs energy reaches the value of the activation energy for the formation of the new phase, $\Delta G = \Delta G^*$, by analogy with the activation energy of chemical reactions in homogeneous systems. The embryo of radius r^* is a critical nucleus that can grow or dissolve with equal probability. In either case the change of the Gibbs energy will decrease. The size of the critical nucleus can be determined by differentiating Eq. (9) with respect to r and by setting the derivative equal to zero:

$$r^* = -\frac{2\sigma}{\Delta \mu_v} = -\frac{2\sigma V_m}{RT \ln S} \tag{10}$$

This equation gives the relationship between the size of the critical nucleus and the

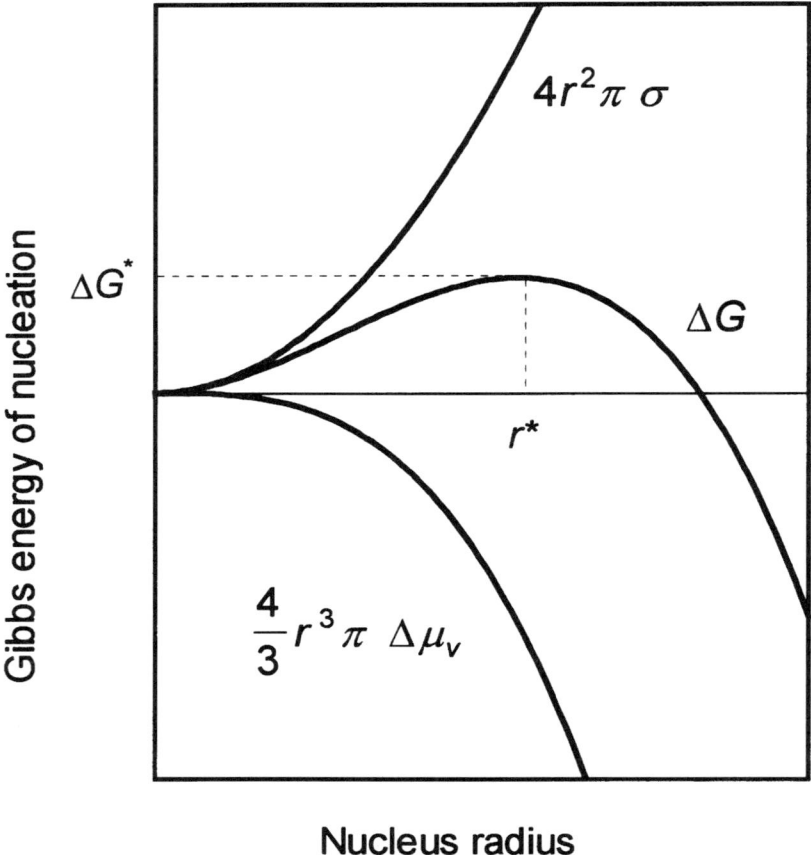

FIG. 3 Change of the Gibbs energy of nucleation as a function of nucleus radius, calculated from Eq. (9).

supersaturation and is known as the Gibbs–Thomson (Kelvin) equation [14,15]. The Gibbs energy for nucleation can be evaluated by introducing Eq. (10) into Eq. (9):

$$\Delta G^* = \frac{16\pi\sigma^3}{3\Delta\mu_v^2} = \frac{16\pi\sigma^3 V_m^2}{3(RT)^2 \ln^2 S} \tag{11}$$

According to the classical model of homogeneous nucleation, the rate at which nuclei form, J, is usually regarded as being controlled by the flux of ions and molecules to the critical nucleus and is often given by the equation [16–18]

$$J = \frac{dN}{dt} = B\exp\left(\frac{-\Delta G^*}{RT}\right) \tag{12}$$

where N is the number density of nuclei, i.e., the number of nuclei produced per volume, and B is the pre-exponential factor related to the frequency of collisions of ions and molecules, which has been estimated theoretically in the range $10^{39\pm3}$ m^{-3} s^{-1} [8,16].

Combination of Eqs. (11) and (12) results in the expression

$$J = B \exp\left(\frac{-16\pi\sigma^3 V_m^2}{3(RT)^3 \ln^2 S}\right) \qquad (13)$$

which shows that the nucleation rate is dependent on supersaturation, S, and interfacial tension, σ.

B. Heterogeneous Nucleation

The formation of a new solid phase governed by the presence of foreign particles in a supersaturated solution is called heterogeneous nucleation. Foreign particles, heteronuclei, may catalyze the nucleation process by reducing the energy barrier to nucleation, so that nucleation takes place at supersaturations that are lower than the critical supersaturation for homogeneous nucleation. In this case, the interfacial tension between heteronuclei and the nucleating solid is smaller than between the solid and the homogeneous solution. Moreover, the better the matching between the two solids, the lower the energy barrier. Experience shows that this matching is more the matter of conformity to atomic distances and lattice type than of chemical similarity [19–21].

The mechanism of heterogeneous nucleation may be conceived as the adsorption of ions or molecules on the impurity surface, and then as either the formation of a two-dimensional nucleus on this surface or the beginning of growth at its dislocation(s) (active sites). Which kind of nucleation will take place depends on the concentration of ions in the interface layer and the catalytic efficiency of the active sites [11].

In analogy to vapor condensation, Walton [8] proposed the following relation to describe heterogeneous nucleation of a solid from solution:

$$\Delta G^*_{het} = \Delta G^* \frac{(1-\cos\theta)^2(2+\cos\theta)}{4} \qquad (14)$$

where θ is the wetting angle of the solid phase by the liquid. In the case of $\theta = 180°$, i.e., when there is no wetting, the Gibbs energies for heterogeneous, ΔG^*_{het}, and that for homogeneous, ΔG^*, nucleations are equal.

Similarly to homogeneous nucleation [Eq. (12)], the rate of heterogeneous nucleation can be estimated from

$$J_{het} = B_{het} \exp\left(\frac{-\Delta G^*_{het}}{RT}\right) \qquad (15)$$

where the pre-exponential factor $B_{het} < B$ [22].

The rate of primary nucleation cannot be measured directly because of the extremely small dimensions of nuclei (not more than a few hundreds of the constituent units), the unpredictable place of their appearance in the system, and their instability. Therefore, in practice, the rate of nucleation is usually estimated on the grounds of measurements of the induction period, t_I, and the number density of particles formed in the system, N, by using the relation

$$J \approx \frac{kN}{t_I} \qquad (16)$$

where k is a constant [7].

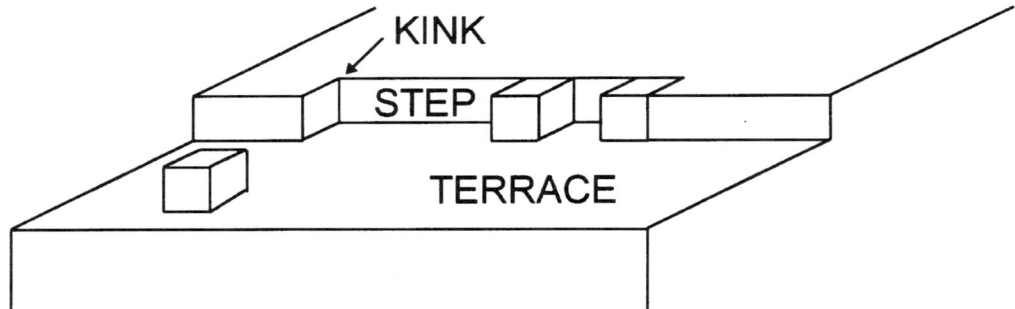

FIG. 4 Schematic drawing of growth sites at a crystal surface.

C. Secondary Nucleation

The growth of primary nuclei to a visible size is sometimes accompanied by the formation of the so-called secondary nuclei [23], thus causing a polydispersed distribution of crystals in the system. There are several ways of formation of secondary nuclei: such nuclei may result from interactions between the already formed crystals and supersaturated solution, collisions of crystals between each other, and collisions between crystals and the stirrer or the reaction vessel. The problem of secondary nucleation is particularly important in the industrial production of solids from solutions, so that mechanisms of secondary nucleation have been discussed in detail [24–26]. This type of nucleation is much more relevant to the precipitation of readily and medium soluble substances than to the precipitation of sparingly soluble ones, since sparingly soluble salts seldom form crystals of such size that secondary nucleation becomes an important process [18,27–29].

IV. CRYSTAL GROWTH

Among the processes included in precipitation, crystal growth is characterized by the most significant changes of solution composition, i.e., the solution concentration of the precipitating phase. Because of the negligible dimensions of the nucleus, as mentioned before, the consumption of matter during the primary nucleation is rather small, while aging processes usually take place under conditions that are close to equilibrium [Gibbs–Thompson equation, Eq. (10)]. At molecular level, crystal growth could be imagined as the continuous addition of one layer of growth units above another. The units arrive from the bulk of a solution at the crystal surface and are fitted to the crystal lattice at growth sites. The growth sites are energetically favored places for incorporation of growth units, which can be described as kinks in steps on the crystal surface (Fig. 4).*
When a growth unit attaches to the surface, it is bonded in one direction to the crystal face. In the attachment to the step or to the kink positions, the bonds in two or three directions, respectively, are involved. The probability of having a growth unit of a typical low molecular sparingly soluble electrolyte in a step position is between 0.7% and 8%,

---

* For this purpose, it is convenient to represent the crystal lattice as a simple cubic, NaCl-like, lattice. Such simplification has no importance in the theoretical treatment of crystal growth and kinetics.

Kinetics and Mechanisms of Crystal Growth

and the probability of having it on the surface is between $2 \cdot 10^{-7}\%$ and 0.004%. This can be shown by a simple calculation based on the increase in surface free energy caused by the addition of growth units to nonkink positions [21].

Although the chemical and physical properties of the precipitate, the number of crystals, and the crystal size distribution in particular, are determined primarily by the mechanism and kinetics of nucleation, the growth kinetics determines how rapidly the solution supersaturation decreases, and thus again the crystal growth influences nucleation. The mechanism of crystal growth could be deduced from the data describing the growth kinetics, that is, by measuring the growth rate as a function of solution concentration and comparing it with the theoretical rate laws based on certain hypotheses. The rate law is an equation in which the rate is expressed as a function of a certain parameter, most usually of concentration of the dissolved substance. The equation is of the general form

Growth rate = constant × function of concentration

There are several ways of expressing (and measuring) the growth rate. When crystals are large enough, the rate can be expressed (and measured) as a linear growth rate of a certain crystal face, v_g, defined as a displacement velocity of that face along its normal. Linear growth rates for different crystallographic faces are usually not equal. The flux density, J, defined as an amount of substance (number of moles) deposited per time on an area, can also be a measure of the growth rate. The flux density is related to the flux of matter, j, by

$$J = \frac{j}{A} \tag{17}$$

and to the linear growth rate by

$$v_g = V_m J = \left(\frac{V_m}{A}\right) j \tag{18}$$

When precipitate consists of a large number of small particles, sizes of about a few micrometers, their linear dimensions could not be determined accurately. In that case, the overall linear growth rate of the crystal, dr/dt, is defined as a time derivative of the radius of a sphere with a volume equivalent to the volume of the crystal. Usually, this is a good approximation as long as the particles are not very aspheric (thin plates, needles, or dendrites).

When crystals grow from solution, a number of elementary processes take place either at some distance from the crystal surface or at the surface/solution interface. Since the transport of growth units (ions, molecules, or segments of flexible linear polymer) from the bulk of solution up to the crystal surface is followed by their transfer to the lattice position, the slowest of the processes involved becomes the bottleneck, thus determining the overall crystal growth rate.* Accordingly, the crystal growth kinetics may be classified as (1) controlled by the transport of matter through solution or (2) con-

---

*Apart from the transport of matter, there are some other processes taking place in the bulk of solution, such as mixing of reactant solutions, chemical reaction, or heat transport, which are not considered in this chapter.

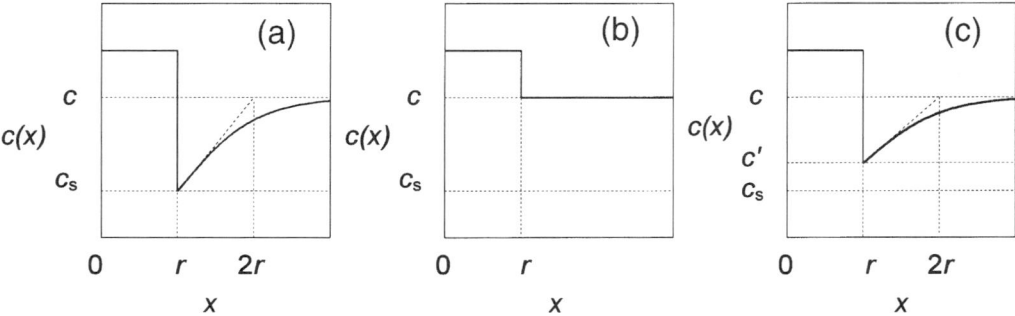

FIG. 5 Concentration profile with respect to distance, x, from the center of a crystal of radius, r, when growth is controlled by (a) transport, (b) surface processes, and (c) both mechanisms. c is the bulk concentration of the liquid phase, c_s is the solubility, and c' is the solution concentration at the crystal surface.

trolled by a reaction at the crystal surface. The concentration in the liquid phase around the crystal, when the growth is controlled by transport, surface processes, or both, is given in Fig. 5.

A. Transport Controlled Growth

Before performing other analyses, we shall assume that crystals are well dispersed and that the suspension is dilute enough so that particles do not influence each other. This means that the average distance between particles is more than 10–20 times their equivalent diameter [11,16]. The transport of growth units through the solution up to the crystal may be influenced by diffusion and/or by convection. When the diffusional concentration field around the particles is not disturbed by their movement relative to the solution, the growth units are transported only by diffusion. This usually holds for particles smaller than 5–10 μm (size depends on the density difference between the solution and the crystal), which sediment sufficiently slowly in a gently stirred solution. A schematic presentation of the solute concentration, $c(x)$, as a function of the distance from the center of the growing crystal, x, is shown in Fig. 5a for the case of diffusion controlled growth. Since the surface processes proceed more rapidly than diffusion, the concentration of solute at the crystal surface ($x = r$, r is the crystal radius) is equal to the solubility, c_s. The total driving force for transport of solute molecules (and for crystal growth) is thus proportional to the difference between the bulk concentration and the concentration at the surface ($c - c_s$). The surface concentration should be distinguished from the concentration in the adsorption or in the electrical double layers.

The rate law describing the diffusion controlled kinetics of crystal growth is derived from Fick's first law applied to the diffusion of growth units toward a spherical sink. Since the diffusion of growth units toward a sink is much faster than the advancement of the crystal layers, a steady state diffusion field around the sink could be assumed [16,30]. The latter approximation (diffusion of growth units toward the spherical sink) holds good for most practical cases; the mathematical treatment of diffusion toward a body of any other shape is much more difficult. In addition, the precipitated particles are often very small, so information about their habits is inaccurate. A flux of matter, dn/dt, diffusing through the area A perpendicular to the x-axis is

$$\frac{dn}{dt} = D \cdot A \cdot \frac{dc}{dx} \tag{19}$$

where D is the diffusion coefficient and dc/dx is the concentration gradient in the neighborhood of the surface equal to

$$\frac{dc}{dx} = \frac{(c - c_s)}{r} \tag{20}$$

(See Fig. 5a, the tangent at the curve at $x = r$.)

Deposition of an amount of matter, dn, increases the volume of the spherical crystal by dr:

$$dV = 4r^2 \pi \, dr = V_m \, dn \tag{21}$$

For a sphere $A = 4r^2\pi$ and from Eqs. (19), (20), and (21) it follows that the rate of diffusion controlled growth is directly proportional to the absolute supersaturation and inversely proportional to the particle size:

$$\frac{dr}{dt} = D \cdot V_m \frac{(c - c_s)}{r} \tag{22}$$

This equation is valid for the case when the solute concentration, c, or the solubility, c_s, can be described by one parameter, as for simple one-molecular compounds. For electrolytes or double salts, the expression becomes more complex, because the diffusion of different ions is characterized by different diffusion coefficients. Thus, for an electrolyte, AB, having the diffusion coefficients D_A and D_B, and the solubility product K_{sp}, the growth rate is defined as

$$\frac{dr}{dt} = \frac{V_m}{2r} \left[c_A D_A + c_B D_B - \sqrt{(c_A D_A - c_B D_B)^2 + 4 D_A D_B K_{sp}} \right] \tag{23}$$

Since the diffusion coefficients normally do not differ much, $D_A \approx D_B = D$, the expression could be simplified to

$$\frac{dr}{dt} = \frac{D \cdot V_m}{2r} \left[c_A + c_B - \sqrt{(c_A - c_b)^2 + 4 K_{sp}} \right] \tag{24}$$

The hypothesis of diffusion controlled growth can be checked by performing several relatively simple tests. For example, it could be checked whether the rate of growth will increase with increasing stirring rate. If so, the rate is diffusion controlled. This test is meaningful only for crystals large enough not to go with the stream under stirring and for very intensive stirring. Furthermore, the activation energy could be calculated from the experimental data and compared with the theoretical dependence of D on temperature. Finally, the growth rate could be calculated from easily accessible data (such as r and Δc, or r, c_A, c_B, and K_{sp}), while D could be either measured independently or calculated from the molar ionic electric conductivities using the Nernst equation

$$D_i = \frac{RT}{F^2} \cdot \frac{\lambda_i}{z_i^2} \tag{25}$$

where D_i is the ionic diffusion coefficient, R the gas constant, T the thermodynamic temperature, F the Faraday constant, λ_i the molar ionic conductivity, and z_i the ionic charge number. The values of the growth rate thus obtained could be compared with empirical results. Usually, even very rough estimates of growth rate will be sufficient, since the rates controlled by a surface reaction are at least 10 times lower.

As mentioned above, the relative movement of the liquid to the growing particles (due to stirring or sedimentation) could increase the growth rate when diffusion is the rate determining step. The effect is due to the disturbance of the stationary diffusion field around the particles, thus supplying the solution of higher concentration closer to its surface. It was observed that the increase in growth rate depends mostly on the size of the particles, or the difference between the particle densities and the surrounding solution, and on the intensity of agitation. Nielsen [30] derived an expression for the size of particles at which convection began to influence significantly the diffusion controlled growth rate (it was arbitrarily assumed that the significant influence was that of an increase of 50%):

$$r_{CD} = \left(\frac{9D\eta}{2a \, |\Delta\rho|} \right)^{1/3} \tag{26}$$

For quiescent and moderately stirred suspensions, the acceleration of gravity $a = g = 9.8$ m s^{-2}, $\Delta\rho = \rho_o - \rho$, is the difference between the solution and the particle density, and η is the dynamic viscosity. Figure 6 shows r_{CD} calculated for particles of various densities in aqueous solution at 20°C and at different stirring intensities ($D = 1 \cdot 10^{-9}$ m^2 s^{-1}, $\rho_o = 1.00$ g cm^{-3}, $\eta = 1.00 \cdot 10^{-3}$ Pa s).

Quantitatively, the influence of sedimentation in the field of gravity or centrifuge and the influence of stirring at a moderate rate can be expressed by introducing a factor F by which the growth rate increases, into the expression for the diffusion controlled rate [Eq. (22)]:

$$\frac{dr}{dt} = D \cdot V_m \frac{(c - c_s)}{r} \cdot F \equiv D \cdot V_m \frac{(c - c_s)}{\delta} \tag{27}$$

This expression is based on a primitive model, the so-called unstirred layer theory. This model assumes that the liquid closer to the surface than some distance δ is immobile and the concentration at the distance δ is the bulk concentration. In the above expression, δ is the thickness of the diffusion (unstirred) layer, $\delta = r/F$. For small crystals, carried with the bulk, $\delta = r$, while for large crystals (10 μm $< r <$ 1 mm), growing in an aqueous solution at ambient temperature, δ could be approximated by

$$\delta \approx \frac{r}{(1 + Pe^*)^{0.285}} \tag{28}$$

where Pe^* is the Peclet number for mass transference,

$$Pe^* = \frac{2r^3 g \Delta\rho}{9\eta D} \equiv \frac{r^3}{r_{CD}^3} \tag{29}$$

For particles larger than about 1 mm, δ becomes proportional to $r^{0.15}$.

If the liquid motion around the particles is highly turbulent, the growth rate can increase above the values of the convective diffusion controlled rate and has been found to depend on the rate of stirring. As could be seen from the equation describing an extreme case of large particles growing at intensive stirring, the rate is exclusively convection con-

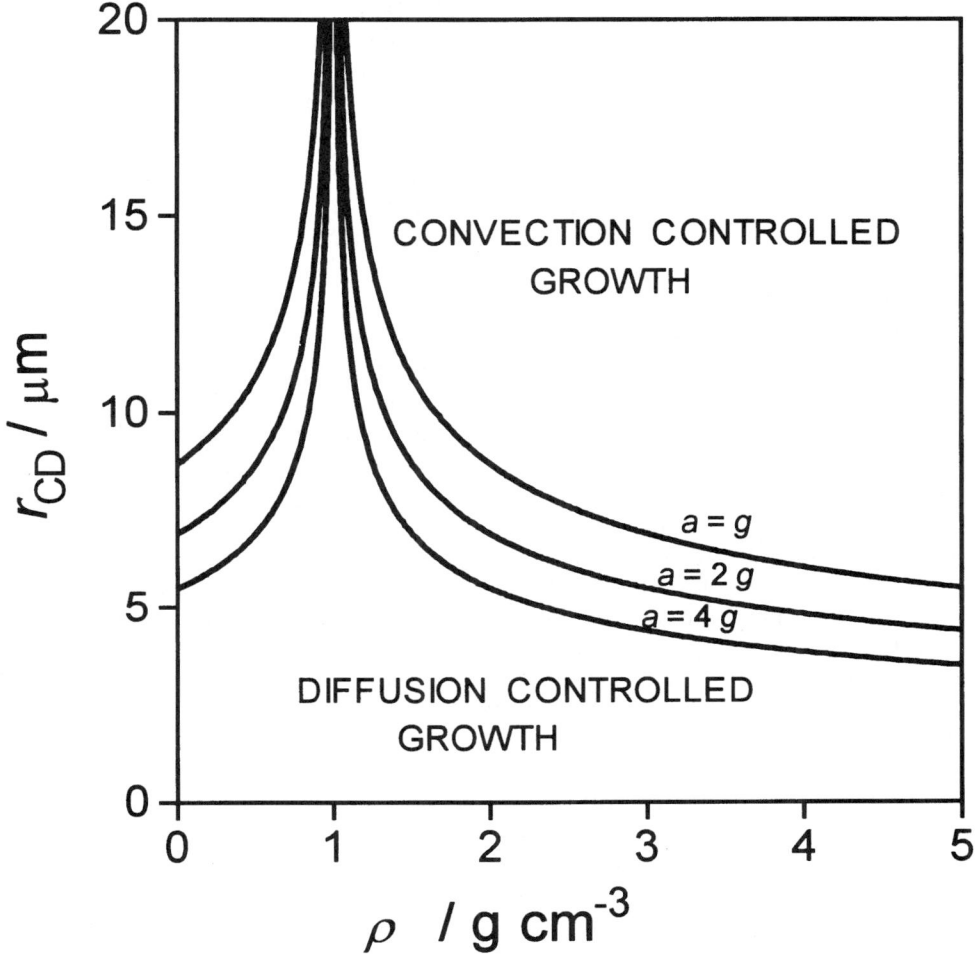

FIG. 6 Size of particles at which convection significantly influences the growth rate as a function of particle density, calculated according to Eq. (26) for $\eta = 1 \cdot 10^{-3}$ Pa s, $D = 1 \cdot 10^{-9}$ m^2 s^{-1}, $\rho_o = 1$ g cm^{-3}, and different stirring intensities.

trolled (the diffusion coefficient, D, is not included in the expression):

$$\frac{dr}{dt} = 1.26 \left(\frac{\rho}{\rho_o}\right)^{1/3} L^{-1/3} \cdot V_m \cdot U \cdot (c - c_s) \tag{30}$$

where L is the container diameter and U is the overall velocity of the liquid.

B. Surface Reaction Controlled Growth

Figure 5b shows a concentration profile of the solute when the overall growth rate is controlled by a certain process at the crystal surface. Since in this case the surface processes are slower than the transport of the solute up to the surface, the concentration of the solute is approximately equal in the whole volume of the solution surrounding the crystals (practically, there is no concentration gradient after establishment of the steady state conditions). Among the processes taking place at the surface of the growing crystals, the predominant role may be played by one (or more) of the following:

The adsorption of growth units on the crystal surface
The migration of growth units across terraces or steps
The dehydration of growth units (ions)
The formation of two-dimensional surface nuclei
The integration of growth units into growth sites (kinks in surface step)

1. Adsorption

For a number of electrolyte crystals, the growth rate has been found to be a linear function of supersaturation [21,31,32]. The Gibbs activation energies of one group of these salts* ($\Delta G^{\neq} \approx 45$ kJ mol^{-1}), calculated from the corresponding rate constants, are about three times the activation energy of the diffusionally controlled growth kinetics (also the linear function of supersaturation). The observed rate law could be explained by assuming that the rate determining step is a surface process or, more specifically, the transition of ions from the bulk of solution to the adsorption layer of crystals. That is, the growth units (ions in this specific case) are attracted by the electric charges on the surface. In order to get close to the surface, they have to release some of their hydration water and to penetrate the hydration layer around the crystal. Since the water molecules leaving the ions should pass through the activated state, any of these steps may be rate determining.† It should be noted that the water molecules are much more strongly attracted by cations than by anions.

The rate law describing the adsorption of ions as the slowest process during the crystal growth may be evaluated from the expression for the flux of ions entering the adsorption layer. It may be assumed that the concentration of ions just outside the crystal surface is equal to the bulk concentration, c, and that the transfer of ions takes place through jumps from a layer of thickness d (the length of a jump is approximated to be equal to the diameter of the water molecule, $d = 0.3$ nm). Under this assumption, the flux is given by the product of the amount of solute in this layer over the surface area, A ($n = Adc$), and the jumping frequency, v_{ad}. Since, by definition, the adsorption is the slowest process and the adsorption layer is at the same time in equilibrium with the crystal ($K_{ad} = c_{ad}/c_s = v_{ad}/v_{ds}$), the net flux is given as the difference between the forward and the reverse flux:

$$j = j_+ - j_- = Adcv_{ad} - AdK_{ad}c_s v_{ds} = Ad(cv_{ad} - c_s K_{ad} v_{ds}) \qquad (31)$$

As $v_g = (V_m/A)j$, the linear growth rate for the adsorption controlled growth is obtained as

$$v_g = (V_m/A)Ad(cv_{ad} - AdK_{ad}c_s v_{ds}) = V_m d v_{ad} c_s (S - 1) = k_1 (S - 1) \qquad (32)$$

In order to test the theory, Nielsen [21,31] compared the values of the Gibbs activation energy for the release of the water molecule with the Gibbs activation energies of adsorp-

---

* Other common features of these salts are the relatively high solubilities or a large amount of water of crystallization [31].
† The forces stabilizing the hydration shell of ions are similar to the forces of the hydration layer of crystals.

tion obtained from experimental results. The release of the water molecules from the hydration shell of cations is related to dehydration frequencies of some cations by the Eyring equation,

$$v_{dh} = \left(\frac{kT}{h}\right)\exp\left(\frac{-\Delta G_{dh}^{\neq}}{kT}\right) \tag{33}$$

The values obtained for $\Delta G_{ad}^{\neq}$ were similar to each other, and somewhat larger than the corresponding $\Delta G_{dh}^{\neq}$, thus indicating the applicability of the theory. More details, as well as the treatment of adsorption specificity, electroneutrality of the adsorption layer, and estimates of K_{ad}, are given in Refs. 21, 31, and 33.

An additional classification of surface processes could be related to the type of crystal surfaces, which may be smooth or stepped or rough at the molecular level. As mentioned before, a kink is the active place for the addition of growth units. In the case of a smooth surface, the appearance of a new step takes place thorough the formation of the surface nucleus. Such a crystal, having a smooth surface, is just an idealized pattern. In real systems, the arrangement of growth units in a three-dimensional pattern is often disturbed by the incorporation of trace amounts of impurities causing the formation of dislocations. The dislocation terminates at crystal surfaces, thus giving a source for the step. Finally, if an interfacial tension of a certain substance is so small that it does not stretch the interface into a plane, the surface becomes molecularly rough. Since there is no need for nucleation on such a surface, and because the concentration of active sites is very high, the growth may become very rapid. The concentration of kinks may be expressed as an "entropy factor," ε [11]. ε is a function of interfacial tension, σ, and the height of a step for a given crystal face, d, and can be calculated using the expression proposed by Bourne and Davey [34],

$$\varepsilon = \frac{4d^2\sigma}{kT} \tag{34}$$

At low values of ε (high concentration of kinks) the integration of growth units could not be the rate determining step and therefore the linear growth rate, given by Eq. (32), reaches the possible maximum value, i.e., that the rate is controlled by the adsorption process. The described pattern of growth is likely to occur in solidification from a melt, but probably never in an aqueous solution [3].

2. Surface Nucleation

The mechanism of formation of a surface nucleus is in a way similar to the formation of a three-dimensional nucleus: the growth units adsorbed on the crystal surface move randomly around and collide mutually. After a series of successful collisions, a stable surface nucleus may form and become the step source. In analogy to the formation of a three-dimensional nucleus, the change of the Gibbs energy of a two-dimensional nucleus is

$$\Delta G' = -N'\phi + P'\sigma_h \tag{35}$$

where N' is the number of growth units forming the surface nucleus, ϕ is the affinity, σ_h is the edge free energy of the step ($\sigma_h = \sigma d$), and P' is the perimeter of the surface nucleus having the area A' and the height d. The height of the nucleus is given by $d = \sqrt{a} = \sqrt[3]{v}$, where a is the surface area per ion and v is the molecular volume. Accordingly, the corresponding change of the Gibbs energy of the critical two-dimensional nucleus is

$$\Delta G'^* = \frac{\beta' \sigma_h^2 d^2}{\phi} = \frac{\beta' \sigma^2 v^{4/3}}{\phi} \tag{36}$$

and the rate of two-dimensional nucleation is

$$J' = \frac{D_s}{d^4} \exp\left(\frac{-\Delta G'^*}{kT}\right) \tag{37}$$

where D_s is the surface diffusion coefficient.

Under appropriate conditions (e.g., for crystals smaller than about 0.1 μm, the surface free of dislocations, and a relatively low supersaturation), the formation of the surface nucleus may be the rate determining process for the crystal growth. In this case, the overall crystal growth rate depends on both the rate of surface nucleation and the rate of two-dimensional growth of the surface nucleus. When a surface nucleus spreads before another nucleus is formed (so-called *mononuclear layer mechanism*), the linear growth rate is given by the product of the available surface area, A, the surface nucleation rate, J', and the thickness of the layer, d:

$$v_g = AdJ' = Ad\frac{D_s}{d^4}\exp\left(\frac{-\Delta G'^*}{kT}\right) = \frac{6r^2 D_s}{d^3}\exp\left(-\frac{\beta'\sigma^2 v^{4/3}}{(kT)^2 v \cdot \ln S}\right) \tag{38}$$

In this expression, β' is the geometric factor depending on the area and the perimeter of the surface nucleus and v is the number of ions in a molecule [11]. It is assumed that the growing crystal always has the same shape; in this particular case, the crystal is a cube with the edge of length r ($A = 6r^2$).

An important feature of the mononuclear mechanism is the dependence of the growth rate on the size [A or r in Eq. (38)]. Thus, assuming that the concentration during the early stage of precipitation is constant (which is a good approximation for crystals smaller than 0.1 μm), Eq. (38) can also be written as

$$\frac{dr}{dt} = \text{constant} \times r^2 \tag{39}$$

By solving this equation one obtains the following relation between r and t:

$$r = \frac{k}{t_I - t} \tag{40}$$

Figure 7 shows a schematic presentation of the relation between the crystal size and the time during the mononuclear layer growth given by Eq. (40). Before the radius becomes infinite at t_I, some other mechanism will take over the rate control.

With increasing crystal size and/or supersaturation, the probability of having more than one nucleus at the surface also increases. In this way, a new layer is formed by intergrowth of nuclei.* The rate of crystal growth following this pattern, the so-called *polynuclear layer mechanism*, does not depend on the crystal size, because the crystal surface is covered by nuclei all the time of the growth process. The theory of polynuclear

---

*Although the resulting layer originates from several nuclei, it is as perfect as a layer formed during mononuclear growth.

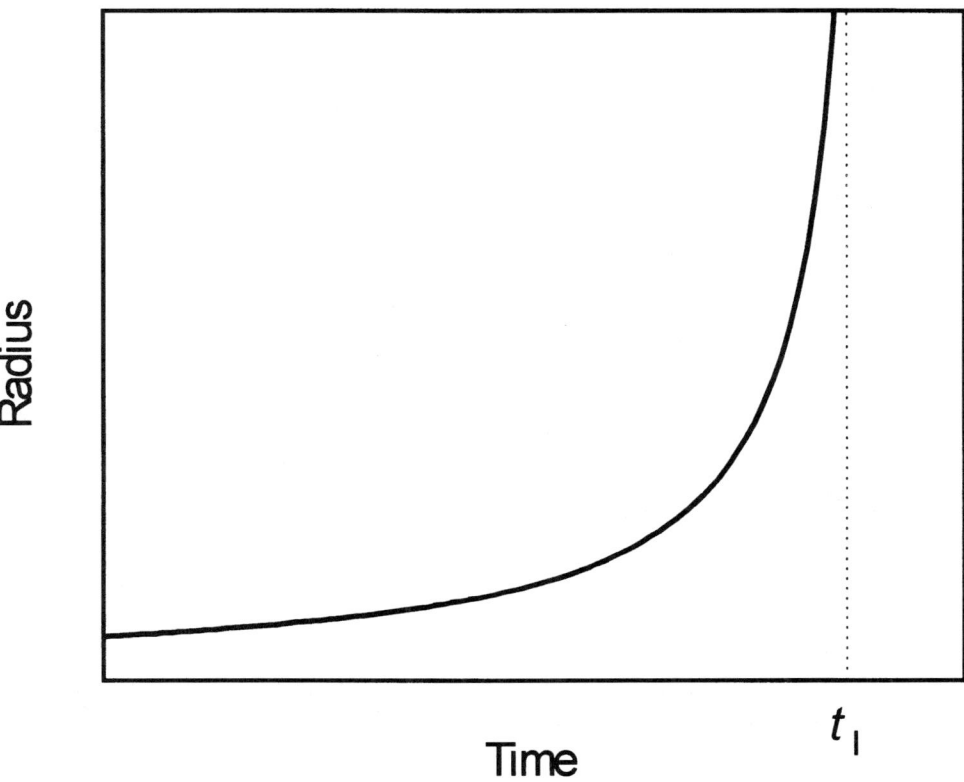

FIG. 7 Schematic representation of crystal size as a function of time for the mononuclear layer growth mechanism, according to Eq. (40).

growth was originally developed for the growth from a melt [35] and was subsequently adopted for the growth of electrolytes from an aqueous solution [31]. The deduction is rather complicated and is based on the assumption that the crystal surface is covered by circular surface nuclei of thickness d and by constituent ions that are in equilibrium with the solution. The linear growth rate is inversely proportional to the time, τ, that passes before the crystal surface of area A is covered by a new layer

$$v_g = \frac{d}{\tau} = d\left(\frac{\pi J' v_\infty^2}{3}\right)^{1/3} \tag{41}$$

where J' is the rate of surface nucleation and v_∞ is the net lateral velocity of growth of the surface island. By inserting the corresponding expressions for J' and v_∞, one obtains the polynuclear rate law:

$$v_g = k_e S^{7/6}(S-1)^{2/3}(\ln S)^{1/6}\exp\left(\frac{-K_e}{\ln S}\right) \tag{42}$$

where

$$k_e = 2dv_{in}(K_{ad}c_s V_m)^{4/3}\exp\left(\frac{-\gamma}{kT}\right) \tag{43}$$

v_{in} and γ being, respectively, the integration frequency and the edge energy, and

$$K_e = \frac{\pi \gamma^2}{3k^2 T^2} \qquad (44)$$

The pre-exponential factor, $F(S) \equiv S^{7/6}(S-1)^{2/3}(\ln S)^{1/6}$, may take several forms [11] and can also be approximated by

$$F(S) \approx G(S) \equiv S^{7/6}(S-1)^{5/6} \qquad (45)$$

for low values of S and by $S^{11/6}$ for large values of S.

For a limited range of supersaturations, the growth rate governed by a mononuclear or a polynuclear layer mechanism can be approximated by the power law of the form

$$v_g = k_g(S-1)^n \qquad (46)$$

where n is the kinetic order of the growth, which can take almost any positive value, depending on the supersaturation range, the physical parameters of the precipitating phase, or even on the way the supersaturation is expressed [11,21]

3. Screw Dislocation

It has been found in practice that crystals grow continuously at supersaturations ($S - 1 \leq 0.01$) much lower than those necessary for the information of a surface nucleus. The explanation of such behavior lies in the fact that real crystals usually have defects in their structure, called dislocations. A frequent type of dislocation is the screw dislocation, schematically shown in Fig. 8a. Crystals having such an "error" in their structure may grow infinitely, with no need for surface nucleation, since by addition of growth units into the step, the step will not disappear but will wind into a spiral instead. This revolutionary concept was introduced and developed by Burton et al. [36] for crystal growth from vapor at low supersaturations and was adopted later on for growth in aqueous solutions [31,33,37].

The structure of the crystal surface with a spiral step (see Fig. 8b) is primarily characterized by its height, d (equal to the diameter of the growth units), the average distance between the kinks along the step, x_o, and the distance between two steps, y_o:

$$x_o = dS^{-1/2}\exp\left(\frac{\gamma}{kT}\right) \qquad (47)$$

$$y_o = \frac{19d\gamma}{kT \ln S} \qquad (48)$$

From the above expressions and using the typical values of the edge energy (work), γ, for electrolytes precipitating in aqueous solutions, it could be easily evaluated that the kink distance at low supersaturation is shorter than the step distance ($x_o < y_o$ whenever $1 < S < 1.6$) [31]. The surface concentration of the kinks (kink density on a crystal face) is given by

$$\frac{1}{x_o y_o} = \frac{S^{1/2} \ln S}{19d^2(\gamma/kT)\exp(\gamma/kT)} \qquad (49)$$

For a range of relative supersaturations up to 10, $S^{1/2} \ln S \approx S - 1$, the above expression

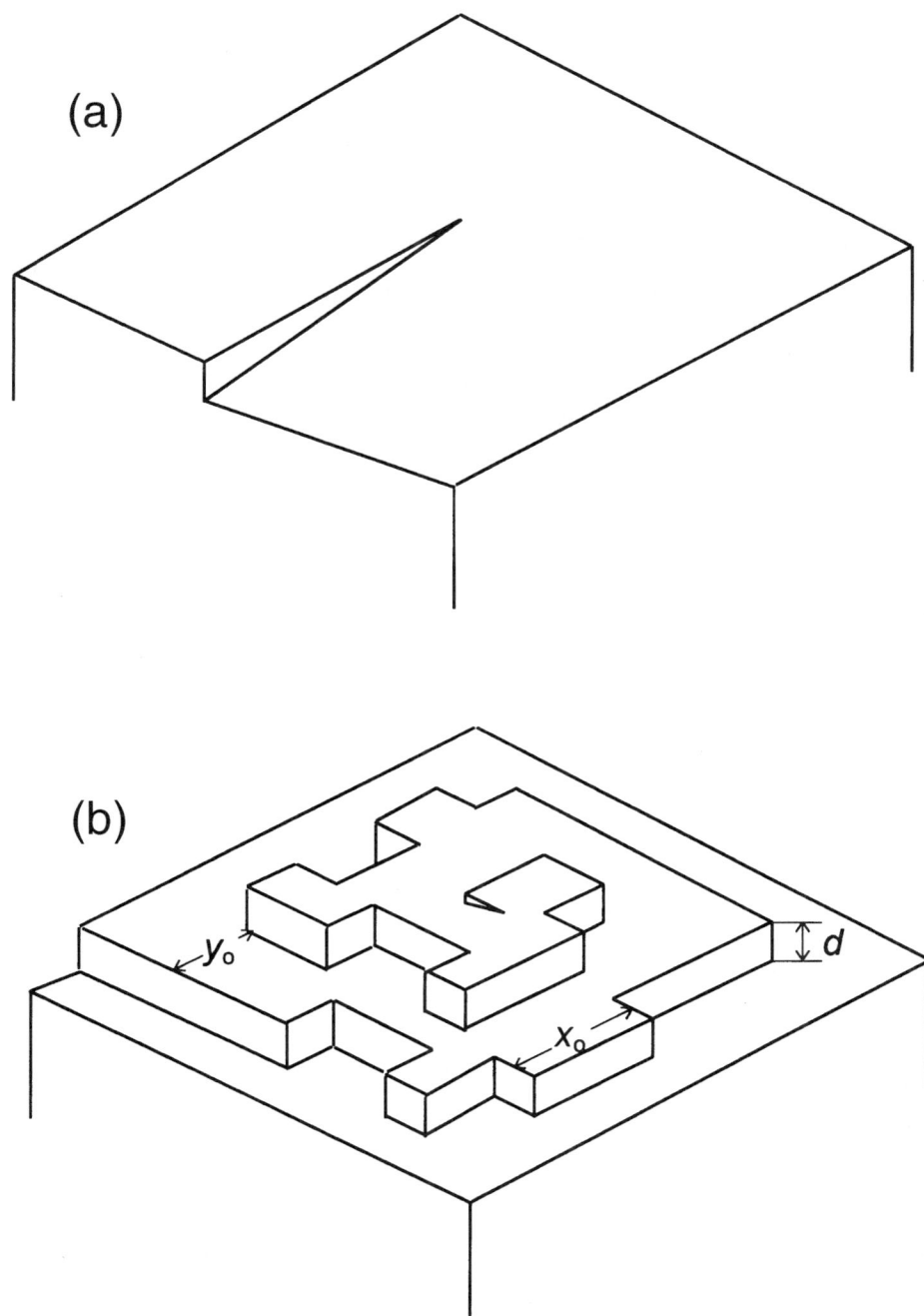

FIG. 8 (a) Slip on a crystal face that produces screw dislocation, and (b) growth step due to a single screw dislocation. Height of spiral step, d, kink distance, x_o, and step distance, y_o, are indicated.

can be approximated by

$$\frac{1}{x_o y_o} \approx \frac{S-1}{19d^2(\gamma/kT)\exp(\gamma/kT)} \tag{50}$$

By analyzing the growth kinetics of crystals having screw dislocations, the parabolic rate law has been found satisfactory for a wide range of supersaturations. Thus, at very low supersaturations, and when adsorption and integration proceed relatively rapidly, diffusion at the surface could be assumed to be the rate determining step. As shown above, for $S < 1.6$, the step distance is larger than the kink distance, and if y_o is at the same time much larger than the length of an average jump of the growth unit, the crystal step may become a continuous sink for growth units. Therefore the linear growth rate of the crystal face is proportional to the lateral growth rate of the step (also visualized as an "adsorption" along the step). The lateral growth rate is, again, proportional to $S - 1$ [see Eq. (32)] and to the step density, y_o^{-1} [Eq. (48)]:

$$v_g \sim k_2(S-1)\ln S \tag{51}$$

For S close to unity, $\ln S$ may be approximated by $S - 1$ and the parabolic rate law is obtained:

$$v_g \sim k_2(S-1)^2 \tag{52}$$

with k_2 being the (parabolic) rate constant.

At supersaturations far from the range of validity of this approximation, the parabolic rate law can be explained by assuming that the rate determining step is the integration of the growth units into the kinks. In this case, the rate is given by the product of the net flux of the growth units (ions) per kink and the density of the kinks at the surface:

$$v_g \sim k_2(x_o y_o)^{-1} j \sim k_2 S^{1/2} \ln S(S-1) \tag{53}$$

For the interval $1 < S < 30$, the above function of S is proportional to $(S-1)^2$.

In order to derive a theoretical expression for the parabolic rate constant, k_2, Nielsen applied the Burton–Cabrera–Frank theory for the growth of crystals in electrolyte solutions [33]. Thus, for the growth rate controlled by the integration of ions from the adsorption layer, he obtained

$$v_g = \frac{2v_{in}d^3 K_{ad} V_m c_s (S-1)}{x_o y_o} \approx \frac{0.1 v_{in} d K_{ad} V_m c_s}{(\gamma/kT)\exp(\gamma/kT)}(S-1)^2 \tag{54}$$

Since the integration jump actually consists of two simultaneous processes, the dehydration of a cation and its diffusion into the lattice position, the integration frequency, v_{in}, and the corresponding activation energy of integration, $\Delta G_{in}^{\neq}$, could be

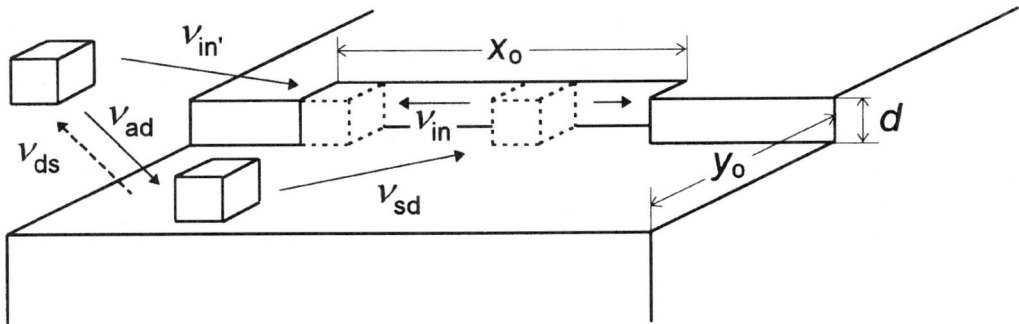

FIG. 9 Possible routes of migration of growth units. Height of step, d, average kink, x_0, and step, y_0, distances are shown.

calculated from the empirically obtained values of the rate constant k_2 by using the Eyring equation [Eq. (33)]. Correlating these $\Delta G_{in}^{\neq}$ values with the values of the activation energy for the removal of a water molecule from the hydration shell of the cation, $\Delta G_w^{\neq}$, Nielsen [21,31,33] found both of them to be equal for monovalent cations and $\Delta G_{in}^{\neq}$ to be about 50% higher for divalent cations.* These results corroborate the Reich–Kahlweit hypothesis [38] that dehydration of cations determines the integration into the lattice.

Nielsen also considered the possibility of surface diffusion or of local diffusion at kinks and the possibility that the integration directly from the solution may be the rate determining step. In all these cases, the quadratic dependence of the growth rate on the relative supersaturation was found to be a better approximation to the exact equation than any other simple mathematical expression. Figure 9 shows the possible routes for migration of growth units on the crystal surface and in the surrounding solution. The growth units can be adsorbed or desorbed, can diffuse on the surface or can be integrated into a kink position. All these processes are comparable, since the elementary event is the same: the jump of a growth unit. The most probable route is determined by the relative proportions of x_0 and y_0, and by the relations between the frequencies of the adsorption/desorption (v_{ad}, v_{ds}), the surface diffusion (v_{sd}), and the integration jump of an adsorbed (v_{in}) or a nonadsorbed (dissolved) ion (v_{in}').

C. Growth Controlled by Different Mechanisms

It has already been mentioned that the mechanism governing the overall rate of growth sometimes depends on parameters that change during the propagation of the process. Thus, for the mechanisms in which the growth rate is size dependent, such as mononuclear or diffusion controlled growth, some other mechanisms would take over after the particles have reached a certain size. Similarly, in a closed system, supersaturation decreases with time, and the nucleation controlled mechanisms would probably be replaced by some other. It should be emphasized again that when two processes are parallel, such as growth on screw dislocations and surface nucleation controlled growth, their rates are added, while two consecutive processes "share" the same driving force. In practice, different elementary mechanisms simultaneously influence the growth rate. The mechanism that

---

*Water molecules are more strongly bonded to divalent than to monovalent cations.

dominates over the others is called the rate determining one. The change from one rate determining mechanism to another is not abrupt; it takes place gradually. A simple example for the interplay between the two consecutive processes is the volume diffusion and the growth on a screw dislocation. In this case, the total driving force is divided into the driving force for diffusion $(c - c')$ and the driving force for surface reaction $(c' - c_s)$, as shown schematically in Fig. 5c. Since the growth rates of both processes are equal, c' can be eliminated by solving the respective double equation:

$$v_g = k_d(S - S') = k_2(S' - 1)^2 \tag{55}$$

in which $k_d = DV_m c_s/r$ and $S' = c'/c_s$. The compound rate law can be written as

$$v_g = \frac{2k_d k_2 (S - 1)^2}{2k_2(S - 1) + k_d(1 + \sqrt{1 + 4(S - 1)k_2/k_d})} \tag{56}$$

At low supersaturations this rate law can be approximated by the parabolic rate law and at high supersaturations by the linear rate law.

V. CRYSTAL DISSOLUTION

In general, there is no conceptual difference between crystal growth and crystal dissolution except for the directions of these processes. Formally, dissolution can be considered as growth in undersaturated solution, and most of the elementary processes mentioned above are the same, with net fluxes of molecules or ions having an opposite direction. An additional difference between these two processes is the fact that during dissolution the constituting units are more accessible to the solvent molecules, those at corners and edges of crystals in particular. Consequently, there is no need for processes equivalent to surface nucleation* or growth at a spiral step, so the rate determining step of dissolution is very often the diffusion of growth units away from the crystal surface. For many electrolytes in nearly the same range of relative supersaturation and undersaturation, growth was found to be controlled by a surface process and dissolution by diffusion.†

The rate laws describing the kinetics of crystal dissolution are similar to those for crystal growth [Eqs. (22), (42), (54)], the only difference being the way of expressing the driving force. Thus, instead of $(S - 1)^2$ for parabolic growth, $(1 - S)^2$ [39] or $(K_{sp}^{1/2} - \Pi^{1/2})^2$ [40] can be used for dissolution and, similarly, $(S - 1)^{5/6}$ and $\ln S$ can be replaced by $(1 - S)^{5/6}$ and $-\ln S$, respectively, for dissolution controlled by surface nucleation. Accordingly, v_g is substituted by $-v_g$, thus indicating the reduction in the crystal linear dimensions. For diffusion controlled dissolution, the relative supersaturation is expressed as $c_s - c$ [41,42].

---

*A process equivalent to surface nucleation is the formation of holes in the top layer of the crystal.
†A practical aspect of this is the fact that a saturated solution is achieved much faster by crystal dissolution than by growth from a supersaturated solution.

Another typical property that should be considered when analyzing the diffusion controlled kinetics of dissolution is the relation between the amounts of solvent and solute. Thus, if the amount of solute is smaller than the amount that could be dissolved under given conditions, i.e., the amount of solute is insufficient to produce a saturated solution, the solute completely disappears from the system at a well-defined time. If the solute is in excess, the process is characterized by a constant half-time [7,30].

VI. INFLUENCE OF IMPURITIES

Any ion or molecule other than the constituent component of the precipitating compound, including the solvent used, is considered an impurity. If the impurity, except the solvent, is added purposely to a system, it is called an additive, in distinction from a variety of impurities naturally existing in all systems. Even a negligible amount of foreign substances may strongly affect the kinetics of precipitation processes, the size and shape of particles, the chemical composition or crystal modification of the precipitated compound, and many other properties of the system and the solid phase.

On the ground of the nature of their action, two basic types of impurities can be distinguished: (1) those that incorporate into the crystal lattice and may cause lattice disturbances as well as a reduction of the overall crystal growth rate and (2) those that adsorb onto the crystal surface and affect the rate of growth of a particular face or of all faces present in the growing crystal. The first group refers primarily to inorganic ions and metal complexes, the influence of which is due to the electrostatic interaction between crystal ions and impurity species, some of which may easily enter the crystal lattice and substitute for the constituent ion, forming solid solutions [1,43–46]. These impurity species are rather small (up to 2–3 atomic dimensions) and usually effective in concentrations above 10^{-5} mol dm^{-3}. The second group of impurities are substances such as polyelectrolytes, multifunctional long-chained molecules, and smaller organic compounds, the effect of which is based on the formation of bonds between the functional groups or negatively charged ligands and the cations at the crystal surface. Such molecules adsorb on specific sites at a certain crystal face(s), thus inhibiting the propagation of regular growth steps across its surface and slowing down the growth in the direction perpendicular to this face. The slower the growth rate in the given direction, the larger the surface area of the crystal face perpendicular to this direction. In this way, the relative growth rates of the crystal faces are affected and concomitantly so are their relative surface areas. As the shape of a crystal is specified by its faces, the adsorption of impurities may cause a change in crystal morphology. The change is more likely when impurities adsorb selectively and irreversibly, remaining immobile on the crystal face terraces. When the adsorption is nonselective and reversible, the impurity being mobile and able to choose among specific surface sites (steps, kinks, or terraces) at the crystal–solution interface, the change of crystal morphology does not have to be pronounced. In the first case, the crystal face grows by penetration of the growth steps in between the adsorbed additive molecules. A consequence of this is the bending of the steps and the slowing down of their advancement. The growth rate of the crystal face can be completely suppressed when the additive concentration on the crystal surface becomes such that the distance between the additive molecules is smaller than the surface nucleus of the growing crystal. Very low concentrations of these additives (sometimes about 10^{-8} mol dm^{-3}) may show such an effect [47–49]. Distinct from the selective adsorption, the nonselectively adsorbing impurities sometimes show an interesting phenomenon: the dependence of the effect of the crystal–impurity interaction on the relative concentrations of the growing crystal and the impurity. Thus, for instance, the reversal of trends in the linear growth rates of certain crystal faces [50] or in enhancement of nucleation [51] can be observed by increasing the impurity concentration in the same system. This means that under certain circumstances in a single system, the same impurity can act either as a promotor or as an inhibitor of a precipitation process; the ability of an impurity to act differently in various systems is also well known.

The most effective of the second group of impurities on the crystallization of sparingly soluble salts turned out to be polysaccharides, proteins, and soluble polymers. These were found to affect most of the precipitation processes [49,52–56], the consequences of which being the habit modification of crystals [57], a change in polymorph type [58–60], a reduction in symmetry because of the formation of solid solution sectors inside the host crystal [61,62], a change of the crystal number and size distribution [51], etc., or a combination of several of these effects.

Apart from the two groups of impurities mentioned above, there are other groups of foreign substances showing different mechanisms of action in influencing the precipitation processes or the shape of crystals formed. The so-called tailor-made additives can be regarded as a special type of the second group of impurities. These are made up of huge molecules, molecular mass of about 10^4, having a site with the molecular structure similar to that of substrate molecules, so adsorbing preferentially on the stereochemically compatible crystal faces [63,64]. As a result, the growth normal to these faces is affected. Surface active agents (surfactants) can be regarded as another special kind of additives which act through adsorption at the solid–liquid interface and aggregation in the bulk solution, or through both. Their molecular structure allows a lot of energetically favorable conformations, so that the kinetics of precipitation processes can be greatly influenced, thereby leading to changes in the crystal habit [65–68]. In addition, the influence of some impurities is due to the formation of complexes with the constituent ions present in the bulk solution, which reduces the concentration and consequently the supersaturation available for crystal growth and hence the rate of this process [69,70].

There is no general principle by which the influence of a certain impurity on a particular precipitation system can be predicted. Still, various models have been proposed for the interpretation of mechanisms involved in the change of crystal habit [63,71–74] and in the inhibition or promotion of crystal growth [75–78] as influenced by impurities. In this section only a few of the proposed models that describe the effect of impurities on the growth kinetics are considered. The distinction between these models can be made on the basis of the adsorption mechanism as a consequence of the impurity concentration and the type of the crystal face involved, i.e., the adsorption on a smooth or a rough face, both at either (a) low or (b) high impurity concentrations. In contrast to smooth faces, which in general require an impurity capable of adsorbing on the surface terrace, the adsorption on faces with a relatively high density of kinks gives many more possible sites for impurity adsorption. Therefore, and because of the nature of deposition on such a face, the concentration of impurity required to inhibit the growth rate is much higher than that for the smooth one.

(a) For low impurity concentrations, the adsorption may take place at any type of specific sites on the crystal face. Thus the adsorption with no preference for any of the specific surface sites, kinks, steps, or terraces, is characteristic of the reversible adsorption of mobile impurity molecules. However, the case of preferential adsorption on the surface terrace is related to the mechanism of irreversible adsorption of immobile impurity molecules.

The fraction of surface sites occupied by the reversibly adsorbed impurity molecules, θ_{imp}, can be expressed by a modified adsorption isotherm, for instance by the modified Langmuir isotherm [54,79]:

$$\theta_{\text{imp}} = \frac{K \cdot c_{\text{imp}}}{K \cdot c_{\text{imp}} + 1} \tag{57}$$

where c_{imp} is the concentration of impurity in the solution and K is the constant. The advancement of the growth steps in the presence of impurity is influenced by the different rates of deposition of the crystal constituent ions at the sites occupied and unoccupied by the impurity. Assuming that the rate of deposition at the occupied sites decreases with increasing covered surface and finally becomes zero, the rate of growth of the crystal face becomes

$$v_{\text{imp}} = v_g (1 - \theta_{\text{imp}}) \tag{58}$$

where v_g is the growth rate without impurity evaluated at the same supersaturation and crystal shape.

Combination of Eqs. (57) and (58) yields

$$\frac{v_g}{v_{\text{imp}}} = K \cdot c_{\text{imp}} + 1 \tag{59}$$

i.e., the relative decrease in growth rate or, in other words, the effectiveness of the impurity.

When the impurity adsorbs irreversibly on the surface terrace, an entirely different mechanism of adsorption is involved, described by Cabrera and Vermilyea [71]. In this case, the velocity of growth steps decreases according to

$$v_r = v_\infty \left(1 - \frac{r^*}{r}\right) \tag{60}$$

where v_r is the velocity of a step with curvature of radius r, v_∞ is the velocity of a straight step, and r^* is the radius of the critical nucleus.

(b) At high impurity concentrations the decrease in crystal growth rate is a result of the formation of either a two-dimensional adsorption layer [75] or three-dimensional clusters [75,78] of the impurity.

The two-dimensional adsorption layer can form only if there is a structural fit between the growing crystal face and the forming layer, the mechanism being similar to that of the formation of an epitaxial or a thin film layer. The growth kinetics of the crystal face without impurity relative to the growth kinetics obtained with impurity at the same supersaturation gives the effectiveness of the adsorbed impurity.

The reduction in growth rate of a crystal face caused by the formation of impurity clusters, randomly localized on the crystal surface, can be described by the Cabrera–Vermilyea mechanism of adsorption [71], considering the participation of two kinds of immobile impurities involved in the process. This is because impurity clusters act as an additional barrier to the growth of the crystal face in regard to the impurity molecules which are also adsorbed.

The effect of impurity on the crystal growth kinetics in precipitation processes, in which the growth of crystals is often presented as the overall crystal growth rate, can be expressed as the ratio between the overall crystal growth rate obtained in the absence, dr/dt, and in the presence, $(dr/dt)_{imp}$, of impurity [80]:

$$e = \frac{dr/dt}{(dr/dt)_{imp}} \tag{61}$$

This holds only if both growth rates are determined at the same supersaturation and shape of crystals.

More about the present understanding of this subject can be found in some recent reviews and books [11,81,82].

VII. AGING

Once formed, solid particles in contact with mother liquor undergo further physical and chemical changes. All these changes result from the tendency of the system to establish equilibrium, thus minimizing the overall Gibbs energy. According to Eq. (7), this will be reached by decreasing the solid–liquid interface to the least possible area, since the contribution of the interfacial tension to the Gibbs energy is proportional to the interfacial area. The processes by which the solid–liquid interface decreases are generally described as "aging" and this includes flocculation* (aggregation, coagulation, agglomeration), growth of larger particles at the expense of smaller ones (Ostwald ripening), and transformation of particles from a metastable into a stable modification.

A. Flocculation

Particles present in a suspension collide with each other, which may end in the formation of larger size entities, or particles may separate again. All successful collisions result in a decrease in the total number of particles and consequently in a reduction of the interfacial area. The flocculation of particles due to thermal (Brownian) motion is described by the relation [83,84]

$$N_i = \frac{N_o(t/t_{1/2})^{i-1}}{(1+t/t_{1/2})^{i+1}} \tag{62}$$

where N_i is the number density of particles at time t consisting of i primary particles, N_o is the initial number density, and $t_{1/2}$ is the time required to reduce the initial number of particles to one-half and is given by

$$t_{1/2} = \frac{3\eta}{4kTN_o} \tag{63}$$

where η is the viscosity. From Eq. (63) an approximate half-time for the flocculation of

---

*The terminology for these processes has not yet been coordinated.

particles initiated by homogeneous and heterogeneous nucleation can be evaluated. For instance, in an aqueous system at the temperature of 300 K with $N_o \approx 10^{17}$ m^{-3} and $N_o \approx 10^{12}$ m^{-3}, $t_{1/2}$ changes from approximately 1 s, or less for larger N_o, to several hours, for smaller N_o. This means that the flocculation of nuclei formed by homogeneous nucleation probably starts during the nucleation process or immediately after it [52,85], whereas flocculation in heterogeneously nucleated systems is not likely to occur in the initial stages of particle formation.

B. Ostwald Ripening

Ripening is another process by which a heterogeneous system, consisting of a polydispersed solid phase in contact with the mother liquor, decreases its interfacial area. This is achieved by dissolution of small particles and simultaneous growth of large ones. Since small particles have a larger surface area per molecular unit and thus a higher Gibbs energy than large particles [Eq. (7)], the processes leading to equilibrium (by decreasing the interfacial area and minimizing the Gibbs energy) occur spontaneously. The equilibrium in such a system can be established only when all crystals reach a uniform size [86,87] or when eventually the solid phase is united in one large crystal. Ostwald [88] was the first to explain this phenomenon using the Gibbs–Thomson equation [Eq.(10)]. From this equation, which shows the dependence of the size of a crystal on supersaturation, it follows that a solution can be in equilibrium only with a precipitate consisting of particles of the same size. In the case of polydispersed precipitate, this solution will be undersaturated with respect to particles smaller than those being in equilibrium and supersaturated with respect to larger ones. Consequently, smaller particles will dissolve and larger particles will grow.

Ostwald ripening has been observed in many experimental studies [89,90], and a number of papers have also been devoted to the theory of Ostwald ripening [86,91–94]. Nevertheless, the proposed theoretical solutions are hardly applicable; these treatments of the theory completely neglect flocculation as a possible process coexisting with the ripening.

C. Transformation of Metastable Phases

It may be said that flocculation and Ostwald ripening are processes that probably always take place in the early stages of particle formation. The transformation process occurs when a metastable modification(s) of the solid phase is formed initially and only later transforms into a thermodynamically stable modification. The fact that a solid phase can exist in several modifications from which only one is thermodynamically stable under given conditions was recognized by Ostwald [95]. According to his "law of stages," the formation of the stable modification is preceded by the formation of less stable ones, i.e., by more soluble modifications. Priority in the sequence of appearance of possible modifications is thus determined by their solubilities and by the relative rates of nucleation and crystal growth for these modifications [91]. This is corroborated by Eq. (10), according to which the radius of the critical nucleus, r^*, is proportional to the interfacial tension, σ. It follows that at a given supersaturation the modification with lower σ, i.e., more soluble and less stable modification, will create a smaller and thus thermodynamically more stable critical nucleus, which is energetically more convenient. Since an identical deduction can be derived from the expression for the rate of nucleation [Eq. (13)], one may say that the modification that corresponds to the nucleus with the lowest σ will first precipitate from the supersaturated solution.

There are a variety of forms in which solid phases may exist, such as hydrates solvates, polymorphs, and amorphous solids. Under suitable conditions, a substance may exist in several different modifications. A well-known example is calcium carbonate, which forms three anhydrous crystalline polymorphs (vaterite, aragonite, and calcite), two hydrates (hexahydrate and monohydrate), and an amorphous hydrated modification [96–98]. At high supersaturations, at which homogeneous nucleation takes place, and in particular when highly hydrated cations are involved, the formation of amorphous hydrated phases [99–102] and metastable higher hydrates [40,58,103,104] is to be expected. Such behavior in the precipitation of certain solids can be linked with the fact that at very high initial supersaturations the critical nucleus may be very small, even smaller than the unit cell of the crystal into which it develops [18]. As such small nuclei grow by flocculation rather than by building the crystal lattice of the constituent units from the surrounding solution, the formation of hydrated amorphous and poorly crystalline structures is likely in the initial stage of precipitation.

There are two possible ways of transition from one solid phase to another. In the first, the metastable phase transforms into the thermodynamically stable phase through an internal rearrangement of ions, atoms, or molecules (solid-state transition) [105–107]. In the second, the metastable phase dissolves in the solution with which it is in contact, and simultaneously an independent nucleation and growth of the stable phase from the same solution occur (solution mediated transformation) [42,108–111]. The overall rate of this type of transformation is determined by measuring the fractions of the solid phases present in the system during the process, as well as the rates of dissolution of the metastable phase and of the growth of the stable one.

A detailed analysis of aging is beyond the scope of this chapter; the reader is referred to the literature quoted above and also to Refs. 7, 11, and 16.

VIII. EXPERIMENTAL APPROACH

In the experimental study of a precipitation system, consideration should be given to both the qualitative and the quantitative information pertaining to the characteristic properties of the system. These properties are determined by parameters such as supersaturation, pH, ionic strength, ratio of ionic activities between reactants and foreign ions or molecules, temperature, and time of reaction.

The understanding of equilibrium conditions is the first requirement in approaching a precipitation system. Therefore it is helpful to construct the appropriate solubility and/or equilibrium phase diagrams [112–114] from the available literature [115] or from the newly measured data. Such diagrams give, respectively, the pH/concentration and temperature/solution composition regions of stable solid phases and possible unstable precursor phases that are likely to form in the system.

Additional valuable information can be achieved by consulting a precipitation diagram of the system. From the precipitation diagrams one can see the dependence of some properties characteristic of the system (e.g., turbidity, pH) or of the precipitate (e.g., morphology, modification of precipitate, etc.) on the initial reactant concentrations (activities) [85,116–118]. In order to obtain such a diagram of a two-component system, a series of precipitation curves has to be made in a wide range of concentrations at constant temperature. This is possible by a stepwise change of the concentration of one reactant at the constant concentration of the other, and by observation of the chosen characteristic(s) at a predetermined time. The time of observation has to be long enough so that a metastable or a stable equilibrium at a particular temperature can be reached.

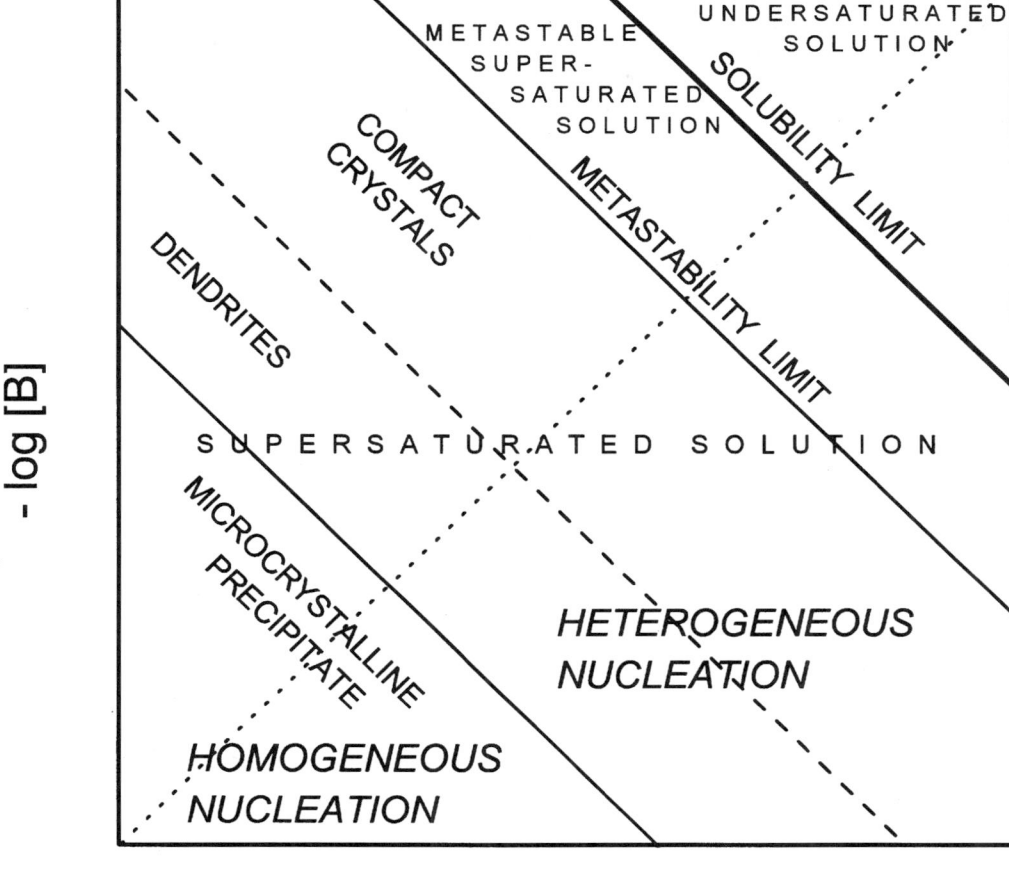

FIG. 10 Schematic representation of a precipitation diagram for a sparingly soluble electrolyte AB. Constituent ions are assumed to be present in the solution only as simple ions A and B (or A^{z+} and B^{z-}).

Thus by plotting the initial concentrations (activities) of one reactant against the initial concentrations (activities) of the other and by connecting the equal values of the observed properties, regions are formed within which the precipitates of similar characteristic (i.e., morphology or chemical composition) are grouped. To characterize a more complicated system, the construction of more than one precipitation diagram may be required. Figure 10 shows a schematic precipitation diagram for the binary electrolyte AB with different stability regions separated by straight lines and perpendicular to the line corresponding to the equimolar concentration ratio of reactants. The construction of this diagram is made under the assumption that the formation of ion pairs and complexes in the solution is negligible, that the chemical composition of the solid phase is independent of the reactant ratio in solution, and also that the nucleation and growth of the solid phase depend only on supersaturation. The initial reactant concentrations (or the corresponding activities), expressed as molarity, are plotted as negative logarithms. The well-defined solubility limit separates the region of undersaturation, where no solid phase could be formed, from the region of supersaturation. The first area next to the solubility line refers to the metastable supersaturated solution where the formation of a solid phase without seeding the solution is not likely within an appreciable observation time. The value of supersaturation corresponding to the line separating the areas of the solution metastability and heterogeneous nucleation is the metastability limit. At observation times long enough, the area of metastability may disappear, and this line can merge into the solubility boundary. In the other two regions, spontaneous nucleation occurs in which the formation of compact crystals and dendrites indicates the mechanism of heterogeneous nucleation, while the formation of numerous small, rather monodispersed particles refers to the prevalently homogeneous nucleation. Thus precipitation diagrams provide qualitative information on the influence that the composition of the solution has on the properties of precipitates and on the mechanisms of nucleation or even on crystal growth [16].

Quantitative information on the rate and mechanism of the precipitate formation can be obtained only by measuring chemical or physical properties of the crystallizing solid and of the associated solution during the reaction time. These kinetic measurements can be made and classified in different ways. Thus, for instance, we can distinguish between the measurements made on single crystals [119–122] and on the overall crystal population in a system [6,32,41,123–125]. Each of these methods has its advantages and drawbacks, and the choice of application depends on the results expected. In the former case, the changes in surface steps or crystal face are usually observed by a microscope and recorded over time, and the supersaturation of the surrounding solution is adjusted and controlled mostly by variation of the solution temperature. In this way, the growth rate of a crystal face is measured as the face advancement in the direction perpendicular to this face, and the linear face growth rate, v_g, is determined according to the relation given by Eq. (18).

The growth rate of a single crystal face hardly represents the growth rate of the crystal as a whole, because these rates often differ for different crystal faces. Thus for the growth rate averaged over the whole crystal, the growth rates and the areas of all faces on a crystal have to be known [126]. Single crystal measurements are favorable in studying the crystal habit modification and the crystal growth process in relation to fundamental theories [36,127], but these can be made accurately only if crystal are large enough. But some recently applied techniques (atomic force microscopy, laser Raman spectroscopy, synchrotron radiation, etc.) give a new dimension in the in situ investigation of surface morphology and kinetics of growth and dissolution of different crystal faces [128–130]. The study of the early stages of the growth of newly formed surface nuclei as well as atomic-scale information on the crystal–solution interface layer is now possible. Figure 11 shows the elementary step spirals and surface nucleation on terraces observed by atomic force microscopy.

For small crystals, kinetic measurements have to be performed on a large number of crystals and are based on the determination of the amount of the solid phase formed in the system as a function of time (changes is mass or size of crystals) and/or of the solution concentration (activity) as a consequence of the deposition of solute into the crystals. The data thus obtained are represented by the "progress curves," i.e., as the time dependence of the chosen characteristic property that determines the progress of the process. The choice of the property depends on the experimental system under investigation and on the information required for the study. In order to obtain the changes in solution concentration, measurements of pH, solution conductivity, scattered or transmitted light intensity, and electrochemical potential are often used. Since the recorded property is

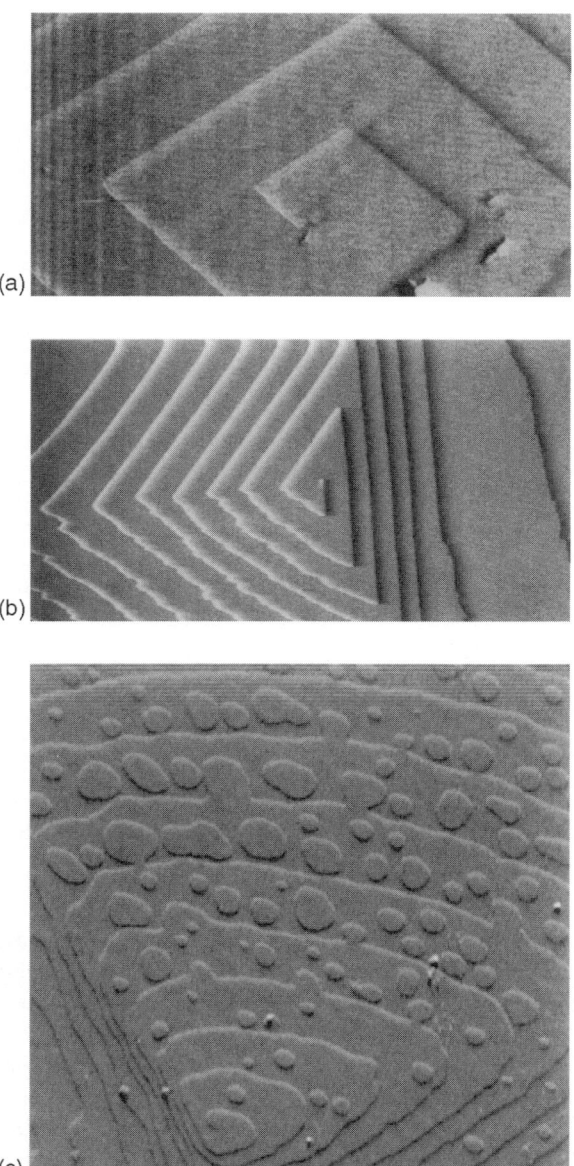

FIG. 11 Atomic force microscopy images of (a) 4 × 2 μm area on the {10$\bar{1}$4} crystal face of calcite showing the first turn of a spiral defect with a step height of 0.31 nm, (b) 6 × 3 μm scanned area of a spiral dislocation on the {010} face of brushite, showing the spreading of the step in three crystallographic directions, and (c) 8 × 8 μm scanned area of the {101} face of KH_2PO_4, showing the steps in three distinct crystallographic directions each with different speeds and terrace widths. Spreading islands, caused by surface nucleation, can be seen on the broadest of these terraces. Crystal surfaces were investigated in situ (a,b) and ex situ (c). (Courtesy of C. Orme, T. Land, and J. J. De Yoreo, Chemistry and Material Science Department, Lawrence Livermore National Laboratory, Livermore, CA, USA.)

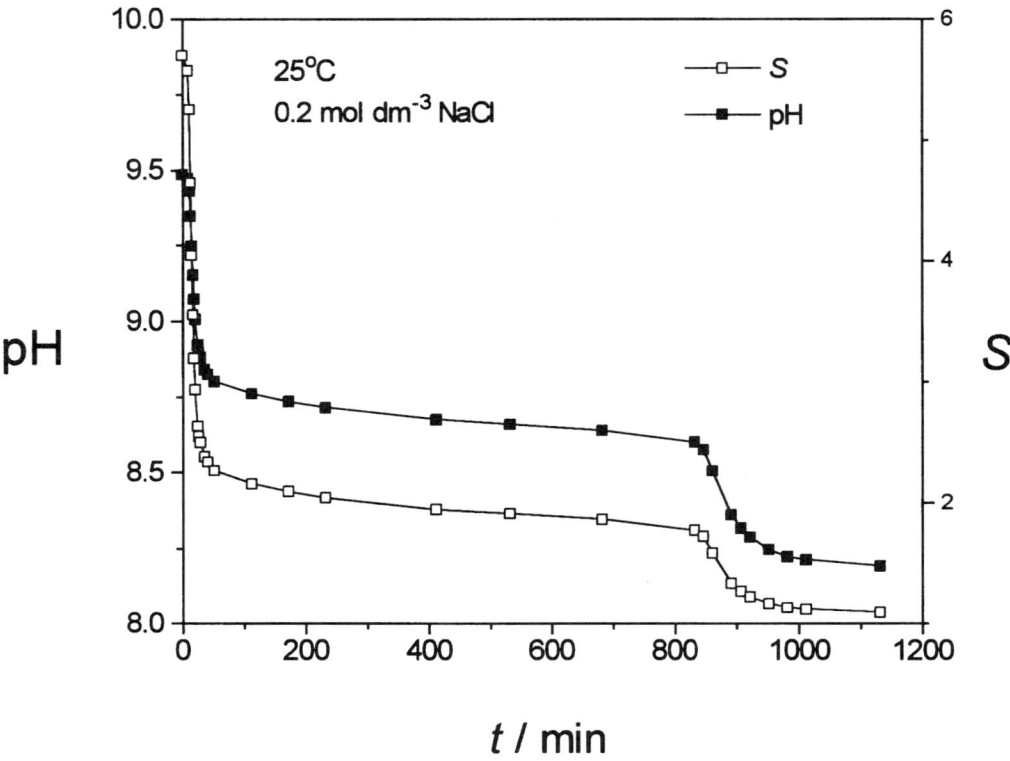

FIG. 12 Time dependence of pH (progress curve) and the corresponding saturation ratio (desaturation curve) for the solution mediated transformation of the unstable calcium carbonate polymorphic modification, vaterite, into the stable modification, calcite.

proportional to the solution supersaturation, the time dependence of supersaturation can be calculated (Fig. 12). The changes of the solid phase are usually obtained by determining the crystal number or the crystal size distribution in the course of the process. From these data the overall crystal growth rates can be calculated and the growth mechanisms deduced using the theoretical background outlined in the preceding parts of this chapter. It should also be mentioned that the chemical and physical characterization of the solid phase formed is necessary for a correct interpretation of the results derived from the progress curves. For this purpose, a number of instrumental techniques are available, such as electron and x-ray diffraction, IR spectroscopy, thermogravimetric analysis, and many others, as well as conventional chemical analysis. The main advantages of measuring a large number of crystals are that the growth rates obtained are independent of individual crystal properties and, owing to the large crystal surface area available, are much less influenced by the impurities present in the system. Sometimes, also, the operation of experiments is much easier. Besides, the information on some secondary processes, such as recrystallization and agglomeration, cannot be obtained by single crystal measurements. This also applies to the crystal size distribution and the influence of the mechanical treatment (e.g., stirring) on precipitation processes and crystal growth in particular. Indeed, the measurements of a large number of crystals are less reproducible, and the accuracy of the measurements is sometimes lower, especially in spontaneously nucleated systems. Likewise, the overall crystal growth rates derived from these measurements are not so good in testing theories of crystal growth. In order to overcome some of the difficulties, well-defined seed crystals can be used in inoculating the reactant solution (applicable only at very low supersaturations; region of metastable solution in Fig. 10), or the experiment can be performed at constant supersaturation. Using these methods, it is possible to obtain highly reproducible results in the study of crystal growth kinetics.

For a detailed description of the experimental methods and techniques used in growth rate measurements and precipitation processes in general, the reader is referred to some recent publications [11,126].

SYMBOLS

A	cationic species
A	surface area
a	activity; acceleration; surface area per ion
B	anionic species
B	pre-exponential factor in Eq. (12)
c	concentration
c'	concentration of solute at crystal surface
c_s	solubility
D	diffusion coefficient
d	diameter; height of surface nucleus
e	effectiveness of impurity
F	Faraday constant; pre-exponential factor defined by Eq. (45)
f	geometric factor
ΔG	change of Gibbs energy
$\Delta G^{\neq}$	Gibbs activation energy
ΔG^*	Gibbs energy for formation of critical nucleus
ΔG^*_{het}	Gibbs energy for heterogeneous nucleation
g	acceleration of gravity
h	Planck constant
I	ionic strength of solution
J	rate of nucleation; flux density
j	flux of matter
K	equilibrium constant
K_{sp}	thermodynamic solubility product
k	rate constant; Boltzmann constant
L	diameter of container
N	number density = number of particles per volume
n	amount of matter; kinetic order
P	perimeter
Pe^*	Perclet number of mass transfer
R	gas constant
r	radius; length of cube edge
r^*	radius of critical nucleus
S	saturation ratio
S^*	critical supersaturation
S_m	metastable limit supersaturation
T	absolute temperature
t	time

t_d	time required for achievement of steady state of nuclei distribution
t_g	time necessary for critical nucleus to grow to detectable size
t_I	induction period
t_n	time for the formation of critical nucleus
$t_{1/2}$	half-time
U	overall liquid velocity
V	volume
V_m	molar volume
v	molecular volume [Eq. (36)]
v_g	linear growth rate
v_∞	spreading step velocity
x	distance from a center
x_o	average distance between kinks along a step
y_o	average distance between two consecutive steps
z	ionic charge number
α	number of cations
β	number of anions
β	shape factor [Eqs. (36) and (38)]
γ	activity coefficient; edge energy
Δ	difference
δ	thickness of diffusion layer
ε	entropy factor
η	dynamic viscosity
θ	wetting angle; fraction of surface occupied by impurity
λ_i	molar ionic conductivity
μ	chemical potential
μ^o	standard chemical potential
$\Delta\mu_v$	Gibbs energy change per volume
v	number of ions in a molecule [Eq. (38)]
v'_{in}	integration frequency of nonadsorbed (dissolved) ions
Π	ionic product in general
ρ	particle density
ρ_o	solution density
σ	relative supersaturation; surface energy (interfacial tension)
σ_h	edge free energy of a step
τ	time needed for formation of a new crystal surface layer
ϕ	affinity

Subscripts

ad	adsorption
d	diffusion
dh	dehydration
ds	desorption
e	exponential rate law
g	growth
het	heterogeneous
i	ith component; ionic
imp	impurity
in	integration

m	molar
s	surface
sd	surface diffusion
w	related to water molecule
+	forward
−	reverse
o	initial
1	linear rate law
2	parabolic rate law

Superscripts

o	standard state
′	two-dimensional (surface) nucleus
*	quantity related to critical nucleus

REFERENCES

1. J. W. Mullin, *Crystallization*, 3d ed., Butterworth-Heinemann, Oxford, 1992.
2. P. Bennema. J. Crystal Growth *24/25*:76 (1974).
3. A. E. Nielsen. Croat. Chem. Acta *42*:319 (1970).
4. A. L. Boskey and A. S. Posner. J. Phys. Chem. *80*:40 (1976).
5. A. E. Nielsen. Faraday Disc. Chem. Soc. *61*:153 (1976).
6. O. Söhnel and J. W. Mullin. J. Crystal Growth *60*:239 (1982).
7. A. E. Nielsen, in *Treatise on Analytical chemistry*, 2d ed., Part 1, Vol. 3 (I. M. Kolthoff and P. J. Elving, eds.), John Wiley, New York, 1983, pp. 269–347.
8. A. G. Walton, in *Nucleation* (A. C. Zettlemoyer, ed.), Marcel Dekker, New York, 1969, pp. 225–307.
9. J. B. Zeldovich. Acta Physicochem. URSS *18*:17 (1943).
10. J. Frenkel, *Kinetic Theory of Liquids*, Oxford University Press, Oxford, 1946.
11. O. Söhnel and J. Garside, *Precipitation: Basic Principles and Industrial Applications*, Butterworth-Heinemann, Oxford, 1992.
12. R. Becker and W. Döring. Ann. Phys. *24*:719 (1935).
13. M. Volmer, *Kinetik der Phasenbildung*, Steinkopff, Dresden, 1939.
14. J. W. Gibbs, *Scientific Papers*, 1906 (The collected works of J. Willard Gibbs), Yale University Press, New Haven, CT, 1957.
15. W. Thomson. Proc. Roy. Soc. Edinburgh *7*:63 (1870).
16. A. E. Nielsen, *Kinetics of Precipitation*, Pergamon Press, Oxford, 1964.
17. E. V. Khamskii, *Crystallization from Solution*, Nauka, Leningrad, 1967; Consultants Bureau, New York, 1969.
18. A. G. Walton, *The Formation and Properties of Precipitates*, Interscience, New York, 1967.
19. J. B. Newkirk and D. Turnbull. J. Appl. Phys. *26*:579 (1955).
20. P. G. Koutsoukos and G. H. Nancollas. J. Crystal Growth *53*:10 (1981).
21. A. E. Nielsen and J. Christoffersen, in *Biological Mineralization and Demineralization* (G. H. Nancollas, ed.), Springer Verlag, Berlin, 1982.
22. D. Turnbull and B. Vonnegut. Ind. Eng. Chem. *44*:1292 (1952).
23. R. F. Strickland-Constable and R. E. A. Mason. Nature *197*:897 (1963).
24. G. D. Botsaris, in *Industrial Crystallization* (J. W. Mullin, ed.), Plenum Press, New York, 1976, pp. 3–22.
25. J. Garside and R. Davey. Chem. Eng. Sci. *4*:393 (1980).

26. J. Nývlt, O. Söhnel, M. Matuchová, and M. Broul, *The Kinetics of Industrial Crystallization*, Elsevier, Amsterdam, 1985.
27. Lj. Brečević and J. Garside. Chem. Eng. Sci. *36*:867 (1981).
28. N. Brown. J. Crystal Growth *16*:163 (1972).
29. B. Tomažič, R. Mohanty, M. Tadros, and J. Estrin. J. Crystal Growth *75*:339 (1986).
30. A. E. Nielsen. Croat. Chem. Acta *53*:255 (1980).
31. A. E. Nielsen. J. Crystal Growth *67*:289 (1984).
32. A. E. Nielsen and J. M. Toft. J. Crystal Growth *67*:278 (1984).
33. A. E. Nielsen. Pure Appl. Chem. *53*:2025 (1981).
34. J. R. Bourne and R. J. Davey. J. Crystal Growth *36*:278 (1976).
35. W. B. Hillig. Acta Met. *14*:1868 (1966).
36. W. K. Burton, N. Cabrera, and F. C. Frank. Phil. Trans. Roy. Soc. London Ser. A *243*:299 (1951).
37. P. Bennema. J. Crystal Growth *1*:278 (1967).
38. R. Reich and M. Kahlweit. Ber. Bunsengesellschaft *72*:66 (1968).
39. A. E. Nielsen and N. D. Altintas. J. Crystal Growth *69*:213 (1984).
40. D. Kralj and Lj. Brečević. Colloids Surfaces A: Physicochem. Eng. Asp. *96*:287 (1995).
41. D. Kralj, Lj. Brečević, and A. E. Nielsen. J. Crystal Growth *143*:269 (1994).
42. D. Kralj, Lj. Brečević, and J. Kontrec. J. Crystal Growth *177*:248 (1997).
43. K. Sangwal, *Etching of Crystals: Theory, Experiment and Application*, North-Holland, Amsterdam, 1987.
44. N. T. Barrett, G. M. Lamble, K. J. Roberts, J. N. Sherwood, G. N. Greaves, R. J. Davey, R. J. Oldman, and D. Jones. J. Crystal Growth *94*:689 (1989).
45. A. Gutjahr, H. Dabringhaus, and R. Lacmann. J. Crystal Growth *158*:310 (1996).
46. Lj. Brečević, V. Nöthig-Laslo, D. Kralj, and S. Popvić. J. Chem. Soc. Faraday Trans. *92*:1017 (1996).
47. S. Sarig and M. Raphael. J. Crystal Growth *16*:203 (1972).
48. M. E. Tadros and I. Mayes. J. Colloid Interface Sci. *72*:245 (1979).
49. M. C. van der Leeden, D. Kashiev, and G. M. van Rosmalen. J. Crystal Growth *130*:221 (1993).
50. E. K. Kirkova and R. D. Nikolaeva. Krist. Tech. *8*:463 (1973).
51. N. Eidelman, R. Azoury, and S. Sarig. J. Crystal Growth *74*:1 (1986).
52. Lj. Brečević, V. Hlady, and H. Füredi-Milhofer. Colloids Surfaces *28*:301 (1987).
53. M. C. Cafe and I. D. Robb. J. Colloid Interface Sci. *86*:411 (1982).
54. H. J. Meyer. J. Crystal Growth *66*:639 (1984).
55. G. H. Nancollas and J. A. Budz. J. Dental Res. *69*:1678 (1990).
56. C. S. Sikes, M. L. Yeung, and A. P. Wheeler, in *Surface Reactive Peptides and Polymers, Discovery and Commercialization* (C. S. Sikes and A. P. Wheeler, eds.), American Chemical Society, Washington, D. C., 1991, ACS Symposium Series 444, p. 50.
57. J. V. Garcia-Ramos and P. Carmona. J. Crystal Growth *57*:336 (1982).
58. A. M. Cody and R. D. Cody. J. Crystal Growth *135*:235 (1994).
59. N. Wada, M. Okazaki, and S. Tachikawa. J. Crystal Growth *132*:115 (1993).
60. G. Falini, M. Gazzano, and A. Ripamonti. J. Crystal Growth *137*:577 (1994).
61. M. Vaida, L. J. W. Shimon, Y. Weisinger-Lewin, F. Frolow, M. Lahav, L. Leiserowitz, and R. K. McMullan. Science *241*:1475 (1988).
62. Y. Weisinger-Lewin, F. Frolow, R. McMullan, T. F. Koetzle, M. Lahav, and L. Leiserowitz. J. Am. Chem. Soc. *111*:1035 (1989).
63. Z. Berkivitch-Yellin, J. van Mil, L. Addadi, M. Idelson, M. Lahav, and L. Leiserowitz. J. Am. Chem. Soc. *107*:3111 (1985).
64. H. A. Lowenstam and S. Weiner, *On Biomineralization*, Oxford University Press, New York, 1989.
65. A. S. Michaels and Jr. F. W. Tausch. J. Phys. Chem. *65*:1730 (1961).
66. A. D. Randolph and A. D. Puri. AIChE J. *27*:92 (1981).
67. A. König and H. H. Emons. Cryst. Res. Technol. *23*:319 (1988).

68. N. Hiquily, J. P. Couderc, and C. Laguérie. Chem. Eng. J. *30*:1 (1985).
69. W. P. Brandse, G. M. van Rosmalen, and G. Brouwer. J. Inorg. Nucl. Chem. *39*:2007 (1977).
70. R. W. Peters and T.-K. Chang, in *Industrial Crystallization 84* (S. J. Jančić and E. J. de Jong, eds.), Elsevier, Amsterdam, 1984, p. 67.
71. N. Cabrera and D. A. Vermilyea, in *Growth and Perfection of Crystals* (R. H. Doremus, B. W. Roberts, and D. Turnbull, eds.), John Wiley, New York, 1958, p. 393.
72. G. W. Sears. J. Chem. Phys. *29*:1045 (1958).
73. G. M. Bliznakov, in *Adsorption et Croissance Cristalline* (R. Kern, ed.), CNRS, Paris, 1965, p. 291.
74. R. J. Davey. J. Crystal Growth *34*:109 (1974).
75. B. Simon and R. Boistelle. J. Crystal Growth *52*:779 (1981).
76. G. M. van Rosmalen and P. Bennema. J. Crystal Growth *99*:1053 (1990).
77. K. Sangwal. J. Crystal Growth *128*:1236 (1993).
78. I. Owczarek and K. Sangwal. J. Crystal Growth *102*:574 (1990).
79. R. J. Davey and J. W. Mullin. J. Crystal Growth *26*:45 (1974).
80. G. M. van Rosmalen and M. C. van der Leeden. Cryst. Res. Technol. *17*:627 (1982).
81. K. Sangwal. Prog. Crystal Growth Charact. *32*:3 (1996).
82. H. Füredi-Milhofer and S. Sarig. Prog. Crystal Growth Charact. *32*:45 (1996).
83. M. von Smoluchowski. Z. physik. Chem. *92*:129 (1917).
84. D. L. Swift and S. K. Friedländer. J. Colloid Sci. *19*:621 (1964).
85. B. Težak, E. Matijević, K. F. Schulz, J. Kratohvil, M. Mirnik, and V. B. Vouk. Disc. Faraday Soc. *18*:63 (1954).
86. C. Wagner. Z. Elektrochemie *65*:581 (1961).
87. M. Kahlweit. Z. physik. Chem. N.F. *36*:292 (1963).
88. W. Ostwald. Z. physik. Chem. *34*:295 (1900).
89. S. Sarig and O. Ginio. J. Phys. Chem. *80*:256 (1976).
90. O. Söhnel. Cryst. Res. Technol. *16*:651 (1981).
91. W. J. Dunning, in *Particle Growth in Suspensions* (A. L. Smith, ed.), Academic Press, London, 1973, p. 3.
92. M. Kahlweit. Adv. Colloid Interface Sci. *5*:1 (1975).
93. M. Kahlweit. Faraday Discuss. Chem. Soc. *61*:48 (1976).
94. P. Voorhees. J. Stat. Phys. *38*:231 (1985).
95. W. Ostwald, *Grundriss der allgemeinen Chemie*, W. Engelmann, Leipzig, 1899.
96. J. Pelouse. Ann. Chim. Physique *48*:301 (1831); Compt. Rend. (Paris) *60*:429 (1860).
97. G. Dorfmüller. Deut. Zuckerind. *63*:1217 (1938).
98. R. Brooks, L. M. Clark, and E. F. Thurston. Phil. Trans. Roy. Soc. London A *243*:145 (1950).
99. J. D. Termine and A. S. Posner. Arch. Biochem. *140*:307 (1970).
100. Lj. Brečević and H. Füredi-Milhofer. Calc. Tiss. Res. *10*:82 (1972).
101. F. Abbona, H. E. Lundager Madsen, and R. Boistelle. J. Crystal Growth *74*:581 (1986).
102. Lj. Brečević and A. E. Nielsen. J. Crystal Growth *98*:504 (1989).
103. G. C. Bye and K. S. W. Sing, in *Particle Growth in Suspensions* (A. L. Smith, ed.), Academic Press, London, 1973, p. 29.
104. Lj. Brečević, D. Kralj, and J. Garside. J. Crystal Growth *97*:460 (1989).
105. Lj. Brečević and H. Füredi-Milhofer, in *Industrial Crystallization* (J. W. Mullin, ed.), Plenum Press, New York, 1976, p. 277.
106. C. H. Bamford and C. F. H. Tipper, *Comprehensive Chemical Kinetics*, Vol. 22, *Reactions in the Solid State*, Elsevier, Amsterdam, 1980.
107. P. T. Cardew, R. J. Davey, and A. J. Ruddick. J. Chem. Soc. Faraday Trans. *80*:659 (1984).
108. A. L. Boskey and A. S. Posner. J. Phys. Chem. *77*:2313 (1973).
109. P. T. Cardew and R. J. Davey. Proc. Roy. Soc. London *A 398*:415 (1985).
110. Lj. Brečević, D. Škrtić, and J. Garside. J. Crystal Growth *74*:399 (1986).
111. R. J. Davey, A. J. Ruddick, P. D. Guy, B. Mitchell, S. J. Maginn, and L. A. Polywka. J. Phys., D: Appl. Phys. *24*:176 (1991).
112. W. Stumm and J. J. Morgan, *Aquatic Chemistry*, 2d ed., John Wiley, New York, 1981.

113. J. Rose, *Dynamic Physical Chemistry*, Pitman and Sons, London, 1961.
114. J. Hulliger. Angew. Chem. *106*:151 (1994).
115. L. G. Sillén, *Stability Constants of Metal-Ion Complexes*, Special Publication, Chemical Society, London, 1964, 1971.
116. B. Težak. Disc. Faraday Soc. *42*:175 (1966).
117. H. Füredi, in Ref. 18, p. 188.
118. H. Füredi-Milhofer, B. Purgarić, Lj. Brečević, and N. Pavković. Calc. Tiss. Res. *8*:142 (1971).
119. J. W. Mullin and A. Amatavivadhana. J. Appl. Chem. *17*:151 (1967).
120. R. Janssen-van-Rosmalen, W. H. van der Linden, E. Dobbinga, and D. Visser. Krist. Tech. *13*:17 (1978).
121. H. Langer and H. Offermann. J. Crystal Growth *60*:389 (1982).
122. K. Tsukamoto. J. Crystal Growth *61*:199 (1983).
123. D. Kralj, Lj. Brečević, and A. E. Nielsen. J. Crystal Growth *104*:793 (1990).
124. P. P. Koutsoukos and C. G. Kontoyannis. J. Chem. Soc. Faraday Trans. *80*:1181 (1984).
125. Lj. Brečević and A. E. Nielsen. Acta Chem. Scand. *47*:668 (1993).
126. E. J. de Jong, J. Garside, M. Kind, A. Mersmann, J. Nývlt, H. Offermann, and J. Pohlisch, in *Measurement of Crystal Growth Rates* (J. Garside, A. Mersmann, and J. Nývlt, eds.), European Federation of Chemical Engineering, Working Party on Crystallization, Druckhaus Deutsch, München, 1990.
127. J. P. van der Eerden, P. Bennema, and T. A. Cherepanova. Prog. Crystal Growth Charact. *1*:219 (1978).
128. D. Cunningham, R. J. Davey, K. J. Roberts, J. N. Sherwood, and T. Shripathi. J. Crystal Growth *99*:1065 (1990).
129. C. Hall and D. C. Cullen. AIChE J. *42*:232 (1996).
130. H. H. Teng, P. M. Dove, C. A. Orme, and J. J. De Yoreo. Science *282*:724 (1998).

13
Dissolution of Calcium Hydroxyapatite

PHILIPPE GRAMAIN Hétérochimie Moléculaire et Macromoléculaire, CNRS—Ecole Nationale Supérieure de Chimie, Montpellier, France

PHILIPPE SCHAAD Department of Chemistry, Institut Universitaire de Technologie, Illkirch-Graffenstaden, France

I.	Introduction	475
II.	Hydroxyapatite: Formation, Morphology, and Structure	476
III.	Hydroxyapatite Dissolution	479
	A. Ionic crystal dissolution	479
	B. Dissolution reaction of HAP	480
	C. Dissolution-rate experiments	481
	D. Kinetics behavior	483
	E. Models of dissolution	484
	F. Dissolution of carbonated apatites and fluorapatites	494
	G. Conclusion	495
IV.	Inhibition of Dissolution by Adsorbed Compounds	496
	A. Adsorption and inhibition of dissolution	496
	B. Inhibition by surfactants	497
	C. Inhibition by adsorbed ions	497
	D. Inhibition by low molecular weight organic compounds	502
	E. Inhibition by proteins	502
	F. Inhibition by synthetic macromolecules	505
	G. Conclusion	506
	References	508

I. INTRODUCTION

Apatites form an important series of minerals that occur as minor constituents of many igneous rocks. They are also present in most metamorphic rocks, especially crystalline limestones. Less well crystallized deposits referred to generally as rock phosphate or phosphorite occur also in large sediments of several marine shelves, some of which were formed by the reaction between phosphate solutions from guano and calcareous rock or precipitated from sea water. The rock phosphates provide most of the world's phosphorus supply for industrial applications including water treatment, fertilizers, detergents, insecticides, and other chemical industries.

Calcium hydroxyapatite [$Ca_5(PO_4)_3OH$] or HAP is a salt of the apatite family slightly water soluble that can easily be substituted by ions like F^-, Cl^-, Ba^{2+}, Sr^{2+}, Na^+, $AsPO_4^{3-}$, and CO_3^{2-}. These substitutions are very often accompanied by none stoichiometry together with vacancies appearing in certain lattice sites. This explains also why HAP is never found as a mineral in a pure form. A review of structural and chemical aspects of apatites may be found in the recent book by J. C. Elliott [1].

For vertebrates, HAP or carbonate containing HAP constitutes the main reference constituent for biological hard tissues like bone and teeth, and their chemical behaviors are models for the study of important biological phenomena such as biological mineralization and demineralization, carious lesion formation, and bone remodeling. HAP is also often used for the development of various dense or porous bioceramics and bioactive composites with polymers for bone replacement, coating metal implants, and prostheses.

Considering the main role of this mineral, the knowledge of HAP growth and dissolution mechanisms are of great importance, and interpretations of studies on complex phenomena like carious lesion formation cannot be achieved when growth and dissolution processes are not well understood. This justifies the important number of physicochemical studies published in this domain. In spite of the considerable progress achieved, important problems remain that give an excellent justification for the name hydroxyapatite, derived from a Greek noun meaning "deceiver" and given by Alfred Warner, who noticed its likeness with precious stones.

This chapter will mainly deal with the physicochemical aspects of HAP dissolution, which in many aspects is in close relationship with crystal growth. The purpose is to present an up-to-date account of the subject, with particular emphasis upon the interfacial aspects and the proposed dissolution models.

II. HYDROXYAPATITE: FORMATION, MORPHOLOGY, AND STRUCTURE

HAP is one of the thermodynamically stable phases defined in the ternary system $Ca(OH)_2$-H_3PO_4-H_2O [2]. The equilibrium diagram based on the activities of the predominant calcium and phosphorous species in solution and the pH as the independent variables is shown in Fig. 1, while Table 1 gives the corresponding chemical reactions.

Above pH 4, the formation of HAP is more kinetically than thermodynamically driven, and saturation must be well controlled in order to avoid the formation of other calcium phosphate phases such as amorphous calcium phosphate (ACP), octacalcium phosphate (OCP), or dicalcium diphosphate (DCPD). In spontaneous precipitation experiments at relatively high supersaturation when the solution becomes very unstable with respect to all phases in the near neutral or alkaline part, ACP is first precipitated and then converts into HAP via OCP. At intermediate supersaturation, formation of HAP can occur without prior formation of ACP but still seems to go via other phases like DCPD or OCP. In practice, very tiny crystals lower than 1 μm in size are formed with considerable surface area in the order of 10–150 $m^2 \cdot g^{-1}$. Larger hexagonal prism shaped monocrystals can be grown by specific methods like hydrothermal boiling. The two most frequently observed morphologies with synthetic and natural apatites are needles and plates [3]. The platelike habit probably constitutes a signature for passage through DCPD or OCP phase.

Crystallographic data show that HAP occurs as hexagonal crystals with structure defined in the space group $P6_3/m$, although a monoclinic form with space group $P2_1/b$ can also happen. In Fig. 2, the three axes *a*, *b*, and *c* are given as well as the most

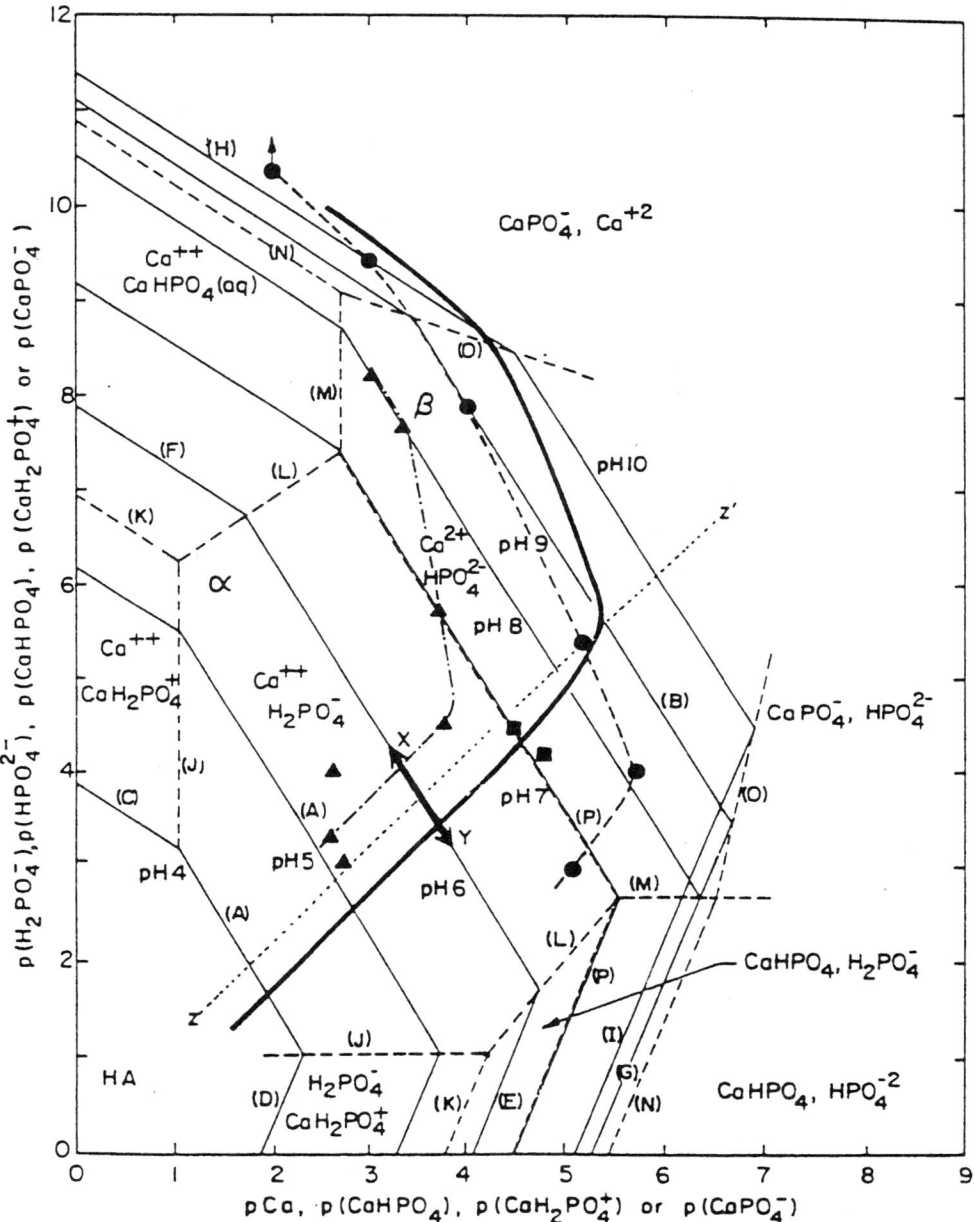

FIG. 1 Equilibrium diagram for hydroxyapatite. Solid lines are the pH-invariant lines on the solubility surface. Broken lines separate the stability domain of various solution species. Heavy solid line is the calculated line of zero charge. (From Ref. 2.)

TABLE 1 List of Chemical Reactions in the Hydroxyapatite–Aqueous Solution System. Each of the Reactions Is Plotted in Fig. 1 and Is Marked by the Respective Letter

A: $Ca_{10}(PO_4)_6(OH)_2 + 14H^+ = 10Ca^{2+} + 6H_2PO_4^- + 2H_2O$;
$pCa + 0.6p(H_2PO_4) - 1.4\ pH = -2.66$

B: $Ca_{10}(PO_4)_6(OH)_2 + 8H^+ = 10Ca^{2+} + 6HPO_4^{2-} + 2H_2O$;
$pCa + 0.6\ p(HPO)_4 - 0.8\ pH = 1.54$

C: $Ca_{10}(PO_4)_6(OH)_2 + 14H^+ = 6CaH_2PO_4^+ + 4Ca^{2+} + 2H_2O$;
$p(CaH_2PO_4) + 0.667\ pCa - 2.333\ pH = -5.51$

D: $Ca_{10}(PO_4)_6(OH)_2 + 4H_2PO_4^- + 14H^+ = 10CaH_2PO_4^+ + 2H_2O$;
$p(CaH_2PO_4) - 1.4\ pH - 0.4\ p(H_2PO_4) = 3.74$

E: $Ca_{10}(PO_4)_6(OH)_2 + 4H_2PO_4^- + 4H^+ = 10CaHPO_4(aq) + 2H_2O$;
$p(CaHPO_4) - 0.4\ p(H_2PO_4) - 0.4\ pH = 1.64$

F: $Ca_{10}(PO_4)_6(OH)_2 + 8H^+ = 6CaHPO_4(aq) + 4Ca^{2+} + 2H_2O$;
$p(CaHPO_4) + 0.667\ pCa - 1.333\ pH = -0.133$

G: $Ca_{10}(PO_4)_6(OH)_2 + 8H^+ + 4HPO_4^{2-} = 10CaHPO_4 + 2H_2O$;
$p(CaHPO_4) - 0.8\ pH - 0.4\ p(HPO_4) = -1.16$

H: $Ca_{10}(PO_4)_6(OH)_2 + 2H^+ = 4Ca^{2+} + 6CaPO_4^- + 2H_2O$;
$p(CaPO_4) + 0.667\ pCa - 0.333\ pH = 8.107$

I: $Ca_{10}(PO_4)_6(OH)_2 + 4HPO_4^{2-} = 10CaPO_4^- + 2H^+ + 2H_2O$;
$p(CaPO_4) + 0.2\ pH - 0.4\ p(HPO_4) = 7.08$

J: $CaH_2PO_4^+ = Ca^{2+} + H_2PO_4^-$

K: $CaH_2PO_4^+ = CaHPO_4(aq) + H^+$

L: $CaHPO_4(aq) + H^+ = Ca^{2+} + H_2PO_4^-$

M: $CaHPO_4(aq) = Ca^{2+} + HPO_4^{2-}$

N: $CaHPO_4(aq) = CaPO_4^- + H^+$

O: $Ca\ PO_4^- + H^+ = Ca^{2+} + HPO_4^{2-}$

P: $H_2PO_4^- = H^+ + HPO_4^{2-}$

Source: Ref. 2.

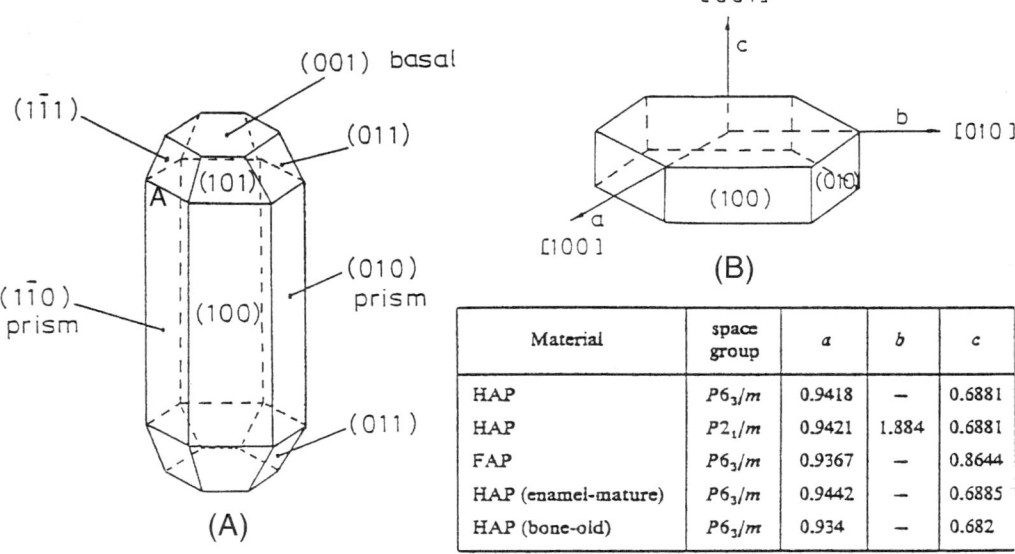

FIG. 2 (A) The most frequently observed crystal faces in apatites. (B) The a-, b-, and c-axes in a platelike apatite crystal. Lattice parameters of HAP and FAP in nm are given in the table. (From Ref. 3.)

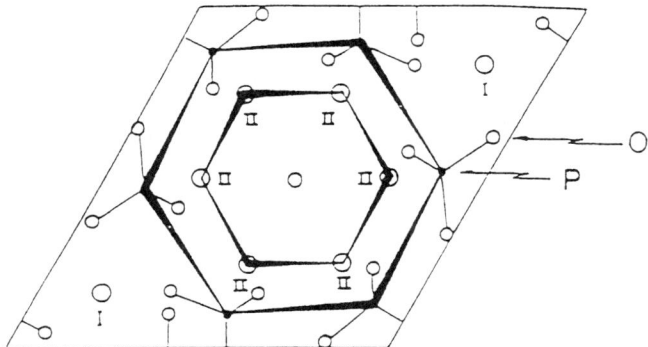

FIG. 3 Crystal structure of apatite projected on (001). The anions are not all shown. The calcium ions are indicated by I and II, respectively. (From Ref. 3.)

frequently observed faces. Fig. 3 shows the positions of the calcium, phosphate, and OH⁻ ions in the unit cell. However, impurities may strongly influence the crystal shape and properties of single crystals. This is in particular the case of carbonate and fluor anions, which constitute two major impurities of biological importance.

An analysis of HAP preparations and crystal morphologies and structures can be found in Ref. 1 and in the short review by J. Arends and W. L. Jongebloed [3].

III. HYDROXYAPATITE DISSOLUTION

A. Ionic Crystal Dissolution

The dissolution of crystals in liquids is usually assumed to occur in two consecutive steps: continuous escape of the dissolving species from the surface, and their subsequent diffusion into the main body of the fluid. However, numerous examples exist demonstrating a much more complex situation with, for example, the formation at the interface of a new phase limiting an easy access of the reactants. The rate of dissolution can be controlled by surface disengagement processes, by the transport of ions through the new phase formed at the interface, by transport through the adjacent diffusion layer, or by a combination of these processes. In any case, the elucidation of the molecular events occurring in each step is of prime importance for reaching a better understanding and a good control of the growth and dissolution processes.

The rate law governing the dissolution process of ionic crystals has the general form [4]

$$R_d = K(C_s - C)^n \tag{1}$$

where R_d is the dissolution rate, C_s the salt solubility, C the salt concentration, and K a constant depending in particular on the surface area. n defines the reaction order. Equation (1) implies that the basic driving force for dissolution is the deviation from the solubility equilibrium. Often one defines by $S = C/C_s$ the degree of saturation of the system.

One of the difficulties of dissolution studies based on the kinetics of dissolution is the knowledge of the exact nature of the rate controlling process in the experimental conditions used. A first order reaction ($n=1$) is generally indicative of a process controlled by a diffusion step for which the rate is described by the Nernst theory. In a diffusion controlled process, the rate depends on the hydrodynamics, and the constant K is given by DS_A/δ, in which D is the diffusion coefficient in the solid adjacent Nernst layer of thickness δ and S_A the surface area. Higher n values demonstrate a surface controlled process for which the concentration near the crystal surface is the same as in the bulk. While an order $n \leq 2$ is generally indicative of a mononuclear or spiral dislocation mechanism, n values higher than 2 reveal the existence of a polynuclear mechanism. The interpretation of the data may however be complicated by a change of order of reaction with experimental parameters such as stirring, temperature, or undersaturation of the solution. Proposition of a model always requires an hypothesis on the rate controlling process, and the justifications are not always convincing.

B. Dissolution Reaction of HAP

The dissolution reaction of HAP can be written as

$$(n_c - 1)H_2O + Ca_5(PO_4)_3OH <=> 5Ca^{2+} + 3x_1PO_4^{3-} + 3x_2HPO_4^{2-} + 3x_3H_2PO_4^-$$
$$+ 3x_4H_3PO_4 + n_cOH^- \quad (2)$$

where n_c is the number of protons consumed when 1 equivalent of HAP is dissolved, and x_i ($i=1-4$) defines the mole fractions of the different phosphate forms present in the solution. Clearly n_c and the x_i's are related by the demand of stoichiometry. The three phosphate ion equilibria are

$$H_3PO_4 + H_2O <=> H_2PO_4^- + H_3O^+ \quad pK_1 = 2.19 \quad (2a)$$
$$H_2PO_4^- + H_2O <=> HPO_4^{2-} + H_3O^+ \quad pK_2 = 7.18 \quad (2b)$$
$$HPO_4^{2-} + H_2O <=> PO_4^{3-} + H_3O^+ \quad pK_3 = 12.3 \quad (2c)$$

If a congruent dissolution is assumed, the ratio $S_C = Ca/P$ of liberated calcium and phosphate ions corresponds to the stoichiometry of HAP ($S_C = 5/3$), and $R_c = n_c/5$ gives the number of protons consumed when on calcium is released. According to Eq. (2), one obtains

$$R_c = \frac{1 + 3(x_2 + 2x_3 + 3x_4)}{5} \quad (3)$$

It is clear that R_c depends on the pH of the solution through the phosphate equilibria defined by Eqs. (2a–c).

The stoichiometry related to the proton consumption during the HAP dissolution process was shown, by many studies done at different pH and in different solution compositions, to correspond well to Eq. (2). Moreover, when pure and well equilibrated HAP is used in order to avoid the presence of adsorbed species such as calcium or phosphate ions [5–7], a congruent dissolution was found, even when the dissolution experiments were realized in the presence of nonstoichiometric initial calcium and phosphate concentrations. All the results are consistent within experimental error with the composition of HAP as the dissolving phase.

Since the deviation from the solubility equilibrium of HAP is the basic driving force for dissolution, many theoretical and experimental studies on the solubility behavior of HAP have been published during the years 1925–1980 [8]. For pure HAP, it is now fairly well established that a well-defined thermodynamic solubility product K_{sp} exists. K_{sp} is the value at saturation of the ionic product K_{IP} of HAP:

$$K_{sp} = [Ca^{2+}]^5 [PO_4^{3-}]^3 [OH^-] \qquad (4)$$

where brackets indicate activities of ions in equilibrium with the pure HAP solid phase. Tabulation of literature values for HAP and enamel mostly determined at 25°C in stoichiometric or nonstoichiometric conditions gives $pK_{sp} = 57.5 \pm 2.5$ [5,6,9,10]. Unlike many sparingly soluble ionic salts, the HAP solubility decreases with temperature ($pK_{sp} = 59.5$ at 37°C and in the presence of 8×10^{-2} M KCl).

It follows that Eq. (1) can be expressed as

$$R_d = \mathcal{K} \left[1 - \left(\frac{K_{IP}}{K_{SP}} \right)^{1/9} \right]^n \qquad (5)$$

In the HAP literature, the bracketed expression $[1 - (K_{IP}/K_{SP})^{1/9}]$ is often referred to as σ and the dissolution rate as the ionic flux J_i (preferentially in mol · s^{-1} · m^{-2}).

C. Dissolution-Rate Experiments

If we except the dissolution studies related to carious formation and monocrystal dissolution, all the experimental studies of the dissolution mechanism of HAP are based on dissolution-rate experiments in well-defined systems. This pre-supposes either the study of the kinetics of loss of weight of the sample when HAP is added in aqueous solutions or the determination and analysis of the evolution with time of the different species involved in Eq. (2). Experiments are often realized at constant pH using an acid buffer or pH-stat equipment. The use of inorganic acids as dissolving agents seems preferable, since organic acids often used under experimental conditions relevant to the dental caries processes may have calcium chelating and HAP surface affinity properties, making more difficult the interpretation of the results. The preferred analytical method is the continuous or semicontinuous recording of proton uptake and calcium release. Phosphate release is determined by using specific electrodes, atomic adsorption spectrophotometry, or calorimetric methods. During the dissolution process, the solution sees its ionic concentration increase up to the attainment of the solubility equilibrium. In order to avoid this permanent change, which may complicate the interpretation of the experiments, Tomson and Nancollas [11] proposed an interesting constant composition method that allows us to maintain constant the undersaturation of the dissolving solution and the chemical potentials of the solution species.

Currently the HAP material most used in dissolution-rate experiments is synthetic HAP crystals of different sizes or enamel pieces dispersed in solution. The use of commercial HAP particles is interesting, since they are easily available, allow good reproducibility, and offer the possibility of particle sieving. HAP microsuspensions or seeds can easily be prepared and have often been used, although aggregation problems may arise [12,13]. The use of rotating disks or pellets of compressed HAP is interesting, since a quantitative evaluation of the thickness of the aqueous boundary layer becomes possible [7,14,15]. However, compressed HAP presents an undefined porosity that complicates the interpretation of the results.

Examination of HAP samples by microscopy and/or by microradiography is also useful, since it allows us to evaluate the morphological effects of acidic attack [2,16–19]. Such observations allow us in particular to control the size, the size dispersity, and the good dispersion of the HAP samples before and during the dissolution process.

1. Influence of HAP Origin

Depending on the origin and method of preparation, HAP materials may have very different physical forms, textures, and surface states. The choice of HAP samples in dissolution studies is of great importance, and it is essential to emphasize that the preparation and handling of this crystalline material may have an appreciable influence on the dissolution kinetics.

Natural HAP such as bovine or human enamel is evidently the preferred materials to study the dissolution process leading to caries formation. However, since HAP possesses the property of easily incorporating into its crystalline lattice a large number of foreign ions, enamel contains various impurities able to affect the dissolution behavior. In addition, its porous structure and its crystalline prismatic orientation depend on the origin of the sample. For these reasons, synthetic HAP, which is a dense material easily prepared with a good reproducibility and purity by various methods, has been proposed as a model for the dissolution behavior of enamel. To justify this choice, a number of studies have been devoted to comparing the behavior of HAP and human and bovine enamel. It is quite remarkable that in spite of the large morphology and origin differences existing between these compounds, recent results have demonstrated comparable dissolution behavior [20–22].

2. Influence of Sample Conditioning

Due to its ionic composition, the HAP interface is charged, and the nature of the surface prior to dissolution is a matter of conjecture. Numerous electrochemical studies of apatite surfaces [1,8,23] suggest that the surface charges and surface potentials are determined by the adsorption of charged ionic species, especially by ions present in the apatite lattice. Information on the surface state can be obtained by the determination of the point of zero charge (p.z.c.) of HAP, which can be calculated [2] and estimated through potentiometric titration or electrophoretic methods. The p.z.c. values reported by various authors show significant variations that can be explained by the differences in both sample origins and experimental conditions. Values around 7 are generally found for synthetic apatites [8,24,25], indicating that the net surface charge is positive in the pH domain concerned by dissolution studies.

Large amounts of phosphate and calcium ions can be adsorbed onto the surface and regulate the attainment of the solubility equilibrium [5,6]. Beside structural ions, HAP is known to adsorb strongly various compounds such as surfactants, organic acids, proteins, and polymers, and numerous studies have demonstrated that the equilibrium properties and the dissolution behavior can thus be markedly altered. In particular, it has been shown that aged slurries dissolve at approximately one-half the rate of similar crystals treated for only two hours [21].

For these reasons and thus for all experiments in which interfacial reactions are considered, sample conditioning and equilibration in saturated conditions are essential before dissolution experiments, as shown in Fig. 4 [26].

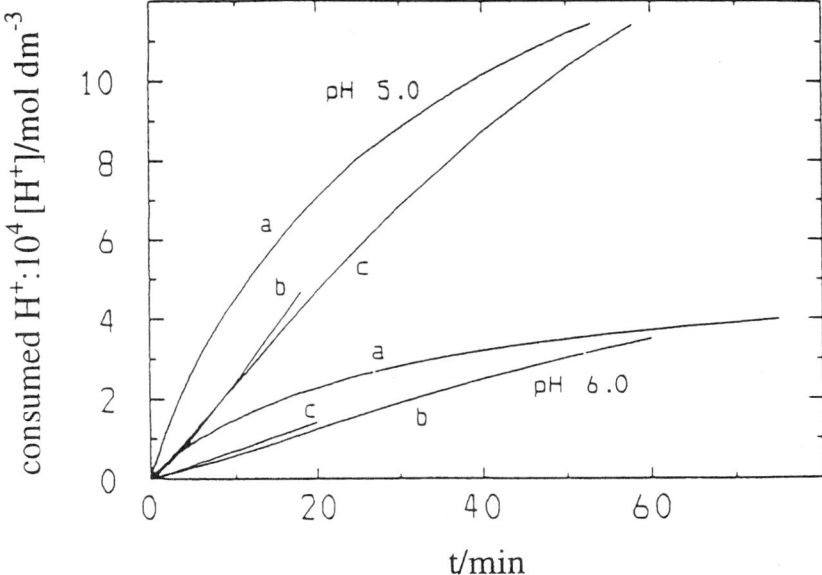

FIG. 4 Influence of conditioning on the HAP dissolution at two pHs and at a stirring speed of 1860 rpm. Proton uptake kinetics for 10 mg of HAP (165–200 μm). (a) Equilibrated powder at pH 7, (b) dry powder directly added to the dissolving solution, and (c) wet powder previously mixed with 1 mL saturated solution at pH 7. (From Ref. 26.)

D. Kinetics Behavior

1. Rate Controlling Process

The size and shape of the HAP samples used for kinetics studies constitute an important parameters, since they may determine the rate controlling process of dissolution. When HAP crystals with a size larger than about 0.1 μm are used, studies show that the rate is dependent on stirring over the whole range of pH values [27]. Most often, a direct proportionality between flux and stirring speed is observed, indicating that the rate is essentially diffusion controlled, although a significant contribution from the surface reaction to the overall kinetics cannot be excluded for certain experimental conditions [7]. With microcrystals of size ≤0.1 μm, the size is so small that they do not move relative to the solution despite stirring, and the rate becoming independent of the fluid dynamics is surface controlled [13,28].

2. Rate Behavior

Whatever the experimental conditions and the controlling process, the rate of dissolution of HAP at constant pH is fast initially; then it declines continuously and falls to a very low level. In all cases, the decline is much faster than that based on a simple first-order law supposing a steady state. While all the mathematical expressions of first-, second-, third-order, or simultaneous two parallel first-order equations are able to fit the experimental data over varying periods of time, none would fit all the data in every case and under all conditions [17]. Depending on the range of time considered, the order of the reaction n was found to vary with dissolution time and saturation degree from 5 or more to 1. To illustrate this behavior, Fig. 5 shows examples of kinetics obtained with HAP powder (165–200 μm) at three pHs with different initial Ca/P ratios, and represented according to Eq. (1).

However, at the end of dissolution and in all conditions of pH and initial stoichiometry, saturation with respect to HAP was always observed, confirming the solubility product as the main driving force.

This decrease in the rate of dissolution with the extent of reaction can undoubtedly be related to the gradual changes of the dissolving surface properties. There is little doubt that a rapid initial adjustment of surface composition occurs when crystals are introduced into the dissolving media [5,6]. This behavior, which is reinforced in the presence of impurities or adsorbed molecules acting as dissolution inhibitors, is probably the key phenomenon for the understanding of the dissolution mechanism [29].

3. pH Dependence of the Rate

HAP dissolution is strongly affected by the acid concentration, but it has been well demonstrated that H^+ diffusion from the bulk to the solid interface is not the controlling step. More precisely, the rate is 100 to 3000 times lower than expected from an H^+ diffusive process, which demonstrates that there is no interfacial H^+ gradient. This conclusion is also confirmed by the observation that the rates determined from the initial slopes of kinetics in diffusion or surface controlled systems are not linearly dependent on H^+ concentration. An $(H^+)^m$ dependence was observed by many authors, with m values varying in the range 0.57–0.81 and depending on the experimental conditions and degree of conversion [26–28,30]. Examples of pH dependence are given in Fig. 6 for diffusion controlled conditions.

E. Models of Dissolution

The experimental data accumulated and in particular the important decrease of the reaction with time together with the pH dependence mentioned above, well demonstrate the occurrence of a complex mechanism. Complementary and concurrent dissolution models have been proposed: a two-site model proposed by Higuchi and coworkers (1976–1978), a model based on a two-dimensional nucleation theory proposed by Smith et al. (1977) and developed by Christoffersen et al (1978), and a permselective interfacial model developed by Gramain and coworkers (1991–1993). These models rationalize the unusual solute concentration dependence observed experimentally by assuming that the thermodynamic solubility C_s is not respected (Higuchi), or that the kinetic rate constant K in Eq. (1) includes a saturation dependent interfacial resistance (Gramain), or that the process is controlled by the rates of disengagement of ions from the surface (Smith and Christoffersen). In the following, the basic points of each model are recalled. It must be emphasized that these models are mostly proposed on the basis of data that are assumed to be obtained in steady-state conditions. This implies that the times of analysis are long in comparison with the observed phenomena. Recent results question this hypothesis [10].

1. The Two-Site Model

Following the first mechanistic approach by Gray [31] proposing that acid dissolution of block dental enamel may be solution diffusion controlled, Higuchi and his collaborators proposed a diffusive model based on experiments realized in weak acid buffers, essentially at pH 4.5 and using low concentrations of calcium and phosphate common ions in the

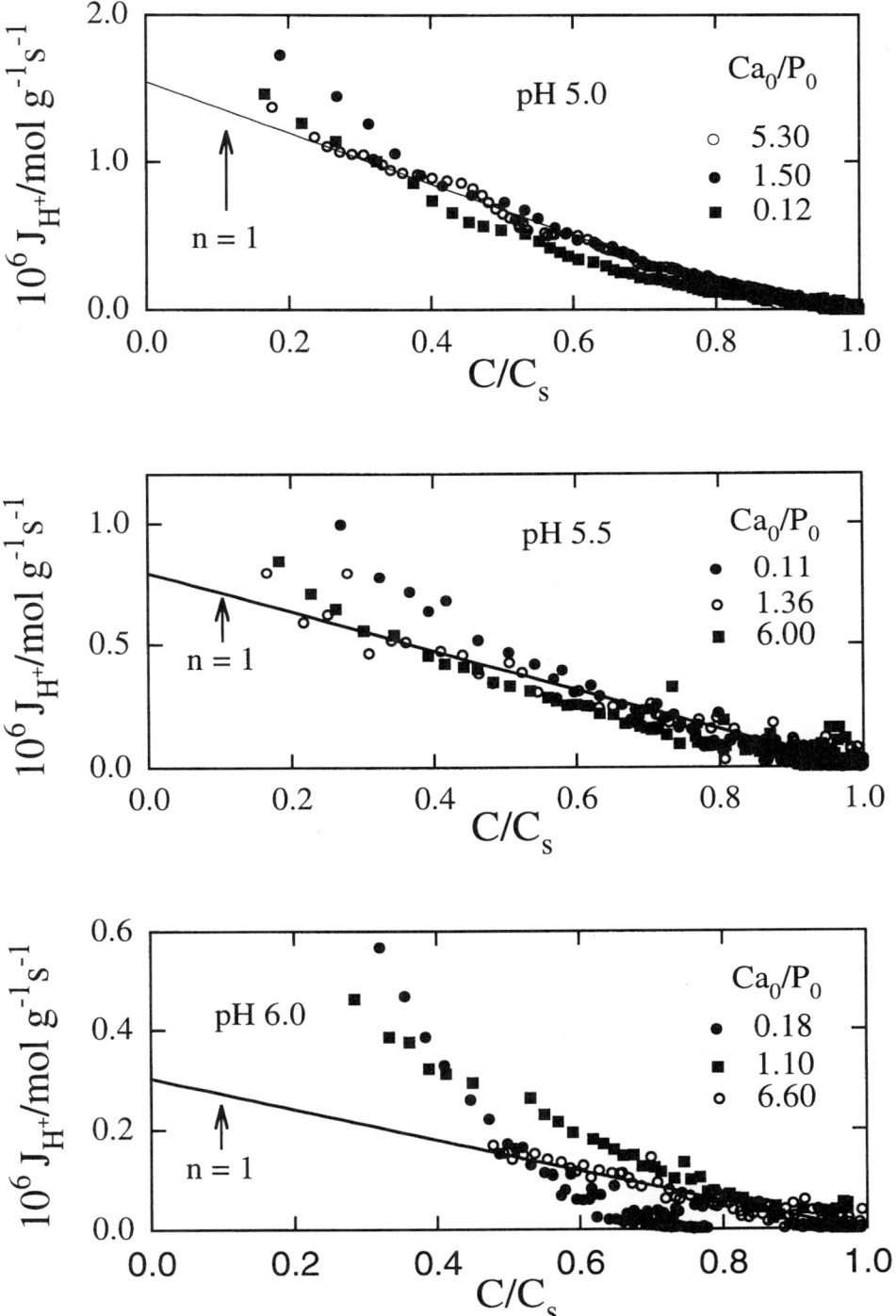

FIG. 5 Dissolution kinetics of HAP powder at three pHs expressed as the proton uptake vs. the saturation degree according to Eq. (1) with $n=1$. The powder was equilibrated at pH 6.4, 6.7, and 6.9, and in the presence of different Ca_0/P_0 ratios before dissolution at pH 5, 5.5, and 6.0 respectively. (From Ref. 10.)

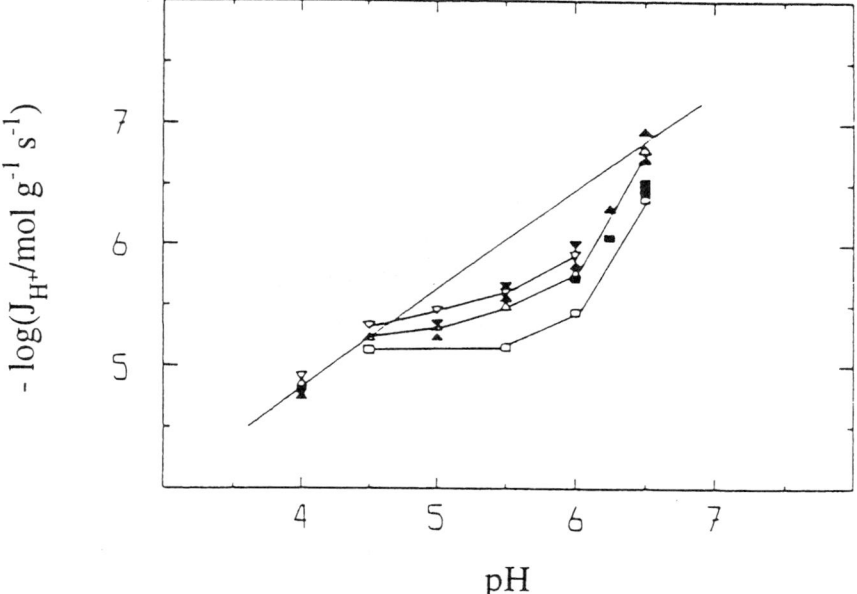

FIG. 6 Examples of pH dependence of the fluxes of consumed protons (initial dissolution rates). Experiments performed with equilibrated HAP (165–200 μm) for different calcium activities in the bulk solution: (□) 5.85×10^{-5} M; (△) 8.24×10^{-5} M; (▽) 1.125×10^{-4} M. Open symbols, stirring speed 1000 rpm; closed symbols, stirring speed 1860 rpm. The full line corresponds to a slope of 0.81. (From Ref. 26.)

bulk. They show that the initial rates of dissolution are well described by a simple first-order equation if an apparent solubility product of about 10^{-62} was considered, a value much lower than the actual thermodynamic solubility product of HAP ($K_{SP} \approx 10^{-58}$). According to the authors, this value corresponds to the solubility of a specific site on HAP surface governing the kinetics of dissolution and called site 2 in the following studies. Further experiments, using a rotating disk technique with porous compacted pellets of synthetic HAP, as already used by Linge and Nancollas [32], confirm the model at low agitation speeds [7]. However, at higher speeds and for certain values of the bulk concentrations, a break in the dissolution rate vs. rotation curve was observed [33,34]. Since such behavior cannot be explained by the single-site model, the authors interpreted this result by the presence on HAP of a second site of dissolution (called site 1) with a much smaller value of the kinetic constant K and a larger apparent solubility of 10^{-60}. The rate of dissolution inside the disk is described by two first-order surface kinetics:

$$Jp = K_1(C'_{S1} - C) + K_2(C'_{S2} - C) \tag{6}$$

where C'_{S1} and C'_{S2} are the apparent solubilities associated with the two crystalline sites, and K_1 and K_2 are the respective first-order kinetic rate constants. C is the concentration of HAP in the aqueous phases. However, this two-sites model was questioned, since the existence of the two sites of dissolution was deduced from a curve-fitting procedure [35].

In order to minimize the participation of the diffusion step in the Nernst layer, further experiments were realized at pH 4.5 with HAP microsuspensions of more or less aggregated power issued from different preparations [13,36]. Various solution compositions covering wide ranges of partial saturation and calcium–phosphate ratios were studied. Although a quantitative assessment of the properties of site 1 was not possible, interpretation of the data based on the initial slopes of the kinetics confirmed the properties of site 2 with a best pK_{sp} value of 64±0.5. The model of demineralization was further reexamined using a rotating disk of bovine enamel at pH 4.5 in sink conditions and in the presence of different buffer types and various amounts of calcium and phosphate ions in solution [15,37]. The results confirmed the role of surface dissolution kinetics and were interpreted with only one dissolution site with an apparent solubility of pK_{sp}=59.4±0.5. This lower value compared to those obtained previously was possibly attributed to the presence of carbonate in the bovine enamel.

In summary, the proposed model based on various experiments carried out in 0% saturated or moderate partial saturation conditions assumes that HAP crystal dissolution occurs at one or two sites, each characterized by an apparent solubility different from the well-established thermodynamic solubility of HAP. It includes one or two first-order rate constants characterizing the crystal surface–solution reaction. Under partial saturation conditions (0–16%), the dissolution rate is presumably governed by site 2, while above 16% of partial saturation, the dissolution comes from site 1 associated with the larger apparent solubility. The fact that site 1 has a smaller apparent dissolution rate constant than site 2 in spite of its larger solubility is believed to reflect a smaller population of these sites. According to the authors, the site responsible for dissolution under partial saturation conditions is very likely the site that is important in the clinical development of carious lesions [18,33]. One of the sites (site 1 under partial saturated solutions) would be associated with dissolution along the c-axis dislocations of HAP.

2. The Surface Nucleation Models

(a). The Restriction Model. Smith et al. [17] reported a study on the kinetics of HAP dissolution in surface controlled conditions using HAP powder with an average size lower than 1 μm and the pH-stat technique. A semiempirical expression fitting well the obtained data in the pH domain 4.5–6.5 was proposed:

$$-\frac{dw}{dt} = k_r(w_S - w_n)\exp(-bw) \tag{7}$$

where w represents the added acid at t, w_S is the amount of acid necessary to dissolve the calcium and phosphate to reach saturation, and k_r and b are constants. Eq. (7), referred to as the restricted model, is able to account for the fact that the decline in the dissolution rate for HAP is much faster than that based on a simple first-order law.

In this model, the interpretation of the experimental results is based on the description of the dissolution of the crystal surface in terms of the probability n of escape of a particle from the surface per unit time [38]. The statement is that any HAP sample consists of a population of crystals of different sizes and carrying on their surface defects such as kinks. The ions of the crystal surface at these defect sites have various probabilities for escape into solution. The distribution of the values of escape probability varies with the form and orientation of the kinks, and the dissolution takes place at sites where n is higher than at other parts of the crystal surface. The sites with the highest n values will dissolve first and so on, leading to a continuous decrease of the dissolution rate. In Eq. (7), the value of b is related to the distribution of the values of n and increases as the distribution widens. The value of k_r is related to the mean escape probability, increasing with the mean probability increases.

The important effect of calcium ion concentration on the k_r and b values is interpreted as the lowering of the mean value of n as the concentration of calcium ions increases. At the molecular level, this expresses the probability for H^+ ions to displace the calcium ions from the sites, a probability that depends on the H^+ and calcium ion concentration. The process is assumed to be slow enough to be unaffected by diffusion through the surface aqueous film.

(b). The Polynucleation Model. To reach a better understanding of the molecular events that take place at the crystal surface during growth and dissolution, Christoffersen et al. developed researches based on the use of polydisperse HAP microcrystals for which the diffusion or convective process from the interface region to the bulk region is not expected to be rate controlling as long as the activation energy for a building unit to leave the edge of a dissolution nucleus is high enough. It is then supposed that at least in a certain range of saturation and pH, the surface process can be studied without interference from transport processes.

From experimental results obtained at pH 6.6–7.2, without inert salt and in the presence of stoichiometric amounts of calcium and phosphate ions in the range $0.1 < C/C_s < 0.7$, the overall rate of dissolution of HAP at constant pH was found to be described by the empirical expression [39,28,19]

$$J = K_J m_0 \left(\frac{m}{m_0}\right)^q \left(1 - \frac{C}{C_s}\right)^p \tag{8}$$

where J is the rate expressed as the time-derivative of the amount of substance dissolved, m_0 is the crystal mass of crystals used in an experiment, m is the mass of crystals present when the rate is determined, C_s is the molar solubility of HAP at the appropriate pH value, and C is the stoichiometric concentration of HAP in the aqueous phase. p and q are constants in the restricted domain studied. The rate constant K_J was found to depend on the hydrogen ion activity according to the law $K_J = 1.0 \times 10^{-2} (H^+)^{0.57}$. This dependence was interpreted as a catalytic effect of the hydrogen ions of the aqueous phase reacting with PO_4^{3-} ions on the HAP surface [40]. In this interpretation, the lateral growth rate of a dissolution nucleus (hole) is proportional to the mole fraction of surface phosphate sites occupied by HPO_4^{-2} ions.

Experiments of Budz and Nancollas [29] showed, however, that q and p are variable and depend on the degree of insaturation. q takes the constant value of 0.06 ± 0.2 found by Christoffersen only in the 20–60% reaction domain, while p was found to depend on m/m_0 and pH, and to vary between 2.9 and 4.7 for pH ranging from 6.6 and 7.2. This is clearly understood since Eq. (8) includes the empirical function $K_J(m/m_0)^q$, which incorporates all changes in crystal specific area, shapes, surface microstructure, and interfacial properties during dissolution. Separate investigations of the effect of each of these factors expected to be saturation dependent become difficult.

For nonstoichiometric solution compositions ($Ca/PO_4 \neq 1.67$), Eq. (8) was found insufficient to describe the kinetics [35]. The rate at constant pH was shown to depend more on the concentration of calcium than on the total concentration of phosphate. An empirical function of the type $[Ca]^a[PO_4]$ was proposed with a value close to 1.67 near saturation and close to 3 for very dilute solutions, showing that at low values of C/C_s, Ca^{2+} ions have a stronger retarding effect than phosphates. This effect was again interpreted as the catalytic effect of H^+ ions, since as the calcium concentration at the surface increases, the mole fraction of surface phosphate sites decreases. According to the authors, this demonstrates that the rate is not solely given by the ionic product and that specific effects of the ions are important. This supposes a description of the rate with at least three concentration parameters, i.e., H^+ and calcium and phosphate concentrations.

The kinetic results obtained in stoichiometric solutions have been described on the basis of the nucleation models for crystal growth and dissolution [19] by combining equations for the rate of nucleation [41] and the rate of nuclei spreading over a crystal surface [42]. Among possible mechanisms like mononuclear, spiral, or transport controlled dissolution, the polynuclear mechanism was chosen since it best fitted the experimental results. In this mechanism, nuclei are formed on the crystal surface and spread over the surface with a finite lateral velocity; they grow laterally and intergrow. Such a mechanism furnishes a possible explanation for the hole formation found during HAP crystal dissolution in which dislocations are present. The linear growth perpendicular to the crystal surface has been expressed as [43]:

$$\frac{dr}{dt} = K' v_+^{1/3} v^{2/3} b^{1/6} \exp\left[-\frac{a^4 s^2}{3(kT)^2 b}\right] \quad (9)$$

where r is the linear dimension of the crystal, K' the rate constant depending on pH, v_+ the lateral rate of growth of a dissolution nucleus when no back reaction takes place, and v the lateral rate of growth of a dissolution nucleus. b is the dimensionless affinity for dissolution for a mean ion ($b = -Dm/kT$, where Dm is the mean increase of the chemical potential for the dissolution of one ion), s is the surface free energy of HAP, and a the linear size of a growth unit. This leads us to express the overall rate of dissolution as

$$J = K'' m_0 F\left(\frac{m}{m_0}\right) v_+^{1/3} v^{2/3} b^{1/6} \exp\left(-\frac{a}{b}\right) \quad (10)$$

Taking into account the proposed H^+-ion catalyzed process, the overall rate can then be expressed as

$$J = K X_{HP} m_0 F\left(\frac{m}{m_0}\right)\left(1 - \frac{C}{C_s}\right)^{2/3} b^{1/6} \exp\left(-\frac{a}{b}\right) \quad (11)$$

with K the rate constant and X_{HP} the mole fraction of phosphate groups HPO_4^{2-} formed at the HAP crystal surface. Experimental results lead to a value of the surface free energy of about 45 mJ/m^2, a much lower value than the one experimentally found and than the theoretical value of HAP growth (240 mJ/m^2) [44]. Acidity constant values for the HPO_4^{2-} surface complexes were estimated to be of the order of 10^{-7} mol/L at pH 7 and 10^{-6} mol/L at pH 5.

Considering the important role of calcium ions demonstrated by the behavior of nonstoichiometric solutions at low pH values, the model was reanalyzed by assuming that calcium ions participate deeply in the rate determining step [45]. It is then suggested that the lateral growth of surface nuclei is controlled by the frequency for a calcium ion to enter in a growth site by performing a diffusion jump into a kink and being simultaneously partly dehydrated.

The effects of dissolution inhibition of various cations (Al^{3+}, Cr^{3+}, Fe^{3+}, Sn^{2+}) [46,47] and other compounds [48,49] were interpreted using the polynuclear model and Langmuir type adsorption isotherms. With negatively charged compounds forming strong bonds towards calcium ions at the surface, the model leads us to assume that the frequency of hole nucleation is strongly reduced in an area around each inhibitor unit.

The dissolution process should be practically nulled when the size of a critical dissolution nucleus is comparable to the surface area occupied by one inhibitor unit adsorbed to the crystal surface.

Although further data are needed to extend the proposed description to the whole range of dissolution reactions, and although this model includes adjustable parameters that cannot be determined experimentally, the proposed approach furnishes an interesting description of the chemical events that may occur at the HAP surface.

3. The Semipermeable Interface Model

Considering that the slowdown of HAP dissolution kinetics is observed in both surface and diffusion controlled experiments and that the experimental approach in diffusion controlled conditions may give direct evidence on the phenomena occurring at the interface and on the role of each determining parameter, Gramain and coworkers analyzed the dissolution process of HAP in diffusion controlled conditions. Using sieved commercial platelike HAP crystals or powder of human enamel, sized in the range of 165–200 μm or 100–160 μm, the kinetics of dissolution were studied in the pH range 3.7 to 6.9 in the whole saturation range, and in stoichiometric and nonstoichiometric conditions. It was shown that proton diffusion was not the key factor of the dissolution process and that no interfacial H^+ gradient exists. A nonlinear pH dependence of dissolution with $(H^+)^{0.64}$ for synthetic HAP [26] and $(H^+)^{0.81}$ laws for human enamel [30] confirmed the previous results. Since H^+ concentration and the HAP solubility gradients were unable fully to describe the dissolution rates, the authors assumed that the controlling step was the calcium diffusion at the interface. The key role played by calcium was justified by the well-documented specific calcium–HAP interactions [50] controlling the HAP surface potential [23].

The specific role of calcium ions during the dissolution process was demonstrated by the experimental observation that during nearly all the dissolution process, important amounts of calcium ions are continuously adsorbed onto the HAP surface, these amounts becoming smaller with the progress of the reaction [10,26]. It was shown that this accumulation depends on various factors such as the sample conditioning, the pH, the undersaturation degree, and the Ca/P ratio in solution. The accumulation period is longer as the calcium in solution is lower and constitutes a classical behavior for adsorption processes. The maximum amounts of accumulated calcium range from 1.5 to 13.4×10^{-10} mol/cm^2, amounts that are comparable with the value of 6.9×10^{-10} mol/cm^2 estimated from the data of Shimabayashi and Nakagaki [50]. Such amounts are 5 to 50 times higher than the value 2.6×10^{-11} mol/cm^2 estimated for a monolayer adsorption. This result suggests either multilayer adsorption or an important increase of the number of adsorption sites during the dissolution, both processes being favored by the increase with time of the calcium concentration in the bulk.

It was further demonstrated that the HAP interface was saturated and even slightly supersaturated for calcium with respect to the solubility product of HAP [22]. This result is in contradiction with all the results showing that in diffusion controlled conditions, a simple first-order kinetic law based on the HAP solubility gradient is unable to describe the dissolution process under all conditions. According to the authors, this contradiction can be leveled off by considering the specific role of calcium ions at the solid interface. A dissolution model schematized in Fig. 7 was proposed in which the calcium saturated interface generates a restriction of diffusion towards the bulk so that the interfacial diffusion becomes much slower than the solid dissolution reaction.

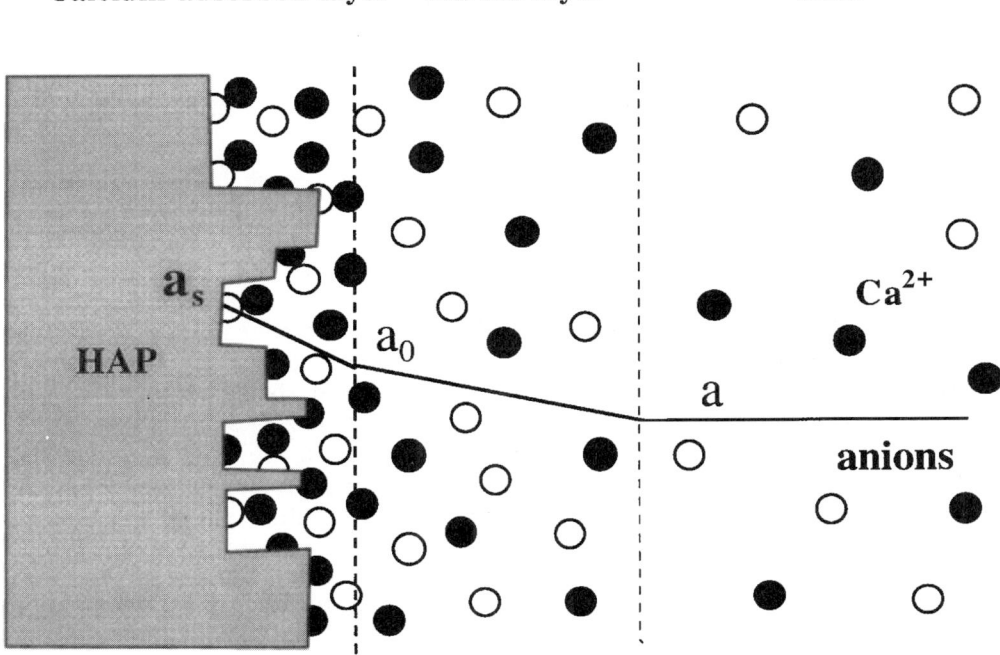

FIG. 7 Schematic picture of the concentration profile of calcium ions through a boundary layer including the calcium adsorbed layer and the Nernst layers as proposed in the semipermeable interface model. A quasi–steady state with a linear concentration profile gradient is assumed for calcium ions.

Since it is well demonstrated that the HAP dissolution is always congruent, the dissolution rate was described by considering the calcium species. It follows that the transport through this interfacial layer is given at constant pH by

$$R_{Ca} = P_{Ca}(a_S - a_0) \tag{12}$$

with

$$-R_H = R_C R_{Ca} = \frac{5}{3} R_C R_P \tag{13}$$

where a_S is the calcium activity at the solid saturated interface, a_0 the calcium activity at the interface between the surface layer and the Nernst layer, and P_{Ca} the mass transfer coefficient characterizing the low permeability of the layer [22]. R_H, R_{Ca}, and R_P are respectively the coupled fluxes of consumed protons and the released calcium and phosphate ions.

The transport rate for calcium through the Nernst layer is simply described by

$$R_{Ca} = P_{Ca}^o(a_0 - a) \tag{14}$$

with a the calcium activity in the bulk and P_{Ca}^o the classical mass transfer coefficient in the absence of chemical reaction. Since at steady state, both fluxes must be equal, the rate of consumed protons is described by

$$R_H = -R_c \frac{P_{Ca}^o}{1+k}(a_S - a) \qquad (15)$$

with

$$k = \frac{P_{Ca}^o}{P_{Ca}} \qquad (16)$$

k^{-1} characterizes the reduced permeability of the surface layer with respect to the permeability in the Nernst layer. $1/(1+k)$ is defined as the rate reduction factor (RRF).

It was shown that this relation well describes the experimental results in stoichiometric and nonstoichiometric conditions [10,22]. Constant from pH 3.7 to pH 5 with k values of about 2000, the RRF increases at higher pH values ($k = 300$ at pH ≈ 7), illustrating a relative acceleration of the dissolution process. The RRF value depends on sample conditioning and on the adsorbed calcium amount. The evolution of RRF during the dissolution up to equilibrium shows an exponential type decay up to a quasi-plateau. According to the authors, this behavior is in close relation with the kinetics of calcium interfacial adsorption leading to the formation of the calcium-rich layer and explains well the fast initial rate and the continuous decrease up to saturation. This shows that a true steady state of the dissolution process is never obtained except perhaps when the RRF plateau is attained. At this pseudo–steady state of dissolution, the RRF is found to be only pH dependent but not on the composition of the initial solution [10].

To summarize, the hypothesis of calcium-rich layer formation, and the nonattainment of steady state characterized by continuous decrease of RRF, explain well the important variation of the apparent order n in Eqs. (1) or (5) observed by various authors during the dissolution process and depending strongly on the experimental conditions.

The restriction of diffusion in the interfacial layer is explained by considering the electrostatic interactions generated by the adsorbed calcium ions acting as an ionic semipermeable interfacial barrier against ionic diffusion. The hypothesis was supported by the well documented permselective properties of HAP due to calcium adsorption [51], and by the pH dependence of the RRF increasing above pH 5 where divalent phosphate ions are mainly present. This behavior should result from the well-known dependence of the Donnan exclusion with the valence of the diffusing counterions as in semipermeable membranes with fixed cations. In this interpretation, the adherent layer should be formed by the eroded solid surface with calcium adsorbed on preferential sites and soaked in a HAP saturated or supersaturated solution.

To describe this approach, a quantitative model was proposed describing the adherent layer by two concurrent approaches, both supposing a steady state, since analytical solutions of the equations are only simple in such situation.

The first approach was based on the scaled particle theory of hard disk [52]. Adsorbed calcium ions constitute the hard disks covering a part of the surface area and limiting, by an electrostatic screening effect, the accessibility to the surface as schematically represented in Fig. 8. The covered fraction is described by a monolayer Langmuir adsorption isotherm. The model leads to Eq. (15) with

$$\frac{1}{1+k} = \frac{S_{A0}}{S_A} \qquad (17)$$

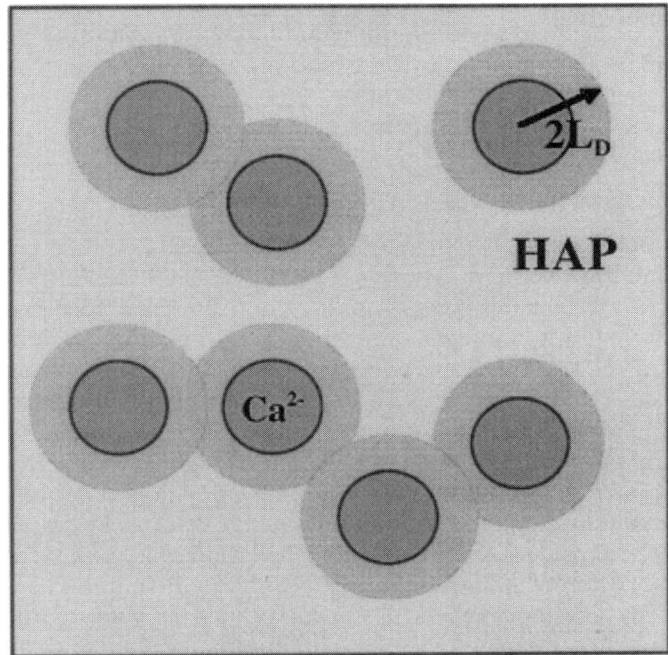

FIG. 8 Schematic representation of the adsorbed calcium ions on the HAP surface. Calcium ions are assumed to be hard disks of effective radius $2L_D$. (From Ref. 52.)

where S_A denotes the total surface area of HAP and S_{A0} the available area for hydrogen attack. Comparison of the pH dependence of RRF obtained in the initial dissolution times with S_{A0}/S_A leads to a value of 1.8×10^5 M^{-1} for k with one adsorption site occupying an area of about 6.4 nm^2.

The second approach uses the current theory of ionic transport through a permselective membrane with the advantage of describing a three-dimensional layer [53]. The overall flux of the ionic species through the layer is described by the Nernst–Planck equations. A linear Freundlich adsorption isotherm for calcium was considered, since the analytical solution of the equations is simple only in the case. The model leads to an equation identical to Eq. (15) with

$$k = \frac{2(X_\infty b)^n}{(1+n)} \left(\frac{P_{Ca}^0}{P_{Ca}}\right) \tag{18}$$

and

$$n = \frac{2}{-z_p} \tag{19}$$

in which X_∞ is the number of adsorption sites per cubic centimeter, b is a constant related to the binding energy, and pH dependent z_p is the charge of one equivalent anionic phosphate group. This relation well accounts for the stronger barrier action at low pH values where the phosphate valence is lower.

According to this model, the description of the HAP–solution interface is as follows. The dominant ionic species of phosphate exposed on the surface may be assumed to be HPO_4^{2-} at pH 6–7 or $H_2PO_4^-$ at pH 3–6. They constitute the most probable sites of adsorption for calcium together with other appropriate sites such as OH^- sites, dislocations, or surface defects. In the presence of H^+ ions and in the pH domain between 3.0 and 6.0, an equilibrated ion exchange reaction occurs:

$$\underline{H_2PO_4^-, H^+} + Ca^{2+} <=> \underline{H_2PO_4^-, Ca^{2+}} + H^+ <=> Ca^{2+} + H_2PO_4^- + H_2O \qquad (20)$$

where the underlined species indicate the solid surface species. Increasing the calcium solution concentration promotes calcium adsorption on sites and liberates adsorbed H^+ for further attack. Since dissolution takes place preferentially at sites such as kinks and defects where the binding of the surface atoms to the surrounding and underlying crystal is the weakest, the process leads to an erosion of the surface with formation of atomic holes increasing continuously the surface rugosity. The rough surface adsorbs more calcium ions and forms a positive barrier against free calcium diffusion. This restricted diffusion generates an interfacial saturation/supersaturation with possible formation of complexes and ion pairs. The equilibria of Eq. (20) are then displaced towards remineralization. In this scheme, the proposed semipermeable layer includes the eroded part of the surface soaked in the saturated solution. Depending on the dissolution conditions, the adherent solution may become under- or supersaturated as soon as the diffusion in the Nernst layer becomes less or more rapid than the solid dissolution process.

In this model, the adsorbed calcium ions play a protective or inhibiting role. As a consequence, any defect on the HAP surface where calcium adsorption is prevented will be a preferred site for accelerated attack. The specific role of calcium ions and the surface equilibrium undersaturation/supersaturation proposed could be of importance in connection with the formation of subsurface lesions in caries formation [54].

The inhibiting effect of various acidic polymers able to bind calcium ions has been studied and interpreted according to this model, assuming that adsorbed acidic polymers increase the ionic capacity of the interface and reduce its permeability [55,56]. The model has been used to interpret the dissolution inhibition of fluoride ions present in solution [57].

F. Dissolution of Carbonated Apatites and Fluorapatites

1. Carbonated HAP

Carbonate ions constitute very important impurities, since they are nearly always present in apatite samples. In the apatite structure, carbonate ions replace either OH^- or PO_4^{3-} groups, and although its distribution in the crystal is not known, an anisotropic distribution is probable [3,58]. From a biological point of view, they are supposed to play a crucial role in the caries process and bone resorption and apposition. Natural human enamel contains 2–3 wt% of carbonate ions, and for this reason carbonated HAP seems to be a better model than pure HAP.

The presence of carbonate in HAP influences to a significant extent its acid reactivity, and fully carbonated HAP (CAP) exhibits higher dissolution rates than pure HAP [21,29]. With partly carbonated HAP, the inhomogenous distribution of carbonate ions [6] and the presence of small crystallite aggregates often found in the preparations [59], have been proposed to explain this behavior. However, the most probable reason is the presence of defects in the crystal lattice of carbonated apatite leading to a change of surface properties during the dissolution process.

Researches on the dissolution of single crystals whose size in synthetic and biological materials varies from 10^2 to 10^8 nm have been realized [3]. In contrast with single crystals of pure HAP or with a low carbonate content, the solubility in acids of high carbonate containing monocrystals is very fast and shows anisotropic initial dissolution with removal of the crystal core and hole formation on the basal planes. This anisotropic dissolution has been explained by the presence of defects where the material is highly strained with lower surface energy. The origin of these defects would be surface screw dislocations [3] or the growth process via octacalcium phosphate [60].

2. Fluorapatites

Despite the biological and industrial importance of fluorapatite $Ca_{10}(PO_4)_6F_2$ (FAP), there are surprisingly few articles concerning the growth and dissolution kinetics of FAP in aqueous solutions. With a pK_{sp} of 59.75 ± 0.75 for the half-cell [61], FAP is less soluble than HAP and dissolves appreciably slower than HAP [62]. As a consequence, the dissolution of mineralized FAP crystals was observed largely controlled by the precipitated FAP phase [62]. In surface controlled conditions at pH 4.5–6.0, and in the undersaturation range $0.6 < C/C_S < 0.8$, the rate of dissolution was found to be proportional to $(1 - C/C_S)^2$ in contrast to the behavior of HAP for which the order was in the range of 3 to 5 in the same conditions [63]. It was concluded that the dissolution of FAP follows a spiral dislocation mechanism rather than the polynuclear mechanism proposed for HAP. However, according to Christoffersen et al. [61], who found a reaction order in the range of 3.0 to 3.3 in nearly the same conditions, a polynuclear mechanism describes well the results with a surface tension of 42 mJ/m^2 comparable to that obtained for HAP.

The rate of growth of FAP has also been studied [45,61,64], and the order of the reaction was found to vary from 1.25 to 5.3. Such a large variation may be indicative of a change of mechanism with experimental conditions. It was deduced to be a tension surface of 120 mJ/m^2, a three times higher value than for dissolution, a difference that was attributed to the presence of defects on the dissolving FAP [45].

Fluorhydroxyapatite (FHAP) in which a varying proportion of the OH$^-$ ions is replaced by F$^-$ ions has been mainly studied under conditions relevant to the dental caries process, since it was suggested that this mineral may be used as fluor reservoir for caries prevention. It was recently shown that FHAP dissolves nonstoichiometrically with respect to F$^-$ release with a F$^-$ deficit in solution [65]. It was suggested that an F-enriched mineral is reprecipitated rather than pure FAP.

G. Conclusion

All the HAP dissolution and growth studies demonstrate the complex behavior of the HAP interface. The complexity of the problem results in large part in that the HAP surface properties are changing as the dissolution reaction proceeds, leading to a decrease of active dissolution sites and possibly of surface perfection. The recent studies show that a description of the dissolution process in terms of solubility product is quite insufficient to describe the process under all saturation conditions and that specific and determining influences of component ions such as calcium are to be considered and included in the models. For this, a better knowledge of the intimate HAP surface composition in aqueous solutions is necessary. It is hoped that future experiments can be designed to demonstrate clearly the specific role of each component ion and of surface defects. From such approaches, it is probable that new ideas for the study of carious processes [54] will be suggested.

IV. INHIBITION OF DISSOLUTION BY ADSORBED COMPOUNDS

A. Adsorption and Inhibition of Dissolution

The effect of contaminants on the rate of mass transfer between the crystal–liquid interface has aroused considerable interest, and numerous studies have been devoted to HAP surface interactions with various inorganic and organic compounds present in the surrounding solutions because they can affect dissolution, nucleation, and growth processes. The well-known classical example is the use of fluoride ions to reduce carious lesion formation. We have seen that the interfacial properties of HAP, a crystal with a highly energetic and ionic surface, determine its dissolution behavior. Adsorption at such a charged surface involves both electrostatic and specific chemical forces with interfacial ion exchanges. Any modification of the surface and in the surrounding solution composition is likely to alter the HAP behavior by affecting the interfacial composition and the free surface energy, a parameter particularly difficult to determine with this type of surface [66,67]. In this context, adsorption phenomena are determinant, and a considerable number of works devoted to the adsorption of ions, surfactants, proteins, and synthetic polymers onto HAP have been published. Many of these works have shown a relationship between inhibition of HAP dissolution or crystal growth and the adsorption properties of inhibitors. Among the proposed compounds that have been shown to act efficiently as inhibitors of dissolution, it is possible to distinguish between two types of action in agreement with the experimental evidence or the proposed interpretations. Following the adsorption step, the HAP interface can be modified either by a surface precipitation phenomenon or by a surface reaction process. Surface precipitation consists in the nucleation and growth of a stoichiometric precipitate of the adsorbed species by a complexing ion at the interface. It often leads to a modification of the solubility properties of HAP. The surface reaction involves chemical interactions between the adsorbate species leading to ion exchange, to electrostatic bonding, or to high-energy bond formation such as a covalent bond (chemisorption). Surface reactions can strongly affect the electrochemical properties and accessibility of the interface and thus modify the free surface energy and the dissolution or growth behavior.

HAP surface charge is an important parameter affecting and controlling the adsorption properties. It has been studied by electrokinetic and potentiometric titration, and predicting methods exist [8,23]. The pH at which the solid possesses no net surface charge is defined as the point of zero charge (p.z.c.). In the case of HAP, the determining ions of the surface charge and the p.z.c. are the lattice ions, but any other species present in solution can affect the interfacial charge and thus the interfacial behavior. Due to the important possibilities and easiness of ion-exchange and ion-substitution reactions, the electrochemical properties of HAP not only are a complex function of pH and of the concentration of constituent and contaminant species but also are strongly dependent upon the method of preparation, cleaning, storing, and conditioning of the solid. However, all studies agree in finding for synthetic HAP a p.z.c. in the range of pH 7–8. Under all pH conditions, phosphate addition makes the apatite surface more negatively charged, and calcium addition renders it more positively charged [2,23,25].

Although in vitro and in vivo researches on dissolution inhibitors or remineralization accelerators are considerable in the dental and bone domains, we will focus in the following on the analysis of studies related to the physicochemical aspects concerning mainly synthetic HAP. Unfortunately, quantification and comparison of the inhibition effect of all studied molecules are difficult in the absence of a precise description of the experimental conditions and in the absence of consensus regarding the data representation.

B. Inhibition by Surfactants

The HAP surface modification by surfactant adsorption is well illustrated by the adsorption of ionic surfactants such as sodium dodecylsulfate or long-chain alkyl ammonium. Both types of surfactants are bound by electrostatic interactions and ion exchanges with the surface-active ions [23,68–70]. As a consequence of the mode of attachment, and depending on the concentration, the hydrocarbon chains either protrude to the aqueous solution or lie on the surface, partially modifying the HAP surface to a hydrophobic one. This property is used to improve the separation of proteins and polypeptides in HAP HPL chromatography [71] and to promote the adsorption of polymers [72]. Studies on the effect of dodecylamine hydrochloride on HAP dissolution show that the rate decreases with increasing surfactant concentration, a behavior that is strongly dependent on the initial HAP heat treatment [68,73].

C. Inhibition by Adsorbed Ions

1. Fluoride Ion

The analysis of the interactions of fluoride ions with HAP surfaces has been the object of extensive studies since it was demonstrated that fluoride ions are effective for the prevention of dental caries and increase both resistance against demineralization and remineralization rates for in vitro and in vivo conditions [74–78]. The knowledge of the mechanism by which the fluoride ion acts is of prime importance in order to define the best form of application for caries prevention, which still remains an acute problem. It is now well established that the interfacial fluoride location is determinant, and that the transformation of only a few layers of the HAP crystal can dominate the chemical reactivity of the apatite phase. This explains why F-treated apatite or enamel is affected so greatly at fluoride contents of 1–2,000 ppm, well below the level 37,000 ppm in pure FAP.

The chemical aspects of the uptake of fluoride by HAP and the different proposed models have been already reviewed [8,77,78]. Fluoride may react with apatite in at least four different ways to form FAP [Eqs. (21) and (24)], FHAP [Eq. (23)], or calcium fluoride [Eq. (21)]:

$$Ca_{10}(PO_4)_6(OH)_2 + 2F^- \rightarrow Ca_{10}(PO_4)_6F_2 + 2OH^- \tag{21}$$

$$Ca_{10}(PO_4)_6(OH)_2 + 2F^- \rightarrow 10CaF_2 + 6PO_4^{3-} + 2OH^- \tag{22}$$

$$10Ca^{2+} + 6PO_4^{3-} + xF^- + (2-x)OH^- \rightarrow Ca_{10}(PO_4)_6F_x(OH)_{2-x} \tag{23}$$

$$10Ca^{2+} + 6PO_4^{3-} + 2F \rightarrow Ca_{10}(PO_4)_6F_2 \tag{24}$$

Such reactions are thermodynamically favored since the solubility product of pure FAP ($K_{sp} = 7.1 \times 10^{-61}$) is two orders of magnitude lower than that of HAP, and the solubilities

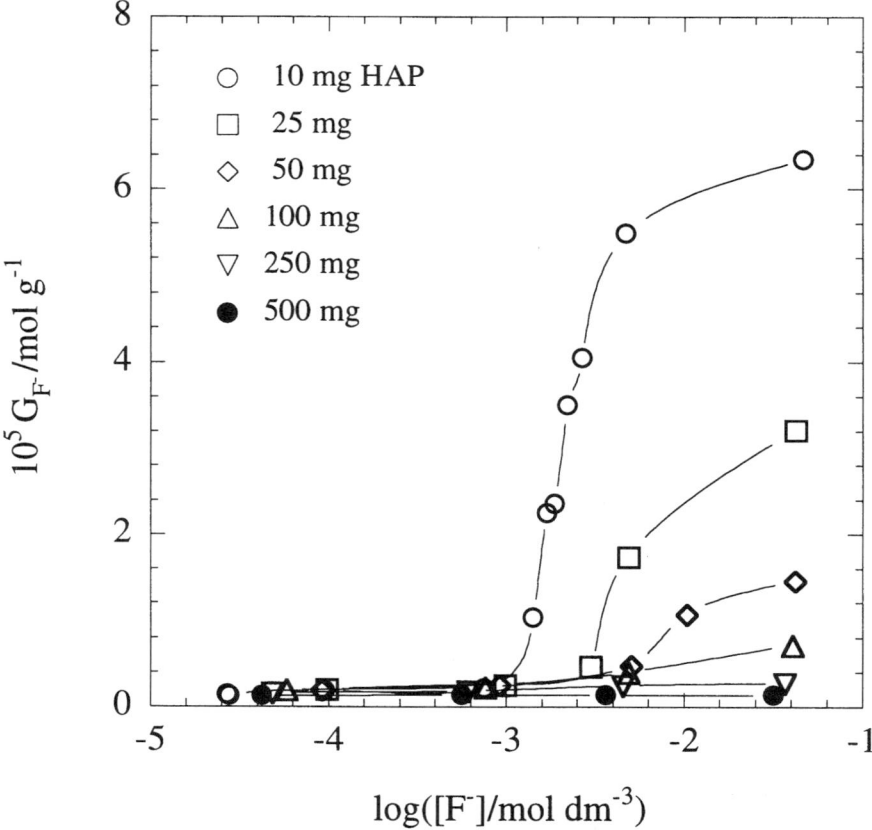

FIG. 9 Fluoride adsorption isotherms vs. fluoride concentration at equilibrium for different amounts of HAP platelets (pH range, 7.8–8.2). (From Ref. 79.)

of FHAPs are intermediate between HAP and FAP. CaF_2, the most soluble phase ($K_{sp} = 3.47 \times 10^{-11}$) [79] is found at high fluoride concentrations. Interpretation of the experimental results dealing with the possible formation of FAP or FHAP phases is however difficult because of the importance of kinetic factors. While kinetics of fluoride adsorption are rapid [79–81], the kinetically controlled formation of new phases is generally a slow phenomenon.

Studies demonstrated that for acidic or neutral pH conditions, the fluoride adsorption isotherms onto HAP showed two plateau domains [79,81–83] as illustrated in Fig. 9. The first plateau obtained for fluoride concentrations lower than about 6.0×10^{-4} M corresponds to the classical Langmuir plateau and occurs at a surface concentration of about 2.0 to 5.0×10^{-6} mol m^{-2}, a value comparable with the number of OH^- sites at the crystal surface. NMR experiments show the probable formation of fluorapatite with 40 to 80% substitution of fluoride for hydroxide ions at pH 7 [84]. The second plateau obtained for higher fluoride concentrations corresponds to the deposit of calcium and fluor ions as shown in Fig. 10. This leads to the formation of CaF_2 promoted by the increasing calcium concentration and the decreasing pH.

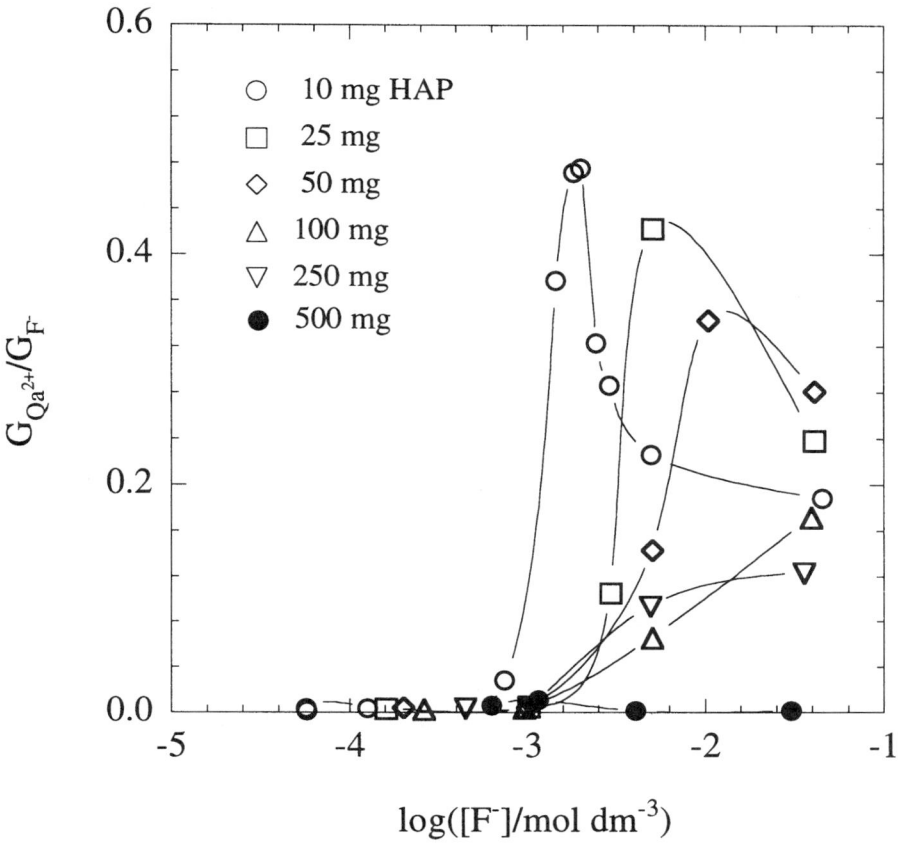

FIG. 10 Ratios of the amount of calcium adsorbed to the amount of fluoride adsorbed vs. fluoride concentration at equilibrium for different amounts of HAP platelets (pH range, 7.8–8.2). (From Ref. 79.)

This is in agreement with the changes in surface charge of apatites accompanying fluoridation [23,82] and making HAP more positive in the acidic and neutral region as expected for CaF_2 formation. This CaF_2 deposit may have an important role in caries prevention, since CaF_2 has been shown to promote the nucleation and mineralization of FHAP [85]. Ambiguity arises from the exact nature of the interfacial layer formed in the first plateau and from the mechanistic approach. This ambiguity demonstrates that the phenomenon is extremely sensitive to experimental conditions and to kinetic factors. Fluoride-treated samples undergo a solid-state transformation over time, which results in the appearance of fluorapatite [84]. Interpretations in terms of lattice destruction with dissolution/precipitation or simple surface exchange have been proposed. It seems obvious that the two mechanisms occur depending on time and experimental conditions. It may be concluded that the reaction dynamics of fluoride at the apatite interface is considerably more complex than the simple reactions of ion exchange, crystal growth, and recrystallization and involves considerable changes in the interfacial double layer of the crystals.

Although the works devoted to the effects of fluoride ions in caries prevention are voluminous and the solubility-reducing effect of fluoride ions for enamel or HAP have been extensively described, physicochemical studies on HAP dissolution in the presence of fluoride ions are rather scarce. It was early demonstrated that inhibition of dissolution increased nonlinearly with fluoride concentration and that the dissolution rates became very slow at high F^- concentration [86]. As a consequence of the fluoride adsorption dynamics, the inhibitory effect of fluoride is dependent on time and degree of HAP saturation, since the diffusion of fluoride ions in the crystal surface may occur during the dissolution [80]. The process may modify the composition of the mineral under attack. Thus the dissolution kinetics (short time) cannot pretend to describe what can happen at longer contact with fluoride. In particular, if at high fluoride concentrations, CaF_2 precipitation may occur during the dissolution reaction, the precipitation of FAP at lower fluoride concentrations is for kinetic reasons much less probable. However, the surface modifications, even if they concern only the first layer of the apatite crystal, are sufficient to induce a dramatic change in dissolution behavior.

A recent study demonstrates that fluoride ions in solution affect the dissolution kinetics in two different ways [87]. The presence of this ion at the interface and in solution is sufficient to reach surface concentrations comparable to fluoridated apatites and to reduce the apatite solubility. The lower FHAP or FAP equilibrium is reached earlier. However, it was also observed that the presence of fluor accelerates the initial dissolution rates as shown in Fig. 11. According to the authors, fluoride adsorption on calcium sites decreases the interfacial positive charge and induces a higher interfacial permeability.

Finally, it is interesting to mention that the cation associated with the fluoride anion plays a role in fluorine–HAP interactions [88] as far as it is also adsorbed onto the surface. This is well understood, since surface-active cations can affect not only the interfacial charge but also the morphology of the formed phases such as CaF_2.

2. Cations

Various studies have shown that divalent (Sr^{2+}, Sn^{2+}, Zn^{2+}, Cd^{2+}, and Fe^{2+}) and trivalent (Al^{3+}, Cr^{3+}, Fe^{3+}, and La^{3+}) cations are strongly adsorbed onto HAP with often important ion exchanges for Ca^{2+}. This property is well illustrated by studies showing that HAP is an efficient cation exchanger [89]. However, most studies have been limited to the clinical aspect, and little is known about the chemical reaction mechanism of these ions with HAP, so that the possible inhibition effects are often difficult to interpret. Although cation adsorption is largely determined by its charge, and within a given charge, by its hydration enthalpy, difficulties arise from the fact that reaction at the HAP surface and in solution may be complicated and quite different according to the nature of the cation. Ion exchange, complex formation, specific adsorption, and precipitation can occur at the HAP surface. True ion exchange may occur only when the size of the cation fits the site occupied by calcium ions, while precipitation will depend on the solubility product of the formed phase. Inhibition of dissolution may be due either to the change of interfacial charge and surface energy or to the decrease of HAP surface accessibility. The published examples seem to demonstrate that in many cases, the inhibition or growth reduction occurs through a surface deposition of a complex few soluble phase easily formed, notably in the presence of phosphate or OH^- ions (in part of oxide, hydroxide, and phosphate), making the system difficult to analyze and leading to a dissolution mechanism rather surface diffusion controlled. This seems to be the case of ions like Cd^{2+} and Mg^{2+} [90], Al^{3+} [46,47,91], Fe^{2+} and Fe^{3+} [47,92], and Pb^{2+} [89]. In particular, it has been demonstrated that La^{3+} ions react with phosphate ions to form insoluble $LaPO_4$ precipitate,

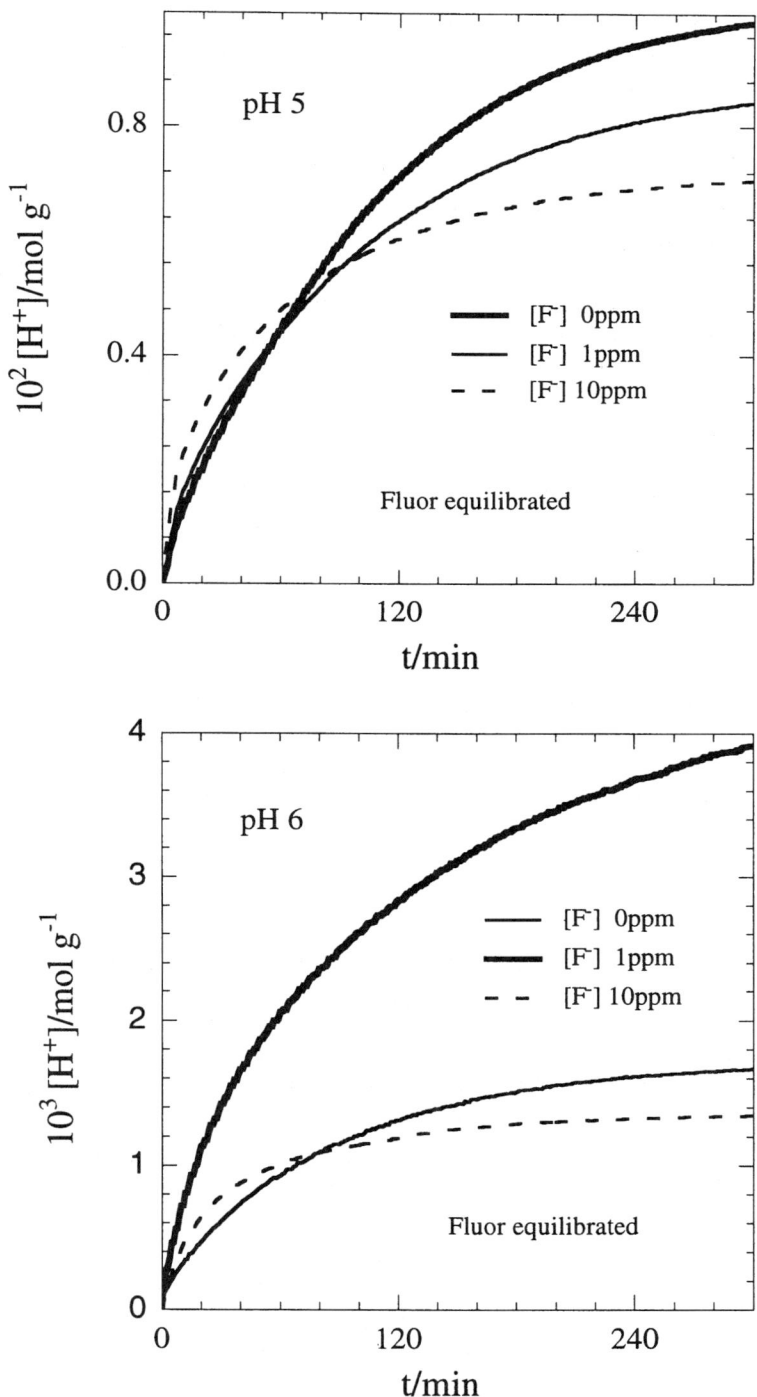

FIG. 11 Dissolution kinetics of HAP powder (165–200 μm) in the presence of fluoride ions. The powder was equilibrated in the fluoride solutions during 3 h before acid addition. For 10 ppm fluoride, CaF_2 is deposited. (From Ref. 87.)

inducing a noncongruent dissolution with a concomitant release of protons. No inhibition effect could be observed suggesting the formation of a non-HAP-adherent $LaPO_4$ precipitate [93].

D. Inhibition by Low Molecular Weight Organic Compounds

Very numerous works describe the adsorption of small organic molecules including amino acids onto HAP. It is particularly interesting to remark that nearly all the molecules showing strong adsorption properties contain carboxylic or phosphoacid groups in their chemical structures, both groups known to interact strongly with calcium ions. In the acidic domain, where the HAP net charge is positive due to a calcium-rich interface, adsorption sites can either be the calcium sites on the solid surface or the adsorbed calcium ions in interaction with surface phosphate groups. The determining role of calcium ions in these adsorption processes is well illustrated by considering that pyrophosphates are adsorbed onto HAP even at basic pH, where the net charge of HAP is highly negative [94]. In the absence of dissolution, the adsorption process is often associated with important interfacial ionic exchanges and ionic rearrangement in the Stern layer.

Various of the adsorbed small organic molecules affect the rate of HAP dissolution. However, interpretation of the inhibition effect may be complicated, since acidic compounds, when also present in solution at sufficient concentration, interact with calcium ions and decrease the calcium activity in solution. As a consequence, as shown with EDTA [95] and citric acid [43,96], acidic compounds act as a calcium pump, increasing the rate of dissolution and affecting the congruency of the HAP equilibrium [97].

Several phosphate-containing compounds and polycarboxylic acids have the ability to act as mineralization inhibitors [12,98]. The phosphocitrate molecule seems of particular interest, since when associated with citrate, the HAP crystal growth rate is decreased by 99% as shown in Fig. 12.

E. Inhibition by Proteins

The interaction of HAP with biomacromolecules has attracted many works considering the use of HAP in chromatographic separation of polypeptides and proteins and the biological importance of the process, since the surfaces of crystals of biological apatites are continuously interacting with salivary or blood proteins. Neutral, acidic, and basic proteins interact significantly with the HAP surface, and it is believed that proteins adsorbed onto enamel surfaces contribute to tooth protection. Elucidation of the role of these proteins requires the knowledge of their adsorption characteristics together with the characterization of their possible effects on the dissolution/mineralization process. Such studies are unfortunately rather scarce.

1. Salivary Proteins

The presence of the so-called acquired enamel pellicle mainly constituted of acidic and polysaccharide-rich proteins induces several processes, such as enamel remineralization or demineralization, and is able to reduce the dissolution rates of carbonated HAP or enamel [99]. In order better to understand their role, the effects of several important salivary proteins have been studied.

Acidic proteins such as proline-rich and phosphoserine-containing proteins [97,100,101], statherin [100,102], and cystatins [103,104] show high adsorption affinity for HAP surfaces. These salivary proteins are active inhibitors of mineralization [104,105].

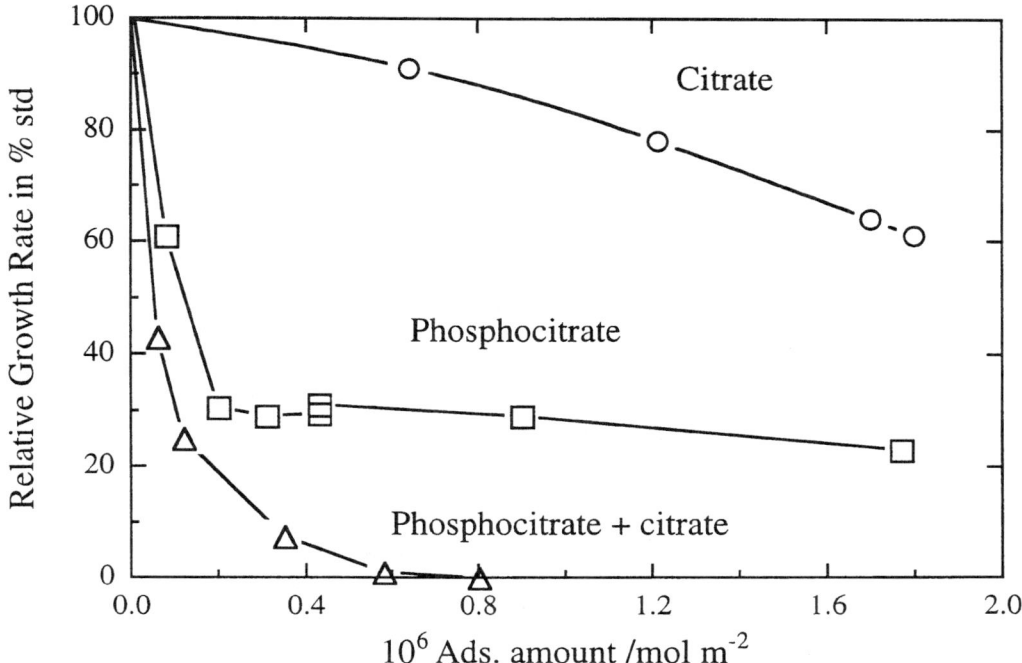

FIG. 12 Effects on HAP crystal growth of adsorbed citrate (○), phosphocitrate (□), and both molecules (△). The surface concentrations of the mixed additives (△) are shown as a function of adsorbed phosphocitrate, and the additional adsorbed citrate amounts are respectively from the left, 1.00, 0.91, 0.88, 0.76, 0.64, 0.56 × 10^{-6} mol · m^{-2}. (From Ref. 12.)

Results suggest that statherin is anchored to the surface through negatively charged residues in the N-terminal groups [102] and that the adsorption is reversible [100]. Salivary cystatins and statherin decrease the HAP dissolution rate by about 30% at pH 6.0, a lower inhibition effect than with albumin and amylase [106]. This is shown in Fig. 13, where it is observed that in constant composition conditions, the inhibition increases markedly with the extent of dissolution. Such a behavior was often observed with proteins, and it is likely that more segments of the molecules adsorb when new surface sites are made available. Glycoaminoglycans such as chondroitin sulfate and hyaluronate were found to promote the incongruent dissolution of HAP with phosphate release and binding of calcium [107].

2. Other Proteins

Among the acidic proteins, albumin, present at high concentration in saliva, seems of particular interest. The adsorption isotherm of albumins onto HAP was found to be quasi-irreversible and of Langmuir type [108,109] or of stepwise nature with possible changes of conformation induced by the surface [110]. The adsorbed amount of the order of 1 to 2×10^{-3} g · m^{-2} was found to decrease with increasing pH, as observed also with other types of substrates. Kinetic data show a fast initial adsorption followed by two slower regimes probably due to conformational changes [111]. Many works suggest that

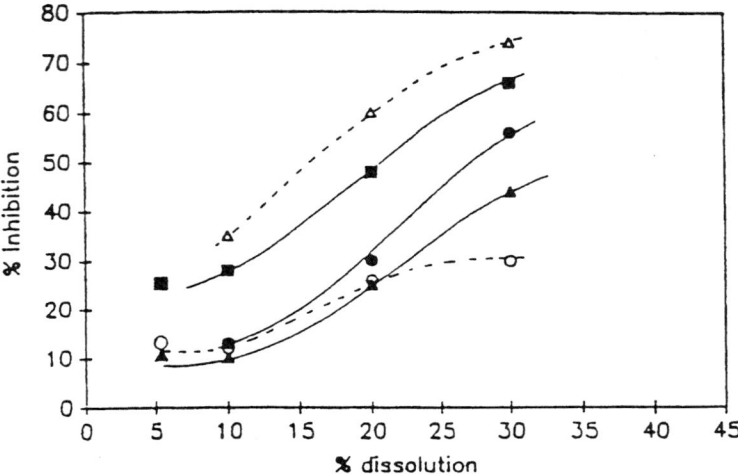

FIG. 13 Inhibition (%) plotted against extent of HAP seed dissolution at pH 6.0 and $C/C_s = 0.41$ (the surface concentration of the proteins was 2.0×10^{-8} mol · m^{-2}): ●, cystatin SN; ▲, cystatin S; △, amylase; ○, statherin; ■, HSA. (From Ref. 106.)

albumin is an active inhibitor of HAP demineralization [106,108,112]. The high affinity of the molecule for HAP and its important size leading to a high surface coverage may in part explain this behavior. However, albumin is also known to bind calcium ions. This property, which obviously is responsible for the inhibition power, complicates the results interpretation, since when present also in solution, albumin decreases the calcium activity and acts as a calcium pump [113] as is observed with other low molecular weight carboxylic compounds.

Phosphoproteins (phosvitin and caseins) markedly reduce the rate of HAP dissolution, and the reduction increases with increasing numbers of phosphoserine residues of the protein [114]. The quasi-irreversibility of adsorption of phosphoproteins compared with the more reversible character of proteins containing carboxylic groups is possibly due to changes in the secondary structure of the macromolecule, and it seems to be one of the factors explaining the efficiency of phosphoproteins as inhibitors [100]. More recently, it was shown that β-lactoglobulin, with an isoelectric pH similar to that of albumin, was also able strongly to reduce the growth and dissolution rate of HAP at pH 7.4 in the concentration range of 2–800 ppm [115]. This is illustrated in Fig. 14.

Adsorption onto HAP of lysozyme, a basic protein, has been well described [116] and occurs in spite of the positive net charge of the HAP surface. Adsorption is favored by high concentrations of phosphate ions, which suggests that adsorption sites are adsorbed phosphate. However, no dissolution inhibition was observed, which has been explained by the absence of affinity of this protein for calcium ions and by its limited coverage of the HAP surface [117]. The low molecular weight, globular shape, rigidity, and stability of lysozyme lead to a small number of well-localized NH anchoring points.

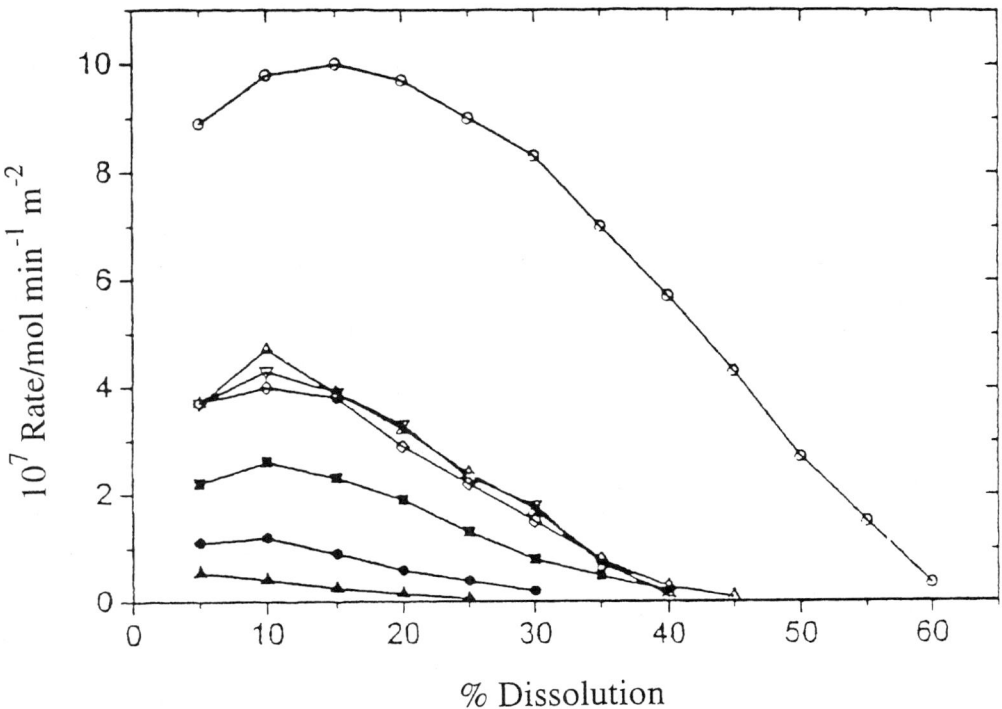

FIG. 14 Dissolution rate of HAP seeds against % extent of reaction in the presence of β-lactoglobulin: 0 ppm (○): 5 ppm (△); 10 ppm (▽); 20 ppm; (◇); 200 ppm; (■); 500 ppm (●); 800 ppm (▲). (From Ref. 115.)

F. Inhibition by Synthetic Macromolecules

1. Polypeptides

In order to have a better understanding of the interactions between proteins and HAP, some investigators have studied the adsorption of positively and negatively charged homopolypeptides onto HAP and enamel [100,118–121]. Only few studies concern the effect of such adsorption on the dissolution of HAP [121].

Polyglutamic (PG) and polyaspartic (PA) acids demonstrate considerable differences of adsorption behavior (reversibility, amounts adsorbed, plateau position). No adsorption of PG on bovine enamel and HAP [100,118,119] was found in contrast with more recent results [113,120]. These differences are possibly explained by the strong tendency for PG to form aggregates [113] and by the difference in conformation of both polypeptides, i.e., a random coil conformation for PA and a helical structure with possible helix–coil conversion for PG. Solid ^{13}C-NMR experiments realized with basic poly-L-lysine and poly-L-glutamic acid at pH 7.5 and 5.5, respectively, in order to have the opposite net charge on HAP, suggest that both polypeptides adsorb in a flat relatively extended conformation through strong electrostatic interactions of side-chain functional groups with the surface [120].

The effect of the adsorption of polylysine (PL) hydrobromide [121], PG, and PA on the HAP dissolution cannot be analyzed without taking into account the ion exchanger behavior of the polypeptides. While PL is bound to phosphate ions, PA and PG strongly complex calcium ions both on the surface and in solution with the exchange of one proton for one calcium ion. This complexation induces not only a nonstoichiometric dissolution but also an acceleration of the dissolution due to the release of protons by the acidic polypeptides. Dissolution inhibition by PA or PG is, however, well demonstrated when no polypeptides are present in solution [113].

2. Other Synthetic Macromolecules

The adsorption properties and interactions of synthetic polymers with HAP or enamel surfaces are particularly interesting to study, since polymers have potential applications in the industrial processes of mineral treatment as well as in dentistry. Polymers are interesting adsorbates, since their chemical structures may be infinitely varied, offering the unique possibility of introducing various specific groups or segments. The great number of segments able to be adsorbed often leads to strong and pseudoirreversible adsorption. They generally adsorb in more or less flat and swollen coil conformations with trains in contact with the surface and swollen loops and tails [122]. However, the adsorption of water-soluble polymers and particularly of polyelectrolytes is strongly affected by important variables such as molecular weight, degree of ionization, substrate charge density, and ionic strength of the solution [123]. These three last parameters are generally not constant during HAP dissolution or remineralization, and this may complicate the interpretation of the results.

Among the relatively few synthetic polymers studied, it is not astonishing to note that polymers containing carboxylic or phosphonic groups and neutral polymers with strong hydrogen bonding power and polarizable groups demonstrate the best adsorption properties towards HAP. Nonionic polymers such as polyethylene oxide, polyvinyl alcohol, or polyoxazoline show no noticeable adsorption onto HAP [56]. Neutral polyacrylamide (Paam) [55,124], dextran, and carboxymethyl cellulose [125] show strong interactions probably generated by dipolar interactions and hydrogen bonding between the polymers and the hydroxyl groups of HAP. Paam demonstrates a moderate power of dissolution inhibition that follows the characteristics of the adsorption layer. The inhibition increases with the molecular weight and the degree of hydrolysis of the polymer, both factors promoting the increase of adsorbed amount and the thickness of the adsorbed layer. Cationic polymers such as quaternized poly(4-vinylpyridine) [126] interact obviously through interactions with surface phosphate or hydroxyl groups.

However, the most studied polymers and the best structures are the anionic polymers known to show strong interactions with calcium ions. Polyacrylic acid [56], hydrolyzed Paam [55,125], and polyphosphonates [127,128] are the most interesting structures. In particular, a copolymer of rather low molecular weight (Mw = 18,800) and containing acrylic and phosphonic acid groups demonstrates a high efficiency as shown in Fig. 15 [56]. As for fluor adsorption, it is noted that the initial rates are strongly accelerated, a behavior not observed with hydrolyzed Paam.

The efficiency of phosphorylated [129–131] and sulfate-containing [132] polymeric substrates as active centers for nucleation of HAP was also demonstrated.

G. Conclusion

All the results demonstrate a straight relationship between the adsorption characteristics of the molecules onto HAP and the inhibition or nucleation effect. A good knowledge of the nature of the adsorbed layer and of the kinetics of its formation is then essential to interpret its role on the HAP dissolution process.

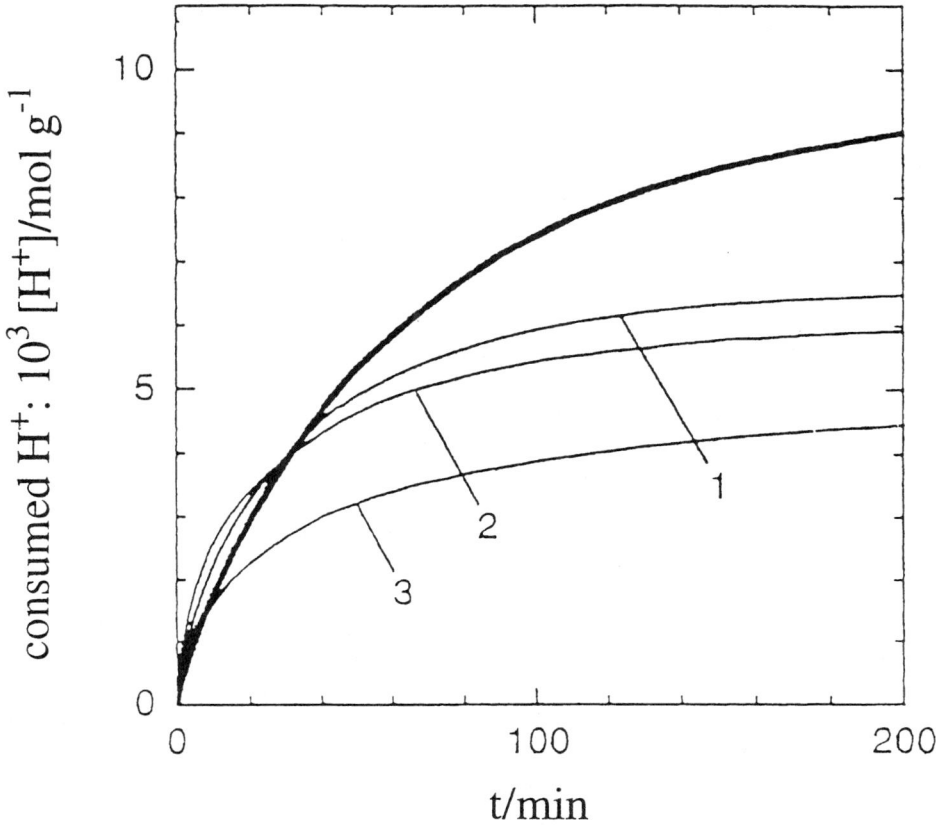

FIG. 15 Proton uptake kinetics of 10 mg of HAP (165–200 μm) for dissolution at pH 5 and at a stirring speed of 1860 rpm, and in the presence of adsorbed poly(phosphoethyl methacrylate-co-acrylic acid). Thick line without polymer; thin lines, with polymer at initial solution concentrations: (1) 0.0025% (w/w); (2) 0.005% (w/w); (3) 0.01% (w/w). (From Ref. 56.)

The role of the layer formed by surface precipitation seems easily understood, since the reduction of the available surface and the change of the nature of the interface leading to a change of the solubility product of HAP are the main parameters. However, the kinetic aspects involving the phase formation and subsequent transformation may be determining. Progress has to be made in quantifying the kinetic effects.

The surface reaction process involving low molecular weight compounds and macromolecules does not involve a change of the HAP solubility product. The adsorption of macromolecules is controlled not only by thermodynamic and kinetic factors but also by electrostatic interactions. Interpretation of the inhibition effect needs a clear knowledge not only of the conformation of the adsorbed molecule and of the type and number of occupied sites but also of the adsorption kinetics. Since the HAP surface is continuously eroded, ionic removal and site occupancy by inhibitors are in constant competition. The rates of both phenomena have to be established and compared. A full knowledge of the conformation of the adsorbate becomes necessary, and the constant evolution of the electrostatic interactions at the interface during the dissolution process may induce dramatic changes of conformation. At each time, the charged surface may locally act as a both attractive and repulsive wall for one particular chemical group of the macromolecule. Moreover, the presence in solution of divalent cations such calcium may induce inter- and/or intramolecular interactions able strongly to affect the conformation of the adsorbed layer. This leads to a complicated situation that can limit the number of bonds to the surface and change the equilibrium between surface-held trains, the free-hanging loops and the tails. The short-distance interactions may be quite important and induce protein denaturation and polymer precipitation. The result is that various parameters have to be considered, i.e., the change of electrostatic interactions at the interface, the rate of ion removal and site occupancy, the conformational change of the adsorbates, and the restriction of diffusion at the adsorbed interfacial layer.

An interesting tentative model relating adsorption constants determined from kinetic experiments to adsorption constants determined from equilibrium measurements has been proposed [133,134]. The model is based on the assumption that the rate controlling process is a surface process involving a critical phenomenon such as a surface nucleation. An inhibitor forming strong bonds towards ions in the HAP surface would cause the frequency of hole nucleation to be drastically reduced in an area around each inhibitor unit. Although it is clear that a simple description in terms of Langmuir parameters is quite insufficient to describe the interfacial changes around the dissolution sites, this approach has the advantage of providing some quantitative analysis relating adsorption and dissolution. The simple HAP dissolution model proposed by Mafé et al. [52] proposing to simulate the occupied sites by hard spheres seems also well adapted to describe the process involving superficial phase formation and LMW compound adsorption. However, the results of the adsorption of synthetic macromolecules demonstrate that inhibition increases with molecular weight and then with thickness of the adsorbed layer. This supposes that at least part of the inhibition is diffusion controlled. Interpretation in terms of a purely steric effect seems quite insufficient, and electrostatic interactions increasing the interfacial resistance to ion diffusion may play a dominant role as proposed in the semipermeable layer model.

Development of quantitative models describing the inhibition effect requires further information about the dissolution mechanism itself and about the detailed description of the HAP interface. Improved surface-characterization techniques are needed. The use of techniques such as high-resolution nuclear magnetic resonance (NMR) spectroscopy [84,135], electron paramagnetic resonance (EPR) spectroscopy [136], surface force apparatus [137,138], small angle neutron scattering [139,140], and ellipsometry [141] provide valuable possibilities for the characterization of HAP surfaces in relation to inhibition and nucleation phenomenon.

ACKNOWLEDGMENT

The authors would like thank Dr. J. C. Voegel and Dr. F. Poumier-Gorce for helping in preparing this chapter.

REFERENCES

1. J. C. Elliott, in *Structure and Chemistry of the Apatites and Other Calcium Orthophosphates. Studies in Inorganic Chemistry 18*, Elsevier, Amsterdam, 1994, p. 160.
2. S. Chandler and D. W. Fuerstenau. J. Colloid Interface Sci. *70*:506 (1979).
3. J. Arends and W. L. Jongebloed. Recueil. J. Royal Netherlands Chem. Soc. *100*:3 (1981).

4. H. G. Linge. J. Colloid Interface Sci. *14*:239 (1981).
5. Ph. Gramain, J. C. Voegel, M. Gumpper, and J. M. Thomann. J. Colloid Interface Sci. *118*:148 (1987).
6. J. C. Voegel, Ph. Gramain, M. Gumpper, and J. M. Thomann. J. Crystal Growth *83*:89 (1987).
7. M. S. Wu, W. I. Higuchi, J. L. Fox, and M. Friedman. J. Dent. Res. *55*:496 (1976).
8. S. Chandler and D. W. Fuerstenau, in *Adsorption on and Surface Chemistry of Hydroxyapatite* (D. N. Misra, ed.), Plenum Press, New York, 1984, pp. 29–50.
9. P. Somasundaran, J. Ofori Amankonah, and K. P. Ananthapadmabhan. Colloids Surfaces *15*:309 (1985).
10. Ph. Schaad, F. Poumier, J. C. Voegel, and Ph. Gramain. Colloids and Surfaces *121*:217 (1997).
11. M. B. Tomson and G. H. Nancollas. Science *200*:1059 (1978).
12. M. Johnsson, C. F. Richardson, J. D. Salis, and G. H. Nancollas. Calcif. Tissue *49*:134 (1991).
13. W. I. Higuchi, E. Y. Cesar, P. W. Cho, and J. L. Fox. J. Pharmaceutical Sci. *73*:146–153 (1984).
14. W. White and G. H. Nancollas. J. Dent. Res. *56*:524 (1977).
15. M. V. Patel, J. L. Fox, and W. I. Higuchi. J. Dent. Res. *66*:1418 (1987).
16. W. C. Chen and G. H. Nancollas. J. Dent. Res. *65*:663 (1986).
17. A. N. Smith, A. M. Posner, and J. P. Quirk. J. Colloid Interface Sci. *62*:475 (1977).
18. E. N. Griffith, A. Katdare, J. L. Fox, and W. I. Higuchi. J. Colloid Interface Sci. *67*:331 (1978).
19. J. Christoffersen. J. Crystal Growth *49*:29 (1980).
20. F. C. M. Driessens, H. M. Theuns, J. M. P. M. Borggreven, and J. W. E. Van Dijk. Caries Res. *20*:103 (1986).
21. J. A. Budz, M. Lore, and G. H. Nancollas. Adv. Dent. Res. *1*:314 (1987).
22. J. M. Thomann, J. C. Voegel, and Ph. Gramain. Colloids Surfaces *54*:145 (1991).
23. P. Somasundaran and Y. H. C. Wang, in *Adsorption on and Surface Chemistry of Hydroxyapatite* (D. N. Misra, ed.), Plenum Press, New York, 1984, pp. 129–149.
24. A. Barroug, J. Lemaitre, and P. G. Rouxhet. Colloids Surfaces *37*:339 (1989).
25. F. Poumier, Ph. Schaad, Y. Haïkel, J. C. Voegel, and Ph. Gramain. Colloids Surfaces, B: Biointerfaces *7*:1 (1996).
26. J. M. Thomann, J. C. Voegel, and Ph. Gramain. Calcif. Tissue Int. *46*:121 (1990).
27. Ph. Gramain, J. M. Thomann, M. Gumpper, and J. C. Voegel. J. Colloid Interface Sci. *2*:128 (1989).
28. J. Christoffersen, M. R. Christoffersen, and N. Kjaergaard. J. Crystal Growth *43*:501 (1978).
29. J. A. Budz and G. H. Nancollas. J. Crystal Growth *91*:490 (1988).
30. J. M. Thomann, J. C. Voegel, M. Gumpper, and Ph. Gramain. J. Colloid Interface Sci. *2*:132 (1989).
31. J. A. Gray. J. Dental Res. *41*:633 (1962).
32. H. G. Linge and G. H. Nancollas. Calc. Tiss. Res. *12*:193 (1973).
33. J. L. Fox, W. I. Higuchi, M. B. Fawzi, and M. S. Wu. J. Colloid Interface Sci. *67*:312 (1978).
34. M. B. Fawzi, J. L. Fox, M. G. Dedhiya, W. I. Higuchi, and J. J. Hefferren. J. Colloid Interface Sci. *67*:304 (1978).
35. J. Christoffersen and M. R. Christoffersen. J. Crystal. Growth *47*:671 (1979).
36. W. I. Higuchi, P. W. Cho, J. L. Fox, and K. Yamamoto. J. Colloid Interface Sci. *110*:453 (1986).
37. M. V. Patel, J. L. Fox, and W. I. Higuchi. J. Dent. Res. *66*:1425 (1987).
38. I. Gallily and S. K. Friedlander. J. Chem. Phys. *42*:1503 (1965).
39. J. Christoffersen and M. R. Christoffersen. J. Crystal Growth *35*:79 (1976).
40. J. Christoffersen and M. R. Christoffersen. J. Crystal Growth *57*:21 (1982).
41. W. B. Hillig. Acta Metallurg. *14*:1868 (1966).
42. G. H. Gilmer and P. Bennema. J. Appl. Physics *43*:1347 (1972).
43. J. Christoffersen, M. R. Christoffersen, and J. Arends. Croatica Chemica Acta *56*:769 (1983).
44. J. Christoffersen and M. R. Christoffersen. J. Crystal Growth *35*:79 (1976).

45. J. Christoffersen, M. R. Christoffersen, and T. Johansen. J. Crystal Growth *163*:304 (1996).
46. M. R. Christoffersen and J. Christoffersen. Calcif Tissue Int. *37*:673 (1985).
47. M. R. Christoffersen, H. C. Thyregod, and J. Christoffersen. Calcif Tissue Int. *41*:27 (1987).
48. J. Christoffersen and M. R. Christoffersen. J. Crystal Growth *53*:42 (1981).
49. J. Christoffersen and M. R. Christoffersen. J. Crystal Growth *62*:254 (1983).
50. S. Shimabayashi and M. Nakagaki. Chem. Pharm. Bull. *231*:2976 (1983).
51. M. S. Tung. J. Dent. Res. *55*:D77 (1976).
52. S. Mafé, J. A. Manzanares, H. Reiss, J. M. Thomann, and Ph. Gramain. J. Phys. Chem. *2*:96 (1992).
53. J. M. Thomann, J. C. Voegel, and Ph. Gramain. J. Colloid Interface Sci. *157*:369 (1993).
54. J. Arends and J. Christoffersen. J. Dent Res. *65*:2 (1986).
55. Ph. Schaad, J. M. Thomann, J. C. Voegel, and Ph. Gramain. J. Colloid Interface Sci. *164*:291 (1994).
56. Ph. Schaad, J. M. Thomann, J. C. Voegel, and Ph. Gramain. Colloids Surfaces A Phys. Eng. Aspects *83*:285 (1994).
57. P. Gasser, Y. Haïkel, J. C. Voegel, and Ph. Gramain. J. Mat. Sci.: Mat. Medecine *6*:105 (1995).
58. R. A. Young and W. E. Brown, in *Biological Mineralization and Demineralization* (G. H. Nancollas, ed.), Springer-Verlag, Berlin, 1982, pp. 37–77.
59. D. G. A. Nelson. J. Dent. Res. *60*:1621 (1981).
60. D. G. A. Nelson, J. C. Barry, C. P. Shields, R. Glena, and J. D. B. Featherstone. J. Colloid Interface Sci. *130*:467 (1989).
61. J. Christoffersen, M. R. Christoffersen, and T. Johansen. J. Crystal Growth *163*:295 (1996).
62. D. J. Crommelin, W. I. Higuchi, J. L. Fox, P. J. Spooner, and A. V. Katdare. Caries Res. *17*:289 (1983).
63. K. O. A. Chin and G. H. Nancollas. Langmuir *7*:2175 (1991).
64. Z. Amjad, P. G. Kouttsoukos, and G. H. Nancollas. J. Colloid Interface Sci. *81*:394 (1981).
65. E. I. F. Pearce, N. Guha-Chowdhury, Y. Iwami, and T. W. Cutress. Caries Res. *29*:130 (1995).
66. H. J. Busscher, H. P. De Jong, and J. Arends. Materials Chem. Phys. *17*:553 (1987).
67. S. Wu, in *Polymer Interface and Adhesion*, Marcel Dekker, New York, 1982.
68. P. W. Cho, J. L. Fox, W. I. Higuchi, and P. Pithayanukul, in *Adsorption on and Surface Chemistry of Hydroxyapatite* (D. N. Misra, ed.), Plenum Press, New York, 1984, pp. 51–70.
69. S. Shimabayashi, H. Tanaka, and M. Nakagaki. Chem. Pharm. Bull. *37*:2514 (1989).
70. S. Shimabayashi, H. Tanaka, and M. Nakagaki. Chem. Pharm. Bull. *375*:4687 (1987).
71. Y. Watanabe, T. Okuno, K. Ishigaki, and T. Tagaki. Anal. Biochem. *202*:268 (1992).
72. S. Shimabayashi, Y. Yoshida, K. Arima, and T. Uno. Phosphorous Research Bull. *4*:89 (1994).
73. M. Otsuka, J. Wong, W. L. Higuchi, D. C. Cheng, K. Yamamoto, and J. L. Fox. Colloids Surfaces *26*:79 (1987).
74. J. Arends and J. Christoffersen. J. Dent. Res. *69*:601 (1990).
75. J. M. ten Cate. J. Dent. Res. *69*:614 (1990).
76. J. D. B. Featherstone, R. Glena, M. Shariati, and C. P. Shields. J. Dent. Res. *69*:620 (1990).
77. H. C. Margolis and E. C. Moreno. J. Dent. Res. *69*:606 (1990).
78. D. J. White and G. H. Nancollas. J. Dent. Res. *69*:587 (1990).
79. P. Gasser, J. C. Voegel, and Ph. Gramain. Colloids Surfaces A *74*:275 (1993).
80. M. R. Christoffersen, J. Christoffersen, and J. Arends. J. Crystal Growth *67*:107 (1984).
81. P. Gasser, Y. Haïkel, J. C. Voegel, and Ph. Gramain. Colloids Surfaces A *88*:157 (1994).
82. J. Lin, S. Raghavan, and D. W. Fuerstenau. Colloids Surfaces *3*:357 (1981).
83. D. J. White, W. C. Chen, and G. H. Nancollas. Caries Res. *22*:11 (1988).
84. J. P. Yesinowski, R. A. Wolfgang, and M. J. Mobley, in *Adsorption on and Surface Chemistry of Hydroxyapatite* (D. N. Misra, ed.), Plenum Press, New York, 1984, pp. 151–175.
85. S. Chandler, C. C. Chiao, and D. W. Fuerstenau. J. Dent. Res. *61*:403 (1982).
86. A. Ludwig, S. C. Dave, W. I. Higuchi, J. L. Fox, and A. Katdare. Inter. J. Pharmaceutics *316*:1 (1983).

87. P. Gasser, J. C. Voegel, and Ph. Gramain. J. Colloid Interface Sci. *168*:465 (1994).
88. V. Caslavska and H. Duschner. Caries Res. *25*:27 (1991).
89. T. Suzuki, K. Ishigaki, and M. Miyake. J. Chem. Soc. Faraday Trans. I. *80*:3157 (1984).
90. E. Dalas and P. Koutsoukos. J. Chem. Soc. Faraday Trans. I. *85*:3159 (1989).
91. Y. Tanizawa, K. Sawamura, and T. Suzuki. J. Chem. Soc. Faraday Trans. *86*:4025 (1990).
92. Y. Tanizawa, K. Sawamura, and T. Suzuki. J. Chem. Soc. Faraday Trans. *86*:1071 (1990).
93. Ph. Schaad, Ph. Gramain, F. Gorce, and J. C. Voegel. J. Chem. Soc. *90*:3405 (1994).
94. T. V. Vasudevan, P. Somasundaran, C. L. Howie-Meyers, D. L. Elliott, and K. P. Ananthapadmanabhan. Langmuir *10*:320 (1994).
95. A. Arbel, I. Katz, and S. Sarig. J. Crystal Growth. *110*:733 (1991).
96. D. N. Misra. J. Dent. Res. *75*:1418 (1996).
97. H. Tanaka, K. Miyajima, M. Nakagaki, and S. Shimabayashi. Colloid Polymer Sci. *269*:161 (1991).
98. T. Aoba and E. C. Moreno. J. Colloid Interface Sci. *106*:110 (1985).
99. M. B. Kautsky and J. D. B. Featherstone. Caries Res. *27*:373 (1993).
100. E. C. Moreno, M. Kresak, and D. I. Hay. Calcf. Tissue Int. *36*:48 (1984).
101. R. F. Troxler, G. D. Offner, T. Xu, J. C. Vanderspek, and F. G. Oppenheim. J. Dent. Res. *69*:2 (1990).
102. M. Johnsson, J. W. Perich, E. C. Reynolds, and G. H. Nancollas. J. Colloid Interface Sci. *160*:179 (1993).
103. J. P. Shomers, L. A. Tabak, M. J. Levine, I. D. Mandel, and D. I. Hay. J. Dent. Res. *61*:397 (1982).
104. M. Johnsson, C. F. Richardson, E. J. Bergey, M. J. Levine, and G. H. Nancollas. Arch. Oral Biol. *36*:631 (1991).
105. T. Aoba, E. C. Moreno, and D. I. Hay. Calcif. Tissue Int. *36*:651 (1984).
106. K. O. A. Chin, M. Johnsson, E. J. Bergey, M. J. Levine, and G. H. Nancollas. Colloids Surfaces *78*:229 (1993).
107. H. Tanaka, T. Arai, K. Miyajima, S. Shimabayashi, and M. Nakagaki. Colloids Surfaces *37*:357 (1989).
108. M. R. Christoffersen, J. Christoffersen, P. Ibsen, and H. Ipsen. Colloids Surfaces *18*:1 (1986).
109. J. C. Voegel, S. Behr, M. J. Mura, J. D. Aptel, A. Schmitt, and E. F. Bres. Colloids Surfaces *40*:307 (1989).
110. V. Hlady and H. Füredi-Milhofer. J. Colloid Interface Sci. *69*:460 (1979).
111. M. J. Mura-Galelli, J. C. Voegel, S. Behr, E. F. Bres, and P. Schaaf. Proc. Natl. Acad. Sci. *88*:5557 (1991).
112. J. Arends, J. Schuthof, and J. Christoffersen. Caries Res. *20*:337 (1986).
113. F. Poumier, Ph. Schaad, J. C. Voegel, and Ph. Gramain. Unpublished results.
114. E. C. Reynolds, P. F. Riley, and E. Storey. Calcif. Tissue Int. *34*:52 (1982).
115. K. D. Daskalakis, T. A. Fuierer, J. Tan, and G. H. Nancollas. Colloids Surfaces A *96*:135 (1995).
116. A. Barroug, J. Lemaitre, and P. G. Rouxhet. Colloids Surfaces *37*:339 (1989).
117. F. Poumier, Ph. Schaad, Y. Haïkel, J. C. Voegel and Ph. Gramain. Colloids Surfaces *7*:1 (1996).
118. A. C. Juriaanse, J. Arends, and J. J. Ten Bosch. J. Colloid Interface Sci. *76*:212 (1980).
119. A. C. Juriaanse, J. Arends, and J. J. Ten Bosch. J. Colloid Interface Sci. *76*:220 (1980).
120. V. L. Fernandez, J. A. Reimer, and M. M. Denn. J. Am. Chem. Soc. *114*:9634 (1992).
121. H. Tanaka, Y. Nuno, S. Irie, and S. Shimomura. Talanta *39*:893 (1992).
122. G. L. Fleer, M. A. Cohen Stuart, J. M. H. M. Scheutjens, T. Cosgrove, and B. Vincent, eds., *Polymers at Interfaces*, Chapman and Hall, London, 1993.
123. B. J. Fontana, in *The Chemistry of Biosurfaces* (M. L. Hair, ed.), Marcel Dekker, New York, 1971.
124. Y. Pradip, A. Altia, and J. W. Fuerstenau. Colloid Polym. Sci. *258*:1343 (1980).
125. E. I. F. Pearce. Calcif. Tissue Int. *33*:395 (1981).
126. T. Bartels and J. Arends. J. Polym. Sci. *19*:127 (1981).

127. T. Bartels and J. Arends. Caries Res. *13*:218 (1979).
128. T. Bartels, J. Schuthof, and J. Arends. J. Dentistry *7*:221 (1979).
129. E. Dalas, J. Kallitsis, and P. G. Koutsoukos. Colloids Surfaces *53*:197 (1991).
130. E. Dalas, J. K. Kallitsis, and P. G. Koutsoukos. Am. Chem. Soc. *7*:1822 (1991).
131. S. Shimabayashi, N. Hashimoto, and T. Uno. Phosphorus Res. Bull. *3*:7 (1993).
132. S. Shimabayashi, H. Kawamura, and T. Uno. Phosphorus Res. Bull. *3*:1 (1993).
133. J. Christoffersen and M. R. Christoffersen. J. Crystal Growth *53*:42 (1981).
134. J. Christoffersen, M. R. Christoffersen, S. B. Christensen, and G. H. Nancollas. J. Crystal Growth *62*:254 (1983).
135. T. Cosgrove and P. C. Griffiths. Adv. Colloid Interface Sci. *42*:175 (1992).
136. H. Hommel. Adv. Colloid Interface Sci. *54*:209 (1995).
137. J. Klein, in *Liquids at Interface* (J. Charvolin, J. F. Joanny, and J. Zinn-Jostin, eds.), North-Holland, Amsterdam, 1980, p. 239.
138. P. F. Luckham. Adv. Colloid Interface Sci. *34*:191 (1991).
139. L. Auvray and J. P. Cotton. Macromolecules *20*:202 (1987).
140. M. A. Cohen-Stuart, T. Cosgrove, and B. Vincent. Adv. Colloid Interface Sci. *24*:143 (1986).
141. M. Kawaguchi and A. Takahashi. Adv. Colloid Interface Sci. *37*:219 (1992).

14

Interfacial Chemistry of Dissolving Metal Oxide Particles: Dissolution by Organic Acids

J. ADRIÁN SALFITY, ALBERTO E. REGAZZONI, and MIGUEL A. BLESA Department of Chemistry, Atomic Energy Commission, Buenos Aires, Argentina

I.	Introduction	513
II.	Brief Review of the Relevant Aqueous Chemistry of Metal Ions and Organic Acids	514
	A. The stability of carboxylate complexes in solution	514
	B. Kinetics of metal ion complexation by carboxylates	516
	C. The structure of dissolved complexes	518
	D. Electron-transfer reactions in solution	519
III.	The Interaction of Carboxylates with Metal Oxide Surfaces Immersed in Water	520
IV.	Nonequilibrated Interfaces in Mineral and Organic Acids	529
	A. Kinetics of surface complexation	529
	B. Kinetics of metal oxide dissolution	530
	C. Dissolution in the presence of carboxylic acids	533
	References	538

I. INTRODUCTION

The surface complexation approach has been much used to describe chemisorption of ions and molecules. Mainly, equilibrated systems have been analyzed, and the analogy with homogeneous coordination chemistry has been stressed. By now, an impressive wealth of knowledge is available that supports this equilibrium description, even accepting the approximate nature of the concept of surface complex (for earlier reviews of the subject, see Ref. 1–4). Within the subjects described by the surface complexation approach, the adsorption of carboxylic and phenolic acids onto metal oxides takes a prominent place. Because of the wide usage of synthetic organic acids, and the widespread occurrence of natural acids, the environmental relevance of the question is large. The mobilization of heavy metal ions in the environment is heavily influenced by surface complexation reactions of this type. Also, in several fields of technology, the interaction is of importance: for instance, in the procedures to decontaminate nuclear reactors, or to clean steam generators. Finally, the biological relevance of the subject is also high.

TABLE 1 Selected Stability Constants for $ML^{(n-m)-}$ Metal Ion–Carboxylate Complexes; the Data Are for 298 K and Ionic Strength $I = 1$ mol dm^{-3}, Except for Some for Which (a) $I = 0.1$ mol dm^{-3} and (b) $I = 0.4$ mol dm^{-3}

Ligand		log (K_1/dm^3 mol^{-1})	
Name	Number of binding groups	Ni(II) complex	Fe(III) complex
Formate	1	0.46	3.1
Oxalate	2	3.7	7.58
Citrate	3	5.35a	11.2a
Benzoate	1	0.55b	—
Phthalate	2	1.57	—
Glycine	2	5.65	10.0
NTA	4	11.50a	15.9a
EDTA	6	18.4a	23.9
Picolinate	2	6.6	—
2,6-Dipicolinate	3	6.6	—

Source: Refs. 5–8.

In this review, we shall address essentially two issues. First, we shall discuss the similarities between homogeneous and heterogeneous (superficial) complexation reactions by organic acids. For this purpose, we shall briefly review the homogeneous complexation kinetics and equilibrium, and later we shall present the state of knowledge of heterogeneous complexation by carboxylic and other organic acids, stressing the dynamics rather than the equilibrium.

Second, we shall explore the cases in which interfacial equilibrium is not achieved. In particular, we shall discuss the description of a dissolving interface in a stationary kinetic regime. The lack of hard experimental evidence about the state of charge of these dissolving interfaces (due in part to the lack of adequate experimental methods, and in part to the need to replace the equilibrium description by a kinetic approach) seriously limits the state of knowledge. Thus this section is largely tentative and includes information about oxides immersed in mineral acids, in the hope that it can provide a basis for analysis in future of the more complex systems containing carboxylic acids. The interfacial chemistry in these latter systems is particularly dynamic, and the assumption of equilibrium is in many cases unjustified.

II. BRIEF REVIEW OF THE RELEVANT AQUEOUS CHEMISTRY OF METAL IONS AND ORGANIC ACIDS

A. The Stability of Carboxylate Complexes in Solution

Table 1 lists the stability constants, K_1, for some selected typical carboxylate complexes. The given values are for 1:1 (metal:ligand) complexes, irrespective of the number of coordinative bonds involved in the complex. The obvious conclusion is that the affinity

TABLE 2 Ratios of Overall Stability Constants for Pairs of Isocharged Complexes of Two Ligands of Different Chelating Abilities, and 1 mol dm^{-3} Concentration, Bound to the Same Metal Ion

Pair of complexes		
Fe–L$_n$	Fe–L	log [β_n(Fe–L$_n$)/β_1(Fe–L)]
Fe(CH$_3$COO)$_2^+$	Fe(ox)$^+$	-1.36
Fe(HCOO)$_3$	Fe(cit)	-2.9
Fe(ox)$_2^-$	Fe(EDTA)$^-$	-11.3

Source: Refs. 5–8.

TABLE 3 Overall Stability Constants for Some ML$_3$ Complexes Formed by Phenolate and Hydroxocarboxylate Ligands at 298 K and 0.1 mol dm^{-3} Ionic Strength

Ligand	log (β_3/dm^9 mol^{-3})		
	Ni(II)	Fe(III)	Ti(IV)
3,4-Dihydroxybenzoate	16.9	42.4	58.6
Catecholate	—	44.9	61.6

Source: Refs. 6–8.

is largely enhanced by chelation; in fact, the underlying main factor is the anionic charge. Polycarboxylates bind the cations as polyvalent anions, and the electrostatic contribution to the bond energy dominates other factors. However, this electrostatic factor is not alone responsible for the stability of chelates. Table 2 compares 1 : n metal : carboxylate overall stability constants, for various n values and constant charge of the resulting complex. The chelate effect here dominates the stability trends. The number of ligands involved in forming chelated complexes is lower, and the ratio of complexed to noncomplexed metal ions in diluted media is therefore even larger in the chelates.

For our purposes, ligands with two different coordinating groups are especially important: hydroxo- and aminocarboxylates. The first group includes salicylate, and humic and fulvic acids, among others: the acidity of the hydroxide group of aromatic (phenolic) alcohols is adequate for metal complexation. The stability of these complexes is especially high for small cations with a high charge, such as Ti^{4+}. Table 3 gives some stability constants for phenolic and aromatic hydroxocarboxylates. In the case of amino acids, the group includes EDTA and NTA (see Table 1) and a wide variety of other important ligands.

A very important conclusion for our purposes is that the Lewis basicity of the R–COO$^-$ group is rather low, as compared to the basicity of water, and not very stable M–OOC–R complexes result in aqueous media. By contrast, the Brønstead basicity is larger, and low pH values further arrest metal complexation through ligand protonation. The picture varies dramatically when chelates may form. Figure 1 compares the speciation of solutions containing Fe(III) and 0.01 mol dm^{-3} ligand, as a function of pH, for L = formate and oxalate.

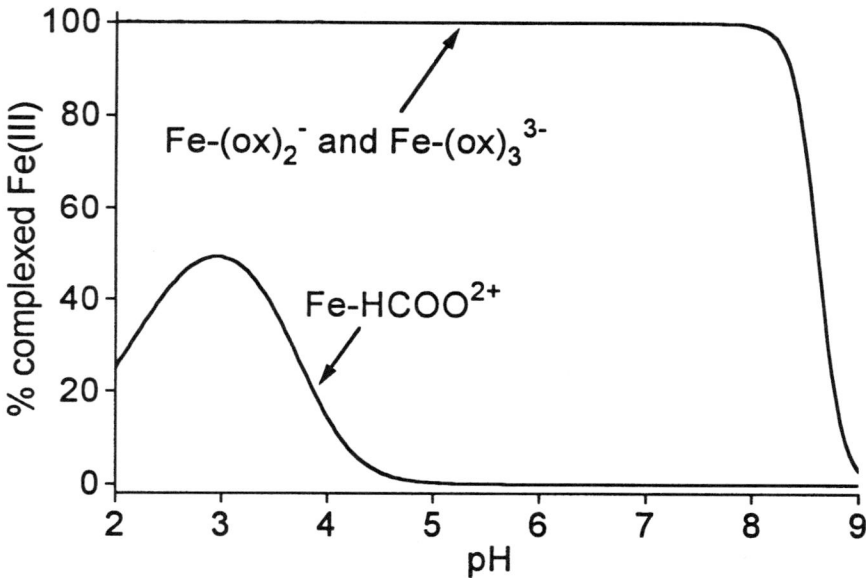

FIG. 1 Speciation, as a function of pH, of solutions containing Fe(III) and excess amounts of formate and oxalate (0.01 mol dm^{-3} ligand concentration), at 298 K and 1 mol dm^{-3} ionic strength.

B. Kinetics of Metal Ion Complexation by Carboxylates

In the case of monodentate complexes, the Eigen–Wilkens mechanism describes the substitution of ligands for water in the first coordination sphere:

$$M-OH_2^{m+} + L^{n-} \rightleftharpoons M-OH_2^{m+}\cdots L^{n-} \qquad K_{os} \qquad (1)$$

$$M-OH_2^{m+}\cdots L^{n-} \rightarrow M-L^{(m-n)+} + H_2O \qquad k_1 \qquad (2)$$

The values of the ion-pair association constants are essentially determined by the charges of the reactants; several equations, such as the Davis equation [Eq. (3)] have been used to estimate activity coefficients, f, as a function of ionic strength, I (expressed in mol dm^{-3}), and thus calculate K_{os}.

$$\log f = -Az^2\left(\frac{\sqrt{I}}{1+\sqrt{I}} - 0.2I\right) \qquad (3)$$

In the case that $L^{n-} = H_2O$, Eq. (2) represents the rate of water exchange, which is usually the upper limit for the rates of substitution. The rates of water exchange span many orders of magnitude, as shown in Fig. 2. By far the most important factor is the electrostatic interaction between the charged metal cation and the water dipole, that limits the rate of M–OH$_2$ bond breaking. In general, for cations of low charge, the rates of water exchange, and of substitution as a whole, are high. This group includes the hydrolyzed metal ions, e.g., $M^{III}(OH)_2(H_2O)_4^+$, that are much more labile than the fully hydrated cations.

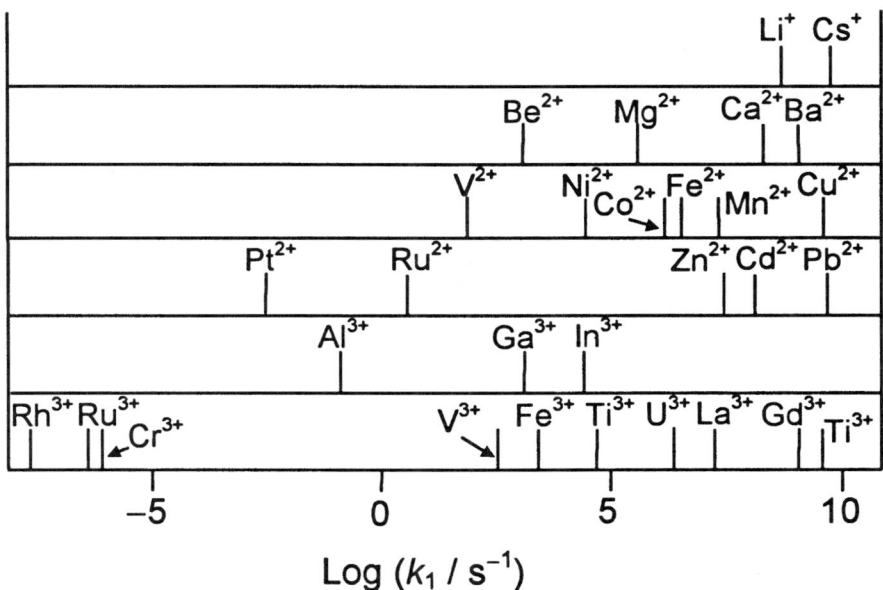

FIG. 2 First-order rate constants, k_1 [cf. Eq. (2)], for water exchange around fully hydrated metal ions at 298 K. (Data from Ref. 9.)

In the general case of substitution of one ligand by another, the rate-limiting step can be influenced by both the outgoing and the incoming ligand. All situations in between the two extreme cases of dissociative (D) and associative (A) mechanisms can be found in practice. For divalent metal cations, the mechanism is usually D or I_d (the latter implies stoichiometric participation of L^{n-} in the activated state), with little influence of the incoming ligand on the enthalpy of activation ($\Delta H^{\ddagger}$). For trivalent metal cations, the activation is in many cases associative, I_a, and $\Delta H^{\ddagger}$ depends on the nature of the incoming anion.

In the case of polydentate ligands, a zippering-like mechanism is involved; the rate-determining step may be associated with any stage of the various water substitutions. In general, the slowest step is the first one, meaning that after the first water molecule is replaced, the zippering-in of the ligand is fast. Because of the principle of microscopic reversibility, the rate-determining step in the rupture of a metal–polycarboxylate complex is also associated with the last M–OOC bond breakage, all previous unzippering processes behaving as unfavorable pre-equilibria. Because of the thermodynamic stability of the complexes, this reverse reaction can proceed in practice only if another ligand, capable of substituting for the first, is present in solution. In this case, the first unzippering may be followed (or even assisted) by the entrance of the second ligand, the following concerted unzippering–zippering reactions being faster. Table 4 collects some kinetic data for substitution reactions in EDTA and related complexes. The first four entries illustrate the range of reaction rates spanned in the substitution of one fully deprotonated ligand by another. The rate enhancement brought about in acid media by the protonation of complexed and free ligands is illustrated by the two calcium entries. On top of an important electrostatic effect, the unzippering of the leaving ligand is assisted by protons. Finally,

TABLE 4 Second-Order Rates of Substitution, k_2, of One Ligand for Another on Metal Complexes. $HCAL^2$ = dianion of 1-(hydroxy, 4-methyl, 2-phenylazo), 2-Naphthol, 4-Sulfonate

Metal complex	Incoming ligand	$k_2/\text{mol}^{-1}\,\text{dm}^3\,\text{s}^{-1}$
$Ca(EDTA)^{2-}$	$EDTA^{4-}$	74
$Ni(EDTA)^{2-}$	$EDTA^{4-}$	1.95×10^{-3}
$Cu(EDTA)^{2-}$	$EDTA^{4-}$	0.174
$Cd(EDTA)^{2-}$	$EDTA^{4-}$	~130
$Ca(EDTAH)^{-}$	HEDTA	3.0×10^5
$Co(EDTA)^{2-}$	$HCAL^{2-}$	7.01
$Co(NTA)^{-}$	$HCAL^{2-}$	2.6×10^5

Source: Ref. 10.

the last two entries show the large differences in the rates of substitution by the same ligand on cobalt–EDTA and cobalt–NTA (nitrilotriacetic acid) complexes. Again, both the electrostatic effect and the degree of chelation of the original complex determine the trend.

The surface complexes we are going to discuss are essentially mixed ligand complexes, in which several ligands are oxo groups, and one, two, or even three positions are substituted by water and other ligands. In order to understand the reactivity of the oxo bonds, it is therefore important to understand the effect of a substituent on adjacent or remote coordination positions. The effect of a given group on the rate of substitution of another ligand placed either cis or trans to the group (in octahedral coordination) has been extensively studied in solution chemistry. In a dissociative mechanism, electronic inductive effects, capable of weakening adjacent bonds, are responsible for the labilization of these coordination positions. Recently, Casey et al. have studied in detail the relationship between the reactivities of dissolved and surface Ni(II) complexes [11–13]. In solution, it is found that the rate of water exchange is very sensitive to the nature of the other ligands present in the coordination sphere. Figure 3 shows that the rate of water exchange in a series of complexes $NiL(H_2O)_n^{2+}$ increases as n decreases, i.e., as the number of chelating bonds formed by L increases. Furthermore, because of the relationship described above between chelation and stability, the rate of water exchange increases as the stability of the complex increases. We shall see below that there is a close parallelism in heterogeneous complexation chemistry.

C. The Structure of Dissolved Complexes

Because of their high stability constants and fast complexation kinetics, in general solutions contain the most stable complexes. Through the usual equilibrium techniques (such as potentiometric titration), the ratio M : L can be derived; however, it is less easy to explore the structure of dissolved complexes. Even now, some ambiguity remains as to the composition of the coordination sphere around several metal–EDTA complexes; the participation of water, and the number of coordination positions occupied by the ligand, cannot be derived from equilibrium data.

For 1 : 1 M : EDTA complexes, in the case M = Co(II), Ni(II), quinquedentate species (one free carboxylate) with one bound water molecule, have been postulated from NMR data [14]. Sexadentate structures have also been reported, and it has been suggested that ionic strength may produce a change from one complex to the other [15]. Trivalent

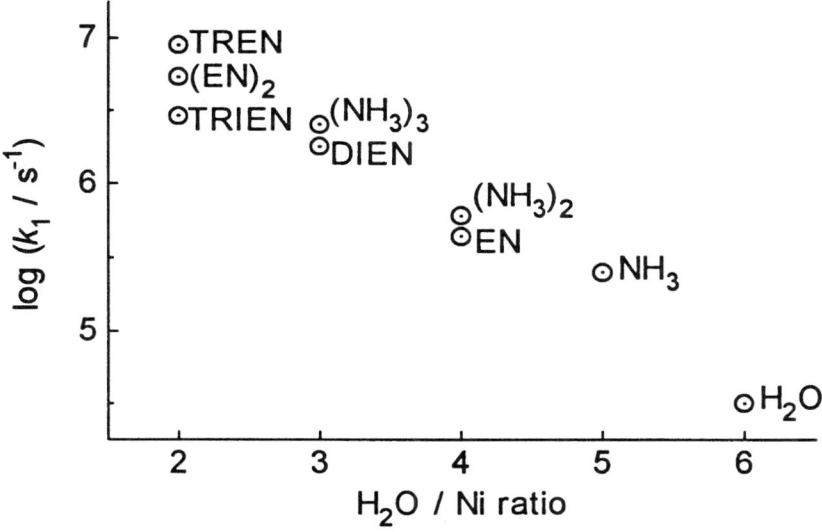

FIG. 3 Rate coefficients for exchange of a single water molecule from around octahedral Ni(II)-ligand complexes, as a function of the number of water molecules in the inner coordination sphere. Ligand abbreviations are EN = ethylenediamine $[NH_2(CH_2)_2NH_2]$; DIEN = diethylenetriamine $[NH_2(CH_2)_2NH(CH_2)_2NH_2]$; TRIEN = $[NH_2(CH_2)_2NH(CH_2)_2NH(CH_2)_2NH_2]$; TREN = $[N((CH_2)_2NH_2)_3]$. (Redrawn from Ref. 12.)

first-row transition metal ions form sexadentate complexes [16], and larger metal ions [Ln(III)] form sexadentate complexes with a coordination number of 10, through the involvement of up to four water molecules [17]. In specific cases of heavier ions, like Pb(II) or Pd(II), bidentate EDTA complexes are formed [16]. A very important feature is the high stability of mononuclear complexes, and the low stability of polynuclear species in which the oligocarboxylates might act as a bridge. Flexible ligands wrap around one metal center, and polynuclear species form only when the coordinating groups of rigid ligands cannot sterically bind the same metal ion; this case is well illustrated by the different coordination chemistry of dipicolinate (2,6-pyridinedicarboxylate) and cinchomeronate (2,4-pyridinedicarboxylate) [18,19].

D. Electron-Transfer Reactions in Solution

Formation of dimeric complexes through ligand bridges has far-reaching consequences in the mechanism of charge transfer between metal centers. The paradigmatic mixed-valence Creutz–Taube complex, pyrazine-bridged binuclear ruthenium (II, III) ammine [20], has been the basis of development of modern ideas about electron transfer in solution. These complexes are especially relevant in the interface metal oxide–aqueous solution when the fluid medium contains carboxylates and metal ions. The intervalence transfer absorption band, typical of these mixed complexes, contains information about the degree of localization of the electron in either metal center, and on the rate of electron transfer between them.

III. THE INTERACTION OF CARBOXYLATES WITH METAL OXIDE SURFACES IMMERSED IN WATER

The subject has been studied through a variety of techniques. The earlier studies resorted to measurements of adsorption isotherms and electrokinetic mobilities in equilibrated systems [21]. The assumption of equilibrium is validated in a time scale that usually spans from several minutes (minimum) to several hours (maximum); for shorter times, the system may be still evolving, whilst for longer times usually secondary drifts, associated with ill-defined further changes, are observed.

In equilibrated systems, the Gibbs adsorption energy for inner-sphere complexation may be derived from the measured adsorption isotherms. The dependence on ligand concentration and pH is used to derive the number of different surface complexes that may form, as well as their stability constants and stoichiometric composition. The surface site density is usually treated as an adjustable parameter that may differ for the various types of surface complexes.

In general, the surface excess Γ is much larger than the surface concentration derived from the diffuse-layer charge, σ_d/F, where $F = 96.500$ C mol^{-1}. Even with this limitation in mind, electrophoretic mobilities yield valuable complementary information about the nature of the interaction between carboxylic acids and the surface. For instance, in an early paper [21], Kallay and Matijević modelled the adsorption of oxalic and citric acids onto hematite through the combined analysis of adsorption isotherms and electrophoretic mobilities. Their basic assumption were

1. The frequently reported change with pH of the apparent maximum adsorption density was an artifact resulting from lateral interactions between charged species.
2. The constancy of the hematite isoelectric point, at pH 2.7, in the oxalic acid concentration range of 1 to 5×10^{-3} mol dm^{-3}, implied that the operative adsorption potential (which is larger in magnitude than ζ) also remained constant.

Kallay and Matijević concluded that the acidity of surface oxalic acid is high enough to guarantee that it exits only as the dianion $C_2O_4^{2-}$. The enhancement in acidity brought about by complexation to metal ions is well known in solution chemistry. The result is also well in line with the spectroscopic characterization of the adsorption of oxalate onto various metal oxides (see below). The modelling of citric acid adsorption [21] suggested that fully deprotonated, and monoprotonated, dinegative anions exist in the surface. Again, the coherent picture of proton release by surface complexation results. In the case of iminodiacetic acid, $HN(CH_2COOH)_2$ [22], the results suggest that only one proton is lost upon adsorption, at least for concentrations within the range 1×10^{-3} to 3×10^{-2} mol dm^{-3}. Thus a pendant free COOH seems to be involved, and probably a bidentate surface complex, involving N and the other carboxylate, is formed.

The more strongly chemisorbing carboxylates are polyprotic and multidentate, and the variety of surface complexes that may result is very large. For instance, in our earlier work [23], two consecutive surface complexation stages were suggested by the shape of the adsorption isotherm of EDTA onto magnetite. This result is coherent with the findings of Torres et al. [22] for iminodiacetic acid; the change in conformation with increasing concentration has important consequences on the reactivity of surface species, as discussed below.

In a recent work on the adsorption of EDTA on goethite [24], it was possible to derive two sets of values for n and m in Eq. (4), $n = 2, m = 1$, and $n = 1, m = 0$, corresponding to equilibria (5) and (6):

$$n(\equiv \text{FeOH}) + Y^{4-} + (n+m)H^+ \rightleftharpoons \equiv \text{Fe}_n YH_m^{(4-n-m)-} + nH_2O \qquad (4)$$

$$2(\equiv \text{FeOH}) + Y^{4-} + 3H^+ \rightleftharpoons \equiv \text{Fe}_2 YH^- + 2H_2O \qquad \log K = 30.36 \qquad (5)$$

$$\equiv \text{FeOH} + Y^{4-} + H^+ \rightleftharpoons \equiv \text{FeY}^{3-} + H_2O \qquad \log K = 14.99 \qquad (6)$$

where Y^{4-} stands for the fully deprotonated EDTA anion. The stability constants were derived using a generalized two-layer model for the interfacial region [25].

These two modes of adsorption represent in fact the successive formation of two complexes, the second step being

$$\equiv \text{Fe}_2 YH^- + Y^4 \rightleftharpoons 2(\equiv \text{FeY}^{3-}) + H^+ \qquad \log K = -0.38 \qquad (7)$$

$\equiv \text{FeY}^{3-}$ contributes appreciably to the surface speciation only at high pH values, as expected from Eq. (6). The degree of protonation of $\equiv \text{Fe}_2 YH^-$ suggests that at least one of the carboxylate groups is not bound to Fe; the reverse inference, that no free groups are present in $\equiv \text{FeY}^{3-}$, cannot be made because deprotonation may be simply due to electrostatic effects. Note that the predominant species of free EDTA in the spanned pH range is $H_2 Y^{2-}$; further deprotonation upon complexation is again indicated by the experimental data. An important and distinct feature of surface EDTA complexes is the occurrence of dimeric binuclear complexes, in which EDTA links two surface ions. Similar bridged complexes may form by interaction of one surface ion, EDTA, and a second metal ion taken from the solution [26,27]. By contrast, we mentioned that EDTA does not form this type of complex in solution. The reasons behind the difference are two: the packing of surface metal ions precludes the wrapping around of EDTA, thus leaving some coordination groups available to interact with a second metal ion; also, the charge and potential buildup in the surface eventually determines the protonation of EDTA, and the breakage of some $\equiv$M–OOC–R bonds.

Figures 4 and 5 show adsorption isotherms of salicylic acid onto titanium dioxide, and the derived surface speciation diagrams; Table 5 shows the surface complexation equilibria and the stability constants [28]. The identification of surface ions with one available ligation site (denoted as $\equiv$vTi) and two ligation sites ($\equiv$ivTi) is coincident with findings from EXAFS measurements [29,30].

It is clear that good coordinative groups for ions in solution are also good complexing agents for the same ions in the surface of oxides. Thus carboxylates and phenolates, especially when chelating, form strong surface complexes with titanium dioxide and with Ti(IV) in solution (see for example, [31,32]). Vasudevan and Stone [33,34] found that the stability of surface complexes formed on TiO_2, Fe_2O_3, FeOOH, and Al_2O_3 followed the trend catechol > 2-aminophenol ≫ o-phenylendiamine, and that the trend for catechol surface complexes was TiO_2 > Fe(III) (hydr)oxides > Al_2O_3. For different TiO_2 samples, significant differences in stability were found, indicating that the detailed surface topology was an important factor. Stone and coworkers have carried out a systematic study of the behavior of carboxylates and phosphorothioates in the surface of various minerals [35–42]. Particularly interesting was the possibility of demonstrating that phenyl picolinate ester forms a surface chelate on TiO_2 or FeOOH, as shown in Fig. 6. As early as 1981, Matijević and coworkers [43] reported that the affinity toward hydrous

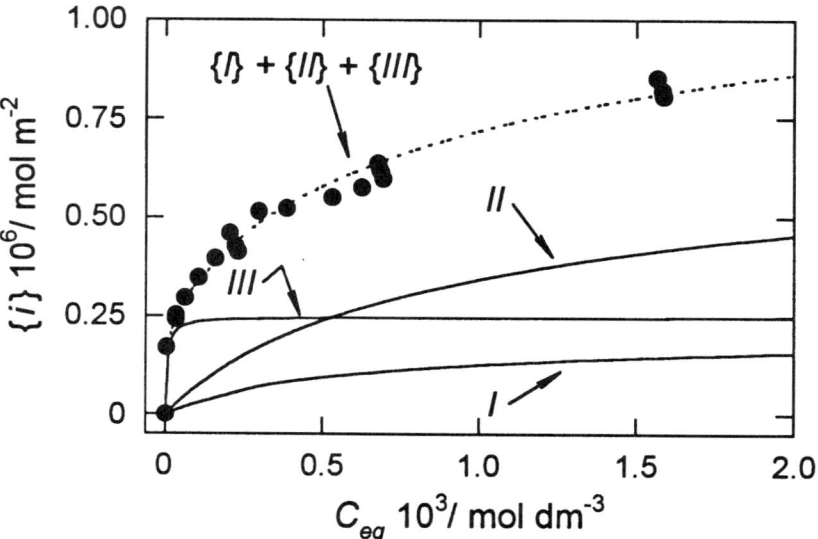

FIG. 4 Speciation of surface Ti(IV)-salicylate complexes as a function of salicylic acid concentration at pH 3.6 and 298 K. The sum $\{I\}+\{II\}+\{III\}$ equals the surface excess Γ; symbols are experimental Γ values. The structures of the complexes are shown in Table 5. (Redrawn from Ref. 28.)

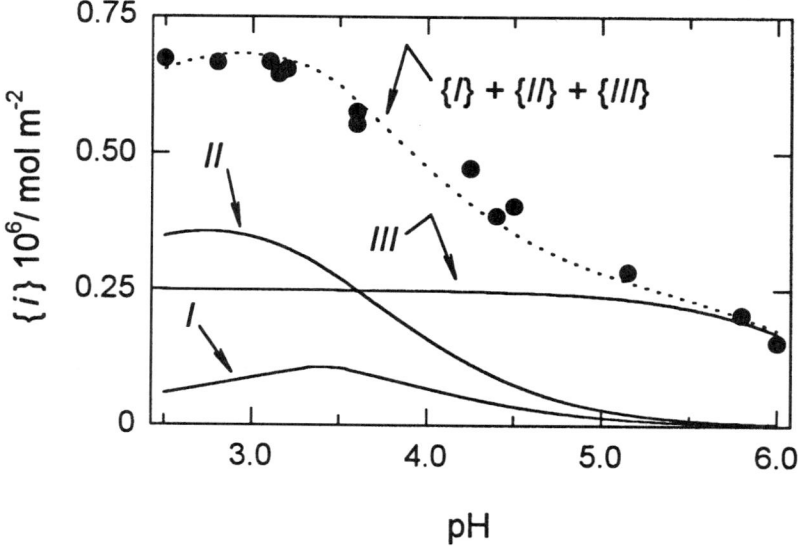

FIG. 5 Speciation of surface Ti(IV)-salicylate complexes as a function of pH at 298 K. The sum $\{I\}+\{II\}+\{III\}$ equals Γ, and symbols are experimental Γ values; the structures of the complexes are shown in Table 5. Other conditions are total salicylic acid concentration, 8.0×10^{-4} mol dm^{-3}; surface-to-volume ratio, 514 m^2 dm^{-3}. (Redrawn from Ref. 28.)

Interfacial Chemistry of Dissolving Metal Oxide Particles

TABLE 5 Surface Complexation Equilibria and Their Corresponding Constants for Adsorption of Salicylate on Titanium Dioxide (cf. Figs. 4 and 5); the Small Roman Numeral Preceding Ti Indicates Its Number of Bonds to Lattice Oxo Anions

[Reaction (I)] $\equiv$vTi—OH$^{1/3-}$ + $\equiv$OH$^{1/3+}$ + salicylic acid $\rightleftharpoons$ [$\equiv$vTi—O—C(=O)—C$_6$H$_4$—OH···O$\equiv$]$^{-}$ + H$_3$O$^+$ $pK_I = 3.3$

[Reaction (II)] $2\ \equiv$vTi—OH$^{1/3-}$ + salicylic acid $\rightleftharpoons$ [$\equiv$vTi—O—C(=O)—C$_6$H$_4$—O—Ti$\equiv$v]$^{2/3-}$ + 2 H$_2$O $pK_{II} = -2.9$

[Reaction (III)] [$\equiv$ivTi(OH)$_2$]$^{2/3-}$ + salicylic acid $\rightleftharpoons$ [$\equiv$ivTi(O—C(=O)—C$_6$H$_4$—O)]$^{2/3-}$ + 2 H$_2$O $pK_{III} = 6.5$

Source: Ref. 28.

chromium(III) followed the trend dipicolinate > picolinate > nicotinate. This trend can be explained as due to the change in chelation along the series, tridentate, bidentate, and monodentate.

Casey and coworkers [11–13] have explored the surface complexation chemistry of bunsenite (NiO), and they were able to demonstrate that the kinetics of interaction of surface nickel ions with the complexing agents was governed by the same factors that apply to homogeneous solution.

The Gibbs energies for metal complexation, either in solution or on the surface of metal oxides, can be solved into an electrostatic contribution, a solvation term, and an intrinsic ("chemical") energy [44]:

$$\Delta G_{ads} = \Delta G_{coul} + \Delta G_{solv} + \Delta G_{chem} \tag{7a}$$

The electrostatic contribution in the heterogeneous case differs from the ion-pairing energy (work term) used in homogeneous kinetics, in the sense that the local macropotential ψ replaces the point-charge interactions. The solvation energy is also different, involving changes both in the inner coordination sphere and in the continuum

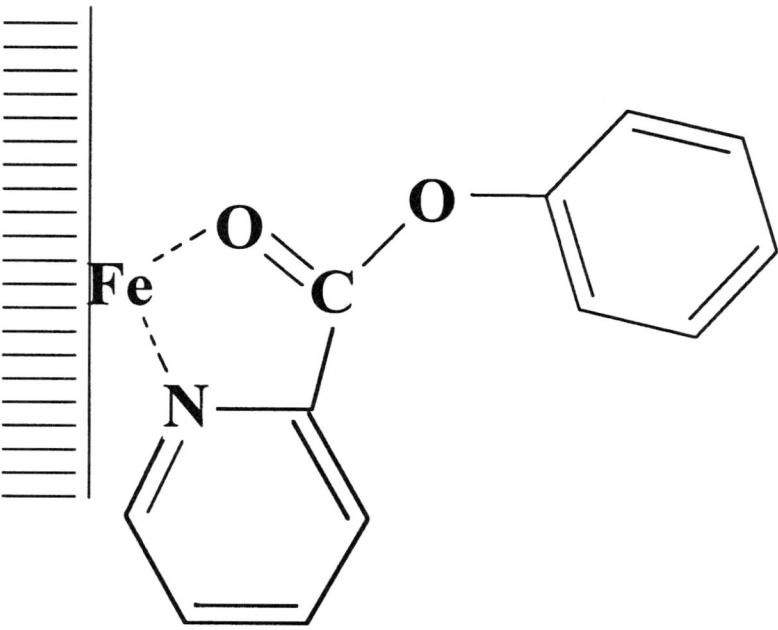

FIG. 6 Sketch of surface coordination by chemisorbed phenyl isonicotinate. (Redrawn from Ref. 42.)

properties of water adjacent to the solid surface. The difference in homogeneous and heterogeneous solvation contribution may change the trends of stability of series of homologous ligands, depending on the hydrophylic/hydrophobic nature of each ligand. It should be noted that the solvation contribution is highly sensitive to the state of charge of the reactants and reaction products, and this state of charge may differ drastically in both cases. For the adsorption of chromate onto titanium dioxide, Huang et al. [45] have derived the values $\Delta G_{solv} = 3 \times 10^{-3}$ to 1.3×10^{-2} kJ/mol, $\Delta G_{coul} = -0.8$ to -1.7 kJ/mol, and $\Delta G_{chem} = -3$ to -10 kJ/mol. In this case, the coulombic contribution is important, and the solvation contribution is small. Solvation is expectedly much more important in the case of adsorption of cations, including H$^+$, and also for large neutral organic molecules. In the context of this review, the case of the interaction of photosensitizer dyes with titanium dioxide, especially ruthenium-bipyridyl derivatives, is important. Even in this case however, the best-behaving sensitizers possess functional groups that render them at the same time more hydrophilic and better complexants. 2,2'-bipy-4,4'-dicarboxylate has been reported to bridge Ru(II) (bound to the bipy end of the ligand) to surface Ti(IV) (bound to the carboxylate groups) [46].

As in the case of homogeneous media, the equilibrium measurements described above do not give information about the structure of the surface species. Several spectroscopic and other techniques have been used for structural characterization, although in general the outcome is a fingerprint for a given species, that renders some gross features of the complexes, rather than a sharp, detailed picture of them. For instance, chelated versus nonchelated species can sometimes be distinguished by the vibrational spectra, and the distances between a surface and an adsorbed metal ion $M \cdots M'$, measured by EXAFS, may be used to differentiate inner- and outer-sphere complexes. Fourier-transform infrared (FTIR) and visible diffuse reflectance have yielded until now the most valuable information about the nature of the chemisorption of organic anions. When the spectrum of surface species at the solid–water interface is sought, a most serious problem is that the signal from the adjoining solution volume is in general orders of magnitude larger than that due to surface species, and the shifts are usually small enough to prevent resolution. The attenuated total reflection FTIR (ATR-FTIR) method has largely solved this problem, because the thickness probed by the evanescent radiation is of the order of 1 μm for a 45° incidence angle. For a list of ATR-FTIR as well as ex-situ infrared studies of the adsorption of organic ligands onto metal oxide surfaces, see [4].

The chemisorption of simple carboxylates has been studied in some detail by this technique. Formation of oxalate surface complexes on TiO_2 [47] and on hydrous chromium oxides [48] has been demonstrated; data collected as a function of ligand concentration and pH could be interpreted as due to the formation of three different surface complexes, the spectra of which could be derived by factor analysis of the spectra collected at different pH values. The mathematical procedure requires that a chemical model of the interaction be assumed; Hug and Sulzberger [47] assumed that the three different complexes are formed independently.

In the case of salicylate adsorption onto TiO_2, two adsorption modes are identified [49]. Two different chemical models were analyzed:

1. Independent adsorption:

$$x(\equiv S) + C_6H_4(OH)COO^- + nH^+ \rightleftharpoons \equiv S_x(C_6H_4(OH)COO)H_n^{(n-1)+} \tag{8}$$

$$y(\equiv S + C_6H_4(OH)COO^- + mH^+ \rightleftharpoons \equiv S_y(C_6H_4(OH)COO)H_m^{(m-1)+} \tag{9}$$

2. Successive adsorption:

$$x(\equiv S) + C_6H_4(OH)COO^- + nH^+ \rightleftharpoons \equiv S_x(C_6H_4(OH)COO)H_n^{(n-1)+} \tag{10}$$

$$\equiv S_x(C_6H_4(OH)COO)H_n^{(n-1)+} + C_6H_4(OH)COO^- + mH^+ \rightleftharpoons$$

$$\equiv S_x(C_6H_4(OH)COO)_2 H_{n+m}^{(n+m-2)+} \tag{11}$$

Models 1 and 2 cannot be distinguished by the goodness of the fitting: Fig. 7 shows the results of modeling the spectra on the basis of the two schemes. In both cases it is assumed that complexation is described by a simple equilibrium constant, unaffected by surface potentials (Langmuirian model). Biber and Stumm [50], on the other hand, found out that the surface binding modes of salicylate onto several oxides and oxohydroxides differ depending on the nature of the metal ion and the chemical composition of the (hydr)oxide.

In the case of 4-chlorocatechol adsorbed on TiO_2, ATR-FTIR studies suggest the formation of a single inner-sphere surface complex of the type $\equiv Ti–OC_6H_3(Cl)O–Ti\equiv$, and a nonspecific mode, perhaps an outer-sphere complex [51]. In turn, catechol adsorption isotherms suggest two adsorption modes, involving a dimer similar to the one

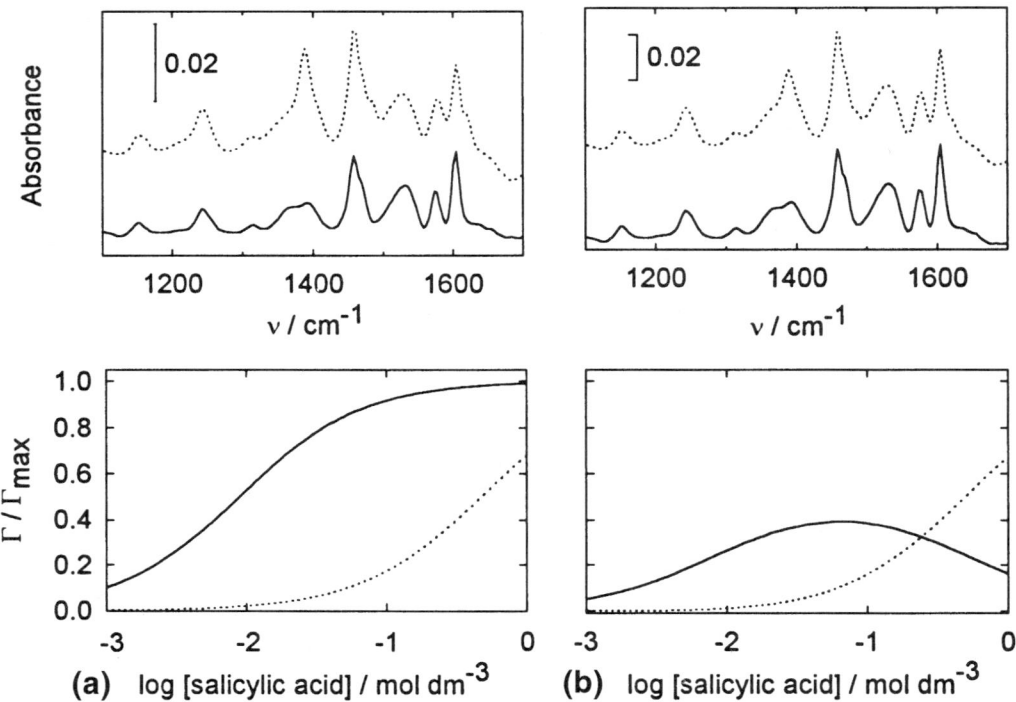

FIG. 7 Modeling of ATR-FTIR spectra of salicylate-Ti(IV) surface complexes. Both derived spectra and calculated Langmuir isotherms (in linear-log scale) are shown for the two adsorption models of salicylate onto TiO_2 (see also text). (a) Two-site, independent adsorption; (b) one-site, successive adsorption. The best fitted apparent stability constants at pH 4.0 are (a) log $K_8 = 5.05$, log $K_9 = 3.32$; (b) log $K_{10} = 4.75$, log $K_{11} = 3.61$; subscripts refer to equation numbers in the text. Experimental conditions: 298 K, pH 4.0, 0.01 mol dm^{-3} ionic strength. (From Ref. 49.)

described above, and a chelated mononuclear complex [52]; Fig. 8 shows the relative contribution of both species ({II} and {I}, respectively) as a function of pH in the absence of added electrolyte.

The strong chemisorption of catechol does not operate on TiO_2 alone; Stone and Morgan [53] demonstrated that catechol complexes were formed more readily than hydroquinone complexes on manganese(III) oxides. Consequently, the reductive dissolution rates (see below) did not follow the trend of the redox potentials, and catechol was more effective than hydroquinone (cf. however with the results obtained on goethite [54]). Inner-sphere complexation, with formation of chelated species, also takes place when catechol chemisorbs on alumina [55], and probably also on iron oxides [56].

The FTIR spectrum of a hydrous chromium(III) oxide exposed to a solution containing oxalate (0.1 mol dm^{-3}) at pH 3.6 demonstrates that chelated surface complexes are formed [48]. The main peaks in Fig. 9b (1710, 1680, 1410, and 1260 cm^{-1}) can be attributed to the $\equiv Cr(C_2O_4)$ moiety; the same bands are observed in the solids $K_3[Cr(C_2O_4)_3]$ and $K[Cr(C_2O_4)_2(H_2O)_2]$ [57], as well as in the aqueous complex, shown in Fig. 9c. In the complex, both carboxylate groups are bound, either to the same surface

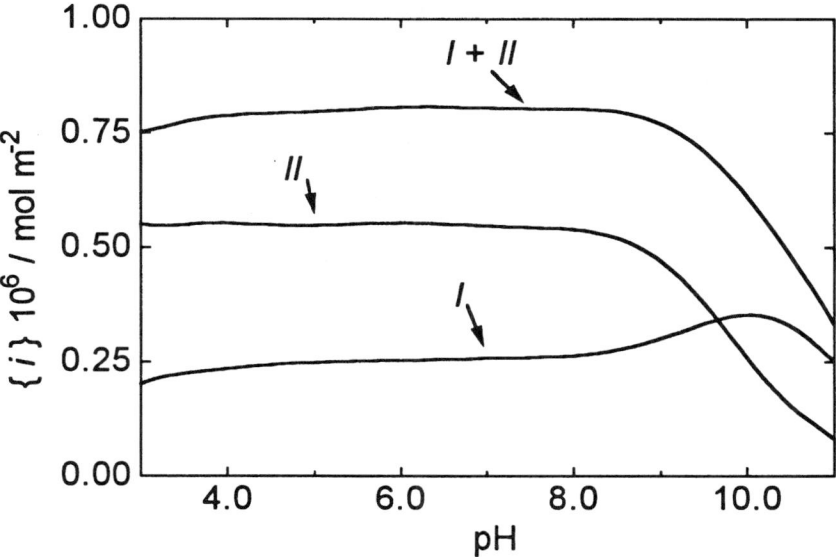

FIG. 8 Speciation of surface Ti(IV)-catecholate complexes as a function of pH at 298 K, in the absence of added electrolyte; total catechol concentration is 1.0×10^{-3} mol dm^{-3}, and surface-to-volume ratio is 822.4 m^2 dm^{-3}. The sum $\{I\} + \{II\}$ equals the surface excess Γ. The structures of the complexes are as follows: I is a chelated mononuclear, and II is a bridged binuclear complex. (Redrawn from Ref. 52.)

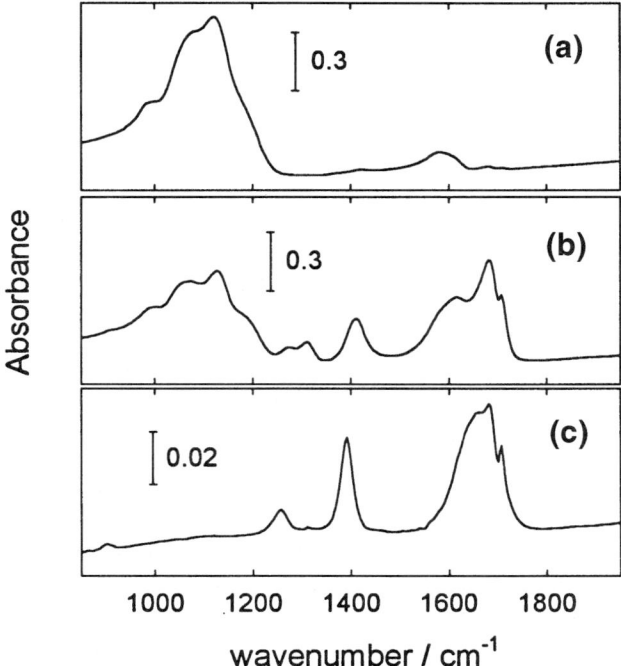

FIG. 9 ATR-FTIR spectra of hydrous chromium(III) oxide before (a) and after (b) exposure to a 0.1 mol dm^{-3} oxalate solution at pH 3.6. Also shown (c) is a differential spectrum of 0.01 mol dm^{-3} dissolved Cr(C$_2$O$_4$)$_3^{3-}$ in 0.1 mol dm^{-3} excess oxalate, at pH 4.9. (Data from Ref. 48.)

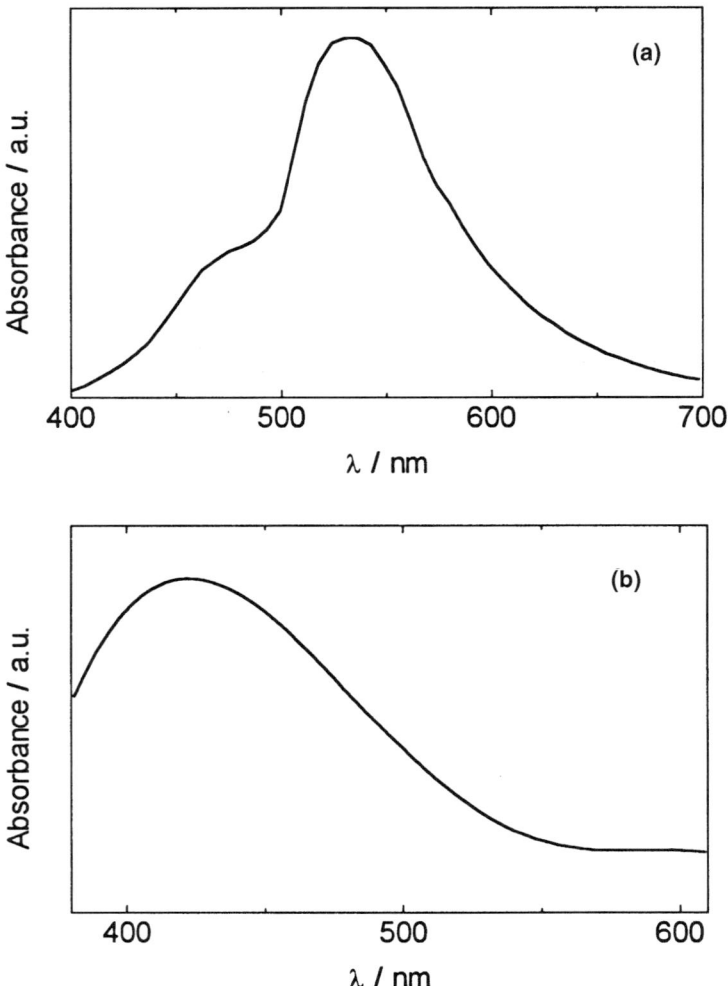

FIG. 10 Diffuse reflectance UV-visible spectra of (a) iron(III)-thiocyanate and (b) titanium(IV)-catecholate surface complexes. (Redrawn from Refs. 58 and 52, respectively.)

Cr(III) ion or to two adjacent ones. Comparison of Figs. 9a and 9b shows some decrease of the bands due to hydroxylated surface sites, as expected if substitution of oxalate for OH takes place.

UV-visible absorption and fluorescence spectroscopies have also been employed to characterize surface complexes formed by organic ligands on metal oxides; for further details, see [4]. Because of its impact on the reactivity of surface complexes, the ligand-to-metal charge transfer bands are especially interesting. An early example was the band observed upon adsorption of thiocyanate onto iron oxides [58], and cathecol onto titanium dioxide [52] (Fig. 10). These bands are closely linked to similar transitions observed in homogeneous solution.

Not much is known about the role played by the solvent and the inner-sphere reorganization energies in heterogeneous systems; the information available up to now is too limited to permit a fundamental analysis.

From the point of view of the structure of the double layer, equilibrated adsorption of carboxylic and other organic acids may be viewed as a derivatized surface that alters the acid–base properties and hence the whole potential/distance profile at each pH, temperature, and ionic strength. Because the surface excess Γ is much larger than the surface charge, the description amounts to accepting the existence of at least two types of acid–base sites, $\equiv$M–OH and $\equiv$M–L. In practice, the number is usually larger, but the resulting complications are essentially due to more complex algebra and to the lack of enough experimental information to reach a quantitative model. At high degrees of coverage, the surface should be viewed as a metal carboxylate, and the acid–base properties shall be governed by factors akin to those operating in solution: type and availability of pendant acid–base groups in partially coordinated ligands. The unraveling of the surface coordination modes remains a task largely unfulfilled.

The basic assumption of equilibrated surfaces becomes unwarranted in many cases when carboxylic acids are present. The agressivity of carboxylic acids towards metal ions is well established, but the concepts to be used to describe the double layer associated to chemically nonequilibrated surfaces are not clearly established. In what follows, we shall review the state of knowledge in the description of nonequilibrated interfaces. Because of the lack of information, much of the discussion is based in systems that involve simple mineral acids.

IV. NONEQUILIBRATED INTERFACES IN MINERAL AND ORGANIC ACIDS

From a basic point of view, the metal oxide–water interface should be described in terms of the thermodynamics and kinetics of ionic and molecular exchange between the solid and the surrounding medium. The basic models that have been developed for these purposes are, respectively, the modified Nernstian approach to describe surface charge and potential [59], and the Vermilyea model of ion exchange in dissolving ionic solids [60]. Both descriptions must be complemented with an adequate model of the double layer region.

The equilibrium description is well established. In the case of insoluble metal oxides, it is based on the assumption that metal ions are not equilibrated in the solid and the liquid phase; the only potential-determining ions are H^+. The standard equilibrium surface complexation description has been used with some success to describe the kinetics of dissolution in mineral acids, in which metal ions are exchanged between the metal oxide and the aqueous solution. Conceptually, however, a dissolving interface should not be described only in terms of proton adsorption, the basic assumption of the usual equilibrium models [59], that H^+ is the only potential-determining ion, is no longer valid. In fact, it is easier to visualize dissolution kinetics in the extreme of solubility equilibrium. Only in cases in which all surface complexation and protolysis reactions are rapidly equilibrated can the (slow) rates of dissolution be easily modelled in terms of surface complexes.

A. Kinetics of Surface Complexation

The dynamics of adsorption exchange are still very obscure, and in some cases the assumption of equilibrated surfaces during typical measurements is unwarranted. Some attempts have been made to measure rates of adsorption through the use of pressure-jump methods [61,62]. Chromate and arsenate adsorption were found to take place in two steps, that in the simplest case may be envisaged as the successive formation of two different surface complexes. Both reactions are fast, equilibrium being reached in less than 1 s [62]. On the other hand, a recent scanning-tunneling-microscopy study of the adsorption of sulfate onto goethite [63] demonstrated that the surface is not static, within the time scale of the experiment, distinct changes in the surface were apparent, probably associated with the movement of sulfate.

An analogy with solution chemistry supports the idea that chemisorption may be fast, unless limited by mass transfer. The overall charge acquired by the surface in any conceivable condition is rather low: σ_0 is equivalent to only a minor fraction of the total metal ion surface density; in these conditions, the $\equiv$M–OH$_2$ bonds are expectedly labile.

B. Kinetics of Metal Oxide Dissolution

We shall describe briefly the standard electrochemical description of oxide dissolution. The surface of a metal oxide MO in equilibrium with a solution of M^{2+} (and H^+) is undergoing constant ionic exchange with the solution, the rates of cation exchange are not in general equal to the rates of anion exchange; furthermore, they depend on solution variables frequently in opposite ways. The basic equations of electrochemical kinetics, as applied to this case, are

$$R_c^+ = k_c^+ N_c = k_c^{o+} N_c \exp\left(\frac{\alpha_c z_c e\psi}{kT}\right) \tag{12}$$

$$R_c^- = k_c^- = k_c^{o-} \exp\left(\frac{-\alpha_c z_c e\psi}{kT}\right) \tag{13}$$

$$R_a^+ = k_a^+ N_a = k_a^{o+} N_a \exp\left(\frac{-\alpha_a z_a e\psi}{kT}\right) \tag{14}$$

$$R_a^- = k_a^- = k_a^{o-} \exp\left(\frac{\alpha_a z_a e\psi}{kT}\right) \tag{15}$$

In these equations, R represents rate, the + or − superscripts correspond to dissolution and ionic deposition, respectively, and the subscripts c and a identify cation and anion; the k's are specific rate constants, N is the surface density of anions or cations, α the electrochemical transfer coefficient, z the charge number (without sign) of the ion being transferred, e the electron charge, ψ the surface potential, k the Boltzmann constant, and T the temperature. The k^o's are the rate constants for $\psi = 0$, and are therefore potential-independent values; they are, however, undefined (up to now) functions of solution variables.

In principle, the kinetics of ionic exchange, either in equilibrium or away from it, are described by this set of equations; it is only required to know the values of the four k^0 constants and of ψ as a function of solution composition. The surface potential values under solubility equilibrium conditions is simply given by

$$\psi = \left(\frac{kT}{z_c e}\right) \ln\left(\frac{a_c}{a_c^o}\right) = \left(\frac{kT}{z_a e}\right) \ln\left(\frac{a_a^o}{a_a}\right) \tag{16}$$

where a represents activity, and a_c^o and a_a^o are the corresponding values in the point of zero charge (pzc) conditions, that are interrelated through the solubility product K_s:

$$a_c^o a_a^o = K_s \tag{17}$$

Introducing Eq. (16) into Eqs. (12)–(15), very simple rate equations are obtained:

$$R_c^+ = k_c^{o+} N_c \left(\frac{a_c}{a_c^o}\right)^{\alpha_c} \tag{18}$$

$$R_c^- = k_c^{o-} \left(\frac{a_c}{a_c^o}\right)^{-\alpha_c} \tag{19}$$

$$R_a^+ = k_a^{o+} N_a \left(\frac{a_a}{a_a^o}\right)^{-\alpha_a} \tag{20}$$

$$R_a^- = k_a^{o-} \left(\frac{a_a}{a_a^o}\right)^{\alpha_a} \tag{21}$$

If the Nernst equation (16) applies, Eqs. (18) to (21) describe the rates of ionic exchange in equilibrium. The simplest further assumptions that may be made are

$$N_c \approx N_a = 0.5 N_s \tag{22}$$

$$k_c^{o+} = \text{constant} \tag{23}$$

$$k_c^{o-} = k_c^{*-} \left(\frac{a_c}{a_c^o}\right) \tag{24}$$

$$k_a^{o+} = k_a^{*+} \left(\frac{a_H}{a_H^o}\right) \tag{25}$$

$$k_a^{o-} = k_a^{*-} \left(\frac{a_a}{a_a^o}\right)\left(\frac{a_H^o}{a_H}\right) \tag{26}$$

Equation (22) simply states that the deviation from electroneutrality on the surface is small, and roughly one half of the N_s surface sites are occupied by each type of ion. The well-known deviation from ideal behavior of insoluble metal oxides is due to the noncompliance with this relation. The surface is dominated by neutral species, and in the pzc,

$$N_c = N_a << 0.5 N_s$$

Equation (23) assumes that the rate constant of M^{2+} release is not affected by the solution composition; Eq. (24) states that the rate constant of M^{2+} deposition is first-order on cation activity. Equations (25) and (26), where a_H represents activity of protons, are written on the assumption that the oxide ion is equilibrated in solution as OH^-, and that hydroxide transfer from the surface to the solution proceeds via attack by H^+ followed by OH^- transfer.

In a dissolving interface, in some cases a steady state may be reached, in which the surface composition does not change appreciably with time. In these cases, electrophoretic mobilities would reflect this kinetic condition rather than the equilibrium ζ-potential values. A simple, limiting case corresponds to the *freely dissolving condition* [60]. In this condition, the deposition rates are negligibly small, and the surface composition is solely determined by the rates of transfer of surface anions and cations. Thus the surface potential ψ_f itself is kinetically determined, and the steady-state condition yields values that are a function of pH:

$$\psi_f = \frac{RT}{(\alpha_c z_c - \alpha_a z_a)F} \ln\left(\frac{N_a k_a^+ a_H}{N_c k_c^+}\right) \tag{27}$$

Equation (27) should be compared with the equilibrium potential value.

The Vermilyea model has been used to derive the expected pH dependency of the dissolution rates. For this subject the reader may consult the bibliography (see, e.g., [3]); here we are mainly interested in the kinetic steady-state potential ψ_f.

Diggle [64] has modified the model by replacing the expression for ψ derived under the assumption of potential determination by constituent ions. His approach is close to the surface complexation picture, which provides a simple way to understand how dissolution influences the acquisition of charge by the interface and hence the surface acid–base properties. The dissolution of vanadium pentoxide in mineral acid proceeds according to the following mechanism [65]:

$$\equiv \text{VO}^- \xleftrightarrow{(K_{a2}^s)^{-1}, \text{H}^+} \equiv \text{VOH} \xleftrightarrow{(K_{a1}^s)^{-1}, \text{H}^+} \equiv \text{VOH}_2^+ \tag{28}$$

$$\equiv \text{VOH}_2^+ \xrightarrow{k_o} \text{VO}_2^+ \tag{29}$$

This mechanism is much used to describe dissolution of metal oxides from surfaces that are equilibrated with respect to protolytic reactions. The rate law is

$$R^o = \frac{R_{\max}^o K_s[\text{H}^+]}{1 + K_s[\text{H}^+]} \tag{30}$$

However, if the rate constant of $\equiv\text{VOH}_2^+$ deprotonation is not much larger than k_o, K_s in Eq. (30) is not the true reciprocal acidity constant $(K_{a1}^s)^{-1}$. Calling k^f and k^r the rate constant for surface protonation and deprotonation, $(K_{a1}^s)^{-1} = k^f/k^r$, the steady-state value for K_s is

$$K_s = \frac{k^f}{k^r + k_o} \tag{31}$$

which becomes $(K_{a1}^s)^{-1}$ only if $k_o \ll k^r$.

Thus, according to the surface complexation description, the acidity of highly reactive surfaces is determined by steady-state considerations. Both the intrinsic acidity of the surface and the reactivity towards dissolution of the protonated surface complexes determine the proton surface excess.

Below, we shall discuss the pH dependence of the electrophoretic mobility of ilmenite particles; the results may in fact be understood in terms of the potential developed in a dissolving surface.

This description is valid provided that dissolution is not so fast as to perturb the double-layer structure; diffusion limited processes are therefore excluded (see [66]: the dissolution of small MgO particles in strongly acidic solution has been reported to proceed in such a way that it is not possible to assume that the ionic distribution in the double layer is equilibrated).

C. Dissolution in the Presence of Carboxylic Acids

Carboxylic acids are aggressive towards many metal oxides, and there is a very large wealth of information on the kinetics of dissolution. In what follows, we shall present a steady-state description of the interface of an oxide being dissolved by a carboxylic acid. We shall first assume that dissolution is a first-order process on a single surface complex, which is formed in a pre-equilibrium step:

$$n(\equiv \text{MOH}) + m\text{Y}^{z-} + (p+n)\text{H}^+ \rightleftarrows \equiv \text{M}_n\text{Y}_m\text{H}_p^{(mz-p)-} + n\text{H}_2\text{O} \tag{32}$$

$$\equiv \text{M}_n\text{Y}_m\text{H}_p^{(mz-p)-} \rightarrow \text{dissolution} \tag{33}$$

$$\sigma_0 = F[\{\equiv \text{MOH}_2^+\} - (mz-p)\{\equiv \text{M}_n\text{Y}_m\text{H}_p^{(mz-p)-}\}] \tag{34}$$

Equation (34) assumes that the surface plane is the locus of the adsorbed ligand; any difference with the plane of proton adsorption is ignored. The steady-state condition is given by

$$\{\equiv \text{M}_n\text{Y}_m\text{H}_p^{(mz-p)-}\} = \frac{k_{32}}{k_{-32} + k_{33}}\{\equiv \text{MOH}\}^n[\text{Y}^{z-}]^m[\text{H}^+]^p \tag{35}$$

where the k's are the specific rate constants for Eqs. (32) and (33). This description differs from equilibrium in the k_{33} term in the denominator; the usual surface complexation approach is thereby modified, and substantial alteration of the surface charge may result when k_{33} becomes comparable to k_{-32}. Combining the above equations with the acidity equilibrium for $\equiv\text{MOH}_2^+$ (K_{a1}), the following expression for the surface charge results:

$$\sigma_0 = F N_s \frac{[\text{H}^+]/K_{a1} - (mz-p)k_{32}/(k_{-32}+k_{33})[\text{Y}^{z-}]^m[\text{H}^+]^p}{1 + [\text{H}^+]/K_{a1} + k_{32}/(k_{-32}+k_{33})[\text{Y}^{z-}]^m[\text{H}^+]^p} \tag{36}$$

Dissolution experiments are often carried out at high ligand concentration; in the absence of dissolution, these conditions would lead to total surface coverage, and the surface charge is essentially determined by the negative charge $(mz-p)$ of the surface complex, which in turn may also depend on surface coverage. The almost constant negative surface charge densities (σ_p) typically found in a wide pH range (that spans rather acidic conditions, see Fig. 11) [67] are in agreement with this interpretation. The constancy in the isoelectric point found at large ligand concentrations [21] may result from the acid–base properties of the ligand (changes in $mz-p$) and from the desorption brought about by ligand protonation. Insofar as k_{33} remains appreciably smaller than k_{-32}, this picture does not change. However, as soon as $k_{33} \approx k_{-32}$, the protolytic equilibrium may introduce positive charge in the surface; the degree of surface protonation and surface complexation (and hence the surface charge) shall result from the dynamics of all the processes involved. In oxides dissolving at fast rates, the steep rise in σ_p shown in Fig. 11 may be due to this kinetic effect. Another good example is provided by the dissolution of vanadium pentoxide in oxalic acid [65]; several different vanadium–oxalate surface complexes mediate dissolution, at characteristic rates, and the description of the steady-state interface requires that all the involved rate constants be known.

If the oxidation state of the metal ion does not change upon phase transfer, Eq. (33) describes a "ligand-assisted" process. Different ligands dissolve a given metal oxide at rates that depend on the surface affinity (k_{32}/k_{-32}) and on the reactivity of the oxo bonds

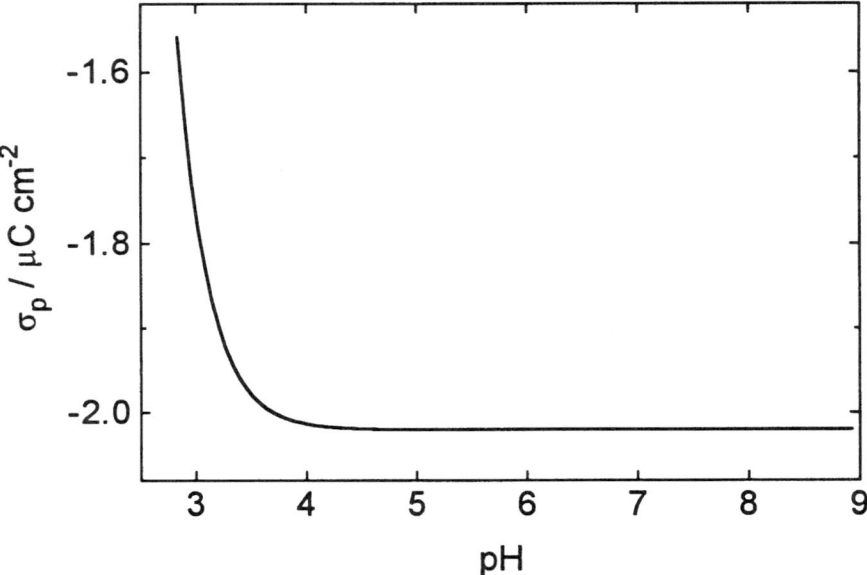

FIG. 11 pH dependence of the surface charge up to the ζ plane, σ_p, for magnetite particles immersed in 2.4×10^{-3} mol dm^{-3} oxalic acid solutions, at 303 K and 2.8×10^{-2} mol dm^{-3} ionic strength. (Redrawn from Ref. 67.)

in the surface complex (k_{33}). A detailed analysis of the chemical factors that influence these figures and hence the steady-state surface condition can be found in Refs. 3, 4, and 12.

In many cases however, redox-mediated dissolution is faster than ligand assisted dissolution, and dominates the overall rate. A very simple scheme for this process is given by Eqs. (37)–(38):

$$\equiv M^{III}X^- + H_2O \rightleftharpoons \ \equiv M^{II}OH_2 + X^\bullet \tag{37}$$

$$\equiv M^{II}OH_2 \rightarrow \text{dissolution} \tag{38}$$

Under steady-state conditions and saturation coverage of $\equiv M^{III}$ sites, the ratio of M^{II} and M^{III} in the surface is given by

$$\frac{\{\equiv M^{II}\}}{\{\equiv M^{III}\}} = \frac{k_{37}}{k_{-37}[X^\bullet] + k_{38}} \tag{39}$$

The more general expression should embody the surface complexation reaction (32).

A very important point refers to the extent of reversibility of interfacial electron transfer. As written, Eq. (37) is reversible if the oxidized species $X^\bullet$ is not scavenged by some other reductant. Carboxylic acids are in general two-electron reductants, that upon oxidation may release carbon dioxide; the simplest case is oxalic acid. In many cases, Eq. (37) represents an unfavorable equilibrium, that is followed by an irreversible further oxidation of $X^\bullet$; the kinetic scheme becomes therefore slightly more complicated.

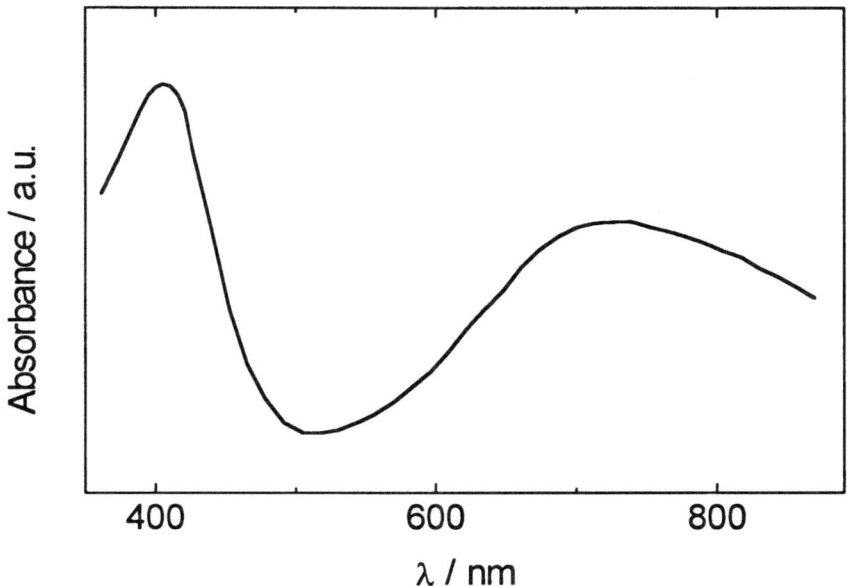

FIG. 12 Visible absorption spectrum of the surface species formed by adsorption of $Fe^{II}(CN)_5(isonicotinate)^{4-}$ onto iron oxide. (Redrawn from Ref. 75.)

The stoichiometry of the dissolution of iron oxides by carboxylic acids shifts from that corresponding to a ligand-assisted acid process to that of reductive dissolution when the temperature increases. Reported systems include oxalic [68], iminodiacetic, and ethylenediaminetetraacetic acids [69,70]. As in the case of ligand-assisted dissolution, both the dissolution rate and the steady-state surface composition depend on the stability of the surface complex [Eq. (32)] and on the rate of the irreversible process; in this case, this latter reaction is the charge-transfer reaction described by Eqs. (37) and (38). Thus the rates are affected also by the redox properties of the carboxylate; oxalate turns out to be especially powerful.

A very important group of reductants to enact oxide dissolution is that of carboxylate complexes of low-valence metal ions [3,23,67,68,71–73]. The dissolved metal ions usually adsorb through the formation of a bridged mixed-valence dimer with the surface metal ions. A chelating agent, acting in this case as a bridge, is responsible for this *ligand-like adsorption* [74]. By analogy with solution chemistry, UV-visible spectroscopy is a potentially important tool to characterize these mixed-valence dimers, when intervalence charge-transfer bands can be detected. Coadsorption of isonicotinate and Fe(II) onto iron(III) oxides provides a good example. Figure 12 shows the visible absorption spectrum of the surface species formed by adsorption of $Fe(CN)_5(isonic)^{4-}$ onto iron oxide [75]. The band centered at around 735 nm may be identified as the intervalence charge-transfer band, also seen in dissolved dimers, in the surface complex $\equiv Fe^{III} - O_2CC_5H_4H - Fe^{II}(CN)_5^{4-}$.

The acid–base properties, and hence the surface charge, shall depend on the surface ratio $\{\equiv M^{II}\}/\{\equiv M^{III}\}$; note that the inner-sphere electron transfer previous to dissolution [Eq. (37)] can be viewed as a homolytic M–X bond breakage. In the case of outer-sphere electron transfer, a proton must also be transferred to the surface. Normally, the process is kinetically complex, but it is worthwhile to model the simple scheme that describes dissolution in the absence of specific adsorption. Equation (40) replaces Eq. (37), and the steady-state condition is described by Eq. (41):

$$\equiv M^{III}OH + X^- + H^+ \rightleftarrows \equiv M^{II}OH_2 + X^\bullet \qquad (40)$$

$$\frac{\{\equiv M^{II}\}}{\{\equiv M^{III}\}} = \frac{k_{40}[X^-][H^+]}{k_{-40}[X^\bullet] + k_{38}} \qquad (41)$$

If both surface M^{III} and M^{II} reach protolytic equilibrium, the surface should be described as a mixed oxide [76,77]. An increase in the point of zero charge is expected if k_{38} is much smaller than the product $k_{-40}[X^\bullet]$, because M^{II} is as a rule more basic than M^{III}. The effect on the pzc may be large if the ratio in Eq. (41) is larger than 0.1. The expected effect depends on the ΔpK_a values for both surface ions. Conversely, a mixed-valence oxide such as magnetite, because of the lability of Fe^{II}, depending on the value of k_{38}, may rapidly evolve to a purely M^{III} surface.

In practice, and even in the absence of specific adsorption, the observed effects are more complex. For instance, the pzc of magnetite is 6.9 [78], and that of γ-Fe_2O_3 is 7.0. These values are appreciably lower than that of hematite, 8.5 [79]. Oxides that may undergo redox exchange are also subjected to related, and specific, structural changes. In this particular case, it has been postulated that maghemite possesses some H^+ in otherwise vacant Fe^{III} positions. Thus the acquisition of surface protonic charge by maghemite may involve the prior uptake of protons in lattice positions. Surface transformation from magnetite to maghemite leads to the same pH_0 value. As an example of the complexities of the acid–base phenomena in the interface of unstable oxides, we may discuss the pH dependence of the electrophoretic mobilities of ilmenite.

Figure 13 presents the pH dependence of the electrophoretic mobilities of ilmenite; the data were collected using fresh, untreated particles, and particles that had been leached in acid for 72 h at pH 2.0. The profile of the untreated particles may be understood as follows: in the alkaline range (pH > 7), the mobility behaves as expected for ilmenite, $FeTiO_3$, which is stable in these conditions. In the highly acidic media (pH < 3), the behavior approaches that of titanium dioxide; in the intermediate region (7 > pH > 3), the influence of iron leaching during the measurement is obvious. Although there is not actual proof that a stationary state has been reached, the leaching away of Fe^{2+} arrests the buildup of positive charge due to proton adsorption, and a wide plateau with a modest negative mobility is observed. The leached particles, on the other hand, resemble titanium dioxide in the alkaline range, as expected for the properties of the leached layer. This layer is however not very thick (ca. 10 nm according to chemical data and HRTEM observations), and the passivity is broken down in acid. The dissolving interface governs again the behavior in the range 5.5 > pH > 3. Complexing ligands influence the nature of the leached layer in different ways, depending on their affinity for Fe(II) and for Ti(IV). Thus, o-phenanthroline enhances the rate of Fe(II) dissolution but does not dissolve Ti(IV), e.g., the leached layer becomes thicker, and the surface properties shall approximate more to those of titanium dioxide. Catechol, on the other hand, dissolves some of the titanium dioxide of the leached layer, the thickness of which decreases accordingly.

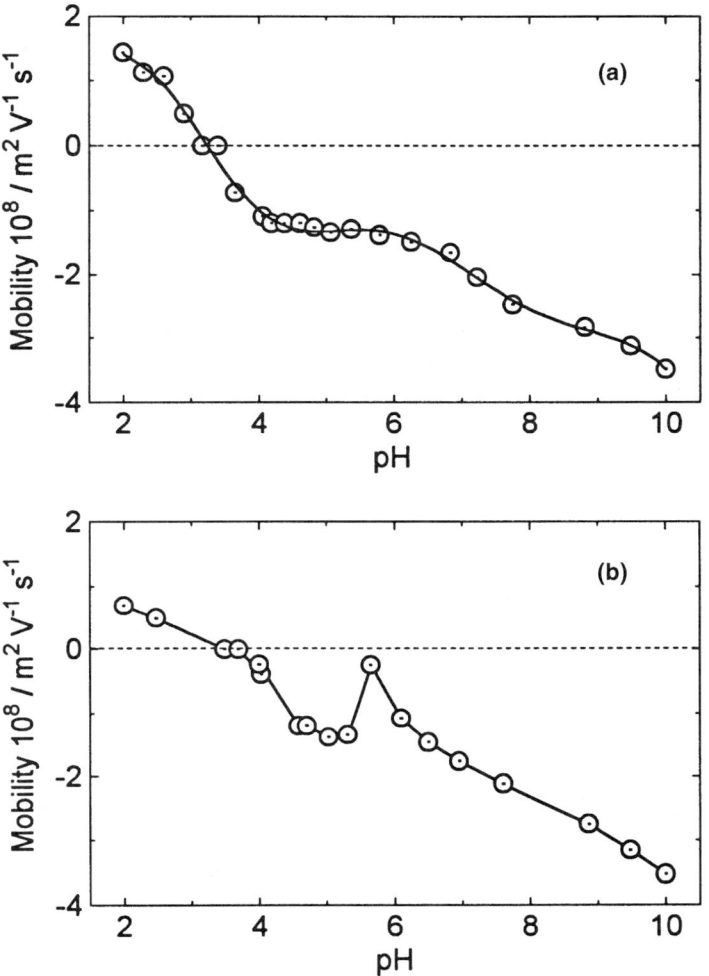

FIG. 13 pH dependence of the electrophoretic mobilities of ilmenite suspended in 0.01 mol dm^{-3} KCl solutions at 298 K. (a) untreated sample; (b) particles previously leached for 72 h at pH 2.0 and 298 K. Circles are experimental values; the lines have been drawn for visual aid only. (Data from Ref. 76.)

A system that has been much studied is that of titanium dioxide, both in mineral acids and in the presence of organic substrates capable of undergoing photocatalytic decomposition. The acid–base properties of titanium dioxide are well characterized [52], and the adsorption isotherms of catechol and salicylic acid (among other ligands) have been determined. In the case of catechol, the electrophoretic mobilities are also available.

Titanium dioxide is a very stable oxide in aqueous media, but in the presence of UV light it undergoes an electronic excitation that may lead to substantial surface changes; for instances, the color turns bluish. A reduced surface is created, and the impact of this phenomenon on the electrical properties of the interface is not known, even though it impacts strongly its photoelectrochemical behavior. In the presence of organic acids and an oxidant (for instance oxygen), illumination brings about a rather fast photocatalytic reaction, whereby the organic adsorbate is oxidized. There is no information available about the impact of these reactions on the double layer properties. The simplest assumption, that proton adsorption/desorption still determines the surface potential, certainly requires stronger experimental validation. Experiments in which electrophoretic mobilities are measured in situ, in the course of an oxidation reaction under light, would provide information important for the understanding of the chemistry of these dynamic interfaces.

ACKNOWLEDGMENTS

The work reported here involved a large number of students and associates, as illustrated in the references. The work was supported by CNEA, CONICET, ANPCYT, Fundación Antorchas, and CYTED. MAB and AER are members of CONICET. JAS acknowledges a fellowship granted by CONICET.

REFERENCES

1. M. A. Anderson and A. J. Rubin, eds., *Adsorption of Inorganics at Solid–Liquid Interfaces*, Ann Arbor Science, Ann Arbor, 1981.
2. M. F. Hochella, Jr., and A. F. White, eds., *Mineral–Water Interface Geochemistry*, Rev. Mineral. 23, Mineralogical Society of America, Washington, D.C., 1990.
3. M. A. Blesa, P. J. Morando, and A. E. Regazzoni, *Chemical Dissolution of Metal Oxides*, CRC Press, Boca Raton, FL, 1994.
4. M. A. Blesa, A. D. Weisz, P. J. Morando, J. A. Salfity, G. E. Magaz, and A. E. Regazzoni. Coord. Chem. Rev., in press.
5. A. E. Martell and R. M. Smith, *Critical Stability Constants, Volume 1: Amino Acids*, Plenum Press, New York, 1974.
6. A. E. Martell and R. M. Smith, *Critical Stability Constants, Volume 3: Other Organic Ligands*, Plenum Press, New York, 1977.
7. A. E. Martell and R. M. Smith, *Critical Stability Constants, Volume 5: First Supplement*, Plenum Press, New York, 1982.
8. A. E. Martell and R. M. Smith, *Critical Stability Constants, Volume 6: Second Supplement*, Plenum Press, New York, 1989.
9. J. Burgess, *Ions in Solution: Basic Principles of Chemical Interactions*, Ellis-Horwood, Chichester, U.K., 1988.
10. D. W. Margerum, G. R. Cayley, D. C. Weatherburn, and G. K. Pagenkopf, in *Coordination Chemistry, Volume 2* (A. E. Martell, ed.), ACS Monograph 174, ACS, Washington, D.C., 1978, pp. 1–220.
11. C. Ludwig, W. H. Casey, and P. A. Rock. Nature 375:44 (1995).
12. W. H. Casey and C. Ludwig, in *Chemical Weathering Rates of Silicate Minerals* (A. F. White and S. L. Brantley, eds.), Rev. Mineral. 31, Mineralogical Society of America, Washington, D.C., 1995, pp. 87–117.
13. C. Ludwig, J.-L. Devidal, and W. H. Casey. Geochim. Cosmochim. Acta 60:213 (1996).
14. D. S. Everhart and R. F. Evilia. Inorg. Chem. 14:2755 (1975).
15. B. Nowack, Ph.D. thesis, ETH Zürich Nr. 11392, 1996.
16. A. A. McConnell, R. H. Nuttall, and D. M. Stalker. Talanta 25:425 (1978).
17. M. D. Lind, B. Lee, and J. L. Hoard. J. Am. Chem. Soc. 87:1611 (1965).
18. E. E. Sileo, M. A. Blesa, G. Rigotti, B. E. Rivero, and E. E. Castellano. Polyhedron 15:4531 (1996).
19. E. E. Sileo, D. Vega, R. Baggio, M. T. Garland, and M. A. Blesa. Aust. J. Chem., 52:205 (1999).

20. C. Creutz and H. Taube. J. Am. Chem. Soc. *91*:3988 (1969).
21. N. Kallay and E. Matijević. Langmuir *1*:195 (1985).
22. R. Torres, N. Kallay, and E. Matijević. Langmuir *4*:706 (1988).
23. M. A. Blesa, E. B. Borghi, A. J. G. Maroto, and A. E. Regazzoni. J. Colloid Interface Sci. *98*:295 (1984).
24. B. Nowack and L. Sigg. J. Colloid Interface Sci. *177*:106 (1996).
25. D. A. Dzombak and F. M. M. Morel, *Surface Complexation Modeling: Hydrous Iron Oxides*, John Wiley, New York, 1990, pp. 306–315.
26. B. Nowack, T. Lutzenkirchen, P. Behra, and L. Sigg. Environ. Sci. Technol. *30*:2397 (1996).
27. M. C. Ballesteros, E. H. Rueda, and M. A. Blesa. J. Colloid Interface Sci. *201*:13 (1998).
28. A. E. Regazzoni, P. Mandelbaum, M. Matsuyoshi, S. Schiller, S. A. Bilmes, and M. A. Blesa. Langmuir *14*:868 (1998).
29. A. Manceau and L. Charlet. J. Colloid Interface Sci. *168*:87 (1994).
30. S. Fendorf, M. J. Eick, P. Grossl, and D. L. Sparks. Environ. Sci. Technol. *31*:315 (1997).
31. R. S. Ramakrishna, V. Paramasigamani, and M. Mahendran. Talanta *22*:523 (1975).
32. B. A. Borgias, S. R. Cooper, Y. B. Koh, and D. N. Raymond. Inorg. Chem. *23*:1009 (1984).
33. D. Vasudevan and A. T. Stone. Environ. Sci. Technol. *30*:1604 (1996).
34. D. Vasudevan and A. T. Stone. J. Colloid Interface Sci. *202*:1 (1998).
35. A. T. Stone. J. Colloid Interface Sci. *127*:429 (1989).
36. A. Torrents and A. T. Stone. Environ. Sci. Technol. *25*:143 (1991).
37. A. Torrents and A. T. Stone. Environ. Sci. Technol. *27*:1060 (1993).
38. A. Torrents and A. T. Stone. J. Soil Sci. Soc. Am. *58*:738 (1994).
39. A. T. Stone and A. Torrents, in *Environmental Impact of Soil Components Interactions* (P. M. Huang, ed.), Lewis, Chelsea, MI, 1995.
40. J. M. Smolen and A. T. Stone. Environ. Sci. Technol. *31*:164 (1997).
41. J. M. Smolen and A. T. Stone. J. Soil Sci. Soc. Am. *62*:636 (1998).
42. A. T. Stone, in *Perspectives in Environmental Chemistry* (D. L. Macalady, ed.), Oxford University Press, New York, 1998, pp. 75–93.
43. C. G. Pope, E. Matijević, and R. Patel. J. Colloid Interface Sci. *80*:74 (1981).
44. R. O. James and T. W. Healy. J. Colloid Interface Sci. *40*:42 (1972).
45. C. H. Weng, J. H. Wang, and C. P. Huang. Wat. Sci. Tech. *35*:55 (1997).
46. N. W. Duffy, K. D. Dobson, K. C. Gordon, B. H. Robinson, and A. J. McQuillan. Chem. Phys. Lett. *266*:451 (1997).
47. S. J. Hug and B. Sulzberger. Langmuir *10*:3587 (1994).
48. L. A. García Rodenas, A. M. Iglesias, A. D. Weisz, P. J. Morando, and M. A. Blesa. Inorg. Chem. *36*:6423 (1997).
49. A. D. Weisz and M. A. Blesa, work in course.
50. M. V. Biber and W. Stumm. Environ. Sci. Technol. *28*:763 (1994).
51. S. T. Martin, J. M. Kesselman, D. S. Park, N. S. Lewis, and M. Hoffman. Environ. Sci. Technol. *30*:2535 (1996).
52. R. Rodríguez, M. A. Blesa, and A. E. Regazzoni. J. Colloid Interface Sci. *177*:122 (1996).
53. A. T. Stone and J. J. Morgan. Environ. Sci. Technol. *18*:617 (1984).
54. J. S. LaKind and A. T. Stone. Geochim. Cosmochim. Acta *53*:961 (1989).
55. R. Kummert and W. Stumm. J. Colloid Interface Sci. *75*:373 (1980).
56. J. A. Davis and J. O. Leckie. Environ. Sci. Technol. *12*:1309 (1978).
57. M. J. Schmelz, T. Miyasana, S. Mizushima, T. H. Lane, and J. V. Quagliano. Spectrochim. Acta *9*:51 (1957).
58. A. E. Regazzoni and M. A. Blesa. Langmuir *7*:473 (1991).
59. Y. G. Bérubé and P. L. de Bruyn. J. Colloid Interface Sci. *27*:305 (1968).
60. D. A. Vermilyea. J. Electrochem. Soc. *113*:1067 (1966).
61. R. D. Astumian, M. Sasaki, T. Yasunaga, and Z. A. Schelly. J. Phys. Chem. *85*:3832 (1981).
62. P. R. Grossl, M. Eick, D. L. Sparks, S. Goldberg, and C. C. Ainsworth. Environ. Sci. Technol. *31*:321 (1997).

63. C. M. Eggleston, S. J. Hug, W. Stumm, B. Sulzberger, and M. dos Santos Afonso. Geochim. Cosmochim. Acta *62*:585 (1998).
64. J. W. Diggle, in *Oxides and Oxide Films*, Volume 2 (J. W. Diggle, ed.), Marcel Dekker, New York, 1973, Ch. 4.
65. V. I. E. Bruyère, P. J. Morando, and M. A. Blesa. J. Colloid Interface Sci. *209*:207 (1999).
66. C. F. Jones, R. L. Segall, R. St. C. Smart, and P. S. Turner. Proc. Roy. Soc. London *A374*:141 (1981).
67. M. A. Blesa, H. A. Marinovich, E. C. Baumgartner, and A. J. G. Maroto. Inorg. Chem. *26*:3713 (1987).
68. M. G. Segall and R. M. Sellers, in *Advances in Inorganic and Bioinorganic Mechanisms*, Vol. 3 (A. G. Sykes, ed.), Academic Press, London, 1984, p. 97.
69. R. Torres, M. A. Blesa, and E. Matijević. J. Colloid Interface Sci. *131*:567 (1989).
70. R. Torres, M. A. Blesa, and E. Matijević. J. Colloid Interface Sci. *134*:475 (1990).
71. M. del V. Hidalgo, N. E. Katz, A. J. G. Maroto, and M. A. Blesa. J. Chem. Soc. Faraday Trans. I *84*:9 (1988).
72. E. B. Borghi, A. E. Regazzoni, A. J. G. Maroto, and M. A. Blesa. J. Colloid Interface Sci. *130*:299 (1989).
73. E. B. Borghi, P. J. Morando, and M. A. Blesa. Langmuir *7*:1652 (1991).
74. P. W. Schindler, in *Mineral–Water Interface Geochemistry*, Rev. Mineral. 23 (M. F. Hochella, Jr., and A. F. White, eds.), Mineralogical Society of America, Washington, D.C., 1990, Ch. 7.
75. M. A. Blesa, V. I. E. Bruyère, A. J. G. Maroto, and A. E. Regazzoni. Anales Asoc. Quím. Arg. *73*:39 (1985).
76. M. A. Blesa, G. Magaz, J. A. Salfity, and A. D. Weisz. Solid State Ionics *101–103*:1235 (1997).
77. G. Magaz, L. A. García Rodenas, P. J. Morando, and M. A. Blesa. Croatica Chem. Acta *71*:917 (1998).
78. A. E. Regazzoni, M. A. Blesa, and A. J. G. Maroto. J. Colloid Interface Sci. *91*:560 (1983).
79. A. E. Regazzoni, M. A. Blesa, and A. J. G. Maroto. J. Colloid Interface Sci. *122*:315 (1988).

15

Ion Transport and Electrical Conductivity in Heterogeneous Systems: The Case of Microemulsions

FEDERICO BORDI Sezione di Fisica Medica, Dipartimento di Medicina Interna, Università di Roma "Tor Vergata," and Istituto Nazionale Fisica della Materia, Unità di Roma 1., Rome, Italy

CESARE CAMETTI Dipartimento di Fisica, Università di Roma "La Sapienza," and Istituto Nazionale Fisica della Materia, Unità di Roma 1., Rome, Italy

I.	Introduction	541
II.	Electrical Conductivity of Typical Colloidal Aqueous Suspensions	543
III.	Electrical Conductivity of Typical Microemulsion Systems	546
IV.	Conclusions	561
	References	562

I. INTRODUCTION

Of the many physical or physicochemical methods that can be used to study colloidal systems, including heterogeneous systems, solid–liquid dispersions, complex fluids, and polyelectrolyte and macromolecular solutions, those involving electrical properties have been prominent. The origin of this specificity lies in the existence of charges at the interfaces disjoining the different media of the dispersed system that lead to intra- and intermolecular interactions that may be stronger and of much longer range than in the case of uncharged interfaces, determining the behavior, at the macroscopic level, of the whole system.

The presence of electrified interfaces as a common feature of dispersed systems influences the thermodynamic, dynamic, and structural properties of the whole system, particularly if the electrostatic potential arising from these charge distributions is not sufficiently screened, as generally occurs in aqueous colloidal dispersions.

Much interest has been focused on electrical conductivity measurements both because of the need to have a fundamental understanding of the ion transport in these systems and because the data carry information about the temporal behavior of the processes (relaxation phenomena) that polarize the electrical double layer, at the interface.

Electrical conductivity (ionic conductivity), being linked to the movement of charges in response to an external electric field, can be broken essentially into a product of three terms: the charge ze of the carriers, the concentration c of the charge carriers, and their mobility u, as the ratio of the average velocity to an applied electric field of unit strength, so that the total effect is obtained as the sum of contributions due to all the carriers in the sample, according to the expression

$$\sigma = \sum_i (|z_i|e) c_i u_i \tag{1}$$

At least in principle, each of these quantities, once appropriately measured, may carry information on the structural and dynamical properties of the system.

In the present account, we will not discuss in detail the phenomenology of the conductive behavior of aqueous colloidal dispersion nor present the numerous theoretical attempts to describe the electrokinetic processes governing the electrical properties of colloidal particles in electrolyte solutions. Instead we shall give a brief review that incorporates the recent development and gives illustrations of representative conductometric properties for a selected heterogeneous system, undergoing a large variety of different structural arrangements. Here, we restrict our discussion to microemulsions (particularly water-in-oil microemulsions) defined as isotropic and homogeneous systems on the macroscopic scale but heterogeneous on the microscopic scale.

Given the large variety of charged interface arrangements occurring in these systems for different temperatures and compositions, microemulsions represent a typical model system where electrical conductivity measurements (more generally frequency dependent dielectric measurements) can be used to investigate electrokinetic properties, especially those relating to electrophoretic mobility and to time dependent relaxation phenomena, in heterogeneous systems.

One of the questions that assumes relevance refers to how changes in the structural arrangements of the system (induced by composition and/or temperature) influence the ionic transport properties (electrical conductivity of the whole system), passing from a "bulk" conductance in diluted systems to a "surface" conductance in long-range correlated systems.

It is noteworthy, however, that the electrical conductivity behavior of typical microemulsion systems (of the type of water-in-oil), both in the single-dispersed droplet regime and in the clusterized and/or organized regime, can be interpreted, at least to a first approximation, by neglecting the polarization of the electrical double layer at the interface. This situation is the opposite to that which occurs in charged colloidal particles dispersed in an aqueous electrolyte solution, where the conductivity behavior strongly depends on the ionic atmosphere distribution at the interface.

The review is organized as follows. Section II will briefly summarize the theoretical treatments that describe the conducting behavior of a dispersed system both from a macroscopic and from a microscopic point of view. After outlining the feature of the average current density in dilute suspensions of identical charged particles, we present the fundamental electrokinetic equations that govern the distribution of ionic species outside a particle migrating in an unbounded solution when an external electric field is applied. In Section III, the conductivity behaviors of a typical water-in-oil microemulsion system are presented, and the importance of different transport mechanisms driven by different charged carriers in the different structures is discussed on the basis of some representative experimental results.

II. ELECTRICAL CONDUCTIVITY OF TYPICAL COLLOIDAL AQUEOUS SUSPENSIONS

Schematically, two different approaches have been developed, a macroscopic approach dealing with the bulk electrical properties of the different media of the dispersed system in the light of the effective medium approximation, and a microscopic approach in which a set of constitutive equations describes the behavior of each charged carrier, and the whole conductivity results by solving, as a small perturbation of the equilibrium, this set of equations subjected to the appropriate boundary conditions, to account for the interface structure.

When a uniform electric field is applied to charged particles suspended in a nonconducting continuous medium or in an electrolyte solution containing n ionic species, the basic equations connecting the macroscopic quantities accessible to the experiment concern the volume-averaged current density

$$<\vec{J}> = \frac{1}{V} \int_V \vec{J}(\vec{r}) \, dv \qquad (2)$$

and the measured electric field

$$\vec{E} = -\frac{1}{V} \int_V \nabla \psi(\vec{r}) \, dv \qquad (3)$$

where $\psi(\vec{r})$ is the electrical potential at position $\vec{r}$ and V is a sufficiently large volume of the system. The usual experimental definition of the electrical conductivity σ is given by the linear relationship.

$$<\vec{J}> = \sigma \vec{E} \qquad (4)$$

valid for statistically homogeneous suspensions, ignoring all boundary effects.

As far as the macroscopic approach is concerned, it is essentially based on the solution of the Laplace equation

$$\nabla^2 \psi(\vec{r}) = 0 \qquad (5)$$

in the far-field condition, which yields the dipolar factor of the colloidal particle that, in turn, can be linked, in the limit of small concentration, to the dielectric and conductometric properties of the whole system by means of the Clausius–Mossotti equation or some other relationships more or less sophisticated, typical of the effective medium approximation [1,2]. Along these lines, a number of successful theories have been elaborated, and a satisfactory agreement with the experimental results has been obtained, regarding both the dielectric and the conductivity properties of a large variety of colloidal systems.

On the other hand, when the interfacial properties of the dispersed particles are important to the characterization of the system (highly charged colloidal particles in electrolyte solution), i.e., when the counterion distribution extends far from the particle surface, the system cannot be further modeled as a heterogeneous system composed of two different media with two well-defined electrical properties. From a formal point of view, this implies that the Laplace equation should be replaced by the Poisson equation

$$\nabla^2 \psi(\vec{r}) = -\frac{4\pi}{\varepsilon} \rho(\vec{r}) \qquad (6)$$

which considers the presence of an ion distribution $\rho(\vec{r})$ in the bulk of each medium. For example, in a microemulsion system, in particular a water-in-oil microemulsion, the ion distribution interests essentially the water core, so that the problem is reduced to the solution, around each water droplet, of this couple of equations,

$$\nabla^2 \psi(\vec{r}) = 0 \qquad r \geq a \tag{7}$$

and

$$\nabla^2 \psi(\vec{r}) = -\frac{4\pi}{\varepsilon} \rho(\vec{r}) \qquad r \leq a \tag{8}$$

where a is a parameter that defines the geometry of the disperse phase and ε the dielectric constant of the continuous medium. In the case of a spherical droplet, a is its radius. Similar equations hold in the case of oil-in-water microemulsions or in general in the case of dielectric particles dispersed in an electrolyte solution.

The problem is more complicated when the overall behavior of the system is heavily due to the interactions between individual aggregates, resulting in the existence of a new phenomenology (i.e., percolation, gelation) governed by scaling laws. In this case, the detailed description of the single mechanism response for the conduction is no longer possible, and a different approach based on an average behavior of the resulting structures must be used. A typical example occurs in water-in-oil microemulsion systems close to percolation. In this case, on increasing the volume fraction, owing to a short-range attractive potential, the water droplets approach each other, causing the formation of large aggregates (clusters), the typical dimension of which diverges at the percolation threshold. The dielectric properties and the conduction mechanisms cannot be treated in detail due to the non trivial geometrical configuration assumed by the system.

In the microscopic approach, the behavior of the system is generally described by a current density $\vec{J}$ written as

$$\vec{J} = \sum_i z_i e \left(N_i \vec{u} - N_i \frac{D_i}{k_B T} \nabla \mu_i \right) \tag{9}$$

where $z_i e$ and N_i are the charge and concentration of ions of species i, $\vec{u}$ the fluid velocity field, D_i the diffusion coefficient of ions of species i, $k_B T$ the thermal energy, and μ_i the electrochemical potential given by

$$\mu_i = \mu_i^0 + k_B T \ln(N_i) + z_i e \psi \tag{10}$$

Eqs. (9) and (10), which govern the fundamental electrokinetic phenomena in electrical transport processes, take into account the convection of the ionic species by the fluids, the diffusion and the electric field induced migration of ions.

The electrical conductivity of a colloidal suspension made up of dispersed nonconducting particles distributed in an aqueous electrolyte solution is influenced by two different effects, i.e., the decrease in conductivity due to the presence of nonconducting particles and the increase in conductivity due to the surface conductivity, arising from the polarization of the ionic atmosphere at the particle–water interface.

According to the standard description, each colloidal particle is surrounded by an ionic diffuse layer whose charge distribution is governed by the Poisson–Boltzmann equation. Application of an external electric field modifies the equilibrium distribution inducing a different balance between hydrodynamic and electrical forces. This situation is generally treated as a small perturbation of the equilibrium state. A general expression for the electrical conductivity of a dilute suspension of spherical colloidal dielectric particles in a z-valent electrolyte solution has been derived by De Lacey and White [3], following a previous work by O'Brien [4]. The total increment in the electrical conductivity, in the weak perturbation limit, has been found by solving the fundamental equations (electrokinetic equations) here:

Charge distribution (Poisson equation)

$$\nabla^2 \psi(\vec{r}, t) = -\frac{4\pi}{\varepsilon} \rho(\vec{r}) = -\frac{4\pi}{\varepsilon} \sum_{i=1}^{N} z_i e n_i(\vec{r}, t) \qquad (11)$$

Conservation of momentum

$$\eta \nabla^2 u - \nabla p - \rho \nabla \psi = \rho \frac{\partial \vec{u}}{\partial t} \qquad (12)$$

Flux of ith ion species (Nernst–Planck equation)

$$n_i(\vec{v}_i - \vec{u}) = -\Lambda_i (e z_i n_i \nabla \psi + k_B T \nabla n_i) \qquad (13)$$

Conservation of ion species

$$\nabla(n_i \vec{v}_i) = \frac{\partial n_i}{\partial t} \qquad (14)$$

which connect the quantities that allow a complete description of the system, i.e., the electrical potential $\psi(\vec{r}, t)$, the concentration $n_i(\vec{r}, t)$ and the drift velocity $v_i(\vec{r}, t)$ of the ionic species, and the velocity $u_i(\vec{r}, t)$ and the pressure $p_i(\vec{r}, t)$ of the medium at every position $\vec{r}$ and time t in the system. η is the viscosity of the solvent medium and the Λ_i the ionic mobilities. The electrokinetic equations, i.e., the Poisson equation for electrostatics, the Nernst–Planck equations for the ionic transport, the ionic continuity equations, and the Navier–Stokes equations for solvent transport, have been put forward in order to describe every electrokinetic phenomenon characterized by tangential motion of liquid with respect to an adjacent charged surface.

In the presence of an external electric field, the above quantities are perturbed about their static values

$$u(\vec{r}, t) = u(\vec{r}) \exp(-i\omega t) \qquad (15)$$
$$p(\vec{r}, t) = p^0(\vec{r}) + p(\vec{r}) \exp(-i\omega t) \qquad (16)$$
$$v_i(\vec{r}, t) = v_i(\vec{r}) \exp(-i\omega t) \qquad (17)$$
$$n_i(\vec{r}, t) = n_i^0(\vec{r}) \exp(-i\omega t) \qquad (18)$$
$$\psi(\vec{r}, t) = \psi^0(\vec{r}) + \delta\psi(\vec{r}) \exp(-i\omega t) \qquad (19)$$

Substitution of these expressions into the fundamental equations results in corresponding equations for the perturbation terms. The resulting coupled equations were solved by numerical methods, and the real and imaginary increments in the dielectric constant and electrical conductivity were obtained. The electrical conductivity response of the system is then expanded in powers of the volume fraction ϕ, viz.

$$\sigma = \sigma_0 (1 + \Delta\sigma\phi) + 0(\phi^2) \qquad (20)$$

and the magnitude of the conductivity increment, depending on the ionic strength of the electrolyte solution, the valency of the ionic species, and the ζ-potential, was observed to compare favorably with the experimental findings. De Lacey and White [3] present a solution of the electrokinetic equations in different conditions, and the details of the numerical procedure can be found in their paper.

More recently, using a similar analysis, Liu and Keh [5,6] determined the effective conductivity of a dilute suspension of charged porous particles or composite particles (porous particles surrounded by a porous shell) and dispersed in an electrolyte solution.

These theoretical approaches are strongly dependent on the way of formulating the surface boundary conditions and can be extended to include a surface transport process through the ion migration within an extended region (Stern layer) between the shear envelope and the rigid surface of the colloidal particle [7]. The improved agreement between experiment and theory, once this is done, supports the view that surface conductive processes play an important role in the electrokinetic behavior of a large variety of colloidal systems.

III. ELECTRICAL CONDUCTIVITY OF TYPICAL MICROEMULSION SYSTEMS

In this section, we survey some examples of the application of electrical conductivity methods to characterize the dynamics of processes occurring in microemulsions, considered as typical of highly heterogenous systems. A microemulsion is a macroscopically single-phase fluid, thermodynamically stable, composed of water and oil separated by a monomolecular layer of amphiphilic molecules. These systems have attracted attention and have been intensively investigated not only because of their practical interest in applications as liquid agents in technical processes (oil recovery, biotechnology, drug carriers) [8] but also because of their aggregation patterns, which span a wide variety of morphologies and thus make them very interesting model systems to study complex fluids and in particular short-range organized fluids, showing different behaviors, namely critical miscibility points, loci of percolation points, lamellar phases, and bicontinuous phases.

Here we will refer in particular to an ionic microemulsion system that has been analyzed in a series of investigations by various techniques such as ultracentrifugation [9], static and dynamic light scattering [10–12], small-angle neutron scattering [13], small-angle x-ray scattering [14], and time-resolved luminescence measurements [15].

Water-in-oil microemulsions consist of nanometer-sized spherical droplets with water in the central core surrounded by a layer of surfactant molecules having their hydrophilic head groups facing the water and their hydrophobic tails oriented towards the continuous oil phase. Ionic surfactants [a typical example is sodium bis(2-ethylhexyl)sulfosuccinate (AOT), a two alkyl chain surfactant with a relatively small hydrophilic head group and therefore appropriate for the formation of inverted micelles] can dissociate into anions containing negatively charged head groups SO_3^- at the water–oil interface and positive Na^+ counterions distributed in the water core.

The phase diagram at a constant water-to-surfactant molar ratio $W = [H_2O]/[AOT] = 40.8$ is shown in Fig. 1. This diagram is characterized by a region of the coexistence of liquid microemulsion droplet phases bound by the binodal line. For this microemulsion [16–19], a critical point exists at a temperature $T \approx 40°C$ and a fractional volume $\phi \approx 0.08$, and from the vicinity of that point a percolation line extends up to very high volume fractions ($\phi > 0.80$). In the present system, starting from low to high temperature, the microemulsion, initially composed of inverted micelles (water-in-oil

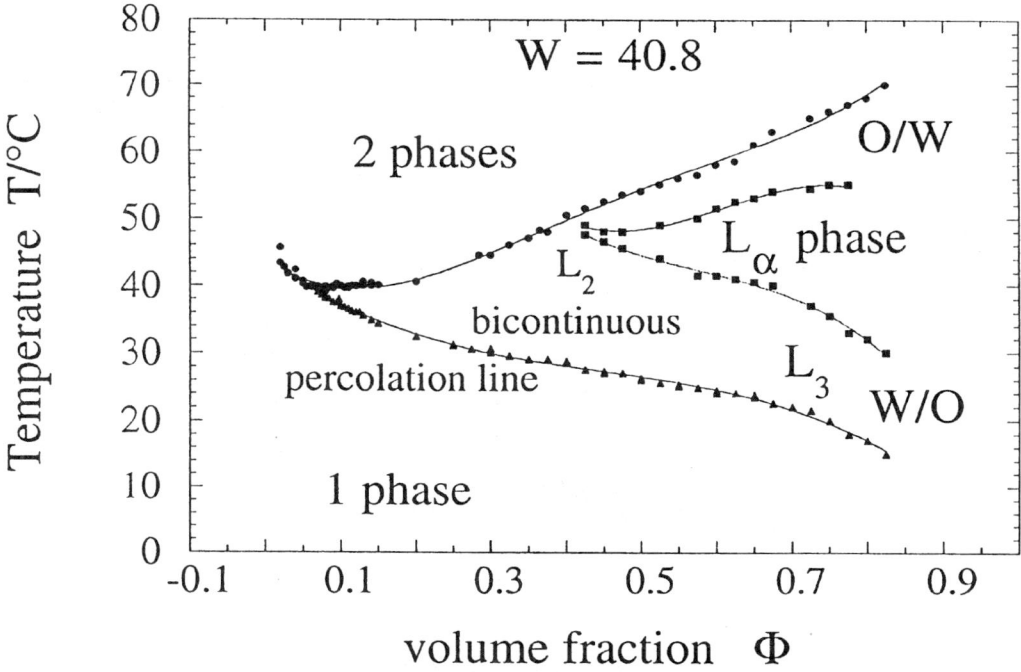

FIG. 1 The phase diagram of water-AOT-n-decane microemulsion at water-to-surfactant molar ratio $W = [H_2O]/[AOT] = 40.8$. In the one-phase region, the system experiences different structural arrangements, depending on temperature and composition. As the temperature is increased, the system undergoes a progressive change from a water-in-oil to an oil-in-water phase, passing through a region where a planar lamellar phase occurs. At higher temperatures, a phase transition takes place.

microemulsion), undergoes a percolation transition, followed by the formation of a bicontinuous structure that evolves, passing through a lamellar phase, from water-in-oil to oil-in-water aggregates, up to the phase transition, at higher temperatures.

We will consider in detail two regions of the phase diagram, (1) the region of low and intermediate water concentration, where the microemulsion consists of surfactant-coated water droplets distributed in an oil phase that undergo Brownian motion, and (2) a region, at higher concentration, where a long-range organized structure, described as a lamellar phase, appears. For both these regions, we will review the current theories developed to take into account the transport process (of ionic nature) occurring in dependence or the different structural organizations of the microemulsion.

The typical electrical conductivity behavior of a water-in-oil microemulsion is shown in Fig. 2 as a function of temperature, for different compositions (volume fractions ϕ of water droplets from 0.30 to 0.60). The logarithm of the microemulsion conductivity exhibits the characteristic sigmoidal behavior as the temperature is increased. This increase of conductivity over more than four orders of magnitude is indicative of percolation. The percolation temperature T_p is defined as the point at which a dramatic change is slope in the conductivity as a function of temperature occurs (as shown in the inset of Fig. 3, where the maximum in the $d/dT[\ln(\sigma)]$ vs. T indicates the percolation temperature).

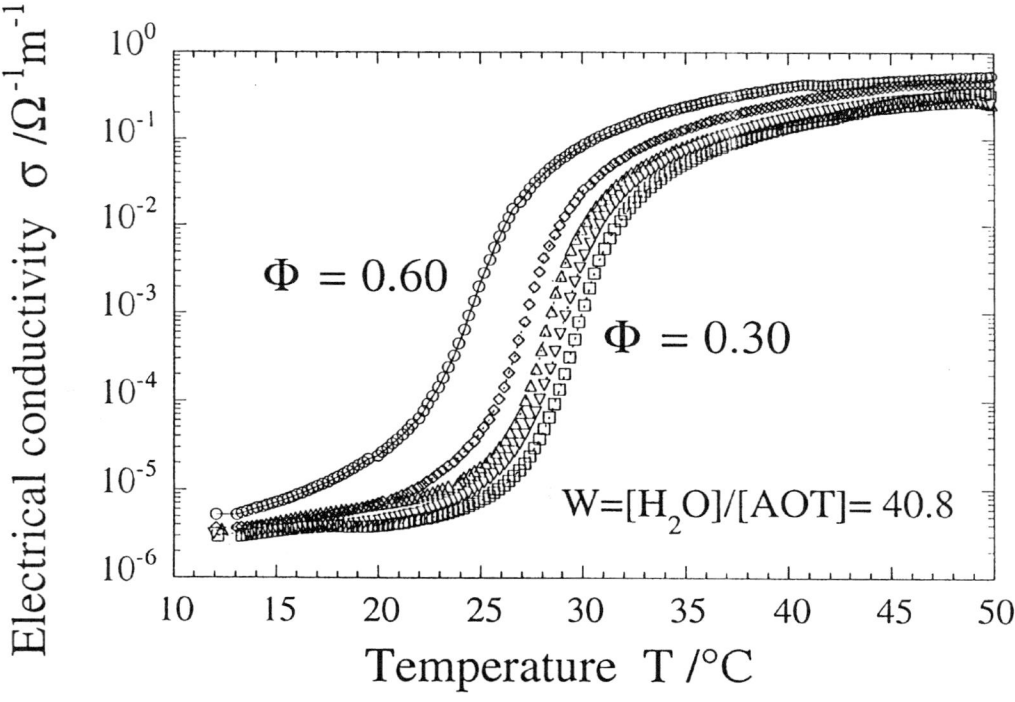

FIG. 2 The electrical conductivity σ (in a semilog plot) of water-AOT-n-decane microemulsion as a function of temperature for different values of the fractional volume ϕ, from $\phi = 0.30$ to $\phi = 0.60$, at the water-to-surfactant molar ratio $W = 40.8$. The curves display the typical sigmoidal behavior, indicating the occurrence of percolation.

A strong correlation exists between a transport property such as electrical conductivity and the different morphologies that microemulsions undergo, depending on composition and temperature. That this type of correlation indeed exists is convincingly exemplified by the observed behavior illustrated in Fig. 3, where the electrical conductivity is shown as a function of temperature, for the particular system water-AOT-n-decane at the fractional volume $\phi = 0.60$ and water-to-surfactant molar ratio $W = [H_2O]/[AOT] = 40.8$. As can be seen, for each structural change occurring in the system, a corresponding change in the electrical conductivity can be evidenced. In Fig. 3, the system, starting from low to high temperatures, undergoes a percolation transition (at the temperature T_p) followed by the formation of a bicontinuous structure that evolves, passing through a lamellar phase, from water-in-oil to oil-in-water structure, up to the phase separation (at a temperature higher than $T = 55°C$).

There are, essentially, two different conducting regimes, below and above the percolation threshold, based on different mechanisms, at the microscopic level. Well below percolation, at low temperatures and low volume fractions, the electrical conductivity (of the order of 10^{-6} Ω^{-1} m^{-1} but several times higher than the oil phase, 10^{-8}–10^{-10} Ω^{-1} m^{-1}) has been explained by the migration of statistically charged water droplets, in terms of the charge fluctuation model. This process, driven by spontaneous thermal fluctuations, provides exchange of the water droplet content when droplets

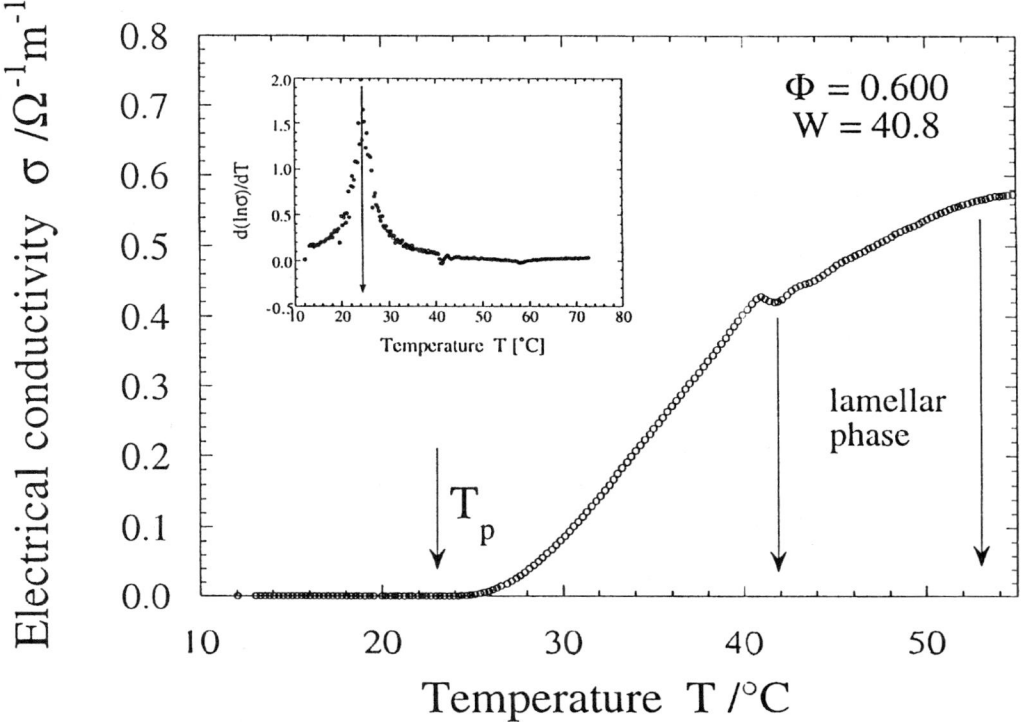

FIG. 3 A typical conductivity behavior (in linear plot) of water-AOT-n-decane microemulsion as a function of temperature at a volume fraction $\phi = 0.60$ and water-to-surfactant molar ratio $W = 40.8$. Deviations occur at the percolation temperature T_p (the inset shows the maximum of $d/dT \ln(\sigma)$ vs. T) and at the transition from bicontinuous water-in-oil structure to lamellar phase (the temperature interval of the lamellar phase is marked by arrows).

approach each other, fuse to form a short-lived droplet dimer, and then redisperse. At higher temperatures and higher volume fractions, close to percolation, water droplets come in close contact, and charge carriers (surfactant anions or surfactant counterions) propagate by hopping [20] between droplets or move [21,22] within transient water channels, arising in droplet clusters upon opening of the surfactant monolayer.

The two mechanisms deeply differ and involve different structural entities, the first being related to the dynamics of the single water droplet (or to small-size water clusters interacting via short-range attractive potential), the latter (heating induced or composition induced percolation) associated with a cooperative phenomenon induced by the deformation of the surfactant interface upon clustering.

We will review and discuss in detail the electrical conductivity behavior of a particular microemulsion (but the results can be generalized to a generic ionic, water-in-oil microemulsion) in both the two above regimes.

The general conduction mechanism of the charge fluctuation model is described in terms of exchange between the water cores of two adjacent droplets on the basis of two elementary steps associated with the temporary formation of a transient aggregate (coalescence process) and its subsequent breakdown into two separate droplets (decoalescence process), according to the reaction

$$n_0 + n_0 \xrightleftharpoons[k_r]{k_d, k_r} (2n_0) \xrightleftharpoons[k_{-1}]{k_1, k_{-1}} <n^+> + <n^-> \qquad (21)$$

where the rate constants k_d, k_r and k_1, k_{-1} refer to the dimer formation and to those dissociations that are effective in the charging process. Eicke and coworkers [23] derived for the reduced conductivity σ/ϕ (conductivity normalized to the fractional volume ϕ of the dispersed water droplets) the equation

$$\frac{\sigma}{\phi} = \frac{\varepsilon_0 \varepsilon k_B T}{2\pi \eta R_d^3} \qquad (22)$$

where ε and η are the permittivity and the viscosity of the oil phase, $k_B T$ the thermal energy, R_d the hydrodynamic radius of the surfactant-coated water droplets, and ε_0 the dielectric constant of free space. Along the same derivation, Kallay and Chittofrati [24] distinguish between the Born radius R_c, which corresponds to the water core where counterions are confined, and the Stokes radius R_d, which defines the size of the moving droplet. These radii differ by the thickness δ of the surfactant tail, $R_d = R_c + \delta$. As a consequence, Eq. (22) is replaced by

$$\frac{\sigma}{\phi} = \frac{\varepsilon_0 \varepsilon k_B T (R_d - \delta)}{2\pi \eta R_d^4} \qquad (23)$$

which predicts the conductivity maximum as a function of the droplet size, as observed experimentally. Figure 4 shows the reduced conductivity σ/ϕ as a function of the hydrodynamic droplet radius R_d for a particular microemulsion composed of water, AOT, and isooctanol. The experimental data, taken from Ref. 23, are compared with the predicted values according to the charge fluctuation model [Eq. (23)]. As can be seen, the agreement is rather satisfactory, especially because the model does not contain any adjustable parameter. Moreover, an analysis based on the Boltzmann statistics [25] confirms the validity of the fluctuation model in describing the conductometric behavior of these systems.

The charge fluctuation model has been further improved by Hall [26], who proposed a more general expression that includes Eq. (22) as a limiting case. This expression reads

$$\frac{\sigma}{\phi} = \frac{e^2}{8\pi^2 \eta R_d^4} \frac{\sum_{z=-\infty}^{z=\infty} z^2 \exp(-z^2 e^2 / 8\pi \varepsilon_0 \varepsilon k_B T R_d)}{\sum_{z=-\infty}^{z=\infty} \exp(-z^2 e^2 / 8\pi \varepsilon_0 \varepsilon k_B T R_d)} \qquad (24)$$

Equation (24) passes through a maximum and then decreases as the inverse droplet volume. In general, only small deviations from electroneutrality are expected in order to justify the observed behavior of the reduced conductivity in a one-phase region of the phase diagram, well below the percolation threshold. A comparison between the predicted conductivity [Eq. (24)] and the observed values in the case of water-AOT-n-decane microemulsion with 10^{-2} M NaCl and without salt addition is shown in Fig. 5 as a function of the fractional volume ϕ. A good agreement is achieved with $z = \pm 1$ (deviation from electroneutrality is confined to a single elementary charge) and with a hydrodynamic radius of $R_d = 80$ Å and $R_d = 60$ Å in the absence and in the presence of added salt, respectively. As the fractional volume is increased, when a clusterization process takes place, deviation from a linear dependence of the conductivity occurs, and a different transport mechanism prevails.

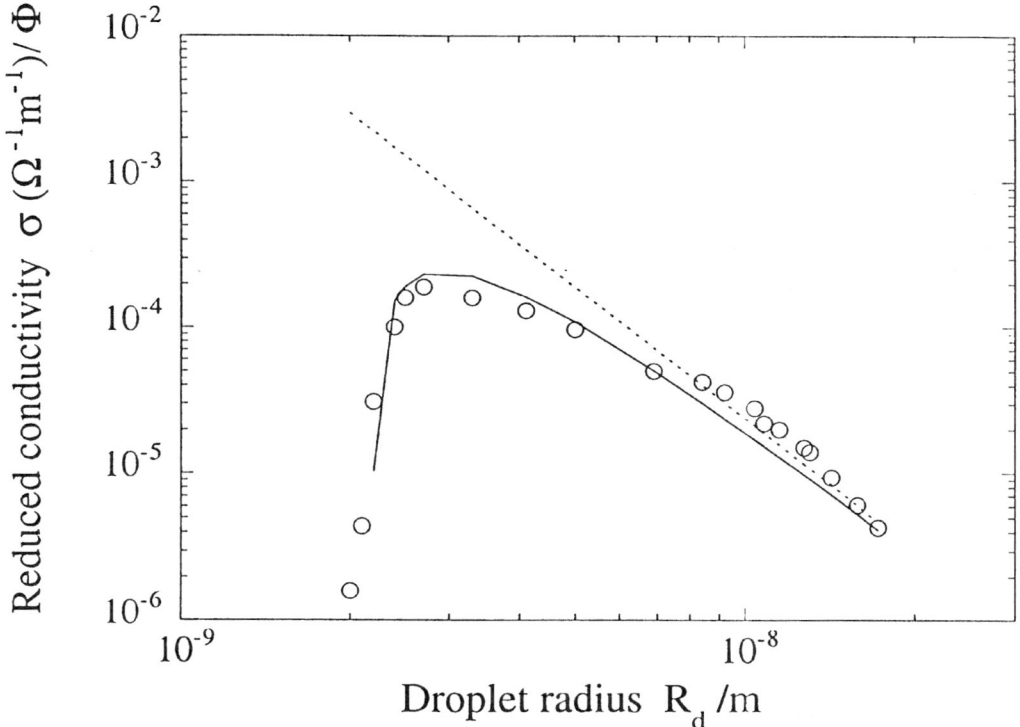

FIG. 4 The reduced conductivity σ/ϕ of a dilute water-AOT-isooctane microemulsion as a function of the radius of the water droplets as determined by dynamic light scattering. Experimental data were taken from Ref. 23. The dashed line was calculated on the basis of the original charge fluctuation model [Eq. (22)] and the full line according to Eq. (23), distinguishing between the hydrodynamic droplet radius R_d and the radius of the water core ($R_c = R_d - \delta$). The following values were used: $T = 298$ K, $\varepsilon = 1.96$, $\eta = 0.47 cP$, $\delta = 2.1$ nm.

A further refinement of the charge fluctuation model, based on a more rigorous thermodynamic approach, has been developed by Halle and Bjorling [27,28] by introducing a further contribution to the chemical potential deriving from short-range droplet interactions and by describing the microemulsion as a "conventional" electrolyte on the basis of the restricted primitive model [29]. Within this framework, the distribution of the net droplet charges is given by

$$P(z) = \frac{\exp(-z^2\alpha)}{\Sigma_z \exp(-z^2\alpha)} \qquad (25)$$

where the coupling parameter α can be written as [30]

$$\alpha = \frac{l_B}{2}\left(\frac{1}{R_c} - \frac{a}{R_d}\right) - \lambda \qquad (26)$$

with l_B the Bjerrum length and a and λ two parameters depending on the Coulombic interactions and on short-range interactions, respectively. The final expression for the

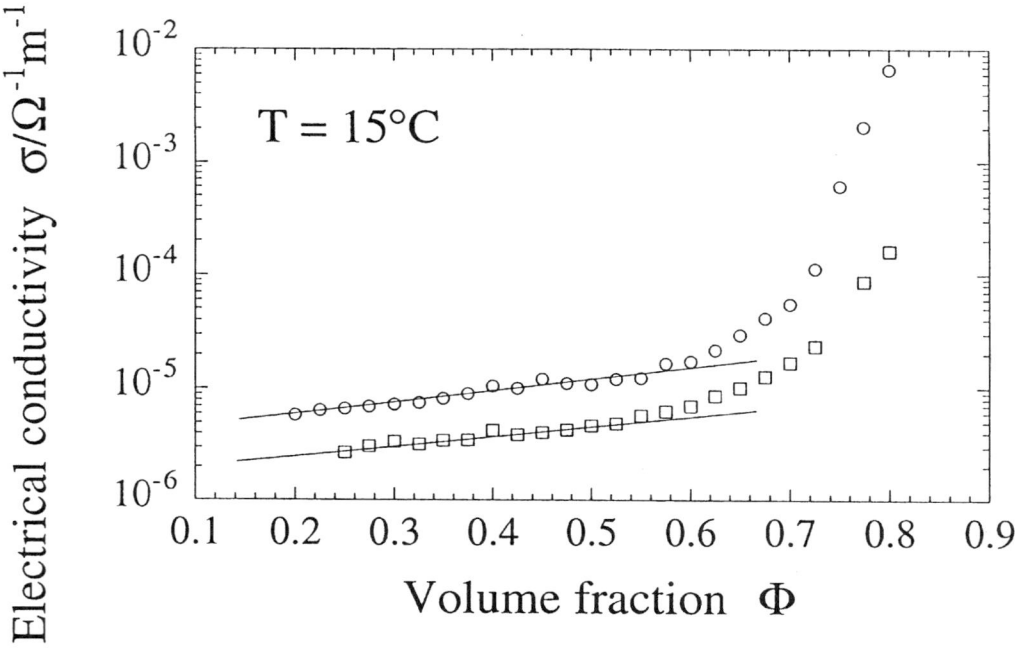

FIG. 5 The electrical conductivity σ as a function of the fractional volume ϕ for a water-AOT-decane microemulsion at the temperature of $T = 15°C$ in the presence of added salt (10^{-2} M NaCl) (○) and in absence of added salt (□). Full lines are the calculated values according to Eq. (24) on the basis of the charge fluctuation model. Deviation from a linear dependence occurs at higher volume fractions, where the percolation threshold is approached.

electrical conductivity (limited to a single elementary charge $z = \pm 1$) is given by

$$\frac{\sigma}{\phi} = \frac{e^2}{8\pi\eta R_d^4} \frac{2\exp[-(e^2/8\pi\varepsilon_0\varepsilon k_B T)(1/R_c - a/R_d) + \lambda]}{1 + 2\exp[-(e^2/8\pi\varepsilon_0\varepsilon k_B T)(1/R_c - a/R_d) + \lambda]} \qquad (27)$$

The effect of the distribution of counterions within the water droplets and their association with surfactant charged groups has been also considered by Tomić and Kallay [31] on the basis of the Poisson–Boltzmann equation. Their analysis confirms that the Born energy, except for extremely strong association, is a good approximation of the energy contribution responsible for charging the water core. The influence of a short-range interdroplet interaction potential has been recently investigated [30] in water-AOT-oil microemulsions prepared by varying the chemical characteristics of the oil phase (three alkanes, i.e., n-pentane, n-heptane, n-decane, and carbon tetrachloride). The overall interactions due to the mutual interpenetration of the surfactant tails and to the Coulombic potential are taken into account by means of the parameters λ and a, respectively. The effect of the different oil phases on the reduced conductivity is shown in Figs. 6 and 7 as a function of the water-to-surfactant molar ratio $W = [H_2O]/[AOT]$. Figure 6 shows, at the selected temperature of $T = 30°C$, the effect of the different alkanes, where the conductivity passes through a maximum, and that due to carbon tetrachloride, where an approximately linear increase is evidenced. The comparison with the expected values [Eq.

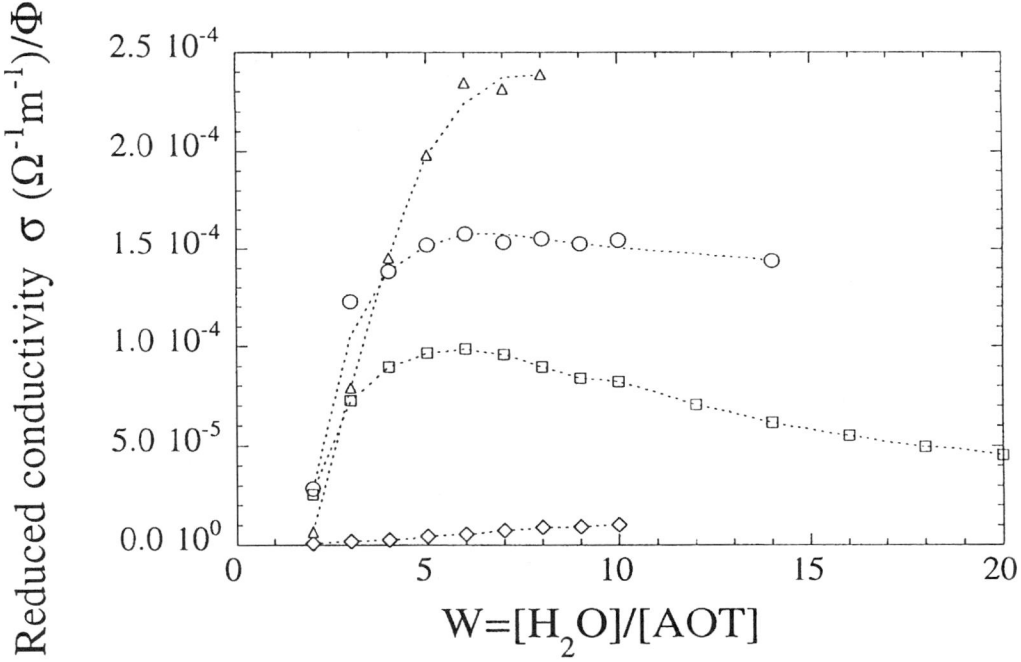

FIG. 6 The reduced conductivity σ/ϕ of water-in-oil microemulsion composed of different oil phases as a function of water-to-surfactant molar ratio $W=[H_2O]/[AOT]$, at the temperature of $T=30°C$. ($\triangle$) water-AOT-n-decane; ($\bigcirc$) water-AOT-n-pentane; ($\square$) water-AOT-n-heptane; ($\diamond$) water-AOT-carbon tetrachloride.

(27)] suggests that the well depth in the short-range attractive potential increases with the length of the molecule solvent chains in agreement with an interdroplet potential [32] due to the penetration of the surfactant tails, over short distances. The effect of temperature in the case of a water-AOT-n-decane microemulsion is shown in Fig. 7. Estimates of activation energies [33,34] for electrical conduction (of the order of 25–40 kJ/mol) support the mechanism of Na^+ counterion transport through water channels in the clusters of water droplets upon collision (rather than hopping of AOT anions from one droplet to another).

At higher temperatures and higher volume fraction, when water droplets come in close contact, a different transport mechanism prevails, owing to the transition from a conductivity mechanism due to the motion of charge water droplet (forced by an external electric field) to a conductivity regime dominated by the motion of charge carriers (surfactant anions or counterions) within connected structures, resulting from droplet clusterization. This process has been analyzed as a percolation process [20,35–37] where the conductance, after the threshold volume fraction or temperature, rises sharply in a narrow range and thereafter remains more or less unchanged. The phenomenon of percolation in microemulsions has been the subject, through the study of the electrical conductivity, of a large number of investigations. For a water-in-oil microemulsion, composed of water droplets undergoing Brownian motion uniformly distributed in a continuous oil phase, as the volume fraction ϕ of the dispersed phase is increased, a very pronounced increase, over several orders of magnitude, of the low-frequency electrical

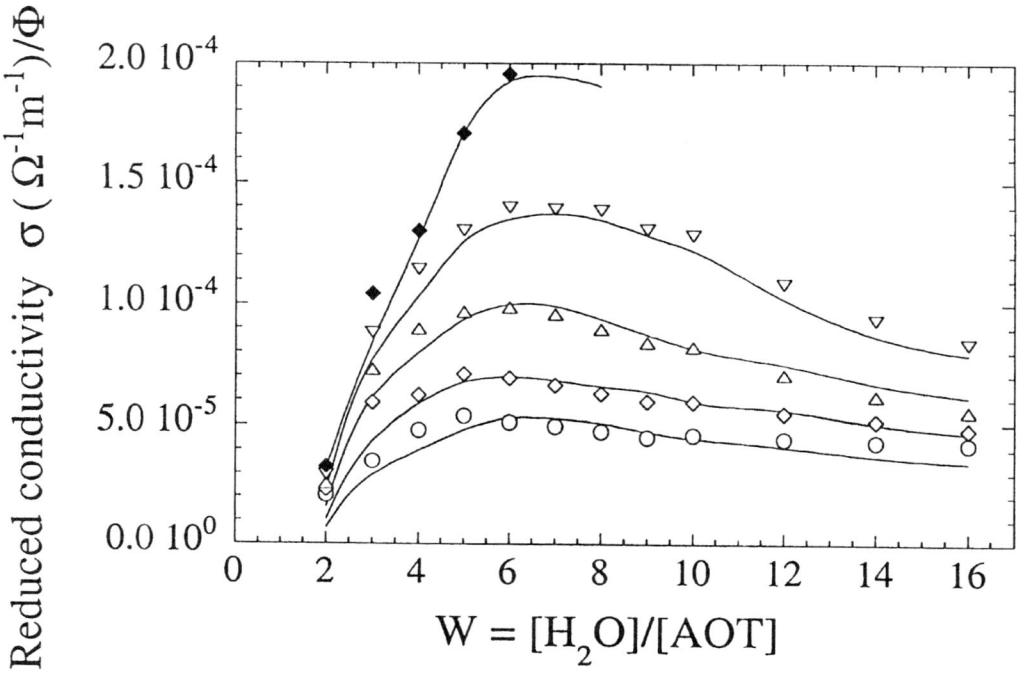

FIG. 7 The reduced conductivity σ/ϕ of water-AOT-n-decane microemulsion as a function of water-to-surfactant molar ratio $W = [H_2O]/[AOT]$ at different temperatures: (○) $T = 10°C$; (◇) $T = 20°C$; (△) $T = 30°C$; (▽) $T = 40°C$ (◆) $T = 50°C$.

conductivity occurs. This effect has been described as percolation phenomena [17,38–40] taking place when the water droplets approach each other sufficiently closely to give rise to a charge transfer between water droplets. This phenomenon is due to the transition from a conductivity mechanism associated with Brownian diffusion of the charge droplets to a conductivity regime dominated by the motion of charged carriers within a connected cluster of water droplets, extended on large-scale length. The connectivity of water droplets over macroscopic length scales probed by the electrical conductivity measurements is achieved by different possible mechanisms, i.e., charge hopping, transient merging of different droplets, or formation of stable water channels, each of them able to produce an enormous increase of the conductivity. There exist two views about the conducting mechanisms in microemulsion close percolation. The first supports the opening of the surfactant layers of the water droplets to form water channels, thus favoring the counterion (Na^+ ions) diffusion [41,42], whereas the second invokes a hopping mechanism [43,44] of the surfactant anions from one droplet to another, resulting in a transport of the surfactant anions.

The usual description of the conductivity behavior near percolation of a mixture of particles of electrical conductivity σ_p dispersed in a continuous medium of electrical conductivity σ_m is based on the scaling law [45]:

$$\sigma = \sigma_p |\phi - \phi_c|^\mu f\left(\frac{\sigma_m}{\sigma_p} |\phi - \phi_c|^{-(\mu+s)}\right) \tag{28}$$

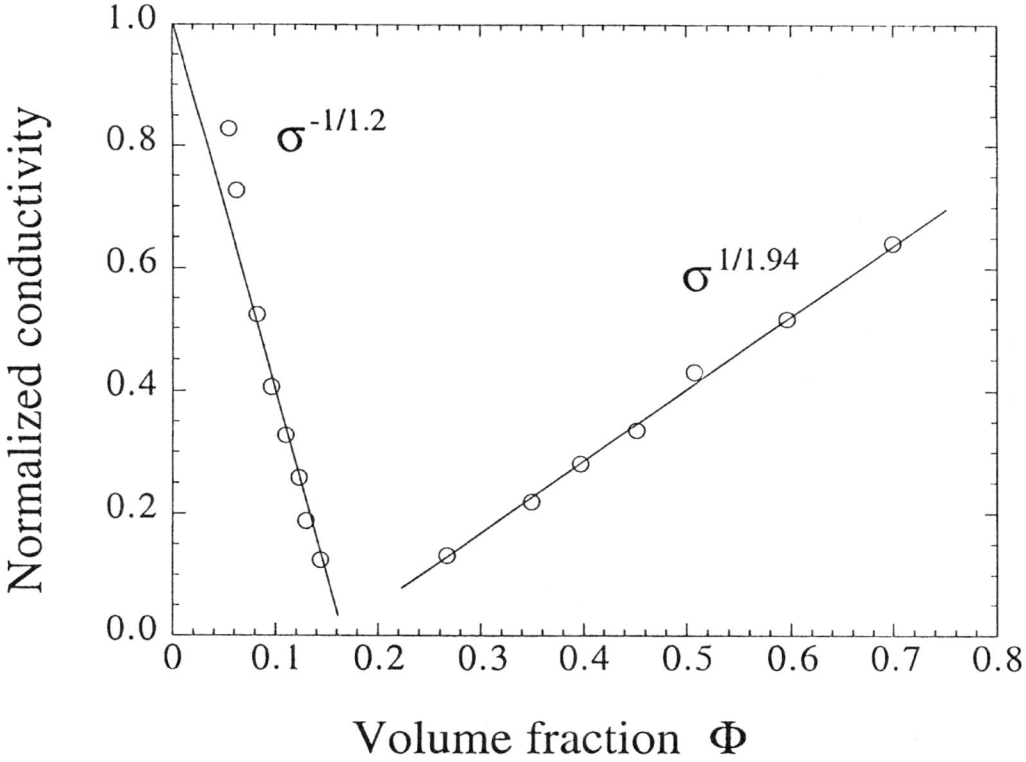

FIG. 8 The plot shows the conductivity of water-AOT-undecane microemulsion at water-to-surfactant molar ratio $W = 8$ and at a temperature of $T = 15°C$. The data are analyzed according to the relationships $\sigma^{1/\mu}$ and $\sigma^{-1/s}$ (normalized to 1 for $\phi = 0$ and 1) as a function of the volume fraction ϕ, above and below percolation, respectively. (Data taken from Ref. 61.)

where ϕ_c is the percolation threshold and s and μ two scaling exponents. In the vicinity of percolation, but not very close to it, Eq. (28) reduces to the asymptotic relations

$$\sigma \approx (\phi - \phi_c)^\mu \qquad \phi > \phi_c \qquad (29)$$
$$\sigma \approx (\phi_c - \phi)^{-s} \qquad \phi < \phi_c \qquad (30)$$

Experiments carried out on different microemulsion systems (of the type water-in-oil, $\sigma_p > \sigma_m$) have shown that the exponents have the values $\mu = 1.94$ and $s = 1.2$, suggesting that dynamic percolation occurs [20] (for static percolation a value $s = 0.7$ is expected).

Typical examples of the electrical conductivity behavior of different microemulsion systems (with very different components, water or waterless systems, ternary or quaternary systems) are shown in Figs. 8 to 10.

Most generally, experimental values of the critical exponents obtained by locating the percolation threshold as a function of temperature or volume fraction for AOT-water-oil microemulsion formulated with different types of oil (cyclohexane, dodecane, undecane, decane, isooctane, etc.) lie in the interval $1.1 < s < 1.6$ and $1.6 < \mu < 2.2$ [17,39,46,38,47–53], the values for the critical index s being in agreement with a dynamic percolation picture.

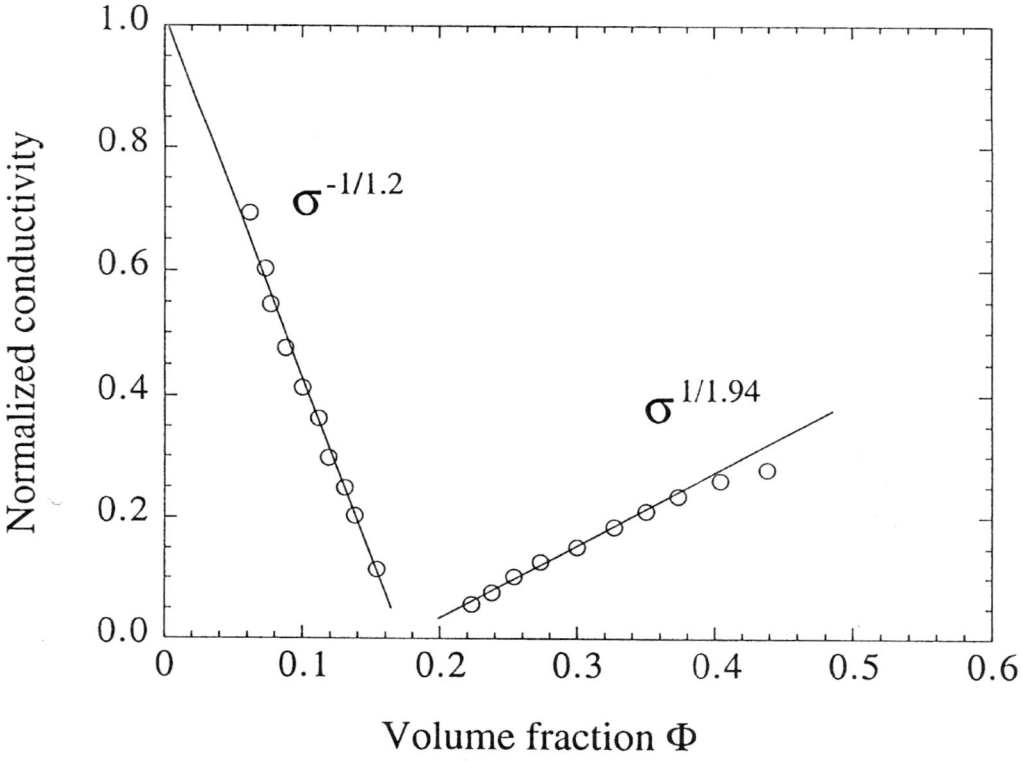

FIG. 9 The plot shows the conductivity of a quaternary system formulated with water, sodium dodecyl sulfate, butanol, and toluene at the temperature of $T = 20°C$. The data are analyzed according to the relationships $\sigma^{1/\mu}$ and $\sigma^{-1/s}$ (normalized to 1 for $\phi = 0$ and 1) as a function of the volume fraction ϕ, above and below percolation, respectively. (Data taken from Ref. 61.)

Above percolation, as the temperature (or the water concentration) is further increased, the topology of the oil and water domains varies, and the bicontinuous water-in-oil structure gives place to a two-dimensional ordered structure (lamellar phase), before the inverted bicontinuous oil-in-water structure holds. Within this region, the structure of the microemulsion is made of alternate sequences of oil and water layers coated with a monomolecular layer of surfactant molecules, whose polar head groups are oriented towards the water phase. In this case, the ionic transport is essentially due to the motion of Na^+ ions derived from the ionization of the surfactant molecules within the water layer, and the electrical conductivity reduces to a "bulk" phenomenon within this region.

If the microemulsion is modeled as a set of randomly oriented multilamellae, the overall conductivity of the system in the effective medium theory approximation can be written as [54]

$$\sigma = \frac{1}{3} \frac{\sigma_l \sigma_w}{\sigma_w(1 - \phi_w) + \sigma_l \phi_w} + \frac{2}{3}[\sigma_l(1 - \phi_w) + \sigma_w \phi_w] \qquad (31)$$

where σ_l and σ_w are the electrical conductivities of the surfactant-coated oil layer and the water layer, respectively, and ϕ_w the volume fraction of the water in the lamellar phase.

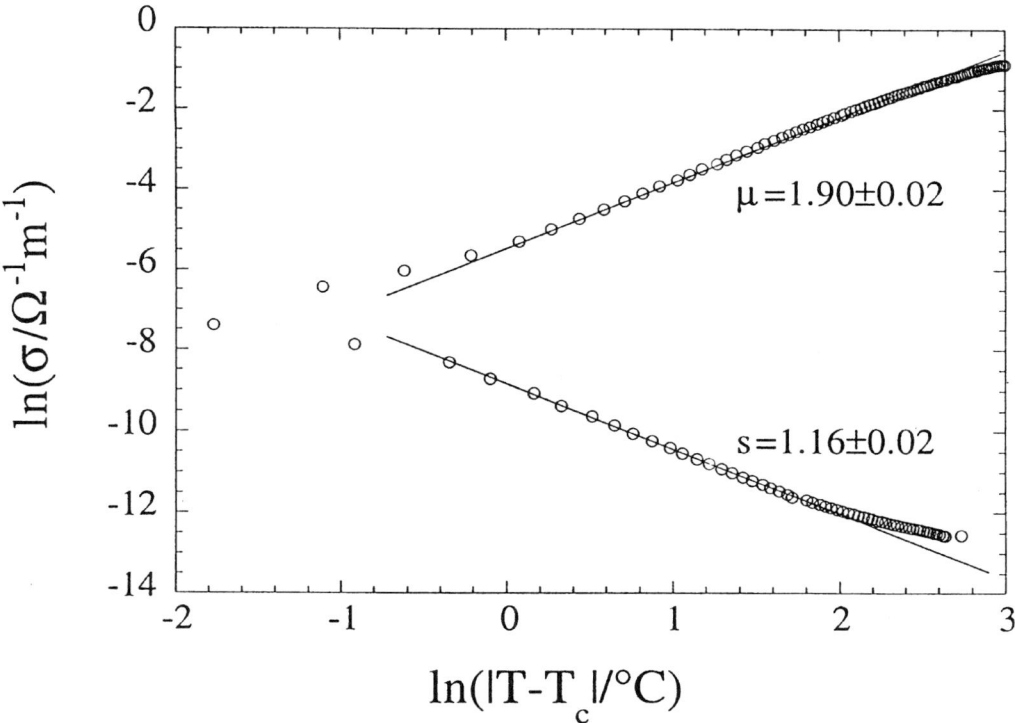

FIG. 10 A typical plot of $\ln(\sigma)$ vs. $\ln(T - T_p)$ for the water-AOT-n-decane microemulsion at $\phi = 0.525$, in the presence of added salt (10^{-2} M NaCl). The two straight lines represent the power laws above and below percolation with exponents μ and s. Deviations from a power law behavior very close to percolation are evident.

The local asymmetry of the system implies that the conductivity σ_w assumes two extreme limits, when the electric field is perpendicular and parallel to the lamellae, resulting in a different ion concentration that contributes to the electrical conductivity. By introducing a parameter η that takes into account the hindering effect at the charged surfactant–water interface on the ion transport mechanism, in the limit $\sigma_l \ll \sigma_w$, the overall conductivity of the microemulsion in the lamellar phase is given by [54]

$$\sigma = \frac{8}{3} z^2 e^2 \sqrt{2f} \tan\left(\frac{f}{2}\left[\frac{k_D d_w}{2} - \eta\right]\right) \frac{\lambda}{k_D d_w^2 S_0} \tag{32}$$

where k_D is the inverse of the Debye screening length, $S_0 = 2/n_0 d_w$ the average interfacial area occupied by each surfactant molecule, d_w the thickness of the water layer, and λ the equivalent conductance of the ions of charge ze. The parameter f gives the fraction of the ion concentration far from the charged interface to that of the imperturbed one.

Typical conductivity behaviors of different microemulsions at different fractional volumes ϕ are shown in Fig. 11, in the temperature interval where the lamellar phase occurs. The solid line represents the calculated values according to Eq. (32), assuming that each surfactant molecule contributes one Na$^+$ ion to the water layer. The values of the parameter η give a thickness, in Debye length units, of immobilized ions at the surfactant–water interface corresponding to 1.7, 2.9, and 3.3 for microemulsions with

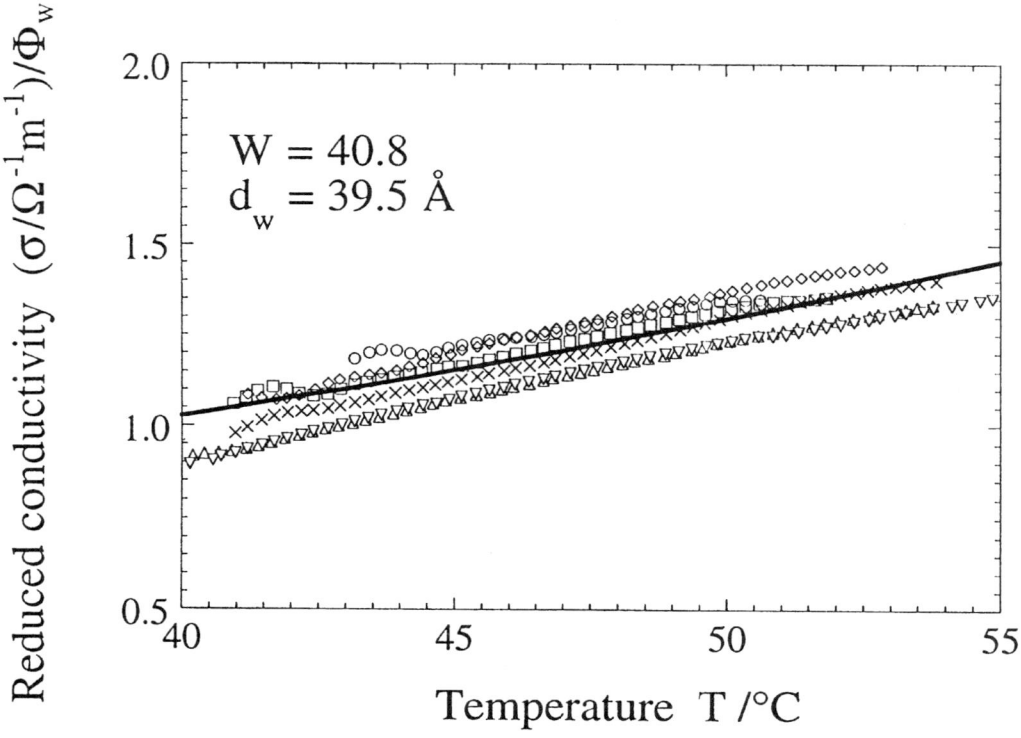

FIG. 11 The normalized conductivity σ/ϕ for a water-AOT-n-decane microemulsion at water-to-surfactant molar ratio $W = 40.8$. Each set of data represents the normalized conductivity for different values of the fractional volume fraction ϕ, where the lamellar phase occurs. The solid line is calculated according to Eq. (32).

water-to-surfactant molar ratios $W = 50, 40.8, 20$, respectively [54]. As can be seen, the agreement with the measured conductivity is remarkably good, for all the volume fractions investigated, in the lamellar phase.

In the percolation region, where, as we have stated above, transient clusters of complex structure are formed, in addition to the relaxation of the single droplets, two further mechanisms having a cooperative nature have been evidenced, the first associated with spatial rearrangement of percolation clusters, with a characteristic time of a few microseconds, and the second associated with the transport of charge carriers along the percolation cluster. A detailed analysis of the influence of these processes on the dielectric (and conductometric) behavior of ionic microemulsions (water-AOT-decane microemulsion, $W = [H_2O]/[AOT] = 26.3$) has been done recently by Feldman et al. [55–57], who found a well-defined hierarchy of the above processes on a time scale expressed as

$$\psi(t) = \psi_1(t) + \psi_2\left(\frac{t}{\tau_c}\right)\psi_3\left(\frac{t}{\tau_r}\right) \approx A_1 \exp\left(\frac{-t}{\tau_1}\right)^{\beta_1} + A_2 \exp\left(\frac{-t}{\tau_2}\right)^{\beta_2} \tag{33}$$

The results presented by these authors show that the phenomenological parameters τ and β, describing the Kohlrausch–Williams–Watts relaxation functions [Eq. (33)], are related to the structure of the system and reflect the cooperative behavior of microemulsion droplets near the percolation threshold, being insensitive to the microscopic details of the charge transport in the system. This is a further example that shows how a different conduction mechanism prevails in different structural arrangements.

We have recently developed a new approach to the dielectric and conductometric behavior of microemulsions close to percolation that gives a more detailed description of the characteristics of this phenomenon [58–60]. The microemulsion droplets, characterized by a complex conductivity $\sigma_p^*(\omega) = \sigma_p + i\omega\varepsilon_0\varepsilon_p$ and dispersed in a continuous oil phase of complex conductivity $\sigma_m^*(\omega) = \sigma_m + i\omega\varepsilon_0\varepsilon_m$, experience a weak short-range attractive potential that causes the formation of large inhomogeneous clusters characterized by a fractal dimension $D = 2.53$. According to the dynamic percolation model of conductivity in microemulsions below the percolation threshold, the conduction mechanism is due to charge migration in the bulk of clusters (charge carriers perform an anomalous diffusion motion). The steps we used are the evaluation of the complex conductivity of a single cluster by means of an effective medium approximation, supplemented by the fractal and dynamical properties mentioned above and the appropriate distribution of cluster size, typical of percolation phenomena.

Using the effective medium approximation, for the complex conductivity of a cluster $\sigma_k^*(\omega) = \sigma_k + i\omega\varepsilon_0\varepsilon_k$ in terms of the analogous quantities for the microemulsion components, we obtain

$$\sigma_k^* = \sigma_m^* \frac{A_k \sigma_p^* + 2\sigma_m^*}{\sigma_p^* + B_k \sigma_m^*} \tag{34}$$

Eq. (34) can be put in the form of a single relaxation process with amplitude Δ_k and relaxation time τ_k:

$$\sigma_k^*(\omega) = \sigma_{k0} + i\omega\varepsilon_o\left(\varepsilon_{k\infty} + \varepsilon_m \frac{\Delta_k}{1 + i\omega\tau_k}\right) \tag{35}$$

In the case of conducting clusters in nonconducting medium ($\sigma_m/\sigma_p \approx 10^{-4}$, $\varepsilon_m/\varepsilon_p \approx 10^{-2}$), these quantities can be easily related to the parameters A_k and B_k through the relations

$$\Delta_k = A_k - \frac{2}{B_k} \tag{36}$$

and

$$\tau_k = \frac{B_k \varepsilon_m}{\sigma_p} \tag{37}$$

The two parameters A_k and B_k must be determined as a function of the number of particles k in a cluster, making reference to the physical mechanism of electrical conductivity in microemulsions below the percolation threshold. Following the stirred percolation introduction some time ago by Lagues [35] and named dynamic percolation by Grest et al [20], the relaxation amplitude, related to the static dipole moment correlation function, is proportional to the mean square charge fluctuation according to the cluster size k and the radius of gyration squared R_k^2:

$$\Delta_k \approx k R_k^2 \approx k(k^{1/D})^2 \tag{38}$$

Since the charge carriers perform diffusive motion on a fractal aggregate, the relaxation time τ_k is governed by anomalous diffusion and is related to the radius of gyration through a power that takes into account the anomaly, i.e.,

$$\tau_k \approx (R_k^2)^{D/\tilde{d}} \tag{39}$$

where $\tilde{d}$ is the so-called spectral exponent. Following the analogy with random-bond percolation, the number of clusters of size k is given by a scaling expression characterized by a power-law behavior and a cut-off cluster size that diverges as the percolation threshold is approached. Consequently, the complex conductivity $\sigma_k^*(\omega)$ is calculated as a sum of properly weighted independent contributions:

$$\sigma_k^*(\omega) = \int_1^\infty c(k)\left(\sigma_{k0} + i\omega\varepsilon_0\varepsilon_{k\infty} + i\omega\frac{\Delta_k}{1+i\omega\tau_k}\right)dk \tag{40}$$

In order to get an explicit expression for the complex conductivity, with a number of clusters of size k characterized by a power-law behavior with index τ and a cut off cluster size k_c, we obtain

$$\sigma_k^*(\omega) = S_N \sigma_p^* B_1 \frac{\sigma_m^*}{\sigma_p^*} \int_1^\infty dk \frac{\exp(-k/k_c)k^{-1/D}}{1+B_1 k^{2/\tilde{d}}(\sigma_m^*/\sigma_p^*)} \tag{41}$$

where the adimensional conductivity scale S_N is given by

$$S_N = N(\tau-2)\left(A_1 - \frac{2}{B_1}\right)\frac{1}{B_1} \tag{42}$$

and A_1 and B_1 are related to the monomer amplitude Δ_1 and the relaxation time τ_1 according to

$$A_1 - \frac{2}{B_1} = \frac{\Delta_1\tau_1(\varepsilon_p/\sigma_p - \varepsilon_m/\sigma_m)}{(\tau_1\varepsilon_p/\sigma_p - 1)(1-\tau_1\varepsilon_m/\sigma_m)^2} \tag{43}$$

and

$$B_1 = \frac{\varepsilon_p}{\varepsilon_m}\left(\frac{\tau_1\varepsilon_p/\sigma_p - 1}{1-\tau_1\varepsilon_m/\sigma_m}\right) \tag{44}$$

N is the total number of monomers initially present in the system.

The final result concerning the static electrical conductivity exactly at the threshold is

$$\sigma = N\left(\frac{3}{D}-1\right)\Delta_1\tau_1\sigma_p\frac{\varepsilon_m}{\varepsilon_p}\frac{(1-\tau_1\varepsilon_m/\sigma_m)(\varepsilon_p/\sigma_m - \varepsilon_m/\sigma_m)}{(\tau_1\varepsilon_p/\sigma_p - 1)^2}\,_2F_1(1,u;1+u;\,) \tag{45}$$

where $_2F_1$ is the Gauss hypergeometric function and $u = \mu/(s+\mu)$. This approach was used to reinterpret experimental data from our laboratory on two microemulsions close to percolation in the presence and in the absence of added salt. Figure 12 shows the two branches of the electrical conductivity below and above percolation for a particular microemulsion formulated with water-AOT-n-decane at two different compositions. As can be seen, at the percolation temperature T_p, the electrical conductivity reaches a finite value, related to the small but finite conductivity of the oil phase. The power laws [Eqs. (29) and (30)] can be observed only in a finite window not too close to T_p. On the contrary, Eq. (41) gives a detailed description over the whole branch of the conduc-

Ion Transport and Electrical Conductivity

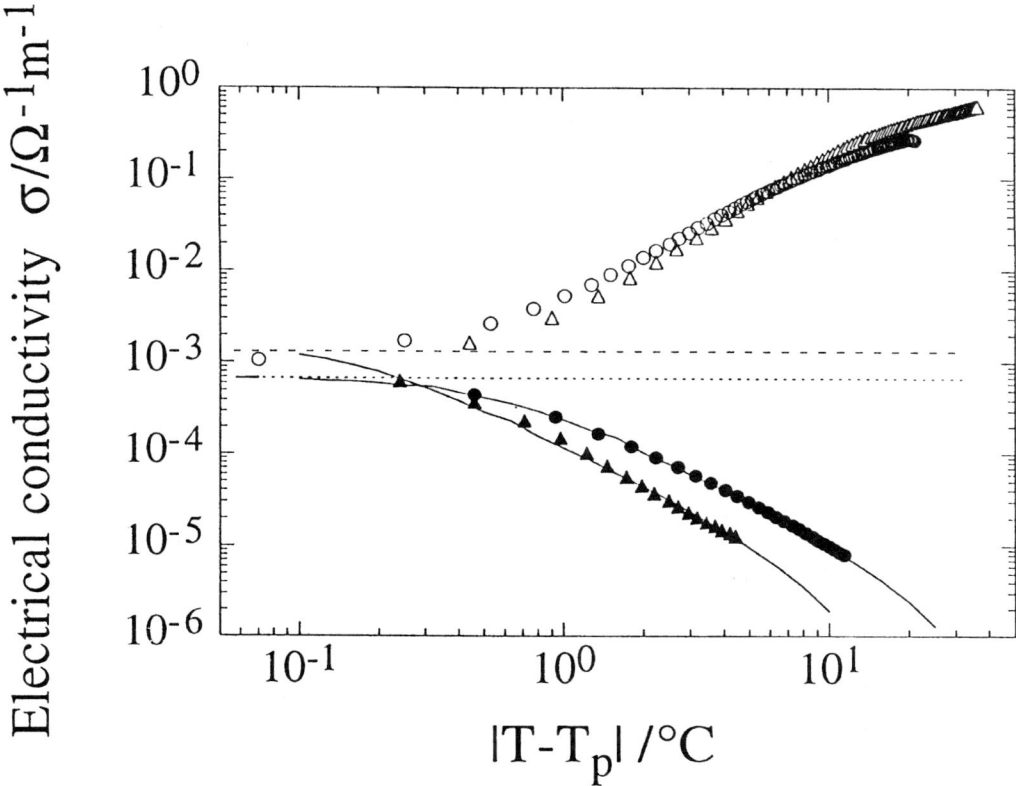

FIG. 12 The two branches of the electrical conductivity for a water-AOT-n-decane microemulsion in the absence and in the presence of added salt (10^{-2} M NaCl). Below percolation: (▲) without added salt, $\phi = 0.40$; (●) with added salt, $\phi = 0.575$. Above percolation: (△) without added salt, $\phi = 0.40$; (○) with added salt, $\phi = 0.575$. The dotted lines indicate the percolation temperature T_p. The full lines, below the percolation, represent the calculated values according to Eq. (40), giving exactly the measured value up to the percolation threshold, Eq. (45).

tivity below percolation, yielding, moreover, exactly the value of σ at $T = T_p$ [Eq. (45)]. The treatment of the resulting aggregates near percolation on the basis of fractal clusters with an appropriate distribution of cluster sizes markedly improves the agreement between theory and experimental data. The above model, giving quantitative expressions for both the conductivity and the permittivity in the entire frequency range of the percolation relaxation phenomena, has been applied in different microemulsion systems of different composition, in different temperature and frequency ranges, with a very good agreement in all cases [59].

IV. CONCLUSIONS

Among various heterogeneous systems, microemulsions have received great attention in the last few years because of their thermodynamics and physicochemical behavior, including spontaneous formation, intricate phase behavior, critical phenomena, and large solubilization of both organic and aqueous phases. Owing to the large interfacial area per unit surface, these systems represent a very interesting model system for the study of a large variety of surface phenomena, in particular those related to the ionic transport induced by an external electric field. Electrical conductivity measurements carried out in typical microemulsion systems have shown that this technique is a powerful tool for the investigation of different conduction regimes (but all of ionic nature) that take place in these systems is dependence on the different morphologies assumed and enhanced by the interfacial properties. We have presented and discussed a comprehensive review of the electrical behavior of a typical microemulsion system in the light of the current theories of highly dispersed charged systems and have shown that a very good agreement is generally achieved between theory and experiments.

REFERENCES

1. H. Ohshima and K. Furusawa, eds., *Electrical Phenomena at Interfaces: Fundamentals, Measurements and Applications*, Marcel Dekker, New York, 1998.
2. S. Takashima, *Electrical Properties of Biopolymers and Membranes*, Adam Higler, Bristol, 1989.
3. E. H. B. De Lacey and L. R. White. J. Chem. Soc. Faraday Trans. 2, 77:2007 (1981).
4. R. W. O'Brien. J. Colloid Interface Sci. 81:234 (1981).
5. Y. C. Liu and H. J. Keh. J. Colloid Interface Sci. 192:375 (1997).
6. Y. C. Liu and H. J. Keh. Langmuir 14:1560 (1998).
7. L. A. Rosen, J. C. Baygents, and D. A. Saville. J. Chem. Phys. 98:4183 (1993).
8. C. Solans and H. Kunieda, eds., *Industrial Applications of Microemulsions*, Marcel Dekker, New York, 1997.
9. B. H. Robinson, D. C. Steyler, and R. D. Tack. J. Chem. Soc. Faraday Trans. 1, 75:481 (1979).
10. R. Zana, ed., *Surfactant Solutions*, Marcel Dekker, New York, 1987.
11. J. Rouch, A. Safouane, P. Tartaglia, and S. H. Chen. J. Chem. Phys. 90:3756 (1989).
12. E. Y. Sheu, S. H. Chen, J. C. Huang, and J. C. Sung. Phys. Rev. $A39$:5867 (1989).
13. M. Kotlarchyk, J. S. Huang, and S. H. Chen. J. Phys. Chem. 89:4382 (1984).
14. C. Cabos and J. Marignan. J. Phys. Lett. 46:L-267 (1985).
15. A. M. Ganz and B. E. Boeger. J. Colloid Interface Sci. 109:504 (1986).
16. M. A. van Dijk. Phys. Rev. Lett. 55:1003 (1985).
17. S. Bhattacharya, J. P. Stokes, M. W. Kim, and J. S. Huang. Phys. Rev. Lett. 55:1884 (1985).
18. A. Ponton, T. K. Bose, and G. Delbos. J. Chem. Phys. 94:6879 (1991).
19. S. H. Chen, J. Rouch, F. Sciortino, and P. Tartaglia. J. Phys.: Condensed Matter 6:10855 (1994).
20. G. S. Grest, I. Webman, S. A. Safran, and A. L. R. Bug. Phys. Rev. $A33$:2842 (1986).
21. A. Jada, J. Lang, and R. Zana. J. Phys. Chem. 93:10 (1989).
22. A. Jada, J. Lang, and R. Zana. J. Phys. Chem. 94:381 (1990).
23. H. F. Eicke, M. Borkovec, and B. Das-Gupta. J. Phys. Chem. 93:314 (1989).
24. N. Kallay and A. Chittofrati. J. Phys. Chem. 94:4755 (1990).
25. N. Kallay, M. Tomić, and A. Chittofrati. Colloid Polymer Sci. 270:194 (1992).
26. D. G. Hall. J. Phys. Chem. 94:492 (1990).
27. B. Halle. Prog. Colloid Polym. Sci. 82:211 (1990).
28. B. Halle and M. Bjorling. J. Phys. Chem. 103:1655 (1995).
29. J. P. Hansen and I. R. McDonald. *Theory of Simple Liquids*, 2d ed. Academic Press, London, 1986.
30. F. Bordi, C. Cametti, A. Di Biasio, and G. Onori. Prog. Colloid Polym. Sci. 110:208 (1998).
31. M. Tomić and N. Kallay. J. Phys. Chem. 96:3874 (1992).
32. J. S. Huang, S. A. Safran, M. W. Kim, and G. S. Grest. Phys. Rev. Lett. 53:592 (1984).
33. F. Bordi and C. Cametti. Colloid Polym. Sci. 276: (1998).
34. F. Bordi, C. Cametti, and A. Di Biasio. Prog. Colloid Polym. Sci. (in press) (1999).

35. M. Lagues. J. Phys. Lett. *49*:33 (1979).
36. S. R. Bisal, P. K. Bhattacharya, and S. P. Moulik. J. Phys. Chem. *94*:350 (1990).
37. D. Stauffer and A. Aharony, *Introduction to Percolation Theory*, Taylor and Francis, London, 1994.
38. M. A. van Dijk. Phys. Rev. Lett. *55*:1003 (1985).
39. M. W. Kim and J. S. Huang. Phys. Rev. *A34*:719 (1986).
40. M. Moha-Ouchane, J. Peyrelasse, and C. Boned. Phys. Rev. *A35*:3027 (1987).
41. E. Dutkiewicz and B. H. Robinson. J. Electroanal. Chem. *11*:251 (1988).
42. L. Mukhofadhyay, P. K. Bhattacharya, and S. P. Moulik. Colloids Surfaces *50*:295 (1990).
43. J. S. Huang, S. A. Safran, M. W. Kim, G. S. Grest, M. Kottarchyk, and N. Quirke. Phys. Rev. Lett. *53*:593 (1984).
44. R. Hilfiker, H. F. Heicke, S. Geiger, and G. Furler. J. Colloid Interface Sci. *105*:378 (1985).
45. Z. Saidi, C. Mathew, J. Peyrelasse, and C. Boned. Phys. Rev. *A42*:872 (1990).
46. A. Ponton and T. K. Bose. J. Chem. Phys. *94*:6879 (1991).
47. M. A. van Dijk, G. Casteleijn, J. G. H. Joosten, and Y. K. Levine. J. Chem. Phys. *85*:626 (1986).
48. C. Cametti, P. Codastefano, A. Di Biasio, P. Tartaglia, and S. H. Chen. Phys. Rev. *A40*:1962 (1989).
49. C. Cametti, P. Codastefano, P. Tartaglia, J. Rouch, and S. H. Chen. Phys. Rev. Lett. *64*:1461 (1990).
50. C. Cametti, P. Codastefano, P. Tartaglia, S. H. Chen, and J. Rouch. Phys. Rev. *A45*:R5358 (1992).
51. H. F. Eicke, R. Hilfiker, and M. Holz. Helv. Chim. Acta *67*:361 (1984).
52. M. T. Clarkson. Phys. Rev. *A37*:2079 (1988).
53. M. T. Clarkson and S. I. Smedley. Phys. Rev. *A37*:2070 (1988).
54. A. Di Biasio, C. Cametti, P. Tartaglia, J. Rouch, and S. H. Chen. Phys. Rev. *E47*:4258 (1993).
55. Y. Feldman, N. Kozlovich, I. Nir, and N. Garti. Colloids Surfaces *128*:47 (1987).
56. Y. Feldman, N. Kozlovich, I. Nir, and N. Garti. Phys. Rev. *E51*:478 (1995).
57. Y. Feldman, N. Kozlovich, Y. Alexandrov, R. Nigmatullin, and Y. Ryabov. Phys. Rev. *E54*:1 (1996).
58. P. Codastefano, F. Sciortino, P. Tartaglia, F. Bordi, and A. Di Biasio. Colloids and Surfaces A *140*:269 (1998).
59. F. Bordi, C. Cametti, J. Rouch, F. Sciortino, and P. Tartaglia. J. Phys.: Condensed Matter *8*:A19 (1996).
60. F. Bordi, C. Cametti, P. Codastefano, F. Sciortino, P. Tartaglia, and J. Rouch. Prog. Colloid Polym. Sci. *100*:170 (1996).
61. C. Boned, J. Peyrelasse, and Z. Saidi. Phys. Rev. *E47*:468 (1993).

16
Electrical Behavior of Inorganic Thin Films in Electrolyte Media

JUANA BENAVENTE and JOSÉ RAMOS-BARRADO Department of Applied Physics, Universidad de Málaga, Málaga, Spain

SEBASTIÁN BRUQUE Department of Inorganic Chemistry, Universidad de Málaga, Málaga, Spain

I.	Introduction	565
II.	Experimental	569
	A. Materials	569
	B. Experimental setup	570
III.	Theoretical Framework	571
	A. Direct current measurements	572
	B. Alternating current measurements	573
IV.	Results and Discussion	579
	A. Direct current results	579
	B. Alternating current results	581
	References	589

I. INTRODUCTION

While organic polymer films or membranes have enjoyed over two decades of commercial success, inorganic films are just making their inroads into a growing number of commercial applications. Most of these inorganic membranes are ceramic mixed oxides or metallic films that typically exhibit stabilities at high temperatures and extreme pH conditions [1,2]. Organoinorganic polymer films are also advantageous when compared to their organic counterparts.

Currently commercial inorganic films are dominated by porous ceramic materials or dense metallic membranes. They have been used commercially or tested successfully in selected liquid-phase industrial production and analytical applications. More remarkable examples are dairy and fruit juice processing, wine and beer clarification, filtration of potable water, wastewater treatment in food and petrochemical processing, paper and textile applications, isotope separation for uranium enrichment, high-purity hydrogen production, and laboratory uses. Depending on the applications requirements, membranes of various ceramic and metallic materials with different pore sizes are used as either microfiltration or ultrafiltration membranes [2].

New and promising developments for higher separation performance have started to emerge: zeolite polycrystal films, gas sensors based on ceramic zirconia or titania materials, and ion-selective electrodes in strong media (oxidizing or high pH). Other goals invoked are the preparation of dielectric films for ionic conductors in batteries, or as microcapacitors [3].

Independently of their applications, an important handicap in the preparation of an inorganic ceramic film is its poor mechanical manageability against the flexibility of organic polymer materials. To bypass this, multiple strategies have been conceived. One of them consists in the association of an inorganic material with a flexible matrix, generally an organic polymer, to yield a composite material with the desired mechanical properties. Another interesting alternative may be the grafting of an organic moiety to an inorganic structure, and the best procedure is to synthesize an ionocovalent inorgano-organic hybrid compound. This is appropriate only if such a compound exhibits a stability close to that of the inorganic compounds together with the flexibility of the organic materials.

Among other materials, there are two classes of inorganic compounds that are proper candidates for the preparation of flexible inorganic films for dielectric purposes in electrolyte media: insoluble layered metal phosphates and metal phosphonates. Many metal acid phosphates can be obtained as layered compounds: $HUO_2PO_4 \cdot 4H_2O$, and the series α-$M^{iv}(HPO_4)_2 \cdot H_2O$ (M^{iv} = Ti, Zr, Hf, Ge, Sn, Pb, ...). Owing to the presence of acid HPO_4^{2-} or $H_2PO_4^{-}$ groups, these compounds exhibit good proton transport [4]. Furthermore, in recent years an extensive chemistry about metal phosphonates has been reported. The phosphonates can be regarded as inorgano-organic polymers in which organic moieties are bridged through phosphorous atoms to a metal–oxygen inorganic matrix. Phosphonates are usually lamellar compounds, but some exist with one-dimensional and three-dimensional structures [5].

The research on phosphonates and lamellar phosphates cited above has been aimed at discovering new proton conductors [6,7], solid electrolytes [8,9], dielectric materials [10], self-assembled films for the Langmuir–Blodgett technique [11], solid-state gas sensors [12], membranes for current rectification [13,14], and electrochemical devices [15]. The physicochemical and mechanical properties of these metal phosphonates and phosphates are strongly dependent upon their crystal structures [16] and hence it is necessary know their frameworks.

On the other hand, inorganic thin films made from precipitates were described several decades ago [17], and several inorganic compounds with a variety of structural features were used: for example, copper cyanoferrates [18], cobalt and nickel sulphides or copper orthophosphate [19–21], or barium sulphate [13,22,23] and zirconium hydrogenphosphate [24]. However, the inorganic solids were not commonly used as films, as the polymeric ones were, due to the difficulty in handling them. In order to solve this problem, the inorganic materials have been deposited either on polymeric [13,22,25] or on alumina porous supports [26].

In this chapter, we will describe the electrical behavior of thin films prepared from different insoluble uranyl phosphates and phosphonates, whose crystal structures have already been determined, when they are in contact with electrolyte solutions at different concentrations.

The hydrogen uranyl phosphate $H_3OUO_2PO_4 \cdot 3H_2O$ (HUP) is a well-known proton conductor [8,9] which can be precipitated from UO_2^{2+} and $H_2PO_4^{-}$ acid solutions. It is a layered tetragonal solid; the layers, made of distorted UO_6 octahedra and PO_4 tetrahedra, are coordinated in a wafflelike square net array (Fig. 1). The layers are displaced sideways

FIG. 1 A view of the a.c. plane of the layer structure of the hydrogen uranyl phenylphosphate (HUP). The water molecules are located in the interlamellar space and they are indicated by the oxygen atoms as shaded circles, the uranium atoms are represented by black circles, the tetrahedra correspond to PO_4 groups, and the octahedra are the UO_6 groups.

relative to each other, and the interlamellar spaces are occupied by the water molecules and the oxonium ions forming $H_9O_4^+$ square clusters [27]. Calcium uranyl phosphate, $Ca(UO_2PO_4)_2 \cdot nH_2O$ or CaUP, and ammonium uranyl phosphate, $NH_4UO_2PO_4 \cdot 3H_2O$ or NUP, have the same framework as HUP, but the oxonium ions are replaced by either Ca^{2+} or NH^{4+} in the interlayer space.

On the other hand, several polymorphs of uranyl phenylphosphonates have been synthesized, and their phase transitions have been recently analyzed [28,29]. The very low soluble compound $[UO_2]_3[HO_3PC_6H_5]_2[O_3PC_6H_5]_2 \cdot H_2O$ presents a complex tubular structure built of a covalently bonded network where the internal faces of nanotubes are

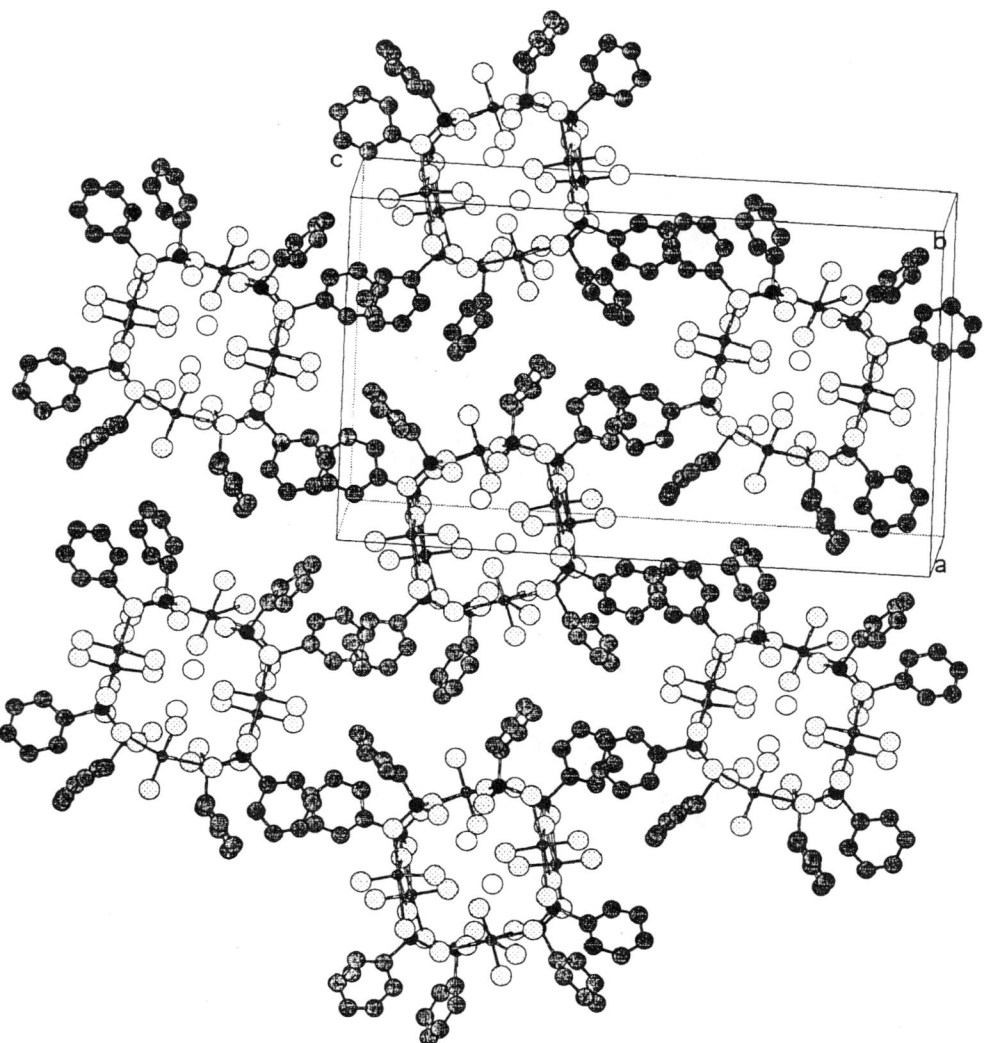

FIG. 2 Projection of the tubular structure of the uranyl phenylphosphonates (UPP) down the b axis, showing unidimensional pores separated by the hydrophobic regions, while the intratubular zones are hydrophilic. (From Ref. 29.)

hydrophylic and the external faces are hydrophobic [29]. The nanotube walls are formed by the inorganic uranium oxide and phosphorus oxide units [UO_7 polyhedra and $PO_3(C)$ tetrahedra]. The organic phenyl rings are linked to the phosphorus and extruded towards the intertubular regions. The full inorgano-organic nanotubes are held together by the Van der Waals forces through the phenyl ring interactions generating hydrophobic intertubular regions (Fig. 2).

II. EXPERIMENTAL

A. Materials

The $H_3OUO_2PO_4 \cdot 3H_2O$ (HUP) solid was prepared by mixing aqueous solutions 1 M and 1.1 M of $UO_2(NO_3)_2$ and H_3PO_4, respectively. The resulting suspension was refluxed for 5 days, and then the yellow polycrystalline solid was washed twice with distilled water and finally with a mixture 1:1 of water and acetone. Calcium uranyl phosphate, $Ca(UO_2PO_4)_2 \cdot 8H_2O$ (CaUP) with autunite structure was synthesized by mixing aqueous solutions 0.5 M of $CaHPO_4$ and $UO_2(NO_3)_2$ in a Ca:U ratio of 1:2. The white resulting suspension was refluxed for 5 days, and then the solid was washed three times with water and stored in a dessicator that contains a controlled ambiance of 82% humidity.

Solid uranyl phosphonates were prepared from phenyl-phosphonic acid and uranyl nitrate solutions in aqueous media with reflux. The precipitation occurs in few minutes, but the suspension is maintained at reflux temperature for 5 days in order to obtain a well crystallised material. Two different P:U molar ratios were used in the synthesis (1:1 or 5:1). The first produces a solid with the stoichiometry $UO_2(O_3PC_6H_5) \cdot 0.7H_2O$ (or UPP-11), but when the molar ratio is 5:1, the composition corresponds to $(UO_2)_3(HO_3PC_6H_5)_2 (O_3PC_6H_5)_2 \cdot H_2O$ (or UPP-51). By refluxing UPP-51 in water, UPP-11 is obtained [28]. The framework of these UPPs are stable up to 330°C; they are very insoluble in water, acetone, and other common solvents. They are also stable and insoluble in strong acid solutions until $c = 0.5$ M, and in basic dilute solutions.

The characterization and identification of active phases in the thin films was carried out by means of x-ray diffraction in a Siemens D-501 automated diffractometer, using monochromated CuKα radiation, i.r. spectroscopy (Perkin-Elmer 883 apparatus), and thermal analysis (on a Rigaku Thermoflex apparatus). The x-ray diffraction study reveals a layered structure with a basal spacing of 8.8 Å for the HUP, while the UPP shows a tubular structure with basal spacing of 14.6 Å for UPP-51 and 19 Å for UPP-11, and good crystallinity; the UPP layer space does not swell and contains the phenyl groups [29].

The inorganic films studied in this chapter were obtained in two different ways:

1. An amount of solid active phase (~ 1 g) was suspended in N,N'-dimethylacetamide (10 mL) and stirred in an ultrasonic bath for 2 h; then 0.5 g of Kynar® (polyvinylidene fluoride) was added. The slurry so obtained was maintained under ultrasonic stirring for 2 h. This mixture was spread on a glass plate, and the solvent was removed in an oven at 85°C. The solvent removal was accomplished in a water vapor saturated atmosphere. A flexible thin film (self-supported film) can finally be easily detached from the plate [30]. By ion exchange of the HUP self-supported films, CaUP and NUP films can be obtained by immersion in Ca^{2+} or NH_4^+ solutions, respectively. The CaUP film contains calcium uranyl phosphate hydrate [$Ca(UO_2PO_4)_2 \cdot nH_2O$] as the main active phase according to the x-ray powder pattern. The basal spacing increases from 8.8 Å (HUP) to 10.1 Å, but the H^+/Ca^{2+} exchange is not complete in these conditions since a small amount of the 8.8 Å phase still remains ($<20\%$). By immersion of the HUP in a 10^{-2} M solution of $NH_4H_2PO_4$ an ion exchange reaction takes place to produce $NH_4UO_2PO_4 \cdot 3H_2O$ (NUP) as the main phase. For the NUP film, the basal spacing is 8.9 Å, which indicates that the H^+/NH_4^+ exchange does not produce structural changes in the

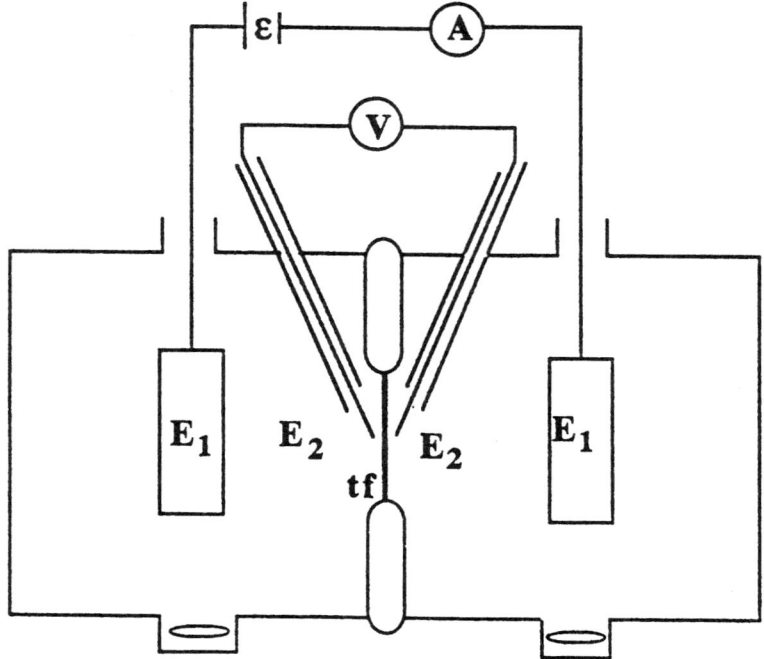

FIG. 3 Experimental setup. tf, thin film; E_1, current supply electrodes; E_2, measuring voltage electrodes.

precipitate; it contains $NH_4UO_2PO_4 \cdot 3H_2O$ as the main active phase, although about 20% of the exchange sites have still not been saturated by NH_4^+ cations, remaining as H^+ sites (HUP), as the chemical analysis of nitrogen reveals.
2. An aqueous suspension of precipitate was filtered through a porous support. In this case, a layer of UPP-51 was deposited on a commercial alumina porous support (Anopore®, 0.2 μm pore size), and a composite UPP+ANP film was obtained.

Solutions of the precipitate generating electrolytes [H_3PO_4 and $UO_2(NO_3)_2$ or $H_2O_3PC_6H_5$ and $UO_2(NO_3)_2$] at different concentrations and at a given temperature, $t = 25.0 \pm 0.3°C$, were used.

B. Experimental Setup

The experimental device used for both d.c. and a.c. measurements is shown in Fig. 3, and it basically consists of two half-cells separated by two methacrylate rings, where the films were placed.

As can be seen in Fig. 3, a four-probe electrode method was used for d.c. measurements: E_1 are the electrodes for the current supply and E_2 are the electrodes for measuring the voltage. Three series of current–voltage measurements for each membrane and concentration were made and, due to the membrane asymmetry, different values of the electrical resistance, depending on the direction of the current, were measured. The final resistance value for each direction and concentration was obtained as the average value of the three corresponding ones, and they are called $R(+)$ and $R(-)$. These measurements were carried out with the films separating two different solutions of the precipitate generating electrolytes at the same concentration.

Impedance spectroscopy measurements (a.c.) for the different membrane/electrolyte systems were carried out in the same measuring cell, but in this case only E_1 electrodes were used. A frequency responses analyzer (Solartron 1255), with 100 frequencies ranging between 10^2 Hz and 10^7 Hz, was used. The experimental data were corrected by software as well as by other parasite capacitances. In order to insure a linear response, the maximum voltage applied to the system only exceeded by 25 mV the corresponding concentration potential. For UPP film, the interface contribution was also studied, and measurements for frequency ranging between 10^{-2} Hz to 10^2 Hz were also carried out. These measurements were made with two different electrolytes at each side of the film and with the same electrolyte at each side of the film, in both cases the solutions having the same concentration.

For the UPP film and the Anopore support, measurements of salt diffusion and concentration potentials with the same electrolyte (but different concentration gradients) in both half cells were also carried out following the procedures indicated in the literature [31–33].

III. THEORETICAL FRAMEWORK

Let us consider an inorganic precipitate film that, interposed between two isothermal media, allows a mass transfer between them. A flow of charged particles, J_i, can be induced either by diffusion or by conduction, according to the Nernst–Planck equation:

$$J_i = -c_i u_i \, \mathrm{grad}\, \mu_i = -u_i\left[RT\left(\frac{da_i}{dx}\right) + v_i c_i\left(\frac{dP}{dx}\right) + z_i F c_i\left(\frac{d\Psi}{dx}\right)\right] \tag{1}$$

where c_i, u_i, a_i, and v_i are the concentration, mobility, activity, and molar partial volume of the ion i, respectively, R and F are the gas and Faraday constants, and T is the thermodynamic temperature. For diluted solutions (concentrations are used instead of activities) and when there is not a pressure gradient, Eq. (1) can be written as

$$J_i = -u_i\left[RT\left(\frac{dc_i}{dx}\right) + z_i F c_i\left(\frac{d\Psi}{dx}\right)\right] \tag{2}$$

If there is no concentration gradient across the system, the flow of charged particles is only induced by the electric field. In this case, the electric current accompanied by the charged particles is expressed by

$$I_i = z_i F J_i = -z_i^2 F^2 c_i\left(\frac{d\Psi}{dx}\right) \tag{3}$$

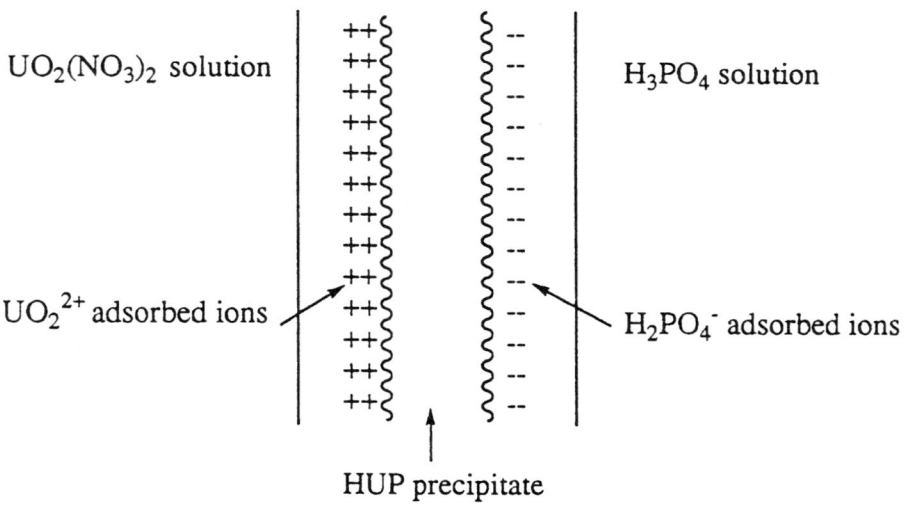

FIG. 4 Bipolar structure of the $UO_2(NO_3)_2/HUP$ film/H_3PO_4 system, due to the adsorption of ions by the HUP film.

If each one of the electrolyte solutions in contact with the film contains one (or more) of the precipitate generating ions, an electrically charged bipolar system will arise, as is shown in Fig. 4 for the $UO_2(NO_3)_2/HUP/H_3PO_4$ system: a surplus of UO^{2+} ions will be adsorbed at one side, while at the other side the adsorbed ions will be $H_2PO_{4^-}$, and two interfaces of opposite electric charge will be thus formed.

A. Direct Current Measurements

A main characteristic of this kind of system is its ability to rectify electric current [22]; this means that they present an asymmetry in the transport of charge depending on the polarity of the external electric field applied to the system, as can be seen in Figs. 5a and b for the $UO_2(NO_3)_2/UPP/H_2O_3PC_6H_5$ system.

1. When the external electric field has the polarity shown in Fig. 5a (hyperpolarizing field), the precipitate generating ions are driven to the film and are stopped there (at the film/electrolyte interface or increasing the precipitate itself).
2. However, for the opposite polarity (depolarizing field) no restriction to the ions' movement exists, as Fig. 5b shows, and the system behaves as an ohmic conductor.

Figure 6 shows a typical current–voltage curve for these kinds of systems where the different resistance of both hyperpolarizing (h) and depolarizing (d) zones can be seen; the transport of charge in the hyperpolarizing zone is mainly attributed to the movement of H^+ and OH^- ions present in the aqueous solutions, and its limit value (I_L) can be obtained either by extrapolating to $\Delta V = 0$ the slope of the straight line for high values of the hyperpolarizing potential [23] or by a Cowan diagram (resistance at each time

Electrical Behavior of Inorganic Thin Films

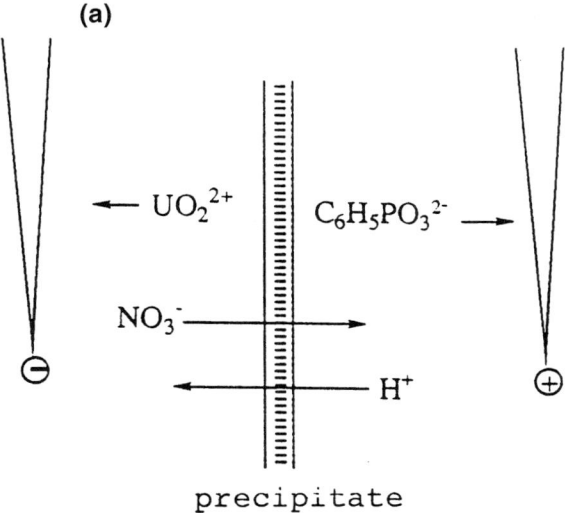

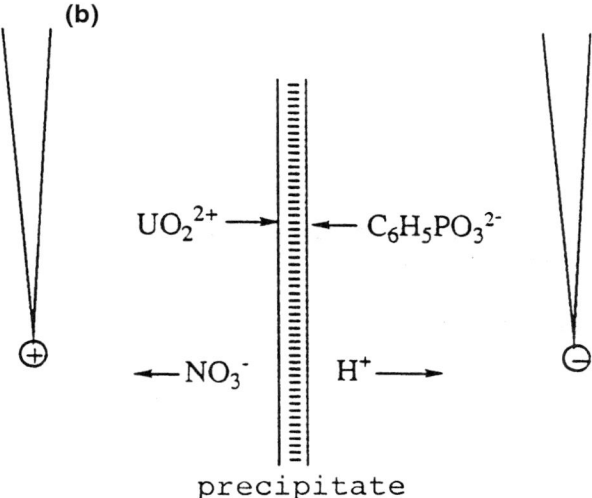

FIG. 5 Different fluxes of ions across the $UO_2(NO_3)_2$/UPP film/$H_2O_3PC_6H_5$ system due to the opposite polarities of the external electric field. (a) hyperpolarizing field; (b) depolarizing field.

instance versus $1/I$ [34]. The $R(h)/R(d)$ ratio is taken as a measure of the film current rectification efficiency, and both $R(h)/R(d)$ and I_L are considered as characteristic values for each system.

B. Alternating Current Measurements

Electrical measurements on solid–liquid interfaces imply the application of an electrical stimulus (a known voltage or current) to the electrodes/electrolyte/film system and the measurement of the response. It is usually considered that the properties of an elec-

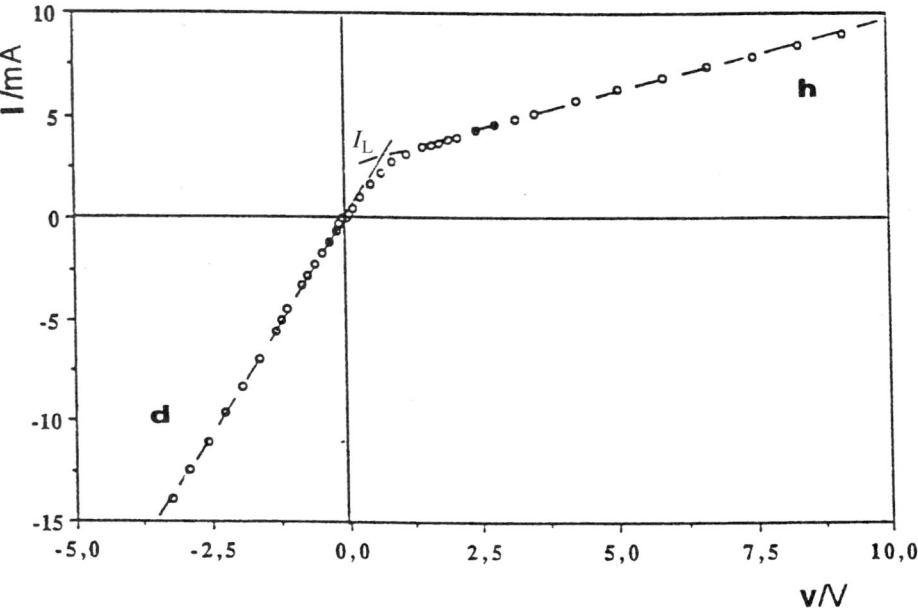

FIG. 6 Characteristic current–voltage curve for the systems with bipolar structure due to the absorbed ions showing the hyperpolarizing (h) and depolarizing (d) regions and limit current (I_L).

trolyte/film system are time invariant, and one of the basic purposes of impedance spectroscopy is to determine these properties, and their dependence on such controllable variables as electrolyte concentration, concentration gradient, valence numbers of ionic species, ionic mobility, boundary condition parameters, and applied static voltage or current bias. Although these systems are nonlinear systems, they can be considered in a linear regime for small external perturbations [35].

When a linear system is perturbed by a small voltage $v(t)$, its response, the electric current $i(t)$, is determined by a differential equation of the nth order in $i(t)$:

$$b_0 \frac{d^n i(t)}{dt^n} + b_1 \frac{d^{n-1} i(t)}{dt^{n-1}} + \cdots + b_n i(t) = a_0 \frac{d^n v(t)}{dt^n} + a_1 \frac{d^{n-1} v(t)}{dt} + \cdots + a_n v(t) \tag{4}$$

or a set of n differential equations of the first order. Hence, if $v(t)$ is a sine wave input,

$$v(t) = V_0 \sin \omega t \tag{5}$$

the current intensity $i(t)$ is also a sine wave,

$$i(t) = I_0 \sin(\omega t + \phi) \tag{6}$$

and a transfer function, the admittance function, can be defined by

$$Y^*(\omega) = |Y(\omega)| e^{j\phi} \tag{7}$$

where $|Y(\omega)| = I_0/V_0$, and $|Y(\omega)|$ and ϕ are the modulus and the phase shift of the admittance function. The impedance function, $Z(\omega)$, is the inverse admittance function,

$$Z^*(\omega) = |Y^*(\omega)|^{-1} \tag{8}$$

If the inverse Fourier transform, $Y(t)$, is considered,

$$Y(t) = \mathfrak{I}^{-1}[Y^*(\omega)] \tag{9}$$

then

$$Y(t) = \int_{-\infty}^{\infty} Y(\theta) V(\theta - t) d\theta \tag{10}$$

and in the frequency domain,

$$I^*(\omega) = Y^*(\omega) V^*(\omega) \tag{11}$$

The impedance $Z^*(\omega)$ of these systems is a complex number that can be represented in either polar coordinate or Cartesian coordinates:

$$Z^*(\omega) = |Z| e^{j\phi} \tag{12}$$

$$Z^*(\omega) = Z' + jZ'' \tag{13}$$

where Z' and Z'' are the real and imaginary part of the impedance. The relationships between these quantities are

$$|Z|^2 = Z'^2 + Z''^2 \tag{14}$$

$$\phi = \arctan \frac{Z''}{Z'} \tag{15}$$

$$Z' = |Z| \cos \phi \quad \text{and} \quad Z'' = |Z| \sin \phi \tag{16}$$

Two types of plots are commonly used to describe these relationships, and they are illustrated in an example for the equivalent circuit of an electrochemical cell. Figure 7a shows the impedance imaginary part vs. log ω, while Fig. 7b shows a representation in the complex plane, which involves plotting the impedance imaginary part against the real part ($-Z''$ vs. Z'). A first approximation of the analysis of the a.c. data may be carried out by the complex plane method: when plotted on a linear scale, the equation for a parallel RC circuit, which can represent an electrochemical cell, gives rise to a semicircle in the complex plane [36]. The semicircle has intercepts on the real axis at $R_\infty(\omega \to \infty)$ and $R_0(\omega \to 0)$, where $(R_\infty - R_0)$ is the resistance of the system. The maximum of the semicircle is 0.5 $(R_\infty - R_0)$ and occurs at a frequency ω such that RC = 1, RC being the single relaxation time [36].

The study of a physical phenomenon often leads to the elaboration of a model. One looks for a model that is a rational representation, sometimes mathematical, of a phenomenon and that contains only the essential features of the real situation. The model at times deals with the very structure of a system but, in certain cases, it can also be useful to elaborate a model of the "input–output" type (equivalent circuit for example), which

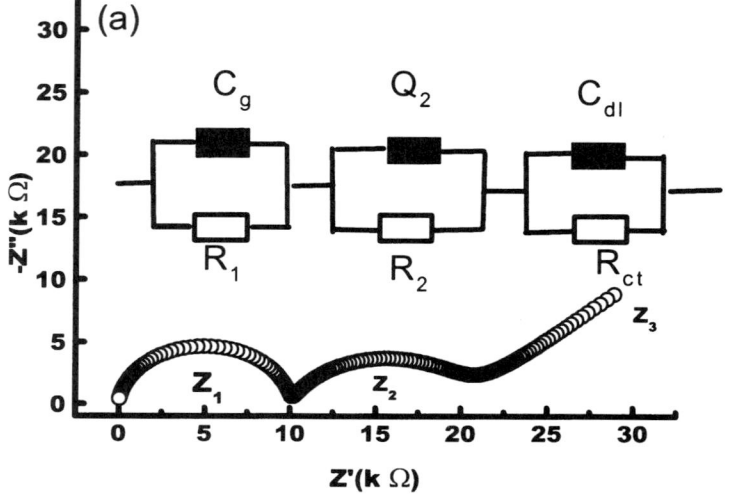

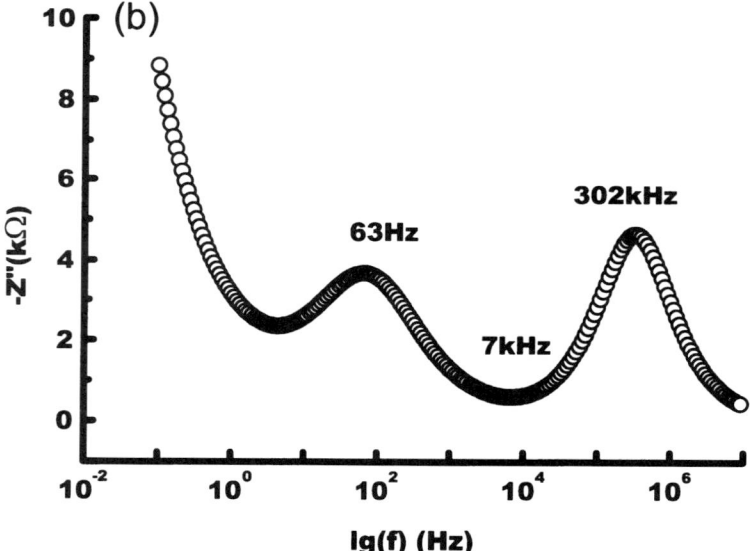

FIG. 7 Typical impedance plots for an electrochemical cell and the associated equivalent circuit. (a) Nyquist plot with the different impedance arcs Z_1, Z_2, and Z_3; (b) Bode plot.

describes the behavior of a system with respect to its environment. Ideal resistors, capacitors, and various distributed circuit elements such as Warburg impedances and constant phase elements compose the equivalent circuit. In such a circuit,

> A resistance represents a conductive path. The resistor might take account of the bulk conductivity, an interfacial conduction process, or even the chemical step associated with the electrode reaction; in general, it implies a long displacement of charge carriers.

The capacitor is associated with a space charge polarization region and it implies a short displacement of the charge carriers.

The distributed elements do not present a simple physical interpretation. We can consider two different distributed elements, both related to the finite spatial extension of a real system: the Warburg impedance, which is associated with diffusion processes and may be obtained from the Fick diffusion equation with appropriate boundary conditions; and the other type, the constant phase element (CPE), which arises because microscopic material properties are themselves often distributed in the bulk and, in the interfacial processes, they take into account the roughs and surface defects in the interfacial boundary.

The elaboration of a model for the electrolyte/interface/film system is derived from the general equations of physics that take into account the nonlinear character of the process involved. The space distribution of the state variables $V(M, t)$ and $c_i(M, t)$, the potential and concentration of species i at point M and time t, is determined by a set of integral–differential equations (Maxwell equations and mass balance equations), subject to the appropriate boundary and initial conditions:

$$\nabla^2 V = -\rho_c + \nabla \cdot \vec{P}$$
$$\frac{\partial c_i}{\partial t} = -\nabla \cdot \vec{J} + \xi_i$$

The first of these $(m+1)$ coupled equations generalizes the Poisson equation, where ρ_c is the electric charge density per unit volume and P the dielectric polarization. The second set of equations generalizes the hydrodynamic equations and applies to the m types of chemical entities involved in the system; c_i is either a surface or a bulk concentration, J_i is the flux, and x_i is a term representing the sources (production and consumption) of the entities i generally arising from chemical and electrochemical reactions.

Such a system as the electrolyte/interface/film, which is an intrinsic–extrinsic conduction system containing mobile positive and negative charge species of arbitrary valence numbers z_p and z_n, arbitrary mobilities u_p and u_n, and quite general boundary conditions, shows, for the most general situation, three connected arcs when the imaginary and real parts of the impedance are represented in a complex plane, which indicates three different conduction–relaxation processes with three different relaxation times [36], as can be seen in Fig. 7.

The Z_1 impedance arc is the high-frequency one. It represents only the electrolyte properties, is always extensive, and shows only a relaxation time characteristic of a Debye behavior. The extensive Z_1 circuit elements are C_g, the geometrical capacitance of the system in the absence of mobile charge, and R_1, its bulk resistance.

The Z_2 impedance arc, the kinetic arc, represents the electrochemical process occurring in the film. These processes may be separated into charge migration (faradic) and charge polarization (nonfaradic) processes, and they can be interpreted, for the most simple and general system, by a parallel combination of a constant phase element (CPE) and a resistor. The CPE impedance can be described by the generalized expression

$$Q_f = A'(j\omega)^{-1} = A'\omega^{-n}\left[\cos\left(\frac{n\pi}{2}\right) - j\sin\left(\frac{n\pi}{2}\right)\right] \tag{17}$$

where A' and n are empirical parameters (A' is the admittance and $0 \leq n \leq 1$). The Z_2 impedance arc, for a certain mobility ratio range for charge carriers, well approximates a depressed semicircle of the Cole–Cole relaxation time distribution type.

The Z_3 impedance arc, the interfacial arc, is made up of just the blocking double layer capacitance in parallel with a resistance, R_{ct} [36]. The double layer interface generally shows a gradient of conductivity, and if it is a constant gradient it can be represented by a parallel combination of a resistance and a capacitance, which is a function of the frequency of the external applied electric field. The complex capacitance can be expressed as [37,38]

$$Q_{dt}^* = C_{dl} + \frac{Y_p}{(j\omega)^n} \tag{18}$$

where

$$C_{dl} = \frac{S(\sigma_2 - \sigma_1)}{d} \frac{B/\omega}{A^2 + B^2} \tag{19}$$

$$Y_p = \frac{S(\sigma_2 - \sigma_1)}{d} \frac{A}{A^2 + B^2} \tag{20}$$

where σ_1 and σ_2 are the electrical conductivities, S and d are the film area and thickness, respectively, while A and B are functions of the frequency, and the electrical conductivity and permittivity.

When σ_1 and σ_2 are the same, we can calculate the limit of the equation, and we obtain

$$\lim_{\sigma_1/\sigma_2 \to 1} C_{dl} = \varepsilon \frac{S}{d} \tag{21}$$

$$\lim_{\sigma_1/\sigma_2 \to 1} Y_p = \sigma_1 \frac{S}{d} \tag{22}$$

Thus, if there is not a gradient of conductivity, and for $n=1$, the double layer interface may be represented by a simple parallel combination of a resistor and a CPE. If there is no electric field gradient, transports of electroactive species to or from the film are governed by the rate diffusion. A solution of the diffusion equation (Fick law) with an appropriate boundary condition (smooth interface) gives rise to a frequency dependent impedance that has the form of a Warburg component. The Z_3 impedance is the low-frequency one, and Z_3 is essentially intensive (an interface quantity) when the frequency of the applied potential is not too low.

Under many conditions, only two of these three arcs may appear, and the melting of arcs into each other can also occur. When both cations and anions pass across the surface at high and low frequency, the finite Warburg region does not exist; moreover, the kinetic arc is present when both cations and anions pass the film surface or when only one passes and its boundary parameter is much lower than 1 [36]. In general, if the characteristic frequencies of two arcs are not very different, they are mixed, and it is difficult to separate them. Because of this, an equivalent circuit with only a depressed arc is the most difficult case to analyze, since this arc may include an average over every relaxation process in the system. However, it is possible to define an equivalent capacitance by means of the relationship [39]

$$C^{eq} = \frac{\tau}{R} = \frac{(R/A)^{1/n}}{R} \tag{23}$$

where τ is the relaxation time and the other parameters have already been defined.

IV. RESULTS AND DISCUSSION

A. Direct Current Results

Figure 8 shows the experimental current–voltage curves (d.c.) for different inorganic films in contact with solutions of the generating electrolytes. The asymmetry of the i–v curves depending on the polarity of the external applied voltage is clearly shown in this picture, but also that a minimum level of generating ions in the solutions, and consequently at the membrane/solution interfaces, should exist in order to obtain an asymmetric behavior with these films.

A comparison of the rectifier effect found with HUP, NUP, and CaUP self-supported films is also shown in Fig. 8a for a given concentration ($c = 0.01$ M); it can be observed that for the depolarizing potential zone the resistance values obtained for the three films are very similar, but significant differences exist for the hyperpolarizing potential zone. The highest rectifier effect appears with the NUP film, while the lowest is presented by the HUP one. This result is attributed to the different transport mechanisms of ions across the films: an almost free exchange of H^+ for the HUP self-supported film, which agrees with the high protonic conductivity presented by this material [12]; but a diffusion of the ions across CaUP and NUP ones. In the latter case, the higher rectification presented by the NUP with respect to the CaUP film is attributed to differences in the matrix structure of each inorganic precipitate film; thus the CaUP has a more open structure than the NUP ($d_L^{CaUP} = 10.1$ Å and $d_L^{NUP} = 8.9$ Å), and it also has more water molecules in the interlayer space, which makes easy the ionic diffusion across the CaUP structure.

Current–voltage curves obtained with the UPP film at different concentrations of the generating electrolytes are shown in Fig. 8b. It can be seen that also in this case concentrations higher than 5×10^{-4} M must be used in order to obtain current rectification.

The fitting of the straight lines shown in Figs. 8a and b allows the determination of $R(+)$, $R(-)$, and I_L values for each film and electrolyte concentration. The $R(+)/R(-)$ ratio represents a measure of the current rectification efficiency, and its value for each thin film and concentration are indicated in Table 1. Values of the limit intensity, I_L, which was determined by the interception of the straight lines obtained for each current polarity [23], are also indicated in Table 1. Differences in I_L values depending on the system considered can be observed: while for the NUP film the limit current values are almost independent of concentration, for the other films the limit current increases when the concentration of the external solution increases, mainly for the HUP thin film, which agrees with the high proton conductivity of the HUP precipitate previously indicated. These results show the importance of protonic diffusion through the HUP film but also that the transport across the NUP film is almost blockaded. Results also indicate that the increase of the concentration of precipitate generating ions in the solution not only affects the transport of charges across the film but also increases the adsorbed layers at the film/solution interfaces, which should be considered for further applications of these kinds of systems, for instance, as sensors in electrolyte media.

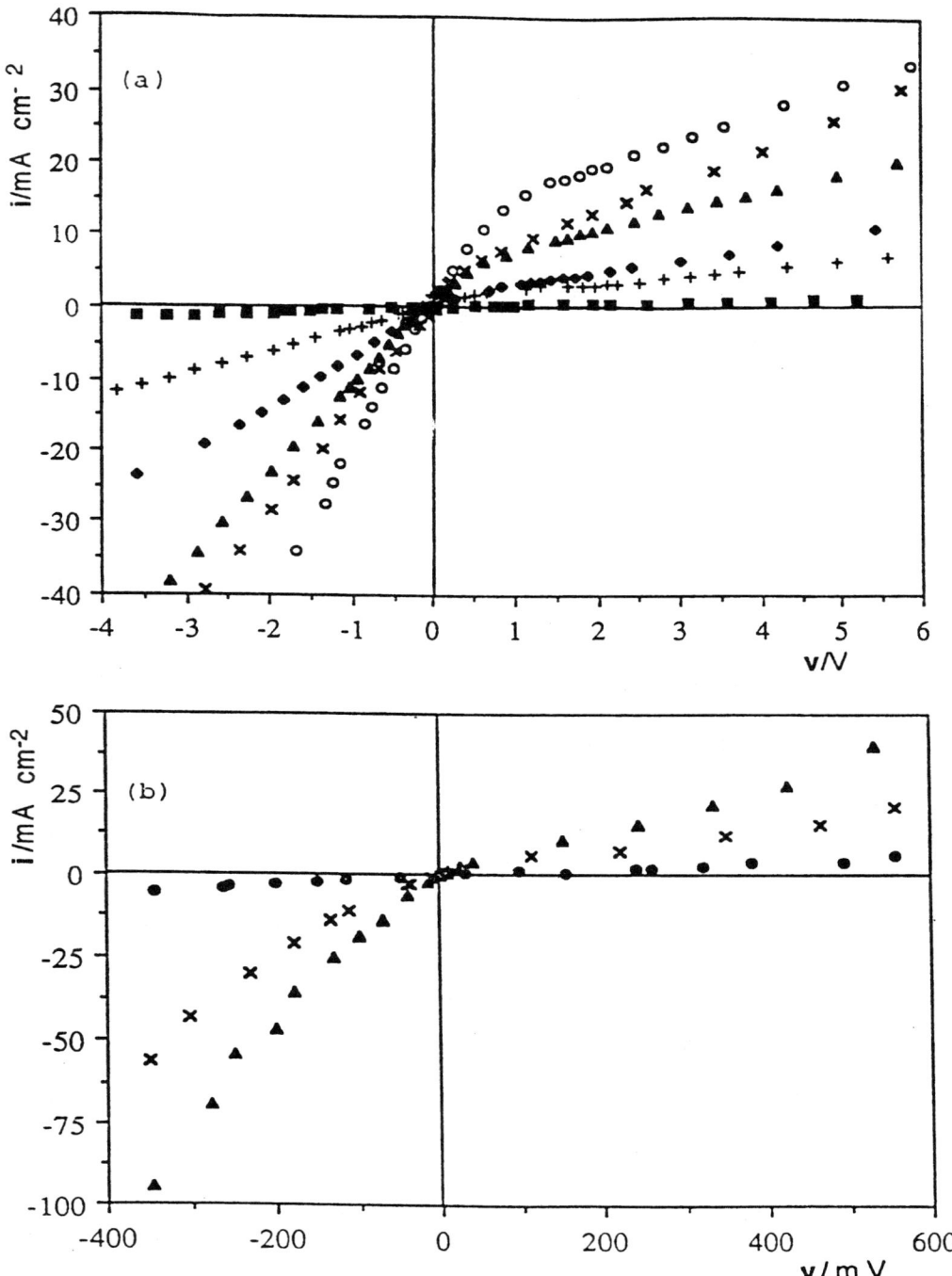

FIG. 8 Current–voltage curves for the different systems studied at different electrolyte concentrations. (a) Self-supported films: HUP (■) 10^{-4} M, (+) 10^{-3} M, (◆) 5×10^{-3} M, and (○) 10^{-2} M; NUP (▲) 10^{-2} M; CaUP (×) 10^{-2} M. (b) UPP film: (●) 5×10^{-4} M, (×) 5×10^{-3} M, and (▲;) 10^{-2} M.

TABLE 1 Resistance Ratio, $r = R(+)/R(-)$, and Limit Intensity Current, I_L, as a Function of the Electrolyte Concentration for HUP, NUP, CaUP, and UPP Thin Films

	HUP		NUP		CaUP		UPP	
C (M)	r	I_L (mA)	r	I_L (mA)	r	I_L (mA)	r	I_L (μA)
0.0001	1.77	0.03	—	—	—	—	1.20	0.36
0.0005	2.39	0.07	—	—	—	—	1.90	0.55
0.001	2.64	0.51	0.76	0.075	2.34	0.19	2.75	0.80
0.005	4.44	1.30	1.86	0.075	3.23	0.77	5.48	2.19
0.01	5.62	2.84	2.45	0.086	4.32	1.24	5.04	5.02

B. Alternating Current Results

Figures 9a and b show the Bode and Nyquist plots for the H_3PO_4/HUP film/$UO_2(NO_3)_2$ system ($c = 5 \times 10^{-3}$ M). From Fig. 9a two relaxation effects can be clearly observed, one at low frequencies (10^3 Hz $\leq f \leq 5 \times 10^5$ Hz), which corresponds to the film, the other at high frequencies (5×10^5 Hz $\leq f \leq 10^7$ Hz), which is due to the electrolyte solution. In Fig. 10 the Nyquist plots for NUP and CaUP films are also shown ($c = 5 \times 10^{-3}$ M). Similar results were obtained with the other concentration studied. The equivalent circuit for the three self-supported films consists of a series association of a capacitor (C_f) and a Warburg impedance (W_f) in parallel with a resistance (R_f).

Figure 11 shows the Nyquist plot for the $UO_2(NO_3)_2$/UPP film/$H_2O_3PC_6H_5$ system at $c = 0.005$ M, for the whole range of frequency (10^{-2} Hz $\leq f \leq 10^7$ Hz). In this case, the equivalent circuit is a series association of three parts (electrolyte, film, and film/electrolyte interface):

1. The electrolyte part, which corresponds to the highest frequencies (4×10^5 Hz $\leq f \leq 10^7$ Hz), consists of a resistance (R_c) in parallel with a capacitor (C_c).
2. The UPP film contribution, for frequency ranging between 6×10^2 Hz and 4×10^5 Hz, consists in a parallel association of a resistance (R_f) and a nonideal capacitor (Q_f).
3. The film/electrolyte interface contribution appears at low frequency (10^{-2} Hz $\leq f \leq 5 \times 10^2$ Hz), and it also consists of a resistance R_{ct} and a nonideal capacitor (Q_{dl}). These two elements are associated with the charge transference resistance and the electrical double layer capacitance.

The experimental data were analyzed using a nonlinear program [40], and the calculated circuit parameters (R_f, C_f, and W_f for the self-supported films and R_f C_f^{eq}, R_{ct}, and C_{dl}^{eq} for the UPP film) for different concentrations are indicated in Tables 2 and 3. In all cases, differences between experimental and calculated values were lower than 8%. Equivalent capacitances were calculated by means of Eq. (22). From the results shown in Table 2, it can be seen that for the three self-supported films (HUP, NUP, and CaUP), W_f and C_f values slightly depend on the thin film and concentration considered. However, for the UPP film (Table 3), it was found that at low concentrations, C_{dl}^{eq} values increase when the concentration increases, but an almost constant value is reached at high concentrations, which can be explained assuming that at a given concentration, around 0.001 M, the electrical double layer built up by the adsorption of the UPP precipitate generating ions at the UPP film/electrolyte interfaces is completely

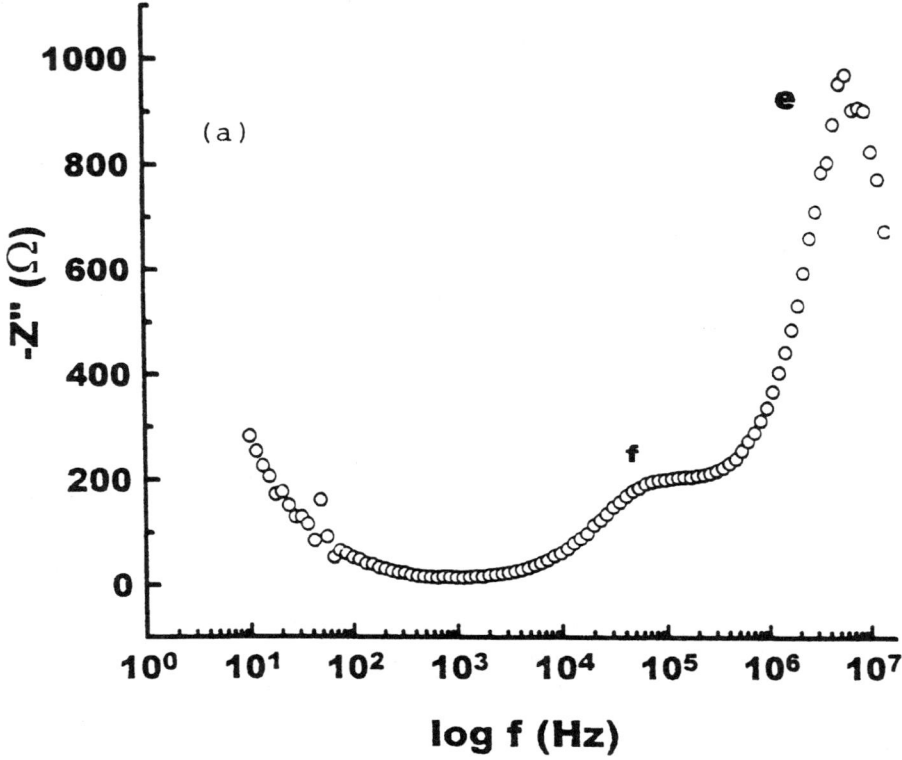

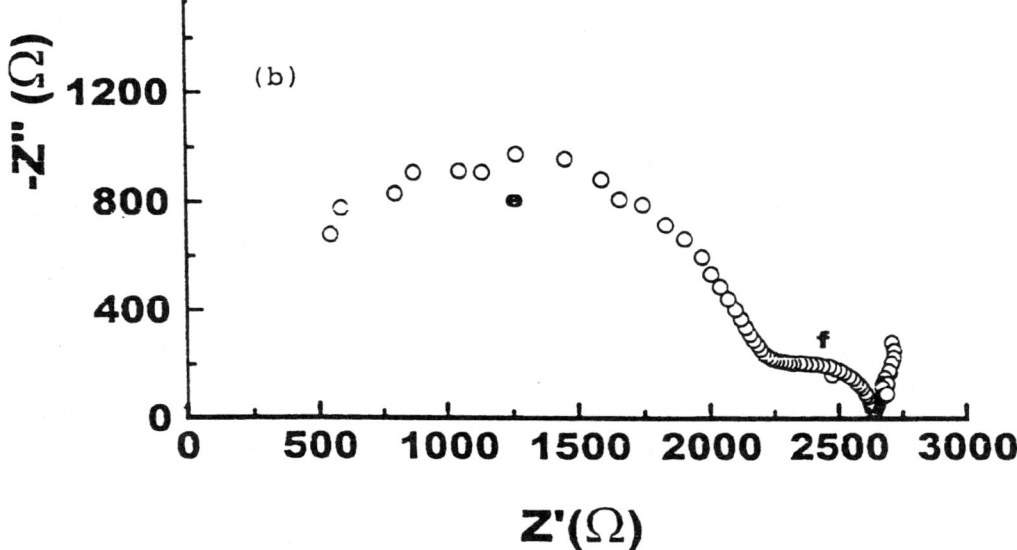

FIG. 9 Impedance plots for the $UO_2(NO_3)_2/HUP/H_3PO_4$ system at $c = 5 \times 10^{-3}$ M, indicating both thin film and electrolyte contributions. (a) $-Z'' = Z_{img}$ vs. f (Bode plot); (b) $-Z_{img} = Z''$ vs. $Z_{real} = Z'$ (Nyquist plot).

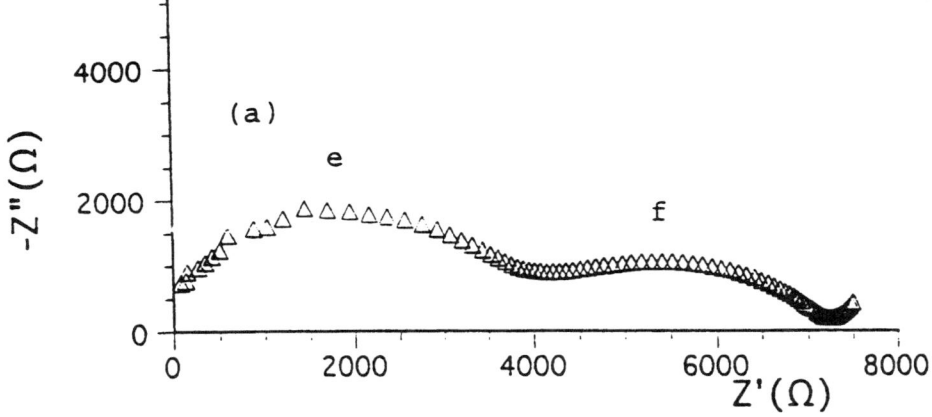

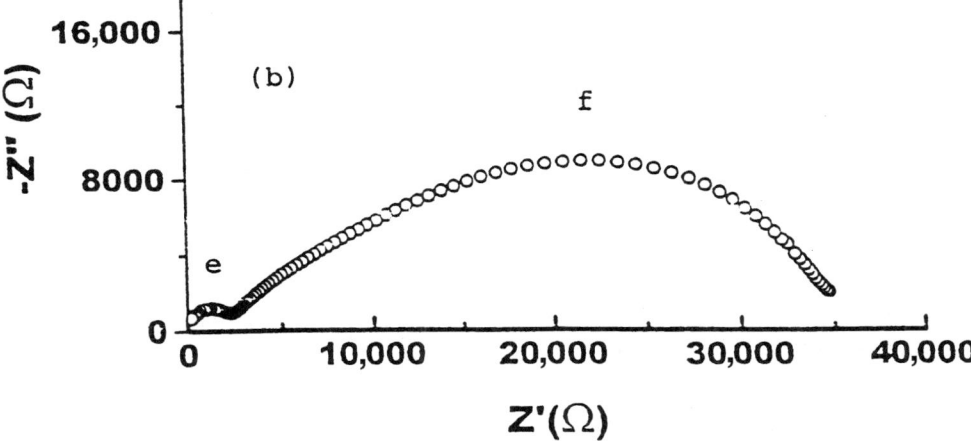

FIG. 10 Nyquist plots for (a) $UO_2(NO_3)_2/NUP/H_3PO_4$ and (b) $UO_2(NO_3)_2/NUP/H_3PO_4$ systems at $c = 5 \times 10^{-3}$ M.

developed and is not greatly affected by an extra amount of ions in the solutions (high concentrations); C_f^{eq} values present a linear increase when the concentration increases, but significant differences are obtained for the resistance values with both film and concentration. With respect to the concentration dependence for the film resistance, results show that R_f values for the different films are strongly dependent on salt concentration, which is mainly due to the electrolyte invasion into the film structure [41,42]; however, for the UPP film, a kind of limit value is obtained at high concentrations. An explanation for this fact will be provided by the formation/compaction of the precipitate layer, which would mainly affect the geometrical parameters of the film.

For the UPP film, the transport of electrolyte solutions across the UPP layer was studied by measuring different characteristic electrochemical parameters for the UPP porous-supported film and the porous support when they separate the same solution of each one of the UPP generating electrolytes ($UO_2(NO_3)_2$ and $H_2O_3PC_6H_5$) or a

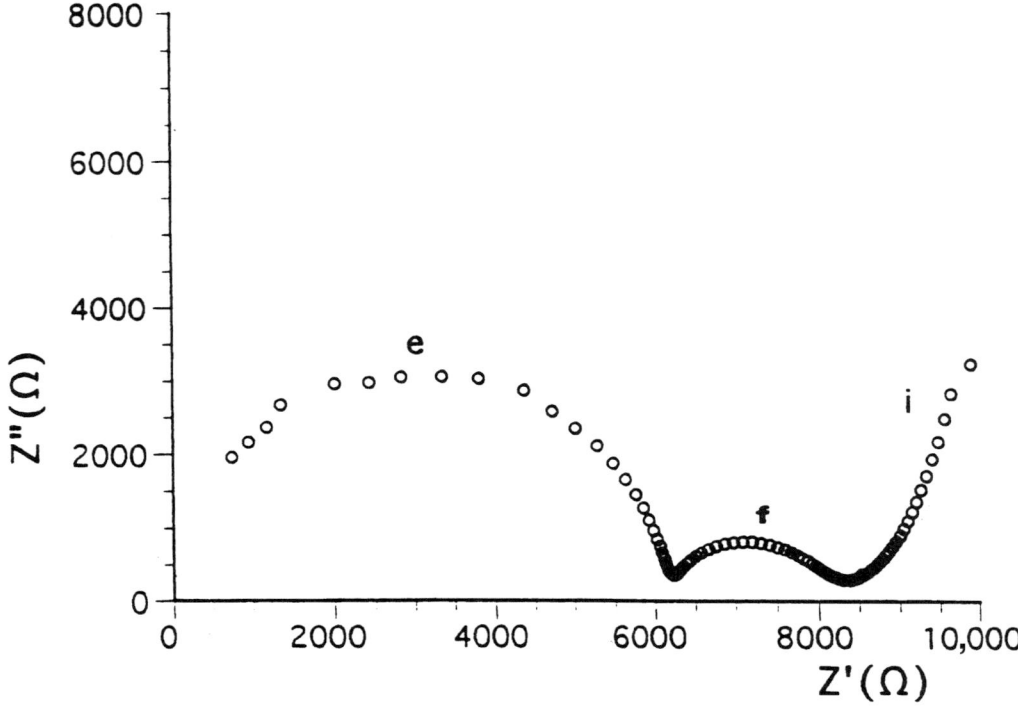

FIG. 11 Nyquist plots for the $UO_2(NO_3)_2/UUP/H_2O_3PC_6H_5$ system at $c = 5 \times 10^{-3}$ M, indicating the different contributions: (e) electrolyte solution; (f) thin film, and (i) film/electrolyte interface.

TABLE 2 Concentration Dependence of the Circuit Parameters: Film Resistance (R_f), Capacitance (C_f), and Warburg Impedance (W) for the Self-Supported Films

	HUP			NUP			CaUP		
C (M)	R_f (Ω)	C_f (nF)	W (Ω^{-1} s$^{0.5}$)	R_f (Ω)	C_f (nF)	W (Ω^{-1} s$^{0.5}$)	R_f (Ω)	C_f (nF)	W (Ω^{-1} s$^{0.5}$)
0.0001	2190	1.4	$0.87 \cdot 10^{-6}$	16240	4.6	$0.34 \cdot 10^{-6}$	93600	11.4	$0.38 \cdot 10^{-6}$
0.0005	1655	2.2	$1.01 \cdot 10^{-6}$	10955	6.9	$0.38 \cdot 10^{-6}$	65120	15.6	$0.49 \cdot 10^{-6}$
0.001	1310	4.3	$1.14 \cdot 10^{-6}$	7620	5.3	$0.42 \cdot 10^{-6}$	36140	18.2	$0.66 \cdot 10^{-6}$
0.005	950	9.1	$2.73 \cdot 10^{-6}$	3140	10.01	$0.60 \cdot 10^{-6}$	18450	22.1	$0.84 \cdot 10^{-6}$
0.01	820	11.6	$4.09 \cdot 10^{-6}$	1734	4.81	$0.72 \cdot 10^{-6}$	8340	17.4	$1.02 \cdot 10^{-6}$

"neutral" electrolyte (NaCl), for different salt concentrations, but an electrokinetic characterization of the material/electrolyte solution interface using UPP colloidal particles was also carried out:

The electrokinetic mobility, v, of UPP colloidal particles was measured by means of a Zeta Sizer IIc (Malvern Instruments) with $UO_2(NO_3)_2$ or $H_2O_3PC_6H_5$ solutions at different concentrations ($10^{-4} < c(M) < 10^{-2}$ M). From v values, assuming the Helmholtz–Smoluchowski equation [43], the zeta potential can be obtained, and its dependence on concentration is shown in Fig. 12. Different signs for the zeta potential of the

TABLE 3 Concentration Dependence of the Circuit Parameters for the UPP Film: Film Resistance Charge (R_f), Equivalent Capacitance (C_f^{eq}), Charge Transfer Resistance (R_{ct}) and Double Layer Equivalent Capacitance (C_{dl}^{eq})

C (M)	R_f (kΩ)	C_f^{eq} (μF)	R_{ct} (MΩ)	C_{dl}^{eq} (μF)
0.0001	8.14	0.14	15.0	12.2
0.0005	4.42	0.16	1.8	26.0
0.001	2.45	0.25	1.4	28.3
0.005	1.25	0.33	1.0	32.2
0.01	0.92	0.45	0.3	33.8

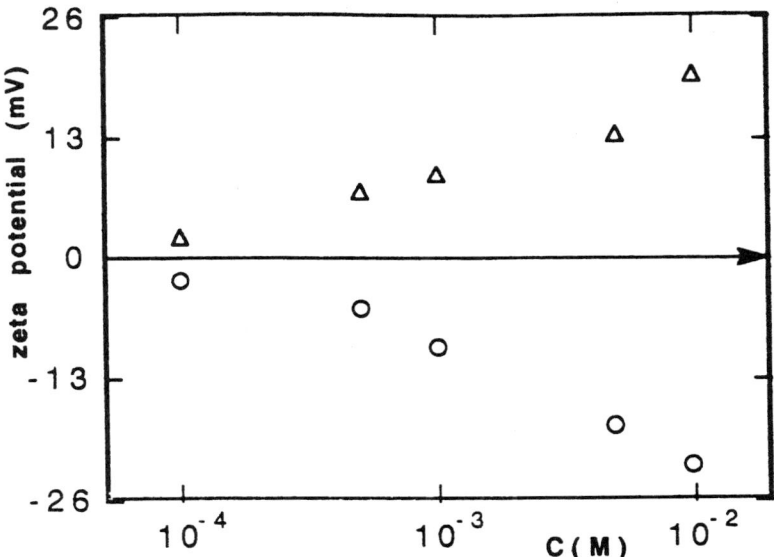

FIG. 12 Zeta potential versus electrolyte concentration for UPP particles: (△) $UO_2(NO_3)_2$, (○) $H_2O_3PC_6H_5$.

UPP particles were obtained depending on the electrolyte, but also an increase in zeta potential values when the concentration of the electrolyte solution increases due to a higher adsorption of ions on the UPP surface.

From impedance measurements, the electrical resistance and equivalent capacitance of the UPP film and the porous support, with the same electrolyte at both sides of the films, were obtained. R_f^{UPP} and R_f^{ANP} values are strongly dependent on electrolyte concentration, and they were fitted to the following expression: $R_f(c) = R_f^* c^{-a}$; values of R_f^* and a are indicated in Table 4, for the three different electrolytes studied. Equivalent capacitance for the UPP film strongly depends on both electrolyte and concentration, and their average values are also shown in Table 4. A comparison of R_f^* and a values determined for the UPP film and the alumina porous support shows important differences in both parameters when

TABLE 4 Electrical Resistance Parameters, R_f^* and a, and Salt Permeability for the UPP film and the Anopore Support in Contact with $UO_2(NO_3)_2$, $H_2O_3PC_6H_5$, and NaCl Solutions

	UPP film			ANP support		
	$UO_2(NO_3)_2$	$H_2O_3PC_6H_5$	NaCl	$UO_2NO_3)_2$	$H_2O_3PC_6H_5$	NaCl
R_f^*	52.0	67.2	8.85	2.69	1.95	5.05
a	-0.655	-0.618	-0.716	-0.957	-0.948	-0.647
$\langle C^{eq}\rangle$ (nF)	7.5±1.9	8.2±2.0	7.3±1.8	—	—	—
P_s (m/s)	$0.82 \cdot 10^{-7}$	$0.6 \cdot 10^{-7}$	$11 \cdot 210^{-7}$	$5.5 \cdot 10^{-6}$	$2.9 \cdot 10^{-6}$	$7.27 \cdot 10^{-6}$

the UPP generating electrolytes are considered, but for the NaCl no significant differences in the electrical resistance of both films were obtained, which indicates that no strong interactions between the UPP material and the "neutral" salt exists.

Diffusion measurements allow the determination of salt permeability across the films. Taking into account Fick's first law (for a quasi-steady state), the following expression can be written [32]:

$$\frac{dc_2}{dt} = \frac{S}{V_0} P_s (c_1 - c_2) = \frac{S}{V_0} P_s \Delta c \qquad (24)$$

where P_S is the salt permeability and S and V_0 are the film area and the volume of the solution at the side of concentration c_2, respectively. Salt permeability through the UPP film and the Anopore support, $P_s(UPP)$ and $P_s(ANP)$, respectively, for the different electrolytes were determined by the fitting of straight lines as shown in Fig. 13. From these results, salt permeability across the UPP layer can be determined [44], and the following values were obtained for each electrolyte:

$UO_2(NO_3)_2$: $\quad P_s^{UPP} = 8.3 \times 10^{-8}$ m/s

$H_2O_3PC_6H_5$: $\quad P_s^{UPP} = 6.2 \times 10^{-8}$ m/s

NaCl: $\quad P_s^{UPP} = 1.3 \times 10^{-6}$ m/s

Concentration potential for both UPP film and Anopore support were also measured at a constant concentration ratio $c_1/c_2 = h = 2$ (10^{-4} M $\leq c_1 \leq 5 \times 10^{-3}$ M) and the experimental values are shown in Fig. 14. From concentration potential, $\Delta\phi$, and film conductance, λ_f, values (or resistance $R_f = 1/\lambda_f$), the ionic permeabilities across the film, P_i, can be determined [45]:

$$P_+/P_- = \exp\left(\frac{F}{RT}\Delta\phi\right) - \frac{h}{1 - h\exp(F/RT)\Delta\phi} \qquad (25)$$

$$P_+ + P_- = \frac{\lambda RT}{F^2 S z_i^2 c_{avg}} \qquad (26)$$

P_+ and P_- are the cation and anion permeability, z_i is the valence of the ions, R and F are the gas and Faraday constants, respectively, and T is the temperature of the system. Conductance values at the average concentration used in the concentration potential measurements [$c_{avg} = (c_1 + c_2)/2$] were obtained by extrapolation of the film resistance–

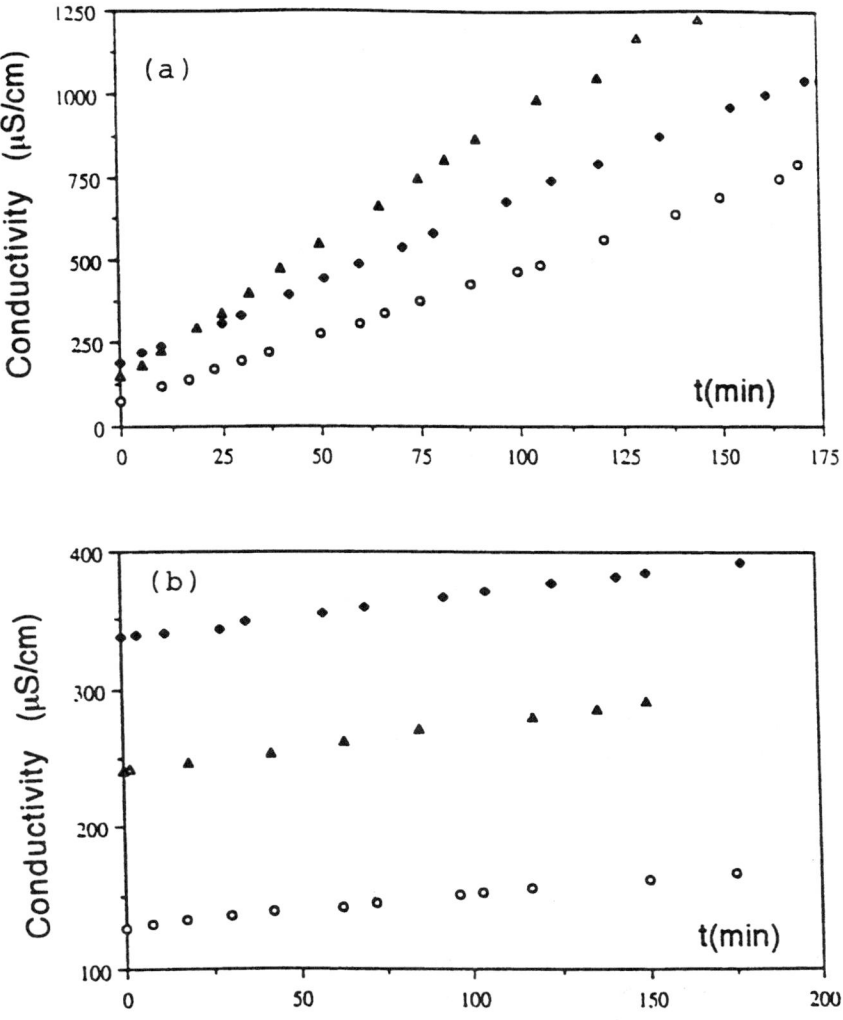

FIG. 13 Salt conductivity versus time in diffusion measurements: (○) NaCl, (▲) $UO_2(NO_3)_2$, (◆) $H_2PO_3C_6H_5$. (a) Anapore support, and (b) UPP film.

concentration dependence results determined by impedance spectroscopy measurements. A comparison of the cationic permeabilities, $P(UO_2^{2+})$ and $P(H^+)$, determined by Eqs. (25) and (26) for the UPP film and the Anapore support, shows an important reduction of the cationic permeability across the UPP film due to the UPP layer; this reduction is around 80% for UO_2^{2+} and 30% for H^+, which agrees with the small size and higher mobility of protons.

From its structure and composition, it was expected that the uranyl phenylphosphonate could act as a good insulator in electrolyte media, but the results obtained for the different characteristic transport parameters determined by electrochemical measurements through a UPP film clearly indicate that ions can move in the material.

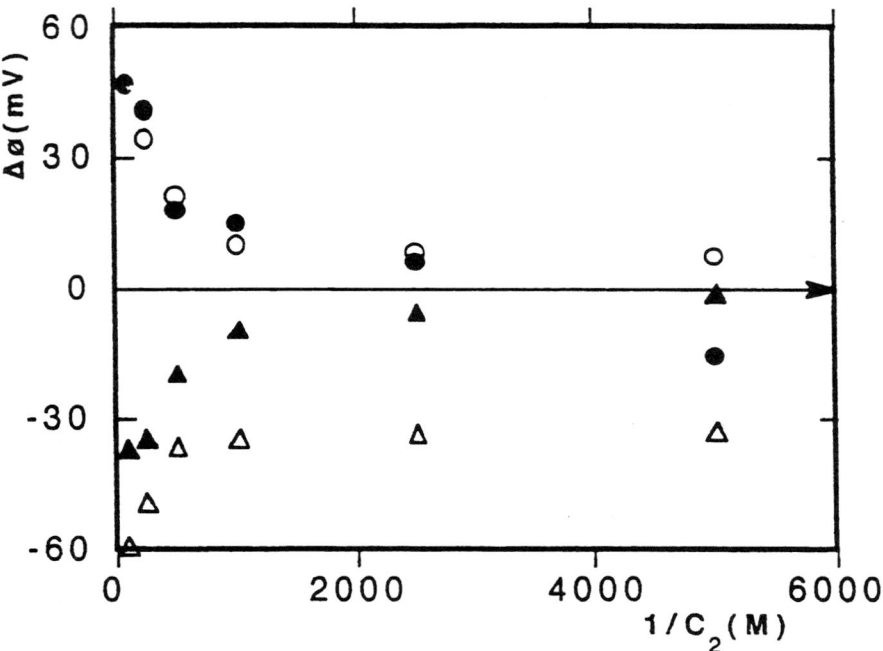

FIG. 14 Concentration potential ($\Delta\phi$) versus electrolyte concentration ($1/c_2$). UPP film: ($\triangle$) $UO_2(NO_3)_2$ and ($\bullet$) $H_2O_3PC_6H_5$. ANP alumina support: ($\triangle$) $UO_2(NO_3)_2$ and ($\bigcirc$) $H_2O_3PC_6H_5$.

These results show how the presence of hydrophobic moieties (phenyl groups) in the layered space of UPP does not prevent the electrolyte invasion, because of which the material becomes permeable to ions.

In this chapter, it was shown how the electrical properties of inorganic thin films in electrolyte media can be evaluated by a multitechnique methodology (alternating and direct current, electrophoretic mobility, etc.) that allows the determination of characteristic parameters for the inorganic materials themselves but also for the solid–liquid interface. The main interest for the study of these kinds of systems is related to the possible application of inorganic films as membranes. In this context, inorganic and inorganic-organic materials like those studied in this chapter can present certain advantages:

1. Good stability in hard media.
2. Selectivity to some ions (mainly, precipitate generating ions) without presenting isolate/hydrophobic character in aqueous media. Their electrical properties (resistance, ionic mobility, etc.) are strongly dependent on the electrolyte solutions.
3. Due to the adsorbed layers at both film/solution interfaces they do not act as "neutral" systems; this means that their electrical responses depend very much on the ions and external conditions (external voltage).

ACKNOWLEDGMENTS

We wish to thank Dr. M. Martinez and Dr. A. Cabeza for their help in the synthesis of the materials and d.c. measurements, and Dr. C. Criado for the preparation and assistance with the computer program for a.c. measurements. We also thank Grupo de Física de Coloides (Departamento de Física Aplicada), Universidad de Granada (Spain), for helping us in the measurements of electrophoretic mobility.

REFERENCES

1. R. R. Bhave, ed., *Inorganic Membranes, Synthesis, Characteristics and Applications*, Van Nostrand Reinhold, 1991.
2. H. P. Hsieh, *Inorganic Membranes for Separation and Reaction*, Membrane Science and Technology Series, Elsevier, 1996.
3. S. Sekido and Y. Ninomiya. Solid State Ionics *324*:153 (1981).
4. M. Casciola, in *Two and Three-Dimensional Inorganic Networks*, Comprehensive Supramolecular Chemistry, Vol. 7, (G. Alberti and T. Bein, eds.), Pergamon Press, 1996.
5. S. Drumel, V. Penicaud, D. Deniaud, and B. Bujoli, Trends in Inorganic Chemistry *4*:13 (1996).
6. E. W. Stein, A. Clearfield, and M. A. Subramanian. Solid State Ionics *83*:113 (1996).
7. G. Alberti and M. Casciola. Solid State Ionics *97*:177 (1997).
8. A. T. Howe, S. H. Sheffield, P. E. Child, and M. G. Shilton. Thin Solid Films *67*:365 (1980).
9. A. T. Howe and M. G. Shilton. J. Solid State Chem. *28*:345–361 (1979).
10. M. Casciola, U. Constantino, S. Fazzini, and G. Tosoratti. Solid State Ionics *8*:27 (1983).
11. H. Byrd, S. Whipps, J. K. Pike, J. Ma, S. E. Nagler, and D. R. Talham. J. Am. Chem. Soc. *116*:295 (1994).
12. G. Alberti, M. Casciola, and M. Palombari. Russ. J. Electrochem. (Transl. from Elektrokhimiya 29, 1433–1441) *29*:1257 (1993).
13. A. K. Kazir-Katchalsky, P. Hirchh-Ayalon, and I. Michelis, Isr. J. Chem. *20*:91 (1984).
14. J. Benavente, A. Cabeza, M. Martinez, and S. Bruque. Solid State Ionics *61*:175 (1993).
15. M. Pham-Thi, Ph. Adet, and G. Velasco. Appl. Phys. Lett. *48*:1348 (1986).
16. A. Clearfield. Progress in Inorganic Chemistry *47*:371 (1998).
17. R. Beutner. J. Phys. Chem. *17*:344 (1913).
18. Z. Freise. Phys. Chem., N. F. *15*:48 (1958).
19. F. A. Siddiqi, N. Lakshminarayanaiah, and M. N. Beg. J. Polym. Sci. *9*:2869 (1971).
20. F. A. Siddiqi, M. N. Beg, S. P. Singn, and A. Haque. Bull Chem. Soc. Jpn. *49*:2869 (1976).
21. M. N. Beg, F. A. Siddiqi, and R. Shyam. Can. J. Chem. *55*:1680 (1977).
22. A. Ayalon. J. Membrane Sci. *20*:91 (1984).
23. G. Bahr and A. Ayalon. J. Membrane Sci. *20*:101 (1984).
24. G. Alberti. Pontif. Acad. Sci. *40*:629 (1986).
25. J. Benavente, J. R. Ramos-Barrado, S. Bruque, and M. Martinez. J. Chem. Soc. Faraday Trans. *90*:3103 (1994).
26. J. Benavente, J. R. Ramos-Barrado, A. Cabeza, S. Bruque, and M. Martinez, 9th CIMTEC, Firenze, 1998.
27. B. Morosin. Phys. Lett. A *65*:53 (1978).
28. M. A. G. Aranda, A. Cabeza, S. Bruque, D. M. Poojary, and A. Clearfield. Inorg. Chem. *37*:1827 (1998).
29. D. M. Poojary, A. Cabeza, M. A. G. Aranda, S. Bruque, and A. Clearfield. Inorg. Chem. *35*:1468 (1996).
30. A. Cabeza, M. Martinez, J. Benavente, and S. Bruque. Solid State Ionics *51*:127 (1992).
31. F. G. Helfferich, *Ion Exchange*, McGraw-Hill, 1962.
32. G. Jonsson and J. Benavente. J. Membrane Sci. *69*:29 (1992).

33. J. Benavente, J. M. García, J. G. de la Campa, and J. de Abajo. J. Membrane Sci. *114*:51 (1996).
34. A. Cabeza, J. Benavente, M. Martinez, and S. Bruque. Key Engineering Materials *61*:473 (1991).
35. C. Gabrielli, Tech. Report No. 004/83, Solartron Instruments, 1980.
36. J. R. Macdonals, *Impedance Spectroscopy*, John Wiley, 1987.
37. T. Hanai, K. Zhao, K. Asaka, and K. Asami. J. Membrane Sci. *64*:153 (1991).
38. J. R. Ramos-Barrado, J. Benavente, S. Bruque, and M. Martinez. J. Colloids Interface Sci. *170*:550 (1995).
39. A. K. Jonscher, *Dielectric Relaxation in Solids*. Chelsea Dielectric Press, London, 1983.
40. B. A. Boukamp. Solid State Ionics *18&19*:136 (1986).
41. J. R. Ramos Barrado, J. Benavente, and A. Heredia. Arch. Biochem. Biophys. *306*:337 (1993).
42. J. Benavente, J. R. Ramos Barrado, M. Martinez, and S. Bruque. J. Applied Electrochem. *25*:68 (1995).
43. R. Hunter, *Zeta Potential in Colloid Science: Principles and Applications*, Academic Press, London, 1981.
44. O. Kedem and A. Katchalsky. Trans. Faraday Soc. *59*:1941 (1963).
45. N. Lakshminarayanaiah, *Transport Phenomena in Membranes*, Academic Press, 1969.

17
Thermosensitive Gels

KIMIKO MAKINO Faculty of Pharmaceutical Sciences, Science University of Tokyo, Tokyo, Japan

I.	Introduction	591
II.	Thermosensitive Properties of Poly(N-Isopropylacrylamide) Hydrogel	592
	A. Preparation of cross-linked poly(N-isopropylacrylamide) hydrogel plate	592
	B. Wettability of a hydrogel surface	592
	C. Equilibrium swelling ratio of a hydrogel	593
	D. Amount of negative fixed charges in a hydrogel	596
	E. Charge distribution in a hydrogel	596
III.	General Aspects of Electrokinetic Properties of a Hydrogel Surface	596
IV.	Changes in Interfacial Electric Properties of Poly(N-Isopropylacrylamide) Hydrogels Depending on Temperature	599
	A. Electro-osmotic velocity on glass surfaces covered with poly(N-isopropylacrylamide) hydrogels	599
	B. Electrophoretic mobility of latex particles covered with poly(N-isopropylacrylamide) hydrogel layers	602
	References	618

I. INTRODUCTION

Considerable research attention has been focused recently on materials that change their structure and properties in response to external stimuli [1–10]. Many kinds of stimuli are available, such as temperature, pressure, pH, electric field. A number of studies have been carried out on poly(N-isopropylacrylamide) hydrogel with a particular emphasis on its thermosensitive properties [11–15] and application to drug delivery systems as well as biomaterials [16–19]. Poly(N-isopropylacrylamide) hydrogel is in swollen state at temperatures lower than its phase transition temperature (LCST, 33°C) and is in a shrunken state above it. At room temperature (below its phase transition temperature), the hydrogel is almost transparent, and it is in a swollen state, while as the temperature rises above 33°C, it becomes opaque, since water molecules are squeezed out of it. This macroscopic change in the hydrogel volume has been explained by Tanaka by four forces that control the hydrogel structure at equilibrium [20].

The surfaces of poly(N-isopropylacrylamide) and its derivatives exhibit reversible hydrophilic/hydrophobic changes depending on temperature. Due to this property, poly(N-isopropylacrylamide) hydrogel has also been applied to bioseparation materials. Cultured cells on poly(N-isopropylacrylamide)-grafted surfaces can be easily recovered by lowering the temperature [19].

Although poly(N-isopropylacrylamide) is itself neutral, the hydrogel carries negative fixed charges, when initiators like potassium peroxodisulfate and ammonium peroxodisulfate are used during the polymerization procedure. With a charged poly(N-isopropylacrylamide) hydrogel, it was found that the interfacial electric properties of poly(N-isopropylacrylamide) hydrogel exhibit a discontinuous change around LCST, as described later [21–25]. A latex particle and a glass surface both covered with poly(N-isopropylacrylamide) hydrogel were prepared, and the temperature-dependent changes in electrokinetic properties of the hydrogel surface were studied. The data were analyzed with an electrokinetic theory for "soft" surfaces developed by Ohshima [26].

II. THERMOSENSITIVE PROPERTIES OF POLY(N-ISOPROPYLACRYLAMIDE) HYDROGEL

Using a flat poly(N-isopropylacrylamide) hydrogel plate, the dependencies of both the swelling ratio of the hydrogel and the wettability of the hydrogel surface on temperature were studied. These measurements give us information about the thermosensitive properties of uniformly polymerized poly(N-isopropylacrylamide) hydrogel with a thickness of a few mm. Also, the total amount of negative charge located in a hydrogel was measured by a potentiometric titration. From the comparison of the total amount of charge in a unit volume of hydrogel with the swelling ratio of the hydrogel at each temperature, the dependence of the charge density in the hydrogel on temperature can be discussed [23–25].

A. Preparation of Cross-Linked Poly(N-isopropylacrylamide) Hydrogel Plate

Cross-linked poly(N-isopropylacrylamide) hydrogel was synthesized using N-isopropylacrylamide (NIPAM), N,N'-methylene-bis-acrylamide (MBAAm), ammonium peroxodisulfate (AP), and N,N,N'N-tetramethylethylenediamine (TEMED). MBAAm, AP, and TEMED were, respectively, used as a cross-linker, a redox initiator, and a coinitiator. A mixture of 0.5 g of NIPAM, 0.0068 g of MBAAm, and 10 μL of TEMED were dissolved in 6 mL of distilled water. This solution was bubbled with dry nitrogen gas for 15 minutes at 4°C. After 0.01 g of AP was added to the solution, it was injected quickly between two Mylar sheets separated with a 0.5 mm-thick Teflon gasket, backed by two glass plates. Polymerization was then performed at room temperature for 6 hours. The hydrogel plates formed were separated from the Mylar sheets and rinsed with distilled water to remove unreacted monomers for 1 week prior to the following measurements.

B. Wettability of a Hydrogel Surface

The dynamic contact angle of the poly(N-isopropylacrylamide) hydrogel surface in distilled water was measured at various temperatures with the Wilhelmy plate technique using an Orientec DCA-20 apparatus. The hydrogel was kept in distilled water at a desired

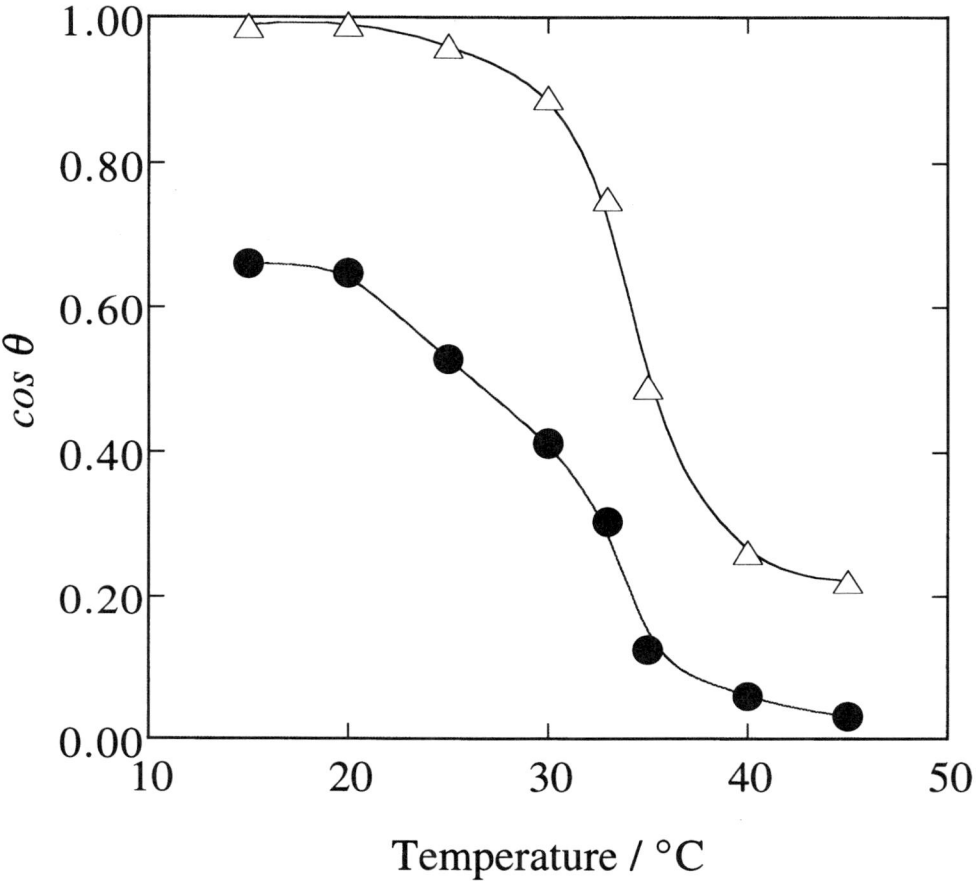

FIG. 1 Dynamic contact angle (θ) of poly(N-isopropylacrylamide) hydrogel surface as a function of temperature: ●, advancing angle; and △, receding angle. (From Ref. 25.)

temperature for 1 day. The dynamic contact angle of poly(N-isopropylacrylamide) surface drastically increases around its phase transition temperature (33°C) as shown in Fig. 1, showing that the hydrogel surface becomes more hydrophobic as the temperature rises.

C. Equilibrium Swelling Ratio of a Hydrogel

The swelling ratio of the hydrogel at equilibrium was measured in a phosphate buffer solution at pH = 7.4 as a function of temperature and the ionic strength of the solution. The hydrogel was immersed in each buffer solution at a desired temperature for 1 day. The sample was removed from the buffer solution and was frequently weighed after being trapped with a filter paper to remove excess water on the surface. The swelling ratio, Ws/Wp, is defined as the ratio of the weight of adsorbed water (Ws) to the weight of the sample at a dry state (Wp). Solutions with different ionic strengths were made by dilution of a buffer solution with the ionic strength of 0.154 mol/dm³.

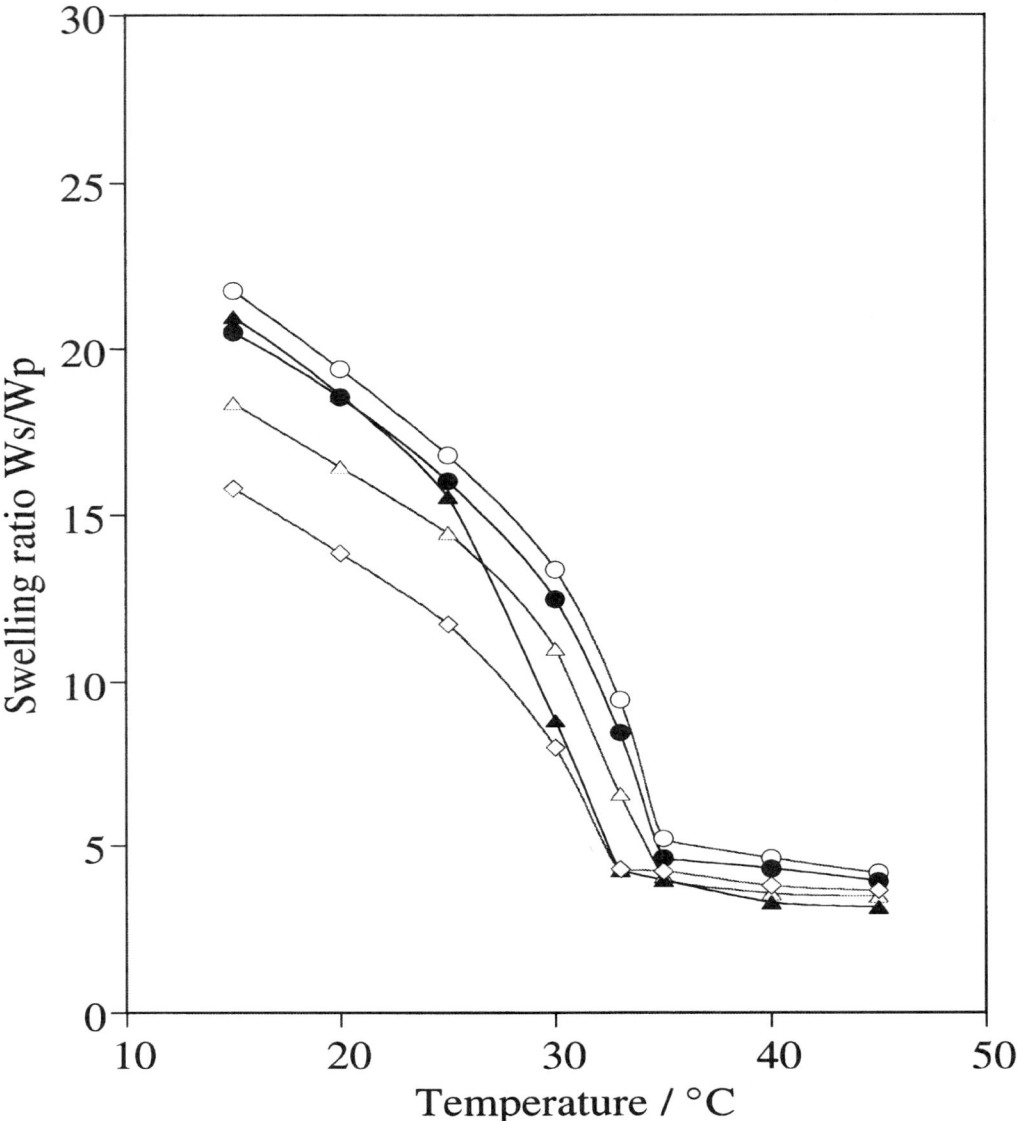

FIG. 2 Volume transition of poly(N-isopropylacrylamide) hydrogel plate in phosphate buffer solution at pH = 7.4 as a function of temperature at various ionic strengths: ○, 0.005 mol/dm^3; ●, 0.02 mol/dm^3; △, 0.04 mol/dm^3; ◇, 0.1 mol/dm^3; and ▲, 0.154 mol/dm^3.

Figure 2 shows the swelling ratio of the poly(N-isopropylacrylamide) gel at equilibrium as a function of temperature in buffer solutions at pH = 7.4 with various ionic strengths. Phase transition is clearly observed around 33°C, below which poly(N-isopropylacrylamide) gel is in a swollen state and above which it is in a shrunken state. Usually, the swelling ratio of a charged hydrogel at equilibrium becomes higher in a solution with lower ionic strength, because of stronger electrostatic repulsion acting among the hydrogel-fixed charges as shown in Fig. 3. Figure 3 shows the swelling ratio of poly(acrylic acid-co-N-isopropylacrylamide) hydrogel plate at various temperatures.

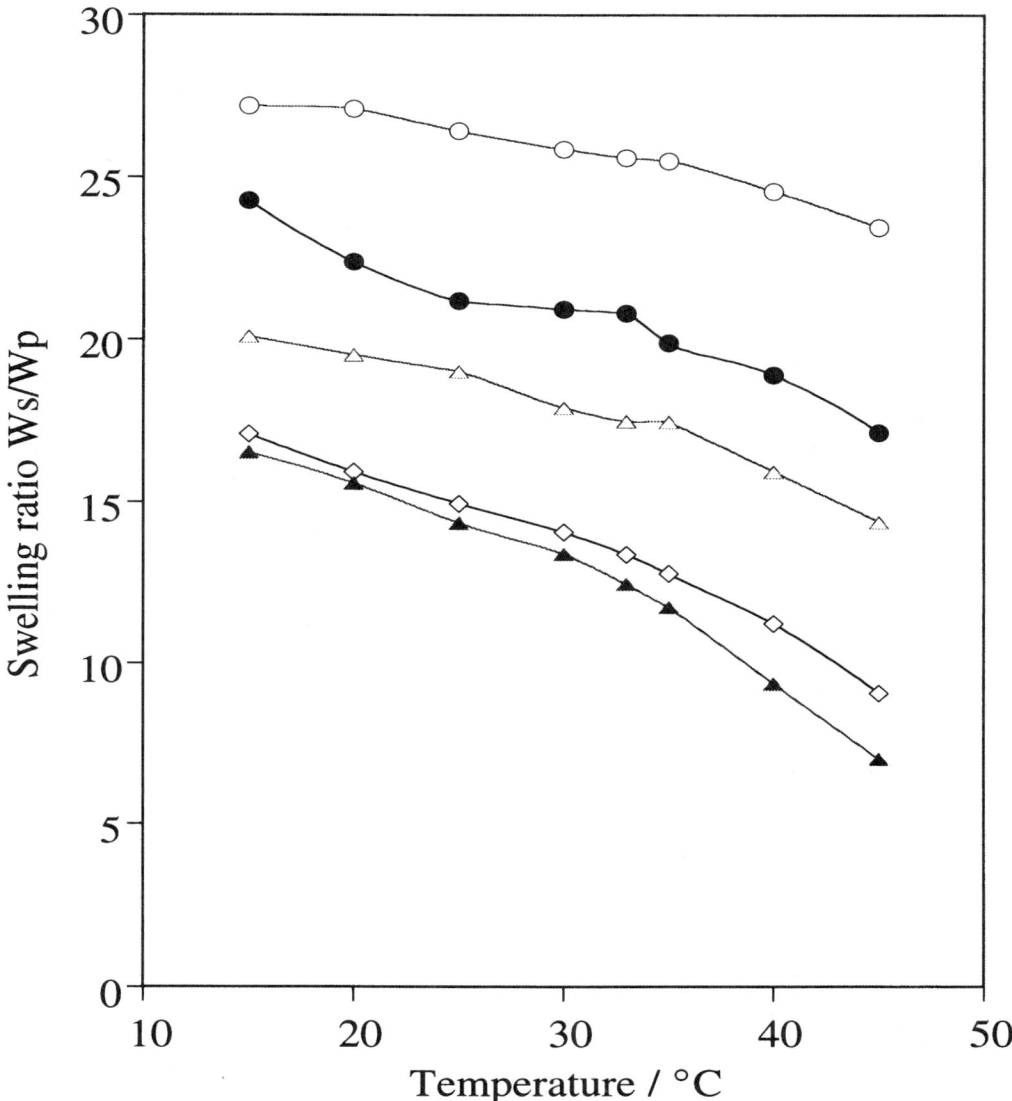

FIG. 3 Volume transition of poly(acrylic acid-*co*-*N*-isopropylacrylamide) hydrogel plate in phosphate buffer solution at pH = 7.4 as a function of temperature at various ionic strengths: ○, 0.005 mol/dm^3; ●, 0.02 mol/dm^3; △, 0.04 mol/dm^3; ◇, 0.1 mol/dm^3; and ▲, 0.154 mol/dm^3.

The hydrogel was prepared from a monomer mixture of 5 mol% of acrylic acid and 95 mol% of *N*-isopropylacrylamide. In a medium with lower ionic strength, the swelling ratio of the hydrogel is higher than in that with higher ionic strength at each temperature. Also, the temperature dependence of the swelling ratio is less clearly observed than that observed in poly(*N*-isopropylacrylamide) hydrogel as shown in Fig. 2. In Fig. 2, the swelling ratio of the poly(*N*-isopropylacrylamide) hydrogel plate exhibits a weak dependence on the ionic strength, since there are relatively small amounts of fixed charges in this hydrogel plate.

D. Amount of Negative Fixed Charges in a Hydrogel

The prepared hydrogel carries sulfonic acid groups, since AP was used as an initiator for polymerization of N-isopropylacrylamide. To determine the fixed negative charges, potentiometric titration was performed as follows. A certain amount of hydrogel was immersed in 0.01 mol/dm^3 NaOH solution to allow complete dissociation of all sulfonic acid groups at the terminal of the polymer chains. Back-titration with 0.01 mol/dm^3 HCl was performed at 25°C, and the residual NaOH concentration in the solution was determined potentiometrically. By the titration, the number of sulfonic acid groups per unit weight of a dry hydrogel was determined to be 10^{-4} mol/g.

E. Charge Distribution in a Hydrogel

It is assumed that the ionized groups of the valence z are distributed at a uniform density of N (or the charge density of zN) in poly(N-isopropylacrylamide) hydrogel, because polymerization proceeds uniformly in a plate, and both the shrinkage and the swelling of the hydrogel are considered to occur uniformly on the plate. In the present case where the ionized groups are sulfonic acid ones, $z = -1$. The charge density zN was calculated from the value of 10^{-4} mol/g obtained from the potentiometric titration and from the values of the swelling ratio. The hydrogel volume was calculated from the value of Ws/Wp at the ionic strength of 0.02 mol/dm^3 in Fig. 2, because the value of Ws/Wp was little affected by the ionic strength. The results are shown in Fig. 4, which shows that negative charge density increases as the temperature is raised and zN changes from -0.008 mol/dm^3 at 25°C to -0.03 mol/dm^3 at 40°C.

III. GENERAL ASPECTS OF ELECTROKINETIC PROPERTIES OF A HYDROGEL SURFACE

Electrokinetic measurements have provided us with fundamental knowledge about surface properties. Electrophoretic mobility measurements are useful for studying the surface properties of colloidal particles in an electrolyte solution. On the other hand, electro-osmotic velocity measurements are advantageous to the study of the properties of a flat surface. Theoretically, these two measurements give us the same information about the surface properties. A polymer hydrogel is a cross-linked polymer network swollen in an aqueous solution. In a soft structure of gels, the motion of polymer networks and the diffusion of ions take place easily. The dependence of the electrophoretic mobility of particles with soft surfaces upon ionic strength of the dispersing medium was well explained via an electrophoretic mobility formula for "soft particles" developed by Ohshima [26]. This formula is applicable to the analysis of the dependence of electro-osmotic velocity on a soft flat surface upon ionic strength of the bulk solution. Both electrokinetic measurements and the explanation of data via the electrokinetic mobility formula for "soft surfaces" give us information about the surface charge density and the "softness" of soft surfaces [27–35]. The mobility formula involves two parameters [see Eq. (1)]: the fixed charge density and a parameter characterizing the "softness" of the polyelectrolyte layer, the latter of which is related to the reciprocal of the frictional coefficient in the polyelectrolyte layer. In this model, ions can penetrate the surface polymer layer, and the bulk solution can flow inside the surface layer. The ion-penetrable surface polymer layer is schematically shown in Fig. 5.

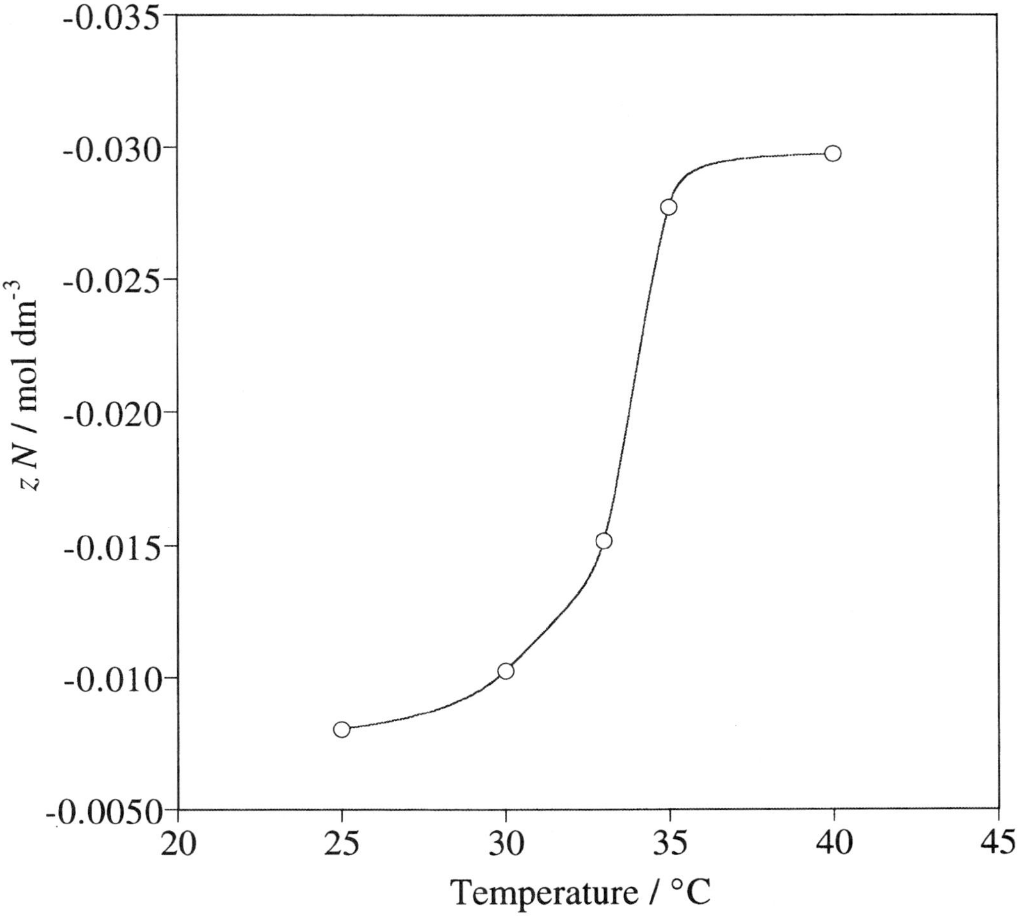

FIG. 4 Charge density, zN [mol/dm^3], as a function of temperature. (From Ref. 23.)

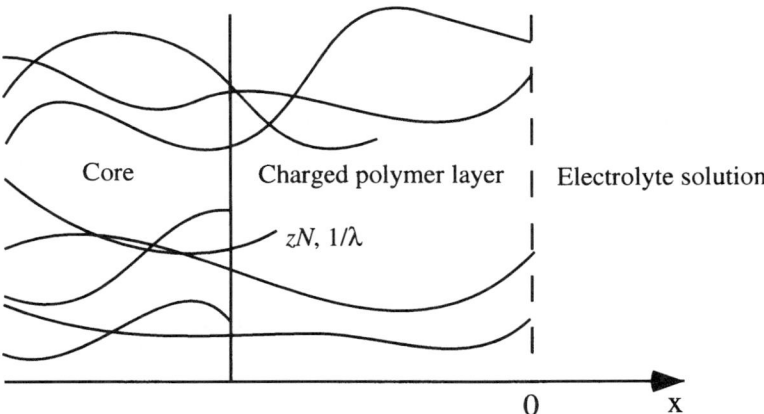

FIG. 5 Schematic representation of ion-penetrable surface polymer layer.

The electrophoretic mobility (μ) data were analyzed by Eq. (1):

$$\mu = \frac{\varepsilon_r \varepsilon_0}{\eta} \frac{\psi_0/\kappa_m + \psi_{DON}/\lambda}{1/\kappa_m + 1/\lambda} + \frac{zeN}{\eta \lambda^2} \tag{1}$$

with

$$\psi_{DON} = \frac{kT}{ve} \ln\left[\frac{zN}{2vn} + \left\{\left(\frac{zN}{2vn}\right)^2 + 1\right\}^{1/2}\right] \tag{2}$$

$$\psi_0 = \frac{kT}{ve}\left(\ln\left[\frac{zN}{2vn} + \left\{\left(\frac{zN}{2vn}\right)^2 + 1\right\}^{1/2}\right]\right.$$
$$\left. + \frac{2vn}{zN}\left[1 - \left\{\left(\frac{zN}{2vn}\right)^2 + 1\right\}^{1/2}\right]\right) \tag{3}$$

$$\lambda = \left(\frac{\gamma}{\eta}\right)^{1/2} \tag{4}$$

$$\kappa_m = \kappa \left[1 + \left(\frac{zN}{2vn}\right)^2\right]^{1/4} \tag{5}$$

$$\kappa = \left(\frac{2ne^2v^2}{\varepsilon_r \varepsilon_0 kT}\right)^{1/2} \tag{6}$$

Here, μ is the electrophoretic mobility of a colloidal particle covered with a layer of polyelectrolytes (or a hydrogel layer), in which ionized groups of valency z are uniformly distributed at a number density of N (m^{-3}). The symbol n (m^{-3}) is used for the bulk (number) concentration of symmetrical electrolytes with valency v in the dispersing medium. If one uses the molar scale, N and n are replaced by 1000 $N_A N$ and 1000 $N_A n$, where N_A is the Avogadro constant. η is the viscosity, γ is the frictional coefficient of the hydrogel layer, ε_r is the relative permittivity of the solution, ε_0 is the permittivity of a vacuum, ψ_{DON} is the Donnan potential of the hydrogel layer, ψ_0 is the potential at the boundary between the hydrogel layer and the surrounding solution, and κ is the Debye–Hückel parameter. ψ_0 is the surface potential of a particle covered with a hydrogel layer, and κ_m can be interpreted as the Debye–Hückel parameter of the hydrogel layer. The parameter λ characterizes the degree of friction exerted on the liquid flow in the hydrogel layer, and zeN represents the density of the fixed charges in the hydrogel layer. The reciprocal of λ, i.e., $1/\lambda$, has the dimension of length and can be considered to be a "softness" parameter, since in the limit $1/\lambda \to 0$, the particle becomes rigid.

On the other hand, the experimental data obtained from the electro-osmosis measurements are electro-osmotic mobility (μ) on a flat hydrogel surface:

$$\mu = -\frac{U_{eo}}{E} \tag{7}$$

where U_{eo} is the electro-osmotic velocity and E is the strength of an external electric field. The electro-osmotic mobility can be analyzed by the same formula for soft particles as given above. The mobility, μ, is analyzed by Eq. (8) [26]:

$$\mu = \frac{\varepsilon_r \varepsilon_0}{\eta} \frac{\psi_0/\kappa_m + \psi_{DON}/\lambda}{1/\kappa_m + 1/\lambda} + \frac{zeN}{\eta \lambda^2} \tag{8}$$

Here, μ in Eq. (8) agrees exactly with the expression for the electrophoretic mobility of a soft particle. By a curve-fitting procedure [29], the fixed charge density, zN, and softness parameter, $1/\lambda$, in a soft surface layer are determined.

IV. CHANGES IN INTERFACIAL ELECTRIC PROPERTIES OF POLY(N-ISOPROPYLACRYLAMIDE) HYDROGELS DEPENDING ON TEMPERATURE

A. Electro-Osmotic Velocity on Glass Surfaces Covered with Poly(N-Isopropylacrylamide) Hydrogels

The electro-osmotic flow of an electrolyte solution with various ionic strengths was measured on the hydrogel surface at various temperatures, with the ELS800 electrophoretic light scattering spectrophotometer (Photal, Otsuka Electronics, Japan). The hydrogel was prepared on the glass surface. The glass surface was pretreated with vinyltrichlorosilane as follows. A slide glass was immersed in a solution of volume fraction 2% (v/v) vinyltrichlorosilane in toluene for 5 min, dried in the air, and washed in ethanol and in distilled water. On the surface of a vinyltrichlorosilane-treated slide glass, N-isopropylacrylamide reacts with vinyl groups on the glass surface so that the prepared hydrogel sheet is immobilized on the glass surface.

The electro-osmotic mobility (μ) on the hydrogel surface immersed in an electrolyte solution was measured at pH = 7.4 as a function of the ionic strength and temperature of the solution. In Fig. 6, the electro-osmotic mobility was plotted against the ionic strength of the solution. We observe that μ ($-U_{eo}/E$) was negative at all temperatures and at all ionic strengths, and that μ becomes less negative as the ionic strength increases, while it does not reach zero even in a solution of very high ionic strength. Also, μ becomes more negative as the temperature increases.

The experimental results, obtained from the electro-osmosis measurements, were analyzed via a theory of electro-osmosis on a "soft" surface [26]. In the present hydrogel plate the fixed charges are distributed uniformly, as mentioned before. The electrolyte ions in the bulk solution phase can penetrate the hydrogel layer, and the electrolyte solution can flow parallel to the surface at some depth from the surface. Therefore the surface of this hydrogel is expected to behave as a "soft surface." Although Eq. (1) assumes symmetrical electrolytes, the electrolytes used are not symmetrical (cations are univalent but anions are not); the value of v is set approximately equal to unity ($v = 1$), since anions are less important for negatively charged latex particles. Also, in the present case, the fixed charges in the hydrogel arise from sulfonic acid groups ($z = -1$) so that ψ_{DON}, ψ_0, and the value of μ become negative.

Equation 1 involves two unknown parameters, N (m^{-3}) and $1/\lambda$ (nm), which represent the fixed charge density in the hydrogel surface and its softness, respectively. By the curve-fitting procedure [29], zN and $1/\lambda$ at each temperature were determined. Figure 6 shows the results of the calculation of μ as a function of the electrolyte concentration via Eq. (1) with solid curves. Each theoretical curve with a pair of zN and $1/\lambda$ revealed a good agreement with the experimental data over a wide range of the ionic strength at each temperature. This means that the surface of poly(N-isopropylacrylamide) hydrogel plate can be considered as a "soft surface" described by Eq. (1) between 25 and 40°C, and the fixed charges are distributed uniformly in the hydrogel at each temperature. Also, this agreement enables us to estimate the values of the unknown parameters N and $1/\lambda$ by a curve-fitting procedure. The best-fit curves (shown as solid lines in Fig. 6) at

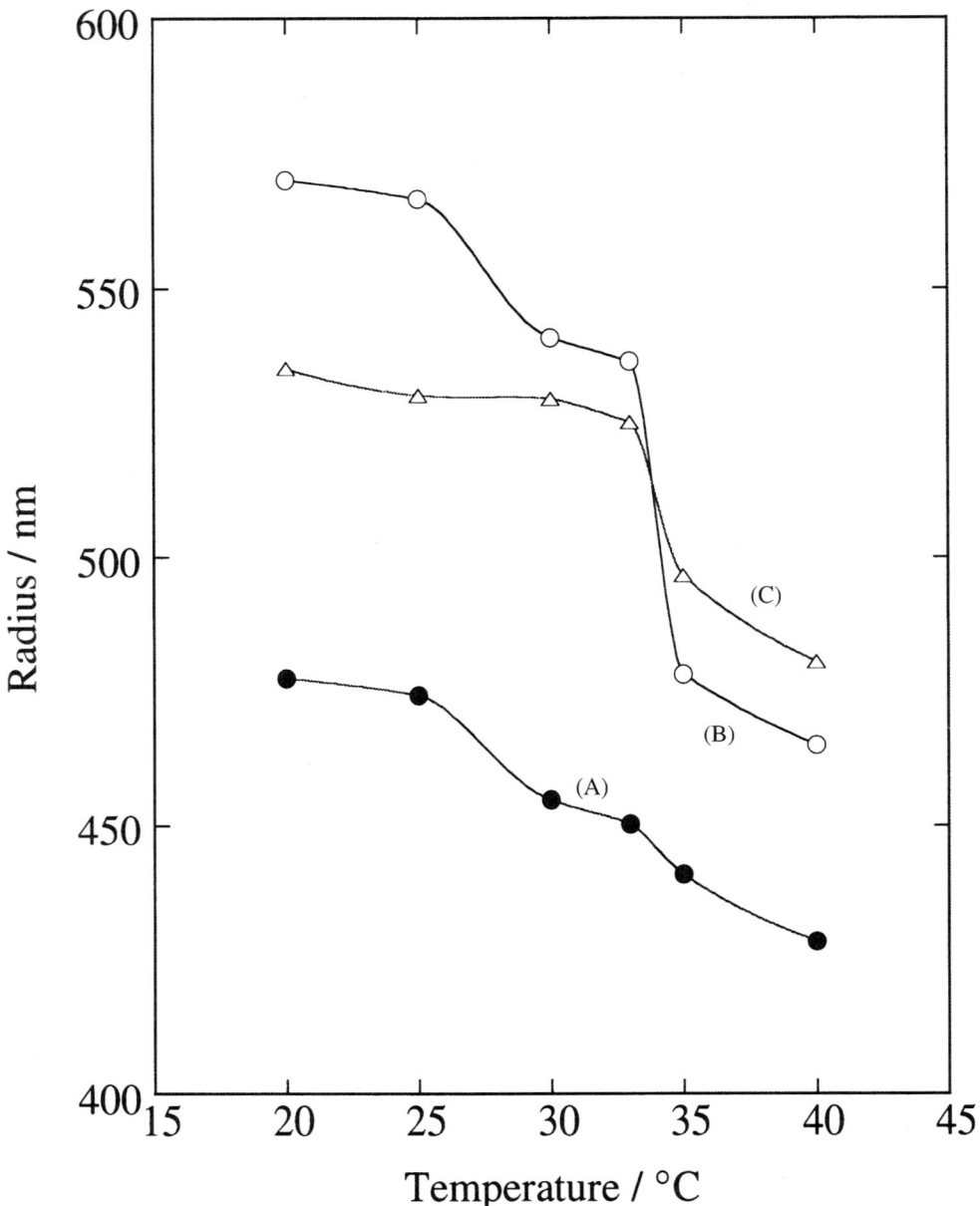

FIG. 6 The negative values of U_{eo}/E against the ionic strength of the bulk solution at several values of temperature. Symbols represent experimental data: ○, 25°C; ●, 30°C; ■, 33°C; △, 35°C; and □, 40°C. Solid curves are theoretical results calculated via Eq. (2) with $zN = -0.008$ mol/dm^3 and $1/\lambda = 3.0$ nm at 25°C (curve 1), $zN = -0.011$ mol/dm^3 and $1/\lambda = 3.0$ nm at 30°C (curve 2), $zN = -0.015$ mol/dm^3 and $1/\lambda = 3.0$ nm at 33°C (curve 3), $zN = -0.028$ mol/dm^3 and $1/\lambda = 2.1$ nm at 35°C (curve 4), and $zN = -0.032$ mol/dm^3 and $1/\lambda = 1.9$ nm at 40°C (curve 5). (From Ref. 23.)

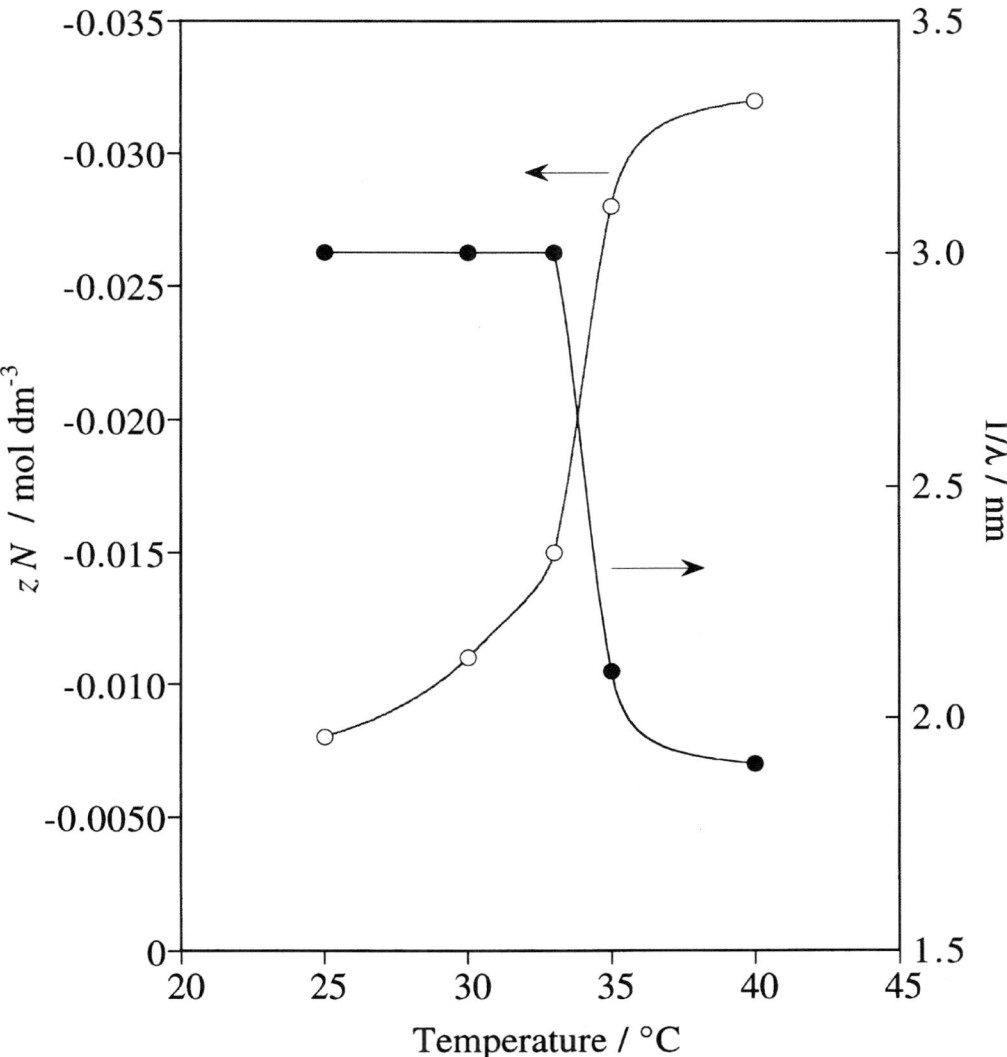

FIG. 7 Charge density, zN (mol/dm^3), and softness parameter, $1/\lambda$ (nm), of poly(N-isopropylacrylamide) surface as a function of temperature. Symbols are for charge density (zN), ○, and softness parameter ($1/\lambda$), ●. (From Ref. 23.)

different temperatures are obtained with $zN = -0.008$ mol/dm^3 and $1/\lambda = 3$ nm at 25°C; $zN = -0.011$ mol/dm^3 and $1/\lambda = 3$ nm at 30°C; $zN = -0.015$ mol/dm^3 and $1/\lambda = 3$ nm at 33°C; $zN = -0.028$ mol/dm^3 and $1/\lambda = 2.1$ nm at 35°C; $zN = -0.032$ mol/dm^3 and $1/\lambda = 1.9$ nm at 40°C, respectively. In the calculation, we used the viscosity η and the relative permittivity ϵ_r of distilled water at each temperature for those of the surrounding media. These values of the charge density (zN) and the softness parameter ($1/\lambda$) of the hydrogel were plotted as a function of temperature in Fig. 7.

As mentioned before, poly(N-isopropylacrylamide) hydrogel is negatively charged due to the residual sulfonic acid groups at the terminal of polymer chains. As shown in Fig. 7, the negative charge density in the hydrogel layer (zN) increases as the temperature rises. This change is most appreciable between 33 and 35°C. Also, the surface of the hydrogel becomes harder ($1/\lambda$ decreases) as the temperature rises. The softness parameter changes abruptly between 33 and 35°C, while it changes a little above 35°C and does not change at all below 33°C. By comparing with the temperature dependence of the change in Ws/Wp between 25 and 33°C (in Fig. 2), it is found that the shrinkage of the hydrogel increases the charge density, but affects little the friction of the liquid flow in the hydrogel surface. At 25°C there are about 100 water molecules around one monomer unit of poly(N-isopropylacrylamide), while at 30°C the number of water molecules located there decreases to 50 molecules, which is calculated from the molecular weight of N-isopropylacrylamide ($M_w = 113.17$) and that of water. By shrinkage of the hydrogel due to the temperature rise from 25 to 33°C, water molecules are squeezed out, making the distance between polymer chains shorter, while there is still enough space for the electrolyte solution to flow in the surface area, and $1/\lambda$ is little changed. As the temperature rises, the hydrogel loses more water molecules. At 35°C one poly(N-isopropylacrylamide) unit is surrounded by only ca. 25 water molecules, and the hydrogel becomes completely opaque. This decrease in the number of water molecules between 33 and 35°C as well as the decrease in the distance between polymer chains causes a discontinuous increase in the frictional coefficient in the hydrogen surface, so that $1/\lambda$ decreases from 3 to 2.1 nm in this temperature difference. Between 35 and 40°C, no appreciable change is observed in Ws/Wp or in zN or $1/\lambda$.

A comparison of Fig. 7 with Fig. 4 shows that the values of charge density (zN) determined from the electro-osmotic measurements agree with those calculated from the volume change of the hydrogel and the charge amount determined by a potentiometric titration.

B. Electrophoretic Mobility of Latex Particles Covered with Poly(N-Isopropylacrylamide) Hydrogel Layers

The electrophoretic mobility and the size of two types of latex particles with a core–shell structure were measured in comparison with those for latex particles with no shell structure. The core particle is negatively charged, while the outer layer composed of poly(N-isopropylacrylamide) hydrogel is negatively charged or uncharged, depending on the kind of initiator used in the polymerization procedure. These three kinds of latex particles were prepared by a surfactant-free emulsion polymerization and seed polymerization, as schematically shown in Fig. 8. It was confirmed that these particles were all monodisperse by a transmittance electron microscopic observation. The influence of the initiator used in the polymerization process upon the structure and the surface properties of the poly(N-isopropylacrylamide) hydrogel layer will be discussed later [21,22].

1. Preparation of Latex Particles

(a) Preparation of Poly(Styrene-co-N-Isopropylacrylamide) Microspheres: Preparation of "Core Particles" (Sample A). A mixture of 18 g of styrene (St) and 2 g of N-isopropylacrylamide (NIPAM) was dissolved in 185 mL of distilled water in a 300 mL round-bottomed flask equipped with a condenser, a nitrogen inlet, and a stirrer.

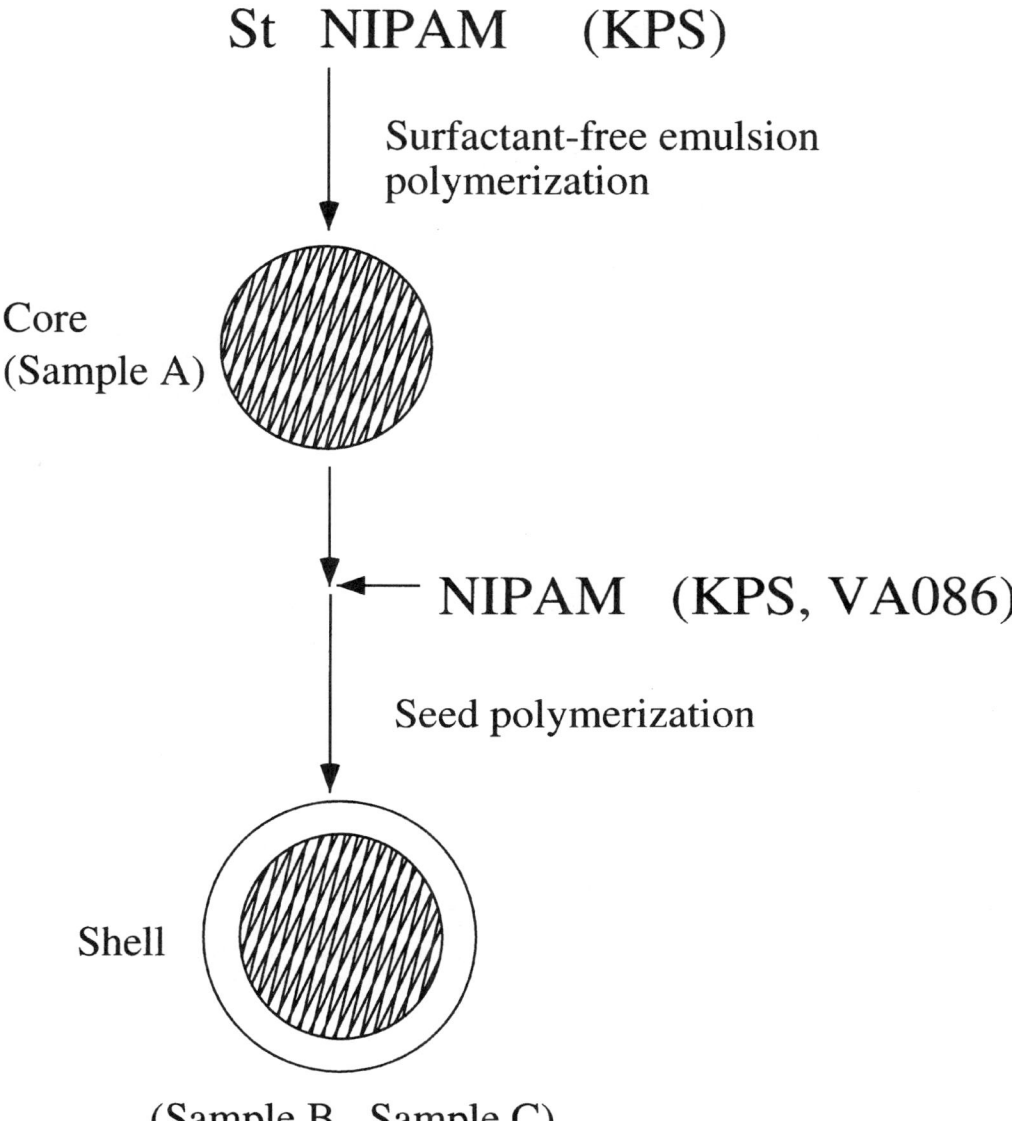

FIG. 8 Schematic representation of latex particle preparation. Sample A is prepared in the first step with KPS as an initiator, and Samples B and C are prepared in the second step from Sample A with the coverage of N-isopropylacrylamide hydrogel layers on it with KPS and VA-086 as an initiator, respectively. Styrene, N-isopropylacrylamide, and methylene-*bis*-acrylamide are abbreviated as St, NIPAM, and MBAAm, respectively. (From Ref. 22.)

Nitrogen was bubbled into the solution, and the mixture was stirred for 30 min at 300 rpm to remove oxygen from the monomeric phase. Polymerization was initiated by adding 15 mL of an aqueous solution containing 0.2 g of potassium peroxodisulfate (KPS) at 70°C. The reaction proceeded for 24 h at 70°C under stirring. The resulting poly(styrene-co-N-isopropylacrylamide) microspheres were dialyzed and purified by repetitive centrifugation, decantation, and redispersion.

(b) Preparation of "Core–Shell Particles" (Samples B and C). Five grams of core particles prepared as mentioned above were dispersed in 195 mL of distilled water, and subsequently 3 g of N-isopropylacrylamide and 0.15 g of N,N'-methylene-*bis*-acrylamide were added to the suspension. Five mL of an aqueous solution containing 0.05 g of KPS and 2,2'-azobis[2-methyl-N-(2-hydroxyethyl) propionamide] (VA-086) were added to the system, and polymerization proceeded for 4 h at 70 and 80°C under stirring at 200 rpm to prepare samples B and C, respectively. The other procedures were carried out in the same manner as those of "core particles." Monodisperse latex particles thus prepared were cleaned by repetitive centrifugation, decantation, and redispersion with distilled water.

2. Changes in Latex Particle Size Depending on Temperature

The hydrodynamic size was determined using photon correlation spectroscopy (Photal, Otsuka Electronics, LPA 3000/3100). Latex particles were redispersed in double distilled water, and their sizes were measured at 20, 25, 30, 33, 35, and 40°C.

Figure 9 shows the observed relationship between the particle radius of latex particles and the temperature of the dispersing medium. A sudden change in the particle radius was observed in sample B around the phase transition temperature of poly(N-isopropylacrylamide), that is, the particle radius altered by 105 nm from 570 nm at 20°C to 465 nm at 40°C, exhibiting an 18% decrease in radius. This particle size change is obvious in a relatively narrow temperature range between 33 and 35°C, as compared with other cases, and the particle radius decreases from 536 nm to 478 nm in this temperature range. Between 20 and 33°C, the particle radius decreases gradually from 570 nm to 536 nm as the temperature rises. Also, between 35 and 40°C, the particle radius decreases from 478 nm to 465 nm. On the other hand, the particle radius in sample A was less sensitive to the temperature, and the particle radius gradually decreases as the temperature rises. The particle radius changes by only 50 nm from 478 nm at 20°C to 428 nm at 40°C. Also, in sample C, the particle radius gradually decreases by 55 nm from 535 nm to 480 nm between 20 and 40°C. As mentioned above, samples B and C contain sample A as the particle core and carry the surface layer of poly(N-isopropylacrylamide) hydrogel on it. By comparing the temperature dependence of the radius in samples A and C, it is clear that the thickness of the poly(N-isopropylacrylamide) hydrogel layer of sample C surface, which covers the particle core (sample A), does not change significantly with temperature, while the particle core size changes. The thickness of the poly(N-isopropylacrylamide) hydrogel layer in sample B, however, decreases by 56 nm as the temperature rises from 92 nm at 20°C to 37 nm at 40°C. Therefore the temperature-dependent change in particle size in sample B is attributed to the changes both in the thickness of the poly(N-isopropylacrylamide) hydrogel layer and in the core particle size. This different behavior of the surface layer composed of poly(N-isopropylacrylamide) hydrogel observed in samples B and C may be considered to be due to the difference in their electrostatic properties. The poly(N-isopropylacrylamide) hydrogel layer at the surface of sample B is negatively

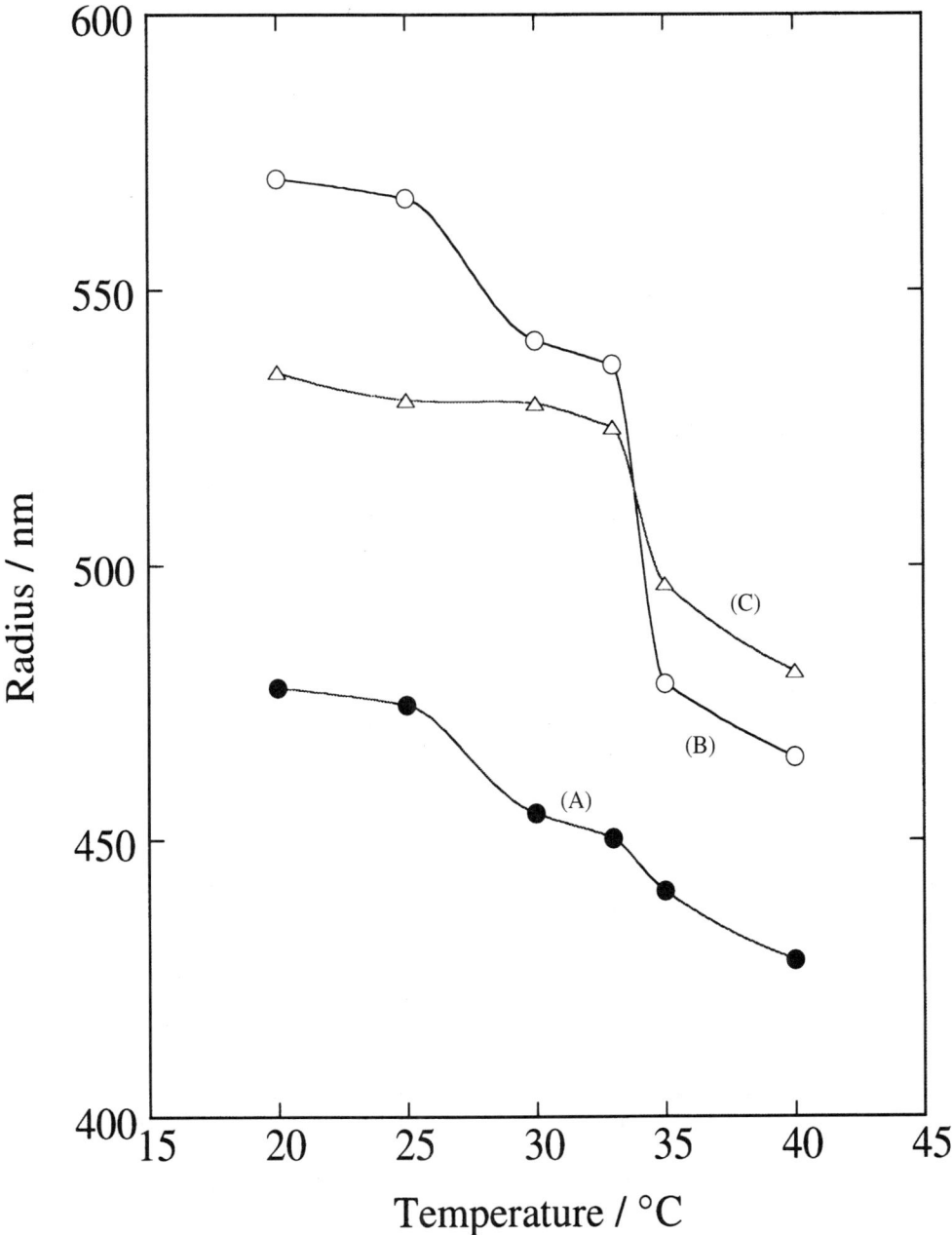

FIG. 9 Particle radius in double distilled water as a function of temperature for sample A (●), sample B (○), and sample C (△). (From Ref. 22.)

charged, but that of sample C is uncharged. In order to estimate the structural difference between these surface layers, the electrophoretic mobility of latex particles has been measured.

3. Electrophoretic Mobility of Latex Particles

The electrophoretic mobility of the latex particles was measured in a phosphate buffer solution at pH = 7.4 with various ionic strengths at five different temperatures, 25, 30, 33, 35, and 40°C, with an automated electrophoresis apparatus Pen Kem system 3000. Before the measurement, the latex particles were redispersed in the phosphate buffer solutions and were kept at the prescribed temperature for 1 h.

The charge density of these particles should change depending on the temperature, because these particles contain poly(N-isopropylacrylamide), which swells at lower temperatures than its LCST, 33°C, and shrinks at higher temperatures.

The core particle (sample A) showed negative electrophoretic mobilities at each temperature and ionic strength as shown in Fig. 10. As the temperature rises, the mobility value becomes more negative at any ionic strength. When the ionic strength is low, the mobility change depending on the temperature is apparent, and as the ionic strength increases, it becomes smaller and the mobility value becomes less negative in the entire temperature region. Especially, at the ionic strength of 0.005 mol/dm^3, the electrophoretic mobility changes from -3.7 to -6.5 μm s^{-1} V^{-1} cm when the temperature rises from 25 to 40°C.

For sample A, the electrophoretic mobility data were analyzed by Eq. (1). Figure 10 displays the mobility calculated as a function of the electrolyte concentration via Eq. (1). Each theoretical curve with a pair of zN and $1/\lambda$ revealed a good agreement with the experimental data over a wide range of ionic strengths at each temperature. This means that sample A shows the electrophoretic behavior of a "soft particle" as described by Eq. (1) between 25 and 40°C.

By the curve-fitting procedure [29], a pair of zN and $1/\lambda$ at each temperature was determined as shown in Figs. 11 and 12. The negative charge density (zN) of sample A increases as the temperature rises. This increase corresponds to the thermosensitive shrinkage of sample A shown in Fig. 9. On the other hand, as seen in Fig. 12, the softness parameter $1/\lambda$ does not change depending on temperature in sample A but keeps a constant value in this temperature range, which is considered to be the effects of poly(styrene) contained in the particle, that is, 90% of the polymer composing the particle is nonthermosensitive poly(styrene).

The electrophoretic mobility of sample B was measured as a function of temperature and the ionic strength of the medium. Figure 13 shows the results. The electrophoretic mobility of sample B changed more drastically than that observed in sample A (Fig. 10). At the ionic strength of 0.005 mol/dm^3, it decreases from 0.25 to -5.0 μm s^{-1} V^{-1} cm for sample B. As is clearly seen in Fig. 13, the electrophoretic mobility varies in the temperature region between 30 and 33°C, corresponding with the phase transition phenomena of poly(N-isopropylacrylamide) gel. Sample B is recognized as a particle with a core–shell structure, in which the negatively charged particle core (sample A) is covered with a negatively charged poly(N-isopropylacrylamide) hydrogel layer. If the surface layer is thick enough and charged, then the charge of the inner core is considered to affect little the electrophoretic behavior of the particle. The thickness of the surface layer of sample B lies between 40 and 90 nm as seen in Fig. 9, which is much thicker than $1/\kappa$. Therefore we estimated zN and $1/\lambda$ for sample B as shown in Fig. 13 with the same model as for sample A. Also, in this case, each

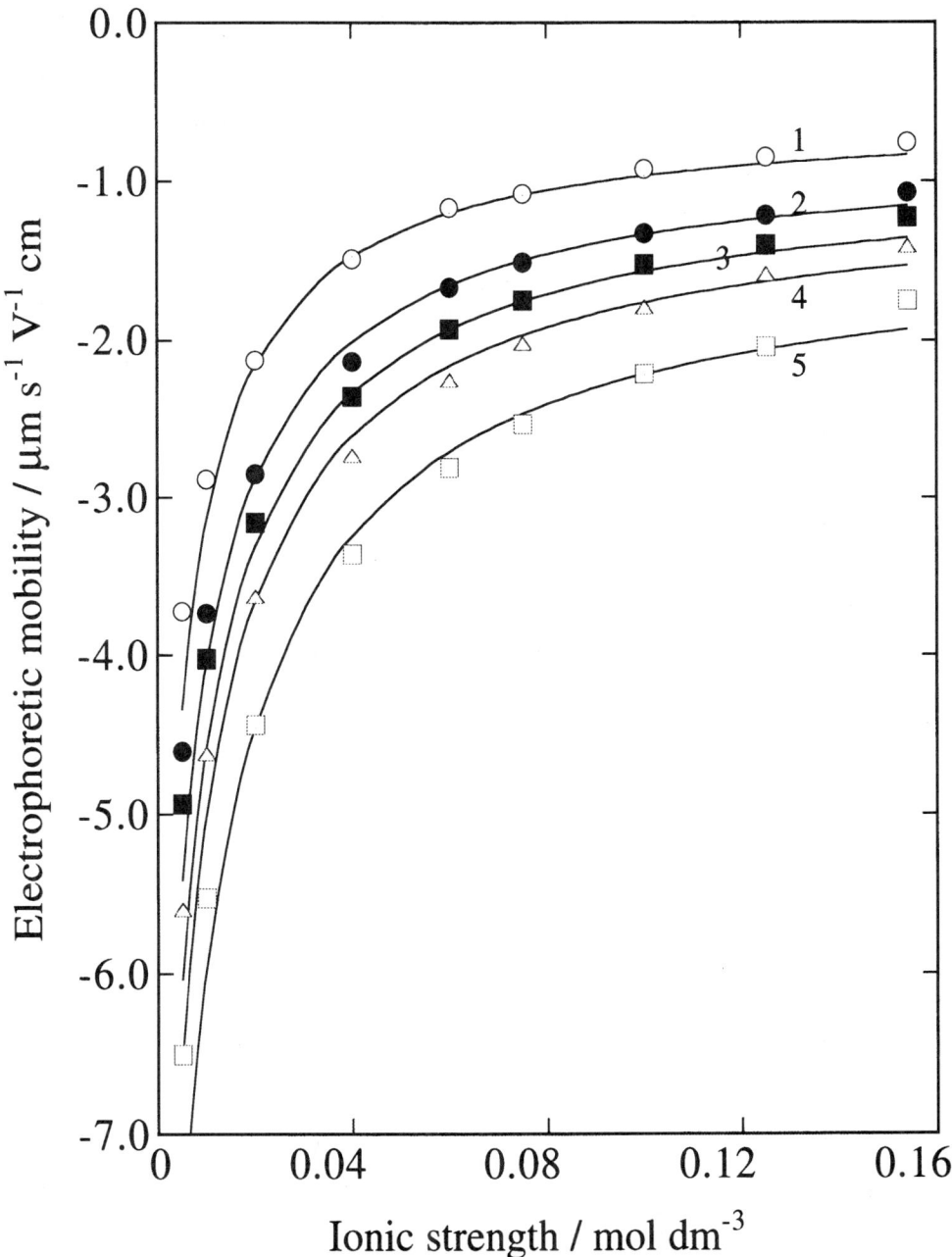

FIG. 10 Electrophoretic mobility of sample A as a function of ionic strength at pH = 7.4 at several values of the temperature. Symbols represent experimental data: ○, 25°C; ●, 30°C; ■, 33°C; △, 35°C; and □, 40°C. Solid curves are theoretical results calculated via Eq. (1) with $zN = -0.052$ mol/dm^3 (curve 1) at 25°C; $zN = -0.065$ mol/dm^3 (curve 2) at 30°C, $zN = -0.072$ mol/dm^3 (curve 3) at 33°C, $zN = -0.078$ mol/dm^3 (curve 4) at 35°C, and $zN = -0.09$ mol/dm^3 (curve 5) at 40°C and with $1/\lambda = 1.0$ nm at all temperatures. (From Ref. 22.)

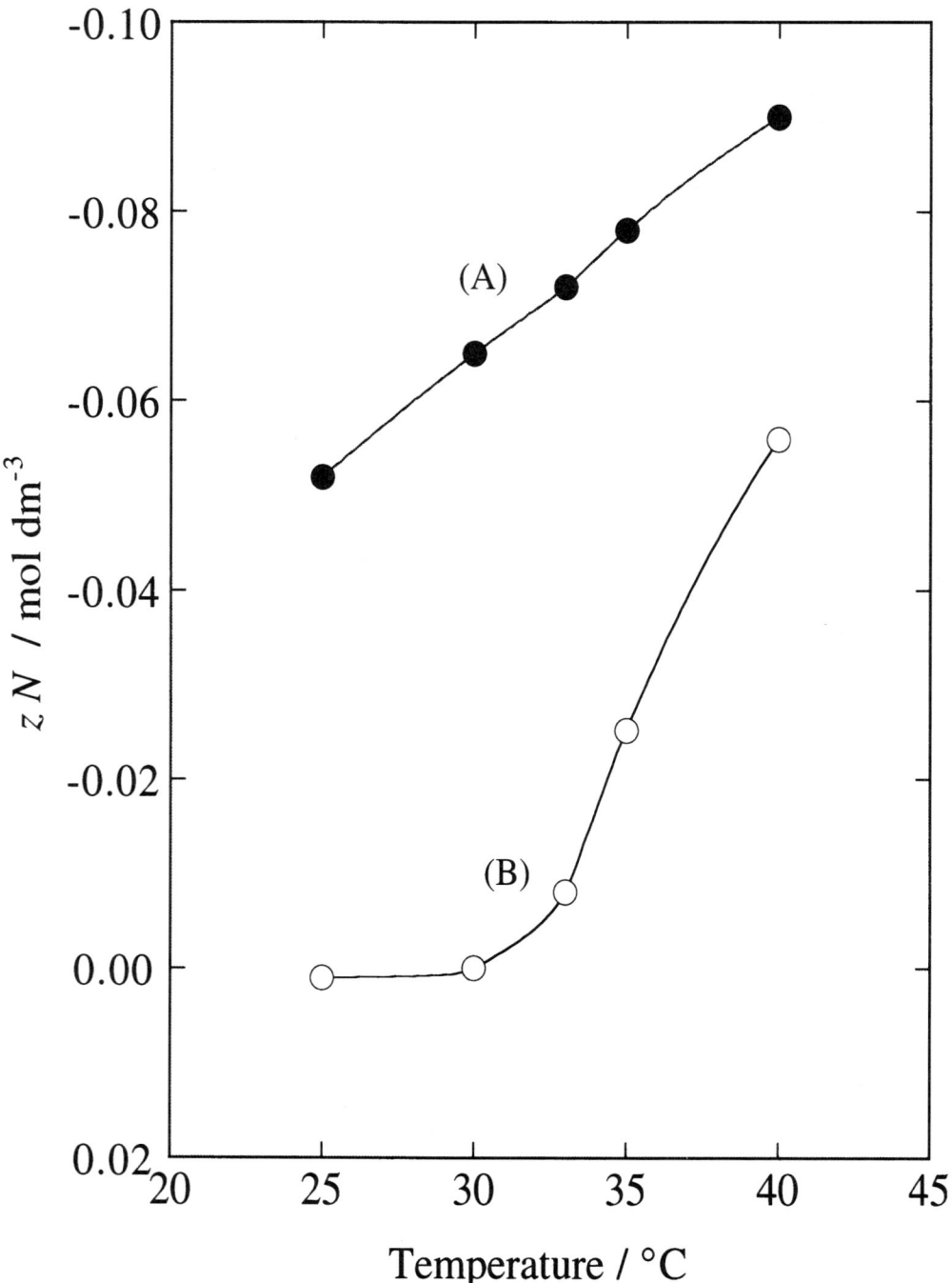

FIG. 11 Charge density zN of samples A and B as a function of temperature. Symbols are for sample A (●) and for sample B (○). (From Ref. 22.)

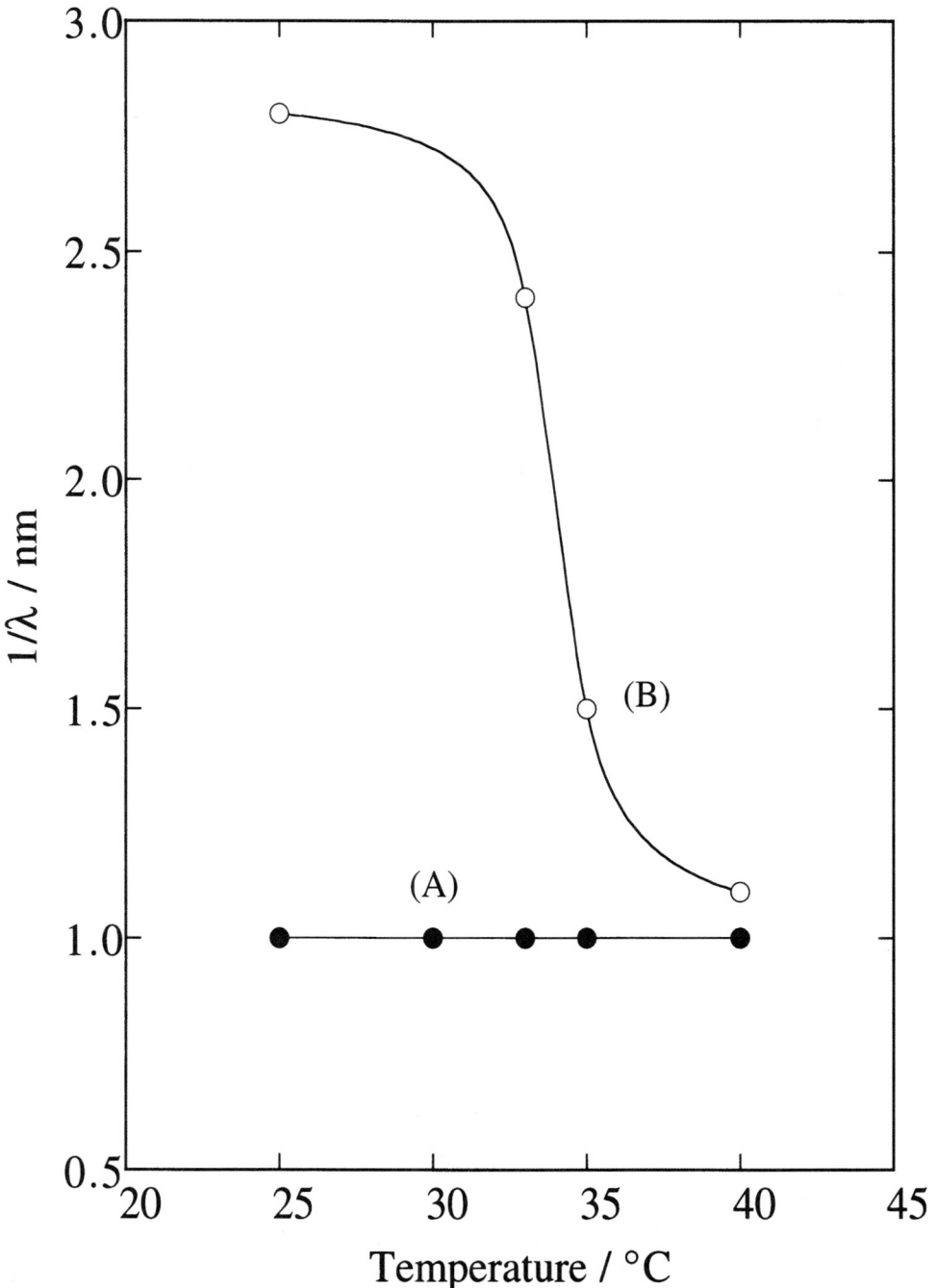

FIG. 12 Softness parameter $1/\lambda$ of samples A and B as a function of temperature. Symbols are for sample A (●) and for sample B (○). (From Ref. 22.)

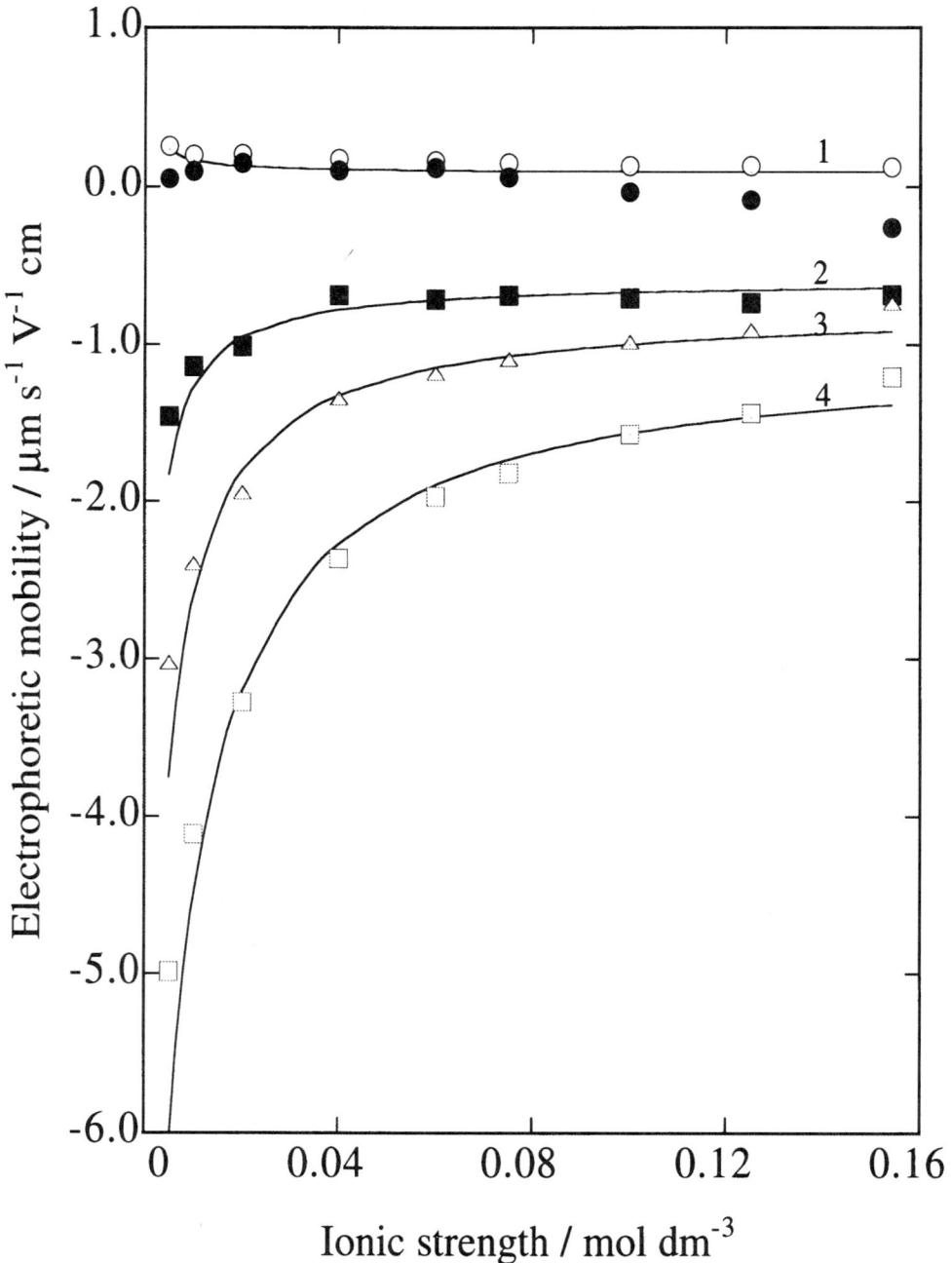

FIG. 13 Electrophoretic mobility of sample B as a function of ionic strength at pH = 7.4 at several values of the temperature. Symbols represent experimental data: ○, 25°C; ●, 30°C; ■, 33°C; △, 35°C; and □, 40°C. Solid curves are theoretical results calculated via Eq. (1) with $zN = 0.001$ mol/dm^3 and $1/\lambda = 2.8$ nm (curve 1) at 25°C, $zN = -0.008$ mol/dm^3 and $1/\lambda = 2.4$ nm (curve 2) at 33°C, $zN = -0.025$ mol/dm^3 and $1/\lambda = 1.5$ nm (curve 3) at 35°C, and $zN = -0.056$ mol/dm^3 and $1/\lambda = 1.1$ nm (curve 4) at 40°C. (From Ref. 22.)

Thermosensitive Gels

theoretical curve calculated via Eq. (1) with a pair of zN and $1/\lambda$ showed a good agreement with experimental data at 25, 33, 35 and 40°C. At 30°C, the electrophoretic mobility changed from a slightly positive to a negative value as the ionic strength increased, which might be caused by the physical adsorption of sodium cations onto the surface layer. It was regarded that zN is zero at 30°C because the electrophoretic mobility is almost zero at any ionic strength; however, $1/\lambda$ was still unknown. The values of zN and $1/\lambda$ obtained from Fig. 13 are plotted in Figs. 11 and 12. Little change in zN is seen from 25 to 30°C, but the negative charge density increases from 0 to -0.056 mol/dm^3 when the temperature increases from 30 to 40°C. Compared with the case in sample A, sample B showed a clear effect of phase transition on zN. Also, a large decrease in $1/\lambda$ is seen between 33 and 35°C, and a smaller decrease is observed from 25 to 33°C and from 35 to 40°C, which shows that sample B becomes harder as the temperature rises; its change is most obvious between 33 and 35°C. These sudden increases of charge density and hardness seen between 33 and 35°C correspond to the decrease in thickness of the shell layer of sample B, which is seen in Fig. 9.

Figure 14 shows the electrophoretic mobility of sample C. The values are affected by both the temperature and the ionic strength of the dispersing media. For analyzing the data of sample C, the two-layer model was essential for the core particle with a charged soft layer and further covered with an uncharged soft layer, as schematically shown in Fig. 15. This model is different from that used for samples A and B (Fig. 5), which is essentially a one-layer model.

Consider a colloidal particle moving with velocity U in a liquid containing a symmetrical electrolyte of valency v and bulk concentration n in an applied electric field E. Imagine that the colloidal particle is composed of a charged-particle core covered with an ion-penetrable surface layer consisting of two sublayers, the outer sublayer (layer 1) of thickness d_1 being composed of uncharged polymers and the inner sublayer (layer 2) of thickness d_2 being composed of charged polymers. In the inner sublayer, fixed charged groups of valency z are distributed at a uniform density N. The particle radius is much larger than $1/\kappa$ (κ being the Debye–Hückel parameter), so that the particle surface can be assumed to be planar and the applied field E to be parallel to the surface. Also we assume that d_1 and d_2 are much greater than $1/\kappa$. We take an x-axis perpendicular to the particle surface with its origin at the front surface of the surface charge layer (Fig. 15). The flow velocity of the liquid $u(x)$ (relative to the particle), which is parallel to the particle surface, is described by the Navier–Stokes equations,

$$\eta \frac{d^2 u}{dx^2} + \rho_{el}(x) E = 0 \qquad x > 0 \qquad (9)$$

$$\eta \frac{d^2 u}{dx^2} - \gamma_1 u(x) + \rho_{el}(x) E = 0 \qquad -d_1 < x < 0 \qquad (10)$$

$$\eta \frac{d^2 u}{dx^2} - \gamma_2 u(x) + \rho_{el}(x) E = 0 \qquad -(d_1 + d_2) < x < -d_1 \qquad (11)$$

where η is the viscosity, γ_1 and γ_2 are the frictional coefficients in sublayers 1 and 2, respectively, and $\rho_{el}(x)$ is the volume charge density resulting from the presence of electrolyte ions at position x. The term $-\gamma u$ arises from the drag force exerted by the polymer segments in the surface layer.

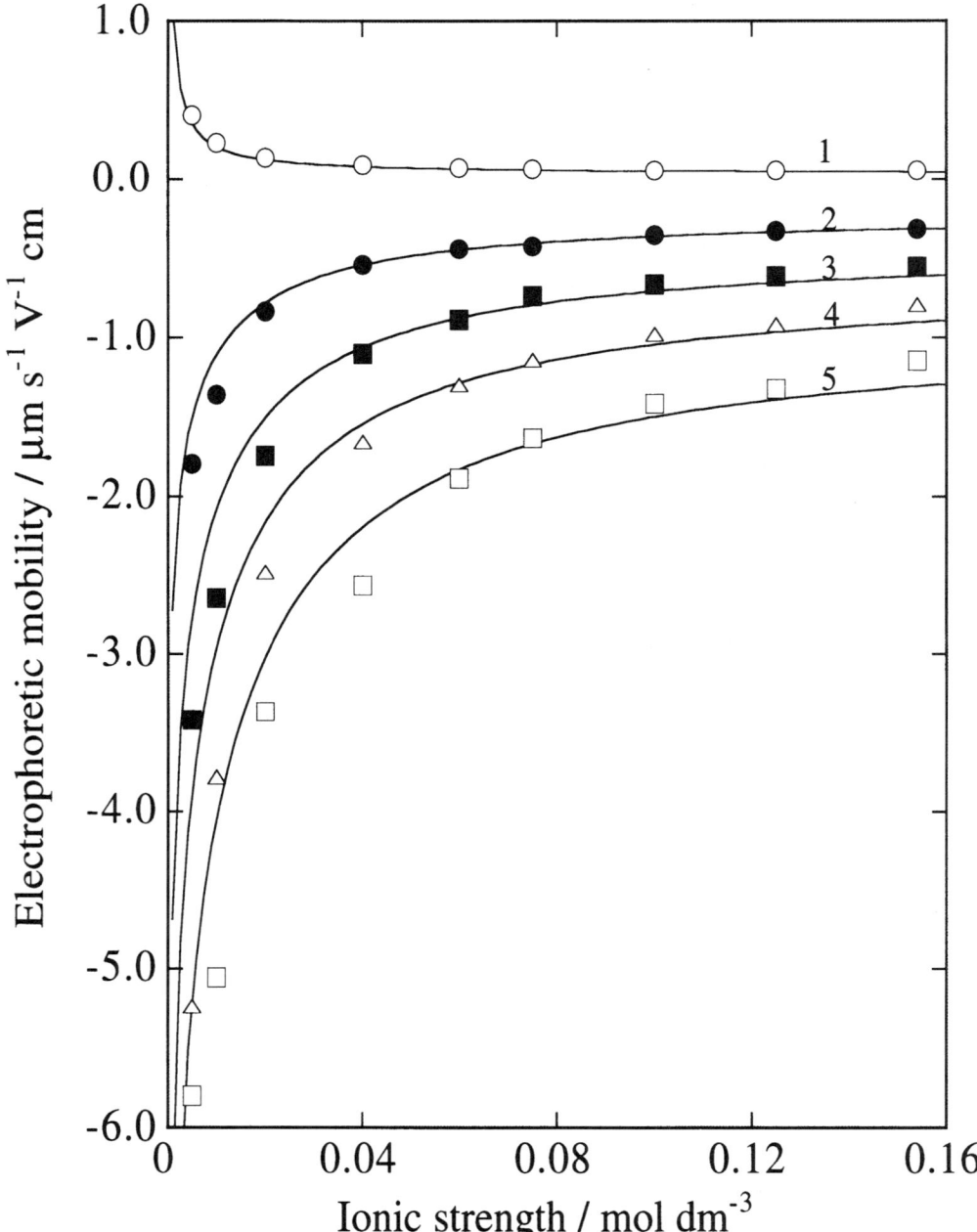

FIG. 14 Electrophoretic mobility of sample C as a function of ionic strength at pH = 7.4 at several values of the temperature. Symbols represent experimental data: ○, 25°C; ●, 30°C; ■, 33°C; △, 35°C; and □, 40°C. Solid curves are theoretical results calculated via Eq. (26) with $zN = 0.01$ mol/dm^3 and $1/\lambda_1 = 26$ nm (curve 1) at 25°C, $zN = -0.065$ mol/dm^3 and $1/\lambda_1 = 26$ nm (curve 2) at 30°C, $zN = -0.072$ mol/dm^3 and $1/\lambda_1 = 36$ nm (curve 3) at 33°C, $zN = -0.078$ mol/dm^3 and $1/\lambda_1 = 46$ nm (curve 4) at 35°C, $zN = -0.09$ mol/dm^3 and $1/\lambda_1 = 54$ nm (curve 5) at 40°C. At each temperature, $1/\lambda_2 = 1.0$ nm was used. (From Ref. 22.)

Two-layer model

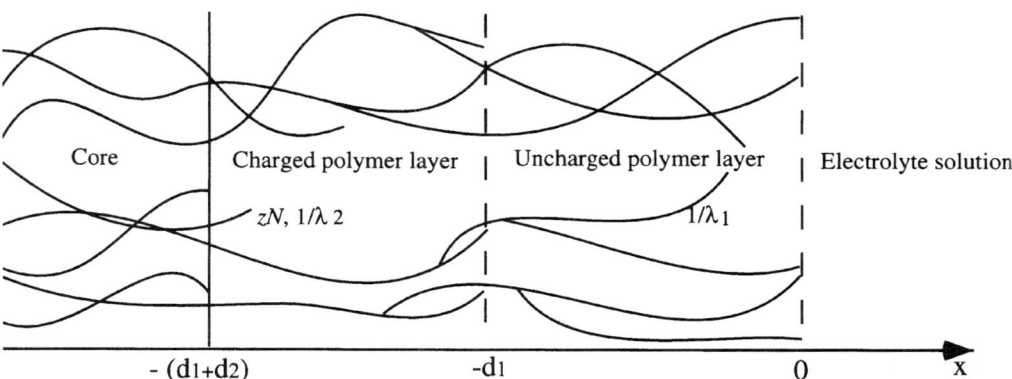

FIG. 15 Schematic representation of the ion-penetrable surface polymer layer 1 of thickness d_1 composed of uncharged polymer and the surface polymer layer 2 of thickness d_2 composed of negatively charged polymer with charge density of zN. The softness parameters for layer 1 and 2 are $1/\lambda_1$ and $1/\lambda_2$, respectively.

The boundary conditions for $u(x)$ are then

$$u(-(d_1 + d_2)) = 0 \tag{12}$$

$$u(-0) = u(+0) \tag{13}$$

$$\left.\frac{du}{dx}\right|_{x=-0} = \left.\frac{du}{dx}\right|_{x=+0} \tag{14}$$

$$u(x) \to = U \quad \text{as} \quad x \to +\infty \tag{15}$$

Equation (12) expresses the boundary condition that the slipping plane, at which $u(x)$ is zero, is assumed to be located at $x = -(d_1 + d_2)$. Equations (13) and (14) are the continuity conditions of $u(x)$ and du/dx at $x = 0$. Equation (15) states that the liquid velocity in the bulk solution phase equals the electrophoretic velocity, but with the opposite sign. In order to calculate the mobility, one needs to know the distribution of the electric potential $\psi(x)$ within the surface layers. The potential distribution $\psi(x)$ is given by the Poisson equations, viz.,

$$\frac{d^2\psi}{dx^2} = -\frac{\rho_{el}(x)}{\varepsilon_r\varepsilon_o} \quad x > 0 \tag{16}$$

$$\frac{d^2\psi}{dx^2} = -\frac{\rho_{el}(x)}{\varepsilon_r\varepsilon_0} \quad -d_1 < x < 0 \tag{17}$$

$$\frac{d^2\psi}{dx^2} = -\frac{\rho_{el}(x) + zeN}{\varepsilon_r\varepsilon_0} \quad -(d_1 + d_2) < x < -d_1 \tag{18}$$

where ε_r is the relative permittivity of the solution, ε_0 is the permittivity of a vacuum, and e is the elementary electric charge.

The boundary conditions for $\psi(x)$ are

$$\psi(-0) = \psi(+0) \tag{19}$$

$$\left.\frac{d\psi}{dx}\right|_{x=-0} = \left.\frac{d\psi}{dx}\right|_{x=+0} \tag{20}$$

$$\left.\frac{d\psi}{dx}\right|_{x=-(d_1+d_2)} = 0 \tag{21}$$

Equations (19) and (20) are the continuity conditions of ψ and $d\psi/dx$ at $x=0$. Equation (21) states that the surface charge of the particle core ($x = -(d_1 + d_2)$) and the electric field within the particle core are both negligible.

On the basis of Eqs. (9)–(11) and (16)–(18) subject to boundary conditions, Eqs. (12)–(15), we derive the following formula for the electrophoretic mobility:

$$\mu = \left(\frac{\cosh \lambda_1 d_1}{\sinh \lambda_2 d_2} + \frac{\lambda_1 \sinh \lambda_1 d_1}{\lambda_2 \cosh \lambda_2 d_2}\right)^{-1}$$

$$\times \left[\frac{\lambda_1 \varepsilon_r \varepsilon_0}{\eta \sinh \lambda_2 d_2} \int_{-d_1}^{0} \psi(x) \sinh \lambda_1 (x + d_1)\, dx \right.$$

$$+ \frac{\lambda_1^2 \varepsilon_r \varepsilon_0}{\eta \lambda_2 \cosh \lambda_2 d_2} \int_{-d_1}^{0} \psi(x) \cosh \lambda_1 (x + d_1)\, dx$$

$$+ \frac{1}{\sinh \lambda_2 d_2 \cosh \lambda_2 d_2} \frac{1}{\eta \lambda_2} \int_{-(d_1+d_2)}^{-d_1} zeN \sinh \lambda_2 (x + d_1 + d_2)\, dx \tag{22}$$

$$+ \frac{1}{\sinh \lambda_2 d_2 \cosh \lambda_2 d_2} \frac{\varepsilon_r \varepsilon_0 \lambda_2}{\eta} \int_{-(d_1+d_2)}^{-d_1} \psi(x) \sinh \lambda_2 (x + d_1 + d_2)\, dx$$

$$\left. + \frac{\varepsilon_r \varepsilon_0 \psi(-(d_1 + d_2))}{\eta \sinh \lambda_2 d_2 \cosh \lambda_2 d_2}\right]$$

with

$$\lambda_1 = \left(\frac{\gamma_1}{\eta}\right)^{1/2} \tag{23}$$

$$\lambda_2 = \left(\frac{\gamma_2}{\eta}\right)^{1/2} \tag{24}$$

The potential distribution in the neutral polymer layer (layer 1) is described by

$$\psi(x) = \frac{2kT}{ve} \ln \frac{1 + \tanh(ve\psi(-d_1)/4kT)\exp(-\kappa(x+d_1))}{1 - \tanh(ve\psi(-d_1)/4kT)\exp(-\kappa(x+d_1))} \qquad -d_1 < x < 0 \tag{25}$$

and the potential for the charged polymer layer region (layer 2) is approximated by

$$\psi(x) = \psi_{DON} + (\psi(-d_1) - \psi_{DON})\exp(\kappa_m(x + d_1)) \qquad -(d_1 + d_2) < x < -d_1 \quad (26)$$

with

$$\psi(-d_1) = \psi_{DON} - \frac{kT}{ve}\tanh\left(\frac{ve\psi_{DON}}{2kT}\right) \quad (27)$$

ψ_{DON} and κ_m given in Eqs. (2) and (5) previously are the Donnan potential and the effective Debye–Hückel parameter of the polymer layer 2. Note that the usual Debye–Hückel parameter κ given in Eq. (6) is involved in Eq. (25), since the polymer layer 1 is uncharged. For the present case in which $\psi(-d_1)$ is very small and d_2 is much larger than $1/\lambda_2$ and $1/\kappa_m$, we have approximately

$$\mu = \left(\cosh\lambda_1 d_1 + \frac{\lambda_1}{\lambda_2}\sinh\lambda_1 d_1\right)^{-1}\left[\frac{\varepsilon_r\varepsilon_0}{\eta}\frac{\lambda_1}{\lambda_1^2 - \kappa^2}\psi(-d_1)\right.$$
$$\times\left\{\left(\kappa + \frac{\lambda_1^2}{\lambda_2}\right)e^{-\kappa d_1}\sinh\lambda_1 d_1 + \left(\lambda_1 + \frac{\lambda_1\kappa}{\lambda_2}\right)(e^{-\kappa d_1}\cosh\lambda_1 d_1 - 1)\right\} \quad (28)$$
$$+\frac{zeN}{\eta\lambda_2^2} + \frac{\varepsilon_r\varepsilon_0}{\eta}\frac{\psi(-d_1)/\kappa_m + \psi_{DON}/\lambda_2}{1/\kappa_m + 1/\lambda_2}\right]$$

Figure 14 shows results of the calculation of the mobility as a function of the electrolyte concentration via the new model, that is, Eq. (28). Experimental data are for sample C. In the calculation, the values of zN and $1/\lambda$ for sample A were used as obtained before as zN and $1/\lambda_2$ in this model at 30, 33, 35 and 40°C, because the core particle of sample C is sample A. At 25°C, we used zN to be 0.01 mol/dm^3, since the electrophoretic mobility measured was unexpectedly slightly positive at all ionic strengths, which might be caused by the cation adsorption onto the surface layer. Each theoretical curve revealed a good agreement with the experimental values over a wide range of the ionic strength except at very low ionic strengths at each temperature.

The calculated zN, $1/\lambda_2$, and $1/\lambda_1$ were plotted in Figs. 16 and 17. The negative charge density (zN) in the inner layer (layer 2) increases in magnitude as the temperature rises, although $1/\lambda_2$ remains constant. It should be noted that $1/\lambda_1$ increases as the temperature rises, showing that the surface layer becomes softer. This is the opposite tendency with sample B and is considered to be caused by the difference in the mechanisms of their temperature-dependent structure change. As was seen in Fig. 9, the shell thickness decreases in sample B by shrinkage as the temperature increases, while it does not change in sample C even by the shrinkage of sample C. From this, it is concluded that in sample C, the uncharged polymer segments in the outer sublayer do not shrink in the direction perpendicular to the particle surface, but approach each other, forming groups on the particle surface, as shown in Fig. 18. This concentration of polymer segments in the surface layer, producing void spaces in the surface layer, may decrease the value of $1/\lambda_1$, making the friction of liquid flow around the particle smaller as the temperature rises. This type of collection is available when the polymer segments

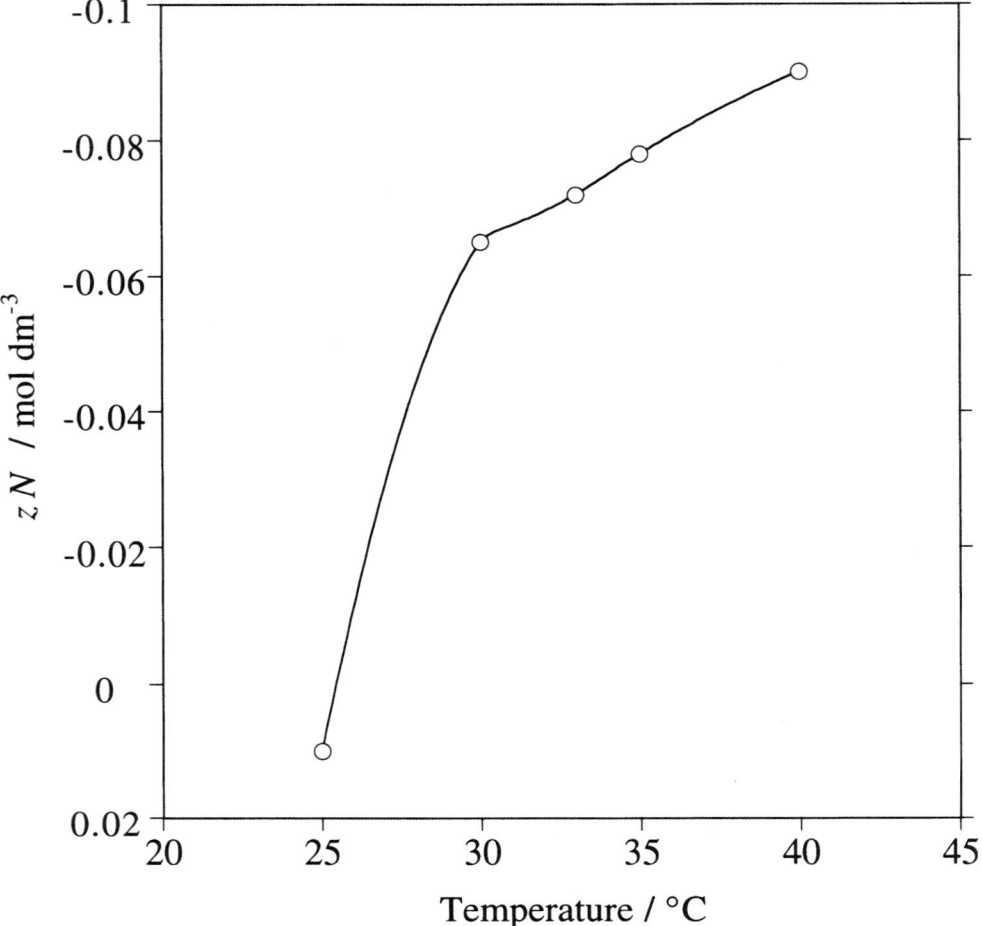

FIG. 16 Change in charge density, zN (mol/dm^3), of sample C as a function of temperature.

have no charges, so that no electric repulsion exists. Therefore, it is considered that this phenomenon is not observed in sample B. Also, this result coincides with the observation that the friction coefficient of the water flow inside the poly(N-isopropylacrylamide) gel layer decreases as the temperature rises when the gel volume is kept constant, which was reported by Tokita and Tanaka [36]. They explain this observation as follows. Some portions of the gel swell while other portions shrink, maintaining constant the total gel volume, and the water passes through the swollen open space, avoiding the shrunk regions.

In conclusion, during the preparation of latex particles, when potassium peroxodisulfate is used as an initiator for the polymerization of N-isopropylacrylamide, the polymer hydrogel layer is negatively charged and shrinks as temperature rises.

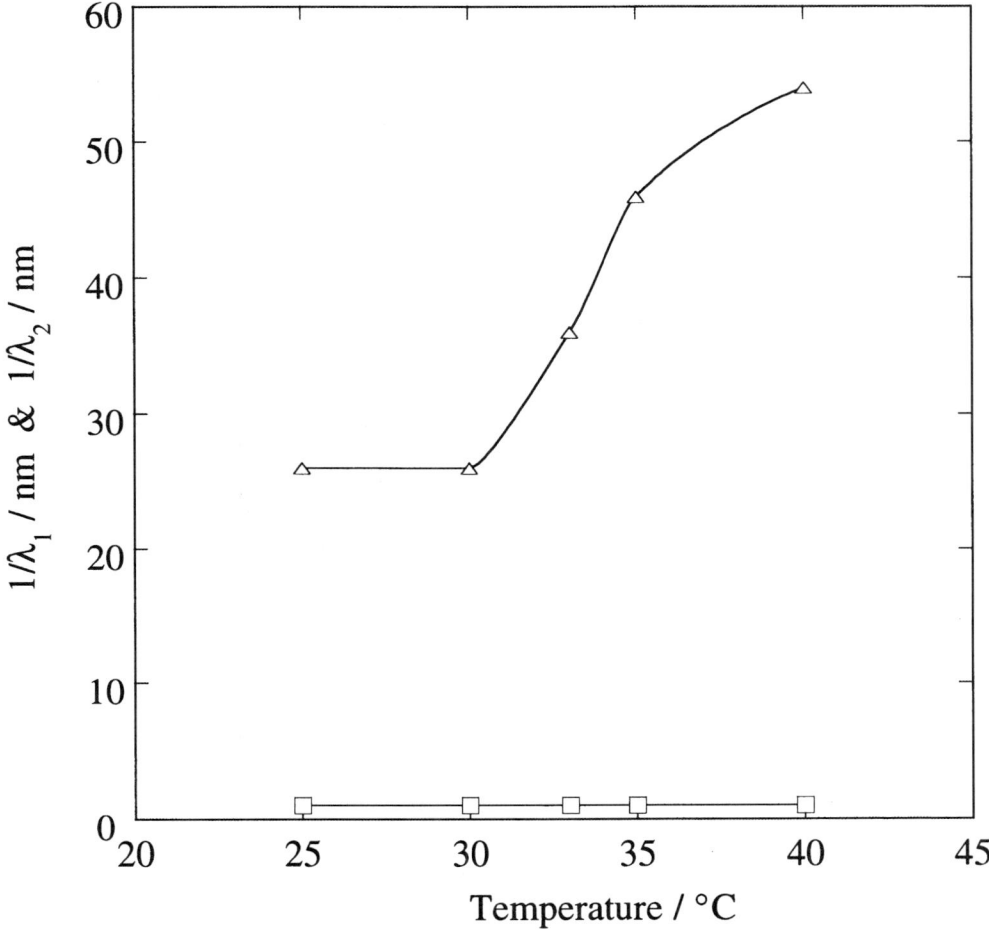

FIG. 17 Change in softness parameters, $1/\lambda_1$ and $1/\lambda_2$ (nm), of sample C as a function of temperature. Symbols represent △, $1/\lambda_1$ and □, $1/\lambda_2$.

Especially, when the shell layer is composed of charged poly(N-isopropylacrylamide), this shrinkage results in an accumulation of poly(N-isopropylacrylamide) segments on the particle core, making the particle surface harder. On the other hand, when VA-086 is used as an initiator for the preparation of the shell layer, the polymer hydrogel layer is uncharged and does not change thickness in the direction perpendicular to the particle surface but accumulates each other, when the temperature rises. Therefore, when the temperature rises, the surface layer becomes softer, because void spaces are created on the particle core by the accumulation of poly(N-isopropylacrylamide) chains.

Also, the temperature dependence of the charge density and of the softness of negatively charged poly(N-isopropylacrylamide)-coated polystyrene latex surfaces (sample B) estimated from the electrophoretic mobility measurements (Fig. 13) is in good agreement with that of glass surfaces covered by poly(N-isopropylacrylamide) hydrogel layers as estimated by electro-osmotic measurements (Fig. 6). That is, the values of

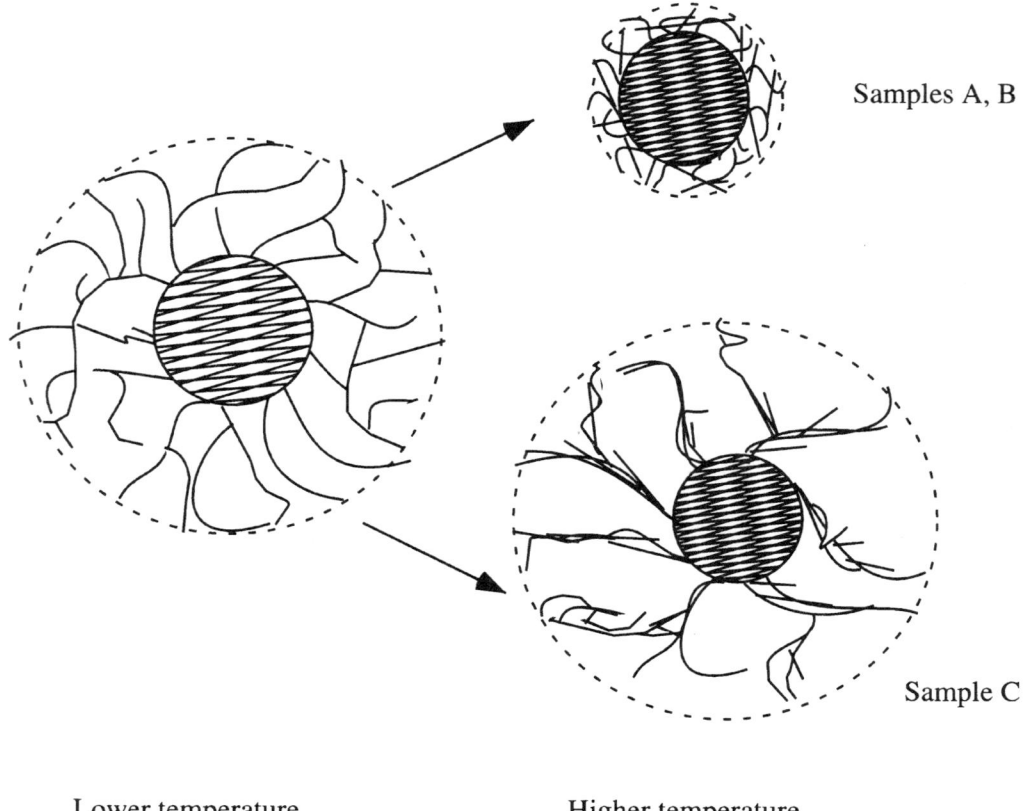

FIG. 18 Schematic representation of the thermosensitive structural changes in samples A, B, and C.

zN and $1/\lambda$ determined by the respective measurements (Figs. 11 and 12 with Fig. 7) are found to be of the same order of magnitude. The numerical discrepancy may be caused by the difference in preparation conditions of the samples.

REFERENCES

1. N. A. Peppas, in *Pulsatile Drug Delivery: Current Applications and Future Trends* (R. Gurny, ed.), Wiss. Verl.-Ges., Stuttgart, 1993, pp. 41–56.
2. J. Heller, in *Pulsatile Drug Delivery: Current Applications and Future Trends* (R. Gurny, ed.), Wiss. Verl.-Ges., Stuttgart, 1993, pp. 57–71.
3. S. W. Kim, E. J. Mack, C. M. Pai, K. Makino, and Y. H. Bae, in *FIP Symposium Proceedings, Munich, FRG* (D. D. Breimer, D. J. A. Crommelin, and K. K. Midha, eds.), Elsevier, Amsterdam, 1989, pp. 3–9.
4. S. W. Kim, C. M. Pai, K. Makino, L. A. Seminoff, D. Brown, J. Gleeson, D. Holmberg, and D. Wilson. J. Controlled Release *11*:193 (1990).
5. K. Makino, E. J. Mack, T. Okano, and S. W. Kim. J. Controlled Release *12*:235 (1990).
6. K. Makino, E. J. Mack, T. Okano, and S. W. Kim. J. Biomaterials, Artificial Cells Artificial Organs *19*:219 (1991).

7. E. Miyauchi, Y. Togawa, K. Makino, H. Ohshima, and T. Kondo. J. Microencapsulation 9:329 (1992).
8. K. Makino, E. Miyauchi, Y. Togawa, H. Ohshima, and T. Kondo, in *ACS Symposium Series 540, Polymers of Biological and Biomedical Significance* (S. W. Shalaby, Y. Ikada, R. Langer, and J. Williams, eds.), ACSBooks, 1994, pp. 314–323.
9. N. Ueda, R. Ohta, Y. Togawa, K. Makino, and T. Kondo. Polymer Gels Networks 3:135 (1995).
10. T. Dobashi, T. Narita, J. Masuda, K. Makino, T. Mogi, H. Ohshima, M. Takenaka, and B. Chu. Langmuir 14:745 (1998).
11. L. D. Taylor and L. D. Cerankowski. J. Polym. Sci. Polym. Chem. 13:2551 (1975).
12. M. Shibayama, T. Tanaka, and C. C. Han. J. Chem. Phys. 97:6829 (1992).
13. H. Kawaguchi, K. Fujimoto, and Y. Mizuhara. Colloid Polym. Sci. 270:53 (1992).
14. K. Mukae, M. Sakurai, S. Sawamura, K. Makino, S. W. Kim, I. Ueda, and K. Shirahama. J. Phys. Chem. 97:737 (1993).
15. K. Mukae, M. Sakurai, S. Sawamura, K. Makino, S. W. Kim, I. Ueda, and K. Shirahama. Colloid and Polym. Sci. 272:655 (1994).
16. A. S. Hoffman, A. Afrassiabi, and L. C. Dong. J. Controlled Release 4:213 (1986).
17. T. Okano, Y. H. Bae, and S. W. Kim. J. Controlled Release 11:255 (1986).
18. T. Okano, N. Yamada, H. Sakai, and Y. Sakurai. J. Biomed. Mater. Res. 27:1243 (1993).
19. A. Kikuchi, M. Okuhara, F. Ogura, Y. Sakurai, and T. Okano, in *Advanced Biomaterials in Biomedical Engineering and Drug Delivery Systems* (N. Ogata, S. W. Kim, J. Feijen, and T. Okano, eds.), Springer-Verlag, Tokyo, 1996, pp. 267–268.
20. T. Tanaka. Sci. Am. 249:124 (1981).
21. H. Ohshima, K. Makino, T. Kato, K. Fujimoto, T. Kondo, and H. Kawaguchi. J. Colloid Interface Sci. 159:512 (1993).
22. K. Makino, H. Ohshima, S. Yamamoto, K. Fujimoto, H. Kawaguchi, and H. Ohshima. J. Colloid Interface Sci. 166:251 (1994).
23. K. Makino, K. Suzuki, Y. Sakurai, T. Okano, and H. Ohshima. J. Colloid Interface Sci. 174:400 (1995).
24. K. Makino, K. Suzuki, Y. Sakurai, T. Okano, and H. Ohshima. Colloids Surfaces A: Physicochemical and Engineering Aspects 103:221 (1995).
25. K. Makino, K. Suzuki, Y. Sakurai, T. Okano, and H. Ohshima, in *Advanced Biomaterials in Biomedical Engineering and Drug Delivery Systems* (N. Ogata, S. W. Kim, J. Feijen, and T. Okano, eds.), Springer-Verlag, Tokyo, 1996, pp. 305–306.
26. H. Ohshima. J. Colloid Interface Sci. 163:474 (1994).
27. K. Makino, in *Electric Phenomena at Interfaces*, 2d ed. (H. Ohshima and K. Furusawa, eds.), Marcel Dekker, New York, 1998, pp. 583–593.
28. K. Makino, T. Taki, M. Ogura, S. Handa, M. Nakajima, T. Kondo, and H. Ohshima. Biophys. Chem. 47:261 (1993).
29. K. Makino, M. Ikekita, T. Kodo, S. Tanuma, and H. Ohshima. Colloid Polym. Sci. 272:487 (1994).
30. Y. Nakano, K. Makino, H. Ohshima, and T. Kondo. Biophys. Chem. 50:249 (1994).
31. T. Mazda, K. Makino, R. Yabe, K. Nakata, K. Fujisawa, and H. Ohshima. Transfusion Medicine 5:43 (1995).
32. T. Mazda, K. Makino, and H. Ohshima. Colloids Surfaces B: Biointerfaces 5:75 (1995).
33. K. Makino, F. Fukai, T. Kawaguchi, and H. Ohshima. Colloids Surfaces B: Biointerfaces 5:221 (1995).
34. K. Makino, F. Fukai, S. Hirata, and H. Ohshima. Colloids Surfaces B: Biointerfaces 7:235 (1996).
35. S. Nagashima, S. Ando, K. Makino, T. Tsukamoto, and H. Ohshima. J. Colloid Interface Sci. 197:377 (1998).
36. M. Tokita and T. Tanaka. Science 253:1121 (1991).

18

Vapor Adsorption Isotherms of Medium-Sized Alcohols on Various Solids and the Behavior of Adsorbed Molecules

AKIRA NONAKA Institute of Applied Physics, University of Tukuba, Tukuba, Japan

I.	General Introduction	621
II.	Behavior of Molecules Adsorbed on the Monolayer: The Gradient Method and the Theory of Surface Area Estimation	622
	A. Introduction	622
	B. Theory of the gradient method	625
	C. Verification of the gradient method	628
III.	New Apparatus for Measuring Small Adsorption at Very Low Vapor Pressure	630
	A. Apparatus	630
	B. Procedure of adsorption measurements using the new apparatus	633
IV.	Multilayer Adsorption of 1-Butanol and 1-Hexanol and the Behavior of Adsorbed Molecules	635
	A. Experimental	636
	B. Results and discussion	638
V.	Conclusions	647
	References	648

I. GENERAL INTRODUCTION

In this paper, adsorption isotherms and adsorption phenomena, especially those related to the dynamic state of the molecules of medium-sized normal alcohol vapors adsorbed in multilayer adsorption above monolayer adsorption on various solid surfaces, are described. In general, since the molecules in multilayer gas adsorption are considered to be in a molecular state analogous to the liquid state at the same temperature and pressure, the dynamic state of the liquid molecules near the liquid/solid interface may be considered similar to the state of the multilayer gas adsorption molecules.

Many reports have been published on the vapor adsorption of lower alcohols on various solid surfaces. However, there are few reports on the vapor adsorption of medium-sized alcohols because of the difficult measurement of low adsorption amounts due to the very low vapor pressure at room temperature. In this report, the apparatus and the method developed by the author recently for the low adsorption measurement at low vapor pressure are outlined in terms of the "gradient method," which is a new method of surface area estimation, developed by the author [1–5]. For analysis of the dynamic state of molecules adsorbed in multilayers on solid surfaces, a fundamental thermodynamic theory formulated for the "method for surface area estimation by the gradient of linear portion of Type II adsorption isotherms (gradient method)" is used. The gradient method theory is considered to be largely correct, because the gradient method has been tested in many experiments using many different solids and a number of adsorbates (e.g., lower alkylbenzenes), which are considered by the theory to be adequate for use in the gradient method. Making use of the simple equations in the gradient method theory, thermodynamic quantities or the dynamic state of the adsorbate molecules in multilayer adsorption on solid surfaces could be estimated. Using the gradient method theory, the dynamic state of the molecules of medium-sized alcohols adsorbed on various solid surfaces is analyzed in this paper. In the first section of this paper, therefore, the principle of the gradient method for surface area estimation will be explained briefly.

II. BEHAVIOR OF MOLECULES ADSORBED ON THE MONOLAYER: THE GRADIENT METHOD AND THE THEORY OF SURFACE AREA ESTIMATION

A. Introduction

For the estimation of the surface area of solids, the BET method [6] has been generally utilized. In this method, nitrogen gas at the liquid nitrogen temperature is generally used as the most reliable adsorbate gas, recommended by most experimenters for the method. Nitrogen, however, is not always adequate for the measurement of a surface area smaller than several square meters, because of the comparatively high vapor pressure at the liquid nitrogen temperature used [7]. In recent years, on the other hand, it has been frequently necessary for very small surfaces, such as platelike and/or nonporous surfaces, to be measured precisely, since many useful functional surfaces having small surface areas have been developed in the industrial field, and one of the most important physicochemical data for these materials may be the exact surface area. Small surface areas, however, may be rather difficult to measure, and it also seems to be difficult to prepare functional solid samples, most of which may be platelike, as large as several square meters in area, which is necessary for the usual surface area measurement. Thus it is desirable that very small surface areas should be precisely and easily measured.

In the BET method, krypton and xenon, which have a comparatively low vapor pressure at liquid nitrogen temperature, have been tested as the adsorbate gases for the measurement of very small surface areas [7]. However, these adsorbate gases are not always fully reliable for surface area estimation, because these gases do not have a constant molecular cross-section like the adsorbate molecule, but different cross-sections depending on the kind of adsorbent solid. Moreover, it is rather difficult to measure precisely the very low vapor pressures of these adsorbates at the liquid nitrogen temperature to be maintained in obtaining the isotherms, due to the problem of "thermal molecular flow" caused by the temperature difference in the gas vessel at a very low measuring pressure [8].

Alternative attempts to measure small surface areas have been made by making use of the vapors of organic materials having a low vapor pressure at room temperature. In this case, however, the BET method for surface area estimation cannot be always applied for the organic adsorbate employed to obtain the proper values for all the kinds of solids, because each of the organic materials used as adsorbate has its own molecular form and components and also its own interaction with the surface specified in each case. Under such complicated conditions for the use of organic vapor adsorbates, the surface area estimation by the customary BET method using a single ordinary molecular cross-section of adsorbate might be essentially inconsistent with the method, and erroneous. This is because the molecular cross-section used in the BET method is usually calculated from the liquid density under the assumption of the spherical form of adsorbate molecules and the closest packing in the liquidlike adsorption layer, which is known generally to be inadequate for organic adsorbates. In practice, in the application of organic vapors in the BET method, the molecular cross-sections estimated experimentally for each kind of surface solid have been used, such an uncertain method being contradictory to the universal use expected for surface area estimation. In these cases, errors due to the surface irregularity and/or contamination specific to each sample would be inevitable for the organic adsorbates, even if the same kind of material were used as the sample. In general, the BET method is considered usually to contain an inevitable error of 20 to 30% in the estimation, even if the most reliable adsorptive gas, nitrogen, rather than organic vapors, is used, although it has been pointed out by many authors that there are many theoretical contradictions in the BET theory [9].

Even on a solid that is considered to have a fairly clean and plane surface, there may be many kinds of fine surface defects, such as micropores, dislocation-kinks, and microcracks. For fine surface defects, adsorbate molecules would be either adsorbed or rejected, depending on the molecular size and the defect size. In the ordinary surface area estimation, therefore, the surface area values obtained may change with the size of the adsorbate molecules used in the experiment. Even in the general use of nitrogen as adsorbate, the more micropores acceptable for nitrogen molecules the surface has, the better the surface area might be estimated, although it is questionable whether the surface area obtained by adding the inner area of such very small pores should be admitted as the true surface area.

To avoid confusion relating to the concept of surface area and its measurement, the definition of "external (outer) surface area" has been introduced [10], where the (external) surface area is measured by the use of a volatile standard gas (e.g., nitrogen) after monolayer preadsorption of a nonvolatile material on the sample [11]. As another estimation method for the external surface area, the t-plot method [12] (or α_s-plot method [13]) is recommended by some authors, where adsorption isotherms of the sample solid are compared with those of the nonporous standard solid to make "comparison plots" [14], though it is considered rather difficult to prepare standard solid materials without surface defects. The gradient method is also one of the estimation methods for the external surface area.

The gradient method (surface area estimation method by the linear portion gradient of Type II adsorption isotherms [15]) was studied, first in order to enable measurement of very small surface areas of solids. The first adsorbates employed for this purpose were inactive vapors of organic materials that have a low vapor pressure at room temperature and no specific attractive forces to the surface of most kinds of solids. This kind of organic adsorbate, however, could not be applied to the ordinary BET method, due to serious errors in the method as mentioned above, even if the adsorption isotherms obtained show typical Type II curves. In this situation, a new method for surface area estimation using gas adsorption was designed to replace the BET method. Observing especially the rather long linear portion of the Type II adsorption isotherms shown by some organic vapors (cf. Fig. 1), where the increase of adsorption is proportional to the increase of the adsorbate pressure (Henry's law, in a broad sense), and considering that the adsorbate molecules

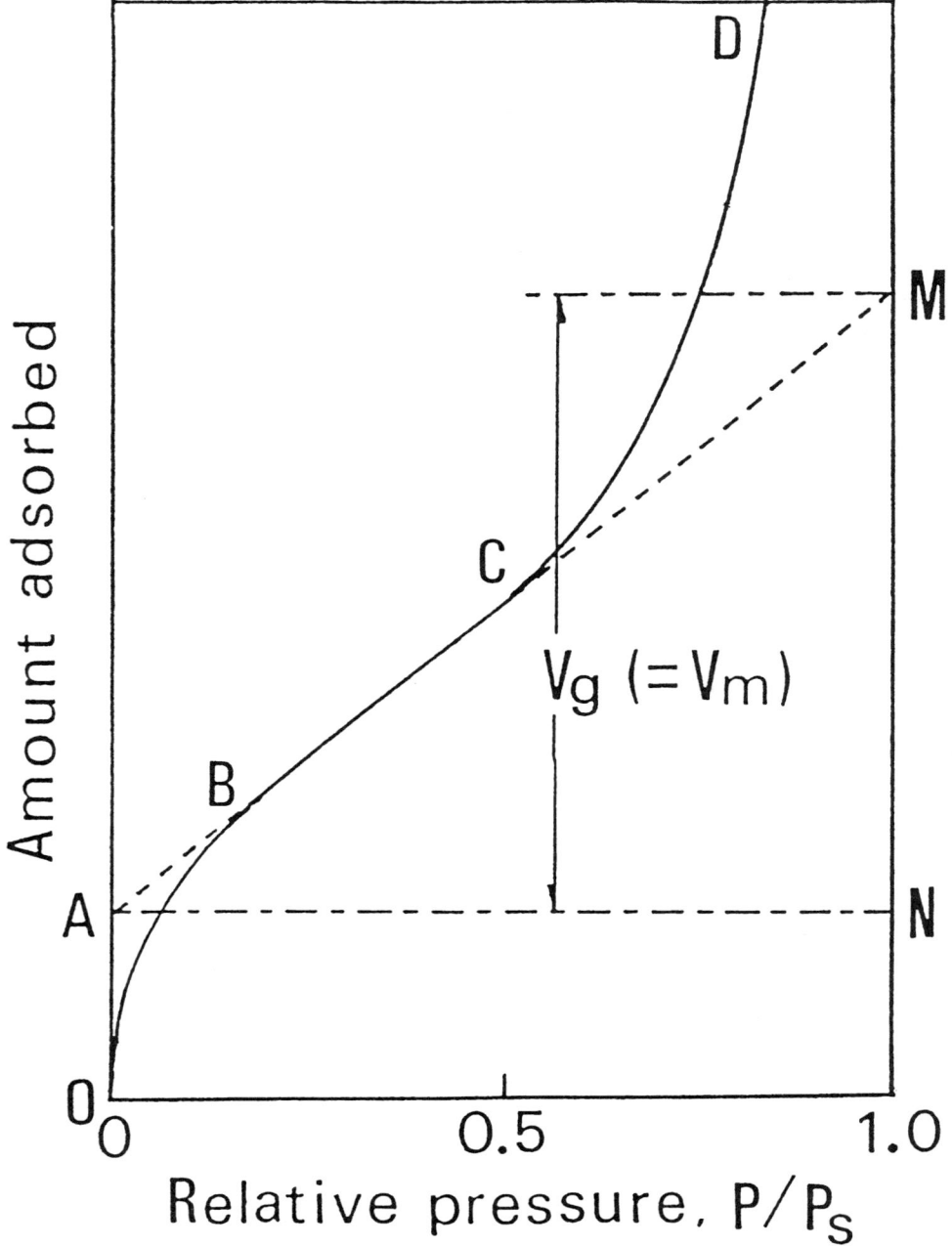

FIG. 1 Gradient method using the Type II adsorption isotherm (OBCD) and the gradient value (Vg) of the linear portion (BC). The monolayer volume (Vm) is equated directly to the gradient value (MN).

in the linear portion on the isotherm would be in a liquid state in most cases [16], a fairly simple theoretical assumption was made: The value of the linear portion gradient in an ideal case should be equal to the monolayer adsorption volume in the use of a nonspecific adsorbate. This concept is shown in Fig. 1. This hypothetical relation was obtained by comparing the nonlocalized adsorbate molecules in multilayer adsorption with the molecules on the corresponding liquid surface, by means of molecular dynamics, or by thermodynamic consideration. In order to test the theoretical conclusion, a large number of experiments were carried out, using a number of comparatively inert organic adsorbates (benzene and lower alkylbenzenes) that might fit the theory, in combination with various kinds of solid samples to investigate the general usefulness of the method. Numerical surface area values obtained experimentally by the gradient method were compared with those provided by the standard BET method, and it was confirmed that both estimated surface area values are in good agreement within the extent of intrinsic errors involved in the BET method. Moreover, geometric areas of some platelike samples having no surface pores were compared with the values obtained by the gradient method to test the reliability of this method.

B. Theory of the Gradient Method

1. Consideration in Terms of the Kinetics of Adsorbed Molecules

We will consider the behavior of the molecules adsorbed as the second layer in equilibrium with the outside adsorbate gas phase at a constant pressure (P) over the monomolecular adsorption layer on a solid surface. Interactions of the adsorbate molecules with the surface solid molecules and with the molecules adsorbed as the first adsorption layer are assumed to be only a simple dispersion force and not any specific attractive force. The molecules adsorbed on the monomolecular adsorption layer are assumed to be nonlocalized and able to exchange their sites with those of the underlayer (the first layer) molecules.

S and P are, respectively, the surface area and the adsorbate gas pressure. The number (N_a) of the molecules that dive onto the surface from the adsorbate gas phase per unit time is assumed to be proportional only to P and S (proportional coefficient, k), irrespective of the number (n) of molecules adsorbed already, making the second layer, over the first adsorption layer, although the assumption is different from that used by Langmuir in his equation. This is because, in this theory, the adsorbed molecules on the monolayer are assumed to be mobile, and all the diving-in molecules could hold on for some time, even if they collide with the molecules already adsorbed on the monolayer, since they have the same sticking energy as the other diving-in molecules. Thus, the simple equation is given as

$$N_a = k \cdot P \cdot S \tag{1}$$

On the other hand, the number (N_d) of molecules that take off from the second layer on the surface, being liberated from the adsorption energy ($\mathcal{E}_2$), per unit time is

$$N_d = k' \cdot n \cdot \exp\left(\frac{-\mathcal{E}_2}{RT}\right) \tag{2}$$

where k' is the proportional coefficient. The number of taking-off molecules from the first layer adsorption has been neglected, because it is assumed to be very small because $\mathcal{E}_1 > \mathcal{E}_2$, and $\exp(-\mathcal{E}_1/RT) \ll \exp(-\mathcal{E}_2/RT)$, where $\mathcal{E}_1$ is the adsorption energy of the first layer molecules. In adsorption equilibrium, $N_a = N_d$.

Considering the liquid surface of this adsorbate material to be in equilibrium with the saturated vapor (the pressure is P_s), on the other hand, the number (N_a^0) of molecules diving onto the surface per unit time is equal to the number (N_d^0) of molecules that take off from the surface per unit time. Then, assuming the same area S and the same proportional constants k and k' as those in the case of adsorption equilibrium on solids,

$$N_a^0 = k \cdot P_s \cdot S \tag{3}$$

and

$$N_d^0 = k' \cdot n_m \cdot \exp\left(\frac{-\mathcal{E}_L}{RT}\right) \tag{4}$$

are obtained, where n_m is the number of the liquid surface molecules on area S, which is equal to the number of molecules of monolayer volume of the solid surface having the area of S, and $\mathcal{E}_L$ is evaporation energy from the liquid surface. From equations (1–4),

$$\frac{P_s \cdot S}{n_m \cdot \exp(-\mathcal{E}_L/RT)} = \frac{P \cdot S}{n \cdot \exp(-\mathcal{E}_2/RT)} \tag{5}$$

or

$$\frac{P}{P_s} = \frac{n}{n_m} \cdot \exp\left(\frac{\mathcal{E}_L - \mathcal{E}_2}{RT}\right) \tag{6}$$

is obtained.

It is well known in gas adsorption on a solid that, when the adsorption shows a typical Type II isotherm, the adsorption energy ($\mathcal{E}_1$) of the molecules adsorbed until the amount adsorbed reaches the "knee (Point B [17])" on the isotherm curve (Fig. 1) is usually fairly large, but the adsorption energy ($\mathcal{E}_2$) of the molecules adsorbed in the linear portion following Point B is so lowered that it is nearly equal to the liquefaction (or vaporization) energy: $\mathcal{E}_L \approx \mathcal{E}_2$ [16]. Therefore Point B is generally considered to be the end of the first monolayer adsorption.

When organic materials (e.g., simple hydrocarbons) that have no specific interactive radicals in the molecule are used as adsorbates, the adsorption isotherm sometimes enters into the linear protein over Point B (quasi–second layer) on the isotherm curve before the amount adsorbed reaches monolayer volume. This is because some solid surfaces may consist of heterogeneous patchwork sections and share some sections on which the adsorption energy would be less than the comparatively large interactive energy between large adsorbate molecules. It may be also taken that another reason for the small initial adsorption below Point B is that the effective cross-section of the long molecules adsorbed directly onto the surface is rather larger than that of the hypothetical spherical shape, which is assumed in the BET theory, owing to flat contact between the surface and the adsorbate molecules. Therefore the value of n, which is the number of molecules in the adsorption phase of the linear portion, used in Eq. (2) and other equations, could not be counted as a definite number, since it is not always known from what adsorption amount the second or quasi–second layer adsorption over Point B began. Using differentials of the amount adsorbed and the adsorbate relative pressure, respectively,

dn and $d(P/P_s)$, expression [6] is changed to the following formation:

$$\frac{dn}{d(P/P_s)} = n_m \cdot \exp\left(\frac{\mathcal{E}_2 - \mathcal{E}_L}{RT}\right) \quad (7)$$

In the case $\mathcal{E}_2$ is nearly equal to $\mathcal{E}_L$, or $\mathcal{E}_2 - \mathcal{E}_L \ll RT$, which has been experimentally proved in most cases, this expression becomes simply

$$\frac{dn}{d(P/P_s)} \doteqdot n_m \quad (8)$$

This shows that the gradient, $dn/d(P/P_s)$, of the linear portion of the Type II adsorption isotherm curve is practically equal to the number of molecules of the monolayer adsorption volume, n_m.

This simple relation has been proven experimentally by using lower alkylbenzenes (toluene, ethylbenzene, n-butylbenzene, and benzene) as adsorbates, which may be well suited to the gradient method theory for a large number of solid samples; the usefulness of the gradient method for surface area estimation has been ascertained together with the theory used. This method has also been proven by other authors using some nonspecific organic adsorbates other than lower alkylbenzenes [18].

In the case where the value of $\mathcal{E}_2 - \mathcal{E}_L$ is small but not zero, for example, in nitrogen or argon adsorption at low adsorption temperatures, it has been clearly shown that the gradient value of the linear portion changes with a change of the adsorption temperature. For nitrogen and argon at the liquid nitrogen temperature, the values of $\mathcal{E}_2 - \mathcal{E}_L$ were estimated, respectively, to be smaller than 400 and 400–800 J/mol for various solid adsorbents [2]. For the small value of $\mathcal{E}_2 - \mathcal{E}_L$, on the other hand, the relation of Eq. (8) may be kept in the measuring condition above room temperature owing to the high value of RT.

For the adsorbates having specific radicals for adsorption in their molecules, Eq. (8) is not always satisfied. This is because the molecules adsorbed on the first monolayer are subject to a specified interactive attraction from the specifically adsorbed first-layer molecules (localized and/or oriented, and so on) on the solid surface and from the solid surface itself, that is, the adsorbate molecules have a coefficient k' in Eq. (2) different from that in Eq. (4).

It should be mentioned that there is a remarkable fact that even the second or quasi–second layer molecules, subject to a strong interaction from polar surface and the underlayer molecules, have an adsorption energy (heat) similar to the liquefaction heat, though the fact was obtained experimentally using medium-sized alcohol adsorbates and some polar solid samples, as will be shown later in this chapter.

2. Thermodynamic Consideration of Adsorbate Molecules

Assuming that the molecules adsorbed on the first monolayer (more exactly, in the linear portion over Point B on the isotherm curve) are in a state similar to that of the molecules on the corresponding liquid surface at the same temperature, the differences in the thermodynamic values of the adsorbate molecules and those of the molecules of the liquid surface will be calculated as follows.

If the chemical potential of the adsorbed molecules, whose number is n, is expressed by μ_a, and that of the molecules in the gas phase in equilibrium with the adsorbed molecules by μ_g, which is equal to μ_a, and the chemical potential of the molecules of the liquid surface, which is equal to that of the molecules in the saturation gas phase above the liquid, is expressed by μ_0, then, by assuming that the adsorbate gas has so low a vapor pressure as to be an ideal gas, the relation in the gas phase is

$$\mu_g = \mu_0 + R \cdot T \cdot \ln\left(\frac{P}{P_s}\right) \tag{9}$$

and for the chemical potential of the molecules adsorbed on the first monolayer,

$$\mu_a = \mu_0 + dH_a + R \cdot T \cdot \ln\left(\frac{n}{n_m}\right) + T \cdot dS_a \tag{10}$$

is obtained. In the above expression, $dH_a = H_2 - H_L$, where H_2 and H_L are, respectively, enthalpies of the adsorbed molecules and the liquid state molecules, is the excess enthalpy of the adsorbed molecules compared to the liquid state molecules. $R \cdot \ln(n/n_m)$ is the entropy change (configurational entropy [19,20]) by the free configuration of the adsorbed molecules on the surface, since the adsorbed molecules are mobile, which is similar to the second term of Eq. (9). dS_a is the excess entropy of the adsorbed molecules, that is, an increase in entropy from the state of liquid surface molecules other than the configurational entropy change, although dS_a may be due to the kinetic restriction of the adsorbed molecules under the various influences of the surrounding molecules adsorbed on the solid surface and the solid surface molecules themselves, if dS_a is nonzero. Using $\mu_g = \mu_a$, from Eqs. (9) and (10),

$$\frac{n/n_m}{P/P_s} = \exp\left(\frac{dS_a}{R}\right) \cdot \exp\left(\frac{dH_a}{RT}\right) \tag{11}$$

is obtained. Using the differentials of n and P, as in the previous section, the expression is rewritten as

$$\frac{dn}{d(P/P_s)} = n_m \cdot \exp\left(\frac{dS_a}{R}\right) \cdot \exp\left(\frac{dH_a}{RT}\right) \tag{12}$$

Using Eq. (12), the value of dS_a, which is a measure showing the degree of kinetic freedom or kinetic restriction of the molecules adsorbed on the first monolayer as compared with the liquid surface molecules, can be estimated from the values of the linear portion gradient of a Type II isotherm $(dn/d(P/P_s))$, the monolayer volume of adsorption (n_m), and the difference of adsorption heat from the liquefaction heat (dH_a = net adsorption heat). Equation (12) is reduced to (8), and it is shown that the gradient method for surface area estimation would be valid, in the special case where $dS_a = 0$ and $dH_a/RT \approx 0$, although the latter condition seems to obtain even for adsorbates having specific adsorptive radicals in their molecules, e.g., medium-sized alcohol adsorbates, as is shown in this chapter.

C. Verification of the Gradient Method

1. Choice of Adsorbates for the Gradient Method

For verification of the gradient method, it was first necessary to choose appropriate adsorbates applicable to the gradient method, because all adsorbates could not always satisfy Eq. (8) and be applied to the gradient method for all sorts of adsorbents. For universal use of the gradient method, adsorbates that have no specific adsorptive effects on almost all adsorbents should be selected as a necessary condition. Moreover, the

TABLE 1 Physical Constants of Alkylbenzenes Used with the Gradient Method

	Toluene	Ethylbenzene	n-Butylbenzene	Benzene
mp (K)	178	178	185	353
bp (K)	383	409	456	278
Ps (Pa/273.2 K)	900	263	21.3	3511
σ_m (nm^2)	0.343	0.377	0.443	0.304

mp, melting point; bp, boiling point; Ps, saturation pressure; σ_m, molecular cross-section.

adsorbates used for this method had to maintain a low vapor pressure at room temperature, because the method was studied for easy estimation of very small surface areas (less than 1 m^2) at ordinary temperature. After full consideration and trial, lower alkylbenzenes (toluene, ethylbenzene, etc.) having comparatively low melting points and high boiling points were chosen as adsorbates. It could be expected that at ordinary temperatures a material having a high boiling point maintains a low vapor pressure and that a material having a low melting point (about $-100°C$) keeps a typical fully liquid state, in which the molecules have no mutual orientation, and full kinetic freedom, which might not always be maintained at a temperature near the melting point. The latter condition might be necessary for the adsorbate used in the gradient method, because in this method the adsorbate molecules adsorbed on the first monolayer should be expected to keep a fully liquid state similar to the bulk liquid, even in the potential field of the active solid surface. Under these considerations, toluene, ethylbenzene, and n-butylbenzene, the last one of which was especially used for samples of very small surface area due to its quite low vapor pressure, were chosen and tested as adsorbates of the gradient method. Then these adsorbates were appraised as effective for the gradient method by comparison with the standard BET/N$_2$ surface areas for many nonporous samples, and/or with geometrical surface areas for some platelike samples having no surface porosity [1–5]. Benzene, other than alkylbenzenes, which has a comparatively high melting point and a low boiling point, was found to be applicable as an adsorbate for the gradient method [1,21]. In studying this adsorbate, it was found that benzene molecules maintain a liquid state even below the solidifying temperature on some solid surfaces to satisfy the condition for the gradient method [21]. The physical constants of these organic adsorbates, which could be used for the gradient method, are listed in Table 1.

From the results of a series of experiments to verify the applicability of the gradient method, it was also established that the gradient method theory would be almost correct for the appropriate adsorbates. Moreover, for the adsorbates inapplicable to the gradient method, excess entropy of the adsorbed molecules, dS_a, the negative value of which expresses a decrease of the degree of kinetic freedom, was estimated from Eq. (12). Analysis of the state, or degree of kinetic freedom, of the adsorbate molecules of middle-sized alcohols on solid surfaces was carried out using Eq. (12), as shown later in this chapter.

2. Solid Samples Used for Verification

It was necessary for the solid samples used for verification of the gradient method to be chosen from those which had, or were expected to have, no micropores and/or specific, or anomalously strong adsorptive sites on the surface, because the "external" surface area estimated by the gradient method might not correspond to the BET surface area, which includes the apparent area raised from micropores and/or specific sites other than the external area, when the solid samples used have such surface complexities. Platelike solid samples used for comparison with their geometrical areas were, or were considered to be, nonporous. Experiments using platelike samples were carried out for evaluation of the gradient method, as independent of the other surface area estimation methods.

3. Results of Verification Experiments

Table 2 shows the comparison of the specific surface areas obtained by the gradient method using various adsorbates with the standard specific surface areas by the BET/N_2 method for various solid samples used. Table 3 shows the comparison of surface areas obtained by the gradient method with the geometrical surface areas of platelike solid samples. From the results shown in Table 2, it is seen that specific areas obtained by both the standard BET method and the gradient method agree fairly well, if it is taken into account that the surface areas by the standard BET method using nitrogen may not always be exact, since the uncertainty may extend to 20% [22], and that the solid samples used may have some residual micropores or defects on the surface. In Table 3, it is shown that there are some samples (e.g., blown glass) whose areas are estimated by the gradient method to be almost equal to the geometrical surface areas, although other platelike samples are estimated to have somewhat larger surface areas than the geometrical ones, since the samples have some residual surface pores on them, because it is quite difficult to prepare perfectly planar samples having no defects on the surface. In the experiments using platelike solid samples, it was found that the relation of various surface-finishing methods with the residual surface defects could be studied, as is also shown in Table 3.

From these experiments on the gradient method, it was concluded that the gradient method is in practice useful and that the theory derived for the method should be valid.

III. NEW APPARATUS FOR MEASURING SMALL ADSORPTION AT VERY LOW VAPOR PRESSURE

For the experiments relating to verification of the gradient method, a new apparatus for measuring quite small amounts of adsorption at a very low adsorbate pressure had to be constructed, because the surface areas of solid samples used for the verification were essentially very small, as was mentioned in the preceding section, and the small amount of adsorption had to be measured accurately under a very low vapor pressure in order to obtain exact adsorption isotherms. The detailed construction of the apparatus, which was essentially a system of volumetry at constant pressure, is given below. It was also used for the experiments of vapor adsorption of medium-sized alcohols in this study without any change. The schematic diagram of the construction is shown in Fig. 2.

A. Apparatus

The whole construction of the new apparatus for adsorption measurement was made of metal and was devised so as to enable easy operation. A gas burette, the most important part of this apparatus, was constructed of a stainless steel bellows and a stainless steel cylinder covering the bellows, the inner space between which functioned as a gas burette having a changeable volume, which could be controlled mechanically from outside the apparatus so as to maintain a constant pressure, for example, when it was used for constant adsorbate pressure operation. The volume change in the gas burette was measured

TABLE 2 Comparison of Specific Surface Areas by Gradient Method with n-Butylbenzene (n-B.B.), Ethylbenzene (E.B.), and Toluene with BET Specific Surface Areas

Sample		S_{BET} (m^2/g)		S_{grad} (m^2/g)			
		N_2 (C_{BET})†	Ar	n-B. B. (S_{grad}/S_{BET})	E. B. (S_{grad}/S_{BET})	Toluene (S_{grad}/S_{BET})	
Rock salt	(S)	0.275 (463)	0.267	0.270 (0.99)	0.273 (0.99)	0.275 (1.00)
Calcite	(S)	0.813 (83)	—	0.799 (0.99)	—	—
	(S)	1.50 (83)	1.22	—	1.53 (1.02)	1.54 (1.03)
MgF$_2$	(S)	0.893 (neg)	—	1.11 (1.24)	—	—
	(S)	1.11 (neg)	0.75	—	1.34 (1.21)	—
K$_2$SO$_4$	(S)	0.269 (135)	—	0.277 (1.03)	0.293 (1.09)	—
K$_2$CrO$_4$	(S)	0.278 (70)	0.212	0.251 (0.90)	—	—
Fluorite	(S)	0.799 (inf)	0.688	—	0.789 (0.99)	—
CaF$_2$	(R)	0.561 (inf)	—	0.462 (0.82)	—	—
	(Colloid)	4.21 (400)	—	—	—	3.40 (0.81)
α-Alumina	(R)	0.697 (112)	0.55	0.707 (1.01)	0.701 (1.00)	0.701 (1.00)
Fe$_2$O$_3$	(R)	10.8 (48)	7.38	10.7 (0.99)	10.50 (0.97)	—
MgO	(R)	6.64 (122)	4.87	8.14 (1.23)	—	—
ZrO$_2$	(R)	22.7 (96)	17.6	18.8 (0.83)	22.3 (0.99)	—
CaCO$_3$	(R)	0.600 (139)	0.498	0.480 (0.80)	0.480 (0.80)	—
CeO$_2$	(R)	20.0 (78)	16.8	37.6/44.7 (hysteres)	—	—
Glass (CPG-10)		10.6 (50.4)	9.87	—	—	12.3 (1.16)
C-22 (Adsorbent)		3.85 (151)	2.98	3.00 (0.78)	3.09 (0.80)	3.09 (0.80)
Silica (Porasil)		4.45 (105)	4.02	—	—	4.24 (0.95)
Au (Colloid)		1.58 (73)	—	1.89 (1.20)	—	1.88 (1.19)
Cu (Powder, R)		1.18 (51)	—	—	1.26 (1.07)	—
Al-foil		0.0896 (39)	0.0770	—	—	0.0844 (0.94)
Mn-Al (Alloy)		0.785 (12.7)	0.126	0.103 (0.82*)	—	—

S: single crystal powdered; R: reagent chemical powder; C-22: diatomaceous-firebrick; Porasil F: pored cristobalite; CPG-10: controlled pored glass.
*Comparing with BET-Ar.
† C-value of BET-equation; neg, negative value.

TABLE 3 Comparison of Gradient Method Surface Areas by n-Butylbenzene Vapor Adsorption with Geometrical Surface Areas

Sample (surface-treatment method)	Geometrical area, S_{geom} (cm^2)	S_{grad}/S_{geom}	S_{grad}/S_{BET}
(1) Fused silica (polished by flame)	72	1.1	(0.95)
(2) Rock salt (cleaved and thermally annealed)	67	1.19	(1.00)
(3) Lithium fluoride (single crystal, cleaved)	64	1.39	(......)
(4) Fluorite (cleaved and polished)	79	1.56	(0.99)
(5) Calcite (cleaved)	49	1.53	(0.99)
(6) Saphire (polished)	72	1.36	(1.00)
(7) Silicate glass (thin plate by blowing)	291	1.03	(1.16)
(8) Aluminum (polycrystal, electropolished)	28	1.3	(0.94)
(9) Silver (polycrystal, rolled)	25	1.36	(......)
(10) Copper (polycrystal, thermally annealed)	100	1.30	(1.07)
(11) Copper (polycrystal, electropolished)	18	1.5	(1.07)
(12) Silicon (single crystal, polished)	66	1.25	(......)
(13) Stainless steel (electropolished)	25	1.3–1.2	(......)

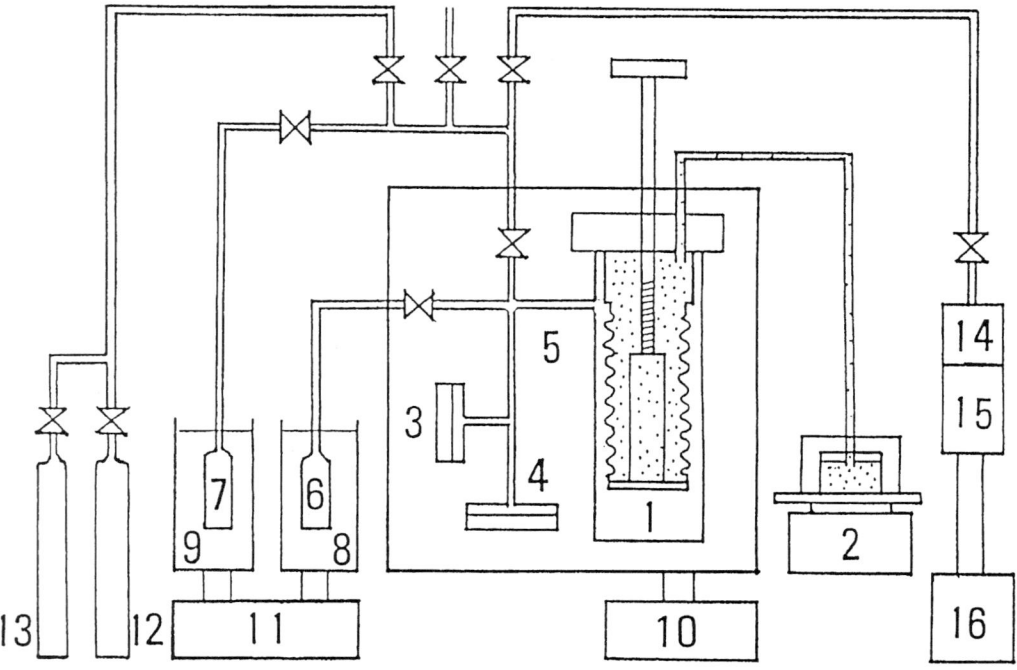

FIG. 2 Schematic drawing of the semimicrovolumetric adsorption measurement apparatus using the constant pressure method. (1) Gas burette having metal bellows for changing the inner volume, (2) electric balance for measuring the volume change of the gas burette, (3), (4) pressure gauges, (5) thermostatted air bath, (6) sample tube, (7) adsorbate liquid boiler, (8), (9) liquid bath thermostatted for the sample tube and adsorbate liquid boiler, (10), (11) thermostats for air and liquid baths, (12), (13) gas supplies for adsorption and dead volume measurement, (14) cold trap with liquid nitrogen, (15) turbo molecular pump, (16) rotary pump.

precisely (±1 μL) by weighing on a digital balance (±1 mg), set outside, the increase or decrease in the liquid (e.g., ethylene glycol) that was fully introduced into the bellows (cf. Fig. 2). An advantage of using this type of gas burette was that no change in the adsorption amount of the adsorbate on the inside surface of the gas burette could be found, although the inside adsorption change of the adsorbate would be induced essentially by the volume change in the use of ordinary gas burettes such as mercury or metal piston gas burettes. This is because the inside surface area of this type of gas burette, making use of bellows, might be slightly changed with a change of the volume by the elastic deformation of a metal bellows. Any small change in adsorption inside the gas burette with the volume change could never be ignored in the experiment where a very low adsorption amount was measured under a very low adsorbate pressure because of a serious error in the result. The changeable volumes of gas burettes in two types of apparatus used were 25 mL and 200 mL maximum, respectively, constructed according to the measurements of small adsorption and of slightly larger adsorption.

In the construction of the apparatus, metal bellows, pipes, valves, gaskets, etc., which were provided commercially for the components of high vacuum equipment, were used, usually after electrolytic polishing, the effect of which was clarified afterwards by the experiments verifying the gradient method itself, that is, by the absolute evaluation experiments for the method by comparing the surface area of platelike samples with the geometric one (cf. Table 3). It was shown by such experiments that electrolyte polishing is very effective in terms of a remarkable reduction of surface porosity; and therefore it improves this kind of apparatus (e.g., reduction of the leading time of measurement by adsorption equilibrium). In particular, the inside surface of the apparatus constructed by welding must be carefully polished electrolytically, because it was proved experimentally by the gradient method that welded metal surfaces generally had a large number of pores in them and a very large surface area. The pressure gauges used in the apparatus were MKS-315-0001 (1 torr maximum, MKS-Japan, Inc., Tokyo) and TJK/806 (25 torr maximum, Sensotec-Sokken, Tokyo); the former is used in the measurement of very low adsorbate vapor pressures.

The main parts of the apparatus, after being assembled, were put into an air bath (25–250°C) for the contaminant gases adsorbed on the inside surface to be baked out at 250°C maximum in high vacuum conditions (cf. Fig. 2).

The evacuation to a high vacuum was carried out with a turbomolecular pump with a cold trap using liquid nitrogen and a rotary pump up to 1×10^{-7} Pa. This was until the maximum pressure of the gas desorbed from the inside surface of the apparatus was held at a pressure less than 1% of the minimum value of the measuring adsorbate gas pressure, in conditions where the pumps were stopped until the adsorption measurement was continued.

An apparatus having the maximum changeable volume of 200 mL was used with a pressure gauge of 3.5 atms maximum in the measurements of the standard BET/N_2 surface areas of the sample solids used.

B. Procedure of Adsorption Measurements Using the New Apparatus

1. Introducing Adsorbate Liquid into the Apparatus

The adsorbate liquid used for vapor adsorption experiments was introduced into the evaporating boiler in the apparatus, filled with helium gas, with anhydrous calcium sulfate, the amount of which was about half of the adsorbate weight. Helium and contaminant gases were pumped out, after the boiler was cooled in a liquid nitrogen bath for the adsorbate liquid to be solidified. Liquefying of the adsorbate during the cessation of pumping, and solidifying of the adsorbate liquid during pumping-out, were repeated several times, in order that contaminant gases other than the adsorbate gas in the apparatus system should be removed thoroughly, until the residual gas pressure at the measurement temperature became constant, which was used as the value of saturation vapor pressure of the adsorbate at the temperature, shown in Tables 1 and 4.

The adsorbate liquids were distilled two or three times with a dehydration reagent before being introduced into the apparatus system.

2. Solid Samples

Solid samples, which were powdery or platelike, were thoroughly washed with pure acetone before being put into the sample tube in the apparatus. Powdery solid samples were wrapped in a 7-μm-thick aluminum foil, the specific surface area of which was known to be 1.04 cm^2/cm^2, lest the sample powder should disperse in the system. The solid samples put into the apparatus were heated at 160 to 250°C to dry *in vacuo* ($\approx 1 \times 10^{-6}$ Pa) for 18 to 24 h before the adsorption measurement.

3. Procedure of Adsorption Measurement

After the gas burette, which was thermostatted at a constant temperature (usually 50°C), and the sample tube, which was maintained at a measuring temperature (usually at normal room temperature, the same as the temperature of adsorbate boiler), were thoroughly evacuated, the adsorbate vapor was introduced into the gas burette from the boiler at measurement pressure after the sample tube was shut off from the gas burette by the connecting valve. A rather long time interval would be necessary for the introduced adsorbate vapor to attain constant measurement pressure to be maintained in the gas burette due to gradual adsorption onto the inside wall of the gas burette. In order to shorten the time for adsorption equilibrium in the first procedure, the adsorbate vapor was introduced at a pressure somewhat higher than measurement pressure; then, after appreciable adsorption onto the inside wall, the introduced vapor was pumped out slowly until the pressure attained measurement pressure. Next, the valve leading from the gas burette to the sample tube was opened for the solid sample to adsorb the adsorbate vapor introduced, although the adsorbate vapor introduced into the sample tube would be consumed by adsorption onto the inside wall of the sample tube, by the filling of the dead space in the sample tube, and by adsorption on the sample solid itself. Then, the decrease of the adsorbate vapor pressure in the gas burette found in this procedure was recovered to the measurement pressure by reducing the gas burette volume, the volume reduction being the measure of consumption of the adsorbate vapor in the sample tube. Since even in the latter procedure, the time necessary for a perfect adsorption equilibrium would be usually fairly long, the adsorbate vapor pressure in the gas burette connected with the sample tube was, first, set to a pressure a little higher than the measuring pressure by pressing the gas burette volume a little more than the proper volume. This must be carried out very carefully in the case where adsorption/desorption hysteresis would appear on the isotherm in such a procedure. After full adsorption at a slightly higher vapor pressure than the proper measurement pressure on the sample and sample tube wall, the gas burette volume was depressed backward for the adsorbate vapor to show precisely the measurement pressure with slight desorption of adsorbate from the sample and sample tube wall. In this type of adsorption experimental procedure, which was made possible only by the use of the new apparatus developed in this study, perfect attainment of the adsorption equilibrium on the sample could be fully ascertained. Therefore the adsorbed amounts shown on the isotherms obtained in this experiment are no transient values but truly reliable adsorption values.

The true amount adsorbed by the solid sample itself can be obtained by subtracting the amount of consumption by filling the dead space of the sample tube and by adsorption onto the sample tube inside wall from the apparent adsorbed amount calculated directly from the decrease of the gas burette volume. The dead space volume of the sample tube was measured separately using the gas burette and an inert gas such as argon or helium. The adsorbed amount onto the inside wall of the sample tube vs. various adsorbate pressures, that is, the sample tube adsorption isotherms, was measured by the same method as mentioned above, in the conditions of no sample in the sample tube. To obtain an adsorption isotherm curve in the process of gradual increase of the adsorbate pressure, the adsorbed amounts of the next stage were obtained by adding the increment of the amount adsorbed at the last stage to the last stage adsorption amount. This procedure was carried out in succession with the increase of adsorbate pressure to complete an isotherm from zero pressure of adsorbate. "Desorption isotherms," which would be obtained in the process of decreasing the adsorbate pressure, were obtained by an entirely inverse process to the process of obtaining "adsorption isotherms." Usually, both adsorption and desorption isotherms were measured in order to examine the accuracy of the adsorption experiment or to detect any existence of adsorption/desorption hysteresis.

Using the new apparatus with the sample tube cooled at liquid nitrogen temperature, adsorption isotherms of krypton, the vapor pressure of which is comparatively low at liquid nitrogen temperature, could be measured strictly without any difficulty.

As an application of the gradient method and the new apparatus, from the total surface area obtained by the BET/Kr method, the external surface area obtained by the gradient method using n-butylbenzene, and the inside volume of mesopores obtained from the size of hysteresis shown in the adsorption/desorption isotherms of both adsorbates used, pore-size analyses were carried out for the mesopores on porous silicon and anodized aluminum, both of which seemed to have uniform-sized cylindrical pores on the surface [23–26]. It was clarified in these reports that simple cylindrical pores having a uniform diameter could have an adsorption/desorption hysteresis, the existence of which used to be considered thermodynamically impossible, and that in the usual analysis of pore-size distribution, both adsorption and desorption isotherm curves should be used as references, though the pore-size analysis was usually carried out using the desorption isotherm only under the assumption of the existence of a set of various sizes of simple cylindrical pores on the surface.

IV. MULTILAYER ADSORPTION OF 1-BUTANOL AND 1-HEXANOL AND THE BEHAVIOR OF ADSORBED MOLECULES

Adsorption studies on various solid samples were carried out to verify the applicability of the gradient method to medium-sized alcohol vapors as adsorbates, and for the thermodynamical research on the kinetic states of the molecules adsorbed on the surface, especially in the case when the gradient method could not be applied, because medium-sized alcohols are similar to lower alkylbenzenes in physical properties, that is, they have a high boiling point and a low melting point, though lower alkylbenzenes can be used effectively in the gradient method [27].

TABLE 4 Melting Points (mp), Boiling Points (bp), Saturation Vapor Pressures (Ps), Molecular Cross Sections (σ_m), and Enthalpy Changes by Melting (ΔH_m) and by Vaporizing (ΔH_v) of 1-Butanol and 1-Hexanol

	1-Butanol	1-Hexanol
mp (K)	183.9	229.2
bp (K)	390.4	430.2
Ps (Pa)	302 (283.16 K)	42.0 (288.16 K)
	657 (293.16 K)	65.6 (293.16 K)
σ_m (nm^2)	0.331	0.383
ΔH_m (kJ/mol)	9.38	15.4
ΔH_v (kJ/mol)	43.2	48.6

A. Experimental

The adsorption experiments were carried out to obtain precise isotherm curves, especially in the multilayer adsorptive portion in the range of 0.1 to 0.9 of relative pressure, using a variety of solid samples that were considered to have a typical nature of the adsorbent for the vapors of medium-sized alcohols.

1. Adsorbate Vapor

1-butanol and 1-hexanol were chosen as typical adsorbates of medium-sized alcohols. Physical constants of the alcohols are shown in Table 4.

Since these alcohols have a comparatively low melting point, it is considered that, at room temperature much higher than the melting point, the molecules of these alcohols in liquid phase may be fully in a liquidlike molecular state having a comparatively large thermal kinetic energy and a large kinetic freedom, presumably owing to insignificant mutual interactive attraction and/or a slight mutual orientation effect compared to the thermal motion, though these molecules may associate with each other to form, temporarily and/or partially, dimers or clusters, due to the interaction between hydroxyl groups in their molecules. It was shown later, however, as an experimental result by comparison of the liquid state molecules with the adsorbed molecules, that such an association of two or more molecules could not be observed at room temperature for the medium-sized alcohols used.

The vapor pressures of these medium-sized alcohols having high boiling points were very low at room temperatures in the experiment in this study, as shown in Table 4.

Molecular cross-sections, σ_m, in the table, which are usually used as the standard cross-section of the adsorbate molecules in the BET method, were calculated from the liquid density by assuming the adsorbate molecules to be spheres in the closest packing in the liquid state. Such a method of calculation for obtaining the molecular cross-section may be valid for the adsorbed molecules in the liquid state at a temperature much higher than the melting point, since the molecules in a typical liquid state are considered to have no molecular association and full diffusional and rotational kinetic freedom. This fact has been verified for the medium-sized alcohols by the application of the gradient method to some solid samples in adsorption experiments using vapors.

Although the values of enthalpy changes by melting, ΔH_m, and by vaporizing, ΔH_v, in Table 4 were obtained from the values of the heat of melting and heat of vaporizing, respectively, the entropy change by melting was calculated from the value of ΔH_m to be used for the estimation of the freedom of adsorbed molecules.

These adsorbate liquids were used after being twice distilled with dehydrating reagent.

2. Solid Sample

Solid samples used for adsorbents were gold, aluminum foil (only for 1-hexanol adsorbate), graphite, sodium chloride, aluminum oxide, and silicate glass. Their nonporous solids were used. These were chosen as solid samples having the characteristic nature of a surface for adsorption; that is, gold is a typical heavy metal that has no oxide film on the surface; aluminum is a light metal that has a hard oxide film on the surface; graphite is a typical half metal having plenty of π-electrons on the surface; sodium chloride is a typical soft ionic solid that presumably has many hydroxide groups on the surface due to surface oxidization by the atmospheric humidity; aluminum oxide (type α) is a representative covalent-bonded, refractory oxide that has a hard hydroxide or its dehydrate film on the surface; silicate glass is a typical noncrystalline material with hydrophilic groups on the surface.

The methods of preparation of these solid samples were as follows: the gold sample was precipitated as fine platelet crystals by reduction of chloroauric acid in aqueous solution and fully dried; the aluminum sample was of chemically pure, 7μm-thick foil and used after being fully washed with dry acetone; the graphite sample was prepared by milling a natural graphite crystal block; the sodium chloride sample was prepared by milling a large single crystal for optical use in a dry atmosphere; the aluminum oxide sample was reagent grade of α-alumina; the silicate glass sample was a macroporous glassy powder for liquid-chromatographic use (GCP-10-3000, Electronucleonics, Inc., New Jersey). All these solid samples were known to have no micropores on the surface. Since the vapor pressures of adsorbate alcohols used in this experiment, especially for 1-hexanol, were rather low, very few solid samples were appropriate for use in volumetry in this experiment. In Table 5, the standard total surface areas and standard specific surface areas of the solid samples used are listed along with the method of estimation of standard surface areas. Although the most common surface area estimation method in the list was the BET method using N_2, the BET method using Kr and the gradient method using n-butylbenzene vapor were employed for a few samples only in cases where the large amount of sample needed for the BET/N_2 method could bot be prepared. For the graphite sample, the surface area obtained by the BET/Kr method was reviewed, because it had been reported that the BET/N_2 method value for graphite would be rather smaller than the true value [28]. The estimation method referred to as N_2/Point-B in Table 5 is the method using the adsorption amount at Point B, at which the linear portion begins on the N_2-adsorption isotherm of Type II, as the monolayer volume (Point B method [17]), in the case where the BET/N_2 method could not be used due to irregular BET plots. These standard surface areas in Table 5 were used for thermodynamic analyses of the molecules adsorbed on solid sample surfaces.

3. Apparatus for Adsorption Experiments

Adsorption experiments of 1-butanol and 1-hexanol vapors on various solid samples were carried out using the new apparatus developed for the measurement of very small adsorption amounts under an extremely low vapor pressure, which was first used for verification

TABLE 5 Standard Total Areas and Specific Areas of Solid Samples Used in the Adsorption Experiments of 1-Butanol and 1-Hexanol Vapors

	Total area (m^2)	Specific area (m^2/g)	Estimation method
	For 1-butanol vapor		
Gold	1.81	0.791	N$_2$/BET
Graphite	1.07	2.81	N$_2$/BET
	(1.23	3.21	Kr/BET[a])
Sodium chloride	3.33	0.253	N$_2$/BET
Aluminum oxide	1.08	0.697	N$_2$/BET
Silicate glass	1.102	6.80	N$_2$/BET
	For 1-hexanol vapor		
Gold	0.103	3.22	n-B.B./GRAD.[b]
Aluminum foil	0.0461	0.132	Kr/BET[a]
Graphite	0.390	2.81	N$_2$/BET
	(0.446	3.21	Kr/BET[a])
Sodium chloride	0.489	0.330	N$_2$/Point B
Aluminum oxide	0.276	0.697	N$_2$/BET
Silicate glass	0.296	6.80	N$_2$/BET

[a] By adoption of Kr, using molecular area of 0.202 nm^2.
[b] Gradient Method using n-butylbenzene vapor.

of the gradient method, as mentioned in detail in Section III. The adsorption isotherm curves obtained with this apparatus were made using exact adsorption amounts that were measured after a fully confirmed adsorption equilibrium, as mentioned in the preceding section. The time necessary for complete equilibrium in adsorption of the medium-sized alcohol vapor on solid samples was not so long as expected for the usual alcohol vapors but similar to that of the adsorption of lower alkylbenzene vapors used in the gradient method. This might be a result of all the solid samples used having no pores on the surface and a very small surface area. All the samples used had no adsorption/desorption hysteresis, and the desorption isotherms agreed perfectly with the adsorption isotherms for all samples (cf. Section III.B.3).

B. Results and Discussion

1. Adsorption Isotherms

The adsorption isotherms of 1-butanol vapor at 10°C and 1-hexanol vapor at 20°C on various solid samples are shown in Fig. 3 and Fig. 4, respectively. The amounts adsorbed shown on the y-axis are the values in mL (STP) per square meter of the standard surface area [/m^2 (std)] shown in Table 5, which are directly proportional to the number of molecules adsorbed per unit area (1 m^2) of the standard surface area for each solid sample. The monolayer volume on unit area is also shown in the figures as Vm.

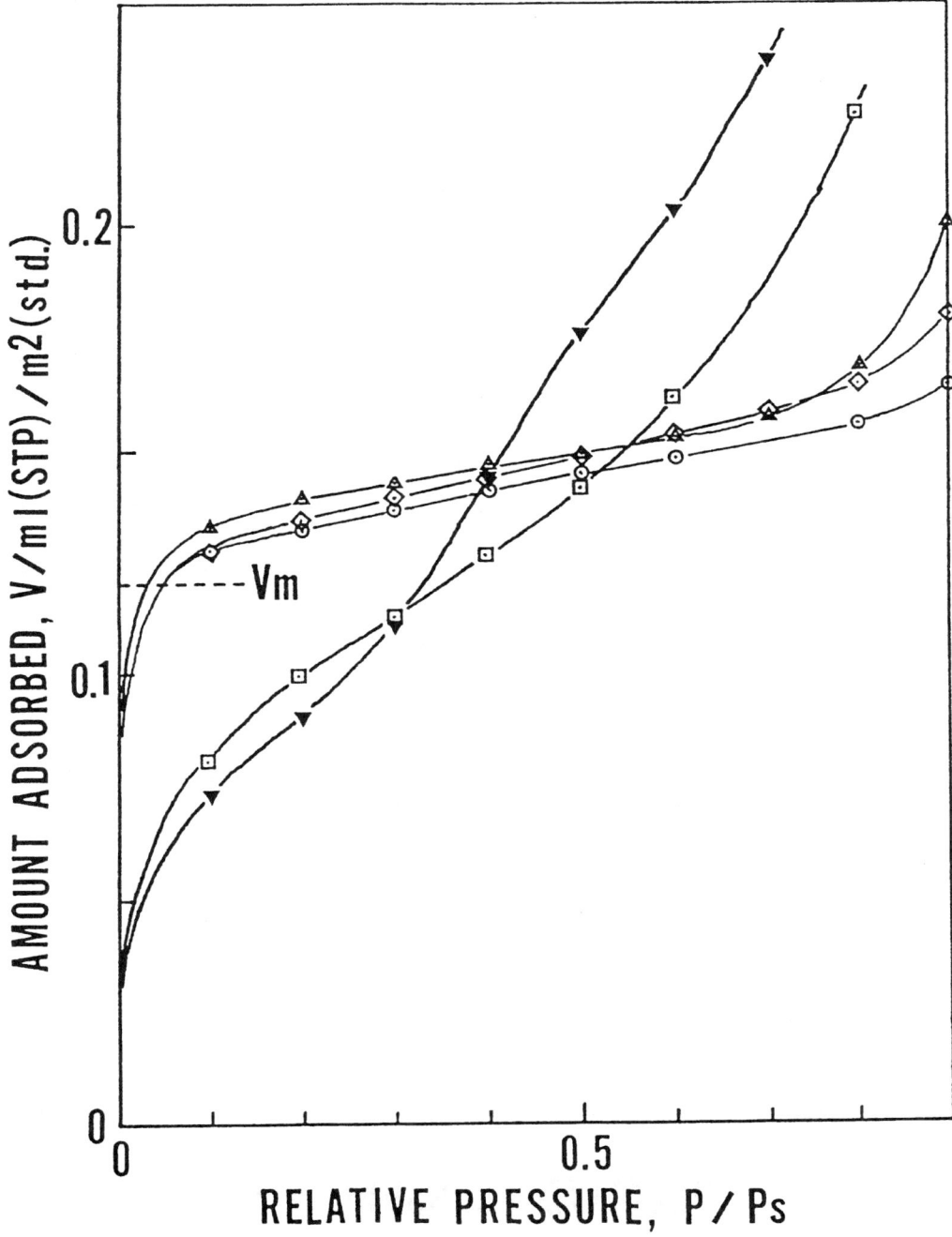

FIG. 3 Adsorption isotherms of 1-butanol vapor on gold (□) graphite (▼), sodium chloride (○), aluminum oxide (△), and silicate glass (◇) samples at 10°C. Vm, monolayer volume of 1-butanol vapor for the standard surface area (per m^2) by BET with N_2 (cf. Table 5).

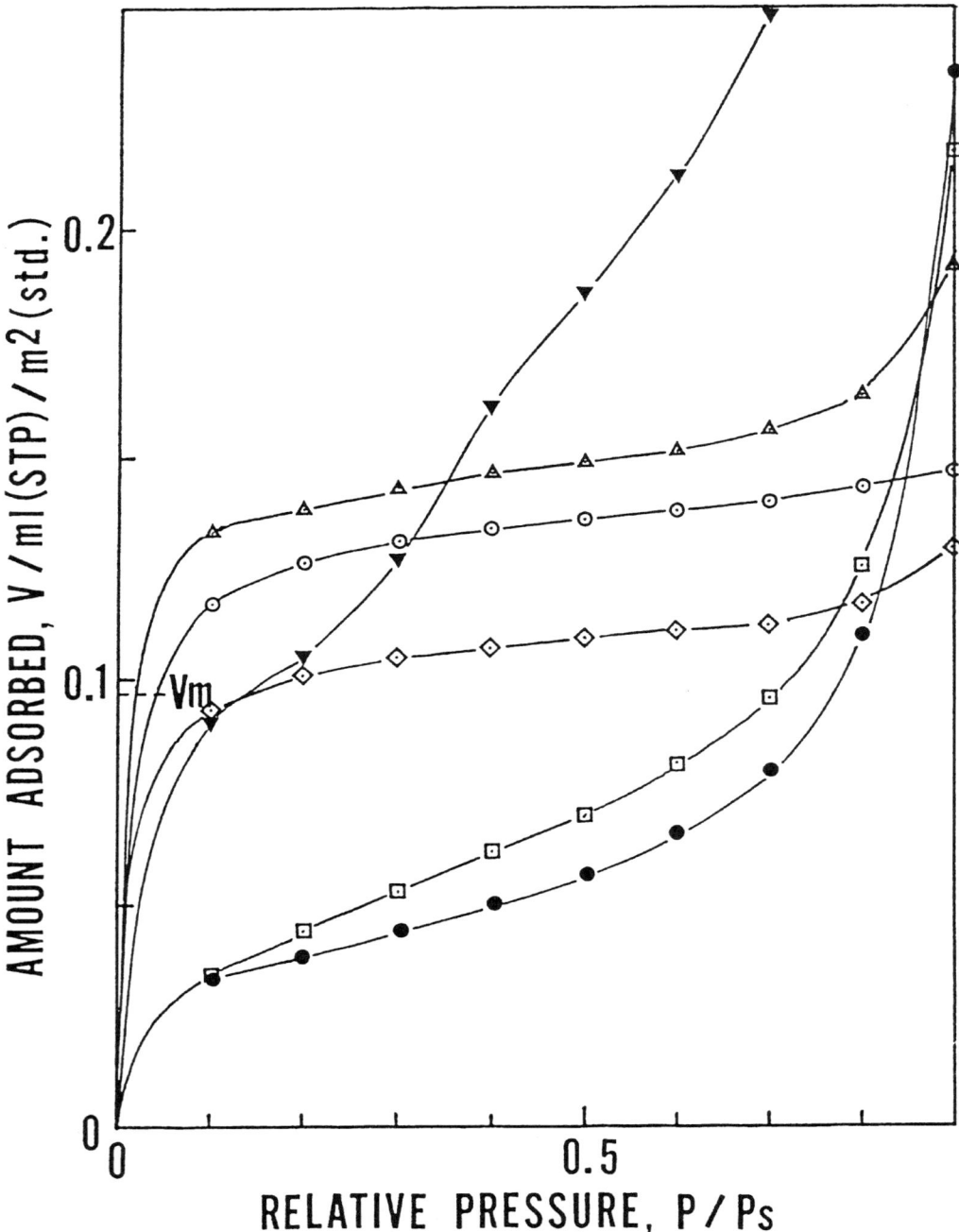

FIG. 4 Adsorption isotherms of 1-hexanol vapor on gold (□) graphite (▼), aluminum foil (●), sodium chloride (☉), aluminum oxide (△), and silicate glass (◇) samples at 20°C. Vm, monolayer volume of 1-hexanol vapor for the standard surface area (per m^2) by BET with N_2 (cf. Table 5).

As seen in Fig. 3 for 1-butanol vapor, the adsorption isotherm curves for sodium chloride, aluminum oxide, and silicate glass samples showed Type II isotherms (in the classification of Brunauer, Deming, Deming, and Teller (BDDT) [15]) fairly similar to a Type I isotherm (in the same classification), which had a long linear portion over Point B, whose gradient was so small that they might be categorized as quasi–Type I isotherms. The amounts adsorbed and the linear portion gradients over Point B for these three solid samples were alike all over the isotherms. On the other hand, the adsorption isotherm curve of 1-butanol vapor on the gold sample showed a typical Type II isotherm. For the graphite sample, the curve was a quasi–Type II isotherm that had a small step in it.

For the adsorbate of 1-hexanol vapor, as shown in Fig. 4, the appearance of the adsorption isotherm curves of sodium chloride, aluminum oxide, and silicate glass samples were similar to those of these solid samples by 1-butanol vapor (quasi–Type I). Although the gradients of the linear portion of these isotherm curves seemed to be similar, the amounts adsorbed per unit area by these solid samples differed considerably all over the isotherms, being unlike the case of 1-butanol vapor adsorption. The difference of the adsorbed amounts per unit area between gold and graphite samples also was much larger than that in the case of 1-butanol vapor, though graphite adsorption had a step in the isotherm similar to that of 1-butanol vapor. The adsorption isotherm of 1-hexanol vapor on aluminum foil was of Type II, which had an initial adsorption amount similar to that on the gold sample but a smaller gradient of the linear portion than that of the gold sample.

2. Isosteric Heat

Adsorption isotherms at different temperatures were measured to obtain isosteric heats of adsorption [29] for all the combinations of adsorbents and adsorbates used, especially in the range of multilayer adsorption (over Point B, in a more exact expression) on the isotherms. Figure 5 shows typical adsorption isotherms of 1-butanol vapor on graphite at temperatures of 10 and 20°C and those of 1-hexanol vapor on sodium chloride at temperatures of 15 and 20°C. As seen from the isotherms in the figure, the adsorption isotherms at different temperatures against relative pressure (vapor pressure relative to saturation vapor pressure of the adsorbate liquid at the measurement temperature) of the adsorbates agreed very well, at least from 0.1 to 0.8 relative pressure, between 10 and 20°C for 1-butanol vapor and between 15 and 20°C for 1-hexanol vapor. From these experimental results and a thermodynamic consideration, it became clear that the isosteric heats of adsorption are perfectly equal to the heats of liquefaction of the adsorbate liquids in the range of multilayer adsorption (over Point B, to be more exact). It may be surprising that the adsorption heats of alcohol molecules are identical with the liquefaction heats on solid samples, such as sodium chloride and aluminum oxide, having hydroxyl and/or its anhydride groups on the surface, although less adsorption heat is expected from the low gradient values shown in the quasi–Type I isotherms, and a higher adsorption heat is expected from the excess entropy decrease of the adsorbed molecules, as will be shown later. The two effects may counteract each other essentially for the adsorption heat so as to agree with the liquefaction heat.

3. Applicability of the BET Method and the Gradient Method

The adsorption isotherms obtained by experiments should be examined in more detail for the behavior of the molecules adsorbed on the solid surface to be investigated. For the solid samples (e.g., sodium chloride, aluminum oxide, silicate glass) showing quasi–Type I isotherm curves both for 1-butanol and 1-hexanol vapors, the BET equation

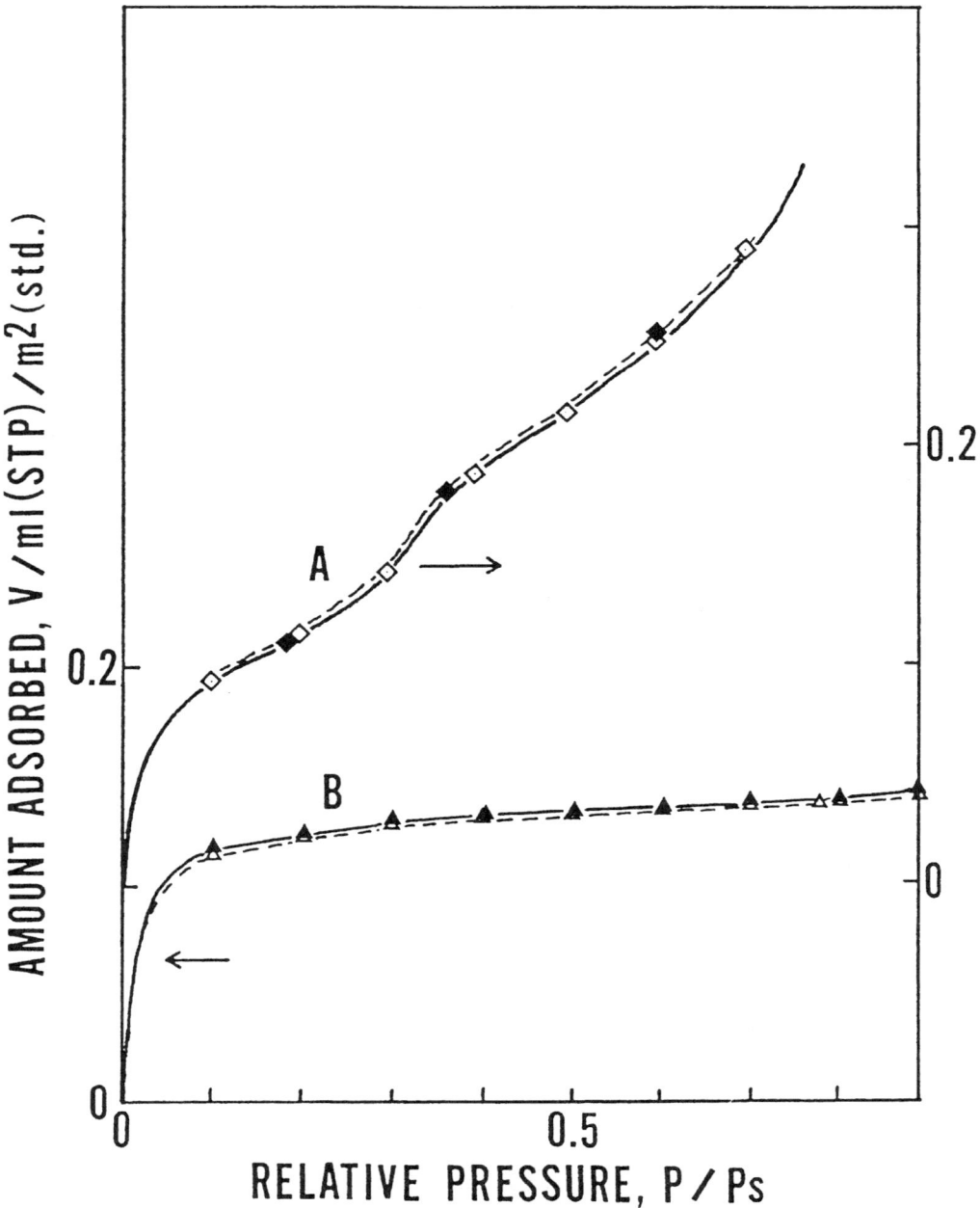

FIG. 5 (A) Adsorption isotherms of 1-butanol vapor on graphite (sample size 1.23 m² at 10°C (◇) and 20°C (◆). (B) Adsorption isotherms of 1-hexanol vapor on sodium chloride (sample size 0.489 m²) at 15°C (△) and 20°C (▲).

could not be applied to the surface area estimation, since hardly any normal liner plots of the isotherms were obtained by the BET equation, showing a negative intersection with the y-axis. On the other hand, since the location of Point B, which is considered to be the point of monolayer completion, was rather clear on these quasi–Type I isotherms, the surface area estimation of these samples really seemed to be easily done by using the adsorption amount at Point B and σ_m, the standard cross-section (Point B method [17]). The surface areas of these samples estimated by the Point B method, however, were not always identical with the standard surface areas by BET/N_2, especially for 1-hexanol vapor adsorbate, showing a large deviation from the standard values on the sodium chloride and aluminum oxide samples (compare the height of Point B on these isotherm curves with the Vm line in Fig. 4), though for these sample slightly larger surface area values than the standard values were obtained by this method from the isotherms of 1-butanol vapor (Fig. 3). On these solid samples, having surface hydroxyl and/or its anhydride groups, alcohol molecules adsorbed as the first layer on the surface might be strongly localized on the adsorption sites by direct interaction of hydroxyl groups of the alcohol molecules with the surface polar groups, possibly by an enhanced dispersion force or hydrogen bonding. This kind of adsorption mode would not meet the assumptions used in the BET theory that adsorbate molecules having a hypothetical cross-section of a "spherical particle" formed temporarily could adsorb on or desorb from the "uniform surface," which has no specific adsorptive sites. The reason the surface areas could not be obtained by the Point B method, on the other hand, is possibly that the medium-sized alcohol molecules adsorbed on the specific adsorptive sites on the surface showing a quasi–Type I isotherm might no longer possess the mean cross-section equal to σ_m, the cross-section calculated from the liquid density by assuming a spherical molecular shape and the closest packing, but the cross-section might depend on the original shape of the adsorbate molecule and on the surface density of adsorptive sites. In the case of 1-butanol adsorbate, however, the surface areas estimated from the adsorption amount of Point B corresponded comparatively well to the standard surface areas. This is because the adsorbate molecules smaller than the 1-hexanol molecules would be adsorbed with a mean shape similar to that of a sphere, due to the violent molecular kinetic motion at the measurement temperature comparatively higher than the melting point, being liberated from the dense polar sites on the solid surface.

The gradient values of the linear portion over Point B (against relative pressure) of quasi–Type I adsorption isotherms shown by solid samples such as sodium chloride, aluminum oxide, and silicate glass were as small as 0.3 to 0.4 of the standard monolayer volume for 1-butanol vapor and as small as 0.2 for 1-hexanol vapor. This is because in these adsorption systems, the molecules adsorbed over Point B, which are considered to be restrained in motion by the specific interaction with the polar solid surface and the partially oriented underlayer molecules, would be no longer in the kinetic or thermodynamic condition required by the gradient method theory. The degree of the restraint in molecular motion will be discussed in the next section.

It seems, on the other hand, that the surface areas of solid samples having a typical Type II adsorption isotherm (e.g., gold and aluminum foil) or a quasi–Type II isotherm (graphite) could be obtained by the BET method procedure. However, the values of the BET surface areas, except for the case of graphite by 1-hexanol vapor, were much lower than the standard values, though such surface area underestimations had been frequently encountered in the use of nonspecific organic vapor adsorbates in the BET method. This shows the general shortcomings of the BET method using the common organic vapors, although in such cases the amount adsorbed at Point B would be anomalously small. The reason the amount adsorbed at Point B, which is usually considered to be the end of the first layer adsorption, is much smaller than the monolayer volume for the alcohol adsorbate used, may be that the interaction energy of a surface having no polar sites, such as a gold surface, with medium-alcohol adsorbate molecules is similar to that with normal, nonspecific chain organic molecules, although a high interaction energy of the imperfect, irregular lattice sites (active sites) on the surface with the adsorbate molecules should be taken into consideration. Thus, after the active sites have been covered by initial adsorption of some adsorbate molecules, the following adsorption may be carried out by the adsorptive energy similar to, or a little higher than, the adsorptive energy needed for the adsorption onto the surface of monolayer adsorption, depending on the nature of the solid surface. Even in the latter case, where the solid may have higher density than the adsorbate molecules, the amount of the first layer adsorption of organic chain molecules on the inactive surface may be smaller than the standard monolayer volume (Vm) estimated using the standard cross-section (σ_m) under the assumption of the mean spherical shape of adsorbate molecules. This is because the molecules in the first layer may be no longer adsorbed in the spherical form but adsorbed by touching all the fractions of the chain adsorbate molecule on the solid surface with a larger coverage area than that of the mean spherical shape. The adsorbing molecules cannot touch each other. This is due to a higher interaction energy of the molecules with a denser solid surface having a higher electron density than that of the adsorbate molecules, since the adsorptive interaction of the residual nonactive surface with the medium-alcohol molecules is a nonspecific, simple dispersion force. Moreover, the first layer adsorbate molecules could occupy the surface area more spaciously by an entropic exclusion effect [30] of the two-dimensional motion of a rather long carbon chain of the medium-alcohol molecules on the surface. Thus the first layer adsorption can end with a rather smaller amount of adsorbate molecules than the monolayer volume estimated as Vm. The difference of the initial adsorption amounts of 1-butanol from 1-hexanol can be explained as the difference of the chain length of the molecules.

In the former case, where the solid surface should be less dense than the adsorbate molecules to have less dispersion force potential, the interaction energy of the solid with the adsorbate molecules might be less than, or similar to, the one between two adsorbate molecules, and the adsorption isotherms would be often found to have a very small initial adsorption so that they seem to be rather Type III isotherms (quasi–Type III isotherms). Incidentally, for the linear portion on the quasi–Type III isotherms, the gradient method would be applicable on such inactive surfaces. In fact, in the adsorption system of sodium chloride or silicate glass with lower alkylbenzene vapors, the quasi–Type III isotherms obtained have been found to obey the theory of the gradient method [2].

On the other hand, the surface area estimated by the gradient method using 1-butanol or 1-hexanol vapor agreed will with the standard surface area for the gold sample, which had a typical Type II isotherm, using the standard values of the cross-section of the adsorbate molecules, σ_m, shown in Table 4, obtained from the liquid density, assuming spherical adsorbate molecules. Considering the applicability of the gradient method for the gold sample, which has no polar sites such as hydroxyl groups or oxygen chemisorbed on the surface, the medium-alcohol molecules adsorbed over Point B on the surface could move freely, at least two-dimensionally and rotationally, keeping a spherical mean molecular shape with the standard adsorptive cross-section, σ_m, following the assumption used in the gradient method theory, as was mentioned in Section II.

Adsorption by graphite is fairly distinct from the others in the isotherm feature in having a small step in it and a remarkably large amount of adsorption in the adsorbate pressure range over the point of the step, especially for 1-hexanol vapor adsorbate, although the steplike isotherms often appear on graphite in the adsorption of inert gases at a comparatively low adsorptive temperature [31,32]. This may be because the graphite sample used in this experiment presumably has many cracklike pores on the surface, developed by the cleavage nature of the layer lattice structure, the width of which is lager than two adsorption layers, though it is considered that the inside of the cleavage pore may have an adsorptive potential somewhat larger than that on the isolated plane surface, due to multiplication of the potential from the two planes facing each other. Therefore, the second layer adsorption on such surfaces facing each other in a cleavage pore will begin at a comparatively lower adsorptive pressure with a steplike increase. In relation to this phenomenon, the author has encountered even in a Type II isotherm that the adsorption on single platelike solid samples increases linearly against the adsorbate pressure in a fairly wide range of relative pressures and does not increase so abruptly as on fine powder solid samples in the comparatively low pressure range. This is because the isolated platelike samples may be unrelated to any enhanced adsorption, or capillary condensation without adsorption/desorption hysteresis, in the fine spaces between powder particles, as is often found in the use of fine powder samples [33].

The linear portion gradient over Point B on the graphite sample is higher than the monolayer volume for both 1-butanol and 1-hexanol vapors. This may be because the gradient is raised by the second layer adsorption, which begins at a comparatively low relative pressure in the cleaved inner surface, though it seems that the gradient value of the shortened linear portion must be essentially identical with the monolayer volume, obeying the gradient method theory.

For the aluminum foil sample, which is considered to have a dense oxide film on the surface, the adsorption isotherm of 1-hexanol vapor showed a typical Type II curve having a fairly long linear portion and a fairly large amount of adsorption at high adsorbate pressure, being different from the other samples having polar groups on the surface. From the normal BET plots, so small a surface area as 0.34 of the standard surface area was obtained, though the gradient surface area was found to be also as small as 0.6 of the standard surface area, which was a medium value among the polar and nonpolar solid samples used.

The numerical results obtained are listed in Table 6.

4. Degree of Kinetic Freedom of the Molecules Adsorbed on the Monolayer: Application of the Gradient Method Theory

As stated in the preceding section, it was proved that, on solids having no surface polar groups, such as gold, even the alcohol molecules adsorbed on the first adsorption layer or over Point B could move around freely with the same kinetic freedom as the molecules on the liquid surface, by the application of the gradient method to such adsorption systems.

On the other hand, the degree of kinetic motion of the alcohol molecules adsorbed over the first adsorption layer or Point B on the solids such as sodium chloride, aluminum oxide, and silicate glass, all of which have dense polar groups on their surfaces, could be clarified by the gradient method theory, although the gradient method was not applicable for surface area estimation to adsorptive systems not showing ordinary Type II isotherms but showing quasi–Type I isotherms.

Expression (12) is simplified by substituting $dH_a = 0$, a result obtained experimentally in this study. Thus we obtain

$$\frac{dn}{d(P/P_s)} = n_m \cdot \exp\left(\frac{dS_a}{R}\right) \qquad (13)$$

TABLE 6 Specific Surface Areas Estimated by the BET or Point B Method (P.B.) and the Gradient Method (GRAD) Using 1-Butanol and 1-Hexanol Vapors and the Ratios Against the Standard Values (Std.)

	BET (m²/g)	Point B (m²/g)	BET or P.B. (Std.)	GRAD (m²/g)	GRAD (Std.)
		1-Butanol vapor			
Gold	0.562	—	0.71	0.803	1.02
Graphite	2.14	—	0.66[a]	4.60	1.43[a]
Sodium chloride	—	0.271	1.08	0.073	0.29
Aluminum oxide	—	0.774	1.11	0.244	0.35
Silicate glass	—	7.77	1.14	2.58	0.38
		1-Hexanol vapor			
Gold	1.21	—	0.38	2.99	0.93
Aluminum foil	0.0446	—	0.338	0.0800	0.61
Graphite	2.64	—	0.82[a]	4.24	1.32[a]
Sodium chloride	—	0.429	1.30	0.0694	0.21
Aluminium oxide	—	0.956	1.37	0.182	0.26
Silicate glass	—	6.55	0.96	1.18	0.17

[a] Ratio against Kr/BET area.

or

$$dS_a = R \cdot \ln\left\{\frac{dn}{d(P/P_s)} \cdot \frac{1}{n_m}\right\} \tag{14}$$

From Eq. (14), dS_a can be easily calculated using the gradient value of the linear portion above Point B, $dn/d(P/P_s)$, and the standard monolayer volume, n_m, where dS_a is the excess entropy of the molecules adsorbed over Point B, the deviation in entropy from that of the molecules on the corresponding liquid surface subtracted by the configurational entropy, $-R \cdot \ln(n/n_m)$. Substituting into $dn/d(P/P_s)$ in (14), 0.3–0.4 n_m for 1-butanol vapor or 0.2 n_m for 1-hexanol vapor, the excess entropy, dS_a, of the alcohol molecules adsorbed over Point B on the surface showing a quasi–Type I isotherm is obtained as -10.0 to -7.6 J/mol · K for 1-butanol vapor or about -13.4 J/mol · K for 1-hexanol vapor. The effective entropy reduction shown by the negative excess entropy (dS_a) could be used for the estimation of the degree of kinetic freedom of the adsorbed molecules, which are considered to be in a quasi-liquid state in which the motion of the molecules may be more or less frozen. Therefore negative excess entropies could be compared with the entropy reduction of the molecules by the phase change from the liquid to the solid state. These values are -31.4 J/mol · K for 1-butanol and -52.5 J/mol · K for 1-hexanol, each of which is calculated from the heat of solidification at 20°C (Table 4). The reduction of entropy by the phase change means the reduction of the degree of kinetic freedom of the molecular motion by the phase change. A comparison showed that the reduction of entropy by the adsorption is 1/3 to 1/4 of the reduction by solidification from the liquid for both 1-butanol and 1-hexanol. This shows that the reduction of kinetic freedom of the molecules adsorbed over Point B is not as high as the reduction of the molecules by solidification. The number of degrees of freedom in the kinetic motion of the liquid alcohol molecules, in which their thermal energy may be quite evenly partitioned, is about six, three of translational motion (involving diffusion of the molecules) or vibration and three of rotational motion or vibration at comparatively high temperatures used in this study. In the adsorption onto the first adsorption layer in which adsorbed molecules stick strongly to the polar adsorption sites on the solid surface, it is considered that the molecules adsorbed as the second layer may be restricted to diffusing into the stiff first layer by exchanging the adsorption sites, and that the adsorbed molecules may be also restricted partially so as to rotate or vibrate around two axes parallel to the surface by a specific attractive potential from the oriented first layer and the polar surface molecules. Some of these restrictions to the molecular motion may result in a reduction in entropy of the molecules adsorbed in the second layer as compared to the liquid molecules, though the reduction is as small as 1/3 to 1/4 of the entropy change in solidification, and most of the freedom of the molecular motion in the liquid state may be left in the molecules adsorbed onto the first layer. On the solid surface having no polar sites and applicable to the gradient method (e.g., $dS_a = 0$), on the other hand, molecules adsorbed as the second layer are exchangeable with the molecules adsorbed roughly and nonlocalized on the surface as the first layer, and the rotational movements other than translational movements may never be affected by the nonpolar surface.

For aluminum samples, which have a medium gradient value of the linear portion on an ordinary Type II isotherm for 1-hexanol, the estimated value of dS_a is -4.1 J/mol · K, this being only 0.08 of the entropy change by solidification of the liquid. The molecules adsorbed on the surface above Point B, at which the adsorbed amount is rather smaller than the standard monolayer volume, can move about on the first layer surface with kinetic freedom to a slightly lesser degree than the molecules in the liquid state. The aluminum oxide film on aluminum foil seems to be fairly thin and somewhat different from the other alumina modifications. The alcohol molecules adsorbed as the first layer on a slightly polar surface would not be so much localized and oriented as unexchangeable with the upper layer adsorbed molecules.

V. CONCLUSIONS

The behavior of the molecules of medium-sized normal alcohols adsorbed from the vapor phase onto various solid surfaces was analyzed in correlation to the form of adsorption isotherms obtained experimentally by the adsorption theories applied. As a theory useful for analysis, the gradient method theory for surface area estimation including the molecular kinetic state of adsorption was used, which was developed by the author in recent years, as described briefly in this paper. Exact data on the small adsorption of the medium-sized alcohols under very low vapor pressures could be obtained using a newly constructed apparatus.

The adsorption isotherms of alcohol vapors obtained at room temperature were categorized into two groups, viz. typical Type II isotherms and those resembling Type I (quasi–Type I) isotherms. The former were obtained from solids possessing no polar groups on the surface and the latter from solids possessing polar group molecules, such as hydroxide and/or oxygen chemisorbed on the surface. Inapplicability of the BET theory for most samples having either type of isotherm showed that the behavior of the adsorbed alcohol molecules used on the surface would not meet the assumption used in the BET theory. For the samples, such as gold, which shows a typical Type II isotherm and applicability of the gradient method, it was evidenced that the molecules adsorbed over Point

B (or in multilayers) could move as freely as the liquid molecules, while for samples having quasi–Type I isotherms and for samples rejecting the use of the gradient method, it was shown by the inapplicability of the gradient method and by the calculation of excess entropy by the gradient method theory that the molecules adsorbed over Point B (or onto the first monolayer adsorption) would be restricted in molecular movement, although the degree of restriction was small.

Many adsorption theories have been presented, most of which could not always describe the adsorptive phenomena for all adsorbent/adsorbate systems, due to the use of oversimplified assumptions on the molecular state of adsorbents and adsorbates in the construction of the theories. From the discrepancy between theory and experimental facts, however, the true behavior of the adsorbed molecules might be clarified to a great extent, if a number of adsorption theories were used in parallel.

REFERENCES

1. A. Nonaka. J. Colloid Interface Sci. *99*:335 (1984).
2. A. Nonaka. J. Colloid Interface Sci. *112*:548 (1986).
3. A. Nonaka. J. Colloid Interface Sci. *117*:355 (1987).
4. A. Nonaka. Colloids Surf. *36*:49 (1989).
5. A. Nonaka. Hyoumen Gijutsu (J. Surf. Finish. Soc. Jpn.) *41*:670 (1990).
6. S. Brunauer, P. H. Emmett, and E. Teller. J. Am. Chem. Soc. *60*:309 (1938).
7. S. J. Gregg and K. S. W. Sing, in *Adsorption, Surface Area and Porosity*, 2d ed., Academic Press, London, 1982, pp. 42–84.
8. A. W. Czanderna and S. P. Wolsky, in *Microweighing in Vaccum and Controlled Environments* (A. W. Czanderna and S. P. Wolsky, eds.), Elsvier Amsterdam/Oxford/New York, 1980, pp. 34, 100, and 149.
9. S. J. Gregg and K. S. W. Sing, in *Adsorption, Surface Area and Porosity*, 2d ed., Academic Press, London, 1982, pp. 72, 102–105.
10. D. A. Payne, K. S. W. Sing, and D. H. Turk. J. Colloid Interface Sci. *43*:287 (1973).
11. S. J. Gregg and K. S. W. Sing, in *Adsorption, Surface Area and Porosity*, 2d ed., Academic Press, London, 1982, pp. 209–214.
12. B. C. Lippens and J. H. de Beor. J. Catalysis *4*:319 (1965).
13. K. S. W. Sing, in *Surface Area Determination* (D. H. Everett and R. H. Ottewill, eds.), Butterworths, London, 1970, p. 98.
14. S. J. Gregg and K. S. W. Sing, in *Adsorption, Surface Area and Porosity*, 2d ed., Academic Press, London, 1982. pp. 214–218.
15. S. Brunauer, L. S. Deming, W. E. Deming, and E. Teller. J. Am. Chem. Soc. *62*:1723 (1940).
16. S. J. Gregg and K. S. W. Sing, in *Adsorption, Surface Area and Porosity*, 2d ed., Academic Press, London, 1982, pp. 58–60.
17. P. H. Emmett and S. Brunauer. J. Am. Chem. Soc. *59*:1553 (1937); S. J. Gregg and K. S. W. Sing, in *Adsorption, Surface Area and Porosity*, 2d ed., Academic Press, London, 1982. pp. 54–56.
18. H. Kowalczyk, G. Rychlicki and A. P. Terzyk. Colloids Surf. *A80*:257 (1993).
19. T. L. Hill, in *An Introduction to Statistical Themodynmics*, Addison-Wesley, Reading, MA, 1960, p. 128.
20. A. W. Adamson, in *Physical Chemistry of Surfaces*, 4th ed., Wiley-Interscience, 1982, p. 527.
21. A. Nonaka. Colloids Surf. *62*:207 (1992).
22. S. J. Gregg and K. S. W. Sing, in *Adsorption, Surface Area and Porosity*, 2d ed., Academic Press, London, 1982, pp. 72, 102–105.
23. A. Nonaka. Hyoumen Gijutsu (J. Surf. Finish. Soc. Jpn.) *43*:84 (1992).
24. A. Nonaka. Hyoumen Gijutsu (J. Surf. Finish. Soc. Jpn.) *46*:74 (1995).
25. A. Nonaka. Hyoumen Gijutsu (J. Surf. Finish. Soc. Jpn.) *46*:84 (1990).

26. A. Nonaka. J. Colloid Interface Sci. *167*:117 (1994).
27. A. Nonaka. J. Colloid Interface Sci. *151*:175 (1995).
28. A. C. Zettlemoyer. J Colloid Interface Sci. *28*:343 (1968).
29. S. J. Gregg and K. S. W. Sing, in *Adsorption, Surface Area and Porosity*, 2d ed., Academic Press, London, 1982, pp. 12–16.
30. T. L. Hill, in *An Introduction to Statistical Themodynmics*, Addison-Wesley, Reading, MA, 1960, p. 225.
31. G. D. Halsey. J. Chem. Phys. *16*:931 (1948).
32. S. J. Gregg and K. S. W. Sing, in *Adsorption, Surface Area and Porosity*, 2d ed., Academic Press London, 1982, pp. 84–89.
33. S. J. Gregg and K. S. W. Sing, in *Adsorption, Surface Area and Porosity*, 2d ed., Academic Press, London, 1982, pp. 115, 160–164.
34. C. F. Prenzlow and G. D. Halsey. J. Phys. Chem. *61*:1158 (1957).

19
Interactions at Colloidal Soil and Sediment Interfaces

BERND DIETER STRUCK and MILAN JOHANN SCHWUGER Institute of Applied Physical Chemistry, Research Centre Jülich, Jülich, Germany

I.	Introduction	651
II.	Major Colloids and Organic Contaminants in Soils and Sediments	652
	A. Colloids	652
	B. Organic compounds	653
III.	Slow Sorption Processes in Soils and Sediments	654
	A. Mechanisms of slow sorption of organic chemicals to natural particles	654
	B. Modeling slow sorption of organic chemicals to natural particles	655
	C. Experimental and model results	665
IV.	Formation of Sorption Hysteresis	674
	A. Mechanisms of the formation of sorption hysteresis	674
	B. Experimental and model results	675
V.	Multicomponent Sorption	677
	A. Organic contaminants	677
	B. Organic contaminants and surfactants	679
	C. Organic contaminants and emulsions	694
VI.	Conclusions	695
	Nomenclature	696
	References	697

I. INTRODUCTION

Hydrophobic organic compounds (HOCs) deposited on soils and sediments play an important role in ecosystem contamination. They result from incineration or combustion processes and from pesticide and detergent deposition and the degradation of the latter.

Sorption of HOCs in soils and sediments is often ruled by slow kinetics due to transport hindrances and energy barriers. The significance of slow sorption is evident for the transport, degradation, and biological activity of organic compounds in the ecosystem.

Also sorption hysteresis has an influence on contaminant transport, provoking a diminution of transport velocity. In this context, experimental artifacts have to be avoided that can result in an apparent sorption hysteresis.

On the other hand, multicomponent sorption can accelerate the transport of individual components by competing adsorption of the components to the same sorption sites unless the competing component increases the hydrophobic properties of the sorbent surface.

A special case of multicomponent sorption is the sorption of HOCs in the presence of surfactants. Surfactants increase the hydrophobicity of the sorbent surface on the one hand and increase the partitioning of HOCs into micelles on the other. Thus, at high surfactant concentrations, soil remediation becomes possible.

Soil remediation by surfactant solutions can be further improved by employing microemulsions, thus enhancing HOC solubility.

Of course, mathematical model approaches support the empirical evaluation of the HOC sorption mechanisms and allow the prediction of transport processes. However, valid model interpretations have to rely on model parameters determined by independent experiments avoiding the determination of multiple model parameters via single curve fitting. Because of the colloidal nature of soil and sediment particles, theoretical and experimental model studies on the interactions of HOCs at the interfaces of natural particles can be performed well by applying the physicochemistry of colloids.

In this context, a number of recent publications on the interactions of hydrophobic organic contaminants at colloidal soil and sediment interfaces have been reviewed and significant results reported.

The study of the interfacial effects of organic contaminants in soils and sediments has to consider a number of colloid properties related to the respective environmental conditions. Basically, the origin and environmental history of the soil and sediment determine these natural colloid properties.

II. MAJOR COLLOIDS AND ORGANIC CONTAMINANTS IN SOILS AND SEDIMENTS

A. Colloids

Soil is formed from the destruction of rocks as a result of weathering, and its formation is affected by numerous factors such as parent rock material, topography, climate, vegetation, and biotic organisms. Soil is a complex three-phase system containing a solid phase and a pore space filled with gas, water, and/or organic liquid. Because of its porous nature, soil has the ability to adsorb large quantities of HOCs, and large amounts of contaminants exist in the soil in the adsorbed phase.

The soil solid phase is a complex and ill-defined matrix. As shown in Fig. 1 it is composed of an inorganic part and an organic part. The inorganic phase is itself divided into crystalline and noncrystalline phases. Important types of crystalline materials are clay and mica minerals, which compose a group in the more general class of silicate minerals. They are virtually ubiquitous in soils and often compose 10% to 70% of a soil solid phase. The particle size of clay minerals is colloidal, often less than 2 μm. The structure of phyllosilicates like clay and mica is that of layers composed of silica tetrahedral sheets and octahedral sheets, centered around either divalent (Mg^{2+}) or trivalent (Al^{3+}) cations.

Soil organic matter is the residue from the partial decomposition of cellulose and other natural organic materials in the soil environment. It also is an ill-defined macromolecule, coiled and branched to an overall length of approximately 50 μm and

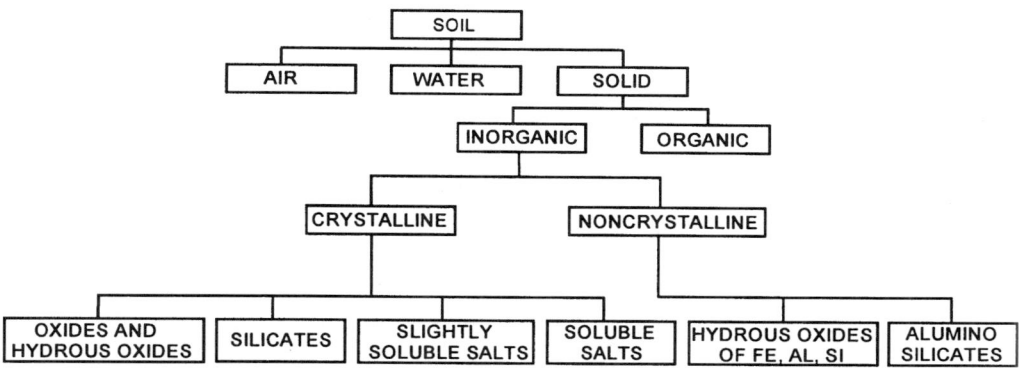

FIG. 1 Scheme of soil composition. (From Ref. 1.)

with a molecular weight of approximately 150,000, but of no definite or commonly agreed molecular formula. Humus has no other meaning than soil organic matter including specifications as humic and fulvic acid [2].

Sediments are also formed from the destruction of rocks. The particles formed are transported by ice, water, and gravity and become deposited e.g. in rivers, lakes, and the sea. Sediments can also be classified by their grain size. The transport of sediments in water can take place in colloidal form. These colloidal particles are mostly composed of siliclastic materials as clay minerals and of organic substances.

B. Organic Compounds

1. Hydrophobic Organic Compounds (HOCs)

Organic contaminants studied result from combustion and incineration processes, as PAHs, from industrial and urban waste, as the chlorinated aliphatic and aromatic hydrocarbons, and from agriculture, as the pesticides.

As representatives of environmentally relevant organic compounds, a number of aromatic hydrocarbons (benzene, toluene, trimethylbenzene, n-propylbenzene, n-butylbenzene, n-hexylbenzene, ethylbenzene, o-xylene, p-xylene, quinoline), polycyclic aromatic hydrocarbons (naphthalene, anthracene, phenanthrene, pyrene), chlorinated aromatic hydrocarbons (chlorobenzene, 1,2-dichlorobenzene, 1,3-dichlorobenzene, 1,4-dichlorobenzene, 1,2,4-trichlorobenzene), polychlorinated aliphatic hydrocarbons (1,2-*trans*-dichloroethene, trichloroethene, tetrachloroethene), the herbicide diuron, the insecticide lindane, and the polyelectrolytes polyacrylic acid and Percol 757 were studied in the publications considered.

2. Surface Active Agents

The surface active agents studied originate from anthropogenic and natural sources, and also by biodegradation. Anthropogenic sources are e.g. detergents and pesticide formulations. Surface active agents enter the environment either directly or e.g. via the overflow of sewage treatment plants or the spreading of sewage sludge. The environmental significance of these surface active substances results from their interactions with toxic substances despite their small concentrations in soils and sediments. Because of their

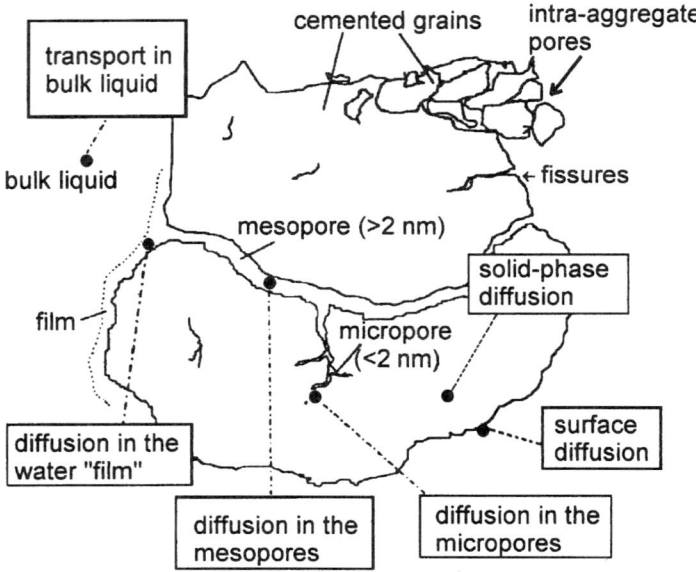

FIG. 2 Diffusion processes in a soil particle. (From Ref. 3.)

environmental compatibility, anionic and nonionic surfactants are used as components of detergents and as formulations of pesticides. Thus the influence of chemically well defined anionic and nonionic surfactants on the sorption behavior of organic chemicals was investigated. Candidates were the anionic surfactants sodium dodecyl sulfate (SDS) and sodium dodecylbenzenesulfonate (SDBS) as well as the nonionic surfactants polyoxyethylene (20) sorbitanemonooleate (Tween 80), polyoxyethylene(23)laurylether (Brij 35), and alkylphenylpolyethyleneglycol (Triton X-100).

III. SLOW SORPTION PROCESSES IN SOILS AND SEDIMENTS

As sorption can be rate-limiting for subsurface transport and biodegradation of HOCs, the characterization of sorption mechanisms is decisive for fate and risk assessment. This concerns especially slow sorption processes, which cause long-term risks and which are difficult to determine.

A. Mechanisms of Slow Sorption of Organic Chemicals to Natural Particles

The use of equilibrium expressions for sorption to natural particles in fate and transport models is often invalid due to slow kinetics. Sorption kinetics are complex and poorly predictable at present. Diffusion limitations appear to play a major role. Contending mechanisms include diffusion through natural organic matter matrices and diffusion through intraparticle nanopores (Fig. 2). These mechanisms probably operate simultaneously, but the relative importance of each in a given system is indeterminate.

Processes causing nonequilibrium can be separated into three groups: physical nonequilibrium, chemical nonequilibrium, and intrasorbent diffusion [4].

Physical nonequilibrium results from a heterogeneous flow domain. Certain regions within the domain may experience minimal or no advective flow. However, diffusional mass transfer between these regions and the advective-flow regions can occur, causing the former to act as sink/source components. When the mass transfer is rate limited, access to the sink/source or immobile regions is constrained, and thus nonequilibrium occurs. The immobile regions in the case of soil and sediment colloids include the internal porosity of soil aggregates and the matrix porosity of fractured media. It is important to realize that physical nonequilibrium affects both nonsorbing and sorbing solutes. However, because of its influence on sorption, physical nonequilibrium affecting nonsorbing solutes is usually also discussed as a sorption-nonequilibrium mechanism.

Chemical nonequilibrium results from a rate-limited reaction between the sorbate and specific sites on the sorbent surface. Generally, relatively large activation energies are associated with these reactions. The sites at which these reactions occur can be associated with mineral surfaces or with organic matter. As opposed to physical nonequilibrium, chemical nonequilibrium affects only sorbing species.

Nonequilibrium caused by intrasorbent diffusion results from rate-limited mass transfer of sorbate from the exterior surface of the sorbent into the interior of the sorbent matrix. Although both intrasorbent diffusion and physical nonequilibrium involve physical mass transfer processes, they may be considered as separate entities for two important reasons.

First, whereas physical nonequilibrium affects both sorbing and nonsorbing species, intrasorbent diffusion influences only sorbing species. Second, the mass transfer process for intrasorbent diffusion is different from that for physical nonequilibrium. For intrasorbent diffusion, the sorbate is in intimate contact with the sorbent matrix essentially at all times. Conversely, physical nonequilibrium is a pore-diffusion process.

Considering organic matter and mineral surfaces as the primary sorbents, intrasorbent diffusion may take two forms, diffusion into the matrix of organic matter, say intraorganic matter diffusion, or diffusion into the interior of mineral particles, say intramineral diffusion.

Organic matter in soils and sediments is commonly pictured as a three-dimensional network of polymer chains, with a relatively open, flexible structure perforated with voids. With such a structure attributed to sorbent organic matter, the diffusive mass transfer of sorbate into the interior of organic matter is conceptually viable. In fact, several researchers have proposed such a mechanism as a potential cause of sorption nonequilibrium.

Intramineral diffusion may occur as diffusion of sorbate into interlamellar regions of clay minerals or as diffusion into microporous sand grains.

Intrasorbent diffusion in the form of intraorganic matter diffusion may predominate in soils that are high in natural organic matter and low in aggregation, while intrasorbent diffusion in the form of sorption-retarded pore diffusion may predominate in soils where the opposite conditions exist [3].

Determination of individual sorption mechanisms can be supported by model calculations following a number of distinct criteria.

B. Modeling Slow Sorption of Organic Chemicals to Natural Particles

1. Criteria for the Derivation of Sorption Mechanisms by Model Approaches

The subsequent criteria should be obeyed in evaluating sorption mechanisms by model approaches:

1. Formulation of the sorption mechanism must be by model equations according to the physicochemical expectation.
2. One must examine whether other mechanisms lead to similar types of model equations.

E.g. the two-region and two-site models are mathematically equivalent in nondimensional form for the linear sorption isotherm case. The parallel and series configurations of the two-site model are equivalent as long as one of the compartments is assumed to be in equilibrium. Hence, under the stated conditions, the models for all three nonequilibrium mechanisms are equivalent. This means that a single bicontinuum model may be used to represent any one of the three mechanisms [4].

Additionally, the mechanism responsible for sorption nonequilibrium in a given case cannot be identified simply by the curve-fitting of a model to experimental data, and therefore

3. Model constants have to be measured by independent experiments.

Furthermore, this also implies that

4. Assumptions reducing the number of model constants have to be formulated clearly.
5. In order to derive the best hypothesis on a sorption mechanism, the verification of the best mechanism can be achieved by minimizing the variance between experimental and model variable values.

A major constraint associated with the investigation of sorption dynamics is the present inability to observe behavior at the microscopic scale. Until techniques for the determination of the relevant microscale parameters are available, it will be necessary to employ indirect approaches to the investigation of sorption dynamics, thus

6. Stochastic distributions for the microscale parameters must be used.

Although indirect approaches may not provide definitive proof of the validity of a hypothesis, they can provide a preponderance of evidence to support or refute a given hypothesis.

Thus deterministic and stochastic model approaches have been made according to the discussed mechanisms of slow sorption of organic chemicals to natural particles. A selected number of these models will be described in the following.

2. Basic Deterministic Models Describing Slow Sorption

Of course, regarding different steps of a sorption mechanism, instantaneous equilibrium steps may be involved, to be described by the classical isotherms as e.g. Henry, Freundlich, Langmuir, Toth, BET.

Models describing reaction kinetic limitations of the sorption mechanism include linear or nonlinear first-order rate equations and are termed first-order kinetic models.

In the case of physical nonequilibria, model descriptions in the field of diffusion concern film diffusion and pore diffusion, and in the case of intrasorbent diffusion, surface diffusion and intraorganic matter diffusion.

Of course, there are models avoiding an interpretation of the diffusion mechanism, only describing the mass transfer numerically. Such a phenomenological model is the first-order mass transfer model.

Most sorption data exhibit a two-stage approach to equilibrium, with a rapid initial rate, which accounts for approximately 20–50% of total sorption, followed by a much slower rate. In this context the two-site model and the two-region models have to be mentioned.

The two-site model is described by Cameron and Klute [5] and has been used to model chemical nonequilibrium. The two-site model in a series configuration may be also employed to model intrasorbent diffusion [6].

Two-region models have ben developed for physical nonequilibrium by Coats and Smith [7] and van Genuchten and Wierenga [8] for nonsorbing and sorbing solutes, respectively. A special case is the bicontinuum model. The use of bicontinuum models is often extended from the modeling of physical nonequilibrium to the simulation of nonequilibrium mechanisms caused by chemical nonequilibrium and intrasorbent diffusion.

As natural sorbents have multifold kinetically effective properties, this complexity can be described by multiclass approaches regarding different single-class properties.

Concerning nomenclature, in the subsequent sections a distinction is made between the distribution coefficient with the dimension L^3/M and the dimensionless partition coefficient.

(a) Description of Basic Deterministic Models for Physical Nonequilibrium, Chemical Nonequilibrium, and Intrasorbent Diffusion

Two-site models. A commonly used nonequilibrium model is the two-site equilibrium/kinetic model. The model is intended to account for the experimental evidence of two different sorption or mass transfer rates according to two different types of sorption site (Fig. 3).

A common case is the two-site first-order kinetic model e.g. employed by Culver et al. [10]. There is instantaneous sorption equilibrium at the equilibrium-type sorption sites e:

$$q_e = f_e K_d C$$

and a mass transfer restriction at the nonequilibrium-type sorption sites t:

$$\frac{\delta q_t}{\delta t} = k_t [q_{te} - q_t]$$

where

$$q_{te} = (1 - f_e) K_d C$$
$$f_e = q_e / (q_e + q_{te})$$
$$1 - f_e = q_{te} / (q_e + q_{te})$$

and

$$q_e + q_{te} = K_d C$$
$$q_{te} = (1 - f_e) K_d C = (1 - f_e)(q_e + q_{te})$$
$$\delta q_t / \delta t = k_t [q_{te} - q_t] = k_t (1 - f_e) - k_t q_t = k_t (q_e + q_{te}) - k_r q_t$$
$$k_f = k_t (1 - f_e)$$
$$k_r = k_t$$

Symbols signify

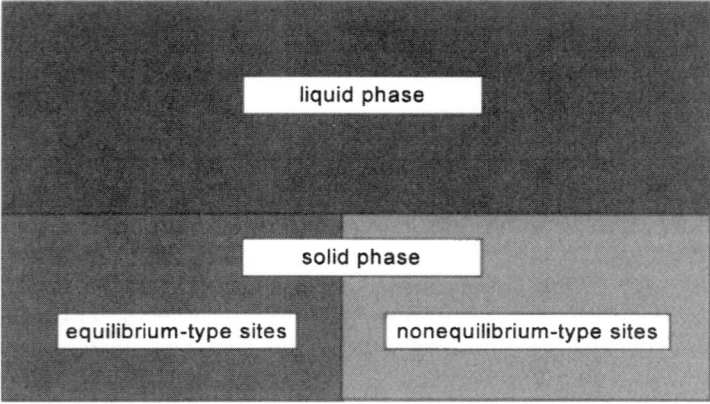

FIG. 3 Schematic of the two-site model. (From Ref. 9.)

C: solution-phase solute concentration
f_e: fraction of sorption sites in equilibrium
K_d: distribution coefficient
k_t: first-order mass transfer rate constant
k_f: forward first-order mass transfer rate constant
k_r: reverse first-order mass transfer rate constant
q_e: amount actually adsorbed at the "equilibrium sites"
q_t: amount actually adsorbed at the "nonequilibrium sites," amount adsorbed in the mass transfer-constrained domain
q_{te}: amount adsorbed at the "nonequilibrium sites" thought to be in equilibrium

The mass transfer restriction can be interpreted in terms of diffusion by defining the rate constant k_r correspondingly. The following equation refers to retarded intraparticle diffusion [11]:

$$k_r = \frac{15\, D_{aq}}{(1 + (\rho/\theta)K_d)\tau R^2}$$

D_{aq}: aqueous diffusion coefficient
θ: volume fraction of internal porosity
R: radius of the spherical solid particles
ρ: sorbent bulk density
τ: tortuosity factor

Two-region models, biocontinuum models. Bicontinuum models (two-region or mobile–immobile liquid-phase models) presume that all of the sorption sites are at equilibrium with the adjacent liquid-phase solute concentration; however, transfer of solute between the mobile (bulk) and immobile (intraparticle) liquid phases is diffusion controlled (Fig. 4).

Brusseau et al. [11] apply the first-order bicontinuum model in a special version in order to describe intraorganic matter diffusion. In this context, the sorbent regions respectively sorbent domains of the bicontinuum model are characterized by their volume

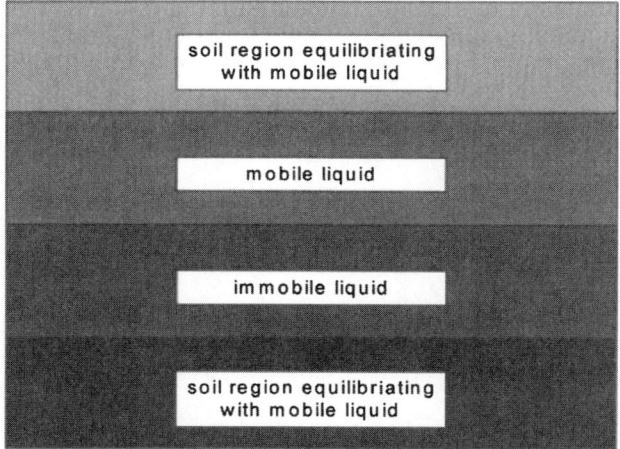

FIG. 4 Soil regions distinguished by the bicontinuum model. (From Ref. 9.)

fractions and not, as usual, by their mass fractions.

$$\theta_e = \frac{V_e}{V_e + V_t}$$
$$1 - \theta_e = \frac{V_t}{V_e + V_t}$$

where

  θ: volume fraction of the respective domain
  e: index "equilibrium domain"
  t: index "rate-limited or diffusion-controlled or mass transfer–constrained domain"
  V: Volume of the respective domain

The instantaneous sorption equilibrium in the mobile domain is

$$q_e = \theta_e K_d C \tag{1}$$

with

  C: bulk solution-phase solution concentration
  q: sorbed-phase concentration in the respective domain

The change in the sorbed-phase concentration q_t of the mass transfer–constrained immobile domain is described by the mass transfer between the equilibrium-constrained and the mass transfer–constrained domain as

$$\frac{\delta C_t}{\delta t} = k_t(C - C_t) \tag{2}$$

  k_t: mass transfer rate constant

followed by the instantaneous sorption equilibrium in the mass transfer–constrained domain

$$q_t = (1 - \theta_e) K_d C_t \tag{3}$$

Defining $k_{t,iomd}$ as the mass transfer rate constant for intraorganic matter diffusion within the mass transfer–constrained domain

$$k_{t,iomd} = k_t(1 - \theta_e) \tag{4}$$

the following is derived considering Eqs. (1–4):

$$\frac{\delta q_t}{\delta t} = k_{t,iomd}\left[\left(\frac{1}{\theta_e}\right)q_e - \left(\frac{1}{1-\theta_e}\right)q_t\right]$$

with

$$k_f = \frac{k_{t,iomd}}{\theta_e}$$

$$k_r = \frac{k_{t,iomd}}{1 - \theta_e}$$

and

$$\frac{\delta q_2}{\delta t} = k_f q_e - k_r q_t$$

According to Brusseau et al. [11], the mass transfer rate constant $k_{t,iomd}$ employed for intraorganic diffusion in polymeric materials can be related to the polymer diffusivity of the sorbate by

$$k_{t,iomd} = \frac{sf\, D_{poly}}{l^2}$$

D_{poly}: diffusion coefficient for the specific sorbate/polymer pair
l: characteristic diffusion length
sf: shape factor

Mass transfer control and degradation. The dual-resistance surface-diffusion model or reactive surface-diffusion model described by Miller and Pedit [12] comprises film diffusion within the boundary layer of the aggregate and surface diffusion within the aggregate. Compound degradation occurs in the bulk solution and within the aggregate. For the solution kinetics in the bulk solution and within the aggregate, two equations were derived.

For the solute concentrations C in the bulk solution,

$$\frac{\delta C}{\delta t} = -\frac{3 k_b m_s}{R V \rho}(C - C_{e,R}) - k_a C$$

is obtained. The first term on the right describes film diffusion and the second term the solute degradation in the bulk solution.

The nomenclature is

$C_{e,R}$: solution-phase solute concentration corresponding to an equilibrium with the solid-phase solute concentration at the exterior of particle
k_a: pseudo-first order solute degradation rate constant for solute loss in the solution phase
k_b: boundary layer mass transfer rate constant
m_s: mass of the solid phase
R: radius of the spherical solid particles
V: volume of the solution phase
ρ: macroscopic particle density of the solid phase

The solid-phase solute concentration q is a function of the radial position r in the aggregate

$$\frac{\delta q}{\delta t} = D_{sd}\left(\frac{\delta^2 q}{\delta r^2} + \frac{2}{r}\frac{\delta q}{\delta r}\right) - k_{a,s}\, q \tag{5}$$

with

$$\delta q/\delta r|_{r=R} = k_b(C - C_{e,R})/D_{sd}\,\rho$$
$$\delta q/\delta r\,|_{r=0} = 0$$

The first term on the right of equation (5) describes the surface diffusion in the aggregate, and the second term solute degradation on the surface of the aggregate. D_{sd} is the surface-diffusion coefficient and $k_{a,s}$ a pseudo-first-order solute degradation rate constant for solute in the solid phase.

(b) Multiclass Models. Multiclass models consider the locally differing physicochemical sorption conditions forming model classes approaching these differences.

A first example is the distributed reactivity model employed by Weber et al. [13]

Distributed reactivity model. This composite model is introduced to characterize intrinsic heterogeneities in the properties and behaviors of soils and sediments and to take up the resulting nonlinearities of sorption isotherms. Contaminant sorption by soils and sediments is described as a multiple reaction phenomenon. The approach is based on the observation that most natural soils and sediments are intrinsically heterogeneous even at the microscopic scale; that is, variable in composition and structure at both interparticle and intraparticle scales.

A variety of different classes of reactions between organic solutes and types of solid surfaces and quasi-solid organic matrices typical of those associated with environmental solids have been identified. The physicochemical and structural features of soils and sediments that lead to energetic differences between or within individual particles are likely to give rise to different combinations of linear and nonlinear sorptions.

Thus a classification was performed distinguishing mass fractions of the sorbent solid phase characterized by different linear or nonlinear sorption isotherms forming a composite isotherm based on the Freundlich model.

The linear isotherm for the ith mass fraction is

$$q_e^i = K_d^i\, C$$

and the nonlinear Freundlich isotherm for the jth mass fraction is

$$q_e^j = K_F^j\, C^{nj}$$

Combination of the linear and nonlinear isotherms results from summarizing the sorbent loadings by

$$q_e = \Sigma_i w_l^i w_i^i + \Sigma_j w_{nl}^j q_e^j$$

deriving the composite isotherm

$$q_e = \Sigma_i w_l^i K_d^i C + \Sigma_j w_{nl}^j K_F^j C^{nj}$$

Mass averaging of the distribution coefficient K_d^i

$$w_l K_d = \Sigma_i w_l^i K_d^i$$

leads to the simplified composite isotherm

$$q_e = w_l K_d C + \Sigma_j w_{nl}^j K_F^j C^{nj}$$

with

$$\Sigma_i w_l^i + \Sigma_j w_{nl}^j = w_l + \Sigma_j w_{nl}^j = 1$$

The nomenclature employed is

K_d^i: distribution coefficient for mass fraction i
K_d: mass-averaged distribution coefficient for the summed linear components
K_F^j: Freundlich constant for mass fraction j
nj: Freundlich coefficient for mass fraction j
q_e: total solute mass sorbed in equilibrium per unit mass of bulk solid for the summed mass fractions in linear and nonlinear equilibrium
q_e^i: solute mass sorbed in equilibrium by mass fraction i per unit mass of that mass fraction
q_e^j: solute mass sorbed in equilibrium by mass fraction j per unit mass of that mass fraction
w_l: summed mass fractions of the solid phase exhibiting linear sorption
w_e^i: mass fraction i of the solid phase exhibiting linear sorption
w_{nl}^j: mass fraction j of the solid exhibiting nonlinear sorption

In the subsequent model the classification is extended to compartments comprising the complete bicontinuum model formulation, distinguishing particles with different model properties.

Multiple particles classes. The model is called the multiple particle class formulation of the first-order mass transfer bicontinuum model and was discussed by Pedit and Miller [14].

It is a first-order mass transfer model, which allows for multiple particle (or sorption site) classes i with equilibrium and diffusion-controlled sorption sites. Physical process interpretations of bicontinuum models, often referred to as two-region or mobile–immobile liquid-phase models, assume, as already mentioned, that all of the sorption sites are at equilibrium with the adjacent liquid-phase solute concentration. However, transfer of solute between the mobile and immobile liquid phases is diffusion controlled. In the model formulation that follows, the mobile and immobile liquid phases are viewed as the bulk and the intraparticle liquid phases, respectively.

Particles are distinguished by mass fraction w^i considering their different properties described by the bicontinuum model.

$$\Sigma_i\, w^i = 1$$

Within the individual particle class the bicontinuum model distinguishes again between equilibrium and diffusion-controlled sorption sites. The kinetics within class i is

$$q_e^i = f_e^i * K_d^i\, C$$
$$\frac{\delta C_t^i}{\delta t} = k_t^i (C - C_t^i) \qquad (6)$$
$$q_t^i = (1 - f_e^i) K_d^i\, C_t^i \qquad (7)$$

Equation (6) and (7) can be combined resulting in

$$\frac{\delta q_t^i}{\delta t} = k_t^i (q_{te}^i - q_t^i)$$

with

$$q_{te}^i = (1 - f_e^i) K_d^i\, C$$

The nomenclature is

- C: bulk solution phase solute concentration
- C_t^i: sorbent phase (mass transfer–constrained domain) solution solute concentration
- f_e^i: fraction of equilibrium-type sorption sites of particle class i
- k_t^i: first-order mass transfer rate constant of particle class i, describing the mass transfer between the bulk and the sorbent-phase solution
- K_d^i: distribution coefficient of particle class i
- q_{te}^i: solid-phase solute concentration of particle class i adsorbed at the "nonequilibrium sites" thought to be in equilibrium with the bulk liquid-phase solute concentration
- q_e^i: solid-phase solute concentration of particle class i in equilibrium with the bulk liquid-phase solute concentration
- q_t^i: solid-phase solute concentration for the class i diffusion-controlled fraction

In the publication of Pedit and Miller [14] the term sorbent-phase includes the solid phase and the solution within the solid-phase pores.

The following model refers to distributed mass transfer rates [10] modeling physical nonequilibrium and intrasorbent diffusion. The model contains no physicochemical presumptions on sorption kinetics.

Distributed mass transfer rate model or one-dimensional mass transfer rate model. The one-site first-order kinetic model is used as a starting point.

$$\frac{\delta q}{\delta t} = k_t [K_d C - q]$$

For the distributed mass transfer rate model different compartments i are introduced referring to different first-order mass transfer rate constants k_t^i and different fractions f^i of sorption sites.

$$\frac{\delta q}{\delta t} = \Sigma_i k_t^i [f^i K_d C - q^i]$$

As a model simplification, the fractions of active sites f^i in each compartment i are assumed to be equal, thus reducing the number of parameters to be determined.

3. Stochastic Models

Stochastic sorption models are formulated to describe microscale variability, which may be interparticle or intraparticle in origin, in terms of both sorption equilibrium and rate properties.

The stochastic model approach demonstrated concerns the stochastic distribution of particles with different bicontinuum model parameter values and in a simplified form the stochastic distribution of mass transfer rates.

(a) Multiple Particle Class Formulation of the First-Order Mass Transfer Model. A batch system in equilibrium does not have a uniform sorbent-phase solute concentration throughout the bulk sample. Instead, the sorbent-phase solute concentration varies due to local variations of the equilibrium distribution relationship. These variations can be viewed as occurring at the interparticle and intraparticle scale. Variations will also exist in the rate of solute transfer to and from the bulk liquid phase, even for particles or collections of sorption sites with the same equilibrium distribution relationship. At the interparticle scale, these variations may occur due to differences in particle size and shape. At the intraparticle scale, these variations may occur due to differences in depth or accessibility of collections of sorption sites within the particle.

Analogously, Culver et al. [10] cite that pore sizes in soil have been found to follow a Γ distribution [15] and that particle size distributions are typically considered to be lognormally distributed [16].

In the model formulation that follows [14], it is assumed that the linear equilibrium model and the first-order mass transfer model apply throughout the bulk sample. The sorbent-phase solute concentration, distribution coefficient, and first-order mass transfer constant are treated as random variables. These variables are to be approached by a stochastic distribution of the mass fractions of particles with different sorbent-phase solute concentrations, distribution coefficients, and mass transfer constants describing the mass transfer between the mobile and immobile liquid regions. The model formulation ignores the presence of equilibrium-type sorption sites, because these sites are represented by large values of the first-order mass transfer constant random variables. The model formulation also ignores direct mass transfer between adjacent collections of sorption sites within a particle.

In this context a bivariate lognormal distribution $\phi(K_d, k_t)$ of the mass fractions of particles is suggested. With the assumptions outlined above, the bulk liquid-phase solute concentration mass balance results [14] from

$$\frac{\delta C}{\delta t} = -\frac{m_s}{V} \int_0^\infty \int_0^\infty \phi(K_d, k_t) k_t (Q_{te} - Q_t) dK_d dk_t$$

$\phi(K_d, k_t)$: mass fraction probability density function
m_s: mass of soil in the reactor
Q_t: diffusion-controlled sorbent-phase solute concentration for a particle or collection of sorption sites
Q_{te}: diffusion-controlled sorbent-phase solute concentration for a particle or collection of sorption sites thought to be in equilibrium with the bulk liquid phase
V: volume of bulk liquid

A model simplification is provided by considering the linear free energy relationship (LFER) approach [4,6,14,17], a relation between K_d and k_t:

$$\ln(k_t) = \alpha_0 + \alpha_1 \ln(K_d)$$

α_0 : intercept
α_1 slope

It allows for the mass fraction probability density function $\phi(K_d, k_t)$ a univariate lognormal distribution, or a univariate Γ distribution $\phi(K_d)$. Models are called the lognormal linear free energy relationship model (LLFER) or the Γ linear free energy relationship model (GLFER).

It follows for the total sorbent-phase/bulk liquid distribution coefficient $K_{d,tot}$ that

$$K_{d,tot} = \Sigma_i (w^i K_d^i) = \int \phi(K_d) \, dK_d$$

K_d^i: sorbent-phase/bulk liquid distribution coefficient of particles class i
w^i: mass fraction of particles class i

Another model simplification assumes K_d to be constant throughout the bulk sample. The bivariate probability density reduces to a univariate probability density function $\phi(k_t)$. The simplified model is called the lognormal rate model (LR) in the case of a lognormal distribution and the Γ model (GR) in the case of a Γ distribution.

In the distributed mass transfer rate model by Culver et al. [10], similarly a continuous probability distribution of the mass transfer rates is used, say a lognormal probability density function or a Γ probability density function for the mass-transfer rate constants k_t^i.

C. Experimental and Model Results

1. Apparent Sorption Equilibria Caused by Centrifugation Errors

Erroneous results on sorption equilibria can have severe consequences for the prediction of transport phenomena in soils and sediments [18]. These results do not only occur due to apparent sorption equilibria caused by slow sorption but e.g. also from centrifugation errors in the course of batch experiments.

Problems in determining the sorbed amount of the adsorbate after the batch experiment arise via centrifugation due to incomplete separation of the solid and liquid phase. Only supercolloidal solids are captured; colloidal solids remain in the supernatant. This means that the adsorbed amount is determined as too low. The possible deviation of true and apparent distribution coefficients is calculated by the two-phase colloidal partitioning model [19].

For the development of the two-phase colloidal partitioning model, the supercolloidal and colloidal solids are assumed to be separable components of the solid phase. Colloidal material is not considered to be partitionable into the supercolloidal phase. Consequently, the experimentally determined apparent distribution coefficient includes only sorbate adsorbed on the supercolloidal solids. Consequently, the experimentally determined apparent distribution coefficient is lower than the distribution coefficient.

A nonlinear relationship between the apparent distribution coefficient K'_d and the distribution coefficient K_d follows.

$$K'_d = \frac{K_d w_{st}}{w_{st} + n(1 - w_{st}) + n(1 - w_{st}) K_d m_t / V} \qquad (8)$$

with

$K_d = q_e / C_e$
$K'_d = q'_e / C'_e$
$m_t = m_{sc} + m_c$
$w_{st} = m_{sc} / m_t$
$q'_e = q_{e,sc} w_{st}$
$n = q_{e,c} / q_{e,sc}$

referring to the nomenclature

- C_e: equilibrium mass concentration of solute in bulk solution
- C'_e: apparent equilibrium mass concentration of solute in bulk solution including colloidal solids
- K_d: overall distribution coefficient
- K'_d: apparent overall distribution coefficient
- m_{sc}: total mass of supercolloidal solids
- m_c: total mass of colloidal solids
- $q_{e,c}$: (mass sorbed in equilibrium by colloidal solids)/(total mass of colloidal solids)
- q_e: (mass sorbed in equilibrium by supercolloidal and colloidal solids)/(total mass of supercolloidal and colloidal solids)
- q'_e: (apparent mass sorbed in equilibrium)/(total mass of supercolloidal and colloidal solids)
- $q_{e,sc}$: (mass sorbed by supercolloidal solids)/(total mass of supercolloidal solids)
- V: volume of solvent

A comparison of experimental and model data is shown in Fig. 5. As expected, results show that the experimentally determined apparent distribution coefficients are lower than the true distribution coefficients calculated according to Eq. (8).

As a typical example of sorption kinetics at natural particles an experimental study on intraorganic matter diffusion will now be discussed.

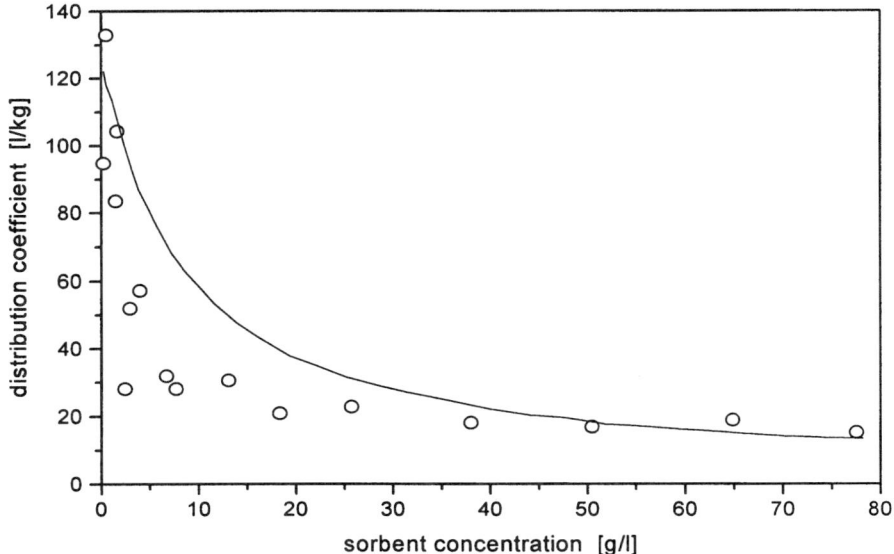

FIG. 5 The distribution coefficient for 1,2,4-trichlorobenzene (TCB) sorption on soil (Ultisol, Typic Hapludult) as a function of sorbent concentration. Experimental results (○), model for two-phase distribution of soil (solid line). Initial TCB concentration 25 mg L^{-1}, background electrolyte 0.01 M $NaNO_3$. (From Ref. 19.)

2. Experimental Study on Intraorganic Matter Diffusion

The results of experiments designed to identify the processes responsible for nonequilibrium sorption of hydrophobic organic chemicals (HOCs) by natural sorbents are reported [11]. The results of experiments performed with natural sorbents were compared to rate data obtained from systems where rate-limited sorption was caused by specific sorbate–sorbent interactions.

This comparison showed that chemical nonequilibrium associated with specific sorbate–sorbent interactions does not significantly contribute to the rate-limited sorption of HOCs by natural sorbents. Transport-related nonequilibrium was also shown not to be a factor for the systems investigated.

Hence attempts were made to interpret the data in terms of two sorption-related, diffusive mass-transfer conceptual models: retarded intraparticle diffusion and intraorganic matter diffusion. The analyses provide strong evidence that intraorganic matter diffusion was responsible for the nonequilibrium sorption exhibited by the systems investigated.

If long-term adsorption/desorption is diffusion controlled, an inverse correlation between a compound's distribution coefficient and the rate constant representing diffusive mass transfer is expected. The larger the distribution coefficient, the slower the diffusive mass transfer, and hence the lower the rate constant [6].

In this context, experimental results from breakthrough curves have been evaluated on the basis of the first-order bicontinuum model referring to retarded intraparticle diffusion and intraorganic matter diffusion as described above in Section III.B. The distribution coefficient K_d and the rate constant k_t were calculated for nonequilibrium sorption of HOCs.

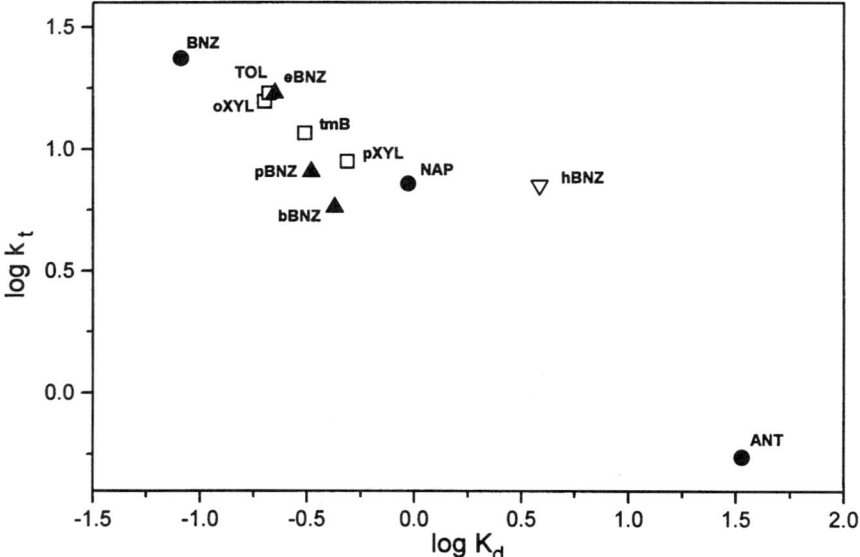

FIG. 6 Influence of methyl group addition on sorption kinetics. BNZ, benzene; NAP, naphthalene; ANT, anthracene; eBNZ, ethylbenzene; pBNZ, *n*-propylbenzene; bBNZ, *n*-butylbenzene; hBNZ, *n*-hexylbenzene; TOL, toluene; oXYL, *o*-xylene; pXYL, *p*-xylene; tmB, trimethylbenzene. (From Ref. 11.)

Experiments applied the miscible-displacement technique. Sorbates employed were pentafluorobenzoic acid, benzene, toluene, *o*-xylene, *p*-xylene, trimethylbenzene, ethylbenzene, *n*-propylbenzene, *n*-butylbenzene, *n*-hexylbenzene, naphthalene, anthracene, chlorobenzene, 1,3-dichlorobenzene, 1,2,4-trichlorobenzene, 1,2-*trans*-dichloroethene, and quinoline. The sorbent was sandy surface soil (Eustis find sand).

The k_t and K_d values determined from the breakthrough curves were analyzed by the well-known linear free energy relationship (LFER) approach [14,17].

The behavior of the molecular structure on diffusive mass transfer exhibited in Fig. 6 can be readily explained by the intraorganic matter diffusion model.

Two major physical differences between organic matter and microporous particles and hence intraorganic matter diffusion and retarded intraparticle diffusion, are apparent. First, the size of the pores or voids associated with organic matter is similar to the size of the sorbate molecules, whereas for porous particles the pores are much larger than the diffusing molecule. Second, while the pore networks for porous particles are fixed and comprised of rigid pores, the pore network associated with organic matter is dynamic and the pores are transient rather than fixed. Accordingly, intraorganic matter diffusion should not be interpreted in terms of fixed-pore models.

The presence of a three- or four-carbon chain (i.e., propylbenzene, butylbenzene) provides an increased opportunity for the molecule to become entangled with polymer chains, which results in an increased constraint on diffusion. However, the presence of additional carbons in the chain (i.e., hexylbenzene) has a negligible effect because the diameter of the mesh required for diffusion of a linear molecule is the same, irrespective of length, up to some critical length [11].

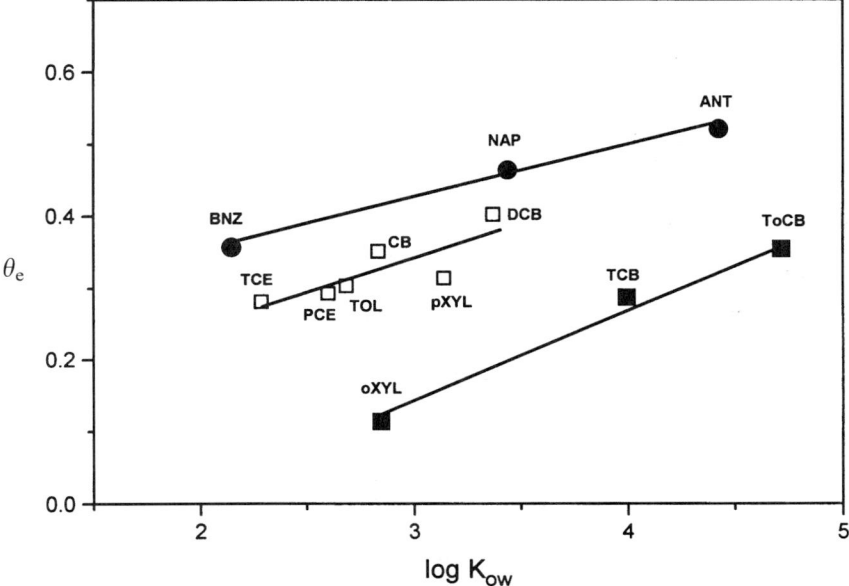

FIG. 7 The volume fraction of instantaneous sorption θ_e as a function of the octanol/water partition coefficient (K_{ow}). (From Ref. 11.)

The influence of molecular structure on the fraction of instantaneous sorption θ_e is also of interest. Values for θ_e are regressed against the log of the octanol/water partition coefficient (K_{ow}) in Fig. 7.

It appears that θ_e increases with log K_{ow}. Since K_{ow} is directly related to the molecular size of the solute, the results shown in Fig. 7 suggest that θ_e varies with molecular size. This behavior is consistent with the intraorganic matter diffusion model.

Recall that θ_e represents the ratio of sorptive volume comprising the instantaneous sorption domain to total sorptive volume, which is the sum of instantaneous and rate-limited domains. Following the concept inherent in the intraorganic matter diffusion model, the instantaneous sorption domain is considered to consist of the near-surface region of the organic matter, while the rate-limited domain is considered to comprise the internal regions. The polymeric mesh characterizing the organic matter matrix may act as a molecular sieve, whereby larger sorbate molecules are excluded from portions of the free volume of the polymer because of size constraints. As a result, the size of the rate-limited domain may vary with sorbate size [11].

This means that θ_e increases because the total sorptive volume decreases for reasons of steric hindrances due to molecular size constraints.

The following models consider the heterogeneity of the natural soils and sediments to be described by distributed sorption parameters.

3. Models Considering Distributed Sorption Parameters

(a) The Distributed Reactivity Model. Experimentally, soil heterogeneity is demonstrated for a number of soils which, on the basis of conventional macroscopic properties, would be considered homogeneous. Soils of the types Ypsilanti, Wagner, and Augusta

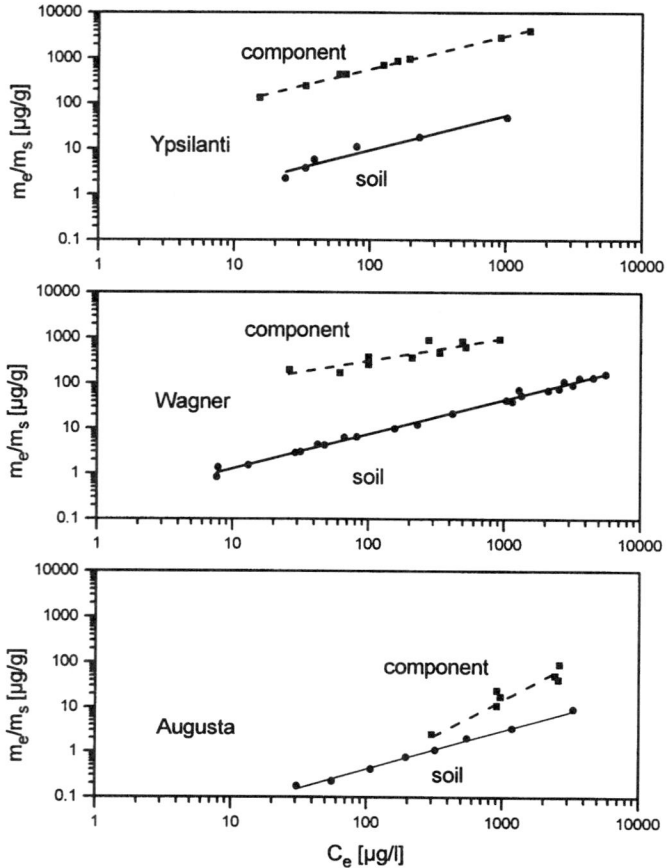

FIG. 8 Sorption of 1,2,4-trichlorobenzene by soils of the type Ypsilanti, Wagner, and Augusta and sorption by an isolated reactive component in each case. (From Ref. 13.)

were collected from subsurface environments in southeastern Michigan and northwestern Ohio. They were collected at depth greater than 1 m. Differing values for the median grain size and the maximum organic carbon content were determined for soil of the type Ypsilanti as 0.31 mm and 1.24%, for the type Wagner as 0.53 mm and 2.49%, and for the type Augusta as 0.13 mm and 0.03%. That such heterogeneities are reflected in sorption reactions, which differ between soils and between different fractions of soil, is demonstrated (viz. Figs. 8 and 9).

In terms of the distributed reactivity model, the significance of particle scale heterogeneity and distributed reactivity is illustrated theoretically. In Fig. 9 the sorption of tetrachloroethene on Wagner soil is calculated by the distributed reactivity model for increasing mass fractions of a nonlinearly sorbing soil component. The sorbing soil fractions were assumed to be of an organic nature. The ratio of the nonlinearly and linearly sorbing organic mass fraction increased by 0, 1/5, 1/2, and 1. The mass of the linearly sorbing fraction was kept constant. The figure shows increasing adsorption of tetrachloroethene with increasing mass fractions of the nonlinearly sorbing soil component.

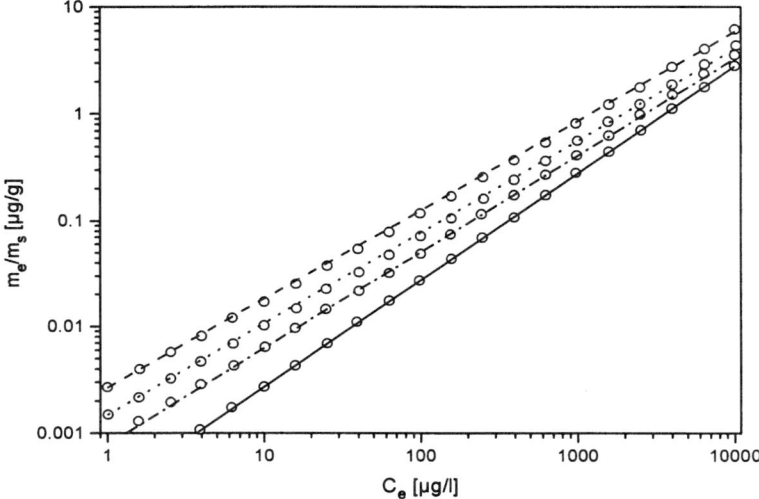

FIG. 9 Isotherms of tetrachloroethene sorption on Wagner soil calculated by the distributed reactivity model (○) for increasing mass ratios w of the sorbing organic soil component keeping the linearly sorbing organic mass constant. Solid line: $w=0$; dashed-dotted line: $w=0.2$; dotted line: $w=0.5$; dashed line: $w=1$. Points have been approximated by single Freundlich fits. (From Ref. 13.)

Figure 9 further illustrates that the points calculated by the distributed reactivity model can also be approximated well by a single Freundlich fit. This confirms again that fitting only does not justify the assumption of a certain sorption mechanism.

(b) Distributed Mass Transfer Rate Model. Model simulations according to the distributed mass transfer rate model using both lognormal and Γ distributions of mass transfer rates were compared to the two-site equilibrium/kinetic model. In all cases, optimal sorption parameters were determined by the best fit to the laboratory data. Results are given in Fig. 10.

Batch experiments at 20°C in continuous-flow stirred tank reactors on desorption of trichloroethene from aquifer material contaminated over a long period were performed [10]. The material was collected at different sites at Picatinny Arsenal, N.J., and was composited including material from a peat layer.

Figure 10 shows for the desorption of trichloroethene from this contaminated aquifer material that the distributed mass-transfer rate model provides significantly improved simulations of aqueous equilibrium concentrations, as compared to the two-site model.

Additionally, use of an apparent distribution coefficient demonstrated that the performance of the two-site model was very sensitive to the value of the distribution coefficient, while the performances of the distributed models were robust over a wide range of distribution coefficients. Furthermore, the desorption studies demonstrated that as the length of the contamination period increases, the simulation capability of the two-site model decreases.

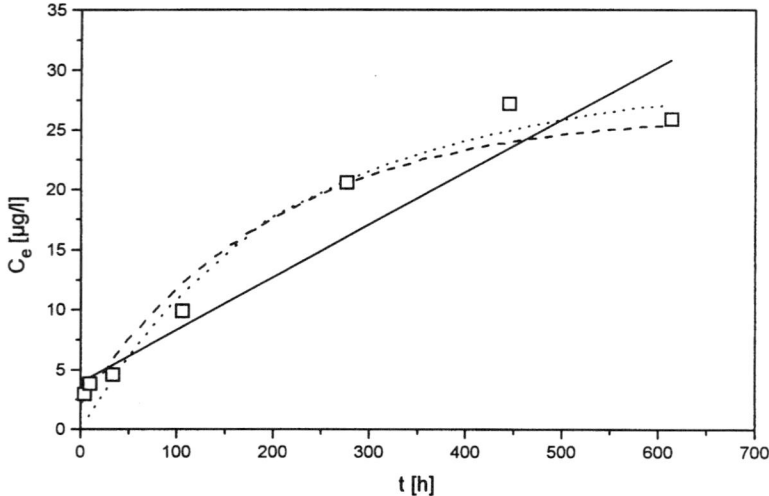

FIG. 10 Aqueous trichloroethane concentrations measured by batch desorption experiments (□) with aquifer material collected at Picatinny Arsenal, N.J. Model calculations were performed by the two-site model (solid line) and by the distributed mass transfer rate model (dashed line) for both the Γ and the lognormal simulations that overlap. The dotted line is obtained by the two-site model employing an apparent K_d determined from the same batch desorption experiment. (From Ref. 10.)

4. Verification of the Sorption Mechanism by Multiple Model Tests

The work of Pedit and Miller [14] responds to deficiencies in current models by formulating two classes of sorption models: an extended set of deterministic models based upon a generalized multiple domain concept, and a class of stochastic models. Stochastic sorption models are formulated to describe microscale variability, which may be interparticle or intraparticle in origin, in terms of both sorption equilibrium and rate properties. The stochastic model formulations treat sorption equilibrium and rate properties as continuously distributed random variables described by lognormal or Γ probability density functions.

Bottle-point rate studies were performed on long-term sorption rates of the herbicide diuron (3-(3,4-dichlorophenyl)-1,1-dimethylurea) on a subsurface material at 23°C. The subsurface material (Wagner, median grain size 0.43 mm, TOC 0.08% [12]) originated from a sand and gravel pit at a depth of about 25 m below the surface of the ground.

The results of a parameter estimation for 16 different rate models applied to the experimental data set are summarized.

The kinetic models discussed are deterministic models characterized as single-particle class, one-parameter model fit (SP-P, SP-S, SP-FO), single-particle class, two-parameter model fit (SP-EP, SP-ES, SP-BC), multiple-particle class, one-parameter model fit (MP-P, MP-S, MP-FO), multiple-particle class, two-parameter model fit (MP-EP, MP-ES, MP-BC), and stochastic models with continuously distributed parameters (LLFER, GLFER, LR, GR).

The abbreviations used have the following significance:

BC: bicontinuum approach
E: instantaneous equilibrium fraction approach
EP: instantaneous equilibrium fraction approach+pore diffusion approach

ES: instantaneous equilibrium fraction approach+surface diffusion approach
FO: first-order mass transfer
GLFER: Γ linear free energy relationship approach
GR: Γ rate approach
LLFER: lognormal linear free energy relationship approach
LR: lognormal rate approach
MP: multiple-particle class
P: pore diffusion approach
S: surface diffusion approach
SP: single-particle class

Sorption rate models were compared calculating the root mean square error (RMSE) for the deviation of the model curves from the experimental kinetic curves.

$$\text{RMSE} = \left[\frac{1}{N} \Sigma_i^N \left(\frac{C_{\exp}^i(t) - C_{\mathrm{mod}}^i(t)}{C_0}\right)^2\right]^{1/2}$$

C_0: initial bulk liquid-phase solute concentration
$C_{\exp}^i(t)$: experimentally predicted bulk liquid-phase solute concentration of sample i, which was collected at time t
$C_{\mathrm{mod}}^i(t)$: model-predicted bulk liquid-phase solute concentration of sample i, which was collected at time t
i: index "number of sample"
N: number of data points
RMSE: root mean square error in units of relative bulk liquid-phase solute concentration
t: time

RMSE values are listed in Table 1.

TABLE 1 Comparison of Deterministic and Stochastic Models Approximating Diuron Sorption on Subsurface Wagner Material Calculating the Root Mean Square Error

MODEL	RMSE
SP-FO	0.163
MP-FO	0.115
SP-BC	0.094
MP-BC	0.063
SP-S	0.059
MP-S	0.046
SP-P	0.045
GR	0.044
SP-ES	0.040
MP-P	0.035
SP-EP	0.033
MP-ES	0.033
MP-EP	0.028
LLFER	0.026
LR	0.026
GLFER	0.024

Source: Ref. 14.

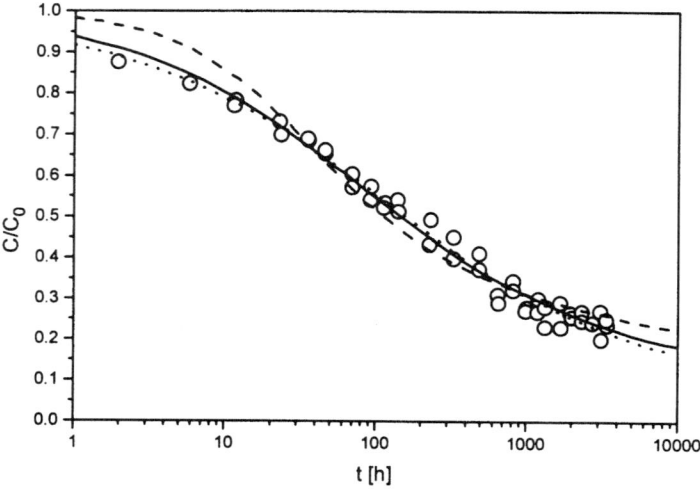

FIG. 11 Experimental data (○) for diuron sorption on subsurface Wagner material. Model approximations were made for models with stochastically distributed parameters: LLFER (solid line); LR (solid line), approximately as LLFER; GLFER (dotted line); GR (dashed line). (From Ref. 14.)

RMSE data comparison shows a range of agreement between individual model fits and experimental data, with several trends evident: diffusion models outperformed first-order models; the addition of an instantaneously sorbing fraction significantly improved model data agreement; multiple-particle class models fit the data more closely than single-particle class models without an increase in the number of fitted parameters, and stochastic first-order models agreed well with the data (viz., Fig. 11).

Of course, model approaches have to consider additionally the sorption hysteresis, since the transport of HOCs in soils and sediments is strongly influenced by sorption hysteresis, mostly leading to an accumulation of HOCs. Basically, the postulation of sorption mechanisms should also include the causes of sorption hysteresis.

IV. FORMATION OF SORPTION HYSTERESIS

A. Mechanisms of the Formation of Sorption Hysteresis

There are a number of reasons for the appearance of sorption hysteresis [20]. Such reasons for anomalous adsorption/desorption behavior are

1. Apparent hysteresis by experimental artifacts
2. True hysteresis, which is time-invariant and repeatable
3. Irreversible adsorption, which is generally associated with some rather permanent change in the adsorbent/adsorbate system

In the case of apparent hysteresis, experimental artifacts may be due to sorption to nonsettling particles, to solute degradation or solute loss, or to the presence of competing solutes. Moreover, regarding short-time batch experiments, slow rates of desorption may lead to apparent sorption equilibria and consequently to apparent hysteresis.

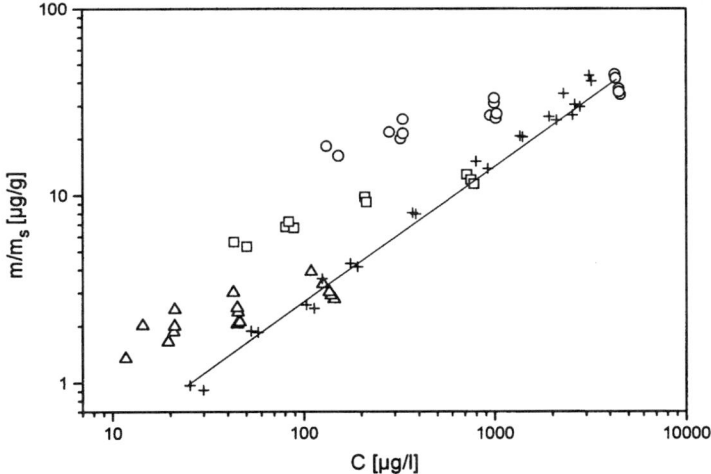

FIG. 12 Pseudo-sorption–desorption equilibrium of lindane on subsurface material from Wagner. Adsorption (+), desorption (○, □, △). (From Ref. 12.)

True hysteresis follows from solute–solid bonds with different sorption energies or site-specific solute–solid bonds.

A special case of true hysteresis is irreversible adsorption by irreversible chemical bonding. However, a pollutant can be also irreversibly bound without chemically reacting with the soil matrix, if some physical alteration takes place. Such physical alterations might include configuration changes to the organic matter as a consequence of changes in ionic strength, pH, or relative humidity. Also, coagulation of either mineral or organic soil particles might contribute.

The results of recent publications on hysteresis are described in the following.

B. Experimental and Model Results

1. Apparent Hysteresis by Solute Degradation or Solute Loss

A reactive surface-diffusion model [12] was used in order to describe apparent sorption–desorption hysteresis and abiotic degradation of lindane in a subsurface material.

Bottle-point rate studies at 23°C were performed to observe sorption–desorption and transformation of the solute lindane in systems that included a subsurface sand material. The subsurface material (Wagner) was from a sand and gravel pit at a depth of about 25 m below ground surface.

Rates of dehydrochlorination were determined for lindane in solution and solid phases. A pseudo-sorption–desorption equilibrium experiment was performed by allowing 720 h for attainment of sorption equilibrium and 384 h for attainment of equilibrium for each of three desorption steps, which is longer than most experiments of a similar nature have been allowed to equilibrate. The pseudoequilibrium data showed marked hysteresis (Fig. 12).

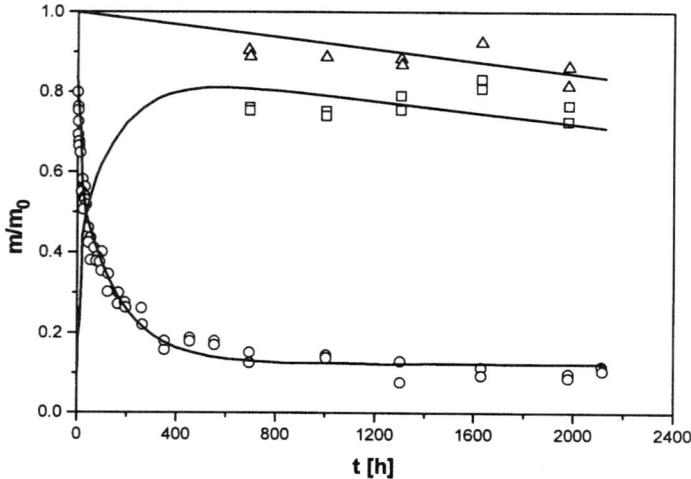

FIG. 13 Kinetics of lindane sorption on subsurface material from Wagner as a function of time and phase. The solute mass is normalized by the initial solute system mass. Total system (△), solid phase (□), liquid phase (○), model calculations (solid lines). (From Ref. 12.)

Results of these experiments were interpreted with a diffusion and reaction model. The model used was a surface-diffusion model, which accounted for solute degradation from the solution and solid phases. It is called the dual resistance surface-diffusion model or reactive surface-diffusion model and was described in Section II.C.3.a. The model predictions of the pseudo-sorption–desorption experiment showed that slow sorption–desorption rates explained most of the apparent hysteresis (cf. Fig. 13).

2. Irreversible Adsorption

The hypothesis that adsorption and desorption are reversible processes has been tested by Kan et al. [20].

The sorption of the organic pollutants naphthalene and phenanthrene was studied in the laboratory using batch reactors at room temperature. Measurements were performed from a few hours to over 2 months. The sorbent was sediment from Johnson Ranch, Lula, Oklahoma (92% sand, 6% silt, 2% clay, about 0.27% organic matter in the sand).

The adsorption experiments were in equilibrium within 1 to 4 days and could be modeled using simple linear isotherms (Fig. 14) with K_d values consistent with published K_{oc} and K_{ow} relationships.

Desorption experiments were conducted with the contaminated sediments by successive dilutions. Desorption experiments varied from 1 day to 5 months, and observed desorption rates were from 1 to 3 orders of magnitude smaller than previously measured or predicted. If equilibrium were obtained during the desorption, typically 82–99% of the adsorbed pollutant would have been desorbed, but generally only 30–50% of the adsorbed pollutant could be desorbed. Consequently, fast adsorption and slow desorption of naphthalene (Fig. 14) and phenanthrene at a sediment demonstrate the presence of a hysteresis.

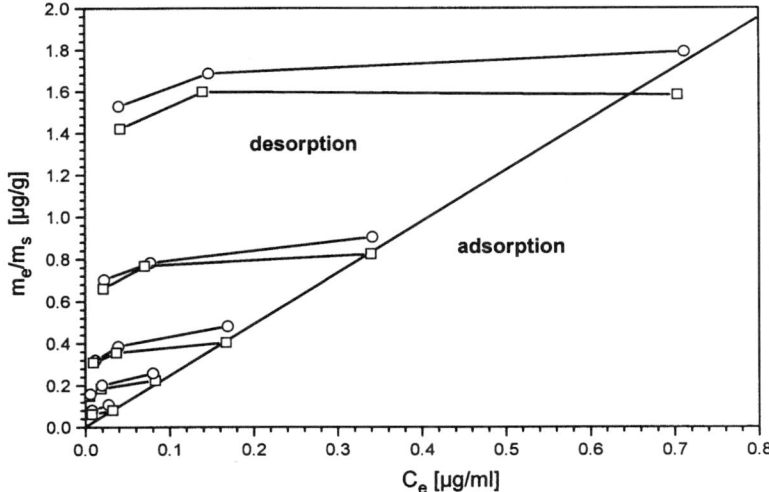

FIG. 14 Adsorption and desorption isotherms of naphthalene on sediments. Adsorption and desorption were continued for either 1 (□) or 7 (○) days. In each desorption step, 80% of the liquid was replaced with water to induce desorption. (From Ref. 20.)

In summary, the adsorption of naphthalene and phenanthrene are fast, whereas a fraction of the adsorbed pollutants desorb very slowly. These desorption results could not be explained by artifacts of the procedure. The difference between the adsorption and the desorption rates is also not explainable by retarded intraparticle diffusion models, where the hindered diffusion should affect the uptake and release similarly. Therefore it is concluded that a substantial fraction of the adsorbed pollutant physically resists desorption or is irreversibly bound.

V. MULTICOMPONENT SORPTION

For reasons of organic contaminants often originating from multicomponent pollutant sources, sorption of organic contaminants has to be described as multicomponent sorption in many cases.

A. Organic Contaminants

1. Mechanisms

The uptake of individual HOCs from aqueous solution by soils that exhibit heterogeneous reactivity is shown to be reduced in the presence of other HOCs. The observed behavior is consistent with expectations for competitive surface adsorption phenomena, being most marked at low solution-phase concentrations and coinciding with single-solute isotherm nonlinearity. Nonlinear sorption and competitive effects appear to relate to components or fractions of soil having different reactivity than that commonly attributed to soil organic matter. In particular, HOC uptake by the organic components of soils deriving from certain sedimentary rocks appears to take place by surface reactions rather than by partitioning to organic matrices [21].

2. Modeling

The ideal adsorbed solution theory is used to predict mixed-solute competitive effects from single-solute sorption isotherms. The sorption of hydrophobic organic compounds by subsurface soils exhibiting the characteristics of distributed reactivity was observed to be influenced by competitive effects in the presence of other HOCs. These effects were modeled using single-solute isotherm parameters and the ideal adsorbed solution theory.

The expression derived for the concentration $C_{e,m}^i$ of solute i in the mixed-solute solution as a function of its single-solute Freundlich isotherm parameters and the sorbed-phase concentrations of all solutes is [21]

$$C_{e,m}^i = \frac{q_{e,m}^i}{q_{e,m}} \left(\frac{q_{e,m}^i + q_{e,m}^j \, n^i/n^j}{K_F^i} \right)^{1/n^i} \tag{9}$$

with

$q_{e,s}^i = K_F^i (C_{e,s}^i)^{n^i}$ single-solute Freundlich isotherm

$q_{e,m} = \Sigma_i \, q_{e,m}^i$ i = 1 to N

$y^i = q_{e,m}^i / q_{e,m}$ i = 1 to N

$\Sigma_i \, y^i = 1$ i = 1 to N

and the nomenclature is

- C: solution-phase solute concentration
- e: index "equilibrium"
- i: index "solute i"
- K_F: Freundlich isotherm parameter
- m: index "multi-solute system"
- n: Freundlich isotherm parameter
- N: number of solutes i
- q: amount sorbed by the soil
- s: index "single-solute system"
- T: absolute temperature
- y^i: sorbed-phase mole fraction of solute i

An equation similar to Eq. (9) can be written for solute j, and the two equations can be solved simultaneously in order to obtain the sorbed-phase concentrations $q_{e,m}^i$, $q_{e,m}^j$ corresponding to any particular equilibrium solution concentrations. Known parameters are the mixed solution-phase equilibrium concentrations and the single-solute Freundlich isotherm parameters.

3. Experimental and Model Results

The accuracy of the predictions of the ideal adsorbed solution theory for bisolute mixtures and different subsoils is influenced by how well the nature of individual single-solute isotherms could be defined. Competitive interactions were found to be most pronounced and the model predictions [21] most accurate for strongly sorbing HOC/soil systems (Fig. 15) exhibiting single-solute isotherm nonlinearity.

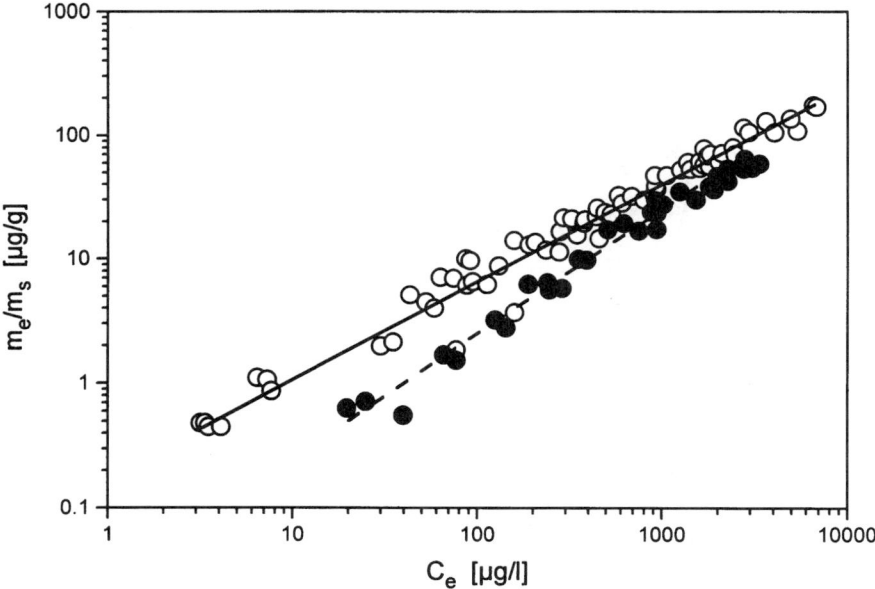

FIG. 15 Experimentally observed tetrachloroethene sorption (○) and tetrachloroethene sorption in the presence of 15 mg/L 1,4-dichlorobenzene (●) by Ann Arbor I subsurface soil. The solid line is the best-fit single-solute isotherm, and the dashed line is the prediction of the ideal adsorbed solution theory using the equilibrium concentration of tetrachloroethene and the initial concentration of 1,4-dichlorobenzene. (From Ref. 21.)

B. Organic Contaminants and Surfactants

1. Mechanisms

Where surfactant and HOCs coexist in soil/water systems there are a number of possible interactions that can occur simultaneously [22,23]:

1. Distribution of surfactant between monomeric, hemimicellar, and micellar forms
2. Competition for active hydrophobic adsorption sites between the surfactant and HOC
3. Partitioning of HOC among soil hydrophobic adsorption sites, surfactant micelles, and hemimicelles

The interaction of HOCs with surfactant monomers is usually very weak and insignificant. At concentrations where micelles and hemimicelles are present, interactions can take place. Sorbed HOCs can be solubilized by free micelles, resulting in mobilization. HOCs in solution are in equilibrium between sorption onto hydrophobic adsorptive sites on the soil, and partitioning into hemimicelles—both resulting in immobilization—and partitioning into free micelles. Whether the HOCs are previously sorbed onto soil or are in solution, partitioning into micelles, hence mobilization, is favored by increasing surfactant concentration.

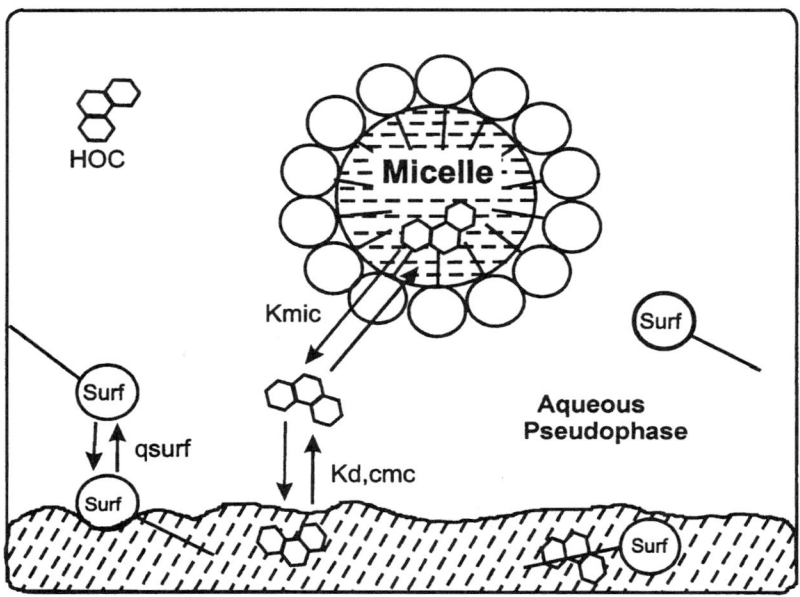

FIG. 16 Partition of a hydrophobic organic compound (HOC) in a soil/aqueous system containing a nonionic surfactant and soil humic matter. (From Ref. 25.)

Although there is evidence that surfactants can affect the fate and behavior of HOCs in soil, the potential for detergent ingredients causing significant effects is limited due to the relatively low concentrations found compared with the critical micelle concentrations (CMC). In addition, the effective CMC in environments such as soils and sediments is generally much higher than in clean water systems. Typical soil concentrations of linear alkylbenzene sulfonate (LAS), the most frequently used surfactant in domestic detergents, are significantly lower than those required to produce micelles in pore water. Therefore it is unlikely that surfactants present in domestic detergents will contribute significantly to the mobilization of HOCs in sludge-amended soil [23].

Surfactant monomers present in the aqueous pseudophase may affect the value of the HOC soil/aqueous pseudophase partition coefficient by enhancing HOC solubility in the aqueous pseudophase; the surfactant sorbed in soil, however, tends to increase the value of the HOC partition coefficient by increasing the fractional organic carbon content of the soil. Conceptually, the fraction of HOC removed from soil and solubilized in bulk solution can be modeled without resorting to fitting coefficients by employing model parameter values obtained from independent experiments and/or estimation techniques. Edwards et al. [24,25] put forward this model describing the effect of nonionic surfactant on the distribution of HOC in a soil–water system.

2. Modeling

In systems consisting of soil and micellar nonionic surfactant in which a HOC exists in dissolved, solubilized, and sorbed forms, but not as a separate solid phase, the HOC is distributed in equilibrium among three separate compartments: a sorbed phase, an aqueous pseudophase, and a micellar pseudophase (Fig. 16).

The partitioning of HOC between the hydrophobic interiors of the nonionic surfactant micelles and the surrounding solution can be characterized by a mole fraction partition coefficient K_{mic}. Nonionic surfactant sorption onto soil may be described by a maximum sorption parameter q_{surf}. Sorbed surfactant molecules tend to increase HOC sorption, and free surfactant monomers in solution tend to decrease HOC sorption by increasing the HOC apparent aqueous solubility S_{cmc} [26]. These effects can be represented by a modified HOC soil/aqueous-pseudophase distribution coefficient $K_{d,cmc}$.

The distribution of HOC between the three compartments can be estimated using a physicochemical model [25], for which model parameter values may be conceptually obtained from independent experiments. As a result, the distribution coefficient $K_{d,tot}$ follows as the ratio of the number of moles of sorbed HOC per gram of solid to the number of moles of HOC either dissolved or solubilized in the bulk solution per liter of solution:

$$K_{d,tot} = \frac{K_{d,cmc}}{1 + K_{mic} V_{m,aq} Y_{mic}} \tag{10}$$

Employment of the model requires the physical values for the parameters K_{mic}, q_{surf}, w_{oc}, Y_{mic}, K_d, S_{aq}, and S_{cmc}, all of which can be independently measured in separate experiments with individuals aqueous, soil, and soil/aqueous systems. The calculation of $K_{d,cmc}$ from the mentioned parameters is provided by the following equations derived according to the mechanism described in Fig. 16.

$$K_{d,cmc} = K_d \frac{S_{aq}}{S_{cmc}} \left(\frac{w_{oc}^*}{w_{oc}}\right)$$

$$w_{oc}^* = w_{oc} + \varepsilon q_{surf} M_{surf} w_{oc,surf}$$

As hydrophobic sorption depends on the organic carbon content of the sorbent, an effectiveness factor ϵ is introduced characterizing the carbon-normalized capacity of sorbed nonionic surfactant to function, relative to humic matter, as a sorbent for HOC. This parameter ϵ is to be estimated from the model calculations.

Summarizing the nomenclature:

K_d: HOC solid/water distribution coefficient
$K_{d,cmc}$: HOC solid/aqueous pseudophase distribution coefficient in the presence of micelles
$K_{d,tot}$: HOC solid/aqueous and micellar solution distribution coefficient
K_{mic}: mole fraction partition coefficient for the distribution of solubilized HOC in the micellar pseudophase and dissolved HOC in the aqueous pseudophase
M_{surf}: molecular weight of the surfactant
q_{surf}: extent of surfactant sorption by adsorption onto, and/or partitioning into, the solids in a system
S_{cmc}: HOC solubility limit in a monomeric surfactant solution at the CMC
S_{aq}: HOC solubility limit in water
$V_{m,aq}$: molar volume of water (L^3/mol)
$w_{oc,surf}$: mass fraction of carbon in the surfactant
w_{oc}: fractional organic carbon content of the solid
w_{oc}^*: effective fractional organic carbon content of a solid after surfactant sorption
Y_{mic}: concentration of surfactant in micellar form in the bulk solution

ε: effectiveness factor characterizing the carbon-normalized capacity of sorbed nonionic surfactant to function, relative to humic matter, as a sorbent for HOC

In the work of Yeom et al. [27] the consideration of coal tar replaces the consideration of humic matter as a sorbent component as published by Edwards et al. [24,25]. An equilibrium model was developed to predict the solubilization of several PAHs from coal-tar contaminated soils for given properties of the soil, surfactant, and component PAHs. In describing the partitioning behavior of individual components in a multicomponent system, it was assumed that the activity coefficients in the aqueous, coal tar, and micellar phases were independent of phase composition and that the total concentration of solutes in the micellar phase was small in comparison to the surfactant concentration. It allows the distribution of PAHs to be predicted given the aqueous solubility S_L^i of pure subcooled PAH at the temperature of the solution, the mass fraction of coal tar in the soil w_{ct}, the coal tar molecular weight M_{ct}, the octanol–water partition coefficient K_{ow} of the PAH, and the concentration of micelles in solution Y_{mic}.

The following equation describing the distribution coefficient for PAH i between the soil/tar and aqueous/micellar pseudophase $K_{d,tot}^i$ was derived on the basis of assumptions made by Yeom et al. [27]. The equation contains only measurable variables.

$$K_{d,tot}^i = \frac{w_{ct}}{(S_L^i M_{ct})(1 + K_{ow} Y_{mic})} \tag{11}$$

The derivation of this equation anticipates the formation of micelles and assumes the partition of PAHs into the tar phase of the soil.

The mole balance of the PAH i in the micellar aqueous solution is

$$C_{tot}^i = C_{aq}^i + C_{mic}^i$$

All concentrations refer to the total solution volume.

C_{tot}^i: total concentration of PAH i in the aqueous and micellar pseudophase
C_{aq}^i: aqueous concentration of PAH i
C_{mic}^i: concentration of PAH i in the micellar pseudophase

Approximate linear sorption isotherms are assumed regarding the aqueous micellar solution

$$K_{d,tot}^i = \frac{q_e^i}{C_{tot}^i}$$

and the aqueous phase

$$K_d^i = \frac{q_e^i}{C_{e,aq}^i} \tag{12}$$

The micelle–water distribution coefficient $K_{d,mic}$ is defined as

$$K_{d,mic} = \frac{C_{e,mic}^i}{C_{e,aq}^i Y_{mic}}$$

It follows that

$$K_{d,tot}^i = \frac{K_d^i}{1 + K_{d,mic} Y_{mic}} \qquad (13)$$

which equation corresponds to Eq. (10).
Further assuming that the coal tar mixture behaves ideally, i.e. Raoult's law

$$C_{aq}^i = x_{ct}^i S_L^i$$

is valid for the aqueous concentration C_{aq}^i of PAH i in equilibrium with coal tar, C_{aq}^i can be introduced into Eq. (12) together with

$$q_e^i = x_{ct}^i \frac{w_{ct}}{M_{ct}} \quad \text{and} \quad ,$$

$$x_{ct}^i = \frac{n_{ct}^i}{n_{ct}}$$

deriving

$$K_d^i = \frac{w_{ct}}{S_L^i M_{ct}}$$

Since $K_{d,mic} \sim K_{ow}$ is a good approximation, Eq. (11) follows by inserting the expressions for K_d^i and $K_{d,mic}$ into Eq. (13).
The additional nomenclature is

- i: index component as PAH
- K_d^i: soil/tar-water distribution coefficient for PAH i
- $K_{d,tot}^i = q_{soil/tar}^i / C_{tot}^i$
- n_{ct}: number of moles of coal tar in the tar phase
- n_{ct}^i: number of moles of PAH i in/on the tar phase
- q_e^i: equilibrium concentration of PAH i in/on the soil
- x_{ct}^i: mole fraction of PAH i in the tar phase

3. Experimental and Model Results

Edwards et al. [25] studied the balance between surfactant sorption and solubilization effects on the sorption of HOCs. Batch experiments on the sorption of phenanthrene were performed on Lincoln fine sand in the presence of the surfactant Triton X-100.

The studied balance depends on a number of factors, principally the surfactant concentration and the nature of the solid sorbent. Sorbed Triton X-100 acts to enhance phenanthrene sorption; not only does the sorbed surfactant directly increase the fractional organic carbon content of the sand but also, on a carbon-normalized basis, the sorbed surfactant is much more effective as a sorbent for phenanthrene than is humic matter.

As shown by batch experiments, phenanthrene sorption in the presence of surfactant micelles is linearly proportional to the concentration of phenanthrene in the bulk solution. Sorption isotherm linearity for HOCs at a constant surfactant dose implies that surfactant effects on HOC distribution can be characterized independently of HOC concentration. The effects of micellar surfactant on the distribution of a HOC in a sediment/aqueous system can be quantified through measurement of the bulk solution HOC concentration in each of a series of batch systems containing the same total amount of HOC, but different amounts of surfactant.

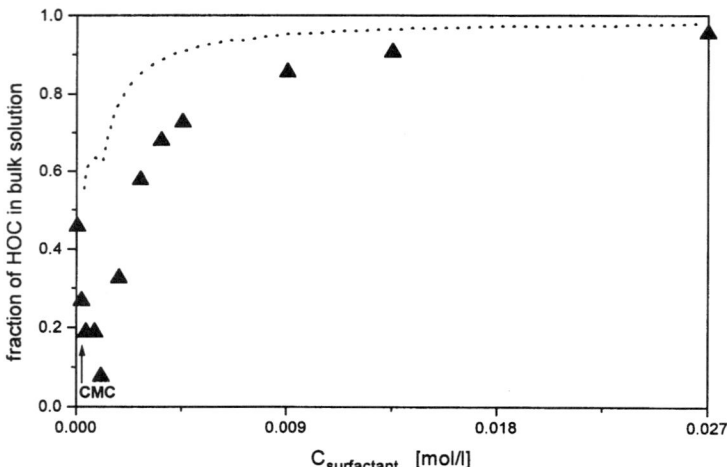

FIG. 17 Experimentally observed phenanthrene sorption by Lincoln fine sand (100 g/L) in the presence of increasing concentrations of Triton X-100 surfactant (▲). Model calculations (dotted line) include a sorption effectiveness factor of 1.0. (From Ref. 25.)

Division of the mass of HOC in the bulk solution by the total mass of HOC gives the fractional mass of HOC in the bulk solution (i.e., in the aqueous and micellar pseudophases) at a given surfactant dose. Figure 17 shows a plot of the fractional mass of phenanthrene in the bulk solution versus the surfactant dose for a number of batch samples containing Lincoln fine sand (organic carbon content 0.053%) and various amounts of Triton X-100 surfactant. The nonionic surfactant can act either to enhance or to inhibit phenanthrene sorption from bulk solution onto fine sand, depending on the bulk solution surfactant concentration.

As the surfactant dose is increased from zero, there is an initial decrease in the fractional mass of phenanthrene in the bulk solution of the system. The effect of sorbed surfactant on enhancing HOC sorption at these low surfactant doses dominates the effect of bulk solution surfactant in enhancing solubilization of HOC. This is true for small supra-CMC values of bulk solution surfactant concentration as well as for sub-CMC values of bulk solution surfactant concentration.

Conversely, Triton X-100 micelles in the bulk solution can greatly enhance the solubilization of phenanthrene and thus its desorption from the sand.

In numbers, the distribution of phenanthrene between the sand and the bulk solution can be characterized by a distribution coefficient that can range in value from less than 0.04 to nearly 10 times that in the absence of surfactant.

Model predictions for this system based on an assumed surfactant effectiveness factor ϵ of unity are shown as a dotted line in Fig. 17. The mass fraction w_{HOC} of HOC in the bulk solution is calculated by

$$w_{HOC} = \frac{A}{A+B}$$
$$A = 1 + K_{mic} V_{m,aq} Y_{mic} \qquad (14)$$
$$B = \frac{K_{d,cmc} m_s}{V}$$

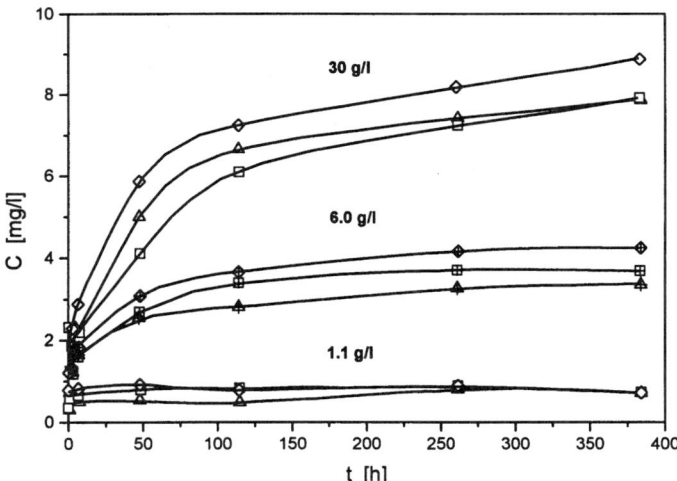

FIG. 18 Solubilization of phenanthrene from a coal-tar contaminated soil (100 g/L) of a manufactured gas plant site by the surfactants Tween 80 (◇), Brij 35 (△), and Triton X-100 (□), employing the dosages 1.1–30 g/L. (From Ref. 27.)

where $K_{d,cmc}$ depends on ϵ. Equation (14) is derived employing Eq. (10).

These model predictions are seen to underpredict the effect of sorbed surfactant on the sorption of HOC, which leads to overpredictions of the mass fraction w_{HOC} of HOC in the bulk solution. Experimental values of w_{HOC} at small supra-CMC surfactant doses are much smaller than those predicted. These results suggest that Triton X-100 surfactant sorbed on Lincoln fine sand in this dose range may actually have an affinity for HOC much greater than that of humic matter on a carbon-normalized basis. The largest discrepancies between experimental data and model calculations using an assumed ϵ value of 1.0 occur over the range of surfactant doses corresponding to sorption of Triton X-100 surfactant in the intermediate region. The much greater affinity for HOCs in this region may be attributed to the development of a nascent hydrophobic surface associated with sorbed surfactant prior to substantive patchy bilayer formation. At the highest surfactant dose tested, about 96% of the phenanthrene is either dissolved or solubilized in the bulk solution in satisfactory correspondence with the model prediction.

Yeom et al. [27] investigated the micellar solubilization of polynuclear aromatic hydrocarbons in coal-tar-contaminated soils. Solubilization of PAHs from a coal-tar-contaminated soil obtained from a manufactured gas plant site was evaluated using nonionic polyoxyethylene surfactants at dosages greater than CMC. Up to 25% of Soxhlet-extractable PAHs could be solubilized at surfactant loadings of 0.3 g/g of soil in 16 days in completely stirred batch reactors (Fig. 18). Longer periods were required to reach equilibrium at higher surfactant dosages.

The equilibrium partitioning of PAHs between the soil/tar and water was obtained in a short period of time, less than 3 h, as reported by Lee et al. [28]. However, with surfactant solutions, equilibration took much longer. Figure 18 shows the solubilization of phenanthrene by three nonionic surfactants. Release of phenanthrene increased with surfactant concentration, but a longer time was required to reach equilibrium at higher concentration. At a dosage of 6 g/L, more than 250 h were required to obtain equilibrium. When the dosage was increased to 30 g/L, equilibrium was not attained even after 380 h.

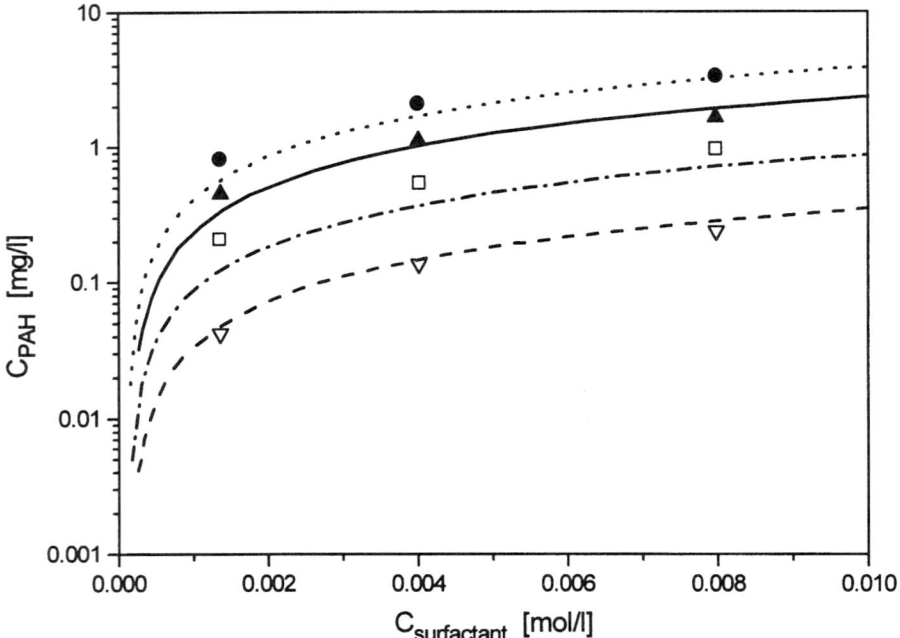

FIG. 19 Measured micellar solubilities C_{PAH} of PAHs from coal-tar-contaminated soil of a manufactured gas plant site in comparison with model predictions. Phenanthrene (●), pyrene (▲), anthracene (□), benzo(a)pyrene (∇); model predictions (lines); surfactant: Triton X-100; soil: 100 g/L. (From Ref. 27.)

Of the three surfactants, Tween 80 was the most efficient in solubilizing phenanthrene. Without surfactant, phenanthrene release was practically zero in terms of the scaling of Fig. 18.

In general, solubilization of coal tar components in surfactant solution can occur by two mechanisms: emulsification and micelle solubilization. In this study, emulsification of coal tar was never observed. Ultracentrifugation of the supernatant of soil-surfactant solution did not reveal separation of an emulsion phase. Therefore, only micellar solubilization was assumed to contribute to enhanced aqueous solubilities of the soil-bound PAHs.

Presumably, mass transfer limitations prevented attainment of equilibrium during the duration (380 h) of solubilization experiments. Seemingly low diffusivities of constituent PAHs in the weathered coal tar matrix imposed mass transfer limitation on solubilization. Nonequilibrium effects become more pronounced as the fractional solubilization of PAHs increases because solubilized PAH must be accessed from deeper within the soil/coal-tar matrix.

The developed model [27] was applied to predict micellar solubilization. In applying the model, most thermodynamic data for the constituent PAHs were obtained from published sources or were measured in the laboratory.

Solubilization of four PAHs from soil of a manufactured gas plant site using Triton X-100 is shown in Fig. 19.

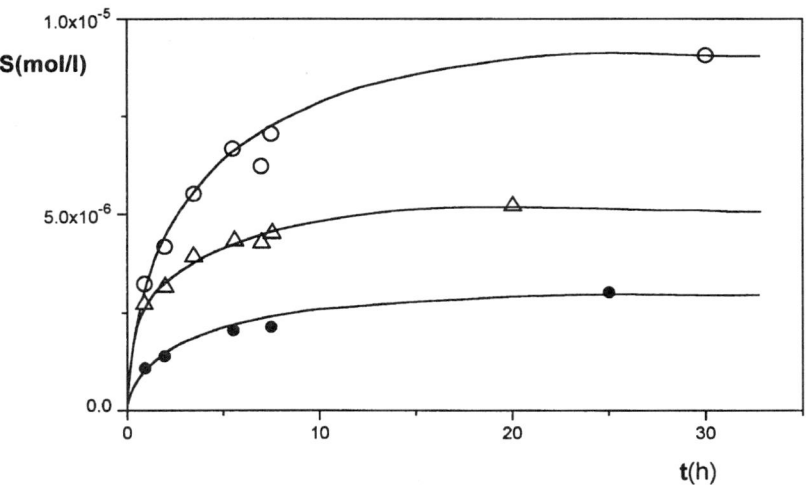

FIG. 20 Shape of the solubilization curves at different surfactant concentrations. ○, -5.0×10^{-3} (mol/L); △, -2.5×10^{-3} (mol/L); ●, -1.0×10^{-3} (mol/L). Surfactant: $C_{12}H_{25}(OCH_2CH_2)_5OH$; temperature: 25 (°C); solubilized substance: Dibenz-(a,h)-anthracene. (From Ref. 26.)

The model predicts solubilization of constituent PAHs reasonably well at low surfactant dosages. At extremely high surfactant dosages, the model failed to predict solubilization reliably.

The solubilization in surfactant micelles is a relatively slow process. The amount of solubilized substance is dependent on the surfactant concentration.

Fig. 20 shows with a model example that the amount of substance solubilized in one time unit increases with rising surfactant concentration. This is easy to understand, because more micelles are formed for solubilization at high surfactant concentrations. However, that does not mean a priori that the kinetics is influenced. If the amount of solubilized substance at time t is related to the amount of solubilized substance at a state of equilibrium, the three curves in Fig. 20 will be reduced to one curve (Fig. 21).

In this respect the solubilization kinetics is independent of the surfactant concentration. These results were verified by Schwuger with different contaminants, e.g. pyrene, pyrelene, anthracene, and naphthalene [26].

In contrast to the surfactant concentration, which has no influence on the solubilization kinetics, the molecule structures of the solubilized substances and the surfactant are very important for the kinetics. In general the kinetics decreases with increasing molecular size of the solubilized substance. In the case of nonionic surfactants, the temperature dependence plays an important role. If the half-life, i.e., the time that is necessary to solubilize half of the amount at a state of equilibrium, is interpreted as a measure of the kinetics, a very interesting curve will be obtained. The half-life runs through a minimum dependent on the temperature (Fig. 22).

In the range of the minimum the half-life is some minutes and increases to 10 and 20 h at higher and lower temperatures, respectively. The minimum is a few degrees beyond the cloud point, i.e., a little bit beyond the phase separation temperature. In this respect the kinetics of the processes is strongly dependent on the number of ethoxylate units in the hydrophilic part of the nonionic surfactant and the respective temperature.

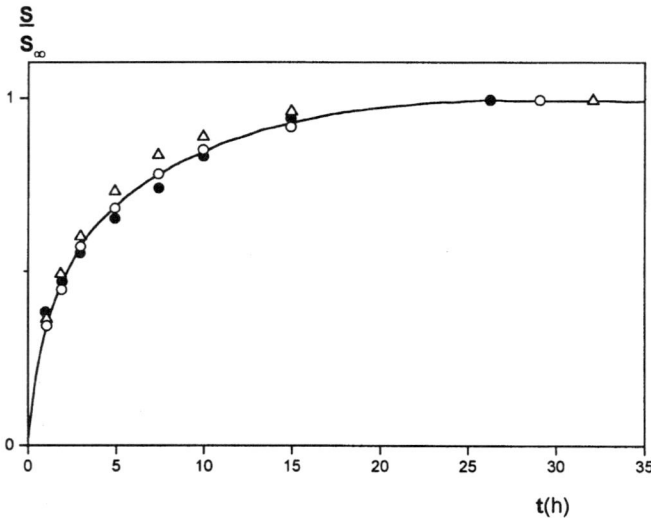

FIG. 21 Influence of the surfactant concentration on the solubilization kinetics. Concentrations of surfactants: ●, -1.0×10^{-3} (mol/L); △, -2.5×10^{-3} (mol/L); ○, -5.0×10^{-3} (mol/L). Temperature: 25 (°C); solubilized substance: Dibenz-(a,h)-anthracene; surfactant: $C_{12}H_{25}(OCH_2CH_2)_5OH$. (From Ref. 26.)

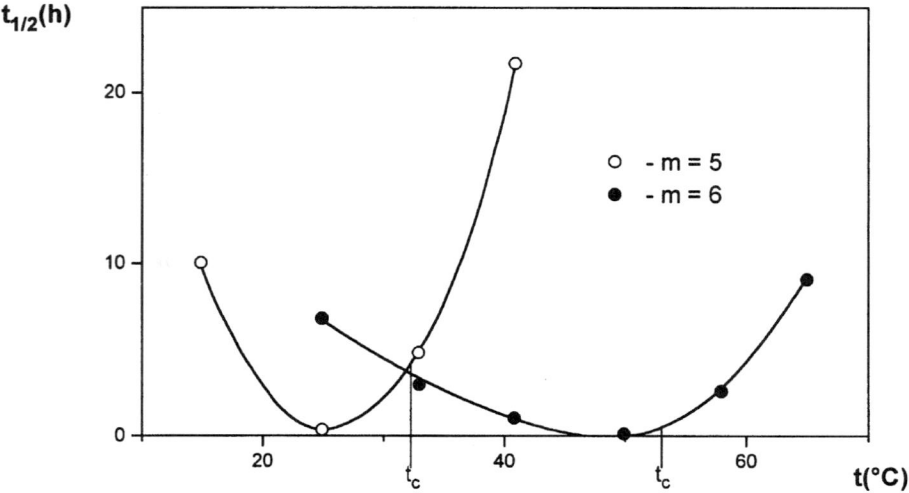

FIG. 22 Dependence of the half-life on the temperature. Solubilized substance: Dibenz-(a,h)-anthracene; surfactant: $C_{12}H_{25}(OCH_2CH_2)_mOH$; concentration: 2.5×10^{-3} (mol/L). (From Ref. 26.)

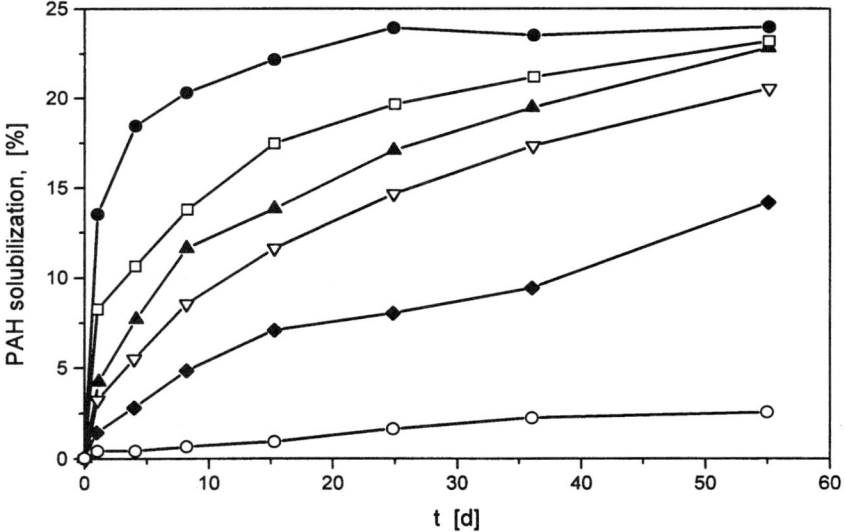

FIG. 23 Fractional release of PAHs from coal-tar-contaminated soil of a manufactured gas plant site in the presence of Tenax. Naphthalene (●), fluorene (□), phenanthrene (▲), anthracene (∇), pyrene (◆), benzo(a)pyrene (○); 0.08 g Tenax/g soil. (From Ref. 29.)

Yeom et al. [29] extended the study of kinetic aspects of solubilization of soil-bound polycyclic aromatic hydrocarbons by applying Tenax resin (polyphenyleneoxide) in the bulk solution in order to accelerate desorption kinetics of PAHs. The presence of Tenax resin causes a steep concentration gradient at the soil surface by adsorbing the bulk PAHs at the Tenax resin and keeping the bulk concentration near zero. Kinetics were described by a modified radial diffusion model. The paper again concerns soil-bound PAH dissolution from the soil of a manufactured gas plant site.

In Fig. 23 solubilizations of various PAHs in the presence of Tenax resin are plotted as a function of time.

The release of PAHs from the soil of the manufactured gas plant site exhibited a nonequilibrium behavior typical of such soils. Except for naphthalene, none of the measured PAHs appeared to have reached equilibrium after 55 days. Naphthalene desorption reached a plateau in about 25 days, having attained a value close to its equilibrium value of 26.5% based on its Tenax–water distribution coefficient ($K_{d,Ten}$ of 45.1 L/g. In comparison, the desorption of fluorene after 55 days was less than half its equilibrium value of 58%, corresponding to its $K_{d,Ten}$ of 131.8 L/g. The desorption of phenanthrene was merely one-third of its equilibrium value of 68% based on a $K_{d,Ten}$ of 270 L/g. Only 2.6% of benzo[a]pyrene adsorbed to Tenax after 55 days. An inverse correlation between the number of rings in the PAH molecule and the extent of release is manifest.

In Fig. 24 the dissolution of phenanthrene from tar-contaminated soil by surfactants is compared with that by water in the presence of Tenax but no surfactant.

Surfactants significantly enhanced the rate of PAH solubilization, the amount solubilized being dependent upon the structure of the surfactant. However, comparison of the rate of phenanthrene release in the presence of surfactant to that in the presence of Tenax (Fig. 24) revealed that phenanthrene release in the presence of Tenax was much

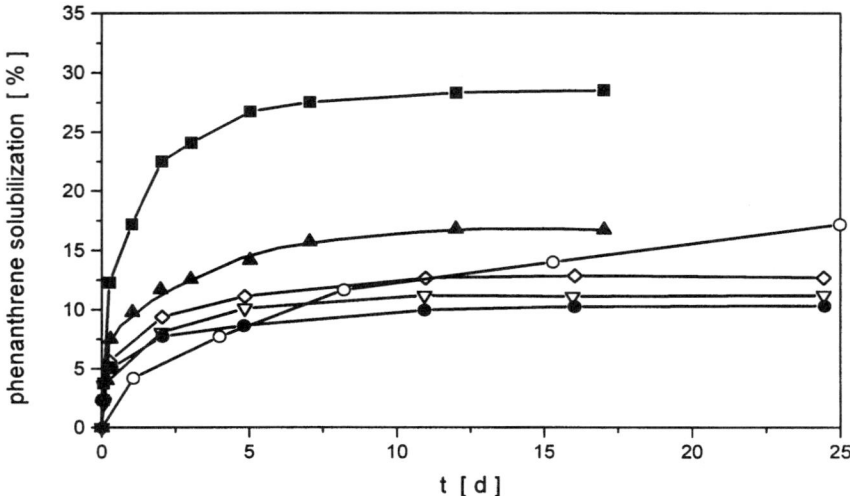

FIG. 24 Phenanthrene solubilization from coal-tar-contaminated soil of a manufactured gas plant site in the presence of surfactants or Tenax. $C_{12}E_4$ (■, 5.5 g/L), $C_{12}E_{10}$ (▲, 5.5 g/L), $C_{12}E_{23}$ (●, 6 g/L), TX-100 (▽, 6 g/L), Tween 80 (◇, 6 g/L), Tenax (○, 8 g/L). (From Ref. 29.)

slower, despite the greater adsorption capacity of Tenax. Therefore enhancement of PAH solubilization by surfactant cannot be attributed entirely to the concentration gradient created by the partitioning of PAH into micelles.

Consequently, in explaining the dissolution of individual PAHs from coal tar mixtures by micellar surfactant solutions, these three steps should be considered:

1. Matrix diffusion of PAH molecules within the mixture
2. Mass transfer at the interface (film diffusion)
3. Partitioning into the micellar phase

Given that step 3 is extremely fast [30], steps 1 and/or 2 must be rate limiting.

In this context, Grimberg et al. [31] suggested that the diffusion of micelles through the hydrodynamic layer (step 2) is rate limiting in the dissolution of phenanthrene crystals by surfactants. Equation (15) was proposed to describe the surfactant-enhanced dissolution of pure crystalline phenanthrene:

$$r_{\text{diss}} = \frac{\delta C}{\delta t} = k_b a (C_{\text{sat}} - C) \tag{15}$$

where r_{diss} is the dissolution rate of crystalline phenanthrene, k_b is the observed mass transfer rate constant, a is the surface area of the solid phenanthrene divided by the volume of the liquid, C_{sat} is the total saturation concentration of phenanthrene in the bulk liquid comprising the aqueous and micellar pseudophases, and C is the actual phenanthrene concentration in this bulk liquid. The constant k_b includes parallel transfer of aqueous and micellar phenanthrene away from the interface. Individual phenanthrene molecules are much smaller than the surfactant micelles, and hence their transport is faster. Thus k_b is inversely related to surfactant concentration. However, a surfactant would increase the overall dissolution rate by significantly increasing C_{sat}, which far outweighs the effect of the smaller k_b of micellar phenanthrene. Analogously, surfactant would have a similar effect on the transport of individual PAHs from the tar–water interface.

Furthermore, interfacial film diffusion failed to describe solubilization of organic constituents in the weathered manufactured gas plant soil used in this study. Thus matrix diffusion appears to be of paramount importance in controlling solubilization of PAHs in the manufactured gas plant soil.

Therefore it is postulated that the enhancement of PAH release by surfactants occurs by two mechanisms:

1. Micellar solubilization by increased concentration gradient at the soil/tar-water interface
2. Sorption and penetration of surfactant molecules, causing intrasorbent swelling of the soil–tar matrix and increased matrix diffusivity of PAHs

Additionally, a modified radial diffusion model was used to evaluate the relative importance of these mechanisms in surfactant washing of contaminated soils. It was concluded that surfactant-enhanced PAH release from the test soil takes place mainly by increasing matrix diffusivities, while increase in solubility by partitioning of PAHs into the micellar pseudophase played a secondary role.

DiVincenzo and Dentel [19] examined the 1,2,4-trichlorobenzene (TCB) sorption–desorption on soil in the presence of sodium dodecyl sulfate (SDS) and/or a cationic polyelectrolyte (Percol 757). Anionic surfactants and cationic polymers were chosen as two classes of anthropogenic chemicals that are released into the environment in significant amounts. In particular, this occurs in the land disposal of sludges, potentially altering the terrestrial mobility of other contaminants such as hydrophobic organics.

Batch experiments on adsorption and desorption were performed at 23°C for 24 h in the dark. The soil was a Sassafras fine sandy loam (Ultisol, Typic Hapludult) collected from the Ap horizon at the Agriculture Experimental Station, University of Delaware. The soil contained 2% organic matter.

Experimental results showed that additions of SDS and/or polyelectrolyte led to significant changes in TCB sorption. A scheme of the derived TCB sorption mechanism for increasing surfactant concentration and constant polyelectrolyte concentration is shown in Fig. 25.

At low surfactant concentrations, the cationic polymer consumes the anionic surfactant due to charge–charge attraction, with residual polymer remaining in solution at the lowest initial surfactant concentrations. Accordingly, no change in the TCB distribution coefficient is seen with an SDS addition of 0.5 g/L.

As the charge equivalence point is approached, the formed surfactant–polymer complex becomes less charged and increasingly hydrophobic. Cooperative binding of the surfactant leads to micellelike surfactant aggregates wrapped by polyelectrolyte, which provides the surfactants' counterions and serves to flocculate the overall structure [32,33]. In a sense, this phenomenon may be viewed as a means of increasing the organic content of the soil, with the familiar result of greater affinity for hydrophobic organics such as TCB.

Further additions of anionic surfactant then provide a surfactant monomer concentration in solution that eventually attains a level equivalent to the CMC. This may be called the point of saturation, differentiated from the CMC because the amount of surfactant required to initiate micelle formation in this system must also include an initial amount for the preferred interaction with polymer. Any surfactant added above the point of saturation may be assumed to form micelles because all the other, more favorable interactions have been completed.

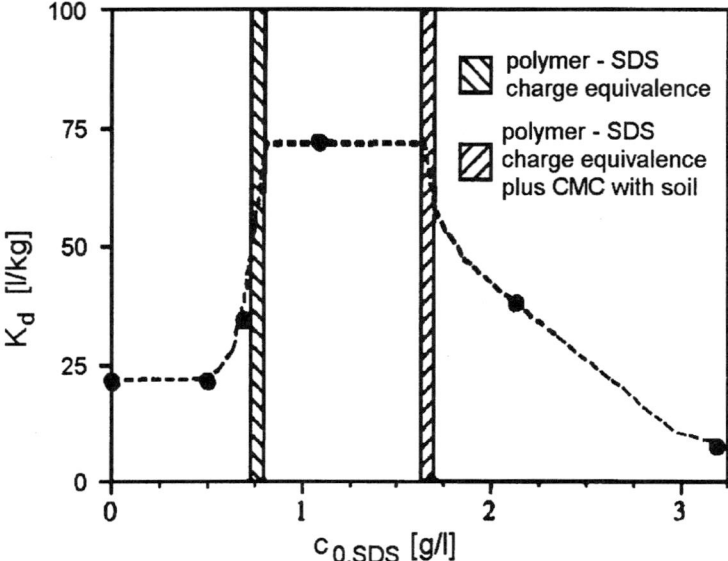

FIG. 25 Distribution coefficients of 1,2,4-trichlorobenzene at Sassafras fine sandy loam in the presence of constant polymer concentration and varying sodium dodecyl sulfate (SDS) concentrations. Also shown is the general trend in the distribution coefficient according to the hypothesized sequence of interactions. (From Ref. 19.)

In parallel the phenomenon of TCB sorption on a high level continues at surfactant concentrations beyond the charge equivalence point, adding surfactant only to the free concentration in solution. Obviously, SDS in this form does not significantly affect TCB uptake. Thus the high K_d is maintained until the point of saturation (Fig. 25).

The increment in surfactant concentration between the charge equivalence point and the point of saturation should equal the CMC in the soil solution; for these results, this would be 0.90 g L^{-1} of SDS, and the total SDS concentration needed to reach the saturation point is thus between 1.63 and 1.70 g L^{-1}.

Once the saturation point is surpassed, the HOC can be solubilized due to micelle formation, and the distribution coefficient decreases. For SDS, this was seen with the 2.13 and 3.19 g L^{-1} isotherms. At sufficiently high surfactant concentrations, the extent of TCB adsorption is significantly less than without any surfactant present.

The alternative case of negatively charged polyelectrolyte and surfactant was studied by Sastry et al. [34].

Batch studies were made on the adsorption of polyacrylic acid (PAA) and sodium dodecylbenzenesulfonate (SDBS) on Na-kaolinite at room temperature.

For the adsorption from a mixture of PAA and SDBS, the results show the chronological importance of the mode of addition. Successive or simultaneous additions of both components indicate a competition between the similarly charged PAA and SDBS for the same binding sites on the kaolinite surface.

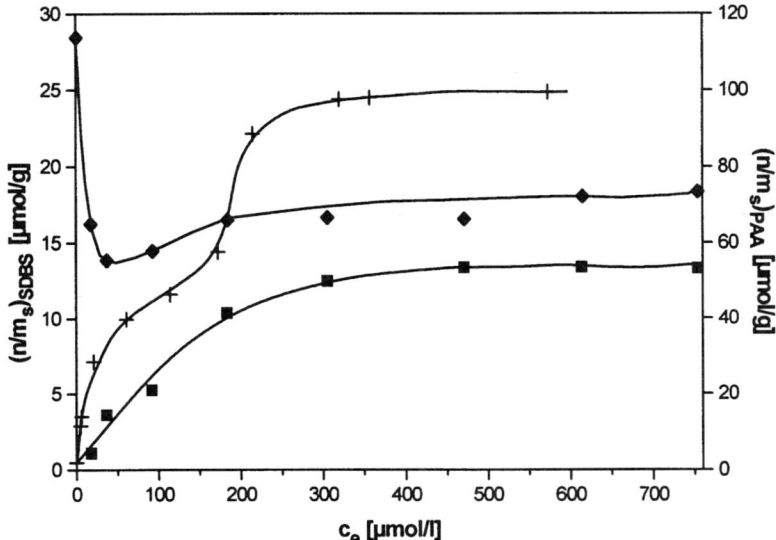

FIG. 26 Individual adsorption of sodium dodecylbenzenesulfonate (SDBS) and simultaneous adsorption of SDBS and of polyacrylic acid (PAA) at pH 4–4.5 on Na-kaolinite. The total SDBS concentration varies; the total PAA concentration is kept constant at 300 mg/L SDBS (+), SDBS in the presence of PAA (■), PAA in the presence of SDBS (◆). C_e: equilibrium SDBS concentration. (From Ref. 34.)

The displacement of a preadsorbed polymer with a surfactant or vice versa allows the relative binding strength to be compared. Attempts have been made for the PAA–SDBS mixtures in both acidic and alkaline media. The displacer SDBS or PAA sequentially replaced part of the supernatant PAA–SDBS mixture.

The results show that added SDBS could not displace the PAA. On the contrary, enhanced adsorption of PAA is observed at acidic and alkaline pH, as is adsorbed SDBS in the presence of PAA. If PAA is added as a displacer only at acidic pH, PAA desorbs SDBS by about 20% for a molar ratio range up to 2.

The sharing of the binding sites for PAA and SDBS in the case of a simultaneous adsorption can be demonstrated in Fig. 26 from the adsorption isotherms of SDBS (both individual and simultaneous with PAA) along with changes in amounts of PAA adsorbed at acidic pH.

The measured individual surfactant adsorption isotherm shows a typical two-stage process. The result is explained by a first stage of adsorption resulting in a monolayer coverage on the surface of kaolinite and by a second stage resulting in the formation of small surface aggregates of three or four surfactant molecules. Simultaneous adsorption of SDBS and PAA demonstrates that in the presence of PAA the adsorbed amount of SDBS diminishes in comparison with the individual SDBS adsorption. Also the adsorbed amount of PAA diminishes with rising concentration of the second component SDBS.

Thus the irreversible character of the blocking effect of both preadsorbed components at the kaolinite surface supposes low-exchange kinetics at the kaolinite surface. Effective simultaneous adsorptions of both components are observed only when they are added in solution at the same time.

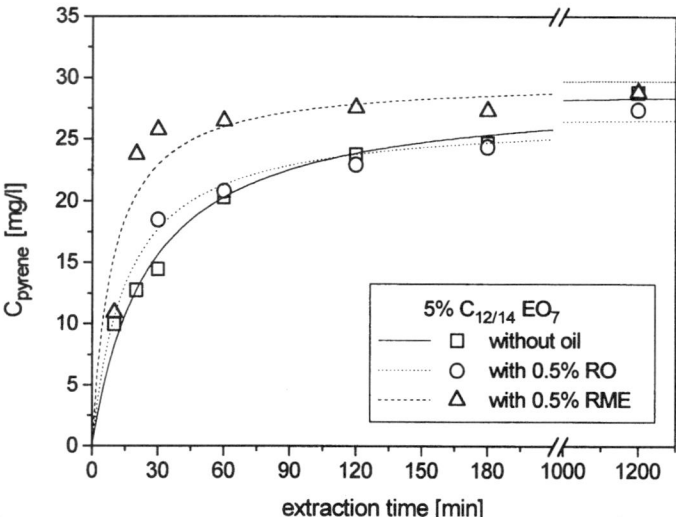

FIG. 27 Extraction kinetics for pyrene with different extraction solutions containing $C_{12/14}EO_7$. (From Ref. 35.)

C. Organic Contaminants and Emulsions

1. Mechanisms

Emulsions and microemulsions intensify the advantages of surfactant solutions, such as decreasing interfacial tension and increasing wettability of soil [35]. Emulsions and especially microemulsions are known to be excellent solvents for polar and nonpolar organic substances. As they are formed by water, oil, and surfactant, they exhibit sites for the solubilization of ions or dipoles as well as for hydrocarbons or amphiphiles. The important difference between soil washing with surfactant solutions and the emulsion or microemulsion treatment is the additional solubility of the pollutant in the oil component, which further enhances uptake [36–38]. Therefore this extraction technique is applicable in particular to soils containing high amounts of silt and clay. Due to their low interfacial tension, microemulsions show excellent wetting behavior for the silt fraction of a soil. As it is inevitable that a small amount of microemulsion remains on the soil particles, microemulsions applied in soil remediation must be composed of biodegradable oils and surfactants.

2. Experimental Results

Apart from equilibrium solubility, the kinetics of PAH transport from the soil surface into the microemulsion is important. The low interfacial tension of a microemulsion should cause good wetting behavior for the soil, so the removal of the contaminants and the following solubilization should occur within a short time. To check this assumption, several kinetic measurements were performed. Real contaminated soil was extracted with different surfactant solutions ($C_{12/14}EO_7$ and $C_{12/14}EO_{10}$) and microemulsions. The soil/extraction solution ratio was set at 1 : 10. Fig. 27 shows the extraction results for pyrene for several solutions with $C_{12/14}EO_2$ as a function of extraction time.

Empirically, this extraction kinetics can be conveniently fitted by first-order kinetics according to

$$C(t) = C_e[1 - \exp(-k_{\text{eff}} t)]$$

Here $C(t)$ is the concentration of pyrene in the extraction solution at time t, C_e the equilibrium pyrene concentration in the extraction solution, and k_{eff} is an effective rate constant.

For the systems with $C_{12/14}EO_7$, the addition of rape oil (RO) and especially of rape oil methyl ester (RME) leads to a pronounced acceleration of the extraction kinetics. The equilibrium pyrene concentration is increased only in the system with rape oil methyl ester. The kinetics of the systems with $C_{12/14}EO_{10}$ are not influenced significantly by added oil.

VI. CONCLUSIONS

Finally, the following conclusions on HOC sorption can be drawn.

For HOCs, physical nonequilibrium and/or intraorganic matter diffusion are probably the major contributors to sorption nonideality in natural colloidal systems. Chemical nonequilibrium may be important under certain conditions as in the case of low organic carbon content of the sorbent or in the case of ionic as well as highly polar sorbates.

Physical nonequilibrium is caused by diffusive mass transfer resistances associated with immobile regions such as intra-aggregate porosity. Several model formulations exist that can be used to represent this system. Use of these models for predictive purposes is partially limited by the inability to determine parameter values independently. Among the parameters most difficult to evaluate are intra-aggregate porosity and aggregate radius. Stochastic approaches are a means of estimating these parameters. A strong argument can be made for the occurrence of intraorganic matter diffusion, which may be responsible or partially responsible for rate-limited sorption–desorption in some cases.

Relative to nonequilibrium, nonlinearity and nonsingularity may not be significant contributors to HOC-sorption nonideality. However, nonlinearity or nonsingularity may be important under certain conditions. Experiments to assess nonsingularity must be carefully designed, considering the many causes of apparent hysteresis. It appears that in some cases nonsingularity may result from rate-limited desorption caused by constraints on diffusional mass transfer. Thus physically caused nonsingularity may be considered an effect of sorption/desorption nonequilibrium [6].

Where surfactant and HOCs coexist in soil–water systems there are a number of possible interactions considering the distribution of surfactant between monomeric, hemimicellar, and micellar forms. There is the partitioning of HOC among soil hydrophobic adsorption sites, surfactant micelles, and hemimicelles, and the competition for active hydrophobic adsorption sites between the surfactant and HOC.

The interaction of HOCs with surfactant monomers is usually very weak and insignificant. At concentrations where micelles and hemimicelles are present, interactions can take place. Sorbed HOCs can be solubilized by free micelles, resulting in mobilization. HOCs in solution are in equilibria between sorption onto hydrophobic adsorptive sites on the soil, partitioning into hemimicelles—both resulting in immobilization—and partitioning into free micelles. Whether the HOCs are previously sorbed onto soil or are in solution, partitioning into micelles, hence mobilization, is favored by increasing surfactant concentration. Sorbed surfactant molecules tend to increase HOC sorption onto soil by increasing its fractional organic carbon content, and free surfactant tends to decrease sorption by increasing the apparent aqueous solubility of the HOC [23].

However, the effective CMC in environments required to produce micelles in pore water of soils and sediments is generally much higher than in clean water systems, and the determined surfactant concentrations are much lower. Therefore it is unlikely that anthropogenic surfactants only will contribute significantly to the mobilization of HOCs in soils and sediments, but in combination with biosurfactants their influence might be very pronounced. On the other hand, in the case of soil remediation by surfactant solutions, the above results on HOC partitioning are to be observed.

An improvement of soil remediation by surfactants is to be expected by further decreasing interfacial tension and increasing wettability of soil or sediment with the aid of microemulsions as a washing solution. Successful experiments are proceeding in this field [35].

NOMENCLATURE

a:	surface area/volume (L^2/L^3)
C:	solution–phase solute concentration (mol/L^3 or M/L^3)
D:	diffusion coefficient (L^2/T)
ϵ:	effectiveness factor characterizing the carbon-normalized capacity of sorbed nonionic surfactant to function, relative to humic matter, as a sorbent for HOC (dimensionless)
e:	index "equilibrium"
f:	fraction of sorption sites (dimensionless)
i:	index "component" or "domain" or "mass fraction" or "particle class" or "reaction" or "sample" or "solute"
K:	partition coefficient (dimensionless)
$k_{a,s}$:	pseudo-first-order solute degradation rate constant for solute loss in the solid phase (1/T)
k_a:	pseudo-first-order solute degradation rate constant for solute loss in the solution phase (1/T)
k_b:	boundary layer mass transfer rate constant (L/T)
K_d:	distribution coefficient (L^3/M)
k_{eff}:	effective rate constant (1/T)
k_f:	forward first-order mass transfer rate constant (1/T)
K_F:	Freundlich isotherm parameter (L^3/M)
K_{ow}:	octanol–water partition coefficient (dimensionless)
k_r:	reverse first-order mass transfer rate constant (1/T)
k_t:	first-order mass transfer rate constant (1/T)
l:	characteristic diffusion length (L)
m:	mass (M) or index "multi-solute system"
m_e:	solute mass sorbed in equilibrium at the solid phase (M)
m_s:	mass of the solid phase (M)
M:	molecular weight (M/mol)
N:	total number
n:	Freundlich isotherm parameter (dimensionless) or number of moles
θ:	volume fraction (L^3/L^3)
q:	solid-phase adsorbate concentration (mol/M or M/M)

Q: sorbent-phase [14] solute concentration (mol/M or M/M)
ρ: particle density or solid phase bulk density (M/L^3)
r: radial position in the aggregate (L)
R: radius of the spherical solid particle (L)
r_{diss}: dissolution rate (M/(L^{3}T))
s: index "soil" or "sediment" or "solid phase" or "single-solute system"
S: solubility (mol/L^3)
sf: shape factor (dimensionless)
S_L^i: aqueous solubility of pure subcooled PAH i at the temperature of the solution (mol/L^3)
S_∞: equilibrium solubility (mol/L^3)
t: time (T) or index "mass transfer"
t_c: cloud point (°C)
$t_{1/2}$: half-life (T)
τ: tortuosity factor (dimensionless)
T: absolute temperature (K)
V: bulk solution volume (L^3)
V^i: volume of domain i (L^3)
$V_{\text{m,aq}}$: molar volume of water (L^3/mol)
w: mass fraction (M/M)
x: mole fraction (mol/mol)
y: mole fraction (mol/mol)
Y_{mic}: concentration of surfactant in micellar form in the bulk solution (M/L^3)
Annotation: L signifies "liter" in the text but "length" in this nomenclature section

REFERENCES

1. G. Uehara and G. Gillman, *The Mineralogy, Chemistry, and Physics of Tropical Soils with Variable Charge Clays*, Westview Press, Boulder, Colorado, 1981, p. 15.
2. C. Thibaud-Erkey, J. F. Campagnolo, and A. Akgerman. Separation and Purification Methods *24*(2):129 (1995).
3. J. J. Pignatello and B. Xing. Environ. Sci. Technol. *30*(1):1 (1996).
4. M. L. Brusseau and P. S. C. Rao. Chemosphere *18*(9/10):1691 (1989).
5. D. R. Cameron and A. Klute. Water Resour. Res. *13*:183 (1977).
6. M. L. Brusseau and P. S. C. Rao. Critical Reviews in Environmental Control *19*(1):33 (1989).
7. K. H. Coats and B. D. Smith. J. Soc. Pet. Eng. *4*:73 (1964).
8. M. Th. van Genuchten and P. Wierenga. Soil Sci. Soc. Am. J. *40*:473 (1976).
9. M. T. van Genuchten and R. J. Wagenet. Soil Sci. Soc. Am. J. *53*:1303 (1989).
10. T. B. Culver, S. P. Hallisey, D. Sahoo, J. J. Deitsch, and J. A. Smith. Environ. Sci. Technol. *31*(6):1581 (1997).
11. M. L. Brusseau, R. E. Jessup, and P. S. C. Rao. Environ. Sci. Technol. *25*(1):134 (1991).
12. C. T. Miller and J. A. Pedit. Environ. Sci. Technol. *26*(7):1417 (1992).
13. W. J. Weber, Jr., P. M. McGinley, and L. E. Katz. Environ. Sci. Technol. *26*(10): 1955 (1992).
14. J. A. Pedit and C. T. Miller. Environ. Sci. Technol. *28*(12):2094 (1994).
15. W. Brutsaert. Soil Sci. *101*:85 (1966).
16. R. A. Freeze. Water Resource. Res. *11*:725 (1975).
17. J. Shorter, *Correlation Analysis in Organic Chemistry: An Introduction to Linear Free-Energy Relationships*, Clarendon Press, Oxford, 1973.
18. B. D. Struck, I. Sistemich, and R. Pelzer. Progr. Colloid Polym. Sci. *95*:119 (1994).

19. J. P. DiVincenzo and S. K. Dentel. J. Environ. Qual. *25*:1193 (1996).
20. A. T. Kan, G. Fu, and M. B. Tomson. Environ. Sci. Technol. *28*(5):859 (1994).
21. P. M. McGinley, L. E. Katz, and W. J. Weber, Jr. Environ. Sci. Technol. *27*(8):1524 (1993).
22. E. Klumpp, B. D. Struck, and M. J. Schwuger. Nachr. Chem. Tech. Lab. *40*(4):428 (1992).
23. S. D. Haigh. Sci. Total Environ. *185*:161 (1996).
24. D. A. Edwards, Z. Liu, and R. G. Luthy. J. Environ. Eng. *120*(1):5 (1994).
25. D. A. Edwards, Z. Adeel, and R. G. Luthy. Environ. Sci. Technol. *28*(8):1550 (1994).
26. M. J. Schwuger. Kolloid-Z. u. Z. Polymere *250*:703 (1972).
27. I. T. Yeom, M. M. Ghosh, C. D. Cox, and K. G. Robinson. Environ. Sci. Technol. *29*(12):3015 (1995).
28. L. S. Lee, P. S. C. Rao, and I. Okuda. Environ. Sci. Technol. *26*:2110 (1992).
29. I. T. Yeom, M. M. Ghosh, and C. D. Cox. Environ. Sci. Technol. *30*(5):1589 (1996).
30. M. H. Geheln and F. C. DeSchryver. Chem. Rev. *93*:199 (1993).
31. S. J. Grimberg, J. Nagel, and M. D. Aitken. Environ. Sci. Technol. *29*:1480 (1995).
32. K. Thalberg, B. Lindman, and K. Bergfeldt. Langmuir *7*:2893 (1991).
33. D. C. Duffy, P. B. Davies, and A. M. Creeth. Langmuir *11*:2931 (1995).
34. N. V. Sastry, J.-M. Sequaris, and M. J. Schwuger. J. Colloid Interface Sci. *171*:224 (1995).
35. K. Bonkhoff, M. J. Schwuger; and G. Subklew, in *Industrial Applications of Microemulsions* (C. Solans and H. Kunieda, eds.), Surfactant Science Series 66, Chapter 17, Marcel Dekker, New York, 1997, pp. 355–374.
36. K. Stickdorn and M. J. Schwuger. Tenside Surfactants Detergents *31*(4):218 (1994).
37. M. J. Schwuger and K. Stickdorn. Chem. Rev. *95*:849 (1995).
38. K. Mönig, F.-H. Haegel, and M. J. Schwuger. Tenside Surf. Det. *33*(3):228 (1996).

20

Application of Surfactants in Liquid Printing Inks

RYSZARD SPRYCHA and RAMASAMY KRISHNAN Technical Center, Sun Chemical Corporation, Carlstadt, New Jersey

I.	Introduction	699
II.	Printing Processes Using Liquid Inks	700
	A. Flexography	701
	B. Gravure	702
	C. Ink-jet printing	702
	D. Liquid toners	702
III.	Effect of Surfactants on Some Processes Essential for Printing	704
	A. Surface free energy of liquids	704
	B. Surface free energy of solids	710
	C. Adhesion	713
	D. Wetting, spreading, and leveling of liquids on solids	714
	E. Stability of dispersions	717
	F. Adsorption of surfactants at the solid–liquid interfaces	722
	G. Polymer/surfactant interactions	725
	References	732

I. INTRODUCTION

Most printing inks used in major printing processes such as lithography, gravure, and flexography contain organic solvents. Some of these solvents, e.g., toluene, are toxic and can be harmful to humans. Both environmental and workplace safety considerations exert growing pressure on the printing industry to limit the use of toxic organic solvents. Due to this pressure, the share of water-based liquid printing inks in packaging printing (including corrugated) has achieved a respectable level of about 50%. In newspaper printing the share is estimated at about 10 to 15% of the total. To prepare for a possible ban on the use of toxic organic solvents, the printing industry is exploring the viability of water-based technology.

The challenge is enormous, as high print quality and low cost printing both have to be met. High-quality printing with solvent-based inks has been achieved over years of research and pressroom practice. Using water-based inks requires a total reformulation of the existing inks because of contrasting properties of the solvents used, i.e., water vs. toluene.

One of the factors important for good printability of liquid inks is low surface tension; low surface tension is indispensable for good ink spreading. Solvent borne inks have inherently low surface tension because of the solvents used (e.g., toluene, alcohol, etc.). Water, on the other hand, has a high surface tension ~ 72 dyn/cm, but the surface tension of water-based inks can be lowered by the addition of the appropriate surfactants, usually nonionic surfactants that ensure low dynamic surface tension (DST). Though the main purpose of adding surfactants to water-based inks is the lowering of their surface tension, the effect of surfactant presence in the system is much more complex.

It is well known from the literature that surfactant molecules show a high tendency to adsorb at the water/air interface and to self-associate in aqueous solutions [1-5]. If species other than surfactant are present in the system (polymer molecules, pigment particles, etc.), the surfactant molecules may self-associate, adsorb at the pigment surface, or interact with polymer molecules, depending on which process will be energetically more favorable. A knowledge of the interactions between surfactant molecules and other ingredients of water-based inks is interesting from both a theoretical and a practical point of view, as it allows for a better understanding of the system and subsequent optimization of its properties. In addition, it can help to change the existing situation, where a large number of commercial products containing complex mixtures of surfactants, polymers, pigments, and other additives are formulated only on an empirical trial-and-error basis.

The purpose of this chapter is to discuss the importance of surfactants in the graphic arts industry and their effect on processes essential for good printability. For readers not familiar with printing processes, a quick overview of major printing processes is presented in Section II. Application of surfactants in liquid inks and some fundamental processes related to printing with liquid inks, such as dynamic surface tension of surfactant solutions; wetting, spreading and leveling of liquids on solids; adhesion; adsorption of surfactants at the solid/liquid interface; stability of dispersions; and polymer/surfactant interactions, are discussed in Section III.

Printing is a very dynamic process—high speed gravure or flexography printing presses can print at speeds up to 3,000 feet per minute (~ 15 m/s). Therefore, to relate the experimental data on surface tension, for example, to the printing process, the data has to be acquired under dynamic conditions. Unfortunately, most of the literature data refers to equilibrium conditions.

II. PRINTING PROCESSES USING LIQUID INKS

Generally, printing processes can be divided into two groups: impact printing, where a significant force (pressure) has to be applied to transfer the ink from the image area to the substrate, and nonimpact printing, where such pressure is not necessary. Among major impact printing processes, gravure and flexography use liquid inks, while lithography and letterpress use paste inks. Nonimpact printing processes like ink-jet printers use low-viscosity liquid inks, while in electrophotography both liquid and dry toners are used. The fundamentals on printing processes using liquid inks are given below with particular emphasis on gravure and flexographic processes. A detailed description of different printing processes can be found elsewhere [6-10].

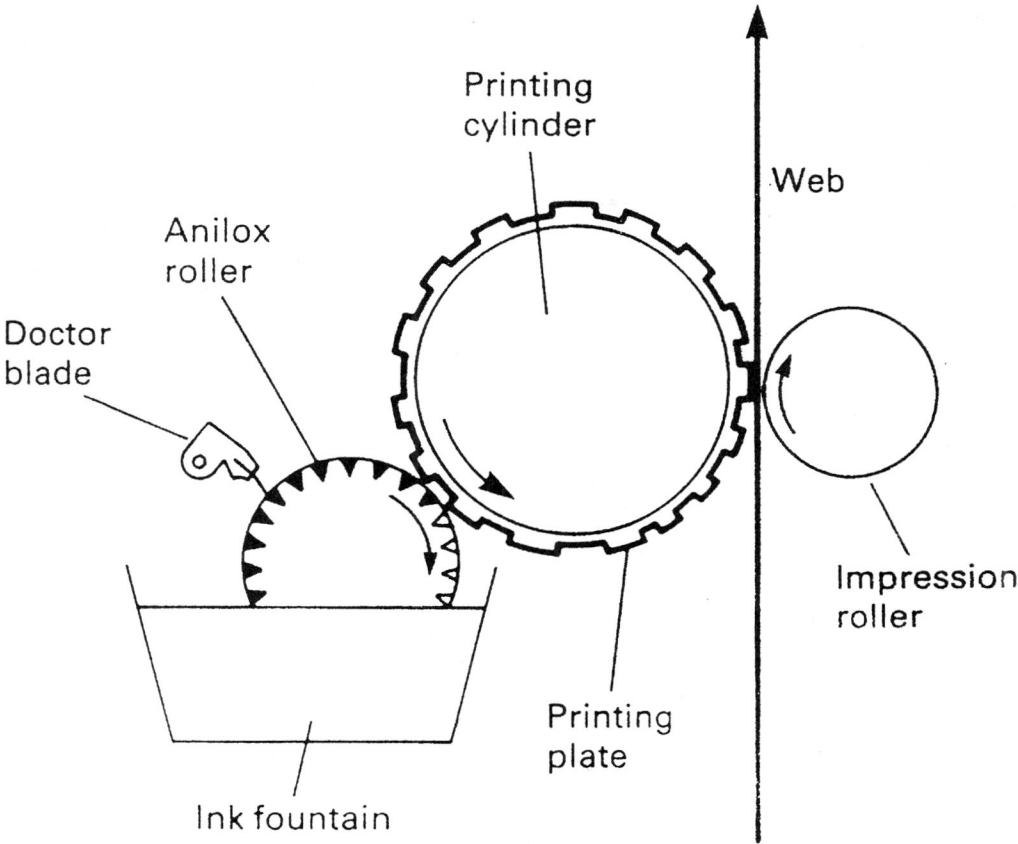

FIG. 1 Schematic picture of a flexographic printing unit. (From Ref. 6 with kind permission from Kluwer Academic Publishers.)

A. Flexography

Flexography is a relief printing process, i.e., the image area is raised above the nonimage area. A flexographic printing unit is presented schematically in Fig. 1. Liquid ink (solvent or water-based) from the ink fountain is transferred to the anilox roller, which contains small cells on its surface. These cells can be engraved in ceramic or metal and may have different sizes to carry the appropriate amount of ink to the printing cylinder. Any excess of the ink on the anilox roller is removed by a doctor blade. Ink from the anilox cells is then transferred to the raised image areas of the printing plate. During the printing process, ink from the image areas of the plate is transferred to the substrate in the nip between printing cylinder and impression roller. The pressure in the nip is relatively low (so-called "kissing" pressure).

The plate cylinder is made of steel and the printing plates are mounted onto this cylinder using clamps and adhesives. The most popular plates are those made of photopolymers. During the development stage, the light-sensitive photopolymer is exposed to UV light through a negative film. The photopolymer solidifies only in irradiated image areas. The polymer can be easily removed during plate processing from the nonimage areas. Flexographic printing is commonly used in packaging printing (plastic films and paper), and some newspapers are also printed with flexographic inks.

B. Gravure

In the gravure printing process, the image area is placed below the plate surface in the form of tiny cells engraved into a metal cylinder. A gravure printing unit is presented schematically in Fig. 2. The printing cylinder rotates in an ink fountain in liquid (solvent or water-based) ink. A doctor blade removes the excess ink from the cylinder so that only the ink in the cells can be transported to the nip (between the printing cylinder and the impression roller) where ink transfer from the cells to the substrate takes place.

The gravure cylinder is made of steel plated with copper (relatively thick layer). The image is engraved in the copper layer mostly by the electromechanical method, and finally the cylinder is plated with a thin chromium layer to protect the image areas against wear during printing. The pressure used in the nip is much higher than that in flexographic printing. Gravure is traditionally used for long-run printing of magazines, plastic films, and metal foils for the packaging industry.

C. Ink-Jet Printing

Ink-jet printing is a nonimpact computer-driven process in which tiny ink droplets are directly deposited on the substrate to form an image. The process involves small droplets of ink being ejected from a fine nozzle by applying high-pressure pulses. Different printers are available on the market using continuous ink-jet or drop-on-demand processes. In the continuous ink-jet process tiny droplets of ink are generated in a steady process and special electronic circuits will direct some droplets onto the substrate to form an image and others to the drain to be recirculated. Drop-on-demand systems generate ink droplets only if they are needed. Examples of such designs are bubble-jet printers or printers using piezoelectric crystals to generate the pressure impulse. The printing head consists usually of a number of nozzles.

Ink-jet inks are mainly dye-based liquids of low viscosity. The making of pigment-based ink-jet inks is more difficult, but in recent years such inks were formulated and are available on the market. Ink-jet printing is much slower than flexography or gravure processes (only about 200–400 fpm i.e., 1–2 m/s) and is limited by the width of the substrates used. Many computer printers use this technology to print black-and-white or color images.

D. Liquid Toners

In printing processes using liquid toners the image is at first created as a pattern of static charges on a photoconductive plate. The latent image is then developed in liquid toner, which consists of polymers and colloidal pigment particles dispersed in very low conductivity liquids, e.g., hydrocarbons. Pigment particles are stabilized by adsorption of so-called "charge control agents" on their surface [11–13]. Charged pigment particles are transported by electrophoresis to the charged sites of the plate (latent image) immersed in liquid toner. Pigment particles deposited on the plate are subsequently transferred to the substrate to form the image. Black and white as well as color images can be produced by this technique, but printing speed is relatively low (300–600 fpm i.e., 1.5–3 m/s). The method is used commercially to print mainly small jobs.

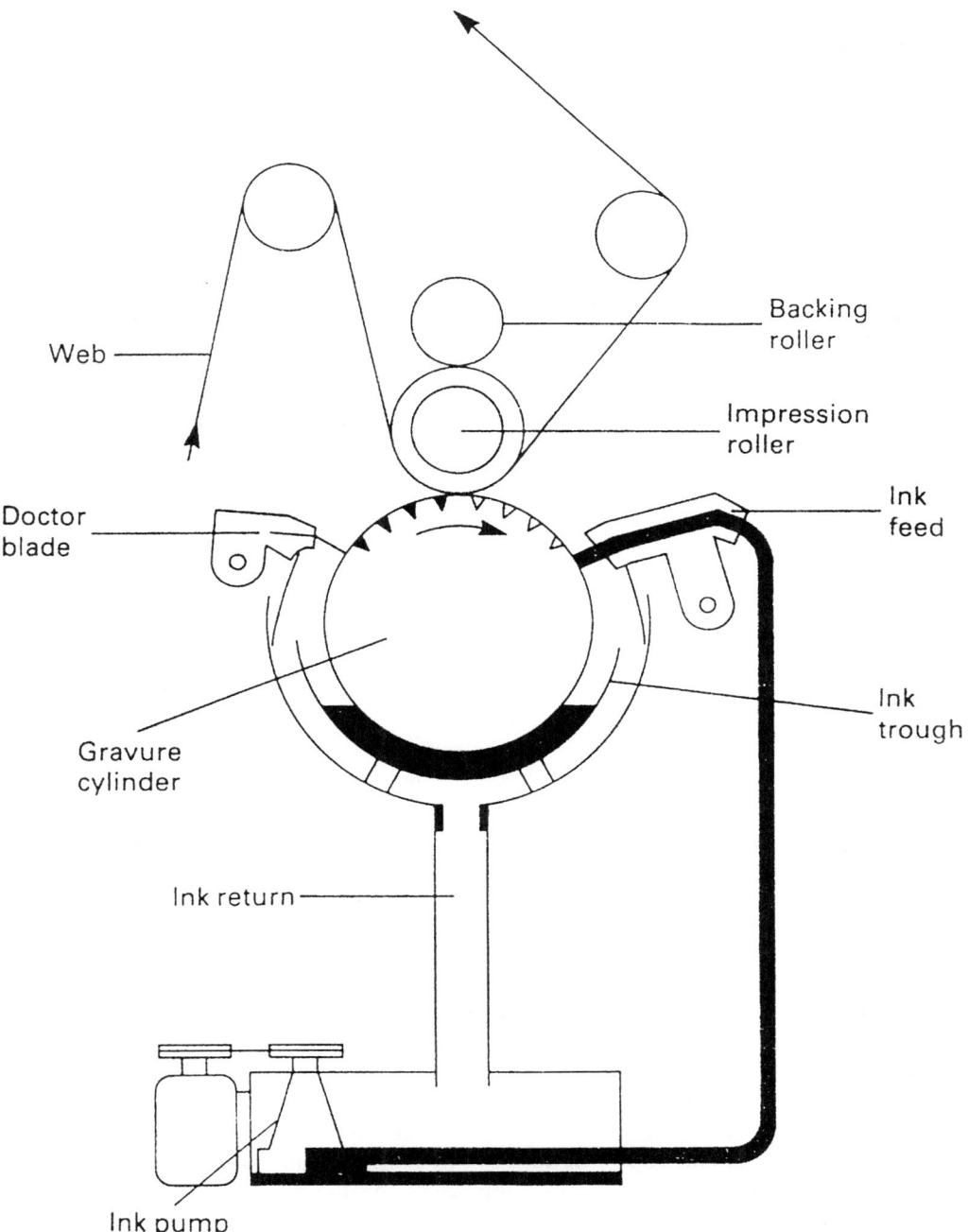

FIG. 2 Schematic picture of a gravure printing unit. (From Ref. 6 with kind permission from Kluwer Academic Publishers.)

III. EFFECT OF SURFACTANTS ON SOME PROCESSES ESSENTIAL FOR PRINTING

Printing with liquid inks is a typical example of the application of surface and colloid chemistry, since the surface properties of printed substrates have to be known to select the appropriate ink to achieve the best print quality. As well as the substrate, the surface properties of pigment particles are extremely important in making stable high-performance inks.

Liquid printing ink, which is a highly concentrated colloidal system, consists of solvent (hydrocarbons, alcohols, water, etc.), appropriate resins that are soluble or dispersible in a given solvent, colloidal-size pigment particles used as colorants, and additives (surfactants, rheology modifiers, defoamers, waxes, etc.). The level of different ink components ranges from 50–70% for solvents, 15–35% for resins, 5–15% for pigments, and 0–4% for additives. Though ink additives are added to the ink in smaller amounts than basic ink ingredients, they can have a profound impact on the printing process. One example of additives widely used in printing inks and coatings is surfactants. In this chapter the fundamentals of the most important phenomena from a printing point of view will be addressed, and the effect of surfactants on these processes will be discussed.

A. Surface Free Energy of Liquids

The surface free energy of liquids (also called surface tension or interfacial tension at the liquid/air interface) results from different energy states between molecules in the bulk of the liquid and molecules at the air/liquid interface. Forces acting in the bulk on molecules are perfectly balanced, and hence the net force is zero. However, surface molecules experience more interactions from other molecules in the bulk of the liquid than from the air phase. Therefore a net force will be exerted on the surface molecules. This force, which is perpendicular to the liquid surface and directed toward the liquid, will tend to minimize the surface area of the liquid. To create a new surface, work therefore has to be done against the surface tension. Thus the value of the surface tension of liquids is expressed as an amount of work needed to create one unit area of new surface, in erg/cm^2 (dyne/cm) or J/m^2 (N/m). The surface tension of pure liquids under constant external conditions depends on their chemistry.

Surfactant molecules, due to their specific hydrophylic–lypophilic structure, show a high tendency to adsorb at the air/water interface. The presence of adsorbed surfactant molecules in the interfacial region will change the surface tension of surfactant solution, which will be lower than the surface tension of water. A typical relationship between the surface tension of surfactant solution and equilibrium surfactant concentration is presented in Fig. 3. As seen from Fig. 3, the surface tension of surfactant solution decreases with increasing surfactant concentration, but above a certain concentration it remains constant—the critical micelle concentration (cmc). At the cmc the adsorption density of surfactant molecules at the air/water interface reaches a maximum (plateau). Above the cmc, surfactant molecules will form micelles in water, since no more space is available for surfactant adsorption at the air/water interface. Because micelles are surface inactive (the outer shell of a micelle consists of polar or ionic hydrophilic groups), the surface tension of surfactant solution stays unchanged.

The type of surfactant also influences the value of the cmc. For nonionic surfactants, interactions between hydrocarbon chains of surfactant molecules are not disturbed by any forces trying to prevent micelle formation. However, for ionic surfactants, repulsive electrostatic interactions between charged groups will act against attractive forces between

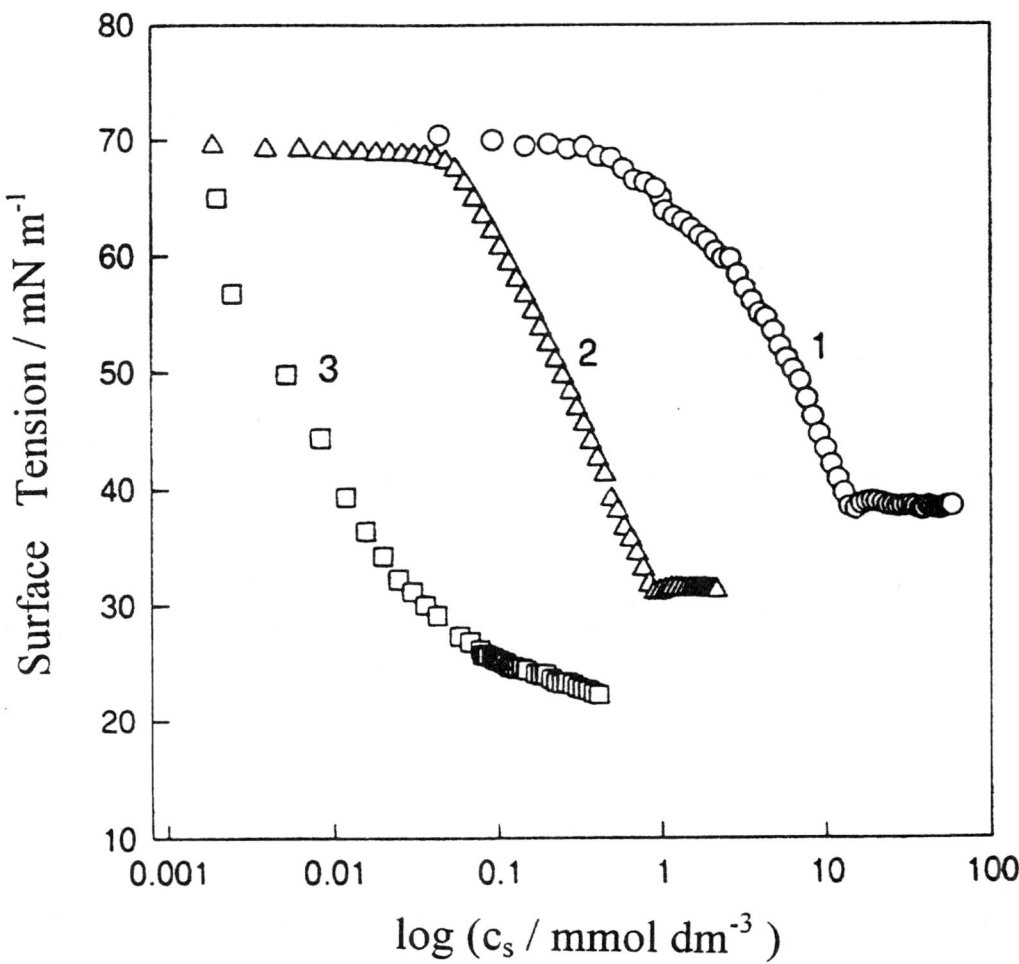

FIG. 3 Surface tension vs. concentration of cationic surfactant in water. 1, dodecyltrimethylammonium bromide; 2, dimeric 1,2-bis(dodecyldimethylammonio)ethane dibromide; 3, trimeric methyldodecylbis(2-dimethyldodecylammonio)ethyl ammonium dibromide. (Reprinted with permission from Ref. 20—Langmuir, Vol. 12(16), August 1996. Copyright © 1996 American Chemical Society.)

hydrocarbon chains. Thus higher concentrations of ionic surfactant will be required to form micelles in solution. Indeed, the typical values of cmc for nonionic surfactants are between 10^{-5} and 10^{-4} mol dm^{-3}, while the equivalent cmc values for ionic surfactants are 10^{-3} to 10^{-1} mol dm^{-3} [1,5,14–20]. Because of an abrupt change in value of surface tension of surfactant solution at the cmc, surface tension measurements are a convenient way to determine the values of the cmc for different surfactants.

Surfactant adsorption density at the air/water interface can be determined from the slope of the γ_l vs. log C_s plot in the Gibbs equation [1,2,14]:

$$\Gamma_s = -\frac{1}{2.3RT}\left(\frac{\partial \gamma_l}{\partial \log C_s}\right)_T \tag{1}$$

where Γ_s is the surface excess concentration of surfactant, R is the gas constant, T is the temperature, γ_l is the surface tension of a solution, and C_s is the surfactant concentration. This equation can be used for diluted solutions of nonionic surfactants and for 1 : 1 ionic surfactants in the presence of a large excess of indifferent electrolyte containing the same counterion as the surfactant. In the absence of an indifferent electrolyte for 1 : 1 ionic surfactant, its dissociation has to be taken into account [1]. Thus Eq. (1) has the form

$$\Gamma_s = -\frac{1}{4.6RT}\left(\frac{\partial \gamma_l}{\partial \log C_s}\right)_T \quad (2)$$

As mentioned above, the lowest surface tension is reached at the cmc, and its value depends on the surfactant type. For most hydrocarbon surfactants, the surface tension value at the cmc ranges from 30 to 40 mN/m, and only with fluorosurfactants can one achieve surface tension values in the range 17–20 mN/m.

The surface tension of surfactant solutions can be measured directly by a number of methods. Some of them are used to measure surface tension under equilibrium conditions and others under dynamic conditions. Time plays a very important role in these surface tension measurements. The transfer of surfactant molecules or ions to a newly formed surface during the measurement is controlled by the diffusion process; diffusivity of different surfactants varies considerably depending on their chemical structure. Some surfactants, which can achieve very low equilibrium (static) surface tension, are very poor performers under dynamic conditions. A number of the methods to determine surface tension used most frequently by ink markers are discussed briefly below. A more detailed description of these methods can be found elsewhere [2,15,16].

1. Surface Tension Measurements Under Static Conditions

(a) Du Noüy Ring. In this method a microbalance is used to measure the force required to detach a small metal (platinum) ring immersed in a liquid from the air/liquid interface. As a first approximation, the force required for detachment is equal to

$$F = mg + 4\pi r \gamma_l \quad (3)$$

where m is the mass of the ring, g is the acceleration due to gravity, and r is the ring radius.

Because the surface tension acts along the ring perimeter on both sides ($2\times 2\pi r$), a factor of 2 was used in Eq. (3) to calculate the contribution from the surface tension. For more accurate measurements (using a Du Noüy ring), corrections to Eq. (3) have to be introduced due to a small amount of liquid remaining on the ring after detachment [2,15,16]. The Du Noüy ring method is widely used to measure the static surface tension of liquids. Care must be taken to keep the ring free of contaminants. During detachment the ring should be perfectly horizontal. This method may be inaccurate when used for systems that attain equilibrium very slowly or when liquid viscosity is high.

(b) Wilhelmy Plate. This method uses a microbalance to measure the force exerted on the plate after the plate edge touches the surface of the liquid. By analogy to the Du Noüy ring method, if the contact angle of a liquid is zero, the force is [2,16]

$$F = mg + 2(x+y)\gamma_l \quad (4)$$

where x and y are the length and width, respectively, of the plate edge in contact with the liquid. Thus the surface tension can be calculated directly from Eq. (4). In general, for a finite value of contact angle, Eq. (4) has the form

$$F = mg + 2(x+y)\gamma_l \cos\theta \tag{5}$$

where θ is the contact angle of the liquid on the plate. A commercial instrument using this technique is available on the market (DCAA, by Cahn, USA).

(c) Drop Weight Method. This is a very simple and accurate method to determine surface tension under equilibrium conditions. Small drops of liquid are formed at the tip of capillary tube, and their weight is measured using a balance. The weight of a single drop is given by the equation [2,16]

$$mg = 2\pi r_c \gamma_l f \tag{6}$$

where r_c is the radius of the capillary and f is a correction factor needed because part of the droplet remains attached to the capillary during the detachment process. The factor f is a function of the capillary radius and the droplet volume. The surface tension of the liquid can be calculated directly from Eq. (6).

2. Surface Tension Measurements Under Dynamic Conditions

The term "dynamic surface tension" was first introduced to describe the time effect on the surface tension of surfactant solutions at constant surfactant bulk concentration. Rosen et al. [21–25] distinguished four characteristic regions on the DST vs. log t plot for aqueous surfactant solution—see Fig. 4. They also found that the following empirical equation describes the effect of time on DST:

$$\frac{\gamma_{l0} - \gamma_{lt}}{\gamma_{lt} - \gamma_{lm}} = \left(\frac{t}{t^*}\right)^n \tag{7}$$

where: γ_0 is the surface tension of pure water, γ_t is the surface tension of surfactant solution at the surface age (t), γ_m is the so-called mesoequilibrium surface tension of surfactant solution (γ_t changes insignificantly with t), t is the time, and t^* and n are constants dependent on the surfactant concentration and its structure.

An example of a plot of DST vs. surface lifetime for different surfactant concentrations is presented in Fig. 5. As observed, the DST decreases with increasing surfactant concentration, but for constant surfactant concentration, DST increases with decreasing lifetime of the surface. The latter phenomenon is due to surfactant properties and depends on how fast surfactant molecules can migrate and adsorb at the interface. Though the curves have similar shapes for different surfactants, the time scale may vary significantly. Details on DST of surfactant solutions can be found in the literature [21–30].

As mentioned previously, high-speed printing is a very dynamic process. In gravure printing, for example, the time required for the ink to travel from the doctor blade, where the new surface is formed, to the nip, where ink is transferred to the substrate, is about 10–40 milliseconds, depending on the press speed. In order to correlate the surface tension measurements of the printing ink to the real on-press situation, the measurements should be performed over the relevant time scale. Hence the lifetime of a surface newly formed by the instrument should also be 10–40 ms. Thus the first three regions represented in Fig. 4 are very important for high-speed printing. Presently, the term DST is used to describe the value of surface tension of any liquid measured under nonequilibrium conditions. Commercially available instruments can be used to perform such measurements.

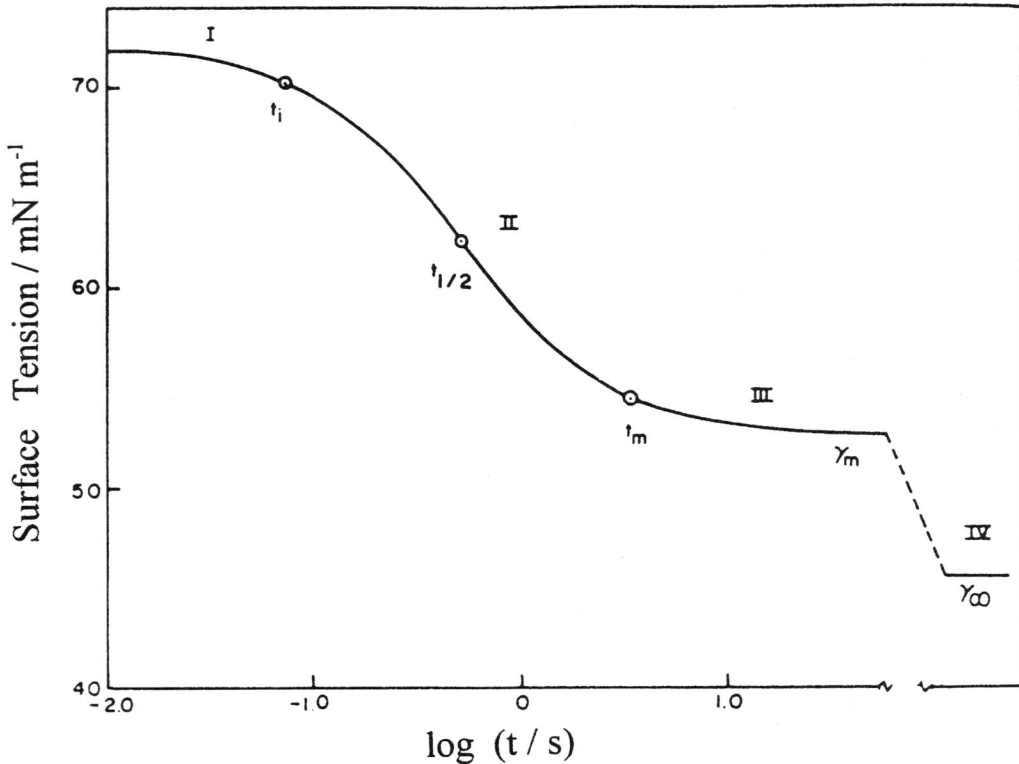

FIG. 4 Schematic plot of dynamic surface tension vs. log t for surfactant solution in water. Region I, induction; region II, rapid fall; region III, mesoequilibrium; region IV, equilibrium. (From Ref. 21. Copyright © 1988 Academic Press, Inc.)

(a) Maximum Bubble Pressure Method. The idea of measuring surface tension using a maximum bubble pressure (MBP) technique is based on measurements of pressure required to generate a bubble at the tip of a small capillary. To generate the bubble at the tip of a capillary immersed in liquid, work has to be done against the surface tension of the liquid and the hydrostatic pressure (dependent on the immersion depth), so that the excess external pressure is

$$\Delta P = \rho g h_i + \frac{2\gamma_l}{r_c} \tag{8}$$

where ρ denotes the liquid density and h_i the immersion depth.

As results from the mechanism of bubble growth [2,31–33], its radius reaches a minimum when it is equal to the radius of the capillary. At the same time, according to Eq. (8), the value of the pressure reaches a maximum (the immersion depth is constant). This maximum value of pressure is recorded by the instrument and used to calculate the value of the surface tension under a given experimental condition. Such an instrument, which

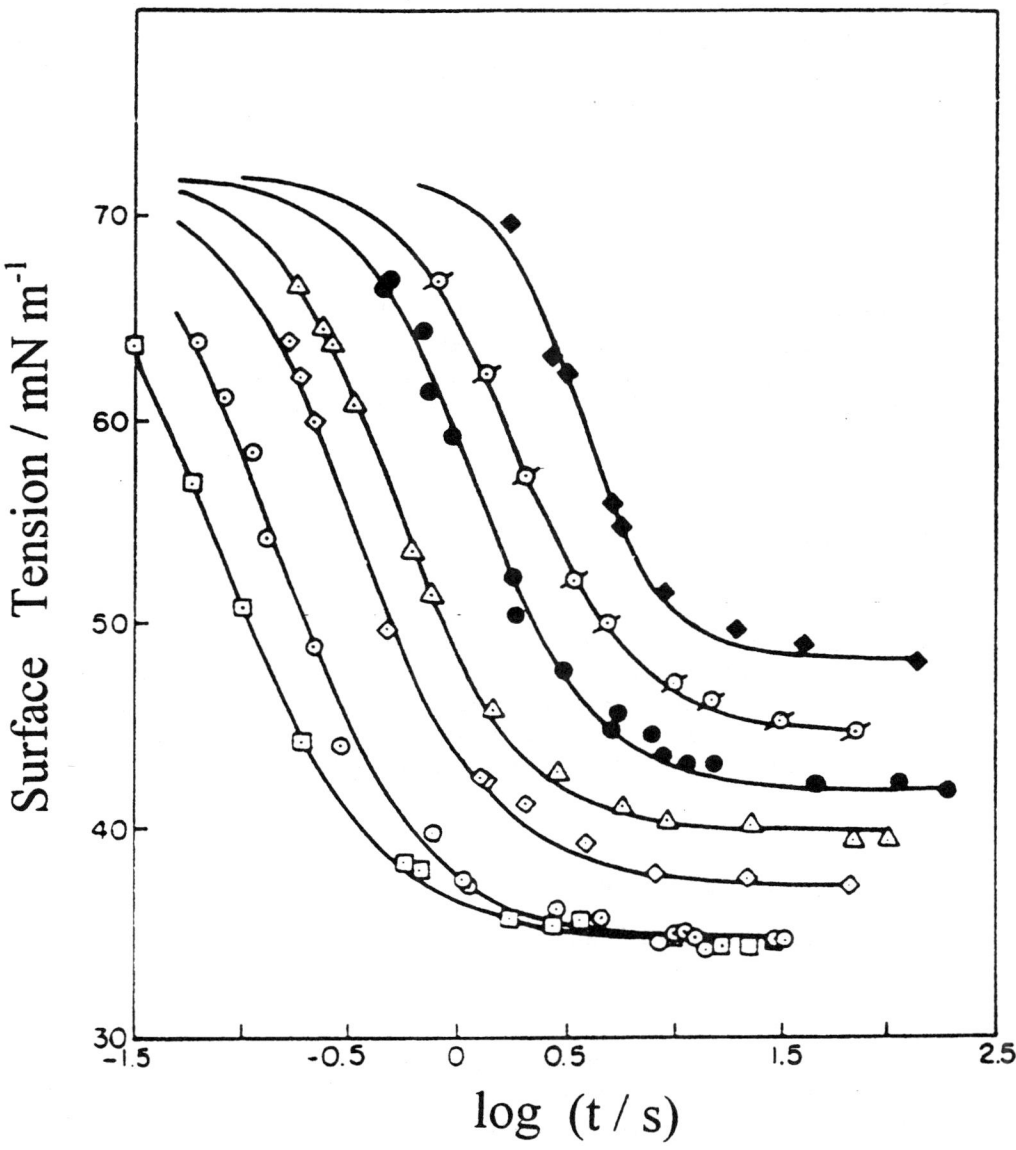

FIG. 5 DST vs. log t for different concentrations of $C_{12}BMG$ (*N*-dodecyl-*N*-benzyl-*N*-methylglycine) surfactant in water (pH = 9.0; T = 298 K). Log C_5 equal to -2.992 (□), -3.224 (⊙), -3.410 (⬥), -3.525 (△), -3.701 (●), -3.826 (⊙), -4.108 (◆). Symbols—experimental points. Solid lines—computer fit. (From Ref. 21. Copyright © 1988 Academic Press, Inc.)

is based on this description, is the BP-2, by Krüss, USA. It can measure the surface tension for surface lifetimes as low as 5 milliseconds. Another instrument, which satisfies the same description (MPT1, by Lauda, Germany), can measure the dynamic surface tension of surfactant solutions in the submillisecond range [34].

(b) Differential Maximum Bubble Pressure (DMBP) Method. This method is a modification of the MBP technique. The bubbles are generated (at constant gas flow) from two capillaries of different radii immersed in a liquid, and the pressure difference between the two capillaries is measured. By analogy to Eq. (8), the pressure difference is

$$\Delta P = \left(\rho g h_i + \frac{2\gamma_{l1}}{r_{c1}}\right) - \left(\rho g h_i + \frac{2\gamma_{l2}}{r_{c2}}\right) \tag{9}$$

Sensadyne 6000 and 9000 instruments, by Chem-Dyne Research Corp., USA, are based on this description, but they use simplified theoretical equations to calculate the DST. Assuming that $\gamma_{l1} = \gamma_{l2} = \gamma_l$ (this condition is not fulfilled for surfactant solutions because lifetimes of bubbles at two orifices bubbling at different rates are different), Eq. (9) can be simplified to

$$\Delta P = \frac{2\gamma_l}{r_{c1}} - \frac{2\gamma_l}{r_{c2}} \tag{10}$$

Schramm [35] showed that the use of simplified theoretical equations might be a source of significant errors. In most cases, however, measurements are accurate enough for practical applications. A very detailed description of methods presented above as well as other methods to measure surface tension (capillary rise, sessile and pendant drops, growing drop, oscillating jet) can be found elsewhere [2,15,16,36–44].

B. Surface Free Energy of Solids

The terms "surface free energy" and "surface tension" are commonly used interchangeably in the literature, as well as in this paper, in application to liquids and solids. However, it should be emphasized that using these terms interchangeably for solids is not entirely correct [2,14,16].

In contrast to liquids, where surface tensions can be easily measured, there is no direct method to measure the surface free energy of solids. Therefore an indirect method must be used to determine the surface energy of solids. The most common technique is to measure the contact angles of various liquids (of known characteristics) on the respective solid sample. Measuring contact angles is easy, but interpretation and calculation of the surface energy of the solid is far from simple. A knowledge of the surface energy of printing substrates is very important for printers because it affects such processes as substrate wetting by ink and ink spreading and adhesion.

The shape of a liquid droplet deposited on a solid is determined by the balance of forces acting at three different interfaces: solid/air—γ_s interfacial tension, liquid/air—γ_l interfacial tension, and solid/liquid—γ_{sl} interfacial tension. At equilibrium, the relationship between the above interfacial tensions and the contact angle can be expressed by Young's equation [1,2,14,16].

$$\gamma_l \cos\theta = \gamma_s - \gamma_{sl} \tag{11}$$

Application of Surfactants in Liquid Printing Inks

Though Young's equation refers to ideal systems, it is widely used in practice for real systems. Because only γ_l and θ can be measured independently, a few different approaches were proposed to eliminate γ_{sl} from Eq. (11).

1. Critical Surface Tension

By collecting a large number of contact angle measurements of homologous series of liquids on different solids, Zisman et al. [45–49] observed the following relationship between the contact angle and the surface tension of the liquid:

$$\cos\theta = 1 + a(\gamma_c - \gamma_l) \tag{12}$$

where: a is a constant for a given series of liquids and γ_c is the critical surface tension of the solid. The value of γ_c can be evaluated by extrapolation of the plot $\cos\theta$ vs. γ_l to $\cos\theta = 1$ ($\theta = 0$). The assumption made in this approach is that $\gamma_{sl} \rightarrow \gamma_s$. The value of γ_c was assumed to be a measure of the surface free energy of the solid.

The idea of "dyne solutions" used in the graphic arts industry to evaluate surface energy of plastic films for printing purposes is based on this description. This method, though widely used in practice, has some limitations. The value of γ_c for a given solid depends on the properties of the test liquids used [47]. The method is limited to systems where hydrogen bond and acid–base interactions at the solid/liquid interface are absent, i.e., to hydrophobic surfaces only. Moreover, Zisman's method cannot be applied to evaluate the surface energy of solids when using solutions of different surface tensions (e.g., alcohol–water mixtures) instead of pure liquids [16,50,51].

2. Two-Liquid Method

Early theories [52–56] assumed that the surface free energy of a solid could be split into a number of components originating from different types of intermolecular interactions. For practical purposes the total energy was simplified to the sum of two components, dispersive γ_s^p: and nondispersive ("polar") γ_s^p:

$$\gamma_s = \gamma_s^d + \gamma_s^p \tag{13}$$

It was assumed that the dispersive component is due to dispersive London interactions between molecules, while γ^p results from such interactions as dipole–dipole, dipole-induced dipole, and hydrogen bonding. To eliminate γ_{sl} from Eq. (11), this value can be expressed by using γ_l and γ_s.

(a) Geometric Mean Approach. It was proposed [52–55] that the interfacial tension γ_{sl} be expressed in the form

$$\gamma_{sl} = \gamma_l + \gamma_s - 2(\gamma_l^d \gamma_s^d)^{1/2} - 2(\gamma_l^p \gamma_s^p)^{1/2} \tag{14}$$

By substituting Eq. (14) into Eq. (11) and rearranging, one obtains for liquids 1 and 2,

$$\gamma_{l1}(1 + \cos\theta_1) = 2(\gamma_{l1}^d \gamma_s^d)^{1/2} + 2(\gamma_{l1}^p \gamma_s^p)^{1/2} \tag{15}$$

$$\gamma_{l2}(1 + \cos\theta_2) = 2(\gamma_{l2}^d \gamma_s^d)^{1/2} + 2(\gamma_{l2}^p \gamma_s^p)^{1/2} \tag{16}$$

Eqs. (15,16) are the basis for the calculation of the surface energy of solids using contact angle data of two liquids for which γ_l^d and γ_l^p components are known. By solving Eqs. (15,16), γ_s^d and γ_s^p components can be calculated and then the total energy γ_s from Eq. (13).

(b) Harmonic Mean Approach. Another approach to eliminate γ_{sl} from Eq. (11) was the use of the harmonic mean equation proposed by Wu [57–60]

$$\gamma_{sl} = \gamma_s + \gamma_l - 4\frac{\gamma_1^d \gamma_s^d}{\gamma_1^d + \gamma_s^d} - 4\frac{\gamma_1^p \gamma_s^p}{\gamma_1^p + \gamma_s^p} \tag{17}$$

By substituting Eq. (17) into Young's equation and rearranging one obtains Eqs. (18) and (19) for two different liquids:

$$\gamma_{l1}(1 + \cos\theta_1) = 4\frac{\gamma_{l1}^d \gamma_s^d}{\gamma_{l1}^d + \gamma_s^d} + 4\frac{\gamma_{l1}^p \gamma_s^p}{\gamma_{l1}^p + \gamma_s^p} \tag{18}$$

$$\gamma_{l2}(1 + \cos\theta_2) = 4\frac{\gamma_{l2}^d \gamma_s^d}{\gamma_{l2}^d + \gamma_s^d} + 4\frac{\gamma_{l2}^p \gamma_s^p}{\gamma_{l2}^p + \gamma_s^p} \tag{19}$$

By analogy to the "geometric mean" method, using two liquids of known γ_l^d and γ_l^p allows for the calculation of γ_s^d and γ_s^p components from Eqs. (18,19).

Methods a and b, which use two test liquids to evaluate the surface energy of solids, have been criticized and presently are assumed to be invalid [50,51]. In spite of these criticisms, for practical purposes, both methods are still widely used to evaluate the surface energies of different substrates [61–70].

3. Acid–Base Approach

The modern approach to the evaluation of the surface energy of solids was initiated by Fowkes [61,71,72] and developed by Good and Van Oss [73–81]. According to the acid–base theory [51,75–81] the total surface energy consists of two dominant terms:

$$\gamma_s = \gamma_s^{LW} + \gamma_s^{AB} \tag{20}$$

where γ^{LW}, the Lifshitz–Van der Waals component, is the combined contribution due to all electromagnetic interactions, and γ^{AB} is the energy component due to acid–base interactions.

The approach uses the electron pair donor–acceptor (Lewis) theory of acids. To calculate the surface energy of solids using acid–base theory, the contact angles of three liquids of known γ^{LW}, γ^+ (acid), and γ^- (base) components on a given solid surface have to be known. Then the surface energy components of the solid can be calculated from the following set of equations [51,76,80]:

$$\gamma_{l1}(1 + \cos\theta_1) = 2\left[\left(\gamma_s^{LW}\gamma_{l1}^{LW}\right)^{1/2} + \left(\gamma_s^+ \gamma_{l1}^-\right)^{1/2} + \left(\gamma_s^- \gamma_{l1}^+\right)^{1/2}\right] \tag{21}$$

$$\gamma_{l2}(1 + \cos\theta_2) = 2\left[\left(\gamma_s^{LW}\gamma_{l2}^{LW}\right)^{1/2} + \left(\gamma_s^+ \gamma_{l2}^-\right)^{1/2} + \left(\gamma_s^- \gamma_{l2}^+\right)^{1/2}\right] \tag{22}$$

$$\gamma_{l3}(1 + \cos\theta_3) = 2\left[\left(\gamma_s^{LW}\gamma_{l3}^{LW}\right)^{1/2} + \left(\gamma_s^+ \gamma_{l3}^-\right)^{1/2} + \left(\gamma_s^- \gamma_{l3}^+\right)^{1/2}\right] \tag{23}$$

where: γ^+ is the acid parameter of surface free energy and γ^- the base parameter of surface free energy. It is recommended that three liquids be selected, two of which should be polar (e.g., water, glycerol). The correct selection of liquids is important because it may happen that unreliable results for γ_s^+ and γ_s^-, due to mathematical difficulties during the solving of Eqs. (21–23), can be obtained [51,82].

Evaluation of the surface energy of some substrates (e.g., papers) used in the graphic arts industry can be difficult due to specific substrate properties. Contact angles of liquids on paper, especially absorptive papers, are not easy to measure, and the measurements can be questioned. On the other hand, there is no better method to evaluate the surface energies of substrates used in the printing industry than via contact angle measurements of well-characterized liquids. One has to be aware, however, that this is an indirect method of measurement and that the accuracy, when using difficult substrates, can sometimes be unsatisfactory. There are a large number of published papers dealing with the determination of the surface energy of solids [45–90]. Detailed information on this subject can be found in several excellent review papers [51,75,81].

C. Adhesion

A knowledge of adhesion plays a very important role in many practical applications. Strong adhesion may be a desired phenomenon for some applications like adhesives, glues, lamination, printing, or an unwanted phenomenon for others, e.g., cleaning, laundry, particle removal. For printers it is extremely important that the printed ink film adhere firmly to the substrate. Adhesion depends strongly on the properties of both substrate and ink. Different theories have been proposed to explain adhesion, but the two more generally accepted are mechanical adhesion and chemical adhesion [2,14].

In the former case, adhesion is due to the mechanical interlocking of tiny asperities at the interface. This mechanism is generally important for rough and porous surfaces. For example, liquid ink may penetrate into large irregular pores of uncoated paper and upon drying the ink film is "mechanically held" by the paper even if intermolecular adhesional forces acting across the ink/paper interface are relatively weak. In the latter case, adhesion is due to the chemical bonding that takes place at the interface. This type of adhesion is especially important for smooth surfaces such as plastic films, heavy-coated papers, and coated boards, which are used in the printing industry.

Thermodynamically adhesion is defined as the amount of work needed to separate two unit areas of surfaces by overcoming the intermolecular interactions across the interface. At the solid/liquid interface the work of adhesion is defined by the equation [1,2,14,16,51]

$$W_a = \gamma_s + \gamma_l - \gamma_{sl} \tag{24}$$

As seen from Eq. (24) for maximum adhesion both γ_s and γ_l should be as high as possible and the value of γ_{sl} should be as low as possible. If the liquid forms a droplet of finite contact angle on the surface of the solid, the work of adhesion is [1,2,14]

$$W_a = \gamma_l(1 + \cos\theta) \tag{25}$$

Thus, using different approaches to the interfacial tensions, the work of adhesion can be defined by Eqs. (15,16), (18,19) or (21–23), respectively, because the right-hand side of Eq. (25) is the same as the left-hand sides of the aforementioned equations. Eq. (25) is valid only for $\theta > 0$. For $\theta = 0$ one obtains

$$W_a = 2\gamma_l = W_c \tag{26}$$

and under this condition the work of adhesion (W_a) is equal to the work of cohesion (W_c). In practice, it is impossible to evaluate the energy of adhesion from Eq. (25) if the surface of the solid is completely wetted by a liquid ($\theta = 0$), because then the work of adhesion is equal to or greater than the work of cohesion [1,91]:

$$W_a \geq 2\gamma_l \tag{27}$$

If both solid and liquid are monopolar surfaces, i.e., $\gamma_s^+ = \gamma_l^+ = 0$ (base–base) or $\gamma_s^- = \gamma_l^- = 0$ (acid–acid), Eq. (21) takes the form

$$\gamma_l(1 + \cos\theta) = 2(\gamma_s^{LW}\gamma_l^{LW})^{1/2} \tag{28}$$

In such a case the adhesion is only due to Lifshitz–Van der Waals interactions between two surfaces. For base–acid or acid–base surfaces, i.e., if $\gamma_s^+ = 0$; $\gamma_s^- > 0$ and $\gamma_l^+ > 0$; $\gamma_l^- = 0$ or $\gamma_s^+ > 0$; $\gamma_s^- = 0$ and $\gamma_l^+ = 0$, $\gamma_l^- > 0$ then only one of the right-hand side terms in Eqs. (21–23) will be equal to zero. The same is true when a monopolar surface interacts with a bipolar surface [51].

The work of adhesion defined by Eqs. (24 and 25) is ideal, and it cannot be met in practical systems. In practice the adhesion is usually a few times weaker than the ideal adhesion. Calculations show [14] that even if dispersion interactions only operate at the interface, one should obtain a strong adhesive bond. Reality proves that such "joints" are not satisfactory. Therefore, for strong adhesion, chemical interactions (hydrogen bond, covalent bond, etc.) have to be involved in the process. This is the reason that the force of adhesion of ink to the substrate depends strongly on the properties of both the ink and the substrate.

As results from Eq. (24) for strong adhesion, the surface tension of the solid, γ_s, should be high. In many cases, printing on plastic films is troublesome because the surface energy of such films is too low (usually in the range 20–30 mNm^{-1}). Generally, the surface energy of polymers can be increased by special treatment of their surface.

Chemical treatment with strong acids/strong bases, strong solvents, though quite effective, is seldom used due to environmental issues. The most popular processes of surface modification of polymers are those that use plasma treatment. In the printing industry, corona treatment is the most widely used technique because it operates in air at atmospheric pressure [92,93]. The main effect of using corona treatment is surface oxidation and formation of functional surface groups such as –C=O or –COOH. Dependent on the extent of treatment, the surface energy of treated polymers can be increased from 20–30 mNm^{-1} to 35–55 mNm^{-1} [64]. In some cases, corona treatment cannot be used because of the sensitivity of the product to corona discharge, e.g., electronic circuits, and in these circumstances flame treatment (H_2; O_2) of the surface is used before printing [94].

Surfactants can affect the process of adhesion via their effect on γ_l and γ_{sl} values. According to Eqs. (24 and 25) the best conditions for strong adhesion are when γ_l is high and $\gamma_{sl} \approx 0$ (or $\theta \approx 0°$). A detailed description of different adhesion theories can be found elsewhere [14,95].

D. Wetting, Spreading, and Leveling of Liquids on Solids

Wetting is the phenomenon of displacing one fluid from the surface of the solid substrate by another fluid. During printing, air is removed from the surface of the substrate and replaced with printing ink. Rosen [1] distinguished three types of wetting.

1. *Wetting by spreading* takes place when, for instance, small droplets of liquid in contact with a substrate spread due to intermolecular interactions across the solid/liquid interface. This type of substrate wetting is partially involved in gravure and ink-jet printing. To characterize the spreading tendency of a given liquid on a given substrate,

the so-called spreading coefficient (S_{ls}) can be calculated according to the formula [1,14]

$$S_{ls} = \gamma_s - (\gamma_{sl} + \gamma_l) \tag{29}$$

If $S_{ls} > 0$, spreading will occur spontaneously, and for $S_{ls} < 0$, no spontaneous spreading is observed. Equation (29) can be easily applied to the system when one liquid spreads on the surface of another liquid, because all interfacial tensions can be measured directly. For a solid/liquid system, the spreading coefficient has to be evaluated indirectly by contact angle measurements. Here the spreading coefficient can be defined as [1,14]

$$S_{ls} = \gamma_l(\cos\theta - 1) \tag{30}$$

For $\theta > 0°$ the value of the spreading coefficient is smaller than zero, $S_{ls} < 0$, because (cos $\theta - 1$) is negative. The spreading coefficient will be equal to or greater than zero only if $\theta = 0°$, i.e., liquid will wet the substrate completely.

2. *Wetting by adhesion* takes place when a liquid contacts a solid substrate and adheres to it. This type of substrate wetting is the most important in the printing industry and dominates all major printing processes such as lithography, letterpress, flexography, and gravure. The process of adhesional wetting is described by the work of adhesion defined by Eqs. (24,25). According to Eq. (25), W_a can never be negative because cos $\theta \leq \pm 1$. The difference between the work of adhesion, Eq. (25), and the work of cohesion, Eq. (26), equals the spreading coefficient [1]:

$$W_a - W_c = S_{ls} \tag{31}$$

If $W_a > W_c$, then $S_{ls} > 0$ ($\theta = 0°$) and complete wetting takes place spontaneously. If $W_a < W_c$, then $\theta > 0°$ and the liquid will form droplets of finite contact angle.

3. *Wetting by immersion* takes place when a solid is immersed completely in the liquid. This mechanism of wetting is important, for instance, for pigment dispersion manufacture. The difference ($\gamma_s - \gamma_{sl}$) is used to characterize immersional wetting [1]. The value of ($\gamma_s - \gamma_{sl}$) can be determined from Young's equation, Eq. (11). If $90° > \theta > 0°$, then ($\gamma_s - \gamma_{sl}$) is greater than zero. If $\theta > 90°$, then ($\gamma_s - \gamma_{sl}$) < 0, and therefore work has to be done to immerse the solid in the liquid. In the former case, wetting is spontaneous. If $\theta \leq 0°$, the value of ($\gamma_s - \gamma_{sl}$) cannot be determined from Eq. (11).

In water-based formulations, surfactants are added to improve substrate wetting. By changing γ_l and γ_{sl}, addition of surfactant to the system may change the spreading coefficient so that $S_{ls} > 0$. Though it is commonly accepted in industry that surfactants will help to wet solid substrates, under certain circumstances their addition to the system will have the entirely opposite effect on wettability. For instance, in wetting of porous material the pressure (ΔP) causing movement of the liquid into the capillary can be expressed as [1,2,14]

$$\Delta P = \frac{2\gamma_l \cdot \cos\theta}{r_c} \tag{32}$$

By substituting Eq. (11) into Eq. (32) one obtains

$$\Delta P = \frac{2(\gamma_s - \gamma_{sl})}{r_c} \tag{33}$$

As seen from Eqs. (32) and (33), if $\theta > 0°$ the value of ΔP depends on ($\gamma_s - \gamma_{sl}$). If surfactant addition does not affect the value of ($\gamma_s - \gamma_{sl}$) by lowering γ_{sl}, due to adsorption at the solid/liquid interface, then ΔP remains constant, and any change in γ_l is counterbalanced by increasing the value of cos θ. For liquids for which $\theta = 0°$ (cos $\theta = 1$) one obtains from Eq. (32)

$$\Delta P = \frac{2\gamma_l}{r_c} \qquad (34)$$

As seen from Eq. (34), adding surfactant and lowering γ_l will only decrease the tendency of a liquid to enter the capillary. This phenomenon is important when printing with water-based inks on porous (uncoated papers, corrugated) substrates.

To change the interfacial tensions, γ_l or γ_{sl}, surfactant molecules must adsorb at a given interface. In some cases, configuration of surfactant molecules in the interfacial region may strongly affect the wetting behavior. The orientation of surfactant molecules at the solid/liquid interface will be addressed in more detail in Section F.

The orientation of surfactant molecules at the solid/liquid interface, in such a way that the hydrophilic (ionic or strongly polar) segment of the surfactant is bonded to the solid surface while the hydrophobic segment is directed towards water, takes place mainly when ionic surfactants adsorb on the oppositely charged solid surface. Such adsorption makes the solid surface more hydrophobic and causes an increase in the γ_{sl} value and a decrease of the spreading coefficient. The area covered with a monolayer of adsorbed surfactant will be hydrophobic and hence very difficult to wet because of the low surface energy. This phenomenon is strongly desired in the mineral ore processing industry, using the flotation process [96], but is unwanted in the printing industry, where printing on already low-energy substrates is very common [64,68]. A detailed description of wetting and spreading in different systems can be found elsewhere [95–103].

The brief discussion presented above on wetting and spreading under equilibrium conditions has limited applicability to printing, which is a very dynamic process. Wetting of papers by liquids and spreading of liquids on paper under dynamic conditions have been studied by some researchers in recent years. Because a paper surface is rough and porous, liquid will penetrate into the irregular capillaries formed by paper fibers and/or coating. A number of papers have been published on wetting and penetration of liquids into capillaries and porous substrates [104–109]. In many cases the wettability of paper was judged by the amount of liquid transferred to the paper under given dynamic conditions using a Bristow wheel [110–113] or other techniques [114–116]. The existence of "wetting delay" from such measurements was postulated by some researchers [112,113] but refuted by others [110]. In our opinion, the so-called wetting delay, judged by the amount of liquid absorbed by paper, is in fact "sorption delay" [117].

The adhesion forces will require a certain amount of time to develop across the solid/liquid interface upon the contact of liquid with the paper surface. However, the time responsible for this real wetting delay would be very short, much less than 1 millisecond. Excellent ink transfer occurring in the nip of the printing press evidences this. The dwelling time on high-speed presses is about 1–3 milliseconds. To reduce wetting delay, surfactants can be used during paper manufacture. It has been shown that the application of the appropriate surfactant to a paper by spraying it onto the paper surface prior to the calendering process can significantly reduce the wetting delay and increase the rate of water absorption [118].

In wetting by spreading, topography of the substrate surface plays an important role. The surface roughness generally enhances the spreading of liquids when $\theta < 90°$ and inhibits spreading if $\theta > 90°$ [1,2,16]. However, the presence of mechanical barriers and sharp edges has a detrimental effect on liquid spreading and results in different values of macro and micro contact angles [119–122]. The effect of surface roughness on liquid spreading is rather complex and under certain conditions can be responsible for super-spreading (complete wetting) or super-repellency (complete nonwetting). Such phenomena have been observed and are described in the literature [123]. Uncoated papers, corrugated, and boards are examples of substrates whose surface is rough and contains a large number of micromechanical barriers and sharp edges that can have a negative effect on ink spreading.

In some printed areas, where a large amount of liquid ink is applied during printing, a continuous wet ink film is formed. The surface of such a film is irregular and will tend to smooth during the process of leveling. Leveling is a process of eliminating surface irregularities of a continuous liquid film under the influence of the liquid's surface tension. Leveling is an important step in obtaining a smooth and uniform ink film. This process is especially critical in the coating industry because in most cases a smooth coating is paramount. The actual process of leveling can be described by the equation [124]

$$l_s = \frac{16\Pi^4 h^3 \gamma_l \ln(a_1/a_0)}{3\lambda^3 \eta} \tag{35}$$

where l_s is the leveling speed, h the average film thickness, a_t and a_o are the final and initial amplitudes, respectively, λ is the wavelength, and η is the viscosity of the liquid.

Printers and ink makers often use the term leveling to describe ink spreading on the substrate. It should be emphasized that spreading and leveling are two totally different processes. These processes are schematically presented in Fig. 6. Though the surface tension of an ink or coating γ_l plays a very important role in both processes, the effect of surface tension on ink spreading, Eq. (29), is opposite to its effect on leveling, Eq. (35). To achieve good spreading, the surface tension of the ink should be as low as possible. For good leveling, the surface tension should be as high as possible. In practical applications, surfactant has to be carefully selected, and a compromise usually has to be reached regarding the γ_l value to achieve optimum leveling and spreading.

E. Stability of Dispersions

Generally, colloidal dispersions can be divided into two classes, lyophobic and lyophilic colloids [16,125]. All types of printing inks are examples of complex concentrated colloidal dispersions. Pigments and very often polymers are the dispersed phase. For good printability and color development, stability of such dispersions is a key issue. Stability can be achieved in different ways, by e.g. electrostatic stabilization of highly charged particles or steric stabilization by an adsorbed layer of polymer/surfactant on the particle surface. The most stable systems are dispersions stabilized by both mechanisms, i.e., electrosteric stabilization [2,14,16].

Stability of a colloidal dispersion is determined by the interactions between the particles as well as the interactions between particles and solution components. The classical DLVO theory of the stability of hydrophobic colloidal dispersions [1,2,14,16,125] assumes a balance between two types of forces, attractive forces, due to dispersive interactions between particles, and repulsive forces, which result from the overlapping of the electrical double layers. These decide about the stability of a given colloidal system. Attractive forces are universal forces acting between molecules and have an electromagnetic origin [1,2,14,16,125]. The source of repulsive forces is the existence of an electrical double layer at the particle/solution interface. The mechanisms by which the electrical

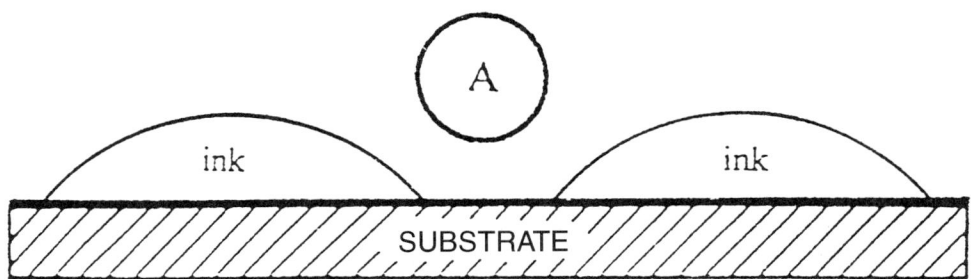

SPREADING

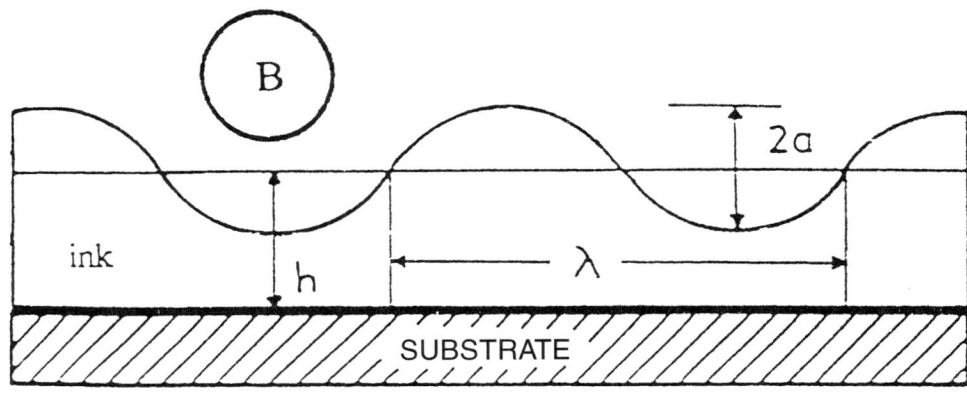

LEVELING

FIG. 6 Schematic presentation of liquid spreading (A) and leveling (B) on solids.

charge is formed at the particle surface may be different depending on the system. A more detailed description of the electrical charge and double layer origin in different systems can be found in Chapters 4 and 6 or in other papers [126–138].

Consider the interaction between two charge spheres of the same size (radius r_s); then the free energy of interaction due to attractive London dispersive forces can be represented by the equation [2,14,16]

$$\Delta G_a = -\frac{r_s A_{121}}{12d} \tag{36}$$

where A_{121} is the Hamaker constant [1,2,14,16] of particle 1 dispersed in liquid 2 and d is the distance between particles ($d \ll r_s$).

Application of Surfactants in Liquid Printing Inks

The energy of repulsive interaction between double layers on the particles at small distances of separation can be expressed as [2,14,16]

$$\Delta G_r = \frac{64 n \pi r_s k T \beta^2}{\kappa^2} \exp(-\kappa d) \tag{37}$$

The value of β is

$$\beta = \frac{\exp(ze\Psi/2kT) - 1}{\exp(ze\Psi/2kT) + 1} \tag{38}$$

where z is the ion charge, e the electron charge, k the Boltzman constant, T the temperature, Ψ the surface potential, and κ is

$$\kappa = \left(\frac{2nz^2 e^2}{\varepsilon \varepsilon_0 k T}\right)^{1/2} \tag{39}$$

where n is the number of ions of each type in the unit volume of the bulk solution, ϵ is the relative permittivity of the solvent, and ϵ_0 the permittivity of free space. The value of $1/\kappa$ is called the electrical double layer thickness. In fact this is a distance, x, from the surface at which the value of the potential Ψ_x is Ψ/e, where e is the base of the natural logarithms. Thus the total energy of interaction, $\Delta G_t = \Delta G_r + \Delta G_a$, is

$$\Delta G_t = \frac{64 n \pi r_s k T \beta^2}{\kappa^2} \exp(-\kappa d) - \frac{r_s A_{121}}{12d} \tag{40}$$

Eq. (40) describes how the energy of interaction between two spheres dispersed in liquid changes with the distance of separation, particle size, surface potential, and electrolyte concentration. Expressions for the energy of interactions for different geometries—two flat plates, flat plat and sphere, etc.—and detailed discussion on this subject can be found elsewhere [2,14,16].

The schematic plot of ΔG_a, ΔG_r, and ΔG_t vs. separation distance for two interacting spheres dispersed in liquid is presented in Fig. 7. The maximum value on the ΔG_t vs. distance plot represents the electrostatic barrier that keeps two particles apart and prevents coagulation. To stabilize the dispersion in this way, this electrostatic barrier has to be high enough. DLVO theory clearly illustrates the importance of electrical surface charges in the stabilizing of lyophobic colloidal dispersions. Though the DLVO theory of stability is simplified, it can roughly predict the effect of electrolyte concentration on the dispersion stability via the Schultze–Hardy rule [2,14,16,125]. In addition, it can explain the phenomenon of weak reversible flocculation at the "secondary minimum" [1,14,16]—see Fig. 7. The existence of this minimum is very important from a practical point of view. Particles flocculated in the secondary minimum can be very easily redispersed, but the structure formed is strong enough to prevent sedimentation and keep the dispersion (paint, printing ink, etc.) stable for a long time.

DLVO theory fails to predict the stability of lyophilic colloidal dispersions and, in some cases, the effect of surfactant addition on the stability of lyophobic systems [1,2,14]. For instance, if an ionic surfactant is added gradually to a dispersion containing particles of opposite charge, a decrease in dispersion stability is first observed due to a decreasing electrostatic (zeta) potential—see Fig. 8. At a certain surfactant concentration, the zeta potential is reduced to zero. Upon addition of more surfactant to the system, the zeta potential increases again with (opposite sign), and the dispersion is stable again due to ionic surfactant adsorption onto the particles precovered with a layer of surfactant,

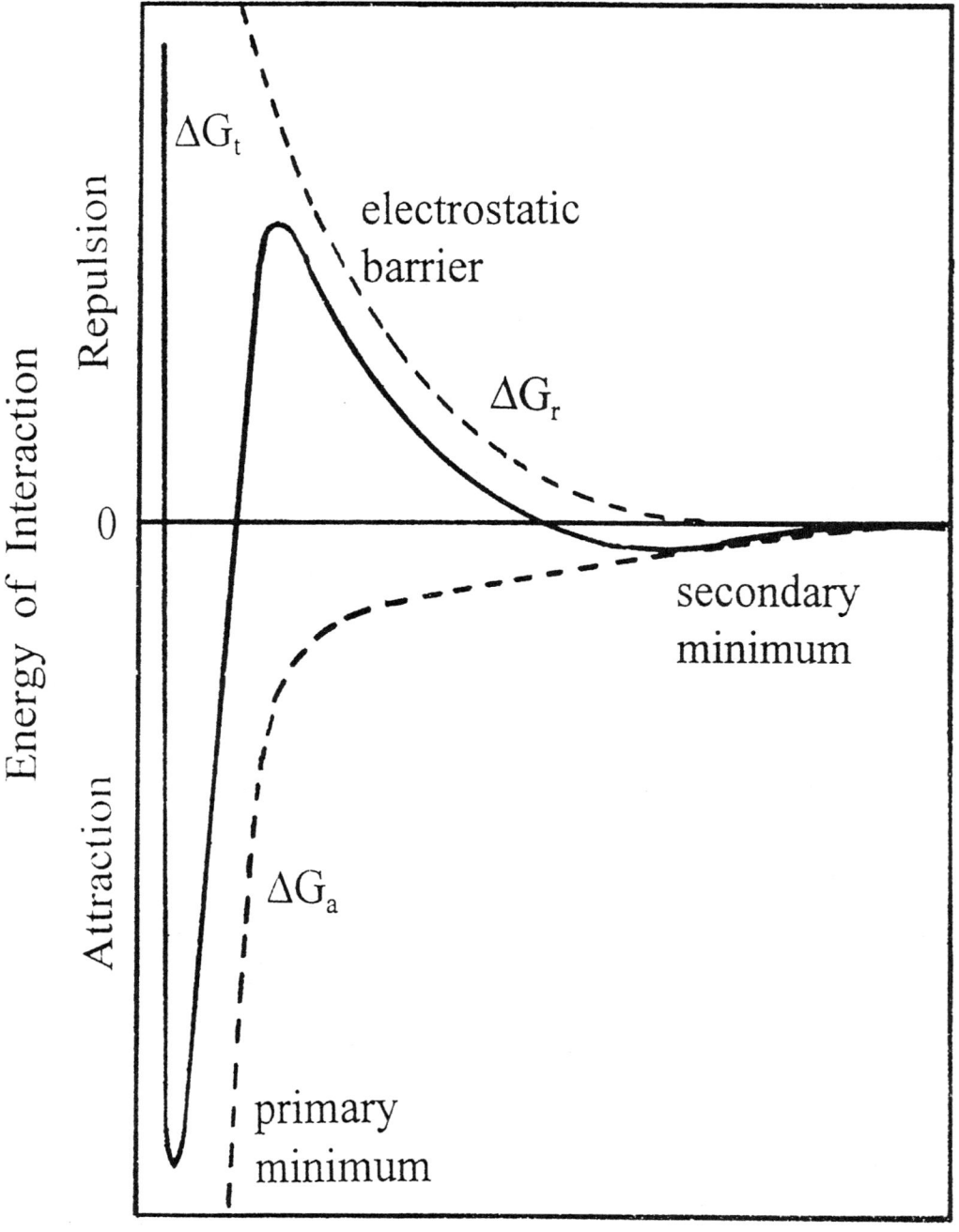

FIG. 7 Schematic presentation of attractive (ΔG_a), repulsive (ΔG_r), and total (ΔG_t) energy of interaction between two spherical particles dispersed in liquid as a function of separation distance—DLVO theory.

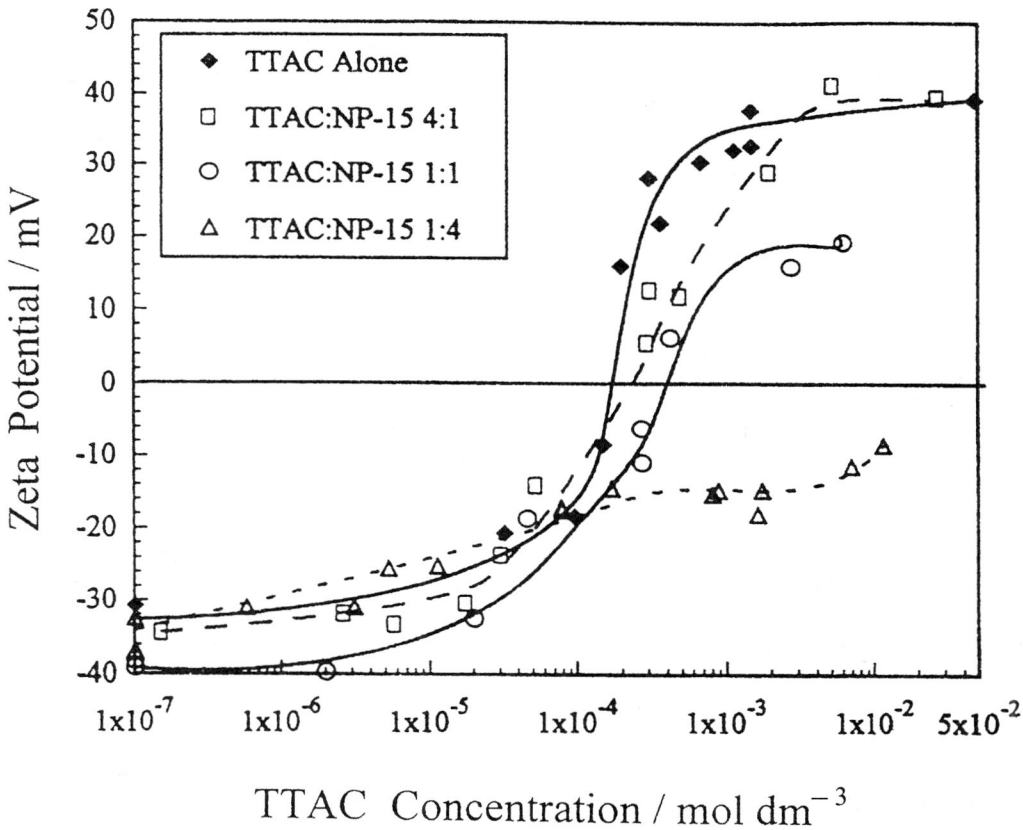

FIG. 8 Zeta potential of alumina particles as a function of cationic surfactant (tetradecyl trimethyl ammonium chloride, TTAC) concentration in the presence and absence of nonionic surfactant (pentadecylethoxylated nonylphenol, NP-15). Electrolyte solution, 0.03 mol dm^{-3} NaCl, (pH = 10). (From Ref. 149. Copyright © 1996 Academic Press, Inc.)

where molecules are oriented with ionic groups towards the particle and hydrophobic groups extended into the solution. A second layer of surfactant molecules with their ionic groups oriented towards the solution will be responsible for the particles' "charge reversal" [1,2,96]. The above example shows that surfactants may stabilize or destabilize the dispersion depending on their concentration in the system. Polymeric surfactants can also flocculate the dispersion by a bridging mechanism [1,2,14]. The above processes have wide practical applications.

Adsorption of polymeric ionic surfactants or polyoxyethylenated nonionic surfactants at the solid/liquid interface may create a layer of close-packed surfactant molecules on the particle surface and give the dispersion good stability even if the electrostatic barrier is very low or nonexistent. Such a mechanism is called steric stabilization [1,2,14,16,125,139,141]. Sterically stabilized aqueous and nonaqueous dispersions are widely used in practice. The stability of such dispersions depends strongly on solvent properties. Usually the solvent should show a high affinity towards this part of the adsorbed surfactant molecule, which extends into the solution. If the solvent interacts strongly with segments adsorbed directly on the particle surface, the bonds will be weakened and flocculation may result. Similarly, this may also occur if addition of the solvent causes thinning or a collapse of the protective layer of surfactant molecules, which extend into solution. More detailed information on the stability of dispersions, including concentrated systems, can be found in the literature [1,2,14,16,96,139–145]. As demonstrated above, the effect of surfactant on dispersion stability is very complex and depends on surfactant type as well as the mechanism of adsorption.

F. Adsorption of Surfactants at the Solid–Liquid Interfaces

As surfactants are commonly used, alone or together with some resins, in the printing ink industry to make stable dispersions of pigments with extremely different surface properties (for instance, very hydrophobic surfaces of phthalocyanine blue or carbon black and hydrophilic surfaces of titania or other pigments), careful selection of a surfactant for the respective pigment is important. Adsorption of surfactants at the solid/liquid interface is a complex process depending on the properties of surfactants as well as the solid surface and solution itself. In aqueous solutions, temperature, pH, and the presence of other additives may influence the properties of both surfactant and solid and thus the mechanism of adsorption. For oppositely charged species, e.g., anionic surfactant and positively charged solid surface, one can expect that electrostatic forces will dominate the interactions between surfactant ion and solid surface. For other systems the problem is more complex.

According to Rosen [1], different types of interactions can be involved in the adsorption of surfactants at the solid–liquid interface, namely ion exchange, ion pairing, hydrogen bonding, dispersion forces, and hydrophobic bonding. Nonelectrostatic interactions may be solely responsible for adsorption or may be only supplementary forces. The brief discussion on adsorption of surfactants at the solid–liquid interface presented in this paper is based chiefly on the properties of the solid surface.

1. Hydrophilic Charged Surfaces

An example of such a surface is titania pigment, whose surface is charged in aqueous solution, with the sign and value of the surface charge being dependent on pH. Only at the point of zero charge is the net surface charge equal to zero [126,127,130,131]. Adsorption of ionic surfactants on an oppositely charged surface is mainly due to electrostatic interactions. However, the adsorption process will not necessarily be terminated when the entire surface charge is counterbalanced by surfactant adsorption. Now, interactions between hydrophobic hydrocarbon chains of the surfactant, adsorbed onto the surface, and surfactant in the solution will promote further adsorption and cause charge reversal of the particle [146,147]. Because of this, the surfactant adsorption isotherm will be S-shaped [1,145–150] as presented in Fig. 9.

Detailed studies on the mechanism of adsorption of cationic surfactant on a mica surface in water have been published recently by Chen et al. [151]. They found that surfactant cations as well as counterions were involved in the initial stage of the adsorption process. Orientation of the surfactant chain at the interface depends on the surfactant surface coverage and surfactant concentration in the solution, as presented in Fig. 10. For charged surfaces, the pH of the solution will have a large impact on the extent of ionic surfactant adsorption because both the surface charge of the solid surface and the solubility of surfactant can be affected by the solution pH.

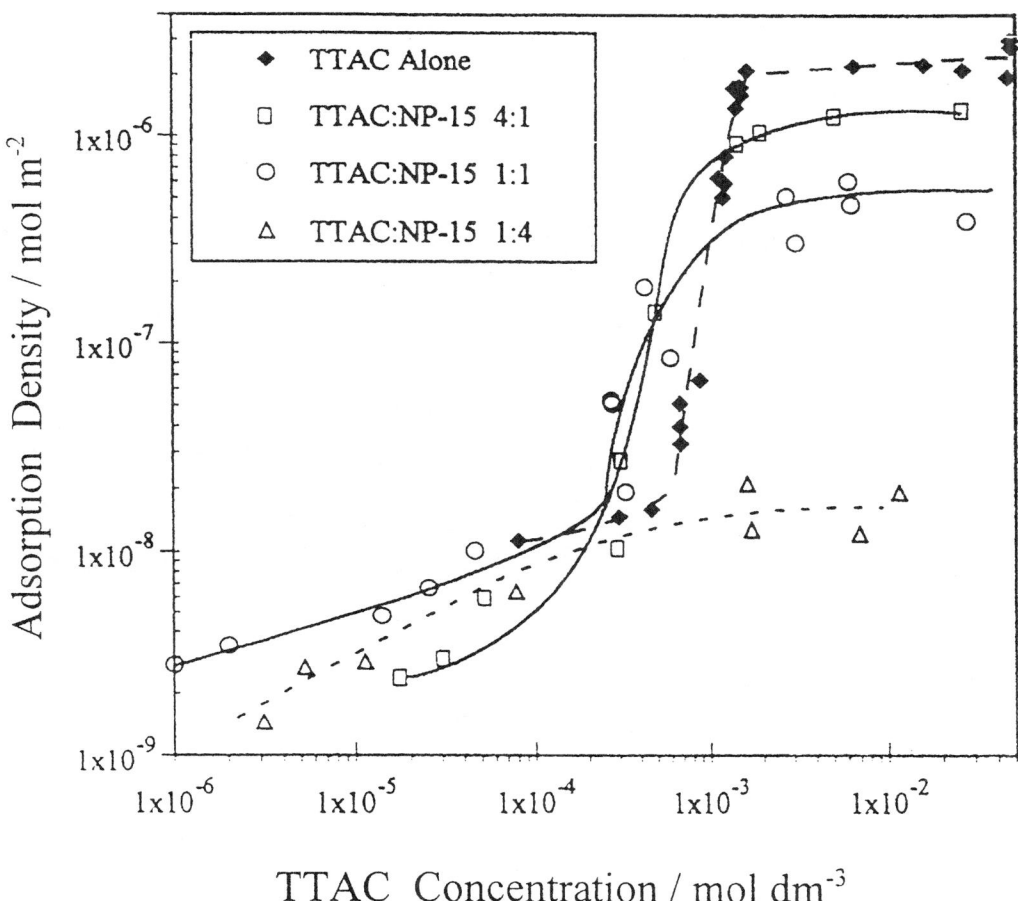

FIG. 9 Adsorption density of cationic surfactant (tetradecyl trimethyl ammonium chloride, TTAC) onto negatively charged alumina surface (pH = 10; NaCl concentration 0.03 mol dm^{-3}) vs. surfactant concentration, in the presence and absence of nonionic surfactant (pentadecylethoxylated nonylphenol, NP-15). (From Ref. 149. Copyright 1996 © Academic Press, Inc.)

Nonionic surfactants can be adsorbed at the charged surface, but in this case electrostatic interactions are not important. Instead, hydrogen bonds can be formed between surface hydroxyl groups (–MOH) and oxygen atoms in polyethoxylated nonionic surfactants [1,152]. The actual mechanism of adsorption of nonionic surfactants seem to be much more complex than that proposed for charged surface and ionic surfactants.

2. Hydrophilic Uncharged Surfaces

The adsorption of surfactants on hydrophilic, uncharged surfaces such as polyesters is determined by nonelectrostatic interactions. Dispersion forces and hydrogen bonding play a dominant role in such systems [1].

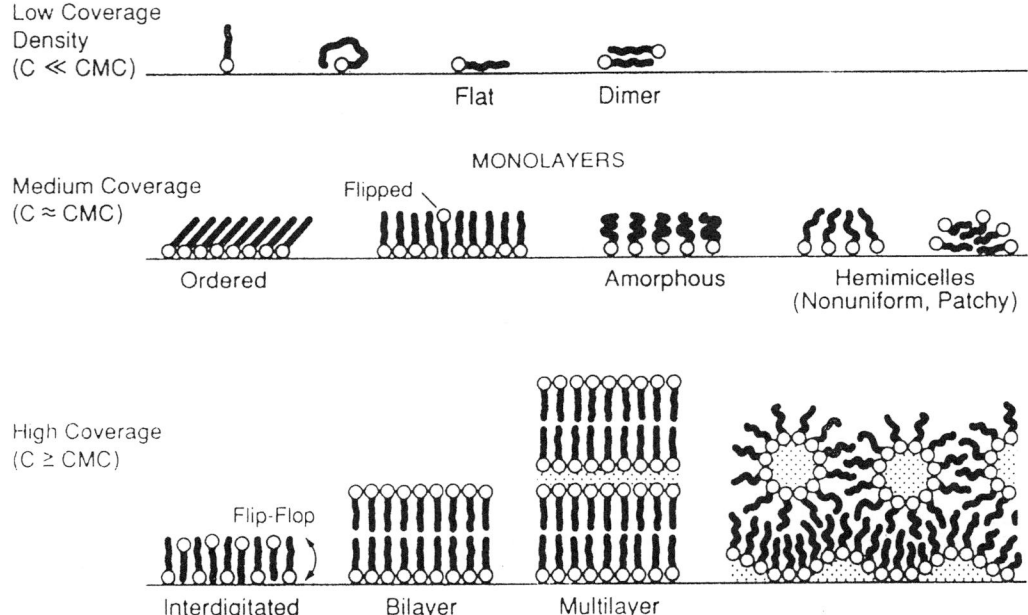

FIG. 10 Surfactant molecule orientation at the solid–liquid interface (ionic surfactant on an oppositely charged solid) at different surface coverage and surfactant concentration. (From Ref. 151. Copyright © 1992 Academic Press, Inc.)

3. Hydrophobic Surfaces

A number of pigments used in the printing ink industry are hydrophobic in nature. Examples are phthalocyanine blue and nonoxidized carbon blacks. Therefore to make stable dispersions of such pigments in water requires a change in their surface properties. One of the ways to make the surface of these pigments hydrophilic is through adsorption of the appropriate surfactants. Adsorption of surfactants on hydrophobic surfaces is mainly due to dispersion forces acting between the solid surface and the hydrocarbon chain of the surfactant. In contrast to hydrophilic charged surfaces, the adsorption isotherm has a Langmuirian shape [1,153–155] as presented in Fig. 11. For such systems, when the maximum adsorption density is reached, surfactant hydrocarbon chains are oriented more or less perpendicular to the surface and polar or ionic groups are oriented towards water. Such orientation of surfactant molecules (ions) will make the particle surface hydrophilic, and no more surfactant will be adsorbed.

Adsorption of surfactants at the solid–liquid interface is a very complex process. Though this has been studied extensively by many researchers, using different techniques [1,147,154], the adsorption process is still not well understood. The adsorption process of ionic surfactants on charged surfaces, where electrostatic interactions dominate, is better comprehended. The adsorption of nonionic surfactants, driven mainly by dispersion and hydrogen bond interactions between surfactant molecules (ions) and solid surfaces, is much more complex and difficult to understand. For more information on the adsorption of surfactants at the solid–liquid interface and the effect of surfactants on the stability

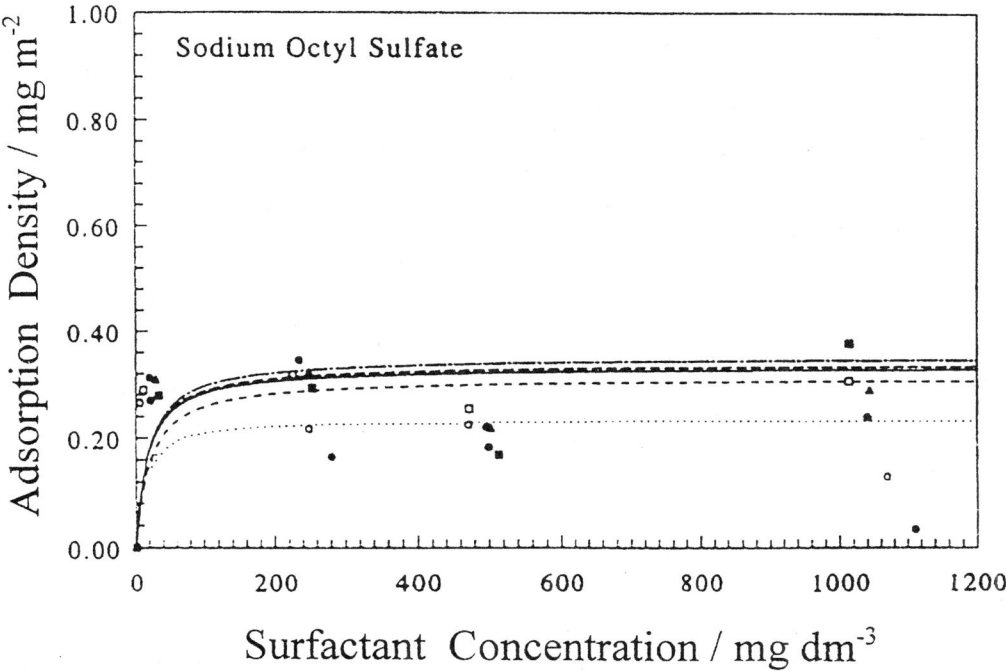

FIG. 11 Adsorption density of anionic surfactant (sodium octyl sulphate) onto hydrophobic negatively charged polystyrene. Surface charge density, -0.034 (●), -0.041 (■), -0.044 (▲), -0.055 (○), and -0.068 (□) C m^{-2}, respectively. (From Ref. 155. Copyright © 1996 Academic Press, Inc.)

of colloidal dispersions, the reader is referred to review papers and books [1,14,139–147,153,154]. A specific case of "adsorption" is the interaction between surfactant and polymer in the solution.

G. Polymer/Surfactant Interactions

Surfactant/polymer mixtures are widely used in many practical systems such as detergents, coatings, and inks, to mention a few. Water-based printing ink is a good example of such a mixture. It consists of pigments, polymer (solution or dispersion), additives (surfactants, defoamers, etc.), and water. The main purpose of adding surfactant to water-based ink is to lower the high (72 mN m^{-1}) surface tension of water. As observed from Fig. 3, the surfactant solution reaches the lowest value of surface tension at the critical micelle concentration. Nonionic surfactants are most generally used in water-based printing inks. One of the reasons is the lower foaming tendency of inks containing nonionic surfactant vs. those containing ionic surfactants. As mentioned previously, the cmc values for nonionic surfactants range from 10^{-5} to 10^{-4} mol dm^{-3}. Assuming there are no interactions between surfactant molecules with any components of the printing ink, such a small amount of surfactant added to the ink should ensure the lowest possible surface tension, for a given surfactant. In practice, the amount of surfactant used in water-based inks to achieve the required level of surface tension is much higher than its cmc. This is because surfactant interacts with other ink ingredients forming inactive or much less active complexes than the surfactant itself. Therefore a knowledge of the interactions between surfactants and polymers is very important for ink manufacturers. It can help to better understand the system and possibly reduce the cost of the ink by optimizing the formulation process.

Interactions of surfactants with polymers may be considered in terms of surfactant adsorption on "adsorption sites" of the polymer chain. Thus the interaction of a cationic surfactant with an anionic polymer is similar to the interactions between a cationic surfactant and a hydrophilic negatively charged solid, where electrostatic forces are responsible for the interactions. Different experimental techniques such as surface tension, conductivity, viscosity, NMR, and calorimetry [156–160] can be used to study polymer/surfactant interactions. For ink makers the effect of surfactant/polymer interactions on surface tension and viscosity of polymer/surfactant mixtures seems to be the most interesting. A very brief discussion on the interaction of surfactants with polymers in different systems is presented below.

1. Ionic Surfactant–Ionic Polymer

The strongest interaction between a surfactant and a polymer is observed when the surfactant and the polymer have opposite charges. Due to the very strong electrostatic interactions, polymer/surfactant interaction can be observed even at very low concentrations of surfactant and polymer. This concentration is very often much lower than the cmc of a given surfactant where self-association is observed in a polymer-free solution [156,158,161]. The effect of the interaction between oppositely charged surfactants and polymers on the surface tension of the polymer/surfactant mixture is shown in Fig. 12. As presented in Fig. 12, the polymer itself is very weakly surface active. The addition of oppositely charged surfactant resulted in polymer/surfactant complexes that were much more surface active than the surfactant itself. At higher concentrations of surfactant, precipitation was observed, as indicated by arrows in Fig. 12.

If surfactants and polymers have like charges, one would expect no interaction between them, because of the strong electrostatic repulsion between like ions. It has been observed, however, in some systems that strong hydrophobic interactions between two species can overcome electrostatic repulsion and cause association [162–164]. Interactions observed for such systems are much weaker than those for oppositely charged species. An example of such interactions is presented in Fig. 13, where the surface tension of hydrophobically modified cationic cellulose polymer solution vs. cationic surfactant concentration is plotted.

2. Ionic Surfactant–Nonionic Polymer

For such systems the electrostatic interaction is not operative, and hydrophobic interactions between surfactant and polymer will play the dominant role. This description is, however, too simple, because substantial differences in interactions between cationic and anionic surfactants were noted. For reasons not clearly understood, the interactions between anionic surfactants and nonionic polymers are much stronger than the interactions between cationic surfactants and nonionic polymers [156,157, 165,166].

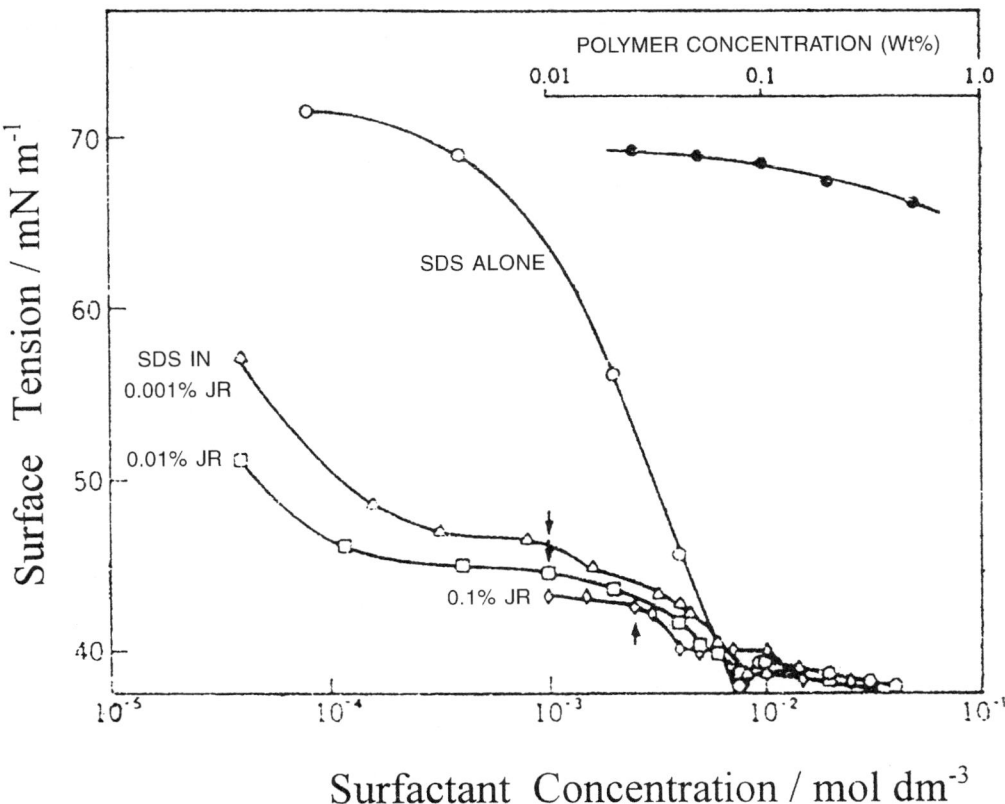

FIG. 12 Surface tension of cationic polymer solution (JR400, a quarternary nitrogen substituted cellulose ether of molecular weight around 500,000) vs. anionic surfactant (sodium dodecyl sulphate) concentration. Arrows indicate the region where precipitation was observed. Surface tension vs. polymer concentration (wt%) plot is presented for comparison. (From Ref. 161. Copyright © 1976 Academic Press, Inc.)

3. Nonionic Surfactant–Ionic Polymer

Among commercially used nonionic surfactants polyoxyethylenated surfactants are the most common. It was found that hydrophobic polymers such as polyacrylic acid interacted with nonionic polyoxyethylene alkyl ethers by forming hydrogen bonds between the acid and the oxygen atom in the ether group of the surfactant [165–168]. Ionization of polymer, by increasing pH, decreased the polymer/surfactant interactions [168,169].

Ionic polymer/nonionic surfactant systems are widely used in the printing ink industry. Some resins, e.g., acrylics and maleics, are made water soluble or dispersible by neutralization of their carboxylic groups with ammonia (anionic polymers). Such resins will decompose upon drying to give the ink film water resistance. Nonionic surfactants that are commonly used in water based formulations are Surfynol® surfactants (by Air Products), which are based on acetylenic diol chemistry. These surfactants represent a group of surface-active agents that provide low dynamic surface tension and good defoaming and wetting characteristics [170,171]. The unique properties of acetylenic diol

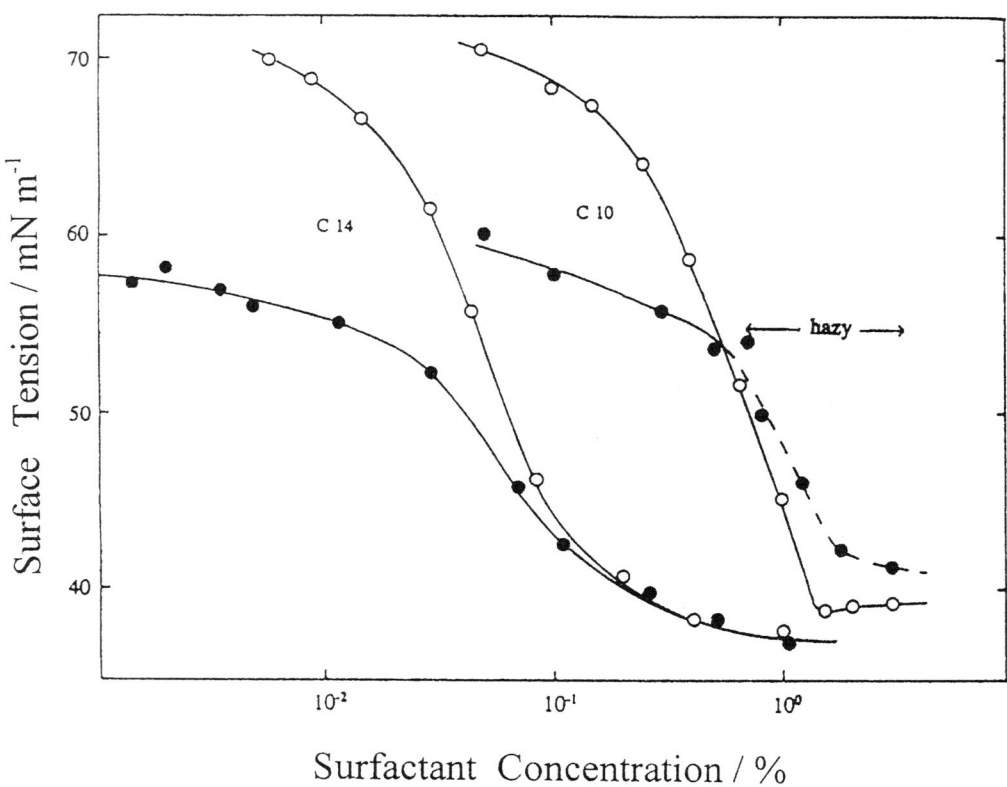

FIG. 13 Surface tension of cationic polymer solution vs. concentration of cationic surfactant. Polymer, chloride salt of dodecyldimethyl (2-hydroxypropylene-3-oxy) ammonium derivative of hydroxyethyl cellulose of molecular weight about 100,000. Surfactants, C10TAB (decyltrimethylammonium bromide) and C14TAB (tetradecyltrimethylammonium bromide). Open symbols, surfactant alone. Solid symbols, in the presence of 1% polymer. (Reprinted with permission from Ref. 163—Langmuir, Vol. 8(5), May 1992. Copyright © 1992 American Chemical Society.)

surfactants result from the chemical structure of the molecule itself—see Fig. 14. A C≡C triple bond and two hydroxyl groups in the center of the hydrocarbon chain are responsible for the high hydrophilicity of this segment. Branched alkyl groups attached to the hydrophilic central segment are strongly hydrophobic. Because of their chemical structure, these surfactants do not form micelles. An acetylenic diol molecule is strongly hydrophobic—the HLB value [1,14,16] is equal to 1. The HLB number for this surfactant can be increased by the addition of ethylene oxide between carbon atoms and –OH groups.

Though the aforementioned systems are widely used in practice, very little literature data exists about the interactions between acetylenic diol surfactants and anionic polymers [172–174]. When strongly hydrophilic polymers (polyacrylates of MW from 1,200 to 30,000) were used, no interactions between acetylenic diol surfactant and polymer were observed [174]. Hydrophobization of anionic polymers by introducing styrene segments into their molecules resulted in strong interactions between styrenated acrylics, or maleics, and acetylenic diol surfactants. The interactions depended on the degree of polymer hydrophobization and the degree of surfactant ethoxylation [172–174]. It was concluded

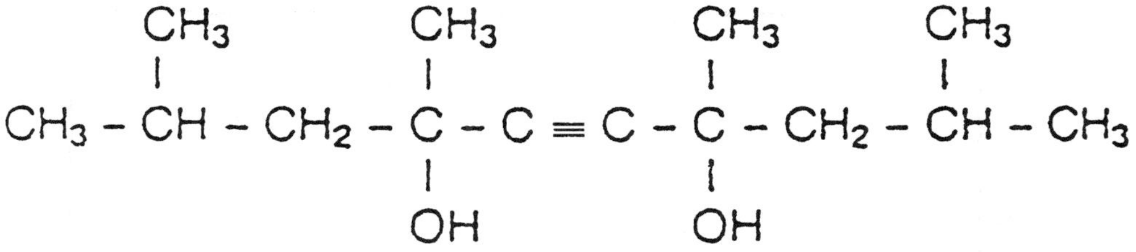

$$CH_3-CH(CH_3)-CH_2-C(CH_3)(OH)-C\equiv C-C(CH_3)(OH)-CH_2-CH(CH_3)-CH_3$$

FIG. 14 Chemical structure of acetylenic diol surfactant (2,4,7,9-tetra methyl-5decyne-4,7diol), Surfynol® 104.

that hydrophobic and hydrogen bonds were responsible for interactions between acrylics (maleics) and acetylenic diol surfactants. Complexes formed between polymer and surfactant in such systems were surface inactive (or had substantially reduced activity). Therefore the DST of polymer/surfactant mixtures was lower than the DST of surfactant solution at the same concentration. An example of the relationship of DST vs. surfactant dose for styrenated acrylic and partially ethoxylated acetylenic diol surfactant is presented in Fig. 15. The data presented in this figure helps to explain why ink formulators have to add much more surfactant to the ink (compared to its cmc or solubility in water) to achieve a low DST value. As noted from Fig. 15 for higher polymer concentrations, a significant "consumption" of surfactant by polymer is responsible for this phenomenon.

Complexes formed between surfactant and polymer may or may not have an effect on the viscosity of such mixtures. The effect of polymer/surfactant interactions on viscosity in the styrenated acrylic (maleic)/acetylenic diol surfactant systems was complex and dependent on polymer and surfactant concentrations, the degree of polymer hydrophobization, and the extent of surfactant ethoxylation [172–174]. A plot of polymer solution viscosity as a function of the amount of surfactant added to the system is presented in Fig. 16. As observed, polymer/surfactant interactions affected the viscosity only at high polymer concentrations. At 20% solids, the curve shows a tendency to plateau at surfactant doses higher than 15,000 mg. At this dose all "adsorption sites" in the polymer structure available for surfactant "adsorption" are occupied and no more complex formation is possible. Indeed, above that dose the presence of free undissolved surfactant was observed in this system [174].

4. Nonionic Surfactant–Nonionic Polymer

Hydrophilic nonionic polymers such as PVP or PVA do not interact with nonionic polyoxyethylenated surfactants [165,166,175]. If, however, the polymer contains hydrophobic segments in its molecule, nonionic surfactant can interact with such polymers due to hydrophobic interactions between surfactant and polymer hydrophobic segments [165,176]. Generally, interactions between nonionic polymers and surfactants are weak and result from hydrophobic interaction and hydrogen bonding.

In summary, interactions between surfactants and polymers are very complex. They are affected by many factors, such as temperature, inert electrolyte concentration, surfactant structure, length of hydrocarbon chain of surfactant, surfactant and polymer charge, molecular weight of polymer, concentrations of both species, and polymer structure and hydrophobicity [156–159,165,166]. Based on the literature on polymer/surfactant

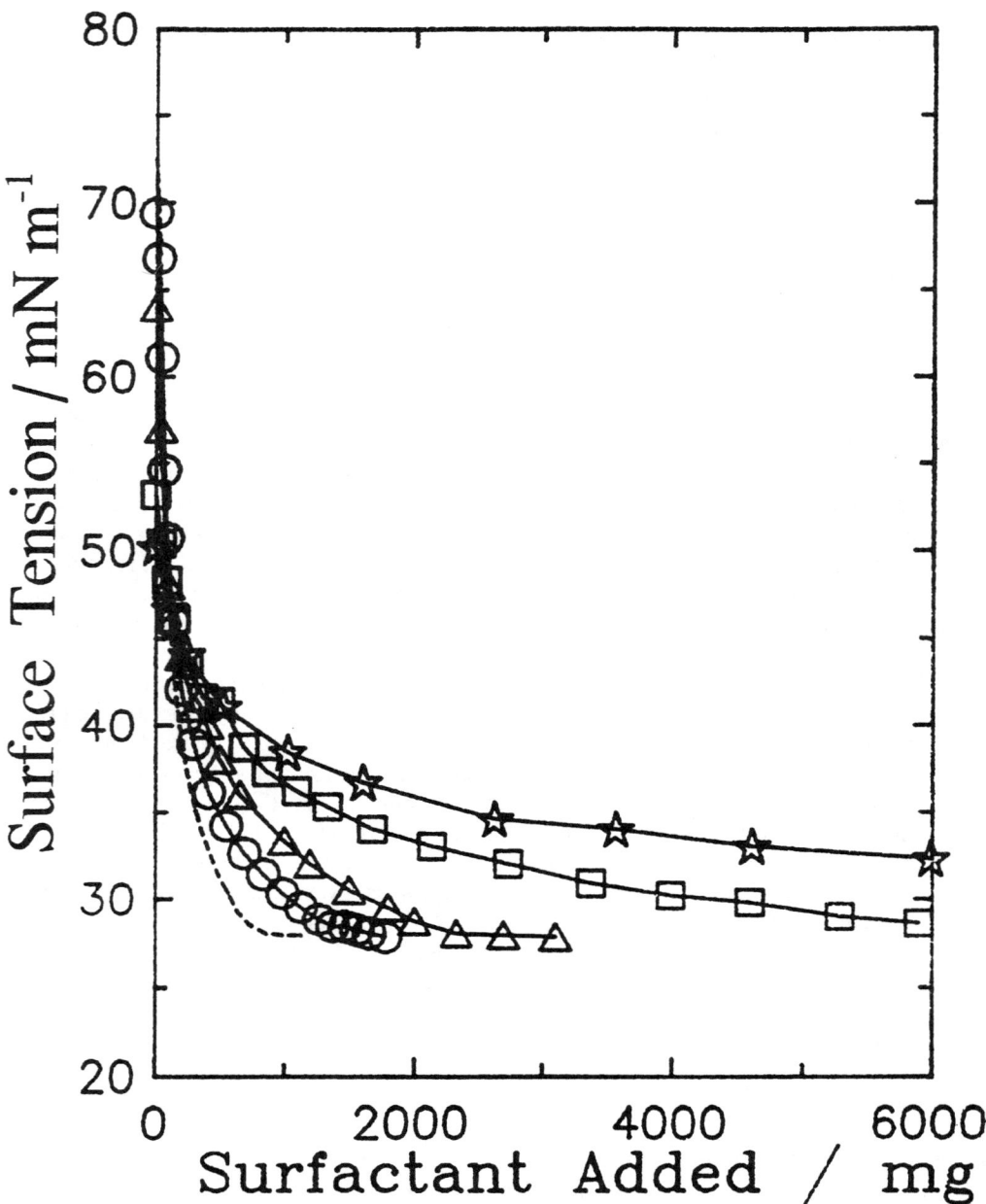

FIG. 15 Dynamic surface tension vs. surfactant dose for styrenated polyacrylic polymer (Joncryl® 63)/acetylenic diol surfactant (Surfynol® SE-F) system. Amount of polymer solution, 350 g. (From Ref. 174, courtesy of TAGA.)

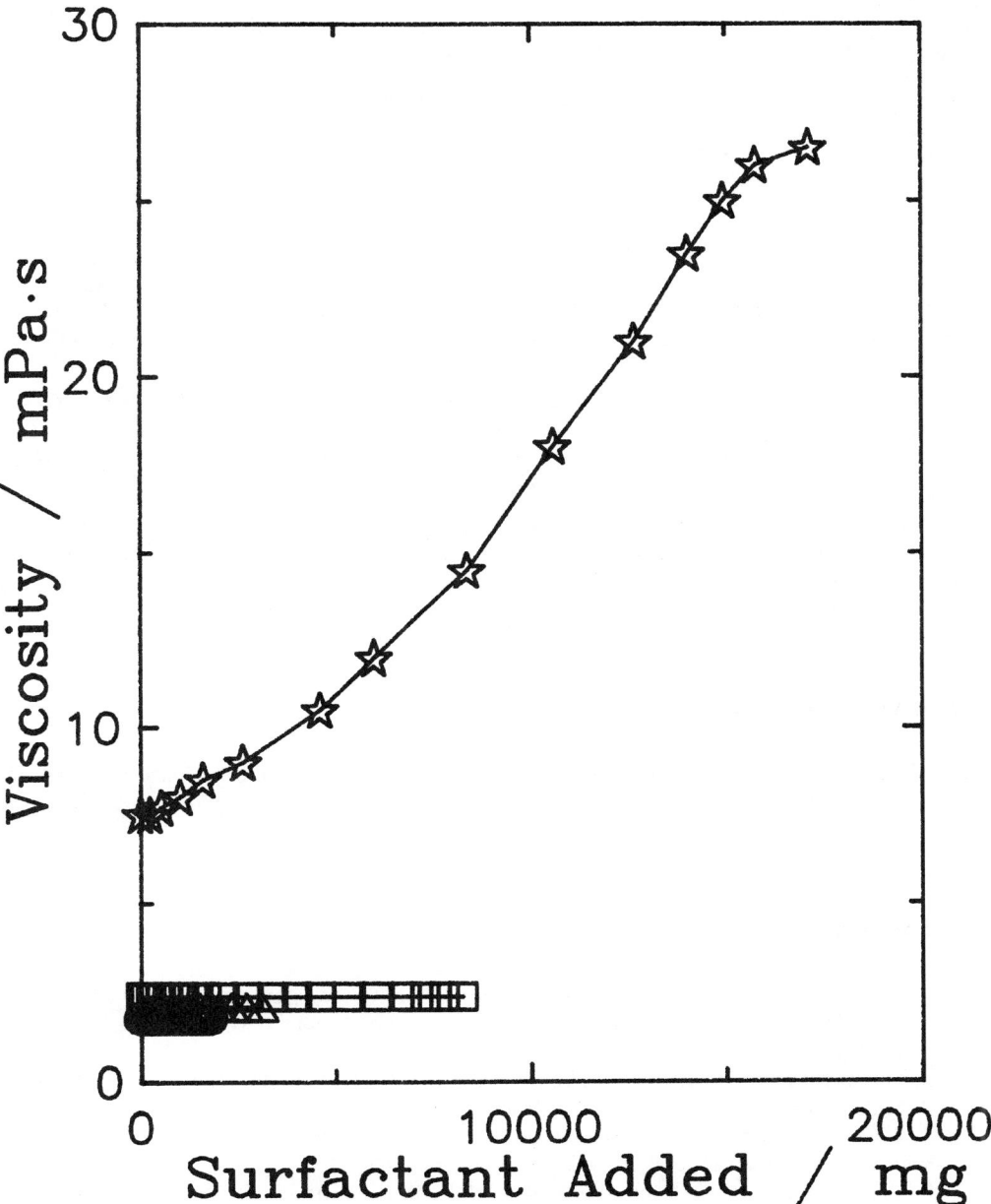

FIG. 16 Viscosity of polymer solution vs. surfactant dose for styrenated polyacrylic polymer (Joncryl® 63)/acetylenic diol surfactant (Surfynol® SE-F) system. Amount of polymer solution, 350 g. (From Ref. 174, courtesy of TAGA.)

TABLE 1 Reactivity of Surfactants

Polymer type	Order of surfactant reactivity
Anionic	Cationic ≫ nonionic ≫ anionic
Cationic	Anionic ≫ nonionic ≫ cationic
Nonionic	Anionic > cationic ≫ nonionic

interactions, some general conclusions can be drawn regarding surfactant affinity toward complex formation with different types of polymers [166]. They are presented in the form of a reactivity table (Table 1).

For brevity, discussion on polymer/surfactant interactions presented in this paper has been intentionally oversimplified. Detailed information on polymer/surfactant interactions in different systems including different theoretical approaches to this subject can be found elsewhere [156–159,165,166].

Polymer/surfactant interactions are important for ink makers. Having thorough knowledge of the effect of such interactions on ink properties means that a better understanding of printing inks will be possible. Careful selection of polymers and surfactants during ink manufacture may help to optimize cost and ink performance.

ACKNOWLEDGMENTS

The authors would like to thank Dr Graeme Thompson for helpful discussions during the preparation of this manuscript.

REFERENCES

1. M. J. Rosen, *Surfactants and Interfacial Phenomena*, Wiley-Interscience, New York, 1989.
2. A. W. Adamson and A. P. Gast, *Physical Chemistry of Surfaces*, Wiley-Interscience, New York, 1997.
3. B. Lindman, in *Surfactants* (Th. F. Tadros, ed.), Academic Press, London, 1984, pp. 83–109.
4. R. H. Ottewill, in *Surfactants* (Th. F. Tadros, ed.), Academic Press, London, 1984, pp. 1–18.
5. K. Meguro, M. Ueno, and K. Esumi, in *Nonionic Surfactants—Physical Chemistry* (M. J. Schick, ed.), Marcel Dekker, New York, 1987, pp. 109–231.
6. J. W. Birkenshaw, in *The Printing Ink Manual* (R. H. Leach and R. J. Pierce, eds), Blueprint, an Imprint of Chapman and Hall, London, 1993, pp. 14–85.
7. J. McPhee, *Fundamentals of Lithographic Printing*, GATF Press, Pittsburgh, 1998.
8. *Flexography Principles and Practices*, Foundation of Flexographic Technical Association, Ronkonkoma, New York, 1991.
9. *Gravure—Process and Technology*, Gravure Association of America, 1991.
10. *Printing Fundamentals* (A. Glassman, ed), TAPPI Press, Atlanta, 1985.
11. M. D. Croucher, S. Drappel, J. Duff, K. Lok, and R. W. Wong. Colloids Surfaces *11*:303 (1984).
12. V. Novotny and M. L. Hair. J. Colloid Interface Sci. *71*:273 (1979).
13. R. E. Kornbrekke, I. D. Morrison, and T. Oja. Langmuir *8*:1211 (1992).
14. D. Myers, *Surfaces, Interfaces, and Colloids—Principles and Applications*, VCH, 1991.

15. P. Becher, *Emulsions—Theory and Practice*, Reinhold, New York, 1965, pp. 381–400.
16. S. Ross and I. D. Morrison, *Colloidal Systems and Interfaces*, Wiley-Interscience, New York, 1988, pp. 113–118.
17. H. Akasu, M. Ueno, and K. Meguro. J. Am. Oil Chem. Soc. *51*:519 (1974).
18. Y. Takasawa, M. Ueno, and K. Meguro. J. Colloid Interface Sci. *78*:207 (1980).
19. K. Meguro, Y. Takasawa, N. Karakoshi, Y. Tabata, and M. Ueno. J. Colloid Interface Sci. *83*:50 (1981).
20. K. Esumi, K. Taguma, and Y. Koide. Langmuir *12*:4039 (1996).
21. X. Y. Hua and M. J. Rosen. J. Colloid Interface Sci. *124*:652 (1988).
22. T. Gao and M. J. Rosen. J. Colloid Interface Sci. *172*:242 (1995).
23. X. Y. Hua and M. J. Rosen. J. Colloid Interface Sci. *141*:180 (1991).
24. T. Gao and M. J. Rosen. J. Am. Oil Chem. Soc. *71*:771 (1994).
25. M. J. Rosen and L. W. Song. J. Colloid Interface Sci. *179*:261 (1996).
26. S. G. Woolfrey, G. M. Banzon, and M. J. Groves. J. Colloid Interface Sci. *112*:583 (1986).
27. L. L. Schramm, R. G. Smith, and J. A. Stone. Colloids Surfaces *11*:247 (1984).
28. K. J. Mysels. Langmuir *2*:423 (1986).
29. V. B. Fainerman. Colloids Surfaces *62*:333 (1992).
30. L. Filippov. J. Colloid Interface Sci. *164*:471 (1994).
31. P. R. Garret and D. R. Ward. J. Colloid Interface Sci. *132*:475 (1989).
32. R. L. Bendure. J. Colloid Interface Sci. *35*:238 (1971).
33. K. J. Mysels. Colloids Surfaces *43*:241 (1990).
34. V. B. Fainerman and R. Miller. J. Colloid Interface Sci. *175*:118 (1995).
35. L. L. Schramm. J. Colloid Interface Sci. *133*:534 (1989).
36. A. V. Makievski, V. B. Fainerman, and P. Joos. J. Colloid Interface Sci. *166*:6 (1994).
37. P. Joos, J. P. Fang, and G. Serrien. J. Colloid Interface Sci. *151*:144 (1992).
38. J. Kloubek. J. Colloid Interface Sci. *41*:1 (1972).
39. J. Kloubek. J. Colloid Interface Sci. *41*:7 (1972).
40. J. Kloubek. J. Colloid Interface Sci. *41*:17 (1972).
41. K. J. Mysels. Langmuir *5*:442 (1989).
42. K. J. Mysels. Langmuir *2*:428 (1986).
43. X. Zhang, M. T. Harris, and O. A. Basaran. J. Colloid Interface Sci. *168*:47 (1994).
44. U. Hofmeier, V. V. Yaminsky, and H. K. Christenson. J. Colloid Interface Sci. *174*:199 (1995).
45. M. K. Bernett and W. A. Zisman. J. Phys. Chem. *64*:1292 (1960).
46. H. W. Fox and W. A. Zisman. J. Colloid Interface Sci. *7*:428 (1952).
47. W. A. Zisman, in *Contact Angle, Wettability and Adhesion* (R. F. Gould, ed.), Adv. Chem. Ser. No. 43, ACS, Washington, D.C., 1964, p. 1.
48. W. A. Zisman, J. Paint. Technol. *44*;42 (1972).
49. E. G. Shafrin and W. A. Zisman, J. Phys. Chem. *71*:1309 (1967).
50. J. Kloubek. Adv. Colloid Interface Sci. *38*:99 (1992).
51. R. J. Good. J. Adhesion Sci. Technol. *6*:1269 (1992).
52. F. M. Fowkes. J. Phys. Chem. *66*:382 (1962).
53. F. M. Fowkes. J. Adhesion. *4*:153 (1972).
54. D. K. Owens and R. C. Wendt. J. Appl. Polym. Sci. *13*:1741 (1969).
55. D. H. Kaelble and E. H. Cirlin. J. Polym. Sci. *9*:363 (1971).
56. L. A. Girifalco and R. J. Good. J. Phys. Chem. *61*:304 (1957).
57. S. Wu. J. Polym. Sci. *C34*:19 (1971).
58. S. Wu. J. Adhesion. *5*:39 (1973).
59. S. Wu and K. J. Brzozowski. J. Colloid Interface Sci. *37*:686 (1971).
60. S. Wu. J. Colloid Interface Sci. *31*:153 (1969).
61. F. M. Fowkes, M. B. Kaczinski, and D. W. Dwight. Langmuir *7*:2464 (1991).
62. B. Miller, in *Surface Characteristics of Fibers and Textiles* (M. J. Schick, ed.), Marcel Dekker, New York, 1977, pp. 417–445.
63. J. W. Swanson, in *The Sizing of Paper* (W. F. Reynolds, ed.), TAPPI Press, 1989, pp. 133–154.

64. R. M. Podhajny, in *Surface Phenomena and Fine Particles in Water-Based Coatings and Printing Technology* (M. K. Sharma and F. J. Micale, eds.), Plenum Press, New York, 1991, pp. 41–58.
65. J. Borch, J. Adhesion Sci. Technol. *5*:523 (1991).
66. D. Kumar and S. N. Sirivastava, in *Surface Phenomena and Fine Particles in Water-Based Coatings and Printing Technology* (M. K. Sharma and F. J. Micale, eds.), Plenum Press, New York, 1991, pp. 299–307.
67. F. M. Etzler, M. Buche, J. F. Bobalek, and M. A. Weiss, in *TAPPI Proceedings* (*Papermakers Conference*), 1995, pp. 383–394.
68. R. W. Bassemir and R. Krishnan. Flexo *15*:31 (1990).
69. H. Al-Turaif, W. N. Unertl, and P. Lepoutre. J. Adhesion Sci. Technol. *9*:801 (1995).
70. Y. Huang, D. J. Gardener, M. Chen, and C. J. Biermann. J. Adhesion Sci. Technol. *9*:1403 (1995).
71. F. M. Fowkes, in *Contact Angle, Wettability and Adhesion* (R. F. Gould, ed.), Adv. Chem. Ser. 43, ACS, Washington, D.C., 1964, p. 99.
72. F. M. Fowkes. Ind. Eng. Chem. *56*:40 (1964).
73. R. J. Good and E. Elbing. Ind. Eng. Chem. *62*;54 (1970).
74. R. J. Good. J. Colloid Interface Sci. *59*:398 (1977).
75. C. J. van Oss, M. K. Chaudhury, and R. J. Good. Chem. Rev. *88*:927 (1988).
76. R. J. Good, M. K. Chaudhury, and C. J. van Oss, in *Fundamentals of Adhesion* (L. H. Lee, ed.), Plenum Press, New York, 1991, pp. 153–172.
77. C. J. van Oss, R. J. Good, and M. K. Chaudhury. J. Colloid Interface Sci. *11*:378 (1986).
78. C. J. van Oss and R. J. Good. J. Dispersion Sci. Technol. *9*:355 (1988).
79. C. J. van Oss and R. J. Good. J. Dispersion Sci. Technol. *12*:273 (1991).
80. C. J. van Oss, R. J. Good, and M. K. Chaudhury. Langmuir *4*:884 (1988).
81. C. J. van Oss, M. K. Chaudhury, and R. J. Good. Adv. Colloid Interface Sci. *28*:35 (1987).
82. A. Hollander. J. Colloid Interface Sci. *169*:493 (1995).
83. F. M. Fowkes. J. Adhesion Sci. Technol. *4*:669 (1990).
84. P. M. Costanzo, R. F. Giese, and C. J. van Oss. J. Adhesion Sci. Technol. *4*:267 (1990).
85. R. J. Good, N. R. Sirivastava, M. Islam, H. T. L. Huang, and C. J. van Oss. J. Adhesion Sci. Technol. *4*:607 (1990).
86. M. D. Vrbanac and J. C. Berg. J. Adhesion Sci. Technol. *4*:255 (1990).
87. K. J. Huttinger, S. Hohmanen-Wien, and G. Krekel. J. Adhesion Sci. Technol. *6*:317 (1992).
88. B. Janczuk, M. L. Kerkeb, T. Bialopiotrowicz, and F. Gonzalez-Caballero. J. Colloid Interface Sci. *151*:333 (1992).
89. B. Janczuk, E. Chibowski, J. M. Bruque, M. L. Kerkeb, and F. Gonzalez Caballero. J. Colloid Interface Sci. *159*:421 (1993).
90. E. Chibowski, M. L. Kerkeb, and F. Gonzalez-Caballero. J. Colloid Interface Sci. *155*:444 (1993).
91. J. C. Berg. Nordic Pulp and Paper Research Journal *1*:75 (1993).
92. E. M. Liston, L. Martinu, and M. R. Wertheimer. J. Adhesion Sci. Technol. *7*:1091 (1993).
93. M. Goldman, A. Goldman, and R. S. Sigmond. Pure Appl. Chem. *57*:1353 (1985).
94. R. Auerbach, A. Reitnauer, and C. Williams. Markem Corporation Technical Bulletin, 1993.
95. D. H. Kaelble, *Physical Chemistry of Adhesion*, Wiley-Interscience, New York, 1971.
96. F. Leja, *Surface Chemistry of Froth Flotation*, Plenum Press, New York, 1983.
97. B. Janczuk, M. L. Kerkeb, F. Gonzalez-Caballero, and E. Chibowski. Langmuir *10*:1012 (1994).
98. V. V. Yaminsky, P. M. Claesson, and J. C. Eriksson. J. Colloid Interface Sci. *161*:91 (1993).
99. K. S. Birdi and D. T. Vu. J. Adhesion Sci. Technol. *7*:485 (1993).
100. S. L. Holmes-Farley, C. D. Bain, and G. M. Whitesides. Langmuir *4*:921 (1988).
101. G. M. Whitesides, H. A. Biebuyck, J. P. Falker, and K. L. Prime. J. Adhesion Sci. Technol. *5*:57 (1991).
102. N. Mourougou-Candoni, B. Prunet-Foch, F. Legay, M. Vignes-Adler, and K. Wong. J. Colloid Interface Sci. *192*:129 (1997).

103. J. C. Berg, in *Wettability* (J. C. Berg, ed.), Marcel Dekker, 1993, pp. 75–147.
104. M. Denisuk, B. J. J. Zelinski, N. J. Kreidl, and D. R. Uhlmann. J. Colloid Interface Sci. *168*:142 (1994).
105. A. Marmur. J. Colloid Interface Sci. *123*:161 (1988).
106. A. Marmur. J. Colloid Interface Sci. *124*:301 (1988).
107. A. Marmur. J. Colloid Interface Sci. *122*:209 (1988).
108. D. Danino and A. Marmur. J. Colloid Interface Sci. *166*:245 (1994).
109. A. Marmur, in *Modern Approach to Wettability: Theory and Applications* (M. E. Schrader and G. Loeb, eds.), Plenum Press, New York, 1992, pp. 327–358.
110. P. Salminen, Research Report, Abo Akademii, Finland, 1988.
111. M. B. Lyne and J. Aspler. Tappi. *65*:98 (1982).
112. J. S. Aspler, S. Davis, and M. B. Lyne. Tappi *67*:128 (1984).
113. P. Lepoutre, M. Inoue, and J. Aspler. Tappi J. *68*:86 (1985).
114. B. L. Anderson and B. G. Higgins, *Tappi Proceedings (Coating Conference)*, 1985, pp. 47–54.
115. Y. L. Pan, S. Kuga, and M. Usuda. Tappi J. *71*:119 (1988).
116. J. F. Oliver, in *Colloids and Surfaces in Reprographic Technology* (M. Hair and M. D. Croucher, eds.), ACS, Washington D.C., 1982, pp. 435–453.
117. R. Sprycha and J. N. Hruzewicz, *TAGA Proceedings*, 1994, pp. 433–458.
118. J. S. Aspler and M. B. Lyne. *Tappi J.* *67*:96 (1984).
119. C. Huh and S. G. Mason. *J. Colloid Interface Sci.* *60*:11 (1977).
120. J. F. Oliver and S. G. Mason. *J. Colloid Interface Sci.* *60*:480 (1977).
121. J. F. Oliver, C. Huh, and S. G. Mason. *J. Colloid Interface Sci.* *59*:568 (1977).
122. J. Drelich, J. D. Miller, and R. J. Good. *J. Colloid Interface Sci.* *179*:37 (1996).
123. T. Onda, S. Shibuichi, N. Satoh, and K. Tsujii. *Langmuir 12*:2125 (1996).
124. C. H. Chan and S. Venkatraman, in *Coating Technology Handbook*, Marcel Dekker, New York, 1991, pp. 19–35.
125. R. J. Hunter, *Introduction to Modern Colloid Science* Oxford University Press, Oxford, 1993.
126. R. J. Hunter, *Zeta Potential in Colloid Science*, Academic Press, New York, 1981.
127. G. A. Parks. *Chem. Rev. 65*:177 (1965).
128. M. A. Blesa and N. Kallay. *Adv. Colloid Interface Sci. 28*:111 (1988).
129. J. A. Davis, R. O. James, and J. O. Leckie. *J. Colloid Interface Sci. 63*:480 (1978).
130. R. Sprycha. *J. Colloid Interface Sci. 127*:1 (1989).
131. R. Sprycha. *J. Colloid Interface Sci. 127*:12 (1989).
132. W. Janusz. *J. Colloid Interface Sci. 145*:119 (1991).
133. L. K. Koopal, W. H. van Riemsdijk, and M. G. Roofey. *J. Colloid Interface Sci. 118*:117 (1987).
134. N. Kallay and M. Tomić. *Langmuir 4*:559 (1988).
135. M. Tomić and N. Kallay. *Langmuir 4*:565 (1988).
136. P. W. Schindler and W. Stumm, in *Aquatic Surface Chemistry*, John Wiley, New York, 1987, p. 83.
137. N. Kallay, R. Sprycha, M. Tomić, S. Žalac, and Z. Trobić. *Croatica Chemica Acta 63*:467 (1990).
138. N. Kallay, D. Babić, and E. Matijević. *Colloids Surfaces 19*:375 (1986).
139. D. B. Hough and L. Thompson, in *Non-ionic Surfactants—Physical Chemistry* (M. J. Schick, ed.), Marcel Dekker, New York, 1987, pp. 601–676.
140. Th. F. Tadros, in *Surfactants* (Th. F. Tadros, ed.), Academic Press, London, 1984, pp. 197–220.
141. R. I. Hancock, in *Surfactants* (Th. F. Tadros, ed.), Academic Press, London, 1984, pp. 287–321.
142. D. J. Walbridge, in *Solid/Liquid Dispersions* (Th. F. Tadros, ed.), Academic Press, 1987, pp. 17–62.
143. B. H. Bijsterbosch, in *Solid/Liquid Dispersions* (Th. F. Tadros, ed.), Academic Press, 1987, pp. 91–110.

144. B. Vincent, in *Solid/Liquid Dispersions* (Th. F. Tadros, ed.), Academic Press, 1987, pp. 149–162.
145. R. Aveyard, in *Surfactants* (Th. F. Tadros, ed.), Academic Press, London, 1984, pp. 153–174.
146. P. Somasundaran, T. W. Healy, and D. W. Fuerstenau. *J. Phys. Chem.* 68:3562 (1964).
147. D. B. Hough and H. M. Rendall, in *Adsorption from Solution at the Solid/Liquid Interface* (G. D. Parfitt and C. H. Rochester, eds.), Academic Press, London, 1983, pp. 247–320.
148. T. Wakamatsu and D. W. Fuerstenau. *Transactions AIME* 254:123 (1973).
149. L. Huang, C. Maltesh, and P. Somasundaran. *J. Colloid Interface Sci.* 177:222 (1996).
150. D. W. Fuerstenau. *Pure Appl. Chem.* 24:135 (1970).
151. Y. L. Chen, S. Chen, C. Frank, and J. Israelachvili. *J. Colloid Interface Sci* 153:244 (1992).
152. H. Rupprecht and H. Liebl. *Kolloid Z. Z. Polym.* 250:719 (1972).
153. F. G. Greenwood, G. D. Parfitt, N. H. Picton, and D. G. Wharton, in *Adsorption from Aqueous Solution*, Adv. Chem. Ser. 79 (W. J. Weber and E. Matijevic, eds.), ACS, Washington, D.C., 1968, pp. 135–144.
154. J. S. Clunie and B. T. Ingram, in *Adsorption from Solution at the Solid/Liquid Interface* (G. D. Parfitt and C. H. Rochester, eds.), Academic Press, London, 1983, pp. 105–152.
155. C. E. Hoeft and R. L. Zollars. *J. Colloid Interface Sci.* 177:171 (1996).
156. E. D. Goddard, in *Interactions of Surfactants with Polymers and Proteins* (E. D. Goddard and K. P. Ananthapadmanabhan, eds.), CRC Press, Boca Raton, FL, 1993, pp. 123–169.
157. E. D. Goddard, in *Interactions of Surfactants with Polymers and Proteins* (E. D. Goddard and K. P. Ananthapadmanabhan, eds.), CRC Press, Boca Raton, FL, 1993, pp. 171–202.
158. B. Lindman and K. Thalberg, in *Interactions of Surfactants with Polymers and Proteins* (E. D. Goddard and K. P. Ananthapadmanabhan, eds.), CRC Press, Boca Raton, FL, 1993, pp. 203–276.
159. K. P. Ananthapadmanabhan, in *Interactions of Surfactants with Polymers and Proteins* (E. D. Goddard and K. P. Ananthapadmanabhan, eds.), CRC Press, Boca Raton, FL, 1993, pp. 319–366.
160. F. M. Winnik, in *Interactions of Surfactants with Polymers and Proteins* (E. D. Goddard and K. P. Ananthapadmanabhan, eds.), CRC Press, Boca Raton, FL, 1993, pp. 367–394.
161. E. D. Goddard and R. B. Hannan. *J. Colloid Interface Sci.* 55:73 (1976).
162. I. Iliopoulos, T. K. Wang, and R. Audebert. *Langmuir* 7:617 (1991).
163. E. D. Goddard and P. S. Leung. *Langmuir* 8:1499 (1992).
164. M. J. McGlade, F. J. Randall, and N. Tcheurakdijan. *Macromolecules* 20:1782 (1987).
165. S. Saito, in *Non-ionic Surfactants—Physical Chemistry* (M. J. Schick, ed.), Marcel Dekker, New York, 1987, pp. 881–926.
166. E. D. Goddard. *J. Am. Oil Chem. Soc.* 71:1 (1994).
167. G. D. Jaycox, R. Sinta, and J. Smid. *J. Polymer. Sci. Polymer Chem. Ed.* 20:1629 (1982).
168. S. Saito and T. Taniguchi. *J. Colloid Interface Sci.* 44:114 (1973).
169. T. Ikawa, K. Abe, K. Honda, and E. Tsuchida, *J. Polymer Sci. Polymer Chem. Ed.* 13:505 (1975).
170. W. Dougherty. *Adhesive Age* 32:26 (1989).
171. S. W. Medina and M. N. Sutowich. *American Ink Maker* 72:32 (1994).
172. R. Krishnan and R. Sprycha, *Ninth Int. Conference on Surface and Colloid Science*, 6–12 July, Sofia, Bulgaria, 1997, Abstract. 617. E5.
173. R. Krishnan and R. Sprycha. *Colloids Surfaces*, 149:355 (1999).
174. R. Sprycha and R. Krishnan. *TAGA Proceedings*, 1998, pp. 585–599.
175. M. J. Schwuger and H. Lange. *Proc. Fifth Int. Congr. Surface Activity* 2:955 (1968).
176. S. Saito, *Koll. Z. Z. Polymer.* 226:10 (1968).

Index

Adhesion, 713–714
Adsorption:
    from binary mixtures onto solids, 85–90
    of 1-butanol and 1-hexanol, 635–647
    of cadmium ions on goethite, 267–270
    calorimetry, 215
    of carboxylates on metal oxides, 520–529
    of cationic surfactants on hydrophilic silica, 131–144
        effect of surface energetic heterogeneity, 131–133
        conceptualization and theoretical model, 133–144
    comparison of the model with experimental data, 144–153
    of counterions, 196–205
    enthalpy of, 215–216
        direct calorimetric measurement, 422
        salt concentration effect on, 427–430
    heat of, 116–124
    of heavy metals, 267–270
    of ions, heat of, 91–106
    isotherms, 621–647
    isotherms forming different surface complexes, 109–116
    of large (organic) ions, 259–267
    at low vapor pressure, 630–635
        apparatus, 630–633
        procedure, 633–635
    of medium sized alcohols on various solids, 622–647
    methods, 406–407
        modeling, 412
    multilayer, 635–647
    of proteins, 406–416

[Adsorption]
    of salicylic acid on hematite and titania, 256–257
    of simple ions, 106–109
    specific, 205–217
    thermodynamics of, 83–85
Adsorption/desorption:
    of alkali metal ions, 391–394
    of amino acids, 394–395
    of anions, 372–380
    of charged proteins, 395–399
    of metal ions, 380–386
    of potential determining ions, 371–372
Adsorption/desorption models:
    multi step, 357–363
    single step, 352–357
        ion-exchange type, 355–357
        Langmuir type, 352–355
Alumina, 297–298, 301, 305, 308
Amino acids, adsorption/desorption of, 394–395
Anatase, 296–297
Apparatus for measuring small adsorption at low vapor pressure, 630–633
Atomic force microscopy (AFM), 207
Attenuated total reflection spectroscopy (ATR), 57, 525

Behavior of adsorbed molecules, 622–625
    the gradient method, 625–630
    theory of surface area estimation, 622–630
Biomaterials:
    application of isothermal titration microcalorimetry, 423–430
    behavior at interface, 420–422

Brönsted acids and bases, 21–22
Bulk water, 37–43
  experimental techniques, 38–41
  spectroscopic studies, 41–43
  theoretical studies, 73–75

Cadmium, 267–270
Calorimetry:
  of immersion, 314–317, 324–331
  of interfacial reactions, 236–246
  of wetting, 314–317
Carboxylate complexes in solution:
  electron-transfer reactions, 519
  stability, 514–516
  structure of dissolved, 518–519
Charge fluctuation model, 550
Colloid stability studies, 44–54
Colloidal suspension, conductivity, 543–546
Complex formation, 32
Complexation constants, determination of, 186–205
Conductivity:
  of colloidal suspensions, 543–546
  of microemulsion systems, 546–561
Constants:
  complexation, 186–205
  equilibrium, 22–26
  surface ionization, 186–205
Counterions:
  adsorption of, 196–205
  association, 230–232
Crystal dissolution, 458–459
Crystal growth, 444
  aging, 462–464
  controlled by different mechanisms, 457–458
  experimental approach, 464–469
  influence of impurities, 459–462
  surface reaction controlled, 449–450
    adsorption controlled, 449–451
    screw dislocation controlled, 454–457
    surface nucleation controlled, 451–454
  transport controlled, 446–449

Dielectric spectroscopy, 57
Dissolution:
  inhibition of:
    by adsorbed ions, 497–502
    influence of adsorption, 496–497
    by low molecular weight organic compounds, 502
    by proteins, 502–505
    by synthetic macromolecules, 505–506
    by surfactants, 497

[Dissolution]
  of metal oxides, 530–532
    in the presence of carboxylic acids, 533–538
Double layer model:
  with counterion association, 180–186
  without counterion association, 169–180
    constant capacitance model, 170–171
    diffuse layer model, 171–173
    MUSIC model, 178–180
    1-pK model, 174–177
    1-pK multisite approach, 177–178
    Stern model, 174

Electrical interfacial layer, 250–254
  in nonaqueous solvents, 273–308
Electrokinetic, (see also Zeta-potential):
  data, 191–196
    interpretation of equilibria on the basis of, 249–270
  mobility of UPP particles, 584–585
  properties of poly(N-isopropylacrylamide) hydrogel, 599–619
Electron paramagnetic resonance (EPR), 210
Electroosmotic velocity, of poly(N-isopropylacrylamide) hydrogel, 599–602
Electrophoretic mobility, 278–280
  of poly(N-isopropylacrylamide) hydrogel, 601–619
Enthalpy:
  of adsorption, 215–216
  electrostatic contribution, 241–246
  of immersion, 317–319
    case of alkanes, 329–330
    case of polar liquids, 330–331
    case of water, 326–329
  of wetting, 315–317
Equilibria:
  in electrolyte solutions, 1–32
Equilibrium constants, 22–26
Extended x-ray absorption fine structure (EXAFS), 209, 525

Fluctuation model, 550

Goethite, 267–270, 308

Heat:
  of immersion, 91–106
  of ion adsorption, 91–106
Hematite, 256–257, 298–299, 301, 306

Hydration:
    hydrophobic, 62–65
    the role of dissolved gases, 65–66
    spectroscopic studies of, 57–62
Hydron transfer reactions, 21–32
Hydrophobic forces:
    long range, 66–69
    short range, 66–69
Hydroxyapatite:
    dissolution, 479–484
        ionic crystal, 479–480
        of HAP, 480–481
        kinetics of, 483–484
        models, 484–494
        of carbonated apatites and fluoroapatites, 494–495
        rate of, 481–483
    formation, 476–479
    morphology, 476–479
    structure, 476–479

Immersion phenomena, 313–346
Inorganic thin films, electrical behavior, 565–589
Interactions:
    polymer-surfactant, 725–732
    protein-surface, 405–406
Interface:
    liquid–liquid, 420–421
    solid–liquid, 421–422
Interfacial water interparticle and intermolecular forces, role of, 43–57
Interfacial water, role of electric, magnetic, electromagnetic and sonic field, 69–73
Inverse gas chromatography (IGC), 323–324
Ionic interactions, 5–21
    advanced theories, 11–14
        higher level theories, 13
        Mayer's virial expansion of osmotic pressure, 11
        mean spherical approximation, 12
        numerical simulations, 14
    ionic cloud models, 6–11
        Debye–Hückel model, 6–11
        primitive models, 6
    ion association, 15–21
        Bjerrum's theory, 15
        Fuoss's revision, 16
        second Fuoss model, 17–19

Kinetics:
    of crystal growth in aqueous systems, 434–471

[Kinetics]
    experimental methods:
        concentration jump, 370–371
        electric field jump, 368–370
        pressure jump, 363–368
    of hydroxyapatite dissolution, 483–484
    ionic adsorption/desorption, 351–400
    of metal ion complexation by carboxylates, 516–518
    of metal oxide dissolution, 530–532
    of protein adsorption, 407–416
        modeling, 410–415
    of surface complexation, 529–530

Langmuir isotherm, 254–255, 352–355
    extended, 255

Magnesia, 306
Mechanism, of crystal growth in aqueous systems, 434–471
Microcalorimetry, isothermal titration, 422
Microemulsions, conductivity of, 546–561
Mixtures, 2–3
MUSIC model, 108, 178–180

Niobia, 306
NMR, 57
Nucleation, 439–444
    heterogeneous, 443
    homogenous, 440–443
    secondary, 444

Organic contaminants:
    in sediments, 652–654
    in soils, 652–654

pH:
    definition, 26–32
    interpretation, 26–32
    scales, 282–285
1-pK model, 174–177, 230
1-pK multisite approach, 177–178
2-pK model, 108
Point of zero charge, 278
    temperature dependence, 233–235
Poly(N-isopropylacrylamide) hydrogel:
    amount of negative fixed charges, 596
    charge distribution, 596
    electrokinetic properties, 596–599

[Poly(N-isopropylacrylamide) hydrogel]
    equilibrium swelling ratio, 593–595
    interfacial electric properties, 599–602
    preparation, 592
    properties, 592
    wettability, 592–593
Printing inks, 700–703
Printing processes, 700–703
    using liquid inks, 700–703
        flexography, 701–702
        gravure, 702
        ink–jet, 702
        liquid toners, 702–703
Proteins:
    adsorption of, 406–416
    adsorption kinetics, 407–416
    interactions with surface, 405–406

Raman spectroscopy, 39
Rheological studies, 44–54
Rutile, 296

Salicylic acid, 256–257
Second-harmonic generation spectroscopy (SHG), 57
Sediments:
    organic contaminants in, 652–654
    sorption processes in, 654–674
Silica, 290–296, 300–304, 308
Soil:
    organic contaminants in, 652–654
    sorption processes in, 654–674
Solution, 3–4
    ionic, 4–5
    stability, 436–439
Sorption (see also Adsorption):
    hysteresis:
        experimental and model results, 675–677
        formation of, 674–677
    multicomponent, 677–696
        organic contaminants, 677–680
    processes:
        comparison of experimental and model results, 665–674
        mechanism, 654–655
        modeling, 655–665
        in sediments, 654–674
        in soils, 654–674
Specific adsorption, 205–217
Spectroscopic studies:
    of colloid hydration, 57–60
    of hydration in biological systems, 60–62

Stability:
    of carboxylate complexes in solution, 514–516
    of dispersions, 717–722
    in mixed and nonaqueous solvents, 307–308
Stoichiometry, of chemical reactions, 226–227
Sum-frequency generation (SFG), 57
Surface charge, 187–191, 289–299
    creation at the solid–liquid interface, 165–168
        ionic crystal (AgI)/electrolyte interface, 166–168
        mechanism, 165
        metal/electrolyte interface, 165–166
        metal oxide(hydroxide)/electrolyte interface, 168
        semiconductor/electrolyte interface, 166
Surface charging (see also Surface reactions):
    enthalpy of, 233–246, 258–259
    mechanism of, 228–230
    in nonaqueous media, 274–278
Surface complexation model, 163–217, 257–270, 513–514
Surface energy of solids, 320–324, 339–343
Surface enthalpy, 339–343
Surface force measurements, 54–57
Surface free energy:
    of liquids, 704–710
    of solids, 704–710
Surface ionization constants, determination of, 186–205
Surface potential, 236–239
Surface reactions (see also Surface charge):
    extent of, 239–240
    stoichiometry of, 226–227
Surface tension, 337–339, 343–345
    measurements, 706–710
Surfactants:
    adsorption at the solid–liquid interface, 722–732
    effect on some processes essential for printing, 704–732

Thermodynamic quantities, electrostatic effect, 240–241
Thin films:
    electrical behavior, 571–589
    inorganic, 565–571
Time-resolved laser induced fluorescence spectroscopy (TRLFS), 207
Titania, 256–257, 300–301, 304
Total internal reflection fluorescence spectroscopy (TIRF), 406
Triple layer model, 108–116, 119–124, 251

Index

Vibration-rotation tunneling spectroscopy, 42

Water clusters, 37–43
Wetting, of liquids on solids, 714–717

X-ray absorption spectroscopy (XAS), 209

Ytria, 301

Zeta-potentials (*see also* Electrokinetics):
  in mixed solvents, 299–301
  in nonaqueous solvents, 301–307
Zinc oxide, 307